CHEMISTRY AND OUR WORLD

CHEMISTRY AND OUR WORLD

CHARLES G. GEBELEIN
PROFESSOR OF CHEMISTRY
YOUNGSTOWN STATE UNIVERSITY
YOUNGSTOWN, OHIO

WCB Wm. C. Brown Publishers
Dubuque, IA Bogota Boston Buenos Aires Caracas Chicago Guilford, CT
London Madrid Mexico City Seoul Singapore Sydney Taipei Tokyo Toronto

Project Team
Editor *Colin H. Wheatley*
Developmental Editor *Brittany J. Rossman*
Marketing Manager *Patrick E. Reidy*
Photo Editor *Janice Hancock*
Advertising Coordinator *Wayne W. Siegert*
Permissions Coordinator *Mavis M. Oeth*
Publishing Services Coordinator *Julie Avery Kennedy*

Wm. C. Brown Publishers

President and Chief Executive Officer *Beverly Kolz*
Vice President, Director of Editorial *Kevin Kane*
Vice President, Sales and Market Expansion *Virginia S. Moffat*
Vice President, Director of Production *Colleen A. Yonda*
Director of Marketing *Craig S. Marty*
National Sales Manager *Douglas J. DiNardo*
Advertising Manager *Janelle Keeffer*
Production Editorial Manager *Renée Menne*
Publishing Services Manager *Karen J. Slaght*
Royalty/Permissions Manager *Connie Allendorf*

A Times Mirror Company

Copyedited by *Anne Cody*

Cover photo © Astrid & Hanns-Frieder Michler / SPL / Photo Researchers, Inc.
Photo Research by Shirley Lanners
Publishing Services by Graphic World Publishing Services
Composition by Graphic World, Inc.
Illustrations by Wilderness Graphics

The credits section for this book begins on page 597 and
is considered an extension of the copyright page.

Library of Congress Catalog Card Number: 95-83494

ISBN 0-697-16574-4

Printed in the United States of America by Times Mirror Higher Education Group, Inc.,
2460 Kerper Boulevard, Dubuque, IA 52001

10 9 8 7 6 5 4 3 2 1

This book is dedicated to my wife Clare, whose encouragement and assistance helped "birth" this text. Our children Clare, Jean, Charles, Joyce, and Keith also shared in this effort.

BRIEF CONTENTS

CONTENTS

CHAPTER 4

ORDER IN CHEMISTRY: THE PERIODIC TABLE AND ATOMIC STRUCTURE 83

CHAPTER 5

ENERGY, THE NUCLEUS, AND NUCLEAR CHEMISTRY 111

CHAPTER 6

CHEMICAL BONDING—ALL-PURPOSE MOLECULAR GLUE 139

CHAPTER 7

CHEMICAL REACTIONS 169

CHAPTER 8

SOLUTIONS, ACIDS, AND BASES 191

CHAPTER 9

ORGANIC CHEMISTRY—THE WONDERFUL WORLD OF CARBON 209

CHAPTER 10

FUNCTIONAL ORGANIC CHEMICALS 243

CHAPTER 11

POLYMERS—THE MIGHTY MOLECULES 271

CHAPTER 12

CHEMISTRY AND OUR WATER SUPPLY 303

CHAPTER 13

SOAPS, DETERGENTS, AND CLEANING 335

CHAPTER 14

THE CHEMISTRY OF LIFE 357

CHAPTER 15

FOOD AND AGRICULTURAL CHEMISTRY 399

CHAPTER 16

COSMETIC CHEMISTRY—THE SCIENCE OF BEAUTY 435

CHAPTER 17

CHEMISTRY IN HEALTH AND MEDICINE 457

CHAPTER 18

AIR POLLUTION, ENERGY, AND FUELS 491

CHAPTER 19

CHEMISTRY AND OUR ATMOSPHERE 523

APPENDICES

PREFACE

TO THE INSTRUCTOR

Philosophy

Teaching science to non-science majors is extremely critical to society and our future, perhaps as critical as teaching science to science majors. Science majors don't need to be convinced that the study of science is worthy, important, and interesting. The scientists of tomorrow already understand the interrelated nature of science—be it chemistry, physics, biology, or geology—and the world we live in. The crucial part is instructing non-science students about this nature. Teaching science to non-science majors can, unfortunately, be extremely difficult and frustrating. The reason the majority even take a non-science majors science class—such as a liberal arts chemistry class—is that by decree of the school, they must take such a class to graduate. This predisposes students to believe they will not enjoy chemistry, much less actually learn it and use it in the larger world. My answer to this challenging problem is the book which you have in your hands. *Chemistry and Our World* strives to teach students that chemistry is indeed relevant and interesting to us as individuals and to our understanding of the changing world around us.

Chemistry can indeed be presented in a relevant fashion to those students for whom this may be their only science course. To that end, *Chemistry and Our World* is carefully sculpted to meet the needs of this specific group of students. The more mathematical and theoretical aspects of chemistry are de-emphasized, and the more practical, or applied aspects are accentuated. There is no need to overwhelm non-science students with the theoretical details of chemistry, but what they do need is a basic understanding of how chemistry affects everyday life. In learning how to apply rational or scientific thinking methods to their lives, their future careers—in whatever field—will benefit.

Content

This text combines five important features.

1. Chemistry is something done by people for people. This book describes the people who developed the science we know as chemistry without becoming a "History of Science" text.
2. Chemists and other scientists think about problems and their solutions using the scientific method. We introduce this concept in chapter 1 with a simple, everyday example to help students start to become accustomed to thinking in this way.
3. The companion problem of risks versus benefits is also introduced in chapter 1 and used frequently throughout the text. Methods that are used to resolve these dilemmas are also considered.
4. Most of the mathematics in this book can be mastered with a knowledge of simple arithmetic. The basics of using the metric system are also included for two reasons: (1) It is standard practice for most areas of science to regularly use the metric system in measurements and calculations, and (2) the majority of the world uses it. As the world truly becomes a global village, it will become increasingly more important to understand the metric system to be able to communicate intelligently with others.
5. Applied chemistry is emphasized in this book. Many examples of applied chemistry will show students how they *already* use chemistry and reveal why chemistry is vitally important to them.

Many non-science chemistry courses are only one semester, or the equivalent of one semester in length. As is the case in almost all courses, there is more material in this text than can be conveniently covered in the available time. You, as the instructor, are free to choose what material you find most relevant and important to your students. This is best done by choosing material within chapters rather than totally eliminating chapters. Some further suggestions are given in the Instructor's Manual. If your course happens to be a full year in length, you have the opportunity to expand on textual material by using the Readings of Interest section and by providing additional materials as appropriate.

Pedagogy

Recognizing that the study of chemistry may be quite frightening to many students in our audience, we have designed the text to be as "friendly" as possible to students.

1. *Chapter Outline*. Each chapter starts with an Outline and a list of the Key Ideas presented in the chapter.
2. *In-Chapter Practice Problems*. Numerous in-chapter Practice Problems help students learn to solve the various problems presented in the text.
3. *Experience This*. Numerous hands-on Experience This studies allow students to see some of the places where chemistry enters into their lives.

4. *Tables*. Several strategically-placed tables provide easy reference to important information, helping students access large amounts of material in an efficient way.
5. *End-of-Chapter Summary*. The Summary concludes each chapter and reviews the information covered.
6. *Key Terms list*. The **boldface** terms throughout the chapters are listed in the Key Terms list located after the summary. These terms are also included in the Glossary.
7. *Readings of Interest*. This section lists many articles, usually from non-technical sources, that provide further insights into chapter materials, and often on other related topics.
8. *Problems and Questions*. A wide selection of problems for students to work out follow each chapter.
9. *Critical Thinking Problems*. These problems may require using certain Readings of Interest or boxed readings to formulate an answer. Some of these questions are of the opinion type and have different answers based on one's perspective and knowledge.
10. *Glossary*. The end-of-book Glossary provides the definition or explanation of important terms used in the book.
11. *Answers to Select Problems and Questions, and Critical Thinking Problems*. This section is located after the Glossary.
12. *Appendices*. In keeping with our desire not to overload students with mathematical information regarding the study of chemistry, some specialized material is condensed into appendices for ease of review. The appendices found in this book include: Appendix A: The Metric System, Appendix B: Exponentials and Scientific Notation, and Appendix C: Equations and the Mathematics of Chemistry.

SUPPLEMENTAL MATERIALS

1. *Instructor's Manual*. The Instructor's Manual provides details about this book and suggestions for using it effectively. Answers are supplied for the even numbered Problems and Questions and also some suggestions for the even numbered Critical Thinking Problems. The printed Test Item File, a list of color transparencies, and additional transparency masters are also included.
2. *Student Study Guide/Solutions Manual*. The Student Study Guide presents students with a variety of exercises, self-tests, and explanations to guide comprehension of the material covered in the text. Answers to the odd-numbered end-of-chapter questions are also included in the Student Study Guide.
3. *Transparencies*. A set of 100 color transparencies is available to help instructors coordinate lectures with key illustrations and tables from the text.
4. *Customized Transparency Service*. For adopters interested in receiving acetates of text figures not included in the standard transparency package or transparency masters, a select number of acetates will be custom-made upon request. Contact your local Wm. C. Brown Publishers sales representative for more information.
5. *MicroTest*. This computerized classroom management system/service contains a database of text questions, reproducible student self-quizzes, and a grade-recording program. Disks are available for DOS, Windows, and Macintosh platform and require no programming ability. If a computer is not available, instructors can choose questions from the Test Item File, and phone or FAX in their request for a printed exam, which will be returned within 48 hours.
6. *Exploring Chemistry in Today's World*. Written by Kathy L. Tyner of Southwestern Community College, this laboratory manual provides experiments suitable for use when a laboratory course is associated with the *Chemistry and Our World* textbook.
7. *Laboratory Manual Instructor's Guide*. This helpful prep guide contains the hints the author has learned over the years to ensure success in the laboratory.
8. *How to Study Science*. Written by Fred Drewes of Suffolk County Community College, this excellent workbook offers students helpful suggestions for meeting the considerable challenges of a science course. Tips are offered on how to take notes and how to get the most out of laboratories, as well as how to overcome science anxiety. The book's unique design helps to stir critical-thinking skills, while facilitating careful note taking on the part of the student.
9. *Doing Chemistry Videodisc*. This critically acclaimed image database contains 136 experiments and demonstrations. It can be used as a pre-lab demonstration of equipment setup, laboratory techniques, and safety precautions. It can also be used as a substitute for lab experiences for which time or equipment is not available. Contact your local Wm. C. Brown Publishers sales representative for more details.
10. *Video Tapes*. Narrated by Ken Hughes of the University of Wisconsin-Oshkosh, the tapes provide six hours of laboratory demonstrations. Many of the demonstrations are of high interest experiments, too expensive or dangerous to be performed in the typical freshman laboratory. Contact your local Wm. C. Brown Publishers sales representative for more details.
11. *Redox Videodisc*. Produced by the American Chemical Society. Redox reactions are among the most important and ubiquitous in chemistry. This unique videodisc is intended to provide students with images of redox reactions and lab equipment that are not usually available for hands-on experiments.

Acknowledgments

A portion of the author's royalties will be donated to the *Amazon Basin Benevolent Association, Inc. (A.B.B.A.)*, a non-profit organization. A.B.B.A.'s goal is to establish floating hospital/dental facilities that will provide low cost (or free) medical and dental care to the natives of the upper Amazon Basin, mainly in Peru. Further details about A.B.B.A. can be obtained by contacting the President/Director Carl E. Schell at the Amazon Basin Benevolent Association, Inc., 417 Canal Street, New Smyrna Beach, Florida, 32168.

I would be seriously remiss if I did not express my thanks to the editorial and production staff at Wm. C. Brown Publishers, without whom this book would still be in type-written pages on my shelf. Special thanks go to Craig Marty, who believed in this project in the early days. Thanks are also due to my cadre of Developmental Editors: Elizabeth Sievers, Robert Fenchel, John Berns (who handled most of the development), Daryl Bruflodt, and Brittany Rossman. Julie A. Kennedy expertly oversaw the production process and kept me up to date and on track throughout the book's last eight months. Additionally, I thank the reviewers who read the manuscript in its varying stages of readiness. Their input has been of enormous value to me, and I thank each one of them:

Raymond F. Bogucki, *University of Hartford*
Sheldon H. Cohen, *Washburn University*
John C. Ford, *Indiana University of Pennsylvania*
Brian R. McGuire, *Northeast Missouri State University*
Donald Halenz, *Pacific Union College*
Larry Kirk, *California State University*
Robert A. Demers, *Massasoit Community College*
George F. Uhlig, *College of Eastern Utah*
William Huggins, *Pensacola Junior College*
Richard W. Cordell, *Heidelberg College*
Richard F. Jones, *Sinclair Community College*
Ralph Shaw, *Southeastern Louisiana University*
Sal Russo, *Western Washington University*
Joanna H. Fribush, *North Adams State College*
Sharmaine S. Cady, *East Stroudsburg University*
Linda Munchausen, *Southeastern Louisiana University*
Elizabeth Wallace, *Western Oklahoma State College*
John B. Holden, Jr, *Mankato State University*
Russell G. Baughman, *Northeast Missouri State University*
Jerry Easdon, *College of the Ozarks*
Rhonda Scott-Ennis, *University of Wisconsin-River Falls*
Mary Johnson, *Loras College*
Lynne Cannon, *University of Iowa*
Richard Treptow, *Chicago State University*
Vaughan Pultz, *Northeast Missouri State University*

Charles G. Gebelein

To the Student

This textbook was written for non-science students. The emphasis is more on the applied, or everyday, use of chemistry than what is typically found in chemistry textbooks. The aim of this is to *enlighten you*, not convert you to the sciences. All people in our society need to understand the foundations of science, not just scientists.

Why is this important? There is little doubt that our society has become increasingly technology-oriented, and that chemistry plays a major role in this development. Even so, approximately 94 percent of today's college students take no chemistry courses. The perception most non-science students have of chemistry is usually a combination of the following:

1. Chemistry is too difficult.
2. Chemistry requires too much math.
3. Chemistry is uninteresting—even boring!
4. Chemistry has little or no relevance to my life or career.
5. Chemistry is, at best, dangerous or maybe even hostile to people.
6. Chemistry has real value only to scientists.

If you relate to the above list, this book is for you! During your studies of chemistry you may find chemistry to be relevant, useful, and possibly even interesting. You will discover that while some simple math (actually arithmetic) is needed, chemistry need not be too mathematical, and can even be fun.

This book will not make you a chemist, or even any type of scientist. That's not the intention. But it can help you become a better informed person who can contribute to society and pursue a successful career in your chosen field. Since everything in life and business involves chemistry, you will be better prepared for today and tomorrow's world.

Here are a few helpful, general suggestions on how to successfully negotiate the raging rapids of chemistry. The best way to learn chemistry, or just about any subject for that matter, is to read the text material *prior* to its presentation in class. Then take notes in class, paying special attention to what your instructor emphasizes. Most instructors base their examinations on those topics on which they focus in class. Answer all the assigned problems. If you do not know how to arrive at the answers, ask for assistance. Additionally, you might purchase the Student Study Guide and work at learning the solutions yourself. This will eventually become your pattern for the remainder of your life. If previous examinations are posted, be certain to practice with them, but don't make these previous exams your sole source of learning. Remember that few instructors will directly copy their old exams to create the new ones.

Good luck in your chemistry course! May your academic pursuits be fruitful, and meaningful to your journey through life!

Charles G. Gebelein

C H A P T E R 1

INTRODUCTION

OUTLINE

Food for Thought—How Chemistry Affects Human Life
Science and Technology
The Risk Dilemma
The Scientific Method
Scientific Measurements

BOX 1.1 Tupper: A Chemist and an Entrepreneur
BOX 1.2 Risking the Cure
BOX 1.3 Useful and Useless Hypotheses
BOX 1.4 The Dynamic Nature of Theory
BOX 1.5 Cold Fusion Confusion

KEY IDEAS

1. The Nature of Chemistry
2. The Difference Between Science and Technology
3. Examples of Benefits that Come from Chemistry
4. How Scientists Balance Risk Against Benefits
5. How the Scientific Method Works
6. Scientific Measurements

"All our knowledge has its origins in our perceptions."
LEONARDO DA VINCI (1452–1519)

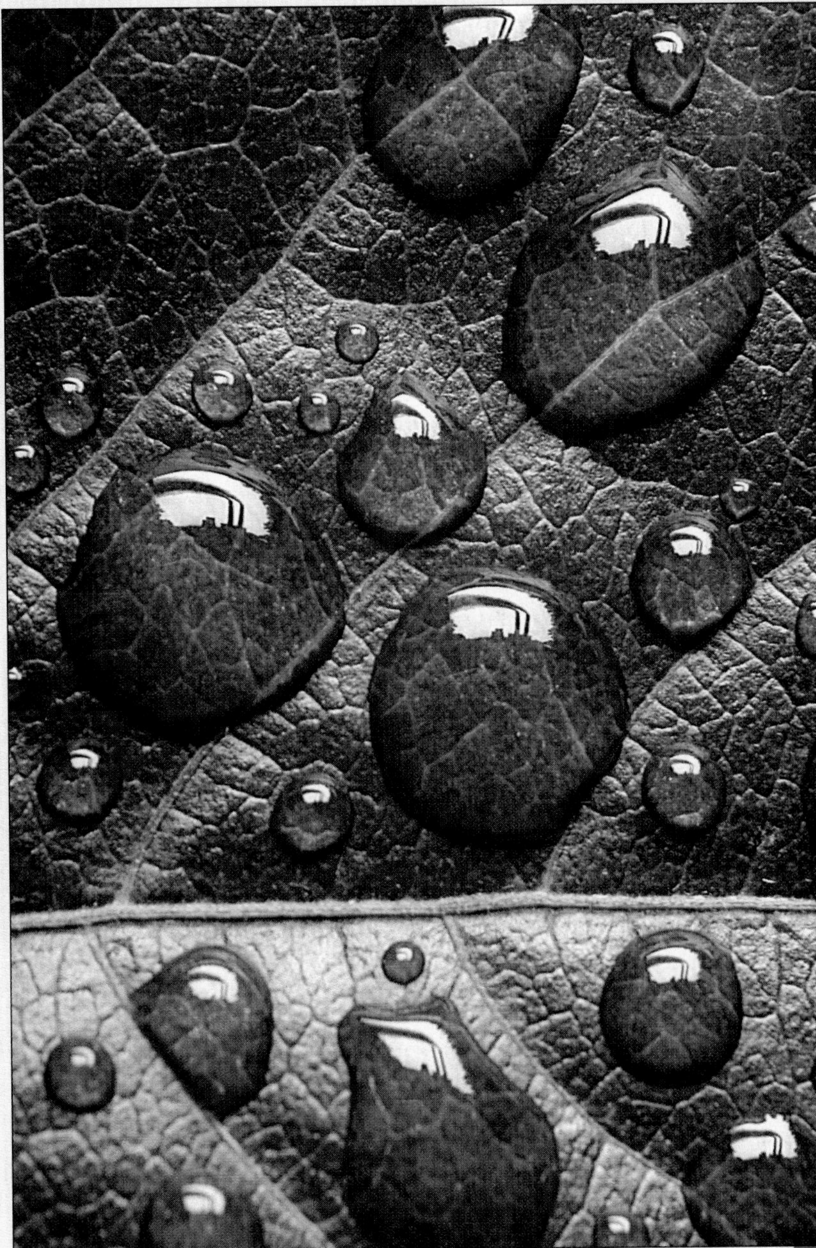

Smokestacks belching smoke, reflected in raindrops on a leaf.

Science is the systematic branch of knowledge wherein people develop, organize, and confirm facts or theories by observation, testing, and experimentation. In short, science is the systematic study of the physical world. **Chemistry** is the branch of science that studies matter, energy, and their interactions. The most interesting aspects of chemistry often apply to our daily lives. This chapter will discuss how chemistry affects our lives and will examine the thought process we apply to chemistry and to science in general: the scientific method.

FOOD FOR THOUGHT—HOW CHEMISTRY AFFECTS HUMAN LIFE

Chemicals are the materials that make up the physical world. Everything you encounter is composed of chemicals, including clothes, soaps, meats, cosmetics, vegetables, eggs, paper, pens, books, and light bulbs. For example, an egg is a complex collection of chemicals. A hard shell, a colorless egg white, and a bright yellow yolk comprise the egg. When we examine the egg under a microscope, we can see additional details. And if we examine the egg *chemically*, we discover each part contains many different kinds of materials.

Chemistry affects basic life functions, new technologies, and products that enhance and even save human lives. Everyone breathes and eats; these familiar actions involve chemistry. When we breathe, a mixture of oxygen and nitrogen dissolves in our blood and moves throughout our body. As the oxygen travels to the various parts of our bodies, it enters individual cells, where it becomes involved in a series of chemical reactions. Carbon dioxide leaves these cells to return through the blood to the lungs and eventually reenters the air. When we eat, we nourish our bodies with food. Most people know that part of this food gives us energy and another portion converts into substances our body can store. Both of these functions, like breathing, involve chemical substances and chemical reactions. Thus, chemistry and chemical reactions form the basis of life.

Even familiar actions such as breathing and eating involve chemistry.

However, chemistry is not limited to life functions. Computers, television, radios, recorders, video games, medical monitors, and countless other devices in our society owe their existence to chemical advances—for example, to the ability to prepare highly purified silicon and germanium. Many of these advances occurred relatively recently; TV traces its brief history back to the mid-1930s, and radio to the early years of this century. The original equipment was bulky and used huge vacuum tubes and large resistors and transformers laboriously soldered onto circuit boards. Today's radios are more sensitive, with circuitry smaller than a dime; small color TV sets can fit into pockets or purses. In 1950, computers occupied hundreds of cubic feet and took hours to perform tasks that portable laptops can now perform in seconds. All these improvements depend, at least in part, on new discoveries and refinements in the field of chemistry.

Future innovations may harness solar power via high-efficiency solar cells that will replace today's less efficient devices. One solar cell innovation can split water into oxygen and hydrogen. This could eventually provide a plentiful, pollution-free fuel for automobiles and heating, yielding only water as a waste product. Chemists and biotechnologists are also developing new sources of petroleum-like fuels from renewable resources.

The cosmetic and personal hygiene product industries, which produce combined annual U.S. sales greater than $60 billion, owe their existence to chemistry. Beauty may be only skin deep, but

Many modern devices owe their invention and refinement at least in part to chemical advances.

Future chemical advances may help provide alternate fuels.

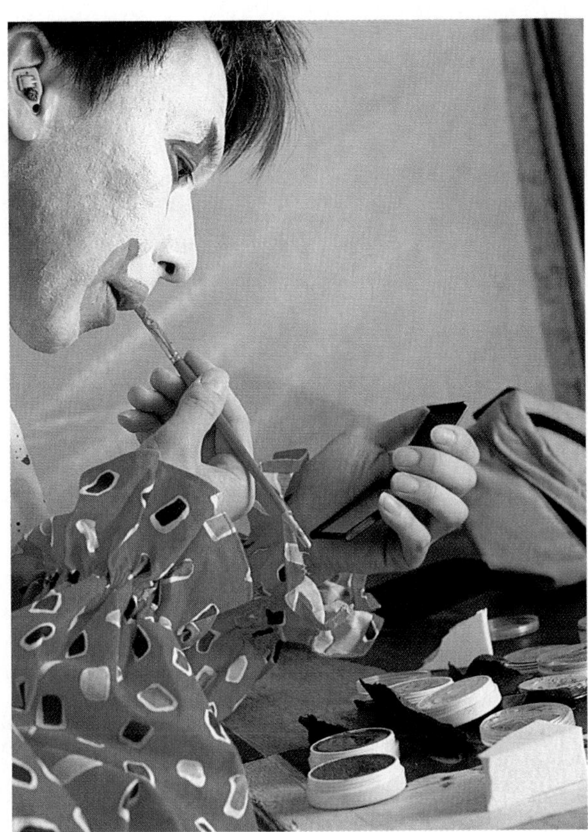

The profitable cosmetic and personal hygiene industries depend on chemistry for their very existence.

the American public evidently believes that the psychic benefits of an improved appearance are enormous. The garment industry also depends on chemistry to enhance our lives. In fact, our entire society and lifestyle depend on chemicals.

But beyond enhancing human life, chemistry actually extends and even saves human life. At the turn of the twentieth century, infectious diseases such as polio, influenza, diphtheria, tuberculosis, and smallpox caused 70 percent of all deaths in the United States. Today, only 2 to 5 percent of U.S. deaths are caused by infectious diseases. Because they developed *chemical* anti-infectious drugs, and because they studied the health effects of sanitation and nutrition, chemists and other scientists helped facilitate this dramatic reversal. Currently, the major causes of death are cancer, cardiovascular disease, and diabetes. Medical practitioners have made major advances in treating these ailments during the past twenty years, including administering new chemical drug agents and using new medical devices made from plastics, metals, or ceramics—all chemical substances. New chemical advances will undoubtedly continue to help in the fight against human disease.

SCIENCE AND TECHNOLOGY

We live in a technological society, and we encounter the products of technology daily. All of these technologies developed because of breakthroughs in scientific research. How can we differentiate between technology and science? **Science** develops a body of knowledge by organizing and verifying facts and theories through observations, tests, and experiments. **Technology,** or applied science, applies that body of scientific knowledge to some practical use.

Examples of applied science, or technology, abound in modern society: television, computers, plastics, the petroleum industry, automobiles, and electric power are integral to our everyday lives. In each case, scientific knowledge allowed the invention of new devices; many involve chemistry in a variety of ways (box 1.1).

Examples of pure science are just as numerous, but less familiar to most of us. Many scientists working in "pure science" (science that has no specific practical application) perform research mainly to satisfy their curiosity and to acquire new knowledge. "Pure" chemistry led scientists to analyze the chemical structure of water, develop atomic theory, and construct the periodic table. Of course, this research may lend itself to practical applications later: If the chemical analysis of water enabled a chemist to deter-

BOX 1.1

TUPPER: A CHEMIST AND AN ENTREPRENEUR

Plastics are such common materials that we most likely take them for granted. But chemistry enabled the invention of plastics, and chemistry helped innovators fit plastics to new uses. One of the best known plastic items in our society—Tupperware® containers—were the brainchild of a self-taught chemist.

Earl Tupper (figure 1) was born to a poor family in Berlin, New Hampshire. He graduated from high school in 1926, and almost everything else he learned was self-taught. Tupper encountered plastics before World War II while working as a chemist in a DuPont chemical plant. He wanted to experiment with some of these new materials and develop new uses for them. Unable to purchase plastics once the war started, he obtained a black, hard, smelly sample of polyethylene waste material and developed a process to purify it. Next, he created an injection molding machine to make new molded items.

Figure 1 Earl Tupper, the inventor of Tupperware.

At the time Tupper was working with his invention, plastic, as a practical, commercially useful material, was just in its infancy. The few plastic products on the market were of poor quality and broke easily. Tupper saw a practical use for his plastics; he could produce plastic ware that would be more flexible and less breakable than current products. Tupper's other innovation was his virtually airtight seal.

In 1946, Tupper began selling his first kitchen products using the new polyethylene material. However, the Tupper products often languished on store shelves because consumers were unaware of how to work the seal. Eventually, Brownie Wise, one of the first Tupperware distributors, developed the idea of having the now-famous Tupperware parties to demonstrate the superior features of Tupper's plastic.

The business became an instant success and reached sales of $25,000,000 by 1954. Tupper sold his company in 1958 and retired to Costa Rica, where he died in 1983. But the products inspired by his chemical experimentation still fill kitchen cupboards around the world.

mine whether our drinking water is safe, he or she would be applying that knowledge to a practical purpose. Still, new and important information that could lead to a momentous technological development could theoretically remain hidden in a dusty notebook for decades. Happily, most scientists publish their research results in journals, making this information accessible to others hoping to fulfill specific tasks.

Sometimes the line between science and technology blurs. The rapidly emerging area of biotechnology poses many problems in distinguishing between science and technology. As its name implies, biotechnology applies biological knowledge to meet specific technological challenges. But this multifaceted discipline has led to an abundance of new scientific information. Similarly, industrial scientists may glean new information in their quest to develop some commercial application. This information, which could be categorized as pure science, is often published in journals specializing in pure science for the benefit of other researchers. Many scientists work in biotechnology, medicine, and other fields both to satisfy their curiosity *and* to develop practical new products or techniques.

PRACTICE PROBLEM 1.1

Which of the following represent pure science? Which are technology, or applied chemistry?

- **a.** Constructing the periodic table
- **b.** Creating color photographic pictures
- **c.** Determining the chemical composition of water
- **d.** Developing a drug to treat cancer
- **e.** Determining the structure of DNA molecules
- **f.** Creating a new plastic suited for use in automobile bodies

ANSWER Examples *a, c,* and *e* are essentially science for its own sake, or pure science. Questions *b, d,* and *f* are goal oriented, or technological. It is interesting to note that a scientist working on *e* might discover knowledge useful to a research chemist working on *d*. DNA is the substance that makes up genes, and some cancers arise from DNA abnormalities. Anticancer drugs often come from chemicals very similar to DNA material.

THE RISK DILEMMA

Although technology can improve our lives, some of the byproducts of technology can be annoying or even harmful. Further technological advances may improve or solve the problem, but these advances often involve substantial costs. For example, consider video games. The process that produces the materials needed to make microchips for the games (or for any computer) also produces enormous quantities of hydrochloric acid as a waste product (figure 1.1). Some of this waste can be refined for use in other applications, but a microchip producer is unlikely to want to take the extra steps needed to produce quality hydrochloric acid; it is simpler and less costly to treat it as waste. Assuming another company does upgrade this acid, they still must ship the crude chemical on highways or railroads and risk spills. About five billion pounds of hydrochloric acid are produced annually for use in food processing, chemical intermediates, swimming pool treatments, metal cleaning, and ore reduction. How can we weigh the benefits of such technologies and products against the risks of producing huge quantities of hydrochloric acid?

Figure 1.1 The risk dilemma. Does the enjoyment these games generate warrant the risks of large amounts of hydrochloric acid during microchip manufacture?

PRACTICE PROBLEM 1.2

Decide whether each of the following chemical items provides a benefit, causes a problem, or both. If possible, give a reason for your answer.

- **a.** Herbicides (for killing weeds)
- **b.** A drug to treat AIDS
- **c.** Microwaved food
- **d.** Radiation-sterilized food
- **e.** Moldy or rancid food
- **f.** Food additives that prevent mold or rancidity
- **g.** Bottled spring water
- **h.** Soft drinks that contain an artificial sweetener

ANSWER Most scientists agree that items *b* and *f* provide benefits. The rest are at least somewhat controversial; for example, *a* and *h* might be considered to provide benefits but they also may cause problems. Most people use *c* or *g* without fear, but no one is fond of *e*, with the exception of cheese. Item *d* remains controversial—though the process offers proven benefits, the public is still skeptical.

Benefit–Risk Ratios

One way to try to balance risks and benefits is to assign numerical values to the risks and benefits of each chemical, process, and waste. Take hydrochloric acid as an example. This chemical is harmful when ingested, which may seem surprising since it occurs naturally in the stomach; it is also irritating to the skin and eyes and is corrosive. Still, few people incur life-threatening injuries from this acid, and hardly anyone dies from it. Conservatively, the danger of death or severe injury from hydrochloric acid exposure is estimated to be less than 1 in 1,000,000; this means that if 1,000,000 people are exposed, 999,999 are likely to suffer only mild effects or none at all. Therefore, hydrochloric acid's risk factor to humans is 1 in 1,000,000. On the other hand, most people receive some benefit from the applications of this chemical. If we conservatively estimate that 80 percent of the population benefits, 800,000 out of every 1,000,000 would benefit from the technologies that use hydrochloric acid.

The **benefit–risk ratio** compares these factors to determine whether benefits outweigh risks. The larger the ratio, the more benefits and the fewer risks the material or process is likely to have. Assuming our estimates for hydrochloric acid were reasonable, the benefits far outweigh the risks (keep in mind that we are considering only direct benefits and risks to humans, not other environmental considerations):

$$\frac{\textbf{benefit}}{\textbf{risk}} = \frac{800{,}000}{1} = 800{,}000 \textbf{ benefit–risk ratio}$$

Of course, not all chemicals have high benefit–risk values. Chlorofluorocarbons (or CFCs) seldom directly cause death or severe injury, but they are presumed to be the major culprits in destroying the ozone layer. (This is discussed in chapter 19.) Ozone loss is believed by some to produce an increase in the incidence of skin cancer, a potentially fatal disease. Risk calculation for CFCs is difficult to estimate because no one is certain how many additional people might contract skin cancer if the ozone layer continued to vanish. Some scientists estimate a risk factor of 15 percent, or 150,000 people per 1,000,000. On the other hand, CFCs have many benefits because they are used in air conditioning and refrigeration. Scientists estimate the benefit factor at 80 percent, or 800,000 people per 1,000,000. We can now determine an estimated benefit–risk ratio for chlorofluorocarbons:

$$\frac{\textbf{benefit}}{\textbf{risk}} = \frac{800{,}000}{150{,}000} = 5.3 \textbf{ benefit–risk ratio}$$

These benefit–risk numbers suggest that it is far more risky to use CFCs than hydrochloric acid. Calculations can aid society in decision making, but only people can make the final decision. Despite the low benefit–risk ratio, people might decide that CFCs are necessary for refrigeration unless an effective substitute exists. The decision requires us to balance the presumed skin cancer danger against potential illness from food spoilage and a variety of economic and sociological factors. Sometimes, as box 1.2 indicates, it is exceedingly difficult to balance a complex array of benefits and risks.

THE SCIENTIFIC METHOD

Just as benefit–risk ratios give us a systematic method for evaluating the effects of chemicals and processes, the **scientific method** gives scientists a systematic approach to gathering scientific data, testing that data, and drawing conclusions from it. The scientific method is important because it underpins all scientific research and the way we interpret that research. Other disciplines, such as business and philosophy, also make use of the scientific method.

Sherlock Holmes's famed deductive powers are one example of the scientific method in action. The scientific method takes

Sherlock Holmes (Basil Rathbone): master of the deductive scientific method.

BOX 1.2

RISKING THE CURE

Although the concept of a benefit-risk ratio is useful, it is sometimes very difficult to quantify the complex factors that affect benefits and risks. For example, consider the decision whether to use DDT (dichloro-diphenyl-trichloroethane), a potent, broad-spectrum insecticide that can kill many different types of insects (figure 1). Pesticides have saved countless human

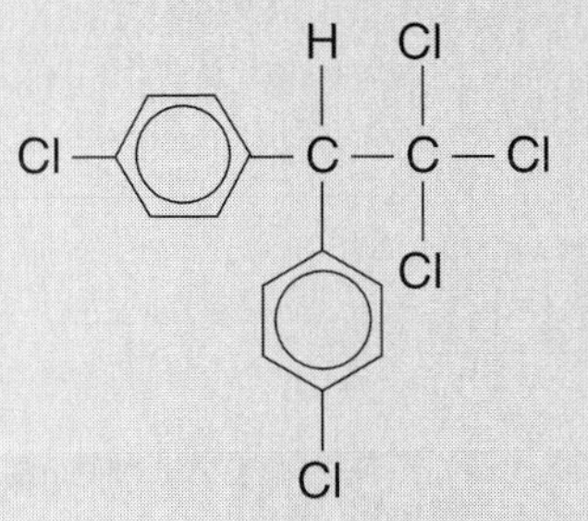

Figure 1 The structure of DDT.

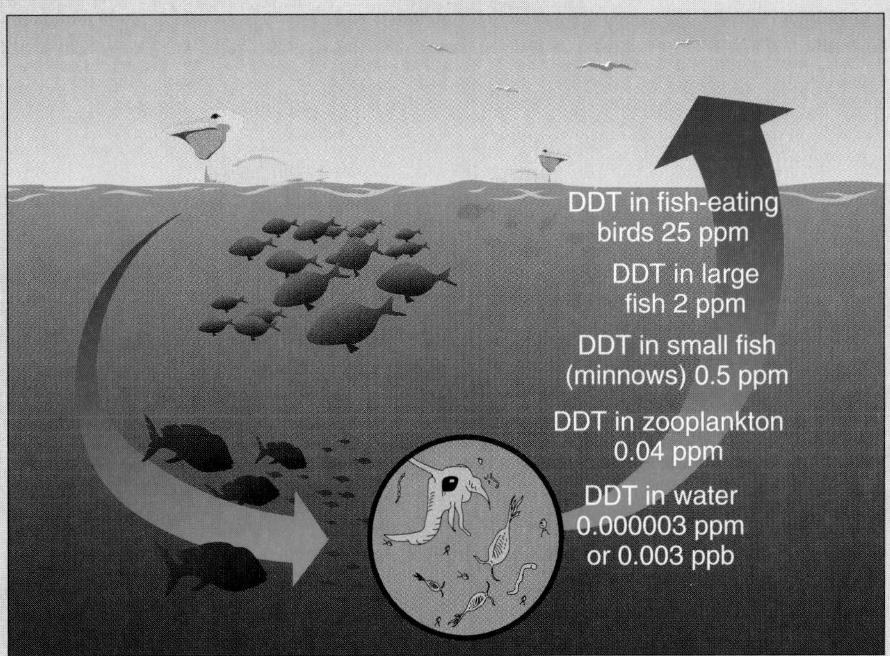

Figure 2 How DDT enters the food chain. Note that the levels of DDT increase as the food chain proceeds upward.

lives; DDT alone, according to World Health Organization estimates, has saved at least 25 million people. DDT was invented by Paul Müller (Swiss, 1899–1965) in 1939; he received the 1948 Nobel Prize in Physiology and Medicine for his research. After testing during World War II, the insecticide won prompt acceptance, and entered the consumer market after the war. It proved exceedingly effective against disease-carrying insects.

However, problems developed as DDT became widely used. DDT enters the food chain when fish eat insects that contain traces of DDT. The insecticide deposits in the fatty tissues of the fish. When bigger fish eat the smaller, DDT-infested fish, the DDT concentrates in even larger amounts in the larger fish. Finally birds, animals, or people eat these larger fish, and the DDT deposits in significant amounts in their fatty tissues (figure 2).

DDT is apparently harmless in the short run to humans, but high DDT levels affect calcium adsorption in certain birds. Poor calcium adsorption causes their eggs to develop thin, breakable shells, decreasing the number of eggs that produce offspring. Eventually, several bird species faced extinction due to the widespread use of DDT, including the bald eagle and the peregrine falcon. This raised questions about the risks of using DDT: No one knew whether similar problems might eventually affect humans—or whether DDT might cause birth defects, cancer, or other diseases if deposits remained in the human body for long periods of time.

To compound the problem, scientists discovered that DDT does not biodegrade quickly. This was initially considered an advantage, but later appeared to be problematic. DDT has an environmental half-life of about fifteen years; this means that more than 6 percent of the DDT used in a given year would still be around 60 years later. During this time, it could continue to enter the food chain and harm wildlife and, possibly, humans. Finally, the benefits DDT afforded began to wane as some insects developed resistance to the insecticide. The United States and most industrialized nations reluctantly banned the "miracle" insecticide in 1978.

Whether DDT is ultimately harmful to humans remains the subject of much debate and a little research. A brief, preliminary study reported early in 1993 (M.S. Wolff, *J. National Cancer Institute*, April 1993), claims to link DDT with breast cancer. However, this study followed only 58 women with breast cancer and 171 women without breast cancer as controls. Generally, medical studies require many more cases. Moreover, many previous studies failed to find a link between DDT and any specific form of cancer. More research might help resolve the DDT dilemma.

We are left with a question—do the benefits of using DDT outweigh the risks?

(Continued)

BOX 1.2—Continued

The *anopheles* mosquito, a tropical insect that carries malaria-causing parasites, has infected millions of people with fatal malaria over the past century, and DDT is the most effective insecticide for mosquito eradication. In Sri Lanka alone, the use of DDT reduced the number of malaria cases from about 2.8 million in 1946 to only 17 cases in 1963! After officials ceased spraying DDT in 1964 in the belief that malaria was eradicated, the incidence of the disease rose to over a million cases by 1968.

All of these factors must be weighed in any DDT benefit–risk equation. If large numbers of people are dying from malaria, is DDT use justified? DDT is still used to control this deadly disease in most Third World countries. Imagine you are a citizen in a country where malaria is prevalent. What would be your verdict on DDT?

logical steps that help scientists avoid faulty conclusions; these steps include observing, predicting, testing, and concluding. When any one of these steps is omitted, the result may be inaccurate interpretations.

To discuss the scientific method in a coherent manner, we must define the terms *facts, data, laws, hypotheses, theories,* and *experiments*. Some of these terms have a slightly different connotation in science than in ordinary conversation.

Facts and Data

The term *fact* is common in daily discourse. In science, a **fact** is a statement that scientists can demonstrate to be true. If an observation cannot be verified, it is not considered factual. Sometimes scientists assume that an unproven idea is true for the purposes of further research; thus, scientists assumed the existence of atoms many decades before they could actually observe individual atoms. Thus, not all current scientific information is rigorously true. Some information is proven as facts; some of it scientists treat as fact because it fits into the general framework of scientific knowledge. **Data** is the raw information scientists gather through observation and measurement; as such, data are considered factual.

Laws

In science, a **law** describes a set of events that have been observed to occur with unvarying uniformity. A law simply describes what always happens—it makes no assumptions about *why* an event occurs. Scientific laws are assumed to hold true everywhere in the universe; the law of gravity is a familiar example. We observe this law operating invariably in our daily existence: rulers fall to the earth, stones fall to the earth, books fall to the earth, and so forth. Our presumption is that all physical objects are pulled toward the earth unless something counteracts the force of gravity. (A helium-filled balloon rises because helium is lighter than air—but the helium balloon is still subject to gravity.) Scientists are confident the law of gravity holds true in places they have never visited. In the Star Trek television programs and films, when the crew of the Starship *Enterprise* boldly travels to the outermost reaches of the universe, they expect the law of gravity to apply (figure 1.2).

It is important for you to make a clear distinction between a scientific law and a legislated law, such as a traffic speed limit. A sign along a roadway may notify you of a specific speed limit law, but you can easily violate this law if you choose to. In contrast, you cannot violate the law of gravity at will. Even if you yearned to walk from a second story window and tread empty space, it is not possible without using some other force or expenditure of energy to counteract gravity. A scientific law describes immutable phenomena and has nothing to do with people's desires or beliefs.

Figure 1.2 Star Trek's *Enterprise* boldly goes where scientific laws remain the same.

BOX 1.3

USEFUL AND USELESS HYPOTHESES

To be of real value, a hypothesis must make testable predictions. This point was illustrated by American physicist P. W. Bridgman (1882–1961). In the late 1920s, Bridgman defined the distinction between useful and useless hypotheses. He won the Nobel Prize in 1946 for some of his research.

To understand the concept of a useless hypothesis, consider a problem Bridgman posed. Suppose you have formulated a hypothesis that the entire universe has doubled in size since 1920, but you are unable to test this hypothesis because all your rulers (being part of the universe) have doubled in size. A book measuring 6 inches in length before 1920 is now 12 inches long, but since your 12-inch ruler is now really 24 inches long, it still measures the 12-inch book as half its length. Consequently, you still find the length of the book to be 6 inches. The basic assumptions allow no way to confirm or refute the hypothesis. Such hypotheses do emerge in science, but they ultimately fade into limbo because they are useless.

Hypotheses

While a scientific law describes observations but makes no assumptions about those observations, a **hypothesis** is a tentative assumption that fits or explains a narrow set of observations or events. Scientists formulate hypotheses as a basis for further reasoning or investigation; they then design experiments to test their hypotheses. A well-designed test or experiment will help determine whether a hypothesis is consistent with known facts and whether it makes sense in explaining how (or why) these facts fit together.

When a scientist first constructs a hypothesis, it is essentially an educated guess. If subsequent testing supports the guess, it becomes a useful hypothesis (box 1.3). Scientists can then proceed with further research on the assumption that the hypothesis is probably true. Nevertheless, a hypothesis by its very definition remains unproven.

Theories

A **theory** is also a potential explanation for a set of phenomena. A theory is similar to a hypothesis in that it involves assumptions, but it differs in that a theory is usually based on a broader range of information. In other words, a theory is an educated explanation for an integrated set of ideas or facts. An example is the *hypothesis* that humans evolved from nonhuman ancestors. This assumption is one part of the general *theory* of evolution, which assumes that all living organisms emerge from simpler life forms.

Like hypotheses, theories are working assumptions; they generally contain some unprovable elements. In spite of this shortcoming, theories are useful in helping us make sense of our observations of the physical world. As scientists test theories and gather data, they often modify their theories.

Good theories and hypotheses tie together lots of information, facts, and ideas, enabling scientists to arrive at new conclusions and to add to our scientific knowledge. Tests and experiments either disprove theories, because the theory contradicts known facts, or lend support to them. When a theory is untestable, scientists have reservations abut its validity. A scientist believes you cannot determine whether a theory is really true if you cannot test it.

Experiments

An **experiment** is an operation conducted under controlled conditions that is intended to test a hypothesis or observation. Experiments frequently yield new and valuable data that help scientists refine hypotheses or develop new ones. Many industrial or consumer products were developed after considerable experimentation. For example, computers, video games, and other electronic devices owe their existence to the development of transistors and semiconductors. These inventions came about because scientists tested and retested their theories about how matter works.

General Features of the Scientific Method

The scientific method integrates facts, data, observations, hypotheses, theories, laws, predictions, and experiments (figure 1.3). It is a cyclic process that may never reach a definite endpoint. The scientific method normally starts when a scientist accumulates *facts, data,* or *observations* that arouse his or her curiosity. If these observations and facts invariably hold true, the scientist may formulate a scientific law to describe them. This law would make no assumptions about *why* these events always occur;

it would simply describe them as invariably true until further tests prove otherwise. On the other hand, if the scientist wants to tie together a wide variety of facts or observations and make assumptions about *why* they occur, he or she devises a theory. Finally, if the scientist wants to focus on a narrow range of observations and make assumptions about why they occur, he or she formulates a hypothesis. The scientist then makes predictions based on his or her hypothesis or theory.

Once the scientist devises a theory or proposes a hypothesis, he or she must try to design *experiments* that will test the hypothesis or theory and confirm or disprove the predictions from it. If the experiments disprove or only partially confirm the predictions, the scientist must revise the hypothesis or theory, or propose a new one and test further. If the experiments confirm the predictions, the scientist may accept the hypothesis or theory unless or until later tests and data challenge them. This cycle of observing, predicting, testing, and interpreting continues as the hypothesis or theory is refined.

Thus, even when the scientist proposes a hypothesis or theory, the cycle of scientific inquiry continues. Scientists make predictions based on their hypothesis or theory, test their predic-

EXPERIENCE THIS

Design experiments to test each of the following hypotheses.

1. An old adage hypothesizes that a watched pot never boils. Is this statement scientifically accurate? What type of experiment might help you test its validity?
2. Some people believe they have a good chance to become rich playing the lottery. How might you test the validity of this assumption? What types of data might be useful to you?
3. An old tradition holds that woolly bears, a kind of caterpillar, forecast the harshness of a coming winter—the harsher the winter, the wider the caterpillar's dark brown band. How can you test this hypothesis? How long would it take to gather reliable data?
4. Astrologists claim they can predict future events from the relative positions of stars. How can you test this claim?

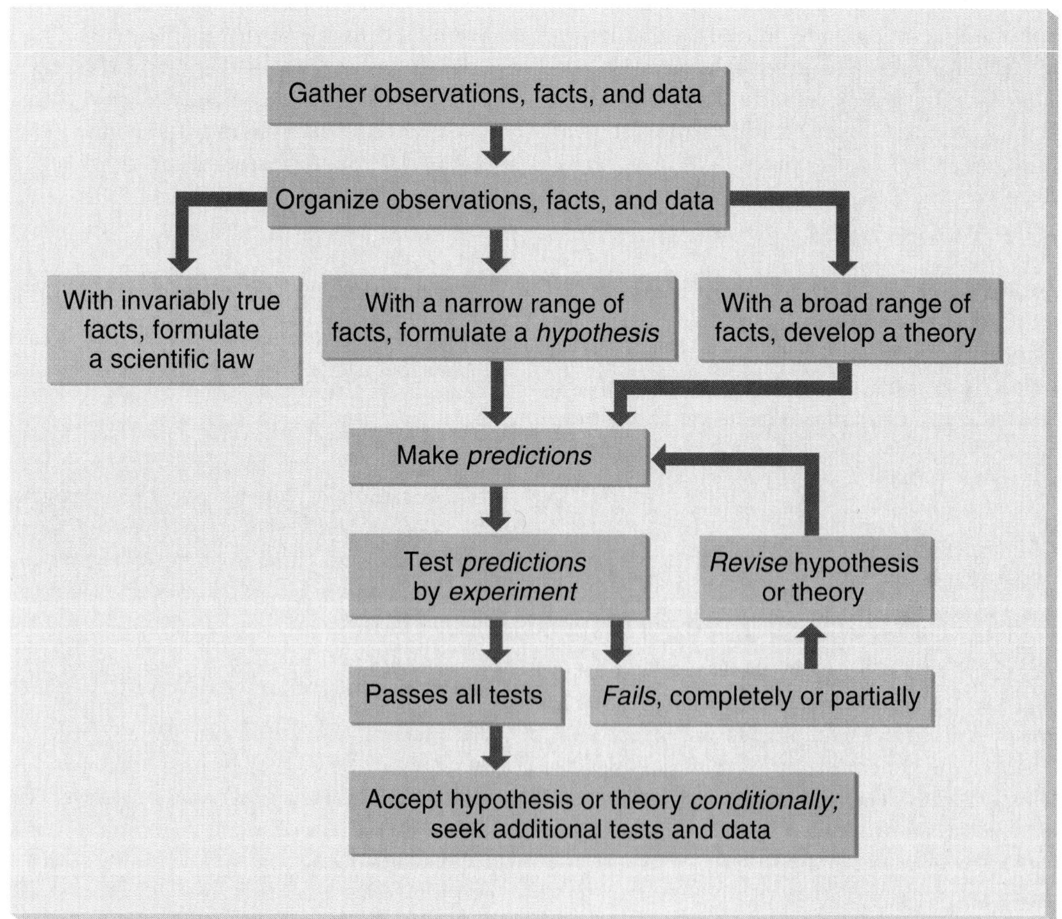

Figure 1.3 An outline of the scientific method.

tions, and confirm or revise their assumptions. This continues in a virtually endless cycle as new scientific knowledge accumulates (box 1.4).

The Scientific Method in Action

You can best learn to use the scientific method by applying it to an everyday occurrence—the evaporation of water. Why does water evaporate? Let's attempt to answer this question by making observations on water evaporation and then setting up hypotheses to explain what is happening. We will follow the pattern outlined in figure 1.3 as closely as possible.

Imagine we place a specific volume of water in an open container. After a few days, we observe that the container holds less water. We have now made a key observation: water vanishes from an open container. We repeat this procedure and observe the

BOX 1.4

THE DYNAMIC NATURE OF THEORY

Once a hypothesis or a theory generates predictions, scientists devise tests to determine whether these predictions are valid. Many times, this takes years; in fact, the testing and refining of a theory is a continuing procedure. A good example of this continuous process is the evolution of gravitational theory. The theory underwent considerable change from Newton's law to Einstein's theories—and it continues to undergo investigation and refinement today.

Isaac Newton (English, 1642–1727) proposed the law of gravity over time, beginning in 1665 and ending in 1687. Newton claimed the inception of the idea came when he watched an apple fall to the ground (figure 1); some scientists feel Newton invented the story many years later. In any event, the law explains the interactions of matter and a force called gravity.

Figure 1 Did Isaac Newton really formulate his law of gravity after observing the fall of an apple from a tree?

Over the next two centuries, scientists continually tested and confirmed Newton's theory. Eventually, between 1905 and 1915, Albert Einstein formulated his gravitational theories, usually called the general and special theories of relativity. Einstein's theory claimed that light is affected by gravity, while Newton's theory did not consider this possibility. Einstein's theory also predicted different planetary orbits than Newton's. Many years passed before scientists devised and ran enough tests to give them confidence in Einstein's theory; it is still tested today during eclipses, with space probes, and with complicated mathematical theorems. The theory of relativity has so far passed these tests and is still considered useful, but future predictions and test results may contradict it. (For example, the other major physical theory of the twentieth century—quantum theory—is not completely compatible with the theory of relativity.) As scientists perform more experiments and gain more data, their theories about gravitational forces will undoubtedly continue to change.

same result, so we decide to formulate some hypotheses to try to explain our observation.

Eventually we settle on two hypotheses. In our first hypothesis, we assume that all liquids, including water, consist of tiny spheres of liquid too small to see. (Remember, hypotheses are based on assumptions.) We further assume that these unseen balls are in constant motion in the container. Every once in a while, some of them actually "jump" out of the bulk of the liquid and enter the air. The fact that we do not normally see small spheres of water moving around and jumping out of a container is not relevant to our hypothesis; we *assume* it happens. We will call the tiny sphere model Hypothesis 1.

In our second hypothesis, we *assume* that the water disappears because invisible creatures drink it. We will further *assume* that these creatures behave like creatures we can see unless we discover some additional facts that contradict this. We will call the invisible creature model Hypothesis 2.

Both hypotheses explain the observation that water disappears from an open container. We might lean toward Hypothesis 2, since we have observed far more animals drinking water than tiny balls jumping out of a container; but before we arrive at any conclusions, we need to make predictions from each hypothesis and test our predictions.

After examining our hypotheses, we predict that placing the water in a closed container will prevent it from vanishing. We reason that the tiny spheres of Hypothesis 1 cannot move through the lid and that the invisible creatures of Hypothesis 2 cannot drink water from a sealed container. Accordingly, we set up an experiment to test our prediction: we fill a container with water and seal it shut. Days later, we discover the prediction was correct—water does not evaporate from a closed container. We now have a second observation or piece of data that seems to fit either hypothesis.

We decide to make another prediction: that increasing amounts of water will disappear from an open container over time. Again, this prediction might support either hypothesis—one tiny sphere of liquid after another might jump into the air until few are left, or an invisible creature might return again and again to quench its thirst. We test the prediction by leaving an open container of water out for several days, measuring the remaining volume of water each day; we find that, indeed, more and more water disappears as each day passes. So far, neither of our hypotheses has been contradicted by experiment, so we do not need to revise or discard either one.

We could continue the prediction-experiment-revision cycle indefinitely, testing the effects of temperature, surface area, and air movement over the surface of the liquid. Some experiments would support our predictions, lending strength to one or both hypotheses; others might contradict our assumptions and cause us to revise our original ideas. We might learn, for example, that water evaporates more rapidly when heated. This could lead us to modify both hypotheses—we might assume that heat makes the tiny spheres of Hypothesis 1 move faster, so that during a given time period, more would jump into the air; and we might assume that the invisible creatures of Hypothesis 2 prefer warm drinking water to cold. However, we might also now start to suspect that Hypothesis 2 may be wholly or partially invalid, since we know that other animals (as well as humans) generally prefer cool drinking water.

Eventually, we might decide to expand our experimentation to other liquids. Perhaps we hypothesize that evaporation is common to all liquids, and to test that prediction, we test open containers of pentane (a chemical found in gasoline) and ethanol (the alcohol in alcoholic beverages). Sure enough, they show the same general behavior as water, though they evaporate at different rates; pentane evaporates fastest, then ethanol, then water. Now we need to reexamine our hypotheses in light of this new data.

The new data concerning pentane and ethanol seem to fit well with Hypothesis 1. If liquids consist of tiny spheres in constant

motion, the pentane and ethanol spheres would likely "jump" into the air just as water spheres do. But our new data appear to contradict Hypothesis 2. Pentane is poisonous to virtually all animals. Is it plausible to assume our hypothetical invisible creatures would lap up this chemical more rapidly than water? It doesn't seem so. We may not want to discard Hypothesis 2 yet, but the newer data do not seem to support the idea that invisible creatures are responsible for evaporation.

Although Hypothesis 2 held up longer than you may have expected, with enough experimentation, we eventually accumulated data that began to call the hypothesis into question. This demonstrates the value of the scientific method—its cyclical nature, involving repeated testing and data gathering, eventually refutes and weeds out poorly conceived hypotheses and helps refine those that are viable. Box 1.5 discusses a recent application of the scientific method to a process known as cold fusion.

BOX 1.5

COLD FUSION CONFUSION

A recent example of a problem in which chemists are using the scientific method involves cold fusion. Fusion is a potential method for producing atomic energy by forcing atoms to combine into larger atoms; this is the process by which the sun generates its energy. Fission, or splitting atoms, is the other way of gaining energy from the atom. These methods are covered in more detail in chapter 5.

On March 23, 1989, Stanley Pons and Martin Fleishmann announced an astounding discovery: they claimed the fusion of hydrogen atoms to form helium atoms, at *room temperature*, in simple laboratory equipment (figure 1). If valid, this discovery would be among the greatest of the century, because fusion could produce enormous quantities of energy, freeing humanity from dependence on fossil fuels. Scientists greeted the claim with a combination of enthusiasm and caution because they wanted proof that Pons and Fleishmann had really discovered a completely new, almost pollution-free, energy-generating source.

Virtually all previous fusion research involved coming as close as possible to the very high temperatures of the sun, approximately 35,000,000 degrees Fahrenheit. Naturally this was a difficult feat; all known materials melt or are destroyed by such great heat. Scientists achieved limited success using strong magnetic fields to contain the atoms to be fused, but no one had been able to make the hot fusion process produce energy on a continuing basis.

Figure 1 Stanley Pons and Martin Fleishmann with their cold fusion apparatus.

After Pons's and Fleishmann's initial news release, scientists worldwide quickly attempted to reproduce their results. Several laboratories confirmed some of their claims, but other laboratories could not. The American Chemical Society arranged a special cold fusion symposium in Dallas, Texas which more than three thousand chemists, including the author, attended. The air was tinged with excitement as Pons explained their research. Other scientists presented supporting theoretical papers. Sudden news releases even arrived, stating that one laboratory or another had just confirmed the cold fusion research—probably the "hottest" was a telegram from chemists at the University of Moscow claiming that they were able to duplicate Pons's and Fleishmann's results. Many left the meeting feeling they were eyewitnesses to a major historical event.

Afterwards, some reports suggested the cold fusion process did not work nearly as well as claimed. Negative results began to accumulate, the initial enthusiasm waned, and skepticism and confusion set in. By mid-1994, the scientific community had divided into three camps on cold fusion: (1) those who avidly maintained the process worked; (2) those who disbelieved and were totally hostile; and (3) most scientists, who remained uncertain. Some, like the Japanese, believe they will develop a practical cold fusion process within the next twenty-five years.

There is more than just energy at stake in this research. Hundreds of scientists have invested their careers in the hot fusion process. If cold fusion replaces hot fusion, these people would likely be out of work; worse, their professional reputations might seem blemished. But the fact is that there may well be a need for both processes. Because of the extremely high temperatures required, hot fusion will probably be limited to stationary power plants, much like present electric plants. Cold fusion might not be subject to this limitation and could possibly be used in individual homes or to power motor vehicles. The scientific method will eventually resolve this riddle, but currently there is more confusion than fusion.

EXPERIENCE THIS

Listen to the weather reports for seven days and answer the following questions.

1. Record the barometric pressures each day for a week, and note whether the following day is clear or stormy. Is there any correlation between barometric pressure and the weather? What hypothesis could you propose from your observations? What experiments could you design to test your hypothesis?
2. Begin to make a daily record of the temperatures in your locality, or in another city of your choice, obtaining the information from TV or radio reports or from the newspaper. Record the date, the high temperature, and the low temperature for the same city every day for several weeks. Retain this information to use when you study chapter 19.

SCIENTIFIC MEASUREMENTS

To collect accurate data and make valid predictions, scientists must be able to make accurate scientific measurements. In nonscientific disciplines, the United States normally uses the English measurement system—a system rarely used elsewhere, including in England. But to take scientific measurements, scientists around the world use the metric system. (Additional information on the metric system appears in Appendix A, including conversion factors between metric and English units.)

The metric system is easy to use. Metric is based on decimal (1,10,100) factors, instead of the varying English factors such as 3, 12, or 16. In addition, metric involves only three basic units you will need to learn—**meter** (length), **gram** (weight), and **liter** (volume). Prefixes, including **micro-, milli-, centi-, kilo-,** and **mega-,** designate varying sizes or amounts (table 1.1). Combinations of these prefixes and units cover almost all measurements discussed in this text.

These prefixes are common in your daily life. A cent is 0.01 dollar (or a centi-dollar). A mil (a unit of taxation) is 0.001, or 1/1000 of a dollar. Electricity is purchased in kilowatts, or 1,000-watt units. A million dollars ($1,000,000) is called a megabuck, and machinists use a micrometer to measure millionths of an inch.

It is useful to have a rough idea of how the metric units compare with our more common English units. A liter is slightly larger than a quart, and a gallon contains 3.75 liters. An inch is about 2.5 centimeters. A meter is about 3.37 inches longer than a yard, but a kilometer is much shorter than a mile: a 100-km trip is about 60 miles. This means that when you drive at a speed of 65 miles per hour, you are going 104.6 km/hour, which certainly sounds faster. A gram weighs very little; a nickel weighs about 5 grams and a pre-1975 penny weighs about 3 grams. A 150-pound person weighs 68.04 kg. If this person is 5 foot 10 inches tall, he or she is also 177.8 cm or 1.778 meters.

Average Values

Part of the challenge of making precise scientific measurements is that it is rare to find identical values when measuring similar items. If you weighed each of two dozen eggs, you would be unlikely to find two eggs that weigh exactly the same. Each weight might be precise, but none is precisely a "typical egg." To adjust for this discrepancy, you might try to determine the average value for the weight or mass of the eggs. You could do so by weighing each egg, adding the weights, and dividing the sum by the number of eggs. You could also determine the average by weighing all the eggs at one time and then dividing by the number of eggs.

Eggs and other items are rarely identical in weight, length, or volume. This variability causes experimental error that can affect

TABLE 1.1 Common metric system units for length, volume, and weight, as used in chemistry.

PREFIX	FACTOR	LENGTH	VOLUME	WEIGHT
micro	0.000001	micrometer (μm)	microliter (μL)	microgram (μg)
milli	0.001	millimeter (mm)	milliliter (mL)	milligram (mg)
centi	0.01	centimeter (cm)	centiliter (cL)	centigram (cg)
	1.0	meter (m)	liter (L)	gram (g)
kilo	1,000.0	kilometer (km)	kiloliter (kL)	kilogram (kg)
mega	1,000,000.0	megameter (Mm)	megaliter (ML)	megagram (Mg)

chemical reactions. Because variability due to experimental error is an inevitable aspect of experimentation, scientists rely on average values to estimate what is "typical" or unusual for an item or event. Like these scientists, you need to recognize this limitation and reserve judgment until a number of measurements and interpretations become available.

PRACTICE PROBLEM 1.3

Halloween is almost here, so you buy several boxes of Reese's Peanut Butter Cups for trick-or-treaters. Because you want to ensure that everyone receives the same size treat, you borrow a balance (scale) from a friendly chemist and patiently weigh the first ten pieces of candy you remove from the box. You find they have masses (or weights) of 17.2, 17.0, 17.0, 17.1, 17.0, 17.2, 17.0, 17.1, 17,0, and 17.1 grams. What is the average mass for a "typical" peanut butter cup? Do you think the candy company is doing a good job of producing candy of uniform sizes?

ANSWER The total mass of the ten peanut butter cups is 170.7 grams. The "average" peanut butter cup is thus 170.7 ÷ 10 = 17.07 grams, or slightly less than 17.1 grams. This is a remarkably consistent achievement for a candy company that manufactures thousands of units daily.

SUMMARY

Chemistry is the science that studies matter, energy, and their interactions. Chemical elements make up the physical world, and chemistry affects our bodies, new technologies, and products that improve human life. Sometimes is is difficult to draw the line between science and technology; basically, science develops knowledge for its own sake, and technology applies that knowledge to practical purposes. It can also be difficult to determine whether a particular technology's benefits outweigh its risks. Benefit–risk ratios help quantify those considerations.

Just as benefit–risk ratios give us a system for evaluating the effects of specific technologies, the scientific method gives scientists a systematic way to collect facts and data, propose hypotheses and theories to account for that data, test their hypotheses and theories, and draw conclusions. Successful use of the scientific method depends on careful scientific measurements. Scientists the world over use the metric system to measure weight, length, volume, and other factors. Often they depend on average values to estimate "typical" measurements, but scientists are careful to recognize the experimental error this introduces and to base their conclusions on repeated tests and measurements.

KEY TERMS

benefit–risk ratio 6
centi 14
chemistry 2
data 8
experiment 9
fact 8
gram 14
hypothesis 9
kilo 14
law 8
liter 14
mega 14
meter 14
micro 14
milli 14
science 2
scientific method 6
technology 3
theory 9

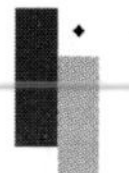

READINGS OF INTEREST

Your local newspaper may have a science or health page, especially on Sunday. Use the scientific method to evaluate scientific reports in the news.

"Trends shaping the world." *The Futurist*, September-October 1991, pp. 11–21. Lists fifty trends which should shape the world as we enter the next century—many of which come directly from chemistry.

Gibson, W. David. "Art or science? Art conservation is a mix of both." *Today's Chemist at Work*, December 1993, pp. 40–44. How museums protect old paintings.

"The new alchemy," *and* "Creating chips an atom at a time." *Business Week*, July 29, 1991, pp. 48–55. A business-oriented view of new wonders emerging from chemistry, including computer chips and insulation.

"Wonder cures." *U.S. News and World Report*, September 23, 1991, pp. 68–71. An analysis of some currently popular, nonstandard medical treatments.

"Food news blues." *In Health Magazine*, November 1991, pp. 40–45. How to tell sense from nonsense in articles about food that appear in your newspaper.

"The wonderland of Lewis Carroll." *National Geographic*, 179(1991) pp. 100–129. This piece offers insight into this fantasy writer who was also a scientist.

Ahearne, John. "What science can and cannot do." *Technology Review*, October 1993, pp. 70–71. Science does not always lead to economic benefits.

Thayer, Ann M. "Justifying technology's value challenges industry R&D managers." *Chemical & Engineering News*, February 6, 1995, pp. 10–14. The public demands that R&D managers justify their research—not always an easy task.

American Chemical Society. *"Chemical risk: A primer"* (12 pages) and *"Chemical risk: Personal decisions"* (16 pages). Free information pamphlets available from the American Chemical Society, Dept. of Government Relations and Science Policy, 1155 Sixteenth Street N.W., Washington, DC 20036.

Lynch, J.; Lewis, S. C.; Goldman, J. H.; and Karlin, A. S. "Exaggerating risk." *Chemtech*, October 1993, pp. 47–53. We must differentiate between risk assessment, which is a scientific procedure, and risk management, which consists of policy-making decisions.

Clarke, David. "The elusive middle ground in environmental policy." *Issues in Science and Technology*, Spring 1995, pp. 63–70. This article discusses the need for risk regulation reform.

"Cold fusion." *Smithsonian*, May 1991. A brief history of the cold fusion controversy.

"Cold fusion." *Science News*, June 22, 1991, pp. 392–93. An update on the cold fusion controversy up to mid-1991.

Rousseau, Denis L. "Case studies in pathological science." *American Scientist*, January/February 1992, pp. 54–63. The loss of objectivity can lead to false conclusions. One of the cases studied in this article is cold fusion.

Pais, Abraham. 1994. *"Einstein lived here."* New York: Oxford University Press. A biography of Einstein written for the nonscientist. This book tells about his wives and children, but it contains no math!

French, A. P. 1979. *"Einstein, a centenary volume."* Cambridge, Mass.: Harvard University Press. This book contains many short articles about Albert Einstein, plus samples of his writings.

Gamow, George. 1973. "Mr. *Tompkins in paperback*." Cambridge, England: Cambridge University Press. This is probably the best and simplest book for readers who want more information on relativity.

"Antiscience trends in the U.S.S.R." *Scientific American*, August 1991, pp. 32–38. An analysis of trends that are far too similar to some taking hold in the United States today.

Gross, Paul R., and Levitt, Norman. "Knocking science for fun and profit." *Skeptical Inquirer*, May/June 1995, pp. 38–42. Science doesn't always work the way some critics claim it should. Their proposed reforms would change science into something else altogether.

Koertge, Noretta. "How feminism is now alienating women from science." *Skeptical Inquirer*, May/June 1995, pp. 42–43. This author claims that some feminists, rather than encouraging women to learn technical subjects, are asserting that logic is a tool of male domination.

Reese, Ken. "Metrication—Should we or shouldn't we? Does it matter?" *Today's Chemist at Work*, June 1994, pp. 49–51. A discussion of the pros and cons of adopting the metric system.

McCurdy, Patrick P. "I'm just mild about metric." *Today's Chemist at Work*, June 1994, p. 60. Some scientists really like the metric system; others are less enthused.

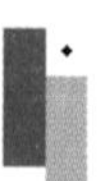

PROBLEMS AND QUESTIONS

1. Define chemistry, and give a few examples of events or items chemists might study.

2. Briefly define the following terms as they apply to chemistry:

experiment	science
fact	technology
hypothesis	theory
law	

3. List at least ten technological benefits chemistry provides. How many of these do you, or a member of your family, use on a regular basis?

4. Five different chemicals yielded the following benefit–risk ratios: 92,500; 6,300; 15.5; 1.3, and 0.5. Which material is the safest to use, and which is the least safe? In *your* opinion, which of these five chemicals are safe enough for use in our society? Now consider the least safe material again. Would you be willing to put this chemical to use if it were an effective anticancer drug?

5. Outline the scientific method.

6. List the main metric units for length, volume and weight.

7. List the main prefixes used in the metric system.

8. Five nickels weigh 5.23, 5.31, 5.27, 5.33 and 5.26 g, respectively. What is the average weight of the coins? What can you conclude from this data about the uniformity of the coins?

9. Five quarters have diameters of 23.5, 23.9, 23.1, 23.7, and 23.6 mm, respectively. What is the average diameter of the quarters? What can you conclude from this data about the uniformity of quarters?

10. In Practice Problem 1.3, you learned that a "typical" Reese's Peanut Butter Cup weighs 17.1 grams. How much does it weigh in ounces?

HINT You can find the conversion factor in Appendix 1.

CRITICAL THINKING PROBLEMS

1. Suppose you could actually repeal the law of gravity. What effect would this have on your life?

2. What is the benefit–risk ratio for high school (or college) athletic programs? Is the risk justified? Why?

HINT Compare the number of accidents and deaths to the attendance at the sports events.

3. What is the benefit–risk ratio for driving a car in the United States? Do you think driving is a safe activity?

HINT Compare the average number of miles driven against the number of deaths and injuries. (According to risk analysis experts, driving a motor vehicle is the most risky activity in this country, ranking well above flying in an airplane or living near a nuclear power plant.)

4. Think about the risks and benefits of using Tupperware. How can you dispose of used Tupperware? Is there any risk in this? How much energy is required to make the plastic—is it excessive compared to the benefits? For that matter, what are the benefits of Tupperware? (Presumably, many benefits must exist if Americans buy over a billion dollars worth of Tupperware each year.)

5. Many tons of chemical X are manufactured every year. This chemical is a liquid usually encountered as a 5 to 50 percent solution in water. (In other words, half or more of the mixture is water.) The main uses of chemical X are as an ingredient in beverages or as a solvent (dissolving fluid) for other materials. About 60 percent of the people in the United States drink the solutions of chemical X and claim some benefit. Approximately 80 percent of the people derive a benefit from chemical X's solvent properties.

On the risk side, chemical X is highly flammable; its use as a solvent causes many fires, injuring people and causing millions of dollars of property damage. When ingested, chemical X impairs body control, especially careful, precise movements, and also impairs judgment. Thousands of industrial and automotive accidents are attributed to the toxic effects of chemical X every year. In fact, chemical X affects nearly everyone, either directly through injury or death or indirectly through increased taxes and insurance rates. We conservatively estimate chemical X is a risk factor for at least 90 percent of the people in the United States.

Calculate the benefit-risk ratio for chemical X. Do you think this chemical should be widely used in our society? Would you personally use this chemical? What common name does chemical X go by, and does this change your answers?

6. Apply the scientific method to the question: why do leaves change color in the fall? Your two hypotheses are: (1) an invisible being named Jack Frost paints the leaves each year; and (2) chemical reactions, caused by changes in the amount of light that strikes the leaves, destroys the green chlorophyll pigment and permits us to see the red and yellow leaf pigments. Make a list of facts or data you can use to test each hypothesis; past observations may be included. Can you design an experiment to test each hypothesis?

7. Read the article "Wonder Cures" in the September 23, 1991 issue of *U.S. News & World Report*, pages 68–77. With the scientific method in mind, evaluate the potential validity of each of these unusual medical treatments.

8. A guest on a TV talk show claims he was healed from a major medical problem by eating ten or more grapefruit daily. The person looks healthy. What should you conclude?

9. Your newspaper reports that a study at a leading medical school examined 100 patients and showed that eating grapefruit has no effect on any severe medical affliction, although it might be good for your general health because it contains some vitamins. What should you conclude about this report? How does the information in the newspaper compare to the information reported by the talk show guest in question 8?

10. Weather prediction is based on theories. Monitor these predictions for a few days and see how well the theories work. Do these theories need modifications?

11. Often, a few scientists continue to research a problem that most other scientists believe is already solved. Why?

12. Why must most scientists also be good laboratory technicians?

13. Much of modern life centers either on the urban or suburban areas. Science and technology are used in either location. Is there anything better that might be possible, compared to the present situations? Two starting point references are: (1) T. W. Wagar, *Technology Review*, April, 1993, pp. 51–59, "*Tomorrow and Tomorrow and Tomorrow*," and (2) R. C. Post, *Invention & Technology*, Spring, 1993, pp. 16–24, "*The Frailties and Beauties of Technological Creativity*."

14. Risk and benefit assessment is important in science, business, and management. There is a trend toward doing risk assessment in the political arena. Is this the same as scientific risk assessment? What are the limitations of risk assessment in politics? Can it be used effectively? Read at least two of the following articles and make your own tentative conclusion: (1) Emily T. Smith, John Careyu, and Mary Beth Regan, Business Week, March 13, 1995, pp. 96–98, "*Voodoo Regulation*," (2) Ronald Begley, *Chemical Week*, Jan. 19, 1994, pp. 24–27, "*Risk-Based Policy Could Finally Be on Its Way*," and (3) M. Granger Morgan, *Scientific American*, July, 1993, pp. 32–41, "*Risk Analysis and Management*."

CHAPTER 2

MATTER, STATES OF MATTER, AND TEMPERATURE

"*Science moves, but slowly, slowly, creeping on from point to point.*"
ALFRED LORD TENNYSON (1809-92)

OUTLINE

The Classification of Matter
The States of Matter
Temperature and Energy
The Gas State
The Liquid State
The Solid State
Evaporation, Vapor Pressure, and Boiling Points

BOX 2.1 Weighing and Floating: Mass, Weight, and Density
BOX 2.2 Creating Thermometer Scales
BOX 2.3 Air Pressure and Helium Density—What Happens to Helium-Filled Balloons?
BOX 2.4 Steam Engines—The Pressure Behind the Power
BOX 2.5 A Few Drops of Water

An eruption of Old Faithful in winter shows steam, water, and snow—the three forms of water.

KEY IDEAS

1. How Matter is Classified
2. The Difference Between Homogeneous and Heterogeneous Materials
3. The Difference Between Elements and Compounds
4. The Difference Between Atoms and Molecules
5. The Significance of Heating Curves
6. The Characteristic Properties for Each State of Matter
7. The Effect of Pressure on Gas Volume (Boyle's Law)
8. The Effect of Temperature on Gas Volume (Charles' Law)
9. The Basic Features of the Combined Gas Law
10. The Basic Features of the Kinetic Molecular Theory of Gases
11. The Difference Between Amorphous and Crystalline Solids
12. How Vapor Pressure Relates to Boiling Point

Take a look around the room in which you are seated. What do you see? At first glance, you may perceive your surroundings as groupings of objects of different sizes, shapes, weights, and textures. You may label the objects with different names, such as chair, book, lamp, or table. But a scientist might classify these objects as different types of matter. In scientific terminology, **matter** is anything that occupies space and has mass (weight). Chemists group matter into several categories for better identification. In this chapter, we will seek to understand this classification of matter and to learn how it relates to physical and chemical changes.

Matter, which is made up of chemical elements, exists in three main forms—solid, liquid, or gas. The air we breathe, the food we eat, and the homes we live in are all composed of matter—indeed, we ourselves are composed of matter. *Matter* is thus such an all-inclusive term that it needs further breakdown or classification. Figure 2.1 outlines the major system of matter classification; subsequent figures elaborate on particular sections of this diagram.

THE CLASSIFICATION OF MATTER

As figure 2.2 indicates, matter may be broadly classified as either homogeneous or heterogeneous. **Homogeneous matter** is relatively uniform in appearance, composition, and properties, whereas **heterogeneous matter** is nonuniform. A casual glance at your surroundings will confirm that some materials appear uniform while others do not; air, clean water, and soft drinks are homogeneous, while dirt, rocks, clothing, this book, and our own bodies are heterogeneous mixtures of materials. Appearance is often revealing: if the object does not appear uniform, it cannot be homogeneous.

Homogeneous substances may exist in either of two forms: as a pure substance or as a solution (figure 2.3). A **pure substance** has a definite and constant composition. A **solution** is a homogeneous mixture of two or more pure substances with a composition that is neither definite nor constant. We will learn more about pure substances and solutions later in the chapter.

Pure substances consist of only one kind of substance and cannot be separated into simpler components by any *physical* means. Figure 2.4 depicts two different types of pure substances—elements and compounds. **Elements** are pure substances composed of a single kind of matter. They cannot be separated into other, simpler components by either *physical* or *chemical* methods. Currently, there are 112 known elements, and each has specific physical and chemical properties that make it different from all the others. You seldom see most elements in their pure forms, but two elements you might see are copper and silver. **Compounds** are pure substances formed when two or more elements bond chemi-

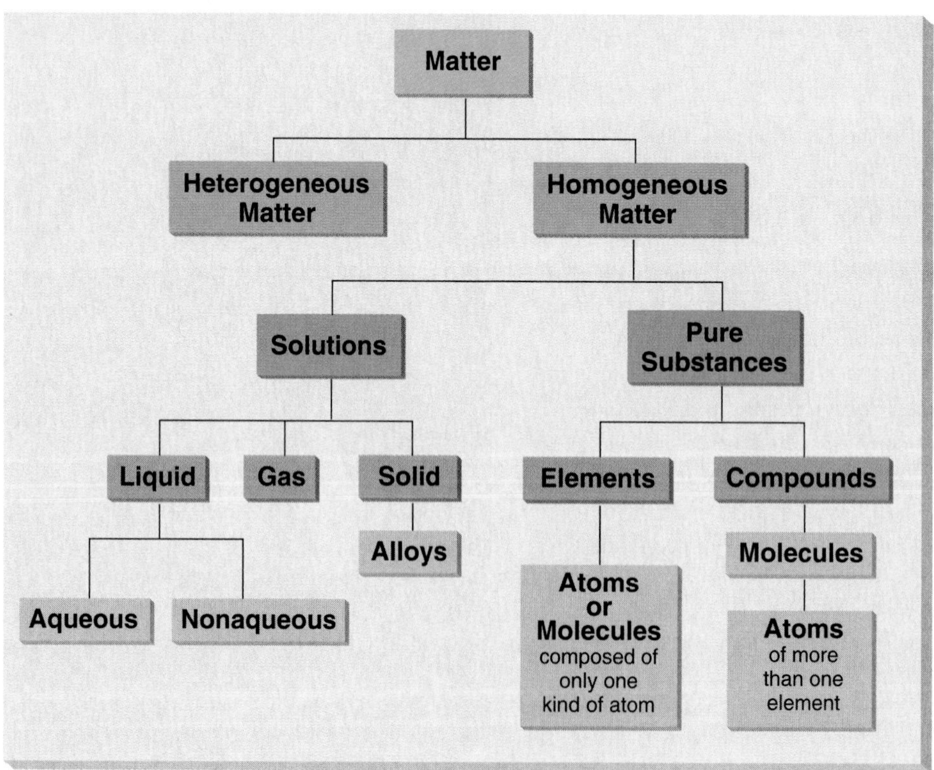

Figure 2.1 The classification of matter.

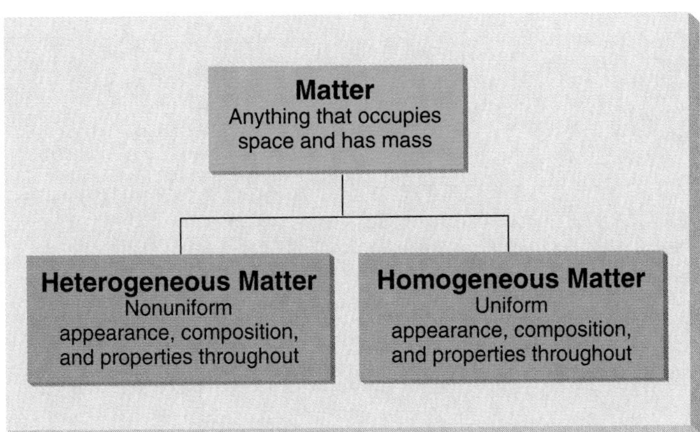

Figure 2.2 Matter may be broadly subdivided into heterogeneous and homogeneous materials.

cally. Compounds always have a definite and constant composition; this means they contain elements in fixed proportions. Compounds cannot be separated into other components by *physical* methods, but they can be separated into different elements or simpler compounds by *chemical* methods. Billions of compounds exist. Table salt, a common example, consists of sodium and chlorine atoms chemically bonded; table salt always contains exactly the same proportions of sodium and chlorine atoms.

Atoms are defined as the basic building blocks of all matter. This is true regardless of whether the matter is homogeneous or heterogeneous—in other words, all matter is ultimately composed of atoms. This also means that all matter is ultimately composed of pure substances; in a heterogeneous mixture, several pure substances mix to create a nonuniform blend, and in a homogeneous mixture, several pure substances combine to form a solution.

Although elements cannot be subdivided into other kinds of matter, you can make a sample of a particular element smaller. Take a 1.0-gram sample of the element gold and divide it into increasingly smaller masses of 0.5 g, 0.25 g, 0.125 g, and so on. As this process continues, the element is still gold. Eventually, you can no longer subdivide the gold; a tiny particle, an atom, remains. If an atom of gold were subdivided, it would no longer be gold. The same is true for any other element. Until recently, scientists assumed that atoms exist, but now they can visualize real atoms using a special technique called a scanning-tunneling microscope.

Molecules are collections of atoms bound together. An element always consists of a single kind of atom, but these atoms are often bound together into molecules. For example, the element oxygen usually consists of two oxygen atoms attached together. Compounds, on the other hand, are made up of molecules containing at least two different elements or kinds of atoms. Water, for example, consists of molecules made up of two hydrogen atoms bonded to one oxygen atom. Compounds can be broken down into elements or simpler compounds by *chemical* methods, which usually require energy. Their smallest subcomponents are molecules; if you break these molecules down to their constituent atoms, the compound no longer exists.

Solutions are homogeneous mixtures of at least two components that appear uniform in composition. Solutions may be further subdivided into liquids, gases, or solids depending on the nature of the main component (figure 2.5). Any mixture, including a solution, can be separated into its components by *physical* actions (filtering, boiling, and so forth).

Liquid solutions may be **aqueous** or **nonaqueous solutions.** Aqueous solutions always contain water as the main ingredient, or solvent; soft drinks or tea are typical examples of aqueous solutions. Nonaqueous solutions contain a liquid other than water as the main component; a common example is gasoline or perfume.

Air, a mixture of the gases oxygen, nitrogen, and carbon dioxide, is the most common gas solution. Other examples include the gases used for cooking and heating.

The most common solid solutions are **alloys,** uniform mixtures of two or more solid substances, at least one of which is a metal.

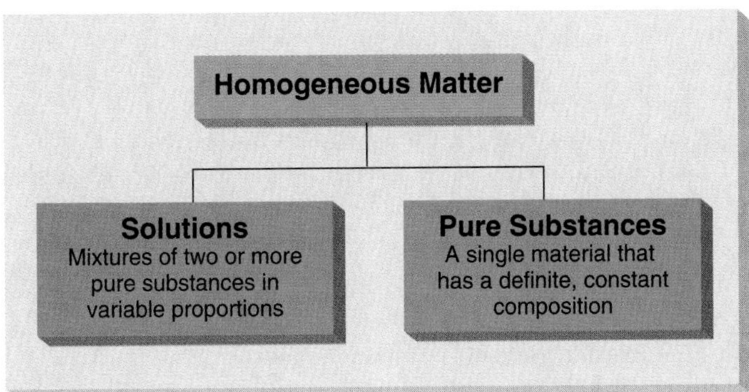

Figure 2.3 Homogeneous matter may exist as either a solution or a pure substance.

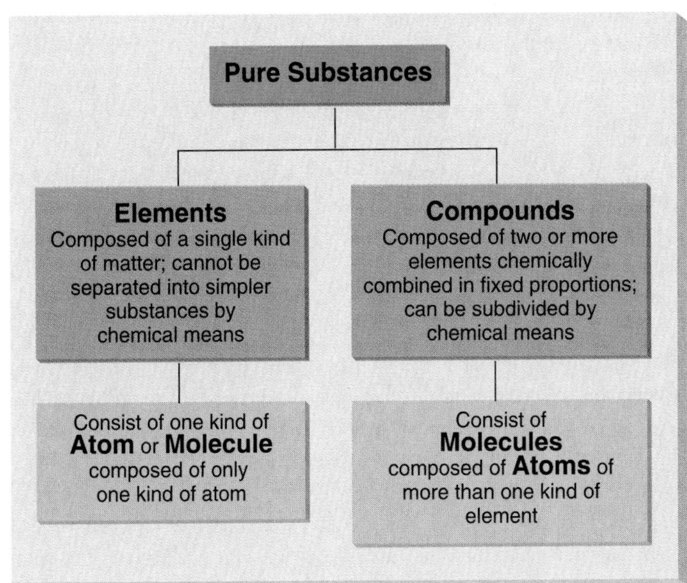

Figure 2.4 Pure substances may be either elements (one type of atom) or compounds (more than one element or type of atom).

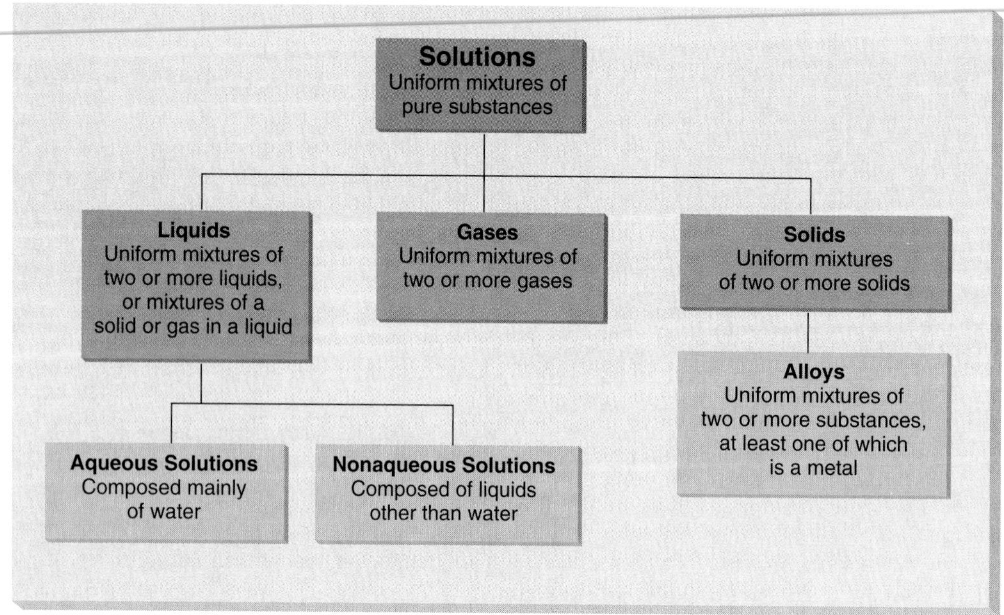

Figure 2.5 Solutions may be liquid (aqueous or nonaqueous), gas, or solid.

Alloys are usually formed by melting the constituents together and allowing them to cool and become solid. Steel is a solution of several metals and carbon in iron; brass is a solution of zinc in copper; solder is an alloy of tin and lead. Dental amalgam, used to fill cavities, is an alloy of mercury and another metal, usually silver.

THE STATES OF MATTER

You are probably aware that water exists as a liquid, a solid (ice), and a gas (steam). These different forms are called states of matter. Three stable states of matter exist on earth. Any type of matter can exist in liquid, solid, or gaseous form, and one form will convert to another when heat is applied or removed.

Figure 2.6 shows a water heating curve that tracks the effects of temperature on the states of matter. The diagram depicts five stages. At stage *a*, water consists of solid ice. At stage *b*, as the temperature increases, the water becomes partially liquid; water begins to melt during this stage. The water temperature plateaus until enough heat is applied to melt all the ice. At stage *c*, the water is entirely liquid as the water absorbs heat energy. Stage *d* is characterized by both liquid and gaseous water; water boils at this temperature, but the water temperature again remains constant, even as more heat is applied, as long as any liquid water remains. At stage *e*, only gaseous water (steam) remains; the steam becomes even warmer as more heat energy is applied. The temperature at stage *b* is the melting, fusion, or freezing point, and the temperature at stage *d* is the boiling, vaporization, or condensation point. For water, the values at stages *b* and *d*, when the temperature stabilizes and the water converts to a different state, are 0°C and 100°C, respectively.

EXPERIENCE THIS

Scientists make models of molecules using balls to represent individual atoms and sticks to represent the bonds between them. You can do the same using gumdrops (or a similar soft candy) and toothpicks. (Styrofoam balls also work.) Make models for the following substances.

a. Helium, which has only 1 atom.
b. Oxygen, which has 2 atoms.
c. Ozone, which has 3 oxygen atoms. Is there more than one way to arrange the atoms in ozone?

Make models of each of the following compounds, using different colored gumdrops for each kind of atom (for example, white for sodium and green for chlorine).

a. Sodium chloride (ordinary table salt), which contains 1 sodium atom and 1 chlorine atom. Is there more than one way to connect the atoms in this model? *Hint:* Don't be confused by the fact that you can place the models in different positions on a table. Make a pair of models and notice that one matches the other when pivoted.
b. Carbon dioxide, which consists of 1 carbon atom and 2 oxygen atoms. Is there more than one way to connect the atoms in this model? *Hint:* Again, make pairs of models and see if they line up the same way.

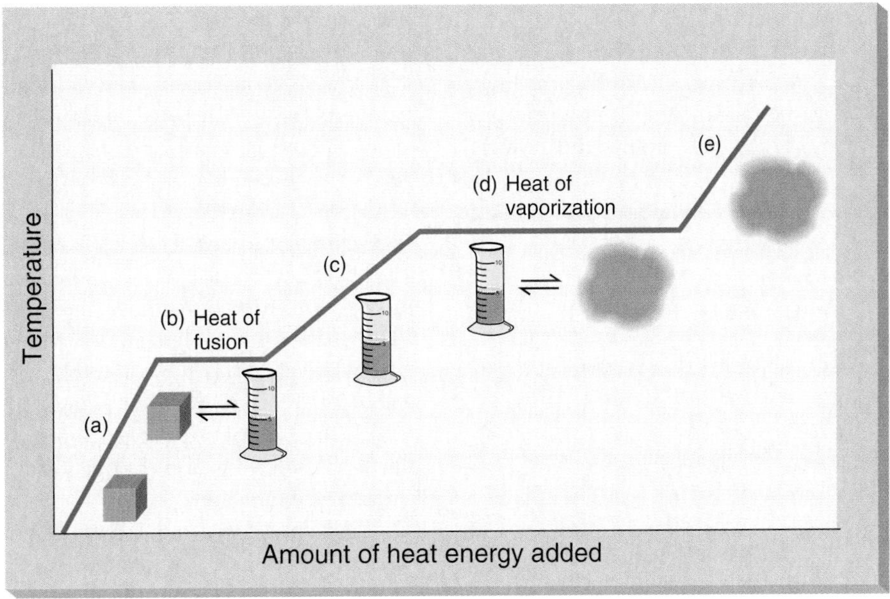

Figure 2.6 A heating diagram for water shows how heat converts water from solid to liquid to gas.

SOURCE: From Jacqueline I. Kroschwitz, et al., *Chemistry: A First Course*, 3d ed. Copyright © 1995 Times Mirror Higher Education Group, Inc., Dubuque, Iowa. All Rights Reserved. Reprinted by permission.

PRACTICE PROBLEM 2.1

Which of the following mixtures are homogeneous? Which are heterogeneous?

a. Sand on a beach
b. Chocolate chip ice cream
c. A glass of fresh, unprocessed cow's milk
d. A glass of freshly made Kool Aid
e. A chocolate bar (with no nuts or other added ingredients)
f. Clear seawater

ANSWER Items *a*, *b*, and *c* are heterogeneous; items *d*, *e*, and *f* are homogeneous. A typical sandy beach consists of shells and other objects embedded in the sand. Even when no shells are present, the beach usually consists of grains of sand surrounded by water. Chocolate chip ice cream *(b)* contains chips embedded in ice cream—two different kinds of substances. Fresh, raw milk *(c)* is not as obviously heterogeneous, but fresh milk separates into a fatty layer (cream) and a pale colored liquid; milk is homogenized (that is, made to appear homogeneous) to hinder this separation. Kool Aid *(d)* is a solution and, therefore, homogeneous. Although the chocolate bar *(e)* and seawater *(f)* both contain several ingredients; they are blended into a uniform mixture. Homogeneous matter can consist of several different substances, as long as the final blend is uniform. When the final blend is nonuniform, the matter is heterogeneous.

PRACTICE PROBLEM 2.2

Which of the following pure substances are elements, and which are compounds? Is the smallest subcomponent of each a molecule or an atom?

a. Table salt, which contains sodium and chlorine
b. Table sugar, which contains carbon, hydrogen, and oxygen
c. Chlorine gas, which contains only chlorine
d. Water, which contains hydrogen and oxygen
e. A rare silver coin, which contains only silver
f. A marble statue, which contains calcium, carbon, and oxygen

ANSWER Only *c* and *e* are elements. The remainder contain at least two elements and are thus compounds. All the examples except silver *(e)* are made up of molecules as their smallest subunit; many elements—including chlorine gas *(c)*—occur naturally as molecules, and compounds always exist as molecules.

PRACTICE PROBLEM 2.3

Which of the following are solutions (that is, homogeneous mixtures)?

a. Chocolate chip ice cream
b. Stainless steel
c. A glass of Kool Aid
d. A soft drink
e. The gas used for cooking or heating
f. A pair of blue jeans

ANSWER Remember that solutions must appear uniform. Thus, *b, c, d,* and *e* are solutions. Chocolate chip ice cream *(a)* is heterogeneous and cannot, therefore, be a solution—you can see the chips embedded within the ice cream. Stainless steel *(b)* is an alloy (solid solution) of iron, carbon and other metals. Kool Aid *(c)* and soft drinks *(d)* are aqueous solutions. Cooking or heating gas *(e)* is a solution of gases. Blue jeans *(f)* might be the solution to your clothing problems, but they are heterogeneous matter.

Pure materials follow a heating curve similar to water's, although the precise temperatures and plateau stages differ. Boiling and melting temperatures are important physical properties that chemists can use to determine the identity and purity of a substance. Pure iron, copper, and ethanol follow this type of heating curve.

Other pure materials, such as iodine or carbon dioxide, do not convert through all three states of matter under ordinary conditions. Iodine lacks a liquid state; the solid converts directly into gas, a process called **sublimation.** (The reverse process—when a gas converts directly to a solid—is called deposition.) Carbon dioxide normally exists as a gas or as solid carbon dioxide, commonly called dry ice. It, too, sublimes directly into the gas. The liquid form exists only under high pressures. Yet other pure materials, such as some plastics, decompose when heated and do not follow a typical heating curve. Heating diagrams represent physical changes; decomposition is a chemical change that makes typical heating curves irrelevant. Impure mixtures, such as coal or dirt, and solutions also do not follow the type of heating curve shown in figure 2.6.

Each state of matter has definite properties that differ from the others, as table 2.1 summarizes. A solid has a definite shape, but the shape of a liquid or gas can vary. Solids and liquids have definite volumes, but gases vary in volume. **Density** is defined as the mass (M) per unit volume (V), measured as grams per milliliter, or g/mL:

$$d = M/V$$

Density is constant if measured at the same temperature and pressure. Table 2.2 lists typical density values. As the table indicates, solids have the highest densities, with liquids close behind; gases are in a distant third place. Density is measurable for homogeneous or heterogeneous mixtures as well as for pure substances. Wood, a heterogeneous mixture, shows density values ranging from about 0.4 to 0.6 g/mL. Box 2.1 discusses density, as well as mass and weight, in more detail.

Most materials expand when heated. Solids expand very little when heated; liquids expand only slightly more. Gases, on the other hand, expand greatly when heated. This is why a hot air balloon rises. The total air mass remains constant, but the volume increases greatly as the air is heated. The density of the air in the

TABLE 2.1 Characteristics of the states of matter.

Property	STATE OF MATTER Solid	Liquid	Gas
Volume	definite	definite	variable
Shape	definite	variable	variable
Expansion during heating	very slight	slight	great
Compressibility	none	slight	great
Density	high	medium	low

TABLE 2.2 Typical density values in g/mL. Solid and liquid densities are measured at 20°C; Gas densities are at 0°C.

SOLIDS 0.5 to 20 g/mL		LIQUIDS 0.5 to 14 g/mL		GASES 0.00009 to 0.003 g/mL	
lithium	0.53	gasoline	0.68	hydrogen	0.00009
ice	0.92	ethanol	0.79	helium	0.00018
sugar (table)	1.59	benzene	0.88	methane	0.00071
magnesium	1.74	cottonseed oil	0.93	ammonia	0.00077
sulfur	2.07	water	1.00	neon	0.00090
salt (table)	2.16	glycerine	1.26	nitrogen	0.00125
aluminum	2.7	hydrogen peroxide	1.44	carbon monoxide	0.00125
iron	7.86	chloroform	1.49	air	0.00129
copper	8.93	carbon tetrachloride	1.60	oxygen	0.00143
silver	10.50	sulfuric acid	1.84	argon	0.00178
lead	11.38	bromoform	2.89	carbon dioxide	0.00196
gold	19.3	mercury	13.55	chlorine	0.00317

BOX 2.1

WEIGHING AND FLOATING: MASS, WEIGHT, AND DENSITY

Most scientists refer more often to *mass* than *weight*. Although mass is not the same as weight, we can usually use these terms interchangeably when describing everyday events, as long as we restrict this use to events on the planet Earth. However, these terms have different scientific meanings. The mass of a specific sample is always constant, but the weight is the result of its interaction with gravity and is not necessarily constant. For example, when astronauts walked on the moon, their mass remained the same even though their weight was much less. A person weighing 100 kg (220 pounds) on the earth weighs only 16.7 kg (37 pounds) on the moon, even though they have not lost any mass. This happens because the moon's gravity is not as strong as the earth's.

Whether a solid or liquid material sinks in or floats on a particular liquid does not depend on the mass (or weight) of the pair of substances. Instead, the two important factors are: (1) whether or not the material dissolves in the liquid, and (2) whether the material is more or less dense than the liquid. A mixture of gases always mixes completely, which is equivalent to the gases dissolving completely in each other. If a solid or a liquid does not dissolve in another liquid, it either sinks to the bottom (if it is denser than the liquid) or floats on top (if it is less dense). Thus, gasoline (density = 0.68 g/mL) floats on water (d = 1.00 g/mL). Similarly, wood has a density ranging between 0.4 and 0.6 g/mL; it floats on water because its density is lower.

Table 2.2 in the text lists the densities of several solids, liquids, and gases. Copper (d = 8.93) sinks in water (d = 1.00) because it has a higher density. In the same manner, liquid carbon tetrachloride (d = 1.60) sinks to the bottom of a container of water. Ice (d = 0.92) floats in water, but sinks to the bottom of a container filled with pure ethanol (d = 0.79). On the other hand, magnesium metal (d = 1.74) sinks in water, but floats on bromoform (d = 2.89).

balloon falls below that of the surrounding air, and the balloon rises.

Compressibility (table 2.1) is a material's ability to accommodate pressure by squeezing into a smaller volume. Solids compress very little, if at all; liquids compress slightly; gases compress greatly. This is why a hydraulic (liquid) brake system works well when the tubing and cylinders contain only liquid, but poorly when a small amount of air is introduced. When the system contains only liquid, pressure on the brake pedal transmits through the largely incompressible fluid and the car stops. When air enters the brake lines, the air compresses, the pedal pressure is not transmitted through the fluid, and the car does not stop.

PRACTICE PROBLEM 2.4

Which of the following materials would be likely to follow a heating curve resembling the heating curve of water?

a. Brass (zinc and copper)

b. Aluminum

c. Nitrogen (purified from the air)

d. Ethanol

e. Gasoline

ANSWER Brass *(a)* is an alloy (a solution), and gasoline *(e)* is a solution; solutions do not show normal heating curves. Aluminum *(b)*, like most metals, will follow a normal heating diagram, as will ethanol *(d)*. Even though nitrogen *(c)* is a gas at room temperature (it makes up about 80 percent of our air), it forms a liquid and a solid at very cold temperatures and follows a normal heating curve.

PRACTICE PROBLEM 2.5

Would it be easier to go hot air ballooning on a cold day or a hot day? Why?

ANSWER The colder day would make it easier because air density increases at lower temperatures. Thus, you would not have to heat the air in the balloon as much to decrease its density below the surrounding air to get it to rise up, up, and away.

Extensive and Intensive Properties

The properties of matter may be either extensive or intensive. An **extensive property** depends on how much material is present; an **intensive property** is independent of the amount of material.

PRACTICE PROBLEM 2.6

Suppose you find that a lump of gold has a volume of 5.0 mL and a mass of 96.6 g. What is the density of gold?

ANSWER

Density is mass divided by volume; thus, you divide as follows:

$$\frac{96.6 \text{ g}}{5.00 \text{ mL}} = 19.3 \text{ g/mL for the density of gold.}$$

Mass and volume depend on the amount of material present and are thus extensive; in other words, a quart of water has a different mass and volume than a gallon of water. Melting and boiling points do not depend on how much matter is present and are therefore intensive; the boiling point is the same for a quart of water as for a gallon.

All the properties in table 2.1 except volume are intensive. Shape does not depend on how large an object is. You can have a small ball or a large one, but both have the same shape. Density is the ratio of two extensive properties (mass and volume) but is itself an intensive property, and compressibility is also intensive—it does not change just because the amount of material changes.

The Differences between the Three States of Matter

The model in figure 2.7a helps explain the changes shown in figure 2.6 by illustrating the molecular differences between solid, liquid, and gas. In the solid state, atoms or molecules are packed tightly together; a given number of atoms fits into a minimal volume. In the liquid state, more empty space exists between the atoms; the same number of atoms must now occupy a larger total volume than they did in the solid state. The gas state contains the most empty space and thus requires the greatest volume for the same number of atoms.

Heat causes atoms to move. In the solid state, the atoms in the middle initially have no room to move. Only the atoms at the edges can move away from the others, and this means that only a small volume increase occurs along the edges of a solid as it begins to heat. Once the matter is liquid, the atoms have more space and can move more freely than in the solid, so the substance undergoes a greater heat expansion. Finally, the atoms in the gas state can move with almost complete freedom because so much empty space surrounds each of them. As heat increases, then, the atoms spread apart, the volume of the substance expands, and the substance moves through the different states of matter.

Figure 2.7 also helps explain compressibility. When matter is compressed, the atoms are squeezed together. As the figure shows, solids and liquids provide almost no room for the atoms to move; this means very little compressibility is possible. On the other hand, compressibility poses no problem in the gas state. The

Figure 2.7 A ball model for the three states of matter. The model shows that organization and density decline from solid to liquid to gas. The dandelion, shown above, provides an analogy for these three states of matter. In the yellow flower stage, the petals are close together, similar to the solid model. As the flowers turn to seed, more empty space accrues between petals, similar to atoms or in molecules in the liquid state. Finally, as the seeds blow away, only a few remain attached to the plant. Between them is much empty space, similar to molecules or atoms in the gas state.

atoms are so far apart that there is plenty of room to press them in together. This is why gases are more compressible than liquids or solids.

TEMPERATURE AND ENERGY

Temperature and heat are related but not identical concepts. Temperature is a measure of the *average* kinetic energy, or energy of motion, in a sample. This is an *intensive property*, independent of sample size. Heat represents the *total* amount of kinetic energy, or energy of motion, in a material. Thus, heat is an *extensive property*, dependent on the sample size.

Figure 2.8 illustrates the difference between heat and temperature by comparing 60°C water samples in 100 mL and 1,000 mL sizes. The temperatures of the two samples are identical—60° C. But since the molecules move at the same average speed in both containers, the total amount of kinetic energy, or heat, is greater in the 1,000 mL sample because it contains more molecules.

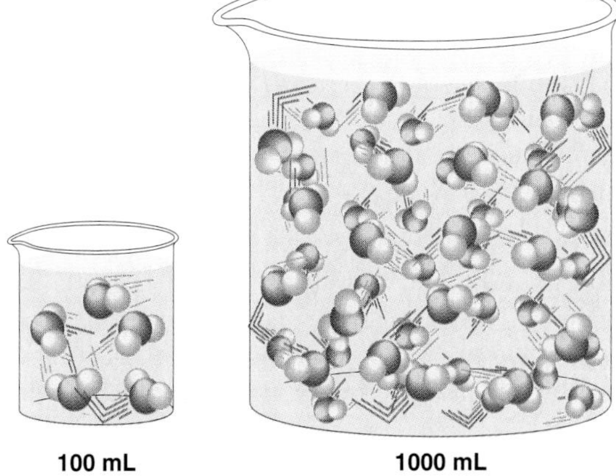

Figure 2.8 Two water samples at 60°C. They share the same temperature, but the larger sample contains more heat because it contains more molecules. Only a tiny fraction of the molecules are shown.

PRACTICE PROBLEM 2.7

Which do you think would be easier (that is, require less energy):
(a) to convert a liquid to a solid or (b) to convert a liquid to a gas?

ANSWER Moving from order to disorder is always easier (requires less energy). Converting a liquid into the solid state requires the atoms to fit into a more ordered arrangement, and that requires extra effort (energy). By contrast, atoms moving from liquid to gas state become even less orderly. This requires less energy expenditure.

This is much like straightening up your room. It is easier to let the room become a complete mess (gaslike) than to put every item in its exact place (solidlike). Putting the room in order requires you to exert more energy than messing the room. Conversion *b* thus requires less energy than conversion *a*.

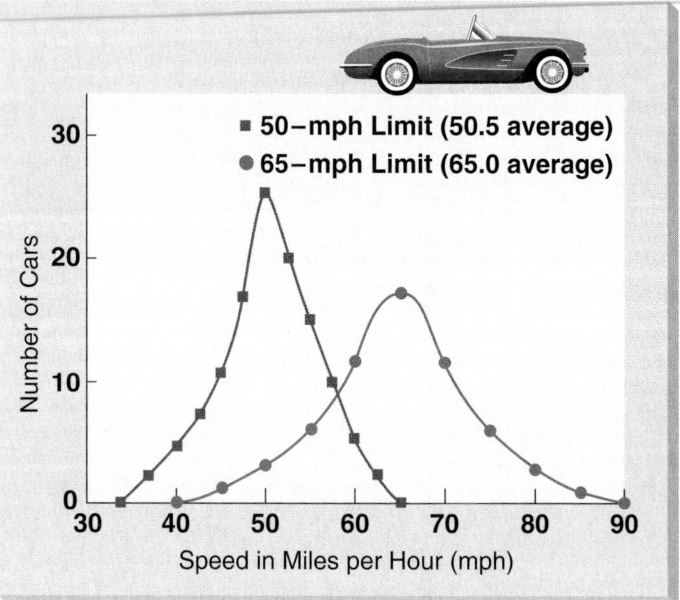

Figure 2.9 Car speed distributions on an expressway demonstrate a chemical principle: the higher the temperature, the faster molecules move and the broader a range they cover.

Molecules move faster and cover a broader range at higher temperatures. Figure 2.9 depicts an analogy to demonstrate this relationship. Imagine a group of cars traveling on an expressway. As the group approaches an area marked with a 50 mph speed limit, each car adjusts its speed. Some cars speed above the 50 mph limit; others travel below the limit. On the average, they move at 50.5 mph, but their speeds range between 35 and 62 mph (a 27-mph range).

Now the cars approach a new section of the highway marked with a 65 mph limit, so they speed up. Their average speed is now 65 mph, but their speeds range from 45 to 85 mph, a 40-mph difference. In the higher-speed area, they move faster and cover a broader range than in the lower-speed area.

A similar situation would exist for gas molecules at 0°C and 1,000°C: the molecules move faster and over a broader range at the higher temperature. Changes in kinetic energy are important factors in many phenomena, such as gas pressure and chemical reaction rates.

Thermometer Scales

The concept of temperature variation dates from antiquity, although early measurements were based solely on the subjective sense of touch. Touch is only capable of detecting temperature differences between objects; it cannot provide quantitative temperature data. For example, you may be able to tell if a person has a fever by placing a hand on their forehead, but exactly how high is the fever? That question requires quantitative data; our senses can mislead us, but a thermometer can give us that data.

Gabriel Fahrenheit (German, 1686–1736) devised the Fahrenheit scale, the oldest common temperature scale, in about 1714. (Prior to this, Galileo Galilei had invented an open tube thermometer in about 1593.) One of Fahrenheit's innovations was to use mercury, which made the liquid column easy to see. Previously, people had had difficulty sealing mercury-containing tubes.

Fahrenheit established his temperature scale by selecting reference points he could reproduce. He designated a freezing ice-snow-salt-water mixture as 0°F because this was the lowest temperature he could create in his workshop. Several stories purport to explain his selection of the reference point for 100°F; my favorite is that he selected the body temperature of a cow, a measurement that interested him and others in the agricultural com-

Lord Kelvin.

BOX 2.2

CREATING THERMOMETER SCALES

How do scientists such as Gabriel Fahrenheit and Anders Celsius create new concepts such as temperature scales? Many times, new discoveries stem from a scientist's personal interests or observations.

A cow's normal body temperature is the same as a human's—about 98.6°F. Yet some evidence indicates that Fahrenheit based his primary reference point for 100°F on the body temperature of a cow. If this story is true, the Fahrenheit scale was based on a sick cow! Some have claimed that Fahrenheit chose his value of 100°F so that the normal human body temperature would be 98.6°F, but this cannot be the case. While most people were aware that there was a normal human body temperature, its value was not determined until *after* the development of the thermometer scales. Instead, it seems that Fahrenheit based his scale on a measurement of interest to him.

Anders Celsius was a Swedish astronomer who had an interest in determining exactly how cold the chilly Swedish nights were as he walked to his observatory. Celsius wanted to know how warmly to dress and whether the metal telescope parts would be too cold to use in gazing at the vastness of the heavens. The Celsius scale grew out of these practical concerns.

As you can see, scientists often have the same kinds of problems nonscientists do. Sometimes, scientists view those problems as an impetus to study new ideas and invent new solutions.

munity (box 2.2). Starting with these primary reference points, Fahrenheit determined values of 32°F for the freezing point and 212°F for the boiling point of water. All subsequent thermometers were designed using the secondary reference points of 32°F and 212°F.

Anders Celsius (Swedish astronomer, 1701–1744) developed a different thermometer scale in 1742. Celsius designated the freezing and boiling points of water as 0°C and 100°C, respectively. His scale was originally called the centigrade scale because centi- means 1/100th part, and the reference points span 100 degrees. The Celsius scale has been adopted worldwide and is the official metric temperature scale.

The third temperature scale was the invention of Englishman Lord Kelvin (born William Thompson, 1824–1907). Lord Kelvin devised his scale in 1848. It is identical to the Celsius scale, but all temperatures are 273 degrees higher. Thus, the freezing point of water is 273 degrees K, the boiling point 373 degrees K. This thermometer scale was derived from observations of gases. We will discuss it in more detail later in this chapter.

Fahrenheit and Celsius temperatures may be converted one to the other using certain equations:

$$°F = 1.8\ (°C) + 32°F$$

$$°C = 0.56\ (°F - 32°F)$$

Celsius temperatures may be converted to Kelvin using the following equation:

$$K = (°C) + 273$$ (degree symbols are not used with degrees Kelvin)

The relationships between the three scales are outlined in figure 2.10.

SOURCE: By permission of Johnny Hart and Creators Syndicate, Inc.

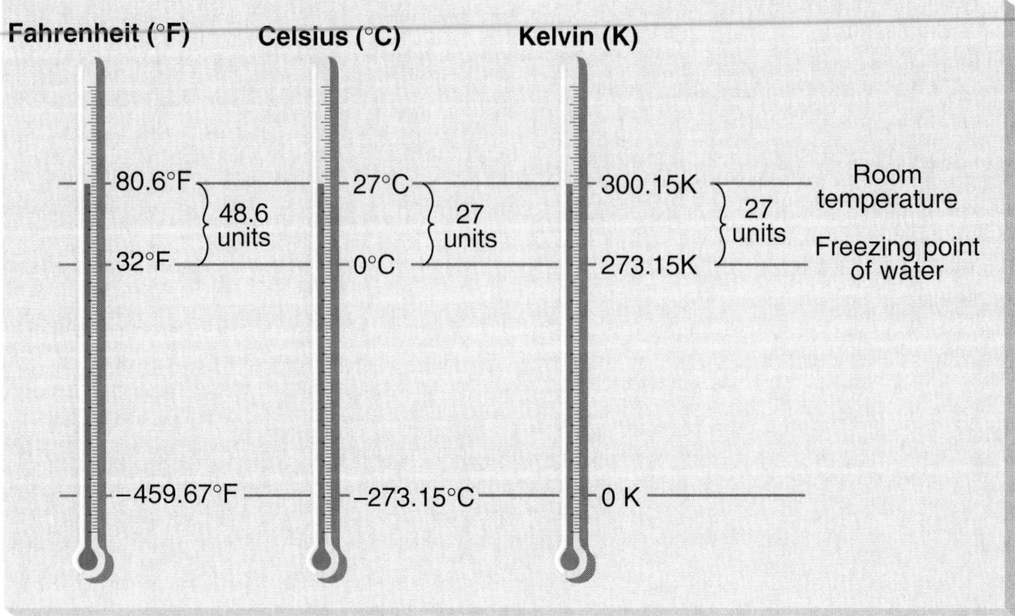

Figure 2.10 Comparison of the Fahrenheit, Celsius, and Kelvin (Absolute) thermometer scales.

PRACTICE PROBLEM 2.8

Convert each of the following temperatures.

a. If the thermometer reads 42°C, what is the temperature in °F?

ANSWER Apply the equation °F = 1.8 (°C) + 32°F. Multiply 42 (the °C value) by 1.8 to get 75.6; add 32 to yield the correct answer: 107.6°F.

b. If the thermometer reads 82.5°F, what is the temperature in °C?

ANSWER Apply the equation °C = 0.56 (°F – 32°F). First, subtract 32 from 82.5 to get 50.5. Then multiply 50.5 by 0.56 to yield the answer: 28.3°C.

c. If the thermometer reads 150°C, what is the temperature in degrees Kelvin?

ANSWER Apply the equation K = (°C) + 273. Simply add 150 and 273; the answer is 423 K.

THE GAS STATE

We live at the bottom of an ocean of gas called the atmosphere. Scientists have studied the gas state in great detail, and people use many gases daily, including those used in tires, aerosol cans, heating, and cooking. Gases have several specific properties:

1. They completely fill any container
2. They are highly compressible
3. They have low densities
4. They mix readily with other gases
5. They exert a constant pressure in all directions on the walls of a container

You can observe some of these properties using a balloon. If you blow up a small, round balloon without stretching it completely, you can see that the air expands to fill the entire balloon. If the balloon is small, you can compress or squeeze it in your hands and observe that all sides seem to have the same resistance to squeezing; that is, the gas exerts the same pressure on all sides. Finally, you will find that a similar balloon filled with the same volume of water is much heavier than an air-filled balloon. This demonstrates that the density of the gas (air) is much lower than the density of water. (Box 2.3 discusses the density of air versus the density of helium.)

Atmospheric Pressure and Barometers

Centuries ago, people realized air had weight. Ancient texts, including the Bible, allude to this fact. In modern times, we measure this weight, or the pressure air exerts in the atmosphere, using a **barometer.** Evangelista Torricelli (Italian, 1608–1647) invented the barometer in 1643.

Figure 2.11 shows a diagram for a simple barometer consisting

BOX 2.3

AIR PRESSURE AND HELIUM DENSITY—WHAT HAPPENS TO HELIUM-FILLED BALLOONS

Did you ever play with helium-filled balloons as a child? Perhaps you watched your balloon drift into the sky after you let go of the string and wondered why it floated. Helium-filled balloons rise in air because helium gas is less dense than air. But what happens as the balloon continues to rise? Does it eventually drift out of the earth's atmosphere and into space?

If you were to travel upwards, as your balloon did, you would soon realize that the air pressure decreases with each passing mile. This is why airplane cabins are pressurized for flights several miles above the earth. A balloon, however, does not travel in such a chamber. As it ascends, the air pressure and density decreases, and the internal pressure of the helium gas eventually becomes greater that the air pressure outside the balloon. The balloon expands, then finally bursts, due to this pressure. The fragments drift back to earth, probably unnoticed. Such is the fate of all helium balloons that drift away from their owners.

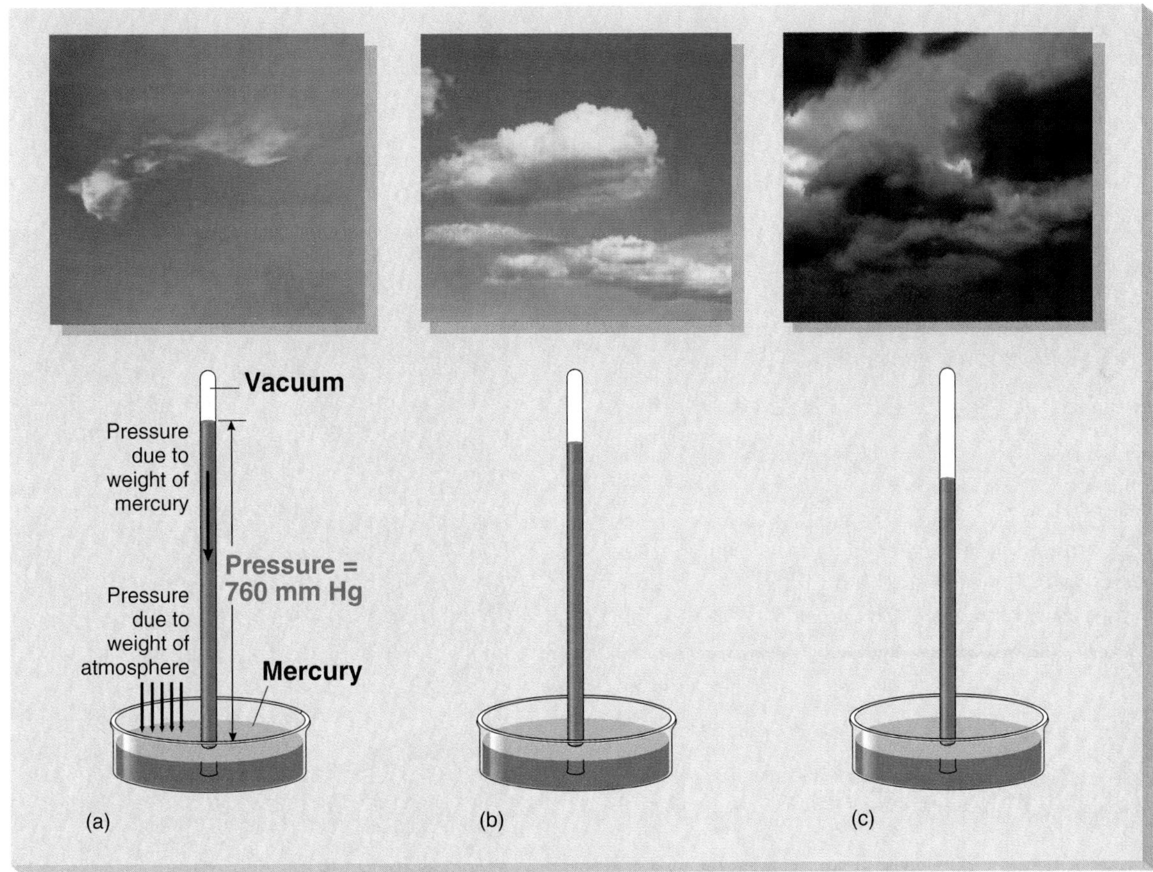

Figure 2.11 A simple barometer illustrating **(a)** high, **(b)** normal, and **(c)** low atmospheric pressure. Notice the weather patterns developing above each measurement. This is why a barometer is useful in weather prediction; the cloud patterns are closely related to the atmospheric pressure changes the barometer measures.

of a narrow tube closed at one end. This tube is placed in a liquid—usually mercury, though any liquid that does not evaporate rapidly works. A mercury barometer is less than 1 meter high, while a water barometer is at least 435 inches tall (over 11 meters).

Air pressure is considered normal when the mercury column is 760 mm (30 inches) high; this pressure measurement is called one standard atmosphere. Barometric pressure may also be measured in torr units, named after Torricelli. Torr units are expressed in mm Hg (Hg is the symbol for mercury). Although 760 mm Hg is considered "normal," the average or "normal" atmospheric pressure in some regions differs from this value. For example, a high elevation will normally show a lower atmospheric pressure, while a low elevation will generally have a higher value.

As atmospheric pressure increases, the added pressure forces additional mercury up the barometer tube. Thus, values above 760 mm Hg characterize areas of high pressure, while values below 760 mm Hg indicate areas of low pressure. These measures of air pressure are useful in weather prediction; nearly all weathercasts refer to high- and low-pressure zones. High pressure implies clear weather, while low pressure indicates possible rain or snow.

PRACTICE PROBLEM 2.9

What is the value in mm Hg for a barometric pressure of 29.75 inches?

ANSWER To answer this, merely convert the inches into millimeters. There are 25.35 mm in an inch. Therefore:

$$29.75 \text{ inches} \times 25.35 = 754.2 \text{ mm Hg, or } 754.2 \text{ torr}$$

Is this an example of low or high pressure? What kind of weather might this reading predict?

ANSWER A barometric reading of 29.75 is below normal, so this is an example of low pressure. Most likely, poor weather is on the way, unless the barometric pressure is rising.

EXPERIENCE THIS

Obtain daily barometer readings from newspaper, radio, or television weather reports for a week. Compare them against the weather each day. What correspondence, if any, did you see between barometric pressure and weather conditions?

Boyle's Law—Gases and Pressure

Robert Boyle (English, 1627–91) was a dominant figure in the early development of chemistry. He amassed a wealth of valuable data that established chemistry as a science. Part of this data resulted in **Boyle's Law,** which describes the relationship between the pressure and volume of a gas. This law can be expressed in several ways:

$$PV = k \quad \text{or} \quad P = \frac{k}{V} \quad \text{or} \quad V = \frac{k}{P}$$

In all these forms, P is the gas pressure, V is the gas volume, and k is a constant number.

This law, which Boyle first presented in 1660, appears to be the first mathematical relationship developed in chemistry. Probably the easiest way to envision Boyle's Law is to imagine a cylinder of gas under different pressures as varying masses are placed on it, as in figure 2.12. For this particular gas, the volume is 2.0 liters, when the gas pressure is 15 pounds per square inch; 1.0 liter when the pressure is 30 pounds per square inch; and 0.5 liters when the pressure is 60 pounds per square inch. If we apply Boyle's Law, we can discover a constant, k, for this gas:

$$PV = k$$
$$15 \times 2.0 = 30$$
$$30 \times 1.0 = 30$$
$$60 \times 0.5 = 30$$

Thus, k = 30 for this gas.

Robert Boyle.

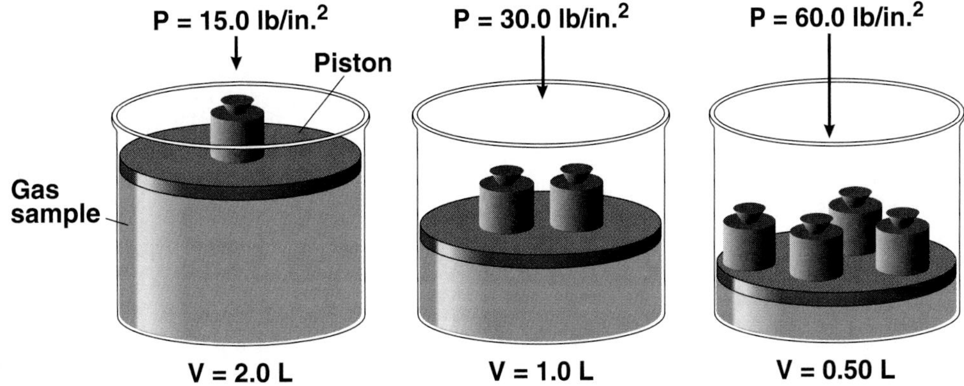

Figure 2.12 According to Boyle's Law, the volume of a gas decreases as more mass, or pressure, is applied.

SOURCE: From Jacqueline I. Kroschwitz, et al., *Chemistry: A First Course*, 3d ed.

A typical Boyle's pressure-volume relationship appears in table 2.3. First, note that the product (P × V) is a constant—k = 10. Next, notice that the volume *decreases* as the pressure *increases*, an inverse relationship.

Another way to describe an inverse relationship is to compare the pressure to the reciprocal of the volume (1/V). If you plot P against 1/V, as shown in figure 2.13, a straight line results. It is more difficult to visualize reciprocal volume because you always see volume directly. But it is easy to remember that volume decreases as pressure increases.

TABLE 2.3 The pressure-volume relationship for an ideal gas at 25°C.

P (atm)	V (liters)	PV = k
0.1	100.00	10.00
0.5	20.00	10.00
1.0	10.00	10.00
1.5	6.67	10.00
2.0	5.00	10.00
2.5	4.00	10.00
3.0	3.33	10.00
5.0	2.00	10.00

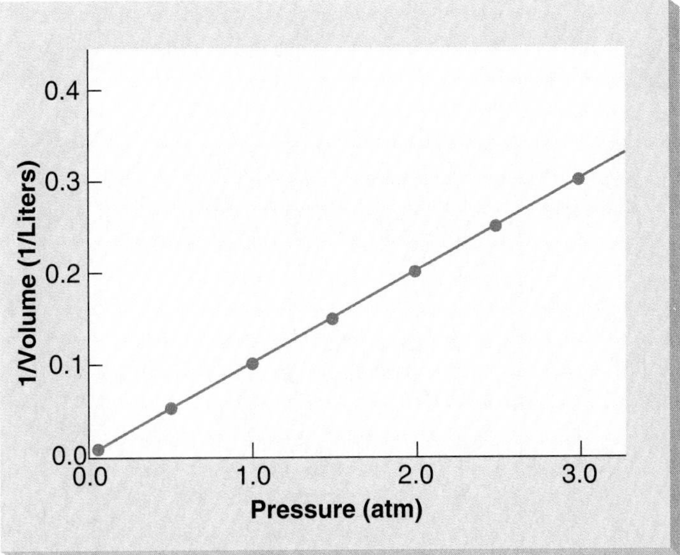

Figure 2.13 The reciprocal gas volume plotted against the pressure: as pressure increases, volume decreases.

EXPERIENCE THIS

Obtain a 12-inch piece of rigid plastic pipe wide enough to pass your hand through. (The pipe should not have holes in the sides.) Next construct a disc made from plastic, metal, or cardboard with a diameter just slightly smaller than the pipe's diameter. Partially inflate a 10-inch elongated balloon and carefully insert it into the pipe; if necessary, inflate the balloon further so that it touches the sides of the pipe, and then push the balloon to the bottom while resting the pipe on a table or another smooth, hard surface. Finally, place the disc over the top of the balloon. Gradually push the disc evenly down the pipe, noticing how much resistance the balloon offers. This illustrates Boyle's Law—the resistance represents the pressure increase as the volume in the balloon decreases.

PRACTICE PROBLEM 2.10

Suppose you doubled the pressure on a 10-L gas sample without adding more gas. What would happen to the volume?

ANSWER When you double the pressure, you halve the volume. If the initial pressure is P, the doubled pressure is 2P. According to Boyle's Law, $PV = k$; therefore, $2P \times V/2 = k$. In this case, the volume is 10 L, so $V/2 = 5$ L.

Suppose you reduced the pressure to only one-tenth of the starting value. What is your volume now?

ANSWER Boyle's Law always works the same way. Once again, $PV = k$; therefore, $P/10 \times 10V = k$. The volume is 10 L, so 10×10 L $= 100$ L. The pressure has reduced tenfold, and the volume has increased tenfold. This demonstrates the reciprocal relationship between pressure and volume.

Remember that Boyle's Law works both ways—when the volume changes, the pressure also changes inversely. Assume your gas sample is at an initial pressure of 1 atmosphere, or 760 torr. If you reduce the volume of your 10 L gas sample to only 5 L, without removing gas, what does this do to the gas pressure?

ANSWER When you halve the volume, you double the pressure. Since $PV = k$, we know that $P \times 10 = k$. This means that $2P \times 5 = k$. When the volume is reduced to 5 L, the pressure increases to 2×760 torr, or 1,520 torr (2 atmosphere).

Jacques Charles.

Charles' Law—Gases and Temperature

In 1787, Jacques Charles, (1746–1823), a French scientist interested in hot air ballooning, developed a second gas law now called **Charles' Law.** Charles' Law describes the interaction of gas volume and temperature:

$$V = k'T \quad \text{or} \quad \frac{V}{T} = k'$$

In this equation, V is the volume, T is the temperature in degrees K, and k′ is a constant (not the same number as in Boyle's Law). Since most Charles' Law applications involve changes in temperature or volume, the following equation is often more useful:

$$\frac{V_1}{T_1} = \frac{V_2}{T_2}$$

This equation considers what happens when the original temperature or volume changes.

Table 2.4 shows typical Charles' Law data; the same data is plotted in figure 2.14. The direct relationship between temperature and volume is evident in both places. When the temperature increases, volume increases as well. In fact, this principle underlies the use of steam as a source of power (box 2.4).

TABLE 2.4 The effect of absolute temperature on gas volume.

TEMPERATURE (K)	VOLUME (liters)
400	14.65
373	13.66
350	12.82
313	11.47
298	10.92
278	10.18
273	10.00
250	9.16
175	6.41
150	5.49
100	3.66

BOX 2.4

STEAM ENGINES—THE PRESSURE BEHIND THE POWER

Steam (the gaseous form of water) is perhaps the most common gas other than air. For centuries, inventors tried to utilize steam for human use; though their progress was slow, these scientists succeeded in harnessing the power of steam.

The story of steam begins almost two thousand years ago. Somewhere between the years 60 and 69 A.D., Hero (Greek, 20 A.D.–?) built a toy that used jets of steam to turn a kettle. However, Hero apparently could not think of any practical use for his toy, and nothing further developed from his studies. In fact, over 1,630 years elapsed before any significant advances occurred in the use of steam power.

During the 1600s, scientists expanded on Hero's early work. In 1690, Denis Papin (French, 1647–1712) became the first person to use steam pressure to move a piston. In 1698, Papin built an engine propelled by the pressure of steam rather than atmospheric pressure. Thomas Savery (English, 1650–1715) built a steam-powered pump in 1698 that was used to pump water out of coal mines; this patented device, known as the Miner's Friend, became the first practical steam-powered machine.

Inventors in the 1700s continued to refine these early machines and expand the use of steam. Thomas Newcomen (English, 1663–1729) devised the first practical steam engine in 1712, the year Papin died. In 1765, James Watt (Scottish, 1736–1819) refined Newcomen's ideas by building a steam engine that separated the condenser from the cylinder so that the steam acted directly on the piston. This produced about six times as much energy as the Newcomen engine. Later, Watt continued his work by experimenting with other forms of steam-powered engines.

Meanwhile, in 1769, Joseph Cugnot (French, 1725–1804) built a steam carriage that could carry four people and travel as fast as 2.25 miles per hour; this was probably the first true automobile. William Murdock (Scottish, 1754–1839) built a working model of a steam-powered carriage in 1784, and not to be outdone, Oliver Evans (American, 1755–1819), patented a steam-propelled land vehicle in 1789.

Other inventors experimented with steam power for other purposes. In 1776, John Wilkinson (English, 1728–1808) increased the efficiency of a blast furnace (used in iron production) by using a steam engine to blow air into the furnace. Twenty-four years later, two dozen English iron production facilities relied on steam-driven blast furnaces. In 1789, the first steam-powered cotton factory was built in Manchester, England. The use of steam rapidly spread to other industries; the London Times was being printed on a steam-driven cylinder press by 1814.

Perhaps the most famous use of steam power was Robert Fulton's (American, 1765–1815) invention of the steamboat in 1803. Although others had experimented with paddle wheel steamboats, Fulton's *Clermont* (figure 1) is considered the first practical and economic version. Interestingly, the first steam-powered warship was built in the United States in 1815, the year of Fulton's death.

As the nineteenth century progressed, steam power became the energy source for railroad locomotives and generators. In 1804, Richard Trevithick (English, 1771–1833) developed a steam locomotive that ran on iron rails and successfully hauled ten tons of iron a distance of ten miles. George Stephenson (English, 1781–1848) followed suit in 1825 with a steam locomotive capable of hauling 30 tons at speeds faster than horse-drawn systems could travel. Within a few years, the steam locomotive would become a major means of transportation and shipping. One might wonder whether the United States would have developed as rapidly if this "iron horse" had not been created (figure 2).

Figure 1 Robert Fulton's *Clermont* was the first practical and economical steamboat.

Figure 2 One of the mighty "iron horses" that criss-crossed the United States on rails.

In 1884, Charles Algernon Parsons (English, 1854–1931) designed and installed the first steam turbine generator for producing electric power. Today, over a century later, steam still powers the turbines that generate electricity. Though early turbines used wood, and then coal or oil, as fuel to create the steam, many generators now use a modern twist: nuclear power heats the steam that generates electric power.

By the year 1900, only 210 years after Papin resurrected steam power from a long period in limbo, steam-powered engines had become the mainstay of industry and transportation. Some innovations came and went, such as the Stanley Steamer automobile; others, such as steam-powered trains and power plants, remain in operation in many parts of the world today. All put a simple principle to use: an increase in temperature creates an increase in gas volume, and as a gas expands, it can be put to work.

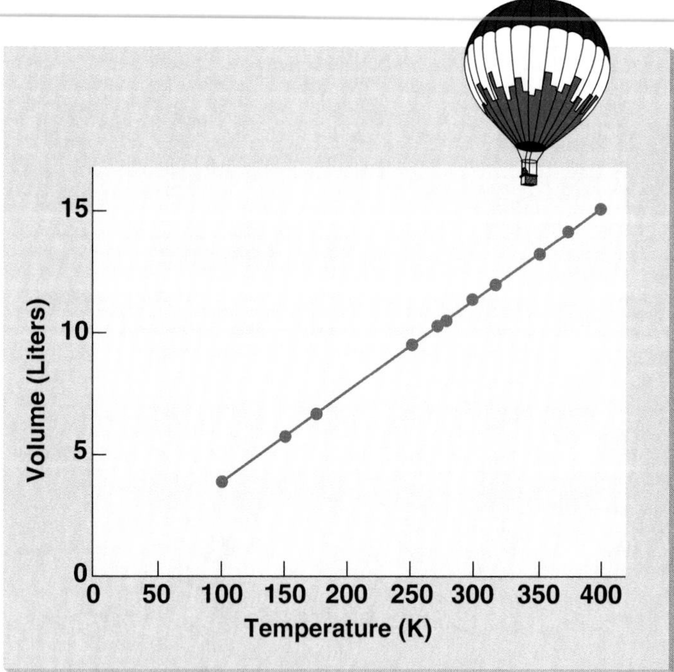

Figure 2.14 The direct effect of temperature, in K, on gas volume: as temperature increases, volume increases.

Refrigerators also apply Charles' Law to cool their contents. A gas, usually a chlorofluorocarbon or CFC, is allowed to expand in a sealed area, drawing heat from within the refrigerator and thus cooling the air within. CFCs are the best available method for this use. Although other gases, such as poisonous ammonia, work in refrigeration, they do not work as well or as safely.

EXPERIENCE THIS

Before your next long car trip (at least 25 miles at 55 to 65 mph), check and record the pressure in each tire. As soon as you complete your trip, immediately recheck and record the new tire pressures, and cautiously touch the tire sidewalls. They should be warm, possibly even hot. The temperature increases because the tires generate heat as they rub against the road. However, unlike balloons, tires are designed to prevent large volume increases—the volume of a tire cannot increase beyond a certain limit. This means that as the temperature rises, and the volume of air inside the tire remains the same, the pressure inside the tire must increase according to Charles' and Boyle's Laws. How much pressure did you measure just after you completed the trip? Did this reflect an increase?

EXPERIENCE THIS

Demonstrate Charles' Law for yourself. Obtain and partially inflate six round balloons to about the same size. Measure their diameters as accurately as possible. Place one balloon in a refrigerator and another in a freezer for a few hours. Remove and quickly remeasure each balloon's diameter. Determine the temperature of your refrigerator (most are about 40°F or 5°C) and the freezer portion (usually −10°F or −23°C).

Meanwhile, carefully place another balloon in a washable pan and put it in an oven with a pilot light. (If your oven is not gas, or lacks a pilot light, you could heat it slightly, but then turn it off; if the oven is too hot, the balloons will explode or the rubber will melt.) Watch carefully. After you see the balloon has expanded, remove it and quickly measure its new diameter. Determine the oven temperature with a household thermometer. (*Caution:* Never turn the oven on to heat with a balloon inside! Make sure it is off before you put the balloon in.)

Repeat this experiment with the other three balloons, and then tabulate all your data in a form similar to table 2.4. (If you'd like, you can convert the diameters to volumes by multiplying the diameters by 0.5236.) Did your balloons become *larger* as the temperature *increased* and *smaller* as the temperature *decreased?* If you examined a greater temperature range, you would see even larger changes.

PRACTICE PROBLEM 2.11

If the temperature of a 10-liter gas sample increases from 25°C to 50°C, what change occurs in the volume?

ANSWER Don't be fooled—the temperature did not simply double, as you might think at first. Charles' Law requires the use of K, not °C; you must first convert the temperatures by adding 273 to each. Thus, the initial temperature is 298 K (25 + 273 = 298), and the final temperature is 323 K (50 + 273 = 323). Now we can plug these numbers into the equation for Charles' Law: $V_1/T_1 = V_2/T_2$. Therefore,

$$\frac{10\text{ L}}{298\text{ K}} = \frac{V_2}{323\text{ K}}$$

If we multiply each side by 323 K,

$$\frac{10\text{ L }(323\text{ K})}{298\text{ K}} = V_2,\text{ or } \frac{3230}{298} = V_2$$

Therefore, V_2 = 10.8 liters. The volume increased by about 8 percent (8 percent of 10 is 0.8).

If a gas increases in volume from 10 liters at 25°C to 20 liters, what change occurred in the temperature?

ANSWER Obviously, if the volume increased, you expect an increase in temperature. Begin by converting the temperature to K, as before: $25 + 273 = 298$ K. Now solve for the unknown final temperature value, T_2, in the equation $V_1/T_1 = V_2/T_2$:

$$\frac{10\text{ L}}{298\text{ K}} = \frac{20\text{ L}}{T_2}$$

If we divide each side by 20 liters (recall that to divide a fraction, you invert it and multiply):

$$\frac{(298\text{ K})\ (20\text{ L})}{10\text{ L}} = T_2,\ \text{or}\ \frac{5960}{10} = T_2,\ \text{or}\ 596\text{ K} = T_2$$

Finally, we convert 596 K to °C by subtracting 273:

$$596 - 273 = 323°\text{C}$$

This is quite an increase from 25°C, which is about 77°F, or close to room temperature.

The Combined Gas Law

We have learned that when gas pressure increases, volume decreases (Boyle's Law) and that when temperature increases, volume increases (Charles' Law). Often, both temperature and pressure of a gas increase at the same time. What happens to the volume then? The Combined Gas Law answers this question by describing the interaction of pressure, temperature, and volume:

$$\frac{PV}{T} = k''$$

In this equation, P is pressure, T is temperature in degrees Kelvin (K), V is volume, and k″ is another constant.

An equivalent equation is generally a more useful expression:

$$\frac{P_1V_1}{T_1} = k'' = \frac{P_2V_2}{T_2}$$

The subscripts refer to different combinations of pressure, temperature, and volume which should maintain the same proportions and therefore yield the same value for the constant k″. The quantities P, V, and T are called *parameters*, a mathematical term meaning values that vary under different circumstances.

PRACTICE PROBLEM 2.12

Suppose V_1 is 10.0 L, T_1 is 300 K, and P_1 is 1.0 atmosphere (atm). When T_2 is increased to 400 K and P_2 increases to 3.0 atm, what happens to the volume?

ANSWER According to the Combined Gas Law, $P_1V_1/T_1 = P_2V_2/T_2$. Therefore,

$$\frac{(1.00\text{ atm})\ (10.0\text{ L})}{300\text{ K}} = \frac{(3.00\text{ atm})\ (V_2)}{400\text{ K}}$$

$$\frac{10}{300} = \frac{3V_2}{400}$$

$$0.0333 = 0.0075\ V_2$$

$$\frac{0.0333}{0.0075} = V_2$$

$$4.44\text{ L} = V_2$$

The volume decreases from 10.0 L to 4.44 L.

The Kinetic Molecular Theory of Gases

The Kinetic Molecular Theory of Gases explains the behavior of gases in general terms. Fortunately, the theory is not as complex as the name suggests. We can summarize it with six simple assumptions:

1. Gases are composed of tiny particles called molecules or atoms.
2. Gas molecules always move in straight lines.
3. No energy is lost when these particles hit each other or the container walls.
4. Little, if any, attraction exists between gas particles.
5. The space between gas particles is very large compared to the size of the particles. Likewise, the total volume of the particles is negligible compared to the total volume the gas occupies.
6. Temperature measures the average kinetic energy (energy of motion) for the gas particles.

Assumptions 1, 4, and 5 reiterate what we learned from the ball model in figure 2.7—that gas molecules tend to spread apart, with wide spaces between individual particles. This makes the gas state shapeless and accounts for gases' high compressibility. Assumptions 2 and 3 explain why gases fill their containers and exert pressure on the walls: the particles move straight out toward the walls and hit them (figure 2.15). We will see assumption 6 in action later.

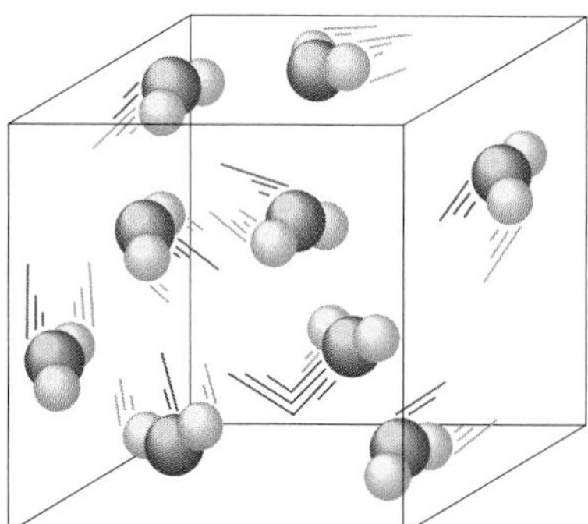

Figure 2.15 Diagram of the Kinetic Molecular Theory of Gases. Pressure arises as molecules strike the walls of the container.
SOURCE: From Jacqueline I. Kroschwitz, et al., *Chemistry: A First Course*, 3d ed. Copyright © 1995 Times Mirror Higher Education Group, Inc., Dubuque, Iowa. All Rights Reserved. Reprinted by permission.

Absolute Zero

We noted earlier that the Kelvin temperature scale was based on observations of gases. Reexamine figure 2.14. What would happen if the gas sample continues to cool? Eventually, as the line moves downward, the volume would decrease to 0.0 L at 0.0 degrees K. This temperature (−273°C) is thus called *absolute zero;* as matter approaches this temperature, it seems to also approach zero volume—to almost vanish altogether.

In reality, the volume does not totally vanish at absolute zero; the molecules become extremely compact. However, scientists have researched this super cold region for several practical reasons. For example, scientists first observed superconductivity (the ability of some materials to conduct electricity without showing resistance to its flow) near zero degrees K.

THE LIQUID STATE

Liquid molecules are less orderly than molecules in a solid and less disorderly than those in a gas. Likewise, most liquid physical properties are intermediate between the values for solids and gases.

Many chemical reactions occur in the liquid state, often in solution, and practically all the chemical reactions that keep us alive take place in aqueous solutions Unsurprisingly, then, the historical development of chemistry rested in large part on the study of water-based reactions. Even today, chemistry often concerns itself with reactions in aqueous solutions. Box 2.5 explores this liquid further.

Liquids have several unique properties. Molecules below the surface of a liquid exert an inward pull on surface molecules, creating **surface tension.** We can measure this force by placing a capillary (small-diameter tube) into the liquid, as shown in figure 2.16. You've seen water rise up in a straw due to surface tension; the effect is greater with a smaller-diameter tube. The liquid ascends the tube until the surface tension is counterbalanced by the mass of the liquid. In effect, then, surface tension "pulls" the liquid up the tube until the force of gravity exerts an equal pull downward. This balance of opposing forces creates an **equilibrium.**

Surface tension has practical applications that we put to use every day. Soap and dishwashing detergents, for example, contain chemicals called surface-active agents that reduce surface tension and permit water to displace the dirt. Straws take advantage of surface tension to move liquids up against the force of gravity.

Another property of some liquids is immiscibility. **Immiscible liquids** do not dissolve in each other; instead, the top liquid spreads over the lower one because liquids always flow and spread.

BOX 2.5

A FEW DROPS OF WATER

Water (H_2O) is not a typical liquid. Its high boiling point and unusual density characteristics give it unique properties.

First, water's boiling point is nearly 190° higher than expected due to the strength of its hydrogen bonds (to be discussed in detail in chapter 6). These hydrogen bonds also give water a second unique quality—they make solid water (ice) less dense than liquid water. Usually, the solid form of a material sinks to the bottom of the liquid form because the solid is denser. But because water's hydrogen bonds force the molecules into an open hexagonal structure, rather than the close-packed arrangement typical of a solid, ice is less dense than water and thus floats.

Finally, the density of water is higher than most common liquids, such as those in petroleum. Oil spills are disastrous partly because of this–the lighter crude oil floats on the surface of the water rather than sinking to the bottom.

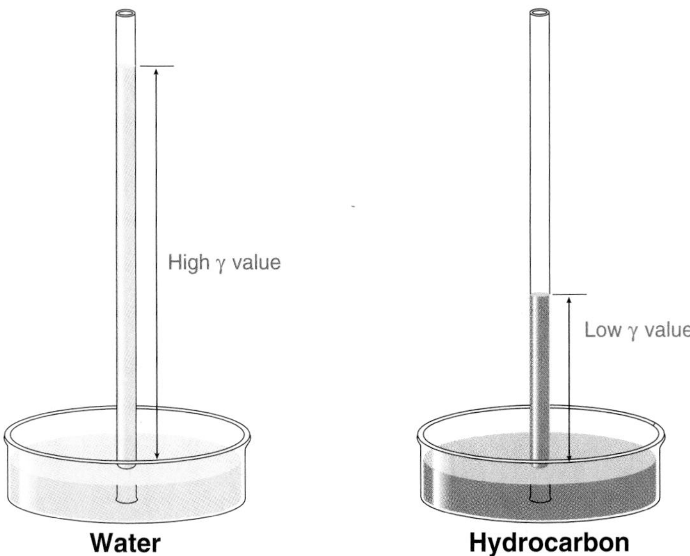

Figure 2.16 Surface tension, γ, is measured by capillary action. The greater a liquid's surface tension, the higher it rises in the capillary tube.

Figure 2.17 Oil spreading across a water surface demonstrates immiscibility.

The spreading is controlled by two main factors—density and wetting. Chemists can demonstrate immiscibility by pouring an oil, such as a cooking oil or kerosene, on water in a dish or tray (figure 2.17). An oil spreads over only part of the water surface and then stops, its edge normally curving down to meet the water surface. The mass of the upper liquid forces it to spread out to its lowest level, but the lack of wetting prevents it from completely covering the water surface.

An **emulsion** is a suspension or dispersion of one immiscible liquid in another, usually in the form of droplets. Emulsions tend to separate into separate immiscible layers unless an emulsifying agent, usually a surface-active agent such as soap, is present. The most common emulsions contain oil dispersed in water and are therefore called O/W emulsions. You can find several O/W emulsions in home refrigerators including milk, mayonnaise, and salad dressings. These foods contain surface-active agents to maintain emulsion and prevent separation into immiscible layers, but these agents differ from soap, which tastes bitter and causes diarrhea. The edible emulsions instead contain a natural emulsifier, such as lecithin.

Water can also emulsify in oil to create W/O emulsions. When your automobile engine oil picks up moisture, it forms a W/O emulsion that leaves a tan, milky coating on the refill cap. (If you see this coating, change your oil, or rust might form inside the engine.) Margarine usually has some water dispersed in a W/O emulsion to give it a more pleasant texture and taste; water-soluble flavoring and coloring agents are also dispersed within this W/O emulsion. The new "light" margarines contain more dispersed water than the traditional versions, making them lower in fat and calories.

EXPERIENCE THIS

Place about one part cooking oil or kerosene and two parts water in a jar. The lower-density oil will float on top of the water. Shake the mix. What happens to the oil? It should break up into tiny droplets. Next, stop shaking the jar and let it rest for a few minutes. What happens to the oil now? Presumably, the droplets coalesce (merge together) and the liquids separate into two distinct, immiscible layers.

You can emulsify the oil and water by adding a small amount of ordinary soap or dishwashing detergent. Shake it again to separate the oil into droplets, then let it rest again. What has happened to the droplets this time? Do they remain emulsified (dispersed) in the water rather than coalescing?

Oil-soap-water emulsions can last for many months without separation because the soap forms a coating around each oil droplet, preventing coalescence.

Dispersions are similar to emulsions but contain scattered bits of solids in a liquid, rather than drops of one immiscible liquid in another. Poison ivy lotions and latex paints are dispersions. A mist or fog is a suspension or dispersion of a liquid in a gas.

PRACTICE PROBLEM 2.13

Which of the following materials are emulsions? Which are dispersions?

- **a.** A milk shake
- **b.** Kool Aid
- **c.** French dressing
- **d.** Thick tomato juice
- **e.** A soft drink
- **f.** Vanilla ice cream
- **g.** Interior latex wall paint

ANSWER Examples *a, c,* and *f* are emulsions; they all contain fat (oil) dispersed in liquids. Examples *d* and *g* are dispersions (assuming the tomato juice contains pulp); they contain solids dispersed in liquids. Examples *b* and *e* are solutions, because they contain homogeneous mixtures of dissolved ingredients.

THE SOLID STATE

Solids may be either amorphous or crystalline in structure. **Amorphous solids** consist of molecules in no organized arrangement. These solids do not form crystals—their atomic arrangement is intermediate between that of a disordered liquid and a highly organized crystalline solid. Amorphous solids also lack the definite melting points characteristic of crystalline solids. Typical examples include glass, rubber, and most plastics.

Crystalline solids consist of regularly repeated, organized patterns of molecules. This creates a distinct crystal shape, as in table salt or sugar. Crystalline solids occur in four types: metallic, molecular, ionic, and macromolecular.

The structure of a solid determines many of its physical properties and qualities. The atoms in a metal are arranged in a crystal lattice, or geometric pattern, with each atom independent of the others rather than bonded in molecules. **Metals** have a wide range of melting points and are insoluble in essentially all liquids. They conduct both electricity and heat. Metals are also malleable—you can beat or press them into a shape—and ductile, which means you can stretch them into wires or thin films. These properties make metals versatile—we form steel into cups or press it into tire rims, draw copper out into threadlike wires, and wrap leftover food in aluminum foil.

Molecular solids are nonmetallic molecules arranged in a lattice. Molecular solids melt at relatively low temperatures and are poor conductors of both heat and electricity; they are used as insulators for this reason. While molecular solids form nonaqueous solutions, most are insoluble in water. Typical examples include table sugar (sucrose), iodine, carbon dioxide (dry ice), and most solid organic compounds. Of these examples, only sugar dissolves well in water.

Ionic solids form lattice structures, but ions rather than atoms occupy the lattice points. An **ion** is an electrically charged atom or group of atoms. Ionic solids have high melting points, are often water soluble, and conduct electricity when molten or in water solution. Typical examples include table salt, saltpeter (potassium nitrate), and magnesium sulfate (Epsom salts).

Macromolecular solids are a more complex form of matter. The prefix macro means giant, and macromolecular solids consist of many thousands of atoms connected into long chains or rings. These materials have very high melting points and are insoluble in essentially all liquids. Macromolecular solids are either nonconductors or semiconductors (that is, materials that partially conduct electricity). Typical examples include sand and diamond.

PRACTICE PROBLEM 2.14

Which of the following materials are amorphous solids?

- **a.** Plastic plumbing pipe
- **b.** A rubber tire
- **c.** A cut-crystal glass goblet
- **d.** A Tupperware container
- **e.** Copper plumbing pipe

ANSWER Examples *a, b, c,* and *d* are amorphous solids; copper *(e)*, like most metals, is a crystalline solid. (The term *crystal* in example *c* describes how the goblet's surface is cut and has nothing to do with the arrangement of the molecules.)

EVAPORATION, VAPOR PRESSURE, AND BOILING POINTS

Thus far we have considered matter, states of matter, temperature, and pressure as separate entities. In reality, these forces and states of matter constantly interact. Evaporation, vapor pressure, and boiling point illustrate common interactions between these forces and the phases of matter.

Evaporation occurs as liquid molecules or atoms convert to the gas state. In an open container, some of these molecules "float" out of the liquid and into the air; air currents transport them elsewhere so that the liquid seems to gradually disappear (as we observed in Chapter 1, in the section entitled "Scientific Method in Action."). In a closed container, at constant temperature, molecules still "float" out of the liquid and into the air enclosed in the container. But because they cannot escape the container, some of these gas molecules condense within it; they then become liquid once again and drop back down into the liquid. When the numbers of molecules evaporating and condensing are roughly equal, equilibrium exists. (Remember that equilibrium is a balance of opposing forces.) While equilibrium looks like a state of rest, it isn't; the molecules are in constant motion and constantly changing. Equilibrium is a fairly common condition in nature.

PRACTICE PROBLEM 2.15

Our sense of odor detects chemical in the gas phase. Why can we smell liquids such as perfumes and solids such as soaps or solid air fresheners?

ANSWER The materials either evaporate from the liquid state or sublime from the solid so that some of the molecules travel through the air to our noses.

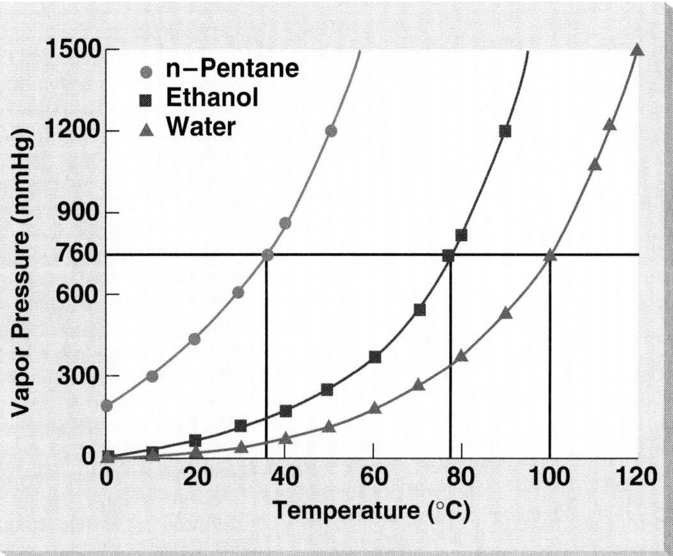

Figure 2.18 Vapor pressure of three liquids at various temperatures. When the vapor pressure equals atmospheric pressure, the liquid is at boiling point (note the vertical lines).

Vapor pressure is the equilibrium pressure created by the gas form of a liquid when it evaporates in a closed container. The normal pressure of the air in a closed container is 760 torr. As the liquid evaporates, its molecules contribute to the total gas pressure inside the container; that portion of the gas pressure created by the liquid's vapor, the vapor pressure, is also measured in torr (mm Hg). Vapor pressure depends on the temperature and the nature of the liquid. The pressure always increases as the temperature rises (direct relationship) but decreases as a liquid becomes more polar (inverse relationship).

Polarity is a measure of how molecules interact with each other; the interaction increases with increasing polarity. Figure 2.18 shows that the vapor pressure of each of three liquids increases as the temperature increases. But because water is the most polar, the temperature must increase more to produce the same increase in vapor pressure. Ethanol has the next greatest polarity, n-pentane the least.

A liquid's normal boiling point is the temperature at which the liquid's vapor pressure equals the normal atmospheric pressure (the horizontal line at 760 mm Hg in figure 2.18). The vapor pressure continues to increase after the temperature rises beyond the boiling point. However, recall that the temperature of the substance plateaus at the boiling point (figure 2.6); it does not rise until all the liquid converts into the gaseous state. Also, note that an increase or decrease in pressure can change a liquid's boiling point.

Pressure Cookers and Vacuum Distillers

Pressure cooking applies the relationship between pressure and temperature to a practical purpose. In a closed container, boiling does not occur until the vapor pressure equals the total gas pressure. If the applied pressure in the container is 1,075 mm Hg, water will not boil until it reaches a temperature of 110°C (figure 2.18). When you seal water in a strong walled vessel and heat it until the water partially vaporizes, the water vapor pressure adds to the air pressure already within the container, increasing the total pressure. Food cooks in this device at 110°C rather than 100°C. In general, chemical reactions occur twice as fast for each 10°C increase in temperature. This means potatoes in a pressure cooker cook in fifteen minutes instead of thirty.

This principle also works in reverse. If the total pressure is reduced to 526 mm Hg, water boils at 90°C, and potatoes require about an hour to cook instead of thirty minutes. This explains why cooking takes longer at higher altitudes, where the atmospheric pressure is lower. It thus takes longer to cook a baked potato on top of Mt. Everest than at the bottom of the Grand Canyon.

If the pressure is reduced inside a sealed vessel, the liquid still boils when the vapor pressure equals the internal pressure. Thus, water boils at 0.0°C if the pressure is only 5 mm Hg. This ability to change the boiling point enables a purification technique called **vacuum distillation.** One advantage of this technique is that a chemist can distill a liquid which decomposes at its normal boiling point by distilling it in a vacuum. Vacuum distillation requires less heat than atmospheric pressure distillation, but energy is needed to reduce the pressure in the vacuum—there is no "free lunch" in chemistry.

PRACTICE PROBLEM 2.16

Food dehydration, one method of preserving foods for later use, removes the moisture from food. Moisture can be removed by boiling, but some foods change in quality when heated. How could you avoid these undesired changes?

ANSWER You can dehydrate food without appreciably changing the quality by dehydrating at a lower temperature. By sealing the food in a low-pressure vacuum, you lower the boiling point of the water in the food. You can then remove the moisture at a low temperature. (Freeze drying employs this method to dehydrate food.)

SUMMARY

Everything on earth—air, food, water, our own bodies—is made up of matter. *Matter* is therefore such a general term that we must classify different types of matter to discuss it.

Broadly speaking, matter may be either homogeneous (uniform) or heterogeneous (nonuniform). Homogeneous matter may be either solutions (mixtures) or pure substances (elements or compounds). Solutions may be solids (such as alloys), liquids (nonaqueous or aqueous), or gases. In fact, any type of matter exists in one of these three states.

Each state of matter has specific characteristics. Gases, for example, have low densities, are highly compressible, and mix readily. Boyle's Law states that an increase in pressure diminishes gas volume (an inverse relationship), while Charles' Law establishes that an increase in temperature increases gas volume (a direct relationship). The Combined Gas Law describes the combined effects of pressure and temperature, while the Kinetic Molecular Theory explains the general behavior of gases.

Liquid molecules are less orderly than solid molecules and more orderly than gas molecules. Similarly, liquids are denser than gases but not as dense as solids, and this makes them intermediate in compressibility as well. Their in-between status gives them unique properties, including surface tension, and in some cases, immiscibility. The quality of immiscibility allows liquids to form emulsions when mixed.

Solids may be highly organized crystal structures or disordered amorphous structures. Crystalline solids may be metallic, molecular, ionic, or macromolecular; each is characterized by different properties. Solids are the most dense and the least compressible of the three states of matter.

Evaporation, vapor pressure, and boiling point illustrate the interactions of pressure, temperature, and the three states of matter. Evaporation occurs when molecules convert from liquid to gas and are transported away in the air; vapor pressure is created as a liquid evaporates in a closed container. The boiling point is the temperature at which a liquid's vapor pressure equals the surrounding air pressure. An increase or decrease in pressure can change a liquid's boiling point, but energy is required to change the pressure.

KEY TERMS

READINGS OF INTEREST

General

Your local newspaper probably has a science or health page that reports the latest scientific news. Be certain to examine these reports using critical thinking.

Dagani, Ron. "Ancient fullerenes." *Chemical & Engineering News*, August 1, 1994, pp. 4–5. Deposits containing fullerenes, laid down about 1.85 billion years ago, have been disovered.

New Techniques

Leutwyler, Kristin. 1994. "Chiller thriller." *Scientific American*, 270:24–25. A lab in Finland claims it has achieved temperatures near or below absolute zero.

Halsey, Thomas C., and Martin, James E. 1993. "Electrorheological fluids." *Scientific American*, 269:58–64. These liquids solidify instantly when exposed to an electric field. They could have some novel uses.

Klingenberg, David J. 1993. "Making fluids into solids with magnets." *Scientific American*, 269:112–113. Describes a simple iron filing/corn oil mixture that solidifies in a magnetic field and that can then be stacked up on one side of a container. To create the mixture, you must have two 100-pound bar magnets (about $12 each).

Seeing Atoms

West, Paul. "Growing large by seeing small things." *Chemtech*, January 1994, pp. 47–51. The scanning tunneling microscope, which can visualize atoms, has some commercial applications.

Gross, Neil; Smith, Emily; and Carey, John. "Window on the world of atoms." *Business Week*, August 30, 1993, pp. 62–64. An easy-to-read article about how the scanning tunneling microscope works and some of its applications.

"Sizing up atoms with electron holograms." *Science News*, September 28, 1991, p. 199. This article describes how scientists are continuing to look even more closely at the smallest bits of matter.

Trefil, James. 1990. "Seeing atoms." *Discover*, June 6, 1990, pp. 54–60. This article tells how scientists can now "see" individual atoms and molecules. Includes many interesting illustrations.

Scientific Method

Sardella, Dennis J. 1992. "An experiment in thinking scientifically: Using pennies and good sense." *J. Chem. Ed.*, 69:933. This experiment requires pennies of varying mint dates and a simple balance to weigh them. Do all pennies weigh the same? If not, why? What could make pennies different from each other?

Wynn, Charles M. 1991. "Does theory ever become fact?" *J. Chem. Ed.*, 69:741. This one-page article explains why and how theories become facts.

Sauls, Frederick C. 1991. "Why does popcorn pop? An introduction to the scientific method." *J. Chem. Ed.* 68:415–16. A good, simple everyday example of the scientific method in action.

Metals, Chips, and Alloys

"Land of bronze." *Discover*, December 1991, pp. 63–66. An article discussing the alloy bronze and the "bronze age."

"The new gold rush." *U.S. News & World Report*, October 28, 1991, pp. 44–54. Modern day gold prospecting.

"Copper-alloy metallurgy in ancient Peru." *Scientific American*, July 1991, pp. 80–86. A South American civilization that flourished more than a thousand years ago used copper alloys.

Boraiko, Allen A. 1982. "The chip: Electronic mini-marvel." *National Geographic*, 162:421–57. Microchips are everywhere—in our watches, kitchen devices, computers, and more. This article discusses how they are made and some unusual uses for them, such as in artificial limbs and in robots.

Johnston, Moira. 1982. "California's Silicon Valley." *National Geographic*, 162: 459–77. While it's no longer the exclusive microchip metropolis it once was, the Silicon Valley is part of a fascinating tale of how new technologies reshape our lives and our futures.

The Gas State and Vapor Pressure

Harby, Karla. "Ballooning to safety." *Discover*, December 1993, p. 44. Hot air balloons could be used to evacuate people from high rise buildings.

PROBLEMS AND QUESTIONS

1. List some common examples of mixtures.

2. List some common pure materials.

3. Briefly define each of the following terms and give an example of each:

a. heterogeneous matter
b. homogeneous matter
c. solution
d. compound
e. element
f. molecule
g. atom
h. alloy

4. Describe the three states of matter, and list the important physical characteristics of each.

5. How do intensive and extensive properties differ?

6. Which of the following are extensive and which are intensive properties: Why?

- **a.** mass
- **b.** volume
- **c.** shape
- **d.** density
- **e.** compressibility
- **f.** melting and boiling points

7. What is the density of a sample when a 6.0 g sample occupies 2.5 ml?

8. Draw a ball model to illustrate each state of matter.

9. What is the difference between a direct and an inverse relationship?

10. What is the difference between heat and temperature?

11. Convert the following Celsius (°C) temperatures into Fahrenheit (°F).

- **a.** 47°C
- **b.** 37°C
- **c.** 65°C

12. Convert the following Fahrenheit (°F) temperatures into Celsius (°C).

- **a.** 83°F
- **b.** −26°F
- **c.** 105°F

13. Convert the following Kelvin temperatures into °C.

- **a.** 407 K
- **b.** 43 K
- **c.** 1,500 K

14. Describe Boyle's Law and show the equation associated with it.

15. Describe Charles' Law and show the equation associated with it.

16. If we increase the pressure on a 10.0-liter sample of air from 1.0 to 2.5 atmospheres, what will the final volume be?

17. If we reduce the pressure on a 15.0-liter air sample from 1.0 to 0.67 atmosphere, what will the final volume be?

18. If we raise the temperature of a 10.0-liter air sample from 25°C to 50°C, what will the final volume be?

19. If we lower the temperature of a 20.0-liter air sample from 20°C to −50°C, what will the final volume be?

20. Outline the six basic assumptions of the Kinetic Molecular Theory of Gases.

21. Define each of the following terms:

- **a.** surface tension
- **b.** emulsion
- **c.** dispersion
- **d.** immiscibility

22. List some examples of emulsions and dispersions.

23. What is vapor pressure? How does it relate to the boiling point of a liquid?

24. Why do pressure cookers cook food faster?

25. Define each of the following terms:

- **a.** crystalline
- **b.** amorphous
- **c.** sublimation
- **d.** evaporation

26. Give some examples of each of the following types of solids:

- **a.** crystalline solid
- **b.** amorphous solid
- **c.** metal
- **d.** molecular solid
- **e.** ionic solid
- **f.** macromolecular solid

CRITICAL THINKING PROBLEMS

1. Is it easier or more difficult to smell a substance when it is cold?

HINT Both sublimation and evaporation decrease as temperature decreases. To smell anything, you must detect some chemical in the air.

2. How does caffeine intake affect the human body in terms of shakiness and wakefulness? Is this a direct or inverse relationship?

3. What effect does ethanol intake (in alcoholic beverages) have on an individual's motor vehicle control? Is this a direct or inverse relationship?

4. Does blowing up a balloon illustrate Boyle's Law, Charles' Law, neither, or both?

HINT You add more gas as you blow up a balloon.

5. Is a tire blow-out an illustration of one of the gas laws? If so, which one? If not, why?

6. If the weather forecast reports a barometric pressure of 29.31 inches, what is the pressure in torr (mm Hg)? in atmospheres? Is this a high or low pressure? How is the humidity affected by high- and low-pressure weather fronts?

7. Why is the atmospheric pressure normally lower at high altitudes than at sea level? In what ways might this affect humans?

8. Mass and weight are closely related yet not identical concepts. Which does a barometer measure—the mass of the air or the weight of the air?

9. Many solids that appear uniform on casual inspection are actually composed of several materials. A mixture of table salt (chemical name: sodium chloride) with ordinary sugar (chemical name: sucrose) looks homogeneous to the human eye. But if you examine this mixture with a magnifying glass, you see two different crystal shapes, and if you dissolve it in water you obtain a homogeneous solution that tastes both sweet and salty. The salt–sugar mixture is heterogeneous, but the water solution is homogeneous. Why? How might you determine whether a salt and white sand mixture is homogeneous or heterogeneous?

HINT Do salt and sand both dissolve in water?

10. Extensive and intensive characteristics exist outside of science. Consider the following descriptions and decide which relate to extensive characteristics and which relate to intensive characteristics. Explain why you made each choice.

HINT Remember that extensive characteristics depend on quantity, and thus can change when the quantity changes, but intensive ones do not; quantity may involve time or energy as well as physical resources.

- **a.** Congress's ability to pass new laws
- **b.** Congress's ability to fund a new project
- **c.** Your ability to buy a new piece of exercise equipment
- **d.** Your ability to use a piece of exercise equipment
- **e.** The color of your eyes
- **f.** Your height

11. Why is Boyle's Law a law instead of a theory or a hypothesis?

12. Could you heat your house by expanding a gas?

HINT Refrigerators, air conditioners, and heat pumps transfer heat; but how will you expand the volume without applying heat?

13. The density of ice is lower than the density of liquid water. What if the density of ice was greater? Would life exist on earth?

14. Gasoline is a liquid solution whose vapor is burned to power car engines. If gasoline had the exact same composition year round, how would it affect the ease of starting your car?

HINT Assume the gasoline starts the car perfectly in warm summer weather.

15. Nanotechnology, the use of very small devices, is likely to become very important in our future. What role does chemistry play in this new technology? You may need to "read between the lines" to answer this question. How are these devices made and where does chemistry enter into the process? Two useful background articles are: (1) Gary Stix, *Scientific American*, Nov., 1992, pp. 106–117, "*Micron Machinations*," (2) Christie R. K. Marrian, Elizabeth A. Dobisz, and Orest J. Glembocki, *R&D Magazine*, Feb., 1992, pp. 123–126, "*Nanofabrication: How Small Can Devices Get?*" Some semi-popular books exist on this subject.

CHAPTER 3

THE COMPOSITION OF MATTER

OUTLINE

KEY IDEAS

1. How the Discipline of Chemistry Developed, and the Main Periods in its Development
2. The Major Discoveries Chemists Made During Each Period
3. How Chemists Developed the Theory of Oxidation, the Law of Definite Proportions, and the Law of Conservation of Mass
4. The Essential Features of Dalton's Atomic Theory
5. How the Atomic Theory Applies to the Structure of Matter
6. The Differences between Elements, Allotropes, and Compounds
7. How Chemists Write Chemical Reactions or Equations
8. How to Calculate the Molecular Mass or Weight for a Compound
9. The Basic Features of Four Analytical Techniques Used in Chemistry: Distillation, Recrystallization, Chromatography, and Spectroscopy
10. How Scientists Handle Very Large or Very Small Numbers

"Men love to wonder, and that is the seed of science."
RALPH WALDO EMERSON (1803–82)

Alchemy laboratory, 14th–15th century.

Although atoms make up all matter, the concept of the atom took centuries to fully develop. Likewise, chemistry as a science did not suddenly appear in full bloom; scientists spent centuries exploring chemical principles. Some early ideas about matter were not only vague—they were completely wrong.

THE DEVELOPMENT OF ATOMIC THEORY

Atomic Theory in the Early Greek Civilization

Although the early Greeks were the first to conceive of atoms, atoms were never at the forefront of Greek thought. In approximately 450 B.C., Empedocles (492–435 B.C.) divided matter into four elements: air, earth, water, and fire. About twenty years later, Democritus (470–375 B.C.) proposed that matter consisted of tiny, indivisible entities called atoms. Although several other philosophers, called atomists, adhered to the idea that matter was ultimately indivisible, Socrates, Plato, and Aristotle—none of whom were atomists—dominated the mainstream of Greek philosophy.

Greek architecture.

Socrates was almost completely disinterested in the physical realm. In 407 B.C., Plato became a student of Socrates, and about twenty years later (390 B.C.), he founded his own philosophy school. Plato redefined Empedocles's elements; he believed all four elements were different-shaped structures of a primary substance he called hyle. Accordingly, Plato taught, fire was a tetrahedron, air an octagon, earth a cube, and water an icosahedron (twenty-sided structure). However, Plato rejected the atomic theory, and his ideas regarding matter are not considered important today.

Plato's most famous pupil was Aristotle. From about 340 to 331 B.C., Aristotle taught that space is always filled with matter. However, Aristotle rejected the atomic theory and believed matter could subdivide indefinitely. Although many aspects of Aristotelian philosophy are still appreciated today, his theories regarding matter are not considered valid. Aristotle did, however, make some contributions to science. For example, he classified 500 animal species into eight classes in approximately 350 to 341 B.C.

After Epicurus (341–270 B.C.) founded the Epicurean philosophical school in about 300 B.C., the atomic theory revived briefly. The Epicureans considered atoms a fundamental unit of matter. Epicurus ascribed weight, shape, and size to the atoms. Later, from about 286 to 268 B.C., Strato seemed to advocate the existence of atoms; he conjectured that air was composed of particles in a void, and that these particles could crowd closer together or pull further apart. Carus (98–55 B.C.), another Greek philosopher, wrote about atoms in about 65 B.C. After this, however, the concept of atoms seemed to fade from Greek philosophy. Some of the ideas, especially Strato's, were carried on to Alexandria, Egypt, where scientists practiced more applied science. It was here that people first attempted chemical experimentation, but they experimented little, if at all, with atomic theory.

Atomic Theory in the Age of Alchemy

Alchemy, a precursor of modern chemistry, was a laboratory-based attempt to produce gold from some less expensive material, such as lead or even dirt. Western alchemy arose in Alexandria, possibly as early as 100 B.C.; this Hellenistic (Greek) phase lasted until about 600 A.D. After that, most western alchemists were Islamic until about 1100.

Eastern or Chinese alchemy began before 175 B.C. and lasted until at least 1000 A.D. Since the Arabs had considerable contact with the Chinese via the Silk Road (a trade route between China and the Mediterranean), some aspects of Chinese alchemy probably permeated the Islamic version, and from there spread slowly into western alchemy. The main thrust of the Age of Alchemy probably occurred from about 300 to 1200 A.D.

Hellenistic alchemy sprang from Aristotle's writings. The Greek philosopher believed one element could transform into another by exchanging qualities. For example, Aristotle theorized

SOURCE: © 1995 Sydney Harris.

that metals form when exhalations become entrapped in dry earth. The metals then "mature" in the ground. Aristotle reasoned that this process might continue until the baser metals transform into gold. In a sense, alchemy's objective was to speed the maturation process artificially.

Alchemists thus experimented with the transmutation of metals (for example, changing lead into gold); they searched for a philosopher's stone they believed would make transmutation possible. Alchemists also sought an alkahest (a liquid that would dissolve anything) and a panacea (a remedy for all diseases). Aristotle believed that each of the four elements had two essential qualities: earth, dry and cold; air, wet and hot; fire, hot and dry; water, wet and cold. Earth could therefore transform into fire if its cold quality changed into hot, and air could transform into water if its hot quality changed into cold.

The idea of transmuting lead into gold was a logical outgrowth of these beliefs. If one element could transform into another, then an inexpensive element such as lead could transmute into gold. Some dishonest alchemists painted materials to look like gold and traveled from village to village selling their fakery to unwary people. Alchemists caught perpetrating such a hoax were often executed.

During the Age of Alchemy, no one explored atomic theory. Still, alchemists invented useful experimental apparatuses, such as distillation equipment, and discovered many important chemicals, including sulfuric and nitric acids. These items proved useful in later periods and laid the foundations for chemical study.

Atomic Theory in the Iatrochemistry Age

The Iatrochemistry Age (the Greek prefix *iatro* means medicine), which placed chemistry in the service of medicine, followed the age of alchemy and lasted until about 1700. Although the goals of iatrochemistry sounded noble, much of the work during this period was outright quackery. Dishonest iatrochemists who preyed on human suffering peddled fake potions for all sorts of ailments. Even honest iatrochemists had insufficient knowledge to develop effective medicines.

Chemical analysis began during this era (especially from about 1500 to 1700) as some people tried to free chemistry from the mysticism of alchemy. From 1620 to 1644, Belgian scientist Jan van Helmont (1579–1644), the father of physiological chemistry, conducted experiments that indicated that matter was never lost nor created during chemical reactions. At about the same time, Francis Bacon (1561–1626), an English scientist and philosopher, revived the atomic theory in his 1620 *Novum Organum*. Bacon believed atoms were the primary particles of any material, and that they differed only in size, not substance. He called them *corpuscula*, which means small bodies.

German scientist Daniel Sennert (1572–1637) advanced essentially the same theory. He proposed the idea of *atoms corpuscula* in 1619. Sennert believed that these particles represented the last degree of subdivision and the first degree of physical composition for all matter. He stressed the idea that a chemical substance was made up of the same types of atoms, no matter what physical transformations the substance underwent. However, Sennett still conceived of elements in the same way Empedocles did—he believed the quartet of earth, air, water, and fire accounted for all matter.

Atomic Theory in the Independent Chemistry Age

Iatrochemistry paved the way for the modern Age of Independent Chemistry, which began about 1650 A.D. During this time, chemistry developed into a science free from the influence of both alchemy and medicine. Two scientists named Jungius and Boyle set the cornerstones in this foundation for change. Joachim Jungius (German, 1587–1657) first defined an element in the modern sense: a unitary substance that cannot decompose into other substances during physical or chemical changes. (An element can change, however, during a nuclear reaction; as we will learn in chapter 5.) Englishman Robert Boyle (1627–1691) vigorously opposed both alchemy and iatrochemistry, stating that:

> . . . the chemists view their task as the preparation of medicines and . . . transmutation of metals. I have tried to deal with chemistry from a quite different viewpoint. If men had the progress of true science more at heart . . . they would render the greatest possible service . . . by devoting all their efforts to . . . experiments and making observations and not by proclaiming any theories without having tested the relevant phenomena.

In other words, Boyle was advocating the development of the scientific method discussed in chapter 1.

Robert Boyle was the dominant figure at the beginning of the Independent Chemistry Age. Boyle was a staunch proponent of corpuscular theory; he claimed that corpuscles differed in the qualities of size, shape, and motion. He believed these corpuscles moved in the solid state and that this motion increased as temperature increased. He adopted Jungius's definition of elements and noted that elements could not decompose chemically without changing their basic nature. In sum, Boyle amassed a wealth of valuable data that established chemistry as a science. (We discussed Boyle's Law in chapter 2.)

Once chemistry began to flourish as a science, several chemists made discoveries or developed theories that eventually culminated in Dalton's atomic theory. Among these foundational discoveries and ideas were the phlogiston theory, the Law of Definite Proportions, the Law of Conservation of Mass, and stoichiometry.

The Phlogiston Theory

By the late 1700s, industrial chemistry was making a major impact on everyday life. Items such as soap, glass, paints, and dyed textiles were produced in quantity for the first time in human history. Still, no major theories combined the pieces of chemical knowledge together or explained why substances behave as they do. An early attempt to integrate chemistry-related observations and ideas resulted in the phlogiston theory, which tried to explain what happens to substances during **combustion** (burning).

Combustion raised questions that intrigued eighteenth- and nineteenth-century scientists. For example, chemists could see that burning produced different results with different substances; why? When wood burned, almost all the mass vanished into smoke and soot. Metals and other materials behaved differently—copper gained weight and left a crumbly residue called a *calx* behind when heated to a high temperature in air, whereas limestone left a calx but lost weight. Smelting copper ore (heating it till it melted or liquified) produced metallic copper, but the copper weighed less than the ore. These facts can be expressed as chemical reactions:

REACTANTS		PRODUCTS
Wood (100 g)	$\xrightarrow{\text{Burning}}$	Ashes (trace)
Limestone	$\xrightarrow[\text{in air}]{\text{Heated}}$	Quicklime
Copper ore	$\xrightarrow[\text{in air}]{\text{Heated}}$	Copper Calx
Copper ore	$\xrightarrow[\text{in wood fire}]{\text{Heated}}$	Copper

(Notice that chemists always write the reactants—the materials that react—on the left of the arrow and the products of the reaction on the right.)

Georg Ernest Stahl (1666–1734), a German scientist, developed a theory to explain the varying effects of combustion. In his **phlogiston theory,** Stahl proposed that a substance called phlogiston was transferred to or from a material during combustion. Stahl believed that when wood burned, it lost its phlogiston. He reasoned that when copper was heated in open air (calcination), it also lost phlogiston (even though it gained weight). During smelting, Stahl thought, the phlogiston transferred from the wood to the ore as the ore purified into metal. This meant that smelting (heating to the melting point) and calcination (heating without melting, leaving a dry residue) were opposite reactions in regard to phlogiston flow–calcination caused copper to lose phlogiston, smelting caused a gain in phlogiston. Limestone, a rock with high mineral content, reacted differently from copper: it gained phlogiston from calcination, but lost weight anyway.

Stahl also believed that phlogiston made certain materials behave as caustics, substances capable of corroding or destroying animal tissue. When he heated limestone, soda, or potash to high temperatures, the resulting quicklime, caustic soda (lye) or caustic potash (potash lye) all behaved as caustics. Stahl believed that this was because they picked up phlogiston from the fire. If these materials were exposed to air for a long time, said Stahl, the phlogiston leaked out and they lost their caustic properties.

The phlogiston theory tied together the chemical operations of burning, smelting, calcination, and caustic behavior. It gained widespread acceptance because it explained so many different phenomena with a single concept—the transfer of phlogiston. Yet, the theory had serious flaws, some quite obvious. For example, why are calxes heavier than the original metal if the metal loses phlogiston? If minerals become caustic because they gain phlogiston, why isn't wood caustic before it loses phlogiston? Other scientists proposed solutions, including a negative weight for phlogiston, but none of the solutions were satisfactory. However, in spite of its inconsistencies, the phlogiston concept continued as the main theory in chemistry for more than a century because no better theory arose.

The Law of Definite Proportions

Eventually, two advances punctured the phlogiston theory and moved scientists closer to a comprehensive atomic theory. English aristocrat and scientist Henry Cavendish (1731–1810) was responsible for both. First, Cavendish developed a method for making quantitative measurements. Although Cavendish, like the other scientists of his day, adhered to the phlogiston theory, his careful measurements of weights and volumes enabled him to quantify the substances he studied and, eventually, to disprove the theory.

Cavendish made his second breakthrough after German chemist Karl Wilhelm Scheele (1742–1786) and English chemist Joseph Priestley (1733–1804) discovered oxygen in 1771 and 1774, respectively. Cavendish soon discerned that 20.84 percent of air is oxygen. In 1785, he determined that 423 volumes of

Henry Cavendish.

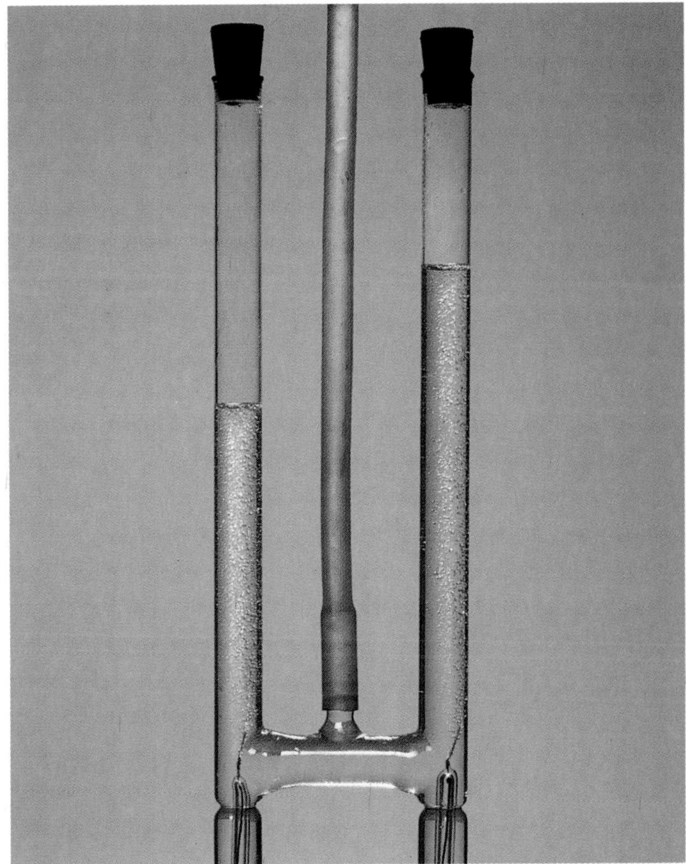

Figure 3.1 The electrolytic decomposition of water. Note that the volume of hydrogen gas, on the left, is twice the volume of the oxygen gas.

hydrogen combined with 1,000 volumes of air (which contained 20.84 percent, or 210 volumes, of oxygen) to produce water. This meant that hydrogen and oxygen combined in a 2:1 ratio to produce water. In 1789, the electrolytic decomposition of water (that is, decomposition caused when electricity passes through a material) confirmed Cavendish's discovery (figure 3.1). His synthesis of water demolished the ancient belief that water is a single element, and his measurements of the components of air and water formed a foundation for Joseph Proust (French, 1754–1826) to later propose the **Law of Definite Proportions,**—the idea that any compound always contains the same proportions of the same elements.

Oxidation and the Law of Conservation of Mass

Antoine Laurent Lavoisier (French, 1743—94) built on the research of Cavendish, Priestley, and others, devised a new theory to replace the phlogiston theory as early as 1789. Still, many scientists still believed the phlogiston theory well into the 1830s, nearly fifty years later.

Nearly all scientists at that time recognized that, during open-air combustion, wood loses weight while metals gain weight. When Lavoisier experimented by burning substances in closed containers, he made an intriguing discovery: the mass of the container's contents was the same before and after combustion. Lavoisier realized that when substances burn, they are not gaining or losing some vague internal material such as phlogiston; instead, they gain or lose a substance from the air. When the air is considered along with the substance, the total mass of the substance *plus the air* remains the same, even when burning changes the nature of the substance. Lavoisier believed that the substance in the air that interacted with a material during combustion was the colorless, odorless gas Priestley called "dephlogisticated air." Lavoisier renamed it oxygen and called his theory the **oxidation theory.**

Lavoisier's oxidation theory answered some of the troublesome questions the phlogiston theory posed. If burning wood released oxygen into the air, the ashes left after combustion would have less mass. If metal combined with oxygen during combustion, the product would have greater mass than the original substance. This would account for wood's weight loss and a metal's weight gain after combustion.

If we express the oxidation theory in the form of a modern chemical equation, the equation for wood might appear as follows:

$C_6H_{10}O_5$	+	O_2	$\rightarrow$	CO_2	+	H_2O	+	Ashes
(100 g)		(119 g)		(163 g)		(56 g)		(trace)
Wood		Oxygen		Carbon Dioxide		Water		

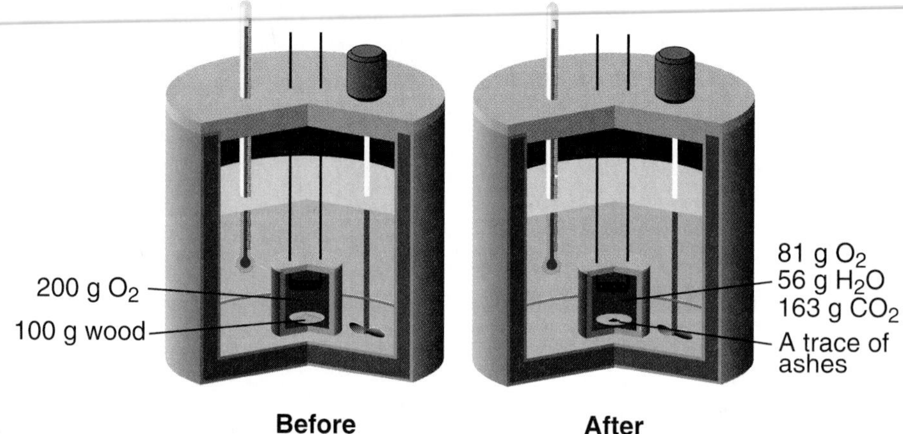

Figure 3.2 An enclosed chamber for burning wood. Total material is 300 g before and after the combustion reaction, but the nature of the materials is not the same.

If this reaction were run in an enclosed chamber (figure 3.2), using 300 g total mass, we would observe that the total mass in the reaction chamber remained unchanged after the reaction. We would also observe that the combustion reaction only involved 219 g. The 300 g initial total mass included 81 g excess oxygen, which remained unreacted. The unreacted oxygen, carbon dioxide, and water are all in the gas phase; this is why early chemists failed to recognize their roles in the reaction.

We can construct a similar equation for a metal (in this case, iron):

Fe	+	O_2	$\rightarrow$	Fe_2O_3
(100 g)		(188 g)		(288 g)
Iron		Oxygen		Iron oxide

In this case, oxygen reacts with the iron to form iron oxide. Again, this reaction could be run in an enclosed chamber, as shown in Figure 3.3, with a grand total of 300 g, including 12 g excess oxygen. Although to early scientists the weight of the metal appeared to increase, the total reacting mass remains 288 g and the total mass remains 300 g.

Lavoisier's discoveries marked a turning point in the study of chemistry (box 3.1). His studies revealed four main types of information: (1) The total mass remained the same during a chemical reaction, showing that matter was conserved; (2) each substance required a specific amount of oxygen, illustrating the Law of Definite Proportions; (3) the amount (mass) of product was different for each set of reactants; and (4) excess reactant (oxygen in this case) remained unreacted when the reaction was completed. Lavoisier summarized his ideas in the **Law of Conservation of Mass,** which states that matter is neither created nor destroyed in a chemical reaction. This law still underpins our understanding of chemistry over two centuries later.

Stoichiometry or Quantitative Chemistry

The use of quantitative measurements increased as chemistry matured. German chemist Jeremiah Benjamin Richter (1762–1807) was a staunch advocate of recognizing the role of mathematics in chemistry. He developed **stoichiometry,** a branch of chemistry based on the idea that quantitative qualities such as mass are a factor in chemical reactions. Richter presented his

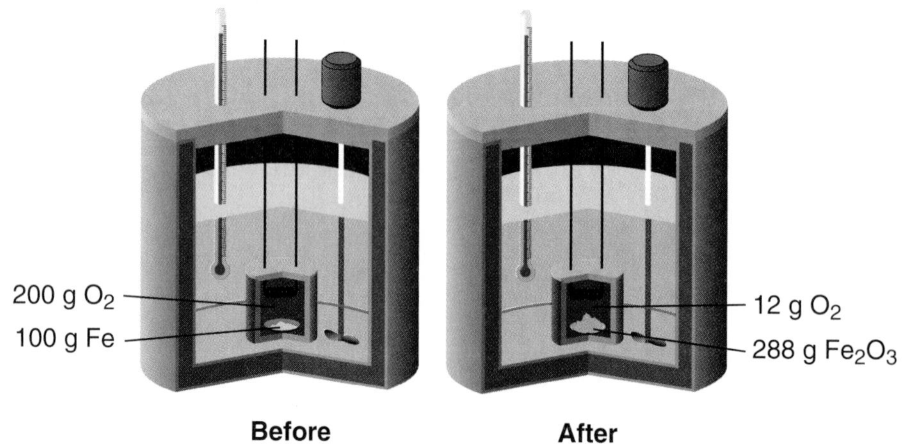

Figure 3.3 An enclosed chamber for burning iron. Total material is 300 g before and after combustion reaction, but the nature of the materials is not the same.

BOX 3.1

LAVOISIER: THE FATHER OF CHEMISTRY

Born the son of a rich lawyer in 1743, Antoine Laurent Lavoisier was one on the first chemists to base his research on systematic experimentation and precise measurement, integral features of scientific endeavor today. Lavoisier entered an arranged marriage with Marie Anne Pierrette Paulze in 1771, when Marie was fourteen years old. Lavoisier's bride was the daughter of a wealthy tax collector, and Lavoisier invested much of his money into a private firm that collected taxes for the French government. He received an income of over 100,000 francs per year from this venture, and this money supported his research. Marie worked alongside her husband in the laboratory and rendered most of the drawings that appeared in his papers (figure 1). Unfortunately, the French Revolution leaders disliked anyone associated with the previous government, especially tax collectors; revolutionaries guillotined the scientist in 1794.

Figure 1 Antoine and Marie Lavoisier working in the laboratory.

Although Lavoisier was a brilliant scientist, his scientific life was not without controversy; on several occasions he claimed credit for another person's research. Nonetheless, Lavoisier's legitimate contributions to the budding field of chemistry are impressive: he originated the Law of Conservation of Mass as well as the oxidation theory, which holds that oxygen reactions cause combustion. He was also one of the first to use systematic names for compounds. But perhaps Lavoisier's most notable accomplishment was the development of accurate mass measurements, which enabled him to perform the research that undergirded all his other achievements. Among the objects confiscated from his laboratory after the French Revolution were 13,000 items of glassware and more than 250 physical instruments, including a trio of precision balances used to weigh chemicals. The "Father of Chemistry" bequeathed to his successors an all-important legacy: the ability to measure substances precisely at the molecular level.

PRACTICE PROBLEM 3.1

Some early chemists claimed that the oxygen in Lavoisier's theory was essentially the same substance as the phlogiston in Stahl's theory. What, if anything, is the difference between the two theories? Are oxygen and phlogiston essentially the same substance?

ANSWER Those who argued that oxygen and phlogiston were the same substance neglected the experimental part of the scientific method. Lavoisier and others had observed (measured) oxygen; no one ever observed phlogiston. Moreover, Stahl claimed phlogiston moved out of the fire and into the reactants during combustion; Lavoisier recognized that oxygen is a reactant, and that heat (fire) simply drives the reaction.

ideas in *On the Modern Objects of Chemistry*, an eleven-volume work published between 1791 and 1802. He proposed that each element and compound has a specific, unvarying mass compared to all other elements and compounds. German chemist Ernst Fischer (1754–1831) adopted this idea to set a system of values for known elements and compounds in 1802, assigning sulfuric acid a value of 100 as the base mass value, and French chemist Claude Berthollet (1748–1822) incorporated these equivalent weights into his 1803 *Essay on Chemical Statics*. These early attempts to quantify chemistry would eventually complement the Law of Conservation of Mass—if you know that mass doesn't change after a chemical reaction, you can mathematically balance the numbers on each side of an equation that represents it.

Richter also tried to develop a systematic table of the elements by arranging them horizontally in order according to mass so that those with similar properties were grouped together in vertical columns. Because elements with like properties appear at regular intervals or periods in such a chart (for example, every eighth element might be what we call an alkali element), this arrangement is called a periodic table. Unfortunately, Richter filled a gap in his chart with a chemical that later turned out to be not an element, but the compound calcium phosphate. The resulting ridicule devastated his attempts to create a periodic chart; chemists had to wait until Mendeleev constructed one in 1869.

Richter never directly associated his ideas of equivalent mass and stoichiometry with the concept of atoms, which would have made these ideas easier to understand; instead, he interpreted his ideas using the phlogiston theory. He lived in poverty throughout his short life and died the year atomic theory became widely known.

DALTON'S ATOMIC THEORY

The research efforts of dozens of chemists culminated in the theory English chemist John Dalton (1766–1844) proposed in 1803. **Dalton's atomic theory** was a landmark that led to all future progress in the field. Merely seven people attended Dalton's October 21, 1803 lecture when he presented the rudiments of his atomic theory and proposed relative masses, based on his theory, for six elements (hydrogen, carbon, oxygen, nitrogen, sulfur, and phosphorus). Dalton's *A New System of Chemistry*, published in 1805, expounded his atomic theory in detail. Box 3.2 tells more about Dalton's life and work.

John Dalton.

Assumptions Behind the Theory

The beauty of Dalton's theory was its ability to integrate many diverse scientific facts and laws into a comprehensive theory of the nature of matter. Dalton based the theory on the following assumptions:

1. Elements consist of tiny particles called atoms. Atoms are the indivisible building blocks of all matter.
2. All atoms of a specific element are identical, and they always have the same unique atomic mass.
3. Atoms of unlike elements differ in mass and volume.
4. Atoms cannot be destroyed or changed during chemical reactions.
5. Atoms of one element combine with atoms of other elements to form molecules, the basic units of compounds.
6. The ratio of atoms in a molecule is constant and can be expressed in simple whole numbers.

These assumptions supported the laws earlier chemists had discovered. The Law of the Conservation of Mass makes sense if atoms are neither destroyed nor changed during chemical reactions; if you cannot destroy the building units, they remain—and retain the same mass—even if arranged into a different form. Similarly, if atoms always combine in the same ratios to form molecules, we can understand why the Law of Definite Proportions operates as it does.

Problems with the Theory

Although Dalton's atomic theory became the basis for our understanding of modern chemistry, some of his assumptions about atoms were not completely accurate. For example, assumption 1 indicates that atoms are indivisible, but scientists later discovered that atoms can subdivide into smaller units called electrons, protons, and neutrons. But although these subatomic particles do have mass and occupy space, they do not constitute matter in the usual chemical sense. Electrons, for example, are the fundamental units of electricity; you cannot make a compound from electricity. Likewise, neutrons and protons do not form chemical compounds; atoms containing neutrons and protons do. Chemists therefore argue that subatomic particles are not matter in the chemical sense, and they approach Dalton's theory as being functionally correct.

Dalton's atomic theory also claims that *all* atoms of an element are identical in atomic mass and size (assumption 2). This is not completely true, either. Many years after Dalton proposed the atomic theory, chemists discovered **isotopes,** atoms of the same element with slightly different atomic mass values. Oxygen occurs as isotopes with atomic masses of 16 and 18 g/mole; hydrogen

BOX 3.2

THE COLORBLIND CHEMIST

John Dalton's list of scientific accomplishments is long and impressive—he proposed the atomic theory of matter, discovered the Law of Multiple Proportions and the Law of Partial Gas Pressure, published one of the first atomic weight tables, and amassed more than 200,000 weather records in his study of meteorology. But Dalton (English, 1766–1844) also made history by becoming the first scientist to describe a common affliction he himself suffered from: colorblindness.

Colorblindness (or Daltonism, as it is still known in England) is a hereditary condition that affects about 5 percent of all males and 0.5 percent of all females. Usually a colorblind individual can perceive some colors; in the most common form of the condition, the person cannot distinguish between red and green (figure 1). In rare cases (about 0.0025 percent of the population), the person is totally colorblind and perceives color only as shades of black, white, and gray. Dalton was red-green colorblind.

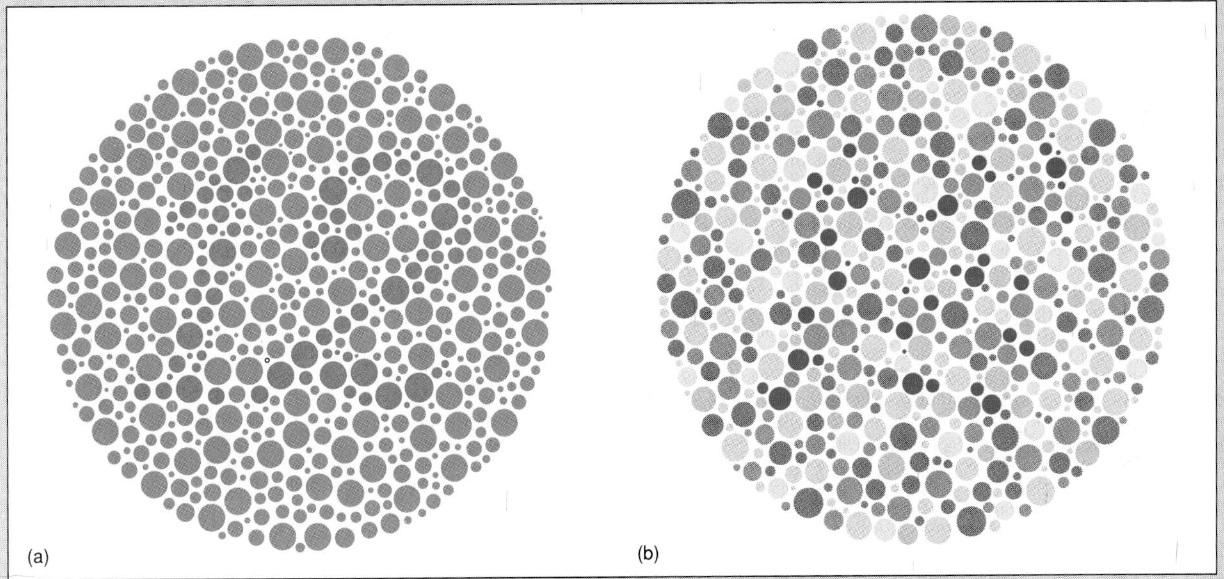

Figure 1 A colorblindness test pattern.
SOURCE: The above has been reproduced from Ishihara's Tests for Colour Blindness published by KANEHARA & CO., LTD, Tokyo, Japan, but tests for colorblindness cannot be conducted with this material. For accurate testing, the original plates must be used.

One story about Dalton's inability to discriminate color relates that the brilliant scientist, at age 66, was to be presented before King William IV. His backers wanted him to wear a scarlet robe to represent the colors of Oxford University, where he had obtained his doctorate; but Dalton, a devout Quaker, preferred modest, plain attire. However, since Dalton was red-green colorblind, the red robe appeared gray to him—so to his supporters' delight, he wore the garment they had chosen.

Color perception arises in certain cells called cones (because of their shape) that are located in the retina. There are three kinds of cones: one for detecting red, one for green, and one for blue. When the red-detecting cones are missing, the perceived color is skewed toward the green. When the green-detecting ones are absent, the perceived color is skewed toward the red. The absence of blue-detecting cones is rare. The other cells used in seeing are called rods (again, because of their shape). Rods translate the power (or intensity) of light into monochromatic (single-color) values of black and white).

No cure exists for colorblindness, but the condition does not seem to handicap those who have it—most learn to compensate from an early age. In fact, people with colorblindness have succeeded in professions that would seem to require color perception. For example, some art historians believe that artist James Whistler, whose most famous work was his portrait of his mother, may have been colorblind; his paintings are filled with somber shades of brown, gray, and black. Similarly, many would argue that a chemist must be able to distinguish chemicals by color. But John Dalton proved that colorblindness need not hamper a chemist in the least.

occurs as isotopes weighing 1, 2, and 3 g/mole. Elements consist of mixtures of isotopes, but the chemical properties of each isotope are essentially the same. (Chapters 4 and 5 consider isotopes in greater detail.)

Dalton himself discovered the final problem with his assumptions. Assumption 6 states that the ratio of the atoms in a particular molecule is constant. For many years, chemists also assumed that this meant any two elements always combine in the same ratios to form the same molecule—in other words, two elements can only combine in a single combination. This also proved incorrect. Hydrogen and oxygen can combine in a 2:1 ratio to form water, or H_2O, and water (as assumption 6 states) always occurs in this ratio. But these same two elements can also combine in a 2:2 ratio to form hydrogen peroxide (H_2O_2).

PRACTICE PROBLEM 3.2

What happens if the number of atoms in a compound changes? For example, suppose water (H_2O) gains another oxygen atom to form H_2O_2?

ANSWER The higher percentage of oxygen forms a new substance with different properties. The new compound is hydrogen peroxide, a very reactive compound. Rinsing your hair with it would turn your hair white, and you could die if you drank hydrogen peroxide rather than water.

The Law of Multiple Proportions

Dalton proposed the **Law of Multiple Proportions** in 1804, just one year after he announced his atomic theory. This law applies when the same elements can combine to produce two or more compounds. It states that the masses of one element that can combine with a given mass of a second element will always occur in ratios of small whole numbers. For example, carbon dioxide, CO_2, and carbon monoxide, CO, both contain only carbon and oxygen, though they are unique compounds with distinctive properties (table 3.1). CO contains these two elements in a 1:1 ratio, CO_2 in a 1:2 ratio. If we compare how many units of oxygen can combine with one unit of carbon in these compounds, the ratio is 1:1 or 1:2, in keeping with the law. We would not see huge masses of oxygen combined with one carbon atom.

Dalton originally developed this law by studying hydrocarbons—compounds containing only hydrogen and carbon. He showed that one of them, methane, had a C:H ratio of 1:4, while another, ethane, occurred in a 1:2 ratio. The ratio of hydrogen atoms to one atom of carbon in these two compounds is 4:2, which can be reduced to 2:1. Again, we would not see huge variations in the number of hydrogen atoms bonding to a single carbon atom.

Dalton's atomic theory did not gain immediate acceptance, but eventually gained hold as experimental and quantitative data accumulated to support it. His theory represented a major turning point in the scientific understanding of matter. Let's explore how it applies to our modern conceptions of chemistry and matter.

TABLE 3.1 Comparison of carbon monoxide and carbon dioxide.

	CARBON MONOXIDE	CARBON DIOXIDE
Formula	CO	CO_2
Molecular mass (grams)	28	44
Melting point (°C)	−207	Sublimes −78.5
Density (g/mL)	1.247	1.974
Combustibility	Purple flame	Does not burn
Toxicity	Highly toxic	Nontoxic

APPLYING ATOMIC THEORY TO THE STRUCTURE OF MATTER

Elements—Atoms of One Substance

An *element* is defined as a pure substance composed of only one kind of atom. Originally, this concept met with skepticism because many scientists adhered to the ancient ideas of Aristotle and Empedocles; they believed that all matter was either air, earth, water, or fire. As chemists developed data on the relative weights of elements and compounds, and as atomic theory became more widely accepted, it became obvious that the simple theory of four elements was no longer useful and they abandoned it.

Prior to the mid-1600s, chemists were aware of very few pure elements. Nine elements were known from antiquity (figure 3.4): the metals gold, silver, copper, tin, lead, mercury, and iron, and the nonmetals sulfur and carbon. Early chemists were also familiar with impure antimony and zinc (pure zinc was unavailable until the sixteenth century), and the German scholar Magnus (1193–1280) isolated arsenic about 1250. Alchemists knew of many fairly pure compounds, including salt, saltpeter (potassium nitrate), sulfuric acid, nitric acid, carbon dioxide, and silver nitrate. They also were aware of perhaps a hundred minerals but these minerals were not identified as particular elements or compounds and were not always pure. With such limited knowledge, it's not surprising that early chemists did not discern the relationship between elements and compounds.

A flurry of experimental activity from the late 1600s through the early 1700s resulted in the discovery of 19 new elements and thousands of compounds. But the greatest advances came in the 1800s with the discovery of an additional 52 elements. Several factors contributed to this sudden explosion of chemical knowledge: (1) Dalton's atomic theory sparked interest in isolating dif-

Figure 3.4 A portion of the gold coffin cover from Egyptian king Tutankhamen's tomb, ca. 1360 B.C. Gold is one of the nine elements known since antiquity. The others are: silver, copper, tin, lead, mercury, iron, sulfur, and carbon.

ferent types of atoms; (2) chemists developed better analytical and separation techniques; and (3) the ability to harness electricity allowed scientists to discover certain elements (such as aluminum and sodium) that are very difficult to isolate without electricity. During the 20th century, 28 more elements, some artificially created in nuclear reactions, brought the total to 111 known elements. The first 92 elements (except technetium, number 43) in a periodic table—up through uranium—occur in nature; the others were created. In 1941, American chemists Glenn T. Seaborg (1912–) and Edwin M. McMillan (1907–1991) created plutonium, the first transuranium element (an element with an atomic mass greater than uranium) via a nuclear reaction. Seaborg and McMillan shared the 1951 Nobel prize in chemistry for this research (see box 3.3). Seaborg went on to spearhead the research that created several other transuranium elements.

Once scientists understood the nature of the elements and the atom, they discovered that subdividing atoms caused an element to lose the properties that define matter. Thus, although atoms are not indivisible, as Dalton originally believed, they are the smallest particles that can function as ordinary matter. Dalton's assumption that atoms of unlike elements differ in mass and volume paved the way to discover new elements and is still accepted as a working definition of an element nearly two centuries later.

Allotropes

Elements may exist as isotopes or allotropes. While isotopes are atoms of the same element with slightly different atomic mass values, **allotropes** are different forms the atoms of one element may combine in. Oxygen, for example, normally exists in the atmosphere in paired atoms (O_2); ozone is an allotrope of oxygen that contains three oxygen atoms (O_3) instead of the normal two. Ozone is helpful in the upper atmosphere because it screens out excess solar ultraviolet light, but it is a dangerous pollutant in ground-level smog. Similarly, elemental sulfur usually exists as an eight-member ring, but an allotrope called plastic sulfur forms a chain of several thousand sulfur atoms. Many other elements have allotropic forms, including manganese, selenium, tin, uranium, and plutonium.

Carbon allotropes are more common and familiar than the other examples; they include carbon black, graphite, diamond, and buckyballs (figure 3.5). Table 3.2 summarizes their properties.

TABLE 3.2 Properties of the carbon allotropes.

PROPERTY	CARBON BLACK	GRAPHITE	DIAMOND	BUCKYBALLS
Form	Powder	Gray solid	Crystal	Hollow cages
Density (g/cm^3)	1.8–2.1	1.5–2.3	3.5	Variable
Preparation	Pyrolysis	Pyrolysis	High °C and pressure	Pyrolysis
Hardness	Low	Medium	High	Varies
Thermal conductivity	Low	High	—	Varies
Electrical conductivity	Low	High	—	Varies
Number of other carbon atoms attached to each carbon	2	3	4	Varies (usually 3)

BOX 3.3

THE NOBEL PRIZE

Most scientists consider the Nobel Prize the most prestigious aware a scientist can receive. But who was Alfred Nobel, and how did this namesake award originate?

Alfred Bernhard Nobel, chemist and engineer, was born in Stockholm, Sweden on October 21, 1833, and died in San Remo, Italy on December 10, 1896. Nobel made his fortune in explosives, as his father had. In 1863, Nobel developed a detonator based on the chemical compound mercury fulminate, which explodes upon impact. This detonator made it possible to use liquid nitroglycerin as a charge in explosives.

Even after an 1864 explosion leveled his factory and killed five people, including his younger brother, Nobel continued his experiments with nitroglycerin. He patented his most famous invention, dynamite, in 1867. Dynamite was prepared from nitroglycerin absorbed in a porous material. It enabled the safe use of nitroglycerin and made Nobel both famous and wealthy. In 1875, Nobel invented an even more powerful explosive he called "blasting gelatine," and in 1887, he introduced ballistite, a nitroglycerin explosive that does not produce smoke. Nobel's innovations made many major civil engineering feats possible, such as blasting through rock to create the Corinth Canal and the St. Gotthard Tunnel.

Prior to Nobel's work, black gunpowder was the primary explosive used in military warfare. Nobel hoped his inventions would decrease the demand for gunpowder and reduce the possibility of war. Ironically, while Nobel's work produced more powerful explosives that proved a boon to civil engineers, it also made military weaponry even more powerful and deadly.

Alfred Nobel.

Still, Nobel's interest in furthering peace, along with his commitment to science and affinity for fine literature, led him to leave much of his fortune to a fund established in his will. The interest the fund generates is awarded annually to individuals who have made outstanding contributions in the fields of science, literature, and peace.

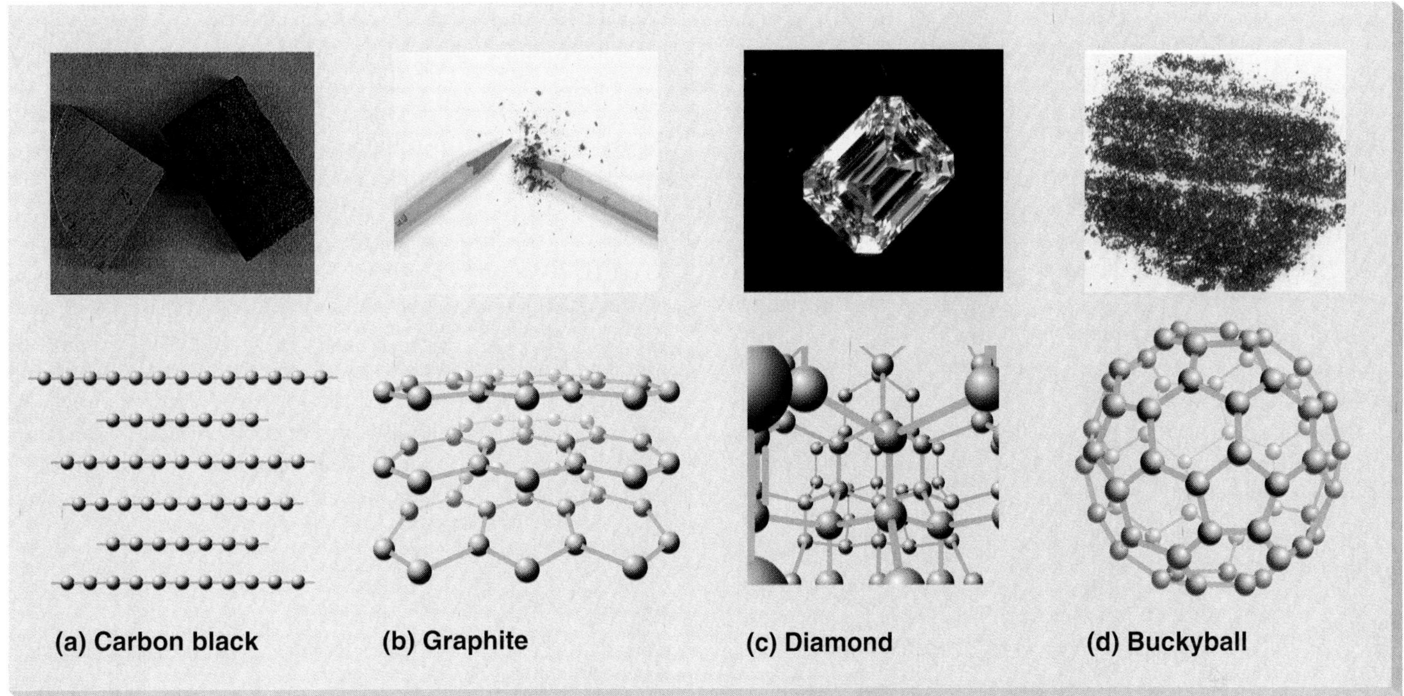

Figure 3.5 The structures of the allotrophic forms of carbon: **(a)** carbon black, **(b)** graphite, **(c)** diamond, **(d)** buckyballs. Only partial structures are shown for the first three allotropes because the actual molecules contain many thousand carbon atoms. Buckyballs, on the other hand, have specific numbers of atoms.

Carbon black is normally impure; it is often formed from the incomplete combustion of natural gas or petroleum. It exists as chains or strings of variable numbers of carbon atoms, each connected to exactly two others (figure 3.5a). Carbon black is used in tires, paints, inks, batteries, and drinking water purification. Preparing carbon black through the pyrolysis of scrap tires could put millions of unwanted tires to use each year if developed successfully on a commercial scale. **Pyrolysis** is the decomposition of one substance into another by applying heat.

Graphite occurs naturally in Madagascar, Sri Lanka, Mexico, Korea, the former Soviet Union, Austria, and China. Its carbon atoms connect into sheets of hexagons, with each C atom connected to exactly three others (figure 3.5b). "Lead" (actually graphite) pencils, electrodes, and generator and motor brushes contain graphite; it is also used as a moderator in nuclear reactors, and in electrolytic cells. Graphite fibers are used in heating pads, protective clothing, space craft, compressor blades, and in aircraft or building composites. Fibers that contain carbon (for example, cotton and rayon) undergo pyrolysis to produce graphite fibers.

Diamonds are not only gemstones (figure 3.6), but are also a valuable industrial material. Diamond atoms are arranged in an elaborate three-dimensional network that links each carbon with four others (figure 3.5c); this structure gives the diamond its extreme strength and hardness. Diamonds occur naturally in South Africa, Brazil, Venezuela, India, Borneo, and in Arkansas.

Figure 3.6 Grace Kelly shows off her diamond engagement ring during the filming of "High Society." She received this ring from Prince Rainier, and this was her last movie before retiring to become Princess Grace of Monaco.

They may be formed synthetically in an electric furnace at high temperatures and pressures. Along with being used as gems, they are used in oil-well bits, grinding, and cutting.

Buckyballs, now called fullerenes (figure 3.5d), are fairly new and are produced synthetically. These hollow cages of 32 or more carbon atoms were first characterized in 1985 and made in bulk quantities in 1990. No major applications exist, as yet, for buckyballs. They belong to a class of carbon molecules called fullerenes (named after Buckminster Fuller because their shapes resemble his geodesic domes). They possess some useful and intriguing electronic properties that enable them to function as insulators, conductors, semiconductors, and superconductors, depending on the arrangements of other atoms. This is an astounding find, since the properties of insulation (preventing the transmission of electricity) and conduction (facilitating the transmission of electricity) oppose each other.

PRACTICE PROBLEM 3.3

Using figure 3.5 as a guide, draw the sulfur allotropes (with the symbol S) for elemental sulfur and plastic sulfur (described earlier in this section).

ANSWER Your drawings should show one allotrope as an eight-member ring and the other as a long chain of S atoms connected together (like the C atoms in carbon black).

Molecules—Atoms Bonded Together

If atoms are the fundamental building blocks of elements, molecules are the basic building blocks of compounds. A **molecule** is two or more atoms bonded together. Some molecules contain atoms of just one element; if the element normally exists as a molecule in its most common form, it is called a molecular element. Examples of molecular elements include the gases hydrogen, oxygen, nitrogen, fluorine, and chlorine; liquid bromine; and the solids carbon, phosphorus, sulfur, and iodine. (Most of the remaining elements exist as atoms rather than molecules, and most are metals.) Other molecules contain atoms of two or more elements; these molecules are compounds. Some familiar compounds include water (H_2O, or two hydrogen atoms bonded to one oxygen atom in each molecule), and salt (NaCl, one atom each of sodium—Na—and chlorine bonded together).

Compounds—Elements Bonded Together

A *compound* is a pure substance made up of atoms of two or more elements bonded together in molecules. A molecule is the smallest subunit of a compound that retains the compound's chemical identity. In other words, if you break apart the molecule into separate atoms, the identity of the substance changes. Any molecule of a pure substance contains the same definite numbers of each type of atom; this is the Law of Definite Proportions, a fundamental property of all compounds. Each compound also has its own unique properties that differentiate it from all other compounds. Purity is discussed further in box 3.4.

We can see how these principles apply by again considering water (H_2O). Water can break down into its constituent atoms, hydrogen and oxygen, but neither hydrogen nor oxygen alone has the same chemical identity as the two together—neither one is water. Water always consists of two hydrogen atoms and one oxygen atom in each molecule; the proportion never varies. And water has unique properties no other compound has–for example, its high capacity for retaining heat helps keep the temperature on earth (and within organisms) much more stable than other liquids would.

PRACTICE PROBLEM 3.4

There are obviously many more compounds than elements. Why?

ANSWER Although the number of elements is limited, they can combine into a great variety of different compounds. Each compound has its own specific set of physical and chemical properties, so compounds allow much more chemical diversity than elements alone.

PRACTICE PROBLEM 3.5

Molecules are the basic subunits for both allotropes and compounds. How do compounds and allotropes differ? Does the total number of atoms in the molecule make a difference?

ANSWER Allotropes contain molecules made up of atoms of only a single element, while compounds contain atoms of at least two different kinds of elements. The total number of atoms in a molecule has no bearing on whether a molecule is an allotrope or a compound. Both types of materials could contain as few as two atoms or as many as several thousand. The difference lies solely in how many different kinds of atoms are present—one (allotrope) or more (compound).

BOX 3.4

PROOF AND PURITY

It is not always easy to determine whether a substance is pure. Solutions, like pure substances, are homogeneous matter and they sometimes look like pure materials; alloys may also appear to be pure. Virtually no progress occurred on the molecular concept until chemists developed the idea of pure compounds.

One familiar example where the question of purity arises is in ethanol solutions. Ethanol is the intoxicating chemical in alcoholic beverages. It has a moderately strong odor, which makes it relatively easy to detect in water-ethanol mixtures, but many other odors can easily mask the odor of ethanol. For example, the mixed drink called a Bloody Mary consists of vodka and tomato juice. The odor of the tomato juice overpowers the alcohol's odor, making it difficult to detect its presence. Even when the human nose can detect the alcohol odor, it can seldom determine just how much of the chemical is present.

Ethanol solutions have been known since ancient times, and some early scientists thought they were pure materials. After gunpowder was invented, a new test method was developed to assay the amount of ethanol in solution. Gunpowder soaked in a 100 proof (50 percent ethanol) solution would still burn, while gunpowder soaked in samples of less than 100 proof ethanol would not sustain burning. (The term *proof* describes the ethanol content of an alcoholic beverage. The proof value is double the percentage of ethanol in the solution; thus, 100 proof alcohol—for example, some vodkas—contains 50 percent ethanol. While strong as a drink, 100 proof alcohol is hardly pure in the chemical sense.) Prior to such tests, scientists sometime mistakenly believed ethanol solutions were pure alcohol.

Such mistakes, though unintentional, caused many problems as scientists worked toward understanding the Law of Definite Composition. Alcohol-water solutions do not have a constant composition. If a chemist was not careful, he or she could make a serious error in an analysis of such a solution. Berthollet made such an error, and as a result, he doubted that compounds have a constant composition. Alcohol offers "proof" that "purity" is not always an easily discernible quality.

The Law of Definite Proportions Applied to Compounds

The Law of Definite Proportions, also called the Law of Constant Proportions, the Law of Definite Composition, or the Law of Constant Composition, was possibly the most important law developed as the Independent Chemistry Age began. The law states that every compound has a specific, unvarying chemical composition; that atoms of two or more elements always combine in the same proportions to form molecules of any given compound. The Law of Definite Proportions is best understood by studying some specific compounds in closer detail.

Water

The chemical composition of pure water is always the same. Elemental analysis shows that water is exactly 11.19 percent hydrogen and 88.81 percent oxygen by weight. (We will discuss atomic weights later in the chapter.) The mass ratio of O:H in water is thus 88.81:11.19. We can reduce the ratio, making it easier to comprehend, by dividing each side of the ratio by the smaller number:

$$88.81 \div 11.19 = 7.94$$

$$11.19 \div 11.19 = 1.00$$

The O:H mass ratio is therefore 7.94:1.00, or about 8:1. The Law of Definite Proportions tells us that water always contains hydrogen and oxygen in the same proportions: 2:1 in number of atoms, about 1:8 by weight or mass.

Calcium Carbonate

Although less familiar than water, calcium carbonate is still a common compound found in limestone, marble, calcite, and chalk; it is also used in antacid tablets. Elemental analysis shows calcium carbonate always contains the elements calcium (Ca), carbon (C), and oxygen (O) in exactly the same proportions. Thus, a 100.00 g sample contains exactly 40.05 g Ca, 12.00 g C, and 47.95 g O (figure 3.7). If we divide each mass by the smallest value (12.00), we can reduce the Ca:C:O ratio to the lowest common terms:

Calcium	$40.05 \div 12.00 = 3.33$
Carbon	$12.00 \div 12.00 = 1.00$
Oxygen	$47.95 \div 12.00 = 4.00$

The Ca:C:O mass ratio in calcium carbonate is thus 3.33:1.00:4.00. Any pure calcium carbonate sample—whether

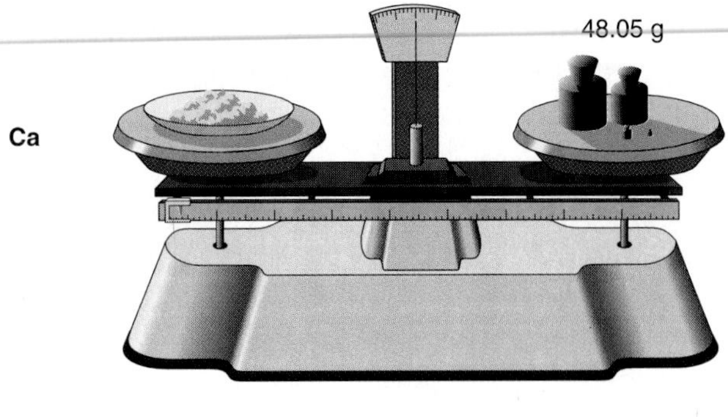

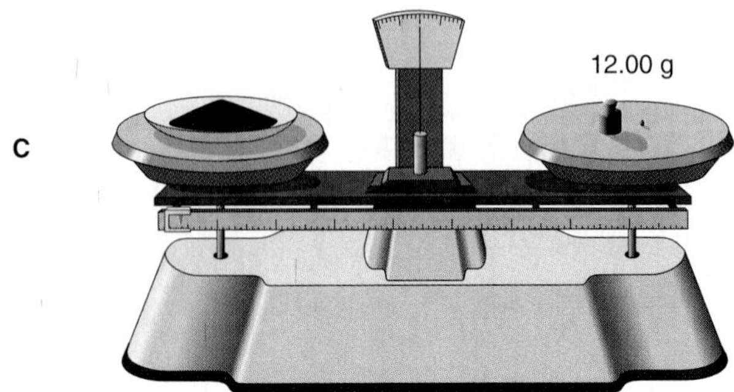

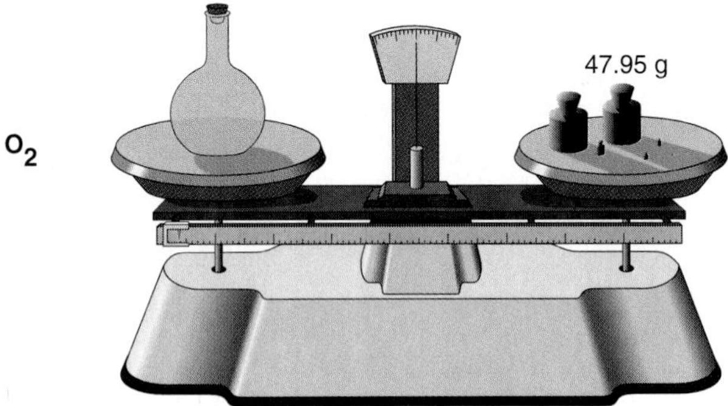

Figure 3.7 The composition of calcium carbonate.

from limestone, chalk, or an antacid tablet—will contain these three elements in exactly these proportions by mass or weight.

We can't round 3.33 off to a whole number because it isn't close enough to either 3.00 or 4.00. However, we can convert the numbers into whole digits by multiplying each value by three:

Ca: $3.33 \times 3 = 10$

C: $1.00 \times 3 = 3$

O: $4.00 \times 3 = 12$

This gives a Ca:C:O mass ratio of 10:3:12 for calcium carbonate. It still tells us nothing about the number of atoms of each element in the calcium carbonate molecule; the chemical formula for calcium carbonate is $CaCO_3$.

PRACTICE PROBLEM 3.6

This section shows that the Ca:C:O mass ratio in calcium carbonate is either 3.33:1.00:4.00 or 10:3:12. Is there a difference between these ratios? If so, which is correct?

ANSWER Both sets of numbers show the same ratio of atomic masses; they are both correct. Because the second ratio contains whole numbers, however, it is the preferable form.

Sodium Sulfate

The German iatrochemist Johann Rudolf Glauber first prepared sodium sulfate in the mid-1650s, and this white, water-soluble, crystalline solid is still called Glauber's salt in his honor. Chemical analysis shows that, by weight, it contains 32.37 percent sodium (Na), 22.57 percent sulfur (S), and 45.06 percent oxygen (O). If we divide these percentages by 22.57 percent, the smallest value, we find the Na:S:O mass ratio:

Sodium $32.37 \div 22.57 = 1.43$

Sulfur $22.57 \div 22.57 = 1.00$

Oxygen $45.06 \div 22.57 = 2.00$

The Na:S:O mass ratio in sodium sulfate is therefore 1.43:1.00:2.00. We can convert this ratio to whole numbers by multiplying each number by seven:

Na: $1.43 \times 7 = 10$

S: $1.00 \times 7 = 7$

O: $2.00 \times 7 = 14$

This gives sodium sulfate a Na:S:O ratio of 10:7:14.

The chemical formula for sodium sulfate is Na_2SO_4 (two sodium atoms, one sulfur, and four oxygen). Note that the Na:S:O mass ratio (10:7:14) is not the same as the atom ratio (2:1:4). Still, whether we are measuring mass or number of atoms, each ratio will always be the same for any molecule of sodium sulfate, as the Law of Definite Proportions predicts. In other words, the mass ratio will always be 10:7:14, the atom ratio always 2:1:4.

PRACTICE PROBLEM 3.7

Steel is manufactured by adding carbon to iron. The amount of carbon added to a given amount of ion varies over a small range. What does this tell us about carbon steel? Is it a compound?

ANSWER Steel is not a chemical compound; it is an alloy, a mixture of two substances. Compounds always have a constant composition; alloys do not.

Atomic Weights or Masses

Once chemists were able to measure the relative weights of elements in a compound, it was only a matter of time until they determined the number of atoms of each element in a molecule of that compound, or the compound's chemical formula. To do so, they first had to determine each element's **atomic weight** or **mass**—that is, how much one atom of that element weighs compared to atoms of other elements. The concept of atomic mass applied atomic theory to the quantitative measurements nineteenth-century chemists were accumulating for various compounds.

To understand this concept, imagine we have a compound called "fruit." "Fruit" always consists of two substances: apples and grapes. When we measure the relative weights of these two substances, we find that apples always comprise just over 88 g in a 100 g sample of "fruit," grapes just over 11 g; a ratio of 88:11 or 8:1. Now we want to determine how many apples and grapes combine to create "fruit." But to do so, we must know how much apples weigh relative to grapes. If an apple weighs 8 times as much as a grape, we can then assume that one apple combines with one grape (an 8:1 ratio by weight) to form a basic unit of "fruit." But if an apple weighs 16 times as much (so that an apple weighs 16 units, and a grape weighs 1), one apple will combine with two grapes (16:2 = 8:1) to form a basic unit of fruit. And if an apple weighs only 4 times as much as a grape, it will take two apples combining with one grape to produce the constant 8:1 weight ratio we measure for "fruit." The weight or mass of each component thus determines the "formula" for this compound.

Amedeo Avogadro.

In the same way, early chemists began trying to determine how many atoms of each element make up one molecule of a given compound. They began by measuring the relative weights or masses of each element in the compound (just as we measured the relative weights of apples and grapes). For example, Dalton knew that the mass ratio of oxygen to hydrogen in the compound water was 8:1—a water sample is 88.81 percent oxygen by weight, and 11.19 percent hydrogen. But to determine *how many atoms* of oxygen and hydrogen combine to produce that 8:1 ratio, we need to know how much oxygen weighs compared to hydrogen. Just as we discovered with apples and grapes, if an oxygen atom weighs 4 times as much as a hydrogen atom, we would need two oxygen atoms and one hydrogen atom to produce a water molecule; if oxygen weighs 8 times as much, one atom of each element; and if oxygen weighs 16 times as much, one oxygen atom and two hydrogen atoms to produce a molecule of water.

Dalton hypothesized that the simplest formula—in this case, one atom of each element, or HO—was always the most likely. However, an Italian chemist named Amedeo Avogadro (1776–1856) proposed an idea that helped correct Dalton's error. Avogadro was aware that while the relative *weights* of oxygen and hydrogen in water occur in an 8:1 ratio, the relative *gas volumes* of oxygen and hydrogen formed during water electrolysis are always in a ratio of 1:2 (see figure 3.1). **Avogadro's hypothesis** assumed that if two gases are at the same temperature and pressure, equal volumes of each gas contain equal numbers of atoms. Accordingly, if one gas has twice the volume, it must contain twice the number of atoms as the second gas. Therefore, if water breaks down into twice as much hydrogen as oxygen when measured by volume, water must contain twice as many hydrogen atoms as oxygen atoms. This would make the correct chemical formula for water H_2O. It would also imply that an oxygen atom weighs 16 times as much as a hydrogen atom (so that one oxygen atom and two hydrogen atoms would produce a 16:2, or 8:1, weight ratio). A modern table of atomic masses reveals that one oxygen atom does indeed weigh 16 times as much as one hydrogen atom. Obviously, Avogadro's hypothesis put chemists on the right track as they tried to determine relative atomic masses or weights for each element.

Once Avogadro's hypothesis became known, researchers could determine relative weights of different elements by weighing the same volume of gas for each element. If the hypothesis was correct, that volume would always contain the same number of atoms, so if the total weight of a gas was three times the weight of another, each individual atom would also weigh three times as much as an atom of the lighter gas. Chemists carefully weighed

TABLE 3.3 Densities and mass ratios for three gases.

GAS NAME	DENSITY (g/L)	MASS RATIO (amu)
Hydrogen	0.089	1.0
Nitrogen	1.247	14.0
Oxygen	1.426	16.0

exactly 1.0 liter of each gas to determine the weight or mass ratios. Because hydrogen had the smallest mass or weight, they initially chose it as the reference point for the relative atomic mass scale, assigning hydrogen a mass value of 1 and determining others by comparing their weights to hydrogen's. As table 3.3 indicates, a 1.0 liter sample of nitrogen weighs 14 times as much as a 1.0 liter sample of hydrogen, so its mass value is 14. Similarly, oxygen weighs 16 times as much as hydrogen, so its atomic mass is 16. Chemists measure atomic mass in an arbitrary unit called an **atomic mass unit** or **amu.** Hydrogen's mass is 1 amu, nitrogen's 14, and oxygen's 16 amu.

Although chemists in the early 1800s used hydrogen as the basis for the atomic mass scale, they later changed the base to oxygen = 16.00 g/mole. Scientists now use a carbon isotope with a mass of 12.00 g/mole as the base value for an amu. They measure g/mole as atomic mass units, or amu.

Moles

Because atoms and molecules are incomprehensibly small, no one could possibly count out or weigh a single one. Chemists therefore use a quantity called a mole (from the Greek for "heap" or "pile") to measure numbers of atoms or molecules. A **mole** is equal to 6.02×10^{23} atoms or molecules—in other words, about 602,000,000,000,000,000,000,000 particles! (We will explain how to handle such large numbers later in this chapter.) When chemists refer to moles, then, they are grouping incredibly large numbers of atoms or molecules. We do the same on a smaller scale when we refer to tons of steel, dozens of doughnuts, or billions of dollars.

We can also define a mole in terms of mass or volume. Avogadro originally defined a mole as enough molecules to give a hydrogen sample a mass of 2.0 grams. (Since hydrogen molecules contain two atoms, a mole of hydrogen *atoms* would weigh 1.0 g.) The number eventually proved to be 6.02×10^{23}; chemists call this **Avogadro's number** in honor of the Italian scientist who first conceived of its existence.

Nitrogen and oxygen, like hydrogen, commonly form molecules containing two atoms; a mole of nitrogen molecules weighs 28 grams (14 times a mole of hydrogen molecules), and a mole of oxygen molecules weighs 32 grams. At standard temperature and pressure, one mole of any gas occupies a volume of 22.4 liters and contains the same number of particles (6.02×10^{23}). Thus, a mole always contains 6.02×10^{23} atoms or molecules and always fills a gas volume of 22.4 liters; but the weight of a mole varies depending on the atomic weight of the element.

Avogadro's work gave chemists the breakthrough they needed to determine relative atomic masses. By weighing exactly one liter of gas and comparing the result to the weight of one liter of hydrogen, they could now determine the atomic mass of any element. And, as we have seen, determining the atomic weights of the elements in a compound was the key to deducing the compound's chemical formula.

PRACTICE PROBLEM 3.8

Why is Avogadro's hypothesis a hypothesis rather than a theory?

ANSWER Avogadro's hypothesis includes only one assumption about one situation. Theories (for example, the atomic theory) utilize several assumptions and attempt to explain a group of observations. This does not mean that Avogadro's accomplishment was unimportant; it was crucial in advancing our knowledge of chemistry and supporting the atomic theory's view of matter!

PRACTICE PROBLEM 3.9

A mole is a unit of measurement, but it can also be a small animal. The mass of our planet is roughly 6×10^{24} g. If every live mole weighs about 100 g each, could you collect an Avogadro mole of mole animals if you converted all the matter on our planet into these animals?

ANSWER To have a mole of animal moles, you need 6×10^{23} animals. Since each animal weighs 100 g, the total mass of a mole of moles would be $6 \times 10^{23} \times 100$ g. The exponential expression 10^{23} means 10 multiplied by itself 23 times; the simple way to think of this is as a 1 followed by 23 zeros. Thus, if we multiply $6 \times 10^{23} \times 100$, we will add two zeros to the product—making the total mass 6×10^{25} g. Since the mass of the entire Earth is only 6×10^{24} g, we could not obtain enough matter from the Earth to create a mole of moles. In fact, we would need to convert the mass of *ten* Earths—$6 \times 10^{24} \times 10 = 6 \times 10^{25}$! While this problem is unlikely to crop up for any scientist, it illustrates how huge Avogadro's number really is.

Chemical Formulas

Once chemists had recorded relative weights and discerned atomic masses for the elements, they could deduce the **chemical formulas** for compounds, or the exact number of each type of atom in the compound. Figure 3.8 shows models of sodium chloride (table salt), water, and glucose (a simple sugar). We can trace the steps chemists took to discover the chemical formulas of the water, calcium carbonate, and sodium sulfate that we discussed earlier.

As we have previously discussed, a 100.00 g sample of calcium carbonate contains 40.05 g calcium, 12.00 g carbon, and 47.95 g oxygen. The atomic masses for these three elements are 40, 12, and 16, respectively. (Remember, this means that calcium weighs 40 times as much as hydrogen, carbon 12 times as much, and oxygen 16 times as much.) We can determine how many atoms of each element make up a calcium carbonate molecule by dividing the relative weight of each element by its atomic mass value:

Calcium $40.05 \div 40.0 = 1.0$ Ca atoms

Carbon $12.00 \div 12.0 = 1.0$ C atoms

Oxygen $47.95 \div 16.0 = 3.0$ O atoms

The chemical formula for calcium carbonate is thus $CaCO_3$—one calcium atom, one carbon atom, and three oxygen atoms combine to form a single molecule of calcium carbonate.

We can determine the formula for sodium sulfate in the same way. A 100 g sample of sodium sulfate consists of 32.37 percent sodium (Na), 22.57 percent sulfur (S), and 45.06 percent oxygen (O). The atomic masses for these three elements are 23 (Na), 32 (S), and 16 (O). If we divide each relative weight by the element's atomic mass value, we find the following:

Sodium $32.37 \div 23.0 = 1.41$ Na atoms

Sulfur $22.57 \div 32.0 = 0.70$ S atoms

Oxygen $45.06 \div 16.0 = 2.82$ O atoms

Obviously, molecules don't contain fragments of atoms. To convert these figures to whole numbers, we can divide each by the smallest number, 0.70:

$1.41 \div 0.70 = 2.0$ Na atoms

$0.70 \div 0.70 = 1.0$ S atoms

$2.82 \div 0.70 = 4.0$ O atoms

The chemical formulas for sodium sulfate is thus Na_2SO_4.

Finally, let's follow the same procedure for water (relative weights are 88.81 percent oxygen, 11.19 percent hydrogen; atomic masses are 16 for oxygen and 1 for hydrogen):

Oxygen $88.81 \div 16 = 5.55$ O atoms

Hydrogen $11.19 \div 1 = 11.19$ H atoms

To convert to whole numbers, divide each by the smaller number (5.55)

$5.55 \div 5.55 = 1$ O atom

$11.19 \div 5.55 = 2$ H atoms

The chemical formula for water is thus H_2O.

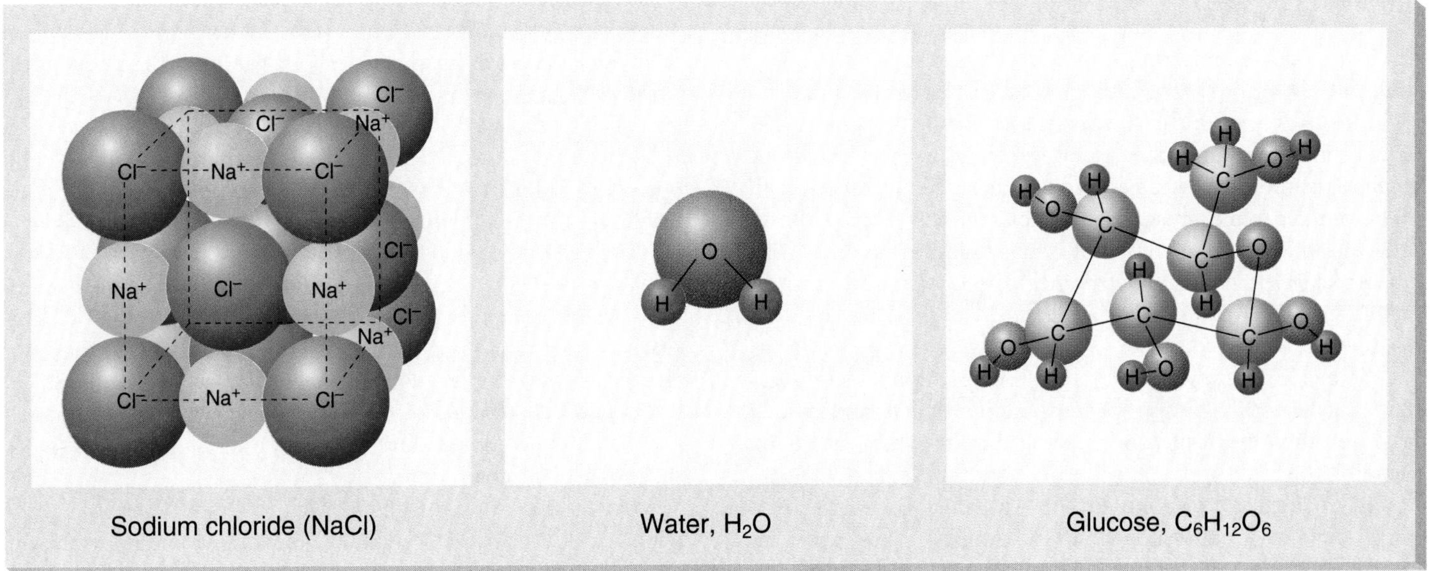

Figure 3.8 Models of the sodium chloride (table salt), water, and glucose (a simple sugar) molecules.

PRACTICE PROBLEM 3.10

Methane gas, used to heat homes and cook food, contains 74.83 percent carbon and 25.17 percent hydrogen by weight. What is the simplest formula for this compound?

ANSWER First, divide the relative weight percentages by the atomic masses of C (12) and H (1).

Carbon	74.83 ÷ 12 = 6.23 C atoms
Hydrogen	25.17 ÷ 1 = 25.17 H atoms

Next, to convert to whole numbers, divide each number by the smaller of the two, 6.23.

Carbon	6.23 ÷ 6.23 = 1.0 C atoms
Hydrogen	25.17 ÷ 6.23 = 4.0 H atoms

Methane thus has the formula CH_4.

INTRODUCTION TO CHEMICAL REACTIONS

When two or more elements interact to create or transform a compound, a **chemical reaction** has occurred. Once scientists determined the exact chemical formulas of a variety of compounds, they were able to learn how and why elements and compounds interact as they do.

Our understanding of chemical reactions rests on most of the foundational discoveries chemists made in the eighteenth and nineteenth centuries. Chemical reactions are stoichiometric in nature—that is, they depend on precise mathematical and quantitative factors such as atomic mass. Such reactions always obey the Laws of Definite Proportions, Multiple Proportions, and Conservation of Mass. For the most part, they follow the atomic theory's assumptions. On a theoretical level, this makes chemistry a predictable, logical science; when we know which elements will interact in a chemical reaction, we should be able to predict how they will interact—the end result of the reaction—as well. On a practical level, chemistry's adherence to precise, clearly defined properties and laws makes it easy for us to summarize chemical reactions in equation form.

Chemical equations always follow the same format: the **reactants,** the "before" substances, appear on the left side, and the end **products,** the "after" substances the reaction produces, appear on the right. An arrow represents the reaction. Soon after they began to analyze chemical reactions in this way, chemists noticed that some chemical compounds behaved similarly to other compounds. For example, lithium (Li), sodium (Na), and potassium (K) react with chlorine (Cl), which occurs as a molecule made up of two atoms, to form similar compounds:

$$2\ Li + Cl_2 \rightarrow 2\ LiCl$$

$$2\ Na + Cl_2 \rightarrow 2\ NaCl$$

$$2\ K + Cl_2 \rightarrow 2\ KCl$$

Lithium, sodium, and potassium are silvery metals, and the products in these three equations are salts that all taste somewhat like the second product—sodium chloride (NaCl), table salt. In each case, the chemical reaction is rapid and **exothermic** (liberates much heat). These equations thus tell us that chemical reactions involving chlorine tend to have certain qualities.

We can replace the common reactant, chlorine, with bromine (Br) or iodine (I) in a similar reaction:

$$2\ Na + Br_2 \rightarrow 2\ NaBr$$

$$2\ Na + I_2 \rightarrow 2\ NaI$$

In these chemical reactions, sodium bromide (NaBr) and sodium iodide (NaI) are the products. These compounds also have a salty taste similar to sodium chloride, and these reactions, like the previous set, are rapid and exothermic. We might predict, from these reactions, that chlorine, bromine, and iodine have similar chemical properties. They do—although they are a gas, a liquid, and a solid, respectively.

These five equations demonstrate several properties of chemical reactions and of the equations that express them. First, they are all balanced equations—there are the same numbers of each type of atom on each side of the arrow. All chemical equations must balance; if the numbers of atoms are not equal on each side of the arrow, it implies that some atoms appear or disappear during the reaction, and this would contradict the Law of Conservation of Mass. In this book, we will use the inspection method to balance equations; in other words, we will examine each side of our equation to make sure it has the same total numbers of each type of atom. Balanced equations demonstrate the stoichiometric nature of chemistry, as well as the Law of Conservation of Mass.

Second, these equations include both compounds (two or more elements bonded together) and elements (in singular form or bonded together in molecules). In the first three equations, chlorine begins as a diatomic (two-atom) molecule, its elemental form. But it reacts to form new compounds that each contain one chlorine atom per molecule. (Again, keep in mind that the total number of chlorine atoms remains the same before and after the reaction—the atoms just form bonds with different atoms as a result of the reactions.)

Third, the equations illustrate several atomic theory assumptions. While compounds break down and reform during a chemical reaction, atoms remain intact. They are neither destroyed nor changed—they simply combine in new ways with other atoms to form molecules. The ratio of the atoms in a particular compound is constant—no matter how much Na and Cl_2 you begin with, they will always combine in a 1:1 ratio as NaCl.

Strictly speaking, substances such as NaCl are ionic compounds, not true molecules. Ionic compounds contain electrically charged atoms. In this book, ionic compounds are called molecules because the distinction is not important in writing chemical reactions. It is important in other areas of chemistry, however.

PRACTICE PROBLEM 3.11

Let's try writing some chemical equations.

a. Barium (Ba reacts with chlorine (Cl_2) to form barium chloride ($BaCl_2$). Write this reaction as an equation.

ANSWER First, arrange the reactants to the left of the reaction arrow and the products to the right.

$$Ba + Cl_2 \rightarrow BaCl_2$$

Now count the numbers of atoms of each element on each side of the arrow. Each side includes one Ba and two Cl; this equation is balanced.

b. Potassium oxide (K_2O) is produced when oxygen (O_2) reacts with potassium (K). Write this reaction in the form of an equation.

ANSWER This one is a bit more challenging. First, put reactants and product on the proper side of the arrow. (Remember, reactants always appear on the left, products on the right.)

$$K + O_2 \rightarrow K_2O$$

Now count the number of atoms of each element on each side of the equation. The left side shows one K atom; the right side, two; and the left shows two O atoms, the right side, one. Obviously this does not balance. How do we remedy this situation?

Begin with the K atoms. If we use two free K atoms as reactants, we will end up with the K atoms balanced:

$$2\ K + O_2 \rightarrow K_2O$$

However, the number of O atoms is still not balanced. Let's put a two in front of the product, K_2O, to account for two O atoms. We now have the following equation:

$$2\ K + O_2 \rightarrow 2\ K_2O$$

This balances the O atoms, but changes the number of K atoms so that they no longer balance—we now have two on the left and four on the right. Don't give up yet; several corrections are sometimes necessary to balance an equation.

Let's try doubling the number of K atoms on the left so that the reactant has as many as the product:

$$4\ K + O_2 \rightarrow 2\ K_2O$$

We can now count four K atoms and two O atoms on each side of the equation. Breathe a sigh of relief! The equation is now balanced—and we understand precisely how potassium and oxygen combine to form potassium oxide.

Moles, Atomic Mass, and Molecular Mass (Weight)

As we have already learned, the chemical mole may be defined in several ways. The primary definition is that a mole is a collection of 6.02×10^{23} atoms or molecules, a sum known as Avogadro's number. A mole is also the mass of a pure substance that contains an Avogadro's number of atoms or molecules; we express this mass value in grams per mole or amu. (This unit can also be called a dalton in honor of John Dalton.) A mole can thus measure numbers or mass of atoms, or numbers or mass of molecules. When it measures the former, we refer to a substance's atomic mass; when it measures the latter, molecular mass.

Atomic mass is the mass of an Avogadro's number of *atoms;* **molecular mass** is the mass of an Avogadro's number of molecules obtained by weighing 6.0×10^{23} molecules of a particular substance. When you know the chemical formula of a compound, you can calculate molecular mass by multiplying the number of each kind of atom by that atom's atomic mass, then adding the products (table 3.4). This value is also called the molecular weight or molar mass.

We can find the molecular mass for any compound by applying this formula. Let's determine the molecular mass of calcium carbonate, $CaCO_3$. We begin by determining the atomic mass for each atom in a $CaCO_3$ molecule and multiplying it by the number of that type of atom:

$$\begin{aligned} \text{Ca:}\quad & 1 \text{ atom} \times 40.08 \text{ g/mole} = 40.08 \text{ g/mole} \\ \text{C:}\quad & 1 \text{ atom} \times 12.01 \text{ g/mole} = 12.01 \text{ g/mole} \\ \text{O:}\quad & 3 \text{ atoms} \times 16.00 \text{ g/mole} = \underline{48.00 \text{ g/mole}} \\ & \qquad\qquad\qquad\qquad\qquad 100.09 \text{ g/mole} \end{aligned}$$

TABLE 3.4 The molecular mass of water.

ATOM	TOTAL NUMBER OF ATOMS IN MOLECULE	ATOM'S MASS (g/mole)	NUMBER OF ATOMS × ATOMIC MASS (g/mole)
H	2	1.01	2.02
O	1	16.00	16.00
			18.02

Total molecular mass for H_2O = 18.02 g/mole

The molecular mass of calcium carbonate is 100.09 g/mole; in other words, a mole (6.02×10^{23}) of calcium carbonate molecules weighs 100.09 g.

Sodium sulfate's formula is Na_2SO_4. We can calculate its molecular mass by multiplying the atomic mass by number of atoms for each type of atom:

Na: 2 atoms × 22.99 g/mole = 45.98 g/mole

S: 1 atom × 32.06 g/mole = 32.06 g/mole

O: 4 atoms × 16.00 g/mole = 64.00 g/mole

The molecular mass of sodium sulfate is 142.04 g/mole; in other words, a mole of sodium sulfate molecules weighs 142.04 g.

Sucrose (ordinary table sugar) has a chemical formula of $C_{12}H_{22}O_{11}$. This is a larger molecule than the previous three (12 + 22 + 11 = 45 total atoms in one molecule), but the procedure for determining molecular mass remains the same. Even though one mole of sucrose contains 45 (6.0×10^{23}) atoms, we can still determine molecular mass by multiplying number of atoms times atomic mass and adding the products:

C: 12 atoms × 12.01 g/mole = 144.12 g/mole

H: 22 atoms × 1.0 g/mole = 22.22 g/mole

O: 11 atoms × 16.00 g/mole = 176.00 g/mole

342.24 g/mole

The molecular mass of sucrose is 342.24 g/mole; one mole of sucrose molecules weighs 342.12 g.

As we have just seen, moles vary greatly in mass. A mole of water weighs only 18.00 grams, but a mole of sugar weighs 342.24 grams. (To visualize the difference, let's look at this another way. A gram of water is about 1 mL and contains about 20 drops. Thus, a mole of water contains about 360 drops and occupies a volume of 18 mL—about two-thirds of a liquid ounce. A mole of sucrose, by contrast, weighs about three-fourths of a pound and occupies about 215 mL.)

The mole is chemistry's primary unit of measurement because reactions occur on a molecular level, but chemists cannot run reactions a molecule at a time—an individual molecule is too small. We can calculate the mass of one water molecule as follows:

$$18.0 \text{ g/mole} \div 6.0 \times 10^{23} \text{ molecules/mole} = 3.0 \times 10^{-23} \text{ g/molecule water}$$

This quantity—a decimal point followed by 23 zeros and then a 3, or 0.000,000,000,000,000,000,000,003—is far too small to weigh with any device ordinarily available in the laboratory. (We will discuss handling such small numbers later in the chapter.) Although much larger, an individual sucrose molecule weighs just 5.7×10^{-22} grams (a decimal point, 22 zeros, then a 57). However, if a pair of molecules react in a certain way, a mole of each of these molecules should react the same way. Chemists can easily weigh moles, or even fractions or multiples of moles, depending on the amount of product desired.

TABLE 3.5 The NaCl reaction: The same reaction with differing amounts of reactants and products.

2 Na	+	Cl_2	→	2 NaCl
2 atoms	+	1 molecule	=	2 molecules
2 moles Na	+	1 mole Cl_2	→	2 moles NaCl
45.98 g		70.90 g	=	116.88 g
0.02 moles Na	+	0.01 moles Cl_2	→	0.02 moles NaCl
0.46 g	+	0.71 g	=	1.17 g

In one of the reactions we looked at earlier, two Na atoms react with one Cl_2 molecule to produce two NaCl molecules. We can also express this reaction in terms of moles—two moles of Na react with one mole of Cl_2 to produce two moles of NaCl. Similarly, 45.98 g Na reacts with 70.90 g Cl_2 to produce 116.88 g NaCl. Table 3.5 summarizes the different ways to view this reaction.

The reaction that occurs when Na and Cl_2 combine to form NaCl occurs very rapidly and with explosive violence. For safety reasons, most chemists prefer using smaller amounts in the laboratory; for example, 0.02 moles Na and 0.01 moles Cl_2 to yield 0.02 moles NaCl. Commercial-scale reactions are run using many moles of reactants. The only difference is how many atoms or molecules actually react at a given time; the calculations that explain the reaction are always the same.

PRACTICE PROBLEM 3.12

Determine the molecular mass for each of the following compounds. Use these atomic masses: H = 1.01; C = 12.01; N = 14.01; O = 16.00; Na = 22.99.

a. Benzene has a formula of C_6H_6. What is its molecular mass?

ANSWER The formula contains six C atoms and six H atoms. Begin by multiplying the number of each kind of atom by the atomic mass of that atom; then add the results to determine the molecular mass.

C: 6 atoms × 12.01 amu = 72.06 amu

H: 6 atoms × 1.01 amu = 6.06 amu

78.12 amu

Total = 78.12 amu (or 78.12 g/mole or daltons)

b. Sodium nitrate's chemical formula is $NaNO_3$. What is its molecular mass?

ANSWER As before, multiply the numbers of each kind of atom by its atomic mass, and then add the results to determine the molecular mass.

Na:	1 atom ×	22.99 amu =	22.99 amu
N:	1 atom ×	14.01 amu =	14.01 amu
O:	3 atoms ×	16.00 amu =	48.00 amu
			85.00 amu

Total = 85.00 amu, g/mole, or daltons

INTRODUCTION TO CHEMICAL ANALYSIS

Synthesis, Separation, and Analysis

Synthesis involves *building* compounds from atoms or from other compounds. Figure 3.9 shows how two compounds in solution are used to synthesize a plastic. This particular process is called the "nylon rope trick" because it results in a "rope" of nylon. The two starting materials, a solid and a liquid, are placed in a pair of immiscible solutions, and then the nylon is drawn out from these two solutions. Although nylon is better known as a textile, it is also a widely used plastic for gears and cases. Practically all commercially manufactured chemical compounds are synthesized.

Analysis involves determining the identities and amounts of the components in a substance. Analysis might determine the proportions of elements in a compound, or it might deduce the proportions of various compounds in a mixture. Figure 3.10 outlines a simple analysis for determining how much solid material (soap or detergent) is present in a typical hair shampoo. We can weigh a sample of the shampoo and then allow the water and any other volatile materials to evaporate; to speed the process, we can use a low-temperature oven. After evaporation is complete, we can weigh the sample again to find out how much solid material is present. Then we can divide the mass of the solid material by

Figure 3.9 The "nylon rope trick," an example of a synthesis.

Figure 3.10 Determining the amount of solid materials in a typical hair shampoo, an example of an analysis.

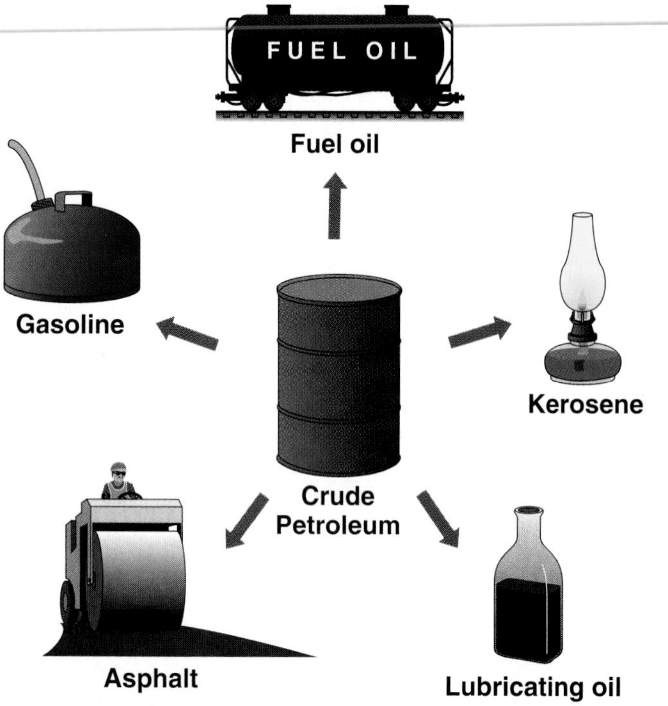

Figure 3.11 An outline showing how some components are separated from a crude petroleum in a refinery. This is an example of a separation and is usually done by distillation.

the mass of the starting mixture and multiply the result by 100 to determine the percentage of solid components in the shampoo. Analysis often involves breaking apart a compound or making a separation occur more readily.

Separations involve separating out pure substances from compounds or impure mixtures. Purification processes, which are closer to analysis than synthesis, are considered separations in this book; separation procedures are used to refine or purify billions of kilograms of chemical products found in nature every year. For example, commercially important materials such as oil, gasoline, sugar, and salt are made by purifying materials existing in nature. Figure 3.11 depicts some of the useful materials separated from crude petroleum at a refinery.

PRACTICE PROBLEM 3.13

Classify each of the following processes as synthesis, analysis, or separation.

a. Refining crude petroleum to produce gasoline

b. Purifying a purple dye from the secretions of a certain kind of mollusk

c. Extracting the anticancer medicine taxol from the yew tree

d. Determining the chemical structure for taxol

e. Making taxol from some other chemicals through a chemical reaction

f. Producing aspirin by triggering a chemical reaction in a substance obtained from the bark of the willow tree

g. Converting ethylene, a petroleum-derived gas, into the plastic called polyethylene through a chemical reaction

ANSWER Processes *a*, *b*, and *c* are separations because they isolate a substance that already exists in a mixture. Example *d* is analysis because it involves determining the chemical identity of a substance. Processes *e*, *f*, and *g* build new compounds from other substances; they are therefore examples of synthesis.

Distillation, Recrystallization, Chromatography, and Spectroscopy

Four techniques help chemists synthesize, analyze, and separate chemical compounds. These techniques are distillation, recrystallization, chromatography, and spectroscopy.

Distillation

Distillation is one of the oldest purification techniques. In this process, a liquid is boiled, and then the vapors pass through a condenser that causes the vapors to return to the liquid state. Although pure substances always have definite boiling points, mixtures usually do not. When a solution of two liquids with widely different boiling points is heated, it is therefore often possible to separate the mixture with a simple distillation apparatus (figure 3.12). One liquid boils and recondenses in a separate container; then, with additional heat, the other does the same. The result is two pure liquids. More elaborate devices contain special condensers and columns to better separate mixtures.

When two liquids have closer boiling points, they may not separate completely in a simple distillation. For example, a mixture of two liquids, ethylbenzene (b.p. 136°C) and pentane (b.p. 36°C) forms a homogeneous solution. The solution can undergo distillation to separate the two chemicals from each other. But when pentane (b.p. 36°C) and benzene (b.p. 80°C) are combined in a distillation apparatus, the result could be a series of fractions like those table 3.6 lists. A **fraction** is a distillate sample that contains both substances and boils at a point between their boiling temperatures. Because the original two substances boil at close temperatures, portions of each substance vaporize and recondense in each fraction; the distillate is not pure.

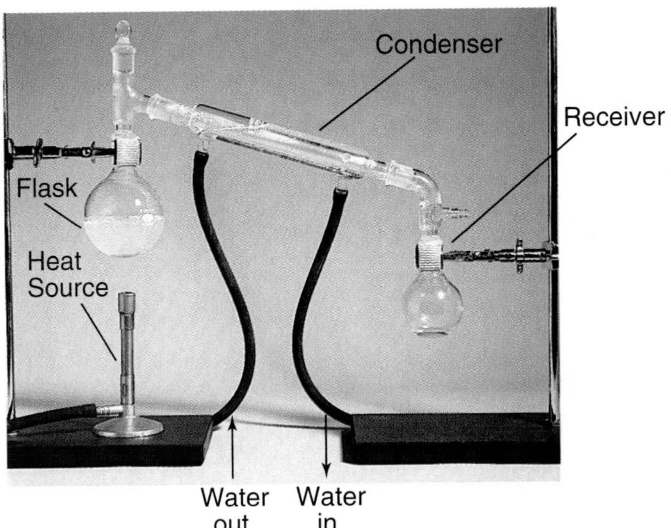

Figure 3.12 A simple distillation apparatus being used to produce pure water from sea water. The sand and the dissolved salt (NaCl) remain in the distillation flask.

In table 3.6, which shows successive samples of the distillate from a 50 percent pentane–50 percent benzene solution, each fraction contains less pentane and more benzene as the temperature increases. This simple distillation partially purified this mixture. A second distillation or a more elaborate apparatus would better separate the chemicals. Many natural solutions consist of compounds with similar boiling points; a series of fractional distillations can separate these mixtures into their components.

Refineries use distillation to separate gasoline, kerosene, and other substances from crude petroleum. A tall vertical column called a refluxing column collects fractions at various temperature ranges.

Liquids containing dissolved solids are easy to purify by distillation because the solids usually do not boil. Seawater may be converted into pure, drinkable water in this manner (a topic covered in more detail in chapter 12).

TABLE 3.6 Fractions collected in a simple distillation of a pentane: benzene solution. The 100 mL sample contains 50 percent of each compound by volume.

Fraction	Boiling Range, °C	Size, mL	PERCENT OF TOTAL VOLUME Pentane	Benzene
A	22–38	35	74	26
B	39–45	13	65	35
C	46–60	5	50	50
D	61–77	12	43	57
E	78–100	35	23	77

Recrystallization

Recrystallization—when a substance dissolves and then crystallizes out of solution—is another technique used to separate or purify mixtures. Recrystallization works because many solids have increased **solubility** when the temperature increases. Solubility is the ability of a solid (the **solute**) to dissolve in a liquid (the **solvent**).

The purification of sugar (sucrose) illustrates recrystallization. Sugar freshly harvested from sugar cane is a brownish color. When it dissolves in warm water, many of the brown-colored solids remain out of solution; they may be removed by filtering or decanting. In **filtering,** the solution is poured through a porous paper or cloth that removes the undissolved solids (figure 3.13). In **decanting** the undissolved solids are allowed to settle to the bottom and the solution is poured off. At this point, the sugar solution is yellowish; most of the solute left in it is sucrose.

Figure 3.13 Photo for a filtration. A solid–liquid mixture is passed through a funnel fitted with a filter paper. The solid is retained on the paper while the liquid passes through and is collected in a beaker.

In the next step in sugar recrystallization, much of the water in the yellowish solution is heated and evaporates, and the more concentrated solution then cools (figure 3.14). **Concentration** measures how much material—in this case, sucrose—a solution contains. As the warm, concentrated sucrose solution cools, much of the sucrose precipitates out of solution, or recrystallizes, because sucrose is more soluble in hot water than in cold. When the temperature decreases so much that the solution contains more sucrose than can remain dissolved; the "excess" sucrose precipitates out of solution and forms crystals. The liquid part of the solution darkens during these steps because the colored impurities become more concentrated as sugar crystals precipitate out. Finally, the purified sucrose is collected by filtration. Figure 3.15 shows a typical sugar refinery; in actual commercial practice, the sugar crystals are recovered from the solution by centrifugation, rather than filtration, because this is a faster procedure. Box 3.5 discusses the American demand for sugar.

Recrystallization may also be used to isolate each of two compounds in a physical mixture. Imagine a mixture of two white, crystalline solids—sugar (sucrose) and table salt (sodium chloride). The crystals are slightly different in shape, and you could separate the sugar out using tweezers and a magnifying glass, but it would require hours of painstaking labor to separate even a small sample.

Recrystallization can separate these two solids more efficiently because sucrose is more soluble in hot water than salt is. In fact, salt has about the same solubility in hot or cold water; while the solubility of sucrose increases with temperature. If a mixture of 50 g of each solid (100 g total) is dissolved in 50 mL hot water (a limited amount of solvent), more sucrose dissolves than salt; the undissolved residue contains more salt than sucrose. This residue may then be filtered off, and the filtered solution evaporated to dryness to obtain a partially purified sucrose. The process is repeated until an adequate separation is achieved.

Figure 3.15 A sugar refinery.

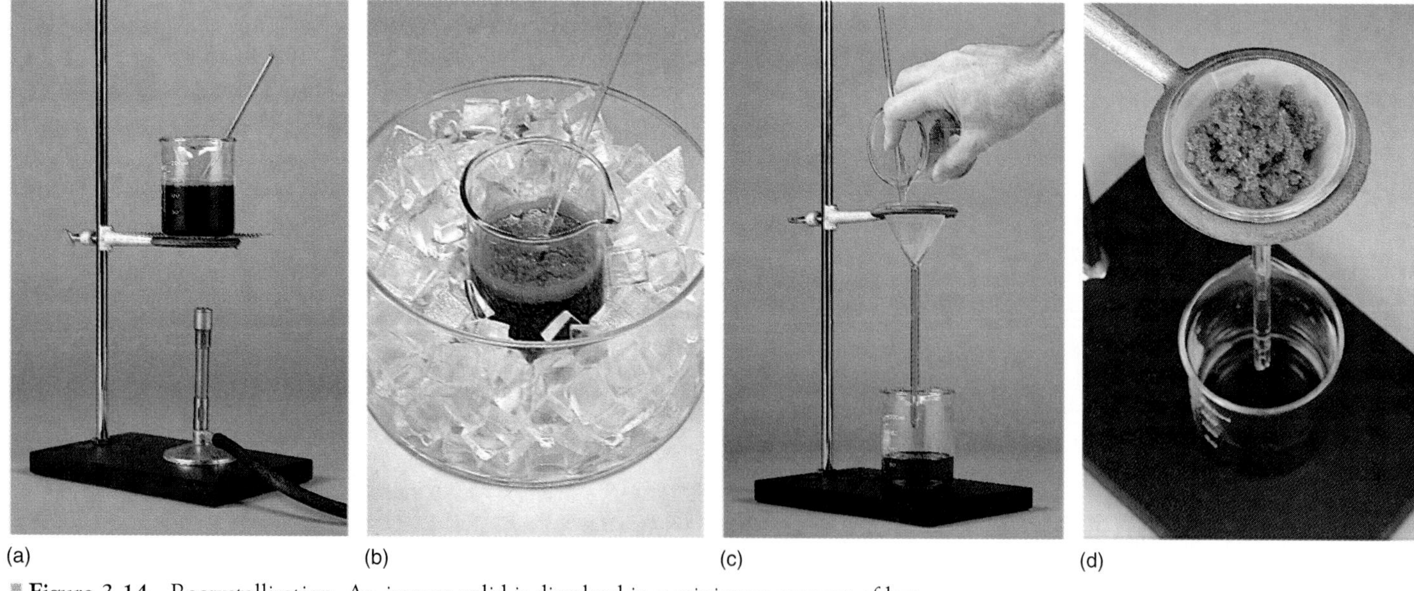

Figure 3.14 Recrystallization. An impure solid is dissolved in a minimum amount of hot solvent in step **(a)**. This solution is cooled in step **(b)**, forming pure crystals. The supernatant liquid is separated by filtration in step **(c)** leaving the pure crystals in step **(d)**.

BOX 3.5

THE AMERICAN SWEET TOOTH

Every year, the United States produces approximately 160 billion pounds of sugar—nearly 700 pounds per person, or about 2 pounds per person per day. Sugar production nearly doubles that of sulfuric acid, the top industrial chemical. Sugar is not only used at the table and in candy (figure 1), it shows up in catsup, salad dressings, bread, pizza, cakes, cookies, ice cream, soft drinks, peanut butter, and jelly.

Figure 1 The American sweet tooth keeps this candy shop in business.

Most dietitians consider sugar a chemical without much food value; in fact, excess sugar intake can cause problems for many people. A diabetic cannot tolerate sugar because his or her body does not produce enough of the enzyme insulin to digest it. The person's blood sugar levels then rise dangerously high. Diabetes, the third leading cause of death in the United States, can lead to many complications, including blindness, heart disease, and even death. Other people suffer from hypoglycemia, a condition in which the sugar level in the blood is too low. Hypoglycemia can cause a lack of energy and other symptoms, but it is rarely a fatal condition.

Most Americans do not suffer from either diabetes or hypoglycemia. For most people, the main consequence of excess sugar consumption is weight increase. Sugar—made up of carbon, hydrogen, and oxygen, all necessary to life—is a much-enjoyed but sometimes harmful part of the American diet.

PRACTICE PROBLEM 3.14

Most grocery stores sell a product called brown sugar. Do some research to find out which material is more pure: brown sugar, or white granulated sugar? Are both materials equally safe for eating and cooking?

ANSWER White sugar is more pure than brown sugar; most brown sugar is made up of white sugar with added molasses. (Molasses is a viscous, dark-colored liquid drained off from raw cane or beet sugar.) Both types of sugar are equally suitable for human consumption, but eating too much of either could add a few kilograms to your weight. Diabetics must avoid all types of sugar because their bodies cannot handle sugar digestion properly.

Chromatography

Chromatography is a separation process in which a fluid (liquid, gas, or solution) flows over a stationary (nonmoving) phase that may be either a solid or a liquid. The basic feature of chromatography is **selective adsorption,** a physical process in which one material "sticks" or adheres to the surface of another (while in absorption, the material actually soaks in). This adsorbed material is later desorbed and collected. Liquid chromatography, the separation of components dissolved in a solvent, is the main example we will consider here. It produces separate bands of substances that often form layers of color.

Michail Semenovich Tswett (1872–1919) invented column chromatography. Tswett was interested in the separation of plant pigments. He noted in his 1901 master's thesis, "I could vividly see differently colored rings when filtering petroleum ether extracts of leaves through Swedish paper." Tswett's extract was the solution formed by mixing ground plant leaves with liquid petroleum ether, a chemical resembling gasoline. Tswett learned he could isolate the individual pigments through this technique, and he published more than a hundred papers on this topic. Not all materials show visible or colored bands, although Tswett's materials did. That is why he called the technique chromatography—the "graphing of colors." When substances are not visible, chemists use other techniques to "see" the substances that make up compounds. Properties such as the ability to conduct electric current or to absorb ultraviolet light allow chemists to differentiate between different substances.

The chromatographic technique works as follows (figure 3.16). A solution containing two or more substances flows through a column, normally a glass tube, that contains a suitable stationary solid phase (for example, sand or powdered sugar). One substance in the solution adheres to this stationary phase better than the

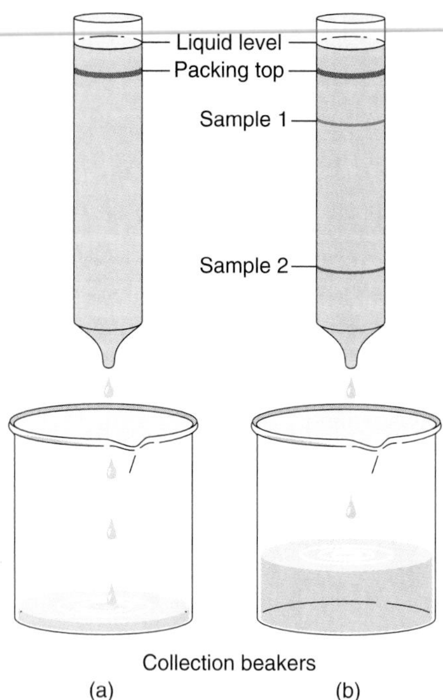

Figure 3.16 Column chromatography. Step **(a)** shows the columns at the start of the process, and step **(b)** shows them after separation. A mixture of samples 1 and 2 was at the top of the column packing in step **(a)**. In step **(b)**, samples 1 and 2 are separated at two locations on the column packing. Each sample is collected as a solution, in the elution solvent, in a beaker at the column bottom.

others. This means the poorer-adhering materials become slightly more concentrated in the solution as it flows down the column. An excellent separation is achieved when the column is sufficiently long and the adsorption difference is adequate. Liquid chromatography is useful for both physical separations and analysis.

Other chromatographic techniques include paper chromatography with a paper stationary phase; gas chromatography (GC), where a gas flows over a stationary phase coated with a nonvolatile liquid; and thin-layer chromatography (TLC), which passes a solution over thin layers of a solid phase attached to a glass plate or other rigid solid.

EXPERIENCE THIS

a. You can observe column chromatography for yourself with a carrot, powdered sugar, and lighter fluid. (Petroleum ether is better, if you can get it, but be careful—it is highly flammable.) Grind the carrot and mix it with the lighter fluid. Pour the resulting solution down a column packed with confectioners sugar. Wash it down with more lighter fluid. (Do not use water, since it will dissolve the sugar.) What happens?

b. You can usually separate ink into colored layers the same way, but paper chromatography works as well and is much easier to do at home. This experiment works best with a liquid ink from a bottle or jar. (You can try ink from a ballpoint pen, but it is likely to prove less satisfactory.) You also need a porous paper, such as blotting paper or a paper towel, a tall, closed jar, and a liquid solvent. Place a small amount of the ink near one end of the paper and hang it into the jar with this spot near the bottom. The jar should contain a small amount of the solvent; do not have your sample spot go in the solvent. As the liquid rises up the paper you should see a series of spots develop. Each spot is a colored substance in the ink.

Spectroscopy

Spectroscopy encompasses several techniques that measure absorbed or emitted energy when materials interact with radiant energy. Radiant energy, known as the electromagnetic spectrum, includes visible light, heat, cosmic rays, X rays, ultraviolet and infrared light, microwaves, and radio waves. All these forms of energy travel in waves. The distance between the wave crests determines the **wavelength** of each type of radiation, the shorter the wavelength, the higher the **frequency** with which the waves pass a given point in one second. (To understand the relationship between wavelength and frequency, visualize closely packed waves moving past a point and stretched-out waves moving past the same point—many more of the closely packed, short-wavelength waves will move past the point in a second, so short wavelengths are associated with higher frequencies.) Also, the shorter the wavelength, and the higher the frequency, the more energy emitted.

When visible white light passes through a prism, the light separates into a spectrum of colors, each representing light of different wavelengths (figure 3.17). In 1814, Joseph von Fraunhofer (Bavarian, 1781–1826) detected dark lines in the usually continuous sunlight spectrum; the lines were later determined to come from elemental helium. In 1861, Gustav Kirchoff (a Russian chemist) and German chemist Robert Bunsen vaporized a solid in a flame and produced an emission line spectrum—a discontinuous spectrum marked by colored lines similar to those Fraunhofer had seen. Eventually, scientists discovered that every element produces a unique emission line spectrum that can be used to identify the element. Figure 3.18 shows several examples of emission line spectra.

Spectroscopy and related technologies are used today in a variety of commercial and analytical techniques and products. Microwaves form the basis for radar and microwave cooking; X rays reveal the insides of our bodies. Heat lamps operate on infrared frequencies, and chemists use infrared and ultraviolet spectroscopy to analyze and identify substances.

High radiation energies may trigger many possible chemical reactions, some of which are potentially harmful to humans or

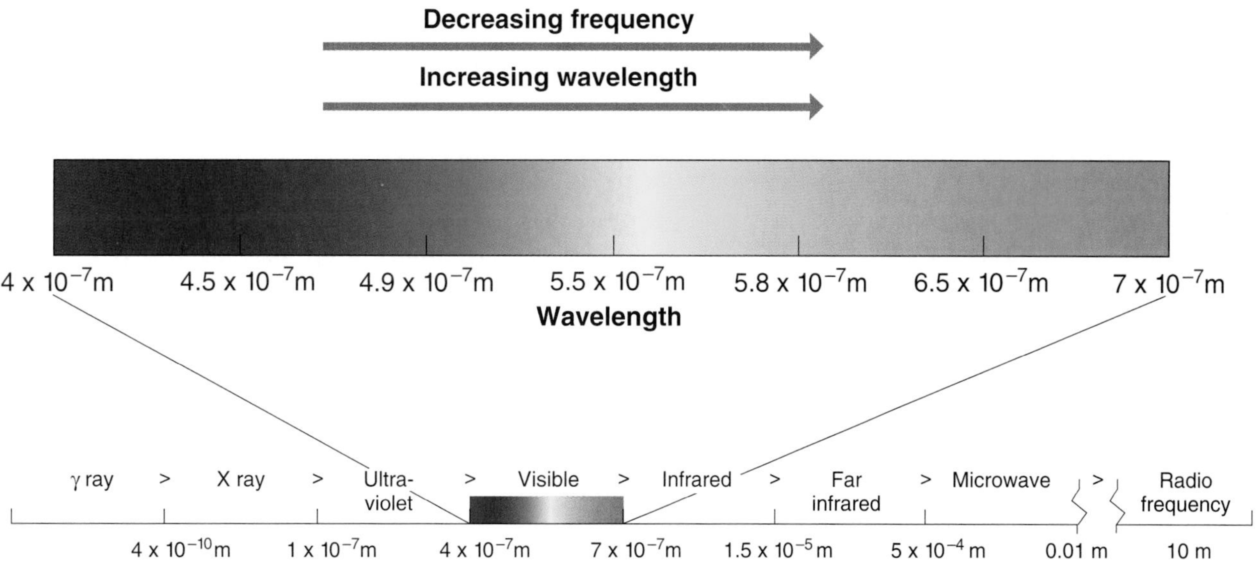

Figure 3.17 The electromagnetic spectrum.

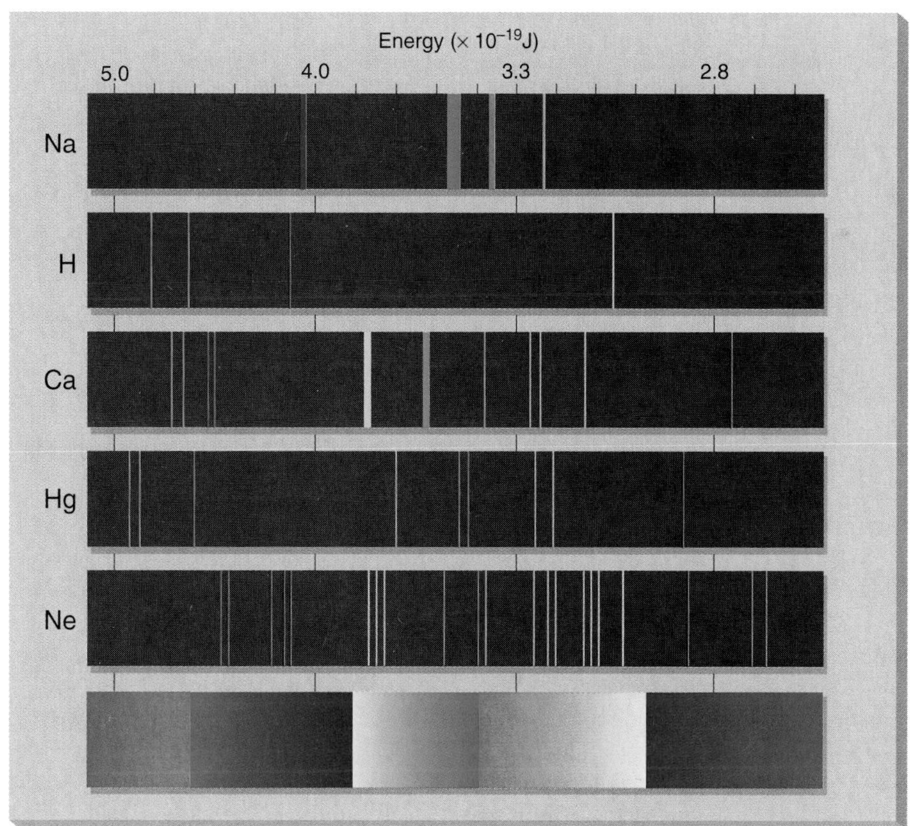

Figure 3.18 Emission line spectra for sodium, hydrogen, calcium, mercury, and neon.
SOURCE: From Jacqueline I. Kroschwitz, et al., *Chemistry: A First Course*, 3d ed.

other life. "Radiation" from a microwave is comparatively safe for cooking, while X rays are potentially harmful. (However, even microwaves could be dangerous. The inside of a microwave oven would be an unsafe place for living creatures.) The interaction between energy and chemistry is constant—in fact, the two are inextricably woven together.

ENERGY AND CHEMISTRY

Energy is crucial to the workings of chemistry; in fact, the development of chemistry as a science parallels the discovery of new forms of energy. Therefore, a brief survey of energy and how it interacts with matter will help us better understand chemical reactions and processes. More information on energy will be presented in chapter 5. For now, note that energy may be classified into five categories: mechanical, thermal, electrical, nuclear, and chemical. These forms of energy may produce either physical or chemical changes in matter.

A Brief History of Energy and Chemistry

The discovery of the elements often depended on the energy sources available. A few elements, such as silver and gold, exist freely (in pure form) in nature; the remainder are combined in various compounds. Early scientists could isolate the free elements through manual labor—a form of *mechanical energy*. Other elements, such as copper and tin, could be isolated from their compounds merely by heating ores in a wood fire, the first human-controlled *thermal energy* source. Years later, when humans discovered that coal produced hotter fires, people could pry iron and a few other elements from ores.

Some elements were impossible to isolate by thermal or mechanical energy. Scientists had to discover and learn to control *electrical energy* before they could isolate large amounts of aluminum or sodium. More recently, *nuclear energy* created a few new elements not found in nature, at least on our planet. The combined application of several energy forms thus helped chemists pull compounds apart and find out what they are made from. In addition, closer examinations of matter with more powerful energy sources such as X rays and particle accelerators revealed amazing details about atomic structure.

Chemical Energy

Much of the energy produced and consumed daily is chemical energy, the energy stored in and released from the chemical bonds between atoms. Sources of chemical energy include common fuels such as wood, coal, and petroleum, the food we eat, and the fat stored in our bodies. Nearly all chemical reactions either require heat (in **endothermic reactions**) or release heat (in **exothermic reactions**). This makes chemical energy the most common energy source on our planet. It is also the energy form most easily stored and transported. Energy and chemistry join together in the bonds between atoms.

PRACTICE PROBLEM 3.15

Is it possible to heat or cool a house using an endothermic (heat-requiring) chemical reaction?

ANSWER You cannot heat a house through an endothermic reaction because such a reaction draws heat out of the surrounding environment. You might theoretically use an endothermic reaction to cool a house, but the practical expense would probably be overwhelming—there are no "free lunches" in chemistry.

Physical and Chemical Changes

The interaction between energy and matter may produce either physical or chemical changes. Physical changes may affect the appearance, structure, or behavior of the material, but they do not affect its chemical or atomic structure. Chemical changes, by contrast, alter the material's structure at the molecular level.

To demonstrate a chemical change, cook an egg by placing it in boiling water. After the egg cooks, the once-liquid inside materials are solid:

$$\text{raw, liquid egg} + \text{heat} \longrightarrow \text{cooked, solid egg}$$

Have you ever wondered why this change occurs? A profound, permanent change has taken place in the atoms of the egg; even after it returns to room temperature (that is, after the heat energy is removed), the egg remains solid:

$$\text{cooked, solid egg} \xrightarrow{\text{cooling}} \text{cooked, solid egg}$$

By contrast, consider what happens when you boil water in a closed container. The water changes into steam when heated, but cools back into a liquid:

$$\text{liquid water} + \text{heat} \longrightarrow \text{steam (water as a gas)}$$

$$\text{steam} \xrightarrow{\text{cooling}} \text{liquid water}$$

Similarly, water changes into solid ice when cooled to its freezing point, but returns to liquid form when warmed.

If we performed a chemical analysis on the egg, we would learn that the egg changed *chemically* when cooked, while the water remained the same chemically when heated or cooled. The egg underwent chemical changes due to *chemical* reactions; the changes in the water were *physical*.

To reverse a chemical reaction, you must run a different chemical reaction that goes in the opposite direction. However, not all chemical reactions run easily in opposite directions; it is not possible to reverse the chemical changes that occurred in boiling an egg. Most (not all) physical changes can be reversed by merely reversing the physical action. Thus, we reverse the physical effects of heating by cooling; ice becomes water when heated, and water returns to ice when cooled.

• • • • • • • • • •

PRACTICE PROBLEM 3.16

Determine whether each of the following examples illustrates chemical or physical changes.

a. Carving a toy from a piece of wood
b. Separating a pile of nuts and bolts
c. Washing your clothing
d. Melting a snowman
e. Mixing a pitcher of Kool-Aid
f. Burning the scraps of wood left from carving a toy
g. Allowing milk to turn sour
h. Cooking a pot roast for dinner
i. Deep-frying sliced potatoes
j. Baking a cake from milk, yeast, salt, sugar, and flour

ANSWER The first five *(a–e)* are physical changes, the last five *(f–j)* are chemical changes. In each of the first five example, no permanent molecular change takes place in the material. Wood chips are still wood, and theoretically, you could reconstruct the original piece by reattaching the carved pieces; the nuts and bolts remain the same when sorted into different piles. Clean clothes may look different, but all you did was remove physically attached dirt; no chemical change occurred in the fabric. The material from the snowman is still water, but it is now in liquid form. Kook-Aid is possibly the most difficult example; if you mixed unsweetened Kool-Aid and added sugar and water, you have mixed the chemicals together and the Kool-Aid and sugar have dissolved. While they now look different, they are separable by a physical process—if the water evaporates, you will still have Kool-Aid and sugar.

The last five are chemical changes because the materials have changed on the molecular level at the end of the process. The wood chips went up in smoke, releasing carbon, hydrogen, and oxygen; the sour milk tastes differently because of chemical changes. The beef changed in color, texture, and taste, and the changes do not reverse when heat is removed. Likewise, the potatoes have changed in ways that do not reverse when they cool. The cake offers the most striking change. From milk, yeast, salt, sugar, and flour, you now have a solid, homogeneous cake with tiny air pockets. No traces of the individual ingredients remain; even after the cake cools back to room temperature, the changes endure.

In the remaining chapters, you will encounter many examples of both chemical and physical changes. Some of these changes will be easy to recognize; others may be more difficult to differentiate. The key is whether a substance still has the same chemical composition at the end of the change. If it does, the change is physical, but if the composition is different, the change is chemical.

HANDLING VERY LARGE AND VERY SMALL NUMBERS

Before becoming deeply involved with atoms and atomic structures, you need to understand how scientists describe very large and very small numbers using exponential and scientific notation. (Additional details appear in Appendix B.)

Many numbers routinely used in chemistry are remarkably big or incredibly tiny. For example, we've already discussed the mole, the quantity of molecules in a designated mass of a compound. This exceptionally large number dwarfs the numbers you usually encounter. A mole of water weighs a mere 18.0 grams (a little more than half an ounce), yet it contains about 602,000,000,000,000,000,000,000 molecules. Each water molecule therefore weighs about 0.000,000,000,000,000,000,000,003 grams. Such very large or very small numbers are difficult to work with. Consequently, scientists use exponential numbers and scientific notation to express them.

An **exponential number** is simply a number raised to a power denoted by a superscript. Thus, 10^2 is ten raised to the second power (or squared), which means ten is multiplied by itself (10×10) to produce 100. The exponential number 10^5, then, means ten is multiplied by itself five times: $10 \times 10 \times 10 \times 10 \times 10 = 100{,}000$. Likewise, a billion is 10^9, obtained by multiplying ten by itself nine times. A trillion is 10^{12}, or the product of ten multiplied by itself twelve times. Notice that with 10, the superscript denotes the number of zeros that follow the 1 in the product: 10^2 is 100 (1 followed by two zeros); 10^5 is 100,000; 10^9 is 1,000,000,000; and so forth.

The numbers million, billion, and trillion are used often, yet people rarely realize their tremendous size. A single stealth bomber costs about 400 million dollars, and our national debt is around 4 trillion dollars, with an interest of over 150 billion dollars annually. More than 1.2 billion people live in China. Other examples crop up daily in the news. To help you grasp the size of these numbers, consider the following example. Suppose you spent exactly 1 dollar a minute for 16 hours a day, or $960 daily. Because there are 365.25 days per year (counting leap years), you would spend $350,640 dollars annually. At this rate, it would take nearly three years to spend a million dollars. A billion is a thousand million; you would need 2,851.9 years to spend a billion dollars! If you had started, for example, to spend a dollar a minute in the year 1 B.C., you would still have the tidy sum of $298,720,000 left to spend from a billion dollars as the year 2000 dawns!

Tiny numbers are also expressed as exponential by writing a negative exponent. This means the number is repeatedly divided rather than multiplied. The quantity one one-thousandth (0.001) is obtained by dividing ten into one three times; in other words, $1 \div 10 = 0.1$; $0.1 \div 10 = 0.01$; and $0.01 \div 10 = 0.001$. This number can be expressed as $1 \div 10^3$ or 10^{-3}, meaning 1 divided by 1,000. A penny is one one-hundredth of a dollar, or 10^{-2} dollars. A millionth of an inch is 0.000001 inches, or 10^{-6} inches. Pollutants are often measured in parts per billion (ppb) by the Environmental Protection Agency (EPA). This means that out of a billion particles of air or water, you are measuring one or more individual particles. One part per billion is 0.000000001, or 10^{-9}.

Exponentials are easiest to handle in powers of ten. Numbers that are not exact powers of ten are written in **scientific notation,** which combines the exponential notation in a power of ten with another number expressed in decimals. Usually this number has one numeral to the left of the decimal point. The number 3,456 cannot be written as a simple exponential of ten; in scientific notation, however, it can be expressed as 3.456×10^3. (If 3.456 is multiplied by 1,000, the result is 3,456—original number.)

Scientific notation offers advantages when working with very large or very small numbers. In our previous example, the money you would have left at the year 2000 if you engaged in a billion-dollar spending spree is 2.99×10^8, or roughly 3×10^8. Even the 4,000,000,000,000-dollar national debt looks more manageable as 4.0×10^{12}. Scientific notation is much easier to write and calculate with for very large and small numbers.

PRACTICE PROBLEM 3.17

The population of the United States is about 250,000,000 people. Write this number in scientific notation.

ANSWER You can generate this number by multiplying 2.5 times a hundred million, which is the equivalent of moving the decimal place eight places to the right. A hundred million is 10^8. Thus, 250,000,000 equals 2.5×10^8.

PRACTICE PROBLEM 3.18

The latest EPA rules require that drinking water contain lead at levels under 15 parts per billion (ppb). Write this number in scientific notation.

ANSWER You can do this several ways, but let's do the easy way first. The number is 15 parts per billion, or 15 divided by a billion. Since a billion is 10^9, this means we divide 15 by this exponential.

$$15 \div 1{,}000{,}000{,}000 = 15 \div 10^9$$

When you divide by an exponential, the quotient is expressed as a product using a negative power. Thus:

$$15 \div 10^9 = 15 \times 10^{-9}$$

Now, since scientific notation generally requires that a single digit be placed to the left of the decimal point, we must convert 15 to 1.5. This number is one-tenth as large as 15, so we can now express 15 ppb as follows:

$$15 \times 10^{-9} = 1.5 \times 10 \times 10^{-9} = 1.5 \times 10^{-8}$$

In scientific notation, 15 ppb is 1.5×10^{-8}.

We can arrive at the same figure by a different method. Begin by writing the number as a division problem:

$$15 \div 1{,}000{,}000{,}000 = 0.000000015$$

What must you multiply this number by to obtain 1.5? The answer is 100,000,000 (one hundred million); in exponential form, 10^8. Since we are actually dividing the number, we will express this by multiplying by 10^{-8}. Thus, 15 ppb is expressed just as before:

$$0.000000015 = 1.5 \times 10^{-8}$$

SUMMARY

Knowledge about chemistry has developed over many centuries, from ancient times through the ages of alchemy, iatrochemistry, and modern independent chemistry. During the modern age, chemists made crucial advances that led from the phlogiston theory to the theory of oxidation and the discovery of the Laws of Definite Proportions and Conservation of Mass. Eventually, Dalton's atomic theory tied all these discoveries together into a comprehensive theory of matter.

The atomic theory explains many observations chemists have made about the structure and behavior of chemical substances. The theory helped scientists identify elements as atoms of one substance, molecules as atoms bonded together, and compounds as different elements bonded into molecules. When chemists applied the Law of Definite Proportions to compounds, they discovered that every compound contains an unvarying proportion of atoms of different elements. Avogadro's hypothesis then assumed that equal volumes of two different gases contain equal numbers of molecules, though their masses differ. These two concepts, taken together, helped chemists rapidly determine the atomic weights for dozens of elements. They also led scientists to determine that a mole of any element contains 6.02×10^{23} molecules or atoms.

Once chemists determined the relative weights and atomic masses of a variety of elements, they could determine chemical formulas for compounds, or the number of atoms of each element in a compound. This information enabled chemists to write chemical reactions. Chemical formulas help us understand how specific chemical reactions occur, as well as reveal the molecular mass or weight for individual compounds.

Chemical analysis involved the synthesis, analysis, and/or separation of chemical compounds. Several techniques help chemists to analyze compounds; some of the most commonly used are distillation, recrystallization, chromatography, and spectroscopy.

Energy and chemistry are integral to one another. The discovery of elements often coincided with the development of new energy sources; in addition, much of the energy produced and consumed every day is chemical energy stored in and released from the bonds between atoms. Nearly all chemical reactions are either endothermic (picks up heat) or exothermic (releases heat).

When scientists study chemical reactions, they are often dealing with either huge or tiny numbers. Scientific notation, which depends on exponential expressions, helps chemists work more effectively with these extreme values.

KEY TERMS

allotrope 57
analysis 69
atomic mass (weight) 67
atomic mass unit (amu) 64
Avogadro's hypothesis 63
Avogadro's number 64
chemical formula 65
chemical reaction 66
chromatography 73
combustion 50
concentration 72
Dalton's atomic theory 54
decanting 71
distillation 70
endothermic reaction 76
exothermic reaction 76
exponential 77
filtering 71
fraction 70
frequency 74
isotope 54
Law of Conservation of Mass 52
Law of Definite Proportions 50
Law of Multiple Proportions 56
mole 64
molecule 60
molecular mass (weight) 67
oxidation theory 51
phlogiston theory 50
product 66
pyrolysis 59
reactant 66
recrystallization 71
scientific notation 78
selective adsorption 73
separation 70
solubility 71
solute 71
solvent 71
spectroscopy 74
stoichiometry 53
synthesis 69
wavelength 74

READINGS OF INTEREST

General

Your local newspaper probably has a science section that reports the latest science "scoops." Apply the scientific methods as you read, and examine the material critically. (These reports are not always correct.)

Fortman, John J. "Pictorial analogies IX. Liquids and their properties." *J. Chem. Ed.*, vol. 70, no. 11, November 1993, pp. 881–882. This article uses a dance party theme to present its model.

Waddell, Thomas G., and Rybolt Thomas R. 1992. "The chemical adventures of Sherlock Holmes: The case of the screaming stepfather." *J. Chem. Ed.*, vol. 69, no. 12, pp. 999–1001.

Brown, T.M., and Dronsfield, A.T. "The phlogiston theory revisited." *Education in Chemistry*, March 1991, pp. 43–45. A brief look back in time to the phlogiston theory, the people who advocated it, and why they did so.

Salzberg, Hugh W, 1991. "*From caveman to chemist.*" Washington, D.C.: American Chemical Society. An excellent overview of the development of chemistry from prehistoric times to the beginning of the twentieth century.

Asimov, Isaac. 1991. "*Atom—Journey across the subatomic cosmos.*" New York: Truman Talley Books. A stylized explanation of atomic structure and how some of the ideas about it developed.

Asimov, Isaac. 1962. "*The search for the elements.*" Greenwich, Conn.: Fawcett Publications. An account of how the elements were discovered.

Important Scientists

Hunt, David M.; Dulai, Kanwajljit S.; Bowmaker, James K. and Molton, John D. "The chemistry of John Dalton's color blindness." *Science*, February 19, 1995, pp. 984–988. This article provides a brief history of Dalton's own account of his colorblindness, plus an intriquing experiment he proposed: examining his eyes after he died to test his hypothesis on why he was colorblind.

Holmes, Frederic L. "Antoine Lavoisier and the conservation of matter." *Chemistry & Engineering News*, September 12, 1994, pp. 38–45. An excellent review of the life and work of one of chemistry's greats.

O'Brien, James F. "What kind of chemist was Sherlock Holmes?" *Chemistry & Industry*, June 7, 1993, pp. 394–398. This fictional character used chemistry in some of his cases; as this article reports, he was a good amateur analytical chemist.

Davies, Mansel. 1991. "Isaac Newton, the alchemist." *J. Chem. Ed.*, vol. 68, no. 9, pp. 726–727. A new and intriguing side of this incredible scientist.

Stock, John T. 1991. "Carl Auer von Welsbach and the development of incandescent gas lighting." *J. Chem. Ed.*, vol. 68, no. 10, pp. 801–803. Gas lights operated in many different ways before the invention of the electric lightbulb. This is the story of one method.

Garmon, Linda. "Happy birthday, Joseph Priestley." *Science News*, March 12, 1983, p. 173. A summary of Priestley's contribution to chemistry written in honor of the 250th anniversary of his birth.

Tiner, John Hudson. 1989. "*Robert Boyle, trailblazer of science.*" Milford, Mich.: Mott Media. A good, easy-to-read biography.

Tiner, John Hudson. 1975. "*Isaac Newton, inventor, scientist, and teacher.*" Milford, Mich.: Mott Media. An easy-to-read account of the life of Newton, one of the greatest scientists of all time.

Duveen, Denis I. 1956. "Lavoisier." *Scientific American*, May, 1956. A brief life story of the man who really got chemistry rolling.

Wilson, Mitchell. 1954. "Priestley." *Scientific American*, October, 1954. A short account of the life of one of the scientists who discovered oxygen.

Allotropes

Aldersey-Williams, Hugh. "The third coming of carbon." *Technology Review*, January 1994, pp. 54–62. A good, well-illustrated explanation of buckyballs.

Baum, Rudy M. "Commercial uses of fullerenes and derivatives slow to develop." *Chemistry & Engineering News*, November 22, 1993, pp. 8–18. Though they have provided excitement for scientists, buckyballs and their kin have yet to make it in the industrial marketplace.

Baum, Rudy M. "Chemists increasingly adept at modifying, manipulating fullerenes." *Chemistry & Engineering News*, September 20, 1993, pp. 31–33. *Fullerenes* is the class name for buckyballs and related substances.

Folger, Tim. "Cages of carbon." *Discover*, September 1993, p. 32. A short explanation of buckyballs.

Boo, W. O. J. 1992. "An introduction to fullerene structures." *J. Chem. Ed.*, vol. 69, no. 8, pp. 605–609. This article is a bit complicated, but it contains excellent drawings of these carbon allotropes.

Beaton, John M. 1992. "A paper-pattern system for the construction of Fullerene molecular models." *J. Chem. Ed.*, vol. 69, no. 8, pp. 610–622. Be the first in your neighborhood to construct your very own fullerene. Imagine the envy. . .

Braun, Robert D. 1992. "Scanning tunneling microscopy of silicon and carbon." *J. Chem. Ed.*, vol. 69, no. 3, pp. A90–A93. Although the article doesn't explain this technique, the photographs are worth seeing; they show atoms directly.

Edelson, Edward. 1991. "Buckyball, the magic molecule." *Popular Science*, August 1991, pp. 52–57, 87. A popular account of a new carbon allotrope.

Analysis

Dickerson, James. "Murders from the past." *Omni*, August 1993, pp. 50–57, 82. Modern chemical analytical techniques can shed new light on past mysteries. For example—was President Zachary Taylor poisoned?

Pollard, Mark. "Tales told by dry bones." *Chemistry & Industry*, May 17, 1993, pp. 359–362. The only remains of many of our ancestors are their bones. Chemical analysis of these bones can provide many insights about these people.

Beilby, Alvin L. 1992. "Art, archaeology, and analytical chemistry: A synthesis of the liberal arts." *J. Chem. Ed.*, vol. 69, no. 6, pp. 437–439. You may have never realized how closely related chemistry is to art or archaeology. Chemistry affects the materials artists use and helps archaeologists determine dates for artifacts and events.

Pilar, Frank L. "Weight-average molecular weights: How to pick a football team." *J. Chem. Ed.*, vol. 69, no. 4, p. 280. For those who really want to see how such averages are calculated.

Thomas, Nicholas C. 1991. "The early history of spectroscopy." *J. Chem. Ed.*, vol. 69, no. 8, pp. 631–634. More historical information on how modern analytical techniques developed.

Suzuki, Chieko, 1991. "Making colorful patterns on paper dyed with red cabbage juice." *J. Chem. Ed.*, vol. 68, no. 7, pp. 588–589. An easy-to-do laboratory experiment using natural dyes.

Gerber, Samuel M., "*Chemistry and crime: From Sherlock Holmes to today's courtroom.*" Washington, D.C.: American Chemical Society. A readable account of how chemical analysis is used to solve crimes.

Ettre, Leslie S. 1978. "Pioneers in chromatography." *American Laboratory*, October 1978, pp. 85–91. How Tsweet discovered chromatography.

Heines, Sister Virginia. 1971. "Chromatography—A history of a parallel development." *Chemtech*, May 1971, pp. 280–291. This article points out the ideas other people had about chromatography at the same time as Tswett.

PROBLEMS AND QUESTIONS

1. List the ages or eras of chemistry, and describe the main ideas chemists had about the atom in each age.

2. What contributions did each of the following people make to the field of chemistry?
   a. Amedeo Avogadro
   b. Robert Boyle
   c. Henry Cavendish
   d. John Dalton
   e. Antoine Lavoisier
   f. Joseph Priestley
   g. Joseph Proust
   h. Jeremiah Richter
   i. Georg Stahl

3. Define the following terms:
   a. allotrope
   b. atomic mass
   c. molecular mass
   d. mole
   e. stoichiometry

4. List the number of elements discovered in each of the following time periods. In which period were elements first synthesized?
   a. before 1000 A.D.
   b. 1000–1599 A.D.
   c. 1600–1799 A.D.
   d. 1800–1899 A.D.
   e. 1900–present

5. In which period were elements first isolated from an ore (rather than found free in nature)?
   a. before 1000 A.D.
   b. 1000–1599 A.D.
   c. 1600–1799 A.D.
   d. 1800–1899 A.D.
   e. 1900–present

6. Convert the following numbers into scientific notation:
   a. 4,600,000
   b. 0.0000073
   c. 831,000
   d. 0.0068
   e. 143.7
   f. 0.042
   g. 29,000
   h. 0.667

7. Convert the following examples of scientific notation into decimal numbers:
   a. 3.2×10^{12}
   b. 4.8×10^{-3}
   c. 7.3×10^{5}
   d. 9.1×10^{-9}
   e. 3.47×10^{4}
   f. 6.02×10^{23}
   g. 3.03×10^{-22}
   h. 5.11×10^{-5}

8. What are the differences between synthesis, separation, and analysis?

9. Why are synthesis, separation, and analysis important in chemistry?

10. Give an example of synthesis.

11. Give an example of separation.

12. Give an example of analysis.

13. Briefly describe the distillation technique.

14. Briefly describe the filtration technique.

15. Briefly describe the recrystallation technique.

16. Briefly describe the liquid chromatography technique.

17. Briefly describe the nature of emission spectroscopy.

18. Briefly summarize how the knowledge and use of different forms of energy were essential to chemistry's development.

19. State the Law of Conservation of Mass.

20. Briefly describe the method Lavoisier used to support the Law of Conservation of Mass.

21. State the Law of Definite Proportions.

22. State the Law of Multiple Proportions.

23. Which of the following pairs of compounds illustrate the Law of Definite Proportions and which illustrate the Law of Multiple Proportions? (*Note:* Some pairs may do both!)
   a. NaCl and NaBr
   b. K_2O and Na_2S
   c. H_2O and H_2O_2
   d. C_2H_6 and C_2H_4
   e. CO and CO_2
   f. $AlCl_3$ and PCl_3

24. What assumptions make up Dalton's atomic theory?

25. Which of Dalton's assumptions about atoms are now known to be at least partially incorrect?

26. What are the essential features of a chemical equation?

27. Write balanced chemical equations for each of the following reactions.
   a. Calcium (Ca) reacts with oxygen (O_2) to form calcium oxide (CaO).
   b. Sulfur (S_8) reacts with oxygen (O_2) to form sulfur dioxide (SO_2).
   c. Nitrogen (N_2) reacts with oxygen (O_2) to form nitrogen dioxide (NO_2).

28. Calculate the molecular mass (molecular weight) for each of the following compounds.

a. NaBr
b. C_3H_8
c. $C_7H_{16}O$
d. $Ca_3(PO_4)_2$
e. AgCl
f. K_2SO_4

29. What are buckyballs? What unusual properties do they have?

CRITICAL THINKING PROBLEMS

1. Many Greek scientists formulated scientific devices or ideas, including the atom concept (Democritus), the mechanical lifting screw and solar mirrors (Archimedes), the astrolabe (Hipparchus), lithotomy or stone cutting (Ammonius), medical diagnosis (Hippocrates), and a primitive steam engine (Hero). With this rich tradition, why did the Greeks fail to develop a technological society? How might history have differed if the Greeks had developed a widespread technology?

2. Does the alchemist's idea of transmuting metals contradict the Law of Conservation of Matter? Why or why not?

3. Some modern forms of mysticism and astrology retain alchemistic beliefs. For example, some people still believe in elixirs that ensure long life. How do these beliefs fit in with the scientific method?

4. Cavendish was the first chemist to perform consistent quantitative measurements, yet he believed in the phlogiston theory. Was this belief reasonable, considering the era he lived in? Would changes have occurred in the history of chemistry if Cavendish had rejected the phlogiston theory?

5. In what ways do chemistry, electricity, and light interact? What benefits, if any, do you derive from this interaction?

HINT Consider how pure elements are separated from ores and how chemicals are analyzed using light. Research how home and commercial alarm systems work to detect intruders.

6. Considering the Greek's limited knowledge of science and their even more limited scientific equipment, what kinds of experiments might they have tried to test the atomic concept?

7. How could someone discover that an alchemist had cheated them by selling fake gold?

8. Fake medical cures are still peddled today. Suppose someone offered such a cure to you or a sick family member. How might you detect the fraud?

HINT Use the scientific method.

9. Why would Boyle's Law support his belief in corpuscles?

HINT Corpuscles are essentially the same as atoms. Why does gas pressure rise—what role do the atoms play in increasing pressure?

10. Why did Cavendish's synthesis of water destroy the ancient idea that water was an element?

11. The compounds benzene and acetylene have different boiling points and chemical properties, yet both contain exactly 92.31 percent carbon and 7.69 percent hydrogen. What can you conclude from this information?

HINT These must be two different chemical compounds even though the ratio of C:H is 1:1 in both cases. How can this be?

12. How is our modern concept of atoms similar to those of the early Greeks, such as Democritus? How is it different from the beliefs of Socrates and his followers?

13. The simple distillation described in table 3.6 assumes you can determine the percentages of pentane and benzene in each fraction. How might you actually do this?

14. How could you detect materials that separated in a chromatographic procedure if they had no color?

15. Have there been limitations to the role women have played in science? If so, are these limitations still present today? Here are some possible background articles: (1) Marguerite Holloway, *Scientific American*, 269 No. 5, Nov., 1993, pp. 94–103, "A Lab of Her Own;" (2) Stephen Goode, *Insight*, Feb. 20, 1995, pp. 15–17, "Feminists Claim Men Do Weird Science," and (3) S. V. Meschel, *J. Chem. Ed.*, 69 (No. 9), Sept., 1992, pp. 723–730, "Teacher Keng's Heritage. A Survey of Chinese Women Scientists."

CHAPTER 4

ORDER IN CHEMISTRY: THE PERIODIC TABLE AND ATOMIC STRUCTURE

OUTLINE

The Development of the Periodic Table
The Discovery of Electrons, Protons, and Neutrons
Atomic Structure
Rutherford's Atom
Bohr's Atom
New Insights into the Atom: The Wave Mechanics Model

BOX 4.1 Marie Curie, A Genius in Any Time
BOX 4.2 A Forensic "Who-Done-It?"
BOX 4.3 Probability and Chemistry

KEY IDEAS

1. The Nature of Chemical Periodicity and How It Was Discovered
2. How Chemists Learned About the Internal Structure of the Atom, Including Electrons, Protons, and Neutrons
3. How the Internal Atomic Structure Relates to Dalton's Atomic Theory
4. The Key Events in Rutherford's Discovery of the Nucleus
5. The Concepts Behind Rutherford's Model of the Atom
6. Spectra and the Development of Bohr's Atomic Model
7. How the Bohr Model Correlates with Mendeleev's Periodic Table
8. How the Wave Mechanics Model Differs from Bohr's Atom

"Science is organized knowledge."
HERBERT SPENCER (1820–1903)

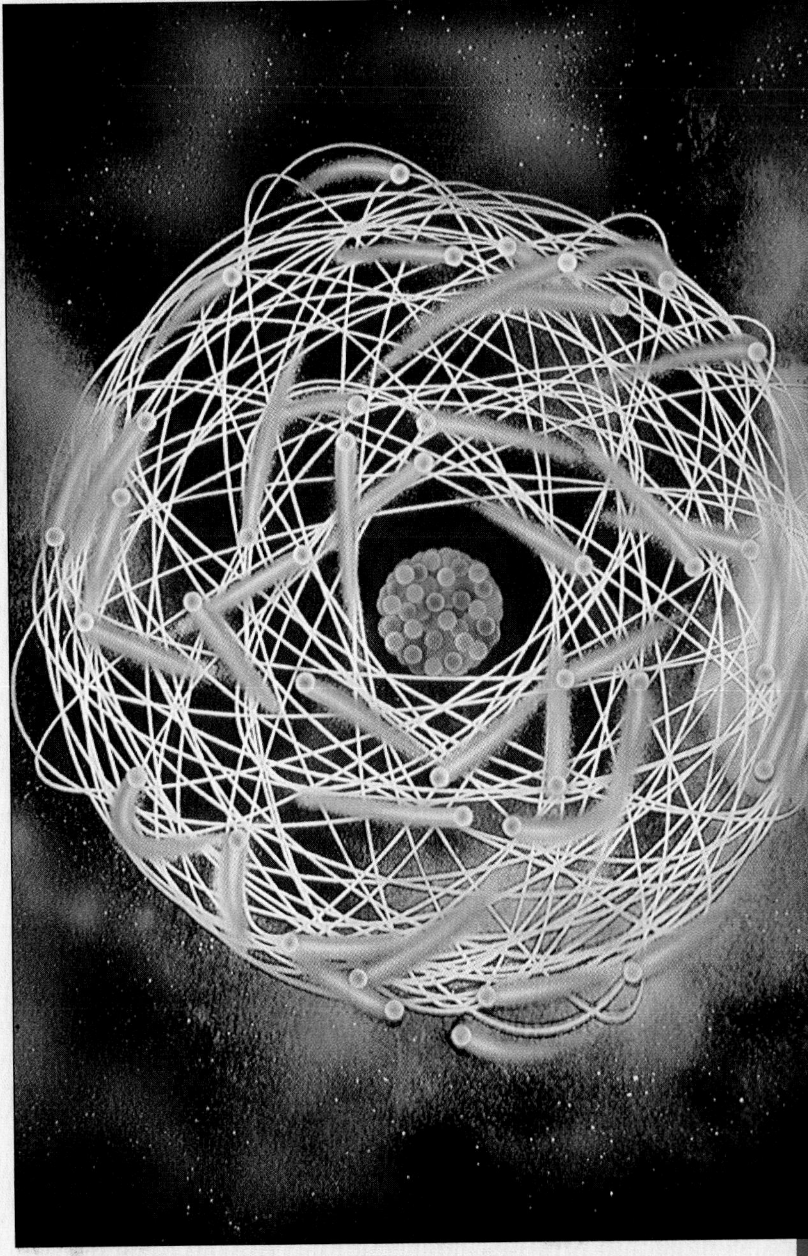

Artist's conception of an atom showing the nucleus enveloped by an outer sheath of orbiting electrons.

In 1800, chemists knew of only about 30 elements. They did not have sufficient data to develop a systematic approach to chemical behavior. After chemists discovered more elements, they noticed that some behaved in similar ways. For example, the metals lithium, sodium, and potassium all react with chlorine to form similar saltlike compounds. Similarly, sodium reacts with chlorine, bromine, or iodine to produce saltlike compounds. As the number of known elements increased, chemists tried to arrange them in chemical families according to shared chemical and physical properties.

SOURCE: © 1995 by Sidney Harris.

Johann Wolfgang Doebereiner.

THE DEVELOPMENT OF THE PERIODIC TABLE

Among the early attempts to organize similar elements into groups were Doebereiner's triads and Newland's octaves.

Chemical Families—Triads and Octaves

Johann Wolfgang Doebereiner (German, 1780–1849) proposed the idea of chemical triads in 1817 and developed the idea further in 1829. Each **triad** consisted of three elements arranged in order of atomic weight. As table 4.1 shows, the middle element in each triad has melting and boiling points that fall between the other two elements. Furthermore, all three elements in each triad form similar compounds.

The first two triads in table 4.1, the metallic triads, form oxides and chlorides with similar chemical formulas. The second two triads, the nonmetallic triads, have virtually the same formulas for the hydride (with hydrogen) and barium compounds. The melting and boiling points decrease as atomic weight increases for the metallic triads, and they increase as atomic weight increases for the nonmetallic triads.

Many chemists did not accept the Doebereiner triad concept because there seemed to be more exceptions than valid examples. Modern chemists know of many more elements than Doebereiner did and can group together many more triads, but the concept does not explain as much about chemical behavior as other systems do.

The next step in organizing chemical families came in 1864 when John A. Newlands (English, 1838–1898) pointed out that certain chemical and physical properties recurred in every seventh element. Newlands arranged the elements in vertical rows called octaves, in keeping with the seven notes in a musical octave. (Mid-1800 scientists believed that music, art, and science form an intricate mosaic; the idea that elements occur in octaves fits well with this belief.) Newlands was aware of only a limited

TABLE 4.1 Properties of some Doebereiner triads.

ELEMENT	SYMBOL	ATOMIC WEIGHT (amu)	MELTING POINT (°C)	BOILING POINT (°C)		
					OXIDE FORMULA	**CHLORIDE FORMULA**
Lithium	Li	7	179	1317	Li_2O	LiCl
Sodium	Na	23	98	892	Na_2O	NaCl
Potassium	K	39	63	770	K_2O	KCl
Calcium	Ca	40	843	1487	CaO	$CaCl_2$
Strontium	Sr	88	769	1384	SrO	$SrCl_2$
Barium	Ba	137	725	1140	BaO	$BaCl_2$
					HYDRIDE FORMULA	**BARIUM FORMULA**
Sulfur	S	32	113	444	H_2S	BaS
Selenium	Se	79	217	685	H_2Se	BaSe
Tellurium	Te	128	452	1390	H_2Te	BaTe
Chlorine	Cl	35	−101	−35	HCl	$BaCl_2$
Bromine	Br	80	−7	59	HBr	$BaBr_2$
Iodine	I	127	114	184	HI	BaI_2

John A. Newlands.

number of elements, and therefore his octaves scheme fared poorly. Table 4.2 shows his first three octaves; each counterpart pair in the first two share similar properties. Unfortunately, the properties of the elements in the third octave failed to match those in the first two octaves. For example, the last two elements in octave 3 are metallic rather than nonmetallic, and their chloride formulas include numbers of chlorine atoms that differ from those of the sixth and seventh elements in the first two octaves.

Newlands still helped advance the understanding of chemical families. His main contribution was to show that larger families exist, and that similar properties occur at fairly regular intervals when elements are arranged in order of atomic weight.

Mendeleev's Marvel—The Periodic Chart

Not long after Newlands proposed octaves, a breakthrough in organizing the elements occurred in two independent locations. In 1869, both Julius Lothar Meyer (German, 1830–1895) and Dmitri Mendeleev (Russian, 1834–1907) almost simultaneously proposed periodic tables. Mendeleev usually receives more credit than Meyer because his system was more complete and was published a few years earlier.

A **periodic table** arranges the elements so that those with similar properties appear at regular intervals or periods. In his table, Mendeleev arranged the 60 known elements in horizontal rows in

TABLE 4.2 Newland's octaves and some of their properties. Most metals in the third octave form several different chlorides; only the major compound is shown here.

OCTAVE 1			OCTAVE 2			OCTAVE 3		
Element	Nature[a]	Chloride Formula	Element	Nature[a]	Chloride Formula	Element	Nature[a]	Chloride Formula
Li	M	LiCl	Na	M	NaCl	K	M	KCl
Be	M	$BeCl_2$	Mg	M	$MgCl_2$	Ca	M	$CaCl_2$
B	N	BCl_3	Al	M	$AlCl_3$	Cr	M	$CrCl_3$
C	N	CCl_4	Si	N	$SiCl_4$	Ti	M	$TiCl_4$
N	N	NCl_3	P	N	PCl_3	Mn	M	$MnCl_2$
O	N	OCl_2	S	N	SCl_2	Fe	M	$FeCl_3$
F	N	FCl	Cl	N	ClCl	Co	M	$CoCl_2$

[a]Metallic (M) or nonmetallic (N)

Dmitri Mendeleev.

order of increasing atomic mass, then lined up the elements so those with similar chemical properties occupied the same vertical columns. However, some elements seemed to be misplaced if he strictly followed both rules. When this happened, Mendeleev matched the chemical properties in the vertical columns, even if the atomic masses were out of order. At the end of this process, Mendeleev had a chart similar to the one in table 4.3.

As table 4.3 shows, Mendeleev had to make two adjustments in his initial scheme. First, he had to reverse the positions of Te and I; tellurium (Te) has a higher mass than iodine (I), but iodine behaves like chlorine (Cl) and bromine (Br), while tellurium behaves like sulfur (S) and selenium (Se). Mendeleev insisted that chemical properties must prevail over atomic mass order, and he arranged Te and I on this basis.

The second adjustment Mendeleev made was to leave two empty spaces in his table. He claimed the holes existed because the elements that filled them were undiscovered when he devised the table, and he boldly predicted the properties of these unknown elements, which he called eka-aluminum and eka-silicon (eka means "next in order"). His colleagues scoffed at these daring claims; how could he know the mass, density, or boiling point of an undiscovered element?

The missing elements, gallium (Ga) and germanium (Ge), were discovered in 1875 and 1886, respectively. Table 4.4 compares Mendeleev's 1869 predictions with the actual properties of these elements. Mendeleev predicted accurate atomic masses and densities and even colors for these unknown elements. Amazingly, his predictions for the oxide and chloride formulas were correct; even more spectacular was his prediction of the eka-silicon chloride boiling point. Remember, this prediction is for an undetermined property of an unprepared compound from an undiscovered element—yet Mendeleev correctly prophesied that the boiling point would be under 100°C! Mendeleev had the fortune to live long enough to see his predictions verified and to see the scientific community accept his periodic table.

The Modern Periodic Table

The periodic table (or periodic chart), has undergone several changes since Mendeleev's time. The discovery of over 45 new

TABLE 4.3 A modified portion of Mendeleev's periodic table, showing atomic weights and modern element symbols.

H 1.0						
Li 6.9	Be 9.0	B 10.8	C 12.0	N 14.0	O 16.0	F 19.0
Na 23.0	Mg 24.3	Al 27.0	Si 28.1	P 31.0	S 32.1	Cl 35.5
K 39.1	Ca 40.1			As 74.9	Se 79.0	Br 79.9
Rb 85.5	Sr 87.6	In 114.8	Sn 118.7	Sb 121.8	Te 127.6	I 126.9

elements and new knowledge of the details of subatomic structure have propelled many of these changes.

General Features of the Periodic Table

Figure 4.1 shows a modern periodic table. Four separate groups of elements appear in large blocks: (1) **main group elements** sometimes called *representative elements*, (2) **noble gases,** (3) **transition elements,** and (4) **inner transition elements.** Each element in the table appears inside a small box with the element's name, **symbol, atomic mass,** and **atomic number** within. (The element names are also listed in an alphabetical order inside the back cover.)

Periods and Families

The horizontal rows of elements, called **periods,** are assigned Arabic numbers from 1 to 7, with 1 marking the top row and 7 marking the bottom. The two blocks of elements to the left and right of the chart, the main group or representative elements, occupy all seven periods. The transition elements only occur in periods 4–7; the inner transition elements occupy only an inner portion of periods 6 and 7.

The vertical columns in the table are called **families** or **groups.** In traditional tables, the eight chemical families in the main group are assigned Roman numerals IA to VIIIA, and transition element families bear Roman numerals IB to VIIIB. (Note that these do not occur in order.) In some recent tables, each group bears an Arabic numeral from 1 to 18. Some families have special names; including the **alkali metals** (family IA), the **alkaline earth metals** (family IIA), the **halogens** (family VIIA), and the **noble gases** (formerly called the inert gases, family VIIIA). The other families are named after the topmost element. Thus, column IIIA is the boron family, IVA is the carbon family, VA is the nitrogen family, and VIA is the oxygen family. All family members react in similar ways to form analogous compounds; in other words, if lithium (Li, family IA) combines with chlorine (Cl) to form LiCl, sodium would also tend to form NaCl, potassium would form KCl, and so on.

TABLE 4.4 Mendeleev's predictions for two unknown elements.

PROPERTY	PREDICTIONS	OBSERVED	PREDICTIONS	OBSERVED
Element	Eka-aluminum	Gallium	Eka-silicon	Germanium
Symbol	E	Ga	Ek	Ge
Atomic weight	68	69.7	72	72.59
Color	gray	gray-white	gray	gray-white
Density (g/mL)		5.9	5.5	5.47
Oxide formula	E_2O_3	Ga_2O_3	EkO_2	GeO_2
Effect of acid on oxide	should dissolve	dissolves	none	none
Chloride formula	ECl_3	$GaCl_3$	$EkCl_4$	$GeCl_4$
Chloride boiling point			<100°C	83°C

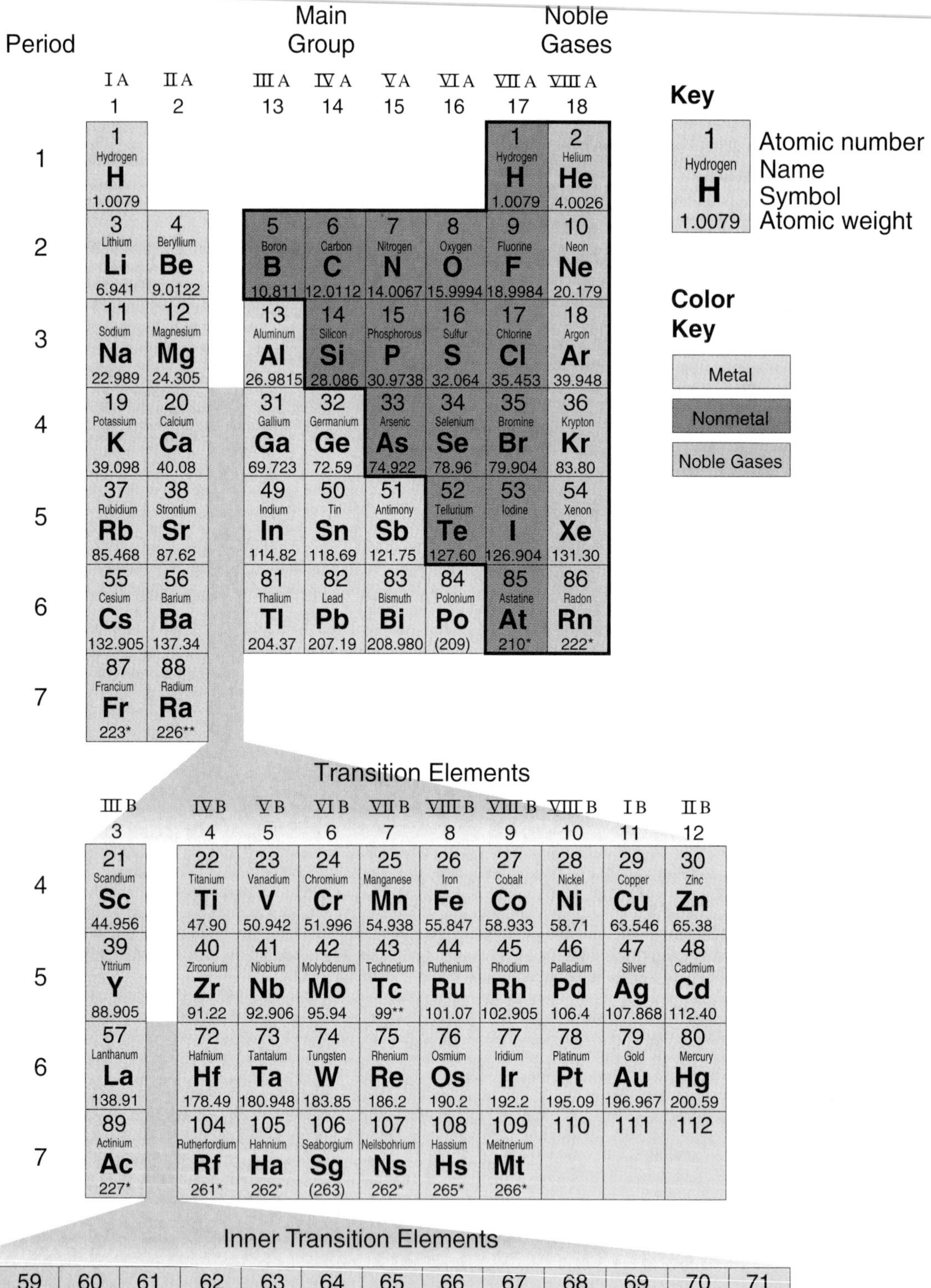

Figure 4.1 The periodic table.

The element hydrogen (H) has always posed a problem in its periodic table placement. Some of its reactions are similar to those of the elements in the alkali metal family (IA), but others resemble those of the elements in the halogen family (VIIA). For this reason, hydrogen is shown on both sides of the table.

PRACTICE PROBLEM 4.1

Which of the following sets of elements belong to the same family? Which belong in the same period? Are there any sets that are in both the same family and the same period?

a. B, Al, In

b. Ca, Ga, Se

c. S, K, Sr

ANSWER Set *a* contains elements in the same family, while set *b* includes elements that appear in the same period. The set *c* elements occupy neither the same family nor the same period, and it isn't possible for a set of elements to be in both the same family and the same period.

Metals, Nonmetals, and Noble Gases

Elements may be either metals, nonmetals, or noble gases. **Metals** are good conductors of electricity and heat, possess luster or shine, are malleable (can be hammered into a shape), and are ductile (can be drawn out into a wire). Most elements are metals, and all metals except mercury (Hg) are solids at normal temperatures. (Mercury is a liquid.) All transition and inner transition elements are metals; thus about 80 percent of all known elements are metals. In compounds, metals exist as positively charged **ions** that form **ionic compounds.**

Nonmetals are poor conductors of electricity or heat and are often brittle in the solid state. The noble gases are often included in the nonmetal group, but noble gases rarely form compounds, while the nonmetals form many. Of the 44 elements in the main group, exactly half lie in the large nonmetal box on the right-hand side of the periodic table. (Hydrogen is counted only in the nonmetal block for this tally). These 22 elements include the 6 noble gases and 16 nonmetals. Aside from the 6 noble gases (family VIIIA), there are 5 other gaseous nonmetals: hydrogen (H), nitrogen (N), oxygen (O), fluorine (F), and chlorine (Cl). One nonmetal—bromine (Br)—is a liquid, and the remaining 10 nonmetals are solids. The other 22 main group elements are all metals. Nonmetals are negatively charged ions when they form ionic compounds; they also form **covalent compounds** in which no charges hold the atoms together.

The **metalloids** are a few elements that lie along the line separating the metal and nonmetal blocks. Under certain conditions, the metalloids have some properties that resemble metals. The metalloids include silicon (Si), germanium (Ge), and arsenic (As). Some of the metalloids are also classed as semiconductors because they are more conductive than nonmetals but less than metals, especially when traces of other elements are present. This unique characteristic enables semiconductors to be used in computers and other electronic devices.

Families and Relatives

The properties of elements in main group (A) families are sometimes similar to the properties of elements in transition (B) families. For example, family IB members behave similarly to the alkali metal family (IA); all alkali metals form chlorides with the general formula MCl, where M is the metal, and silver (Ag, from family IB) forms the chloride AgCl. Likewise, family IIB behaves similarly to alkaline earth family IIA; zinc (Zn) forms the oxide in ZnO and alkaline earths form oxides such as CaO.

However, many differences exist between families IA and IB or IIA and IIB. Copper (IB) forms both a monochloride with one chlorine atom (CuCl) and a dichloride with two ($CuCl_2$), but no alkali metals form dichlorides. Mercury (IIB) forms chlorides with the formulas Hg_2Cl_2 and $HgCl_2$, but no alkaline earth element forms a chloride corresponding to Hg_2Cl_2. Similarly, the alkali metals (IA) are extremely reactive; some even react with water. But family IB members are far less reactive and react slowly, if at all, with water. The relationship between the A and B families, then, is more like cousins than siblings.

PRACTICE PROBLEM 4.2

We discussed some of the similarities and differences between "relatives" such as the elements in families IA and IB. For example, family IA elements are extremely reactive; some react even with water. IB elements, on the other hand, are less reactive and react more slowly (or not at all) with water. Which of these elements might be useful in household plumbing? Which family might be useful in electrical wiring?

ANSWER Family IA elements are completely unsuitable for plumbing or electrical wiring because of their reactivity. It would be impractical to install water pipes that react with water or electrical wires that are destroyed by moisture. Some, but not all, family IB elements are useful for plumbing or wiring. Mercury, (Hg) is unsuitable because at room temperature it is a liquid, and silver (Ag) is too expensive. On the other hand, copper (Cu) is commonly used for both plumbing and wiring.

THE DISCOVERY OF ELECTRONS, PROTONS, AND NEUTRONS

When atoms were considered indivisible, chemists were unable to determine the reasons behind periodicity. If each element was a unique substance occurring in indivisible units called atoms, wouldn't each type of atom also be unique? Then why did similar properties occur at such consistent intervals on a periodic chart?

One key ingredient in understanding the periodic nature of the elements was the discovery of **subatomic particles**—particles smaller than atoms. Ample evidence for such smaller units existed for centuries, but chemists found the idea of subdividing atoms difficult to accept; it seemed to contradict the atomic theory finally established after centuries of struggle.

Still, Dalton's atomic theory could not answer certain questions. One mystery Dalton's theory did not address was the electrical nature of certain types of matter. A second problem arose years after Dalton proposed the theory, when Henri Becquerel discovered radioactivity. How could scientists account for these phenomena in terms of atomic theory?

Electrical Behavior

Scientists were aware of static electricity centuries before Dalton proposed the atomic theory. The ancient Egyptians left records revealing that rubbed amber attracted small objects. Many years later, in the late 1700s, Benjamin Franklin distinguished between two types of electrical charges. He named them positive and negative charges.

Figure 4.2 illustrates the effect of static charges on two small balls made of a material called pith. Initially, when the balls are hung by a hard rubber rod (figure 4.2a), they don't interact with the rod or with each other. If the rod is rubbed with a silk cloth, however, the balls repel one another and the rod (figure 4.2b). The rod acquired a positive charge after being rubbed by the silk; part of the charge transferred to the balls; and because the charges repel one another, rod and balls now repel each other.

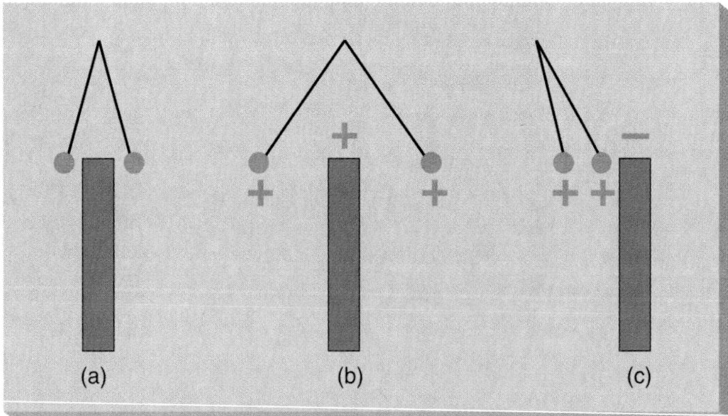

Figure 4.2 Positive and negative electrical charges. **(a)** When two pith balls hang by a hard rubber rod, they don't interact with the rod or each other. **(b)** When the rod is rubbed with silk, it acquires a positive charge, which is then transferred to the balls. The balls become positive, repelling each other and the rod. **(c)** When the rod is subsequently rubbed with wool, it acquires a negative charge. It now attracts the positively charged pith balls. Like charges attract; opposites repel.

If the rod is next rubbed with a piece of wool, a different effect occurs. The rubber rod acquires a negative charge, and opposite charges attract. Thus, the positively charged pith balls are now attracted to the negatively charged rod (figure 4.2c). If the balls also acquired a negative charge, the rod would again repel the balls. Opposite charges attract, while identical charges repel.

The repulsion of like charges and the attraction of unlike charges is a fundamental electrical property, but Franklin's discovery of the two types of charges and their behavior told scientists nothing about the source of the charges. They concluded that rubbing must either add or remove something from matter, because matter consists of atoms and atoms do not contain free-floating bunches of electricity.

Another demonstration of electrical behavior was in the ability of certain types of matter to conduct electricity. For years, scientists knew that some molten compounds and some solutions could conduct electricity. In 1884, Svante Arrhenius (Swedish, 1859–1927) was researching conductivity for his doctoral thesis. Arrhenius proposed that some compounds dissociate into positively and negatively charged particles in solution. Although he eventually proved his thesis through experimentation, scientists initially rejected his ideas because they still believed atoms were indivisible and the electrical charge must come from outside the atom. Eventually, however, Arrhenius's proof that certain compounds break down into charged ions forced chemists to conclude that Dalton's assumptions were not entirely accurate. Where did the electrical charges come from? They appeared to come from the ions themselves.

Radioactivity

A second challenge to Dalton's atomic theory arose in 1896 when Antoine Henri Becquerel (French, 1852–1908) accidentally discovered **radioactivity.** Becquerel was studying fluorescent uranium ore minerals. (Fluorescent materials absorb a light frequency invisible to us and emit a visible frequency; glowing black-light posters are fluorescent.) He recorded his observations photographically, and one day he discovered some unused film was totally exposed, as if subjected to light. Becquerel, an experienced photographer, kept his film tightly wrapped in dark paper to avoid accidental exposure. Why did the film appear to be exposed?

Becquerel discovered that he had stored the film near pitchblende, the uranium ore. He wondered if this ore released something that exposed his film. Systematic studies by Becquerel and Marie Curie (Polish, 1867–1934), his student at the time, proved that the uranium ore did indeed emit an invisible "radiation." Marie Curie later isolated radium, a new element, from the pitchblende ore and the study of radioactivity, or the release of radiant energy from matter, began (see box 4.1).

Becquerel and his coworkers eventually found three different kinds of radiation emanating from pitchblende ore; **alpha (α)** and **beta (β) particles,** and **gamma (γ) radiation.** The 1903 Nobel Prize in physics was awarded to Marie and Pierre Curie (French,

BOX 4.1

MARIE CURIE, A GENIUS IN ANY TIME

Marie Curie, (figure 1), often called Madame Curie, was a woman of unequalled achievements in the history of chemistry and physics. Born in Warsaw, Poland on November 7, 1867, Curie was to become a pivotal figure in the scientific world by the early twentieth century.

Curie's scientific career began inauspiciously: she was refused admission to the University of Warsaw in 1891 because she was a woman. However, she entered the Sorbonne in France that same year and soon earned her master's and doctor's degrees in physics. It was during this period that two significant people entered her life: Pierre Curie, one of her professors, who became her husband in 1895; and eminent physicist Antoine Henri Becquerel. Becquerel introduced the Curies to his discovery of the phosphorescence of uranium; this led to Marie's discovery of radium years later.

Toward the end of the 1890s, Curie's achievements began to accumulate with impressive speed. Along with her husband, Pierre, Marie discovered polonium on July 18, 1898; on December 26, 1898, she discovered radium with the aid of Gustave Bélmont. Curie named the element polonium after her homeland, Poland. Also during that eventful 1898, Marie and Pierre discovered that thorium emitted what they first called "uranium rays." Marie later renamed these rays radioactivity.

In 1899, Marie Curie teamed up with Fritz Geisel and Antoine Henri Becquerel to prove that β-rays, one form of radiation, were actually high-speed electrons. The 1903 Nobel Prize in physics was awarded to Marie and Pierre Curie for their study of radiation and to Antoine Henri Becquerel for work on spontaneous radioactivity.

Figure 1 Marie Curie.

Marie achieved another first in 1906 when she became the first woman professor at the Sorbonne in the entire 650 years of the school's existence. After Pierre died in a traffic accident, Marie was appointed to fill his position. Marie continued her radioactivity research. She published her landmark book, *Traité de radioactivité* (Treatise on radioactivity), in 1910. The following year, Curie won the Nobel Prize in chemistry for her discovery of the elements radium and polonium. Winning two Nobel Prizes is a feat rarely duplicated.

Curie died at the age of 67 from pernicious anemia, a condition probably induced by exposure to radiation. She left behind a rich heritage. Her daughter, Irène Joliot-Curie (French, 1897–1956), collaborated with her husband Frédéric Joliot-Curie to develop artificial radioactive elements—an effort that won them the 1935 Nobel Prize in chemistry. Irène died in 1956 from leukemia; her disease was also probably induced by working with radioactive materials.

The honors and awards Marie Curie earned are almost unmatched by any modern scientist. Along with her two Nobel Prizes, she received the Berthollet Medal (1903), the Davy Medal of the Royal Society of London (1903), the Benjamin Franklin Medal (1921), the Willard Gibbs Medal from the American Chemical Society (1921), and many other awards. Over a hundred institutions, scientific societies, cities (including Warsaw), and countries bestowed honorary titles on Curie. Although she once was determined not to marry, feeling that science was her first and only love, her marriage to Pierre actually furthered the work of both scientists and left an indelible mark on the study of radioactivity.

1859–1906) (for their study of radiation) and Antoine Henri Becquerel (for his work on spontaneous radioactivity).

The discovery of radioactivity raised an obvious question: what is the source of the radiation? This time, scientists could not even hypothesize that rubbing a piece of matter somehow added energy or an electrical charge to it; logically, these rays must come from within the atoms. Although chemists resisted this notion, other scientists were working on projects that would soon show the intricate relationship between matter and energy at the atomic level. Specifically, they would soon discover the electron.

Electrons Charge onto the Scene

Physicists discovered electrons through a device called a cathode ray tube. The **cathode ray tube,** created by William Crookes (English, 1832–1919) in 1875, passed an electric current through a vacuum (figure 4.3). (A perfect vacuum contains no matter.) When a cathode ray tube is in operation, a pump draws gas out of the tube, creating a vacuum, as a high-voltage electrical source sends a current of electricity through the tube. When enough gas is removed, this electrical charge causes the remaining gas to glow. The glowing beam comes from the negatively charged **cathode,** moves through the hole in the metal disk, and flows to the positively charged **anode.** Because the beam seems to emanate from the cathode, Crookes called the beam cathode rays.

In 1897, J. J. Thomson (English, 1856–1940) carried the Crookes tube studies further by applying a second electrical field across the cathode ray tube (figure 4.4). When Thomson set up the second field, the cathode ray beam was deflected toward the second anode or positive plate. Since negative charges are attracted to positive charges, the cathode rays had to consist of negative particles. Thomson received the 1906 Nobel Prize in physics for his investigations of electricity's passage through gases.

In 1874, Irish physicist George Johnstone Stoney (1826–1911) had suggested the name **electrons** for these negatively charged particles. Thomson measured the mass-to-charge ratio for the electrons by varying the voltage of the electrical field, but he was unable to measure electron mass or charge separately. In 1909, Robert A. Millikan (American, 1868–1953) measured the charge of the electron using the technique outlined in figure 4.5. Millikan sprayed tiny oil droplets between two charged plates in an enclosed chamber. As the droplets settled under the influence of gravity, Millikan determined their rate of fall and found that the electrical field caused no change in gravity's effect. Next, Millikan bombarded the oil droplets with X-rays to introduce extra electrons, giving the droplets negative charges. These charged oil droplets moved *upwards,* against gravity, toward the positive anode. Millikan could counterbalance the droplets' movement by varying the electrical field's strength. When the forces of gravity and the electric field were equal, the oil droplets hung suspended in space. By measuring the charge of the electrical field at this point, Millikan could calculate the charge on any given droplet. In 1923, Millikan won the Nobel Prize in physics for his work.

J. J. Thomson.

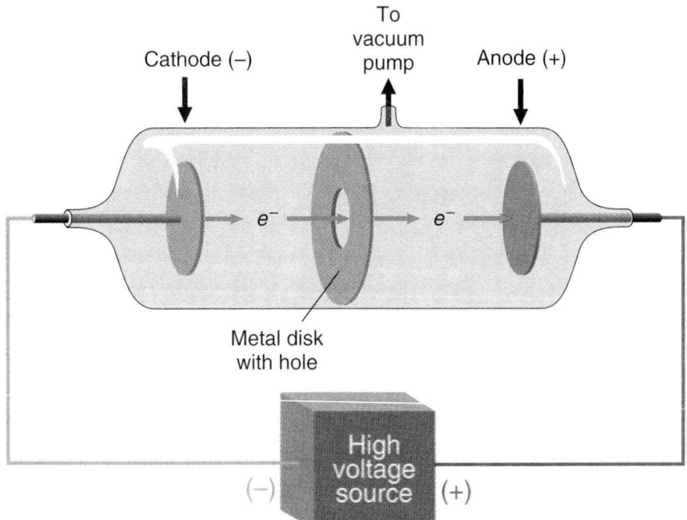

Figure 4.3 A cathode ray tube (Crookes tube). As a pump removes gas from the tube, electrical current passes through and causes the residual gas to glow. Eventually, scientists determined that this glowing beam consists of electrons.

Once Millikan had determined an electron's charge, this information could be combined with Thomson's mass-to-charge ratio to determine the mass of an electron as well. Millikan determined that the electron charge is 1.60×10^{-19} coulombs; we now express this value as −1 e.s.u. (electrostatic units). The mass of an electron is 9.11×10^{-28} grams. Other properties of the electron are summarized in table 4.5.

TABLE 4.5 Subatomic particle properties.

PARTICLE	SYMBOL	CHARGE (e.s.u.)	MASS (grams)	MASS (amu)
Electron	e^-	−1	9.11×10^{-28}	0.00055
Proton	p^+	+1	1.673×10^{-24}	1.0073
Neutron	n	0	1.675×10^{-24}	1.0087

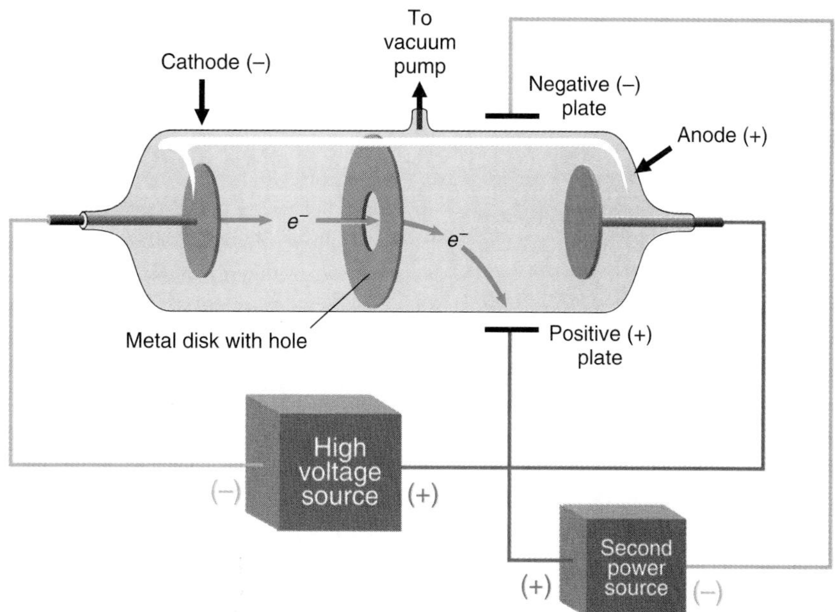

Figure 4.4 The effect when an external electric field is applied to a cathode ray tube. The cathode ray deflects to the positive plate, indicating the ray's negative charge.

Canals of Protons

The discovery of electrons helped account for the negative charges scientists had observed in matter. But what about positive charges?

The first evidence for positive particles came from the canal ray studies Eugen Goldstein (German, 1850–1930) performed in 1886. Goldstein's canal ray apparatus was similar to a cathode ray tube, except the cathode was perforated (figure 4.6). Goldstein discovered that positively charged particles moved toward the cathode as electrons moved toward the anode. He called these positive particles **canal rays.** The rapidly moving canal rays shot through the holes in the cathode to strike the end of the tube.

In 1907, Wilhelm Wien (German, 1864–1928) found that the mass of the canal ray particles varied with the gas used in the tube. The lightest mass occurred when hydrogen gas was used. Wien concluded the canal rays were actually gas atoms that had lost

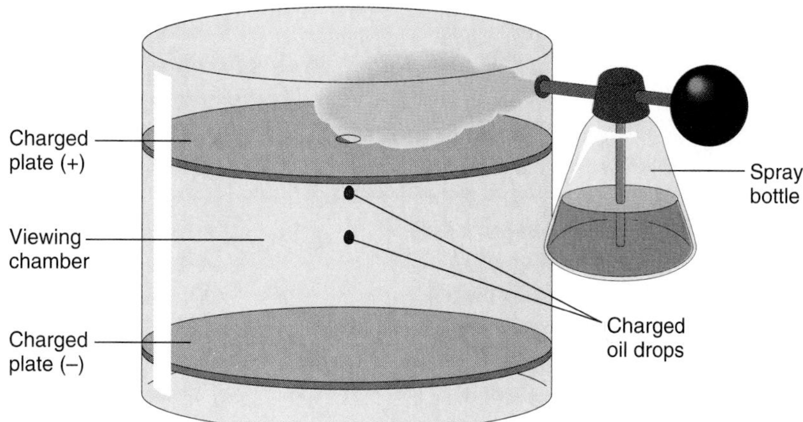

Figure 4.5 Millikan's oil drop experiment. When Millikan sprayed oil droplets into a chamber and bombarded them with X-rays to place a negative charge on them, the charged droplets were attracted to the positive plate. Changing the strength of the electrical field offset the attraction and allowed Millikan to determine the charge of the electrons.

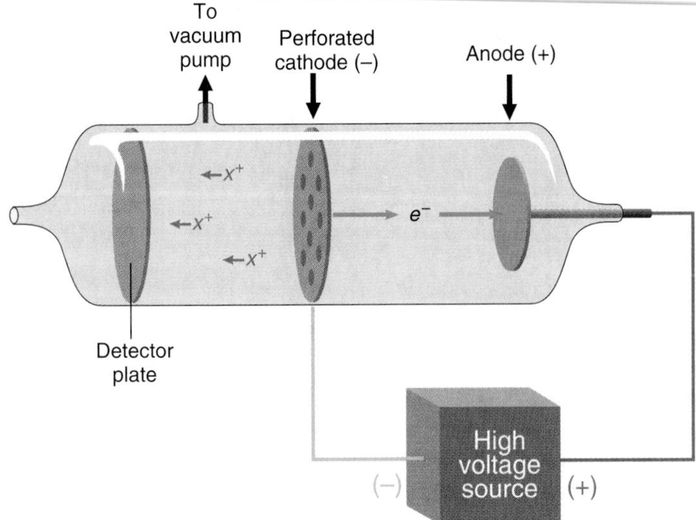

Figure 4.6 Cathode ray tube modified to detect canal rays (X^+).

electrons, giving them a positive charge. When an atom gains or loses an electron it is ionized, and the resulting positively or negatively charged atom is called an **ion.**

Atoms are normally neutral—neither positive nor negative in electrical charge. If losing negatively charged electrons leaves an atom with a positive charge, the atom must contain positively charged particles that normally balance the electrons' charges. Scientists decided that Wien's positively ionized hydrogen atom must contain a positive particle called a **proton** (Greek for first in position or rank). Table 4.5 summarizes the properties of protons.

Neutrons—The "Unattractive" Particle

Scientists realized the atom had to contain exactly the same numbers of electrons and protons, but this resulted in too low a mass for most atoms. Therefore, they concluded that at least one other particle existed that had no charge and weighed about the same as a proton. This particle, the **neutron,** was eventually discovered in 1932 by James Chadwick (English, 1891–1974). The neutron (whose properties are summarized in table 4.5) had the characteristics scientists predicted. Chadwick was awarded the 1935 Nobel Prize in physics for the discovery of the neutron.

The discovery of protons, neutrons, and electrons had several effects. First, it showed that atoms were not the solid, indivisible units Dalton had imagined; they were cloudlike structures made up of differing numbers of subatomic particles, and an individual atom might gain or lose some of those particles under certain circumstances. Second, the discovery of these electrically charged particles helped explain several mysteries; similarities in subatomic structure might account for similar behaviors among elements in a family, and the electrical charges helped explain electrical behavior and radioactivity. Finally, the concept of protons, neutrons, and electrons led to a clearer understanding of atomic structure—an understanding that began with the twentieth century and that continues to shape our view of the atom as the century draws to a close.

ATOMIC STRUCTURE

Once scientists knew that subatomic particles existed, they had to learn how these units assembled into atoms. They now realized that an atom contains protons, neutrons, and electrons. But what was the structure of the atom?

Thomson Cooks a Plum Pudding

In 1898, J.J. Thomson proposed a model for the structure of an atom. Thomson's model pictured a sphere of positive electricity studded with negatively charged electrons (figure 4.7). Thomson

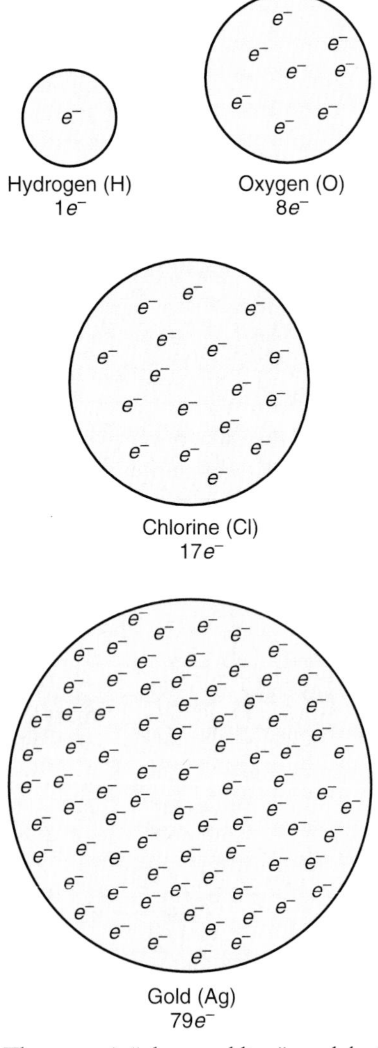

Figure 4.7 Thompson's "plum pudding" model of the atom, with electrons as the plums embedded in a "pudding" of positive electricity.

called his idea the "plum-pudding model" because it resembled the English dessert with plums or raisins (electrons) embedded uniformly throughout the pudding (made up of protons). Modern analogies include cookies with chocolate chips embedded in the dough or ice cream with cherries dispersed throughout.

By the time Thomson proposed his model, scientists were aware that negative and positive charges in an atom must be equal for the atom to be neutral. This meant that if an atom contained just one electron and one proton, as in hydrogen, the electron must be "dispersed" throughout the "proton." If an atom contained ten of each, ten protons would make up the positive sphere, and ten electrons would be embedded uniformly within this sphere.

After the proton proved to be a separate particle, the model was modified. Scientists now believed that electron-proton pairs were uniformly dispersed throughout the volume of the atom. The plums, chips, or cherries became pairs of particles, presumably clinging to each other because of their opposite charges. Although this theory sounded reasonable, no data existed to support it. But Thomson's brilliant student, Ernest Rutherford (born in New Zealand, 1871–1937), wanted to prove his mentor's theory. Instead, he did the opposite; Rutherford showed that Thomson's model could not be correct.

Ernest Rutherford.

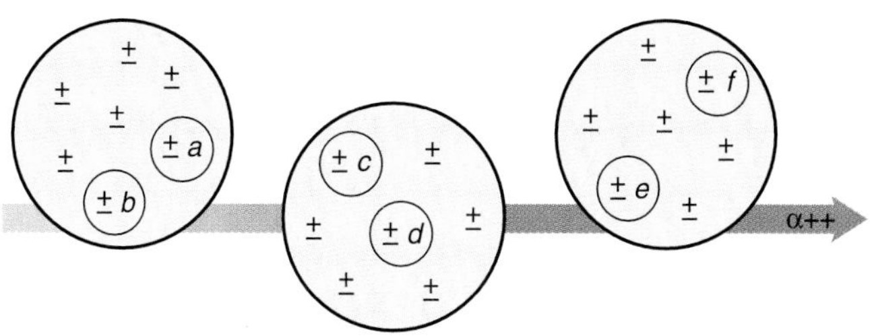

Figure 4.8 The predicted results of Rutherford's alpha particle experiment, assuming Thomson's plum-pudding model of the atom. When Rutherford shot an α-particle through the atom in the metal, he reasoned there would be little deflection since the protons and electrons were uniformly distributed and their charges would cancel each other.

Rutherford Tests the Recipe

Rutherford set out to prove Thomson's model with alpha (α) particles, doubly-positive ionic particles. (In 1906, Rutherford had discovered that these particles were actually helium atoms from which two electrons were removed, giving them a double positive charge, He^{++}). He received the 1908 Nobel Prize in chemistry for this discovery. A year later (1909), Rutherford ran his crucial studies on how these α-particles were scattered by atoms. Rutherford reasoned that if he shot an α-particle through an atom, there should be little electrical deflection since the protons and electrons were uniformly distributed and their charges would cancel each other (figure 4.8). Any slight physical deflection would be counterbalanced by an equal deflection in the opposite direction by another proton in a nearby proton-electron pair. (In these physical collisions, the major effect would be due to the more massive proton, which would repel the positively charged α-particle.)

To test his ideas, Rutherford placed a thin gold foil directly in the path of a narrow α-particle beam (figure 4.9). He then set a

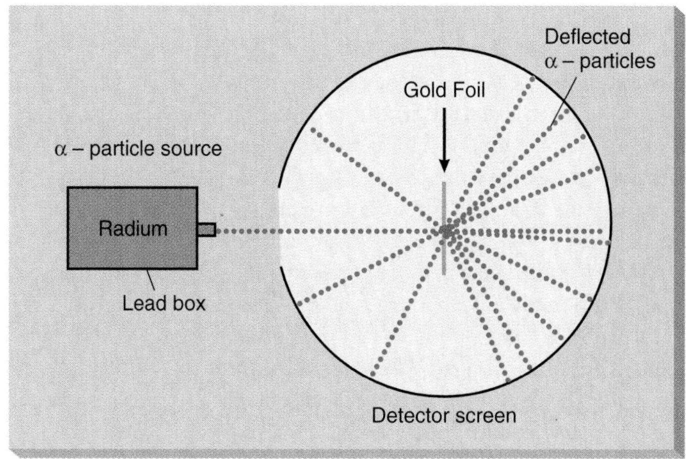

Figure 4.9 Rutherford's α-particle experiment.

detection plate (made of zinc sulfide, which glows when struck by α-particles) all around the thin plate of foil. The alpha particles came from a radioactive material enclosed in a lead box with just one small hole; this would prevent stray radiation from escaping and distorting the results.

Rutherford clearly expected the α-beam to pass undeflected through the gold foil and strike the center of the detection plate. He surrounded the foil with a circular detection plate just to prove that no α-particles were deflected in other directions. Most of the α-particles actually did strike the center of the plate, but some were scattered. A few even bounced directly back toward the α-particle source! This totally unexpected result astonished Rutherford. Years after his landmark experiment, he said, "It was almost as incredible as if you fired a fifteen-inch shell at a piece of tissue paper and it came back and hit you."

It was clear the Thomson plum-pudding model just did not fit the data. Instead of proving Thomson's theory, Rutherford had destroyed it. Sometimes, the scientific method works in unexpected ways.

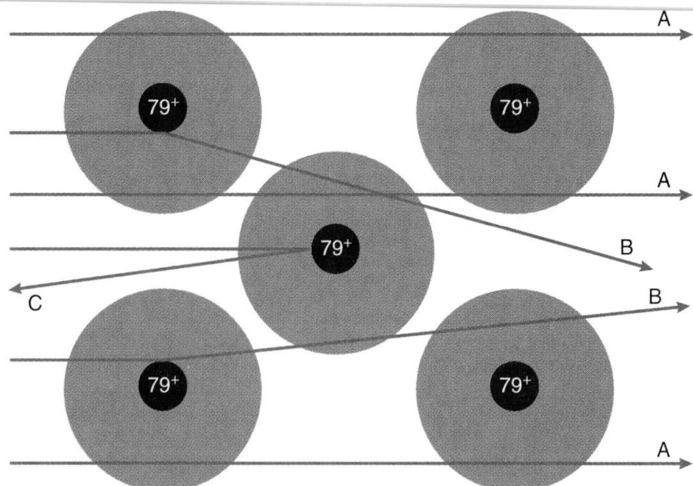

Figure 4.10 Rutherford's α-particle experiment explained in accordance with his new model of the atom. Most α-particles pass through the empty space in the atoms without interference (*path* A). Some come close to a nucleus and are slightly deflected (*path* B); a few directly strike a nucleus and deflect back toward the source (*path* C).

EXPERIENCE THIS

Let's dwell on Rutherford's observation about his experiment for a few moments. Fasten a tissue to an open cardboard frame with tape. Now try throwing small objects as the tissue target. How big (or how heavy) must a missile be to tear the tissue? Will a paper clip go through? a wad of paper? What about a baseball? Rutherford described his experiment as similar to firing a fifteen-inch shell at a tissue and seeing it bounce off. What would happen if you threw a fifteen-pound bowling ball at your tissue target?

RUTHERFORD'S ATOM

By 1911, after completing his initial alpha particle experiments, Rutherford suspected that nearly all the mass of the atom was in a tiny, centralized location he called the **nucleus** (from the Latin *nux*, for "nut"). He believed that *all* the protons and neutrons were densely packed in the nucleus. The electrons remained outside the nucleus and roamed freely in the remaining volume. By 1914 Rutherford's coworkers, Ernest Marsden and Hans Geiger, had amassed additional evidence to support Rutherford's atom model. This new model of the atom represented a dramatic change from previous theories, but it coincided well with Rutherford's startling α-particle rebound data.

Figure 4.10 shows why Rutherford's model was consistent with the α-particle results. The gold atoms were tightly packed in the foil—as is true for virtually all solids—but the atoms themselves were mostly empty space with the atomic mass crowded into a tiny nucleus. (Each nucleus in a gold atom contains 79 protons, as figure 4.10 shows.) Most of the incoming α-particles passed through the empty space, usually in a straight line. Although electrons were moving throughout the empty space, they were too light to affect the α-particles; and since most α-particles missed the atomic nuclei (path A), they passed through without interference and struck the center of the detector plates. Some α-particles came close to a nucleus (path B) and were slightly deflected as the positive charge of the protons repelled them. And on exceedingly rare occasions, an α-particle collided head-on with a nucleus (path C) and ricocheted directly back toward its source.

Rutherford's model not only explained the observed α-particle data, it also fit other data for the atom. Scientists quickly accepted his model and replaced the old theory with the new. Let's examine Rutherford's theory of the structure of an atom more closely.

The Size of the Nucleus

According to Rutherford, the volume of the gold atom was about 10^{12} (one trillion) times larger than the volume of its nucleus, and the atomic radius was 10^4 (ten thousand) times larger than the radius of the nucleus. If you placed a pea (about 1 cm in diameter) in the center of a large football stadium to represent the nucleus in an atom, you would have roughly the same relative proportions as the Rutherford model. The smaller α-particles might correspond to tiny grains of rice. Imagine someone in New York City firing rice grains across the continent, trying to hit a green pea in the middle of the Rose Bowl in Pasadena, California. This approximates what Rutherford was doing in his α-particle experiments.

At first glance, this seems to be a very strange model! Almost all the mass is compressed into a tiny fragment of the total volume of the atom. In fact, the pea analogy does not do justice to the model because peas are much too light. The gold nucleus contains 79 protons and 118 neutrons, which together make up over 99.98 percent of the atom's mass, crowded into 10^{-12} (one trillionth) of the atom's volume. The 79 electrons wander about through almost all the volume but comprise less than 0.2 percent of the atom's mass. Our solar system is considerably more compact, in relative terms, than the atom. Pluto's orbit is about 4,242 times the diameter of the sun; the atom is about 10,000 times the diameter of its nucleus.

Rutherford's model raised or left unanswered several questions. First, why does periodic behavior occur? Rutherford's model still did not explain why elements in the same family share chemical properties. Another unanswered question: How do chemical reactions actually occur, and what role (if any) does the nucleus play? The nucleus would be hidden too deeply in the atom to be of much use in chemical bond formation; moreover, if changes occurred in the nucleus during a reaction, the identity of the atoms would sometimes change—and scientists knew that this does not happen. Some additional structure involving the electrons must exist in the outer volume of the atom and must relate to chemical reactivity.

Rutherford's model also answered—or led to the answers for—some puzzling questions; among them, what relationships exist between numbers of protons and neutrons, the periodic table, and chemical properties and behaviors?

Henry Moseley.

Atomic Numbers (Numbers of Protons)

As we previously noted, the elements in a modern periodic table are arranged in order of atomic number rather than atomic mass. Mendeleev had to reverse some elements in his periodic table in terms of atomic mass to group the chemical properties in element families properly. Tellurium (Te) and iodine (I) are examples; although tellurium has a larger atomic mass, it appears in family VIA, just before iodine in family VIIA. Tellurium's atomic number is 52, iodine's 53.

Where did the atomic numbers originate, and what do they mean? Henry Moseley (1887–1915), an exceptionally brilliant student of Rutherford, performed studies that showed that each kind of atom not only had a unique atomic mass, it also contained a specific number of protons. Moseley assigned an atomic number to each element that reflected the number of protons one atom of that element would contain. Once Moseley assigned these numbers, the Mendeleev inversions made sense; tellurium, number 52, contains 52 protons, and iodine, number 53, contains 53.

At least five Mendeleev inversions appear in our modern periodic table: argon (Ar, number 18) and potassium (K, 19); cobalt (Co, 27) and nickel (Ni, 28); tellurium (Te, 52) and iodine (I, 53); thorium (Th, 90) and protactinium (Pa, 91); and uranium (U, 92) and neptunium (Np, 93). In each case, the element with the smaller number of protons, and thus the smaller atomic number, has the greater atomic mass. Additional Mendeleev inversions may exist for elements 94 and 95 (plutonium, Pu, and americium, Am) as well as for 101 and 102 (mendelevium, Md, and nobelium, No). The atomic masses of these more recently discovered elements have not been determined accurately enough to judge whether inversions exist.

The atomic number identifies an element as well as reveals the number of protons its atom contains; changing the atomic number would change the identity of the atom. If the atomic number of an atom of carbon changed from 6 to either 5 or 7, that atom would change into boron or nitrogen, respectively. But atoms do not transform into other types of elements. Any atom with 6 protons, or an atomic number of 6, must be carbon.

Moseley's atomic number also clarified the role neutrons play. Gold has an atomic weight of 197 but contains only 79 protons. Additional particles must account for the missing 118 amu; electrons are too small, and additional protons would change gold into another element. For twenty-one years after Rutherford proposed his model of the atom in 1911, chemists had to assume the existence of the neutron. We now know that an atom of gold contains 79 protons and 118 neutrons.

Moseley would possibly have received a Nobel Prize for his research, but he was drafted into the British army during World War I and was killed in action in Turkey in 1915, at the height of his career. As with Lavoisier and many other scientists, the political arena superseded the agenda of this particular chemist and of science in general.

PRACTICE PROBLEM 4.3

Determine how many neutrons are in each of the following atoms. Use the atomic numbers and atomic masses given.

ELEMENT	SYMBOL	ATOMIC NUMBER	ATOMIC MASS
a. Carbon	C	6	12
b. Sulfur	S	16	32
c. Iron	Fe	26	56
d. Tin	Sn	50	119
e. Mercury	Hg	80	201

ANSWER To determine the number of neutrons in each atom, subtract the atomic number (number of protons) from the atomic mass (number of protons plus neutrons). This process yields the following answers: *(a)* 6; *(b)* 16; *(c)* 30; *(d)* 69; *(e)* 121. As you can see, the number of neutrons often exceeds the number of protons.

Isotopes (Numbers of Neutrons)

Although it is possible to determine the number of neutrons in an atom if you know the atom's mass and atomic number, some atoms of a single element may have slightly different masses. Atoms with slightly different masses are called **isotopes;** most elements have two or more isotopes (see chapter 3).

The concept of an isotope was developed in 1913 by another Rutherford student named Frederick Soddy (English, 1877–1956). Soddy coined the word *isotope* from a Greek term meaning "same place." Isotopes always have the same number of protons and electrons (and thus are always the same element), but they differ in number of neutrons.

Hydrogen provides a simple example of isotopes. Its normal atomic weight is 1 amu, but two isotopes exist with masses of 2 and 3 amu, respectively. Chemist Harold Urey (American, 1893–1981) discovered deuterium, the 2 amu isotope, in 1934; as a result, he won the Nobel Prize for chemistry in 1934. Marcus Oliphant (Australian, 1901–) discovered tritium, the 3 amu isotope, also in 1934. Each hydrogen isotope has only one electron and one proton, but 0, 1, or 2 neutrons for masses of 1, 2, and 3 amu, respectively.

Another example of an element with isotopes is carbon. The most common carbon isotope weighs 12 amu, but others weigh 13 and 14 amu. All three contain 6 protons and 6 electrons, but ^{13}C (carbon 13) has 7 neutrons and ^{14}C (carbon 14) has 8 neutrons. The current atomic mass values are based on a value of 12.000 for the ^{12}C isotope, but the average mass for all carbon isotopes is actually 12.011 amu because there are many ^{12}C atoms but relatively few ^{13}C and ^{14}C atoms. Similarly, chlorine exists as isotopes with masses of 35 and 37. Both isotopes have 17 protons and electrons, but ^{35}Cl has 18 neutrons and ^{37}Cl has 20.

The discovery of isotopes pointed out another flaw in Dalton's atomic theory, which holds that all atoms of an element have the same mass. Chemists still use Dalton's theory, making a mental note that isotopes have slightly different masses. Dalton's error is understandable when we recall that he did not know about subatomic particles.

PRACTICE PROBLEM 4.4

How many neutrons does a chlorine (Cl) atom contain if its atomic number is 17 and its atomic mass is 35.5?

ANSWER This is a trick question! Subtracting 17 from 35.5 leaves 18.5, but an atom cannot contain half a neutron.

Ordinary chlorine consists of a mixture of two isotopes with masses of 35 (about 75 percent of the atoms) and 37 (about 25 percent of the atoms). This averages out to an atomic mass of 35.5 for the mixture, but each individual atom has a whole number of neutrons (either 18 or 20).

BOHR'S ATOM

The determination of an atomic number for each element and the discovery of isotopes answered many questions about the way chemical families align on the periodic table and why atoms of one element may have slightly differing masses. But scientists still wanted to know how a chemical reaction occurs. If the nucleus, left unchanged deep in the volume of an atom after a reaction, does not participate, do the electrons somehow react in the atom's outer reaches? The clues would come from an apparently unrelated phenomenon—spectra.

Spectra

In 1665, Isaac Newton (1642–1727) discovered that white light separates into a rainbowlike arrangement when it passes through a prism. This separation of light into its components is called a **spectrum** (plural, spectra). Newton's intriguing discovery lay dormant as a scientific tool for nearly two hundred years, until Kirchoff and Bunsen invented the spectroscope in 1861 (see chapter 3).

In a spectroscope, light from a flame (or other source) is passed through a prism, which separates the light into its component colors. After an element absorbs energy from the light source, it is raised from a ground state (the lowest energy possible) to a higher energy state. It then emits light as it loses this energy. This procedure is called emission spectroscopy, and the spectrum is an **emission spectrum.** As we learned in chapter 3, not all spectra are continuous; while white light creates a continuous, smooth spectrum, most emission spectra contain only a few lines of color broken by vast regions of black. (Some typical line spectra appeared in figure 3.18.) Every element has a characteristic line

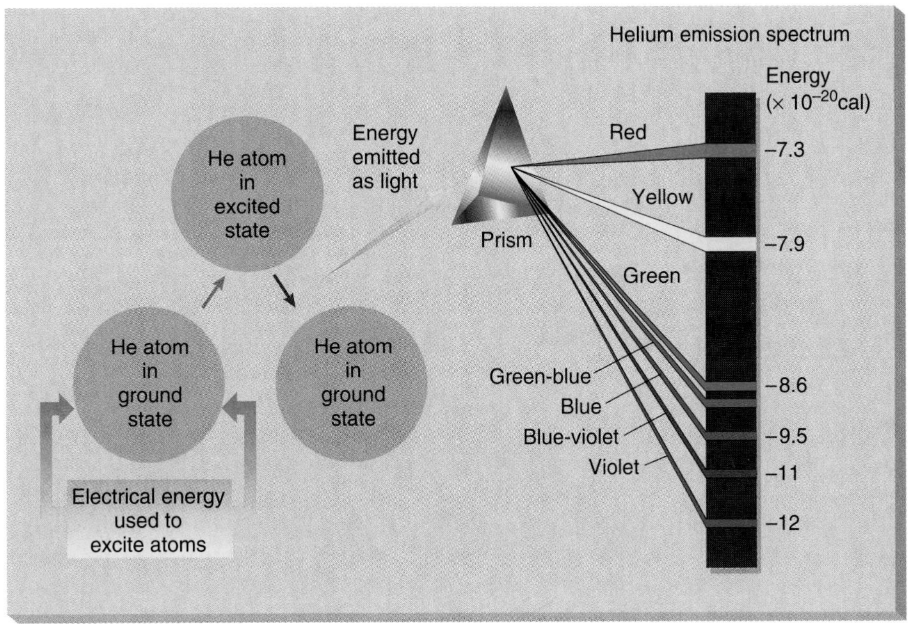

Figure 4.11 A typical emission spectroscope. When an atom absorbs energy, its electrons become excited and jump to a higher excited state. As the electrons return to a lower state, they emit light. Each element produces a unique emission spectrum.

spectrum, and a trained scientist can identify the elements in a sample, or the elements on a distant star, by examining the spectra they produce.

What Causes Emission Spectra?

Emission spectra demonstrate the interaction of matter and energy. When atoms absorb energy, they become "excited." The electrons in the excited atoms move, and these moving electrons emit light. Thus, emission spectra are produced when excited electrons move (figure 4.11).

Recall from chapter 3 that radiant energy is measured in terms of wavelengths (the distance between wave crests) and frequency (how frequently a wave crest passes a given point—in other words, how closely packed the waves are). The shorter the wavelength, the higher the frequency, and the higher the frequency, the more energy emitted. Each line of color in a spectrum has a specific frequency and thus represents a definite amount of energy (figure 4.12). Light has other properties, including amplitude and intensity. **Amplitude** is defined as the range of the wave—its height above and below a reference line down the middle of the wave. **Intensity** is the relative strength of the light; it measures how many photons are in a light beam, not how much energy each photon has. The energy depends on the frequency.

The light in an emission spectrum is released when electrons move; since only certain amounts of energy are released in emission spectroscopy, creating characteristic thin bands of color, the implication is that electron movements are associated with very specific amounts of energy. Niels Bohr (Danish, 1885–1962) offered the first good explanation for this behavior. Box 4.2 describes a very practical spectroscopic application.

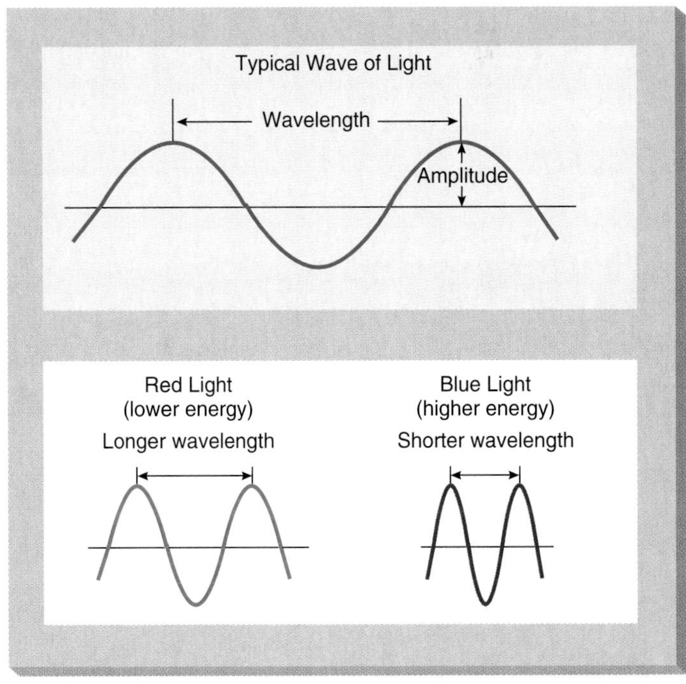

Figure 4.12 Characteristics of different colors of visible light. Each has a different wavelength and thus a different amount of energy associated with it.

BOX 4.2

A FORENSIC "WHO-DONE-IT?"

Modern chemistry uses many different kinds of spectroscopy based on infrared, ultraviolet, and magnetic waves to determine the exact chemical composition of sample materials. Although this may sound complicated, the process has great practical importance. Let's consider an example. A hit-and-run driver kills a pedestrian, and there are no witnesses. How can police catch the driver? Will he or she escape justice?

Chemical techniques may help catch the culprit. Suppose a forensic chemist examines the victim's clothes and finds tiny paint particles from the car embedded between the fibers. Sam, the chemist removes these paint fragments and divides them into two samples. One sample will be used for immediate analysis; the other will be retained in the event that a defense attorney someday wishes to have an independent analysis run. This is standard practice at the lab run by the Federal Bureau of Investigation and many other forensic laboratories.

Next, Sam analyzes the paint fragments spectroscopically. He can compare the results with spectra previously measured for all the paints used on foreign and domestic cars for the past several years. Most likely, his first analysis will be a flame emission spectrograph, since this narrows the possibilities quickly and only requires an infinitesimal amount of material. Sam places a sample in the instrument and ignites it. Within seconds a recording of the spectra appears and a computer compares these spectra with the thousands of automobile paint samples on record.

Next, Sam decides to run an infrared spectral analysis. Infrared is a portion of the electromagnetic spectrum that is lower in frequency (and energy) than visible light (see figure 3.17). This technique gives the researcher some specific information concerning the structure of the materials used to make the paint. Again, a computer compares this data against information stored in its memory, or perhaps accessed through its modem from some other location. The results are definitive. Our forensic chemist discovers that the paint came from a red 1985 Chevrolet Chevette manufactured in Lordstown, Ohio.

The police now know the color and make of car. They check vehicle registration records, and in a short time they learn that only 38 such cars are registered in the immediate vicinity. These vehicles are checked as quickly as possible. In the meantime, detectives interview body shop employees to ascertain whether a red 1985 Chevette was repaired or repainted since the accident. In a few days, the police have likely determined whether the hit-and-run driver was from their locality. If so, they may even make an arrest. If no local car is suspect, the police can transmit the identifying information to other law enforcement agencies to look for the specific car involved.

It may take a few days or weeks, perhaps even longer, but ultimately, the hit-and-run driver is likely to be located and brought to trial. Justice is served—thanks to spectroscopy and forensic chemistry.

Niels Bohr.

Bohr's Theory: Orbiting Electrons

Bohr was another of Rutherford's students. He knew the Rutherford atomic model did not explain each element's unique spectral lines, and in 1913, he devised a new theory to do so. Bohr worked with the hydrogen atom, since it has only one electron and one proton. He reasoned that if he could explain the spectrum for this atom, he might extend his theory to more complicated atoms as well.

Bohr already assumed, from Rutherford's model, that electrons travel around outside the nucleus. He further hypothesized that electrons move in a set of circular **orbits** around the nucleus, similar to the planets circling the sun. Each orbit, at a specific distance from the nucleus, represented a specific **energy level.** The electron was usually in one orbit, but it could jump to a higher energy orbit if it absorbed additional energy from an external source, such as the heat from a flame. The electron could also drop back to a lower energy orbit by releasing part of its energy in the form of emitted light. It could not occupy space between the set orbits.

Max Planck (German 1858–1947) conceived the term **quantum** of light in 1900 to explain the intensity and color of light omitted from hot objects. In 1905, Albert Einstein (German, 1879–1955) had proposed that all light consists of bundles of energy called **photons** or **quanta.** Bohr now suggested that a specific amount of energy, or a quantum, was required to boost an electron from one energy level (or orbit) to the next. As an electron absorbs more quanta (or energy), it can jump to progressively higher orbits.

Figure 4.13 diagrams Bohr's orbits and energy levels for the hydrogen atom. The electron can jump from a lower orbit to a higher orbit, or drop from a higher orbit to any lower orbit. When it drops, the electron releases a specific amount of energy, producing a spectral line at a definite frequency and of a specific color. This frequency corresponds to the energy difference between the two orbits. According to Bohr, the number of possible jumps was limited, since only certain orbits exist; this meant electrons could only produce light in certain, specific frequencies and not a continuous spectrum. Instead, atoms of any element would, according to Bohr, produce a discontinuous line spectrum—which was consistent with the emission spectra actually obtained from different elements.

TABLE 4.6 Comparison of Bohr's predictions with the actual hydrogen spectral lines.

ORBITS INVOLVED	WAVELENGTH (nm) Predicted	WAVELENGTH (nm) Observed	ELECTROMAGNETIC SPECTRUM REGION
$4 \rightarrow 1$	97.3	97.3	ultraviolet
$3 \rightarrow 1$	102.6	102.6	ultraviolet
$2 \rightarrow 1$	121.6	121.7	ultraviolet
$5 \rightarrow 2$	434.3	434.3	visible, blue
$4 \rightarrow 2$	486.5	486.5	visible, green
$3 \rightarrow 2$	656.6	656.7	visible, red
$4 \rightarrow 3$	1876.0	1876.0	infrared

As table 4.6 indicates, the actual hydrogen spectral lines agreed exceedingly well with Bohr's predictions. Bohr published his findings in 1913 and won widespread acceptance for his theory. Like Mendeleev, Bohr predicted beyond known facts. Several of his calculated frequencies did not correspond to known spectral lines for hydrogen; Bohr boldly predicted the existence of these unknown lines, and later research proved him correct. In fact, his predictions were so uncannily accurate that some unfairly suspected he had a secret new spectroscope that measured the "unknown" lines!

When Bohr applied his theory to larger atoms, he was unable to make the outstanding spectral predictions he achieved with hydrogen. Nevertheless, Bohr had achieved a major breakthrough: his electron orbit model finally explained periodic behavior.

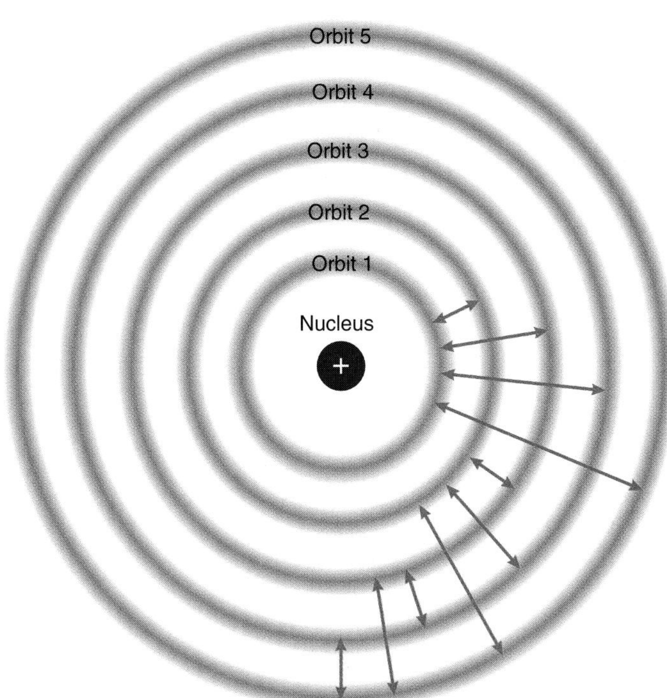

Figure 4.13 Bohr orbit-energy diagram for the hydrogen atom. When an electron absorbs energy, it jumps to a higher energy level, moving to an orbit farther from the atom's nucleus. When an electron drops from a higher level to a lower orbit, it releases the energy it previously absorbed. The more energy an electron absorbs, the higher the energy level it jumps to.

PRACTICE PROBLEM 4.5

Planck, Einstein, and Bohr established the term *quantum* in science. The term has also become common in our society, and we use the basic idea daily. Money is quantized as pennies, nickels, dimes, quarters, half-dollars, dollars, and so on; all monetary transactions operate using money quanta. If you need 17 cents, you must use some combination of money quanta to get it—you cannot use two 8.5-cent units because they do not exist. TV channels are also quantized. Which of the following items are quantized—that is, which exist only in specific quantities rather than in continuous amounts?

a. Postage stamps
b. Stations on the AM radio dial
c. Soft drinks
d. Cookies
e. People's weights
f. A carton of a dozen eggs

ANSWER Postage stamps *(a)* and a dozen eggs *(e)* are clearly quantized. You can buy as many stamps as you wish, but they only come in specific values. Similarly, you can purchase as many dozen eggs as you desire, but a dozen will remain a group of twelve. For all practical purposes, an AM radio dial *(b)* is quantized—though it is one continuous sweep, you will only find stations broadcasting at certain frequencies.

The others are not clearly quantized. You can bake cookies *(c)* in virtually any size, shape, or weight. People's weights (as a group) are also a continuous set of values *(d)*. We express weights in pounds, but you can weigh between 150 and 151 pounds.

Though some items (such as cookies, the radio dial, and people's weights) may seem ambiguous, this demonstrates the long confusion over the wave-particle duality of light. Light is a particle that moves as a wave. Some properties of light depend on one aspect or the other.

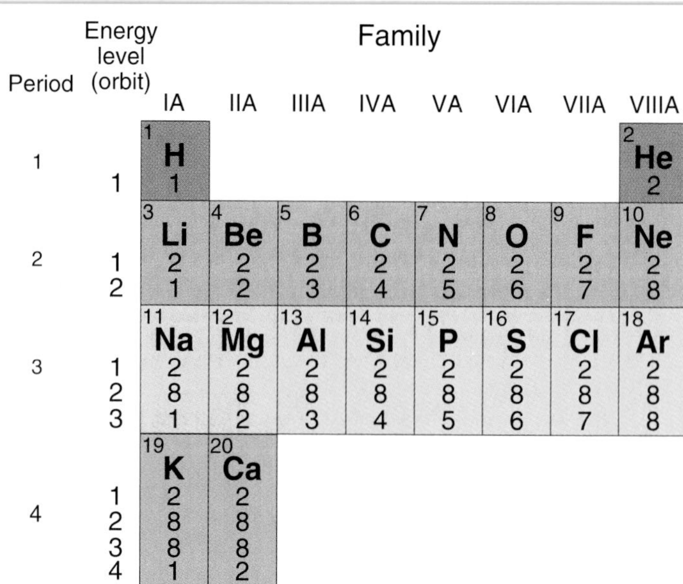

Figure 4.14 The Bohr orbit electron arrangement for the first twenty elements in the periodic table. Although many potential energy levels are available, only those containing at least one electron are shown. This corresponds to the lowest energy state (ground state).

The Bohr Atom and the Periodic Chart

As we move through the elements in the periodic table, each successive element contains one additional proton and one additional electron. Each proton is in place in the nucleus, and each electron orbits around the nucleus. All atoms have a **ground state,** a resting condition when the total energy they contain is at the lowest possible value. In this ground state, the electrons lie in the lowest-energy orbits—those that lie as close as possible to the nucleus. Electrons thus fill up the inner orbits before they appear in the orbits that are further out. Based on information from the periodic chart and his own calculations, Bohr assumed each orbit can contain a limited number of electrons: he stated that the first five orbits can hold 2, 8, 8, 18, and 18 electrons, respectively.

Since orbit number 1 can hold only 2 electrons, the element helium fills up this orbit with its 2 electrons. The next element, sodium, has 3 electrons. The first two can fill orbit 1, but the remaining electron must occupy orbit number 2, at the next higher energy level. Each successive element has an additional electron that fills another opening in orbit 2. Neon, with an atomic number of 10 and therefore 10 electrons completes the second energy level; 2 of its electrons fill orbit 1, and 8 occupy all the available spaces in orbit 2.

Figure 4.14 shows the electron locations for the first twenty elements. This abbreviated periodic table reveals several important ideas. First, the total number of electrons occupying all energy levels always equals the element's atomic number as well as the number of protons in the nucleus (oxygen, for example, whose atomic number is 8, has 8 total electrons and 8 protons). Second, the number of electrons in the outermost energy level is identical to the family number at the top of the chart (with one exception—helium). Aluminum is in family IIIA, and it has 3 electrons in its outer orbit. This means that although each element has a different total number of electrons, all family members have exactly the same number of electrons *in their outermost energy levels*. Third, the period number and the number of energy levels that contain electrons is identical; all the elements in period 3, for example, have electrons in orbits 1, 2, and 3. Finally, the outermost energy levels are completely filled for all members of family VIIIA, the noble gas family.

Mendeleev's Puzzle Solved!

Bohr's orbit electron arrangement, as depicted in figure 4.14, finally provided an explanation for the periodic behavior of the elements. Each number of a chemical family contains the same number of electrons in its outer energy shell, which accounts for the family's shared properties. Every alkali metal family member (IA) has exactly 1 electron in its outermost orbit; every alkaline earth element (IIA) has exactly 2 electrons in its outermost orbit. The same is true for the boron, carbon, nitrogen, halogen, and noble gas families. If electrons are involved in chemical reactions such as forming compounds, the outermost energy levels would come into play because they are the most accessible.

Although Bohr's model would be supplanted by more accurate models as scientists gained new knowledge, the magnitude of Bohr's accomplishment cannot be overstated; Bohr's theory explains periodicity and nearly every feature in chemical bonding. Still, Bohr built his theory on the work of previous scientists. As Isaac Newton wrote to Robert Hooke in 1675, "If I have seen further, it is by standing on the shoulders of giants." Bohr stood on the shoulders of giants such as Mendeleev, Thomson, and Rutherford; he also provided his own shoulders for the many scientists who followed.

Bohr received the Nobel Prize for physics in 1922. During World War II, he fled his native Denmark and settled in the United States, where he was instrumental in developing the atomic bomb. In his later years, Bohr avidly stressed the potential value of peaceful atomic energy applications. In 1957 he received the first Atoms for Peace award.

PRACTICE PROBLEM 4.6

How many total electrons does each of the following elements contain? How many electrons occupy the outermost orbit or shell of each atom?

a. Silicon (Si)

b. Strontium (Sr)

c. Tin (Sn)

d. Selenium (Se)

e. Antimony (Sb)

f. Xenon (Xe)

ANSWER Total electrons and number of electrons in the outermost shell are as follows: *(a)* 14, 4; *(b)* 38, 2; *(c)* 50, 4; *(d)* 34, 6; *(e)* 51, 5; *(f)* 54, 8. Did you notice that the total number of electrons is identical to the atomic number, and the number of outermost electrons is identical to the family number in the periodic table? These are very important patterns to remember.

NEW INSIGHTS IN THE ATOM: THE WAVE MECHANICS MODEL

Although Bohr's model of the atom accounted for many previously unexplained atomic properties and behaviors, research soon demonstrated that the model was not entirely accurate. During the 1920s, several scientists pieced together theories and data that would overturn the Bohr model. These theories included wave mechanics and the uncertainty principle.

Wave Mechanics

Several discoveries shaped the wave mechanics theory. One was the similarity between electrons and light. For centuries, scientists had debated the nature of light. Some believed light exists as waves, continually moving like waves in an ocean; others argued that light consists of particles moving in a stream, like tiny droplets sprayed in a mist. Initially, the wave theory was favored because it was useful in understanding spectroscopy and other phenomena. However, the wave theory failed miserably at explaining the photoelectric effect, the fact that certain metals release electrons when a beam of light shines on them.

The photoelectric effect depends on the frequency of the light (hence, the energy of the photon) and not its intensity. A high-intensity beam of low-frequency light will not emit electrons from a metal, but a low-intensity, high-frequency beam will. The photoelectric effect was best explained by assuming light was also a particle.

Incidentally, you see the photoelectric effect in operation whenever you step past an "electronic eye" in a store and hear a bell sound because you interrupted an invisible light beam. This is a practical application of the quantum theory that Planck, Einstein, and others developed.

Scientists eventually concluded that light is both particles and waves. In other words it is made of quanta or photons (tiny bundles of energy) that behave like waves in their movement. On this basis, it seems reasonable to assume that electrons might behave in a similar manner.

In 1923, Louis de Broglie (French, 1892–1987) startlingly claimed that electrons are also both particles and waves, waves of energy with extremely short wavelengths. Most scientists did not immediately agree with this radical concept, but in 1927 Clifton Davisson and George P. Thomson independently discovered electron diffraction; since diffraction was one of the main properties supporting the wavelike character of light, it also confirmed the electron's wave behavior. (Eventually, acceptance of de Broglie's ideas led Ernst Ruska to develop the electron microscope in 1931. Because electrons have such extremely short wavelengths, they allow scientists to see very small objects such as DNA and viruses, including—as figure 4.15 shows—the AIDS virus.)

In 1924, Sir Jagadis Chandra Bose (Indian, 1858–1937) developed a new, highly mathematical method for handling quanta, and Albert Einstein applied this method to matter. Although at that time only a few people in the entire world understood the mathematics behind Einstein's new treatment of matter, his research had profound effects on how scientists viewed matter from that point onwards. Einstein's main conclusion was that *all* matter exhibits wave properties. Three years later, in 1927, Erwin Schrödinger (Austrian, 1887–1961) published his first paper on wave mechanics, a theoretical system in which matter is described in terms of waves. The wave mechanical theory, or **wave mechanics model,** views electrons as waves of energy that surround the nucleus of the atom, rather than as discrete particles.

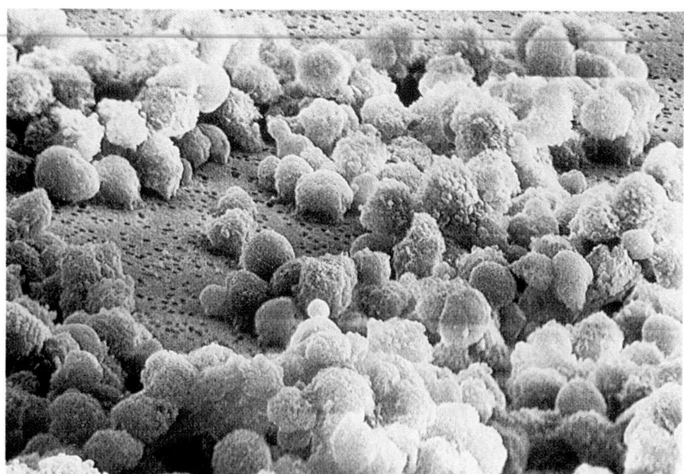

Figure 4.15 A scanning electron micrograph of the AIDS virus at a low magnification.

The Uncertainty Principle

In 1927, the same year Schrödinger published his ideas on wave mechanics, Werner Heisenberg proposed the **uncertainty principle,** which states that it is possible to determine both the momentum and position of a small particle. In essence, this principle says that if scientists know *exactly* where an electron is they cannot tell *exactly* at what speed it is moving, or where it's going; conversely, if they know its exact speed and direction, they cannot pinpoint its location. (This proposal makes common sense—to determine a particle's location, it must be still—without momentum—for at least a split second; and to measure its momentum, it must be moving—not in one particular spot for any length of time.) Heisenberg's deceptively simple theory would profoundly affect the way scientists view the atom.

Goodbye Orbits, Hello Orbitals

The combination of wave mechanics and the uncertainty principle played havoc with the Bohr model of the atom. In Bohr's model, electrons circle the nucleus as planets circle the sun. Just as scientists can determine the position and speed of the planets in their predictable orbits, they should be able to determine the position and speed of the electrons circling the nucleus in their orbits. But wave mechanics and the uncertainty principle combined to assert that this is impossible. Instead, scientists can only predict the probability that an electron, moving as a wave of energy, will be located in a three-dimensional volume called an orbital at any given time. (See box 4.3 for more information on probability.)

An **orbital** is the volume of space around an atom's nucleus in which a particular electron will be located 95 percent of the time. While an orbit is a defined, two-dimensional line surrounding the nucleus, an orbital is a three-dimensional volume with a defined shape. At each energy level (formerly thought of as each successive orbit), one or more orbitals exists. The first energy shell contains one orbital that holds up to 2 electrons; the second shell contains two orbitals that hold up to 8 total electrons; the third shell contains three orbitals that hold up to 18 total electrons; and the pattern continues, with one more orbital at each successive energy level.

The orbitals are known as s, p, d, or f orbitals, with each type representing a different shape. The simplest, the s orbitals, are spherical, like a ball (figure 4.16a). Each p orbital is shaped roughly like a jack in a child's ball-and-jacks game; it consists of three suborbitals, each dumbbell-shaped and set perpendicular to the other two (figure 4.16b). Each p suborbital is identified as p_x, p_y, or p_z, according to its axis. The d and f orbitals are also made up of suborbitals; their shapes are complicated and beyond our concern in this text.

The orbitals become larger at higher energy levels; this is the easiest to picture with the s orbitals for periods 1–7, which form a series of spheres within spheres. For example, for the s orbitals in periods 1–4, imagine a marble within a ping-pong ball, within a softball, within a basketball. The nucleus, with practically all of the mass of the atom, is a tiny speck of dust at the center of these nested balls. The electrons are waves of energy that may be in any location within each orbital. The shape of the orbitals determines the shape of the molecules the atoms will form.

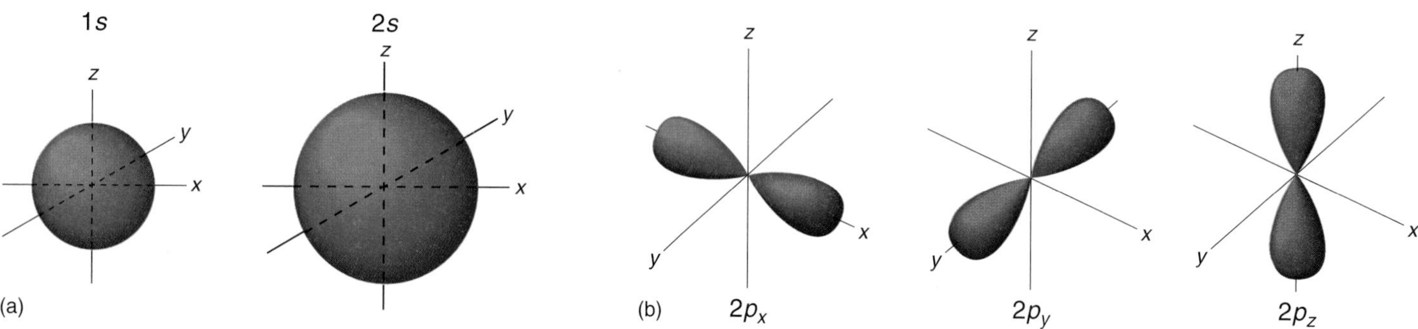

Figure 4.16 Atomic (a) s orbitals and (b) p orbitals.
SOURCE: From Jacqueline I. Kroschwitz, et al., Chemistry: A First Course, 3d ed.

BOX 4.3

PROBABILITY AND CHEMISTRY

Most people have a vague concept of the nature of probability. We hear that the odds of a team winning the championship are five to one; that the chances of winning the lottery with a particular ticket are one in a million; or that the chance of rain tomorrow is 20 percent. Most of us understand that this means the team, perhaps our all-time favorite, might win the coveted prize one time in five tries. Likewise for every million lottery tickets, only one will win the prize. We know that out of five days with conditions like tomorrow's, four will be dry and one will see rain.

As these examples demonstrate, the idea of chance or probability permeates our daily lives. One key feature to remember about probability is that while the odds may be accurate, they do not guarantee a specific result. The scene depicted in figure 1 provides a good demonstration of this idea. The author once had five dogs, and it seemed that three were asleep at any given time. But it was impossible to predict whether the sleeping trio would be Ilsa, Patty, and Champ, with Valentino and Georgie standing watch, or some other combination. And at times, more or fewer than three would be asleep. In the long run, three out of five (60 percent) might be asleep at any given time, but there was no guarantee this would be true at any given moment.

Figure 1 Probability—it gets even worse with more dogs.

What does probability have to do with chemistry? In the past, many scientific laws were expressed in absolute terms. For instance, Boyle's Law states the gas volume will decrease at pressure increases. There was no need to express this as a probability because the law held true *every* time. These laws were essentially chiseled in stone.

As the study of chemistry developed, scientists realized that many phenomena are far less certain. For example, in most cases when a chemical reaction occurs, all the reacting compounds eventually convert into products. Chemists quickly realized that part of the chemicals reacted early, and another portion reacted later. Is it possible to predict which molecules will react first? The answer turned out to be a resounding "no!" Chemists might predict with absolute certainty that 10 percent of the molecules (1 in 10) would react in ten minutes, but they could not possibly determine whether a specific molecule would be in this select group. Instead, they could state that the one particular molecule had a 10 percent chance of reacting within 10 minutes. As it advanced, chemistry dealt less with absolutes and more with probability.

During the early twentieth century, probability grew more important in science as a whole. Perhaps Werner Heisenberg's uncertainty principle was the ultimate application of probability; it stated that it is not possible to determine both the speed and position of a particle. Stated another way, this means you can only predict the probability that a given particle will be in a certain location or traveling at a certain speed.

This assertion flew against long-established "knowledge" that a scientist could indeed measure both the speed and location of an object. After all, when cars run in the Daytona 500, the broadcasters tell us that a car is racing along the track at some extremely fast speed while passing a specific post on the track. But the uncertainty principle claims that this is impossible—and it is correct. Although we believe we are measuring exact speeds and locations, our measurements are actually subject to a small amount of error. The speeds are just too slow and the cars too big for the error to make much practical difference.

Tiny particles such as electrons move at speeds that make race cars seem to stand frozen in time and space. With these tiny particles, the effects of the uncertainty principle do become significant. As a result, chemists now speak of electron movement in terms of probability. They can predict the likelihood that a given electron will occupy a specific orbital; in fact, the definition of an orbital is the volume of space an electron is likely to occupy 95 percent of the time. But they cannot guarantee that the electron will occupy this orbital at any given moment. Put another way, the electron will be within the orbital 19 out of every 20 times you look—but it will not be there one of those 20 times.

What we now have in science, and in the world as a whole, is a mixture of absolute and probable events—which can be either an unpleasant or quite intriguing blend. Nevertheless, we must live with uncertainty both in science and in our everyday affairs. After all, we can live with the prediction that there is a 60 percent chance of rain today and still realize the sun lies somewhere behind the clouds. For most purposes, the simpler scientific theories and laws explain the world sufficiently. In a few cases, such as electron orbitals, a more complex—and less certain—explanation furthers our understanding.

Electrons in Orbitals

Just as in Bohr's model, electrons fill the lower-energy orbitals first. S orbitals fill first because they contain the lowest levels of energy. The alkali (IA) and alkaline earth metals (IIA), plus helium, fill s orbitals with 1 or 2 electrons in their outermost shells. The remaining main group elements (families IIIA–VIIIA) fill in p orbitals in their outer shells. Transition elements (IB–VIIIB) fill in d orbitals in their outer shells, and the inner transition elements fill in f orbitals. Figure 4.17 summarizes the order in which orbitals fill.

PRACTICE PROBLEM 4.7

Which type of orbital (s, p, d, or f) forms the outermost shell containing electrons for each of the following elements?

a. Al **b.** Ar **c.** Au **d.** Ag **e.** Am

f. As **g.** At **h.** Ac **i.** K

ANSWER Locate each symbol in the electron-filling pattern shown in figure 4.17. According to the figure, each outer shell is the following type of orbital: *(a)* p; *(b)* p; *(c)* d; *(d)* d; *(e)* f; *(f)* p; *(g)* p; *(h)* d; *(i)* s.

Figure 4.17 is color-coded to correspond to the types of orbitals being filled. Thus, since the S orbitals can only hold two electrons, only two columns of elements are colored blue. Since there are three p orbitals, which can hold up to two electrons each, six columns of elements are colored green. The five d orbitals can hold a total of ten electrons (two each), so ten columns of elements are orange. Finally, since the seven f orbitals can hold a pair of electrons each, fourteen columns of elements are violet in color. Each individual orbital (or electron) can be identified by its energy level (represented by a number that is the same as the period number) and its shape (s, p, d, or f). Thus we refer to the 3s or 5s orbital, or the 2p or 4f orbital.

The wave mechanics model of the atom has several advantages over Bohr's model. It shows, for example, that the transition and inner transition elements are offsets in the periodic table because different orbital sublevels are filling. It also predicts line spectra more accurately for most elements, and it allows scientists to determine atomic and molecular shapes more accurately—a quality with commercial applications, since many industrial products are manufactured through chemical reactions that depend on the molecules' specific shapes. Finally, as we will learn in future chapters, the wave mechanics model helps explain chemical properties and behaviors with more precision than the Bohr model.

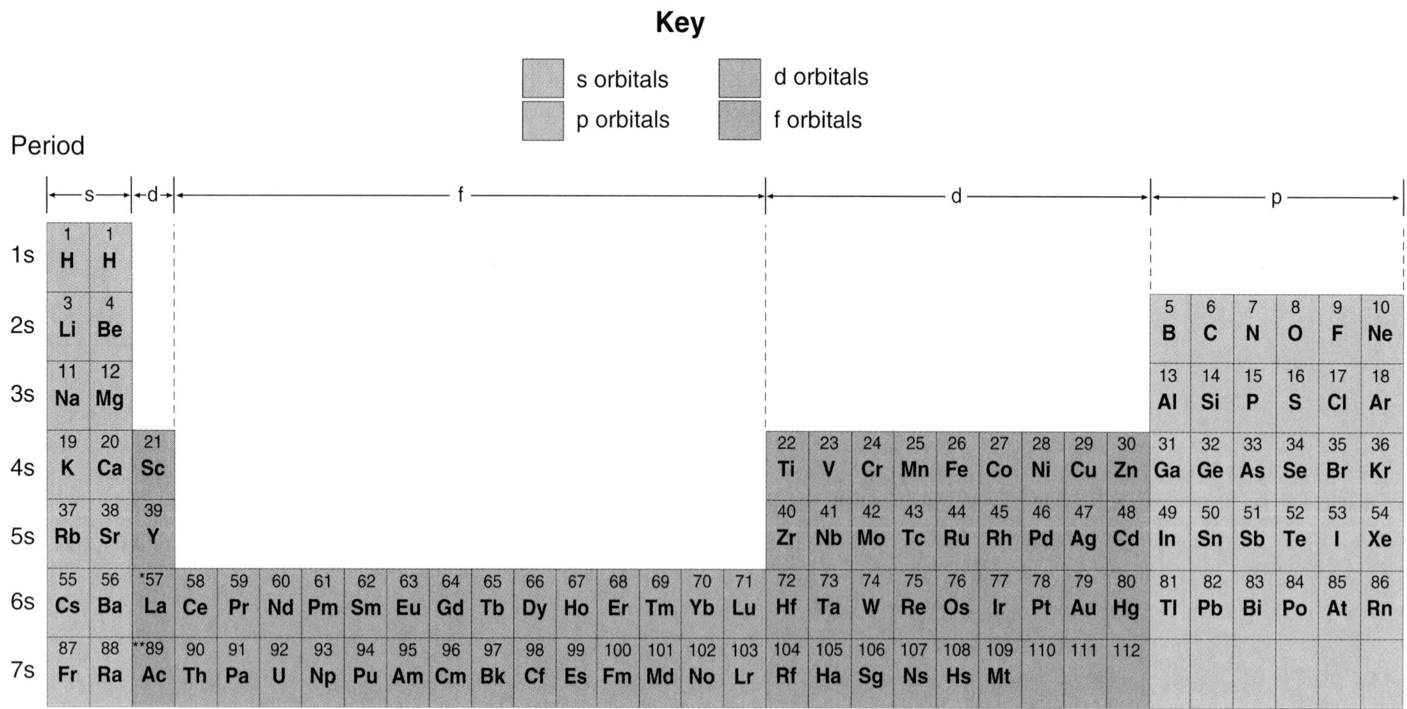

Figure 4.17 A periodic table showing the types of orbitals being filled for each group of elements.

Over the past century, our concept of the atom has changed from a solid, indivisible structure to a cloudlike form made up of tiny, electrically charged subatomic particles. The most accurate models we have picture a dense, infinitesimal nucleus containing nearly the entire atomic mass, surrounded by electrons moving rapidly as waves of energy through three-dimensional orbitals. How will another century of research affect our concept of atomic structure? Only time will tell.

SUMMARY

During the eighteenth and early nineteenth centuries, chemists worked to develop an organized theory that explained the chemical behaviors and properties of the elements. They began by organizing known elements into groups with similar qualities and behaviors. Doebereiner proposed grouping similar elements together in triads; Newlands organized them in octaves of seven elements. Although neither scheme worked, they laid the groundwork for other chemists to experiment with new groupings.

Dmitri Mendeleev was one of the first to propose a comprehensive periodic table, a chart that organized elements in horizontal rows by increasing atomic mass and in vertical rows by shared chemical properties. Today, the modern periodic table is based on Mendeleev's early efforts. Modern tables group the elements into three major categories: main group elements, transition elements, and inner transition elements. Each horizontal row in a modern table is a period, and each vertical row is a family. The elements may also be classified as metals, nonmetals, or noble gases.

Originally, Dalton conceived of the atom as a solid, indivisible unit. But chemists were unable to reconcile this conception with the periodic behavior of the elements. They achieved a breakthrough when several scientists discovered subatomic particles. The subatomic particles helped explain puzzling behaviors such as electrical conductivity and radioactivity.

Electrons, negatively charged subatomic particles, were discovered through Crookes's and Thomson's cathode ray experiments. Goldstein discovered positively charged particles in his canal ray studies; Wien later determined that the masses of these particles varied according to the nature of the gas remaining in the tube. Scientists reasoned that hydrogen, the lightest element, must exist as a single proton. This proton was discovered a few years later by Rutherford. Though chemists long believed that a third, neutral particle existed in the atom, it wasn't until 1932 that Chadwick proved the existence of the neutron.

The discovery of electrons, protons, and neutrons caused scientists to rethink the structure of the atom. Thomson first proposed a new structure; his plum pudding model argued that the atom consists of a sphere of positive charges (protons) studded with negatively charged electrons, like a pudding with raisins mixed in. Rutherford, Thomson's student, set out to prove this model through his alpha particle experiments; instead, to his surprise, he discovered that the model was incorrect. Rutherford proposed a new model, with the mass of the atom (the protons and neutrons) densely packed in a tiny nucleus, and the electrons moving through the rest of the space around it. Rutherford's model led others to determine how many protons and neutrons are present in the nucleus of an atom of each element, which in turn enabled Moseley to assign atomic numbers to the elements.

Rutherford's discovery of the nucleus was a major advance, but chemists still wanted to find out why chemical reactions and periodic behavior occur. Bohr studied hydrogen's emission spectrum and proposed a new theory. What if electrons containing specific amounts of energy orbited the nucleus of an atom? When the electrons absorbed energy, they would jump to higher energy levels, or to orbits farther from the nucleus; conversely, when they dropped to lower energy levels or closer orbits, they would emit energy as light. Bohr's spectra predictions for hydrogen, based on this theory, were uncannily accurate, and chemists quickly accepted his atomic model. It finally explained periodic behavior according to the number of electrons in an atom's outermost energy shell. It also implied that chemical reactivity involves the electrons in the outer shell—two concepts that were extremely significant in our understanding of the atom.

Still, Bohr's model was soon supplanted by a more complex but more accurate revision. Several scientists proposed theories that laid the foundation for Schrödinger's wave mechanical theory. This theory views electrons not as particles, but as waves of energy that move around the nucleus of the atom in three-dimensional orbitals. About the same time, the uncertainty principle lent support to the idea that electrons could not travel in fixed, predictable orbits, but instead moved as waves through a volume of space. According to the wave mechanics model, electrons move in orbitals designated s, p, d, or f, depending on their shape. This model, though difficult to visualize, explains the positions of the transition and inner transition elements in the periodic table, predicts line spectra more accurately, and allows chemists to determine atomic and molecular shapes more accurately. Finally, the wave mechanics model helps explain chemical properties and behaviors with more precision than the Bohr model.

KEY TERMS

alkali metals (IA) 87
alkaline earth metals (IIA) 87
alpha (α) particles 90
amplitude 99
anode 92
atomic mass 87
atomic number 97
beta (β) particles 90
Bohr's atom 102
canal ray 93
cathode 92
cathode ray tube 92
covalent compounds 89
electrons 92
emission spectrum 98
energy levels 100
families 87
gamma (γ) radiation 90
ground state 102
groups 87
halogens (VIIA) 87
inner transition elements 87
intensity 99
ion 89
ionic compounds 89
isotopes 98
main group elements 87
metals 89
metalloids 89
neutrons 94
noble gases (VIIIA) 87
nonmetals 89
nucleus 96
orbital 104
orbits 100
periodic table 85
periods 87
photons 101
protons 94
quantum (plural, quanta) 101
radioactivity 90
Rutherford atom 96
spectrum (plural, spectra) 98
subatomic particles 90
symbol 87
transition elements 87
triad 84
uncertainty principle 104
wave mechanics model of the atom 103

READINGS OF INTEREST

General

Meyerson, Bernard S. 1994. "High-speed silicon-germanium electronics." *Scientific American*, vol. 270, no. 3; pp 62–67. These new electronic devices are faster than the standard silicon versions.

Borra, Ermanno F. 1994. "Liquid mirrors." *Scientific American*, 270, no. 2: 76–81. Mercury can potentially make larger and better astronomical mirrors than glass.

Stix, Gary. 1994. "Material advantage." *Scientific American*, 270, no. 1: 146–147. Silicon-germanium chips provide advantages for computers and other electronics.

Wallich, Paul. 1993. "Bright future." *Scientific American*, vol. 269, no. 6; pp. 22, 24. Porous silicon emits light of different colors depending on the influence of an electrical field.

Waddell, Thomas G., and Rybolt, Thomas R. 1991. "The chemical adventures of Sherlock Holmes: A Christmas story." *J. Chem. Ed.*, 68, no. 12: 1023–1024. Elementary, my dear Watson; this is a simple mystery for students with just a little knowledge of chemistry.

Travis, John. 1991. "Riding the atomic waves." *Science News*, 140: 158–159. An introduction to quantum mechanics and atoms.

Boraiko, Allen A. 1984. "Lasers—A splendid light." *National Geographic*, 165, no. 3: 335–363. This article tells how lasers are made and used in medicine, entertainment, chemistry, and atomic fusion.

The Periodic Table and the Elements

Freemantle, Michael. "Heavy-ion research institute explores limits of periodic table." *Chemical & Engineering News*, March 13, 1995, pp. 35–40. Researchers expect to make element number 112 soon—followed by 113 and 114.

Freemantle, Michael. "Element 111 created by fusing nickel, bismuth." *Chemical & Engineering News*, January 2, 1995, p. 7. Although only three atoms of element 111 were created, scientists were able to determine its radioactive decay sequence.

Dagani, Ron. "Heavy element nomenclature." *Chemical & Engineering News*, November 21, 1994. The American Chemical Society disagrees with other international groups about the names for new elements. Who will prevail?

Carrado, Kathleen A. 1993. "Presenting the fun side of the periodic table." *J. Chem. Ed.*, vol. 70, no. 8; pp. 658–659. Shows a novel periodic chart in which the size of the boxes depict the abundance of the elements.

Bates, Alton J. 1992. "Silicon." *J. Chem. Ed.*, vol. 69, no. 2; p. 99. A quick glimpse at the second most abundant element, some of its uses, and how we get it from the sand on our beaches.

Nelson, P. G. 1991. "Important elements." *J. Chem. Ed.*, vol. 68, no. 9; pp. 732–737. Considers the usage, value, and availability of some of the elements.

Spindel, William, and Ishida, Takanobu. 1991. "Isotope separation." *J. Chem. Ed.*, vol. 68, no. 4; pp. 312–318. This article covers a topic not presented in this book; better students should find it readable.

Banks, Alton. 1991. "What's the use? Magnesium." *J. Chem. Ed.*, vol. 68, no. 3; p. 196. Short but interesting set of facts about element number 12.

Albertovich, Mikhael. 1991. "Dostoevsky and the periodic table (a chemical paradigm)." *Education in Chemistry*, January 1991, p. 32. Compares the use of the periodic table with learning to read in another language. Would you attempt it without first learning some of the basic vocabulary?

Young, Gordon. 1983. "Platinum, the miracle metal." *National Geographic,* vol. 164, no. 5; pp. 686–706. Aside from jewelry, coins, and costly ornaments, platinum finds use in implants, anticancer drugs, and as a catalyst.

White, Peter T. 1974. "Gold, the eternal treasure." *National Geographic,* vol. 145, no. 1; pp. 1–51. This comprehensive article describes gold mining and the myriad of uses for this precious metal.

Atomic Structure

Walton, Harold F. 1992. "The Curie-Becquerel story." *J. Chem. Ed.,* vol. 69, no. 1; pp. 10–15. An interesting historical account of the discoveries this pair made concerning nuclear energy, plus several others. It includes excerpts about Becquerel's own photographic observations leading to the discovery of radioactivity.

Spindel, William, and Ishida, Takanobu. 1991. "Isotope separation." *J. Chem. Ed.,* vol. 68, no. 4; pp. 312–318. This article covers a topic not presented in this book. It is recommended for all interested students.

Boslough, John. 1985. "Worlds within the atom." *National Geographic,* vol. 167, no. 5; pp. 634–663. We have come a long way from the idea of an indivisible atom. This highly readable article will continue your journey past quarks, cyclotrons, tevatrons, and even the big bang theory. You will even see how this can relate to your everyday life through medicine.

Curie, Marie. 1961. "*Radioactive substances.*" New York: Philosophical Library. This is a translation of Marie Curie's presentation to the Faculty of Sciences in Paris; a first-person account of her discoveries.

Curie, Eve. 1939. "*Madam Curie.*" New York: Doubleday, Doran. This is a translation of the biography written by Eve Curie about her mother in 1937, only three years after Madame Curie's death. It is probably still the best biography available on this legendary woman.

PROBLEMS AND QUESTIONS

1. Define the following terms as they are used in this chapter:
- **a.** Canal rays
- **b.** Cathode rays
- **c.** Octave
- **d.** Radioactivity
- **e.** Triad

2. What contributions did each of the following people make to the development of the periodic table?
- **a.** Johann Doebereiner
- **b.** Dmitri Mendeleev
- **c.** Lothar Meyer
- **d.** John Newlands

3. Where are the metals, nonmetals, and noble gases located in the periodic table?

4. How do metals and nonmetals differ?

5. Name the families and the periods in the periodic table.

6. Where are the main group elements located in the periodic table?

7. Where are the alkali metals located in the periodic table?

8. Where are the alkaline earth metals located in the periodic table?

9. Where are the halogens located in the periodic table?

10. Where is the boron (B) family located in the periodic table?

11. Where is the carbon (C) family located in the periodic table?

12. Where is the nitrogen (N) family located in the periodic table?

13. Where is the oxygen (O) family located in the periodic table?

14. What is the main chemical characteristic of the noble gases?

15. Which of the following sets are part of a chemical family?
- **a.** Be, Mg, Ba
- **b.** B, Si, As
- **c.** C, N, O
- **d.** Cl, Br, I
- **e.** Na, Ca, Ra
- **f.** P, As, Sb
- **g.** Li, K, Fr

16. Where are the transition elements and the inner transition elements located in the periodic table?

17. How did the contributions of each of the following people affect Dalton's atomic theory and the concept of the indivisible atom?
- **a.** Svante Arrhenius
- **b.** Antoine Henri Becquerel
- **c.** James Chadwick
- **d.** Marie Curie
- **e.** Eugen Goldstein
- **f.** Robert Millikan
- **g.** Ernest Rutherford
- **h.** J. J. Thomson
- **i.** Wilhelm Wien

18. What, if anything, did electrical behavior and radioactivity imply regarding Dalton's atomic theory?

19. How did Thomson determine the mass-to-charge ratio for the electron?

20. How did Millikan determine the charge of the electron?

21. What were the main features of Thomson's model of the atom?

22. What were the main features of the Rutherford atom, and how does this model differ from the Thomson atom?

23. What is the relative size of the atomic nucleus compared to the volume of the atom?

24. What relationship exists between the atomic number and the number of electrons, protons, and neutrons?

25. What relationship exists between atomic number and atomic mass?

26. What are the main characteristics of each subatomic particle—electron, proton, and neutron?

27. What are isotopes? Give some common examples.

28. What is the difference between an anode, and a cathode?

29. What contributions did each of the following people make to our modern understanding of atomic structure?
   a. Niels Bohr
   b. Louis de Broglie
   c. Albert Einstein
   d. Werner Heisenberg
   e. Erwin Schrödinger

30. What role did emission spectroscopy play in determining the structure of the atom?

31. What do the following terms mean in relation to waves and wave theory?
   a. Amplitude
   b. Frequency
   c. Wavelength

32. Which has more energy, a photon of blue light or a photon of red light?

33. Which has more energy, a photon emitted by a microwave or a photon emitted by an ultraviolet lamp?

34. Describe the Bohr model of the atom. How does this differ from more recent views?

35. What relationship exists between spectral lines and the Bohr orbits?

36. Looking only at a periodic table, tell how many electrons are in the outermost orbit of each of the following atoms:
   a. Si   e. Ar
   b. Mg   f. K
   c. Al   g. P
   d. F   h. O

37. What shapes do s- and p-orbitals take?

38. What are the differences between orbits and orbitals?

39. What kinds of outermost orbitals are being filled in each of the main group, transition, and inner transition elements in the periodic table?

Critical Thinking Problems

1. Microwaves, widely used in our society, are a form of radiation. Why might microwaves be considered safe to use in the home, while gamma rays are considered unsafe?

2. At one time, shoe stores used x-ray devices that enabled people to see their feet (and bones) inside their shoes. This device made shoe fitting easier, but was it safe to use? Why or why not?

3. Why did Rutherford restrict his α-particle studies to solids (gold) rather than liquids or gases?

HINT Consider the differences in each state of matter's properties, as discussed in chapter 2.

4. The periodic table is considered one of the major achievements in chemistry. What other examples of periodic behavior can you think of in other areas of life? Can any be arranged in a similar manner?

HINT think about weather, time intervals, clothing sizes, and so on.

5. In this chapter, you learned that Dalton's atomic theory contained several errors, yet scientists still apply almost all its features to modern chemistry. How can you explain this?

6. What might the consequences be to the study of chemistry if scientists found definite proof that electrons, protons, and/or neutrons could be subdivided into smaller entities?

7. What do you think might have happened to chemical theory if spectroscopy had never been developed?

8. Frequently the news media report complaints about the dangers of electromagnetic (EM) radiation. However, the actual danger relates directly to the amount of energy in the radiation, with the total amount of radiation normally less directly dangerous. Rank the following forms of electromagnetic radiation from least harmful to most dangerous: IR, UV, microwave, γ-rays, AM radiowaves, X rays, visible light.

HINT See figure 3.17 for data on energy levels.

9. How have metalloids helped us in our daily lives? For further background, read the following articles: (1) Bernard S. Meyerson, High-Speed Silicon-Germanium Electronics, *Scientific American* 270:3 (March 1994), pp. 62–67, and (2) Gary Stix, Material Advantage, *Scientific American* 270:1 (January 1994), pp. 146–147.

10. Science does enter into the courtroom, as shown in the O. J. Simpson trial. Should there be limits on how this is done and what is acceptable? Besides DNA profiling and other examples of forensic chemistry, there are expert witnesses and other ways science gets involved in courtroom decisions. Read the following pair of articles on these matters and form your own opinion: (1) Pamela Zurer, *Chemical & Emgineering News*, Oct. 10, 1994, pp. 9–15, "DNA Profiling Fast Becoming Accepted Tool for Identification," and (2) Francisco J. Ayala and Bert Black, *American Scientist*, May/June 1993, pp. 230–239, "Science and the Courts."

CHAPTER 5

ENERGY, THE NUCLEUS, AND NUCLEAR CHEMISTRY

OUTLINE

Potential Versus Kinetic Energy
Forms of Energy
The Basics of Nuclear Energy
Nuclear Applications

BOX 5.1 Jack and Jill: A Rhyme of Potential and Kinetic Energy
BOX 5.2 "Lightning Love"—Chemiluminescence
BOX 5.3 Is EMF Radiation Dangerous?
BOX 5.4 Radiometric Dating
BOX 5.5 Risk Perception Versus Scientific Data
BOX 5.6 Food that Glows in the Dark?

KEY IDEAS

1. The Difference Between Kinetic and Potential Energy
2. How Different Energy Forms—Mechanical, Thermal, Electrical, Light, Chemical, and Nuclear—Convert Into Each Other
3. The Basics of Radioactivity and Nuclear Energy
4. The Nature of a Radioactive Decay Series
5. The Meaning and Uses of a Nuclear Half-Life
6. Some Issues Surrounding the Safety of Nuclear Applications
7. The Difference Between Nuclear Fission and Nuclear Fusion
8. Some Examples of How Nuclear Science is Used in Medicine
9. How Food is Irradiated and Whether Irradiation is Safe
10. The Basic Problems With Nuclear Waste Disposal
11. The Nature of Radon and Whether It Poses a Safety Problem

"Ignorance is never better than knowledge."
ENRICO FERMI (1901–1954)

Lightning in Arizona.

In chapter 3, we briefly examined the subject of energy and the way the development of chemistry paralleled the discovery of more powerful energy sources. For the same reason, scientists' understanding of the forces that hold the atom together paralleled learning about nuclear energy. To appreciate nuclear energy, you must first understand the difference between potential and kinetic energy.

POTENTIAL VERSUS KINETIC ENERGY

Potential energy is the energy an object holds because of its position or internal features; it is stored energy, such as chemical energy, that may be released and converted into kinetic energy to do work. **Kinetic energy,** by contrast, is the energy of motion. Figure 5.1 illustrates the relationship between these two types of energy. In the figure, ball 1 sits at the top of a steep hill, ball 2 is rolling down the precipice, and ball 3 rests at the bottom. Ball 1 has greater potential (stored) energy than ball 3; but since neither ball 1 or ball 3 is moving, neither exhibits any kinetic energy. As ball 2 falls, it possesses both kinetic and potential energy; its potential energy converts into kinetic energy as it moves downhill. The total amount of energy does not change; that is, the sum of the potential and kinetic energy is constant. Ball 1 contains much potential energy and no kinetic energy; ball 2 possesses intermediate amounts of each; and ball 3 possesses little potential energy and no kinetic energy, but represents the end result of a large release of kinetic energy.

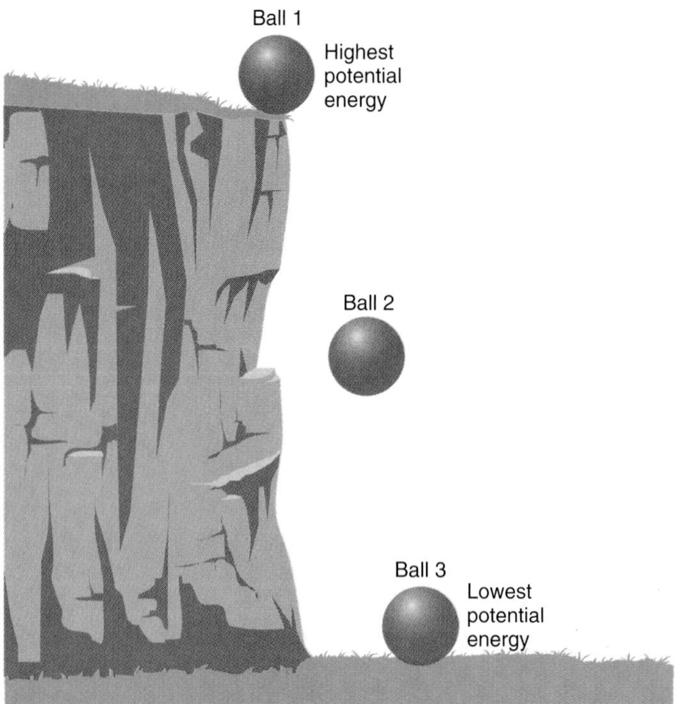

Figure 5.1 Potential versus kinetic energy.

The reverse is also possible if enough kinetic energy is applied to carry the ball back to the top of the hill. In such a transformation, the flow of energy would reverse; as kinetic energy moved the ball uphill, the kinetic energy available would diminish and the potential energy in the ball would increase. Still, the total amount of energy would remain the same. The **Law of Conservation of Energy** (also known as the **First Law of Thermodynamics**) states that *energy cannot be created or destroyed, but it can change form.* Box 5.1 discusses this idea in more detail.

In actuality, energy conversion and conservation do not work as simply as figure 5.1 indicates. For example, the figure does not consider the effects of friction, the resistance between two objects rubbing together. Energy is required to move against this resistance, which means that some of the kinetic energy in ball 2 dissipates as heat due to friction. This energy is not "lost"; it is simply unavailable as kinetic energy to help move the ball. The **Second Law of Thermodynamics** describes this by stating that *whenever energy converts to a different form, some of it always converts to heat energy that is then unavailable to do work.*

PRACTICE PROBLEM 5.1

Sledding and skiing are popular winter activities in the snowbelt. Imagine you have ridden a ski lift up a modest peak and now stand poised at the top of a 1,000-meter downhill run. You dig the poles into the crunchy snow, kick off, and away you go, slaloming around occasional trees on the pathway. Draw this event and note how energy converts from kinetic to potential, or the reverse. What would happen, from an energy standpoint, if you crashed into one of the trees?

ANSWER Your drawing should show that the lift ride changes kinetic energy (the mechanical energy of the ski lift) into potential energy. As you whiz down the slope, you reconvert this potential energy back into kinetic energy. Watch out for that tree; it could convert your kinetic energy into a bunch of broken bones!

FORMS OF ENERGY

Some forms of energy conversion are easy to use. Sunlight converts into heat on a sunny day, and the chemical energy from a flashlight battery changes into a beam of light. Many changes are less obvious, such as sunlight converting into chemical energy as plants grow. But the conversion of energy into other forms is a well-established fact. Some of these forms include mechanical, thermal, electrical, light, chemical, and nuclear energy.

BOX 5.1

JACK AND JILL: A RHYME OF POTENTIAL AND KINETIC ENERGY

Everyone knows the story of Jack and Jill who went up the hill, but did you know that this is also a tale of potential and kinetic energy? Look at scene *a* in figure 1. Jack and Jill both possess their lowest level of potential energy due to position as they begin their climb up the hill to the well. On the way (*b*), they convert kinetic energy into potential energy as they take each step up the slope. But where did this kinetic energy come from? Both Jack and Jill have a reservoir of chemical energy stored inside them; as they walk uphill, some of this chemical potential energy converts into kinetic energy.

It may seem that once Jack and Jill use this kinetic energy to climb the hill, the energy vanishes. However, the law of conservation of energy tells us this can't be true. Instead, both Jack and Jill possess a greater amount of potential energy due to position when they reach the top (*c*). Jack and Jill converted part of their chemical potential energy into kinetic energy of motion, which, in turn, reconverted into potential energy of position as they climbed. This positional potential energy reaches its peak when the duo arrive at the top of the hill and draw the pail of water. The total amount of energy never changed—it just took different forms. Chemical potential energy changed to kinetic energy, which changed again to positional potential energy.

Figure 1 (*a*) Before the uphill climb, Jack and Jill are at their low point in potential energy. (*b*) They exchange some kinetic energy for potential energy as they climb the hill. (*c*) When Jack and Jill reach the top of the hill, they are at their highest level of positional potential energy. (*d*) Jack has fallen to the bottom of the hill, and Jill is on her way down. As they fall, they convert their potential energy back to kinetic energy, expending the kinetic energy in movement. They once again have very little potential energy at the bottom.

Although Jack and Jill demonstrate the conservation of energy, our explanation is admittedly simplified. Was any energy "lost" from the system? Theoretically, no, but in the real world the answer is almost always yes. A little energy probably dispersed into the atmosphere in the form of heat (kinetic energy) and did not return to either Jack or Jill as they climbed. (The pair were understandably perspiring a little from the exertion of the climb.) Likewise, some energy probably dissipated due to friction as their shoes scuffed along the ground. Yet, neither type of energy—heat from perspiration and heat from friction—was truly "lost." It still exists, but is no longer in a useful form. All systems tend to "lose" some energy to heat.

A dramatic change occurs in scene *d*. Jack lies at the bottom of the hill (with a broken crown) and poor Jill is tumbling down after him, head over heels. As she falls, the potential energy she possessed at the top of the slope converts to kinetic energy, and she sometimes slides along the ground and loses energy due to friction. Jack's positional potential energy also decreased very rapidly, with no apparent increase in his chemical potential energy. Where did this "lost" energy go?

Much of it powered Jack's and Jill's flailing movements as they slid and tumbled down the hill. Some of it crushed small plants and insects or sent pebbles skipping downhill as Jack and Jill rolled by. Some dissipated as heat due to their movement and due to friction. All in all, this whole episode was just a frustrating loss of chemical potential energy without even the successful acquisition of a pail of water; the water also spilled as they tumbled downhill.

Mechanical Energy

Mechanical energy, energy used in a machine or mechanism, is kinetic energy. Mechanical energy only works over short distances because it dissipates easily, usually as heat. When mechanical energy is used over long distances, nearly all of the energy converts to heat. We measure efficiency in terms of energy conversion; the more energy available to do work, the greater the efficiency. Mechanical devices usually operate at less than 30 percent efficiency, meaning that 70 percent or more of the energy that operates them is wasted. Some motors are less than 10 percent efficient, which means over 90 percent of the energy is wasted.

Thermal Energy

Thermal energy is the energy of heat, and **thermodynamics** is the study of heat flow (the motion of heat). Heat results when atoms or molecules move, and is thus kinetic energy. Thermodynamics helps chemists improve machine efficiency and predict whether chemical reactions can occur. Machines become more efficient as the difference between the operating and surrounding temperatures increases; the additional thermal energy from the heat can be converted back into mechanical energy to do work.

Modern automobiles operate at warmer temperatures than the older models to take advantage of this fact. Although it is impossible to convert heat completely into mechanical energy because of friction between moving parts, improvements are possible. For example, if a car engine operates at 100°C with an exhaust temperature of 40°C, the maximum efficiency is only 16 percent and the real efficiency is even lower due to friction. By raising the operating temperature to 150°C—50°C above the boiling point of water—the maximum efficiency rises to about 26 percent. However, at this temperature, if the radiator coolant boils away, the engine could be destroyed. Modern cars use antifreeze in both summer and winter to avoid this problem. Antifreeze prevents the coolant from freezing in winter or boiling in summer and rupturing the radiator.

Heat cannot be stored for long periods, nor can it be transmitted over long distances without large losses in available energy. Consider the temperature of coffee you put into a thermos yesterday. As it cooled, heat flowed from the thermos to the cooler surroundings of the room. This is a consequence of the Second Law of Thermodynamics, which, as you recall, states that all systems tend to lose heat energy. Good insulation slows the heat loss but cannot prevent it—there is no "free lunch" in any energy transaction.

Electrical Energy

Electrical energy involves the motion of electrons and is thus another form of kinetic energy. Unlike mechanical or thermal energy, however, electricity may be transmitted over long distances through power lines. Energy losses, chiefly in the form of heat and other types of radiation, still occur, but the losses are smaller than with mechanical or thermal energy. Electrical energy may be converted into other energy forms, such as thermal or mechanical energy, or used directly, as in radio, television, computers, and so on. Even in direct usage, a significant amount of energy dissipates as heat that can cause electrical component breakdown. But sometimes the conversion of electrical energy into heat is the goal—stoves, toasters, and radiant heaters are all designed to produce heat.

Electrical energy is routinely produced from chemical energy (through coal, oil, or wood, for example), mechanical energy (from moving water or wind), and nuclear energy. Electrical energy cannot be effectively stored (the electricity in a battery is actually stored chemical energy). Electricity thus requires an energy conversion on demand.

PRACTICE PROBLEM 5.2

Back to the ski slopes! Everyone knows you can ski better on snow than on a dirt path. Why? Isn't the potential energy due to position the same at the top of our 1,000-meter run in the summer?

ANSWER This is partly a matter of efficiency. The smooth surface of snow imparts very little friction to the surface of the skis, whereas dirt imparts a lot of friction. This friction changes some of the kinetic energy into wasted heat and slows the ride tremendously. It gets worse as you progress, because friction increases with the surface temperature of the skis. (Snow also helps reduce friction by cooling the skis.) You could still ski in the summer if you used a surface that did not result in as much wasted heat. Plastic ski runs cause less friction against ski surfaces and thus allow warm-weather skiing.

PRACTICE PROBLEM 5.3

Let's take one last trip to the ski slopes. This time, we'll look into the question of getting to the mountain in the first place, and how the Second Law of Thermodynamics affects your trip. Suppose you are driving the same car to the slope in summer, at about 85°F (29°C), and in winter, at about 0°F (−18°C). Assume that the driving conditions are the same for each trip, and that you do not run the car heater or air conditioner on either journey. On which trip would your car engine operate more efficiently?

ANSWER Winter wins hands down on the engine efficiency! The temperature is many degrees lower in winter than in summer. This means the heat produced by the engine can move away more effectively, reducing the engine's exhaust temperature and increasing its efficiency. It's nice to know winter driving offers some benefits.

Light Energy

Yet another use for electrical energy is to power incandescent, fluorescent, and neon lightbulbs. In all three cases, the electrical energy "excites" atoms, causing electrons to jump to higher energy orbitals. When the electrons fall back, light is emitted.

In 1879, Thomas A. Edison and Sir Joseph Swann simultaneously developed incandescent bulbs. These bulbs produce light by heating tungsten wires to high temperatures, converting electrical energy to thermal energy to light energy. About 80 percent of the energy is wasted as dissipated heat; fluorescent lamps are much more efficient; using more energy to produce light and less to produce heat.

In fluorescent lamps, the electricity interacts with a trace of mercury vapor, which then emits rays of ultraviolet light. This UV radiation stimulates the atoms of a solid fluorescent coating inside the lamp to glow with visible white light. A. H. Becquerel tried to invent fluorescent lighting in 1867, but Dr. Arthur Compton and his coworkers at General Electric produced the first effective fluorescent lamps in 1934.

Georges Claude invented neon lamps, widely used in advertising, in 1910. In neon lamps, electricity flows through a tube containing traces of neon. The electrical energy excites the gas molecules, which then emit a red-orange light. Other gases produce different colors (figure 5.2).

In June 1992, American Electric Power of Columbus, Ohio and Intersource Technologies of California announced a new type of lighting device. This device, called the E-lamp, generates a radio signal from a magnetic coil that interacts with the same gases used in fluorescent lamps to produce a plasma (an ionized gas). This plasma, also like the gases in a fluorescent lamp, causes a phosphor coating to glow on the inside of the lamp. E-lamps are claimed to be very energy efficient and long-lived. Both features are important, since the initial cost is expected to run between ten and twenty dollars each when they become available for home use.

Light energy (which is kinetic because it is moving photons) may be obtained when certain substances are oxidized—for example, wood. This chemical reaction produces heat or thermal energy, which in turn "excites" other molecules, causing them to emit light. Some chemical reactions emit light without producing much heat; this is called "cold light." A common example of cold light is the light fireflies (lightning bugs) emit. The chemical term for this process is *chemiluminescence* (box 5.2).

Chemical Energy

Many chemical sources of energy exist, although they are not always recognized as such. Several common fuels—wood, coal, oil, and petroleum—contain energy stored in chemical bonds. Our personal energy supply comes from food, which also stores chemical energy, and our bodies store energy as chemicals. In fact, chemical energy is the most commonly used energy source on our planet, and is also the most easily stored and conveniently transported energy form. The potential energy contained in chemical bonds can convert through a chemical reaction to mechanical, thermal, electrical, or light energy. Box 5.3 focuses on yet another type of energy—electromagnetic field radiation.

Figure 5.2 Neon lights brighten the evening sky in Las Vegas.

PRACTICE PROBLEM 5.4

What type of energy do the following materials all contain: wood, straw, gasoline, bituminous coal, peat, anthracite coal, flashlight batteries, heating fuel, kerosene? What types of energy do they produce?

ANSWER All these materials contain chemical energy. Except for flashlight batteries, all these commodities may be burned to produce thermal energy to heat homes, or to produce mechanical or electrical energy to operate machinery. Flashlight batteries convert chemical to electrical energy without burning.

BOX 5.2

"LIGHTNING LOVE"—CHEMILUMINESCENCE

Fireflies, or lightning bugs, could be called "nature's neon signs." The transparent tail of a firefly contains the chemicals luciferin and luciferase. In the presence of oxygen, these chemicals emit a glowing light in the insect's tail. Interestingly, firefly lights do not produce any appreciable amount of heat; therefore, they are an example of cold light (figure 1). The male fireflies respond to the females' light flashes, and this cold light, or chemiluminescence, thus serves to attract mates and maintain the firefly population.

In chemiluminescence, the energy from a chemical reaction is converted directly into light, rather than the more usual heat. Several other combinations of chemicals will produce chemiluminescence, but this light production method is comparatively rare. A few cold light applications do exist, however. You can purchase sticks that contain a luciferin-luciferase combination and use it for emergency lighting; it will glow for several hours. You can also see hundreds of happy children roaming around amusement parks or fairs wearing bracelets or necklaces with a luciferin-luciferase combination enclosed in a transparent flexible plastic tube. These cold light toys glow almost as brightly as the faces of the children wearing them.

Figure 1 Male and female fireflies find each other using their chemiluminescent lanterns.

Nuclear Energy

Nuclear energy marked its beginnings in 1896 with Becquerel's and the Curies' explorations of radioactivity, energy that radiated from certain atoms (chapter 4). In 1905, Einstein claimed that energy and matter are interconvertible, proposing his famous equation, $E = mc^2$. (E = energy, m = matter, and c = velocity of light.) Prior to Einstein's proposal, energy and matter were considered totally different entities. Einstein's recognition that matter could convert into energy ultimately led to the development of atomic or nuclear energy. Nuclear reactions, rather than converting one form of energy into another, convert small amounts of matter into huge amounts of energy.

Initially, the concept of converting mass to energy may seem a violation of the Law of Conservation of Energy and the companion Law of Conservation of Matter; however, what it really shows is that neither law was complete in itself. Scientists have therefore combined these laws into a **Law of Conservation of Matter and Energy** that states that *the sum total of matter and energy remains constant, but one form can convert into the other.* Scientists have not yet converted energy into matter in the laboratory, but this process may have occurred during the formation of the universe.

The most important use of nuclear energy in our society is its conversion into electrical energy, although the medical applications, are also significant. Energy production via nuclear energy is restricted to certain locations because the byproduct radiation is potentially harmful. Moreover, nuclear power is not easily transferred from one place to another unless it is converted into electrical power. Nuclear-powered vehicles (for example, submarines) do exist, but are not yet practical enough for most uses. Most of the remainder of this chapter considers nuclear energy and its uses.

SOURCE: © 1994; reprinted courtesy of Bunny Hoest and *Parade Magazine*.

BOX 5.3

IS EMF RADIATION DANGEROUS?

Electromagnetic fields (EMFs) arise from the nonionizing portion of the electromagnetic spectrum. Ionizing radiation, which can cause atoms to gain or lose electrons, includes cosmic and γ-rays, α- and β-particles, and even ultraviolet (UV) radiation. EMFs are lower in frequency, higher in wavelength, and less energetic than visible light; they arise from the microwave and radiowave regions of the electromagnetic spectrum.

In recent years, EMFs have sparked a controversy: are electromagnetic fields harmful to humans? Whenever electricity is in motion, an electromagnetic field is created. Strong EMFs are produced by some household devices, such as electric razors and electric blankets, and by power lines and radio broadcast antennas. In 1966, Soviet scientists claimed EMFs cause headaches and fatigue and reduce male sexual potency. Although U.S. studies failed to confirm the Soviet results, other concerns arose. Cancer clusters, increased numbers of cancer cases clustered in a relatively small locality, raised further questions about the safety of EMFs.

The first possible EMF cancer cluster was discovered when epidemiologists (scientists who study the occurrence and distribution of diseases) noticed that four people in Guilford, Connecticut who lived along a substation power line were diagnosed with brain cancer between 1968 and 1988. Other clusters, involving similar numbers of people, were reported. Since then, people have claimed that cellular phones, video terminals, televisions, microwave ovens, hair dryers, electric razors, and even radiant heating systems can emit harmful EMFs and, perhaps, cause cancer in humans (figure 1). While these devices do emit EMFs, none of the disease-causing claims have been substantiated to date. In fact, a court case filed in early 1993, claiming cellular phones cause cancer, was thrown out of court in May 1995 for lack of evidence. Other studies examining a possible correlation between EMFs and breast cancer have yielded mixed results.

So far, scientists have been unable to determine exactly how these low-energy fields might cause harm. Laboratory studies are inconclusive, but the most logical proposed mechanism seems to be chemical in nature, involving the human body's reduction in melatonin production under the influence of magnetic fields. Melatonin is a chemical released mainly at night by the pineal gland, located in the lower part of the brain. Many animal studies have shown that melatonin can suppress the growth of breast cancer cells, and possibly other forms of cancer; in fact, melatonin is sometimes used to treat cancer. If EMFs do suppress human melatonin production, perhaps this leaves the body more susceptible to cancer cell growth.

On the other hand, EMFs have some definite positive effects. They promote the healing of bone fractures and can help treat osteonecrosis, a serious condition that, left untreated, almost always results in irreversible deterioration of the hip.

The bottom line is that eliminating all EMF sources is not feasible in our technological society. However, EMF exposure can be minimized. New power lines can be routed to avoid schools and residences. Some electrical devices—for example, electric blankets—can be redesigned to reduce EMF emissions (blankets produced after 1989 are very low in EMFs). Other devices may benefit from shielding if suitable materials exist. And simply moving farther back from your television set or microwave oven, or reducing the amount of time you spend on a cellular phone, reduces exposure to EMFs.

EMF-emitting devices are an integral part of our society and are undoubtedly here to stay. In May 1995, the 45,000-member American Physical Society took a strong stand on the EMF question, claiming they could find no evidence that power line EMFs cause cancer. In its published statement, the Society stated, "More serious environmental problems are neglected for lack of funding and public attention. The burden of cost placed on the American public [in confining EMF from power lines and devices] is incommensurate with the risk, if any." Although the likelihood of danger seems minimal, the debate will probably rage on for many years to come.

Figure 1 Some common devices that emit EMFs.

THE BASICS OF NUCLEAR ENERGY

Nuclear energy is one of the most important energy sources in the modern world. Indeed, it is nuclear energy that powers the sun and that keeps the Earth's interior molten.

Nuclear energy comes from the forces that hold an atom's nucleus together. Because a nuclear reaction splits the nucleus, the protons and neutrons are released, and the atom transmutes into another kind of atom. (Transmutation is the conversion of one element into another.) Rutherford was the first to transmute an atom; in 1919, he bombarded nitrogen atoms with radioactive α-particles and produced oxygen atoms with a leftover proton. Such a change in the nucleus of an atom constitutes a nuclear reaction.

Although transmutation was the alchemist's dream, Rutherford's feat was vastly different from the accomplishment alchemists envisioned. It released a subatomic particle—something early scientists had not conceived of—and radioactive byproducts; and, of course, it resulted in oxygen, not gold.

However, radioactive materials contribute in unique ways that can be as valuable as gold. Nuclear medicine, for example, has saved many lives—far more than the total deaths caused by nuclear energy, including both atomic and hydrogen bombs and inadvertent releases of radioactivity. But no other energy source seems to arouse so much emotion when people weigh its benefits and risks. In the next few sections, we will explore nuclear energy and its applications, briefly considering its advantages and disadvantages.

Isotopes

After Rutherford, Moseley, and Soddy completed their pioneering work on the atom (chapter 4), scientists realized that two numbers characterize each atom: the atomic mass (a) and the atomic number (z). The atomic number represents the number of protons in the atom and is always constant for an element; if this number changes, the element transmutes into another element. The atomic number is also the number of electrons in a neutral atom (that is, an atom that is not an ion).

However, the atomic mass is not necessarily constant for all atoms of a particular element. Isotopes are atoms of the same element that have different atomic masses (see chapter 4). Scientists found it convenient to symbolize different isotopes by using the atomic number, z; the atomic mass, a; and the symbol of the element, Q, in the following configuration:

$$^{a}_{z}Q$$

Thus, a scientist could indicate a particular uranium isotope atom with atomic number 92 and atomic mass of 238 amu (atomic mass units) like this:

$$^{238}_{92}U$$

In a similar manner, the most common isotopes for carbon (C), oxygen (O), potassium (K), and lead (Pb) would be written like this:

$$^{12}_{6}C \quad ^{16}_{8}O \quad ^{39}_{19}K \quad ^{208}_{82}Pb$$

These notations tell us that the most common carbon atoms have 6 protons (and 6 electrons, since the atom is neutral) and a total mass of 12 amu. Since electrons have negligible mass, all the additional mass must come from neutrons; this particular carbon therefore must contain 6 neutrons ($12 - 6 = 6$). But not all isotopes have the same numbers of neutrons. Consider, for example, the following carbon isotopes that contain exactly 6 protons (or they could not be carbon) but 6, 7, or 8 neutrons for isotopes with masses of 12, 13, or 14 amu, respectively:

$$^{12}_{6}C \quad ^{13}_{6}C \quad ^{14}_{6}C$$

PRACTICE PROBLEM 5.5

How many protons, electrons, and neutrons does each of the following isotopes contain?

$$^{51}_{23}V \quad ^{63}_{29}Cu \quad ^{107}_{47}Ag \quad ^{202}_{80}Hg$$

ANSWER Determining the number of protons is easy; the subscript reveals this in every case. Thus, vanadium has 23 protons, copper 29, silver 47, and mercury 80. Since these are neutral atoms, an equal number of electrons and protons are present in each isotope.

Determining the number of neutrons is also easy if you remember that a (the atomic mass) $= n + p$, where $n =$ the number of neutrons and $p =$ the number of protons. That means if you subtract the atomic number (or number of protons) from the atomic mass, the remaining number will equal the number of neutrons. Thus, vanadium has 28 neutrons ($51 - 23 = 28$); copper has 34; silver has 60; and mercury has 122.

But be careful—this procedure does not work with *mixtures* of isotopes. For example, ordinary chlorine is a mixture of 75.4 percent $^{35}_{17}Cl$ and 24.6 percent $^{37}_{17}Cl$, with an average atomic mass of 35.45. Subtracting 17 from 35.45 would result in 18.45 neutrons in a typical chlorine atom, but that is nonsense. Atoms do not contain fractions of neutrons.

PRACTICE PROBLEM 5.6

Now let's try the problem in 5.5 in reverse. Write the symbols with the appropriate superscript/subscript notations for each of the following isotopes, using the numbers of protons (p) and neutrons (n) shown.

Aluminum, 13p, 14n

Bromine, 35p, 44n

Barium, 56p, 82n

Thorium, 90p, 142n

ANSWER Here, all you need to do is to add together the *p* and *n* values to determine the atomic mass. Write this as the superscript and the number of protons as the subscript. Your answers should look like this:

$$^{27}_{13}Al \quad ^{79}_{35}Br \quad ^{138}_{56}Ba \quad ^{232}_{90}Th$$

Radioactivity

Radioactivity is the spontaneous decay of one atomic nucleus into another with the emission of α-, β-, or γ-radiation. Marie and Pierre Curie coined the term in 1898 after they concluded that rays emanating from uranium were characteristic of the element and were not related to prior chemical treatments or to some passing physical state. That same year, Marie Curie and G. C. Schmidt independently discovered that the element thorium also emitted similar rays. Subsequent studies by other scientists showed that the α-ray was actually a doubly charged helium ion ($^4_2He^{2+}$), or a helium nucleus, and that the β-ray was an electron ($_{-1}^{\ 0}e$). Rutherford showed, in 1900, that γ-rays were true electromagnetic radiation.

Since 1896, chemists have discovered dozens of radioactive elements in addition to those Becquerel and the Curies identified. Many of these radioactive elements did not exist in nature, at least not on this planet. The first artificially created radioactive isotope was $^{13}_{7}N$ (a short-lived isotope) that Irène and Frédéric Joliot-Curie (the daughter and son-in-law of Marie and Pierre) prepared in 1934. They did so by bombarding $^{10}_{5}B$ with α-particles, producing $^{13}_{7}N$ and a leftover neutron:

$$^{10}_{5}B + ^{4}_{2}He^{++} \longrightarrow ^{13}_{7}N^{++} + ^{1}_{0}n$$

Notice that the sums of the atomic numbers and the atomic mass numbers balance on each side of the new equation. The total numbers stay the same (in keeping with the Law of Conservation of Matter and Energy), but the subatomic particles are distributed differently. And of course, different numbers of protons mean different elements.

The first synthetic element was technetium, element 43, which filled one of several holes in Mendeleev's periodic chart. Most holes ultimately filled with stable, naturally occurring elements as scientists discovered them, but not the gaps eventually filled by elements 43 and 85. Emilio Segrè (Italian, 1905–1989) created technetium in 1937 and astatine (element 85) in 1940. As near as we can tell, neither element existed on Earth prior to this time; both elements are radioactive with fairly short lifetimes. (Segrè and Owen Chamberlain shared the 1959 Nobel Prize in physics for their 1955 demonstration of the existence of the antiproton.)

Transuranium elements are those that lie beyond uranium in the periodic table (that is, those with atomic numbers greater than 92). In 1940, Philip Abelson and Edwin McMillan (American, 1907-1991) created the first transuranium element, neptunium (element 93). Glen Seaborg continued to explore the new field of synthetic elements vigorously; he helped create plutonium (element 94) in 1941 and numerous others later. (The 1951 Nobel Prize in chemistry went to both Seaborg and McMillan for the discovery of plutonium and research on transuranium elements.) All the elements beyond francium (element 87) are radioactive. Most isotopes of the elements below atomic number 87 are not radioactive, although some have a few radioactive isotopes.

The main characteristic of a radioactive element is that it emits energy in some form: an alpha (α) particle, a beta (β) particle, or gamma (γ) rays, true electromagnetic radiation. One of the earliest discoveries about radioactive elements is that some convert into a different element, then another, and eventually end up a much lighter atomic species. The set of elements involved in this process make up a **radioactive decay series.**

The idea that some atoms could be unstable, decaying into other elements, was inconceivable to Dalton and the early atomic theory proponents. Carried to its logical extreme, it could mean that the entire cosmos was falling to pieces—hardly a comforting thought. But keep in mind that most isotopes of most elements are stable. Stability, not radioactivity, is the hallmark of most atoms and compounds; elements that move through radioactive decay series are more the exception than the rule.

Radioactive Decay Series

Let's look more closely into radioactive decay wherein one element transmutes into another. Consider uranium-238, the most common uranium isotope, which releases an α-particle (^4_2He) as it decays. In this transmutation process, which is written as follows, the uranium releases the α-particle and converts into a new form. (Note that nuclear reaction equations do not normally indicate electrical charges—hence, no 2+ superscript appears with the alpha particle symbol.)

$$^{238}_{92}U \longrightarrow ^{4}_{2}He + ?$$

The main question we must consider now is what belongs in place of the question mark? The Law of Conservation of Mass must apply; if the left side of the equation has a mass of 238 amu, the right side must have the same, and if the left side has 92 protons, the right side must, too. Therefore, we must add 234 amu and 90 protons to the right side of the equation to make the mass equal on each side. We can indicate this as follows, retaining the question mark to represent whichever element uranium has changed into:

$$^{238}_{92}U \longrightarrow ^{4}_{2}He + ^{234}_{90}?$$

Now we have the same mass (4 + 234 = 238) and the same number of protons (90 + 2 = 92) on each side of the arrow. We know that the uranium emitted an α- particle and changed into another atom that has 90 protons and a total mass of 234 (from 90 protons and 144 neutrons). What is this mysterious new atom?

A quick glance at the periodic table reveals the element with atomic number 90 is thorium (Th). Finally, we can write the full equation:

$$^{238}_{92}U \longrightarrow ^{4}_{2}He + ^{234}_{90}Th$$

PRACTICE PROBLEM 5.7

Try your hand at writing the equation that describes a nuclear reaction. Thorium–232 (element 90) loses an α-particle and transmutes into another chemical element. Write the equation for this nuclear reaction.

ANSWER First, write out the known isotope and the α-particle (a helium nucleus) as:

$$^{232}_{90}Th \longrightarrow ^{4}_{2}He + ?$$

Next, determine the mass of the new element (represented by the question mark) and how many protons it has. The mass must be $232 - 4 = 228$, and the number of protons must be $90 - 2 = 88$. Add this new information:

$$^{232}_{90}Th \longrightarrow ^{4}_{2}He + ^{228}_{88}?$$

Finally, determine the new element by consulting the periodic table. The element radium (Ra) is element 88. Enter its symbol in the equation:

$$^{232}_{90}Th \longrightarrow ^{4}_{2}He + ^{228}_{88}Ra$$

This reaction was first observed in 1901 by Rutherford and Soddy. However, they did not realize that the new element was radium until later.

In the late 1890s and early 1900s, chemists did not know about the nucleus of the atom or the true nature of the various forms of radiation. While Rutherford and Soddy proposed in 1903 that radioactive elements transformed into other chemical elements, this was difficult to prove because modern analytical techniques did not yet exist. In fact, in 1904, Sir William Bragg (English, 1862–1942) suggested that the α-particle was a cluster of many thousands of electrons rather than one massive particle.

Scientists have described hundreds of nuclear reactions, but not all nuclear transformations involve only α-particles. Thorium–232 for example, moves through a radioactive decay series involving the release of both α- and β-particles in different steps. First, thorium (element 90) loses an a-particle and transmutes to radium–228 (element 88):

$$^{232}_{90}Th \longrightarrow ^{4}_{2}He + ^{228}_{88}Ra$$

Next, radium–228 decays by releasing a β-particle (an electron). The net effect is to transmute radium into actinium (Ac), element 89, with the same mass (228 amu, because gaining or losing electrons does not affect mass). The net result of any β-emission is an increase of one in the atomic number for the product nucleus, with no change in the atomic mass. Thus, the symbol for a β-particle is $^{0}_{-1}e$, and we can express the radium–228 transmutation in the following equation:

$$^{228}_{88}Ra \longrightarrow ^{228}_{89}Ac + ^{0}_{-1}e$$

After this transmutation, the actinium also loses an electron and converts into thorium (element 90), still retaining a mass of 228:

$$^{228}_{89}Ac \longrightarrow ^{228}_{90}Th + ^{0}_{-1}e$$

The net result of these first three nuclear reactions, then, is that thorium–232 ultimately converted into thorium–228, with the sequential release of an α- and two β-particles. This process would actually continue through several more steps; we can summarize the first five steps in the thorium–232 radioactive decay process as follows:

$$^{232}_{90}Th \xrightarrow{-\alpha} ^{228}_{88}Ra \xrightarrow{-\beta} ^{228}_{89}Ac \xrightarrow{-\beta} ^{228}_{90}Th \xrightarrow{-\alpha} ^{224}_{88}Ra \xrightarrow{-\alpha} ^{220}_{86}Rn$$

Notice that, every time a reaction releases an α-particle ($^{4}_{2}He$), the atomic mass value decreases by 4 and the atomic number by 2. Similarly, every time a reaction releases a β-particle ($^{0}_{-1}e$) the atomic mass remains unchanged, but the atomic number increases by 1.

Other, similar, radioactive decay series start with uranium–238 and uranium–235. In all three series, the sequential decay process continues until a stable lead isotope (Pb, element 82) forms. In the series beginning with thorium–232, the ultimate product is $^{208}_{82}Pb$—lead with a mass of 208.

PRACTICE PROBLEM 5.8

Try writing one more equation describing atomic decay. This time we will look at one that emits a β-particle. Polonium–218 (Po, element 84) loses a β-particle and transmutes into another element. Show what happens.

ANSWER Remember that when an element loses an electron, or β-particle, its atomic number increases by 1 and its mass stays the same. We can begin by writing the equation with the known isotope and the β-particle:

$$^{218}_{84}Po \longrightarrow ^{0}_{-1}e + ?$$

Next, we can determine the new element's mass and atomic number:

$$^{218}_{84}Po \longrightarrow ^{0}_{-1}e + ^{218}_{85}?$$

Now all we need to do is look up the mass and symbol for element 85; it is astatine (At). We can "plug" this symbol into the equation to arrive at our final answer:

$$^{218}_{84}Po \longrightarrow ^{0}_{-1}e + ^{218}_{85}At$$

If you solved Practice Problems 5.7 and 5.8 easily, you may want to consider a career in nuclear physics designing atomic reactors! However, before you can rush out to build your own nuclear reactor at home, a word of caution is vital: we are not covering all possible types of nuclear reactions in this book. Some of those omitted are essential in reactor design, operation, and safety.

Half-Lives

One striking observation about radioactive decay, first noted by Rutherford, is that specific nuclear changes always follow the same timelines. As atoms of a radioactive element transmute into a new element, they change at a steady, predictable rate. Although scientists cannot predict when a particular atom will transmute, they can determine quite precisely the percentage of a sample of the radioactive element that will transmute over a given period of time. Plotting these percentages on a graph produces a characteristic decay curve for each radioactive element.

The basic shapes of several decay curves appear in figures 5.3 through 5.5. Figure 5.3 shows the decay curve for iodine–131, an isotope commonly used in medicine. Figure 5.4 is the decay curve for tritium, a radioactive isotope of hydrogen, and figure 5.5 shows the curve for uranium–238, the most common uranium isotope. Notice that the three curves are all the same shape—in fact, one curve can superimpose on another, if you ignore the horizontal axis values. The numbers on the vertical axis are also the same; in all three cases, they represent the percentage of each isotope remaining after any given time. On the vertical axis, each number is half the value of the one above. On the horizontal axis, the numbers on all three examples increase, but at widely different magnitudes. Figure 5.3 measures decay over days, figure 5.4 over years, and figure 5.5 over billions of years.

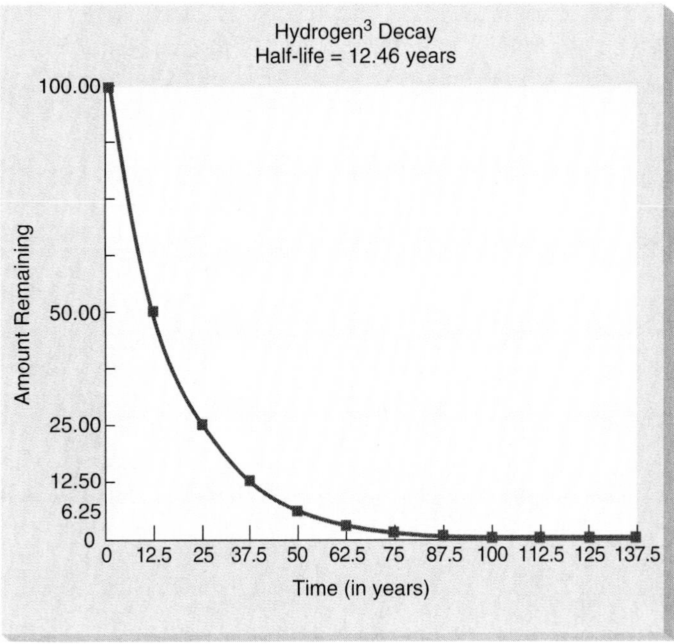

Figure 5.4 The radioactive decay of hydrogen–3 (tritium).

Let's assume that, in each case, we started with a 100-gram sample of the radioactive element. The next dot on the curve marks the point at which 50 g (50 percent of the original) has transmuted into a new element. In figure 5.3, 50 percent of

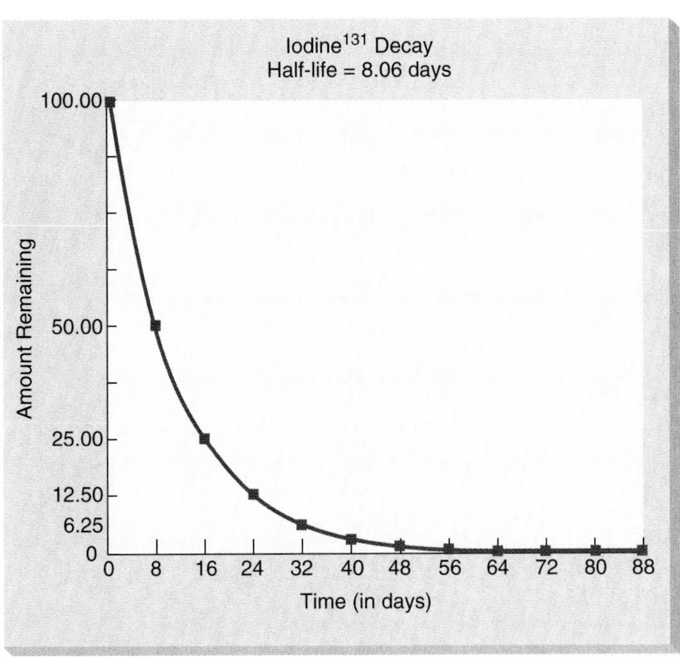

Figure 5.3 The radioactive decay of iodine–131.

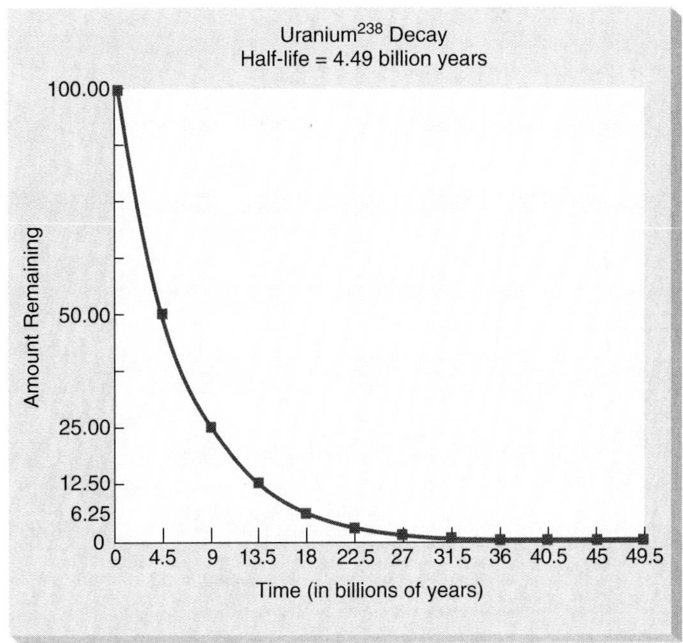

Figure 5.5 The radioactive decay of uranium–238.

the sample has transmuted after 8 days. The next point on the curve in figure 5.3 indicates that half of the remaining 50 g—another 25 g of iodine–131—decays into another element after another 8 days, 16 days after we began measuring the decay. After another 8-day period passes (at 24 days on the horizontal axis), half of the remaining 25 g has decayed, so that we now have only 12.5 g of the original iodine–131 left. This pattern continues: every 8 days, half the remaining sample transmutes into a new element. We therefore know that iodine–131 has a **half-life** of 8 days. In equations, we symbolize half-life as $t_{1/2}$, with *t* representing the time and $t_{1/2}$ representing the amount of time it takes for half of an element's atoms to decay. Scientists apply this knowledge in a variety of technologies, including radiometric dating (box 5.4).

The only real difference between the curves in figures 5.3 to 5.5 is the magnitude of the numbers on the horizontal axis. Half of an iodine–131 sample transmutes in about 8 days, but half of a tritium (or hydrogen–3) sample takes almost 12.5 years to decay. Uranium–238 has the longest half-life of the three; it takes about 4.5 billion years for half of this radioactive isotope to disappear.

The half-life is a characteristic value for each radioactive isotope (or radioisotope for short). It can vary over an incredibly wide time range, from less than a second to billions of years; the isotope $^{8}_{4}Be$ has a half-life of 10^{-16} seconds, while the half-life for $^{115}_{49}In$ is 6×10^{14} years! Whether an element's half-life is very short, very long, or in between, the decay curve is the same shape; the only difference is the time it takes for half of a particular radioisotope to decay.

To predict the effects of radioactivity, scientists must consider the element's half-life, the amount of material involved, the nature of the released rays, and the total exposure to the radioactivity released. Obviously, a large radioactive sample emits more radioactivity than a small one. Similarly, an isotope with a short half-life emits more radioactivity over a given time than an isotope with a longer half-life; a 100-gram sample of iodine–131 emits half its radioactive rays in an 8-day period, while a 100-gram sample of uranium–238 would have emitted only a minute fraction of its radioactivity during the same amount of time.

But this does not necessarily mean an isotope with a short half-life is more dangerous than an isotope with a long half-life. The type of radiation and how much shielding (or protection) is present are also important considerations (Table 5.1). Both α- and β-particles may be trapped by small amounts of shielding; for example, a piece of paper or a layer of skin stops α-particles, and a thin sheet of aluminum foil blocks β-particles. In fact, external α-rays do not pose as much danger to humans as β- or γ-rays because they cannot penetrate the skin, and because the latter two types can cause atoms to ionize, which sometimes leads to biological damage. However, if α-rays enter the body, they are dangerous because these energetic particles can cause many undesirable chemical changes.

Ultimately, γ-rays are the most dangerous form of radiation. They pose the dominant problem in storing radioactive waste because they require massive amounts of heavy shielding to avert damage to humans and other forms of life.

Radioactivity and Safety

Early scientists did not think much about safety in dealing with radioactive materials. But as knowledge accumulated and new technologies developed, it became increasingly important to address the dangers of radioactivity in our society. Today, it is well established that radiation can cause illness or even death. Are radioactive materials safe to use—or can the dangers to humans be controlled? Do the potential benefits of using radioactive materials outweigh possible risks? All processes, natural or human-derived, produce waste products; can we dispose of nuclear waste materials safely?

Scientists generally consider a radioactive material "safe" after twenty half-lives have elapsed. After that time, the total emissions from a sample are presumably too small to cause appreciable biological damage. In the hypothetical 100-gram samples in figures 5.3 to 5.5, the amount of the original radioactive isotope left after twenty half-lives is merely 0.000095 g. For radioactive iodine–131, twenty half-lives have elapsed after 161.2 days, or a little over five months. Twenty half-lives for tritium ($^{3}_{1}H$) have passed after 249 years—longer than the United States has existed as a nation. Uranium–238 does not reach the twenty half-lives mark until almost 90 billion years have elapsed; most experts estimate the universe to be far younger.

TABLE 5.1 The properties of the three forms of radiation.

RADIATION TYPE	PENETRATING POWER	NATURE OF RADIATION	NATURE OF PARTICLE	PARTICLE MASS, amu
α	low—stopped by sheet of paper	positively charged particle	He^{2+} ion	4.0
β	medium—stopped by aluminum foil	negatively charged particle	electron	5.5×10^{-4}
γ	high—Requires thick layer of lead	true radiation	none	none

RADIOMETRIC DATING

Nuclear chemistry can sometimes help scientists determine the approximate age of an object through a procedure called radiometric dating. Willard Libby (American, 1908–1980) introduced the most common form of this process, radiocarbon dating, in 1946. The basic concept of radiometric dating is simple: since each radioisotope decays at a steady rate, the amount of radioisotope present in a sample gives us a measure of its age. All radiometric measurements are made using radioactive elements that occur naturally.

Radiocarbon dating (sometimes known as carbon–14 dating) determines the ages of once-living tissues. Radioactive carbon–14, which forms naturally in the upper atmosphere, reacts with oxygen to form radioactive CO_2. Plants and animals take in a small amount of this CO_2 along with the nonradioactive CO_2 they take in normally by ingestion or respiration. C–14 acquisition ends when the plant or animal dies.

Because the C–14 level in the atmosphere is fairly constant, living organisms contain a constant amount of C–14. However, dead tissues contain lower amounts because part of the C–14 has decayed into nitrogen ($^{14}_{7}N$). Since scientists know that C–14's half-life is 5,730 years, they can determine the approximate age of a sample by measuring the amount of C–14 still present with a device called a Geiger counter (figure 1). Radiocarbon dating has been used to determine the ages of such diverse items as the Dead Sea Scrolls (figure 2), the Shroud of Turin, old trees, and Indian artifacts.

The radioisotope C–14 arises from an upper atmospheric nuclear reaction in which the neutrons come from natural cosmic rays. The reaction results in the production of carbon–14 plus hydrogen:

$$^{14}_{7}N + ^{1}_{0}n \rightarrow ^{14}_{6}C + ^{1}_{1}H$$

The unstable carbon atom breaks back down into nitrogen when it loses an electron:

$$^{14}_{6}C \rightarrow ^{14}_{7}N + ^{0}_{-1}e$$

Naturally, radiocarbon dating is restricted to materials of plant or animal origin, but other radiometric dating methods are used to determine the ages of rocks. Scientists date a rock that contains U–238 ($t_{1/2}$ = 4.5 billion years) by determining the amount of Pb–206 (lead), a daughter radioisotope, present in the rock. When a rock does not contain U–238, chemists can often estimate its age by determining the ratio between radioactive potassium K–40 ($t_{1/2}$ = 1.3 billion years) and its daughter product Ar–40 (argon) within the rock. These techniques, widely used in geology and archeology, have enabled scientists to estimate the age of the oldest rocks on earth—granite from Greenland—at 3.7 billion years. The moon rocks in figure 3 proved to be about the same age. Since the molten materials that comprise our planet had to cool down to form these rocks, scientists believe the Earth itself is about 4.6 billion years old.

Of course, radiometric dating makes certain assumptions. When scientists measure the Pb–206 in a rock sample, they assume the original sample contained no Pb–206 and that no Pb–206 was lost during the rock's existence. But within these assumptions, scientists can ascertain with a fair degree of precision the ages of materials thousands or even billions of years old. Radiometric dating, based on naturally occurring radioactive materials, allows us to date past events more accurately.

Figure 1 A Geiger counter being used to test for radiation.

Figure 3 Rocks obtained from the moon.

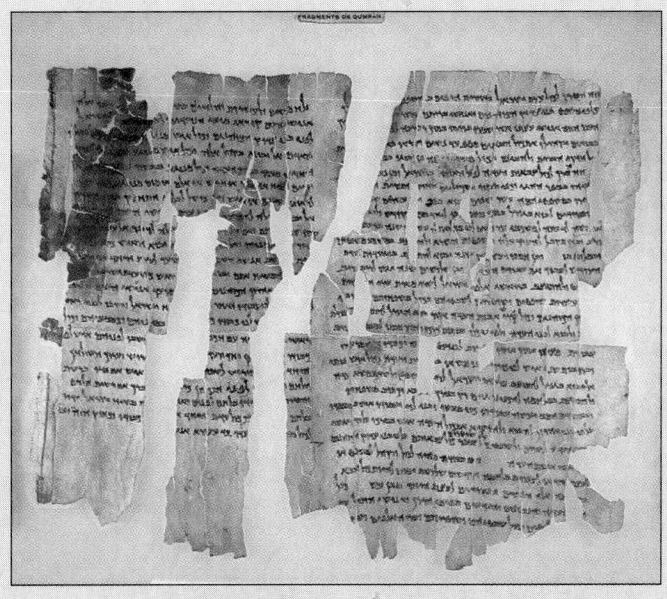

Figure 2 A portion of the Dead Sea Scrolls.

The radioactivity emanating from a sample after twenty half-lives have passed is less than 0.0001 percent of the starting amount; this level is in the range of "background" radiation. We are constantly bombarded with such radiation from natural sources, including low levels of radiation from radon and cosmic rays. In fact, as figure 5.6 shows, 82 percent of all radiation emanates from natural rather than from artificial sources, such as nuclear reactors. Approximately 55 percent of this radiation comes from radon that occurs naturally in the environment, and an additional 8 percent comes from other radioactive elements in the rocks and soils near the Earth's surface. Cosmic rays, which enter our atmosphere from outer space, account for another 8 percent of this background radiation. In actuality, life on planet Earth has always been exposed to radiation, but most of it has not proven especially harmful to life. Artificially derived radiation makes up 18 percent of the total, with about 15 percent deriving from medical applications, including X-rays. The remaining 3 percent comes from nuclear power plants and miscellaneous consumer applications, such as food irradiation. This small slice of the pie, however, is the part that concerns many people.

As you can see from table 5.2, some people (in this survey the generally well-educated members of the league of women voters and college students) perceive nuclear power as the number one risk to human health and safety. By contrast, risk analysis experts, who study safety records of various activities or processes, rate nuclear power twentieth on a list of thirty possible risks—far behind such everyday objects and activities as motor vehicles, smoking, and alcoholic beverages. However, these same experts rank X-rays, nonnuclear electric power, and swimming as much more dangerous than the other survey respondents do. What causes these variations in risk perception?

Several factors enter the picture. First, bad news makes good press, even if the news is slanted or inaccurate. Too often, the public's "knowledge" about a particular material or activity comes from a biased, unscientific report. A second factor is the public's tendency to confuse technologies that may only be peripherally related. For example, many people associate the terror of nuclear war with the much milder risks involved in a nuclear power plant.

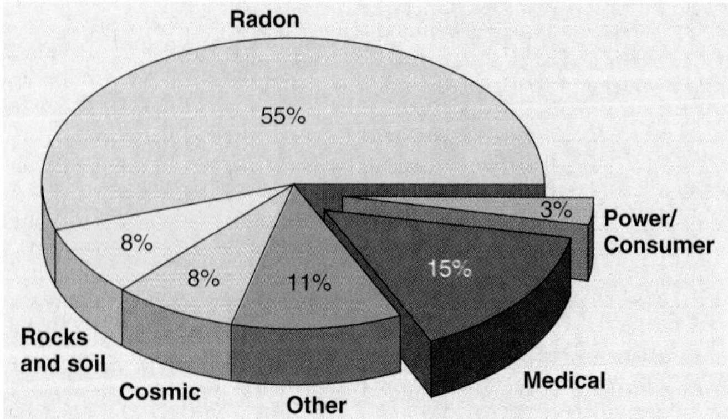

Figure 5.6 The sources of radioactivity. The artificial sources are shown as offsets from the various natural sources.
SOURCE: Data from National Council on Radiation Protection and Measurements.

A common misconception is that a nuclear power plant might explode and devastate a large area, as the atomic bombs did in Hiroshima and Nagasaki. Interestingly, hardly anyone associates the benefits of nuclear medicine with the risks of nuclear power generation *or* nuclear war. All three items (bombs, electric power, and nuclear medicine) depend on nuclear reactions, but the general public, as table 5.1 reveals, tends to exaggerate the risks of nuclear power generation and downplay the risks of nuclear medicine.

Other factors add to the public's misconceptions. The modern mass media often do not have the time or space needed to discuss complicated technologies thoroughly. The newsperson may not always fully understand the scientific details or the ramifications involved. And in an age of skepticism, people will not always believe the experts, even when scientific evidence backs them up. (Box 5.5 examines the subject of risk perception in greater detail.)

The only real way to evaluate the safety of nuclear processes is to examine each, assessing the evidence with a scientific mindset. Let's look now at specific nuclear applications, how they work, what risks they pose, and what benefits they offer.

NUCLEAR APPLICATIONS

Nuclear applications are remarkably important in today's society. How do modern technologies harness energy from the atom? How does this energy power nuclear weapons, nuclear medicine, and food irradiation, and what risks do nuclear waste disposal and natural radiation sources such as radon pose? These are the questions we will explore next.

Harnessing Energy from the Atom

Most atoms are stable entities, so some force must hold them together. This is especially evident when you realize the nucleus consists of a collection of positively charged particles which would normally repel one another—surely some force must hold together this concentration of like charges. As scientists theorized about these forces, they wondered if they could tap the energy of the atom as they did with fossil fuels? This question intrigued many researchers in the early portion of the twentieth century.

Perhaps the concept that most impelled scientists was Einstein's equation $E = mc^2$, which equated energy and matter. Since c represents the speed of light (3×10^{10} cm/sec), the conversion of even minute quantities of matter, multiplied by this huge figure squared, could release incredibly large amounts of energy. This energy source would satisfy human needs for ages to come—at least, so scientists believed at that time.

Nuclear Fission

In 1934, Enrico Fermi (Italian American 1901–54) suggested that slow-moving (or thermal) neutrons could split atoms. This idea sprang partly from a series of studies on the neutron bombardment of uranium–235 (element 92) aimed at producing new elements

TABLE 5.2 Risk perception of thirty activities and technologies by three different groups. The rankings run from number 1 (riskiest) to number 30 (least risky).

ACTIVITY OR TECHNOLOGY	GROUP SURVEYED League of Women Voters	College Students	Risk Analysis Experts
Nuclear power	1	1	20
Motor vehicles	2	5	1
Handguns	3	2	4
Smoking	4	3	2
Motorcycles	5	6	6
Alcoholic beverages	6	7	3
General (private) aviation	7	15	12
Police work	8	8	17
Pesticides	9	4	8
Surgery	10	11	5
Fire fighting	11	10	18
Large construction	12	14	13
Hunting	13	18	23
Spray cans	14	13	26
Mountain climbing	15	22	29
Bicycles	16	24	15
Commercial aviation	17	16	16
Electric power (nonnuclear)	18	19	9
Swimming	19	30	10
Contraceptives	20	9	11
Skiing	21	25	30
X-rays	22	17	7
High school/college football	23	26	27
Railroads	24	23	19
Food preservatives	25	12	14
Food coloring	26	20	21
Power mowers	27	28	28
Prescription antibiotics	28	21	24
Home appliances	29	27	22
Vaccination	30	29	25

SOURCE: Data from Paul Slovic, "Perception of Risk," *Science*, vol. 236, April 1987.

with higher atomic numbers. Along with Emilio Segrè, Fermi observed that, rather than producing just the desired element 93—neptunium—the neutron bombardment actually produced four radioactive species. Ultimately, Fermi detected a barium isotope (element 56) among these products. Where did this come from, he wondered? Could the uranium actually split into smaller atoms?

German scientists Otto Hahn and Fritz Strassman, as well as Austrian Lise Meitner, repeated Fermi's and Segrè's studies, but Hahn was reluctant to call his results "atom splitting" when he published them in January 1939. However, Meitner, also in January 1939, advanced the theory that the uranium broke into smaller atoms when it was bombarded by neutrons. She coined the term **nuclear fission** to describe the process. Later in 1939,

BOX 5.5

RISK PERCEPTION VERSUS SCIENTIFIC DATA

Nuclear chemistry and radiation, as well as the problems with EMFs, raise questions about risk perception. How much risk an operation or process entails and how this risk is determined are specialized issues beyond the scope of this book. Nevertheless, we can acquire a general idea of the risks posed by considering a few simple facts about probability and our perceptions.

Many states now raise money by running lottery games. In one, a person bets on a three-digit number. Since exactly 1,000 three-digit numbers exist from 000 to 999, the chance of guessing the winning number is 1 in 1,000. If the United States ran a national lottery and every individual played it daily, we would expect 250,000 winners each day (1 of every 1,000, based on a total population of 250 million people). Other forms of gambling carry different odds (figure 1).

Figure 1 Gambling devices always carry specific odds for winning and losing.

These odds sound attractive when we think about the lottery, but if we were considering a fatal disease instead, and the chances of dying from it during a given year were 1 in 1,000, we would undoubtedly be horrified at the huge number of deaths (250,000 per year, or 684 per day) this one disease would cause. These odds—1 in 1,000—correspond to 100 deaths per 100,000 population per year, the usual measure for death rates. But the latest available figures (1991) actually listed an overall death rate, from all causes, of 854 per 100,000 population (about 2,135,000 people nationwide). In other words, the actual death rate is about 8.54 *times higher* than our 1 in 1,000 lottery odds! And this is substantially lower than the 1970 death rate (945 per 100,000). Few people worry about the odds of dying in a given year, but the fact is that the risk of dying is 8.54 times higher than the 1 in 1,000 odds of winning a three-digit lottery game.

Why do some odds seem riskier than others, even when the numbers don't bear the risk out? Often, this is a matter of perception. The overall risk of dying is 854 per 100,000, and we hardly give it a thought. The risk of dying of AIDS is slightly less than 10 in 100,000, but we consider AIDS an epidemic. The possible increase in skin cancer deaths due to ozone depletion is projected to be about 4 deaths per 100,000 per year. Assuming the projections are correct, is this an epidemic? Most scientists would say no, but if you were one of the unfortunate people who contracted skin cancer, you might disagree.

Finally, consider heart disease, the number one cause of death in the United States. In 1991, there were 360.3 deaths per 100,000 due to heart disease—900,750 total deaths compared to 22,675 deaths (10 in 100,000) due to AIDS and 10,700 (4 in 100,000) projected deaths due to skin cancer increases caused by ozone depletion. Which disease poses the greatest risk? Which should we be most alarmed about? Despite the numbers, risk perception depends on what interests you.

There is seldom an absolute standard by which to judge risk. The best we can do is to research the data and determine whether a claim is reasonable or not. (The numbers cited here are from the 1993–94 edition of the *American Almanac*.) But social and personal factors always enter into our judgments about risk. This is not necessarily right or wrong—but you should be aware of the effects of perception on risk estimation.

Philip Abelson identified three antimony (Sb), six tellurium (Te) and four iodine (I) isotopes among the products of uranium fission, and John Dunning experimentally confirmed Meitner's uranium fission proposal. A few months later, Frédéric and Irène Joliot-Curie showed that uranium fission could lead to a chain reaction. Hahn received the 1944 Nobel Prize in chemistry for his discovery of atomic fission; unfortunately, Meitner, who had fled to Sweden to escape Hitler's reign, did not share this award.

Figure 5.7 shows a typical chain reaction using U–235 as the starting material and producing tin (Sn) and molybdenum (Mo) isotopes. (The more common uranium isotope, U–238, does not undergo fission.) Note that fission produces more than one neutron for each U–235 atom split by a neutron (it produces two neutrons in the example in figure 5.7). This means the process produces enough neutrons to maintain an ongoing reaction for a long period of time, assuming the neutrons are moving at the correct speed (in the thermal range) and are confined within the sample.

The following equation describes this fission reaction. The subscript and superscript totals, as always, are the same on each side of the arrow $(92 + 0 = 92$ and $42 + 50 = 92$; $235 + 1 =$

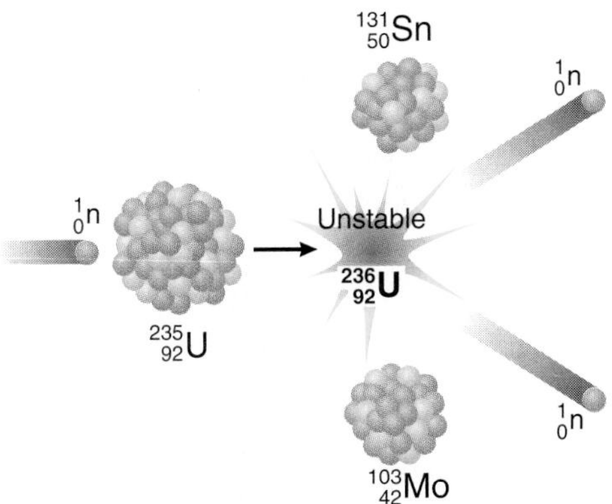

Figure 5.7 A typical fission reaction for uranium–235.

236 and 103 + 131 + 2 = 236). The neutrons are part of this reaction; while we may be tempted to cancel one neutron symbol from each side, we must realize they represent different neutrons. Moreover, the fission reaction will *not* occur unless the U–235 is bombarded with slow-moving (thermal) neutrons. U–235 undergoes radioactive decay when it releases an α-particle, but nuclear fission only occurs when thermal neutrons bombard the U–235 atom.

$$^{235}_{92}U + ^{1}_{0}n \longrightarrow ^{103}_{42}Mo + ^{131}_{50}Sn + 2^{1}_{0}n$$

This fission reaction does not always occur in exactly the same manner. The U–235 may also decay into strontium (Sr, element 38) and xenon (Xe, 54), into barium (Ba, 56) and krypton (Kr, 36), or into any of several other distinctively different combinations. All these reactions include the same total atomic numbers and atomic masses, and all the fission reactions produce at least two new neutrons, which in turn strike other U–235 atoms and perpetuate the reaction.

$$^{235}_{92}U + ^{1}_{0}n \longrightarrow ^{90}_{38}Sr + ^{143}_{54}Xe + 3^{1}_{0}n$$

Figure 5.8 shows how a chain reaction develops. Each neutron from the initial fission reacts with one U–235 atom, which splits and, in this case, releases 3 more neutrons. Each of these 3 neutrons splits another U–235 atom, releasing a total of 9 more neutrons. These 9 neutrons react with 9 more U–235 atoms, splitting them and releasing 27 new neutrons (3 × 9 = 27); the next generation splits another 27 U–235 atoms and forms 81 neutrons; they split 81 U–235 atoms and form 243 neutrons; and so on. In a short time, enormous numbers of neutrons are emitted and an

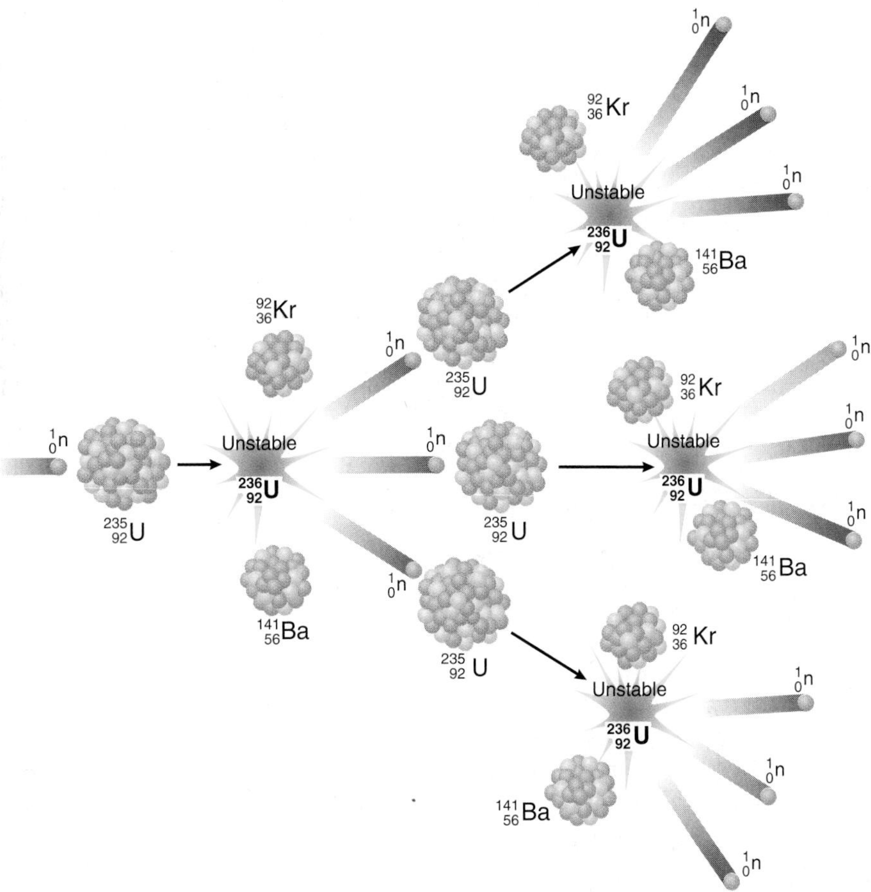

Figure 5.8 Part of a chain reaction from the fission of U–235. Each atom split produces 3 new neutrons.

equally enormous number of atoms split. Each atomic fission releases some heat, and the system rapidly becomes very hot.

The essential feature of any chain reaction is that some substance that is consumed in the reaction is also regenerated as part of the product. In fission, neutrons are consumed, and replacement neutrons are liberated. This regeneration allows a reaction that begins with a single neutron to consume huge numbers of atoms.

Every atom split also produces some heat. Although this heat either is or can be converted into electricity, problems quickly develop if the amount of heat keeps rapidly doubling or tripling. Nuclear reactors are designed to control the heat by "trapping" some of the neutrons and restricting the number of U–235 atoms split. Thermal neutrons themselves are not especially harmful to people, but if just *one* neutron escapes into the fission chamber and splits another U–235 atom, a chain reaction can begin. Successful nuclear power plants are designed to keep the number of fission reactions and the amount of heat they generate controlled and constant (figure 5.9).

Where does the released heat energy come from? The released energy comes from the conversion of part of the subatomic matter into energy. When the *exact* masses for all the emitted subatomic particles are added, they contain slightly less total mass than the original atoms. This small difference, if we measured it, would make up just a fraction of an amu for each atom; for a helium atom, the Einstein equation predicts a mass of 0.0305 amu will convert into energy. In the atom, this mass creates the **binding energy** that ties the subatomic particles together. When fission occurs, this binding energy is released and becomes available for use. The energy released from the fission of a single gram of U–235 is about 10^7 kcal, equivalent to the energy released by burning more than 5,500,000 grams (over 6 tons) coal, or about 14 barrels of oil! By controlling the number of neutrons available to continue the reaction, one can either allow all this energy to be released at once (as in a bomb) or slowly (as in a nuclear power plant).

PRACTICE PROBLEM 5.9

Try your hand at balancing the following equation for U–235 fission producing Kr.

$${}^{235}_{92}U + {}^{1}_{0}n \longrightarrow {}^{92}_{36}Kr + {}^{141}_{56}Ba + ?{}^{1}_{0}n$$

ANSWER First, add the atomic mass values and atomic numbers on each side. The atomic numbers already balance: 92 + 0 = 92, and 36 + 56 = 92. The mass values differ by three units (235 + 1 = 236, but 92 + 141 = 233). The number of neutrons produced, with 1 amu each, must therefore be 3:

$${}^{235}_{92}U + {}^{1}_{0}n \longrightarrow {}^{92}_{36}Kr + {}^{141}_{56}Ba + 3{}^{1}_{0}n$$

Figure 5.9 A nuclear power plant, located in Stedman, Missouri, near Jefferson City which generates electricity for nearby communities.

Nuclear Bombs

In 1942, a team headed by Enrico Fermi built the first nuclear pile (the basis for nuclear power plants) beneath the football stadium at the University of Chicago. Most of this early work on atomic energy was shrouded in a cloak of secrecy as the United States entered World War II. At the urging of Leo Szilard, who confirmed in 1939 that nuclear reactions could be self-sustaining, Albert Einstein wrote his now-famous letter to President Franklin D. Roosevelt. In it, he warned that the Germans were working on a bomb employing nuclear fission, and he suggested that an atomic bomb might be constructed from uranium. This led to a secret research project, called the Manhattan Project, charged with developing an atomic bomb; it operated at 37 different sites in 19 states and Canada with a budget of 2.2 billion dollars. At that time, the Manhattan Project was the largest single enterprise in the history of science.

Entire new cities sprang up in response to this enormous research effort. Scientists isolated U–235 in Oak Ridge, Tennessee (which reached a population of 75,000); in Hanford, Washington (which grew to 60,000), they transmuted U–238 into plutonium (Pu, element 94). In Los Alamos, New Mexico, workers constructed the atomic bombs; they exploded the first on July 16, 1945, in the nearby desert. Shortly thereafter, in August 1945, the United States dropped atomic bombs on Hiroshima and Nagasaki, Japan, marking the end of World War II.

While a nuclear reactor is meant to control a nuclear reaction, a bomb is meant to explode. The challenge the Manhattan Project scientists faced was how to make the reaction multiply rapidly, and still retain enough neutrons and fissionable isotopes to continue to drive the reaction until the heat bursts everything apart. The scientists discovered that a certain minimum amount of fissionable material is required to sustain the reaction: They designated this amount the **critical mass.** If a bomb contained less fissionable matter than this critical mass, the reactants dispersed before an explosion could occur. The plants in Oak Ridge and Hanford worked to produce enough U–235 and plutonium–239 to build bombs containing this critical mass.

Nuclear Fusion

While the atomic bomb relied on nuclear fission—splitting the nucleus—to release energy, scientists soon produced a new explosive that utilizes **nuclear fusion**—the joining of two nuclei. The hydrogen bomb was first exploded in the South Pacific at Eniwetok Atoll on November 6, 1952. A team headed by Edward Teller developed the bomb, which depended on the fusion of hydrogen isotopes to form helium. Hydrogen bombs release much more energy—about four times as much for an equal amount of mass—than atomic bombs. In fact, an atomic bomb was used to trigger the hydrogen bomb by raising the temperature into 100,000°C range necessary for fusion.

Fusion is the nuclear process the sun uses to generate its energy. It does so through a process called the **carbon cycle,** detailed in the following set of equations. (The ${}^{0}_{1}\beta^{+}$ is a positron, a positively charged electron.)

$$ {}^{12}_{6}C + {}^{1}_{1}H \rightarrow {}^{13}_{7}N \rightarrow {}^{13}_{6}C + {}^{0}_{1}\beta^{+} $$
$$ {}^{13}_{6}C + {}^{1}_{1}H \rightarrow {}^{14}_{7}N $$
$$ {}^{14}_{7}N + {}^{1}_{1}H \rightarrow {}^{15}_{8}O \rightarrow {}^{15}_{7}N + {}^{0}_{1}\beta^{+} $$
$$ {}^{15}_{7}N + {}^{1}_{1}H \rightarrow {}^{12}_{6}C + {}^{4}_{2}He $$

The net result of this reaction sequence is that 4 hydrogen atoms convert into 1 helium atom. The seemingly insignificant mass difference between the 4 H atoms and the 1 He atom converts directly into a promethean quantity of energy. The fusion of a single gram of hydrogen releases as much energy as 20 tons (18,000,000 g) of coal. (If you compare this with the fission of U–235, which releases the energy equivalent of 5,500,000 g coal, you can see why fusion is considered a better energy source.)

Several possible reactions exist for the fusion of hydrogen into helium, but all require temperatures between 45 million and 400 million °C. The simplest fusion sequence, which occurs at about 45,000,000°C, is:

$$ {}^{2}_{1}H + {}^{3}_{1}H \rightarrow {}^{4}_{2}He + {}^{1}_{0}n $$
$$ {}^{6}_{3}Li + {}^{1}_{0}n \rightarrow {}^{4}_{2}He + {}^{3}_{1}H $$

Since step 2 in this sequence regenerates ${}^{3}_{1}H$, known as tritium, the reaction can become self-sustaining. Although fewer than 1 in 6,800 hydrogen atoms (0.015 percent) are ${}^{2}_{1}H$, or deuterium, the oceans contain over 10^{13} tons of ${}^{2}_{1}H$. In fusion, this amount could produce more than 10^{27} kcal of energy, which would likely meet all human energy requirements for a billion years.

Obviously, hydrogen bombs are not a practical energy source. Much research has thus gone into the attempt to develop a usable high-temperature fusion reactor. One major problem is creating a container that can withstand the tremendous heat required for fusion; essentially all containment materials vaporize well below these temperatures, so most fusion reactors constructed to date rely on strong magnetic fields for containment.

Small wonder people became excited about the prospects of "cold" fusion when Pons and Fleishmann claimed they had achieved it in 1989 (see box 1.5 in chapter 1). Unfortunately, scientists remain uncertain that a cold fusion process will ever become practical. Most research, including Pons' and Fleishmann's seems to lack the reproducibility essential to the scientific method. While some hold out hope for cold fusion, many scientists believe hot fusion will prove more successful.

To date, the most successful fusion reaction ran self-sustained for about four seconds in December 1993. Whether it employs hot or cold fusion, a practical electricity-generating power plant using fusion seems a good thirty years or more in the future. (We will discuss this topic further in chapter 18.)

Nuclear Medicine

Even before World War II ended, scientists were attempting to apply atomic energy to peaceful uses, and **nuclear medicine** quickly became a prominent research area. Physicians had been using X-rays since the late 1800s, shortly after Roentgen discovered them. If X-rays can diagnose or treat illnesses, scientists reasoned, why not the γ-rays emitted from radioactive materials?

Radioactive materials did prove useful in diagnosing and treating human illness. One of the best known diagnostic uses of radioactivity is to detect thyroid disorders—either hyperthyroidism, when the thyroid gland produces too much hormone, or hypothyroidism, when the gland fails to produce enough. To evaluate thyroid function, physicians administer a potassium iodide solution containing radioactive iodine–131 to the patient. Iodine migrates selectively to a healthy thyroid gland within a few hours, because thyroid hormone picks up iodine. When the patient is examined with a scintillation device that detects radioactive emissions, the doctor can determine how much iodine reached the thyroid gland and treat the patient accordingly.

Figure 5.10 shows some possible results. Part *a* shows a large amount of I–131 in a hyperactive thyroid; *b* shows a normal gland with a normal level of I–131; and *c* shows an enlarged gland that produces very little hormone and picks up almost no I–131. Doctors not only diagnose but treat hyperthyroidism with radioactive iodine. In treatment, the patient is exposed to much larger amounts of I–131 to destroy part of the overactive gland, reducing thyroid activity into the normal range.

Radioisotopes have several other diagnostic applications, including the use of chromium–51 for blood circulation studies; cobalt–59 or cobalt–60 for vitamin B_{12} adsorption evaluation; gadolinium–153 to study bone mineralization and osteoporosis; phosphorus–32 for the detection of skin cancer; iron–59 for red blood cell formation and lifetime studies; tritium (hydrogen–3) to determine the total amount of water in the body; imaging with technetium–99 to study the brain, vascular system, and many

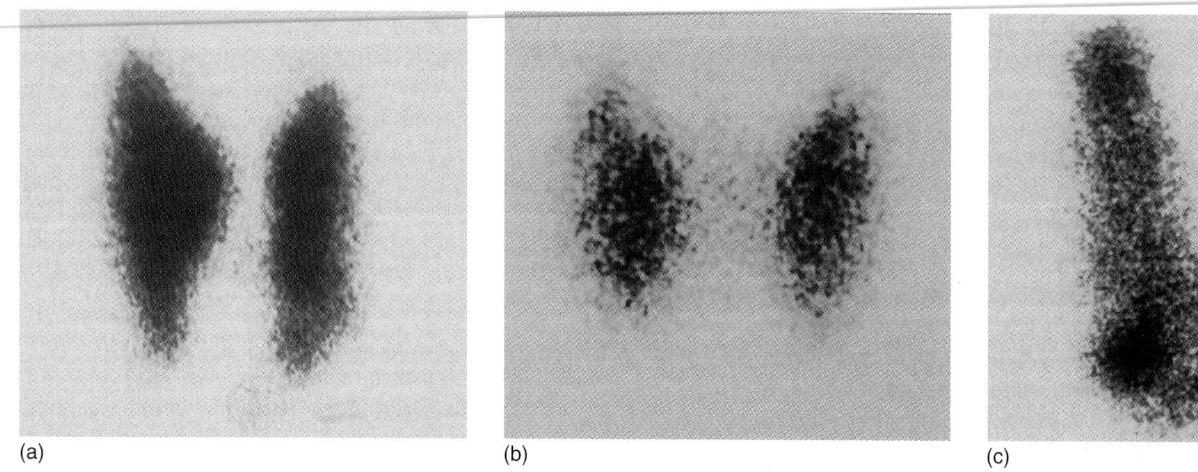

Figure 5.10 Scans of thyroid glands after administration of radioactive iodine–131. **(a)** A thyroid gland that produces too much hormone. **(b)** A normal thyroid gland. **(c)** A tumor-enlarged thyroid that produces very little hormone.

internal organs; and the use of sodium–24 to detect blood vessel obstructions.

Radiation therapy treats cancer in three different ways. Teletherapy, the simplest and oldest method, involves exposing the cancer directly to a beam of radiation. Brachytherapy involves inserting radioisotopes into the region of the body the cancer cells have invaded. Finally, a technique called radiopharmaceutical therapy involves administering radioisotopes through oral or intravenous means. The radioisotopes are attached to another chemical that migrates to the cancerous part of the body. Ideally, this method can be very specific without the invasiveness of the other methods.

It is important to recognize how and why radiotherapy works. Radiation kills cells—both healthy and cancerous. Tumor cells, however, grow and reproduce more quickly than healthy tissue and are more sensitive to radiation damage. Thus, radiation destroys tumor cells more frequently than normal cells. Although radiation treatment often causes nausea and malaise because it affects healthy tissues along with cancer cells, this undesirable side effect usually passes after the treatment ends. Radiation therapy has saved many lives from the ravages of cancer.

Food Irradiation

One of the most controversial uses of radioactivity is **food irradiation.** When the first U.S. commercial food irradiation plant, located in Mulberry, Florida, began shipping irradiated strawberries in January 1992, dozens of people picketed the plant and the stores selling the produce. The antinuclear activists voiced all kinds of fears—even that people who ate the irradiated fruit would glow in the dark! They claimed that irradiation would increase the risk of cancer and birth defects, as well as rob the food of its nutritional value by destroying enzymes and DNA.

On the other side of the issue, Food and Drug Administration (FDA) officials claimed the method was a safe and effective way to prevent food-borne illnesses from infecting fruits, vegetables, poultry, and pork. Although ample evidence backed their claims, consumer acceptance has been slow in the United States (box 5.6). Food irradiation represents another vivid example of the difference between scientific evidence of risk and public perception of it.

Incidentally, nuclear power is not the only victim of nonscientific reporting and resulting consumer fear. Alar, a chemical (damionzide) sprayed on apples to improve ripening, was targeted in a 1989 TV special; the program claimed Alar could potentially poison thousands of children. It was true that small Alar residues clung to some (though not all) apples. But testing showed that a person would need to eat 57,000 apples per day for nearly 70 years in order to ingest enough Alar to cause harm—that is, 40 apples per hour, 24 hours a day, every day of one's life. Although it would be virtually impossible to ingest this much Alar, the chemical was removed from the market and the apple industry suffered for months following the television program. The industry recovered, but some commentators still cite the overblown dangers of Alar exposure as if they were accurate.

Nuclear Waste Disposal

The single most serious problem with nuclear power applications is the disposal of the highly radioactive waste it produces. Waste disposal is a problem associated with most materials and processes in our society, but nuclear waste poses additional complications: danger and time. Most waste materials can poison the environment for several years, but nuclear waste can produce potentially lethal emissions for centuries because some materials have such long half-lives. If the twenty half-life rule is a reliable safety guide, any material with a long $t_{1/2}$ remains active for an exceptionally long time. Two questions arise: (1) Where can we store the materials until they are safe? and (2) How can we safely store them?

Much research has gone into safe storage of nuclear waste. So far, the most effective storage method seems to be the glassification (vitrification) process. In this technique, the radioisotope is incorporated into the chemical structure of glass and made into

BOX 5.6

FOOD THAT GLOWS IN THE DARK?

What is food irradiation? What advantages does it provide? Is the process safe? These are questions of potential importance to most people. The first, at least, is easy to answer. Food irradiation is a process whereby γ-rays bombard foods to extend shelf life and control the growth of potentially disease-causing bacteria and pests. Is it safe? Review the evidence, and decide for yourself.

In food irradiation, the food is placed on a conveying system and passed through a chamber where it is bombarded with γ-rays. The food never comes in direct contact with the radioactive cobalt–60 that produces the γ-rays, nor do the residual γ-rays remain in the food. Alternate radiation sources include X-rays or high-energy electrons; however, most irradiation facilities use Co–60 (figure 1).

This technique is not new; food irradiation has been studied for over forty years. In fact, more than 160 commercial food irradiators operate in 37 countries worldwide. The first country to irradiate food for public use was South Africa, in 1968, but the first large-scale food irradiator went online in 1973 in Japan. In Europe, irradiated foods include poultry, shellfish, eggs, spices, rice, onions, and wine. A few countries (such as Austria, Germany, New Zealand, Singapore, Sudan, Sweden, and Switzerland) ban food irradiation or the importation of irradiated foods, but most countries do not. The United States has lagged behind most of the world—you don't need even five fingers to count the number of full-time commercial irradiation plants in the United States, because only one exists!

This doesn't mean we never eat irradiated foods. The FDA approved irradiation of wheat and flour (to remove insects) in 1963; spices in 1983; pork in 1985 (to control *Trichinella spiralis*); and fresh fruits in 1986 (to delay ripening and retard mold growth).

The FDA also approved irradiation of poultry in 1992, to kill the *Salmonella* bacteria that contaminate almost half of all raw poultry. Still, the food industry hesitates to make use of the technique because of consumer concerns.

Is irradiation safe? Animals fed irradiated foods have never shown any signs of radiation sickness, and no incidents of human radiation sickness have been reported from other countries where food irradiation has been routinely used for decades. The U.S. space program has had over thirty-five years of safety and success with irradiated foods. Medical facilities have sterilized medical instruments with radiation for decades with no harm to patients. No evidence at all exists to date that irradiation makes food unsafe for human consumption.

Moreover, irradiation offers some health advantages. The technique—which leaves no residue—could replace potentially harmful chemical pesticides used on fruits and vegetables. Irradiation could make other foods safer as well. In 1993, on the west coast of the United States, a toxic strain of *E. coli* bacteria contaminated meat supplies and infected many people—some fatally. The outbreak, which affected between 400 and 1,000 people and killed 3, could have been averted if the beef had been irradiated with γ-rays to kill the bacteria. (Since this event, the American Meat Institute has spent $200,000 to study the effects of gamma irradiation on the five

Figure 1 (Above) FOOD TECHnology Service, Inc., a food irradiation plant in Mulberry, Florida. (Right) People safely load food for irradiation.

(Continued)

BOX 5.6—Continued

Figure 2 Irradiated strawberries have a twenty-to-thirty-day longer shelf life and are as nutritious as fresh-picked.

most common bacteria present in beef, including the virulent *E. coli* strain.)

Not all consumers object to irradiated foods. The strawberries irradiated at the Food Technology Service plant in Florida have a twenty-to-thirty-day longer shelf life and seem to have met with consumer acceptance; some prefer the irradiated strawberries over untreated fruit (figure 2). Foods that are irradiated are required to list that fact on the package. Still, despite these precautions and a forty-year record of safety, some people seek a ban on irradiated foods in the United States. Is such a ban in the public's best interest? You be the judge.

beads. The beads are then buried in lead containers. Where to bury these containers remains a problem—most people understandably do not want nuclear waste burial sites near their communities.

Some have suggested that nuclear waste should be disposed of near the places it is formed. With nuclear waste, this is a poor suggestion. Figure 5.11 shows the location of the nuclear reactors in the United States. Most are located near high-population areas.

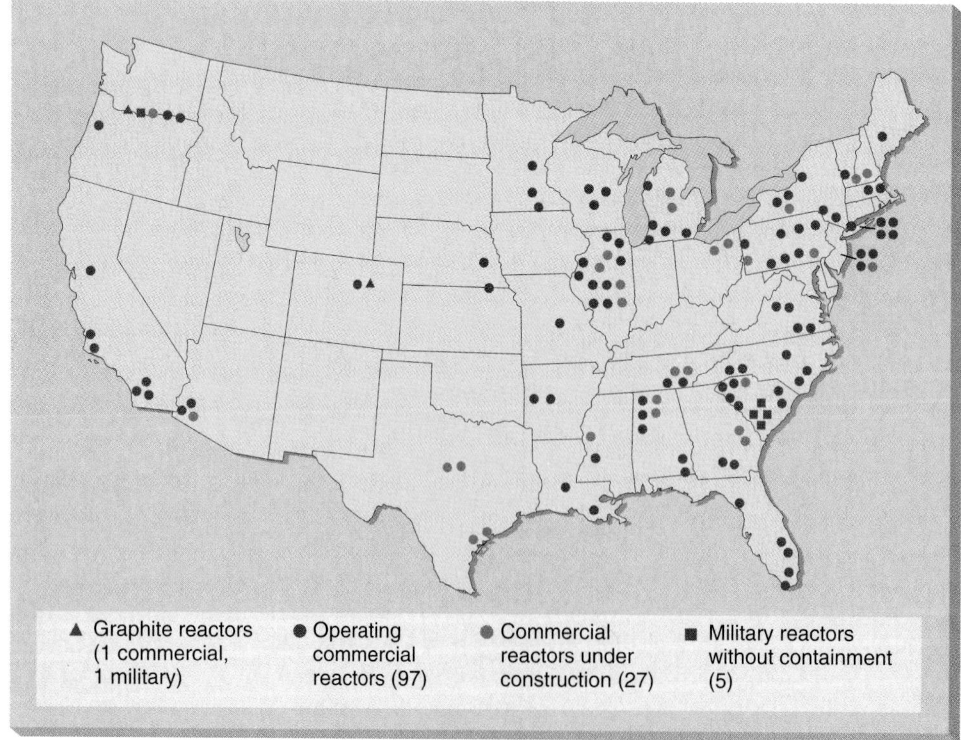

Figure 5.11 The location and status of each U.S. nuclear reactor as of 1990.

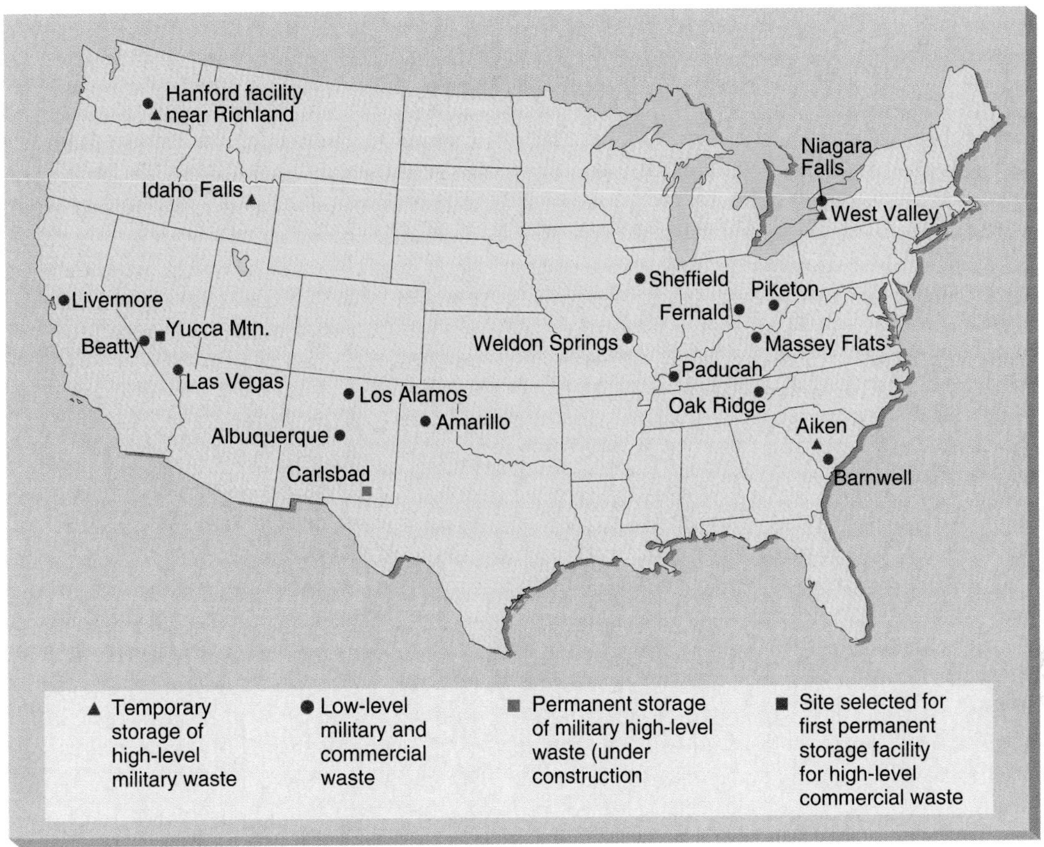

Figure 5.12 Proposed nuclear waste disposal sites.

Others are in regions not conducive to safe long-term storage; locations near large water sources, especially in porous rock or sand, are likely to erode, leading to premature release of the nuclear waste. For example, where might we bury nuclear waste in Florida, a state surrounded by water and composed largely of sand?

Numerous sites have been proposed for nuclear waste storage (figure 5.12). Many of the twenty-two sites shown are near military-related nuclear research centers (Oak Ridge, Hanford, Livermore, Los Alamos, and Aiken) designed for low-level storage or temporary use. The site chosen for long-term, high-level commercial storage is in the Yucca Mountains of Nevada, a relatively unpopulated area. But controversy has clouded this decision. Logically, the sites must be in stable geological formations, free from erosion due to underground or above ground water sources. The sparsely populated western states may be about the only places available.

Radon

Radon accounts for 55 percent of human exposure to radiation (see figure 5.6). This naturally occurring radioactive gas can seep into houses located near uranium-rich soils or rocks. Radon's half-life is 3.8 days, but it decays into radioisotopes called radon daughters; while most have short half-lives, some, such as lead–210 with a half-life of 22 years, last a fairly long time.

Some scientists believe the emissions from radon and its daughters may be responsible for causing many non-smoking-related cases of lung cancer. In 1987, a ten-state EPA survey centered around Pennsylvania claimed 21 percent of the surveyed homes had indoor radon-activity levels that exceeded safety standards. The EPA advised immediate corrective action, either venting the radon out of the dwellings or sealing exposed places that might permit radon into the homes.

Since then, people across the country have tested their homes for radon. Home testing kits are available, but some work poorly or not at all. However, although the radon risk is real and possibly dangerous in some areas, remember that 79 percent of the survey homes were deemed safe. Only in areas high in uranium should a radon (or radon-daughter) problem pose any threat.

SUMMARY

As scientists learned more about atomic structure, they also learned about different forms of energy—including the energy that holds the atomic nucleus together.

Energy may be either potential (stored) or kinetic (energy of motion). Although the energy in a system may convert back and forth from potential to kinetic energy, the First Law of Thermodynamics states that energy can be neither created nor destroyed; it simply converts into other forms. The Second Law of Thermodynamics goes on to assert that when energy converts, some of it tends to convert to heat energy that dissipates into the atmosphere and is then unavailable to do work.

Energy may take many different forms that convert from one to another. Mechanical energy is kinetic energy used in a machine or mechanism. It tends to be relatively inefficient (that is, much of the energy dissipates as heat). Thermal energy is heat energy, and thermodynamics is the study of heat flow. Heat results when atoms or molecules move and is therefore kinetic.

Electrical energy, another form of kinetic energy, involves the motion of electrons. It is a more efficient energy than either mechanical or thermal energy—electricity can travel long distances through power lines without much energy loss. Electricity is converted into light in incandescent, fluorescent, and neon lamps. Light is also kinetic energy because it consists of moving photon particles. Sometimes, a chemical reaction creates "cold light"—light without much heat—or chemiluminescence.

Chemical energy is potential energy stored in chemical bonds. The most common energy source on earth, chemical energy encompasses food energy, the energy in fossil fuels, and the energy stored in our bodies. Finally nuclear energy is energy stored in the nucleus—the energy that holds the nucleus together. When the nucleus splits open or fuses with the nucleus of another atom, a tremendous amount of energy is unleashed.

A nuclear reaction causes an atom to transmute from one element into another. This occurs naturally when a "heavy" isotope (containing more than the usual number of neutrons) breaks down into a more stable form. The decay of such isotopes releases radioactive α- β- and γ-rays. Certain atoms proceed through a radioactive decay series, decaying from one element to another to another as they lose α- and β-particles, until they arrive at a stable form. Each of these series progresses at a steady predictable rate; the amount of time is takes for half of a radioactive sample to decay into another element is called the original element's half-life.

Early scientists did not think much about radioactivity and safety, but the safety of nuclear reactions is a controversial issue today. Often, risk perception is based more on subjective opinion than scientific fact. As a result, the public overestimates some risks and underestimates others.

Nevertheless, several modern technologies put nuclear energy to use. Both nuclear fission (splitting) and fusion (joining) have produced powerful bombs, converting small amounts of matter into tremendous energy. Nuclear medicine employs radioactive matter to diagnose and treat human illness, and food irradiation—though controversial in the United States—extends shelf life and controls potentially harmful bacteria and mold growth. Nuclear reactors churn out electrical energy throughout the United States, but they present a problem—how should we dispose of the highly radioactive waste they produce? So far, most proposals involve burying waste in containers in geologically stable, sparsely populated regions.

Radon accounts for 55 percent of human radiation exposure. Although it poses a health risk in uranium-rich areas, radon is not a great threat to most people. Those at risk may take corrective action to seal or vent radon out of their dwellings.

Human life has always been exposed to natural forms of radioactivity. Future scientists will undoubtedly continue to search for ways to harness the incredible power of the atom while minimizing the risks of radiation.

KEY TERMS

alpha (α) particles 119
beta (β) particles 119
binding energy 128
carbon cycle 129
chemical energy 115
critical mass 129
electrical energy 114
electromagnetic fields (EMFs) 117
first law of thermodynamics 112
food irradiation 130
gamma (γ) radiation 119
half-life 121
kinetic energy 112
law of conservation of energy 112
law of conservation of matter and energy 116
light energy 115
mechanical energy 114
nuclear energy 116
nuclear fission 125
nuclear fusion 129
nuclear medicine 129
potential energy 112
radioactivity 119
radioactive decay series 119
second law of thermodynamics 112
thermal energy 114
thermodynamics 114

READINGS OF INTEREST

General

Nuclear news seems perpetually timely; reports appear in the newspapers or on television almost daily. As you read or listen examine this material critically, using the scientific method. The reports are not always accurate.

Spector, Tami I. "Naming names: A brief biography of women chemists." *Chemical Education*, May 1995, pp. 393–395. Though only eighteen women are listed, one, Arda A. Green (1899–1958), was the first person to isolate luciferase, the chemical that causes fireflies to light up the summer nights.

Hileman, Bette. "Findings point to complexity of health effects of electric, magnetic fields." *Chemical & Engineering News*, July 18, 1994, pp. 27–33. A follow-up of Hileman's earlier article.

Fackelmann, K. A. "Do EMFs pose breast cancer risk?" *Science News*, June 18, 1994, p. 388. The best current data suggest the answer is, "maybe."

Satyamurthy, N.; Namavari, M. Mohammad; and Barrio, Jorge R. "Making ^{18}F radiotracers for medical research." *Chemtech*, March 1994, pp. 25–32. This somewhat technical article describes how this radioisotope can be used to enhance PET (positron emission tomography) scans for the study of Parkinson's and Huntington's diseases and schizophrenia. It also shows how the isotope can be introduced into chemical compounds.

Gunther, Judith Anne. "The food zappers." *Popular Science*, January 1994, pp. 72–77, 86. An easy-to-read evaluation of food irradiation.

Hileman, Bette. "Health effects of electromagnetic fields remain unresolved." *Chemical & Engineering News*, November 8, 1993, pp. 15–29. Possibly the best, and simplest, introduction to this complex topic.

Raloff, Janet. "EMFs run aground." *Science News*, August 21, 1993, pp. 124-127. Although the article deals mainly with underground water pipes and their effects on EMfs, it also provides much general information.

Epperly, Michael W., and Bloomer, William D. "Radiotherapy: The closer the better." *Chemtech*, December 1991, pp. 744–749. More information on how this form of nuclear medicine works.

Background and Historical

Hively, Will. "X-ray dreams." *Discover*, July 1995, pp. 70–79. At present X-ray lasers, which could power microscopes that could peer into living cells and produce three-dimensional holograms, exist but are impractical. Charles Rhodes believes he may have the solution to making practical X-ray lasers.

Hamilton, Joan O'C. "X-rays under fire." *Business Week*, May 22, 1995, pp. 126–128. Are some cases of breast cancer caused by overuse of X-rays? Most experts doubt this, but one doctor feels certain the answer is yes.

Curie, Marie. 1961. *Radioactive substances*. New York: Philosophical Library. An account by one of the founders of nuclear science. Rather difficult reading, but it contains interesting historical insights.

Fermi, Laura. 1961. *The story of atomic energy*. New York: Random House. Good easy-to-read, historical account by the widow of one of the pioneers who worked on atomic energy.

Jungk, Robert. 1958. *Brighter than a thousand suns*. New York: Grove Press. Well-written story of the atomic bomb.

Hot and Cold Fusion

Dagani, Ron. "Latest research on cold fusion to be presented at Anaheim ACS meeting." *Chemical & Engineering News*, March 20, 1995, pp. 22–23. Much research is in progress on this controversial topic.

Storms, Edmund. "Warming up to cold fusion." *Technology Review*, May/June, 1994, pp. 19–29. This article concentrates mainly on positive recent results in cold fusion research.

Silber, Kenneth. "Fusion forces hot reactions." *Insight*, March 14, 1994, pp. 14–16. In spite of the fact that the government has spent billions of dollars researching hot fusion, huge technical problems must be overcome to make it a viable energy source.

Bishop, Jerry E. "It ain't over till it's over . . . cold fusion." *Popular Science*, August 1993, pp. 47-51. Not all scientists reject the possibility of cold fusion. This article presents quotes from both camps in the argument.

Peat, F. David. 1990. *Cold fusion. The making of a scientific controversy*. Chicago: Contemporary Books.

Nuclear Power Plants

Ryan, Megan. "Power move: The nuclear salesman target the third world." *World Watch*, March/April 1994, pp. 30–33. This article questions the United Nations' advocacy of nuclear technology without emphasis on renewable energy, natural gas, or improvement of existing energy systems.

Copulos, Milton R., and Simonson, Scott. "A peace dividend for nuclear power. *Also*, Don't be fuelish with warheads." *Insight*, January 24, 1994, pp. 22–27. With the end of the Cold War, thousands of nuclear weapons became available for possible peacetime use. Should they be buried or used in nuclear power plants? Two views.

Ahearne, John F. "The future of nuclear power." *American Scientist*, January/February 1993, pp. 24–35. Will the United States turn to nuclear power as electricity demand increases, nuclear costs come down, and fears about global warming grow?

Cook, William J. "Nuclear power, act II." *U.S. News & World Report*, May 29, 1989, pp. 52–53. Easy-to-read report on some newer nuclear reactor designs and their continuing promise for power generation with low environmental impact.

Lipkin, Richard. "A safer breed of reactor in sight." *Insight*, January 23, 1989, pp. 52–53. Simplified account of safer reactor in a field that keeps trying to improve.

Nuclear electricity and energy independence. U.S. Council for Energy Awareness, 1776 I Street N.W., Washington, DC 20006. Send for this free brochure that compares the costs and environmental impacts of various energy-generating methods. Worth reading, but obviously slanted in favor of nuclear power.

Nuclear Weapons

Burrows, William E. "Nuclear chaos." *Popular Science*, August 1994, pp. 54–59, 76. A popularized account of the status of nuclear fuels and waste from weapons manufacture.

Hileman, Bette. "U.S. and Russia face urgent decisions on weapons plutonium." *Chemical & Engineering News*, June 13, 1994, pp. 12–25. If these two nations do not exercise sufficient care, plutonium could fall into the hands of terrorists. This article tells how the plutonium is currently stored and includes a map showing major storage locations.

Panofsky, Wolfgang K. H. "Safeguarding the ingredients for making nuclear weapons." *Issues in Science and Technology*, Spring 1994, pp. 67–73. A reasoned plea for more controls.

Budiansky, Stephen. "The dim glow of history." *U.S. News & World Report*, April 18, 1994, pp. 74–75. One section of the Nevada desert is unlikely to become a major tourist attraction.

Budiansky, Stephen; Goode, Erica; and Gest, Ted. "The cold war experiment." *U.S. News & World Report*, January 24, 1994, pp. 32–38. Tells a small part of the story about secret radiation tests conducted on thousands of Americans, often without any informed consent, during the Cold War.

Herzenberg, Caroline L. and Howes, Ruth E. "Women of the Manhattan Project." *Technology Review*, November/December 1993, pp. 32–40. Contrary to the beliefs of some, the development of the atomic bomb was not an all-male project. This tells the story of some of the women scientists who worked on the Manhattan Project.

Nuclear Waste

Wheelwright, Jeff. "For our nuclear wastes, there's gridlock on the road to the dump." *Smithsonian*, May 1995, pp. 39–50. While people argue over the location of the long-term storage of nuclear waste materials, these radioactive wastes pile up in some of the least desirable places.

Pasternak, Douglas; Cary, Peter; and Martinez, Ancel. "Department of horrors." *U.S. News & World Report*, January 24, 1994, pp. 47–52. An article on the 22,824 Department of Energy workers who were injured on the job during the past four and a half years.

Carter, Luther J. "Ending the gridlock on nuclear waste storage." *Issues in Science and Technology*, Fall 1993, pp. 73–79. This article recommends the Yucca Mountain site in Nevada for nuclear waste storage and explains why the author feels it is the best choice.

Illman, Deborah L. "Researchers take up environmental challenge at Hanford." *Chemical & Engineering News*, June 21, 1993, pp. 9–21. Hanford is a plutonium preparation site. What potential pollution problems exist there?

Pasternak, Douglas, and Cary, Peter. "A $200 billion scandal." *U.S. News & World Report*, December 1992, pp. 34–46. An expose on the waste and contamination at U.S. nuclear weapons plants and the costs involved in clean-ups.

Stanglin, Douglas, and Pope, Victoria. "Toxic wasteland." *Also* "Breathing sulfur and eating lead." *Also* "Poisoning Russia's river of plenty." *U.S. News & World Report*, April 13, 1992, pp. 40–46. Three consecutive articles that cover the impact of the Chernobyl disaster and other environmental disasters in the former Soviet Union.

Pasternak, Douglas. "Moscow's dirty nuclear secrets." *U.S. News & World Report*, February 10, 1992, pp. 46–47. Nuclear waste disposal gone mad! Or, one reason you might not want to move to the former Soviet Union.

Kelly, Kevin. "What this town needs is . . . nuclear waste?" *Business Week*, August 19, 1991, p. 24. Why Martinsville, Illinois is seriously considering becoming a nuclear waste site.

Ahearne, John F. "Fixing the nation's nuclear-weapons plants." *Technology Review*, July 1989, pp. 24–29. Post–Chernobyl discussion of some of the nuclear plant problems in the United States.

Sweet, William. "Chernobyl—What really happened?" *Technology Review*, July 1989, pp. 42–52. Very good article on this topic.

Mould, Richard F. 1988. *Chernobyl: The real story*. New York: Pergamon Press. Possibly the most thorough account of this disaster, up to 1988.

Barlett, Donald L., and Steele, James B. 1985. *Forevermore: nuclear waste in America*. New York: W. W. Norton. Rather detailed discussion of nuclear waste disposal, its possible solutions and potential problems.

Radon

Cole, Leonard. "Radon, the silent killer." *Garbage*, Spring 1994, pp. 22–31. A fairly balanced account, with a rebuttal, of the radon problem.

Egginton, Joyce. "The radon menace: An update." *Reader's Digest*, 1989, pp. 141–146.

Lafavore, Michael. "Warning! This house contains radon." *Reader's Digest*, June 1986, pp. 110–114.

PROBLEMS AND QUESTIONS

1. Briefly describe each of the following forms of energy, and give some examples of each.

a. potential energy
b. kinetic energy
c. mechanical energy
d. thermal energy
e. electrical energy
f. light energy
g. nuclear energy
h. chemical energy

2. Diagram the flow of energy (from potential to kinetic) in a waterfall. Would you expect the water temperature to be higher at the top or the bottom of the waterfall?

3. Gasoline-powered generators are sometimes used for emergency back-up power. These devices also charge storage batteries for future electricity utilization. Outline the energy conversion pathways followed for such a use.

4. Explain why a sliding piece of metal comes to a stop, using the Law of Conservation of Energy in your explanation.

5. Relate recycling to the Law of Conservation of Matter and Energy.

6. Could you cool your house (or at least the kitchen) by leaving the refrigerator or freezer door open? Why or why not?

7. What types of energy do you use daily, and how can you reduce the total amount of energy you use in your home?

8. What happens from an energy standpoint when you start a fire by rubbing two sticks together?

9. What is chemiluminescence? Give some common examples of its use.

10. What contributions did each of the following people make to the field of nuclear science?

a. Glenn Seaborg
b. Philip Abelson
c. Edwin McMillan
d. Emilio Segrè
e. Enrico Fermi
f. Otto Hahn
g. Lise Meitner

11. What is a beta (β) particle? What is an alpha (α) particle?

12. What does each of the numbers represent in the symbol ${}^{200}_{80}Hg$?

13. Examine the following list of elements. Determine the numbers of neutrons and protons in each atom, and then answer the questions that follow.

${}^{235}_{91}Pa$ ${}^{235}_{92}U$ ${}^{238}_{92}U$ ${}^{239}_{93}Np$ ${}^{235}_{94}Pu$ ${}^{239}_{94}Pu$

a. Which of these materials are isotopes?
b. What are the materials called that are not isotopes?
c. Which materials have the same mass numbers?
d. Which nuclei have an identical number of protons?
e. Which nuclei have an identical number of neutrons?
f. Which materials have the same number of electrons?

14. Show with an equation what happens when ${}^{239}_{92}U$ decays and emits a β-particle (${}^{0}_{-1}e$).

15. Show the products resulting when ${}^{222}_{86}Rn$ emits an α-particle (${}^{4}_{2}He$).

16. Can a nuclear power plant explode like an atomic bomb? Why or why not?

17. Write an equation for the following sequence of events: ${}^{223}_{87}Fr$ emits a β-particle, and the product then emits an α-particle.

18. What is the difference between nuclear fission and nuclear fusion?

19. List several medical applications for nuclear technology.

20. Uranium–238 has a half-life of 4.5 billion years, while tritium's half-life is only 12.5 years. How many half-lives would it take for the radiation each material emits to drop to a safe level? How many years will this take for each element?

CRITICAL THINKING PROBLEMS

1. Modern automobile engines operate at higher temperatures than older models. Does this increase or decrease gas mileage? Why?

2. Radiation sometimes causes cancer, but it can also be used to treat and cure cancer. Can you explain this?

3. Do EMFs pose a potential health threat? Read box 5.3 and at least one of the following articles before answering this question: K. A. Fackelmann, "Do EMFs Pose Breast Cancer Risk?" *Science News*, June 18, 1994, p. 388; Bette Hileman, "Health Effects of Electromagnetic Fields Remain Unresolved," *Chemical & Engineering News*, November 8, 1993, pp. 15–29; and Bette Hileman, "Findings Point to Complexity of Health Effects of Electric, Magnetic Fields," *Chemical & Engineering News*, July 18, 1994, pp. 27–33.

4. Some reports suggest that the average American changes jobs every seven years. If one particular job is continuously refilled, how many people would theoretically occupy this job in a thirty-five-year period? Now suppose a company initially had 100 positions, and lost half its employees every seven years, but did not fill the positions as they were vacated. How long would it take for the company to drop to only 25 percent of its original work force? What is the "half-life" for jobs at this company?

5. A study conducted by the Office of Occupational Safety and Health Administration (OSHA), reported by the Associated Press during August 1994, estimated annual U.S. deaths due to second-hand smoke at 47,000. Read box 5.5 for background. Then calculate the death rate per 100,000 people, assuming 250 million as the U.S. population. Assuming the OSHA figure is valid, how does it compare to deaths from AIDS, skin cancer, and heart disease?

6. Would you eat irradiated food? Why or why not? Read box 5.6 and the following article for background: Judith Anne Gunther, "The Food Zappers," *Popular Science*, January 1994, pp. 72–77, 86.

7. On March 28, 1979, a meltdown incident occurred at the Three Mile Island nuclear plant near Harrisburg, Pennsylvania. The reactor emitted small amounts of radioactive iodine–131. If the $t_{1/2}$ for I–131 is 8.06 days, how long would it require for the radiation to diminish to a safe level? Later studies have not shown any significant increases in cancer or birth defect rates among the population near TMI. What does this suggest about nuclear safety?

8. Have your opinions about nuclear activity changed after reading this chapter? In what ways?

9. Read any or all of the following three articles on the challenges of maintaining safe storage for the plutonium from nuclear weapons. Then decide what you would do if the storage decisions were yours to make: William E. Burrows, "Nuclear Chaos," *Popular Science*, August 1994, pp. 54–59, 76; Bette Hileman, "U.S. and Russia Face Urgent Decisions on Weapons Plutonium," *Chemical & Engineering News*, June 13, 1994, pp. 12–25; and Wolfgang K. H. Panofsky, "Safeguarding the Ingredients for Making Nuclear Weapons," *Issues in Science and Technology*, Spring 1994, pp. 67–73.

10. The Department of Energy controls nuclear waste disposal in the United States. How well does the DOE do their job? Read the following articles for background: Douglas Pasternak, Peter Cary, and Ancel Martinez, "Department of Horrors," *U.S. News & World Report*, January 24, 1994, pp. 47–52; Luther J. Carter, "Ending the Gridlock on Nuclear Waste Storage," *Issues in Science and Technology*, Fall 1993, pp. 73–79; Deborah L. Illman, "Researchers Take Up Environmental Challenge at Hanford," *Chemical & Engineering News*, June 21, 1993, pp. 9–21; and Douglas Pasternak and Peter Cary, "A $200 Billion Scandal," *U.S. News & World Report*, December 1992, pp. 34–46.

11. Suppose you discovered high radon levels in your home. What would you do? How likely is this to occur where you live?

12. The EPA has published a brochure that claims 14,000 annual deaths occur from radon exposure. This is greater than the combined annual number of deaths from airplane crashes, drownings, and fires. Assuming the EPA figure is valid, what level of risk does this estimate entail? For background information, read box 5.5 and Leonard Cole, Radon, "The Silent Killer," *Garbage*, Spring 1994, pp. 22–31.

13. What would happen if a law (presumably divine) were passed that suddenly outlawed all forms of nuclear fusion and fission?

14. One possible nuclear waste storage area is at the bottom of the oceans. Assuming scientists could design containers that would not rupture for hundreds of years, does this idea seem feasible? What would you recommend if the decision were yours?

HINT Answers will vary, but it is important to consider all possible options. Nuclear waste generally must be stored in a safe place for 400 years or longer. It cannot be stored near high-population areas, nor buried in earthquake-prone regions. And no matter where you decide the best burial place would be, thousands of angry protesters and activists will rail against you. Can you come up with a reasonable option, even under these difficult conditions?

15. Read the following article: Joan O'C. Hamilton, "X-rays Under Fire," *Business Week*, May 22, 1995, pp. 126–28. On the basis of the limited information presented, form a tentative opinion on whether X-rays may have caused some cases of breast cancer in the past. Today's X-rays are at lower doses. Would they be likely to cause breast cancer? Finally, what effect might genetics have on this issue? Many studies imply a strong genetic component for breast cancer.

16. Some recently released documents suggest that workers in nuclear plants were deliberately exposed to radiation so that the effects could be determined. This occurred during wartime, when some believed this information was desperately needed. Was this approach ethical? What alternatives existed?

CHAPTER 6

CHEMICAL BONDING—ALL-PURPOSE MOLECULAR GLUE

OUTLINE

KEY IDEAS

1. The Main Features of Metallic, Ionic, and Covalent Bonding
2. How the Octet Rule Drives Chemical Bonding
3. How to Write a Lewis Electron Dot Structure for Main Group Elements
4. Determining the Valence Number from the Periodic Chart and Using It to Write Correct Formulas for Compounds
5. The Nature of Multiple Covalent Bonds
6. How Electronegativity and Ionization Potential Affect Bond Formation
7. Why Hydrogen Bonds are so Important
8. The Use of the Valence Shell Electron Pair Repulsion (VSEPR) Method

"Blest be the tie that binds."
JOHN FAWCETT (1740–1817)

Wood fire.

Bonding and chemical reactions lie at the very heart of chemistry. Chemists run reactions, synthesize compounds, and analyze substances or processes. Almost all of chemical technology stems from reactions, breaking and reforming chemical bonds, and the compounds that result. Before we can explore chemical synthesis and analysis, we need to learn more about the bonds that hold atoms together. **Chemical bonds** occur in several types: metallic bonds (metal to metal), ionic bonds (metal to nonmetal), covalent bonds (nonmetal to nonmetal), and hydrogen bonds (molecule to molecule).

METALLIC BONDING

About 80 percent of the elements in a periodic table are metals. A **metal** consists of spherical atoms packed into a total volume as compactly as possible to form a regular crystalline solid. Obviously, according to this model, metals have high density. We know they also are hard and malleable, have high boiling points, and conduct heat and electricity. These unique characteristics stem from the way metal atoms bond together.

Scientists believe that the outermost electrons in each metal atom are shared equally by *all* the metal atoms in a sample, rather than being associated with particular atoms. These "free" electrons can then conduct heat and electricity throughout the metal. For example, sodium atoms have a total of 11 electrons, with just 1 in the valence, or outermost, energy level. Scientists think this outer electron can break free from the sodium atom, making it a positively charged ion, and wander freely throughout the entire sample. When several sodium atoms are together, this arrangement produces a collection of sodium cations (Na^+) embedded in a sea of freely moving electrons, e^- (figure 6.1). This **metallic bonding** model is reminiscent of Thomson's plum-pudding model (see chapter 4), except in this case, the "plums" occupy most of the "pudding" volume, with a tiny amount of electron "sauce" poured around them. At no time is any Na^+ ion far away from an electron. The metal atoms are therefore held together by the electrical interaction between the positively charged ions and the negatively charged electrons. Metals do not, therefore, form compounds with other metals; they form alloys, solutions of one metal in another that seldom have definite formulas or structures.

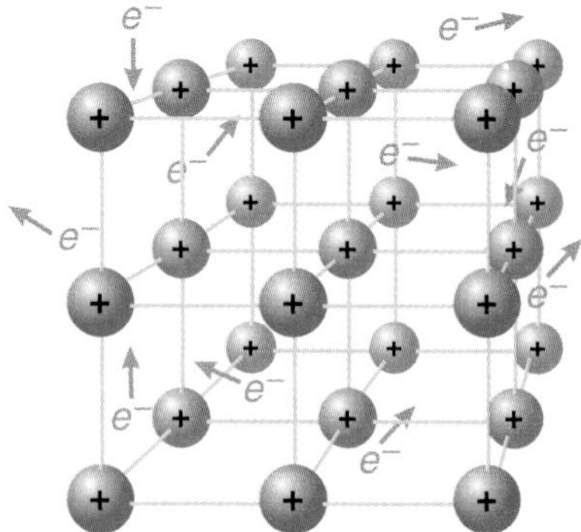

Figure 6.1 A metallic bonding model. The sodium cations are embedded in a sea of electrons that come from the valence (outermost) shell.

Electrical conductivity is a general metal property because all metal atoms have free outermost electrons. Electricity flows through these electrons when one end of a piece of metal is more positively charged than the other.

EXPERIENCE THIS

All metals conduct electricity, but do all metals conduct to the same extent? Construct the apparatus shown here using the lamp portion of a flashlight, three pieces of copper wire, and a dry-cell battery (a typical type of C or D battery). Hook two sections of copper wire to the battery as shown; one should run from the battery to the flashlight lamp and one should run from the other electrode on the battery to the test object (any piece of metal). Now run the third piece of wire from the test object to the lamp. Does it light?

Place different pieces of metal (test objects) between the battery and the lamp, connecting battery, test object, and lamp with the copper wire. Try iron (steel), lead solder, aluminum, gold, and silver. Do they all light the lamp to the same degree of brightness? Which ones best conduct electricity? Summarize your results in a table.

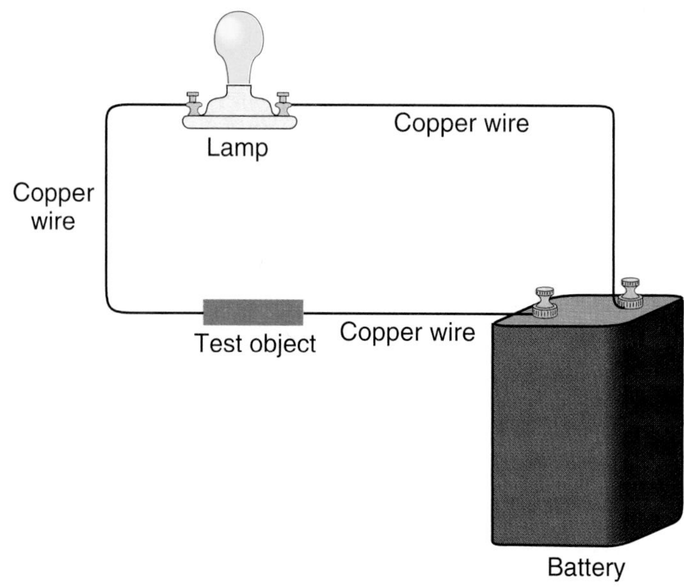

Many metals, such as aluminum and magnesium, have enormous practical importance in our society, but most commercially used metals are transition elements. Table 6.1 shows a few examples of metals and their uses. Let's look at aluminum in more detail.

Aluminum

By mass, aluminum is the third most common element on Earth (after oxygen and silicon) and the most abundant metal in the Earth's crust. Because it does not occur in nature as a free metal, aluminum has found widespread commercial usage only in this century. Previously, this metal was used only in jewelry and specialty items because it was so expensive. In fact, one member of the nobility served a banquet using aluminum dishes instead of the customary gold. Today, anyone can eat from aluminum dishes, or cook in aluminum pots and pans, because the cost of the metal has drastically decreased.

TABLE 6.1 Some important metals and their uses.

METAL	SYMBOL	TYPICAL USES
Aluminum	Al	Soft drink cans, metal foil, construction, wiring
Chromium	Cr	Alloys, plating on metals and plastics
Cobalt	Co	Electroplating, ceramics, lamp filaments
Copper	Cu	Electrical wiring, plumbing, heating, construction
Germanium	Ge	Electronics, semiconductors, alloys
Gold	Au	Precious metal, space vehicles, printed circuits
Iron	Fe	Steel manufacture, construction, magnets
Lithium	Li	Batteries, alloys
Magnesium	Mg	Alloys, space missiles, photography, fireworks
Manganese	Mn	Alloys with iron, aluminum manufacture
Mercury	Hg	Thermometers, batteries, mercury-vapor lamps
Nickel	Ni	Alloys, electroplating, batteries, fuel cells
Platinum	Pt	Precious metal, catalyst, dentistry, surgery
Silver	Ag	Precious metal, photography
Tin	Sn	Tin plate, foil, alloys such as pewter and bronze
Titanium	Ti	Aircraft, corrosion-resistant metal, alloys
Tungsten	W	Alloys, high-speed tools, lightbulb filaments
Zinc	Zn	Plating, galvanizing, alloys such as brass and bronze

Aluminum occurs naturally in several different types of compounds. Some gemstones, such as rubies and sapphires, are aluminum compounds. Both kaolin (china clay) and ordinary clay contain impure aluminum silicates, compounds formed from the three most abundant elements. The main aluminum ore, bauxite, which was discovered in 1820 in Baux, France, is largely aluminum oxide, Al_2O_3 (the ore formed by the combination of aluminum and oxygen). Cryolite, another aluminum ore, is a mixed salt of aluminum fluoride and sodium fluoride with the formula $AlF_3 \cdot 3NaF$.

Chemists have used a variety of processes to isolate aluminum from the compounds that contain it. In 1827, Friedrich Wöhler (German, 1800–1882) isolated a small quantity of aluminum metal by treating aluminum chloride with potassium, a highly reactive, expensive metal. In 1855, Henri-Etienne Sainte-Claire Deville (French, born in the Virgin Islands, 1818–1881) replaced the potassium in Wöhler's process with less expensive sodium, producing kilogram-sized aluminum ingots. These ingots created a sensation when they were displayed at a World Exposition in Paris later that year; but while the cost of aluminum declined a hundredfold when this process was invented, the metal remained expensive. Napoleon III underwrote some of Deville's research and development costs because he believed (correctly) that the lightweight metal had great military potential. However, the Deville process had some undesirable side effects, such as chlorine vapor pollution and unsanitary working conditions. Moreover, the process had produced only about 12,000 kilograms of aluminum by 1855.

Although Deville realized bauxite would be a less expensive aluminum source than aluminum chloride, he was unable to isolate the aluminum from bauxite. Michael Faraday (English, 1791–1867), a pioneer in both electricity and chemistry, had isolated aluminum via electrolysis in 1833, but the process required molten ore, and bauxite's high melting point (2,000+°C) made it economically infeasible. Eventually, in 1886, Hall and Héroult succeeded in extracting aluminum from bauxite (box 6.1). Finally, aluminum was sufficiently cheap and available for widespread human use.

THE OCTET RULE

Compound formation is governed by the **octet rule,** which Gilbert N. Lewis (1875–1946) introduced in 1916. The rule stems from the observation that most atoms tend to end up with an octet (eight) of electrons in their outermost or **valence** shell after forming a compound. Whenever an atom gains, loses, or shares electrons to achieve an octet in its outermost shell during compound formation, the arrangement of electrons (or electronic configuration) in the resulting atom matches the configuration of a nonreactive noble gas atom. This gives the compound unusual

BOX 6.1

HALL, HÉROULT, AND ALUMINUM

In one of those remarkable coincidences that seem to occur in science, both Charles M. Hall (American) and Paul-Louis Toussaint Héroult (French), who each developed a process for isolating aluminum from bauxite ore, were born in 1863 and died in 1914. Hall's interest in aluminum was sparked when he heard Frank F. Jewett, a former student of Wöhler, describe a silvery, lightweight metal that no one had yet found a way to produce cheaply. Years later, after successfully inventing an aluminum-production process, Hall presented an aluminum pellet to Jewett.

Charles M. Hall and Paul-Louis Toussaint Héroult.

Both Hall and Héroult were familiar with the Deville process of aluminum extraction, which required the conversion of bauxite ore (Al_2O_3) to aluminum chloride. Both set out to isolate aluminum directly from bauxite.

The main problem the two scientists faced was to find an energy source powerful enough to rip the aluminum from its oxide ore. Battery power produced only a small electrical current and still required the aluminum chloride conversion. Electrolysis required molten ore, and since bauxite melts at 2,000+°C, the huge amount of energy needed to heat the ore would make the process too expensive. But electrolysis can also work when a compound is in solution. Hall and Héroult independently discovered that bauxite dissolves in cryolite, $AlF_3 \cdot 3NaF$, and that the resulting solution melts below 1,000°C. Hydroelectric plants, first built in 1882, could provide enough power to electrolyze the solution at the lower temperature.

Both researchers also discovered that the electrolysis process required carbon; they used carbon electrodes to pass an electrical current through bauxite dissolved in molten cryolite (figure 1). The chemical reactions that result can be summarized as:

$$Al_2O_3 + 3\,C \rightarrow 2\,Al + 3\,CO$$

The carbon monoxide (CO) burns to form CO_2, and pure aluminum remains.

Hall and Héroult's breakthrough had enormous commercial implications. (Hall founded the company that later grew into the industrial giant ALCOA, formerly Aluminum Corporation of America—the largest producer of aluminum in the world; Héroult, though initially not as successful, eventually sold his process to European manufacturers including Péchiney and Company, now Europe's largest aluminum producer.) The price of aluminum plummeted; the metal now costs about $0.60 per pound, a far cry from the $30,000 per pound it once commanded. The chemical extraction of aluminum from bauxite thus turned a precious metal into an everyday commodity. A typical American uses about 42 pounds

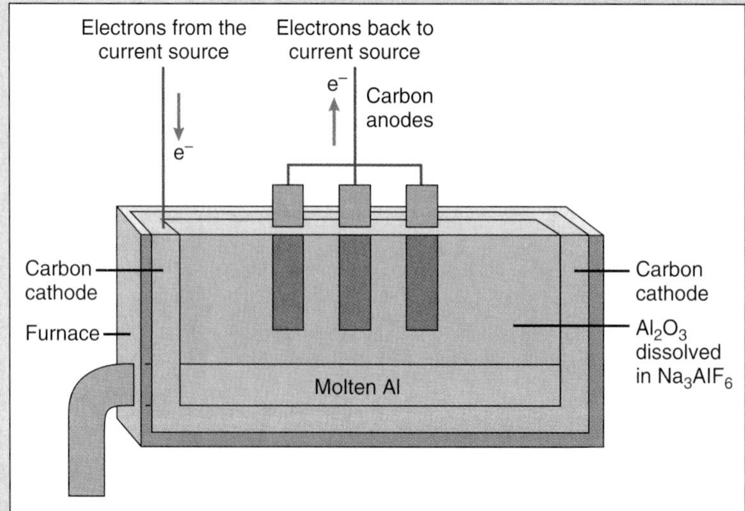

Figure 1 Aluminum manufacture via the Hall process.

(19 kg) of aluminum each year; 31 percent is used in packaging (mostly beverage cans), 22 percent in transportation (mostly automobiles), 19 percent in building materials, 10 percent in electrical wiring, and the remaining 18 percent in miscellaneous applications (figure 2).

In addition, recycling can potentially make aluminum an even more practical, economical metal. Presently, it requires the energy equivalent in electrical power of about 3 ounces of gasoline to produce one 12-ounce soft drink container. Recycling requires just 5 percent as much energy. Although the United States currently recycles only 20 percent of its aluminum, this metal is the most profitable material to recycle; most recycled materials cost more than their virgin counterparts, but recycled aluminum actually nets about $800 profit per ton. Perhaps increased aluminum recycling can offset the losses of recycling other materials and encourage further energy conservation.

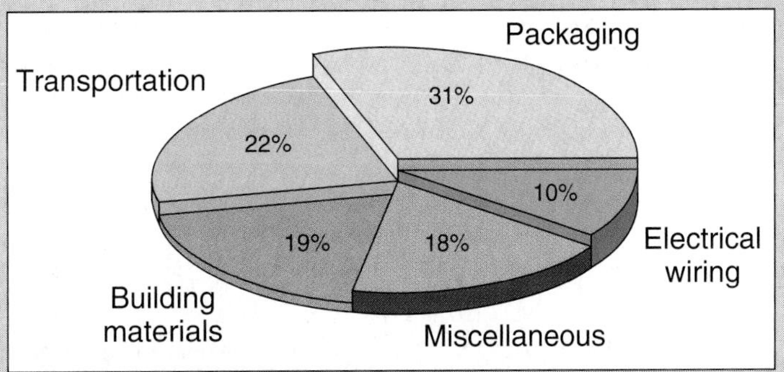

Figure 2 The uses for aluminum metal in the United States. Packaging includes beverage cans and averages about eight pounds per person per year.

stability. It also means that a minimal amount of energy is used to form compounds. (In general, low energy usage confers greater stability on a compound, whereas a greater energy input creates instability.) Most elements tend to adhere to the octet rule. The main exceptions are hydrogen and helium atoms, which fill their outermost shells with only two electrons.

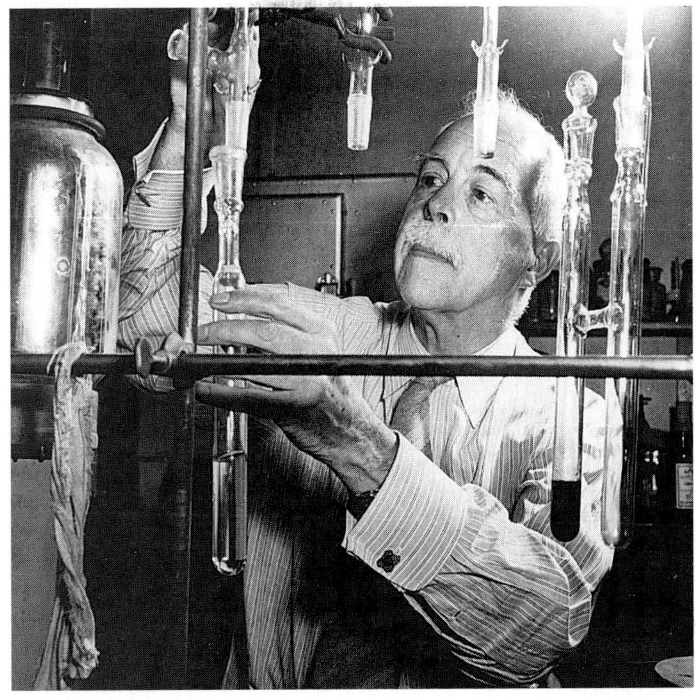

Gilbert N. Lewis.

Octet Bonding Rules

The main group family members (that is, the A-families) follow the octet rule when they form compounds, achieving a noble gas electronic configuration of eight electrons in their final outermost shells. They can achieve a full set of 8 electrons either by sharing electrons in a covalent bond or transferring electrons from one atom to the other in an ionic bond. Compounds are therefore either covalent or ionic. (Metallic bonds do not form compounds; they form alloys.)

The Noble Gases

When atoms follow the octet rule in forming compounds, they emulate the electron configuration of the nearest noble gas. For many decades, chemists called the noble gases (family VIIIA) *inert gases* because these elements did not form any compounds. They attributed this inert behavior to the fact that electrons completely filled the noble gases' outermost, or valence, electron shells. Adding additional electrons to a noble gas requires enormous amounts of energy; it is virtually impossible. Similarly, removing electrons from a noble gas atom requires so much energy it is unlikely to occur.

In 1962, Niel Bartlett showed that one of the "inert gases" formed the compound xenon platinum hexafluoride ($XePtF_6$). Bartlett's startling discovery was initially presented to just seven people, including the speaker and the meeting chairman, at the 141st semiannual American Chemical Society meeting (see box 6.2). Since this landmark event, scientists have discovered many other noble gas compounds, though none have achieved commercial or practical importance. Since these elements proved not

BOX 6.2

THE AMERICAN CHEMICAL SOCIETY

The American Chemical Society is a professional organization for chemists and related scientists. Founded in 1876, the ACS is headquartered in Washington, DC and currently counts nearly 150,000 members worldwide. The ACS is divided into many specialized subunits, called either Divisions or Secretariats, to meet the needs of its members and to disseminate information about various aspects of chemistry. This is partially accomplished through two national meetings each year, which attract multitudes of suppliers (figure 1) and 10,000 to 20,000 attendees. At the national meetings, several thousand papers are presented on the latest developments in chemistry.

The ACS also has local sections and student affiliate groups that meet on a regular basis. The society publishes scientific journals and also aims publications at elementary and high school students in addition to college students and chemists. The weekly *Chemical and Engineering News (C&EN)* carries news about chemists and chemistry and helps maintain contact between members and other interested persons. Its featured articles help chemists keep up to date in a rapidly changing field of knowledge.

Figure 1 The chemical exposition at the August 1994, American Chemical Society national meeting in Washington, D.C.

truly inert, chemists renamed them "noble," referring to them as the nobility among elements.

Although the noble gases are not completely inert, chemists still consider them nonreactive. Just as a completely filled valence shell makes the noble gases extremely stable, a completed electron octet confers unusual stability on a cation, anion, or neutral atom in a compound. The simplest way to show these filled shells is with Gilbert Lewis's electron dot structures.

Lewis Electron Dot Structures

Lewis devised a simple method for depicting the octet rule by drawing dots around the element's symbol to represent the electrons in the outermost shell of the atom. Such a depiction is called a **Lewis electron dot structure.** Figure 6.2 shows the electron dot structures for each element in the first three periods of the periodic chart. Only the outer shell electrons are shown.

Note that an individual electron is placed on each side of the symbol before any electrons are paired. This leaves a maximum number of unpaired electrons to become involved in bond formation. Note also that the numbers above the main group families correspond to the number of outermost electrons (family IIIA has three valence electrons, family VIIA has seven, and so on). These electron numbers determine a related value called the valence number for each element.

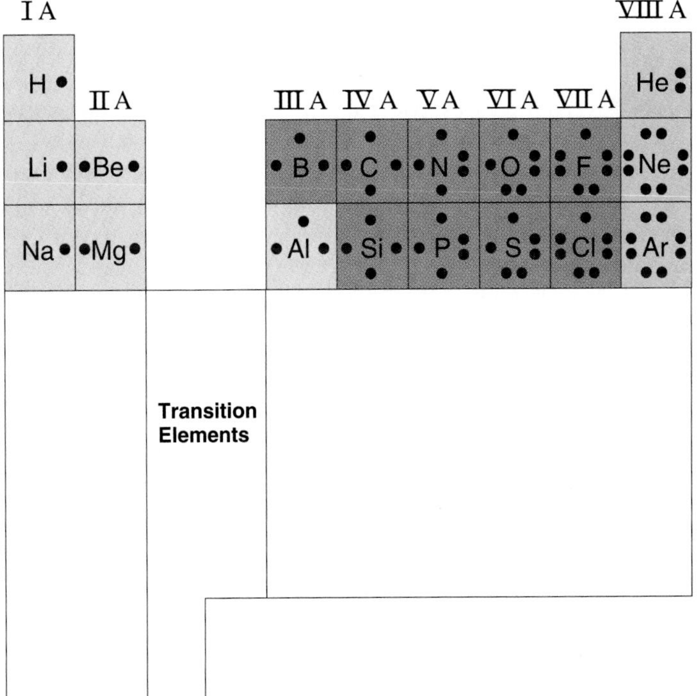

Figure 6.2 Lewis electron dot structures for the first three periods. An electron is placed on each side of the symbol before any electrons are paired. This produces the maximum number of unpaired electrons, leaving them available to form bonds.

VALENCES AND CHEMICAL BONDING

An element's **valence number,** which may vary between 1 and 4, indicates how many electrons an atom must gain or lose to achieve an octet in its outermost orbital (figure 6.3). This number corresponds to the position of a main group element in the periodic chart. Thus, group IA, with 1 valence electron, only needs to lose 1 electron to achieve an octet; group VIIA must gain 1 electron to achieve an octet. An element in group IVA can either gain or lose 4 electrons to achieve its octet. Whether the resulting bond is ionic (gaining or losing electrons) or covalent (sharing electrons) depends on the nature of the elements forming the compound. If one of the elements is a metal, ionic bonds form; if both elements are nonmetals, covalent bonds form.

Valence Number Value	Main Family Group Number	
4	IVA	
3	IIIA	VA
2	IIA	VIA
1	IA	VIIA

Figure 6.3 The relationship between valence number and main group family number.

PRACTICE PROBLEM 6.1

What is the valence number for each of the following elements: B, Mg, P, Se, Sb, Rb, Br, Ar?

ANSWER Locate each element in the periodic chart and note the family number at the top of the vertical column. Then find the valence number for this family from figure 6.3 using this column number. For the above elements, the answers are: B = 3; Mg = 2; P = 3; Se = 2; Sb = 4; Rb = 1; Br = 1; Ar has no valence number because it is in the VIIIA column.

Valence Number Ground Rules

Three valence number ground rules (abbreviated VR) help determine (1) the correct formula for a compound formed between main group elements, and (2) whether this compound is ionic or covalent:

VR-1. In compounds of only two elements that have the same valence number, the compound formula contains one atom of each. For example, in the compound potassium chloride, potassium (K) comes from family IA and chlorine (Cl) from family VIIA. According to figure 6.3, they both have a valence number of 1. Therefore, the compound contains one atom of each; the formula—KCl—reflects this ratio. (Since this compound formed from a metal and a nonmetal, it is an ionic compound. The metal (K) lost an electron—in keeping with the octet rule—while the nonmetal (Cl) gained 1.

VR-2. When two elements in a compound have different valence values, the valence numbers may be multiplied by a valence factor. The **valence factor** for one element is normally the same value as the valence number of the other element; this means that multiplying the two will result in equal valences—numbers of electrons lost or gained—for both types of atoms. The valence factor also tells us the number of atoms required to form a compound—if the first element has a valence number of 1 and the second atom has a valence number of 3, three atoms of the first element will bond into one molecule of the compound. This valence factor appears as a *subscript* to the atom in the chemical formula. (A valence factor of 1 is left unwritten.) Because the valence factor for an atom is identical to the valence number of the other atom, the application of this rule is sometimes called the crossover method.

To understand how VR-2 works, consider the formula for a compound containing the elements calcium (Ca) and bromine (Br). Figure 6.3 shows that the valence numbers are 2 for Ca and 1 for Br (family VIIA). VR-2 says to multiply the Br valence number by a valence factor of 2 (the valence number of Ca) to make the *total* Br valence identical to the total valence of the Ca atom; since multiplying yields 2, we know that 2 electrons will be involved in the transfer. VR-2 also tells us that the final chemical formula contains one Ca and two Br atoms: $CaBr_2$. Since calcium is a metal, this is an ionic compound. The Ca atom lost 2 electrons, while each Br atom gained 1.

VR-3. By custom, when a compound contains a metal and a nonmetal, the formula is written with the symbol of the metal atom first and then the nonmetal symbol. (See the periodic table in figure 4.1 to determine which elements are metals.) Hydrogen behaves as a metal with all nonmetals except for families IVA and VA.

To see rule VR-3 in action, consider the formula of a compound made from sulfur and hydrogen. Figure 6.3 shows that the valences are 2 for S (family VIA) and 1 for H (family IA). After applying VR-2 by multiplying hydrogen's valence number by a factor of 2 (the valence number for S), we know the formula will contain H_2 and S. Since hydrogen is treated as a metal according to VR-3, we place it first in the formula: H_2S. However, the bond in this compound is covalent because both elements are actually nonmetals.

Let's look at one last example of how the valence rules work. What is the formula of the compound formed between the elements aluminum (Al) and oxygen (O)? The periodic chart and figure 6.3 shows us that Al has a valence number of 3, while O has a valence number of 2. VR-2 tells us to multiply the Al valence number by 2 and the O valence number by 3, creating a total valence of 6 for each atom. It also tells us the compound will contain Al_2 and O_3, and VR-3 instructs us to write the formula with aluminum first—Al_2O_3. Each Al atom lost 3 electrons (for a total loss of 6), while each O atom gained 2 electrons (for a total gain of 6) during the formation of this ionic compound.

Generalizations about Bonding

Compound formation involves much more than merely following the correct formulas. Just as three valence rules help determine chemical formulas and the types of bonding that occur when a compound forms, five bonding generalizations (BG) help predict how elements combine and give us a clearer understanding of exactly how compounds form.

BG-1. Metals form ionic bonds with nonmetals; this gives the metal a positive charge making it a cation, and the nonmetal a negative charge, making it an anion. This charge is identical to the atom's valence number.

BG-2. Nonmetals form covalent bonds with each other, with no charge on either atom.

BG-3. Hydrogen generally acts as a nonmetal, creating covalent bonds.

BG-4. Metals normally form metallic bonds with other metals. The formula for a pure metal is simply the symbol of the metal atom.

BG-5. Noble gases (family VIIIA) do not normally form compounds.

One other bonding observation is in order. Ionic compounds (or salts) are sometimes characterized as having formula units instead of molecules. But since formula units behave like normal molecules, the terms are used interchangeably in this book.

Let's look back at some of our earlier examples of compounds, this time applying the bonding generalizations. First, consider KCl. This compound is ionic because it forms from metallic K and nonmetallic Cl. BG-1 tells us that the ions are K^+ and Cl^-.

The second example compound, $CaBr_2$, is also ionic because it forms from Ca, a metal, and Br, a nonmetal. According to BG-1, the calcium becomes a Ca^{2+} cation (because its valence number is 2) and the bromine becomes a Br^- anion (Br's valence number is 1).

The next compound, H_2S, is covalent since both H and S are nonmetals (BG-2 and BG-3). There are no charges on either atom.

The last example, Al_2O_3, involves bonding between a metal and a nonmetal; it is therefore ionic (BG-1). The ions are Al^{3+} cations and O^{2-} anions.

The three VRs and five BGs help us determine almost everything we need to know about bonding and chemical formulas. You need not memorize the BGs because you will learn to apply them by solving the problems at the end of this chapter. In a similar manner, the best way to learn the three VRs is by problem solving.

PRACTICE PROBLEM 6.2

Which of the following compounds are ionic and which are covalent:
BaO, NO_2, Na_2O, $CaCl_2$, H_2Te, ClBr, IrS_2, CS_2?

ANSWER The ionic compounds, consisting of a metal and a nonmetal, are BaO, Na_2O, $CaCl_2$, and IrS_2. (The latter—which includes a transition element—was included to show that these rules are almost completely general, and are not restricted to the main group elements.) The covalent compounds, consisting of two nonmetals, are the remaining ones: NO_2, H_2Te, ClBr, and CS_2.

IONIC BONDS

As you have already learned, **ionic bonds** form between metals and nonmetals. This section explores in more detail how and why ionic bonds form. Ionic compounds are common in nature and industry; table 6.2 lists some typical examples. We can best understand ionic bond formation by examining a specific case following the octet rule.

The Saga of Salt City, or NaF Formation

Although the word *salt* is commonly used to describe ordinary table salt (NaCl), this term applies to many other compounds as well. In chemistry, the word **salt** describes the reaction product between acids and bases (studied further in chapter 8); salts are always ionic compounds.

Let's consider the formation of a salt, sodium fluoride (NaF), from elemental fluorine and sodium. Since elemental fluorine is diatomic (occurring as two atoms) while elemental sodium is a single atom, two Na atoms must react with one F_2 molecule to yield two NaF molecules. (A reaction equation must contain the same numbers of atoms on each side of the reaction arrow.)

$$2\ Na + F_2 \longrightarrow 2\ NaF$$

Sodium, an alkali metal family (IA) member, has 11 protons, 12 neutrons (usually), and 11 electrons. The halogen (VIIA) fluorine has 9 protons, 10 neutrons, and 9 electrons. Sodium has only 1 electron in its outer shell. Fluorine, on the other hand, has 7 electrons in its valence shell.

The next shell inward in sodium, energy level 2, contains a complete octet of electrons, just like the closest noble gas, neon (Ne). If Na could shed its outer shell electron, the new outermost shell—energy level 2—would contain the desired octet (figure 6.4a). The Na atom would still have 11 protons, but only 10 electrons; it thus would become a Na^+ cation with 11 protons, 12 neutrons, and 10 electrons. The outer-shell electron configuration of Na^+ is now identical to the configuration of the Ne atom. It holds a stable octet, but the atom is also charged.

In a similar manner, fluorine can achieve an outermost shell octet by gaining 1 electron (figure 6.4b). But just as borrowing can cause problems between humans, transferring electrons can cause some new, and perhaps unforeseen, results. If the fluorine atom gains an electron, it will have 10 electrons and only 9 protons. The F atom then bears a negative charge and becomes a fluorine anion, F^-. The fluorine atom also has a filled octet identical to the configuration of Ne, but it is now negatively charged.

Experimental evidence supports these ideas. If we measure the initial sizes of the Na and F atoms and compare them with the sizes of the ions in the salt Na^+F^-, we would find that the Na^+ cation is smaller than the Na atom, while the F^- anion is slightly

TABLE 6.2 Common ionic compounds and their uses.

COMPOUND	FORMULA	TYPICAL USES
Aluminum phosphate	$AlPO_4$	Paints, cosmetics, dental cements
Barium nitrate	$Ba(NO_3)_2$	Fireworks, electronics, rat killer
Calcium carbonate	$CaCO_3$	Building materials (limestone, marble, chalk), toothpaste
Copper chloride	$CuCl_2$	Wood preservation, fireworks, dyes
Lead sulfide	PbS	Ceramics, semiconductors
Magnesium sulfate	$MgSO_4$	Cosmetic lotions, medicines
Manganese dioxide	MnO_2	Matches, fireworks, textile dyeing
Potassium bromide	KBr	Medicine, photography, soaps
Silver chloride	AgCl	Photography, medicines, batteries
Sodium bicarbonate	$NaHCO_3$	Baking powder, beverages, ceramics
Sodium chloride	NaCl	Table salt, seasoning, preservation
Sodium fluoride	NaF	Rat poison, fungicide, fluoridation
Sodium hydroxide	NaOH	Soap making, cleaning, paper making
Titanium dioxide	TiO_2	Paint pigment, cosmetics, ceramics
Zinc chromate	$ZnCrO_4$	Pigment, rust-resistant primer

Note: Chapter 8 will discuss the nature of groups of atoms such as phosphates, nitrates, and carbonates. These groups behave as a unit, similar to an individual chlorine or sulfur atom.

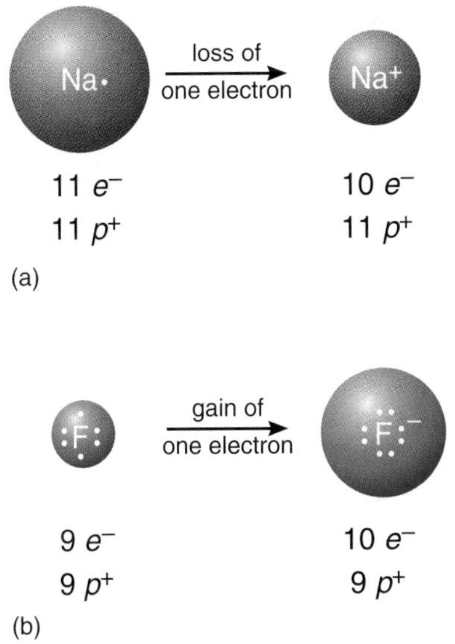

Figure 6.4 **(a)** A sodium atom loses an electron, becoming a cation. **(b)** A fluorine atom gains an electron, becoming an anion. In each case, the final ion has the neon octet. The relative sizes of the Na atom, the Na^+ cation, the F atom, and the F^- anion are shown.

larger than the F atom (figure 6.4). In the Na, the entire third energy level was emptied, making the atom shrink as it formed the Na^+ cation. The F, by contrast, had to swell a little to hold one extra electron and become the F^- anion. Figure 6.5 shows the arrangement of the Na^+ and Cl^- ions in another compound, sodium chloride.

In actual practice, the Na atom transfers its outer-shell electron directly to the F atom (figure 6.6). Both atoms thus achieve the desired octet configuration simultaneously. This electron exchange is the driving force behind the reaction, which releases enormous amounts of heat. In this reaction, the reactants are more valuable and expensive than the product; in fact, NaF is made less expensively from HF and NaOH: HF + NaOH $\rightarrow$ NaF + HOH. Sodium fluoride is a chemical used in fluoridation and fluoridated toothpaste (see chapter 12).

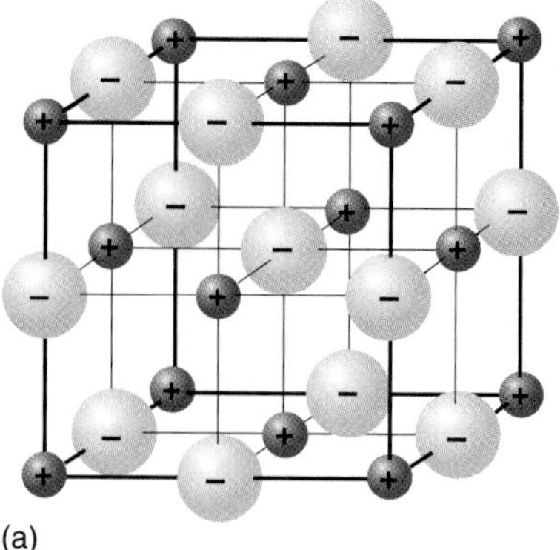

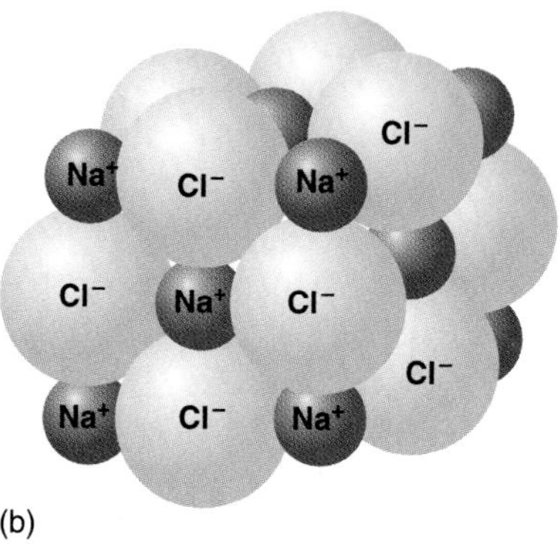

Figure 6.5 **(a)** The arrangement of the Na^+ and Cl^- ions in sodium chloride or table salt. **(b)** The arrangement of the Na^+ and Cl^- ions in sodium chloride or table salt, showing the relative sizes of the ions.

EXPERIENCE THIS

Invent a game that helps you understand electron configurations and ionic bonding. Imagine a sodium atom that dreams of becoming noble, like noble neon, and living a life of ease and leisure. Perhaps name these characters something like Sam Sodium and Nita Neon. Sam Sodium could, under certain conditions, deposit his "extra" outer-shell electron in an electron bank to fulfill this fond dream. This action, like depositing funds in your savings account, means losing immediate use of the electron; but Sam also gains something—the security of having a "noble" octet in his new outermost shell.

Now imagine Flo Fluorine borrowing an electron from the electron bank, filling her outermost energy shell, and achieving an octet like Nita Neon's. In each case, Sam Sodium and Flo Fluorine reach the octet state by becoming ions—or getting a "charge" out of life. In real life, no electron bank exists. Both atoms achieve the desired neon status when Sam Sodium gives an electron to Flo Fluorine.

Ionic Bonding and Multivalent Ions

In figure 6.6, both atoms were monovalent, or had a charge or valence number of 1. Let's examine the more complex situation shown in the following equation, involving aluminum (Al) and oxygen (O). Metallic Al has a valence of 3, and nonmetallic O has a valence of 2. The bond between them is ionic. If we consult the Lewis electron configuration shown in figure 6.2, we see that Al has 3 "extra" electrons in its outer level; if they are lost, Al would achieve an octet like that of neon, the previous noble gas. Similarly, O needs 2 electrons to achieve the Ne octet. We know from VR-2, the crossover rule, that a compound will contain Al_2 (because O's valence number is 2) and O_3 (because Al's valence number is 3).

However, the situation is complicated by the fact that oxygen is diatomic—it normally occurs as O_2. Therefore, we need three O_2 in the reactants to produce two O_3 in the product; and to balance those two O_3, we need two Al_2.

The equation describing this chemical reaction would thus read:

$$\underset{\text{metal}}{4\,Al} + \underset{\text{nonmetal}}{3\,O_2} \rightarrow \underset{\text{ionic compound}}{2\,Al_2^{3+}O_3^{2-}}$$

This equation is balanced with four Al and six O atoms (or ions) on each side of the arrow. Each Al atom must lose 3 elec-

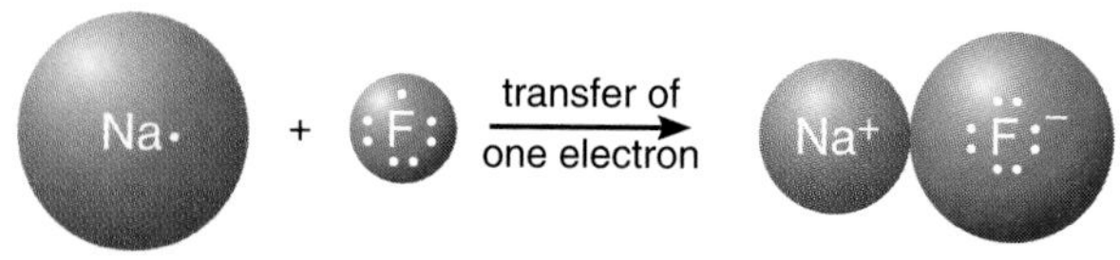

Figure 6.6 The Na atom transfers its electrons to an F atom, forming a Na^+ cation and an F^- anion in a sodium fluoride molecule.

trons and each O atom must acquire 2. This creates a total loss of 12 electrons from the four Al atoms and a gain of 12 electrons by the six O atoms. Figure 6.7 shows this atom-to-atom transfer, using Lewis electron dot structures; all the resulting ions have a neonlike outer shell octet. Multivalent ions, or ions with charges greater than one, form bonds in the same way as monovalent atoms. Note that the Al becomes larger and the O smaller when they become ions.

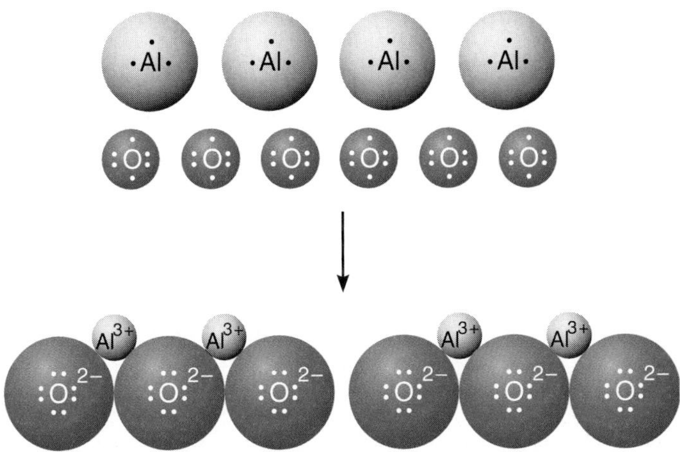

Figure 6.7 Transferring 12 electrons from four Al atoms to six O atoms forms four Al^{3+} cations and six O^{2-} anions in two Al_2O_3 molecules. In the end, each ion has a neon octet. The relative sizes of the atoms and ions are shown.

PRACTICE PROBLEM 6.3

Using the Lewis electron structures shown in figure 6.2, demonstrate how the following reaction takes place:

$$Mg + Br_2 \longrightarrow MgBr_2$$

ANSWER The Mg atom, with 2 outer electrons, transfers 1 electron to each Br atom. Because it loses 2 e^-, Mg becomes a Mg^{2+} cation, while each Br, gaining one e^-, becomes a Br^- anion.

COVALENT BONDS

Covalent bonds, in which atoms share electrons, are the most common type of chemical bond; most chemical compounds essential for life form covalent bonds. This may seem remarkable when you consider that covalent bonding only occurs between nonmetals, which comprise only about 20 percent of all elements. But the vast majority of chemical compounds contain covalent bonds. Table 6.3 lists some common covalent compounds, along with typical uses.

Covalent Bond Formation—or Sharing is Caring

The simplest covalent compound is the hydrogen molecule, H_2, that forms from two individual H atoms. To fill its outermost shell, a hydrogen atom needs 2 electrons; each atom actually has 1 proton and 1 electron. Which hydrogen atom gives up an electron, and which gains an electron in bond formation? Or, to put this another way, which H atom becomes a cation (actually a "naked" proton) and which becomes an anion (figure 6.8a)?

There is no reason why either H atom should lose an electron to the other, nor is there any reason why the other should acquire this electron; the H atoms are identical. Fortunately, an alternative bond exists. Rather than gaining or losing electrons to form ions, the pair can share their 2 electrons and form a covalent bond (figure 6.8b). When each H atom shares its electron, the outer shell is complete, with 2 electrons, like the noble gas helium. (In this case, a duet forms rather than an octet, but the basic principle is the same.) Neither atom actually surrenders or gains another electron. Sharing an electron pair lowers the total energy and gives stability to the bond.

Covalent bonds are not limited to hydrogen atoms. When the diatomic chlorine molecule, Cl_2, forms, two chlorine atoms share a pair of electrons (figure 6.9). This enables each to achieve the desired electron octet of the noble gas argon. Each Cl atom normally has 17 protons, 18 neutrons, and 17 electrons—7 in the valence shell. By sharing, they can each complete the octet and match the Ar configuration.

TABLE 6.3 Common covalent compounds and their uses.

COMPOUND	FORMULA	TYPICAL USES
Acetone	C_3H_6O	Solvent, cosmetics, cleaning solutions
Ammonia	NH_3	Fertilizers, refrigerant, dyes
Butane	C_4H_{10}	Fuel, lighters, propellant, refrigerant
Glycerol	$C_3H_8O_3$	Plastics, soaps, cosmetics
Methane	CH_4	Fuel, byproduct from rice paddies
Phenol	C_6H_6O	Plastics, medicines, dyes
Sucrose	$C_{12}H_{22}O_{11}$	Sweetener (ordinary table sugar), syrups, candy
Water	H_2O	Drinking, solvent, paper making, coolant

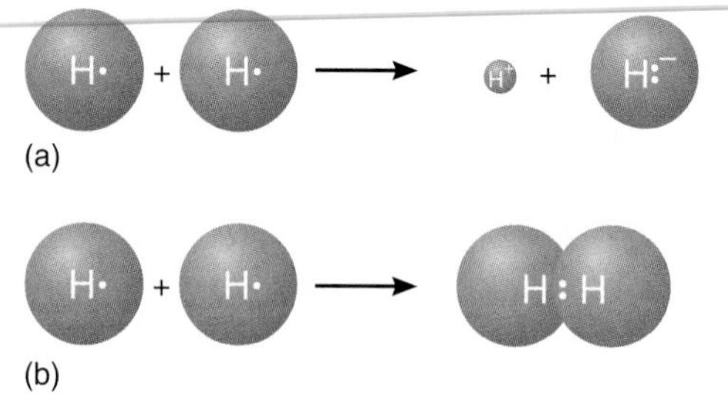

Figure 6.8 Bond formation between two hydrogen atoms. **(a)** If an ionic bond formed, one hydrogen would lose its electron and the other gain it. **(b)** In actuality, the two H atoms share their two electrons in a covalent bond.

In Cl_2, the covalent bond forms as each atom shares 1 electron. The remaining 6 electrons in each atom are not involved in bond formation; they are called nonbonding electrons. Though these nonbonding electrons, shown in the Lewis dot structures, do not bond, they can have an effect on the molecule's properties.

PRACTICE PROBLEM 6.4

Use Lewis dot structures (see figure 6.5) to draw the covalent bond that forms in the bromine molecule, Br_2.

ANSWER Odd as this may seem, the correct answer looks exactly like the structure in figure 6.9, except the element symbol is Br instead of Cl:

:Br· + ·Br: ⟶ :Br:Br:

This is because both chlorine and bromine are halogens, members of family VIIA, and they have the same number of outermost electrons. The only difference—which doesn't appear in the Lewis dot structures—is that each Br atom emulates the krypton (Kr) rather than Ar structure.

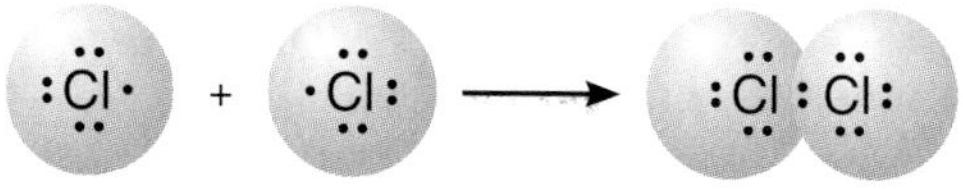

Figure 6.9 Covalent bond formation: two chlorine atoms share electrons.

Multiple Covalent Bonds

Some elements do not achieve a stable electron octet by sharing just one electron pair. The diatomic molecule N_2, for example, shares three electron pairs to achieve a stable octet (figure 6.10). If the pair shared just one electron pair, each atom would end up with only 6 electrons in its valence shell. If they share two pairs of electrons, each N atom in the double bond would have 7 electrons in its outer shell. Each N atom achieves a full octet only by sharing three pairs of electrons and forming a triple bond. N_2 therefore contains a **multiple covalent bond.**

Although N_2 always contains a triple bond, two nitrogen atoms may single or double bond if some of their unpaired electrons have already combined with electrons from other atoms to form covalent bonds. Organic molecules (to be discussed further in chapter 10) frequently contain such formations.

Other main group elements also form double or triple bonds. Carbon and silicon form double or triple bonds in compounds. Figure 6.11 shows the formation of single or double bonds between carbon and oxygen using Lewis electron dot structures. Each final structure in figure 6.11 contains some unpaired carbon electrons. These will eventually pair with electrons from some other atom to form additional covalent bonds. In *a*, the oxygen atom still has one unpaired electron, which will form a covalent bond with some other atom; in *b*, the O atom has no unpaired electrons and will form no more covalent bonds. Covalent carbon compounds, called organic compounds, are more important than any other class. Chapters 9 and 10 will discuss them extensively.

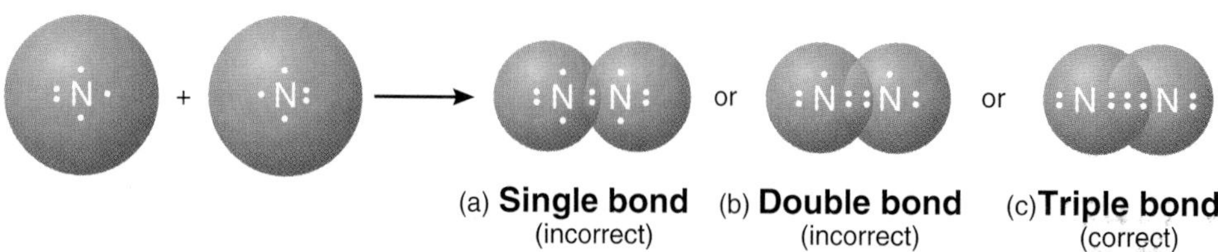

Figure 6.10 The formation of **(a)** single **(b)** double, and **(c)** triple covalent bonds between two nitrogen atoms. The actual structure for the N_2 molecule corresponds to **(c)**; a triple bond forms between a pair of nitrogen atoms. A single or double N—N bond will only form if some of the unpaired electrons on each N atom are already combined in covalent bonds with other atoms.

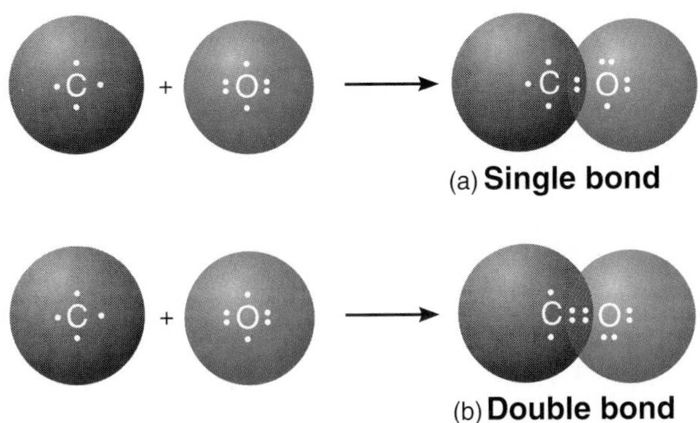

Figure 6.11 The formation of **(a)** a single or **(b)** a double bond between carbon and oxygen. In **(a),** notice that the C atom still has three unpaired electrons, and the O atom still has one. Each of these electrons must eventually combine with an electron from a different atom and form another covalent bond. In **(b),** notice that the C atom still has two unpaired electrons after forming the double bond with the O atom, but both unpaired electrons in the original O atom are now involved in bond formation. The two unpaired electrons on the C atom must each combine with an electron from a different atom and form one more covalent bond, but the O atom cannot form any more covalent bonds because it has no unpaired electrons.

ELECTRONEGATIVITY AND IONIZATION POTENTIAL

We have established that metals form ionic bonds with nonmetals, while nonmetals form covalent bonds when they bond together. Two important questions remain. First, do atoms share the electrons equally in all covalent bonds? And second, why do metals form ionic bonds with nonmetals? The answer to both questions depends on two fundamental properties: electronegativity and ionization potential.

Electronegativity

Although atoms share electrons in covalent bonds, they do not always share an electron pair equally. When the two atoms are identical, they share electrons equally; when the atoms are different elements covalently bonded, the electrons are more attracted toward one atom than the other.

Electronegativity, or EN, measures the attraction of an atom's nucleus toward electrons. Figure 6.12 shows a portion of the periodic chart that contains the EN values for many of the elements. The chart is divided into three sections: metals, nonmetals, and hydrogen, which is separate because it can behave like either a metal or a nonmetal. EN values increase as you move toward the upper righthand corner of the periodic chart and decrease as you move toward the lower lefthand corner. In addition, EN values increase going from left to right across a period (a horizontal row) or bottom to top within a family (vertical row). Fluorine is the most electronegative element. (EN values are unknown for the noble gases.)

As figure 6.12 indicates, nonmetals are more electronegative than metals. An element with low EN does not tend to hold its electrons strongly and will not attract more electrons. These low-EN elements surrender their electrons more readily than others because they have lower EN values. For example, Na gives up its

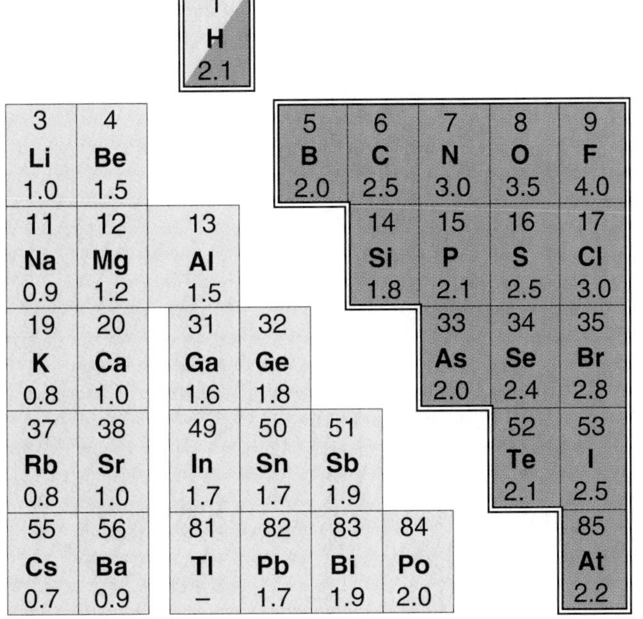

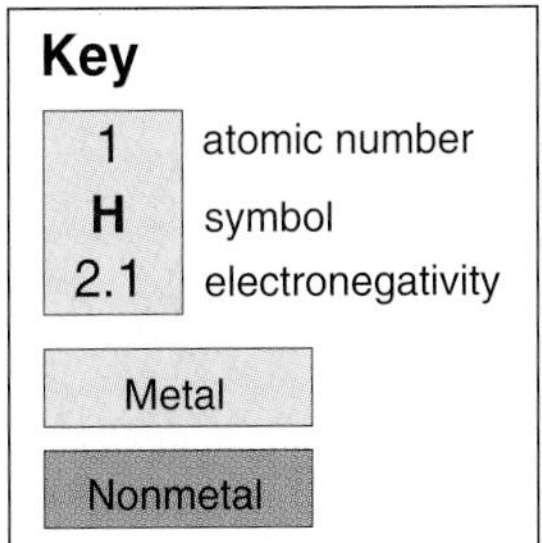

Figure 6.12 The electronegativity values of the main group elements arranged in a modified periodic table.

single outer-shell electron more readily than Mg gives up its two, but both lose electrons and form ionic compounds with nonmetals. On the other hand, Mg surrenders its electrons far more readily than nonmetallic Cl. In a similar manner, some elements attract electrons more than others. F attracts electrons more strongly than Cl, which attracts electrons more than Br or I. Imagine EN as a rating system for an element's ability to "vacuum" up electrons; the higher the EN number, the better the element "sucks up" or attracts electrons.

PRACTICE PROBLEM 6.5

Using figure 6.12, rank the elements from lowest to highest in electron-attracting ability within each of the following sets:

- **a.** Na Mg P Al
- **b.** I Cl Br F
- **c.** Si P O N
- **d.** O Br C Na

ANSWER Lowest to highest: *(a)* Na, Mg, Al, P; *(b)* I, Br, Cl, F; *(c)* Si, P, N, O; *(d)* Na, C, Br, O.

Ionization Potential

Ionization potential (IP) measures the amount of energy required to remove an electron from an atom, making it an ion. In a sense, it is the opposite of electronegativity; while electronegativity measures an atom's ability to attract electrons, ionization potential measures the atom's ability to hang onto electrons even when another atom attracts them. In another sense, however, EN and IP are closely related concepts, since they both measure the atom's ability to attract and hold electrons. In fact, elements with high EN values tend to have high IP values as well.

While electronegativity is a constant value, an atom's ionization potential increases as the atom loses electrons. When one electron is removed, the resulting positive charge amplifies the ion's attraction to negatively charged electrons, making it more difficult to lose another electron. The first two IP values, the amounts of energy needed to remove one or two electrons from an atom, are shown in figure 6.13.

Figure 6.13 divides the elements into four groups: metals, nonmetals, noble gases, and hydrogen. As you can see, for any given element, the energy required to remove an electron increases dramatically as each succeeding electron is removed. Thus, for fluorine, the IP values increase from 17.4 to 35.0 as one or two electrons are removed; this makes sense since fluorine tends to *attract* electrons rather than give them up. Noble gases have exceedingly high IP values—it takes so much energy to dislodge their elec-

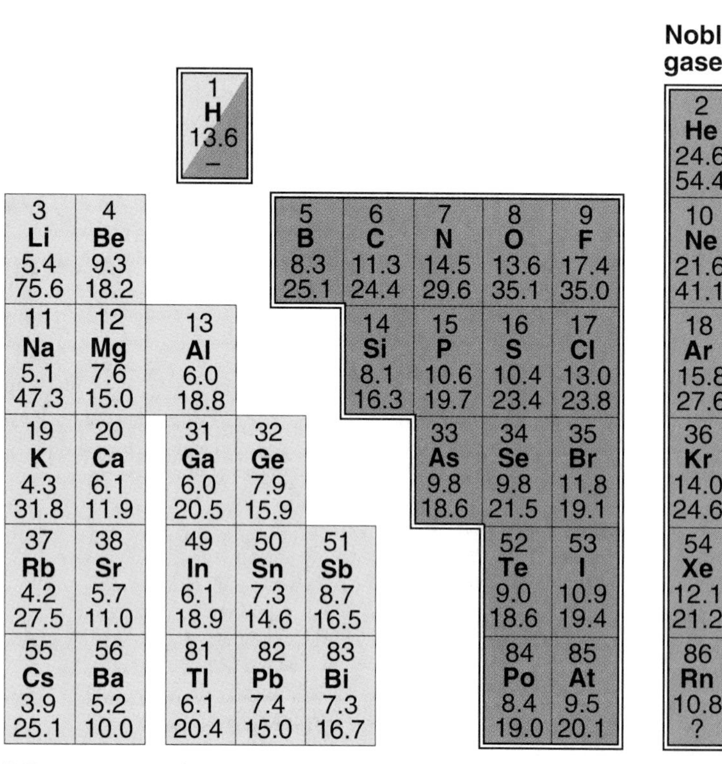

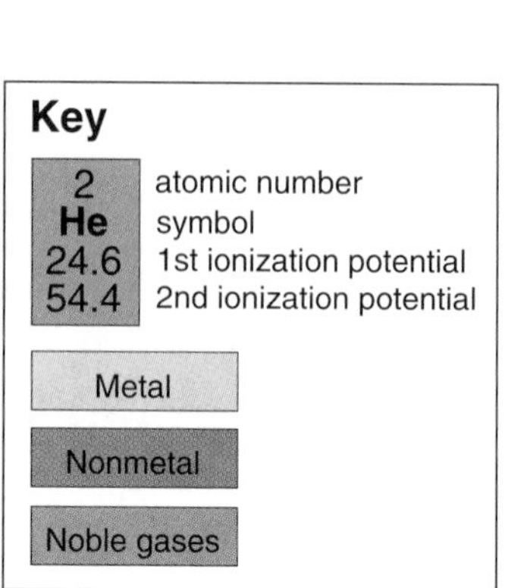

Figure 6.13 The ionization potential values (in electron volts) of the elements arranged in a modified periodic table.

trons that they do not form chemical compounds readily. The IP values therefore confirm earlier assumptions about the stability of the electron octet. As figure 6.13 shows, nonmetals have higher IP values than metals. This means metals lose electrons—and become cations—more easily. This is one reason why metals tend to form ionic rather than covalent bonds.

PRACTICE PROBLEM 6.6

Using figure 6.13, rank the following elements in decreasing order of first IP value: C, Ca, H, Br, Na, O.

ANSWER The rank (from highest to lowest) is O, H, Br, C, Ca, Na.

Ionic versus Covalent Bond Formation

The closer the EN values are for two elements forming a molecule, the more equally they share the electron pair in a covalent bond, since each atom attracts the electrons with equal force. This means that atoms of the same element, and therefore the same EN, share electrons equally. We can expect approximately equal electron sharing in bonds between such element pairs as C—S, H—P, and N—Cl, which have nearly the same EN values. We would not expect equal electron sharing between C—F, or H—O because these elements differ significantly in EN values. In these pairs, the atom with the higher electronegativity would attract the electrons more powerfully.

Ionic bonds form when the difference in EN is great, while covalent bonds form when the difference is small—but the transition is not so sharp that we can always neatly predict which type of bond will form by determining the EN difference. In this book, we assume that metals and nonmetals usually form ionic bonds (since they tend to differ widely in EN values), while nonmetals (since they generally have similar EN values) tend to form covalent bonds. Box 6.3 focuses on one covalently bonded substance—silicon dioxide.

POLARIZED COVALENT BONDS

When the difference between the EN values of two atoms is marked enough, one atom attracts electrons more strongly than the other. Sometimes this results in an ionic bond; sometimes it results in a **polarized covalent bond,** when two atoms *share* electrons unequally because one attracts the electrons more strongly.

In an ionic bond, when an electron transfers completely from one atom to the other, the atoms acquire positive or negative charges and become ions. In a nonpolar covalent bond, the atoms share electrons equally and acquire no charge. A polar covalent bond is intermediate between the two; the atoms share electrons, but since one pulls them more powerfully, that atom acquires a partial negative charge—more than a nonpolar bond, less than an ionic bond. Table 6.4 summarizes these bonds, shows how electronegativity affects bond formation, and also compares the effect of an electrical field on ionic, covalent, and polarized covalent bonds. As box 6.4 points out, some substances contain combinations of these bonds.

Partial Charges

Bonds between chlorine and fluorine atoms (C—F) and between hydrogen and oxygen atoms (H—O) are essentially covalent, but the atoms do not share electrons equally—instead, the more electronegative atom (F or O) attracts a greater allotment of the shared electron pair. Still, since the other atoms (C and H) continue to share rather than give up the electron pair, the F and O atoms acquire only a **partial charge.**

Partial charges, denoted as $\delta+$ or $\delta-$, are expressed as a fraction of a whole charge; this fraction indicates the degree to which the bond is ionic in character. A bond with 50 percent ionic character possesses about half the charge of a single electron. The electron does not split in half; it just lies closer to one atom than the other and thus interacts with one nucleus more than the other. In a 50 percent covalent (or 50 percent ionic) bond, the electron pair lies about three-quarters of the distance toward the more electronegative element, instead of midway between each atom.

TABLE 6.4 Comparison between types of bonds that can form between pairs of elements.

	IONIC BOND	NONPOLAR COVALENT BOND	POLAR COVALENT BOND
Example	Na^+Cl^-	H:H	$H:\ddot{\underset{\cdot\cdot}{O}}:$ with H below O
EN difference	Large	Small	Moderate
Behavior with electricity	Conducts when dissolved or melted	Nonconductor; does not align with electric field	Nonconductor; does align with electric field

BOX 6.3

SILICON DIOXIDE—FROM SAND TO SEMIPRECIOUS STONES

Silicon, a solid nonmetal, is the second most common element on earth (after gaseous oxygen), making up about 25.7 percent of the earth's crust. Silicon is not naturally found in its uncombined state, but covalently bonded silicon dioxide, SiO_2, is very common—it is found as sand on beaches everywhere. Covalently bonded silicon dioxide is present in many other substances as well, including quartz, mica, asbestos, glass, and the gemstones agate, amethyst, and jasper.

QUARTZ AND MICA

In quartz, silicon and oxygen atoms are covalently bonded into a three-dimensional structure similar to a diamond (figure 1). The basic unit is a SiO_4 tetrahedron (a four-sided figure with triangular walls), but the Si:O ratio is 1:2 because each O atom, at the edges of each tetrahedron, is shared by two Si atoms. Mica also consists of SiO_4 tetrahedra, but they are arranged in a flat sheet that cleaves into thin, transparent layers of isinglass, a substance used for windows before glass became common.

High-purity quartz can generate an electric potential when its structure is reshaped; this property, shared by many other crystalline solids, is called the piezoelectric effect. The piezoelectric effect, discovered by Pierre Curie in 1880, is the basis for the operation of quartz watches, noted for their accuracy (figure 2). A quartz crystal can transform mechanical energy into electrical energy, or the reverse. If it is subjected to an applied alternating current with a specific frequency, the crystal vibrates in an exact resonance, creating a standard for measuring time. The piezoelectric effect is also used in police and citizen's band radio transmitters and in pressure-sensing devices.

ASBESTOS

Another silicon dioxide-containing product, asbestos, has excellent thermal insulation properties and is nonflammable. For this reason, asbestos was widely used to insulate furnaces, steam lines, boiler rooms, water pipes, brake linings, and firefighters' fire-resistant clothing. In fact, the U.S government mandated the use of asbestos because it was far superior to other materials for these uses.

Unfortunately, concerns arose about the potential health hazards of asbestos exposure, causing the government to ban asbestos use. These hazards were mainly associated with the inhalation of short fibers about 10 to 50 μm long, which causes a condition called asbestosis when the fibers become embedded in the lung sacs. In some cases, asbestosis can lead to lung cancer or a rare form of cancer that affects the body cavity. Eventually, researchers learned that a relatively uncommon form of asbestos called crocidote was the main culprit in asbestosis. Chrysotile, the more common form, does not cause these health problems; but the governmental restrictions on asbestos failed to distinguish between the forms and banned asbestos of any type.

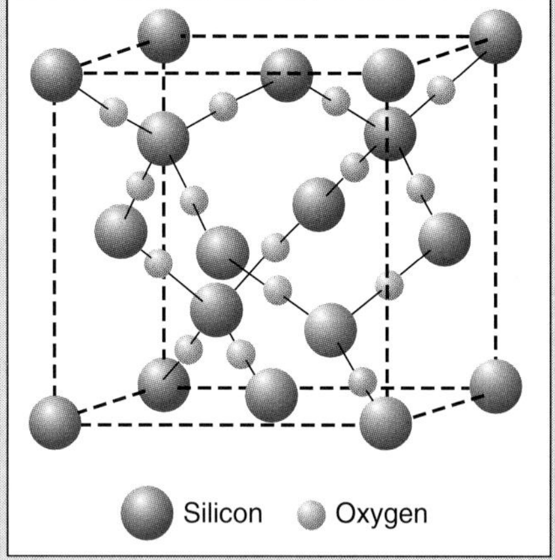

Figure 1 The tetrahedral structure of silicon dioxide.

Figure 2 A quartz crystal watch. Because of quartz's piezoelectric properties, this watch keeps very accurate time.

The government regulations enforced the removal of much of the fire-resistant asbestos from public buildings, such as schools, at an enormous price tag (more than $100 billion dollars). Most risk assessments indicate that this removal was unnecessary. Any type of asbestos, including the crocidote form (which comprises less than 5 percent of the total), is relatively harmless when fibers are firmly in place as insulation. In fact, removal can stir up the fibers and potentially cause more harm than leaving the asbestos in place. Still, it is unlikely we will see a return to widespread asbestos usage in the near future.

GLASS AND GEMSTONES

Glass-making technology is of more recent origin than the ability to create ceramics (see box 6.4). Egyptians made glass beads between 3000 and 2500 B.C., and produced larger glass objects by 1400 to 1350 B.C. Syrians developed glass blowing (that is, forming glass objects by blowing air through molten glass) between 50 and 41 B.C. From then on, humans continually made glass, though not in large-scale production. Glass first appeared in the windows of English private houses in 1180 A.D.

The large-scale glass industry was born in England in 1614, two years after Italian Antonio Neri wrote a manual on glass making. Plate glass, formed by hardening molten glass on the smooth surface of molten tin, was invented in 1688, and mass production of glass began in Venice in 1736. Other places soon followed.

Glass is normally a colorless, transparent solid; any color it contains comes from other materials (figure 3). Cobalt produces blue glass, and copper imparts a red, green, or blue color, depending on the concentration and ionic state of the copper. Gold ions turn glass red, violet, or blue, while iron makes glass either green or yellow. Selenium ions give glass a red color.

The glass normally used as windowpanes and in bottles is called soda-lime glass. It is made by fusing SiO_2, Na_2O, and CaO in a ratio of 7:1:1. This form of glass has low thermal stability—it breaks easily when heat is applied. Soda-lime glass also shatters easily on impact. If B_2O_3 replaces part of the Na_2O and CaO, the resulting borosilicate glass has greater heat and mechanical resistance. Pyrex and Kimax glass ovenware are made from a borosilicate glass.

When glass contains fairly large amounts of PbO (lead oxide), it develops a brilliance and clarity that makes it, when cut into a crystal form, resemble precious stones. This crystal or cut glass is used for special optical applications as well as fine drinking goblets. (Leaded crystal can contaminate liquids unless care is exercised; see chapter 8 for more details.)

The silicate based gemstones, figure 4, derive their colors from traces of other elements, usually combined in oxides. These gemstones are often semiprecious stones, some with commercial uses as well. Garnet gemstones, for example, are used in laser technology.

Figure 3 Glass comes in many colors—but only when other elements are present.

Figure 4 Some silicates are semiprecious gemstones.

BOX 6.4

CLAYS AND CERAMICS

Some common substances consist of atoms held together by both ionic and covalent bonds. Clays are silicate minerals that also contain aluminum ions; they are thus called aluminosilicates. Clays do not have an exact chemical formula, but they consist primarily of mixtures of Al_2O_3 (ionic bonds) and SiO_2 (covalent bonds) with a considerable amount of water, H_2O. A "typical" clay formula might be $Al_2O_3 \cdot 2SiO_2 \cdot 2H_2O$.

Clays are formed when granite and gneiss rocks weather, shedding minute platelets that are easy to mold into shapes when wet. Wet clay can be molded into objects such as bowls, vases, or cups. After the water evaporates, the clay is no longer malleable and retains the new shape indefinitely. Most clays can be remoistened and remolded, however, unless they are heated to form a ceramic.

Ceramics are formed by heating clay, or a similar substance, in a kiln (furnace) to a temperature high enough to partially fuse the ionic and covalent blend. Clay-based ceramics consist of an interlocked SiO_2–Al_2O_3 network; the main bonding is covalent, but the aluminum exists as Al^{3+} ions and part of the oxygen exists as O^{2-} ions. Traditionally, ceramics have been widely used for a wide variety of applications, including porcelain vases, china dishes, and glazed pottery. Glazing involves painting the surface of a ceramic with a water-soluble salt and then baking the object in a high-temperature furnace. The surface acquires a glasslike appearance.

Ceramic technology is one of the oldest known to humans. The earliest examples come from what is now eastern Europe and are dated from 30,000 to 25,000 B.C. Pottery making was well established by 6000 B.C., and clay pots served for cooking food as well as carrying water. The Sumerians invented the potter's wheel around 4000 B.C., making it possible to construct smooth, symmetrical pieces (figure 1).

Figure 1 An example of Sumerian pottery.

Eventually, the Chinese invented a highly developed technology for creating porcelain (a very white, hard, translucent ceramic that was usually glazed). Today, we still call fine dinnerware china (figure 2).

Figure 2 Ming dynasty cup (A.D. 1465–87).

Ceramic technology developed in Europe in the 1700s. In 1708, in Dresden, Germany, ceramics workers discovered the secret of making hard, white porcelain similar to the porcelain made in China. By the mid-1700s, ceramic technology began to spread around the world; today, ceramics boast of a $10 billion annual market.

Modern ceramics are not exclusively aluminosilicates or clays—and many other minerals have been used to prepare such advanced ceramics as those used in the heat-protecting tiles on the Space Shuttle (figure 3) and in experimental automobile engines. Ordinary automobile engines require coolants and lubricants, but ceramic versions can run without them, even when operated at 2,200°F. (Such elevated temperatures increase engine efficiency remarkably as chapter 5 explains.) A car with a ceramic engine is expected to be on the market by the year 2000.

Ceramics lend themselves to many other uses as well. They are harder, lighter, and stiffer than steel and are far more resistant to heat and corrosion. Although most ceramics are very brittle, some newer forms hold up fairly well under adverse circumstances. The Kyocera Corporation in Japan manufactures scissors that rarely need resharpening from a zirconium oxide (ZrO_2) ceramic. Human elbow, knee, hip, and skullcap prostheses (replacement body parts) are made from ceramics. Kyocera is even studying a diesel motor made from silicon nitride (Si_3N_4) that they believe will last five times longer than metal engines, and an American company, Norton, has made ball bearings that do not break under severe stress from Si_3N_4.

Most people have already encountered some advanced ceramics in their daily life. The catalytic converter in an automobile is a ceramic, Corning Ware is found in most kitchens in the United States, and most hardware stores carry drill bits coated with titanium nitride (TiN), which helps maintain the bit's cutting life. Undoubtedly researchers will continue to develop many new uses for these versatile substances.

Figure 3 Replacement of heat-shielding ceramic tiles on Space Shuttle Columbia (left). Ceramic tiles on the nose of the Space Shuttle Orbiter.

In a polar covalent bond, the two atoms reach a state of equilibrium where they each exert a certain amount of pull on the electrons. For example, as the F atom in a C—F bond pulls electrons toward it, the C atom becomes slightly more positive and thus more attractive to the electron pair. The electron pair settles *between* the two atoms, although it resides closer to the F atom. The same happens in an H—O bond. As the O atom attracts the electron pair, the developing δ+ charge that forms on the H atom offsets this attraction, and the O atom ends up with a partial rather than full negative charge (δ−). Essentially, the atoms play a tug-of-war with the electron pair; neither side wins, but one side drags the pair closer. The resulting polarized or polar bonds make the molecule a **dipole,** or a molecule with partial positive and negative charges on each end.

Dipoles and the Dielectric Effect

What happens if a *polarized* bond is placed between such a pair of oppositely charged electrodes? As the δ+ atom is attracted toward the negative cathode, the δ− atom simultaneously is attracted to the positive anode. Since the partially charged atoms are linked together, neither can go any place. Instead, they line up so the δ+ points toward the cathode and the δ− points toward the anode in what is called a **dipole effect** (figure 6.14). With no applied electrical field (figure 6.14a), the atoms are arranged randomly. When an electrical field is applied (figure 6.14b), the partial charges line up with the field, but the atoms cannot move toward one electrode or the other. Instead, these dipoles stay in place and quiver; neither partially charged atom actually moves anywhere. (Table 6.4 compares the effects of electricity on different types of bonds.)

This phenomenon, also called the **dielectric effect,** has important practical ramifications. Bonds with a high partial ionic character have large dielectric constants. Materials with high dielectric constants do not make good insulators because the electric

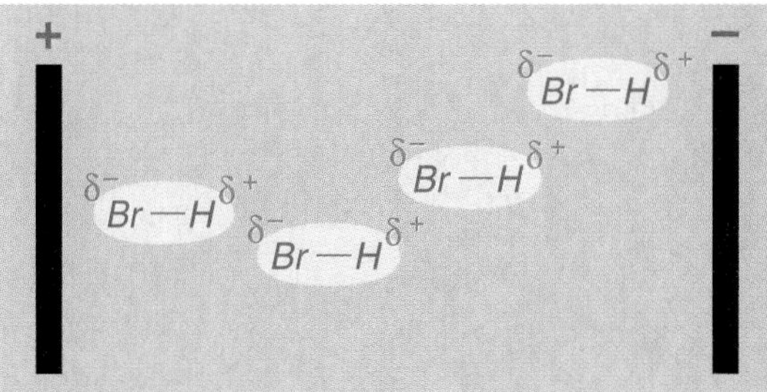

Figure 6.14 The effect of an electric field on a polarized bond. **(a)** No electric field. **(b)** An applied field.

charge "leaks" through. Electrical insulation and tape are therefore made from plastics with a low dipolar character, such as polyethylene, polyvinyl chloride, or rubber. A highly polar molecule with a high dielectric constant such as Dacron, a polyester, does not make good electrical insulation. (Materials that conduct electricity especially well are called superconductors; box 6.5 discusses superconductors more extensively.)

Hydrogen Bonds

Metallic, ionic, and covalent bonds all form between atoms within a molecule. **Hydrogen bonds** are a type of bond that forms between different molecules, most of them containing hydrogen and either fluorine, oxygen, or nitrogen. They form when molecules are strongly polar; since nitrogen, oxygen, and fluorine are much more electronegative than hydrogen, they create a strong dipole when combined with it. The negative end of each dipole (N, O, or F) attracts the positive end of another dipole (H) to form the bond.

Hydrogen bonds have a strong effect on the physical properties of certain compounds. For example, scientists attribute water's unexpectedly high boiling point to hydrogen bonds between water molecules. Figure 6.15 charts the boiling points for compounds containing hydrogen and four family VIA elements. The boiling points of the compounds between hydrogen and sulfur, hydrogen and selenium, and hydrogen and tellurium fall on a straight line, as expected for normal periodic behavior. But the data for water (hydrogen and oxygen) does not fit on this line at all. The point marked *x* on figure 6.15 (at 18 amu, of water) predicts a boiling point for H_2O at about −95°C instead of the actual +100°C. In other words, the expected natural state for water should be a gas instead of a liquid.

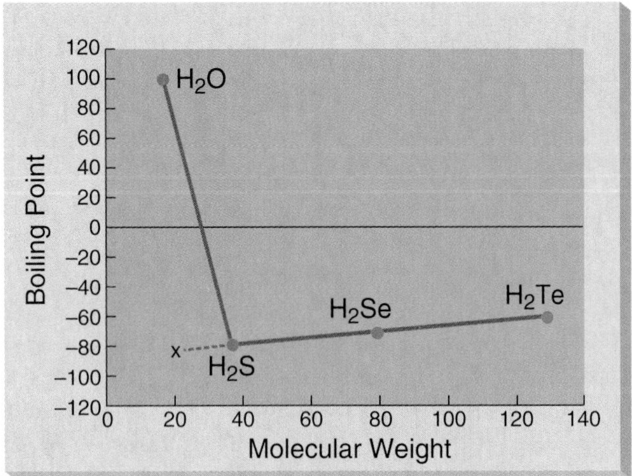

Figure 6.15 Boiling points for family VIA hydrides (hydrogen compounds). The *x* marks the predicted H_2O boiling point.

Why, then, is water's boiling point so much higher than expected? The explanation relates to the intermolecular *hydrogen bonds* that form between water molecules. Although individual hydrogen bonds are not particularly strong (just a fraction of the strength of a covalent bond), they are quite powerful. The individual water molecules form clusters, and for the water to evaporate, the water molecules must detach from these hydrogen-bonded clusters (figure 6.16). Breaking so many hydrogen bonds requires much additional energy, which dramatically increases the amount of heat needed to bring water to the boiling point.

Hydrogen bonds also cause the structure of ice to differ markedly from the structures of other solids (figure 6.16). Notice the hole in the center of each cagelike ice unit. This open structure means fewer H_2O molecules make up a unit volume of solid ice than of liquid water. This gives ice a lower density than liquid water, so that ice floats. All other common solids have higher densities than the corresponding liquid state, so the solids sink.

Hydrogen bonds also form when hydrogen combines with members of the nitrogen family (group VA) and the halides (group VIIA). The H—F molecule occurs as a chain of atoms connected by strong hydrogen bonds (figure 6.17). Because of this, its boiling point of 19°C is about 140° higher than expected. Although the carbon family (group IVA) does not form hydrogen bonds, some polar carbon compounds involve O or N atoms that do form hydrogen bonds (we will discuss this further in chapter 10).

STRUCTURE AND SHAPE

Chemists have known for decades that most molecules have definite shapes, but discovering why has presented a major challenge. Shape has a powerful effect on how one molecule reacts with another, and reactions are the heart and soul of chemistry. Molecular shape also strongly affects solubility, boiling point, and even the properties of compounds used in medicine.

A related mystery chemists had to tackle was why certain compounds form in the first place. Methane, CH4, is the simplest hydrocarbon; but why? Early theories suggested that carbon should only have two valence bonds—in other words, the simplest hydrocarbon should be CH_2 instead of CH_4. The discovery of the octet rule and the invention of Lewis dot structures established a valence of four for carbon because carbon belongs to family IVA. However, these discoveries did not help scientists determine molecular shapes.

Spotty Problems with Electron Dots

The Lewis electron dot structures were never designed to depict molecular shape. This becomes apparent when we draw the formula for water using the Lewis dot structures (figure 6.18). Six possible structures arise because the H atoms and electrons may be placed around the oxygen symbol in several ways. Hydrogen has only one electron, which is always unpaired, but oxygen has six outer electrons—two pairs and two singles. Although figure

BOX 6.5

SUPERCONDUCTORS

Superconductivity allows electricity to flow without any resistance, heat generation, or power loss. The electric current therefore flows through the closed circuit indefinitely. Heike Kamerlingh Onnes (Dutch physicist, 1853–1926) discovered the remarkable phenomenon in 1911. Onnes found superconductivity occurred in mercury metal (Hg) at 4.2K (−269°C). He also discovered how to liquefy helium in 1908, using it to achieve the extremely cold temperatures required for superconductivity; Onnes received the 1913 Nobel Prize in physics for his studies in low-temperature physics and the liquefaction of helium.

Other scientists continued to study superconductivity. In 1957, the trio of John Bardeen (American, 1908–), Leon N. Cooper (American, 1930–), and John R. Schrieffer (American, 1931–) proposed the BCS theory to explain superconductivity. The three believed superconductivity occurs when electrons cannot scatter energy (the usual cause of electrical resistance in conductors). They received the 1972 Nobel Prize in physics for this theory.

Meanwhile, researchers found that the critical temperature, T_c, at which superconductivity occurs in metals climbed slowly as they tested different elements. For example, metallic lead (Pb) becomes superconductive at 7.22K; an alloy of niobium (Nd), aluminum (Al), and germanium (Ge) becomes a superconductor at 20.05K. In 1986, K. Alex Müller and J. Georg Bednorz discovered that a ceramic based on a lanthanum-barium-copper oxide became superconductive at 30K (−243°C); this was a major breakthrough that showed superconductivity is not limited to metals.

In 1987, a team led by Paul C. W. Chu at the university of Houston replaced the lanthanum (La) with yttrium (Y) and achieved superconductivity at 93K, well above the temperature of liquid nitrogen (77K or −196°C). This offered a great advantage since liquid nitrogen costs about $0.10/L, while liquid helium costs $4/L. In addition, liquid nitrogen lasts 60 times longer than liquid helium at the same heat load. These superconductive ceramics were dubbed warm-temperature superconductors. When a magnet is placed above a superconductive ceramic disk cooled with liquid nitrogen, the magnet floats (figure 1).

Figure 1 A magnet suspended above a superconducting ceramic disk cooled with liquid nitrogen.

The critical temperature continued to rise as researchers experimented with superconductors during the late 1980s and early 1990s. In 1993, Chu reported obtaining superconductivity at 161K (−112°C) in a mercury-based ceramic placed under high pressure (230,000 atmospheres). At this T_c, materials can be cooled with freon and ordinary air-conditioning technology. Late in 1993, a French group at CERN (Centre National de la Recherche Scientifique) claimed to have prepared a thin film of a copper oxide-based ceramic with a critical superconductivity temperature of 235K (−38°C).

At the present time, superconductors are used mainly in large and powerful magnets, such as those in particle accelerators and magnetic resonance imaging (MRI) machines used in medicine. The floating magnet in figure 1 may not seem impressive, but it is a forerunner of a magnetically levitated train that could someday whisk passengers across the country at 300 MPH (figure 2). In the future, superconducting power lines may carry electricity between cities without any power loss. Superconductive rings may store electricity for future use, and even motors and automobile engines may be powered by superconducting devices.

Scientists must overcome two major obstacles to make superconductivity a practical technology. First, the superconductivity T_c must be raised to near room temperature or above; a superconducting device will no longer act as a superconductor above the T_c. (Imagine a maglev train losing its superconductivity and dropping onto the track at 300 MPH!) The second problem lies in the nature of the superconducting ceramics: like other ceramics, the superconducting materials are very brittle and prone to fracture. Brittle ceramics do not tend to carry high electrical loads in the first place, and a cracked superconducting magnet does not work well. Much research is currently aimed at overcoming both obstacles, and many scientists believe superconducting magnets will soon operate above room temperature. If this happens, we will see a revolution in power transmission and transportation.

Figure 2 A magnetically levitated (maglev) train in Japan.

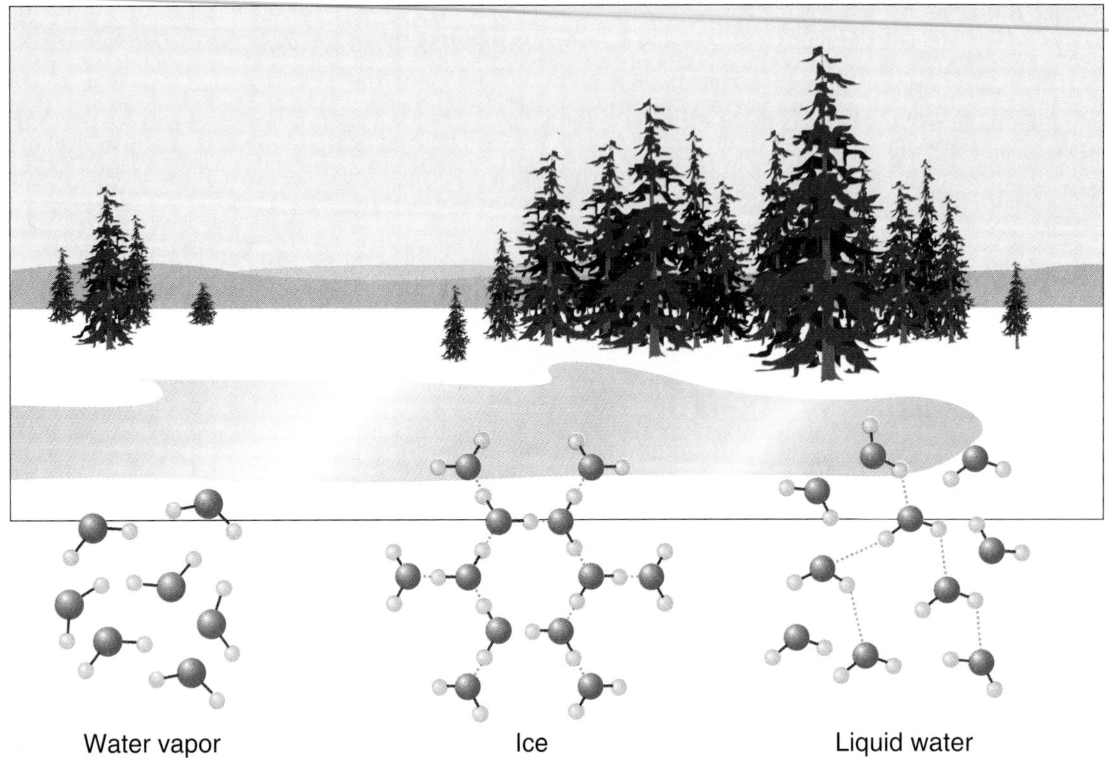

Figure 6.16 Hydrogen bonding in water. Intermolecular hydrogen bonds (represented by dashed lines) hold clusters of molecules together in both the solid and liquid states. The bonds break and the water molecules separate in the gaseous state.

6.2 showed these pairs of electrons in a specific place, they could just as easily be on any of the sides of the O symbol, as figure 6.18 shows. Covalent bonds form when an unpaired electron from one atom combines with an unpaired electron from another atom. Since the unpaired electrons may be on any side of the atoms, the covalent bonds could also form on any side of the O symbol. Figure 6.19 thus shows six possible line structure drawings for the six potential structures, as well; each line represents a bonded pair of electrons.

The wave-mechanics model of the atom (chapter 4) helped chemists take a step forward in determining the shape of a water molecule. According to this theory, orbital shape helps determine molecular shape. In the ground state (when the atoms are at lowest energy), the paired electrons in a water molecule are in the 2s and 2p orbitals, and the single (unpaired) electrons are each in a 2p orbital. Since p-orbitals lie at right angles (90°) to each other, like the corners of a room, it seems covalent bonds in a water molecule should also lie at 90° angles. This means the linear molecules in figure 6.19 (the first two) would not be accurate depictions of a water molecule's shape.

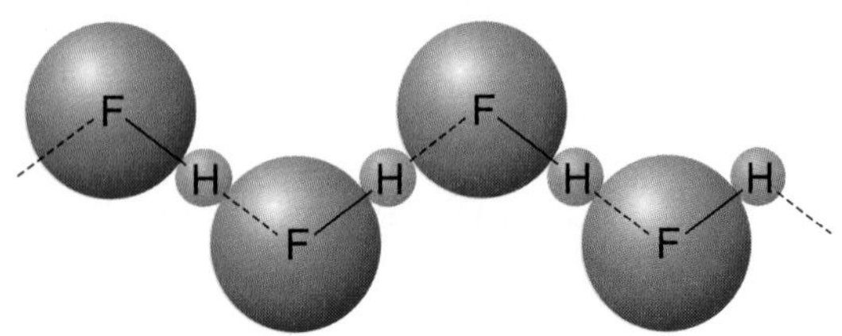

Figure 6.17 Hydrogen bonding in hydrogen fluoride.

	H		H	H	
H:Ö:H	:Ö:	H:Ö:	H:Ö:	:Ö:H	:Ö:H
	H	H			H
	H		H	H	
H–Ö–H	:Ö:	H–Ö:	H–Ö:	:Ö–H	:Ö–H
	H	H			H

Figure 6.18 The six possible Lewis electron dot structures for the water molecule, along with the corresponding line structures showing both pairs of nonbonding electrons.

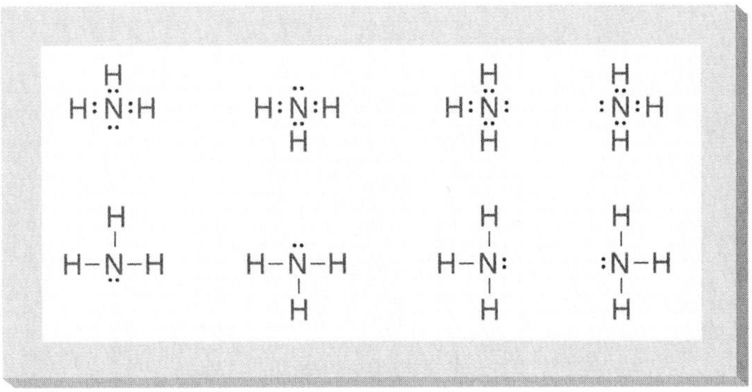

Figure 6.19 The four possible Lewis electron dot structures for the ammonia molecule, along with the corresponding line structures showing the one nonbonding electron pair.

EXPERIENCE THIS

To envision the identical nature of the four right-angled structures in figure 6.18, make gumdrop-and-toothpick models. (The gumdrops will represent the hydrogen and oxygen atoms, and each toothpick will represent the electron pair in each covalent bond.) Use one color gumdrop for the H atoms and another for the O atom. Then arrange the four structures so that the gumdrops line up; you should be able to see that all four structures are actually the same.

Incidentally, did you know you can make your own gumdrops? Place a small amount of powdered, flavored gelatin in a dish and add five to seven drops of water, one at a time; allow the water to seep into the gelatin before adding the next drop. After the last drop soaks into the gelatin, carefully remove the blob, using a knife, and place it on wax paper. Because gelatin contains some protein molecules that swell in water, the gumdrops will "gel" or become firm. The gumdrops you buy include either gelatin, pectin, or some other natural gum. These materials are all natural polymers (to be discussed in chapters 11 and 14).

Still, scientists had not arrived at the correct molecular shape for water. New technologies enabled chemists to measure the *exact* bond angle for the water molecule. They determined that the angle is 104.5°, rather than the predicted 90°. Similar Lewis dot and line structures for the ammonia molecule, NH_3, predict that three N—H bonds will form between electrons in three p-orbitals (figure 6.19). These covalent bonds would thus result in a 90° angle for each H—N—H bond; however, the actual angle is 106.5°.

How could chemists reconcile these measurements with their understanding of molecular shapes? There are several ways to determine the correct shapes. The simplest method is the valence shell electron pair repulsion (VSEPR) technique.

PRACTICE PROBLEM 6.7

Using Figures 6.2, 6.18, and 6.19 as guides, construct the possible Lewis dot drawings for the molecules H_2S and PH_3.

ANSWER Look at the periodic table—S is in the oxygen family (VIA), and P is in the nitrogen family (VA). The six possible structures for H_2S are thus the same as those for H_2O in figure 6.19; the only difference is that the S symbol replaces the O. Likewise, the four apparent structures for PH_3 are just like those for NH_3 in figure 6.20. Just substitute P for every N symbol.

Valence Shell Electron Pair Repulsion (VSEPR)

The basic assumption behind the **valence shell electron pair repulsion (VSEPR)** theory is that the covalently bonding electrons in one orbital repel the electrons in the other orbitals. In addition, the VSEPR method also assumes that nonbonded electron pairs repel other electrons. The final molecular shape is affected as the orbitals repel one another; the end result is that the bonding orbitals arrange themselves as far as possible from each other and from the nonbonding orbitals. VSEPR accounts for the measurements chemists have established for various molecular angles.

To test VSEPR, consider methane, CH_4, which contains four C—H bonds. A Lewis dot structure (figure 6.20) shows that all the electron pairs are bonded, leaving only a single apparent structure. If the VSEPR method holds true, the four C—H bonding orbitals must arrange themselves as far from each other as possible, with the C atom in the center and the four H atoms at the ends of the four bonds. To achieve the greatest distance between all four pairs of bonding electrons and all four H atoms, the molecule must be a tetrahedron with the C atom in the center and the bonds splayed out toward the four apexes (figure 6.21). This is the *only* geometrical form in which all four H atoms are as far from each other as possible.

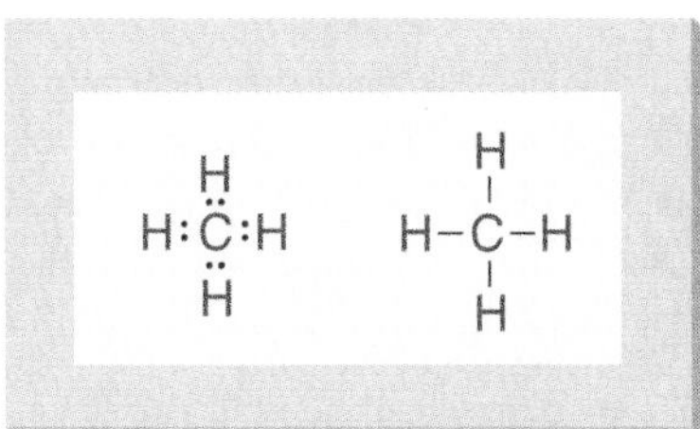

Figure 6.20 The Lewis electron dot structure and corresponding line structure for the methane molecule (CH_4).

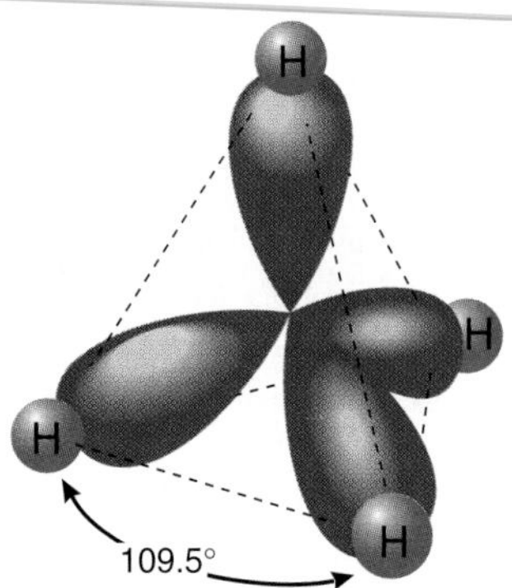

Figure 6.21 The tetrahedral structure of the methane molecule (CH_4).

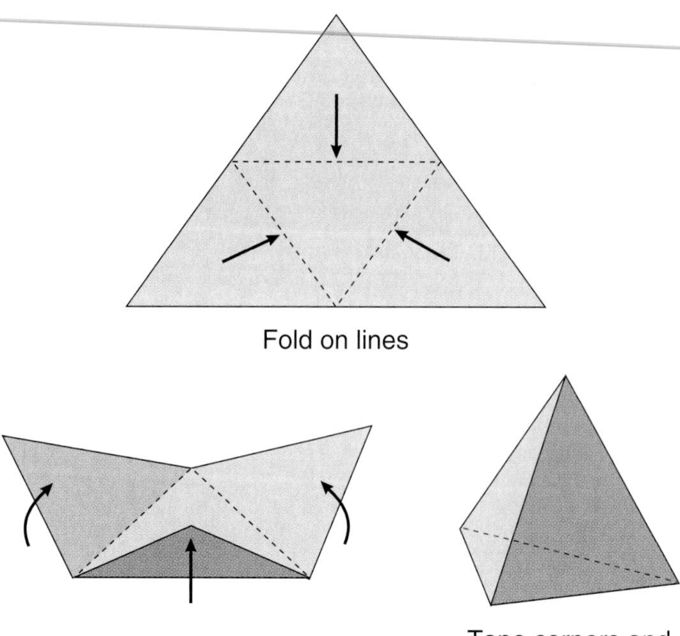

Figure 6.22 How to construct a typical tetrahedron.

Several features distinguish the tetrahedral shape. First, the tetrahedron is symmetrical; it can rest on *any* three corners or edges. Second, the four sides always form equilateral triangles. Third, when viewed from above, the angles between the bottom three corners and the center point appear to be 120°, like a trigonal shape. Fourth, the angle in any H—C—H formation is always 109.5°. Fifth, any attempt to move two atoms further apart causes the other two to come closer to each other. The tetrahedron is the primary shape for the carbon atom and, on the molecular level, the most common shape in the world. Several common molecules form distorted tetrahedra.

EXPERIENCE THIS

Although tetrahedra are not nearly as commonplace in everyday life as squares and cubes, they are sometimes used as disposable creamer containers in restaurants. (In fact, one restaurant supply company named Carbon uses the tetrahedron as its symbol.) You can easily make a tetrahedron from a piece of stiff cardboard (figure 6.22). First, draw a large equilateral triangle (that is, a triangle where all sides are the same length and the angles are all 60°). Next, make a mark at the exact center of each side, and connect these three points with straight lines to form an inner triangle. You now have four small equilateral triangles. Carefully cut out the larger triangle and fold it neatly along the lines of the four small triangles so that the three outer points come together. Finally, tape all the edges together. You have created a tetrahedron.

An alternate tetrahedron model uses gumdrops to represent the individual atoms. Obtain five gumdrops; four should be the same size and color. The fifth should be of a different color and be larger, if possible. Take four toothpicks and connect the four identical gumdrops to the odd one in a three dimensional structure so that each one is as far away from all the others as possible. When your gumdrops are positioned correctly, as in figure 6.23, you will have a good spatial model for methane

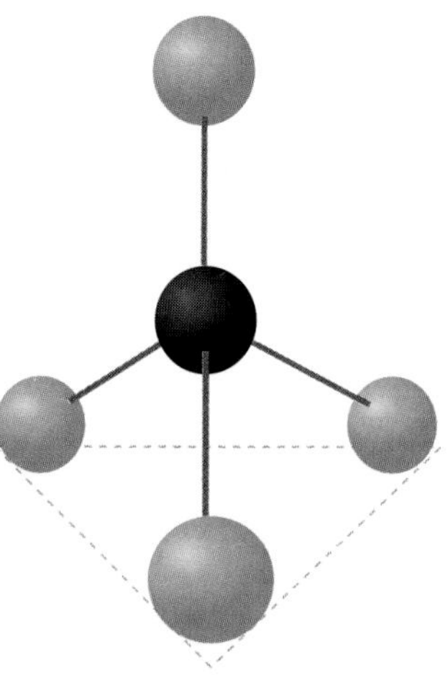

Figure 6.23 A tetrahedron model constructed from gumdrops and toothpicks.

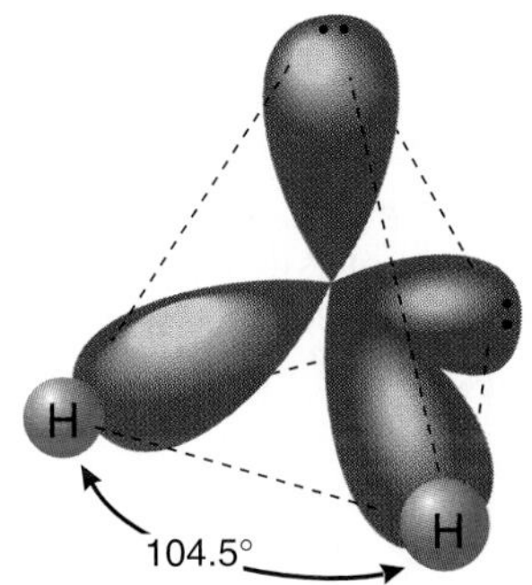

Figure 6.24 The shape of the water molecule.

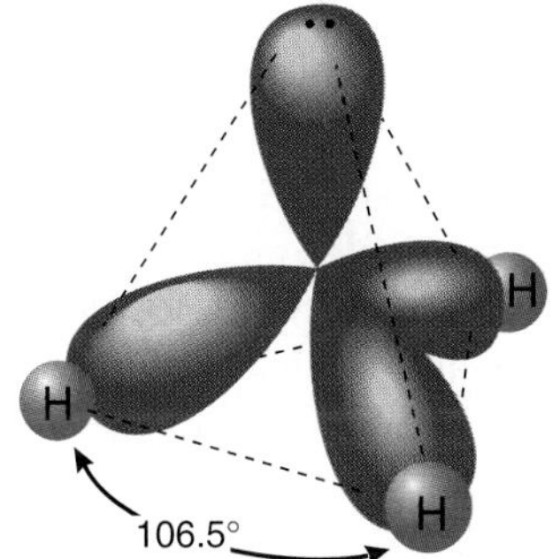

Figure 6.25 The shape of the ammonia molecule (NH_3).

Water and Ammonia Molecules

Let's return to exploring the structure of the water molecule. Since the two original p-orbitals push each other away, the bond angle enlarges from the predicted 90° value to the observed 104.5°. This repulsion occurs between the two electron pairs in the two H—O bonds and the two nonbonding orbitals of the O atom. The final result, as figure 6.24 shows, is a distorted tetrahedral bond (if we consider the electron pairs and the two nonbonding orbitals) or an angular bond (if we consider the atoms). In either case, the H—O—H bond angle is 104.5°.

We can see a similar effect in the ammonia (NH_3) molecule. The bond angles of the three original p-orbitals are enlarged from the 90° value as the three bonded and one nonbonded electron pairs repel one another. The final structure, as figure 6.25 shows, has an H—N—H bond angle of 106.5°. This structure is a distorted tetrahedron if we consider the electron pairs, counting the nonbonding electrons as a fourth "bond," or a trigonal pyramidal shape if we consider the atoms only.

Writing Formulas with Covalent Bonds

Figure 6.26 summarizes the various ways to represent the structures of covalent molecules. The molecular formula provides only the numbers of each atom present in a molecule. The Lewis electron dot structure provides more insight on the bonding and arrangement of the individual atoms, and a line structure conveys most of the same information in a slightly different style. The line structure doesn't show the nonbonding electrons, but they can be added if needed. In many cases, the compact or condensed formula adequately shows the essential information for a molecule. Three-dimensional formulas (shape formulas) convey more information, but they are much more difficult to draw.

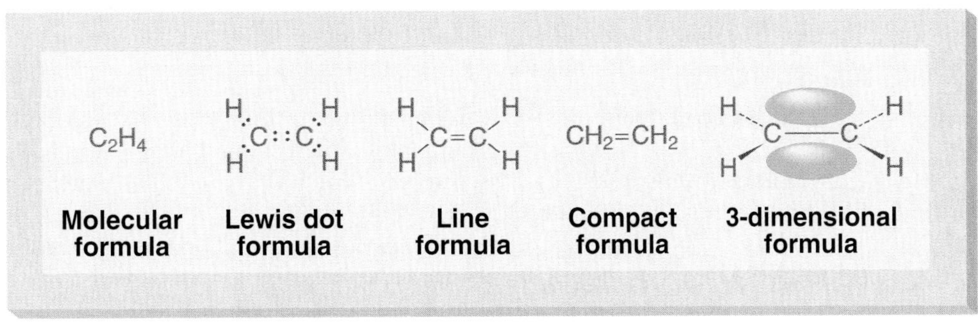

Figure 6.26 The ethylene molecule depicted in five different ways.

SUMMARY

Bonding and chemical reactions form the very heart of chemistry. Chemical bonds occur in several types, including metallic bonds, ionic bonds, covalent bonds, and hydrogen bonds.

Chemists believe that the outermost electrons in each metal atom can move freely among all the atoms in a sample and that this creates the metallic bond that holds the atoms together.

These free electrons can then conduct electricity through the sample.

The octet rule explains the formation of most ionic and covalent compounds. When main group elements form compounds, they tend to end up with 8 electrons in their valence (outermost) shells. This arrangement mirrors the electron configuration of the noble gases, generally making the compound more stable.

An element's valence number, ranging from 1 to 4, indicates how many electrons it must gain or lose to form an octet in its outermost orbital. Whether a compound forms ionic or covalent bonds depends on the nature of the elements that make up the compound. When a metal bonds with a nonmetal, an ionic bond forms. When two nonmetals bond, a covalent bond forms.

Ionic bonds (which transfer electrons) create a charged compound; the metal loses electrons, making it a positively charged cation, and the nonmetal gains electrons, making it a negatively charged anion. Salts have ionic bonds and electrical charges on the atoms.

Covalent bonds (which share electrons) are essential for life. Since the atoms in the bond share electrons, covalent bonds have no electrical charge. They can carry a partial charge when one atom in the bond attracts the shared electrons more than the other does. Some atoms form double and triple covalent bonds, sharing more than one pair of electrons.

Electronegativity and ionization potential are related properties that help explain why elements tend to form either ionic or covalent bonds when they join in a compound. Electronegativity measures an atom's attraction toward electrons. Highly electronegative atoms attract electrons more strongly. Ionization potential measures an atom's tendency to "hold on" to its electrons, or the amount of energy it takes to dislodge electrons from that atom. When the EN values of two elements are very similar, we expect a covalent bond; when EN values differ widely, we expect either an ionic bond or an unequal (polarized) covalent bond to form.

Polarized covalent bonds occur when two atoms share electrons unequally because one attracts the electrons more forcefully. This results in a partial charge; the more attractive element gains a partial negative charge because it pulls the electrons closer, and the other element acquires a partial positive charge because the electrons pull away. The resulting molecule is a dipole.

Hydrogen bonds form between molecules rather than between atoms. Most hydrogen bonds couple hydrogen with a second nitrogen, oxygen, or fluorine. These three elements create strong dipoles with hydrogen; the negative end of each molecule then tends to attract the positive end of another molecule. Hydrogen bonds account for the unusually high boiling points of water and other compounds.

Many molecular properties relate directly to the molecule's shape. The valence shell electron pair repulsion (VSEPR) theory proposes that electrons in one orbital repel electrons in others, so that the orbitals arrange themselves as far as possible from each other. This often results in a tetrahedral shape.

Bonds and chemical formulas may be represented in a variety of ways: by chemical formula, Lewis electron dot structure, line structure, compact formula, or three-dimensional formula. Often, depicting the molecule in graphic style can help a chemist understand more about the bonds that molecule forms and the reactions that create or affect it.

KEY TERMS

chemical bonds 140
covalent bond 149
dielectric effect 157
dipole 157
dipole effect 157
electronegativity 151
hydrogen bond 158
ionic bond 146
ionization potential 152
Lewis electron dot structure 144
metal 140
metallic bonding 140
multiple covalent bond 150
octet rule 141
partial charge 153
polarized covalent bond 153
salt 147
valence 145
valence factor 145
valence number 145
valence shell electron pair repulsion (VSEPR) 161

READINGS OF INTEREST

General

Information on chemical bonding seldom appears in newspapers or on TV. Nevertheless, be on the lookout for occasional news items on the science page.

Fagan, Paul J., and Ward, Michael D. "Building molecular crystals." *Scientific American,* July 1992, pp. 48–54. Crystal growth is only one step beyond bond formation. This article provides an excellent foundation for understanding crystal formation and describes some new potential uses.

Bailey, James. "Uncommon light from common rocks." *Earth,* March 1992, pp. 70–74. Chances are you never realized you could strike two rocks together and make them light up. This fascinating article, complete with photos, on triboluminescence tells why some fairly common rocks have this unusual property.

Aluminum

Teissier-DuCros, André R. "New technology to the rescue for aluminum." *Chemtech*, June 1994, pp. 31–35. New technologies can reduce the production and operating costs of aluminum.

McCosh, Dan. "Aluminum revolution." *Popular Science*, April 1994, pp. 76–78. How new technology can make all-aluminum cars more available.

Banks, Alton J. 1992. "Aluminum." *J. Chem. Ed.*, 69, no. 1, 18. Just a few notes on the uses of aluminum and how it was discovered.

Studt, Tim. "Research maintains aluminum as material of choice." *R & D Magazine*, June 1992, pp. 48–50. Why aluminum is still used for so many different things in our modern society.

Young, John E. "Aluminum's real tab." *World Watch*, March–April 1992, pp. 26–33. Interesting—but one-sided—commentary on the environmental impact of aluminum production. (Also read the letter to the editor in the May–June 1992 issue of this magazine, page 4.)

Theiler, Carl R. 1962. "The discovery of aluminum." In *Men and molecules*. New York: Dodd, Mead. Though somewhat dated, this is a good, brief overview of an important metal, covered in a simple style.

People and Bonding

Borman, Stu. "Gilbert Newton Lewis day proclaimed in Massachusetts." *Chemical & Engineering News*, Nov. 1, 1993, pp. 25–27. This article provides a bit of information about the man behind the electron dot structures.

Ahmad, Wan-Yaacob, and Omar, Siraj. 1992. "Drawing Lewis structures: a step-by-step approach." *J. Chem. Ed.*, 69, no. 10, 791–792. We frequently use the Lewis structures in this textbook. This short article tells you a little more about them.

von Baeyer, Hans. "Atom chasing." *Discover*, July 1992, pp. 42–49. The story of Sam Hurst and his hunt for individual atoms.

Cantor, Geoffrey. 1991. *Michael Faraday: Sandemanian and scientist*. New York: St. Martin's. A much longer account about Faraday that explains much about his philosophy of life.

Asimov, Isaac. 1962. *The search for the elements*. Greenwich, Conn.: Fawcett. Good review, written in a light style and containing lots of information about some of the people who developed modern concepts of bonding.

Kondo, Herbert. "Michael Faraday." *Scientific American*, October 1953; also in *Lives in science*. New York: Simon & Schuster, 1957, pp. 127–140. Short biography of a man best known for his research on electricity, but a pioneering chemist as well.

Glass, Gems, and Ceramics

Irvine, Reed, and Goulden, Joe. "False alarms over asbestos rack up costs." *Insight*, January 24, 1994, pp. 31–32. Malcolm Ross of the U.S. Geological Survey, and formerly president of the Mineralogical Society of America, reports that the most common form of asbestos (chrysotile) poses no health risk.

Ellis, William S. Glass: "Capturing the dance of light." *National Geographic*, December 1993, pp. 37–69. An explanation of how glass is made and what its uses are.

Ryan, Jason; McPhail, David; Rogers, Philip; and Oakley, Victoria. "Glass deterioration in the museum environment." *Chemistry & Industry*, July 5, 1993, pp. 498–501. Older specimens of glass show some degradation, even in museums. This article shows photos of some old glass and the problems it develops.

Harbottle, Garman, and Weigand, Phil C. "Turquoise in pre-Columbian America." *Scientific American*, February 1992, pp. 78–85. The value, significance, and trade of this gemstone discussed in an easy-to-read article.

Fine, Gerald J. 1991. "Glass and glassmaking." *J. Chem. Ed.*, 68, no. 9, 765–768. We use glass every day. This article explores the nature of glass and how it's made.

Kernan, Michael. "Rock hounds never know what they are going to find." *Smithsonian*, September 1991, pp. 60–70. You do not have to be a chemist to collect rocks, but they are definitely chemicals.

Canby, Thomas Y. 1989. "Advanced materials reshaping our lives." *National Geographic*, 176, no. 6, 746–781. A good summary article about the chemical revolution that is changing our lives. Topics covered include prosthetic devices, superconductors, ceramics, polymers, alloys, composites, and metals.

Theiler, Carl R. 1962. "Glass." In *Men and molecules*. New York: Dodd, Mead. This chapter discusses glass, an important material in our modern world, in a simple style.

Superconductivity

Pool, Robert. 1993. "A big step for superconductivity." *Science*, 262, 1816–1817. This summary of the French report of achieving superconductivity at 250K (about −10°F) includes a good graphic showing how the magnitude of the T_c has risen since 1986.

Yam, Philip. 1993. "Trends in superconductivity." *Scientific American*, 269, no. 6, 118–126. Important progress has been made in developing high-temperature superconductors.

Lipkin, R. "Superconductivity possible at 250 Kelvin." *Science News*, December 18, 1993, p. 405. These results were obtained on a thin film of a ceramic.

Lipkin, R. "Super pressures heat up superconductors." *Science News*, October 2, 1993, p. 214. How very high pressures—up to 235,000 atmospheres—enhance superconductivity.

Adrian, Frank J., and Cowan, Dwaine O. "The new superconductors." *Chemical & Engineering News*, December 21, 1992, pp. 24–39. Parts of this article are fairly technical but the beginning is largely historical. Some excellent diagrams appear in this paper.

PROBLEMS AND QUESTIONS

1. Define or give examples of each of the following.
   a. chemical formula
   b. covalent bond
   c. dipole
   d. electron dot structure
   e. hydrogen bond
   f. ionic bond
   g. metallic bond
   h. noble gas
   i. octet rule
   j. partial charge
   k. polarized bond

2. What were the main contributions each of the following people made to chemistry?
   a. Henri Deville
   b. Charles Hall
   c. Paul-Louis Héroult
   d. Gilbert N. Lewis
   e. Friedrich Wöhler

3. Describe metallic bonding. How does it differ from ionic and covalent bonding?

4. Give some examples of materials that bond by metallic bonding. What uses might these materials have?

5. Without looking at figure 6.2, draw the Lewis electron dot structures for the following elements (the atomic number is given in parentheses):
   a. C (6)
   b. N (7)
   c. Na (11)
   d. O (8)
   e. He (2)
   f. Ne (10)
   g. Cl (17)
   h. Ca (20)
   i. Al (13)

6. Tell which of the following compounds are ionic and which are covalent.
   a. NH_3
   b. SO_2
   c. NaBr
   d. $CaCl_2$
   e. SrO
   f. MgF_2

7. Write the correct chemical formulas for the covalent compounds that might form when the following element pairs react:
   a. S + Br
   b. N + Cl
   c. Te + H
   d. P + Cl

8. Write the correct chemical formulas for the ionic compounds that might form when the following element pairs react:
   a. Na + S
   b. Ca + Br
   c. Al + S
   d. Mg + O

9. Write Lewis electron dot formulas for each of the following covalent molecules. (*Note:* It is often easier to draw Lewis dot formulas using different symbols for the electrons in different atoms. For example, you could use • for the electrons in one atom and x for those in another. This helps you keep track of how many total electrons you have, and which atoms they came from.)
   a. S + Br
   b. N + Cl
   c. Te + H
   d. P + Cl

10. Write Lewis electron dot formulas for each of the following ionic molecules.
   a. Na + S
   b. Ca + Br
   c. Al + S
   d. Mg + O

11. Draw the Lewis electron dot formula for each of the following ionic compounds. (Be certain each element obeys the octet rule.)
   a. Na_2O
   b. $CaBr_2$
   c. KF
   d. BaO

12. Draw the Lewis electron dot and line structures for each of the following covalent compounds. (Be certain each element obeys the octet rule.)
   a. C_2H_6
   b. NCl_3
   c. CO_2
   d. H_2CO

13. Define each of the following terms:
   a. linear
   b. tetrahedral
   c. right-angled
   d. valence shell electron pair repulsion (VSEPR)

14. Using gumdrops and toothpicks, construct three-dimensional models for each of the following molecules. Use different colors for the central and attached atoms.
   a. CH_4
   b. CCl_4
   c. H_2O
   d. NH_3
   e. $CH_3—CH_3$

15. How do electronegativity and ionization potential differ? What effect does each have on bond formation and bond type?

16. Why do polarized bonds form? Give some examples of molecules with polarized bonds.

17. Explain how chemists realized that hydrogen bonds existed.

18. What are some effects of hydrogen bonding?

19. Would the bonds between each of the following element pairs be mostly covalent or mostly ionic?

a. N + O
b. Ca + Br
c. Na + O
d. C + H
e. S + O
f. N + Cl

20. Which of the following bonds would you expect to be dipolar?

a. C—F
b. S—O
c. H—Br
d. N—H
e. O—H
f. C—Br

21. What type of bond would be present in each of the following materials?

a. aluminum foil
b. carbon dioxide (CO_2)
c. zinc chloride ($ZnCl_2$)
d. solder (an alloy of tin and lead)
e. calcium oxide (CaO)
f. sulfur dioxide (SO_2)
g. brass (an alloy of copper and zinc)
h. lithium hydride (LiH)
i. carbon tetrachloride (CCl_4)
j. dental amalgam (an alloy of mercury and copper)
k. barium chloride ($BaCl_2$)
l. iodine monochloride (ICl)
m. iron metal

CRITICAL THINKING PROBLEMS

1. Suppose Mendeleev had known about the nonperiodic behavior of the group VIA hydrides. Would he have had more difficulty developing his periodic chart? If he had tried anyway, how might his critics have reacted?

2. Life might never have occurred without hydrogen bonding. Imagine that all hydrogen bonds suddenly disappeared. (All other types of bonds remained.) What effect might this have on life on Earth?

3. Read the articles on aluminum in the Readings of Interest and evaluate the impact of aluminum on the environment and on our society. (Consider the uses of the metal as well as pollution caused by its manufacture.)

4. Glass, a molecule composed mainly of tetrahedral Si—O bonds, has been known since antiquity. In spite of this, scientists have developed new forms of glass in the past century. Read at least two of the articles on glass in the Points of Interest. Based on the information presented in those articles, do you think glass will remain an important material over the next century? (Keep in mind that the main raw material for glass is ordinary sand.)

5. Asbestos has been banned from new construction. Read the information in box 6.3 and the article by Reed Irvine and Joe Goulden in *Insight*, January 24, 1994, pp. 31–32: "False Alarms over Asbestos Rack Up Costs." What is your opinion regarding this ban? What fireproof replacements exist for asbestos?

6. In what ways might practical forms of superconductivity (box 6.5) change our lifestyles?

7. As a noble gas, radon (Rn) does not undergo many chemical reactions. Why is this nonreactive chemical classed as a hazardous material?

8. Consider the consequences of hydrogen bonding in water. One consequence is that ice rises to the top of a lake, river, or stream as it forms. What would happen to aquatic life if hydrogen bonding did not occur? Why?

HINT Aquatic life, except in the tropics, would perish during a cold winter if hydrogen bonding did not occur.

9. It seems likely that our future will have flat TV screens and computer monitors that hang from a wall, smaller laptop computers, and dozens of other electronic devices. Naturally chemistry has been instrumental in developing these technologies. Make a brief summary of the role of chemistry in these areas. Four possible sources are: (1) Michael W. Davidson, *Discover*, Sept., 1993, pp. 72–76, "The Flat Face of Technology;" (2) Steven W. Depp & Webster E. Howard, *Scientific American*, March, 1993, pp. 90–97, "Flat-Panel Displays;" (3) Ron Goldberg, *Popular Science*, Nov., 1993, pp. 100–103, "The Big Squeeze;" and (4) Michael W. Geis & John C. Angus, *Scientific American*, Oct., 1992, pp. 84–89, "Diamond Film Semiconductors."

CHAPTER 7

CHEMICAL REACTIONS

OUTLINE

KEY IDEAS

1. How to Classify Chemical Reactions
2. The Nature of Oxidation-Reduction (Redox) Reactions
3. Some Examples of Redox Reactions
4. The Limits on Chemical Reactivity
5. The Effect of Heat on Chemical Reactions
6. How Chemical Reactivity Affects Reaction Kinetics
7. The Nature and Function of Catalysts
8. The Effect of Surface Area on Chemical Reactions

Science really creates wealth and opportunity which did not exist before.
KARL TAYLOR COMPTON (1887–1954)

Iron powder reacting with oxygen in a bunsen burner flame.

Nobody knows exactly how many different chemical reactions can take place, but the number exceeds several trillion. It is obviously utterly hopeless to attempt to memorize even a small percentage of such a huge number. Fortunately, most chemical reactions follow a few simple patterns, and chemists can predict many unknown combinations by comparing them with similar known reactions. This chapter will examine chemical reactions, how they are classified, and the limits that restrict them.

Types of Chemical Reactions

The most basic reactions involve decomposition, combination, and displacement.

Decomposition Reactions

In **decomposition** reactions, a compound breaks up into two or more separate substances:

$$A{-}B \rightarrow A + B$$

The products of such a reaction could be elements or compounds. Frequently, one of the products is either a gas or water. Often, but not always, the products can be visually distinguished from the reactants. For example:

$$\underset{\text{Red solid}}{2\,HgO} \xrightarrow{\text{heat}} \underset{\text{Silvery metal}}{2\,Hg} + \underset{\text{Gas}}{O_2}$$

This equation and figure 7.1 show the red solid mercuric oxide, HgO, decomposing into elemental mercury—a silvery metallic liquid—and gaseous oxygen. An obvious change has occurred: Mercuric oxide (used as a pigment in some paints and ceramics and an antifouling agent to prevent mold growth) has decomposed into gaseous oxygen and mercury (sometimes called quicksilver), the substance used in thermometers and electrical applications.

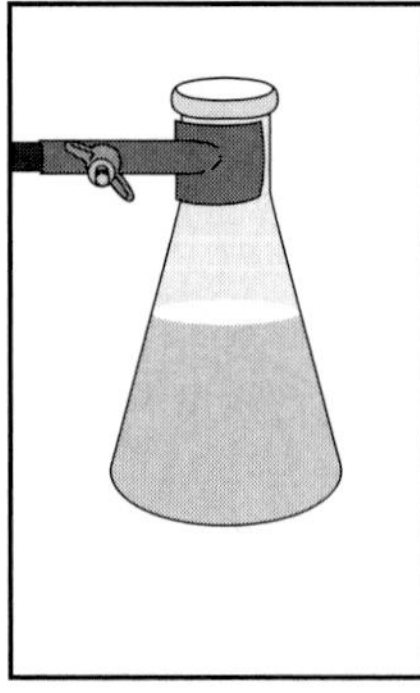

Figure 7.1 The conversion of red mercuric oxide into silvery mercury metal and gaseous oxygen.

The following equation presents a decomposition reaction that is harder to detect because the solid product looks just like the starting reactant. The white solid magnesium carbonate, $MgCO_3$, decomposes to form another white solid, magnesium oxide (MgO), plus carbon dioxide:

$$\underset{\text{White solid}}{MgCO_3} \xrightarrow{\text{heat}} \underset{\text{White solid}}{MgO} + \underset{\text{Gas}}{CO_2}$$

Chemical analysis reveals that these solids have different formulas, and chemical tests can detect the gaseous CO_2. In this reaction, magnesium carbonate (used in white-walled tires, cosmetics, and to promote free flow in table salt) decomposes to magnesium oxide (used in ceramics, pharmaceuticals, paper manufacture, and white electrical insulation) plus carbon dioxide (a colorless, tasteless, odorless gas). Decomposition changes not only appearance, but chemical properties and uses.

Combination Reactions

The following equation shows a schematic diagram for a **combination** reaction. In such a reaction, two substances combine to form a new compound:

$$A + B \rightarrow A{-}B$$

The following two equations represent two actual combination reactions:

$$\underset{\text{Black solid}}{C} + \underset{\text{Colorless gas}}{O_2} \rightarrow \underset{\text{Colorless gas}}{CO_2}$$

$$\underset{\text{Colorless gas}}{H_2} + \underset{\text{Yellow-green gas}}{Cl_2} \rightarrow \underset{\text{Colorless gas}}{2\,HCl}$$

The first equation combines solid carbon and gaseous oxygen, by burning, to form carbon dioxide. The second equation shows how hydrogen, a colorless gas, reacts with chlorine, a poisonous, yellow-green gas, to form hydrogen chloride (also called hydrochloric acid), another colorless gas (figure 7.2). In both examples, a pair of reactants combines to form a single product.

Chlorine is used industrially to prepare other chemicals, including hydrochloric acid, polyvinyl chloride (a plastic), and solvents. Because it kills harmful bacteria and algae, chlorine is also used for water purification. Hydrochloric acid is used in oil wells, food processing, swimming pools, and metal pickling.

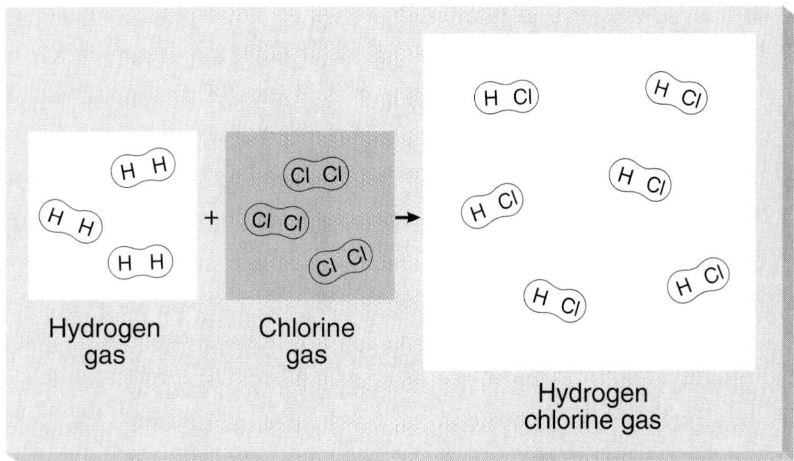

Figure 7.2 The reaction of colorless hydrogen gas with yellow-green chlorine gas to produce colorless hydrogen chloride.

Displacement Reactions

In a **displacement** reaction, one chemical agent displaces another one in a chemical compound. Such a reaction can involve a single pair of substances in **single displacement,** or two pairs of substances in a **double displacement.** When displacement reactions occur, the displaced chemical unit often goes into solution, while the displacing agent comes out of solution as a precipitate or a gas; sometimes the product is a water molecule that remains in solution. These events provide the driving force for the reaction; if they don't occur, no chemical reaction happens.

Single Displacement Reactions

In a single displacement reaction, an uncombined element displaces another element from a compound:

$$A + B{-}D \rightarrow A{-}D + B$$

These types of reactions are sometimes called single replacement reactions. Whether a single displacement reaction occurs depends on the relative reactivity of the chemicals involved; we will consider this further later in this chapter.

Single displacements are usually run in aqueous (water) solutions containing the reactant and product compounds. The abbreviation *aq* indicates this in an equation. The following equation shows the single displacement reaction of elemental, metallic zinc with an aqueous solution of hydrogen chloride (HCl). The products are an aqueous solution of zinc chloride ($ZnCl_2$) plus hydrogen gas (H_2):

Zn	+	$2\ HCl\ (aq)$	$\rightarrow$	$ZnCl_2\ (aq)$	+	H_2
Gray solid		Water solution of gas		Water solution of white solid		Colorless gas

Zinc occurs in several alloys (including brass and bronze) and is used in electroplating, in galvanizing iron, and in batteries. Zinc chloride is used as a catalyst (see chapter 4), a soldering flux, a pigment, a denaturant to give alcohol a bad taste, and in dental cement.

The following equation shows the single displacement reaction of copper metal from cupric sulfate ($CuSO_4$) in an aqueous solution and metallic zinc (Zn). An aqueous solution of zinc sulfate ($ZnSO_4$) forms (figure 7.3).

Zn	+	$CuSO_4\ (aq)$	$\rightarrow$	$ZnSO_4\ (aq)$	+	Cu
Gray solid		Water solution of blue crystals		Water solution of white solid		Reddish metal

Copper sulfate is used as a blue pigment, a pesticide, a wood preservative, and a medicine. Copper metal is used for electrical wires and plumbing. Industrially, copper finds use in chemical and pharmaceutical machinery, alloys (brass, bronze, Monel metal), electroplating, antifouling paints, and copper-coated cooking utensils. Zinc sulfate is used as a fungicide, in ore flotation, and in galvanizing iron.

Double Displacement Reactions

In double displacement reactions (also called double replacement reactions), two sets of atoms (or ions) are displaced. These reactions occur both in solution and the solid state, although more commonly in solution. The following equation shows the two displacements occurring in a double displacement. The atoms or groups of atoms trade position in each compound pair.

$$A{-}B + C{-}D \rightarrow A{-}D + C{-}B$$

In most real examples, one of the reaction products is a water-insoluble precipitate, or a water molecule. When there is no precipitate, the product is a solution containing individual cations and anions that were part of the reactant compounds; dissolving these compounds into constituent ions is *not* a chemical reaction.

Figure 7.3 The reaction of metallic zinc with blue aqueous solution of cupric sulfate to form a colorless zinc sulfate solution and copper metal.

For a chemical reaction to occur, some chemical change must take place.

The following equation shows compounds dissolving into ions—since no change takes place, and no new bonds form, no chemical reaction occurs:

$$Na^+(OH)^- + K^+(NO_3)^- \xrightarrow{\text{dissolve in water}}$$

Water-soluble ionic compounds

$$Na^+ + K^+ + (OH)^- + (NO_3)^-$$

A collection of soluble ions

In other examples, both sets of ionic reactants dissolve in water, but a cation and an anion unite to form a water-insoluble compound. Because these substances bond, a chemical reaction takes place. The following equation shows a true double displacement; the Ag^+ displaces the Na^+ in Na^+Cl^- and the Cl^- displaces the $(NO_3)^-$ in $Ag^+(NO_3)^-$ (two displacements):

$$Na^+Cl^-(aq) + Ag^+(NO_3)^-(aq) \rightarrow$$

Water-soluble compounds

$$Ag^+Cl^- + Na^+(aq) + (NO_3)^-(aq)$$

Insoluble — Soluble ions

Silver nitrate ($AgNO_3$) is used in silver plating, as a germicide, and as an antiseptic to cauterize wounds. Silver chloride ($AgCl$) is used in photography, batteries, glass, and medicine.

EXPERIENCE THIS

Calcium chloride ($CaCl_2$) is available at many stores since it is used to melt ice. Baking soda ($NaHCO_3$) is available in virtually any grocery store. Dissolve roughly equal amounts of each solid in about 100 mL (about 4 ounces) water in separate jars. Next, pour these two solutions together into a third jar. A white precipitate will form. This precipitate is calcium carbonate ($CaCO_3$), found in chalk, limestone, and marble. The reaction also produces hydrogen chloride (HCl) and sodium chloride ($NaCl$—table salt) as byproducts. (You will not be able to detect these two substances since the reaction was run as an aqueous solution.) Write the equation for this reaction. What type of reaction is it?

Save the $CaCO_3$ for use in a future experiment by pouring the solution through a coffee filter. Rinse the $CaCO_3$ with water (to remove any salt) and let it air dry. Store the $CaCO_3$ in a jar or envelope.

Next we will discuss two reactions that produce water as a product. Both reactions include an acid as a reactant (H_2SO_4, sulfuric acid; and HCl, hydrochloric acid). Both acids release an H^+ cation in an aqueous solution. The other reactant in each reaction is a base; this type of reaction bears the special name of neutralization. (Chapter 8 will discuss acids, bases, and neutralization more fully.) In both examples, portions of each compound exchange with the other compound in a double displacement.

In the first reaction, the H^+ cation combines with the OH^- anion to form water (HOH or H_2O), a covalent compound. The loss of the compound's ionic character through this covalent bond formation provides the driving force for the reaction. The formation of other covalent bonds would have the same effect.

$$H_2SO_4 + 2\,NaOH \rightarrow Na_2SO_4 + 2\,HOH$$

Liquid (acid) — Solid (base) — Solid — Liquid water

Figure 7.4 Some examples of magnesium hydroxide. Left to right are powdered $Mg(OH)_2$, an aqueous suspension of $Mg(OH)_2$ called milk of magnesia, and Maalox tablets.

In the preceding equation, both reactants are water-soluble. In the next equation, reactant $Mg(OH)_2$ (magnesium hydroxide) is usually water-insoluble, but the formation of water is sufficiently energetic to overcome the insolubility of the $Mg(OH)_2$ and drive the reaction to completion. The magnesium chloride ($MgCl_2$) product is water-soluble.

$2\ HCl\ (aq)$	+	$Mg(OH)_2$	$\longrightarrow$	$MgCl_2\ (aq)$	+	$2\ HOH$
Water solution of a gas (acid)		White solid (base)		Water solution of white solid		Liquid water

Magnesium hydroxide—$Mg(OH)_2$—mixed in large quantities in a water suspension is called milk of magnesia and is a laxative (figure 7.4). $Mg(OH)_2$ is also used in smaller quantities as an antacid to neutralize stomach acid (hydrochloric acid). Magnesium hydroxide finds nonmedicinal uses in sugar refining and in foods as a drying agent, color retention agent, or weak alkaline material. Magnesium chloride, $MgCl_2$, is used in fire extinguishers, for fire-proofing wood, in ceramics, and in paper manufacturing.

Table 7.1 summarizes the four types of chemical reactions studied thus far in this chapter. We will discuss oxidation-reduction, or redox, reactions in the next section; any redox reaction may also be one of the four types of reactions listed in table 7.1.

TABLE 7.1 Summary of the types of chemical reactions.

REACTION TYPE							
Decomposition	AB			→	A	+	B
Combination	A	+	B	→		AB	
Single displacement	A	+	BC	→	AC	+	B
Double displacement	AB	+	CD	→	AD	+	BC

PRACTICE PROBLEM 7.1

Determine what type of reaction each of the following reactions is—decomposition, combination, single displacement, or double displacement.

a. $NaBr + AgNO_3 \longrightarrow NaNO_3 + AgBr$

b $Mg + PbSO_4 \longrightarrow MgSO_4 + Pb$

c. $2\ AgCl + \text{light} \longrightarrow 2\ Ag + Cl_2$

d. $2\ Al + 3\ Cl_2 \longrightarrow 2\ AlCl_3$

e. $NaHCO_3 + C_2H_3O_2\text{–}H \longrightarrow C_2H_3O_2Na + H_2CO_3 \longrightarrow C_2H_3O_2Na + H_2O + CO_2$

ANSWER Reaction *(a)* is a double displacement; *(b)* single replacement; *(c)* decomposition; and *(d)* combination. Reaction *(e)* is a two-step reaction; the first step is a double displacement, while the second step is a decomposition reaction that occurs almost immediately.

EXPERIENCE THIS

Run the following simple chemical reaction at home. Dissolve some baking soda (sodium bicarbonate, $NaHCO_3$) in water in a tall glass or jar. Add some vinegar ($C_2H_3O_2$—H) to this solution. You will see an immediate release of gas bubbles from the solution. (If you repeat the experiment, adding a small amount of liquid dishwashing detergent before adding the vinegar, the gas will foam as it bubbles out of the container.) The chemical reaction is a double displacement followed by a decomposition; the reaction equation is shown in Practice Problem 7.1*e*.

OXIDATION-REDUCTION REACTIONS

Oxidation-reduction, or **redox reactions** are extremely important in chemistry. They occur in a variety of processes, including iron manufacture, fuel burning, and rusting. Batteries also depend on redox reactions. We will examine three definitions for oxidation-reduction reactions. One important concept underlies all three: These reactions *always* occur in pairs. If an oxidation reaction takes place, a reduction reaction occurs at the same time.

The earliest and simplest definition for oxidation was a reaction in which a compound or element *gained* one or more oxygen atoms. By the same token, in a reduction reaction, a compound would *lose* oxygen atoms. Let's look at some examples of redox

SOURCE: By permission of Johnny Hart and Creators Syndicate, Inc.

based on this simple definition. Perhaps the easiest oxidation example is the rusting of iron:

$$4\ Fe + 3\ O_2 \rightarrow 2\ Fe_2O_3$$

Clearly, in this reaction, iron gains oxygen atoms and is oxidized. Although it is less obvious, oxygen is reduced at the same time (the O_2 molecules lose oxygen atoms to become single atoms, which then combine with the iron). The oxygen becomes part of an ionic compound. When it applies, this definition of oxidation is the easiest one to use.

Metallic iron is a strong, silvery metal, but iron oxide, or rust, is a weak, reddish-brown solid. You might need superhuman strength to bend a thick iron nail, but most people could snap a fully rusted nail between their fingers. Iron oxide, or ferric oxide (rust), does have valuable uses as a pigment in marine paints or metal primers, rubber or plastic items, rouge, and grease paints. It is even used in rocket fuels (box 7.1). Iron oxide is sometimes called turkey red because its color is similar to the comb of this fowl.

Iron applications are much better known. Iron is used in steel and other building materials; it also serves as a catalyst in ammonia synthesis, forms magnets, and is a constituent of the compound hemoglobin in blood, which binds oxygen for transport throughout the body.

EXPERIENCE THIS

Obtain five samples of iron wool (for example, from a scrubbing pad) and five small jars with lids or five test tubes with stoppers. Place a sample of iron wool into each jar and number them. Add nothing to jar 1. To jar 2, add water that was boiled and then cooled to room temperature. Add warm tap water to jar 3. Place a solution of salt water in jar 4. Finally, place some rubbing alcohol in jar 5. Let these jars stand for several weeks, and observe each daily for signs of rust. Try to estimate the percentage of rust in each jar each day. Does the iron rust at the same rate in each jar? What factors does this experiment indicate are important in the rusting process? Could this information be used to prevent rust formation (box 7.2)?

The preparation of iron from iron ore is an example of reduction (figure 7.5). Ferric oxide (Fe_2O_3) loses its oxygen atoms and is reduced to metallic iron (Fe). At the same time, carbon gains two oxygen atoms and is oxidized. Again, the reduction and oxidation reactions are paired—hence the term *redox*.

$$2\ Fe_2O_3 + 3\ C \rightarrow 4\ Fe + 3\ CO_2$$

The second, more inclusive definition of oxidation and reduction involves the gain or loss of electrons rather than oxygen atoms. Consider the equation that describes the rusting of iron:

$$4\ Fe + 3\ O_2 \rightarrow 2\ Fe_2O_3$$

In this equation, the iron atoms (Fe) are neutral, or uncharged. As the reaction occurs, each Fe atom loses 3 electrons to form an

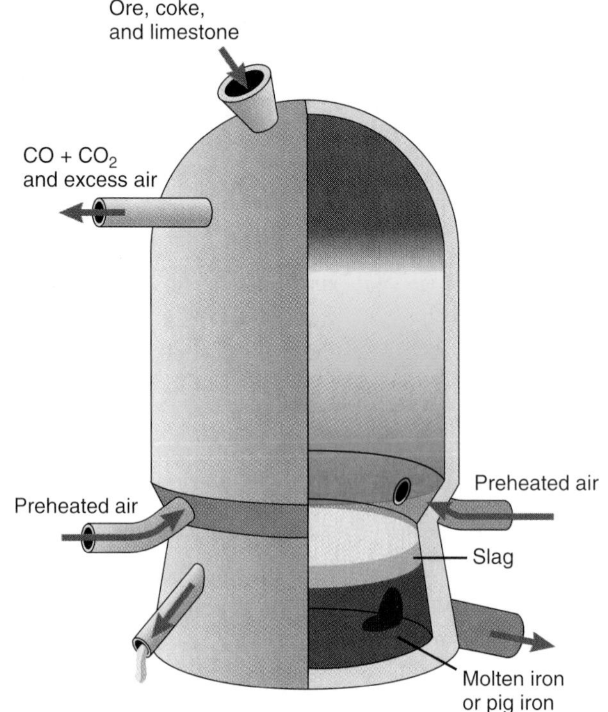

Figure 7.5 The preparation of iron from iron ore using a blast furnace. The ore is reduced and the carbon oxidized during this reaction.

BOX 7.1

THE CHEMISTRY THAT FUELS THE SPACE SHUTTLE

Witnessing a live space shuttle liftoff at the Kennedy Space Center in Florida is one of the most awe-inspiring events a person can experience. The astronauts enter the space craft hours before liftoff, riding an elevator up through the gigantic attached gantry. Though dwarfed by the gantry, the shuttle, its external fuel tanks, and two booster rockets tower more than 184 feet and have a combined mass of 4.4 million pounds (figure 1). The clock gradually counts off the minutes and seconds before this massive object will blast into space. Excitement grows as the gantry pulls away and the countdown nears the critical moment. As the huge clocks tick off the last seconds, the crowd chants 5, 4, 3, 2, 1, blast off! The boosters ignite and the shuttle slowly ascends.

What chemical reactions provide enough energy to lift the shuttle? The two booster rockets, nearly 150 feet long, contain a mixture of aluminum metal (Al) and ammonium perchlorate (NH_4ClO_4) with a trace of iron oxide (Fe_2O_3) catalyst held together in a plastic matrix. Upon ignition, the aluminum metal converts to aluminum oxide (Al_2O_3) and the ammonium perchlorate breaks up into a variety of gases in a decomposition reaction. These gases expand immensely due to the high reaction temperature. A white smoke of powdered Al_2O_3 seeps from the boosters and moves swiftly across the launch pad as the shuttle lifts off. The ground trembles as the booster rocket and shuttle emit nearly seven million pounds of thrust, leaving Earth with a deafening noise that travels more than thirty miles.

Figure 1 The space shuttle during the early stages of liftoff. The external fuel tank is the large red object directly behind the shuttle, and the two white booster rockets are still attached to the fuel tank. These boosters will burst into flame for only about 2.5 minutes before the reactants are consumed and the two boosters descend to the Atlantic Ocean by parachute for reuse in a future flight.

The boosters burn for approximately 2.5 minutes, sending the shuttle high above the earth as the crowd cheers. Once they have finished their task, the boosters separate from the shuttle. Over 378,000 gallons of liquid hydrogen and 139,000 gallons of liquid oxygen from the external fuel tank now power the shuttle. The shuttle continues on its way as the boosters fall to earth.

H_2 and O_2 burn in the external fuel tank in a simple combination reaction that produces water and an enormous amount of heat. This heat expands the volume of the gaseous water immensely, pushing the shuttle farther into space. After about eight minutes, this fuel is consumed and the external fuel tank detaches and falls to earth, breaking into small burning fragments that strike the surface of the Indian Ocean. The 78-foot shuttle may appear small in comparison to the external fuel tank and the booster rockets, yet it has a payload capacity of 65,000 pounds and can carry its crew and many scientific experiments into space at an orbiting speed of 25,000 feet per second, or 17,045 miles per hour.

As the trip nears its end, retrorockets fire to slow down the shuttle and allow it to descend in a controlled flight pattern. The shuttle lands at either Kennedy Space Center or at Edwards Airbase in California. In an adventure filled with potential danger, returning to Earth poses special problems caused by the reentry into the atmosphere. The shuttle orbits between 150 and 180 miles above the earth's surface, and there is virtually no atmosphere that far out. The friction of the air within the atmosphere causes the outside temperature of the falling shuttle to rise above 800°C and even as high as 2,700°C. Most materials catch fire at these temperatures. Chemists have, however, designed special plastics and ceramics (see also box 6.3) to reduce this problem. The ceramics contain clay-stiffened SiO_2 fibers held in place by an epoxy glue (see chapter 10).

Chemistry affects the shuttle in many other ways. Unlike a car traveling across the country, the shuttle cannot stop for meals or restroom breaks. Everything necessary for life must be carried in the shuttle itself, including food, water, and air. Since it would be impractical to constantly replace air and water, they are instead recycled using elaborate chemical-based systems. The astronauts' space suits are also constructed from special fibers to permit the astronauts to venture outside the shuttle in the frigid, airless environment in space.

Although few spectators watching a liftoff are fully aware of the extent to which chemistry is involved, a space flight would be impossible without it. As in any technological field, we are still learning new information regarding space flights. The January 1986 Challenger disaster, in which seven astronauts died, was caused by a faulty O-ring in a fuel tank, a device much like the rubber gasket in a faucet. O-rings are still used in the shuttle, but scientists and engineers have improved the materials, quality control, and inspection routine greatly. Space travel will probably always be a risky venture. But chemistry not only makes it possible—it makes it more efficient, safer, and more comfortable.

BOX 7.2

CORROSION CONTROL

Corrosion is nothing more than the unwanted oxidation of a metal with another substance, often oxygen. Probably the most devastating form of corrosion is the rusting of iron, which leads to many billions of dollars in damages every year. This problem occurs worldwide but is probably most acute in the snow-belt regions of the United States, where salt used on winter roads tends to promote automobile corrosion. Rust also ravages bridges and other structures (figure 1). For these and other equally compelling reasons, the control of corrosion, especially rust, is of paramount importance.

The principles of corrosion control are the same for rust prevention as for the breakdown of other metals. Although we can readily write an equation depicting the conversion of iron into iron oxide ($4\ Fe + 3\ O_2 \longrightarrow 2\ Fe_2O_3$), the actual process is far more complex than the equation reveals. Rust is affected by such factors as the amount of moisture and oxygen present, whether salts, acids, or bases are also present, and the metal's surface characteristics. There are several approaches to rust prevention.

Figure 1 Rust, or iron oxide, destroys both automobile bodies and bridges, costing millions of dollars each year.

Since water always promotes corrosion, one prevention method involves coating the metal to keep water and oxygen off the surface. This coating could be a paint, wax, or grease. This is why manufacturers urge you to keep your car's exterior paint surface clean and waxed: Rust forms quickly on the bare metal surface exposed in a scratch. This rust then spreads underneath the paint and corrodes unexposed metal.

Another surface coating solution involves applying undercoating to reduce rusting problems. Mixtures of waxes can cover any potentially exposed or unpainted metal surfaces. Most undercoating, as the name suggests, covers the underside of the car, but a good rustproofing job also treats the hidden inside surfaces of doors to retard future rusting (figure 2). In a similar application, steel soft drink cans are coated with a lacquer or plastic film to prevent rusting.

Other methods of rust prevention used for iron include chromium coating, tin plating, and galvanizing. Each method uses a second metal to protect iron from rust formation, although the approach differs with each metal.

Chromium plating reduces rust formation because chromium forms a tough, nonreactive, transparent oxide coating on the iron. Since the chromium coating is thin, the metal's bright luster is still evident, which makes this method very popular. However, chromium coating fails drastically if the surface coating becomes damaged. Once the underlying iron is exposed, it is capable of rusting.

Figure 2 Rustproofing an automobile.

Like chromium, tin forms a nonreactive oxide coating on iron. However, if a scratch penetrates the tin surface, the more reactive iron will rust faster than it did with no coating. You can see this by examining crushed tin cans in the trash or recycle bins (figure 3). In other words, tin only protects when the coating is intact.

In galvanizing, a thin layer of zinc coats the iron. Zinc is more reactive than iron and also forms a protective oxide (figure 4). Even if the oxide is scraped off or the zinc is penetrated, the iron will not rust until *all* of the zinc has corroded away. Any iron rust that might form reconverts into iron, at the expense of the zinc, which is oxidized in the process. Galvanizing is used on buckets, some nails, water pipes, and the undersides of cars. Galvanizing imparts a dull gray color to the metal and is not considered very attractive.

Figure 3 Rusting tin-plated cans.

Figure 4 Some galvanized iron objects.

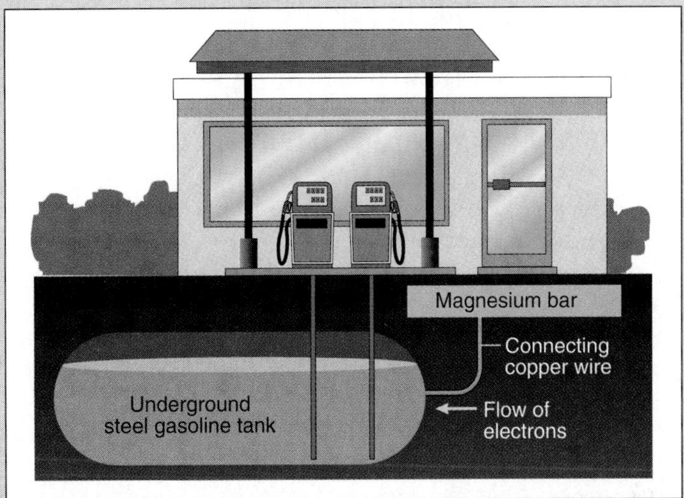

Figure 5 A buried magnesium bar corrodes more rapidly than the iron tank it is attached to by copper wires. This fact enables us to protect underground fuel tanks from corrosion.

SOURCE: From Uno Kask and J. David Rawn, *General Chemistry*. Copyright © 1993 Times Mirror Higher Education Group, Inc., Dubuque, Iowa. All Rights Reserved. Reprinted by permission.

The principle of preventing corrosion by combining an object with a more reactive metal extends beyond surface coating. Buried fuel storage tanks made from iron or steel eventually rust since they are always in moist ground, a prime location for corrosion. Rust formation in such tanks is usually averted by burying a block of highly reactive magnesium metal near the surface and connecting this block to the steel tank with electrical wires (figure 5). When the magnesium corrodes, it is replaced by a new bar, which is much easier than replacing a rusted fuel tank. Fiberglass or plastic tanks are also sometimes used to store fuel, since they do not corrode.

Fe^{3+} cation, and each O atom gains 2 electrons to become an O^{2-} anion. The *loss of an electron* corresponds to oxidation, while the *gain of an electron* corresponds to reduction. Thus, iron is oxidized and oxygen is reduced. (The Fe_2O_3 molecule remains neutral because the 6 positive charges in the two Fe cations offset the 6 negative charges in the three O anions.)

The electron transfer definition of oxidation and reduction applies equally well to the equation that describes the reduction of iron ore to metallic iron:

$$2\ Fe_2O_3 + 3\ C \longrightarrow 4\ Fe + 3\ CO_2$$

In this reaction, the iron atoms in ferric oxide each *gain* 3 electrons, reducing Fe^{3+} to neutral Fe^0 metal. (The ^{0} superscript indicates an uncharged or neutral atom, with zero charge.) Similarly, carbon oxidizes from neutral C^0 to an effective C^{4+} as it forms carbon dioxide, *losing* 4 electrons per carbon atom. In actual fact, carbon dioxide is covalent, with no charge on either the carbon or oxygen atoms. We approximate an ionic compound by writing charges corresponding to the valence numbers, C^{4+} for carbon and O^{2-} for oxygen. This makes the electron transfer method easier to understand.

Just as an equation must balance numbers of atoms of a particular element on each side of the equation, the total number of electrons transferred must be the same for both substances. Thus, the 4 Fe^{3+} atoms gain a total of 12 electrons (3 each) to form 4 Fe^0 atoms, while the three C^0 atoms lose a total of 12 electrons (4 each) to form 3 C^{4+} atoms. The charges in the covalent CO_2 would occur if the molecule were ionic; these implied charges are sometimes called oxidation numbers. Although electron transfer is the more fundamental definition for oxidation and reduction, it is usually more difficult to apply and understand. The important point is to grasp which substance is oxidized and which is reduced.

Earlier in the chapter, we discussed a single displacement reaction that is a redox reaction as well:

$$Zn^0 + 2\ H^+Cl^-(aq) \longrightarrow Zn^{2+}Cl_2^-(aq) + H_2^0$$

When we define oxidation and reduction in terms of oxygen gain or loss, we cannot determine which atoms are oxidized and reduced in this reaction because no oxygen atoms are involved. The electron transfer approach allows us to decide which atoms are oxidized or reduced. Each zinc atom loses 2 electrons and is oxidized to Zn^{2+} to form zinc chloride, an ionic compound. Each hydrogen atom gains 1 electron and is reduced to a neutral elemental hydrogen molecule (H_2^0).

The third redox definition explains oxidation and reduction in terms of the transfer of hydrogen atoms. If a compound *loses* hydrogen atoms, it is oxidized; if it *gains* hydrogen atoms, it is reduced. In the previous equation, the hydrogen chloride (HCl) molecule loses hydrogen atoms as molecular hydrogen (H_2) forms. The H atoms from the HCl gain each other to form the H_2 and are therefore reduced. Because redox reactions always occur in pairs, the zinc must be oxidized (though it doesn't actually lose any hydrogen atoms). This is the same result obtained with the electron transfer approach. The Cl^- does not change during this reaction and is thus neither oxidized nor reduced.

Let's consider one example reaction, testing all three definitions of redox against it. The following equation describes the oxidation of octane, a gasoline component, by oxygen, producing water and carbon dioxide:

$$2\ C_8H_{18} + 25\ O_2 \rightarrow 18\ H_2O + 16\ CO_2$$

Reactions like this occur whenever you burn (oxidize) gasoline in an automobile engine. The O atom changes are easy to track, and they show that the octane is oxidized; it gains oxygen atoms. It is very difficult to track electron transfer in this reaction, however. The hydrogen change method is also easy to apply since the C atoms lose hydrogen and are therefore oxidized. In a similar way, some of the O atoms gain hydrogen and are thus reduced.

The final example involves food digestion—a subject we encountered in chapter 1. The following equation shows the reaction of a simple sugar called glucose to form carbon dioxide and water. In the body, this reaction actually occurs in several steps, each an oxidation-reduction reaction, but these details are unnecessary here. You can see that the carbon in this food substance is oxidized, and the oxygen (from the air) is reduced:

$$\underset{\text{Glucose}}{C_6H_{12}O_6} + 6\ O_2 \rightarrow 6\ H_2O + 6\ CO_2$$

All three redox definitions yield the same results, but some are easier to use in analyzing a particular reaction. For this reason, all three definitions are important. Remember that oxidation involves a *gain* of oxygen, a *loss* of hydrogen, or a *loss* of electrons; reduction involves a *loss* of oxygen, a *gain* of hydrogen, or a *gain* of electrons. Table 7.2 summarizes the three definitions, and box 7.3 provides a tip for remembering them. Box 7.4 discusses a familiar object that operates using redox reactions.

TABLE 7.2 Comparison of the three methods for determining oxidation and reduction.

METHOD	OXIDATION	REDUCTION
1. Oxygen change	Gain of oxygen	Loss of oxygen
2. Electron transfer	Loss of electrons	Gain of electrons
3. Hydrogen change	Loss of hydrogen	Gain of hydrogen

PRACTICE PROBLEM 7.2

Using table 7.2 as a guide, decide which elements are oxidized and reduced in each of the following reactions:

a. $2\ Ag^+Cl^- + \text{light} \rightarrow 2\ Ag + Cl_2$

b. $CH_4 + 2\ O_2 \rightarrow CO_2 + 2\ H_2O$

c. $2\ Al_2O_3 + \text{electricity} \rightarrow 4\ Al + 3\ O_2$

d. $2\ Al + 3\ Ag_2O \rightarrow 6\ Ag^0 + Al_2O_3$

ANSWER *a.* Ag is reduced; Cl is oxidized. *b.* C is oxidized, O is reduced; H remains unchanged. *c.* Al is reduced; O is oxidized. *d.* Al is oxidized; Ag is reduced.

PRACTICE PROBLEM 7.3

Rework Practice Problem 7.2 using the mnemonic device of the Left-Handed LEGO. Do your answers change?

ANSWER Your answers should be exactly the same as in Practice Problem 7.2. The mnemonic would not help much if it provided the wrong answers! Did you find it easier to use the Left-Handed LEGO device? Most students do.

LIMITS ON CHEMICAL REACTIONS

Not everything we can write down in an equation corresponds to a reaction that actually happens. Some simple ground rules and factors govern whether a reaction occurs. This section briefly examines some of these rules and factors.

Limitations on Chemical Reactivity

Nature imposes several limits on chemical reactivity. In an equation we examined previously, you saw zinc displacing hydrogen from hydrogen chloride to form zinc chloride ($ZnCl_2$). However, this reaction does not occur with all metals: For example, copper or gold will not displace hydrogen from hydrogen chloride. Zinc also displaces copper from a copper salt; again this does not occur with all metals. Conversely, zinc will *not* displace magnesium from a magnesium salt. Why?

In general, two types of factors govern reactivity: those that relate to the reactivity of the elements involved, and those that relate to energy. Table 7.3 lists metals in order of reactivity, or

TABLE 7.3 Chemical reactivity ranking.

METAL	SYMBOL
MOST REACTIVE ↑	
Potassium	K
Calcium	Ca
Sodium	Na
Magnesium	Mg
Aluminum	Al
Zinc	Zn
Iron	Fe
Tin	Sn
Lead	Pb
Hydrogen	H
Copper	Cu
Mercury	Hg
Silver	Ag
Gold	Au
LEAST REACTIVE	

tendency to react. More reactive elements usually displace those with lower reactivity. Thus, under normal conditions, calcium displaces tin from a compound, but cannot displace potassium. Table 7.3 permits us to predict which displacement reactions can and cannot occur.

PRACTICE PROBLEM 7.4

Using table 7.3 as a guide, predict which of the following reactions will actually occur:

a. $Zn + CaCl_2 \longrightarrow Ca + ZnCl_2$
b. $6\ Na + Fe_2O_3 \longrightarrow 2\ Fe + 3\ Na_2O$
c. $3\ Sn + Al_2O_3 \longrightarrow 3\ SnO_2 + 4\ Al$
d. $Mg + 2\ HCl \longrightarrow MgCl_2 + H_2$

ANSWER *a.* no; *b.* yes; *c.* no; *d.* yes. In all the reactions that actually occur, the free metal on the left side of the equation is more reactive than the metal it displaces, which becomes the free metal on the right side of the equation.

BOX 7.3

PLAYING WITH A LEFT-HANDED LEGO

How can you remember the definitions for redox reactions? Fortunately, a useful mnemonic can help: **the Left-Handed LEGO!** Most students are familiar with Lego toys, but what, pray tell, is a Left-Handed LEGO? Simply put, the capital letters indicate the conditions for *oxidation:* Lose **H**ydrogen, **L**ose **E**lectrons, **G**ain **O**xygen = **L**eft-**H**anded **LEGO.** Does a material Lose Hydrogen? Does it Lose Electrons? Does the substance Gain Oxygen? In all these cases, the Left-Handed **LEGO** tells us this compound or element is *oxidized*. Obviously, reduction is the opposite: if a substance gains hydrogen, gains electrons, or loses oxygen, it is *reduced*. Even if you never play with actual Lego toys (figure 1), playing with a Left-Handed LEGO will help you determine which atoms are oxidized.

Figure 1 A Lego toy village scene.

BOX 7.4

BATTERIES—REDOX IN ACTION

Contrary to popular belief, batteries do not store electricity. They produce electricity on demand through an oxidation-reduction reaction. No reaction occurs until a circuit is connected to the battery, but then a spontaneous redox reaction occurs and the transferred electrons flow through the circuit. Batteries may be either primary cells, which cannot be recharged, or secondary cells (storage cells), which can be.

The idea of using a chemical source to produce electricity dates back to an observation Luigi Galvani (Italian, 1737–1798) made in 1791: that a frog's legs twitch when an electrical current is applied to them. Galvani discovered that the frog's legs still twitch, seemingly in the absence of electricity, if strips of two different metals are placed on them, implying that chemicals in the frog's body somehow produce or conduct electricity. Eventually, Alessandro Volta (Italian, 1745–1827) invented an electric battery in 1800. Volta's battery consisted of a stack of alternating zinc and silver disks, separated by felt soaked in salty brine. Not long afterwards, others replaced the silver with less expensive copper. This type of battery was what we now call a wet cell.

Over the next century, many researchers produced new or improved batteries. French scientist Georges Leclanché invented a zinc-carbon battery in 1868; this was a precursor to the present dry cell flashlight battery (figure 1). Gaston Planté (French, 1834–1889) invented the first lead-acid storage battery in 1869, and Thomas Alva Edison (American, 1847–1931) developed a nickel-iron alkaline storage battery in 1901. In 1908, a Swedish inventor named Junger invented nickel-cadmium storage cells, a precursor to today's NICADs. Ruben Mallory developed the mercury cell (figure 2) in the 1930s, and H. André of France invented the silver-zinc battery in 1941.

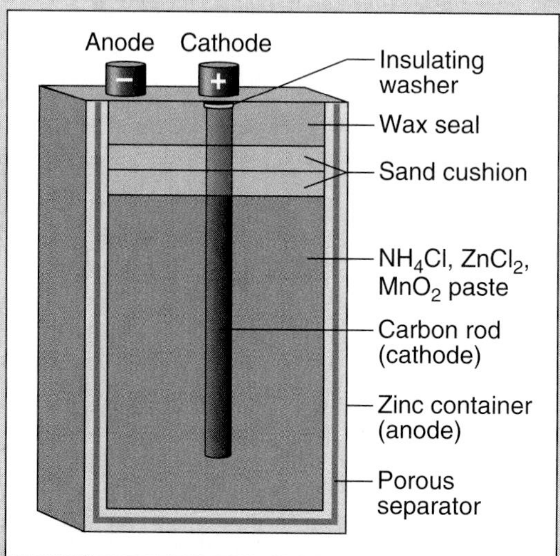

Figure 1 Cross section of a typical dry cell.

The mercury dry cell in figure 2 contains KOH (potassium hydroxide) and $Zn(OH)_2$ (zinc hydroxide). Other mercury cells contain mercury oxide (HgO) and zinc. The net reaction that takes place when a circuit is connected is:

$$Zn^0 + Hg^{2+}O^{2-} \longrightarrow Zn^2 + O^{2-} + Hg^0$$

In the silver oxide cell, the same reaction takes place, but the cathode (positive terminal) is Ag_2O (silver oxide) instead of HgO. The anode (negative terminal) is still zinc:

$$Zn^0 + Ag_2^+O^{2-} \longrightarrow Zn^{2+}O^{2-} + 2Ag^0$$

The most common storage battery is the lead (Pb) storage cell used in automobiles. This battery can be recharged either by the car's generator or alternator or by a battery recharger. Figure 3 shows a cross section of one cell in a lead storage battery (which contains six such cells).

As figure 3 indicates, the lead storage battery uses a cathode of $Pb^{4+}O_2{}^{2-}$ (lead dioxide) and an anode of Pb^0 (lead) immersed in a solution of H_2SO_4 (sulfuric acid). This battery produces current by this net reaction:

$$Pb^0 + Pb^{4+}O_2^{2-} + 4H^+(aq) + 2SO_4^{2-}(aq) \longrightarrow 2Pb^{2+}SO_4^{2-}(aq) + 2H_2O$$

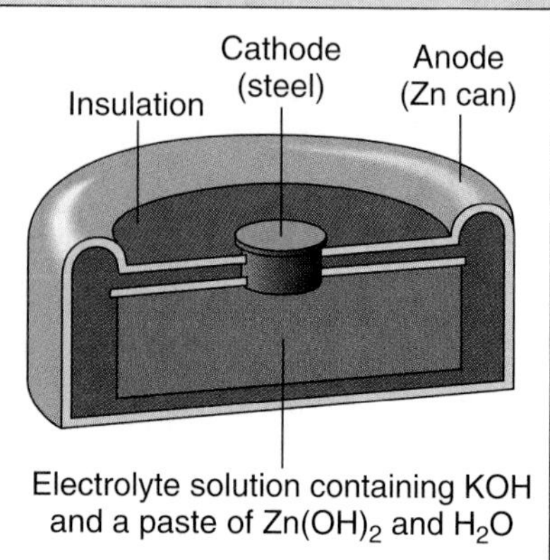

Figure 2 Cross section of a mercury dry cell.

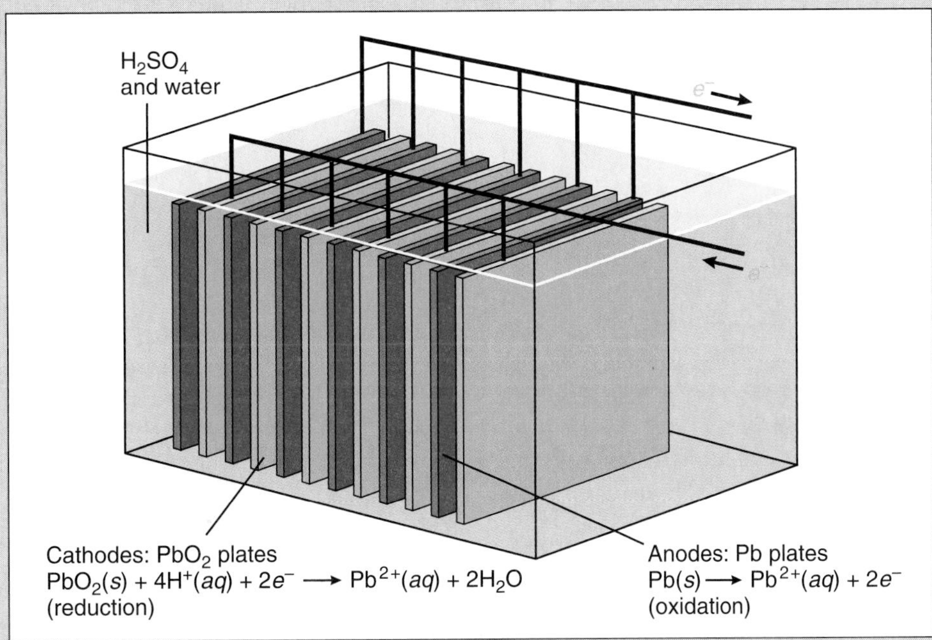

Figure 3 Cross section of one cell in a lead storage battery.
SOURCE: From Jacqueline I. Kroschwitz, et al., *Chemistry: A First Course*, 3d ed. Copyright © 1995 Times Mirror Higher Education Group, Inc., Dubuque, Iowa. All Rights Reserved. Reprinted by permission.

The Pb in both the anode and cathode is converted into soluble $PbSO_4$. If this process continued, the battery would eventually go dead as the redox reaction became exhausted. But the alternator (or generator) forces this reaction to reverse, restoring the Pb in both the anode and cathode. The acidity of the battery decreases during the discharge step, when H_2SO_4 breaks down, and increases during recharging. The voltage fluctuates with the acid concentration.

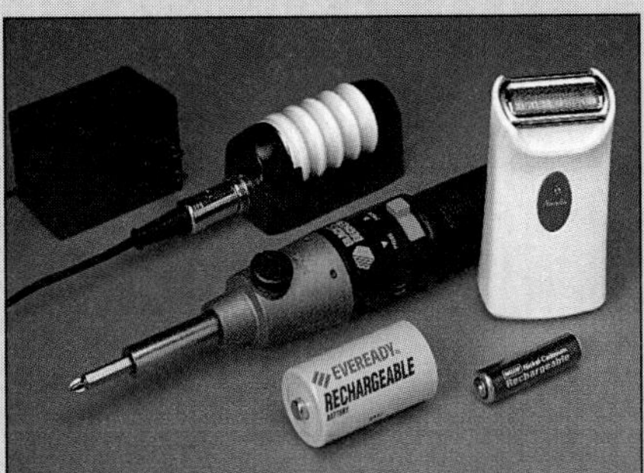

Figure 4 A typical nickel-cadmium, or NICAD, battery.

The other major kind of storage battery is the nickel-cadmium cell, or NICAD (figure 4). NICADs consist of a $Ni^{4+}O_2^{2-}$ (nickel oxide) cathode and a Cd^0 (cadmium metal) anode immersed in an alkaline KOH paste. The overall reaction is:

$$Ni^{4+}O_2^{2-} + Cd^0 + 2H_2O \rightarrow Ni^{2+}(OH)_2^- + Cd^{2+}(OH)_2^-$$

When NICADs are recharged, the reaction is reversed, and the NiO_2 and Cd are restored. NICADs are much lighter than lead storage batteries and also maintain a more constant voltage. For this reason they are used in calculators, cordless power tools, and electric razors. NICADs can last for several years and many recharge cycles before they wear out.

Research on batteries continues. One objective researchers have is to develop an electric car. Various types of rechargeable, lithium-based cells are now common, and a flexible lithium-polymer battery seems to be on the horizon. This lightweight type of plastic battery could become part of an automobile body and serve a dual purpose. Another battery now making gains in the consumer market is a rechargeable zinc-air cell that offers more power than most other kinds of batteries.

EXPERIENCE THIS

Mix a solution containing 1/4 cup baking soda (sodium bicarbonate) and 1/4 cup table salt in 2 quarts water in an enameled or glass pan. Place some aluminum foil and some tarnished silver (perhaps a spoon) in this solution. Let the solution stand. What do you observe? Are these changes due to chemical or physical actions? *Hint:* Check the relative positions of aluminum and silver in table 7.3.

Note: This experiment can also be run in an aluminum pan, but it will lose some aluminum and need scouring afterwards. If you use an aluminum pan, you do not need to add the aluminum foil to the solution.

Thermal Factors that Affect Chemical Reactions

Not only might chemical reactivity limit the likelihood that a particular reaction will occur, but energy-related factors also affect the chances that a reaction will take place. Two thermal factors determine whether a specific chemical reaction occurs **spontaneously,** (that is, runs without additional energy input): enthalpy and entropy. **Enthalpy,** or heat of reaction (ΔH), corresponds to the amount of heat released or consumed by a reaction. An **exothermic reaction** liberates heat, while an **endothermic reaction** requires an input of heat energy to proceed.

Many exothermic reactions run spontaneously, whereas endothermic reactions generally do not. Figure 7.6 depicts the endothermic reaction of nitrogen with oxygen, for example, which occurs only at high temperatures, and the exothermic reaction between hydrogen and oxygen, which occurs readily at ordinary temperatures. Because the nitrogen-oxygen reaction requires heat energy, it is endothermic and unlikely to occur spontaneously.

Entropy, (ΔS), is the degree of randomness or disorder in a system; it generally increases with increases in temperature. Reactions that increase disorder often occur spontaneously. The combination of enthalpy and entropy can be used to predict whether a reaction is likely to occur, even if it is endothermic; but for our purposes, it is sufficient to say that as a general rule, exothermic reactions and reactions that increase in entropy tend to occur spontaneously.

Another important rule of thumb is that reactions run about twice as fast for each 10°C increase in temperature. If a reaction produced 20 g product in 10 minutes at 20°C, then it should produce 40 g product in the same amount of time at 30°C. This thermal effect is extremely important in cooking and food storage. Recall from chapter 2 that food cooks faster in a pressure cooker because the temperature and pressure are higher. On the other hand, reactions run slower when the temperature decreases; thus, only 10 g of product would be expected in 10 minutes at 10°C, or 5 g at 0°C. Refrigeration makes use of this principle because food spoilage is a series of chemical reactions that slow down appreciably at lower temperatures.

EXPERIENCE THIS

Obtain a thermometer that you can place into water without damaging it. Place 100 mL water in an 8-ounce jar or glass and measure its temperature. Now dissolve about 1 ounce baking soda ($NaHCO_3$, sodium bicarbonate) in the water and remeasure the

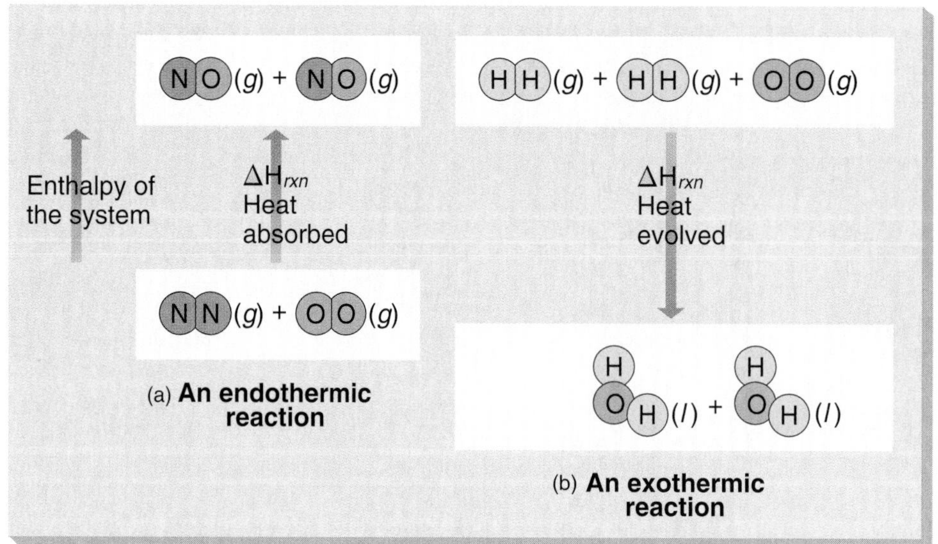

Figure 7.6 Comparison between an **(a)** endothermic (nitrogen and oxygen) and **(b)** exothermic (hydrogen and oxygen) reaction.

temperature. Did the temperature change? Next, measure the temperature of 100 mL vinegar in another jar. Add this vinegar to the baking soda solution, stir, and record the temperature. Did the temperature change again? Repeat the experiment using twice as much baking soda in the same amount of water. Is the resulting reaction exothermic or endothermic? What is the nature of the gas that formed? Is there any way you could trap and collect this gas?

You can study several other concepts using the same basic chemical reaction. (1) Use warm vinegar in one sample of $NaHCO_3$ and water, and cold vinegar in another sample. What effect does this have on the rate of gas production? (2) Set up a pair of containers, each containing $NaHCO_3$ and water. Add diluted vinegar (store strength mixed with five parts of water) to the first sample. Add full-strength vinegar to the other sample. What effect does the concentration of the vinegar play in this reaction?

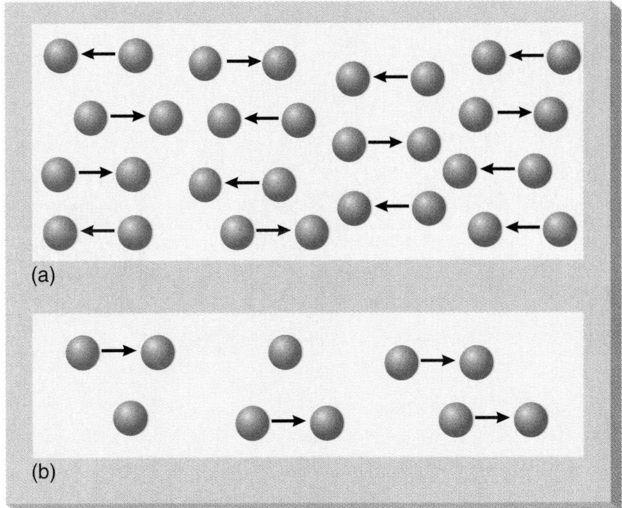

Figure 7.7 The effect of concentration on the number of collisions (indicated by arrows) that occur during a reaction. **(a)** Many particles create many collisions. **(b)** Fewer particles have fewer collisions.

Kinetic Factors that Affect Chemical Reactions

Thermal factors are not the only energy-related conditions that affect chemical reactions; kinetic factors also have an impact. With the exception of decomposition reactions, most chemical reactions involve two or more reactants. For a reaction to happen, the molecules of each substance must contact each other at a high enough energy to permit chemical combination. Additionally, the molecules must strike each other in the proper direction or orientation. Thus, three important kinetic factors affect chemical reactions: (1) frequency of collision, (2) amount of energy, and (3) orientation. These factors are especially important for reactions that involve only covalent compounds. They are much less significant for reactions that involve only ionic compounds.

Frequency of collision, which measures how frequently molecules collide with each other, is regulated by the number of molecules present and the average speed with which they move. When more molecules are present (a higher **concentration**) at constant volume and speed, they collide more frequently and the reaction proceeds faster (figure 7.7). This explains why gas-phase reactions run faster at higher pressure; higher concentrations of atoms or molecules are present. If the number of molecules remains constant, an increase in temperature will make them move faster, and they will collide more often. Thus, increases in the number of molecules present or in the speed they move at increase the frequency of collision as well.

The amount of energy available affects the likelihood of a reaction because covalent chemical reactions require bonds to break and re-form. Energy is required to break the old bonds, and although the formation of new bonds provides some energy, this energy is usually unavailable at the start of the process. For this reason, molecules must collide with sufficient energy, called **activation energy,** to rupture the existing bonds. If this activation energy is lacking, the molecules bounce away from each other like billiard balls. You can picture activation energy as a hill the reactants must climb before they can form the products (figure 7.8). The total energy needed to make a reaction occur is the heat of reaction plus the activation energy.

Figure 7.9 illustrates the importance of **orientation.** This schematic reaction shows an exchange of bonds between one molecule with two red atoms and one with two gray atoms. If these molecules collide in the correct alignment or orientation, they form two mixed-atom molecules. Even if the molecules possess enough energy, this energy is wasted if the collision happens as in orientation *a;* the molecules merely bounce off each other. But if the molecules are in orientation *b* and possess sufficient impact energy, the reaction occurs. The more often the alignment is correct, the faster the reaction proceeds. Therefore, when the percentage of the molecules in the proper alignment doubles from 1 to 2 percent, the reaction rate also doubles.

It is difficult to determine numerical values for each of these three kinetic factors, but understanding each factor does allow chemists to improve reaction rates. Making such an improvement could mean the difference between an economical and profitable reaction and one that would cause a business to lose money (and jobs); fortunately, chemists can manipulate these factors to improve reactions. It is fairly easy to increase concentrations and raise temperatures to provide the necessary collision frequency and activation energy. Improving orientation is harder, but not impossible. Chemists often use catalysts to accomplish this.

Other Factors that Affect Chemical Reactivity

We have learned that chemical reactions are affected by chemical reactivity, temperature, concentration, and orientation. **Catalysis**

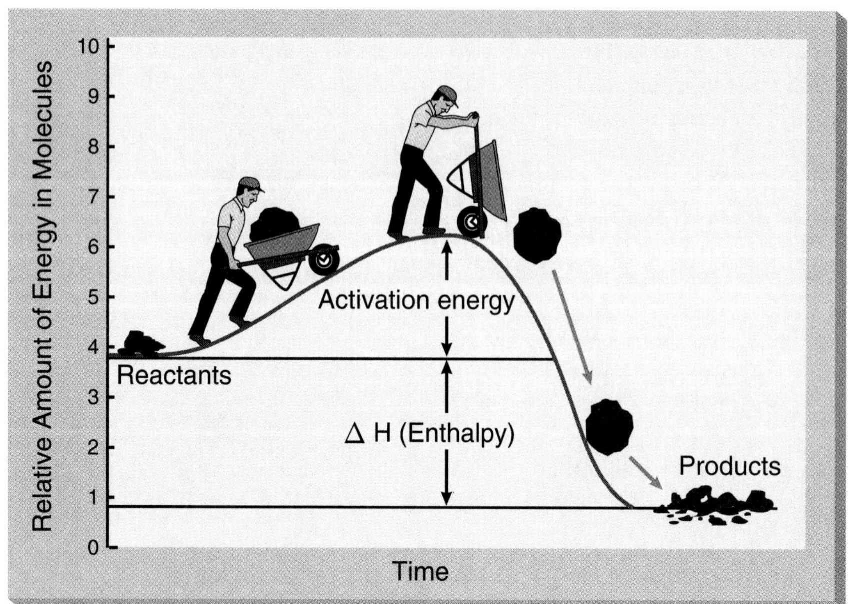

Figure 7.8 The effect of activation energy on a chemical reaction. This energy must be available for the reaction to occur.

SOURCE: From Eldon D. Enger, et al., *Concepts in Biology,* 7th ed. Copyright © 1994 Times Mirror Higher Education Group, Inc., Dubuque, Iowa. All Rights Reserved. Reprinted by permission.

and **surface area** also have an impact on whether a reaction occurs.

Catalysis

A **catalyst** is an agent that makes a reaction run faster, more completely, or at a lower temperature without being consumed in the reaction—in other words, the catalyst still retains its identity when the reaction is complete:

$$A + B + \text{catalyst} \longrightarrow A\text{—}B + \text{catalyst}$$

One way a catalyst speeds up a reaction is by lowering the activation energy. This means the reaction can run faster at a given temperature. The catalyst does this by providing an alternate pathway—a shortcut—for the reaction to follow. Figure 7.10 illustrates this effect. The reaction occurs slowly in the absence of a catalyst because the activation energy is high. The catalyst provides a new reaction route with a lower activation energy, and the decomposition occurs faster. As the figure shows, we could picture a typical reaction path as a hill to climb. At the top, the reactants slide downhill to form the products. The catalyst gives the reactants a shortcut over a lower hill, which requires less energy to climb. This means more molecules transform into product because more energy is available for the actual reaction, rather than being used to climb the hill. Thus, a catalyzed reaction requires less energy per molecule and proceeds more rapidly.

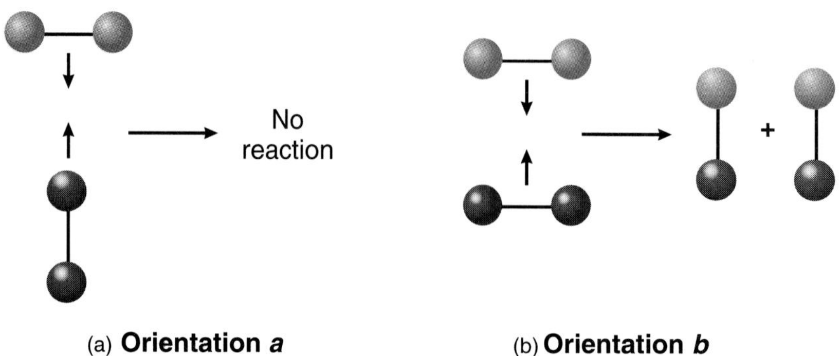

Figure 7.9 Orientation and chemical reaction. **(a)** This double displacement reaction does not occur when the molecules collide with the wrong orientation. **(b)** When the orientation is favorable, the double displacement reaction does occur.

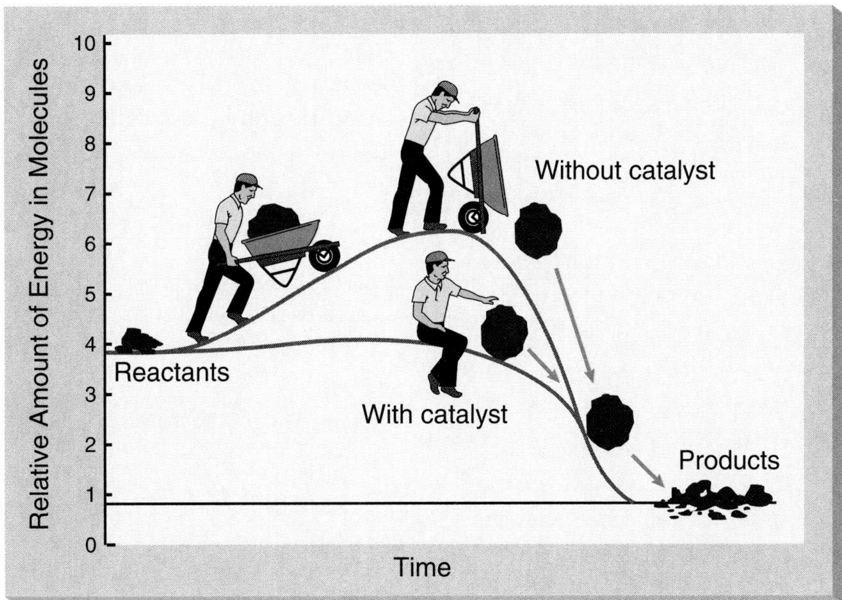

Figure 7.10 A catalyst can provide an alternate reaction pathway that requires less energy (lowering the hill) and makes the reaction proceed faster.
SOURCE: From Eldon D. Enger, et al., *Concepts in Biology*, 7th ed. Copyright © 1994 Times Mirror Higher Education Group, Inc., Dubuque, Iowa. All Rights Reserved. Reprinted by permission.

Sometimes, a catalyst forces the reacting molecules to come together in a high-reactivity orientation. If a catalyst caused all the molecules in a sample to line up as in figure 7.9b, the reaction would run faster and generate more product in a given time interval (figure 7.11). Catalyst shape is critical in orientation catalysis. One example of shape-specific catalysis involves the natural catalysts called enzymes; we will discuss them in chapter 14.

Surface Area

Surface area is another major factor in the rate of certain reactions, such as burning coal, or forming a reversible complex between oxygen and hemoglobin in red blood cells. These reactions occur at the surface of the material, and the speed of reaction depends directly on the total surface area available. When this area increases, the rate increases.

Consider a cube of coal with an edge length of exactly 2.0 cm. Since the cube has six sides, and each side has an area of 4 cm^2 each (2 cm by 2 cm), the total initial surface area for the whole cube is 24 cm^2. However, if we divide each edge in half, we form eight smaller cubes with edge lengths of 1.0 cm. Each new cube side therefore has a surface area of 1 cm^2, and each new cube a total surface area of 6 cm^2. But since there are now eight such cubes, the total surface area of all eight is 48 cm^2. Combustion would proceed twice as fast because the surface area has doubled (figure 7.12a).

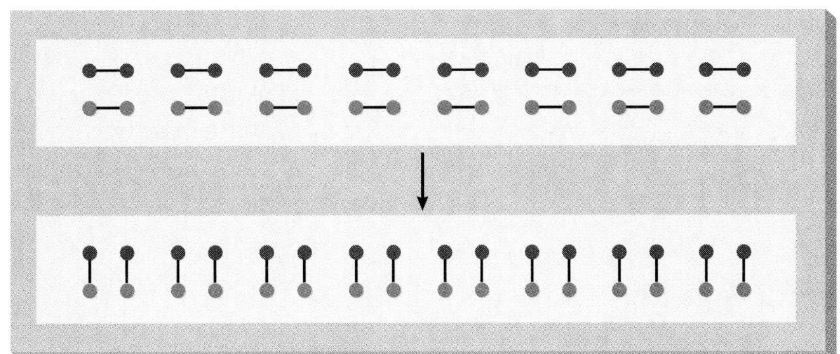

Figure 7.11 A catalyst can change the orientation of reacting molecules and make a reaction proceed faster. The reacting molecules in this double displacement reaction are entrapped in a cavity on the catalyst surface, forcing them into the proper orientation for reaction.

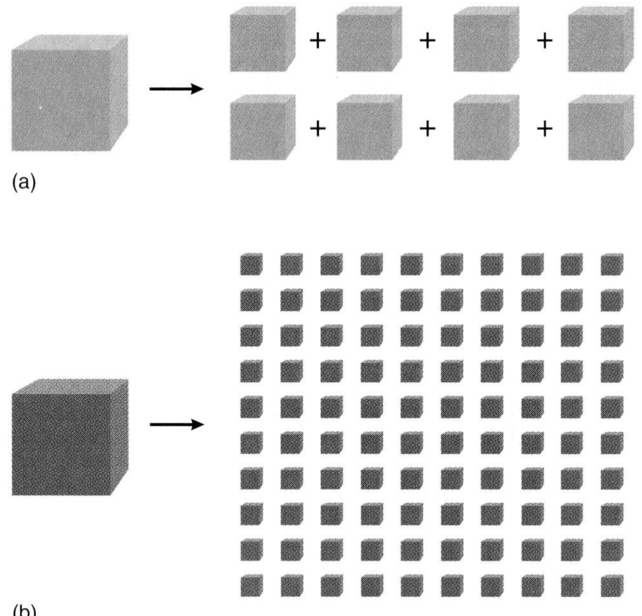

Figure 7.12 Subdividing the edge of a cube increases the surface area and speeds reaction rates. **(a)** Subdividing each edge in half forms 8 new cubes. **(b)** Subdividing each edge into 10 parts creates 1,000 new cubes. Only the front sides of 100 new cubes are shown here.

To take this idea further, consider what happens if we subdivide each edge into 10 pieces 0.2 cm long (figure 7.12b). Although each cube would then have a total area of only 0.24 cm^2, this subdivision would produce 1,000 tiny cubes. The total surface area of all these cubes combined would thus be $1000 \times 0.24 = 240$ cm^2. The reaction would proceed 10 times faster than the original reaction ($240 \div 24 = 10$). This represents both a direct relationship—as surface area increases so does the reaction rate—and an inverse relationship—as particle size decreases, reaction rate increases. In fact, if each cube edge is subdivided into 1,000 parts, the combustion rate increases by a factor of 16,667.

This surface area effect is the reason coal is usually pulverized when used as a commercial fuel. It also explains why grain or dust fires frequently result in explosions. Wheat grain stored in a grain elevator burns fairly slowly. However, if the grain is powdered, the rate of combustion increases dramatically because of the increase in surface area. Since burning is exothermic and produces heat, the surrounding air expands rapidly, and the grain elevator explodes.

EXPERIENCE THIS

Rusting is an oxidation-reduction reaction, but the rate depends on surface area. You can study this reaction by placing some iron objects of different sizes in moist air and examining them daily. (Ideally, you should weigh the samples because iron oxide has a greater mass than iron, but do so only if you have access to an accurate balance.) Although the reaction uses oxygen from the air, it proceeds faster in the presence of moisture. Some samples you could use include iron filings, iron nails of different sizes, a small piece of iron pipe, the fibrous form of iron used in cleaning pads, and some iron washers of various sizes.

One way to increase surface area is to dissolve a substance—when a chemical is dissolved, its surface area increases dramatically as clumps of matter separate into atoms or molecules. Most reactions run faster when a substance is dissolved in some nonreactive material, usually a liquid such as water.

AN OVERVIEW OF CHEMICAL REACTION FACTORS

As we have learned in this chapter, many factors affect chemical reactions. First, and perhaps foremost, reactions are controlled by the chemical components themselves. Some chemicals will react with others, while others will not. In reactions involving metals, the more reactive metal (listed in table 7.3) can displace the less reactive one. The reverse reaction will not occur unless large amounts of energy are available. The reactions that do occur are almost always exothermic, or heat-producing.

Reaction rates and product yields are increased by:

1. Raising the temperature
2. Increasing the concentration
3. Increasing the pressure (if the reaction involves gases)
4. Increasing the time of reaction
5. Adding a catalyst
6. Increasing the surface area

Doing the opposite diminishes reaction rates.

PRACTICE PROBLEM 7.5

Try to predict whether each of the following changes will speed up or slow down a chemical reaction:

a. Lowering the temperature
b. Adding more water to an aqueous solution before starting the reaction
c. Running a reaction between two gases, such as hydrogen and chlorine, at enhanced pressure conditions
d. Converting a log into sawdust and then burning it
e. Adding a catalyst to the reaction mixture

ANSWER The effect on the rate is *a*. a decrease (half the rate for every 10°C temperature reduction); *b*. a decrease (because you lowered the concentration); *c*. an increase (the concentration increased); *d*. an increase (because the surface area was enlarged); *e*. an increase.

SUMMARY

No one knows exactly how many possible chemical reactions can take place, but the number exceeds several trillion. Most follow a few general patterns, including decomposition, combination, and displacement.

In a decomposition reaction, a compound breaks down into its separate components, while in a combination reaction, two separate substances combine to form a compound. In a displacement reaction, one chemical agent displaces another in a compound. A single displacement involves a single pair of substances, whereas a double displacement involves two.

Oxidation-reduction, or redox, reactions are extremely important reactions that always occur in complementary pairs. Redox reactions are defined in one of three ways: in an oxidation reaction, a substance either loses hydrogen, loses electrons, or gains oxygen; and in a reduction reaction, a substance either gains hydrogen, gains electrons, or loses oxygen. Iron's reaction with atmospheric oxygen to form ferric oxide (rust) is an example of an oxidation reaction.

Several factors impose limits on chemical reactions. The chemical reactivity of the potential reactants is the primary limitation. More reactive elements usually displace less reactive ones. Thermal factors also affect reactivity; in general, exothermic (heat-releasing) reactions are more likely to occur spontaneously, as are reactions that increase entropy, or disorder. The kinetic factors that affect reactions include frequency of collision, or how often molecules collide; activation energy; and orientation of the molecules. Finally, catalysis and an increase in surface area may speed a reaction. Chemists can manipulate many of these factors to speed or slow reactions.

KEY TERMS

activation energy 183
catalysis 184
catalyst 184
combination reaction 170
concentration 183
decomposition reaction 170
displacement reaction 171
double displacement 171
endothermic reaction 182
enthalpy (heat of reaction) 182
entropy 182
exothermic reaction 182
frequency of collision factor 183
Left-Handed LEGO 179
orientation 183
oxidation-reduction (redox) reaction 173
single displacement 171
spontaneous reaction 182
surface area 185

READINGS OF INTEREST

General

Newspapers sometimes have articles on various chemical reactions. Look for reports on new medicines, foods and food preparation, and new fuels. All involve chemical reactions.

Stover, Dawn. "Toxic avengers." *Popular Science,* July 1992, pp. 70–74, 93. Not the B-movie, but an intriguing account of microorganisms that love to devour oil spill wastes and other types of pollution. Although this article is light on chemistry, it does discuss how the microorganisms use chemical reactions to consume waste materials.

Theiler, Carl R. 1962. *Men and molecules*. New York: Dodd, Mead. Chapter 4, New Medicines for Mankind. A historical perspective on an important topic in our modern world, covered in a simple style.

Caulfield, H. John. "The wonder of holography." *National Geographic*, vol. 165, no. 3, March 1984, pp. 364–377. An account of how we make all those fascinating three-dimensional images, including a little about the underlying chemistry.

Corrosion

Dhawale, S. W. "Thiosulfate. An interesting sulfur oxoanion that is useful in both medicine and industry—But is implicated in corrosion." *J. Chem. Ed.*, vol. 70, no. 1, January 1993, pp. 12–14. Examples of thiosulfate use include photography, paper making, pharmaceuticals, and medical research.

Space Travel

Carey, John; Payne, Seth; and Schine, Eric. "A small step for man—A tiny one for industry." *Business Week,* August 12, 1993, pp. 46–48. While there is much talk about industry venturing into space, very little actual business has been done so far.

Ostling, Richard N. "Arabia's lost sand castle." *Time,* February 17, 1992, p. 69. The drifting desert sands bury many ancient and modern structures. This article tells how space-age technology helped find the 4,000-year-old city of Ubar again.

Batteries

Krause, Reinhardt. "High energy batteries." *Popular Science,* February 1993, pp. 64–68, 92. More information on modern battery technology.

Freedman, David H. "Batteries included." *Discover,* March 1992, pp. 90–97. This article is mainly on electric batteries in cars.

Glanz, James. "Lithium battery takes to water—And maybe the road." *Science,* vol. 264, May 20, 1994, p. 1084. This article describes a battery that uses a lithium ion water solution instead of the organic liquids used in other versions.

Illman, Deborah L. "Automakers move toward new generation of 'greener' vehicles." *Chemical & Engineering News,* August 1, 1994, pp. 8–16. This article provides a lot of data on the various kinds of batteries now available. It also discusses the emission regulations (federal and state) impelling automakers toward electrically powered cars.

Woodruff, David, Templeman, John, and Sandler, Neal. Assault on batteries. *Business Week,* November 29, 1993, p. 39. Just how enthusiastic are automakers about the electric car?

Thermodynamics

Gibbon, Donald L.; Kennedy, Keith; Reading, Nathan; and Quieroz, Mardsen. "The thermodynamics of home-made ice cream." *J. Chem. Ed.,* vol. 69, no. 8, August 1992, pp. 658–661. Can ice cream form the basis for a chemistry course? This yummy article suggests that we can learn much chemistry just by making some home-made ice cream. A few terms in this article may seem unfamiliar, but you should be able to figure out their meaning.

Steitberger, H. Eric. "How bright are you? Energy and power hands-on activities." *J. Chem. Ed.,* vol. 69, no. 4, April 1992, pp. 307–308. This article considers candles, cashews, and calories and the energy they provide.

Catalysts

Black, Pamela J. "Better living through—Catalysts?" *Business Week,* March 23, 1992. A readable account, aimed at business executives, of some recent advances in catalysis.

Thayer, Ann M. "Catalyst suppliers face changing industry." *Chemical & Engineering News,* March 9, 1992, pp. 27–49. This article considers the use of catalysts in several fields of chemistry, including petroleum refining, plastics, oxidation reactions, and emissions control. *Chemical & Engineering News* runs an article on this topic every year at about the same time.

Mullin, Rick; Chynoweth, Emma; Rotman, David; Morris, Gregory D. L.; Young, Ian; Wood, Andrew; and Kirschner, Elizabeth. "Catalysts." *Chemical Week,* July 1/8, 1992. This is actually a set of six short articles dealing with business aspects of catalysts. Topics include recycling, emissions control, fuel refining, plastics, and medical uses.

Yates, John T. "Surface chemistry." *Chemical & Engineering News,* March 30, 1992. A lot of a catalyst's work occurs on a surface. This article considers catalysis and several other aspects of surface chemistry. (This article is moderately complicated.)

PROBLEMS AND QUESTIONS

1. Briefly define each of the following terms.

a. catalysts
b. chemical equation
c. combination reaction
d. decomposition reaction
e. displacement reaction
f. double displacement
g. enthalpy
h. entropy
i. oxidation
j. oxidizing agent
k. reducing agent
l. reduction
m. single displacement

2. Balance the following chemical reactions and identify the metals and nonmetals in each equation.

a. $CaCl_2 + KOH \longrightarrow Ca(OH)_2 + KCl$
b. $AlBr_3 + Ca(OH)_2 \longrightarrow Al(OH)_3 + CaBr_2$
c. $C_4H_{10} + O_2 \longrightarrow CO_2 + H_2O$
d. $Hg_2S + \text{heat} \longrightarrow Hg + S_8$
e. $Na + O_2 \longrightarrow Na_2O$
f. $Li + HOH \longrightarrow LiOH + H_2$

3. Classify each of the following unbalanced reactions as decomposition, combination, single displacement, or double displacement.

a. $BaCl_2 + NaOH \longrightarrow Ba(OH)_2 + NaCl$
b. $AlCl_3 + Sr(OH)_2 \longrightarrow Al(OH)_3 + SrCl_2$
c. $C_6H_{14} + O_2 \longrightarrow CO_2 + H_2O$

d. Hg_2S + heat → Hg + S_8
e. Li + O_2 → Li_2O
f. K + HOH → KOH + H_2

4. Which of the following are oxidation-reduction reactions? Determine which elements are oxidized and reduced.
a. HCL + NaOH → HOH + NaCl
b. $AlCl_3$ + HOH → $Al(OH)_3$ + HCl
c. $C_2H_4 + O_2 \rightarrow CO_2 + H_2O$
d. PbS + O_2 → $PbSO_4$
e. Rb + O_2 → Rb_2O
f. Cs + HOH → CsOH + H_2

5. Balance each of the following equations (assume the compound formula is correct). Decide whether each reaction would occur spontaneously. (*Hint:* Use table 7.3.)
a. Ca + $ZnCl_2$ → $CaCl_2$ + Zn
b. Mg + Fe_2O_3 → Fe + MgO
c. K + SnO_2 → K_2O + Sn
d. Fe + HgS → Fe_2S_3 + Hg

6. Tell how the following factors affect chemical reactions:
a. catalyst
b. collision energy
c. concentration
d. frequency of collision
e. orientation
f. surface area
g. temperature
h. time of day

7. What happens if the temperature of a chemical reaction is increased by 10°C?

8. What change occurs in the rate of a reaction if the chemical concentrations are doubled?

9. What happens if the temperature of a chemical reaction is decreased by 30°C?

10. How many degrees must a reaction temperature be reduced to make it run only 25 percent as fast?

11. Suppose you burn a wood log that has a surface area of 425 cm^2, then burn sawdust from the same size log. Assume there are 2,000 sawdust particles, each with an area of 2.5 cm^2. Which burns faster, and how many times faster?

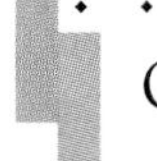

CRITICAL THINKING PROBLEMS

1. Ordinary household flour is made by grinding wheat. Both whole wheat grain and wheat flour burn; which one should burn faster? Why?

2. Hydrogen gas reacts with ethylene ($CH_2{=}CH_2$) to form ethane ($CH_3{-}CH_3$). Assume we obtain 20 g ethane in 30 minutes at 50°C, but then find that the yield increases to 35 g in 30 minutes at 50°C if we add a very small quantity of platinum (Pt) metal to the reaction. How can you explain this result? Suppose we let the reaction run for 60 minutes, with the Pt present. How much ethane should we obtain?

3. When an iron or steel gasoline storage tank is buried underground, it is often connected by wires to a piece of magnesium metal buried nearby. What is the purpose of this Mg metal?

HINT Reexamine table 7.3. If you're still unsure, reread box 7.2.

4. Suppose you are trapped on a desert island, and you discover that your only source of drinkable water is the liquid you can drive out of some rocks by heating them. (Seawater is not drinkable.) Assume you have any type of chemical equipment you may need, including matches and fuel, and a plentiful supply of rocks. However, the release of water from these rocks is much too slow to obtain enough water to sustain life. How could you speed the reaction enough to obtain a sufficient amount of water to survive?

5. Imagine you are still on the deserted island and need food. Fortunately, the surrounding ocean contains lots of fish, though several days often pass by between your catches. Although you have matches and fire, you have no refrigeration or electricity. How can you keep your fish from spoiling between fresh catches?

HINT Recall some of what you learned in chapter 1 testing hypotheses on water evaporation.

6. An oxidizing agent, such as hydrogen peroxide (H_2O_2), is sometimes used as a germicide to kill bacteria in small cuts on the skin. How might this work? Does this type of treatment pose any possible danger to the skin?

7. So-called tin cans are actually iron cans coated with a thin layer of tin. Why does this coating prevent the iron from rusting? When the coating is ruptured, the rusting process usually occurs even faster than it would on uncoated iron. Why?

HINT Reexamine table 7.3.

8. Which method of preventing iron corrosion would last longer, galvanizing or tin coating? Why?

9. Read box 7.1 and the article by John Carey, Seth Payne, and Eric Schine called "A Small Step for Man—A Tiny One for Industry," in *Business Week,* August 12, 1993, pp. 46–48. Why haven't more

space-related industrial applications developed? Do you think industry will venture into space in the future? Why or why not?

10. Read the following two articles dealing with batteries and electric cars: Deborah L. Illman, "Automakers Move Toward New Generation of 'Greener' Vehicles," *Chemical & Engineering News*, August 1, 1994, pp. 8–16; and David Woodruff, John Templeman, and Neal Sandler, "Assault on Batteries," *Business Week*, November 29, 1993, p. 39. If possible, research the electric cars that were used in the early parts of this century. What problems do electric cars raise? Are these types of vehicles likely to increase in popularity in the near future? (Note that most automobile companies have prototypes but no large-scale marketing plans.) How would electric cars fit in with the thrust for increased public transportation?

11. If the following reaction could run in reverse, would it be a safe way to chlorinate a swimming pool?

HINT Think about what happens to the hydrogen.

$$H_2 + Cl_2 \rightarrow 2\ HCl$$

12. Figure 1 in box 7.1 shows a space shuttle launch. Could the decomposition of mercuric oxide be used to power this vessel? What would the limitations of such a reaction be?

13. Some food decomposes by a chemical reaction with oxygen in the air. What type of reaction would you suspect this is? How could you prevent this type of food decomposition?

14. Iron is galvanized by applying a coating of zinc to the surface. How does the zinc prevent the iron from rusting? How long will the zinc keep the iron from rusting?

HINT Look at table 7.3.

15. What relation exists between food spoilage and chemical reactions? How does refrigeration retard food spoilage?

C H A P T E R 8

SOLUTIONS, ACIDS, AND BASES

OUTLINE

KEY IDEAS

1. How to Predict the Water Solubility of Most Ionic Compounds
2. How the Terms *Solute* and *Solvent* Define or Describe Solutions
3. What Concentration Is, and How It Is Measured
4. How Molarity Is Expressed and Calculated
5. How to Use Parts Per Million (ppm) and Parts Per Billion (ppb)
6. The Nature of Acids, Bases, Salts, and Neutralization
7. The Meaning and Importance of the pH Scale

Learn to reason forward and backward on both sides of a question.
THOMAS BLANDI

Vortex in a beaker.

Almost all chemical reactions that occur in nature take place in aqueous solution—in water. Water is not a typical liquid. It is strongly hydrogen bonded and highly polar (see chapter 6). This unusual combination of properties permits water to dissolve more chemicals than any other common liquid. Thus, soft drinks, swimming pools, and living systems contain aqueous solutions.

Along with occurring in aqueous solution, many naturally occurring chemical reactions are acid-base reactions. Like redox reactions, acid-base reactions involve a transfer—but of a proton rather than an electron. Acids and bases react with each other to form water and salts.

In this chapter, we will examine (1) water solutions and solubility in water, and (2) acids and bases. Aqueous solutions you encounter daily include drinking water, milk, soft drinks, juices, tea, and coffee. We swim, bathe, and shower in water containing dissolved salts. We use acidic and basic chemicals every day, and acid rain (a solution of water and chemicals) often makes the news.

SOLUBILITY—THE SOLUTION TO SOLUTIONS

Before we look at what makes up a solution, we must first learn about **solubility,** or the tendency to dissolve. Not all compounds are water-soluble or completely dissolvable. What makes one material soluble and another insoluble is not easy to define; it is, in fact, beyond the scope of this text. However, a simple rule of thumb applies to solubility: Like dissolves like, and unlike substances do not dissolve. For example, a polar compound would most likely dissolve in a polar solvent, but not in a nonpolar solvent.

Water is polar; therefore, according to our rule, polar compounds are expected to dissolve in water. Ionic compounds are extremely polar, and many (although not all) ionic compounds are water-soluble. However, ionic solids do not usually dissolve in a nonpolar solvent such as gasoline (a hydrocarbon mixture). By similar reasoning, nonpolar organic molecules should dissolve in nonpolar organic solvents, but not in water. Thus paraffin wax, a mixture of solid, nonpolar hydrocarbons called alkanes, will dissolve in nonpolar gasoline, but not in water.

Figure 8.1 shows that the water solubility of an ionic compound depends on which ions are present. This chart is divided into four quadrants. The compounds in the upper left quadrant are nearly all water-soluble. The exceptions among the halides and sulfates fall into the upper right quadrant; these compounds contain lead, mercury, or sometimes, barium or silver. By contrast, although a few exceptions (alkali metals and the ammonium ion) are water-soluble and appear in the lower left quadrant, most carbonates, phosphates, hydroxides, and sulfides are water-insoluble, and therefore fall into the lower right quadrant. Oxides are not included in figure 8.1, but they are normally insoluble unless they react with water. Thus, Fe_2O_3 (iron oxide) is unreactive and remains insoluble; Na_2O and other group IA oxides react with water to form soluble products. Most other oxides do not react readily with water.

From figure 8.1, we can make some quick observations on whether a substance will dissolve in water. Note that any compound containing a group I element (the alkali metals), an ammonium ion, acetate, nitrate, or sulfate group is soluble. Almost every compound with a group VII element (the halogens, except fluorine) is also soluble. Practically everything else either has low solubility or is insoluble.

PRACTICE PROBLEM 8.1

Determine which of the following compounds will be water-soluble and which will be insoluble. (HINT: Use figure 8.1 to identify and decide on the solubility of important multiatom groups in each compound. Use the periodic table if you need to find out which elements are alkali [IA] or IIA metals.)

a. $BaCO_3$
b. Na_3PO_4
c. $AgNO_3$
d. $PbSO_4$
e. Na_2S
f. BaS
g. $CaCl_2$
h. Al_2O_3

ANSWER Compounds *b, c, e,* and *g* are soluble; *b* and *e* contain the alkali metal Na, *c* is a nitrate, and *g* contains a IIA metal, and is a chloride. Compound *d,* though a sulfate, is one of the exceptions containing lead (Pb) and is thus insoluble. Compounds *a, f,* and *h* are insoluble. Compound *a* is a carbonate, compound *f* a sulfide, and compound *h* an oxide—all insoluble with few exceptions.

SOLUTIONS AND CONCENTRATION

In chapter 2, we defined solutions as homogeneous mixtures involving any combination of solids, liquids, and gases. Air is a gaseous solution of approximately 19 percent O_2 in 80 percent N_2. Seawater is a liquid solution of about 3.5 percent NaCl, along

Compound Type	Soluble	Insoluble
Alkali metals Ammonium (NH_4^+)	all	none
Nitrate $(NO_3)^-$ Acetates $(C_2H_3O_2)^-$	all	none
Chlorides Cl^-), Bromides (Br^-), Iodides (I^-)	most metals	compounds containing Pb^{2+}, Hg_2^{2+}, Ag^+
Sulfates $(SO_4)^{2-}$	nearly all	compounds containing Pb^{2+}, Hg_2^{2+}, Ba^{2+}
Carbonates $(CO_3)^{2+}$ Phosphates $(PO_4)^{3+}$	alkali metals & NH_4^+	all others
Hydroxides $(OH)^-$	alkali metals	Ca^{2+}, Sr^{2+}, Ba^{2+} moderately soluble; all others insoluble
Sulfides $(S)^{2-}$	NH_4^+, IA & IIA metals	all others

Figure 8.1 Water-solubility "rules of thumb" for common ionic compounds.

with many other ionic compounds, in water. Solder is a solid solution (an alloy) of 33 percent tin and 67 percent lead. The amount of each material present in a solution is fairly important.

Solutions consist of a **solute** and a **solvent,** with the solvent normally the major component. Thus, the solvents (the materials in the previous examples) are nitrogen, water, and lead—a gas, a liquid, and a solid, respectively. The solutes (the dissolved materials) are oxygen, sodium chloride, and tin.

Concentration is a measure of the amount of one material dissolved in another. There are many ways to express concentration, but the most common are percentage (parts per hundred, or pph), molar (moles/liter), parts per million (ppm), and parts per billion (ppb). All methods have advantages in certain cases, but most chemists prefer to express concentration in moles/liter. People are daily concerned about concentration levels in acid rain, water toxins, and other solutions. Many of these examples of concentration will be discussed later in this chapter.

Percentage (pph, or %), the most common concentration scale used in everyday life, expresses how many parts of one substance are in 100 parts total. Thus, solder has 33 parts tin and 67 parts lead in 100 parts total, or is 33 percent tin and 67 percent lead. U.S. coin values are expressed as percentages; one per*cent* of a dollar is one cent, and 10 percent of a dollar is 10 cents. Food labels often list percentages of ingredients. For example, a typical soft drink contains 0.1 percent sodium benzoate as a preservative. Percentages are familiar, useful measurements in chemistry and in other disciplines and activities as well.

PRACTICE PROBLEM 8.2

Let's look at a few problems involving percentage calculations.

a. Suppose you have twelve nickels and eight pennies in your pocket. What percentage of your coins are nickels?

b. A bowl of fruit contains exactly twenty pieces of fruit. The bowl's contents are 60 percent apples, 30 percent bananas, and 10 percent oranges. How many of each kind of fruit are in the bowl?

c. The sucrose molecule has a formula of $C_{12}H_{22}O_{11}$. What percentages of the atoms in this molecule are carbon, hydrogen, and oxygen?

ANSWER *a.* You have 20 total coins in your pocket. To determine the percentage, divide the number of nickels by the total number of coins and multiply the result by 100. Therefore, nickels make up 60 percent of the total number (12 ÷ 20 = .6; .6 × 100 = 60 percent).

b. The bowl contains 12 apples, 6 bananas, and 2 oranges. Sixty percent of 20 is 12 (.60 × 20 = 12), 30 percent is 6 (.30 × 20 = 6), and 10 percent is 2 (.10 × 20 = 2).

c. One molecule of sucrose contains 45 atoms total (12 + 22 + 11 = 45). Divide the numbers for each specific kind of atom by 45 and multiply by 100 to determine the percentage of each. Thus, to find the percentage of carbon atoms: 12 ÷ 45 = .2667; .2667 × 100 = 26.67 percent. To determine the percentage of hydrogen atoms: 22 ÷ 45 = .4889; .4889 × 100 = 48.89 percent. And to determine the percentage of oxygen atoms: 11 ÷ 45 = .2444; .2444 × 100 = 24.44 percent. These three percentages (26.67 + 48.89 + 24.44) add up to 100 percent.

A word of caution: These numbers are the percentages of *atoms* in this molecule and not the *weight percentages* of each element.

Molarity

The calculation of solution **molarity** (moles/liter solution, or M) is another measure of concentration. While parts per hundred (pph) is familiar and easy to use, molarity may take a little practice at first. Determining the molarity of a solution requires three steps. First, the molecular weight of the solute is determined. Next, the number of moles of solute is determined, using this weight. Finally, the moles of solute are divided by the solution volume in liters to yield molarity (M). To understand this process further, examine Practice Problems 8.3 and 8.4. These problems demonstrate everyday chemical applications of molarity.

PRACTICE PROBLEM 8.3

Let's determine the moles/liter for a solution prepared by dissolving 24.8 g KCl (potassium chloride) in enough water to produce 0.5 L total volume.

ANSWER To determine the molarity, we need to take three steps:

1. Determine the molecular weight of the solute in grams/mole.
2. Determine the number of moles of solute by dividing the grams of solute by the grams/mole.
3. Divide the number of moles of solute by the volume in liters to obtain moles/liter, or M.

First, we determine the molecular weight of KCl by consulting the periodic table. Potassium (K) has a molecular weight of 39.1 amu, or grams/mole, and chlorine (Cl) 35.5 amu. The sum of 39.1 + 35.5 = 74.6 amu, or 74.6 grams/mole. The solution contains 24.8 g of KCl, so to determine the moles of KCl, we divide 24.8 g by 74.6 g/mole = 0.33 moles KCl. Finally, we divide this by the volume in liters. Since there is 0.5 L total volume, 0.33 ÷ 0.50 = 0.66 M.

PRACTICE PROBLEM 8.4

Let's practice calculating molarity by determining the molarity of the following solutions.

a. A solution of 48.0 g calcium chloride ($CaCl_2$) in enough water (H_2O) to provide 200 mL total solution

b. A solution of 96.0 g benzene (C_6H_6) in enough hexane (C_6H_{14}) to give 250 mL total solution

ANSWER

a. Refer to the three steps listed in Practice Problem 8.3.

1. Begin by determining the molecular weight of the solute: Ca = 40.1 amu or grams/mole, and Cl = 35.5 grams/mole (so Cl_2 = 71.0 g/mole). Therefore, the molecular weight of the solute, $CaCl_2$, is 40.1 + 71.0 = 111.1 g/mole.
2. Next, determine the number of moles of solute by dividing grams of solute by the molecular weight expressed in g/mole. There are 48.0 g $CaCl_2$ in the solution, so 48.0 ÷ 111.1 = 0.43 moles. This quantity is dissolved in the amount of water necessary to bring the final volume up to 200 mL, which is 0.20 L.
3. To determine moles/liter, or M, divide the moles of solute by the volume in liters. 0.43 ÷ 0.20 = 2.15 M solution of $CaCl_2$.

b. Again, follow the three steps for calculating molarity.

1. Since C = 12.0 g/mole, C_6 = 72.0 g/mole; H = 1.0 g/mole, so H_6 = 6.0 g/mole. 72.0 + 6.0 = 78.0 g/mole. This is the molecular weight of the solute, C_6H_6.
2. The solution contains 96.0 g C_6H_6. 96.0 ÷ 78.0 = 1.23 moles C_6H_6 (benzene).
3. 250 mL = 0.25 L. 1.23 ÷ 0.25 = 4.92 M. This solution is 4.92 M benzene.

Parts per Million

Parts per million (ppm), another measure of concentration, tells how many parts of one substance are present in 1,000,000 (1 million) parts of another. Bituminous coal (soft coal) contains a small amount of sulfur (S) that burns with the coal, combining with oxygen to produce SO_2:

$$S \text{ (in the coal)} + O_2 \rightarrow SO_2$$

If the level of SO_2 in the air is 3 ppm, 3 liters of SO_2 are present in 1,000,000 liters total, or 3 L SO_2 in 999,997 L air. Although this may seem a small amount, air containing this much sulfur would be objectionable to most people, and many plants, fish, and statues. Sulfur dioxide, SO_2, the major pollutant in acid rain, is a noxious material that irritates eyes and mucous membranes. SO_2 pollution may reach this level—3 ppm—near coal-burning power plants and smelters.

A few other examples will illustrate ppm. One ppm of the total U.S. population (about 230,000,000) is 230 people. This may also be expressed as 0.0001 percent of the total population. One penny in $10,000.00 is one ppm; so are 8 average steps of two feet each in a 3,036-mile trek from New York City to San Francisco. Although one ppm is quite small, air pollution levels of just a few ppm may be hazardous. On the other hand, equally small amounts of certain chemicals may have some benefits, as box 8.1 explains.

PRACTICE PROBLEM 8.5

Some people believe that carbon dioxide (CO_2) buildup may result in catastrophic global warming. Carbon dioxide is present in the atmosphere at about 350 ppm. What is this value expressed as a percentage?

ANSWER Since ppm is parts per million, and percentage is parts per hundred (pph), the ratio of ppm/pph is always 1,000,000/100, or 10,000, or 10^4, and the ratio of pph/ppm is 100/1,000,000, 0.0001, or 10^{-4}. This means 350 ppm = 350×10^{-4} or 0.035 percent. Could a quantity this small cause worldwide global warming? The point is debatable.

Parts per Billion

Yet another measure of concentration, **parts per billion (ppb),** describes how many parts of a substance are present in 1,000,000,000 (1 billion) parts total. To envision one ppb, imagine yourself with four friends—1 ppb of the world's population of 5 billion. Or think about one penny in comparison to 10 million dollars ($10,000,000), or about two-thirds of an average two-foot step in comparison to the steps taken walking around the world ten times. Although 1 ppb is obviously extremely tiny, a few chemicals are hazardous even at this level. For example, lead (Pb) is a very poisonous chemical. The Public Health Service recently reduced the permissible lead level in drinking water from 50 ppb to only 15 ppb (0.0000015 percent). Even at seemingly minuscule levels, lead can cause nerve damage and learning disabilities (box 8.2).

PRACTICE PROBLEM 8.6

Determine the value of 15 ppb as a percentage.

ANSWER Since ppb is parts per billion, the ppb/pph ratio is $10^9/100$, or $10^9/10^2$, which is 10^7. The inverse pph/ppb ratio is $10^2/10^9$, or 10^{-7}. Thus, 15 ppb = 15×10^{-7} or 0.0000015 percent.

PRACTICE PROBLEM 8.7

A more common and useful comparison converts ppb to ppm, or the reverse. Convert each of the following examples. Which value—ppm or ppb—seems more useful and easier to work with in each case?

a. Convert 40 ppm to ppb
b. Convert 700 ppb to ppm
c. Convert 7 ppm to ppb
d. Convert 3,000 ppb to ppm
e. Convert 5.0 ppb to ppm
f. Convert 0.0017 ppm to ppb

ANSWER The ratio of ppb/ppm is $10^9/10^6 = 10^3$, or 1000. The inverse ratio of ppm/ppb is $10^6/10^9 = 10^{-3}$, or 0.001. Thus, to convert ppb to ppm, you multiply by 1,000; to convert from ppm to ppb, you divide by 1,000. Therefore:

a. 40 ppm × 1,000 = 40,000 ppb
b. 700 ppb ÷ 1,000 = 0.700 ppm
c. 7 ppm × 1,000 = 7,000 ppb
d. 3,000 ppb ÷ 1,000 = 3.000 ppm
e. 5 ppb ÷ 1,000 = 0.005 ppm
f. 0.0017 ppm × 1,000 = 1.7 ppb

ACIDS AND BASES—THE pH SCALE

Several definitions of acids and bases exist, but the simplest and most common define them in terms of proton transfer. According to this definition, an **acid,** is a compound that can give up or transfer a labile proton (H^+). (A labile proton is one prone to separate from the rest of the molecule.) A **base,** on the other hand, is a compound that can accept a proton.

BOX 8.1

FLUORIDATION OF DRINKING WATER

Although no one debates the removal of lead contaminants from drinking water, the addition of fluoride ions to the water supply has sparked debate for over fifty years. Beginning in 1945, F^- was added to many American drinking water supplies to combat tooth decay. Fluoridated drinking water contains about 1 ppm (range = 0.7–1.2 ppm) F^- ions (from sodium fluoride, NaF, or another fluoride). What are the benefits and risks of fluoridation?

THE SOCIOLOGICAL PERSPECTIVE

Some experts believe that fluoride ions reduce the formation of caries (cavities) in teeth. Early studies indicated reductions of as much as 60 percent; more recent reports claim a caries reduction rate of less than 35 percent. These same studies also indicate fluoride administration is effective in combatting tooth decay only in children under ten years of age.

Fluoride prevents or reduces caries by replacing the OH in tooth enamel. Tooth enamel is composed of the compound calcium hydroxyapatite, $Ca_5(PO_4)_3OH$. The F^- ion replaces some of the OH groups in the compound, forming fluorinated calcium hydroxyapatite, $Ca_5(PO_4)_3OH_{(1-x)}F_x$, where x is between zero and one. Bones, which also consist of calcium hydroxyapatite, take up the F^- ion replacement as well.

But does fluoridated drinking water pose any risks? The *Merck Index* and the *Physician's Desk Reference* note that the ingestion of 0.25–0.45 g NaF can cause severe nausea, vomiting, and diarrhea; ingesting 4 g NaF can be fatal. If you drink eight 8-ounce (200 mL) glasses of water daily, you ingest 250 L water and 0.25 g NaF in 156 days (a little more than five months). At this rate, you would ingest a total of 4.0 g NaF in slightly less than seven years.

Fortunately, healthy individuals excrete most of the F^- ions they ingest daily. (Some people with kidney diseases and other ailments don't excrete it as efficiently and are more susceptible to the toxic effects of NaF.) However, excessive F^- usage can cause a condition known as chronic endemic dental fluorosis in which teeth become mottled and brittle. Skeletal fluorosis, a bone disorder than can cause arthritis-like pain and immobility, may also arise from high F^- intake; and an increase in the occurrence of hip fractures has been directly linked to F^-, even at the 1 ppm level. For most people, however, the benefits of fluoridation seem to outweigh its risks.

THE SCIENTIFIC PERSPECTIVE

Since F^- is a chemical, let's examine this topic from a chemical perspective. Domestic water is used many ways, as figure 1 indicates. Only about 0.6 percent is actually consumed by humans; this means that 99.4 percent of fluoride-treated water is used for nondrinking purposes and has no decay-prohibiting effect. In addition, many studies indicate that F^- application is most effective under age ten, but children under ten consume less than 0.05 percent of the total public water supply. This effectively reduces the F^- treatment to 0.003 percent of the drinking water supply. Furthermore, industrial, agricultural, and car washing uses of water were excluded from consideration—even though they, too, contain F^- ions. Obviously, most of the fluoride added to the water supply provides few or no health benefits or risks.

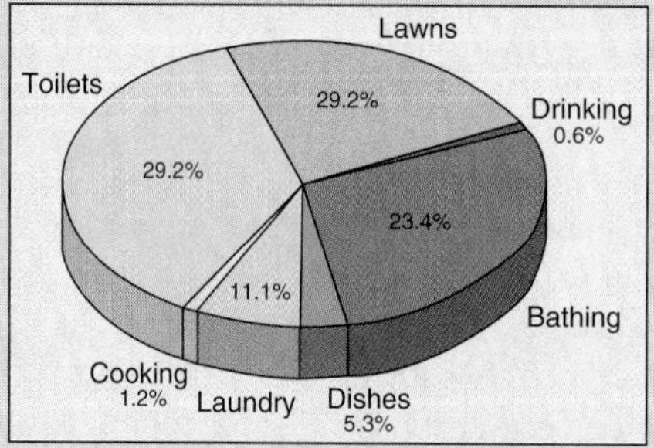

Figure 1 Domestic uses for drinking water.

In addition, although F^- ions do actually bond with the calcium hydroxyapatite that forms tooth enamel, tooth decay is not caused by lack of F^- ions, but rather by streptococcal bacteria that metabolize sugar in the mouth. A low-sugar diet and good dental hygiene are thus important factors in cavity reduction. Also, F^- ions may be obtained naturally from some foods as well as from many brands of toothpaste and from concentrated fluoride solutions a dental patient can swish and spit. There seem to be few scientific reasons for continuing to fluoridate drinking water. (The latest dental technology makes fluoridation nearly obsolete by sealing teeth in plastic to prevent caries.)

And so the debate continues. Antifluoridationists feel their right to choose is violated and their health endangered when the water supply is fluoridated, while fluoridation proponents argue that the benefits of caries prevention outweigh this objection. Unlike the chlorine in drinking water, which is present as dissolved Cl_2 gas that boils off from the water, F^- ions cannot be easily removed. For an excellent review of both sides of the fluoride controversy, see the 1988 article by Bette Hileman listed in the Readings of Interest section at the end of the chapter.

BOX 8.2

GET THE LEAD OUT!

Most heavy metals (that is, metallic elements with an atomic mass that exceeds about 200 amu) are poisonous to humans and animals. Lead, one of these heavy metals, interferes with hemoglobin production in the blood and produces anemia. Lead contamination can also cause mental retardation, learning disabilities, and damage to the nervous system.

The symbol for lead, Pb, comes from the Latin word *plumbum* (lead), which also gives us the words *plumbing* and *plumber*. In the Roman Empire, lead pipes carried water in the cities and lead cooking utensils were used by wealthy Romans. (The poorer peasants did not have running water, and they mainly used clay cooking vessels.) As a result, upper-class Romans may well have suffered lead poisoning that possibly caused brain or central nervous system damage. In fact, 1994 studies of ice cores in Greenland revealed extreme lead pollution caused by smelting during the period from 500 B.C. to 300 A.D. (See the 1994 *Science* article in the Readings of Interest section at the end of the chapter.) Lead poisoning may have contributed to the fall of the Roman Empire.

However, the Romans aren't the only ones who have been exposed to toxic levels of lead. Lead pigments were once common in household and furniture paints—lead compounds produce so many different colors that an artist could compose an entire palette using only lead compounds (table 1). Unfortunately, lead poisoning from paint in older houses causes problems because old paint peels so readily. Small children eat the paint chips; nearly 250,000 become ill every year. In some cases, permanent brain damage occurs. Lead paint cleanup requires professional care and disposal of lead paint fragments, subject to EPA requirements. New paints make use of other types of pigments.

Another source of lead pollution in modern homes still involves the water supply. In many homes, copper water pipes are joined with lead solder; the brass alloy used for plumbing fittings also sometimes contained lead. Small amounts of solder slowly dissolve each day. For this reason, if you have copper pipes, it is wise to run your tap water a few minutes before drinking it, especially if the tap has not been used for a day or more. If you wash your face or hands before drinking the tap water, you can avoid wasting the water as it runs. Newer homes often make use of plastic water pipes made from poly(vinyl chloride), or PVC. These pipes are joined with a small amount of volatile solvent. At this time, the polymer appears to be a safe alternative to the older lead-and-copper systems.

Like the ancient Romans, we still sometimes use glasses and dishes that contain lead; lead improves the appearance and luster of crystal glass. However, crystal glass is another potential source of lead poisoning, especially when used to serve acidic beverages such as wine and soft drinks. Similarly, many ceramic dishes are coated with lead-containing glazes (especially those imported from foreign markets, where the health regulations are not as much at the forefront).

Physicians treat lead poisoning with EDTA or ethylene-diamine-tetra-acetic acid, usually written without hyphens—figure 1. EDTA forms soluble compounds called *chelates* with the lead, permitting the body to excrete it. But far preferable to using EDTA as an antidote is avoiding lead exposure as much as possible. There is no doubt that it is wise to get the lead out of paints, drinking water, and other everyday solutions.

```
HOOC-CH2                  CH2-COOH
        \                /
         N-CH2-CH2-N
        /                \
HOOC-CH2                  CH2-COOH
```

Figure 1 Ethylenediaminetetraacetic acid, EDTA, is a chemical used to treat lead poisoning.

TABLE 1 The colors of various lead compounds.

COMPOUND NAME	COMPOUND FORMULA	COMPOUND COLOR
Lead sulfide	PbS	black
Lead sulfate	$PbSO_4$	blue
Lead molybdate	$PbMoO_4$	yellow
Lead acetate	$Pb(C_2H_3O_2)_2$	white
Lead oxide	PbO	red, brown, yellow, or black in various forms

Bases sometimes have a labile hydroxyl group (OH^-)—an OH ion that separates readily from the rest of the molecule. However, this trait does not define a base because some bases do not have a hydroxyl group. A few bases even generate the OH^- group via a chemical reaction with water. However, even these bases fit the proton-accepting definition.

What Are Acids and Bases?

Acids and bases have been known since antiquity. Ancient texts (for example, the Bible) written as long ago as 1,000 B.C. describe the reaction of vinegar (an acid) with soda (a base). While the ancients obviously did not understand the underlying chemistry, they observed acid-base reactions as everyday occurrences.

Table 8.1 lists the properties of acids and bases. Acid properties are often opposite to base properties, and vice versa; a base is an anti-acid, and an acid is an anti-base. Common medicinal **antacids** are anti-acids, or bases. Solutions of bases are also called **alkaline** solutions.

EXPERIENCE THIS

This experiment will take many months to complete, but you could be the envy of your neighborhood by the time you finish. Plant two hydrangeas about ten feet apart. Occasionally sprinkle some baking soda (a base) around one plant, and apply an acidic fertilizer to the soil around the other. As the flowers appear, the first set will be blue, while the second set pink. The blossoms, like litmus, indicate the acidic or basic quality of the surrounding soil.

Acid-Base Properties

The litmus mentioned in property 1 of table 8.1 is a dye obtained by treating an extract from the lichens *Variolaria lecanora* and *Variolaria rocella* with ammonia and potassium carbonate and then fermenting the product. Because litmus turns red to indicate an acid and blue to indicate a base solution, it is called an **indicator.** As figure 8.2 shows, litmus is one of many indicators. Indicators are used to test the acidity or basicity of soils, swimming pool water, and other solutions.

We routinely exploit the sour taste of acids (property 2 in table 8.1) in cooking. White vinegar, a 5 percent acetic acid solution, is used as a flavoring. The sour tastes of rhubarb and spinach come from small amounts of oxalic acid, and the bitter taste of tonic water comes from a natural base such as quinine. The antacids we use to relieve excess acidity in the stomach are also bases; a flavoring agent, such as peppermint, usually masks their bitter taste.

Acids liberate H_2 gas in an oxidation-reduction (redox) reaction when they react with certain metals (property 3 in table 8.1).

TABLE 8.1 Properties of acids and bases.

ACIDS	PROPERTY	BASES
Turns litmus red	1	Turns litmus blue
Tastes sour	2	Tastes bitter
Reacts with active metals to release hydrogen gas	3	Reacts with skin oils to feel slippery on skin
Reacts with base to form water and a salt	4	Reacts with acid to form water and a salt
Usually gives up a proton	5	Usually accepts a proton

Bases react with fats and oils to produce soaplike compounds. When acids and bases react with one another in equal proportions, the result is a neutralization reaction. The products of such a reaction are neither acids or bases; they do not change the color of litmus paper, have a sour or bitter taste, or display any other acid or base properties.

The following equations represent two typical acid-base neutralization reactions:

$$HNO_3 + KOH \rightarrow KNO_3 + H_2O$$

Nitric acid; Potassium hydroxide (base); Potassium nitrate; Water

$$H_2SO_4 + 2\,NaOH \rightarrow Na_2SO_4 + 2\,H_2O$$

Sulfuric acid; Sodium hydroxide (base); Sodium sulfate; Water

These equations illustrate property 4 in table 8.1. **Neutralization** always involves an acid and a base as reactants, and nearly always

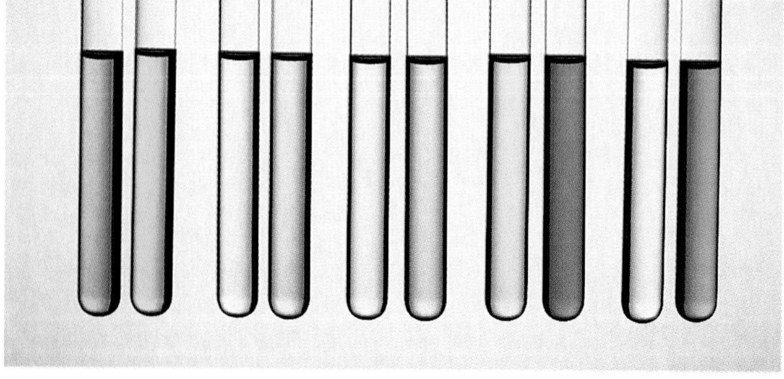

Figure 8.2 How indicators change color with changes in acid or base content. The indicators are (left to right): methyl orange, bromcresol green, litmus, phenol red, and phenolphthalein.

yields salt and water as products; **salt** is a generic term for an ionic compound produced in an acid-base reaction. In these equations, the salts are KNO_3 and Na_2SO_4.

In nearly all acid-base reactions, including neutralizations, a proton (H^+) transfers from the acid to the base (property 5 in table 8.1).

Since most bases include hydroxyl (OH^-) groups, the H^+ and OH^- combine to produce water (H_2O) as one of the products.

Other neutralization reactions—involving bases with no hydroxyl (OH^-) group—do not form water:

$$\underset{\text{Hydrochloric acid}}{HCl} + \underset{\text{Ammonia (base)}}{NH_3} \rightarrow \underset{\text{Ammonium chloride}}{(NH_4)^+Cl^-}$$

The Hydronium Ion

As we have mentioned, acids have labile protons, or hydrogen ions (H^+) they can transfer, and many bases have hydroxide ions (OH^-) that can accept free H^+ ions. Most bases are ionized in the solid state (for example, Na^+OH^- or K^+OH^-), but many acids are not. Hydrogen chloride (HCl), or hydrochloric acid, is a gas with a covalent bond between the proton and chlorine atoms; concentrated sulfuric acid is also essentially nonionic. However, in water, these covalent bonds break apart and produce ions. This is illustrated by the reaction of HCl with a water molecule to form a H_3O^+ group (a **hydronium ion**), and a Cl^- ion:

$$H{-}Cl + H_2O \rightarrow H{-}\overset{\displaystyle H \atop |}{\underset{| \atop \displaystyle H}{O^+}} + Cl^-$$

Hydronium ion

H^+ ions do not normally exist freely in water because they combine with water molecules to form hydronium ions, as this equation shows. Therefore, it is actually the hydronium ion, H_3O^+, that produces the properties characteristic of an acid in water—not the hydrogen ion (H^+). Still, for brevity and for simplicity's sake, we will refer to acids as compounds that transfer protons—H^+.

Acid and Base Strength

Some acids break apart, or dissociate, completely in water to form ions, whereas other acids ionize only partially. Acids that completely or almost completely dissociate into ions in water are called **strong acids;** those that dissociate to a smaller degree are either moderate or **weak acids.** In the same vein, a **strong base** dissociates readily in water; a **weak base** doesn't.

Keep in mind that *strong* and *concentrated* are not interchangeable terms when applied to acids and bases; *strong* refers to the extent to which an acid or base ionizes in water, and *concentration* describes how much of an acidic or basic compound is present in a solution. Table 8.2 shows several acids described as strong (near 100 percent dissociation), moderate (10–80 percent), and weak (less than 10 percent dissociation in water). The more an acid ionizes, the more free protons are present; because the free protons define the properties of an acid, an acid that dissociates completely has acidic properties to an extreme.

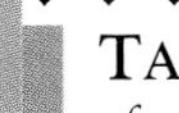

TABLE 8.2 The percentage dissociation (ionization) of some strong, moderate, and weak acids.

ACID	FORMULA	STRENGTH	PERCENT DISSOCIATED
Hydrobromic	HBr	strong	@100
Hydrochloric	HCl	strong	@100
Hydroiodic	HI	strong	@100
Nitric	HNO_3	strong	@100
Perchloric	$HClO_4$	strong	@100
Sulfuric	H_2SO_4	strong	@100
Oxalic	$C_2H_2O_4$	moderate	77
Sulfurous	H_2SO_3	moderate	41
Phosphorus	H_3PO_3	moderate	32
Phosphoric	H_3PO_4	moderate	27
Hydrofluoric	HF	weak	8.2
Nitrous	HNO_2	weak	6.7
Formic	CH_2O_2	weak	4.2
Acetic	$C_2H_4O_2$	weak	1.3
Hydrosulfuric	H_2S	weak	0.1
Hydrocyanic	HCN	weak	0.007
Carbonic	H_2CO_3	weak	@0.0

*Percentage ionized in a 0.1 M solution

Most organic acids are weak; the most familiar example is acetic acid. Only 1.3 percent of acetic acid molecules dissociate in water. An aqueous solution of acetic acid, such as vinegar, contains mainly nonionized acid. The following equation represents the reaction of acetic acid with water:

$$\underset{\text{Acetic acid}}{CH_3{-}COO{-}H} + \underset{\text{Water}}{H_2O} \rightleftharpoons \underset{\text{Acetate anion}}{CH_3{-}COO^-} + \underset{\text{Hydronium cation}}{H_3O^+}$$

This equation is an example of an equilibrium reaction, or a reaction in which two opposite reactions occur at equal rates. Some acetic molecules continue to dissociate, but at the same time, an equal number of hydronium cations combine with acetate anions to re-form acetic acid molecules. The specific molecules dissociating may change, but the total number of ionized molecules remains constant, at about 1.3 percent.

Another common weak acid is carbonic acid, H_2CO_3, found in carbonated beverages. Carbonic acid forms when CO_2 reacts with water:

$$H_2O + CO_2 \rightleftharpoons \underset{\text{Carbonic acid}}{H_2CO_3}$$

While weak acids such as acetic acid and carbonic acid are common ingredients in food, stronger acids such as sulfuric acid would destroy the mouth and esophagus linings.

A similar situation exists for bases. Sodium and potassium hydroxides are fully ionized and completely water soluble. Ammonia (NH_3), the most common weak base, must react with water to generate an OH^- group. Notice in the following equation that water donates an H^+ to ammonia and thus acts as an acid:

$$NH_3 + H_2O \rightleftharpoons NH_4^+ OH^-$$

This is also an equilibrium reaction, since only a few NH_3 molecules react at any given time, and NH_3 re-forms as quickly as it converts to NH_4. Because of this, ammonia is nearly 100 percent undissociated—very few NH_3 molecules react. This makes ammonia a weak base.

Other weak bases, such as calcium hydroxide—$Ca(OH)_2$—and magnesium hydroxide—$Mg(OH)_2$—are used as antacids (box 8.3). Strong bases would destroy body tissues.

EXPERIENCE THIS

Observe an acid-base reaction for yourself. For this experiment, you will need (1) camphor mothballs (not the parachloricide type); (2) several small pieces of marble (chalk will work if you cannot find marble); (3) hydrochloric acid, which is sold as muriatic acid for use in swimming pools; (4) table salt; (5) food coloring; and (6) a tall, clear glass cylinder or vase. Prepare a solution in the glass cylinder by dissolving 25 g (about an ounce) salt and 30 mL muriatic acid (HCl) in 1,000 mL water. (*Caution:* Handle muriatic acid carefully; if you spill any on your hands, wash it off immediately with cold water.) Add food coloring to suit your preference. Now add about 10 g marble or chalk chips and several mothballs. The acid will react with the marble chips (a base) to release CO_2 gas. The mothballs then adsorb the gas, which makes them lighter than the water, and they rise to the surface and discharge CO_2 into the air; after discharging the gas, the mothballs fall back to the bottom. This experiment requires at least thirty minutes to begin to operate well, but it will run for several hours with only an occasional addition of new chips.

Water Dissociation

Not only does water react with acids and bases, it can ionize into H^+ and OH^- ions:

$$H{-}OH \rightleftharpoons H^+ + OH^-$$

When water ionizes, H^+ and OH^- ions must form in equal numbers. One liter of pure, distilled water at 25°C contains 55.56 moles of water. In this quantity, there are only 0.0000001 (or 1×10^{-7}) moles of H^+ or OH^-. Although 10^{-7} moles seems insignificant, these ions have a profound effect on all aqueous chemistry, and especially the chemistry of acids and bases.

Since the concentrations of H^+ and OH^- are identical, pure water is neutral, and the reaction is marked by an equilibrium between products and reactants. The following equation expresses the mathematical relationship between the reactants and products in this equilibrium reaction. The terms within the brackets (for example, $[H^+]$) are the concentrations of the ions in moles/liter (mole/L). K is a constant for the reaction.

$$K = \frac{[H^+][OH^-]}{[H_2O]}$$

If we multiply each side of the equation with the $[H_2O]$ figure, and rearrange it to [HOH] to see the ions more clearly, we obtain a more useful equation that includes the actual ion concentrations. The concentration of the undissociated water does not *significantly* change during ionization; it holds steady at 55.56 moles. This effectively creates a water ionization constant, K_W, with a value of 10^{-14}:

$$K[HOH] = [H^+][OH^-] = [10^{-7}][10^{-7}] = 10^{-14} = K_W$$

The water ionization constant K_W regulates the concentrations of $[H^+]$ and $[OH^-]$; the $[H^+][OH^-]$ product *never* exceeds 10^{-14}. What happens if some H^+ or OH^- ions are added? The concentration of the other ion must *decrease* so that the total product still equals 10^{-14} moles!

Now suppose we add 10^{-4} mole/L HCl. This adds 0.0001 mole/L H^+ and 0.0001 mole/L Cl^- ions to the solution. The Cl^- has no effect on K_W, but the H^+ does. This new amount of H^+ is a thousand times greater than the H^+ concentration in neutral water: the $[OH^-]$ decreases to compensate for the large shift in $[H^+]$, and the water becomes highly acidic. The following equation calculates the change as the concentration of OH^- ions decreases to 1/1000 the amount of neutral water:

$$K_W = 10^{-14} = [H^+][OH^-] = [10^{-4}][OH^-]$$

$$[OH^-] = 10^{-14}/[10^{-4}] = [10^{-10}]$$

(If exponential manipulation seems confusing, review chapter 3 and appendix B.)

EXPERIENCE THIS

Visit the over-the-counter drug section at a grocery or drug store and examine the available selection of antacids, both brand name and generic. List the active anti-acid (base) ingredient in these materials. What other chemicals do the antacids contain?

BOX 8.3

ANTACIDS—OR HOW DO YOU SPELL RELIEF?

One example of an acid-base reaction occurs in the digestive system of anyone who swallows an antacid tablet to relieve indigestion. An antacid neutralizes the excess HCl acid that builds up in the stomach during times of stress or after overeating. Hundreds of brand names and generic antacids produce annual gross sales of over a half billion dollars in the United States alone. Figure 1 shows some common antacids.

Figure 1 Typical antacids.

Sodium bicarbonate—($NaHCO_3$), also known as baking soda—is the oldest antacid; overingestion can make the blood too alkaline (basic). Alka-Seltzer contains $NaHCO_3$, citric acid, and aspirin (for pain relief). The well-known plop-plop, fizz-fizz occurs when the citric acid reacts with some of the $NaHCO_3$ to produce CO_2. When $NaHCO_3$ reaches the stomach, it neutralizes the HCl and liberates CO_2, which may erupt as a belch.

The sparingly water-soluble magnesium and aluminum hydroxides are weak bases found in both name-brand and countless generic antacids. Milk of magnesia, an aqueous suspension of $Mg(OH)_2$, can serve as an antacid or a laxative. Aluminum hydroxide, by contrast, tends to cause constipation if used in excess; Amphogel, used both as an antacid and an antidiarrheal medicine, contains $Al(OH)_3$. Maalox contains both magnesium and aluminum hydroxides, with the expectation that the laxative and constipation effects will cancel each other. (In chapter 17, we will discuss a possible relationship between aluminum salts and Alzheimer's disease.)

Calcium carbonate (chalk) is the base in several antacids, including Tums, which also contains magnesium carbonate. Like sodium bicarbonate, these weak bases react with stomach acid and release CO_2. Rolaids contains aluminum sodium dihydroxy carbonate, $AlNa(OH)_2CO_3$, a complex carbonate. One major problem with aluminum-containing antacids is that they can deplete the body of phosphate. Aluminum combines with phosphate in the body to produce aluminum phosphate ($AlPO_4$). This compound is very insoluble in water and is excreted as solid waste, removing essential phosphate ions.

How do you spell relief? The abundance of brands and chemical formulations give you plenty of options. But whichever you choose, the principle is the same: a base neutralizes the acid in your stomach.

A similar result occurs if we add a base to neutral water. If we add 0.001 mole/L NaOH to pure water, then both 10^{-3} mole/L Na^+ ions and 10^{-3} mole/L OH^- ions are added. The $[Na^+]$ has no effect on K_W, but the $[OH^-]$ level becomes 10,000 times greater than in neutral water. The $[H^+]$ concentration must then decrease to maintain the value of K_W, and the solution becomes strongly basic. In this calculation, the final $[H^+]$ value is 1/10,000 the amount present in pure, neutral water:

$$K_W = 10^{-14} = [H^+][OH^-] = [H^+][10^{-3}]$$

$$[H^+] = 10^{-14}/[10^{-3}] = 10^{-11}$$

The solution is acidic when $[H^+]$ is 10^{-4}, as when we added hydrochloric acid to the water; the solution is basic or alkaline when $[H^+]$ is 10^{-11}, as when we added the base NaOH to pure water. In fact, we can generalize: $[H^+]$ concentrations *above* 10^{-7} yield *acidic* solutions, and $[H^+]$ concentrations *below* 10^{-7} yield *alkaline* (or *basic*) solutions. $[H^+]$ concentrations of exactly 10^{-7} are neutral.

In reality, the equilibrium reaction for the self-ionization of water requires more than one water molecule to avoid the formation of the naked proton, H^+. When a pair of water molecules takes part in the reaction, a proton from one H_2O molecule is transferred to the other to produce H_3O^+ and OH^- ions. These ions can recombine to form water molecules at the same rate at which water dissociates into the ions:

$$H{-}OH + H{-}OH \rightleftharpoons H_3O^+ + OH^-$$

In actuality, the H_3O^+ hydronium ion appears to associate with three other water molecules and is probably more correctly written $H_9O_4^+$—one H^+ associates with four H_2O molecules. The net is still a constant tally of 10^{-7} moles of each ion, H^+ and OH^-. Therefore, in the remainder of this book, we will continue to refer to the H^+ ion rather than the H_3O^+ or $H_9O_4^+$ ion.

PRACTICE PROBLEM 8.8

Which of the following solutions are acidic? Which are basic?

a. 0.00035 mole/L $[H^+]$

b. 10^{-9} M $[H^+]$

c. 0.00067 mole/L $[OH^-]$

ANSWER In each case, convert to an exponent and decide whether the $[H^+]$ exponential value is higher or lower than 10^{-7}. *a.* The $[H^+]$ value is 3.5×10^{-4}; the exponent is -4, which is greater than -7, so the solution is acidic. *b.* The exponent is -9, which is less than -7, so the solution is basic. *c.* The $[OH^-]$ value is 6.7×10^{-4}; this makes the $[H^+]$ value 10^{-10}. The exponent is -10, which is less than -7, so the solution is basic.

Condition	H^+	pH	OH^-	Condition
Acidic	10^0	0	10^{-14}	Acidic
	10^{-1}	1	10^{-13}	
	10^{-2}	2	10^{-12}	
	10^{-3}	3	10^{-11}	
	10^{-4}	4	10^{-10}	
	10^{-5}	5	10^{-9}	
	10^{-6}	6	10^{-8}	
Neutral	10^{-7}	7	10^{-7}	Neutral
Basic	10^{-8}	8	10^{-6}	Basic
	10^{-9}	9	10^{-5}	
	10^{-10}	10	10^{-4}	
	10^{-11}	11	10^{-3}	
	10^{-12}	12	10^{-2}	
	10^{-13}	13	10^{-1}	
	10^{-14}	14	10^0	

Figure 8.3 The relationship between pH and the H^+ or OH^- concentrations in mole/L.

The pH Scale

Rather than using exponential values, a more convenient way to express acidity or basicity is by measuring **pH.** In 1909, S. P. L. Sorensen (Danish, 1868–1939) introduced the concept of pH (from the French *pouvoir hydrogene*—hydrogen power). pH is defined as the value of the $[H^+]$ exponent multiplied by -1. The term *log*, an abbreviation for logarithm, is a mathematical device used to make this conversion:

$$pH = -\log[H^+]$$

We can apply this formula to three hypothetical solutions: (1) pure neutral water, (2) an acidic solution with an $[H^+]$ of 10^{-4}, and (3) a basic solution with an $[H^+]$ of 10^{-11}:

1. $pH = (-1) \times (-7) = 7$ (neutral)
2. $pH = (-1) \times (-4) = 4$ (acid)
3. $pH = (-1) \times (-11) = 11$ (base)

Just as exponential values of $[H^+]$ indicate whether a solution is acidic, neutral, or basic, so do pH values—with a slight difference. Because pH values are positive rather than negative numbers, a pH *below* 7 is *acidic* (rather than *above* 10^{-7}) and a pH *above* 7 is *basic* (rather than below 10^{-7}). A pH of 7 is neutral (figure 8.3).

Remember, since each pH unit change represents an exponential change in $[H^+]$, each pH unit change is actually a tenfold change in $[H^+]$. Thus, an acidic solution with a pH of 3 is ten times as strong an acid as one with a pH of 4.

Figure 8.4 shows the pH values of some common solutions. Fruit juices and fruit are slightly acidic because they contain natural organic acids, such as citric acid. Soft drinks have pH values between 3.0 and 4.0, due mainly to dissolved CO_2 and citric acid. Most cola soft drinks also contain phosphoric acid, which further increases their acidity. (This is why a cola drink will clean rust off of iron, while noncolas will not.) Battery acid, at pH = 0.8 is over a hundred times more acidic than any soft drink on the market. Acid rain, as box 8.4 explains, varies from being moderately to strongly acidic.

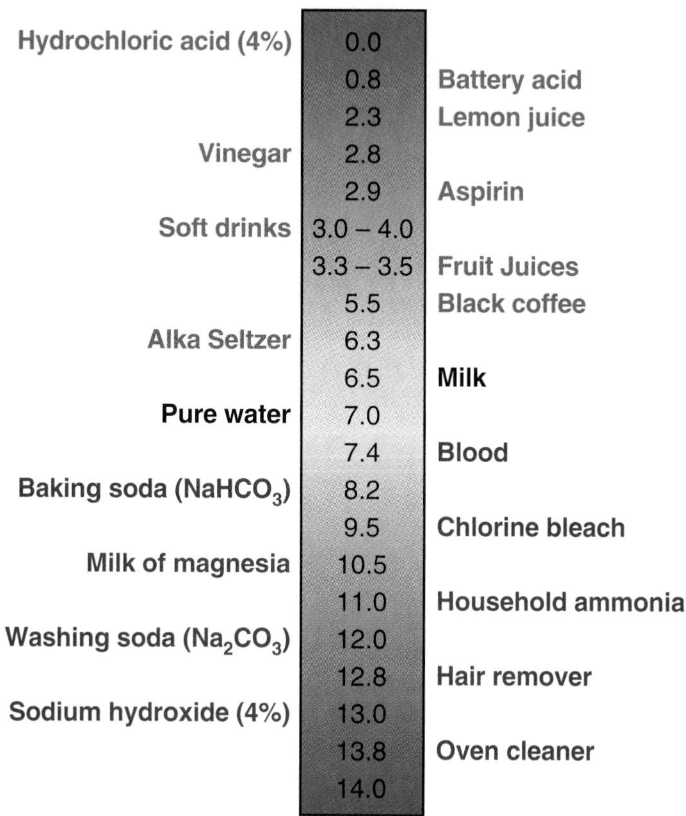

Figure 8.4 pH values for some common solutions and substances.

Acids taste sour, so sweeteners mask the sour taste in a soft drink. Many people also add sweeteners to coffee to reduce its bitter taste.

PRACTICE PROBLEM 8.9

Which of the following pH values are acidic? Which are alkaline (or basic)?

a. 8.3
b. 2.9
c. 6.6
d. 11.3
e. 9.4
f. 5.1
g. 1.8
h. 7.7

ANSWER The pH values below 7.0 are acidic; this includes *b*, *c*, *f*, and *g*. The remaining values for *a*, *d*, *e*, and *h* are all above 7.0 and are thus alkaline or basic.

EXPERIENCE THIS

Here is an easy way you can determine the pH of several different materials. Drain off and collect in a jar the juice from a can of red cabbage. (If this is unavailable, you can cook a piece of red cabbage in water.) This juice will work as an indicator, changing from red when acidic to green when basic. To use this indicator, place the material you want to test in a glass, beaker, or jar. If it is a solid, dissolve it in some water. Then add some of the red cabbage juice and note whether the final color is red (acidic) or green (basic). Try running this test on vinegar, soda water, baking soda, apple juice, household ammonia, aspirin, liquid laundry detergent, antacid tablets, diluted shampoo, oven cleaner, vitamin C, borax, tea, and coffee.

Other natural juices change color with pH and can be used as indicators, including the juice from beets, rhubarb, cherries, hollyhocks, and carrot stems. If you decide to try some of these juices, test them first with vinegar as a typical acid and household ammonia as a typical base to determine the colors of the indicators at different pH values.

BOX 8.4

ACID RAIN

Acid rain is rainwater with a pH value between about 1.8 and 5.5. The normal pH for ordinary rainwater is 5.6 to 6.2; rainwater is slightly acidic because it always contains a small amount of dissolved CO_2 gas that reacts with water to form the weak acid H_2CO_3:

$$H_2O + CO_2 \rightarrow \underset{\text{Carbonic acid}}{H_2CO_3}$$

The acidity in acid rain comes from dissolved acidic substances derived from both natural and human-generated sources. Most living organisms can only briefly tolerate a pH value below 4.0 because living systems operate at pH values near neutral. Thus, acid rain can harm plants, animals, and people.

Acid rain has posed problems for decades. It is a solution of specific strong and moderate acid compounds, including sulfuric, sulfurous, nitric, and nitrous acids, formed as certain gases and dissolved solids react in rainwater. The aqueous reactions of SO_2, SO_3, and NO_2 all produce acids that can create acid rain:

$$SO_2 + H_2O \rightarrow \underset{\text{Sulfurous acid}}{H_2SO_3}$$

$$SO_3 + H_2O \rightarrow \underset{\text{Sulfuric acid}}{H_2SO_4}$$

$$2\,NO_2 + H_2O \rightarrow \underset{\text{Nitric acid}}{HNO_3} + \underset{\text{Nitrous acid}}{HNO_2}$$

Acid rain originates from a variety of sources and has a variety of effects. Figure 1 shows the widespread destruction of vegetation around a smelting plant, which produces SO_2 as a byproduct.

Figure 1 SO_2 emissions and acid rain from a copper smelting plant in Ontario, Canada, destroy surrounding vegetation. The plant is visible in the background.

(Continued)

BOX 8.4—Continued

Smelting is the process of extracting a metal from an ore by heating it in air. Power-generating plants also produce SO_2; scrubbers can remove the SO_2 from some of these sources (see chapter 18). But not all sources are human-controlled. Huge amounts of sulfur oxides are belched into the atmosphere every time a volcano erupts, though fortunately, this is relatively uncommon. Similarly, the main human-made source of nitrogen oxides (NO_2 and others) is automobile emissions, but lightning, which heats the surrounding air several hundred degrees, also produces nitrogen oxides. At such elevated temperatures, N_2 and O_2 combine to form various forms of nitrogen oxides, NO_x.

Acid rain has a deleterious effect on both animate and inanimate objects. It destroys vegetation and severely corrodes metals. In addition, some building materials, such as marble and limestone, react with acids to produce water-soluble products that are then susceptible to erosion. An example is calcium carbonate, $CaCO_3$, the main component in chalk, marble, limestone, and calcite:

$$\underset{\text{Water-insoluble}}{CaCO_3} + H_2SO_4 \rightarrow \underset{\text{Water-soluble}}{CaSO_4} + H_2O + \underset{\text{Gas}}{CO_2}$$

Many classical works of art made from marble show the effects of acid rain. Figure 2 shows the loss of detail in a medieval statue after long acid rain exposure.

Acid rain sometimes reaches the pH range of 1.8 to 2.0, while soft drinks are nearer to pH 3.0, less than one-tenth as acidic. Figure 3 compares the $[H^+]$ concentrations of acid rain with the $[H^+]$ levels in a battery, lemon juice, vinegar, soft drinks, and ordinary rainwater.

Figure 4 shows the extent of acid rain precipitation in the United States. The numbers indicate the pH values of the rain falling in these regions; many regions have values between 4.3 and 5.3. Perhaps the biggest environmental change in the past twenty-five years is that there are about 38 percent more people in the United States. More people means more production and consumption, which increases environmental costs.

Figure 2 Acid rain corrodes a medieval statue. Insoluble $CaCO_3$ in the statue was converted into water-soluble $CaSO_4$ by the acids formed from atmospheric SO_2.

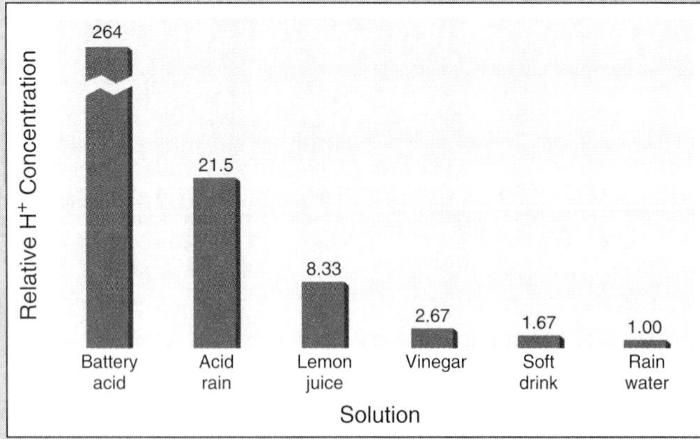

Figure 3 The relative $[H^+]$ concentrations in acid rain and some other acidic solutions compared against ordinary, unpolluted rainwater, which is assigned a value of 1.0.

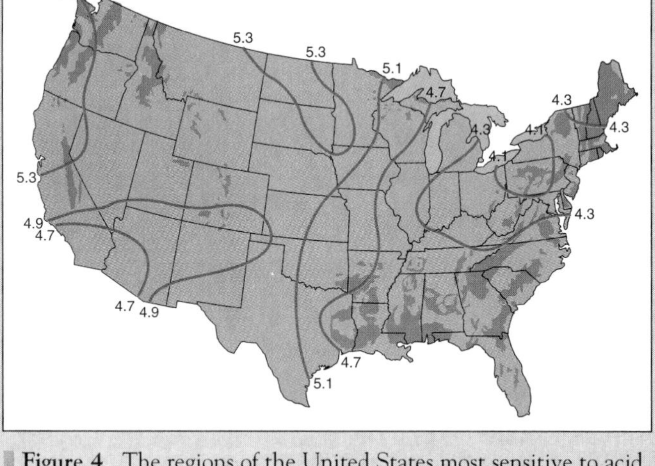

Figure 4 The regions of the United States most sensitive to acid rain. Dark areas show that regions containing lakes are likely to become acidic from the acidic rainfall. The contour lines show the average pH of the rain in each particular region.

SOURCE: From Volker Mohnen, "The Challenge of Acid Rain", illustration by Bob Conrad, August 1988.

Summary

Almost all natural chemical reactions take place in aqueous solution, and nearly all are either redox or acid-base reactions. Not all substances dissolve in water; as a general rule, like dissolves like, but unlike substances don't dissolve. Since water is polar, this means that polar compounds are more likely to dissolve in water, and ionic compounds are likely to be water-soluble.

Solutions are homogeneous mixtures involving any combination of solids, liquids, and gases. Solutions consist of a solute dissolved in a a solvent; concentration is a measure of how much solute is dissolved in the total amount of solution. Parts per hundred (pph), parts per million (ppm), and parts per billion (ppb), and moles/liter are all measures of concentration.

The simplest and most useful definition of acids and bases defines them in terms of proton (H^+) transfer. According to this definition, an acid is a compound that can give up a proton, and a base is a compound that can accept a proton. When acids and bases react in equal proportions, they neutralize one another. Often a neutralization yields a salt and water as products.

Free protons (H^+) do not normally exist in water because they combine with water (H_2O) to form hydronium ions (H_3O^+). Still, the net result is an increase in the H^+ ions present in solution; therefore, this text refers to H^+ rather than H_3O^+ ions.

Strong acids and bases dissociate completely in water to free H^+ and OH^- ions; weak acids and bases dissociate very little. Weak acids occur in many foods, including vinegar, soft drinks, and spinach. Weak bases such as calcium carbonate are used in antacids for the stomach.

Water itself can dissociate into H^+ and OH^- ions. One liter of pure, neutral water contains 10^{-7} mole/L H^+ ions and 10^{-7} mole/L OH^- ions; the product is a constant 10^{-14}. The balance between the H^+ and OH^- ions may shift, but the total amount is a constant. Adding more ions of one type would decrease the number of opposite ions and would result in either an acidic or alkaline (basic) solution.

Acidity and alkalinity may be expressed as exponential values denoting number of H^+ ions in moles/liter, or as pH values. Exponential H^+ values greater than 10^{-7} (for example, 10^{-3}) moles/liter signify an acid; lesser exponential values (for example, 10^{-9}) signify a base. Similarly, positive pH values below 7 signify an acid, while pH values greater than 7 signify a base. A pH of 7 is neutral. Acid rain is moderately to strongly acidic, with a pH value between 1.8 and 5.5.

Key Terms

acid 195
acid rain 203
alkaline 198
antacid 201
base 195
hydronium ion 199
indicator 198
molarity, (moles/L) 194
neutralization 198
parts per billion (ppb) 195
parts per million (ppm) 194
pH 202
salt (generic) 199
solubility 192
solute 193
solvent 193
strong acid 199
strong base 199
weak acid 199
weak base 199

Readings of Interest

Lead Poisoning

Hong, S.; Candelone, J.-P.; Patterson, C. C.; and Boutron, C. F. "Greenland ice evidence of hemispheric lead pollution two millennia ago by Greek and Roman civilizations." *Science*, September 23, 1994, pp. 1841–1843. Lead pollution caused by the ancient Greeks and Romans rivals what we have caused today.

Rosen, John F. "Effects of low levels of lead exposure." *Science*, 256, April 17, 1992, p. 294; Needleman, Herbert L., same issue, pp. 294–295. Two short discussions of the title topic.

Yanke, Charles. "From the cradle to the grave—A lead recycling blueprint." *Purchasing Management Magazine*, February 1992. Many people in industry are very concerned about pollution. This paper from Vulcan Lead tells how they control lead pollution by recycling.

Franklin, Deborah. "Lead, still poison after all these years." *Health*, September/October 1991, pp. 39–49. An easy-to-read article about lead pollution in the home, with several surprising conclusions.

Fluoridation

Hileman, Bette. "Study links fluoridated water and hip fractures." *Chemical & Engineering News*, August 17, 1992, p. 8. This study was done in Brigham City, Utah, where the fluoride level was 1 ppm. The work was originally reported in *J. Am. Med. Assn.*

Coffel, Steve. "The great fluoride fight." *Garbage*, May/June 1992, pp. 32–37. *Garbage* was one of the better magazines in the popular environmental field. This article gives good coverage on the history of the problem and cites the 1989 American Dental Association report that notes only a 25 percent reduction in tooth decay when water is fluoridated. Worth reading.

Donahue, Peggy Jo. "Nip cavities in the bud." *Prevention*, May 1992, pp. 88–90, 135. This article deals primarily with tooth sealants, which is possibly the most effective method for averting caries at this time.

Martin, Brian. 1991. *Scientific knowledge in controversy: The social dynamics of the fluoridation debate*. Albany: University of New York Press. If you think we exaggerated the fact that the scientific method has not been systematically applied to the fluoridation debate, this is the book for you to read.

Oliwenstein, Lori. "Flap over fluoride." *Discover*, July 1989, pp. 34–35. A brief summary of the arguments on both sides of the fight that points out how illogical the whole argument has become.

Hileman, Bette. "The fluoridation controversy." *Chemical & Engineering News*, August 1, 1988, p. 26. This article, cited in the chapter, is probably the most balanced presentation available. Subsequent letters to the editor were evenly divided between those who believed this author sided with antifluoridationists and those who believed this author sided with profluoridationists.

Raloff, Janet. "Rinsing away decay." *Science News*, April 19, 1986, pp. 251–253. Alternate ways to prevent tooth decay instead of fluoridating drinking water systems.

Waldbott, George L. 1978. *Fluoridation, the great dilemma*. Lawrence, Kansas: Coronado Press. This work, written by a medical doctor, is probably still the key book on the topic. It contains background on why some industries like fluoridation, plus hundreds of examples of harm caused by fluoridation.

Acids and Bases

Reese, K.M. "Household bleach and ammonia are bad mixture." *Chemical & Engineering News*, January 16, 1995, p. 72. Why you should never mix these common household chemicals.

Kirschner, Elisabeth M. "Sodium hydroxide demand is booming and supplies are dwindling." *Chemical & Engineering News*, January 9, 1995, pp. 11–13. This base is one of the major inorganic chemicals in our economy. What will happen if it is in short supply?

Thomas, John Meurig. "Solid acid catalysts." *Scientific American*, April 1992, pp. 112–118. This topic was not discussed in this chapter, but it has important industrial applications. This article tells how solid acids can help manufacture other materials.

Foltz-Gray, Dorothy. "The swimming pool solution." *Health*, February/March, 1992, pp. 114–116. While everyone knows chlorine is used in swimming pools, pH is also extremely important.

Acid Rain

Monastersky, R. "Acid precipitation drops in the United States." *Science News*, July 10, 1993. This paper reports the results of a study that showed acid rain levels dropped between 1980 and 1991.

Glanz, James. "Environmental research." *R&D Magazine*, September 28, 1992, pp. 83–86. This article describes the possible use of 450-foot-tall towers to suck the pollutants out of the Los Angeles air. The designer, Melvin Prueitt, believes ninety-five such towers could purify half the air in L.A. each day.

Pennisi, E. "Salt, not acid rain, may mottle Ms. Liberty." *Science News*, August 17, 1991, p. 101. Acid rain cannot be blamed for all the corrosion in the world. Salt can be even worse!

Marx, Wesley. "Environmental countdown: Where we're losing—And winning." *Reader's Digest*, May 1990, pp. 99–105. Contains much information on acid rain.

Raloff, Janet. "Where acids reign." *Science News*, July 22, 1989, pp. 56–58. A not-so-pleasant story about the effects of acid rain on the Bavarian timber forests.

American Chemical Society. 1985. *Acid rain*. This free pamphlet is probably the best single source for acid rain information. Copies can be obtained by writing to the Office of Federal Regulatory Programs, ACS Dept. of Government Relations and Science Policy, 1155 16th Street, NW, Washington, D.C. 20036.

Peterson, Ivars. "Acid rain's political web." *Science News*, July 28, 1984, pp. 58–59. This article points out how the political solutions to the problem of acid rain seem to lag behind research. Not much has changed since 1984.

PROBLEMS AND QUESTIONS

1. Define each of the following terms.
   - a. acid
   - b. acid rain
   - c. antacid
   - d. base
   - e. molarity
   - f. neutralization
   - g. pH
   - h. water ionization constant

2. Calculate the percentage, ppm, and ppb for each component in the following gas mixture.

   Mix A 300 parts O_2
   1500 parts N_2
   100 parts CO_2
   100 parts SO_2

3. Calculate the percentage, ppm, and ppb for each component in the following gas mixture.
   Mix B 400 parts O_2
   2400 parts N_2
   150 parts CO_2
   50 parts SO_2

4. Calculate the percentage and molarity (moles/liter) for each of the following solutions. (*Note:* Water has a density of 1.0 g/mL.)
   a. 126 g KCl in 500 mL water.
   b. 36 g sucrose $[C_{12}H_{22}O_{11}]$ in 1,000 mL water.
   c. 225 g $CaCl_2$ in 500 mL water.
   d. 17 g glucose $[C_6H_{12}O_6]$ in 200 mL water.
   e. 125 g $AgNO_3$ in 500 mL water.
   f. 88 g NH_4NO_3 in 300 mL water.

5. Which of the following compounds are likely to be water-soluble, and which are likely to be insoluble?
   a. $CaCO_3$
   b. NH_4Br
   c. K_2CO_3
   d. $AlPO_4$
   e. $SrCO_3$
   f. $Ca(OH)_2$
   g. Na_2SO_4
   h. $Ba_3(PO_4)_2$
   i. $MgCl_2$
   j. $Al(OH)_3$

6. Where does the lead in our drinking water come from, and how can we eliminate it?

7. Other than lead, what other metal ions pollute our drinking water?

8. What are the advantages and disadvantages of fluoridating drinking water?

9. What is the molarity of a solution containing 6.0 g NaOH in 500 mL total solution? Should the pH of this solution be acidic or basic?

10. What is the molarity of a solution containing 14.6 g NaCl in 250 mL total solution? What pH is expected for this solution? (Careful, this is a trick question!)

11. What is the molarity of a solution containing 9.8 g H_2SO_4 in 1,000 mL total solution? Should this solution be acidic or basic?

12. What is the pH for each of the following solutions?
    a. $[H] = 10^{-4}$ mole/L
    b. $[OH^-] = 10^{-9}$ mole/L
    c. $[H^+] = 10^{-10}$ mole/L
    d. $[OH^-] = 10^{-3}$ mole/L

13. What value is the water ionization constant?

14. List some of the chemicals commonly used as antacids.

15. What are the causes of acid rain?

16. What are some of the effects of acid rain?

17. How can acid rain be prevented?

18. How many years must pass before a billion seconds have elapsed?

CRITICAL THINKING PROBLEMS

1. Many lakes, rivers, and streams suffer from heavy metal pollution as industry dumps metals such as mercury and lead into them. At times, these metals enter the food chain through the fish in these waters. On the other hand, some natural methods may exist to remove these heavy metals from waterways. Would you eat fish harvested from these waters? What would you do to control water pollution caused by heavy metals?

2. Substantial mercury contamination exists in the Florida Everglades. Most of the mercury apparently comes from herbicide and pesticide run-off from surrounding sugar mills. Cleaning up the Everglades sounds like a good idea, but who should pay for it? The profit margin is low in sugar manufacturing, and some smaller companies might go bankrupt if they have to pay. Others might lay off employees or cut their salaries or benefits. On the other hand, some people do live in these polluted regions, and they may be vulnerable to heavy metal poisoning. How would you handle this dilemma? What reasons can you give for your decision?

3. Fluoride occurs naturally in many water supplies. In fact, natural fluoridation spawned the idea of using fluoride to prevent tooth decay. If fluoridation is really dangerous, why isn't anyone leading a campaign to remove naturally occurring fluoride from drinking water?

4. Read the following three articles on fluoridation as well as information in box 8.1. (1) Lori Oliwenstein, "Flap Over Fluoride," *Discover*, July 1989, pp. 34–35; (2) Steve Coffel, "The Great Fluoride Fight," *Garbage*, May/June 1992, pp. 32–37; and (3) Bette Hileman, "Study Links Fluoridated Water and Hip Fractures," *Chemical & Engineering News*, August 17, 1992, p.8. Weigh the available evidence for and against fluoridation. Do *you* think fluoridating drinking water is a good idea?

5. Would a dilute acid or a dilute base better remove chalk dust from a glass-top table? Why?

6. Would an acid or a base better remove iron rust from a glass surface? Why?

7. Would a dilute acid or a dilute base better remove hard water deposits from a countertop? Why?

8. Some people want to ban all hazardous chemicals from home use. Do you think materials with very low (perhaps 0–2) or very high (perhaps 12–14) pH should be banned? If so, why? What types of chemicals would you suggest as replacements?

9. If most of the chemicals that cause acid rain come from natural sources, does it make sense to try to reduce these same chemicals when they are derived from industrial sources?

10. Read the following three articles: (1) Ivars Peterson, "Acid Rain's Political Web," *Science News*, July 28, 1984, pp. 58–59; (2) Janet Raloff "Where Acids Reign," *Science News*, July 22, 1989, pp. 56–58; and (3) R. Monastersky, "Acid Precipitation Drops in the United States," *Science News*, July 10, 1993, p. 22. Acid rain seems to have decreased in the United States but not in Europe. Why? Would it be reasonable to say the United States has solved its acid rain problem? Explain your answer.

11. Carbon dioxide is a natural waste product from people and animals as well as a product of combustion. Is CO_2 a pollutant? Why or why not?

HINT This opinion question has no right or wrong answer, but ponder the following points. Can the term *pollutant* always be defined the same way? For example, if a pollutant is defined as an unnatural material that has entered the environment, then CO_2 is not a pollutant. However, if unusually high levels of any substance constitute pollution, CO_2 could fit this definition—but so could oxygen. Try to decide what a pollutant is in your own mind.

12. The fluoridation controversy is unlikely to be resolved in the near future. Form your own opinion regarding fluoridation of drinking water using the scientific method. Some specific questions to ponder include: (1) Can we determine the benefit/risk ratios in a problem with so many variables? (2) Can the benefits for many outweigh the risk for a few? (3) Is it economical to put F^- ions in the drinking water supply when at least 99.94 percent of this treated water has no positive benefit for our population?

13. Using the information you have learned concerning lead poisoning, state your opinion in response to each of the following questions.

a. Should communities change building codes to reduce lead in drinking water? How?

b. Should older houses and public buildings be retrofitted with lead-free plumbing? If so, who should pay for these changes?

c. Some dishes are coated with lead-containing surface glazes, and some crystal glass contains a high lead level to achieve a special appearance. Could part of this lead be extracted by food or beverages and possibly pose a health hazard?

d. Some older buildings still have lead paint on the walls. Should this be removed by law? If so, who should pay for removal? And where should discarded lead paint scrap be stored?

e. Some environmental advocates complain that artists often use heavy metals, such as lead and cadmium salts, in their pigments. Should this practice be outlawed?

f. Can you form some general guidelines for society to follow concerning lead use and removal? What are the prime considerations in forming such policies?

14. Read box 8.4 and some of the articles on acid rain in the Readings of Interest section. How much of the acid rain problem is caused by natural forces, and how much by industries? How would you handle this problem if it were suddenly thrust into your lap? A study released in January 1994 suggests that many "cures" for acid rain were counterproductive and actually aggravated the problem. What were some of these remedies? Can you think of possible improvements?

CHAPTER 9

ORGANIC CHEMISTRY—THE WONDERFUL WORLD OF CARBON

OUTLINE

"The universe is full of magical things patiently waiting for our wits to grow sharper."
EDEN PHILLPOTS (1862–1960)

Full moon over a petroleum plant.

KEY IDEAS

1. Why More Compounds Containing Carbon Exist Than Compounds Containing Any Other Element
2. The Basic Features of the IUPAC Nomenclature System Used to Name Organic Compounds
3. The General Physical and Chemical Characteristics of Alkanes
4. The Nature and Function of Multiple Bonds in Organic Compounds
5. How Alkanes, Alkenes, Alkynes, and Aromatics Differ in Structure and Reactivity
6. The Nature of Addition and Substitution Reactions
7. The Uses of Hydrocarbons, Including Coal and Petroleum, as Fuels
8. How Chemists Modify Crude Petroleum to Make New Materials
9. The Main Features of Addition Polymerization
10. How Chemists Made Natural Rubber into a Major Industry

Organic chemistry is the chemistry of carbon compounds. Carbon compounds were first isolated from living organisms, and for a long time, scientists believed a vital force (that is, a force derived from a living organism) was required to form them. The term *organic* denoted this special peculiarity. Scientists first classified organic compounds in the 1700s. They called compounds derived from mineral sources *inorganic* because they came from lifeless sources.

Carbon—The Covalent Superstar

In 1828, Friedrich Wöhler (German 1800–1882; figure 9.1) synthesized the organic compound urea from inorganic ammonium cyanate. Prior to Wöhler's synthesis, urea was obtained only from animal or human urine; it was clearly an organic compound living organisms produced. Ammonium cyanate was just as clearly inorganic. How could Wöhler create an organic ("life") compound from an inorganic ("lifeless") one? Wöhler's synthesis of urea can be expressed as:

$$\underset{\text{Ammonium cyanate}}{(NH_4)\,(OCN)} \xrightarrow{\text{Heat}} \underset{\text{Urea}}{H_2N{-}\overset{\overset{\displaystyle O}{\|}}{C}{-}NH_2}$$

Chemists were initially surprised by Wöhler's synthesis, but through the years, others confirmed his claim by applying the scientific method and retesting his results. Soon they succeeded in manufacturing many organic compounds from inorganic compounds without using any "vital force." Today, chemists routinely synthesize organic compounds from inorganic substances, and the term *organic* now applies more broadly to chemistry involving carbon compounds. This immensely important division of chemistry has led to the synthesis of plastics, medicines, fuels, soaps, cosmetics, and clothing.

Figure 9.1 Friedrich Wöhler.

Why is carbon such a prevalent and important element? Because carbon has four valence (outermost) electrons, each carbon atom can form four covalent bonds. This allows carbon atoms to combine with each other and with other elements in an endless variety of ways. No other element can form such long, stable chains, branched chains, and rings, or form so many different compounds. Carbon's unique properties make it not only the most common element in compounds, but also an element essential to life.

This chapter explores a specific group of carbon compounds—hydrocarbons, which contain only carbon and hydrogen atoms. Millions of unique hydrocarbon molecules exist. Many are structurally similar, and chemists must differentiate them by naming hydrocarbons in a systematic way. But two qualities are shared not only by all hydrocarbons, but by other carbon compounds as well: catenation and stability.

Catenation and Stability

Carbon is different from all other elements because of two properties—catenation and stability. **Catenation** describes the ability of an element to combine readily with many other atoms of itself. A carbon atom can combine with hundreds or even thousands of others, forming compounds containing more than 100,000 carbon atoms. (The carbon allotropes discussed in chapter 3 provide an illustration.) This is the first property that sets carbon apart from other elements. The other is the stability of these catenated compounds.

The unique stability of catenated carbon compounds lends these compounds the ability to survive under conditions that destroy other catenated substances. For example, chemists can create a ten-atom chain of either carbon atoms or silicon atoms (figure 9.2). The carbon chain (n-decane) has C—C and C—H covalent bonds; the silicon analog has Si—Si and Si—H covalent bonds. The carbon compound is stable in the presence of water, but the silicon compound breaks apart. Chemists can make many catenated molecules, but those from carbon are useful in unique ways because they are so stable. Noncarbon catenated molecules are often unstable—some even burst into flame on contact with water or oxygen.

Because carbon so easily combines with additional atoms of itself, and because the resulting molecules are highly stable, carbon can form many more different compounds than other ele-

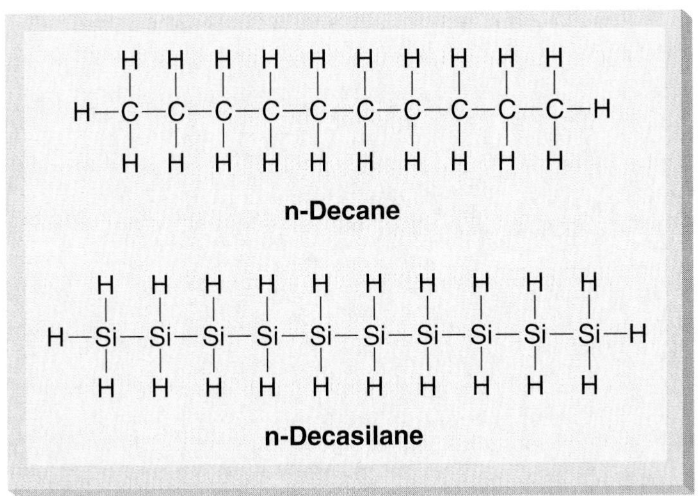

Figure 9.2 Two compounds with catenated covalent bonds that have the same number of atoms but vary in stability; only n-decane is stable in water.

ments can. To classify all these compounds, chemists had to establish a system of nomenclature—a way to name them.

HYDROCARBON NOMENCLATURE—ALKANES

Nomenclature is a system for naming things, in this case, organic molecules. One of the purposes of establishing a nomenclature is to assign specific names to closely related compounds. This is not a simple task; the molecule n-decane (figure 9.2) has just 10 C atoms and 22 H atoms, but there are exactly seventy-five different, correct ways to arrange these 32 atoms into different stable molecules.

Learning seventy-five different names may not seem a difficult chore; everyone knows the name of more than seventy-five people. However, in organic chemistry, the numbers of possible compounds multiply more rapidly than a proverbial pair of rabbits. You can make more than five thousand different molecules from combinations of 15 C atoms and 32 H atoms. With 40 C atoms and 82 H atoms, you could form more than 60 trillion different molecules. You would therefore need more than 60 trillion names to describe all possible molecules with the general formula $C_{40}H_{82}$. To assign names to so many compounds, using common chemical names like benzene and decane, would obviously be a hopeless task.

Chemists solve this problem with the International Union of Pure and Applied Chemistry nomenclature system, or **IUPAC system** for short. The development of the IUPAC system began in 1892 and continues today. Systematic nomenclature helps chemists maintain sanity in the face of such huge numbers of very similar molecules. The IUPAC system relies on a major division of hydrocarbons—the alkanes—to form root names for most other hydrocarbons.

Naming n-Alkanes

Hydrocarbons may be divided into four classes: alkanes, alkenes, alkynes, and aromatic hydrocarbons. The **alkanes** family, also called the paraffins and saturated hydrocarbons, are the most fundamental class of organic compounds. Table 9.1 lists the IUPAC names and structures for the first ten normal or n-alkanes. A "normal" alkane has all its carbon atoms arranged in a continuous chain. It is considered a **saturated hydrocarbon** because it contains the maximum possible number of hydrogen atoms; it is therefore "saturated" with H atoms.

Note that all ten names in table 9.1 have one common feature—they end in the suffix *ane*, as in the family name alk*ane*. This consistent suffix is an important feature of the IUPAC system; any organic compound ending in *ane* is an alkane. The first part of each name also has significance. The term *deca*, for example, is the Greek word for "ten." Thus *dec*ane is an alkane with 10 carbon atoms.

FRANK & ERNEST® by Bob Thaves

SOURCE: FRANK & ERNEST reprinted by permission of NEA, Inc.

TABLE 9.1 The formulas and IUPAC names for the first ten *n*-alkanes ("normal" alkanes).

NAME	FORMULA
methane	CH_4
ethane	$CH_3—CH_3$
n-propane	$CH_3—CH_2—CH_3$
n-butane	$CH_3—CH_2—CH_2—CH_3$
n-pentane	$CH_3—CH_2—CH_2—CH_2—CH_3$
n-hexane	$CH_3—CH_2—CH_2—CH_2—CH_2—CH_3$
n-heptane	$CH_3—CH_2—CH_2—CH_2—CH_2—CH_2—CH_3$
n-octane	$CH_3—CH_2—CH_2—CH_2—CH_2—CH_2—CH_2—CH_3$
n-nonane	$CH_3—CH_2—CH_2—CH_2—CH_2—CH_2—CH_2—CH_2—CH_3$
n-decane	$CH_3—CH_2—CH_2—CH_2—CH_2—CH_2—CH_2—CH_2—CH_2—CH_3$
n-decane	$CH_3—(CH_2)_8—CH_3$

Each of the first four alkanes listed in table 9.1 has a nonnumerical prefix followed by the suffix *ane:* methane, ethane, propane, and butane. All remaining n-alkanes have a Greek numerical prefix signifying the number of C atoms in the molecule. (Table 9.2 lists some Greek prefixes and their Arabic equivalents). Thus, n-octane is a normal alkane with 8 C atoms. Although the compound octanal also contains 8 C atoms, it cannot be an alkane because it has the wrong suffix.

TABLE 9.2 Greek prefixes and the numerical equivalents.

PREFIX	NUMBER
mono-	1
di-	2
tri-	3
tetra-	4
penta-	5
hexa-	6
hepta-	7
octa-	8
nona-	9
deca-	10
dodeca-	12
tetradeca-	14
hexadeca-	16
octadeca-	18

Each alkane class member fits the **class formula** C_nH_{2n+2}, where *n* is the number of C atoms in the molecule. Thus, n-octane has 8 C atoms and $(2n + 2) = (2 \times 8 + 2) = 18$ H atoms. This class formula describes the formula of all alkanes except cycloalkanes (which we will discuss soon). The alkanes thus form a homologous series, a series of compounds differing from each other by exactly one CH_2 (a methylene unit). If you look again at table 9.1, you will see that each successive n-alkane has one additional CH_2 "sandwiched" between the CH_3 units on either end.

PRACTICE PROBLEM 9.1

Memorization can be hard, but it is necessary in organic chemistry. Use tables 9.1 and 9.2 as guides to draw the following n-alkanes in the condensed and compacted forms. (See chapter 6 for definitions of these forms.)

a. n-nonane

b. n-pentane

c. n-octadecane

ANSWER

a. $CH_3CH_2CH_2CH_2CH_2CH_2CH_2CH_2CH_3$
or $CH_3(CH_2)_7CH_3$

b. $CH_3CH_2CH_2CH_2CH_3$, or $CH_3(CH_2)_3CH_3$

c. $CH_2CH_2CH_2CH_2CH_2CH_2CH_2CH_2CH_2CH_2CH_3$
|
$CH_2CH_2CH_2CH_2CH_2CH_2CH_3$, or $CH_3(CH_2)_{16}CH_3$

PRACTICE PROBLEM 9.2

Use tables 9.1 and 9.2 as guides to determine the names for each of the following n-alkanes.

a. $CH_3CH_2CH_2CH_3$
b. $CH_3CH_2CH_2CH_2CH_2CH_2CH_3$
c. $CH_3CH_2CH_2CH_2CH_2CH_2CH_2CH_2CH_2CH_2CH_2CH_3$
d. $CH_2CH_2CH_2CH_2CH_2CH_2CH_2CH_2CH_2CH_2CH_2CH_3$
   |
   $CH_2CH_2CH_2CH_3$

ANSWER

a. n-butane
b. n-heptane
c. n-dodecane
d. n-hexadecane

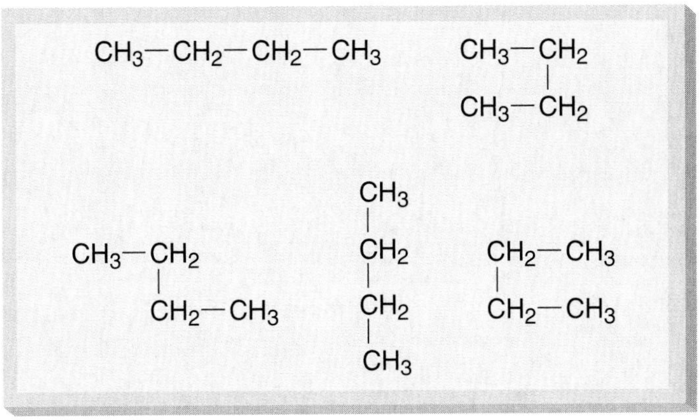

Naming Other Alkanes

Most alkanes are not n-alkanes, or "normal" alkanes. Recall that n-alkanes have all their carbon atoms arranged in a continuous chain. But other alkanes may have their carbon atoms arranged in branched chains or rings. For example, the seventy-five alkanes with the formula $C_{10}H_{22}$ are called **structural isomers,** molecules with the same chemical formula but different structures. Only the n-decane isomer has all 10 carbon atoms in a continuous chain. The remaining seventy-four isomers have shorter carbon chains with other carbons branching from these chains.

The first alkane with an isomer pair is butane, which exists as the molecules n-butane and iso-butane. As figure 9.3 shows, all 4 C atoms in n-butane are connected in one chain, and each is connected to no more than two others, while the central C atom in iso-butane is attached to 3 other C atoms. In fact, n-butane and iso-butane are, despite their identical chemical formulas, different compounds with different properties that can be separated from each other by physical methods. Figure 9.4 shows different views of n-butane; in each view, four C atoms are in a continuous chain and no C atom is attached to more than two others. All of these views correspond to n-butane, and none matches iso-butane.

$CH_3-CH_2-CH_2-CH_3$ — **n-Butane**

$CH_3-C(CH_3)(H)-CH_3$ — **Iso-butane**

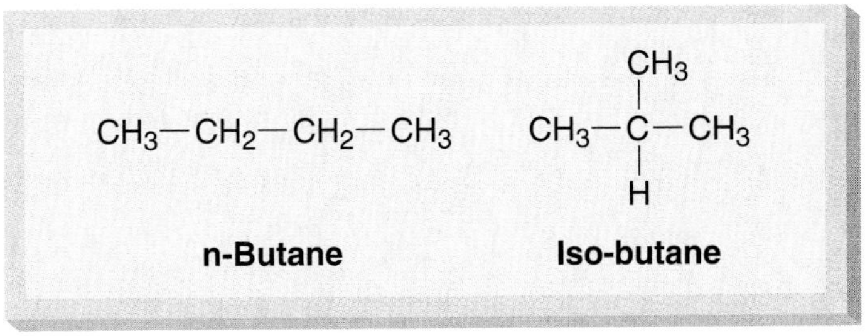

Figure 9.3 Butane isomers.

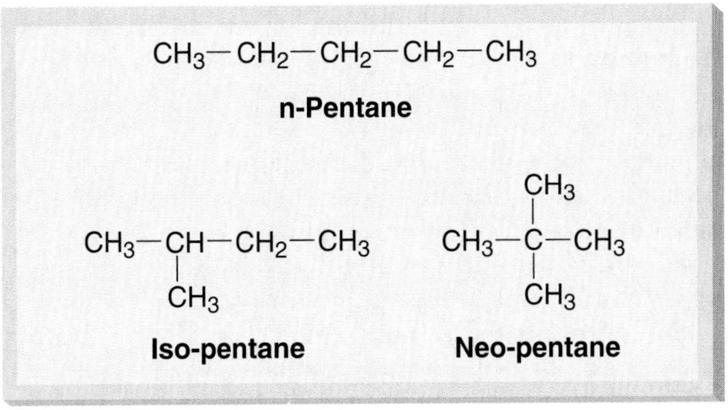

$CH_3-CH_2-CH_2-CH_3$

Figure 9.4 Five views of an n-butane molecule.

The term *iso-* is a prefix with a specific meaning; it signifies an alkane isomer with one C atom attached to the second C atom in the chain. Figure 9.5 shows three pentane molecules—n-pentane, iso-pentane, and a third molecule that also has the formula C_5H_{12} but that is neither an n- or iso-alkane. This new molecule is called *neo*-pentane; *neo-* means new. Notice that the middle C atom in neo-pentane's longest chain of 3 is attached to 4 other C atoms.

So far you have seen a total of five isomers, each with its own specific name. Isomers become more complex as more C atoms are added. Figure 9.6 shows five hexane isomers with the formula C_6H_{14}. Obviously isomer *a* is n-hexane, with its carbon atoms in a continuous series, and *b* is iso-hexane, with one C atom attached to the second C atom in its major chain. But what are the other three? You might call isomer *e* neo-hexane because it features a central C atom attached to 4 others in a structure similar to neo-pentane's, but structures *c* and *d* are unprecedented and require new names. Eventually, naming huge numbers of carbon compounds with their own common chemical names becomes impossible.

$CH_3-CH_2-CH_2-CH_2-CH_3$ — **n-Pentane**

$CH_3-CH(CH_3)-CH_2-CH_3$ — **Iso-pentane**

$CH_3-C(CH_3)_2-CH_3$ — **Neo-pentane**

Figure 9.5 Three pentane isomers.

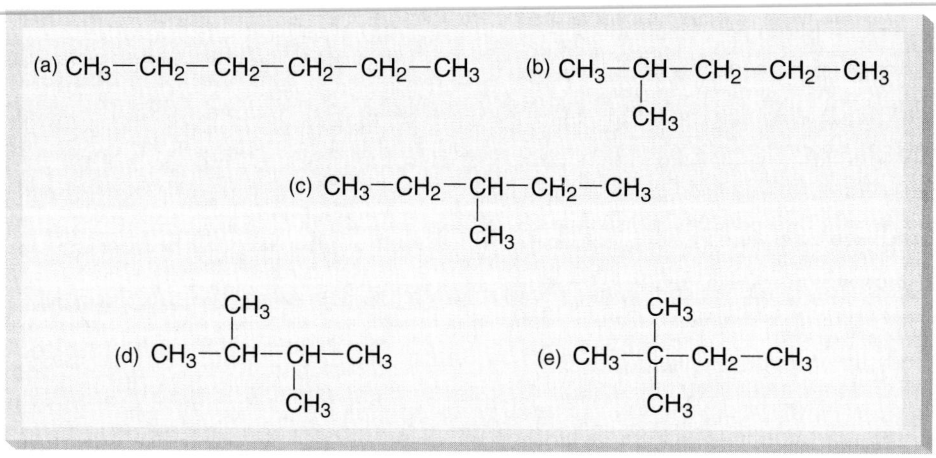

(a) $CH_3-CH_2-CH_2-CH_2-CH_2-CH_3$

(b) $CH_3-CH(CH_3)-CH_2-CH_2-CH_3$

(c) $CH_3-CH_2-CH(CH_3)-CH_2-CH_3$

(d) $CH_3-CH(CH_3)-CH(CH_3)-CH_3$

(e) $CH_3-C(CH_3)_2-CH_2-CH_3$

Figure 9.6 Five hexane isomers.

EXPERIENCE THIS

Make gumdrop-toothpick models for n-butane and iso-butane. For simplicity, leave off the H atoms and assemble only the C atoms. Notice that you cannot match n-butane with iso-butane; only n-butane is a normal alkane with all its C atoms arranged in a series.

Alkyl Groups

The IUPAC system solves the problem of naming organic compounds; chemists no longer assign special common names like neo-pentane to each individual molecule. Instead, the system relies on alkanes to provide root names for classes of hydrocarbons, and on alkyl groups to supply prefixes. An **alkyl group** is an alkane with one hydrogen atom removed. Table 9.3 shows the eight most common alkyl groups; methyl and ethyl are the main groups this book discusses. Note that alkyl groups are a family with the suffix *yl* replacing *ane* in the original alkane name.

Methane has four identical H atoms, and removing any one yields the same methyl group (CH_3). Likewise, removing any of the six identical H atoms from ethane (C_2H_6) will always yield the same ethyl group. Figure 9.7 shows an ethyl group viewed from three different directions. Although the three drawings look different, they are identical. (This is comparable to figure 9.4, which shows several structures for n-butane.)

Propane (C_3H_8) forms two different alkyl groups, depending on which H atom is removed. The removal of any of the 6 H atoms from one of the C atoms on the ends yields the same n-propyl group, but the removal of either H atom from the center C atom produces an isopropyl group (table 9.3). Four different alkyl groups derive from n-butane and iso-butane.

TABLE 9.3 Some common alkyl groups.

NAME	STRUCTURE
methyl	CH_3-
ethyl	CH_3-CH_2-
n-propyl	$CH_3-CH_2-CH_2-$
isopropyl	$CH_3-CH(CH_3)-$
n-butyl	$CH_3-CH_2-CH_2-CH_2-$
sec-butyl	$CH_3-CH_2-CH(CH_3)-$
iso-butyl	$CH_3-CH(CH_3)-CH_2-$
tert-butyl	$CH_3-C(CH_3)_2-$

CH_3-CH_2- (H above and below C, bond to the right) $\quad$ CH_3-CH_2-H (open bond above C, H below) $\quad$ CH_3-CH_2-H (H above C, open bond below)

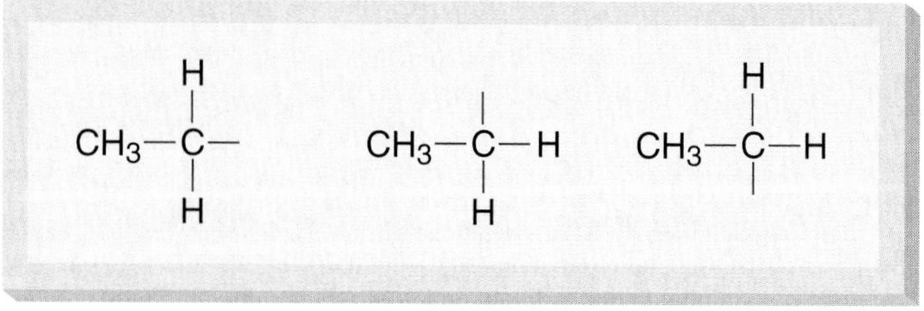

Figure 9.7 Three views of the same ethyl group.

Applying the IUPAC Nomenclature System

The IUPAC system names all alkanes as derivatives of an n-alkane. The root name of a molecule is based on its longest continuous carbon chain; for example, as table 9.1 indicates, if the longest chain contains 4 carbon atoms the root name will be butane, and if the longest chain has 6 C atoms the root name will be hexane. These root names are taken from the n-alkanes. Alkyl group prefixes then provide further information about the isomer. Methylpentane, as tables 9.1 and 9.3 tell us, would have a methyl unit (CH_3) attached to a continuous chain of 5 carbon atoms. Dimethylpentane (table 9.2) would have *two* methyl units attached to a 5-carbon chain. Figure 9.8 shows the five isomeric hexanes shown in figure 9.6 with their IUPAC system names added.

Hexane *a* in figure 9.8 remains n-hexane in the IUPAC system. Hexane *b* has five carbons in its longest continuous chain, making pentane its root IUPAC alkane name, and it has a methyl group (CH_3) hanging from the second carbon of this chain. This gives the molecule the root name methylpentane. Hexane *c* is also a methylpentane; to differentiate it from *b*, the IUPAC system specifies which carbon the methyl group hangs from. Since the methyl group hangs from the second carbon of the 5-carbon chain in hexane *b*, it receives the name 2-methylpentane; hexane *c* is 3-methylpentane, because the methyl group is attached to the third carbon in the longest chain. These two isomers are different compounds with different names.

Hexanes *d* and *e* both contain pairs of methyl groups attached to a 4-carbon butane chain. Molecule *d* is named 2,3-dimethylbutane and *e* is 2,2-dimethylbutane. Notice that the location of *every* methyl group attached to the main chain is specified, even if two are joined to the same carbon atom, as in hexane *e*. Under this system, each isomer has a distinctive name that describes its unique structure.

Just as we looked at the structures in figure 9.8 to determine their IUPAC system names, we can reverse the process—we can begin with an IUPAC name and determine the corresponding molecule's structure. Consider the molecule, 2,7-dimethyloctane. Because the root name is octane, we know the longest chain contains eight carbons. As figure 9.9 shows, we can draw the rest of the structure in four steps. Step 1: Draw a chain of 8 carbon atoms, showing all possible bonds. Step 2: Attach two CH_3— at the second and seventh carbon atoms in the chain. Step 3: Place H atoms on the remaining bonds. Step 4: Condense the formula. The advantage of the IUPAC system is that every chemist gives a molecule the same name and draws the same structure from that name.

PRACTICE PROBLEM 9.3

Name the alkane shown, using the IUPAC system. If you wish, use tables 9.1 and 9.2 to guide you.

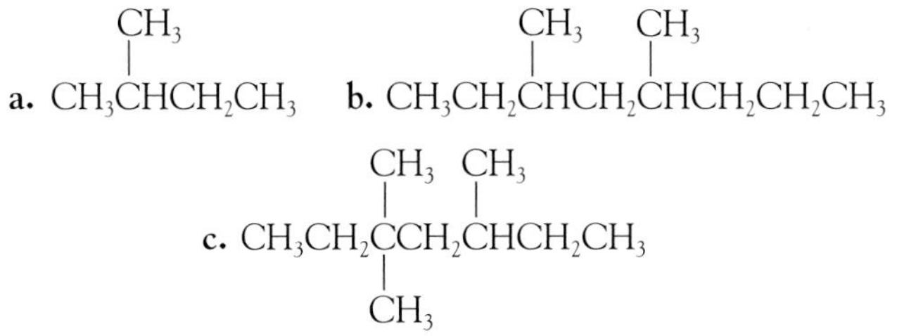

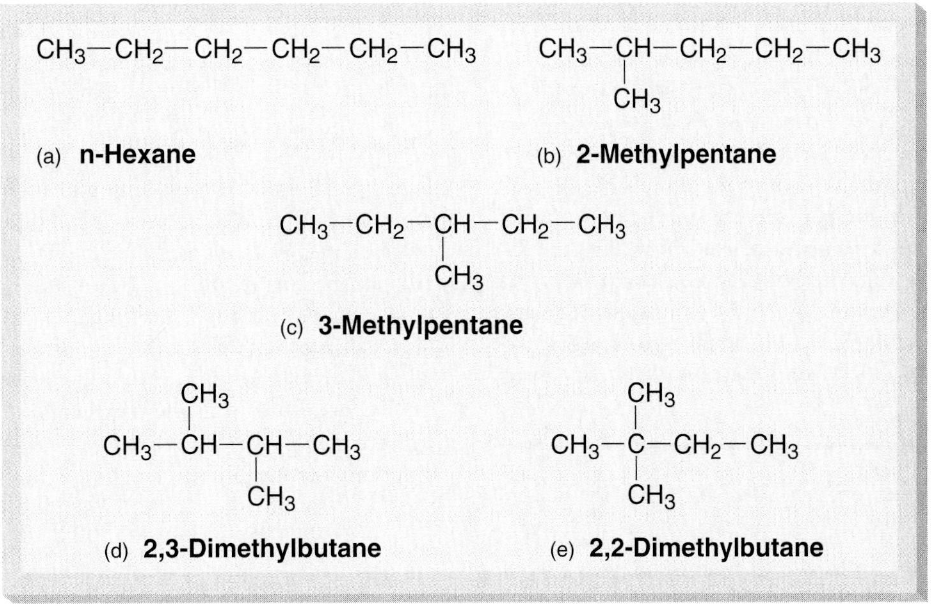

a. $CH_3CH(CH_3)CH_2CH_3$

b. $CH_3CH_2CH(CH_3)CH_2CH(CH_3)CH_2CH_2CH_3$

c. $CH_3CH_2C(CH_3)_2CH_2CH(CH_3)CH_2CH_3$

$CH_3-CH_2-CH_2-CH_2-CH_2-CH_3$

(a) **n-Hexane**

$CH_3-CH(CH_3)-CH_2-CH_2-CH_3$

(b) **2-Methylpentane**

$CH_3-CH_2-CH(CH_3)-CH_2-CH_3$

(c) **3-Methylpentane**

$CH_3-CH(CH_3)-CH(CH_3)-CH_3$

(d) **2,3-Dimethylbutane**

$CH_3-C(CH_3)_2-CH_2-CH_3$

(e) **2,2-Dimethylbutane**

Figure 9.8 Correct IUPAC system names for the five hexane isomers first depicted in figure 9.6.

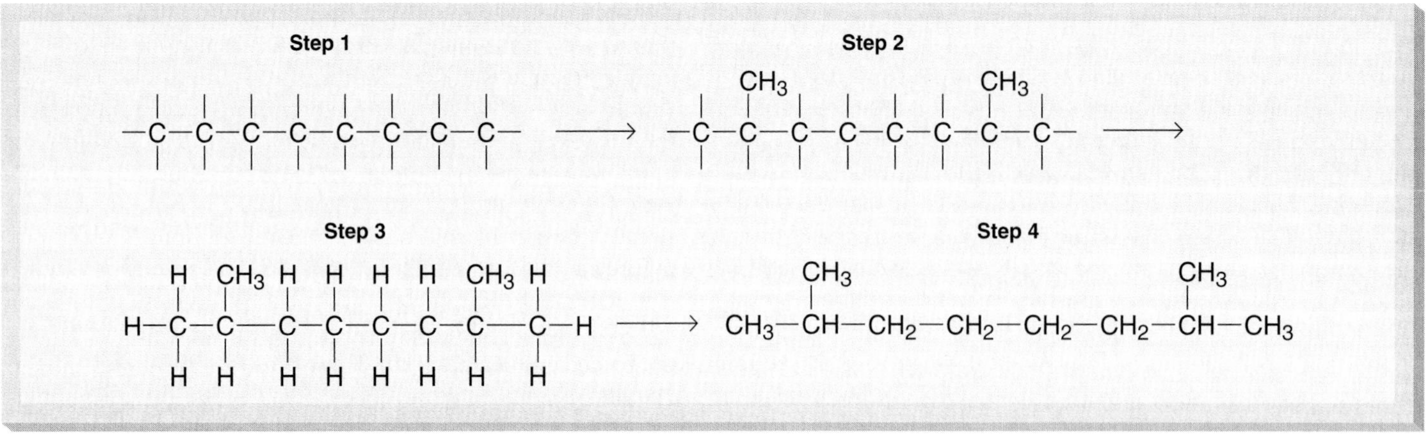

Figure 9.9 The steps for drawing the structure of 2,7-dimethyloctane.

ANSWER

a. 2-methylbutane (2-methyl-butane)
b. 3,5-dimethyloctane (3,5-di-methyl-octane)
c. 3,3,5-trimethylheptane (3,3,5-tri-methyl-heptane)

PRACTICE PROBLEM 9.4

Try the opposite—draw the structure for 2,4,5,7-tetramethylnonane.

ANSWER To determine the structure of this molecule, first break the name into smaller units to see the root and prefixes: 2,4,5,7-tetra-methyl-nonane. Then follow the four-step process. Step 1: The root chain name is nonane, so draw a 9-carbon chain. Step 2: Add the four CH_3—units at the specified sites—carbons 2,4,5, and 7 in the 9-carbon chain. (You can count from either end of the chain. Figure 9.10 shows the molecule; one CH_3—group is placed beneath the 9-carbon chain simply for clarity, to avoid crowding the CH_3 groups in the drawing. (Alternatively, you could draw the horizontal bonds longer and place all groups on the same side of the chain.) In step 3, add H atoms to each open carbon bond. Finally, in Step 4, write the formula in condensed form.

The IUPAC system works for very complex molecules, but additional prefix rules apply. In Chapter 10, we will learn some of the rules for naming complex functional organic molecules.

Cycloalkanes

So far, we have discussed the n-alkanes, which align in chains, and other alkanes whose carbon atoms form branched chains. Carbon forms rings as well as chains, and the simplest ring class is the **cycloalkane** family, class formula C_nH_{2n}. The IUPAC system names these compounds in a manner similar to the way it names open-chain molecules. Figure 9.11 shows some typical cycloalkanes.

As figure 9.11 demonstrates, we can draw cycloalkanes in at least two styles. The first style shows each C atom and all attached H atoms, while the second shows a shorthand geometric symbol, with each apex in the symbol representing a CH_2 unit. Cycloalkanes can also have attached methyl groups, but we will not be exploring these more complex structures. Cycloalkanes are used in fuels and the synthesis of other important organic compounds. For example, cyclohexane is used to prepare the adipic acid used to make nylon, an important fiber.

Summing Up the Alkanes

Figure 9.12 shows the melting and boiling points for n-alkanes containing up to 20 C atoms. As the figure indicates, the melting and boiling points increase with an increase in the number of C atoms. The smaller alkanes (less than 4 carbon atoms) are gases at room temperature, and alkanes with 5 to 16 C atoms are liquids. Alkanes with 17 or more C atoms are waxy solids (hence, alkanes are sometimes known as paraffins).

The main use of alkanes is in fuels, as we will discuss later in this chapter. The following equation shows a typical combustion reaction for n-hexane:

$$2\ C_6H_{14} + 19\ O_2 \rightarrow 12\ CO_2 + 14\ H_2O$$

As the equation shows, alkanes form carbon dioxide and water when they undergo combustion. These exothermic reactions make alkanes an ideal fuel.

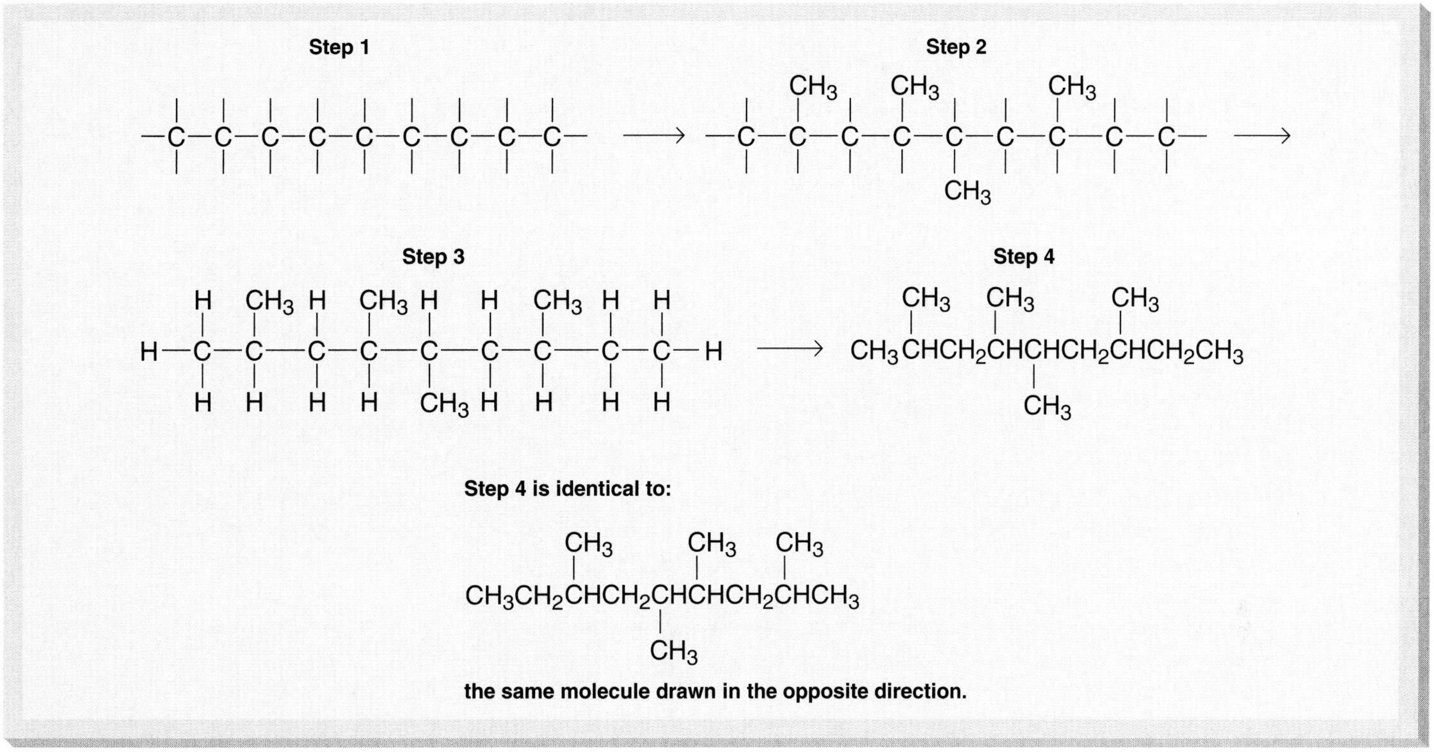

Step 1

—C—C—C—C—C—C—C—C—C— ⟶

Step 2

CH_3 CH_3 CH_3

—C—C—C—C—C—C—C—C—C— ⟶

CH_3

Step 3

H CH_3 H CH_3 H H CH_3 H H

H—C—C—C—C—C—C—C—C—C—H ⟶

H H H H CH_3 H H H H

Step 4

CH_3 CH_3 CH_3

$CH_3CHCH_2CHCHCH_2CHCH_2CH_3$

CH_3

Step 4 is identical to:

CH_3 CH_3 CH_3

$CH_3CH_2CHCH_2CHCHCH_2CHCH_3$

CH_3

the same molecule drawn in the opposite direction.

Figure 9.10 The steps for drawing the structure of 2,4,5,7-tetramethylnonane.

Alkanes participate in very few other important reactions, and most of these involve substituting a H atom with some other atom, such as a Cl atom. A typical substitution reaction occurs between ethane and chlorine:

$$CH_3—CH_3 + Cl_2 \rightarrow CH_3—CH_2—Cl + H—Cl$$

However, since alkanes are largely unreactive, these reactions are relatively uncommon.

In chapter 6, you learned that more than one covalent bond can form between atoms with a valence greater than 2. This is particularly true for carbon, which can form multiple-bonded systems that greatly increase the versatility of hydrocarbons. While alkanes contain only single C—C bonds, the alkenes, alkynes, and aromatic hydrocarbons contain multiple bonds. Let's look now at these other types of hydrocarbons.

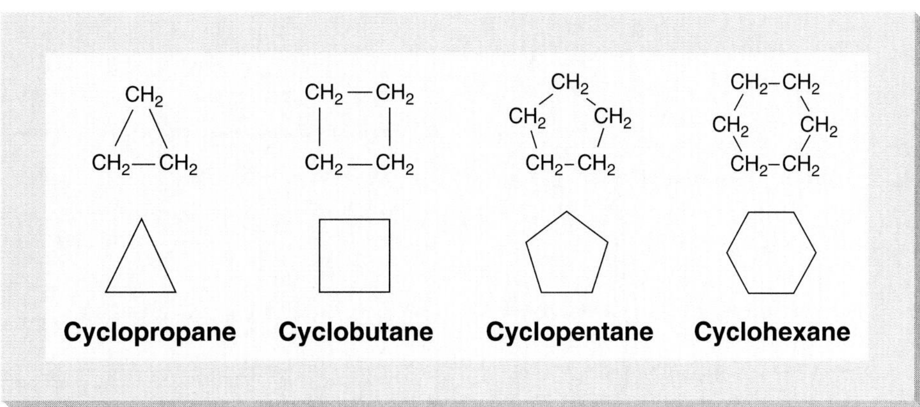

Figure 9.11 Some typical cycloalkanes.

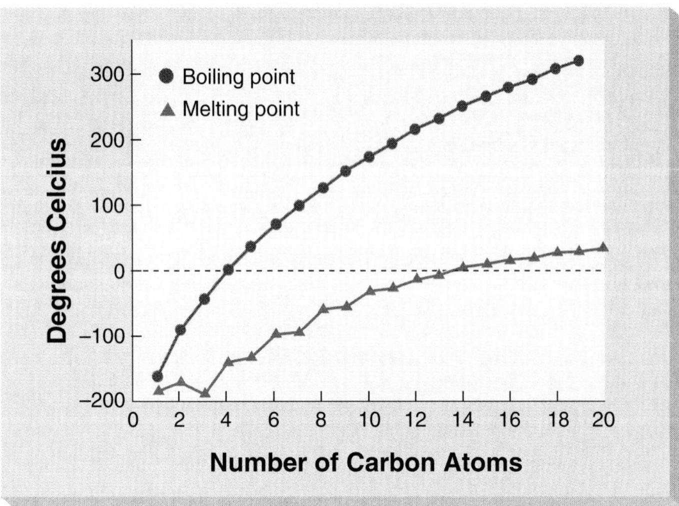

Figure 9.12 Melting and boiling points for the n-alkanes up to $C_{20}H_{42}$.

ALKENES

Alkenes are hydrocarbons that contain a C=C double bond in the continuous carbon-atom chain. They have the class formula C_nH_{2n}. Alkenes are also called olefins and unsaturated hydrocarbons; the latter also applies to alkynes and aromatics. You have likely seen the term *unsaturated* used to refer to edible oils; this means some of the oil molecules contain a double bond. A C=C double bond consists of a **sigma bond (σ-bond)** and a **pi bond (π-bond);** nearly all alkene reactions occur at the more reactive π-bond.

The suffix *ene* indicates that a compound is an alkene, while the prefix refers to the number of C atoms in the longest chain containing the C=C double bond. Two important IUPAC rules apply to alkenes: (1) The root name comes from the longest continuous chain of C atoms *containing a double* C=C *bond,* even if a longer chain is present; and (2) the double bond is assigned the lowest possible number corresponding to the location of the first C atom in the double bond, regardless of other alkyl substituents.

According to these rules, $CH_2{=}CH{-}CH{-}CH{-}CH_3$ is 1-pentene (never 2-pentene or 4-pentene). The name *pentene* comes from the molecule's chain of 5 carbons, with one double bond; the 1 is the lowest number assignable to the C=C because the double bond joins the first carbon to the second. The correct name for $CH_3{-}CH{=}CH{-}CH_2{-}CH_3$ is 2-pentene, never 3-pentene. Similarly, the following molecule is 3-n-propyl-6-methyl-1-heptane; though the longest continuous chain has 8 carbons, the longest chain *containing* a *double bond* has 7, making the root name heptene:

```
CH3—CH2—CH2               CH3
          |               |
CH2=CH—CH—CH2—CH2—CH—CH3
```

The compound is 1-heptene because the double bond joins carbon atom 1 to carbon atom 2. The 3-n-propyl indicates the n-propyl unit attached to the third carbon in the main chain; the 6-methyl indicates the position of the methyl unit, attached to the sixth carbon in the chain. If we flipped the molecule over sideways, we might first believe it to be 2-methyl-5-n-propyl-6-heptene (simply read the condensed formula shown from right to left). However, this would violate one of the IUPAC alkene rules; because the double bond receives the lowest possible number, the molecule is 1-heptene rather than 6-heptene, and the branches' positions are assigned relative to this. More details about the structure and bonding in alkenes and other unsaturated molecules appear in box 9.1.

PRACTICE PROBLEM 9.5

Name the following alkenes, referring to tables 9.1 and 9.2 as needed.

a. $CH_3CH_2CH{=}CHCH_3$

b.
```
CH3CH=CHCH2CHCH3
             |
             CH3
```

c.
```
   CH3            CH3
   |              |
CH3CHCH2CH=CHCHCH3
```

ANSWER

a. 2-pentene. Here, to assign the double bond the lowest possible number, you must start counting from the right.

b. 5-methyl-2-hexene. Because the C=C double bond takes precedence, it is numbered as being joined with carbon atom 2 counting from the left. This makes the methyl group end up on carbon atom 5.

c. 2,6-dimethyl-3-heptene. The C=C begins at carbon 3, which places the two methyl groups at carbon atoms 2 and 6.

PRACTICE PROBLEM 9.6

Now reverse the process; draw the structures that correspond to each of the following names:

a. 4-dodecene

b. 2,4,4-trimethyl-2-octene

ANSWER

a. $CH_3CH_2CH_2CH_2CH_2CH_2CH_2CH{=}CHCH_2CH_2CH_3$, counting from the right. (You can also reverse the condensed formula to show the double bond four carbons from the left.)

b. $CH_3\overset{\displaystyle CH_3}{\overset{|}{C}}{=}CH_2\underset{\displaystyle CH_3}{\underset{|}{\overset{\displaystyle CH_3}{\overset{|}{C}}}}CH_2CH_2CH_2CH_3$, counting from the left. (When a C atom is *both* double bonded *and* bonded to an additional alkyl group, it appears as a C atom.)

Geometrical Isomerization

The C=C double bond in an alkene consists of one σ-bond and one π-bond. If we examined the single bond in an ethane molecule, we would see that all the hydrogen atoms in ethane appear identical no matter which way the molecule is turned, because the two methyl groups can rotate freely around the single bond between the carbon atoms:

```
    H   H
    |   |
H — C — C — H
    |   |
    H   H
```

By contrast, when a C=C double bond links two carbon atoms, the attached groups cannot rotate around the central bond. This leads to a new type of isomerization called geometrical isomerization. **Geometrical isomers** are two molecules with the same chemical formula and structure, but with the atoms arranged differently. Geometrical isomers are different compounds with different properties.

Figure 9.13 shows geometrical isomers of 2-butene. The prefixes cis- and trans- differentiate one geometrical isomer from the other. Notice that each carbon atom is attached to one methyl (CH_3) group and one hydrogen atom. They are geometrical isomers because the attached groups are in different positions; it would be impossible to superimpose trans-2-butene over cis-2-butene and align the same atoms and bonds. (Imagine the C=C bond is like a fence; the CH_3 groups are on the same side of the fence in the cis- isomer and opposite sides in the trans- isomer.) The 2-butene geometrical isomers have different melting and boiling points and slightly different chemical reactivity. After all, they are different compounds.

To have geometrical isomers, two different groups must be attached to each of the C atoms in the C=C bond. Figure 9.14 shows that, despite its C=C double bond, ethylene, known as ethene in the IUPAC system, cannot form a geometrical isomer because each C atom in the double bond has two identical H atoms attached. Similarly, 1-cholorethene (commonly called vinyl chloride) cannot form a geometrical isomer because one C atom in the C=C bond has an identical pair of H atoms attached. 1,2-dichloroethene does form geometrical isomers because each C atom in the C=C bond has two different kinds of atoms or groups attached, an H atom and a Cl atom. In the cis-isomer, both H atoms (or Cl atoms) are on the same side of the C=C bond; in the trans-isomer, the H atoms are on opposite sides.

cis-2-Butene
m.p. = −139°C
b.p. = +4°C

trans-2-Butene
m.p. = −106°C
b.p. = +1°C

Figure 9.13 The two geometrical isomers of 2-butene. Although they have identical chemical formulas, they are different compounds with different melting and boiling points.

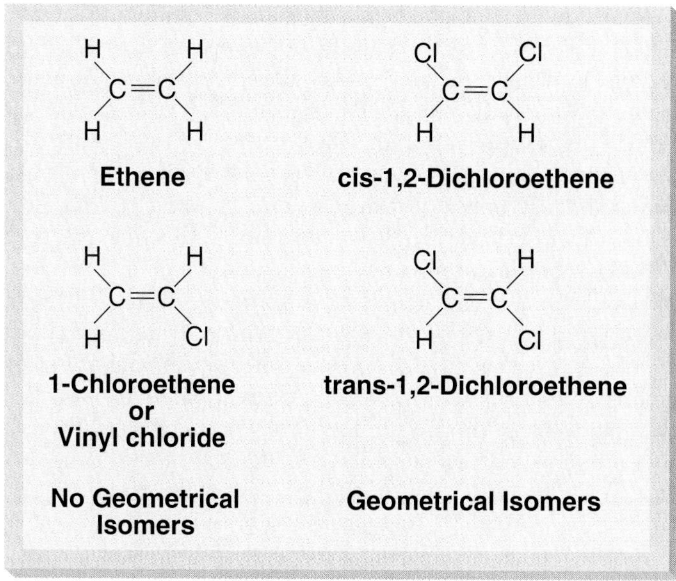

Figure 9.14 Geometrical isomerization.

EXPERIENCE THIS

Use toothpicks and gumdrops to make models of the two geometrical isomers in figure 9.14. Use black gumdrops for the C atoms, fastening them together with a pair of toothpicks arranged so you cannot rotate one gumdrop around the other; this represents the π-bond. Lay this part on a tabletop and stick two more toothpicks in each gumdrop so that all six toothpicks are parallel to the table; ideally, your bond angle is 120°. First make the cis-isomer by placing a pair of white gumdrops (representing H atoms) on the same side of the double bond. Place a pair of green gumdrops (Cl atoms) on the remaining toothpicks. Make the trans-isomer by constructing a second model, this time placing a white pair of gumdrops on opposite sides of the double bond; do likewise with a green pair. After constructing your models, try to line them up so that the white gumdrops on one model are always aligned with the white ones on the other model, and the green gumdrops aligned on top of the other green ones. You will find this impossible; although the models contain the same "atoms," they are geometrically different.

Addition Reactions

Alkenes, with their more reactive π-bonds, are more reactive than alkanes. Their most important reaction is addition, which occurs at the π-bond and produces a product in which the reactants are added to the double bond. (Addition polymers are a special type of addition reaction we will discuss further later in the chapter.) Several different addition reactions may occur.

Ethene (or ethylene) reacts with water to form ethanol; when a trace of an acid catalyst is used, the H—OH adds in an addition reaction to the C=C.

$$\underset{\text{Ethene}}{CH_2{=}CH_2} + \underset{\text{Water}}{H{-}OH} \xrightarrow[\text{An acid}]{H^+} \underset{\text{Ethanol}}{H{-}CH_2{-}CH_2{-}OH}$$

Br_2 adds to the ethene double bond, one Br atom attaching to each C atom like extra links in the chain.

$$\underset{\text{Ethene}}{CH_2{=}CH_2} + \underset{\text{Bromine}}{Br_2} \rightarrow \underset{\text{1,2-Dibromoethane}}{Br{-}CH_2{-}CH_2{-}Br}$$

An alkene can also add H_2 to form an alkane in the process called hydrogenation. A similar process is used to hydrogenate vegetable oils. This changes the unsaturated alkene to a saturated alkane hydrocarbon.

$$\underset{\text{Ethene}}{CH_2{=}CH_2} + \underset{\text{Hydrogen}}{H{-}H} \rightarrow \underset{\text{Ethane}}{H{-}CH_2{-}CH_2{-}H}$$

ALKYNES

Alkynes are hydrocarbons with at least one C≡C triple bond. This triple bond consists of one σ-bond and two π-bonds; it is the key characteristic of the alkyne family, which has a class formula C_nH_{n-2}. Nearly all alkyne chemical reactions, like alkene reactions, occur at π-bonds. Alkynes also undergo addition reactions.

Alkynes add water to form new classes of compounds called either aldehydes or ketones. (These compounds are discussed further in chapter 10.) Acetylene, a gas used in welding that is called ethyne in the IUPAC system, adds water to form an aldehyde (figure 9.15a), but the other alkynes form ketones when adding water (figure 9.15b). Alkynes normally add other reagents, such as bromine, to both π-bonds (figure 9.15c). And alkynes can add one or two hydrogen molecules to form either a cis- or trans-alkene (figure 9.15d) or an n-alkane (figure 9.15e).

An important industrial use for the alkyne acetylene is the synthesis of vinyl acetate when acetic acid reacts with acetylene (figure 9.16a). Vinyl acetate may also be prepared by reacting acetic acid with ethene (figure 9.16b). The reactions shown in figure 9.16 each require a specific catalyst. Both ethene and acetylene are inexpensive compounds obtained from petroleum and are good raw materials for synthesizing the vinyl acetate used to make **polymers** (large organic molecules made of chains of smaller molecules; chapter 11).

The IUPAC names for alkynes are identical to those of alkenes except that they include the suffix *yne*. In the following examples, note that the correct prefix number denoting the C≡C bond in the first example is 3-, not 4-; this is true no matter which direction you write the chemical formula in. The methyl group added to the second example is assigned as 6- (never 2-) because the C≡C group sets the numbering sequence.

$$CH_3{-}CH_2{-}C{\equiv}C{-}CH_2{-}CH_2{-}CH_3$$
3-Heptyne

$$CH_3{-}CH_2{-}C{\equiv}C{-}CH_2{-}\underset{|}{\overset{CH_3}{CH}}{-}CH_3$$
6-Methyl-3-heptyne

PRACTICE PROBLEM 9.7

Name the following two alkynes according to the IUPAC system.

a. $CH_3CH_2CH_2CH_2{-}C{\equiv}C{-}CH_2CH_3$

b. $CH_3CH_2{-}C{\equiv}C{-}CH_2\overset{CH_3}{CH}CH_2CH_3$

ANSWER *a.* There are eight C atoms in the chain, making this an octyne; the C≡C begins at the third carbon, so the compound is 3-octyne. *b.* There are eight C atoms in the chain with the triple bond, so this also is an octyne. Again, it must be 3-octyne because counting it as 5-octyne would assign it a higher number, which violates IUPAC rules. The location of the CH_3-group can be determined once the C≡C location is assigned; it is attached to carbon 6, so the full name is 6-methyl-3-octyne.

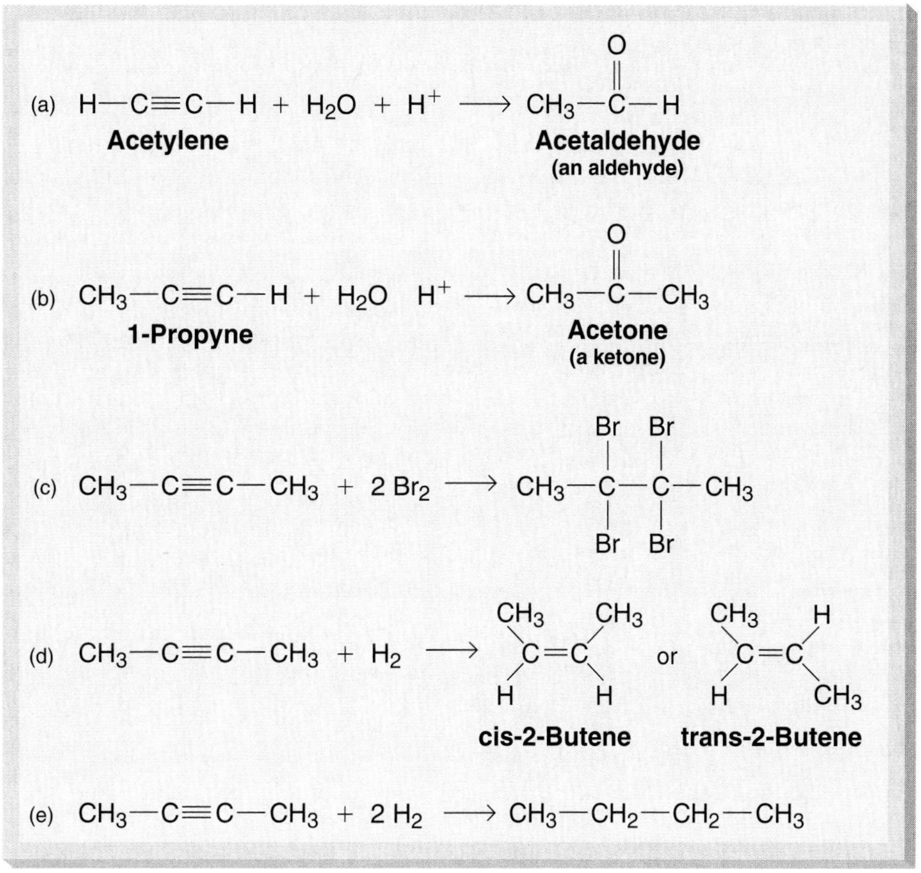

(a) $H{-}C{\equiv}C{-}H + H_2O + H^+ \longrightarrow CH_3{-}\overset{\displaystyle O}{\overset{\|}{C}}{-}H$

Acetylene → **Acetaldehyde** **(an aldehyde)**

(b) $CH_3{-}C{\equiv}C{-}H + H_2O \;\; H^+ \longrightarrow CH_3{-}\overset{\displaystyle O}{\overset{\|}{C}}{-}CH_3$

1-Propyne → **Acetone** **(a ketone)**

(c) $CH_3{-}C{\equiv}C{-}CH_3 + 2\,Br_2 \longrightarrow CH_3{-}CBr_2{-}CBr_2{-}CH_3$

(d) $CH_3{-}C{\equiv}C{-}CH_3 + H_2 \longrightarrow$ (CH₃)(H)C=C(CH₃)(H) or (CH₃)(H)C=C(H)(CH₃)

cis-2-Butene **trans-2-Butene**

(e) $CH_3{-}C{\equiv}C{-}CH_3 + 2\,H_2 \longrightarrow CH_3{-}CH_2{-}CH_2{-}CH_3$

Figure 9.15 Some interesting alkyne addition reactions.

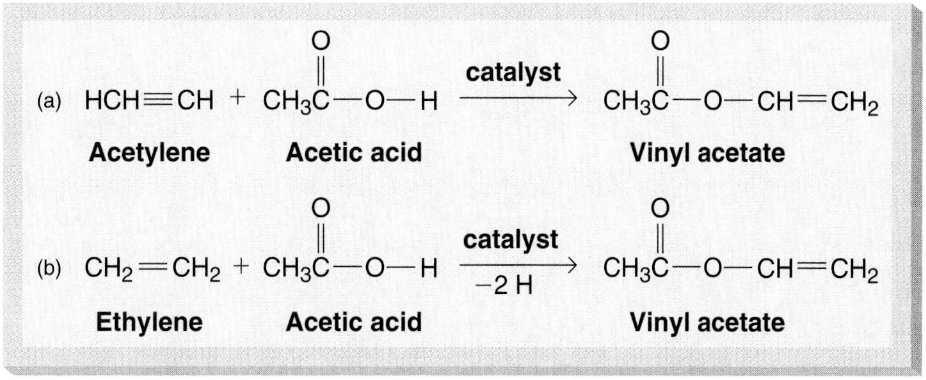

(a) $HCH{\equiv}CH + CH_3\overset{\displaystyle O}{\overset{\|}{C}}{-}O{-}H \xrightarrow{\text{catalyst}} CH_3\overset{\displaystyle O}{\overset{\|}{C}}{-}O{-}CH{=}CH_2$

Acetylene **Acetic acid** **Vinyl acetate**

(b) $CH_2{=}CH_2 + CH_3\overset{\displaystyle O}{\overset{\|}{C}}{-}O{-}H \xrightarrow[-2\,H]{\text{catalyst}} CH_3\overset{\displaystyle O}{\overset{\|}{C}}{-}O{-}CH{=}CH_2$

Ethylene **Acetic acid** **Vinyl acetate**

Figure 9.16 Two different industrial ways to synthesize vinyl acetate.

BOX 9.1

HYBRID ORBITALS

You are probably familiar with hybrids—composites of two different species or types—such as plants, mules, and breeds of dogs. The details of hybrid orbitals involve complex mathematics, but the basic ideas behind them are fairly simple. Remember the Lewis dot structures introduced in chapter 6? These structures assume the electrons are in the ground state, but this is not true at bond formation temperatures. As the temperature increases, the electrons increase their energy levels slightly, unpair, and undergo fundamental changes. In carbon, the s-orbital and p-orbitals "pool" their energy and form up to four new orbitals that share exactly the same bond angles and bond lengths. These are called **hybrid orbitals** or **hybrid bonds,** and the process of forming them is **hybridization.**

Carbon bond hybridization does not always use all three p-orbitals, but it does always use the lower-energy s-orbital. The total number of hybrid orbitals is the sum of the s- and p-orbitals used. Think of mixing three cans of red paint (p-orbitals) with one can of yellow (the s-orbital). If you mix all four cans, you end up with four cans of reddish-orange paint. If you only mix three cans—two red plus the yellow—you end up with three cans of orange-red paint, with one can of red left over. And if you mix only two cans—one red, one yellow—you end up with two cans of orange and two unused cans of red paint. In all cases you start and end with four cans of paint, or four orbitals. But the final number of cans of mixed-color paint or the final number of hybrid orbitals depends on how many cans or orbitals participated in hybridization (figure 1).

As figure 1 shows, when one s-orbital is mixed with three p-orbitals, the resulting four mixed orbitals are called sp^3-orbitals (the superscripts denote how many

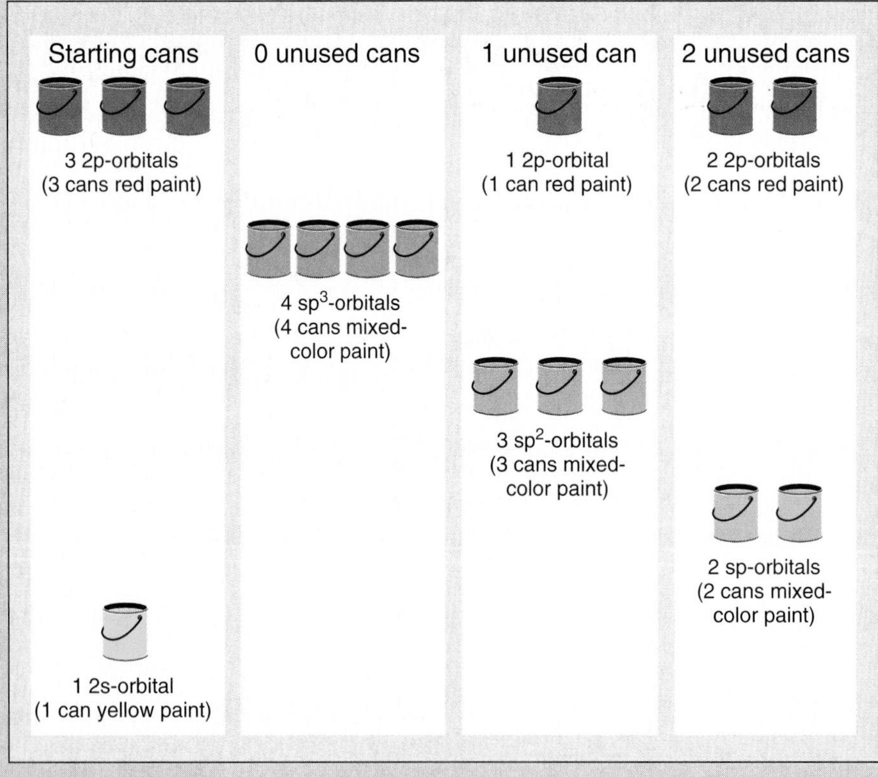

Figure 1 An analogy for forming hybrid orbitals from one s-orbital (yellow can of paint) and up to three p-orbitals (red cans of paint). The less red used, the more yellow the shade of the final orange paint (the greater the s-orbital character of the bond); this corresponds to a lower-energy orbital in the analogy. The leftover cans of red paint (unused p-orbitals) are still at the highest energy levels.

p-orbitals were used to form the hybrid.) The sp^3 hybrids all have 109.5° angles, just like a methane (CH_4) molecule, which is shaped like a tetrahedron. In an ethane (C_2H_6) molecule, all three p-orbitals are also used. The molecule is shaped like two carbon tetrahedrons joined together. A bond forms where two of the sp^3 hybrid orbitals, one from each carbon atom, overlap. The bond connecting the two C atoms is a *sigma bond* (σ-bond). Sigma bonds always form on a straight line between the two atoms. The bond is a single sp^3—sp^3 bond; the ethane σ-bond appears in two different ways in figure 2a. The view on the left shows the bond formed from the overlap of two sp^3 orbitals. The view on the right is a line representation showing bond lengths and angles.

When one s-orbital is mixed with two p-orbitals, three hybrid sp^2-orbitals with 120° bond angles are produced. One p-orbital remains unused. Ethylene, C_2H_4, is the simplest carbon molecule with sp^2-orbitals; its structure appears in figure 2b. One bond forms between the carbon atoms, where the sp^2-orbitals overlap. A second bond forms between the two C atoms from the unused p-orbitals, one on each C atom. This second bond is called a *pi bond* (π-bond), but these π-bonds do not form along a straight line between the

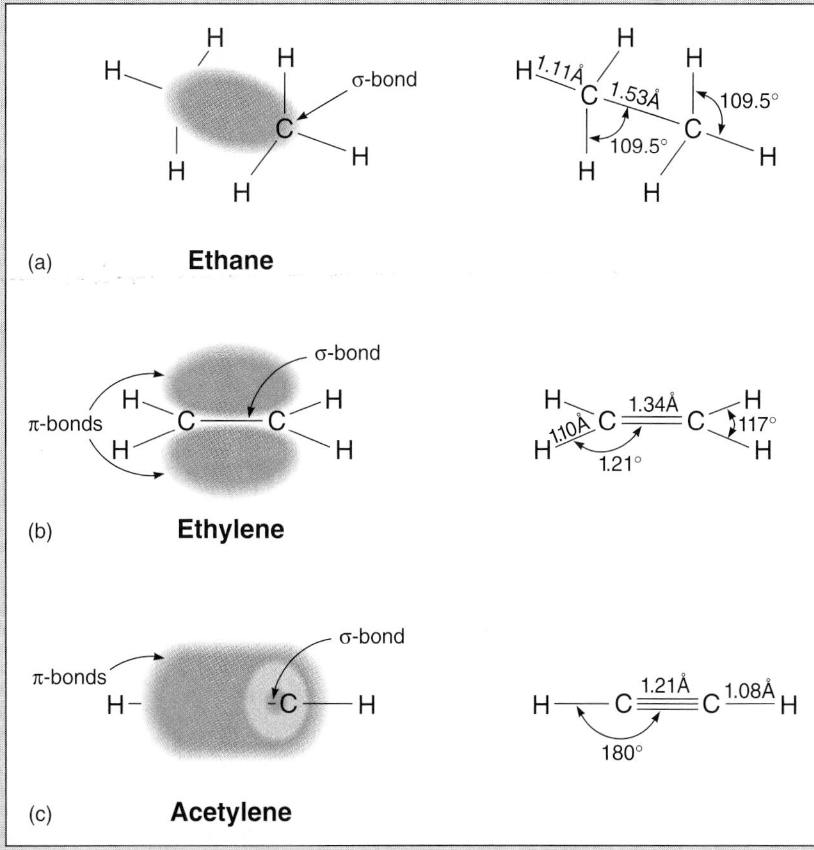

Figure 2 The structures of the ethane, ethylene, and acetylene molecules. The structures on the left depict the sigma (σ) bonds formed by the overlap of sp^3-, sp^2-, or sp-orbitals and pi (π) bonds formed from the overlap of p-orbitals . In ethane **(a)**, the sigma bond is shown as an overlapping cloud; the corresponding sigma bonds in ethylene **(b)** and acetylene **(c)** are shown as lines, but the pi (π) bonds are shown as overlapping clouds from p-orbitals. The structures on the right depict each molecule in the simpler line notation. The bond distances and angles are included in the line drawings.

(Continued)

BOX 9.1—Continued

atoms. Instead, they lie above and below the C atoms like a frame or an electron cloud. The electrons in the π-bonds are more reactive than those in σ-bonds. The carbons are united in a double (C═C) bond; one is sp^2—sp^2, the other p—p.

When one s-orbital is mixed with one p-orbital, two hybrid sp-orbitals form and two p-orbitals remain unused. Acetylene, C_2H_2, is the simplest carbon molecule with sp-orbitals. Its structure appears in figure 2c. The sp-orbitals are linear (that is, bonds from the central atom lie in a straight line) with a bond angle of 180°. The two C atoms are held together by one σ-bond (from two sp-orbitals) and two π-bonds formed from the two unused p-orbitals on each carbon atom. The pair of π-bonds form a cylindrical sheath around the σ-bond. The carbon atoms are thus joined in a triple (C≡C) bond; one is sp—sp, two are p—p.

Since p-orbitals have more energy than s-orbitals, the hybrid orbitals end up with more energy when more p-orbitals are blended into them. The orbital energy increases in this order, s, sp, sp^2, sp^3, p. Hybrid orbitals change the shape and chemical reactivity of the molecules. Carbon maintains its valency of four (and thus its versatility) by utilizing leftover p-orbitals to form bonds.

Two consequences of the different σ-bond and π-bond shapes show up in their strength and chemical reactivity. A σ-bond is always stronger than a π-bond between the same pair of atoms. Imagine each type of bond as consisting of four wooden planks. If the wooden boards are piled up and nailed together, it is extremely difficult to separate one board from another. This is similar to a σ-bond. However, if you construct a square frame from these four boards, joining them together at the corners and using the same number of nails, you can separate one board from the others relatively easily (figure 3). This is similar to a π-bond. The same number of boards and nails were used in both cases, but the strength of the "bond" depends upon its anatomy or geometry. In covalent bonds, the electrons correspond to the nails used to join the boards. When the electrons are joined in a straight line (σ-bond), a stronger bond (about 88 Kcal/mole) results than when the same number of electrons are joined in a framelike π-bond (with a strength of about 70–79 Kcal/mole). In the ethane, and acetylene examples, the hybrid σ-bonds are stronger than the p-orbital π-bonds.

The second consequence of hybrid bonds shows up in chemical reactivity, which involves the valence electrons. In ionic bonds, the valence electrons are located in the outermost shells or orbitals and are thus very accessible. But in covalent bonds, the electron pair nestles between the C atoms in the σ-bond and becomes difficult to reach. Thus, reactions involving these electrons occur with greater difficulty and are slower. The electrons in π-bonds are more exposed than those in σ-bonds. Again, to understand why, consider the strip-versus-frame analogy. All boards and nails are accessible in the frame structure, but this is not so in the strip arrangement. Because the electrons are more accessible, chemical reactions involving π-bonds occur more rapidly than those involving σ-bonds. But they are still slower than ionic reactions.

Differences in bond length are a third consequence of bond geometry. The presence of additional bonds tend to draw the C atoms closer together (table 1). Because more bonds are involved, the total bonding energy between the two C atoms also increases. As table 1 indicates, the triple bond typical in an acetylene molecule is shorter and stronger than those in an ethylene or ethane molecule. The presence of hybrid orbitals thus changes the shape, strength, and chemical reactivity of a molecule.

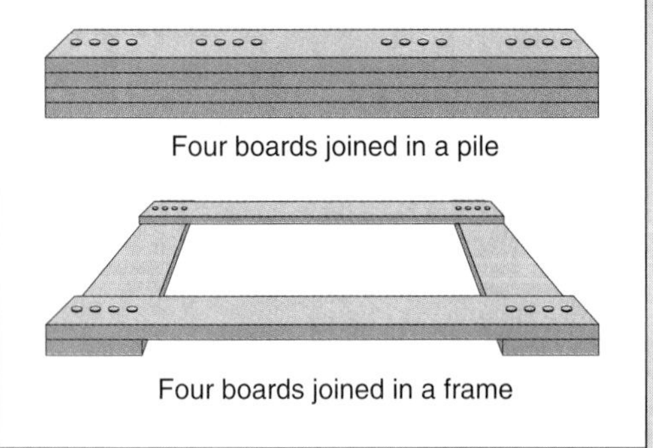

Figure 3 The differences in "bonding" strength between boards nailed together in a pile and boards nailed together as a frame. There are 16 nails in each board arrangement, and all nails pass completely through the boards. The boards in the pile are very difficult to separate, but the boards in the frame can easily be separated.

TABLE 1 The dependence of bond strength and energy on orbital type.

BOND	ORBITALS USED	BOND DISTANCE, Å	TOTAL BOND ENERGY, KCAL/MOLE
C—C	sp^3—sp^3	1.54	80
C═C	sp^2—sp^2 + p—p	1.34	146
C≡C	sp—sp + 2p—p	1.20	200

AROMATICS

Aromatic hydrocarbons are carbon compounds that contain benzene rings. Aromatics acquired their name because many of them have strong and persistent odors. Michael Faraday isolated benzene in a fairly pure condition in 1825. Subsequent work showed its formula was C_6H_6, implying a highly unsaturated molecule, since the class formula for a saturated hydrocarbon is C_nH_{2n+2}—14 hydrogen atoms to 6 carbon atoms. However, benzene's low chemical reactivity seemed to contradict this, since it does not undergo addition reactions as unsaturated alkenes and alkynes do. Eventually, chemists learned that benzene is a cyclic molecule with six CH units in a hexagonal shape. Friedrich August Kekulé (German, 1829–96; figure 9.17) concluded in 1865 that the benzene molecule consisted of a ring of alternate C=C and C—C bonds (figure 9.18). Kekulé claimed his idea came from a dream in which he saw snakes swallowing their tails and rolling around as loops. Whether this story is true is debatable, but Kekulé was the first chemist to propose a benzene structure that accounted for the lack of isomers for certain benzene compounds.

As figure 9.18 shows, when benzene's molecular structure is rotated on a piece of paper, the two forms match after a 60° rotation. The two apparently different benzene molecules in figure 9.18a and b are actually identical because they superimpose each other perfectly after a simple rotation; this always means the molecules are identical. As more benzene-derived compounds were

Figure 9.17 Friedrich August Kekulé (1829–1896) first proposed the cyclic structure for benzene.

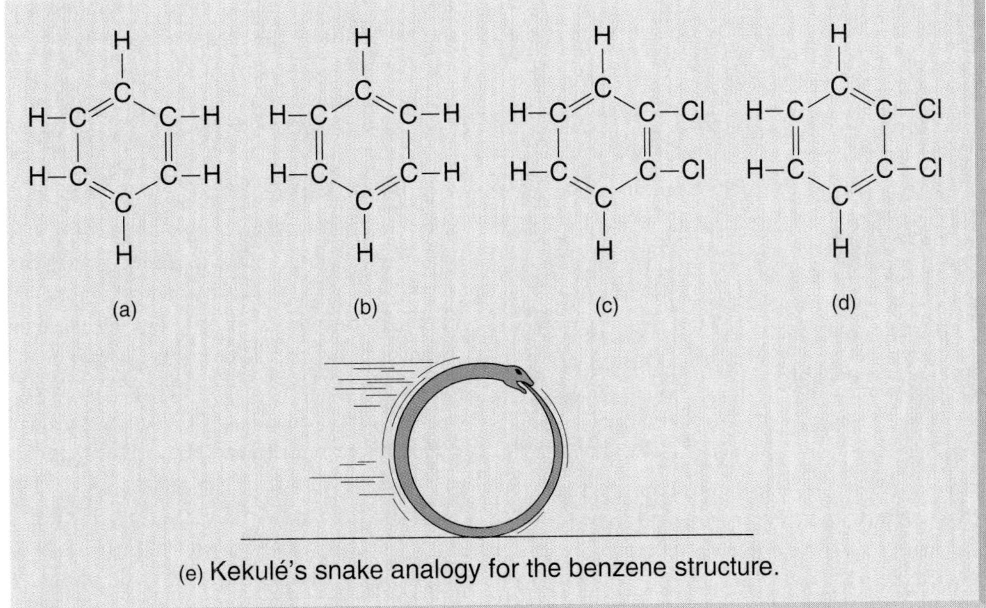

Figure 9.18 The Kekulé structures for the benzene molecule and two potentially isomeric 1,2-dichlorobenzenes. Structures **(a)** and **(b)** are identical, but rotated 60°. Structures **(c)** and **(d)** cannot be made identical by rotation and would be isomers if the bonds in the benzene structure were fixed. Kekulé's snake analogy **(e)** gave him the clue to benzene's cyclic structure.

synthesized, chemists realized that other molecules analogous to the dichlorobenzenes (figure 9.18c, d) should form isomers. This is because distinct compounds are expected if the Cl atoms are located on adjacent carbons of a C═C or a C—C bond; a similar effect produced cis- and trans-isomers in alkenes. We expect isomers because the one structure cannot be perfectly matched to the other by rotation. However, chemists found that the expected isomers did not exist. Why?

Eventually, chemists found that a property called **resonance** explained this lack of isomerization. According to resonance theory, the double and single bonds in the molecule exchange places so that a C═C becomes a C—C, and vice versa. No atoms move in this interchange; only the bonding electrons move. Figure 9.19 illustrates resonance for molecules *c* and *d* in Figure 9.18. As the double and single bonds exchange places by electron movement, structure *c* converts into structure *d*, and vice versa. Because the two structures continually interconvert, they can never be separated from one another; they instead produce a molecule with neither single nor double bonds, but six identical intermediate bonds between the carbon atoms. This means they are isomers in theory only. Since actual isomers have never been found, chemists are confident that true isomerization does not happen. The two structures in figure 9.19 are theoretical "parents" of a hybrid benzene ring that is intermediate in structure but impossible to portray adequately in a diagram.

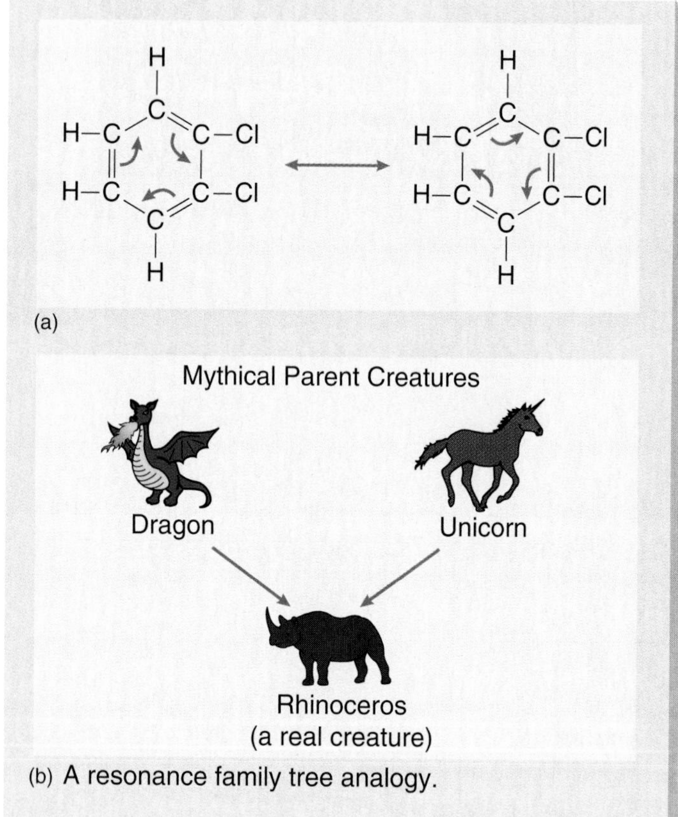

Figure 9.19 The resonance explanation for the lack of isomers in the 1,2-dichlorobenzene molecule. **(a)** The atoms are stationary, but the electrons flow freely around the ring, causing the double and single bonds to continually exchange positions. These two structures correspond to structures **(c)** and **(d)** in figure 9.18. **(b)** The "family tree" illustrates the resonance analogy; a known creature (the molecule with intermediate bonds) is created from a pair of mythical creatures (the depictions of "isomers" with single and double bonds in part **a**).

Several chemists have developed analogies to help explain resonance. American chemist J.D. Roberts depicts a medieval traveler in the heart of Africa who sees a charging rhinoceros which makes a lasting (and pointed!) impression. Upon his return to Europe, he wants to tell everyone in the royal court about this fantastic animal. He describes the rhinoceros as a cross (hybrid) between a scaly dragon and a horned unicorn, two mythical creatures whose forms were understood by all, even though they were never really observed. Like the dragon and unicorn in the analogy, the aromatic molecules depicted in figure 9.19 do not actually exist–their "offspring," the intermediate structure does.

Current views of aromatic structure are based on modern bonding concepts. When C—C, C═C, and C≡C bonds are compared, chemists find that the distance between C atoms becomes smaller as the number of bonds increases (see the table in box 9.1). Each multiple bond consists of a σ-bond and one or two π-bonds; although π-bonds are weaker than σ-bonds, they still pull the carbon nuclei closer together. The six C—C bonds of the benzene ring are intermediate in length between alkane (σ-) and alkene (σ- and π-) bonds, and all six C atoms lie in the same plane.

The σ-bond in an alkene forms from two sp^2-orbitals, and the π-bond forms from two p-orbitals that stick out of the plane of the molecule. These more accessible π-bonds account for both the greater reactivity and the lower strength of an alkene (see the frame analogy in box 9.1). In a typical alkene, such as 3-hexene, only two of the six carbon atoms (carbons 3 and 4) have unused p-orbitals and form a π-bond (figure 9.20a).

Like the alkenes, aromatic compounds have σ-bonds and π-bonds, but aromatics also have an important peculiarity; *every* carbon atom in a benzene molecule has a p-orbital that can form a π-bond (figure 9.20b). In 3-hexene, the C═C must form between carbons 3 and 4, but in the benzene molecule, all C atoms are part of a π-bonding arrangement extending around the entire ring. These atoms have 6 p-lobes extending above and 6 p-lobes extending below the carbon ring. The π-bond electrons circulate through these lobes, producing an electron cloud above and below the ring (figure 9.20b). Figure 9.20c shows how chemists now draw the benzene ring. The points on the hexagon represent the C atoms, and the inscribed circle represents the electron cloud and the shared bonds. All the bonds in benzene are identical and isomers do not occur.

Figure 9.21 shows the names and structures of some common aromatic compounds. These compounds form when other atoms or groups replace the hydrogen atoms attached to the benzene

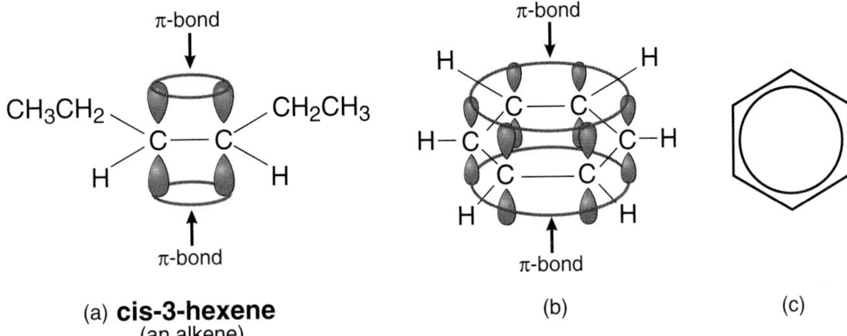

Figure 9.20 Double bond formation in *cis*-3-hexene **(a)** (an alkene) compared against two views of the modern benzene (aromatic) structure. View **(b)** shows the six C atoms with a p-orbital (and thus a π-bond) on each and shows the electron cloud formed from these overlapping p-orbitals with the resulting π-bond above and below the plane of the six C atoms. View **(c)** shows the same structure, viewed from above, with the C atoms represented by the points of the hexagon and the electron cloud by the inscribed circle.

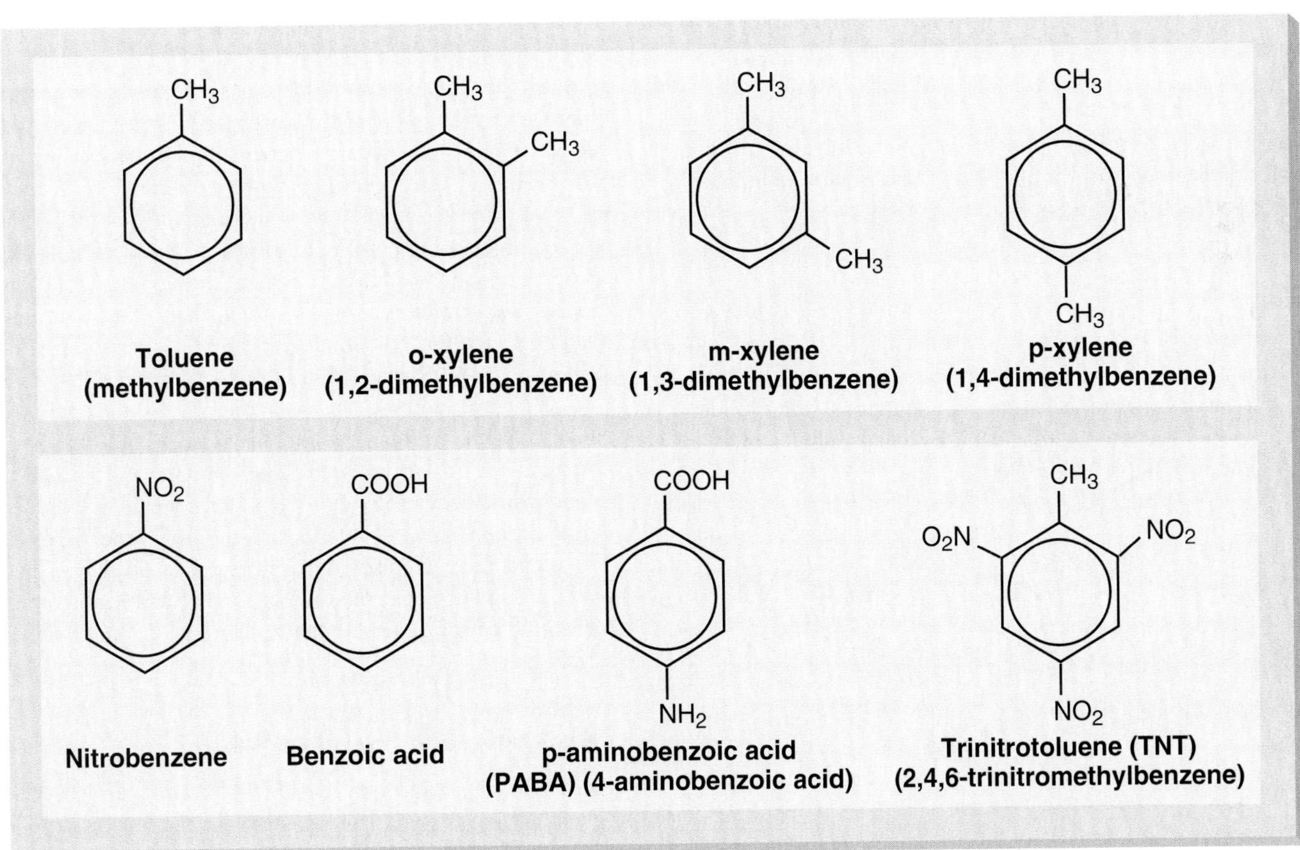

Figure 9.21 The names and structures of some common aromatic compounds. The name shown directly beneath the structure is the common name; the name in parentheses is the IUPAC version.

carbon ring. Ring substituents are assigned numbers or letters when more than one group is present. The letters *o*-, *m*-, and *p*-, representing ortho-, meta-, and para-, indicate substituents at locations 2-, 3-, or 4- on the carbon ring, in addition to position 1. If more than two substituents replace the hydrogens, numbers or common names are used (figure 9.21).

The aromatics in figure 9.21 have a variety of practical uses. Benzene, toluene, and the xylenes are used as solvents, high-octane aviation fuels, and as starting materials in the synthesis of insecticides, medicines, and plastics. Nitrobenzene, an excellent but toxic solvent, is used in polishes and is an important dye intermediate. Benzoic acid is used as a food preservative, and its derivatives are used in plasticizers, perfumes, flavoring agents, and medicines; 4-aminobenzoic acid (or PABA) is used in medicines, nutritional supplements, and ultraviolet sunscreens. Trinitrotoluene (TNT), or 2,4,6-trinitromethylbenzene, is a powerful explosive.

Alkenes react rapidly with bromine, but aromatics do not. This lack of reactivity served as a clue for mid-1800 chemists, telling them that the C═C bonds they presumed to be present in benzene were unusual. Eventually, they learned that aromatic hydrocarbons react by substitution instead of addition. When Br_2 is added to an alkene, it adds to form a dibromo-alkane. But when Br_2 reacts with benzene, one Br atom substitutes for an H atom in the final reaction product. The displaced H atom combines with the other Br to form a second reaction product, HBr:

$$C_6H_6 + Br_2 + \text{catalyst} \rightarrow C_6H_5\text{—}Br + HBr$$

This preservation of the aromatic ring—by substituting an atom or group in place of an H atom—sets the aromatic compounds apart from other hydrocarbons.

PRACTICE PROBLEM 9.8

Using Figure 9.21 as a guide, name the following aromatic compounds. Use both the number and letter systems to identify each. (Remember that each assigned number must be the lowest possible.)

```
a.        H                    b.        NO2
          |                               |
          C                               C
       //   \                          /    \\
  H—C        C—NO2              H—C          C—H
     ||      |                     ||        |
  H—C        C—NO2              H—C          C—H
       \   //                          \    //
          C                               C
          |                               |
          H                              NO2

c.        NO2
          |
          C
       //   \
  H—C        C—H
     |       ||
  H—C        C—NO2
       \\   /
          C
          |
          H
```

ANSWER All three compounds are dinitro- (two-nitro-) derivatives of benzene. Their specific names are: *a.* 1,2-dinitrobenzene, or o-dinitrobenzene; *b,* 1,4-dinitrobenzene, or p-dinitrobenzene; and *c.* 1,3-dinitrobenzene, or m-dinitrobenzene.

FUELS: HYDROCARBONS IN USE

The largest single use for hydrocarbons is as fuels; they release their latent chemical energy in oxidation reactions. The most common hydrocarbon fuels are coal, petroleum, and natural gas. We will compare these substances to wood (box 9.2).

Coal—There's No Fuel Like an Old Fuel

Coal, the most prevalent fossil fuel, is derived from plants. Good grades of coal contain small amounts (1–15 percent) of oxygen and small amounts of bound, but burnable, hydrogen (5–15 percent). Soft (bituminous) coal also has 2–3 percent sulfur; this produces SO_2, considered the chief culprit in acid rain, when bituminous coal is burned. Hard coal (anthracite) contains very little sulfur, but it is harder to mine and is less available for human use.

The need for clean, efficient, easy-to-use fuels has been the stimulus behind a search for ways to create synthetic fuels. Chemists have conducted research on a liquefaction process to convert coal into synfuel, a petroleumlike fuel for automobiles. However, coal is a poorer fuel than petroleum, and it takes extra energy to convert coal into a liquid fuel. Scientists have also worked on creating petroleum substitutes from organic materials, such as garbage, by adding hydrogen. Inevitably, synfuels will cost more than petroleum as long as inexpensive petroleum is readily available. Interest in synfuels rises and falls depending on factors such as economics, world conditions, and potential war.

Coal, along with its use as a fuel, is a source for chemicals called coal-tar derivatives. These substances are obtained by the destructive distillation (pyrolysis, chapter 3) of coal in coke ovens. One ton of coal yields 8.8 gallons of coal tar, which is converted into benzene, toluene, xylene, and other hydrocarbons. These chemicals are used in applications as diverse as medicines, plastics, solvents, and dyes. Coal tar is hydrogenated to form a fuel suitable for residential heating. Crude coal tar is used in road surfacing.

Petroleum—Too Useful to Waste

Petroleum is a high-quality, efficient, and versatile fuel. The petroleum industry has found valuable uses for almost every distillation fraction obtained from crude petroleum (figure 9.22). Table 9.4 lists the boiling ranges of the typical fractions along with the number of C atoms in each type of molecule and its usual physical state. The percentage of each fraction produced from a given volume of petroleum varies with the specific sample of

BOX 9.2

WOOD AS A FUEL

At one time, wood was the main heating and cooking fuel in most of the world. With the discovery of more efficient fuels, such as coal, the use of wood gradually diminished in the more highly developed regions of the world. Third World countries still widely use wood and other vegetation as a fuel, at least partially because of wood's ready availability. In more developed nations, as civilization encroached on the forest areas, wood gradually became less available and had to be transported from remote locations. Coal and similar fuels were easier to transport and often were less expensive.

Wood is not, however, an efficient fuel, nor is it a hydrocarbon. Approximately half of the dry chemical composition of wood is oxygen, as compared with 2 to 10 percent in coal. An efficient fuel must be in a highly reduced form, and wood can never meet that requirement since it is already partially oxidized.

In spite of its limitations, wood fuel has enjoyed a resurgence in popularity in the United States during the past several decades, largely due to the fact that it is a "natural," renewable resource and the supply in some areas is fairly plentiful. In addition, the newer wood-burning stoves are more efficient than the older types. Still, wood has many disadvantages as a fuel. Wood is a dirty fuel that produces soot, even in the newer stoves. Wood fires leave deposits in chimneys that can catch fire and burn down a home. Finally, using wood as a fuel requires cutting down trees, which many consider an ecological and esthetic problem. For these and other reasons, as scientists explore new fuels or develop more efficient ways to use the fuels we have, it seems unlikely that wood will ever rival the more efficient hydrocarbons as a major fuel.

Table 1 summarizes the composition of various fuels, their quality, and their ease of use. The fuel quality increases as you move downward in table 1, corresponding to lower percentages of oxygen in the fuel; only petroleum is a better fuel than anthracite coal. Alcohol fuels (methanol and ethanol) fare poorly by comparison because they contain high levels of oxygen. Oxygenated fuels are already partially oxidized; the better fuels are in a more reduced form. Liquid fuels are usually easier to use than solid, but they are hardly pollution free—alcohols produce toxic aldehyde pollutants on incomplete combustion, and petroleum produces several pollutants, including aldehydes. Virtually all carbon-containing fuels can produce soot.

TABLE 1 Fuel composition, quality, and ease of use.

FUEL MATERIAL	PERCENT CARBON	PERCENT HYDROGEN	PERCENT OXYGEN	FUEL QUALITY	EASE OF USE[1]
Plants (wood)	45	6	49	very poor, dirty	difficult
Methanol	37	13	50	poor, clean[2]	easy
Ethanol	52	13	35	poor, clean[2]	easy
Peat	60	5	33	poor, dirty	difficult
Lignite	69	4	24	fair, dirty[3]	difficult
Bituminous	84	4	9	good, dirty[3]	difficult
Anthracite	95	3	2	very good, dirty	difficult
Petroleum	84	16	0	best, clean[4]	easy
Natural gas	83	17	0	best, very clean[4]	easiest

Notes: [1]Ease of use for general purposes, including automotive.
[2]Produces aldehydes when partially burned.
[3]Contains sulfur and produces some SO_2 when burned.
[4]Can produce soot when burned incompletely, but otherwise clean. Can produce other pollutants by various chemical reactions.

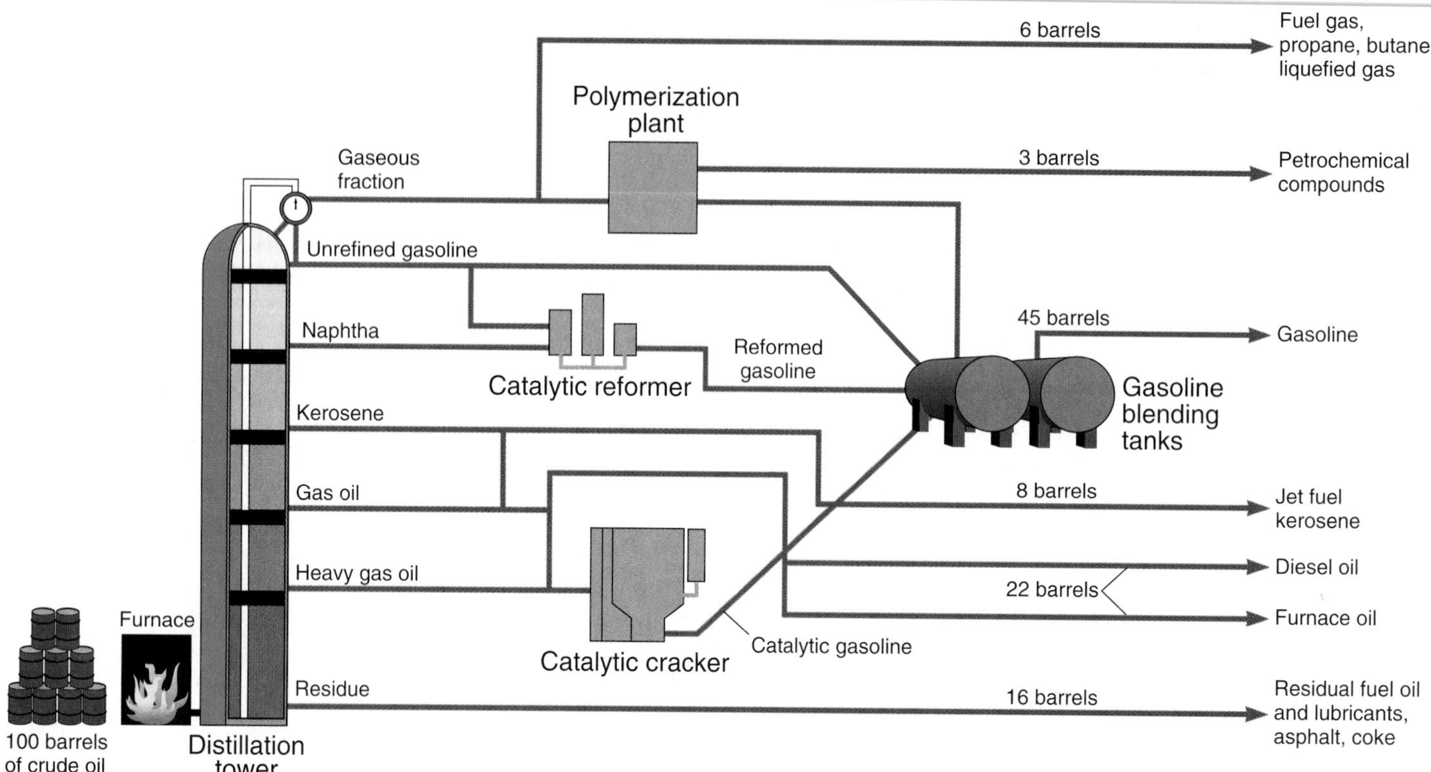

Figure 9.22 Refining crude petroleum in a distillation tower results in a wide variety of important products. A typical barrel of crude petroleum produces only a half barrel of gasoline.
SOURCE: From Eldon D. Enger and Bradley F. Smith, *Environmental Science: A Study of Interrelationships*, 4th ed. Copyright © 1992 Times Mirror Higher Education Group, Inc., Dubuque, Iowa. All Rights Reserved. Reprinted by permission.

crude petroleum. Petrochemicals form the basis for our modern chemical industry; petroleum is used to manufacture not only gasoline and motor oil (box 9.3), but also plastics, medicines, textiles, food, pesticides, and herbicides. When crude oil increases in price, the cost of almost everything else we buy goes up. Gasoline is only a small part of the picture.

TABLE 9.4 Properties of typical petroleum fractions.

FRACTION	NUMBER OF CARBON ATOMS	BOILING RANGE	STATE
Gas	1–4	<40	gas
Gasoline	5–12	40–200	liquid
Kerosene	10–16	180–280	liquid
Diesel	12–20	250–400	liquid
Fuel oil	15–20	250–400	liquid
Lubricating oil	16–24	300+	viscous liquid
Asphalt	20+	350+	semisolid to solid

Octane Numbers and Additives

Gasoline, a petroleum product, is a homogeneous hydrocarbon mixture consisting mainly of C_5—C_{12} alkanes, alkenes, and aromatics. However, not all hydrocarbons burn equally well—for example, branched alkanes burn more rapidly than n-alkanes. When combustion is poor, the engine of an automobile "knocks" and runs roughly.

Chemists devised the **octane number** to evaluate fuel quality. Octane numbers are based on the combustion of n-heptane, a poor-burning fuel assigned an octane number of 0, and 2,2,4-trimethylpentane (sometimes called "isooctane"), an efficient-burning fuel assigned an octane number of 100. To determine a gasoline's octane number, chemists measure the burning characteristics of the gasoline and compare it to the percentage of 2,2,4-trimethylpentane in a mixture of n-heptane and 2,2,4-trimethylpentane that burns at the same rate. Thus, the higher the octane number, the more efficiently the fuel burns. Table 9.5 lists the octane numbers of several hydrocarbons as well as regular and premium grade gasolines.

Certain chemicals and additives promote burning and reduce engine knocking. In 1921, Thomas Midgley, Jr. (American, 1889–1944) discovered that tetraethyl lead was an outstanding antiknock agent for low-grade fuels. This chemical was added to

BOX 9.3

MOTOR OILS

Motor oil functions as a lubricant by clinging to moving engine parts at all temperatures and in all seasons. Motor oil's ability to maintain contact with these parts is related to its **viscosity,** the property of liquids that describes their ability to resist flow. Motor oil viscosity is denoted on the label as 10W, 30W, and so on; the higher numbers denote a more viscous (thick and sticky) oil for use at warmer temperatures.

Different viscosity grades are prepared by blending various oil fractions. Normally, viscosity decreases with increasing temperatures, like the proverbial "molasses in January"; this is an inverse relationship. However, the most effective multigrade motor oils increase in viscosity as the temperature rises—a direct relationship.

Multigrade motor oils have two number designations, such as 10W30, to indicate the range of operating temperatures. The viscosity of a multigrade motor oil increases with increasing temperature in a direct relationship because it contains certain kinds of polymers. When cool, these polymers exist in ball-like coils (figure 1). As the temperature rises, these coils expand, making the oil blend more viscous.

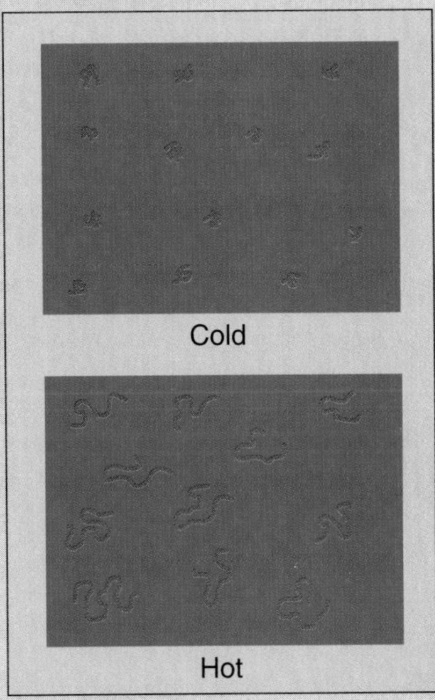

Figure 1 How a multigrade motor oil works. At cold temperatures, the polymers coil; at warmer temperatures they expand, making the oil more viscous.

"leaded" gasolines for decades because it promotes better combustion in poorer fuels. Eventually, lead's toxicity encouraged fuel producers to remove tetraethyl lead from gasolines; the driving force behind that decision was the worry that lead could "poison" the catalysts used in the catalytic converters installed on cars to reduce air pollution. (See chapter 8 to learn other reasons why lead removal was desirable.) Without lead, less expensive gasoline blends with low octane ratings did not burn efficiently. This problem was partially alleviated by engine redesign and partially by blending higher-octane components (such as benzene and toluene) into the fuel, but both approaches increased gasoline prices significantly. (Chapter 10 will discuss some of the oxygenated compounds added to increase the octane number.)

Other additives improve combustion by cleaning the engine of carbon deposits. Poor combustion causes carbon deposits to build up inside the automobile engine; chemicals are sometimes blended into gasoline to help prevent this. Most engine-cleaning products are hydrocarbons. For example, toluene is used in solvent systems to clean carburetors.

EXPERIENCE THIS

From all you learned about nomenclature, is *isooctane* a good name for 2,2,4-trimethylpentane? The next time you fill your car with gas, check the octane numbers at the various pumps. Then make the following calculations, using table 9.5.

a. Determine the percentages of n-heptane and 2,2,4-trimethylpentane needed to match each octane number.

b. Suppose the gasoline was made from an n-heptane and toluene mixture. What percentage of each would be needed to match the octane numbers? *Hint:* The two percentages should always add up to 100%. Assume the octane number of the gasoline you bought is 90. In *a*, you need a blend that is 90 percent 2,2,4-trimethylpentane and 10 percent n-heptane. However, toluene has an octane number of 118. To determine the percentages of toluene and n-heptane in *b*, divide the octane number at the pump (90) by toluene's number (118). Multiply the result by 100 to find the toluene percentage. Thus, in our example: $90 \div 118 = 0.7627 \times 100 = 76.27$ percent. The n-heptane value would therefore be 23.73 percent.

Beating the Seasons

Fuel needs vary according to the season. We use less gasoline in winter than in summer, but we use more fuel oil to heat homes and other buildings in winter. How can we most efficiently produce the fuels in the right proportions as needs and seasons change?

Although petroleum companies could store large portions of each fraction until they are needed, this is not an economical solution. A better approach is to convert one petroleum fraction into another. Two processes exist for this: catalytic cracking and catalytic reforming. Both these processes convert lower-octane fractions into higher-octane compounds. As we learned in chapter 7, a catalyst provides a new reaction pathway that makes a chemical reaction progress faster, more efficiently, or both. Without a catalyst, the energy requirements to drive the conversion reaction are too great.

Chemists developed **catalytic cracking** in the 1930s to split large molecules into smaller ones. Gasoline demand generally exceeds the demand for fuel oil. But if a typical fuel oil molecule with 16 C atoms is split in half, the new product falls into the gasoline range (table 9.4, figure 9.23). However, cracking does not yield just one product. It usually produces a mixture of an alkane and an alkene, with each pair containing a total of 16 C atoms and 34 H atoms. As figure 9.23 shows, several splits are possible; all of those shown are in gasoline range (5 to 12 carbon atoms). The catalytic cracking products have higher octane ratings as well.

Catalytic reforming is another process that converts one hydrocarbon to another, often a lower-octane compound into a higher-octane one. As table 9.5 shows, branched or cyclic hydrocarbons have higher octane numbers than linear ones. A good rule of thumb is that the more highly branched a hydrocarbon, the higher its octane number. Figure 9.24 diagrams three examples of catalytic reforming. Low-octane-number n-hexane can be isomerized into higher-octane-number branched 2-methylpentane (figure 9.24a). The same starting material can also be converted into cyclohexane (figure 9.24b). The process can go still further; the equation in figure 9.24c shows how cyclohexane, the product in part b, can be converted into the aromatic benzene. With each step, the hydrocarbon gains a higher octane number and thus becomes a more efficient fuel.

Combining hydrocarbon molecules is also a possible conversion method. For example, 2-methylpropene (commonly called isobutylene) combines with another molecule of itself to form an alkene, which in turn is hydrogenated to produce 2,2,4-trimethylpentane (isooctane; octane number = 100).

$$2\,CH_2{=}C(CH_3){-}CH_3 \xrightarrow{\text{catalyst}} CH_3C(CH_3)_2CH{=}C(CH_3)CH_3$$

2-Methylpropene → Alkene

$$\xrightarrow[\text{catalyst}]{H_2} CH_3C(CH_3)_2CH_2CH(CH_3)CH_3$$

2,2,4-Trimethylpentane

TABLE 9.5 The octane numbers of various hydrocarbons. The octane numbers for regular and premium grade gasoline are included for comparison.

COMPOUND	OCTANE NUMBER
n-octane	−20
n-heptane	0
2-methylheptane	24
n-hexane	25
n-pentane	62
2-methylpentane	71
Regular grade gasoline	87
2-methylbutane	90
1-pentane	91
Premium grade gasoline	93
2,2,4-trimethylpentane ("isooctane")	100
benzene	105
o-xylene	107
p-xylene	116
2,2,3-trimethylpentane	116
toluene	118
cyclopentane	122

$$CH_3CH_2CH_2CH_2CH_2CH_2CH_2CH_2CH_2CH_2CH_2CH_2CH_2CH_2CH_2CH_3$$

$$\xrightarrow{\text{catalyst, heat, pressure}}$$

$$\underset{\textbf{n-Octane}}{CH_3CH_2CH_2CH_2CH_2CH_2CH_2CH_3} + \underset{\textbf{1-Octene}}{CH_2{=}CHCH_2CH_2CH_2CH_2CH_2CH_3}$$

$$\underset{\textbf{n-Nonane}}{CH_3CH_2CH_2CH_2CH_2CH_2CH_2CH_2CH_3} + \underset{\textbf{1-Heptene}}{CH_2{=}CHCH_2CH_2CH_2CH_2CH_3}$$

$$\underset{\textbf{n-Heptane}}{CH_3CH_2CH_2CH_2CH_2CH_2CH_3} + \underset{\textbf{1-Nonene}}{CH_2{=}CHCH_2CH_2CH_2CH_2CH_2CH_2CH_3}$$

$$\underset{\textbf{n-Decane}}{CH_3CH_2CH_2CH_2CH_2CH_2CH_2CH_2CH_2CH_3} + \underset{\textbf{1-Hexene}}{CH_2{=}CHCH_2CH_2CH_2CH_3}$$

$$\underset{\textbf{n-Hexane}}{CH_3CH_2CH_2CH_2CH_2CH_3} + \underset{\textbf{1-Decene}}{CH_2{=}CHCH_2CH_2CH_2CH_2CH_2CH_2CH_2CH_3}$$

+ Other hydrocarbons

Figure 9.23 Catalytic cracking splits a large molecule into smaller ones. In this example, a fuel oil (an *n*-alkane with 16 carbon atoms) can split into any of several products, all consisting of alkane-alkene mixtures.

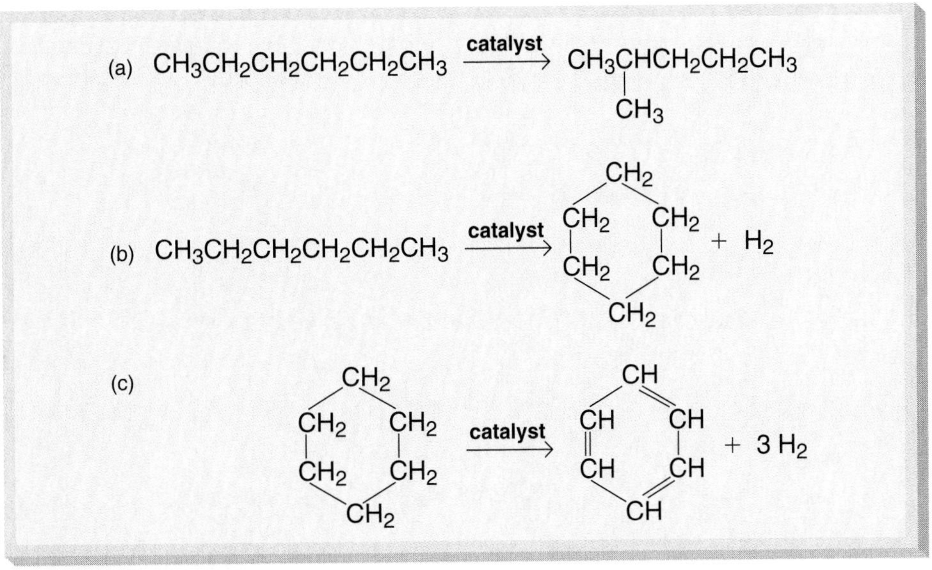

(a) $CH_3CH_2CH_2CH_2CH_2CH_3 \xrightarrow{\text{catalyst}} CH_3\underset{\displaystyle CH_3}{\underset{|}{C}}HCH_2CH_2CH_3$

(b) $CH_3CH_2CH_2CH_2CH_2CH_3 \xrightarrow{\text{catalyst}}$ cyclo-$(CH_2)_6$ (ring of six CH_2 groups) $+ H_2$

(c) cyclo-$(CH_2)_6$ (ring of six CH_2 groups) $\xrightarrow{\text{catalyst}}$ ring of six CH groups with alternating double bonds $+ 3\,H_2$

Figure 9.24 Catalytic reforming. **(a)** Low-octane n-hexane converts to higher-octane 2-methylpentane. **(b)** The same reactant, *n*-hexane, converts to cyclohexane. **(c)** Cyclohexane converts to benzene.

ADDITION POLYMERIZATION

As we have already mentioned, hydrocarbons have a myriad of commercial uses. The largest industrial usage of alkenes is in the manufacture of plastics, textiles, rubbery substances, and related materials that are lumped into a class called polymers. **Polymers** are large organic molecules consisting of repeating smaller units; many polymer-forming reactions are addition reactions. Polymers are everywhere: We cook in plastic-lined vessels, eat from plastic dishes, drive in cars containing sizable quantities of plastics and textiles, and ride on rubber tires.

Chemists use two terms to describe this group of molecules—*macromolecule* and *polymer*. The prefix *macro* means large or long, so a macromolecule is simply a large molecule. Hermann Staudinger (German, 1881–1965), often called the "Father of Polymers," coined the term *macromolecule*. Staudinger (figure 9.25) began his polymer research in 1926 and continued until his death; he was awarded the Nobel Prize in 1953.

Figure 9.25 Hermann Staudinger, the founder of polymer chemistry, won the Nobel Prize in chemistry in 1953.

The word *polymer* comes from two Greek words: *poly*, meaning many, and *meros*, meaning unit; a polymer is therefore a substance composed of many units. The chemical units that form polymers are called **monomers.** The term comes from *mono*, meaning one or single, and *meros*.

Polymerization reactions may be either addition or condensation reactions. Only **addition polymerization** is considered in this chapter; we will examine condensation reactions in chapters 10 and 11.

The most common synthetic polymer is poly(ethylene), or PE, which consists of a long chain of $—CH_2—$ units. Figure 9.26a shows a 15-unit segment with a molecular weight of 210. The entire chain for a typical poly(ethylene) molecule with a molecular weight of 10^6 has 71,428 $—CH_2—$ units. It would require 4,762 lines or 88 pages (at 54 lines per page) to print out the condensed chemical formula for this molecule.

Because polymers can be such large molecules, most chemists use a shorthand notation called the repeat unit to represent a polymer. A **repeat unit** is a collection of atoms regularly occurring in the polymer chain. This repeat unit is enclosed in brackets (figure 9.26b) and marked with a subscript *n*, where *n* is the average number of repeat units in the chain. (A typical polymer sample contains many individual polymer chains of different lengths.) The repeat unit concept is illustrated for poly(ethylene) in figure 9.26c. An *n* value of 35,714 describes a polymer with a molecular weight of 10^6 and more than 71,000 C atoms in its chain.

The Polymerization of Ethylene

In an addition polymerization reaction, a collection of C=C double bonds opens up and reassembles into a long chain of repeat units from the starting monomers. Figure 9.27 shows such a reaction for ten ethylene monomers ($CH_2=CH_2$) forming a 10-

(a) $—CH_2—CH_2—CH_2—CH_2—CH_2—CH_2—CH_2—CH_2—CH_2—CH_2—CH_2—CH_2—CH_2—CH_2—CH_2—$

(b) $\left[\text{Repeat unit}\right]_n$

(c) $\left[CH_2—CH_2\right]_n$

Figure 9.26 Polymers. **(a)** A 15-unit portion of a poly(ethylene) polymer. **(b)** The general notation for a repeat unit. **(c)** The notation for the repeat unit of a poly(ethylene) chain.

$$CH_2{=}CH_2\ \ CH_2{=}CH_2\ \ CH_2{=}CH_2\ \ CH_2{=}CH_2\ \ CH_2{=}CH_2\ \ CH_2{=}CH_2\ \ CH_2{=}CH_2\ \ CH_2{=}CH_2\ \ CH_2{=}CH_2\ \ CH_2{=}CH_2$$
$$\downarrow$$
$$-CH_2-CH_2-CH_2-CH_2-CH_2-CH_2-CH_2-CH_2-CH_2-CH_2-CH_2-CH_2-CH_2-CH_2-CH_2-CH_2-CH_2-CH_2-CH_2-CH_2-$$

Figure 9.27 Ten ethylene monomers ($CH_2{=}CH_2$) join in an addition reaction to form a 10-unit polymer. The repeat unit for this chain is $[CH_2{-}CH_2]_{10}$.

unit portion of a **poly(ethylene)** or **PE** chain. Imagine ten people standing with their hands folded across their chests—this corresponds to a double bond. Now suppose they unfold their arms, extend their hands, and join together with their neighbors to form a 10-person human chain. This is similar to the monomers bonding in a polymerization reaction.

The equations in figure 9.28 show ethylene and vinyl chloride monomers combining through addition polymerizations to form poly(ethylene) (PE) and poly(vinyl chloride) (PVC), respectively. The formulas for the monomer and the repeat unit are identical in addition polymerization. Thus, ethylene and the PE repeat unit both have the formula C_2H_4, although their structural formulas differ. Addition polymerization requires the formation of a high-energy species to make the reaction occur; this is usually an atom or group of atoms with an unpaired electron (that is, a free radical). Not all C=C double bonds polymerize well; this reaction is essentially limited to 1-alkenes and related molecules.

Two major versions of poly(ethylene) are manufactured commercially—low-density poly(ethylene), or LDPE, and high-density poly(ethylene), or HDPE. These two variations are plastic recycling types 1 and 2, respectively. LDPE is prepared by radical polymerization (figure 9.28a), while HDPE is prepared using a special catalytic surface. More poly(ethylene) is manufactured than any other polymeric material; poly(vinyl chloride) comes in second. HDPE consists mainly of linear polymer chains, while LDPE is mainly branched chains (figure 9.29); this gives LDPE its lower density. Linear chains have no irregular side branches. A branched crosslinked polymer has many chains linked together by chemical bonds to form one extremely large molecule. Crosslinked poly(ethylene) is used to manufacture the caps on plastic bottles.

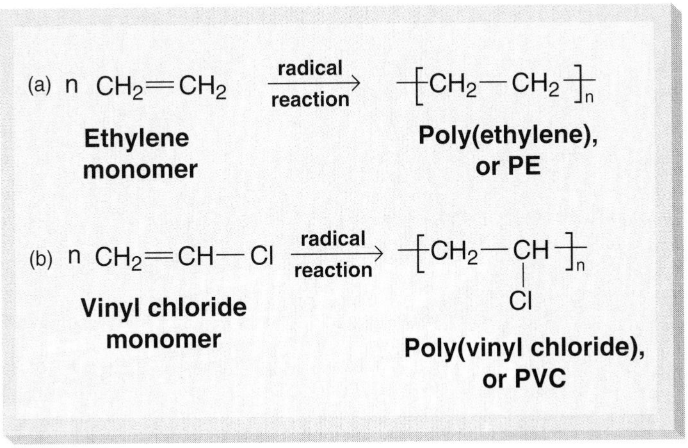

(a) $n\ CH_2{=}CH_2 \xrightarrow[\text{reaction}]{\text{radical}} {-}\!\!\left[CH_2{-}CH_2\right]\!\!{-}_n$

Ethylene monomer — Poly(ethylene), or PE

(b) $n\ CH_2{=}CH{-}Cl \xrightarrow[\text{reaction}]{\text{radical}} {-}\!\!\left[CH_2{-}\underset{\displaystyle Cl}{CH}\right]\!\!{-}_n$

Vinyl chloride monomer — Poly(vinyl chloride), or PVC

Figure 9.28 Addition polymerization reactions producing **(a)** poly(ethylene), or PE, and **(b)** poly(vinyl chloride), or PVC.

PRACTICE PROBLEM 9.9

Many applications exist for various grades of poly(ethylene); most depend on the molecular weight.

a. A grade of poly(ethylene) with a molecular weight of about 2,000 amu has waxlike properties and is used in floor polishes. How many repeat units are in this polymer?

b. Another PE grade is used to make plastic toys. It has an average of 4,800 repeat units. What is its molecular weight?

ANSWER *a*. To find the answer, divide the polymer's molecular weight by the molecular weight of the repeat unit. A PE repeat unit has a molecular weight of 28 ($CH_2 = 14$; $CH_2 - CH_2 = 28$). Thus, $2{,}000 \div 28 = 71.4 = n$. This chain has about 71 units. *b*. To find the molecular weight, multiply 4,800 repeat units by the molecular weight for each repeat unit; $4{,}800 \times 28 = 134{,}400$. This particular PE grade would not work well as a floor polish because it would not flow properly. Nor would the polymer from part *a* make quality plastic toys since it would lack strength.

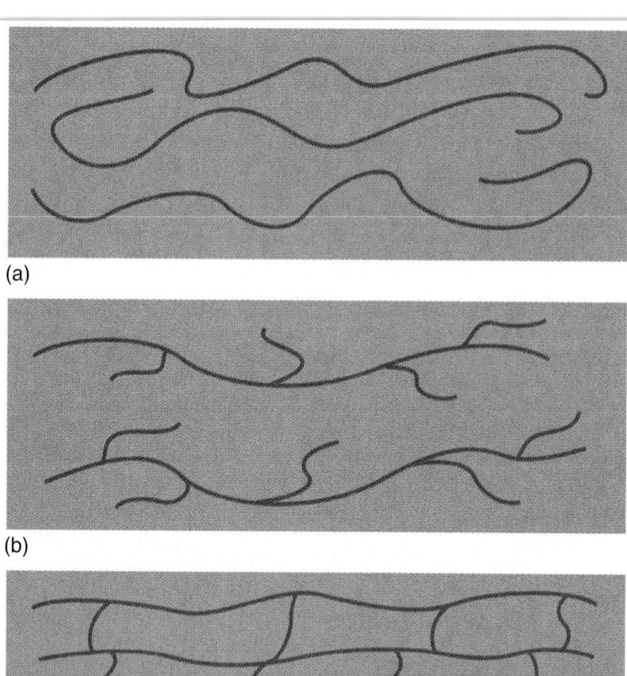

Figure 9.29 Polymer chain structures. **(a)** Three linear polymer chains. **(b)** Two branched polymer chains. **(c)** One crosslinked polymer chain.

DIENES AND RUBBER

Some molecules contain more than one C=C double bond; those with two C=C bonds are called **dienes.** Two important examples of dienes, butadiene and isoprene (figure 9.30), form rubbery polymers. **Elastomers** are polymers with elastic or rubber-like properties (they can stretch and recover). Until comparatively recently, the only known elastomer was natural rubber, known since ancient times. Then chemists developed the first synthetic elastomers in 1937. Usually elastomers are lightly crosslinked in a process called **vulcanization** that improves their strength and elasticity.

$CH_2{=}CH{-}CH{=}CH_2$

Butadiene

$$CH_2{=}\underset{|}{\overset{CH_3}{C}}{-}CH{=}CH_2$$

Isoprene

Figure 9-30 Two examples of dienes.

Natural Rubber

Natural rubber was known to the American Indians, who made balls from the sap of certain trees and even played ball games before Columbus arrived in the New World. The sap is called rubber latex, and it is a water dispersion of solid rubber in assorted proteins and fats. Rubber was originally called by the Indian name *caoutchouc*. Joseph Priestly (English, 1733–1804), one of the chemists who discovered oxygen, coined the name *rubber* when he noted it could rub lead markings off a piece of paper.

Rubber latex is produced by more than 500 plant species, including common milkweed, but the major source is the *Hevea brasiliensis* tree that grows wild in South America and is cultivated in South Asia. Figure 9.31 shows the latex draining from a *Hevea* tree into a bucket. This latex is precipitated by acid treatment to produce slabs of crude rubber that are dried and then carried to a nearby town or village. The crude rubber is then shipped to a rubber mill. Although elastic, crude rubber is quite sticky and difficult to use. Columbus introduced rubber to Spain in the late 1400s, but even 200 years later, there was no real market for this tacky substance. It wasn't until 1839 that Charles Goodyear made a discovery that eventually made rubber a commercially viable product (box 9.4).

Vulcanization

In 1839, Charles Goodyear discovered the vulcanization process. The chemistry underlying the process is shown in figure 9.32; Thomas Hancock (English) named the procedure *vulcanization*

Figure 9-31 Latex flows from the *Hevea* tree into a collection vessel in Malaysia.

BOX 9.4

THE CREATION OF THE RUBBER INDUSTRY

Although rubber's water-resistant qualities have been well known since antiquity, it wasn't until 1823 that Charles Macintosh (Scotland, 1766–1843) invented the waterproof fabric that bears his name in the British raincoat by sandwiching sticky natural rubber between layers of fabric. Unfortunately, natural rubber becomes brittle when cold and sticky when hot—not exactly a convenient combination of properties. The large-scale utilization of rubber did not occur until after Goodyear created the vulcanization process in 1839. Vulcanization eliminated both brittleness and tackiness.

Charles Goodyear (American, 1800–1860; figure 1), believed he could "cure" rubber as tanning cures leather. (Tanning crosslinks the polymers in animal skins and prevents rapid degradation of the dead tissues.) Goodyear experimented by combining rubber with many other chemicals, but success came by accident. Goodyear mixed rubber and sulfur together and accidentally left part of the sample in a metal pan on a hot stove while he went to lunch. When he returned, he discovered the sample had melted and congealed into a mass that was stronger, more elastic, and less tacky than the original rubber. Goodyear put the sample outside overnight and discovered that the rubber retained these properties even in the cold. Goodyear's process was eventually named vulcanization after Vulcan, the Roman god of fire. His laboratory accident eventually enabled the development of a major industry.

Figure 1 Charles Goodyear discovered the vulcanization process, which eventually led to today's rubber industry, in 1839.

Charles Goodyear ran out of money before he could develop vulcanization into a commercial process; he had borrowed heavily and even went to prison for his debts. He died in 1860 leaving about $200,000 in unpaid bills as a legacy for his family. In 1851, Nelson Goodyear converted rubber into hard plastic by using large amounts (about 15 percent) sulfur. He called this plastic Vulcanite and it proved successful in the marketplace.

About 1870, Dr. Benjamin Franklin Goodrich, a successful surgeon-turned-entrepreneur, founded the company that still bears his name. He began producing rubber fire hoses in his Akron, Ohio plant using vulcanization to improve the elasticity. At that time, fire hoses were made from highly rupture-prone leather. Goodrich watched a close friend's home burn to the ground before his rubber fire hoses were marketed. Garden hoses, gaskets, and bottle stoppers followed, and much later, rubber tires arrived on the scene.

The first rubber tires (for bicycles) were solid rubber. Although cushion or balloon tires (air-filled tubes at low pressures) were marketed earlier, pneumatic tires (tires filled with compressed air) were invented in 1888 by John Boyd Dunlop (Scottish, 1840–1921). The Dunlop Rubber Company also produced the first foam rubber in 1929. Almost all of these developments occurred before anyone understood what a polymer was or even understood most of the chemistry you have learned in this book. In spite of this lack of knowledge, the rubber industry grew into a multibillion dollar business.

Eventually, chemists learned that the structure of natural *(Hevea)* rubber was cis-poly(isoprene) (figure 2); the related trans-poly(isoprene) is harvested from balata or Gutta percha trees. These trans-poly(isopropene) analogs are rigid polymers with essentially no elasticity. Gutta percha shows excellent sound-conducting qualities and is used to make stethoscopes; it is also the core material in golf balls. In the mid-1950s, chemists learned to polymerize isoprene monomers into a synthetic polymer virtually identical to natural rubber. Today, rubber is used in hundreds of ways in our everyday lives. But without Goodyear's accidental discovery, rubber might never have become a commercially feasible product.

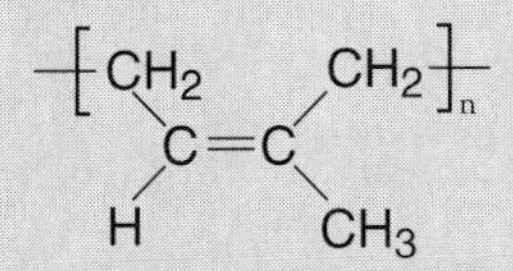

$$\left[CH_2(H)C=C(CH_3)CH_2 \right]_n$$

Natural rubber or cis-Poly(isoprene)

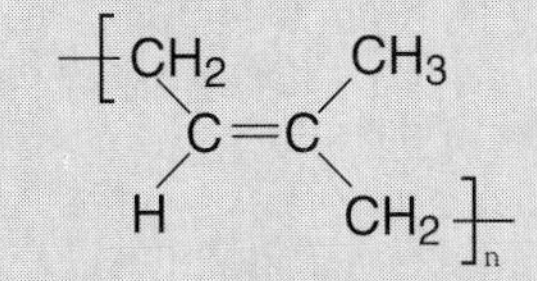

$$\left[CH_2(H)C=C(CH_3)CH_2 \right]_n$$

Gutta percha or trans-Poly(isoprene)

Figure 2 The chemical structures of natural rubber and gutta percha.

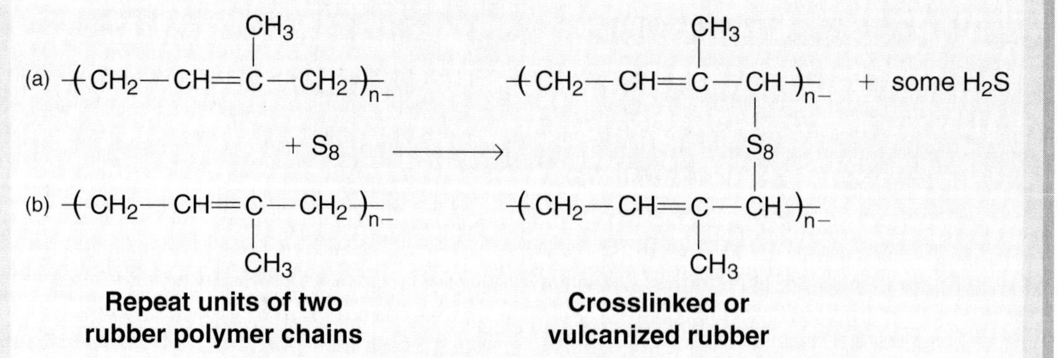

$$(a)\ \text{-}(CH_2\text{—}CH\text{=}C(CH_3)\text{—}CH_2)_{n}\text{-} + (b)\ \text{-}(CH_2\text{—}CH\text{=}C(CH_3)\text{—}CH_2)_{n}\text{-} + S_8 \longrightarrow \text{-}(CH_2\text{—}CH\text{=}C(CH_3)\text{—}CH)_{n}\text{-}\text{—}S_8\text{—}\text{-}(CH_2\text{—}CH\text{=}C(CH_3)\text{—}CH)_{n}\text{-} + \text{some } H_2S$$

Repeat units of two rubber polymer chains — **Crosslinked or vulcanized rubber**

Figure 9-32 Goodyear's vulcanization process.

after Vulcan, the mythical Roman god of fire. Rubber vulcanization normally uses sulfur, although sulfur compounds and some metal oxides are also used. During vulcanization, the individual rubber polymer chains connect through sulfur bridges, converting the substance into a crosslinked structure. This makes the rubber stronger, more elastic, and less sticky.

EXPERIENCE THIS

Hard rubber objects are made from rubber vulcanized with high levels of sulfur (15 percent). Window-shop in some stores and locate some examples of hard rubber. If you have any samples around the house, see if they will dissolve in gasoline. (*Caution:* Gasoline is very flammable!) If possible, make a photographic collection of these objects. Possible examples include: combs, rubber stoppers, balls, and some automobile distributor caps.

SUMMARY

Organic chemistry is the chemistry of carbon compounds. Carbon is unique among the elements. Because they have four valence electrons and can form covalent bonds, carbon atoms combine in an endless variety of chains and rings, a property called catenation. In addition, these combined or catenated molecules are extremely stable. These two properties—catenation and stability—give carbon a unique role in the complex molecules essential to life.

Because carbon can form so many different types of molecules, scientists devised the IUPAC nomenclature, or naming system, for organic molecules. Hydrocarbons, for example, are divided into four classes: alkanes, alkenes, alkynes, and aromatics. Alkanes fit the class formula C_nH_{2n+2}, where *n* is the number of C atoms in the molecule. Some—the normal or n-alkanes—are structured with all their carbon atoms in a continuous chain; the carbon atom in other alkanes may be arranged in branched chains or rings. Most alkanes have several structural isomers, or molecules with the same chemical formula but different structures. Alkanes provide root names for most hydrocarbons and other organic molecules.

Alkenes are hydrocarbons that contain a C=C double bond in their continuous carbon chain. Their class formula is C_nH_{2n}. The suffix *ene* indicates an alkene. Alkenes sometimes have geometrical isomers—molecules with the same formula and structure, but with individual atoms or groups of atoms arranged in different locations at the double bond.

Alkynes are hydrocarbons with at least one C≡C triple bond. Their class formula is C_nH_{n-2}. Both alkenes and alkynes undergo addition reactions.

Aromatic hydrocarbons, such as benzene (C_6H_6), are unsaturated compounds that have unusual properties compared to alkenes. These complex molecules undergo substitution rather than addition reactions, preserving the benzene ring. This makes aromatics unreactive and sets them apart from other hydrocarbons.

The largest single use for hydrocarbons is as fuels; they release their energy in oxidation reactions. Coal, petroleum, and natural gas are commonly used, efficient hydrocarbon fuels. Octane numbers denote automotive fuel quality. Polymers, including rubber, are another major use for hydrocarbons.

KEY TERMS

addition polymerization 234
alkanes 211
alkenes 218
alkyl groups 214
alkynes 220
aromatic hydrocarbons 225
catenation 210
catalytic cracking 232
catalytic reforming 232
class formula 212
cycloalkanes 216
dienes 236
elastomers 236
geometrical isomers 219
hybrid bond 222
hybridization 222
hybrid orbital 222
IUPAC system 211
monomer 234
nomenclature 211
octane number 230
pi (π) bond 218
poly(ethylene) or PE 235
polymer 234
repeat unit 234
resonance 226
saturated hydrocarbon 211
sigma (σ) bond 218
structural isomers 213
viscosity 231
vulcanization 236

READINGS OF INTEREST

General

Organic chemistry and hydrocarbons make the news often. Watch for reports in your newspapers or on television. Look for articles on new fuels or fuel additives, new hydrocarbon plastics, or new uses for older plastics. Watch for reports on how these plastics, as well as petroleum products such as motor oil, can be recycled.

Borman, Stu. "Nineteenth-century chemist Kekulé charged with scientific misconduct." *Chemical & Engineering News*, August 23, 1993, pp. 20–21. The "misconduct" charge suggests that Kekulé's snake dream was a fabrication, but not everyone agrees with that charge.

Breedlove, C.H. "Some illuminating chemistry provided by a camping lantern." *J. Chem. Ed.*, vol. 69, no. 8, August 1992, p. 621. Did you ever wonder about the chemistry that powers a Coleman lantern? If you enjoy camping you will want to read this one!

Cassidy, "David C. Heisenberg, uncertainty and the quantum revolution." *Scientific American*, May 1992, pp. 102–112. Hybridization stems from quantum theory, and Heisenberg was one of the scientists who created this theory. This article provides more of a personal perspective on Heisenberg than a discussion of the theory.

Aihara, Jun-ichi. "Why aromatic compounds are stable." *Scientific American*. March 1992, pp. 62–68. We only ventured slightly into aromatic hydrocarbons. This article discusses more compounds, including some that occur in tobacco smoke and can cause cancer. It even has a photograph of actual benzene molecules (via a scanning microscope).

Merry, Barbara A., and Martinez, Ben. "Rope." *Invention & Technology*, Fall 1991, pp. 38–45. Today, most of our ropes are made from synthetic polymers such as polyolefins or nylon. Rope made from natural materials also consist of polymers, although not hydrocarbons. This article tells how this industry developed and about the techniques still used to make rope.

Smoking

Himelstein, Linda, and Melcher, Richard A. "Did big tobacco's barrister set up a smokescreen?" *Business Week*, September 5, 1994, pp. 68–70. In other words, did the lawyers cover up data that might restrict the use of tobacco?

Brownlee, Shannon. "The smoke next door." *U.S. News & World Report*, June 20, 1994, pp. 66–68. This article discusses secondhand smoke.

Brownlee, Shannon; Roberts, Steven V.; Cooper, Matthew; Goode, Erica; Hetter, Katia; Wright, Andrea; Watson, Traci; and Grant Linda. "Should cigarettes be outlawed?"; "Teens on tobacco"; and "Smoke, flame, and fire" *U.S. News & World Report*, April 18, 1994, pp. 32–47. This is a trio of articles on the topic of tobacco use viewed from several perspectives.

Carey, Benedict. "Saying goodbye to an old flame." *Health*, May/June 1992, pp. 80–85. This old flame is smoking. Many of the problems coming from smoking are due to materials called polynuclear hydrocarbons, which you will examine more closely in chapter 17. This article will give you some useful background information on smoking and its problems.

Fuels

Carpenter, Betsy. "Fighting fire with fire." *U.S. News & World Report*, October 10, 1994, pp. 93–96. What is the best way to control forest fires?

Linden, Eugene. "Chain saws invade Eden." *Time*, August 29, 1994, pp. 58–59. Eden in this case is sparsely populated Guyana in South America, where large amounts of the forest have been cut down for lumber and fuel. How much should be conserved?

Kaufman, Eric N., and Scott, Charles D. "Liquify coal with enzyme catalysts." *Chemtech*, April 1994, pp. 27–32. In many ways, coal would be more useful if it were converted into a liquid fuel. This article discusses a new method that might possibly accomplish this.

Svitil, Kathy. "From brown to black." *Discover*, April 1994, p. 24. A simple outline of how coal forms.

Mullin, Rich; Roberts, Michael; and Morris, Gregory D. L. "U.S. industry confident that improvements signal a trend"; "Continuing low oil prices fuel high output and consumption"; and "Producers plan for global competitiveness." *Chemical Week*, March 1994, pp. 26–36. A trio of articles on the use of petroleum as a fuel and as a source of valuable chemicals.

Shepherd, Mark A. "Searching for oil at sea." *Chemical Computing & Automation*, January 1994, pp. 21–22. Computers play a big role in this search.

Reese, Ken. "Crackers: How refiners learned to wring more gas from crude." *Today's Chemist at Work*, November/December 1993, pp. 59–60. An easy-to-read account about how catalytic cracking was developed.

Stodolsky, Frank, and Santini, Danillo J. "Fueling up with natural gas" *Chemtech*, October 1993, pp. 54–59. What are the best ways to use natural gas as a fuel?

Warner, Barry G. Peat: "Nature compost." *Earth*, March 1992, pp. 44–49. Peat is one of the precursors to coal. While peat is a low-grade fuel, it has many other important uses in agriculture and gardening.

Vartanian, Paul F. "The chemistry of modern petroleum product additives." *J. Chem. Ed.*, vol. 68, no. 12, December 1991, pp. 1015–1020. Petroleum is much more than just gasoline, motor oil, and grease. Most petroleum-based products use an assortment of additives. This readable article tell this intriguing story.

Lucier, Paul. "Petroluem: What is it good for?" *Invention & Technology*, Fall 1991, pp. 56–63. With so many environmental groups asking this very question, this article deserves your attention.

Hapgood, Fred. "The quest for oil" *National Geographic*, vol. 176, no. 2, August 1989, pp. 226–259. Our society runs on petroleum, and this is unlikely to change in the near future. This article tells how oil is found.

Rubber

Pennisi, Elizabeth. "Rubber to the road." *Science News*, vol. 141, March 7, 1992, p. 155. Every year thousands of old tires are discarded. What can we do with them? This brief article tells about recycling them in road asphalt.

Prioli, Carmine. Rubber dentures for the masses. *Invention & Technology*, Fall 1991, pp. 28–37. Among other things, this article tells what made George Washington's smile so appealing.

Theiler, Carl R., 1962. *Men and molecules*. New York: Dodd, Mead. Chapter 8, The rise of man-made rubber. A historical perspective on synthetic rubber that is worth reading.

PROBLEMS AND QUESTIONS

1. What contributions did each of the following people make to chemistry?

- **a.** Kekulé
- **b.** Wöhler
- **c.** Goodyear
- **d.** Goodrich

2. Define the following terms, or give an example of each.

- **a.** addition reaction
- **b.** substitution reaction
- **c.** structural isomerization
- **d.** geometrical isomerization
- **e.** addition polymerization
- **f.** vulcanization
- **g.** resonance
- **h.** hybridization

3. What two factors help carbon form uniquely stable covalent bonds?

4. How do the melting and boiling points of n-alkanes change with the number of carbon atoms present?

5. Identify the class (alkane, alkene, alkyne, or aromatic hydrocarbon) each of the following compounds is in. Name each compound according to the IUPAC system.

a. $CH_3CH_2CH_2CH_2CH_2CH_2CH_3$

b. $CH_3CHCH_2CH_2CH_2CH_3$ (with CH_3 attached to the second carbon)

c. $CH_3CH_2CH_2CH{=}CH_2$

d. $CH_3{-}C{\equiv}C{-}CH_3$

e. $CH_3CH_2(H)C{=}C(H)CH_2CH_3$ (the CH_3CH_2 groups on opposite sides: upper left and upper right; H lower left and lower right)

f. $CH_3{-}CH{-}CH_2{-}CH{=}CH_2$ (with CH_3 attached to the second carbon)

g. $CH_3(CH_2)_3CH(CH_3)_2$

h. $C_6H_5CH_3$

6. Draw the structures for each of the following compounds:
   a. 2,2,4-trimethylpentane
   b. 3,4-dimethylhexane
   c. *cis*-3-heptene
   d. 2-methyl-3-hexene
   e. 1,3-dichlorobenzene
   f. 2-hexyne
   g. n-heptane

7. Name the following alkyl groups:
   a. CH_3—
   b. CH_3CH_2—
   c. $CH_3CH_2CH_2$—

8. Name the following aromatic compounds using both the number and the letter (o-, m-, p-) methods.

a.
```
          H
          |
          C
       //   \
  H—C        C—CH3
     |       ||
  H—C        C—H
       \\   /
          C
          |
         CH3
```

b.
```
            H
            |
            C
         //   \
    H—C         C—NO
       ||       |
 O2N—C          C—H
         \    //
            C
            |
            H
```

c.
```
            H
            |
            C
         /    \\
    H—C         C—H
       ||       |
  CH3—C         C—H
         \    //
            C
            |
           CH3
```

9. Show the products for the following reactions (assume any necessary catalyst is present):
   a. $CH_3—CH_2—CH_3 + O_2 \rightarrow$
   b. $CH_3—CH{=}CH_2 + Br_2 \rightarrow$
   c. $CH_3—C{\equiv}CH + 2\ Cl_2 \rightarrow$
   d. $C_6H_6 + Br_2 \rightarrow$

10. What are the main uses for alkanes?

11. What are the main uses for alkenes?

12. Tell some uses for the following compounds:
   a. hexane
   b. octadecane
   c. cyclohexane
   d. acetylene
   e. ethylene
   f. toluene
   g. nitrobenzene
   h. benzoic acid
   i. p-aminobenzoic acid (PABA)

13. What is a fuel, and how are good fuels distinguished from poor ones?

14. What kind of chemical reaction releases the latent chemical energy in a fuel?

15. How do the following chemicals compare as fuels: wood, gasoline, anthracite coal, bituminous coal, methanol, peat?

16. Describe how the octane number is determined.

17. Why was tetraethyl lead added to gasoline?

18. If one product formed in the catalytic cracking of n-octadecane is n-decane, what is the other compound formed in the same reaction?

19. How does a multigrade motor oil work?

20. What are the differences between low- and high-density poly(ethylene) in (a) preparation and (b) properties?

21. What polymer is produced in the largest quantities each year?

22. What uses does a crosslinked poly(ethylene) have?

23. Who discovered the vulcanization process?

24. What chemicals are used to vulcanize natural rubber?

25. What changes occur in natural rubber when it is vulcanized? Answer this question from the standpoint of both chemical and physical changes.

CRITICAL THINKING PROBLEMS

1. Suppose that the average American car has an oil change every 4,000 miles. Assume that the average car travels 12,000 miles per year, that four quarts of oil are needed for each oil change, and that there are 230,000,000 cars in the United States. How many gallons of used oil would mechanics dump in a month? How does this compare to a major oil spill?

2. Review the reasons why tetraethyl lead was added to gasoline. Was this a good decision, based on the knowledge we had back in 1921? Is this a wise choice for an antiknock additive, based on present knowledge? Several chemicals are now added to increase gasoline's octane level. Suppose that, a decade from now, researchers learn that one of these chemicals causes cancer. What bearing would this have on our decisions today? What bearing would it have ten years from today?

3. Burning vegetable matter, including tobacco, produces hydrocarbons known to cause cancer in humans. The government supports tobacco farming with subsidies and also provides many millions of dollars for cancer research. Are these actions contradictory? What

ethical considerations, if any, do you believe should come into play? For more background, read the following articles: Shannon Brownlee, Steven V. Roberts, Matthew Cooper, Erica Goode, Katia Hetter, Andrea Wright, Traci Watson, and Linda Grant, "Should Cigarettes Be Outlawed?" "Teens on Tobacco," and "Smoke, Flame, and Fire," *U.S. News & World Report,* April 18, 1994, pp. 32–47; Shannon Brownlee, "The Smoke Next door," *U.S. News & World Report,* June 20, 1994, pp. 66–68; Linda Himelstein and Richard A. Melcher, "Did Big Tobacco's Barrister Set Up a Smokescreen?" *Business Week,* September 5, 1994, pp. 68–70; and Benedict Carey, "Saying Goodbye to an Old Flame," *Health,* May/June 1992, pp. 80–85. Based on what you learn, can you justify smoking or selling cigarettes? Is your answer affected by whether you are a smoker?

4. Can you think of any reasons why poly(ethylene) is produced in larger quantities than any other polymer?

HINT Consider economic factors.

5. Read the article by Paul Lucier in *Invention & Technology,* Fall 1991, pp. 56–63, entitled "Petroleum: What is it Good for?" Is is possible to continue using petroleum and still maintain good environmental quality? What are the alternatives to petroleum use in our society? For more information, read the trio of articles by Rich Mullin, Michael Roberts, and Gregory D. L. Morris entitled "U.S. Industry Confident that Improvements Signal a Trend." "Continuing Low Oil Prices Fuel High Output and Consumption," and "Producers Plan for Global Competitiveness," *Chemical Week,* March 1994, pp. 26–36.

6. What can we do with old, used tires? Read the article "Rubber to the Road" by Elizabeth Pennisi in *Science News,* March 7, 1992, p. 155 for one suggestion. Can you think of any other possible uses for this rather stable waste material?

7. Based on the strip-and-window-frame models for multiple bonds in box 9.1, which would you expect to be more reactive—a C—O or a C=O bond?

8. Make gumdrop-and-toothpick models for each of the following covalent compounds. (See figure 9.16 for their chemical formulas and structures.) Use a wad of cotton to represent each π-bond.

a. ethylene
b. acetylene

9. Suppose water was added to the double bond in a $CH_3—CH=CH_2$ molecule. (a) Show the structure(s) for the product(s). (b) Why is more than one product possible? (c) Do you think chemists could control which product actually forms?

HINT *(a),* remember that water adds as H- and HO- units. Either C atom of the double bond could bond with either of these units. For *(b),* try matching the two structures. For *(c),* you might reasonably guess that chemists could control the product; however, this is not always easy to do.

10. How did the discovery that aromatics react by substitution instead of addition help chemists realize that the bonding in these molecules was totally different?

HINT Did anything suggest that ordinary C=C bonds do not always add other groups?

11. Why does the presence of irregular branching cause LDPE to have a lower density than linear HDPE?

HINT Density is a measure of mass/volume; how would branching change the amount of mass in a given volume? Also, compare this with the models for liquids and solids.

12. How can bacteria be used for cleaning up oil spills? Is there a high element of risk in doing this? For background, read: Thomas Y. Canby, *National Geographic,* 184 (No. 2), August, 1993, pp. 36–61. "Bacteria. Teaching Old Bugs New Tricks."

CHAPTER 10

FUNCTIONAL ORGANIC CHEMICALS

OUTLINE

KEY IDEAS

1. The Names and Structures of the Main Functional Groups
2. The Structures, Main Reactions, and Uses of the Hydroxyl Group Classes—Alcohols and Phenols
3. The Structure and Use of Ethers
4. The Structure and Use of Halides
5. The Structures and Uses of the Carbonyl Group Classes—Aldehydes and Ketones
6. How Amines Make Our World More Colorful
7. The Structures and Uses of the More Common Carboxylic Acids
8. How Esters Add Flavor and Perfume to Our Life
9. The Nature of Chemical Equilibrium
10. How Addition Polymerization and Condensation Polymerization Differ
11. Polyesters and Polyamides

"Science is nothing but developed perception . . . minutely articulated."
GEORGE SANTAYANA (1863–1952)

Colorful yarn at market in Solalá, western highlands, Guatemala.

Organic chemistry encompasses not only hydrocarbons, which we studied in chapter 9, but billions of other molecules derived from hydrocarbons as well. These other molecules contain **functional groups,** groups of atoms that replace the hydrogen atoms in a hydrocarbon. These functional groups take part in chemical reactions and help determine how the molecules that contain them function. Billions of organic compounds may be divided into a small number of functional group classes.

FUNCTIONAL GROUPS—NEW DIMENSIONS IN REACTIVITY

Table 10.1 points out three essential lessons about functional groups. First, it lists the names and structures of the major functional groups. Second, it shows that some functional groups look similar; it is important to learn to tell them apart. Third, it demonstrates the differences between the two classes that contain the C=O functional group—aldehydes have at least one H atom attached to the C=O functional group, while ketones have two carbon atoms attached to the C=O. Other functional group classes are examined in the following sections.

Different functional groups have different chemical and physical properties. For example, many functional groups form hydrogen bonds, making the boiling points of the molecules that contain them greater than their parent hydrocarbons. Alkanes, alkenes, and alkynes do not form hydrogen bonds to any extent, and their boiling points are all approximately the same. However, carboxylic acids and alcohols (alkanols), both derived from alkanes, form strong hydrogen bonds, which increase their boiling points compared to the alkanes' (figure 10-1). Similarly, aldehydes are dipoles that have higher boiling points than an alkane with the same number of C atoms.

Figure 10.1 Hydrogen bonding raises the boiling points of carboxylic acids, 1-alkanols (alcohols), and n-aldehydes compared against the boiling points of the n-alkenes.

PRACTICE PROBLEM 10.1

Write the formula for each of the following functional groups. What compound class or classes are formed from each functional group?

a. Hydroxyl
b. Amine
c. Halide
d. Carbonyl
e. Carboxyl

ANSWER

a. R—OH or Ar—OH; alcohol, phenol
b. $R{-}NH_2$; amine
c. R—X; alkyl halide
d. $R{-}\overset{\overset{O}{\|}}{C}{-}H$ or $R{-}\overset{\overset{O}{\|}}{C}{-}R'$; aldehyde, ketone
e. $R{-}\overset{\overset{O}{\|}}{C}{-}OH$ or $R{-}\overset{\overset{O}{\|}}{C}{-}OR'$; carboxylic acid, ester

TABLE 10.1 The major classes of organic compounds according to their functional groups. *R* is an organic group, such as methyl or ethyl. *Ar* is an aromatic (aryl) group, such as phenyl.

FUNCTIONAL GROUP	COMPOUND CLASS	STRUCTURE	EXAMPLE FORMULA	EXAMPLE NAME
Hydroxyl —OH	alcohol (alkanol)	R—OH	CH_3—OH	methanol
	phenol	Ar—OH	C_6H_5—OH	phenol
Oxide —O—	ether	R—O—R Ar—O—Ar (or R)	C_2H_5—O—C_2H_5 C_6H_5—O—C_6H_5	diethyl ether diphenyl ether
Halide —X	alkyl halide	R—X where X = Fl, Cl, Br, I	CH_3CH_2—Cl	chloroethane
Amine —NH_2	amine	R—NH_2	CH_3—NH_2	methylamine
Carbonyl —C=O	aldehyde	R—C(=O)—H	CH_3—C(=O)—H	acetaldehyde
	ketone	R—C(=O)—R′	CH_3—C(=O)—CH_3	acetone
Carboxyl —C(=O)—O—	carboxylic acid	R—C(=O)—OH	CH_3—C(=O)—OH	acetic acid
	ester	R—C(=O)—OR′	CH_3—C(=O)—O—CH_2CH_3	ethyl acetate
Amide —C(=O)—NH_2	amide	R—C(=O)—NH_2	CH_3—C(=O)—NH_2	acetamide

PRACTICE PROBLEM 10.2

Examine figure 10.1 carefully and answer the following questions:

a. What functional groups define each of the classes shown in the figure (except alkanes)?

b. Which functional group undergoes the greatest boiling point increase?

c. Which group (except alkanes) undergoes the smallest boiling point increase?

d. Why does each functional group undergo characteristic changes in boiling point?

ANSWER *a.* The carboxyl functional group defines the carboxylic acids; the hydroxyl group defines the 1-alkanols (alcohols) and the carbonyl group defines the n-aldehydes. *b.* The greatest boiling point increase occurred in the carboxylic acids, which contain the carboxyl functional group. *c.* The smallest increase occurred in the n-aldehydes, which contain the carbonyl group. The 1-alkanols (containing the hydroxyl group) are intermediate. *d.* These b.p. increases are caused by hydrogen bonding and/or dipolar effects; the carboxyl group contains the most hydrogen bonds, followed by the hydroxyl and then the carbonyl.

HYDROXYL GROUPS

Two major classes of organic compounds contain the **hydroxyl (—OH) group** in place of a hydrogen atom—the alcohols and phenols. (Carboxylic acids also contain a —OH group, but we classify them as carboxyls because of their carboxyl group.) Alcohols have their hydroxyl group attached to a carbon chain; in phenols, the —OH group is attached to an aromatic ring. The —OH is covalently bonded in these compounds; do not confuse it with the ionic HO^- in bases (chapter 8).

Alcohols

Alcohols are used widely in chemical reactions and as solvents. **Alcohols** have a general formula of R—OH, where R represents an organic group, such as an alkyl. (Recall that an alkyl is an alkane with one or more hydrogen atoms removed.) The IUPAC name of an alcohol is derived by determining the longest carbon chain that contains the —OH group, then replacing the *ane* in the alkane name with the suffix *ol*. Thus meth*ane* becomes methan*ol* and eth*ane* becomes ethan*ol*. In keeping with IUPAC rules, the carbon atom attached to the —OH group is always assigned the lowest possible number. This means that the compound shown in figure 10.2a is 1-pentanol, never 5-pentanol, and the one in figure 10.2b is 3-hexanol, never 4-hexanol. Similarly, figure 10.2c shows 5-methyl-3-hexanol, not 2-methyl-4-hexanol. Molecules that have more than one —OH group are called diols, triols, and so on. Thus, figure 10.2d shows 2,3-butanediol. (Diols are also called glycols.)

Alcohols are prepared naturally by fermenting carbohydrates (sugars and starches) or synthetically by hydrating alkenes (figure 10.3a). They are key reactants in the preparation of dozens of other compounds and are an extremely versatile class of molecules. In general, alcohols are toxic when ingested. This is partly due to the oxidation reactions shown in figure 10.3b and c, which produce aldehydes and ketones that are more toxic than the original alcohols.

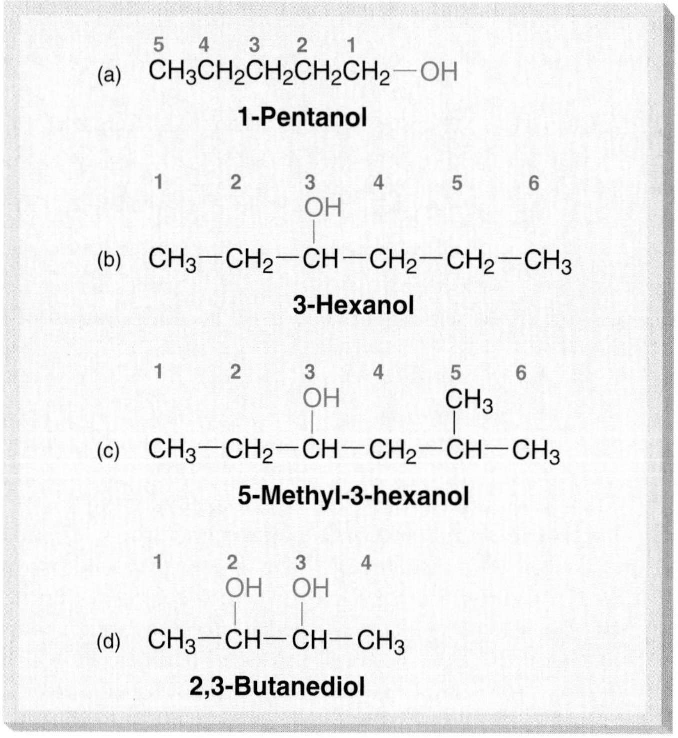

(a) $\overset{5}{C}H_3\overset{4}{C}H_2\overset{3}{C}H_2\overset{2}{C}H_2\overset{1}{C}H_2—OH$

1-Pentanol

(b) $\overset{1}{C}H_3—\overset{2}{C}H_2—\overset{3}{C}H(OH)—\overset{4}{C}H_2—\overset{5}{C}H_2—\overset{6}{C}H_3$

3-Hexanol

(c) $\overset{1}{C}H_3—\overset{2}{C}H_2—\overset{3}{C}H(OH)—\overset{4}{C}H_2—\overset{5}{C}H(CH_3)—\overset{6}{C}H_3$

5-Methyl-3-hexanol

(d) $\overset{1}{C}H_3—\overset{2}{C}H(OH)—\overset{3}{C}H(OH)—\overset{4}{C}H_3$

2,3-Butanediol

Figure 10.2 Examples of the IUPAC names of alcohols and diols. The red numbers above each structure indicate the IUPAC numbering sequence for the carbon atoms in the carbon chain.

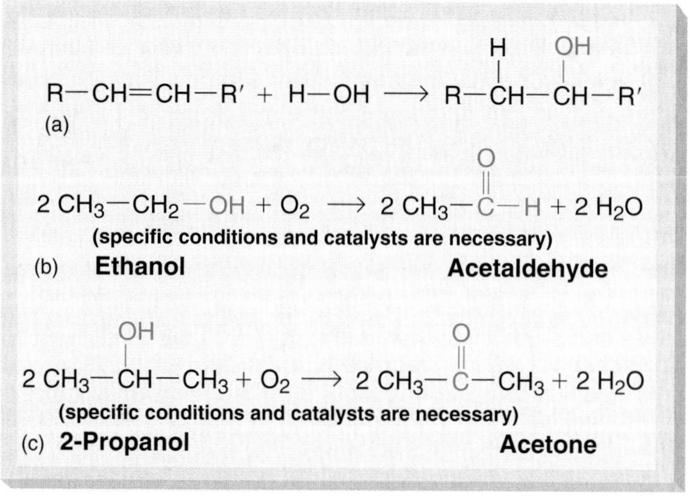

(a) $R—CH{=}CH—R' + H—OH \longrightarrow R—CH(H)—CH(OH)—R'$

(b) $2\ CH_3—CH_2—OH + O_2 \longrightarrow 2\ CH_3—C({=}O)—H + 2\ H_2O$

(specific conditions and catalysts are necessary)

Ethanol **Acetaldehyde**

(c) $2\ CH_3—CH(OH)—CH_3 + O_2 \longrightarrow 2\ CH_3—C({=}O)—CH_3 + 2\ H_2O$

(specific conditions and catalysts are necessary)

2-Propanol **Acetone**

Figure 10.3 **(a)** Hydrating (adding water to) an alkene produces an alcohol. **(b)** and **(c)** Oxidation reactions convert alcohols to aldehydes and ketones.

PRACTICE PROBLEM 10.3

Name the following molecules using the IUPAC system. (Refer back to table 9.1 if you need to refresh your memory on the alkane root names.)

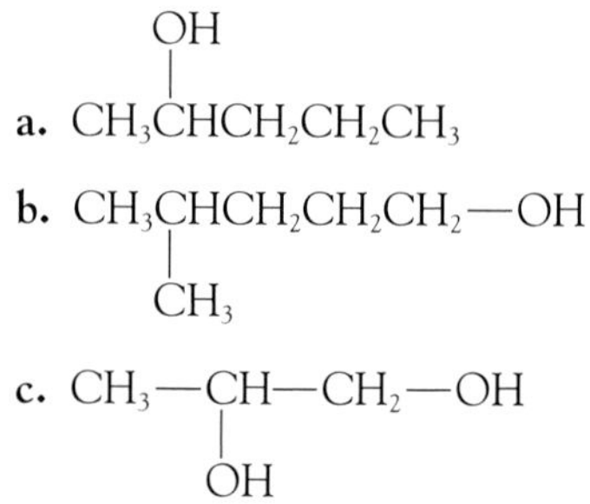

a. $CH_3CH(OH)CH_2CH_2CH_3$

b. $CH_3CH(CH_3)CH_2CH_2CH_2—OH$

c. $CH_3—CH(OH)—CH_2—OH$

ANSWER The location of the hydroxyl functional group always determines the numbering of the C atoms in alcohol compounds. Thus, compound *a* is 2-pentanol—it has five carbons, with the —OH attached to the second. Compound *b* is 4-methyl-1-pentanol. It also has five carbon atoms in its longest chain; it has a methyl group attached to carbon 4 and a hydroxyl group attached to carbon 1. (Remember, the —OH always is attached to the C atom with the lowest possible number.) Compound *c* is a diol (two —OH groups) with three carbon atoms in its chain; the —OH groups are on the first two carbons, so its name is 1,2-propanediol.

Methanol

Methanol (the IUPAC name for an alcohol with its hydroxyl group attached to a single carbon atom) is also called methyl alcohol or wood alcohol. This exceedingly toxic compound can cause blindness, paralysis, or death when ingested. Methanol was first prepared by heating wood in a process called destructive distillation. As wood is heated in a distillation flask in the absence of air to prevent oxidation, its chemical structure is destroyed, and methanol is among the reaction products.

Industrially, methanol is prepared by reacting hydrogen and carbon monoxide at high temperature and pressure:

$$2\ H_2 + CO \xrightarrow[\text{catalyst}]{\text{high pressure}} CH_3{-}OH$$

Methanol can also be produced by partially oxidizing hydrocarbons from natural gas. More than seven billion pounds of methanol are prepared annually.

Methanol has many uses. It serves as a solvent for shellac and other compounds used for coatings, as a temporary antifreeze, and as a gasoline additive or alternative fuel. Methanol is also converted into plasticizers (chemicals used to manufacture plastics) and other important organic compounds.

Ethanol

Ethanol, also called ethyl alcohol or grain alcohol, is much less toxic to humans than the other alcohols. This is fortunate since it is the form of alcohol that occurs in beverages such as beer, wine, and whiskey. However, this does not mean that ethanol is nonpoisonous; a toxic dose is about 8,000 grams for a person weighing about 132 pounds. Ethanol is one of the oldest chemicals used by humans, who have prepared alcoholic beverages from grains for thousands of years. References to alcoholic beverages appear in ancient Babylonian writings and in the Bible. Wine was made in China before 2000 B.C., and fairly pure alcohol was distilled from wine by the twelfth century A.D.

TABLE 10.2 Toxic effects of ethanol on humans.

AMOUNT OF ALCOHOL (mg/100 mL BLOOD)	PERCENTAGE IN BLOOD	EFFECT ON THE PERSON
less than 50	0.00–0.05	slower reflexes
50–150	0.05–0.15[1]	poor coordination
150–200	0.15–0.20	intoxication
300–400	0.30–0.40	unconsciousness
500+	0.50+	possibly fatal

[1]Legal limit for driving in many states.

Table 10.2 summarizes the toxic effects of ethanol. Ethanol is a depressant that significantly impairs a person's reflexes even when it is below toxic levels. A level of 50 mg alcohol in 100 mL blood (0.05 percent), achieved when one ingests as little as two "shots" of whiskey (about 30 mL each) or two 10-ounce beers, is the legal blood-alcohol limit in some states. Recent studies at Penn State University (reported in the Orlando *Sentinel*, January 10, 1993) indicate that a person's ability to drive a motor vehicle is significantly impaired at blood-alcohol levels below 0.10 percent. The drinker could die if the blood level reaches 500+ mg/100 mL (0.5 percent), but most people "pass out" before reaching this concentration. Ethanol is the most commonly used addictive drug in the United States and has cost untold billions of dollars in health care costs, lost productivity, and other types of damage. More than 10 million alcohol addicts live in the United States—about 1 in every 25 people.

Ethanol is prepared commercially from ethene. Figure 10.3a shows how ethene is processed into ethanol; *R* and *R′* in the figure are H atoms. Ethanol may also be prepared by fermenting sugar and starches, as shown in figure 10.4. Diastase, maltase, and zymase are enzymes—natural catalysts (discussed in Chapter 14) whose names always end in the letters *ase*. Along with being used in alcoholic beverages, ethanol is used as an industrial solvent and in medicines. (Many cough syrups contain ethanol.) It is also

$$\underset{\textbf{Starch}}{(C_6H_{10}O_5)_n} \xrightarrow[\text{diastase}]{H_2O} \underset{\textbf{Maltose}}{n\ C_{12}H_{22}O_{11}} \xrightarrow[H_2O]{\text{maltase}} \underset{\textbf{Glucose}}{C_6H_{12}O_6} \xrightarrow{\text{zymase}} \underset{\textbf{Ethanol}}{CH_3{-}CH_2{-}OH} + CO_2$$

Figure 10.4 Ethanol may be produced by fermenting sugars and starches.

an important reagent used in preparing other chemicals. Ethanol is sometimes used as an antifreeze, but the cost is generally too high to make this a common usage. Gasohol is gasoline that contains 10 percent ethanol.

2-Propanol

The compound 2-propanol, also known as isopropyl alcohol, is obtained commercially by hydrating propene. (In figure 10.3a, *R* would be CH_3 and *R′* would be H.) It is a useful solvent for oils, gums, resins, and other coatings and is a chemical intermediate used in the preparation of acetone (figure 10.5). In addition, 2-propanol as a 70 percent aqueous solution is readily available as a rubbing alcohol. More recently, it has become available in a 90+ percent aqueous solution used to clean electronic components and tape heads. 2-propanol is fairly toxic when ingested (box 10.1).

Ethylene Glycol

The IUPAC name for ethylene glycol is 1,2-ethanediol, but the common name is used much more often. Ethylene glycol's major use is as a "permanent" automobile antifreeze and summer coolant (figure 10.6). Its high boiling point (187°C) means this material remains even when water boils away from a mixture of the two. Ethylene glycol is moderately toxic; 100 mL is estimated to be a lethal dose. Since ethylene glycol has a sweet taste, children (and pets) sometimes drink lethal amounts.

Ethylene glycol is used not only as an antifreeze but in brake fluids, cosmetics, printing inks, and lacquers. It is a solvent for polyesters and cellulose esters. Ethylene glycol is made by oxidizing ethylene in air, then adding water (figure 10.7); catalysts are needed to produce the chemical reactions in both steps.

Glycerol

The alcohol known as glycerol is also called glycerin; the correct but rarely used IUPAC name is 1,2,3-propanetriol. This chemical is a clear, colorless, odorless, syrupy liquid with a sweet taste. It is used to sweeten toothpaste as well as in many types of cosmetics (chapter 16). Glycerol is obtained as a byproduct in soap manufacture and is made synthetically from propene. Its boiling point is 290°C, which gives it the potential for use as an antifreeze, but it is far too expensive to use routinely. Glycerol's main uses are in cosmetics, soaps, hydraulic fluids, and printing inks. It is often used as a lubricant.

In addition to its own commercial uses, glycerol also serves as a raw material for producing trinitroglycerin (also called nitroglycerin), a potent, highly shock-and-heat-sensitive explosive. Nitroglycerin is dispersed in carbonaceous materials to form the less sensitive dynamite. Alfred Nobel (Swedish, 1833–96) who later established the Nobel Prize awards, was the first to use nitroglycerin in this way (see box 3.3). Nitroglycerin is used medically to treat angina pectoris, a chest pain caused by coronary problems. The chemical dilates blood vessels, thereby relieving the pain.

Figure 10.6 Four of the many brands of antifreeze based on ethylene glycol. Most mixtures contain other organic molecules, such as amines, to reduce corrosion.

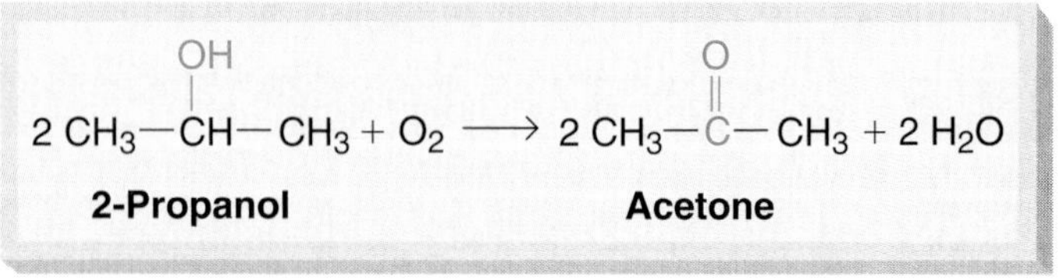

$$2\,CH_3-\underset{}{\overset{OH}{\overset{|}{C}H}}-CH_3 + O_2 \longrightarrow 2\,CH_3-\overset{O}{\overset{\|}{C}}-CH_3 + 2\,H_2O$$

2-Propanol → **Acetone**

Figure 10.5 Isopropyl alcohol, or 2-propanol, is used to produce acetone.

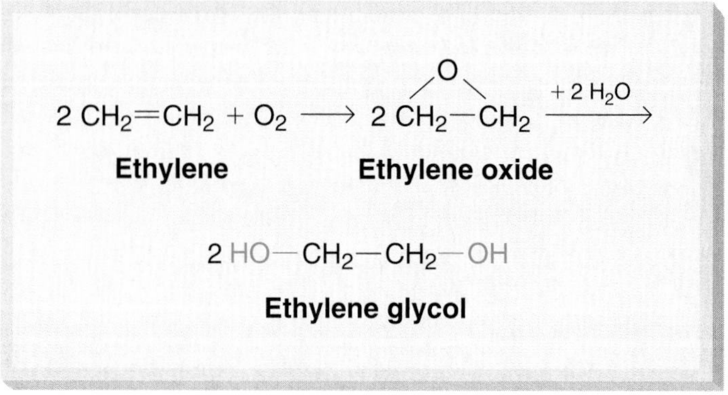

$$2\,CH_2{=}CH_2 + O_2 \longrightarrow 2\,CH_2\overset{O}{-}CH_2 \xrightarrow{+2\,H_2O} 2\,HO-CH_2-CH_2-OH$$

Ethylene → **Ethylene oxide** → **Ethylene glycol**

Figure 10.7 Ethylene glycol is made by (1) oxidizing ethene in air, then (2) adding water. Both steps require catalysts.

BOX 10.1

TOXICITY AND LD_{50} VALUES

Almost all chemicals, natural or synthetic, possess some degree of toxicity. In fact, some of the most poisonous chemicals come from natural sources; for example, South American indigenous people dip their poison darts into a potent toxin they squeeze out from a frog (figure 1). Because the toxicity of various chemicals varies widely, scientists have developed a measurement system called the LD_{50}.

The **LD_{50}** values represent the amount of a chemical that kills 50 percent of the laboratory animals in a toxicity test. These values are usually expressed as mg/kg or g/kg body weight. Sometimes toxic effects do not appear until the animal has been exposed to the drug over most of its life span. The higher the LD_{50} value, the safer the chemical, since it takes more of the drug to produce toxic effects.

Figure 1 This South American frog exudes a deadly poison from its skin. Native people dip darts into this poison and use them to kill.

Table 1 lists the LD_{50} values for some alcohols. Since compounds with lower LD_{50} values are more toxic, 1-propanol (with an LD_{50} value of 1.9) is more toxic than ethanol (with a value of 13.7). Companies developing new drugs must determine the LD_{50} for each potential medicine. Chapters 15 and 17 will explain more about LD_{50} values and their uses.

Since LD_{50} values are determined using small laboratory animals—usually mice or rats—it can be difficult to extrapolate LD_{50} values to human poisoning. Table 2 estimates lethal values of a few chemicals for a 60 kg (132 pound) person. As the table shows, methanol is toxic at a fairly low level—just 3 g (about 60 mL)—because it rapidly oxidizes to formaldehyde in the body. Around 300 g of the ethylene glycol used in permanent antifreeze might cause the death of a 60 kg person, but it would take about 8,000 g of ethanol to produce the same result. Glycerol is even less toxic than ethanol.

TABLE 1 The toxicity of some alcohols and related compounds.

ALCOHOL	STRUCTURE	LD_{50} (g/kg BODY WEIGHT)[1]
methanol	CH_3—OH	0.07[2]
ethanol	CH_3—CH_2—OH	13.7[3]
1-propanol	CH_3—CH_2—CH_2—OH	1.9
2-propanol	CH_3—CH(OH)—CH_3	5.8
1-butanol	CH_3—CH_2—CH_2—CH_2—OH	4.4
ethylene glycol	HO—CH_2—CH_2—OH	5.8
glycerol	HO—CH_2—CH(OH)—CH_2—OH	>20

[1]These values are in g/kg body weight. Other units, such as mg/kg, are sometimes used to measure LD_{50} values.
[2]This value is the LD_{50} for formaldehyde. Methanol oxidizes too rapidly in the body to measure its LD_{50}; it becomes formaldehyde.
[3]This LD_{50} value is for ethanol. The corresponding LD_{50} for acetaldehyde (the ethanol oxidation product) is 1.9 g/kg body weight—making it more toxic than the original ethanol.

TABLE 2 Estimated relationships between LD_{50} values and human lethal doses.

LD_{50} (mg/kg BODY WEIGHT)	PROBABLE HUMAN LETHAL DOSE[1] (ORAL)	ALCOHOL EXAMPLES
<5	<0.3 g	
5–50	0.3–3 g	
50–500	3–30 g	methanol[2] (at lower end)
500–5,000	30–300 g	1-propanol; 1-butanol
5,000–15,000	0.3–9 kg	ethylene glycol (at lower end) ethanol (at upper end)
>15,000	>9 kg	glycerol

[1]Based on a 60 kg (132 pound) body weight.
[2]This LD_{50} value is for formaldehyde. Methanol oxidizes too rapidly in the body to measure its LD_{50}; it becomes formaldehyde.

Phenols

Although **phenols** have a —OH group attached directly to an aromatic ring, they are *not* alcohols, nor are their properties very similar to those of alcohols. Most phenols are solids, whereas most alcohols of similar molecular weight are liquids. Phenols are acidic—the most common, called phenol, is also known as carbolic acid. Most phenols react with and dissolve in aqueous NaOH, a base.

Figure 10.8 shows the synthesis of phenol by the oxidation of isopropylbenzene (also called cumene). This reaction also produces acetone. Although it is toxic to humans when ingested, phenol is used as a topical (surface) antiseptic because it has germicidal properties (that is, it kills germs). Phenol is also used in some over-the-counter cough syrups, presumably due to an analgesic or pain-relieving effect. Phenol's major use is in the preparation of phenolic resins and as a starting material for other plastics and pharmaceuticals. Phenol is also an effective slimicide (it kills slime-producing creatures) and disinfectant (for floors, bathroom and kitchen fixtures, and so forth).

OXIDES

The **oxide** or —O— group is the functional group in the ethers. These compounds are derived from alcohols and phenols.

Ethers and Anesthetics

Ethers are formed when a pair of alcohol molecules or phenols lose a molecule of water between them. Since the two molecules originally contained two —OH groups, the loss of one H_2O molecule leaves a single O atom bonded to the *R* or *Ar* groups. Ethers are named by placing the names of the two alkyl groups in front of the word *ether*. The product in the following equation is thus

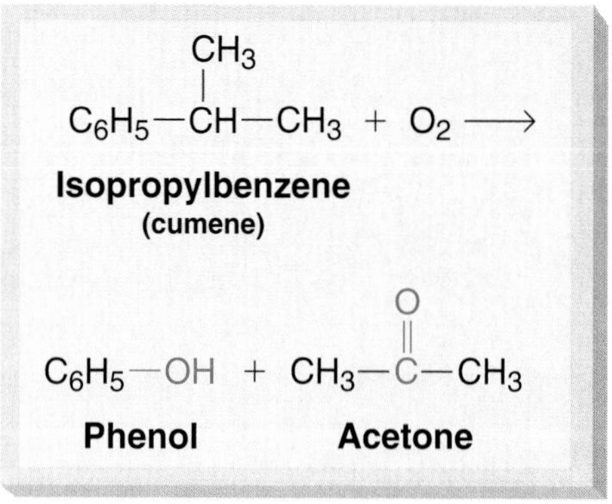

$$C_6H_5{-}CH(CH_3){-}CH_3 + O_2 \longrightarrow$$

Isopropylbenzene (cumene)

$$C_6H_5{-}OH + CH_3{-}C({=}O){-}CH_3$$

Phenol **Acetone**

Figure 10.8 Phenol is synthesized by oxidizing isopropylbenzene (also called cumene).

diethyl ether, since it contains two (di) ethyl groups ($CH_3{-}CH_2{-}$) attached to the oxide group (—O—).

$$\underset{\text{Ethanol}}{CH_3{-}CH_2{-}OH} + \underset{\text{Ethanol}}{HO{-}CH_2{-}CH_3} \rightarrow \underset{\text{Diethyl ether}}{CH_3{-}CH_2{-}O{-}CH_2{-}CH_3} + H_2O$$

If an ether is derived from two different molecules—perhaps methanol and ethanol—the resulting groups are ordered alphabetically. Therefore, the following compound is called ethyl methyl ether:

$$\underset{\text{Methanol}}{CH_3{-}OH} + \underset{\text{Ethanol}}{HO{-}CH_2{-}CH_3} \rightarrow \underset{\text{Ethyl methyl ether}}{CH_3{-}O{-}CH_2{-}CH_3}$$

In 1846, Boston dentist William Thomas Morton (American, 1819–1868) first used diethyl ether (often just called ether) as an effective general anesthetic. Others rapidly adapted it to nondental uses. Because the amount of diethyl ether required to produce toxic effects is much higher than the amount needed to induce unconsciousness, ether is relatively safe. However, diethyl ether has other disadvantages—it produces nausea, is extremely flammable, and is less effective than other agents available. Although it is still used in other countries (for example, China), ether is no longer widely used as a general anesthetic in the United States.

PRACTICE PROBLEM 10.4

What are the names for the following ethers?

a. $CH_3{-}O{-}C_6H_5$
b. $CH_3CH_2{-}O{-}CH_2CH_2CH_3$
c. $CH_3{-}O{-}CH_2CH_3$

ANSWER *a.* Methyl phenyl ether; *b.* ethyl propyl ether; *c.* ethyl methyl ether. Notice that the groups are arranged alphabetically—methyl before phenyl, ethyl before propyl, and ethyl before methyl.

Some alcohols and ethers are used as gasoline additives (box 10.2). Increasing the oxygen content of gasoline helps cut down on CO (carbon monoxide) emissions; since alcohols and ethers contain oxygen, they are added to gasoline to increase its oxygen content. In addition, because they often have higher octane ratings, oxygenated compounds improve the smoothness of combustion.

BOX 10.2

OXYGENATED GASOLINE ADDITIVES

Oxygenated chemicals are now added to gasoline to promote cleaner air, smoother combustion, and higher octane numbers at low cost. Ethanol is the main alcohol used for the purpose; the gasohol introduced in the 1970s contained about 10 percent ethanol. Ethanol's main drawback is that it absorbs water from the atmosphere, which causes corrosion, though some additives can reduce this tendency. Like most alcohols and related chemicals, ethanol has a high octane number (table 1).

Another alcohol additive, methanol, is often used as a gas line antifreeze. Methanol's high octane number also reduces knocking, but it, too, absorbs water and causes severe corrosion. The less soluble 1-butyl alcohol avoids this problem, since it does not absorb water well; in fact, t-butyl alcohol reduces the tendency of gasoline blends to pick up water. A typical blend may contain 2.5 percent methanol, 2.5 percent t-butyl alcohol, and 95 percent gasoline.

Methyl t-butyl ether, MTBE, is added to gasoline to promote smoother burning and to promote hydrocarbon burning. Unresolved questions continue to arise about the toxicity of MTBE. MTBE is used as an additive rather than a gasoline replacement, partly because it is much more expensive.

Although these oxygenated compounds have high octane numbers, a high octane number relates to smoothness of combustion and does not relate directly to the energy content of a fuel. In general, all oxygenated additives contain less energy than gasoline because they are already partially oxidized. In a practical sense, this means that you get less knocking in the engine, but you also get less energy per gallon of fuel. Still, because ethanol and methanol can be made from biomass or other renewable sources (figure 1), they are good options for nonpetroleum fuels.

Although ethanol and methanol add oxygen to gasoline, they are not totally nonpolluting fuels. Both can produce toxic aldehydes on incomplete oxidation and, like all carbon-containing fuels, they produce CO_2. Consumers sometimes express one other complaint: additives, such as ethanol, methanol, and MTBE increase the cost of gasoline without increasing the energy it provides.

TABLE 1 The octane numbers for some oxygenated compounds.

COMPOUND	FORMULA	OCTANE NUMBER
methanol	CH_3—OH	107
ethanol	CH_3CH_2—OH	108
t-butyl alcohol	$(CH_3)_3C$—OH	113
methyl t-butyl ether (MTBE)	CH_3—O—$C(CH_3)_3$	116

Figure 1 A crop of corn is harvested for conversion into ethanol, a renewable fuel source.

HALIDES

In the halides, a halogen atom—fluorine, chlorine, bromine, or iodine—replaces the hydrogen atoms in the parent hydrocarbon. In the IUPAC system, halides derive their name from the original hydrocarbon, with a prefix designating the halogen (fluoro-, chloro-, bromo-, or iodo-). Thus, methane (CH_4) would become chloromethane (CH_3Cl) if a chloride atom replaced one of the H atoms. Many halides also have common names.

Halides have many applications. Chlorofluorocarbons (CFCs) were commonly used in cooling systems and aerosol sprays until relatively recently, when scientists claimed that the buildup of CFCs in the atmosphere was depleting the ozone layer. Other halides, including chloroform, were used as solvents in laboratories and dry cleaning establishments, though they were later replaced by safer chemicals. Chloroform was also used as an anesthetic. Still other halides are used in poly(vinyl chloride) or PVC plastics.

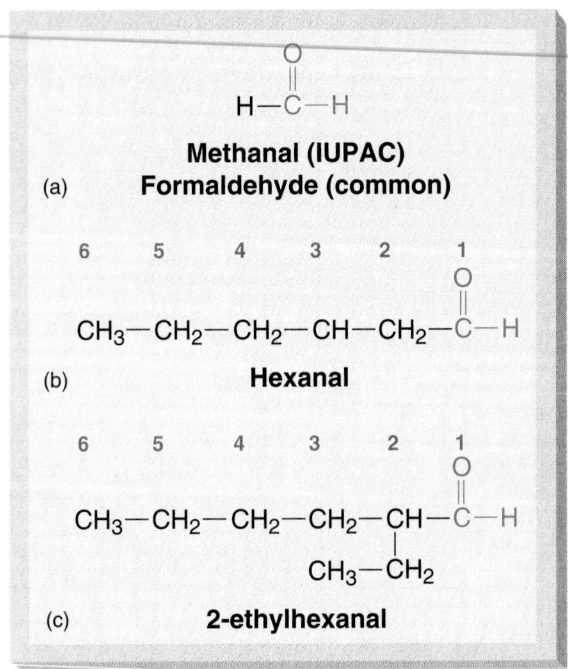

Figure 10.9 Examples of formulas and IUPAC names for aldehydes. Example *a* is better known by its common name, formaldehyde. Red numbers above structures indicate numbering sequence.

CARBONYL GROUPS

Two major organic classes contain the **carbonyl group,** a C=O unit: the aldehydes and the ketones. Carboxylic acids and their derivatives also contain a carbonyl group that forms as part of the carboxyl group. They are considered separately in a later part of the chapter.

Aldehydes

In an **aldehyde,** the C=O group always has one H atom directly attached to it at one end; the other end is either an H atom or an alkyl, alkenyl, or aryl group (see Table 10.1). Aldehydes' IUPAC names are derived by locating the longest carbon chain attached to the C=O group, determining its alkane name, and replacing the *e* in the parent alkane with the suffix *al*. Thus, methane becomes methanal, hexane becomes hexanal (figure 10.9). The carbon in the C=O group is always numbered as carbon 1 in the aldehydes. Even if another chain in the molecule contains more carbon atoms, the name of the compound is always based on the longest carbon chain attached to the C=O group.

Many aldehydes are much better known by their common names; formaldehyde (methanal in IUPAC), the simplest aldehyde, is obtained by the partial oxidation of methanol (figure 10.10a). Formaldehyde is a gas often dissolved in a 40 percent aqueous solution called formalin. It is used as a germicide, a disinfectant, and a preservative for biological tissues (box 10.3).

Another major use of formaldehyde is to prepare polymers such as the phenol-formaldehyde resins (chapter 11). Formaldehyde can also polymerize by itself to form poly(oxymethylene) (figure 10.10b). This compound is marketed as a lightly colored thermoplastic called Delrin, which is used to make household dishes and other molded items. Hermann Staudinger, one of the early leaders in polymer chemistry, first prepared poly(oxymethylene) plastic from formaldehyde.

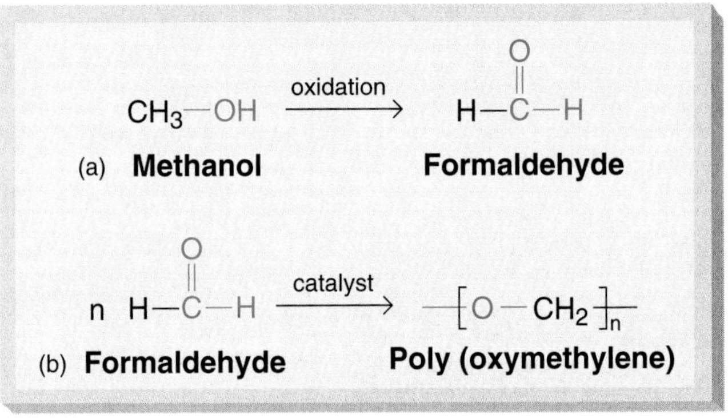

(a) Methanol $CH_3-OH \xrightarrow{\text{oxidation}} H-C(=O)-H$ Formaldehyde

(b) Formaldehyde $n\ H-C(=O)-H \xrightarrow{\text{catalyst}} \left[O-CH_2 \right]_n$ Poly (oxymethylene)

Figure 10.10 **(a)** Formaldehyde (methanal) is obtained by the partial oxidation of methanol. **(b)** Formaldehyde can polymerize to form poly(oxymethylene).

PRACTICE PROBLEM 10.5

Name the following aldehydes by the IUPAC method.

a. $H-C(=O)-CH_2CH_2CH_3$

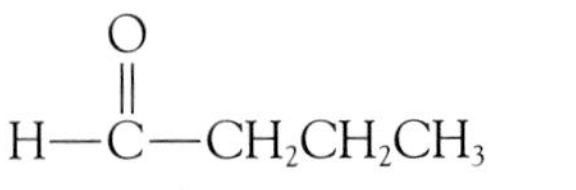

b. $CH_3CH(CH_3)CH_2CH_2-C(=O)-H$

ANSWER Compound *a* is butanal since it has four C atoms. Compound *b* is 4-methyl-pentanal; it has five C atoms, and the methyl group is on the fourth one. (Recall that the aldehyde functional group is always in position 1.) It does not matter which direction these structures are drawn in—the naming rules give the same result either way.

As figure 10.11 shows, some aldehydes are used as food flavoring agents. The primary taste in natural almond, cinnamon, and vanilla flavors are produced by aldehyde compounds. These compounds can now be manufactured synthetically from petroleum, and these synthetic aldehydes are generally regarded as safe for human consumption in the small amounts used for flavoring. Although some people claim that "natural" flavorings are better and safer than the synthetic versions, the natural sources contain traces of other chemicals, and the nature and toxicity of many of these minor chemical components is unknown. Synthetic flavorings may well be less toxic than their natural counterparts.

EXPERIENCE THIS

Does the odor of paint or some other volatile organic chemical bother you? You can mask most odors with a small amount of vanilla. Your nose can detect as little as 0.5 ppm of the aldehydes that create vanilla odor.

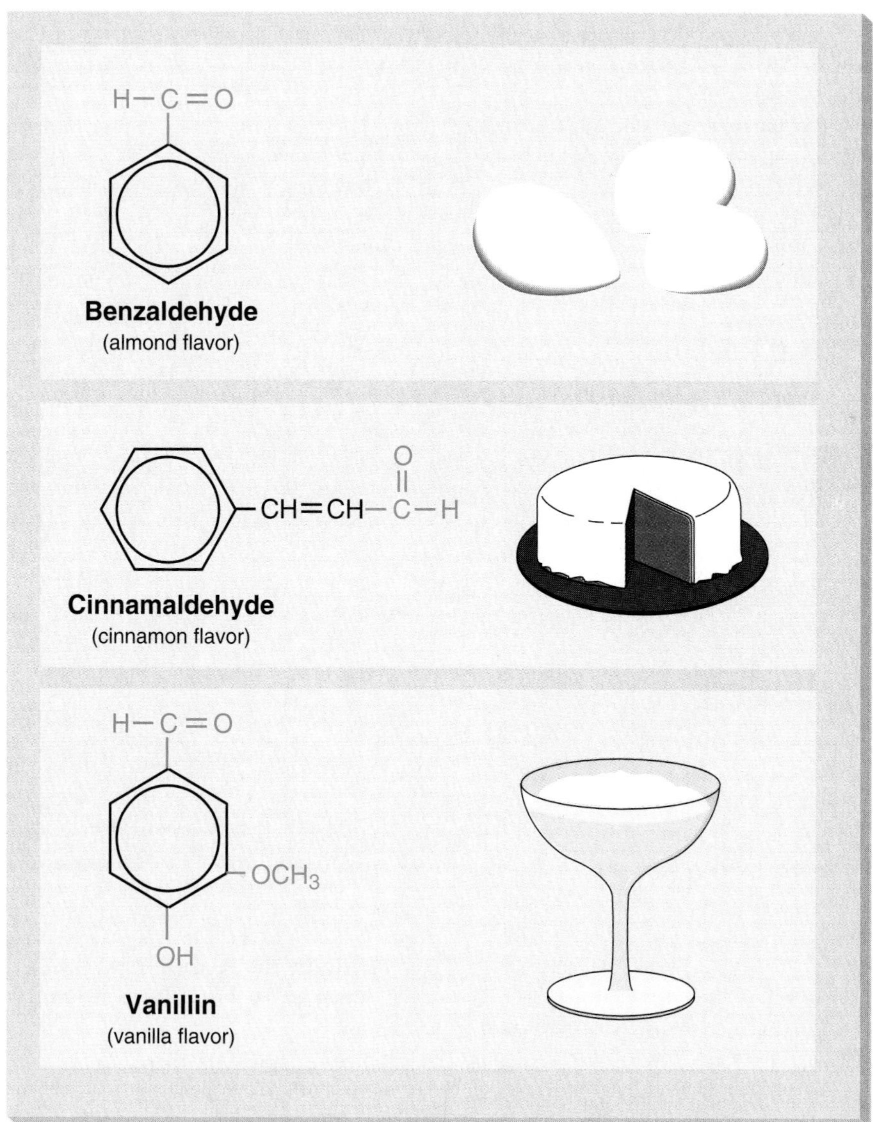

Figure 10.11 Some aldehyde flavoring agents with examples of their uses.

BOX 10.3

PRESERVED LIKE AN EGYPTIAN

One major use of aldehydes is in the process of embalming, or preserving a dead body. Formaldehyde is the main chemical used for embalming, but this was not always true. The earliest descriptions of embalming come from ancient Egypt (figure 1). The Egyptians preserved the deceased in salts and oils, wrapping them in cloth, as a religious ritual. The Old Testament mentions that the patriarch Joseph was "embalmed" and buried in Egypt, and Herodotus (Greek historian, 485–424 B.C.) gave detailed descriptions of the three different embalming practices the Egyptians used around 500 B.C. Other ancient peoples also practiced embalming, including those in Ethiopia, Peru, the Canary Islands, and North America. The Persians immersed their dead in jars of honey and wax for preservation.

The early Christians were essentially disinterested in embalming and focused instead on the spiritual aspects of death. But during the fifteenth century there was a renewal of interest in embalming due, in part, to Leonardo da Vinci's (Italian, 1452–1519) desire to study human anatomy more closely. Da Vinci developed a method of injecting embalming fluids into the veins. These fluids included turpentine, camphor, wine, rosin, and various oils from plants. Miscellaneous materials were used until the early 1900s.

Figure 1 The Sphinx, an Egyptian landmark, was built at a time when Egyptians were already embalming their dead.

Formaldehyde, first used in embalming in the early 1900s, soon became the chemical of choice (figure 10.2a). The use of formaldehyde as a tissue preservative stemmed from a chance observation in 1893 by Ferdinand Blum (German physician, 1865–1959) that a 4 percent aqueous solution made his fingers stiff. Formaldehyde soon became the primary chemical for tissue preservation, partly because it is very inexpensive. Formaldehyde preserves tissues by crosslinking the proteins, which are polymers. (See chapter 9 for more information on crosslinking and chapter 14 for more on proteins.) The exact mechanism of this chemical reaction is unknown, however.

Exposure to formaldehyde presents a variety of health concerns to the living. The chemical is suspected to cause cancer, although this remains to be proven; but contact dermatitis and asthmalike problems caused by formaldehyde are well documented. However, no environmental problems arise from the use of formaldehyde in embalming, according to the New Jersey EPA. The search for a better replacement has met with limited success. Glutaraldehyde (figure 2b), first used in 1955, seems to work better. It also works by crosslinking protein molecules in the body tissues, but it is five to eight times more expensive than formaldehyde.

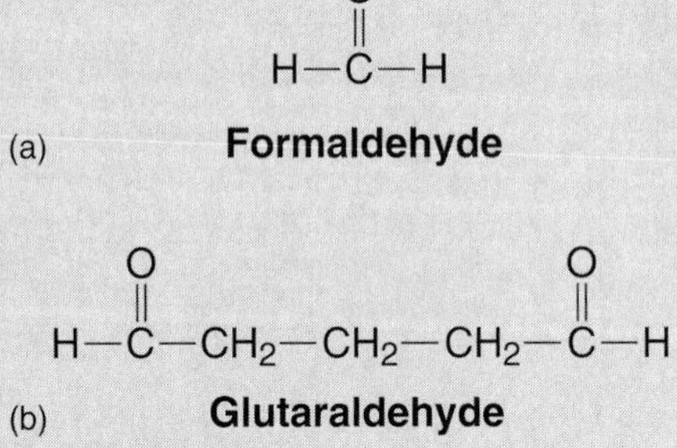

(a) $H-C(=O)-H$ **Formaldehyde**

(b) $H-C(=O)-CH_2-CH_2-CH_2-C(=O)-H$ **Glutaraldehyde**

Figure 2 The structures of formaldehyde and glutaraldehyde.

One problem that arises in embalming is the chemical reactions between formaldehyde and any drug previously used to treat the deceased person. Many of these interactions cause undesirable color changes or marks on the skin. Frequently the embalmer does not know what drugs the deceased person used and cannot avoid these difficulties. This problem seems especially acute with people who die from AIDS and who are usually treated with combinations of drugs. Formaldehyde does destroy the AIDS virus, however, so the risk of transmission during embalming is minimal.

Ketones

In a **ketone,** the C=O group is flanked by any combination of two alkyl or aryl groups (see table 10.1). The IUPAC name is derived from the parent alkane, but the *e* is replaced with *one* (so that propane becomes propanone and heptane becomes heptanone). To determine the alkane root name, the longest chain containing the C=O unit is located and the C=O is assigned the lowest possible number. Figure 10.12 shows three examples of ketones.

Like aldehydes, many ketones are better known by non-IUPAC names. Often, this name is the alkyl or aryl group name followed by the word *ketone*. Thus, figure 10.12b is ethyl n-butyl ketone; propanone (figure 10.12a) is dimethyl ketone (acetone is

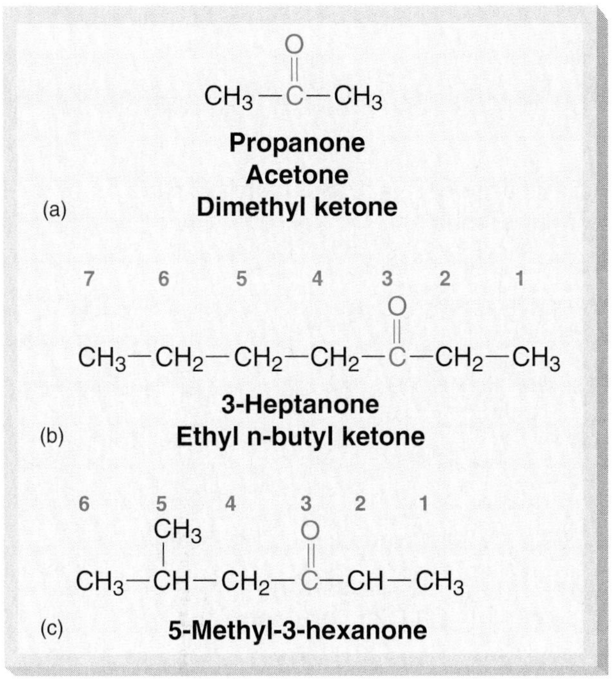

(a)

$$CH_3-\overset{\overset{\large O}{\|}}{C}-CH_3$$

Propanone
Acetone
Dimethyl ketone

(b)

7 6 5 4 3 2 1

$$CH_3-CH_2-CH_2-CH_2-\overset{\overset{\large O}{\|}}{C}-CH_2-CH_3$$

3-Heptanone
Ethyl n-butyl ketone

(c)

6 5 4 3 2 1

$$CH_3-\underset{}{\overset{\overset{CH_3}{|}}{CH}}-CH_2-\overset{\overset{\large O}{\|}}{C}-CH-CH_3$$

5-Methyl-3-hexanone

Figure 10.12 Examples of ketones. The red numbers specify the IUPAC carbon numbering; the common names are listed below the IUPAC names.

its common name). It is unnecessary to specify the number 2 position for the C=O in propanone because there is no other position the C=O group can take in a three C atom chain; a ketone *must* contain two alkyl or aryl groups with a C=O group between.

Acetone is a useful solvent found in paint remover, nail polish, and nail polish remover. Acetone is also used to clean and dry precision equipment parts because it has a low boiling point (56°C) and evaporates easily. Methyl ethyl ketone (2-butanone), or simply MEK, is also used as a cleaning fluid and a solvent in paint removers, lacquers and coatings, and lube oil dewaxing. Diethyl ketone (3-pentanone) is sometimes used as a solvent, but its major uses are in medicine and organic synthesis. Ketones are fairly flammable and require careful handling.

PRACTICE PROBLEM 10.6

Name the following ketones by the IUPAC system. Also try to name them with the ketone name; there are no common names for these compounds. Which method makes it easier to establish the structure for the compound?

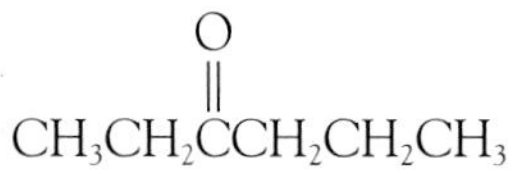

a. 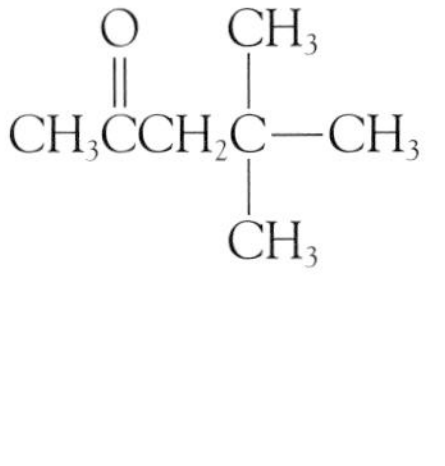 $$CH_3\overset{\overset{O}{\|}}{C}CH_2\underset{\underset{CH_3}{|}}{\overset{\overset{CH_3}{|}}{C}}-CH_3$$

b. $$CH_3CH_2\overset{\overset{O}{\|}}{C}CH_2CH_2CH_3$$

c. $$CH_3CH_2-\overset{\overset{O}{\|}}{C}-CH_2CH_3$$

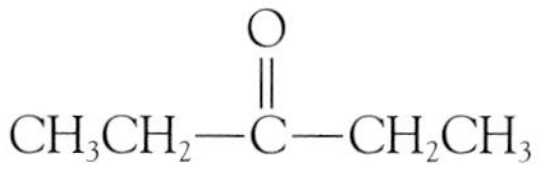

ANSWER The position of the carbonyl group sets the IUPAC numbering sequence in ketones. Thus, compound *a* is 4,4-dimethyl-2-pentanone because it has five carbons in the longest chain, with the C=O in position 2 and two pendant methyl groups at position 4. (Recall that the number must be shown for each methyl group, even when they are attached to the same carbon.) You would have trouble naming compound *a* by the ketone method since you do not know the name of the group on the right side of the carbonyl unit. Compound *b* is 3-hexanone by the IUPAC method and ethyl n-propyl ketone by the ketone method. Compound c is 3-pentanone or diethyl ketone. In most cases, the IUPAC system provides a shorter name and does not raise questions regarding the structure of the compound. Chemists use the IUPAC system most of the time and use common names only for compounds in frequent use.

AMINES

The first thing people notice about **amines,** which contain the $-NH_2$ functional group, is that they usually have a *very* foul odor. Amines are responsible for the odors that emanate from decomposing garbage and dead bodies, and they are one of the main causes of body odor. However, amines are essential to life itself; an important related class called amino acids will be covered in chapter 14.

Classification of Amines

Unlike most other functional groups, amines are subdivided into three distinct categories—primary, secondary, and tertiary—based on the number of organic groups attached to the nitrogen atom. Primary (1°), secondary (2°), and tertiary (3°) amines have 1, 2, and 3 organic groups, respectively. Figure 10.13 illustrates how amines are named. In the IUPAC system, 1° amines are named by locating the carbon atom the $-NH_2$ is attached to and assigning this C atom the lowest possible number on its chain. The name then becomes this number, the prefix *amino* and the alkane name for the chain—for example, 3-aminohexane or 2-aminopropane. Naming 2° and 3° amines is rather complicated because the individual organic groups vary greatly. In practice, many amines are given common names by identifying each organic group attached to the nitrogen atom then placing the names in front of the word *amine*—thus, aminomethane could also be called methyl amine (figure 10.13).

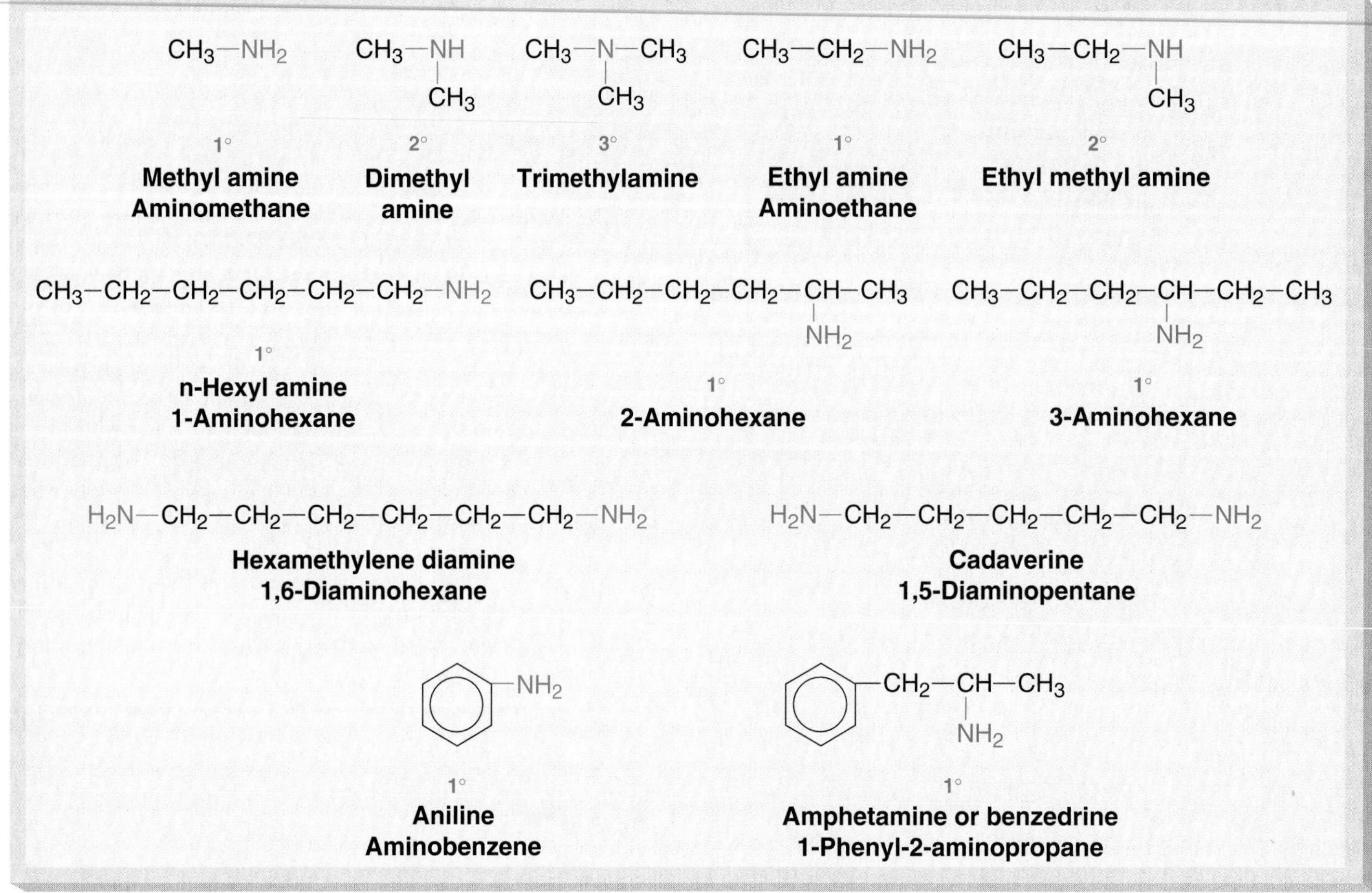

Figure 10.13 Structures and names for some amines. The symbols 1°, 2°, and 3° denote whether the amine is primary, secondary, or tertiary, respectively. Common and IUPAC names appear for primary amines.

PRACTICE PROBLEM 10.7

Name the primary amines in compounds *a* and *b* by the IUPAC system. Name the amines in compounds *c* and *d* by common name only. Are the amines in *c* and *d* secondary or tertiary?

a. $CH_3CH_2CH(NH_2)CH_2CH_2CH_2CH_3$

b. $CH_3CH(CH_3)CH_2CH_2CH(NH_2)CH_3$

c. $CH_3CH_2-NH-CH_2CH_3$

d. $CH_3-N(CH_2CH_3)-CH_3$

ANSWER In amines, the $-NH_2$ group sets the IUPAC numbering sequence. Compound *a* is therefore 3-aminoheptane, since the amino group is on the third of seven C atoms. Determining a common name for *a* is very difficult unless you know the name of the group; this group is too complex and is not used elsewhere in this book, so we won't assign it a common name. Compound *b* is named 5-methyl-2-aminohexane; again, the "common" name is too hard. The IUPAC method is much easier to use with primary (1°) amines.

Compound *c* is the 2° amine called diethyl amine because the NH group is attached to two (di) ethyl groups (CH_3CH_2). Compound *d* is the 3° amine called dimethyl ethyl amine because the N is attached to two methyl groups (CH_3) and one ethyl group. Finally, we know *c* is secondary because the NH group is attached to two organic groups, and *d* is tertiary because the N atom is attached to three organic groups.

BOX 10.4

ANILINE DYES: LIKE THE LILIES OF THE FIELDS

Although it is easy today to enter a clothing store and find a dazzling array of garments in virtually every shade and hue, this was not always so. A few thousand years ago, dyes were comparatively rare and expensive, and dyeing methods and sources were guarded trade secrets, handed down from generation to generation. Chemistry changed this situation dramatically!

Prior to the mid-1800s, the selection of dyes was limited to a few materials from plant or animal sources. Alizarin, a red dye, was extracted from madder plant roots *(Rubia tinctorum)*, and indigo, a blue dye, was extracted from the tropical *Indigofera* plant. Both have been known for at least three thousand years and were expensive. The most valuable dyes, however, were the purple dyes from shellfish and scarlet red dye obtained from some insects. The cost of these dyes was equivalent to the cost of gold or silver, so they were used only on cloth worn by wealthy people. Poorer people wore white or drab colors made from vegetable dyes.

This situation changed radically in 1856 when William Perkin (English, 1838–1907) made an accidental discovery (figure 1). Perkin was trying to synthesize quinine, a drug used to treat malaria, by reacting aniline sulfate with an oxidizing agent. Instead of quinine, he produced a brown material that yielded a rich purple dye solution when extracted with ethanol. This purple dye, called mauve, formed the basis of the synthetic dye industry that made Perkin wealthy. Perkin later synthesized some perfume ingredients and founded another industry. Meanwhile, other researchers, mostly in Germany, developed numerous synthetic dyes that eventually led to the limitless array of colors available today (figure 2).

Figure 1 William Perkin, the Englishman who founded the synthetic dye industry in 1856.

Figure 2 A homemade patchwork quilt displays the colorful chemical beauty of aniline dyes.

Amines are organic bases, as contrasted with organic acids (R—COOH). (We studied acids and bases in chapter 8.) Many drugs, dyes, and food supplements contain amino groups. Figure 10.13 shows the ringlike structure of amphetamine, a powerful central nervous system stimulant commonly called an "upper." Amphetamine is addictive, but is has some legitimate medical uses, for example, in nasal sprays to relieve congestion.

Figure 10.13 shows several other amines. Some contain more than one amino group—for example, an amine with two groups is called a diamine. The diamine 1,6-diaminohexane is used to make nylon-66, a common polymer discussed later in this chapter. Cadaverine, or 1,5-diaminopentane, with one less C atom than 1,6-diaminohexane, is the main compound that causes the stench in decaying fish. Aniline, a highly toxic aromatic amine, is the primary starting material for many important dyes and drugs (box 10.4). Aniline is also used as a catalyst in some chemical reactions, as well as in photography and the synthesis of herbicides and fungicides (chemical agents that kill unwanted plants or fungi).

Amines form hydrogen bonds as the alcohols and organic acids do, but the effect is weaker with nitrogen than with oxygen atoms (figure 10.14). This means that the boiling points of amines do not increase as much as those of the alcohols. The hydrogen-bonding effect is about the same for amines as for aldehydes.

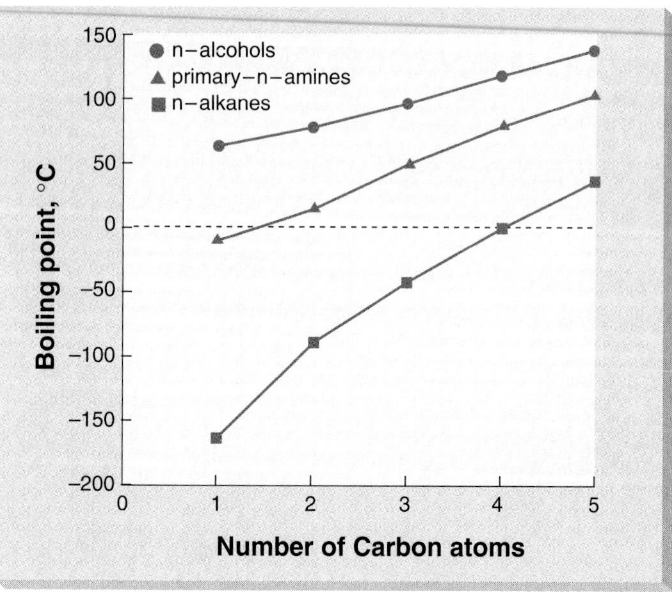

Figure 10.14 The effect of hydrogen bonding on amine boiling points.

ORGANIC ACIDS AND THEIR DERIVATIVES: THE CARBOXYL AND AMIDE GROUPS

The members of the organic acid class—carboxylic acids, esters, and amides—all possess a **carboxyl group,** a composite group consisting of the C=O carbonyl unit plus an O atom attached to the carbonyl unit. The O atom, in turn, may be part of an —OH group or another organic group (see table 10.1). In this sense, the carboxyl group is a blend of a carbonyl group and an alcohol (—OH) or ether (—O—) linkage. The resulting molecule is called a **carboxylic acid** when a hydrogen atom is attached to the oxygen atom, and an **ester,** when an alkyl or aryl group is attached to the oxygen atom. **Amides** are derivatives of the carboxyl group in which the O atom is replaced by a N atom. These groups are shown in figure 10.15.

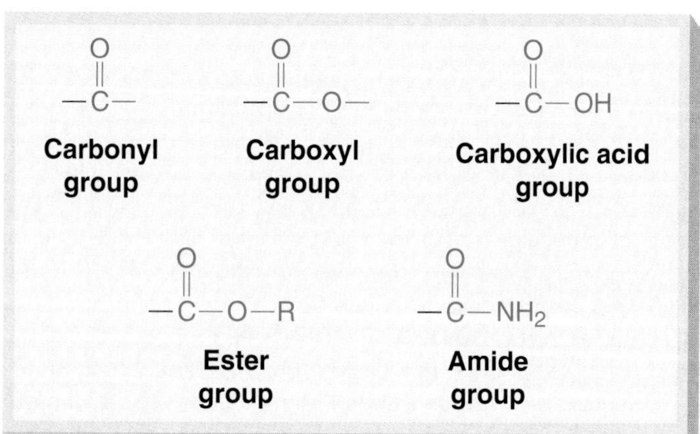

Figure 10.15 The carboxyl unit and its derivatives.

Carboxylic Acid

In the IUPAC system, carboxylic acids are named by replacing the *e* of the alkane name with the composite ending *oic acid.* Some examples, shown in figure 10.16, include formic (methanoic) acid, acetic (ethanoic) acid, butanoic acid, stearic acid, and benzoic acid.

Most carboxylic acids are weak acids that lower the pH of a solution only slightly. Carboxylic acids hydrogen bond strongly (see figure 10.1) and therefore have high boiling points. When there are more than 8 C atoms in a saturated molecule, the carboxylic acid is a solid. Organic acids usually occur in nature as esters or amides, but they sometimes exist in the "free" state, where they have a strong, rather unpleasant odor. Rancid butter, for example, contains the unpleasant-smelling 4-carbon butanoic acid (or butyric acid). The "low-price spread" (margarine) contains longer chain acids and never emits butyric acid if it goes bad. Other acids noted for unpleasant odors are the 6-, 8-, and 10-carbon acids, which go by the common names of caproic, caprylic, and capric acids, respectively, derived from the Latin word *caper,* for goat. Goats achieve their unenviable body odor by exuding these three acids, which is why most communities do not permit goat farms within city or village limits. However, goats can make good pets if you wash the acids away often enough!

PRACTICE PROBLEM 10.8

Name the following carboxylic acids by the IUPAC system.

a. $CH_3CH_2CH_2CH_2\text{—}\overset{\overset{\large O}{\|}}{C}\text{—}OH$

b. $CH_3CH_2\overset{\overset{\large CH_3}{|}}{C}HCH_2\text{—}\overset{\overset{\large O}{\|}}{C}\text{—}OH$

c. $CH_3\overset{\overset{\overset{\overset{\large OH}{|}}{C=O}}{|}}{C}HCH_2CH_2CH_3$

ANSWER The carboxylic acid group sets the numbering of the carbon atoms in the organic acids; this group is almost always located in position 1. Compound *a* has five carbons with the —COOH in position 1, so it is pentanoic acid. Compound *b* has a methyl group attached to the third carbon in the five-carbon chain and is, therefore, 3-methylpentanoic acid. Compound *c* is 2-methylpentanoic acid. This is because the carboxylic acid group is in position 1, with a methyl group (CH_3) attached to carbon 2 on the five-carbon chain. These acids do not have common names.

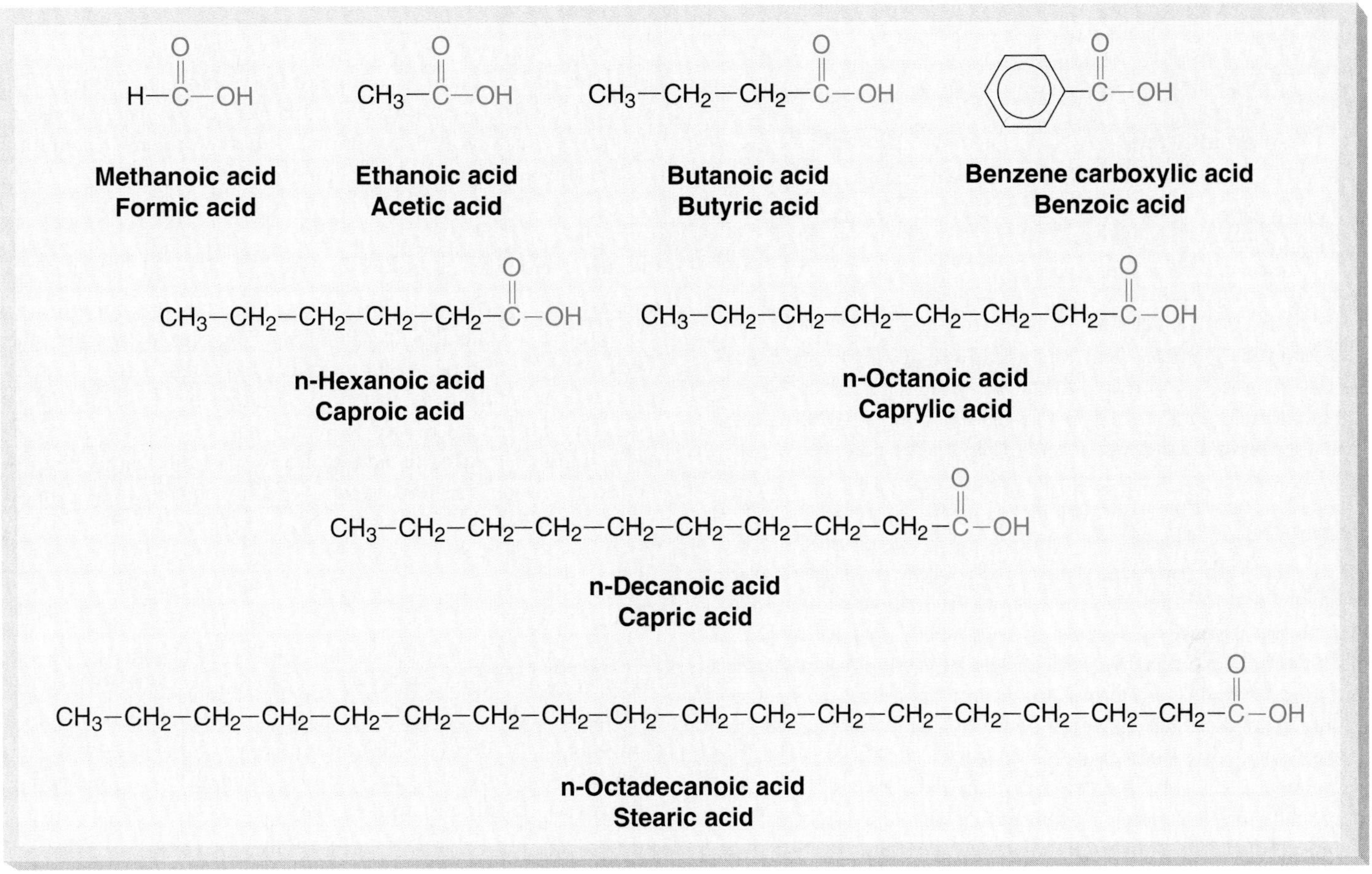

$H-\overset{\overset{O}{\|}}{C}-OH$

Methanoic acid
Formic acid

$CH_3-\overset{\overset{O}{\|}}{C}-OH$

Ethanoic acid
Acetic acid

$CH_3-CH_2-CH_2-\overset{\overset{O}{\|}}{C}-OH$

Butanoic acid
Butyric acid

$C_6H_5-\overset{\overset{O}{\|}}{C}-OH$

Benzene carboxylic acid
Benzoic acid

$CH_3-CH_2-CH_2-CH_2-CH_2-\overset{\overset{O}{\|}}{C}-OH$

n-Hexanoic acid
Caproic acid

$CH_3-CH_2-CH_2-CH_2-CH_2-CH_2-CH_2-\overset{\overset{O}{\|}}{C}-OH$

n-Octanoic acid
Caprylic acid

$CH_3-CH_2-CH_2-CH_2-CH_2-CH_2-CH_2-CH_2-CH_2-\overset{\overset{O}{\|}}{C}-OH$

n-Decanoic acid
Capric acid

$CH_3-CH_2-CH_2-CH_2-CH_2-CH_2-CH_2-CH_2-CH_2-CH_2-CH_2-CH_2-CH_2-CH_2-CH_2-CH_2-CH_2-\overset{\overset{O}{\|}}{C}-OH$

n-Octadecanoic acid
Stearic acid

Figure 10.16 The structures and names of some common organic acids.

EXPERIENCE THIS

Colorless or lightly colored fruit juices, such as lemon, apple, or grapefruit juice, contain carboxylic acids and make good invisible inks. Use a toothpick and one of these juices to write a message on paper; when the juice dries, no trace of the message is visible. However, when you heat the piece of paper over a lightbulb, the message suddenly appears. You can also use diluted milk, sugar water, or soft drinks in the same way. The heat causes the organic compounds in these "invisible inks" to decompose, leaving a trace of carbon behind.

Formic Acid

The name *formic acid* (as figure 10.16 indicates, methanoic acid in the IUPAC nomenclature) is derived from the Latin word *formica*, which means "ant." The acid was first isolated as one of the products formed during the destructive distillation of ants. The sting of an ant bite is due to formic acid (figure 10.17). This acid stimulates our body's pain-sensing nerve endings by lowering the pH in their environment. The bite area swells as water flows in to dilute the formic acid and raise the pH. Because the sting is caused by an acid, the best treatment is to apply a mild base, such as baking soda ($NaHCO_3$). Ants and other insects often release other chemicals when they bite that can cause severe allergic reactions in some people.

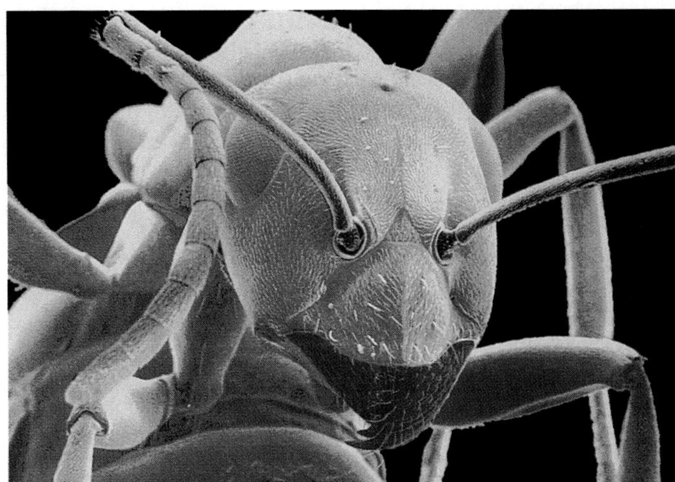

Figure 10.17 Ant bites contain formic acid.

$$2\,CH_3-\overset{O}{\overset{\|}{C}}-H + O_2 \longrightarrow 2\,CH_3-\overset{O}{\overset{\|}{C}}-OH$$

Acetaldehyde **Acetic acid**

Figure 10.18 Acetic acid is synthesized by oxidizing acetaldehyde.

Formic acid bottles bear the warning "avoid skin contact" because this acid is corrosive to the skin and body tissues. Nevertheless, formic acid is used in the manufacture of dyes, insecticides, perfumes, medicines, and plastics.

Acetic Acid

Acetic acid (ethanoic acid in the IUPAC system; as figure 10.16 indicates) is by far the most important carboxylic acid. It is manufactured industrially by oxidizing acetaldehyde (figure 10.18), a raw material obtained by oxidizing ethanol or hydrating acetylene. Acetic acid is also formed in vinegar, as *Acetobacter* bacteria oxidize ethanol. Vinegar, which contains about 5 percent acetic acid in water, has been used for centuries to flavor foods. But glacial acetic acid (100 percent acetic acid) is toxic (LD_{50} = 3.5 g/kg). The first person to synthesize acetic acid directly from chemical elements was Adolph Kolbe (German, 1818–84) in 1845. Acetic acid is used in the manufacture of cellulose acetate, vinyl acetate, drugs, dyes, insecticides, photographic chemicals, and food additives.

PRACTICE PROBLEM 10.9

Just how safe a food is vinegar? Ordinary vinegar is a 5 percent aqueous solution of acetic acid, but 100 percent acetic acid is toxic with an LD_{50} = 3.5 g/kg. How many grams of vinegar would a 100 kg person have to ingest in a short period of time to take in toxic levels of acetic acid? If you assume that 1.0 L vinegar weighs 1,000 g, how many liters of vinegar would this be?

ANSWER The total amount of acetic acid toxic for a 100 kg person is 100 kg × 3.5 g/kg = 350 g. Since vinegar only contains 5 percent acetic acid, the person would need to drink 350 ÷ 0.05 = 7,000 g vinegar. This works out to be about 7 L, or the equivalent of three and a half 2-liter bottles of soft drink. While this might seem a feasible amount to drink, the strong taste of vinegar makes it unlikely anyone would ingest so much over a short period of time.

Butanoic Acid

Although butanoic acid (see figure 10.16) has an obnoxious odor, the esters derived from it are useful in perfumes and flavors. The acid itself is also used in the manufacture of emulsifying agents, pharmaceuticals, and disinfectants. As noted previously, butanoic acid is the culprit behind rancid butter odor.

Stearic Acid

Stearic acid (see figure 10.16), an organic acid containing 18 carbon atoms, is a member of the fatty acid group, a group of acids with 12 to 22 carbon atoms that may be isolated from fats and oils. Stearic acid (m.p., 69.6°C), obtained by reacting beef fat with NaOH, has been used to make soap for many generations (see chapter 13). Stearic acid is also used in lubricants, pharmaceuticals, cosmetics, rubber compounding, coatings, shoe polish, and toothpaste.

Benzoic Acid

Benzoic acid, the best known aromatic acid (figure 10.16), is a solid (m.p. 121°C). Combined with sodium, it has antibacterial properties and is a common food additive for that reason. Sodium benzoate is also used as a rust and mildew inhibitor. Benzoic acid itself is used in the manufacture of perfumes, flavors, medicines, plasticizers, dentifrices, and polymers.

Esters

Another class of compounds, along with carboxylic acids, that contains the carboxyl group is the **esters.** Although acids have unpleasant odors, the esters derived from them usually have a fairly pleasant aroma and are frequently used in perfumes and flavorings. Table 10.3 lists some common esters; in the table, the carboxyl group is shown as —COO. Esters are named by replacing the *ic* or *oic acid* ending with *ate* and then placing the name of the organic group (R in figure 10.15) in front of the name. For example, two of the esters derived form acetic acid are ethyl acetate and propyl acetate.

Esters are synthesized by reacting acids with alcohols, as shown for ethyl acetate in figure 10.19. As the reaction proceeds, a molecule of water is lost between the acetic acid and the ethanol. Usually a small amount of a strong inorganic acid, such as sulfuric acid, is added as a catalyst, or the reaction proceeds extremely slowly. However, even with a catalyst, the reaction shown in figure 10.19 will not run to completion; it reaches an **equilibrium,** or balance. This occurs because as new ester molecules form, other ester molecules react with water and hydrolyze back to acetic acid and ethanol. In ethyl acetate synthesis, this equilibrium occurs when the reaction is about 67 percent completed—in other words, about 33 percent acetic acid and 33 percent ethanol remain unreacted. Chemists use double arrows as in figure 10.19 to indicate an **equilibrium** or **reversible reaction.**

TABLE 10.3 The names, structures, and odors of some esters.

NAME	STRUCTURE	ODOR
ethyl formate	$HCOO-CH_2CH_3$	rum
isobutyl formate	$HCOO-CH_2CH(CH_3)_2$	raspberry
ethyl acetate	$CH_3COO-CH_2CH_3$	floral
propyl acetate	$CH_3COO-CH_2CH_2CH_3$	pear
amyl acetate	$CH_3COO-CH_2CH_2CH_2CH_2CH_3$	banana
isoamyl acetate	$CH_3COO-CH_2CH_2CH(CH_3)_2$	banana
octyl acetate	$CH_3COO-CH_2CH_2CH_2CH_2CH_2CH_2CH_2CH_3$	orange
benzyl acetate	$CH_3COO-CH_2C_6H_5$	jasmine
isobutyl propionate	$CH_3CH_2COO-CH_2CH(CH_3)_2$	apple
methyl butyrate	$CH_3CH_2CH_2COO-CH_3$	rum
ethyl butyrate	$CH_3CH_2CH_2COO-CH_2CH_3$	pineapple
butyl butyrate	$CH_3CH_2CH_2COO-CH_2CH_2CH_2CH_3$	pineapple
amyl butyrate	$CH_3CH_2CH_2COO-CH_2CH_2CH_2CH_2CH_3$	apricot
isoamyl pentanoate	$CH_3CH_2CH_2CH_2COO-CH_2CH_2CH(CH_3)_2$	apple

Esters and Equilibrium Reactions

Figure 10.20 compares an equilibrium reaction to a familiar example—a pair of small, equal-weight children on a seesaw. When both sit balanced on the seesaw with their feet off the ground (figure 10.20a), they are at equilibrium. If one child is given a 5 kg weight, the seesaw tips so that this child touches the ground (figure 10.20b). If the seesaw is long enough, the other child might shimmy away from center to counterbalance the extra weight. Equilibrium is once more achieved (figure 10.20c). But what if the first child hands back the 5 kg weight? The second child must shimmy back in order to rebalance the seesaw. As the weight is added or removed, the children adjust to achieve equilibrium.

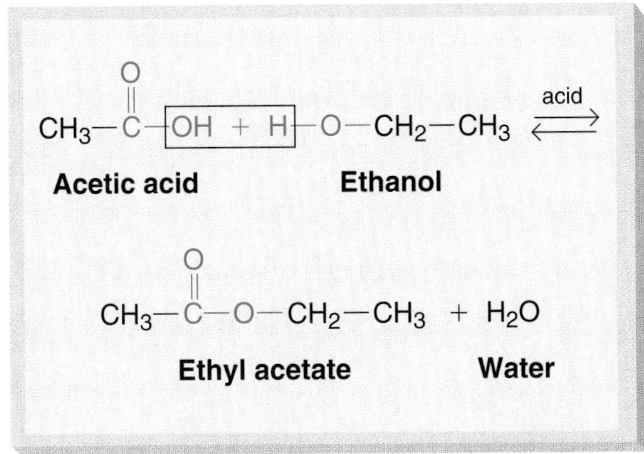

Figure 10.19 Ethyl acetate is synthesized by reacting acetic acid with ethanol. The double arrow indicates that this is an equilibrium (or reversible) reaction.

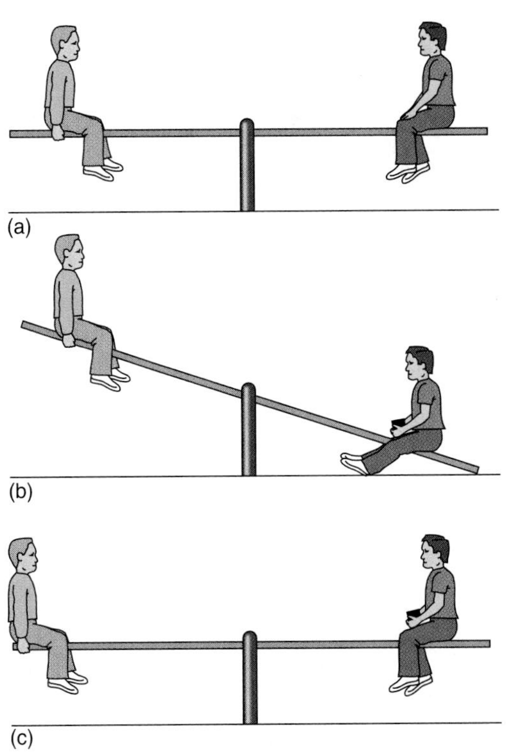

Figure 10.20 **(a)** Two equal-weight children at equilibrium on a seesaw. **(b)** The child on the right has been given a 5 kg weight; equilibrium is disturbed. **(c)** The child on the left shimmies farther back on the seesaw and equilibrium is restored again.

In the same way, a chemical reaction can "rebalance" itself to achieve equilibrium after it has been manipulated or disturbed. If some water (a product) is removed from the reaction shown in figure 10.19, the system immediately seeks to rebalance itself by producing more water. Since both ethyl acetate and water are produced in the reaction, as the system produces more water, it also produces more ethyl acetate. As more water is removed, the reaction continues to reestablish equilibrium until all of the reactants are consumed and converted into ethyl acetate. (This assumes an equal number of acetic acid and ethanol molecules are available as reactants.) The reaction is carried out by boiling (refluxing) the acid-alcohol mixture and removing the chemical (either the ester or the water) with the lowest boiling point by condensing it from the gaseous state into the liquid state and collecting it in another container. This type of reaction—called a **condensation reaction**—always produces a small molecule (typically water or ammonia) as a byproduct. Condensation reactions are important synthesis reactions.

Along with removing product, there is a second way to disturb a chemical equilibrium. If we add more of one reactant—ethanol, for example—the reaction will shift to reestablish equilibrium, removing ethanol by producing more ethyl acetate. If we add enough ethanol, we can "force" almost all of the acetic acid to convert into ethyl acetate. Chemists routinely use these techniques to increase the yield of a reaction. (The yield is the percentage of the desired product—ethyl acetate, in this case.)

Esters are important not only in cosmetics and flavorings but in other applications as well. Beeswax, a mixture of esters such as $C_{25}H_{51}COO—C_{30}H_{61}$, and carnauba wax, another ester, are used in many automobile and furniture polishes. Fats and oils (not petroleum oil) are esters of glycerol with fatty acids; they are important in our diet (chapter 15). The esters aspirin and methyl salicylate are used in medicine as analgesics and anti-inflammatories (chapter 17). Methyl salicylate, also called oil of wintergreen, is the main ingredient in wintergreen flavor. Ethyl acetate is used as a nail polish remover.

PRACTICE PROBLEM 10.10

The hardest problem in naming esters is figuring out which part comes from the acid and forms the basis for the name, and which part comes from the alcohol to form the prefix. The following two examples may help you understand the difference. First, locate the —COO— group; the group attached to the C is the acid and the portion attached to one of the O atoms is the alcohol. Can you determine the names for these two esters? (Refer to figure 10.16 and table 10.3 for help.)

a. $CH_3—\overset{\overset{O}{\|}}{C}—O—CH_2—CH_3$

b. $CH_3—CH_2—CH_2—\overset{\overset{O}{\|}}{C}—O—CH_3$

ANSWER In *a.* the original acid is acetic acid (CH_3COOH). The alcohol is ethanol ($HOCH_2CH_3$). The base name of this compound is therefore acetate, the prefix ethyl: ethyl acetate. In *b.* the acid is butyric acid ($CH_3CH_2CH_2COOH$—see figure 10.17). The alcohol is methanol (CH_3OH). The ester derived from these two compounds and shown here is methyl butyrate.

Amides

Amides are the reaction products of carboxylic acid with ammonia (NH_3) or with ammonia-derived primary or secondary amines containing NH_2 or NH (figure 10.21). Amides may therefore have zero, one, or two alkyl groups attached to the N atom. Amide formation is achieved through a condensation reaction, as the boxed area in figure 10.21 indicates.

Proteins (chapter 14) and nylon polymers contain amide groups. Acetaminophen, an important analgesic, and sulfa drugs are amide-containing pharmaceutical agents (chapter 17). Some amides are used as plasticizers, chemicals that soften plastics.

Condensation Polymerization

Condensation reactions produce several important classes of polymers, including polyesters and polyamides. While addition

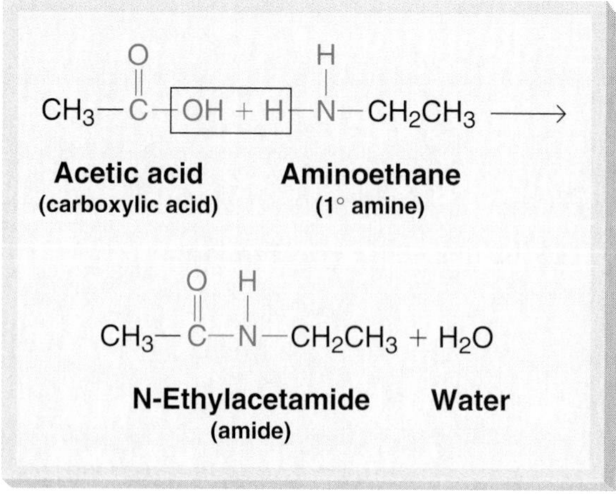

Figure 10.21 Amides form from a carboxylic acid and ammonia or a primary or secondary amine in a condensation reaction.

polymerization and **condensation polymerization** both produce polymers, or large organic molecules formed by chains of smaller molecules, these two types of reactions differ in at least two major ways. First, condensation polymerization produces a byproduct—often water—while addition polymerization does not. Second, most condensation polymers have a noncarbon atom such as oxygen or nitrogen within the polymer chain, while most addition polymers have only C atoms in their polymer chains. In either kind of polymer, the pendant groups hanging from the main polymer chain could contain atoms other than C or H.

While these are the general differences between condensation and addition polymers, there are several important exceptions to the rules. For example, poly(oxymethylene) is an addition polymer containing an O atom—not just C atoms—in its chain (see figure 10.10). And phenolic polymers (chapter 11) are condensation polymers that have only C atoms—no noncarbon atoms—in their polymer chains.

Polyesters (chains of esters) and polyamides (chains of amides) are condensation polymers with a large commercial market. The annual production of polyesters is about 2.087×10^6 metric tons. About 74 percent of these polyesters are used in fiber applications, with the remaining 26 percent used mostly in films and bottles, such as the PET—poly(ethylene terephthalate)—in soft drink bottles (recycle class 4). Poly(ethylene terephthalate) is the most commercially important polyester.

Poly(ethylene Terephthalate)

Poly(ethylene terephthalate) (commonly called **PET**) was first introduced in 1940 and is the major polyester in both plastic and fiber applications. Polyesters resulted from the research of Wallace Hume Carothers, one of the pioneers who created the field of polymer chemistry in the 1920s. Figure 10.22 depicts the synthesis of poly(ethylene terephthalate) from ethylene glycol and methyl terephthalate ester. (The reaction could be run with terephthalic acid, but the acid requires more energy than the ester and also produces lower yields.) Other polyesters are prepared in a similar manner.

PET fiber, marketed under the trade names **Dacron,** Terylene, and Kodel, is the most common synthetic fiber used in the permanent press, wash-and-wear garments that revolutionized clothing maintenance. Most people born after the 1950s are hardly familiar with the tedium of ironing that preceded the advent of these wonder fabrics. Polyesters have captured most of the fiber market in the United States. Less than thirty years ago, King Cotton and the cellulose fibers (rayon) reigned supreme in the marketplace. Although cotton and rayon are making a comeback, today's fiber market is only about 28 percent cotton and cellulose derivations. Some types of cotton are now being cultivated that have a natural brown or green color; they are usable as dye-free fabrics.

The plastic film version of poly(ethylene terephthalate) which goes under the trade name **Mylar,** is unusually tough and has high abrasion resistance and low moisture absorption. After Mylar is coated with an iron oxide coating, it serves as the magnetic recording tape used in video and audio tapes and computer disks. It is unlikely that the magnetic tape industry could have blossomed without a polymer film. While most people tend to think of these fields as electronic, chemistry was essential to their development.

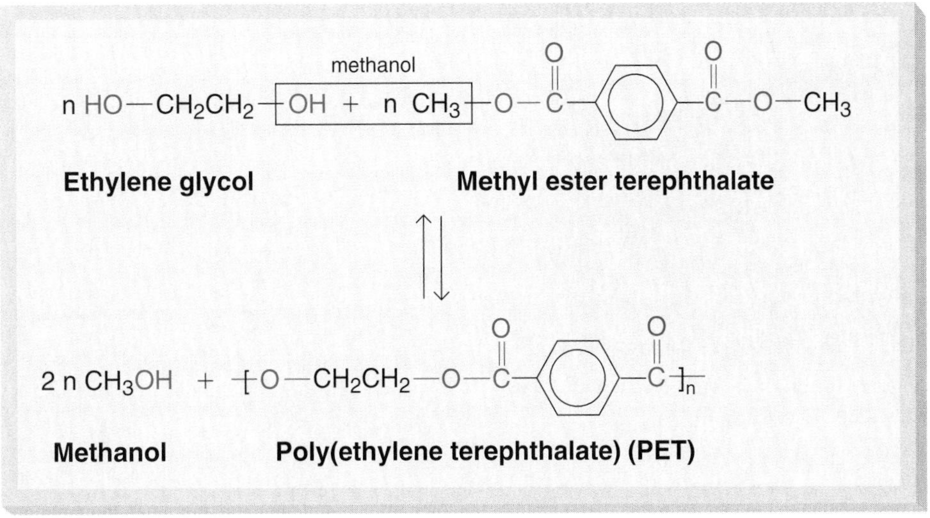

$$n\,HO-CH_2CH_2-OH + n\,CH_3-O-\overset{O}{\overset{\|}{C}}-C_6H_4-\overset{O}{\overset{\|}{C}}-O-CH_3$$

methanol

Ethylene glycol **Methyl ester terephthalate**

$$\rightleftharpoons$$

$$2n\,CH_3OH + \left[O-CH_2CH_2-O-\overset{O}{\overset{\|}{C}}-C_6H_4-\overset{O}{\overset{\|}{C}}\right]_n$$

Methanol **Poly(ethylene terephthalate) (PET)**

Figure 10.22 PET, or poly(ethylene terephthalate), is a commercially important polyester synthesized from ethylene glycol and methyl terephthalate ester. Methanol is formed as a byproduct in a condensation reaction.

PRACTICE PROBLEM 10.11

Figure 10.22 shows an equilibrium reaction. It is easy to shift the equilibrium to form high amounts of the product. What is the key to do this? (*Hint:* What compound is formed as a byproduct in a condensation reaction?)

ANSWER There are several ways to shift the equilibrium point. The most effective is to remove methanol, which has a low boiling point, in a condensation reaction. In fact, this is the main reason for using the methyl ester of terephthalic acid, rather than the acid itself.

Figure 10.23 A wide variety of products are made from Mylar, the trade name for PET plastic film.

Poly(ethylene terephthalate) films are strong enough to be used in greenhouses and other locations where a durable, transparent film is needed. The polyester fibers used in automobile tires are nearly always poly(ethylene terephthalate). Polyester films are also used in home heat-sealing devices. Figure 10.23 features examples of the many uses of this important plastic and fiber.

Poly(hexamethylene Adipamide)

A second commercially important condensation polymer is **poly(hexamethylene adipamide).** Figure 10.24 shows the synthesis of this polyamide, better known as **Nylon 66,** by a condensation reaction between adipic acid and 1,6-diaminohexane. Water is formed as a byproduct.

The term *nylon* is now used in a generic sense to describe several different fibers (box 10.5). Most nylons are made using a reaction similar to the one shown in figure 10.24; some other nylon structures are shown in table 11.1, chapter 11. Nylon, a familiar fabric, has many nonfabric applications as well—it is used to make gears for small motors, monofilament fishing line, and molded plastic items.

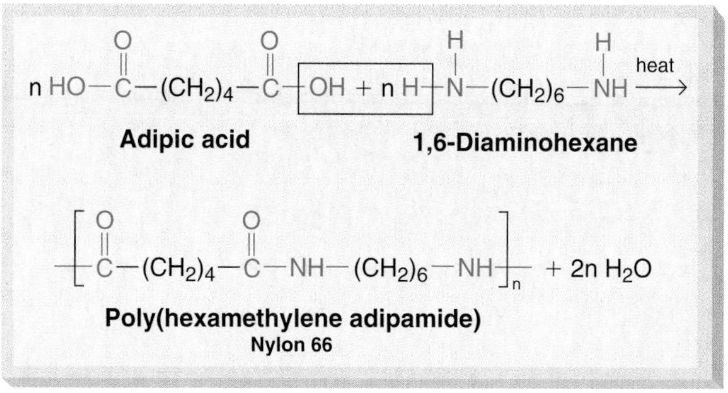

$$n\,HO-\overset{O}{\overset{\|}{C}}-(CH_2)_4-\overset{O}{\overset{\|}{C}}-\boxed{OH + n\,H}-\overset{H}{\overset{|}{N}}-(CH_2)_6-\overset{H}{\overset{|}{N}}H \xrightarrow{heat}$$

Adipic acid **1,6-Diaminohexane**

$$\left[-\overset{O}{\overset{\|}{C}}-(CH_2)_4-\overset{O}{\overset{\|}{C}}-NH-(CH_2)_6-NH-\right]_n + 2n\,H_2O$$

Poly(hexamethylene adipamide)
Nylon 66

Figure 10.24 Nylon 66, or poly(hexamethylene adipamide), is synthesized from adipic acid and 1,6-diaminohexane. Water is formed as a byproduct in a condensation reaction.

BOX 10.5

THE STORY OF NYLON

Nylon, the first of a series of polyamides, was introduced in 1938, two years before polyesters. Like polyesters, polyamides sprang from the research of Wallace Hume Carothers. Nylon is a mimic of silk, a natural polypeptide containing an amide group (chapter 14). Silk fiber is obtained from the cocoons of silk worms, but we cannot raise these insects successfully in the United States because of our climate. Consequently, silk fiber is difficult to obtain whenever anything disturbs the foreign supply—wars, adverse economics, or unsuitable environmental factors, for example. For these reasons, and also because scientists were curious and desired to duplicate or even surpass the inventions of nature, Carothers and many others studied how to synthesize an artificial silk.

The research bore fruit on October 27, 1938 when the DuPont company announced the availability of a new silk substitute, claiming it was "passing in strength and elasticity any previously known textile fibers." DuPont advertised that this new polymer was made from air, water, and coal. This was a half-truth since the chemicals used to make this fiber were obtained from coal tar. Nylon was first displayed at the New York World's Fair in 1939 and first marketed as women's hosiery on May 15, 1940, dubbed "Nylon Day." A wave of near-hysteria swept the country as women waited in long lines to obtain this new miracle hosiery. Stockrooms were rapidly depleted of their limited supplies, nearly leading to riots. Before the year's end, enthusiastic shoppers had snapped up more than 36 million pairs of the "miracle hose."

Today, because nylon is so common and inexpensive, it is difficult to visualize these events. But during the early 1940s, nylon rapidly replaced silk in essentially the entire market. Nylon has also taken over the toothbrush bristle market once monopolized by animal hair (mainly pig bristles). These "improved" toothbrushes are cheaper and definitely more sterile. And, of course, nylon has many other everyday uses, from rainwear to sails for boats (figure 1).

Figure 1 A regatta of boats heading off into the sunset.

SUMMARY

Organic chemistry encompasses not only hydrocarbons, but millions of other molecules derived from them. These other molecules contain functional groups—groups of atoms that react chemically and that help determine how the molecule functions. Billions of organic compounds may be organized into a smaller number of functional group classes.

Different functional groups have different chemical and physical properties. For example, alcohols and carboxylic acids form strong hydrocarbon bonds and thus have higher boiling points than their parent hydrocarbons. Similarly, aldehydes are dipoles, which also increases their boiling points.

The major functional groups are the hydroxyl groups, the oxides, the halides, the carbonyl groups, the amines, the carboxyl groups, and the amides. Two major compound classes include the hydroxyl (—OH) group: the alcohols and phenols. Alcohols have their —OH group attached to a carbon chain. Their IUPAC names, derived from their parent alkanes, end in *ol* rather than *ane*. Alcohols, widely used in chemical reactions and as solvents, include methanol, ethanol (the alcohol in alcoholic beverages), 2-propanol (used in rubbing alcohol), ethylene glycol, and glycerol. All are more or less toxic to humans when ingested.

The other hydroxyl group class, the phenols, have their —OH group attached to an aromatic carbon ring. Most phenols are acidic solids. They are used in cough syrups, topical antiseptics, and other pharmaceuticals, and in the preparation of phenolic resins and plastics.

The oxide (—O—) group is the functional group in the ethers. These compounds are formed when a pair of alcohol or phenol molecules lose a molecule of water and bond with the remaining oxygen atom between them. Diethyl ether was first widely used as general anesthetic. Other alcohols and ethers are used to oxygenate gasoline, making engines run smoother and cleaner.

Halides contain a halogen atom—fluorine, chlorine, bromine, or iodine. They derive their names from their parent hydrocarbon, with the halogen as a prefix (fluoro-, chloro-, bromo-, or iodo-). Chlorofluorocarbons (CFCs) were commonly used in cooling systems and aerosol sprays until recent concerns arose about their effects on the ozone layer in the upper atmosphere. Other halides are used in dry cleaning, laboratories, and in manufacturing poly(vinyl chloride) or PVC plastics.

Carbonyl groups (C=O) define the aldehyde and ketone classes. An aldehyde has one H atom attached to one end of the C=O group, and a hydrogen, alkyl, alkenyl, or aryl group attached to the other end. It derives its name from the parent alkane, replacing the *e* with *al*. Aldehydes are used to preserve biological tissues and to manufacture plastics.

Ketones have a C=O group flanked by two alkyl or aryl groups. The IUPAC name of a ketone comes from the parent alkane, with *one* replacing the *e* at the end of the alkane name. Ketones are used as solvents and cleaning fluids as well as in medicine and organic synthesis.

Amines, containing the $—NH_2$ functional group attached to the carbon atom, usually have a foul odor. They are divided into three categories—primary, secondary, and tertiary—depending on whether 1, 2, or 3 organic groups are attached to the nitrogen atom. Amines cause the stench that emanates from decomposing garbage and dead bodies. Many drugs, dyes, and food supplements contain amino groups.

The carboxyl group is contained in the carboxylic acids and esters. Carboxylic acids contain an O atom and a H atom attached to the carbonyl group (creating a —COOH acid group), while an ester has an O atom and an alkyl or aryl group attached to a carbonyl group. Carboxylic acid names end in *ic* or *oic acid* rather than the *e*, ending from the parent alkane name. Carboxylic acids serve a variety of uses in pharmaceuticals, lubricants, perfumes, flavors, plasticizers, and polymers. They include formic acid, acetic acid, butanoic acid, stearic acid, and benzoic acid—all weak organic acids, some with unpleasant odors.

Esters have pleasant fragrances and are frequently used in perfumes. They are synthesized by reacting acids with alcohols, and their names derive from their parent carboxylic acids, with *ate* replacing *ic* or *oic acid*. Many of the reactions that produce esters are equilibrium (reversible) reactions.

Amides are formed by a condensation reaction between (1) carboxylic acids and (2) ammonia or primary or secondary amines. Proteins and nylon polymers contain amide groups. Amides are also used in pharmaceutical agents and plasticizers.

Condensation polymerization produces polymers, or large organic molecules made up of chains of smaller molecules, by condensing out a byproduct. Condensation polymerization differs from addition polymerization in two ways. First, condensation reactions produce a byproduct, while addition reactions don't. Second, most condensation polymers have a noncarbon atom within the units that make up the polymer chain, while most addition polymers have only carbon atoms in their polymer chains. Condensation polymers include polyesters and polyamides. PET, or poly(ethylene terephthalate), used in soft drink bottles, fibers, and films, is a polyester. Nylon 66, or poly(hexamethylene adipamide), is one of many types of nylon. Nylon is used in textiles, for gears in small motors, in fishing line, and in molded plastics.

KEY TERMS

alcohols 246
aldehydes 252
amides 258
amines 255
carbonyl group 252
carboxyl group 258
carboxylic acid 258
condensation polymerization 263
condensation reaction 262
Dacron 263
equilibrium 261
esters 260
ethers 250
functional groups 244
hydroxyl group 246
ketones 254
LD_{50} 249
Mylar 263
Nylon 66 264
phenols 250
poly(ethylene terephthalate) (PET) 263
poly(hexamethylene adipamide) 264
reversible reaction 261

READINGS OF INTEREST

Alcohols and Related Compounds as Fuels

Goehna, Hermann, and Koenig, Peter. "Producing methanol from CO_2." *Chemtech*, June 1994, pp. 36–39. Why not take the CO_2 accumulating in our atmosphere and convert it into a useful fuel?

Lucas, Allison. "MTBE still facing pressure from ethanol under latest fuel proposal." *Chemical Week*, January 26, 1994. Since ethanol is less expensive than MTBE and has an octane number of 108, this competition is hardly surprising. Ethanol is also easier to make from plant sources.

Anderson, Earl V. "Brazil's program to use ethanol as transportation fuel loses steam." *Chemical & Engineering News,* October 18, 1993, pp. 13–15. Brazil's market for ethanol as a motor vehicle fuel peaked at 28 percent in 1988 and has declined ever since. This paper outlines the historical background on Brazil's use of ethanol and explains why ethanol use is declining.

Anderson, Earl V. "Health studies indicate MTBE is safe gasoline additive." *Chemical & Engineering News,* September 30, 1993, pp. 9–18. With an octane number of 116, MTBE is a promising gasoline additive, but some recent reports claim it might be harmful to people. This article reviews studies that suggest these fears are without foundation.

Driscoll, William L. "Fill 'er up with biomass derivatives." *Technology Review,* August/September 1993, pp. 74–76. A review of the book *Renewable Energy Sources for Fuels and Electricity* by T. B. Johansson, H. Kelly, A. K. N. Reddy, and R. H. Williams (Washington, D.C.: Island Press). It is possible, in principle, to convert biomass (including garbage) into useful fuels such as ethanol. But the road to success seems bumpier than expected.

Boddey, Robert. "'Green' energy from sugar cane." *Chemistry & Industry,* May 17, 1993, pp. 355–358. Ethanol, obtained from sugar cane, is widely used as a motor vehicle fuel in Brazil.

"Process might boost ethanol output from municipal waste." *R&D Magazine,* February 1992, p. 26. This process, developed at Colorado State University to produce ethanol from garbage, may help us recycle waste for use as an ethanol fuel, solvent, or chemical for synthesizing other molecules.

Alcoholism

Goode, Stephen. "Are America's college students majoring in booze?" *Insight,* August 8, 1994, pp. 15–17. This article explores the problem of alcoholism on college campuses.

Blum, Kenneth. 1992. *Alcohol and the addictive brain.* New York: Free Press. This book supports the genetic basis for alcoholism, or at least presents alcoholism as a disease rather than a social problem. Explains one side of a controversial topic.

Bower, B. "Alcoholism: Nurture may often outdo nature." *Science News,* February 1, 1992, p. 69. Which is more responsible for producing alcoholism—genes or the environment? This article leans toward the latter.

"Alcoholism's drain on the economy may be deeper than ever." *Business Week,* December 30, 1991, p. 26. This drain may run as deep as $150 billion by the mid-1990s! Eventually, you will foot this bill.

Bower, B. "Early alcoholism: Crime, depression higher." *Science News,* March 25, 1989, p. 180. If you even vaguely contemplate joining America's 10 million alcoholics, you might want to know what lies in store.

Ziegler, Edward, and Goldfrank, Lewis R. "Knocking back a few." *Reader's Digest,* February 1988, pp. 115–118. Written from the standpoint of an emergency room physician, this condensed article presents a sobering view of the "disease" that afflicts ten million Americans. Easy reading, but not pleasant!

Jackson, Charles. 1944. *The lost weekend.* New York: Noonday Press. This is a classic work of "fiction" about alcoholism that led to the movie of the same name (starring Ray Milland). The book is actually based on Jackson's own battle with alcohol and is still one of the best books on America's alcohol problem. The film version of this book is also available at videotape stores.

Toxic Chemicals

Noble, Charles. "Keeping OSHA's feet to the fire." *Technology Review,* February/March 1992, pp. 43–51. Part of OSHA's business is to appraise toxicity and set toxic limits, but more than a thousand new substances appear in the workplace every year. How can OSHA protect the American worker?

Dyes, Flavors, and Odors

Kiefer, David M. "Organic chemicals' mauve beginning." *Chemical & Engineering News,* August 9, 1993, pp. 22–23. A review of the book *The Rainbow Makers: The Origin of the Synthetic Dyestuffs Industry in Western Europe,* by Anthony S. Travis (Cranbury, N.J.: Lehigh University Press). Contains information about Perkins and others who developed the dye industry.

Dunkel, Tom. "It's time to wake up and smell the rosemary." *Insight,* June 15, 1992, pp. 14–16, 36–38. Spices and flavors add variety to life. This article discusses the possibility that they can also relieve some of the stress of daily living.

Amato, Ivan. "The ascent of odorless chemistry." *Science,* vol. 256, April 17, 1992, pp. 306–308. Many people associate chemistry with odors, often bad ones; but many chemists now run odor-free experiments by computer.

Hirshlag, Jennifer. "Only a rose smells as sweet as a rose." *Chemical Week,* August 24, 1988, pp. 64–66C. Perhaps, but what chemicals give a rose its sweet fragrance? Roses contain at least ten different chemicals which combine to shape specific odors. Each has at least one functional group, but the actual structures of these chemicals are quite different. Why do they all smell like a rose?

Engen, Trygg. "Remembering odors and their names." *American Scientist,* 75, September-October 1987, pp. 497–503. Once an odor enters your memory, you remember it, even if years pass before you smell the same aroma again. What enables us to distinguish odors and why do we remember them so well?

Layman, Patricia L. "Flavors and fragrances industry taking on a new look." *Chemical & Engineering News,* July 20, 1987, pp. 35–38. Along with discussing the industry in general, this article lists the top fifteen firms that hold half the world's market in flavors and fragrances. See how chemistry and business interact.

Nylon

Hounshell, David A., and Smith, John Kenly Jr. "The nylon drama." *Invention & Technology,* Fall 1988, pp. 40–55. An easy-to-read account of the development of nylon, including photographs of

"Nylon Day," tug-of-wars over nylon stockings, and the first nylon football pants, used in 1941 at Notre Dame.

Miscellaneous

McKone, Harold T. "Embalming: A rite involving early chemistry." *Today's Chemist at Work*, April 1994, pp. 68–70. This readable article gives more information on the topic discussed in box 10.3.

Kirschner, Elisabeth. "ACS meeting highlights advance in organic magnets." *Chemical Week*, April 15, 1992, p. 6. Tomorrow's magnet just might not be made from iron or another metal—some functional organic molecules show promising magnetic properties.

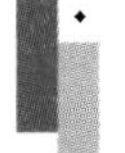

PROBLEMS AND QUESTIONS

1. Briefly define the following terms (as used in chemistry):

a. anesthetic
b. functional group
c. equilibrium
d. LD_{50}
e. fungicide
f. herbicide
g. addition polymerization
h. condensation polymerization

2. What contribution did each of the following people make to chemistry?

a. Carothers
b. Perkin
c. Kolbe

3. Draw an example of a compound that contains each of the following functional groups. (*Hint:* You can obtain almost all of this information from table 10.1).

a. ester
b. amide
c. primary (1°) amine
d. secondary (2°) amine
e. aldehyde
f. ketone
g. carboxylic acid
h. alcohol
i. phenol
j. ether

4. How do the physical properties of organic compounds with functional groups differ from the physical properties of organic compounds without them?

5. Which functional groups engage in the greatest amount of hydrogen bonding?

6. Which are generally more reactive—organic compounds with functional groups or without such groups?

7. Identify the class (for example, alcohol, ketone, or aldehyde) for each of the following compounds. (*Hint:* You will find table 10.1 very useful in determining the class of compound.)

a. $CH_3CH_2CH_2CH_2CH_2{-}OH$

b. $CH_3\underset{\underset{OH}{|}}{C}HCH_2CH_2CH_3$

c. $CH_3CH_2CH_2CH_2{-}\overset{\overset{O}{\|}}{C}{-}H$

d. $CH_3CH_2{-}\overset{\overset{O}{\|}}{C}{-}CH_3$

e. $CH_3CH_2CH_2CH_2{-}NH_2$

f. $CH_3CH_2{-}\overset{\overset{O}{\|}}{C}{-}CH_2CH_2CH_3$

g. $CH_3CH_2{-}NH{-}CH_2CH_3$

h. $CH_3CH{-}\overset{\overset{CH_3}{|}}{C}HCH_2{-}\overset{\overset{O}{\|}}{C}{-}H$

i. $CH_3CH_2CH_2CH_2{-}\overset{\overset{O}{\|}}{C}{-}OH$

j. $CH_3CH_2CH_2{-}\overset{\overset{O}{\|}}{C}{-}O{-}CH_3$

k. $CH_3CH_2{-}\overset{\overset{O}{\|}}{C}{-}NH_2$

l. $CH_3\overset{\overset{CH_3}{|}}{C}HCH_2CH_2{-}\overset{\overset{O}{\|}}{C}{-}OH$

8. Name the following compounds by the IUPAC system. (*Hint:* You may need to consult the individual sections in the chapter to find the IUPAC naming rules.)

a. $CH_3CH_2CH_2CH_2{-}\overset{\overset{O}{\|}}{C}{-}OH$

b. $CH_3CH_2CH_2CH_2{-}\overset{\overset{O}{\|}}{C}{-}CH_3$

c. $CH_3CH_2\underset{\underset{CH_3}{|}}{C}HCH_2CH_2{-}OH$

d. $CH_3CHCH_2CHCH_3$
$\quad\ \ |\qquad\ \ |$
$\quad OH\quad CH_3$

$\qquad\qquad\quad O$
$\qquad\qquad\quad \|$
e. $CH_3CH_2—C—H$

$\qquad\quad CH_3\qquad\qquad\quad O$
$\qquad\quad\ |\qquad\qquad\qquad\ \|$
f. $CH_3CH—CH_2CH_2CH_2—C—H$

$\qquad\qquad\qquad\ O$
$\qquad\qquad\qquad\ \|$
g. $CH_3CH_2CH_2—C—O—CH_2CH_3$

h. $CH_3CHCH_2CH_2—NH_2$
$\qquad\quad |$
$\qquad\ CH_3$

$\qquad\qquad\qquad\ O$
$\qquad\qquad\qquad\ \|$
i. $CH_3CH_2CH_2—C—CH_2CH_3$

j. $CH_3CH_2CH_2—NH—CH_2CH_2CH_3$

$\qquad\qquad CH_3\qquad\qquad O$
$\qquad\qquad\ |\qquad\qquad\quad \|$
k. $CH_3CH_2CHCH_2CH_2—C—OH$

$\qquad\qquad\qquad\ O$
$\qquad\qquad\qquad\ \|$
l. $CH_3CH_2CH_2—C—NH_2$

9. Draw the structures for each of the following compounds:
   a. butanoic acid
   b. 2-butanol
   c. ethyl butanoate
   d. 2-methyl-3-pentanol
   e. iso-propyl amine
   f. methyl ethyl ketone
   g. hexanal
   h. 3-hexanone
   i. 2-aminobutane
   j. methyl ethyl ether

10. Write the reaction equations for the formation of each of the following compounds from the specified reactants.
   a. butyl acetate, from acetic acid and 1-butanol
   b. N-methyl acetamide, from methylamine and acetic acid
   c. propionic acid from propanol
   d. butanal from 1-butanol

11. What are some of the uses for each of the following alcohols?
   a. methanol
   b. ethanol
   c. 2-propanol
   d. ethylene glycol
   e. glycerol

12. Name some examples of the phenol class of compounds and list some of their uses.

13. What are anesthetics?

14. Name a chemical used as an anesthetic now or in the past.

15. What was the first chemical widely used as a general anesthetic?

16. What are some uses for each of the following aldehydes?
   a. formaldehyde
   b. benzaldehyde
   c. cinnamaldehyde
   d. vanillin

17. What are some of the uses for each of the following ketones?
   a. acetone
   b. methyl ethyl ketone
   c. 3-pentanone

18. Are the carboxylic acids strong or weak acids?

19. What compound causes the rancid odor in spoiled butter? Can this odor also occur in margarine?

20. What chemical compounds cause goats to have a strong odor? Can you think of a way to neutralize the odor?

21. What is the main chemical that produces pain in ant bites?

22. Why does our skin swell up after an ant bite?

23. Name some uses for each of the following organic acids.
   a. formic acid
   b. acetic acid
   c. butanoic acid
   d. stearic acid
   e. benzoic acid

24. Give some examples of condensation polymers.

25. What are the main differences between condensation and addition polymers?

26. Show how to synthesize each of the following polymers. (*Hint:* Pattern your reactions after those in the chapter.)
   a. a poly(ester) from adipic acid and ethylene glycol
   b. a poly(amide) from adipic acid and 1,4-diaminobutane

27. Where are you likely to encounter "free" amines in your daily life?

28. What are the distinguishing features of primary, secondary, and tertiary amines? How do these amine classes differ from each other?

29. What are some of the uses for each of the following amines?
   a. aniline
   b. amphetamine
   c. 1,6-diaminohexane

30. What flavor would each of the following esters have? (*Hint:* Consult table 10.3).
   a. octyl acetate
   b. butyl butyrate
   c. ethyl acetate
   d. amyl acetate
   e. benzyl acetate

31. Aside from flavorings, what are some other uses of esters?

32. What other chemicals, in addition to esters, are commonly used as flavoring agents?

CRITICAL THINKING PROBLEMS

1. The use of ethanol as a gasoline additive or replacement runs in cycles in this country. Read a few of the following five articles and form your own opinion on ethanol gasoline: (1) Allison Lucas, "MTBE Still Facing Pressure from Ethanol Under Latest Fuel Proposal," *Chemical Week*, January 26, 1994, p. 9; (2) Robert Boddey, "'Green' Energy from Sugar Cane," *Chemical & Industry*, May 17, 1993, pp. 355–358; (3) Earl V. Anderson, "Brazil's Program to Use Ethanol as Transportation Fuel Loses Steam," *Chemical & Engineering News*, October 18, 1993, pp. 13–15; (4) William L. Driscoll, "Fill 'Er Up with Biomass Derivatives," *Technology Review*, August/September 1993, pp. 74–76; (5) "Process Might Boost Ethanol Output from Municipal Waste," *R&D Magazine*, February 1992, p. 26.

2. Methyl t-butyl ether, or MTBE for short, has a high octane number but is having a difficult time establishing itself in the fuel market. Read the following two articles: (1) Allison Lucas, "MTBE Still Facing Pressure from Ethanol Under Latest Fuel Proposal," and (2) Earl V. Anderson, "Health Studies Indicate MTBE is Safe Gasoline Additive," *C&EN*, September 20, 1993, pp. 9–18. What are the objections to using MTBE? Is this chemical likely to make a major impact on the motor fuel market? Cost considerations aside, could MTBE be used as a total replacement for gasoline?

3. Describe four different ways the equilibrium point can be shifted in a reaction such as the formation of an ester from acetic acid and ethanol.

4. Alcoholism is claimed to be the number one drug problem in this country. Read the following three articles and form your own opinion on this much-debated topic: (1) Stephen Goode, "Are America's College Students Majoring in Booze?" *Insight*, August 8, 1994, pp. 15–17; (2) "Alcoholism's Drain on the Economy May Be Deeper than Ever," *Business Week*, December 30, 1991, p. 26; (3) B. Bower, "Early Alcoholism: Crime, Depression Higher," *Science News*, March 25, 1989, p. 180.

5. Read the article by B. Bower in *Science News*, February 1, 1992, p. 69, called "Alcoholism: Nurture May Often Outdo Nature." Is there really good evidence that a person's genetic makeup affects alcoholism more than their environment?

6. Some of the most deadly poisons are derived from amine compounds, yet other amines are essential for life. How can you explain that?

7. The article "It's Time to Wake Up and Smell the Rosemary," by Tom Dunkel in *Insight*, June 15, 1992, pp. 14–16, 36–38, discusses the possibility that spices and flavoring agents might relieve some of the stresses we encounter in daily living. Read this article and decide whether Dunkel's assertion seems reasonable in light of applying the scientific method.

8. Read the article entitled "Keeping OSHA's Feet to the Fire," by Charles Noble in *Technology Review*, February/March 1992, pp. 43–51. Suppose you work for OSHA and can determine policy. How would you cope with the task of evaluating the safety of over a thousand new compounds per year, plus a backlog of many thousands more from previous years?

9. Campaigns are frequently aimed at eliminating all potentially toxic or hazardous chemicals from the home. Review this chapter and make a list of the chemicals with specific home uses. How many might be classed as potentially toxic or hazardous? Would it be easy to find a safe substitute for each of them?

10. What chemicals cited in this chapter normally occur in our food as part of the food itself?

11. What chemicals listed in this chapter are used in personal hygiene?

12. Alcohols react in the human body to produce, among other things, aldehydes. Do you think the higher toxicity of the aldehyde from methanol accounts for the fact that this alcohol is more toxic to humans than ethanol?

HINT Compare the LD_{50} values for methanol (formaldehyde) and ethanol (acetaldehyde) in table 1 in box 10.1.

13. Gasohol was popular in the United States for many years, especially in the 1970s when shortages of crude oil occurred, but it was never widely accepted, partly because ethanol costs more than gasoline in the United States. In Brazil, gasoline is several times more expensive than in the United States, and ethanol is much more widely used as an automobile fuel. However, ethanol absorbs water, causes engine corrosion, and does not provide as much power per gallon as gasoline since it is already partially oxidized. Redesigning engines to operate on ethanol helps, but you cannot overrule the laws of thermodynamics that make gasoline a more efficient fuel. Finally, incomplete ethanol oxidation produces acetaldehyde, a toxic chemical. Based on these facts, what do you think is the future for ethanol as an automobile fuel? Would you use it in your car, even if it cost more? Explain.

14. Some people recommend coffee, which contains the stimulant caffeine, as a treatment for an alcoholic hangover, but this does not really work well. Can you explain why?

15. There are four different ways to shift the equilibrium point in the reaction shown in figure 10.19. List them. Which is the most practical, in your opinion?

16. If you compare the reactions in figures 10.19 and 10.21, you will notice that only the one in figure 10.19 has a double arrow. What does this indicate?

17. Amines are organic bases. Will the base shown in figure 10.24 react with the adipic acid in an acid-base reaction? What effect, if any, might this have on the reaction?

CHAPTER 11

POLYMERS—THE MIGHTY MOLECULES

OUTLINE

Classification of Polymers
Addition Polymers
Condensation Polymers
Copolymers—Adding More Versatility
Elastomeric Polymers
Changing the Properties of Plastics
Paints and Coatings

BOX 11.1 Silicones and Implants
BOX 11.2 Polymer Recycling and Disposal

KEY IDEAS

1. How Chemists Classify Polymers
2. The Names of the Five Most Produced Polymers
3. The Main Uses for the More Important Thermoplastics
4. The Names and Uses for Major Examples of Thermosetting Resins
5. What Copolymers are and Why They are Important
6. How Chemists Modify the Properties of Polymers
7. The Basic Nature and Composition of Paints and Coatings
8. How to Recycle Polymers

"What man's mind can create, man's character can control"
THOMAS ALVA EDISON (1847–1931)

Sprays of fiber optic cables transmitting colored light.

We have already learned much about polymers, large molecules consisting of repeated units of a smaller molecule called a monomer. We know that the structure of a polymer may be linear, branched, ring-shaped, or crosslinked. We know that polymers form in addition and condensation reactions. All of these concepts are examples of pure chemistry—they describe the structure, function, and behavior of polymers.

In this chapter, we will turn to the applications of synthetic polymers. Synthetic polymers—better known as plastics, textiles and rubbers—have totally changed modern life. They are essential to our homes, our businesses, our automobiles, our clothing—sometimes, as in the artificial heart, to our very lives.

Table 11.1 lists the names, classes, starting monomers, repeat units, and **T_g (glass transition temperature)** for several common synthetic polymers. The T_g—the temperature at which a polymer becomes soft or flexible rather than rigid—is a key polymer property. It helps make synthetic polymers the most versatile of substances, suited to a virtually endless variety of uses.

Classification of Polymers

Polymers may be broadly classified as either natural or synthetic. Natural polymers include molecules essential to life, such as proteins and DNA, as well as such varied substances as natural rubber and diamonds. Synthetic polymers are synthesized in the laboratory. This chapter especially explores these synthesized polymers.

Synthetic polymers themselves can be further classified according to thermal characteristics, physical structure, use, and method of preparation. Let's examine each of these classification methods in the following sections.

Thermal Behavior: Thermoplastic versus Thermosetting

The most important industrial classification of polymers depends solely on their behavior when heated. A **thermoplastic,** which is made up of branched or linear molecules, softens every time it is heated above its T_g. Therefore, it can be remolded repeatedly. A **thermosetting polymer** is made up of crosslinked molecules. These crosslinks form as the reaction proceeds and harden during the heating process. Thermosetting polymers do not soften when cooled and cannot be remolded by applying heat, though they can be reshaped by machining or other physical means. Over 80 percent of all industrial polymers are thermoplastics, including poly(ethylene), poly(vinyl chloride), poly(ethylene terephthalate) (for example, Dacron or Mylar), the nylons, and the acrylates. Typical thermosetting polymers include the phenolic resins (made from phenol and formaldehyde), urea-formaldehyde polymers, and melamine formaldehyde polymers (for example, Melmac). Figure 11.1 shows how we put some of these plastics to use in our daily lives.

Figure 11.1 Polymers are everywhere you look.

Structure: Linear, Branched, or Crosslinked

Another classification scheme used to categorize polymers is physical structure. The terms **linear, branched,** and **crosslinked** refer to the arrangement (morphology) of the polymer chains (see figure 9.29). Linear polymers consist of long chains tangled together much like cooked spaghetti. Branched polymers are similar, but they have small side-branches at irregular intervals. In a crosslinked polymer, the chains are interconnected into a network. The physical structure of the polymer determines various properties. For example, linear and branched polymers are generally soluble in solvents and flow under the influence of pressure or heat. Crosslinked polymers are insoluble in essentially all solvents and do not flow under pressure or heat. Most linear or branched polymers are also thermoplastics, while thermosetting polymers are crosslinked. An oligomer is a small polymer consisting of only a few units.

EXPERIENCE THIS

You can make a plastic by boiling milk (low fat milk works best) with vinegar. After the curd forms, boil the milk for about ten minutes. Then separate the solid polymer, washing it repeatedly with water until the rinse water is clear. This elastic polymer, called casein, was at one time widely used in coatings and adhesives.

Because it is not very stable, the casein polymer will decompose after a while. Decomposition can be preventing by mixing the polymer with formaldehyde solution; the resulting hard plastic was once used to make buttons. However, casein has largely been replaced with more stable synthetic plastics.

Use: Plastics, Fibers and Elastomers

The terms *plastic, fiber,* (or *textile*), and *elastomer* (or *rubber*) refer to the end use of a polymer. These three uses provide the classification method most nonchemists use to describe polymers.

The first term, *plastic*, means "capable of being molded." **Plastics** are either thermoplastic or thermosetting polymers, and are molded at some point during manufacturing. Although common solid organic compounds such as sugar exist as crystals, most common polymers have little or no crystallinity. Instead, they are **amorphous;** they have no fixed atomic lattice structure.

Figure 11.2 contains schematic diagrams of an amorphous polymer, a slightly crystalline polymer, and a highly crystalline polymer. Amorphous polymers include poly(vinyl chloride) and poly(styrene). Semicrystalline polymers include low-density poly(ethylene) and poly(vinyl alcohols), and crystalline polymers include Nylon 66 and poly(ethylene terephthalate), or PET. All are capable of being molded. Herman Mark (Austrian-American, 1896–1992) developed the theory of polymer crystallinity. Mark also developed the theory of rubber elasticity and many of the commercial processes for synthetic polymers.

Herman Mark

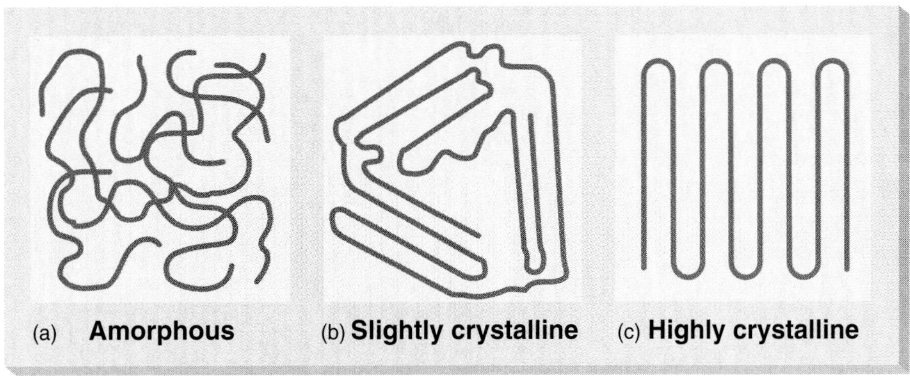

Figure 11.2 Schematic diagrams for: **(a),** a linear, amorphous polymer; **(b),** a slightly crystalline polymer; **(c),** a highly crystalline polymer.

TABLE 11.1 Names, classes, monomers, repeat units, and T_g of some common polymers. The classes are addition (A), condensation (C), and semi-inorganic (I).

NAME, TRADE NAME, ABBREVIATION	CLASS	MONOMER	REPEAT UNIT	T_g (°C)
Poly(ethylene) PE	A	$CH_2{=}CH_2$	$-[CH_2CH_2]_n-$	−125
Poly(propylene) PP	A	$CH_2{=}CH{-}CH_3$	$-[CH_2CH(CH_3)]_n-$	0
Poly(vinyl chloride) PVC	A	$CH_2{=}CH{-}Cl$	$-[CH_2CH(Cl)]_n-$	80
Poly(vinylidene chloride) Saran	A	$CH_2{=}CCl_2$	$-[CH_2CCl_2]_n-$	−17 190*
Poly(styrene) PSt	A	$CH_2{=}CH{-}C_6H_5$	$-[CH_2CH(C_6H_5)]_n-$	105
Poly(tetrafluoroethylene) Teflon PTFE	A	$CF_2{=}CF_2$	$-[CF_2CF_2]_n-$	127
Poly(vinyl acetate) PVAc	A	$CH_2{=}CH{-}OC({=}O){-}CH_3$	$-[CH_2CH(O{-}C({=}O){-}CH_3)]_n-$	28
Poly(vinyl alcohol) PVAl	A	None	$-[CH_2CH(OH)]_n-$	85
Poly(ethyl acrylate) PEA	A	$CH_2{=}CH{-}C({=}O){-}O{-}CH_2CH_3$	$-[CH_2CH(C({=}O){-}O{-}CH_2CH_3)]_n-$	−24
Poly(methyl methacrylate) Plexiglas, Lucite PMMA	A	$CH_2{=}C(CH_3){-}C({=}O){-}O{-}CH_3$	$-[CH_2{-}C(CH_3)(C({=}O){-}O{-}CH_3)]_n-$	105
Poly(acrylonitrile) Orlon, Dynel PAN	A	$CH_2{=}CH{-}C{\equiv}N$	$-[CH_2CH(C{\equiv}N)]_n-$	105

TABLE 11.1 (Continued)

NAME, TRADE NAME, ABBREVIATION	CLASS	MONOMER	REPEAT UNIT	T_g (°C)
Poly(amide) Nylon 66	C	$H_2N-(CH_2)_6-NH_2$ + $HOOC-(CH_2)_4-COOH$	$[-NH-(CH_2)_6-NH-C(=O)(CH_2)_4C(=O)-]_n$	57 265*
Poly(amide) Nylon 610	C	$H_2N-(CH_2)_6-NH_2$ + $HOOC-(CH_2)_8-COOH$	$[-NH-(CH_2)_6-NH-C(=O)(CH_2)_8C(=O)-]_n$	228*
Poly(amide) Nylon 6	C	ε—caprolactam	$[-NH-C(=O)-(CH_2)_5-]_n$	215*
Poly(carbonate) Lexan	C	$COCl_2$ + bis-phenol A	$[-O-C(=O)-O-C_6H_4-C(CH_3)_2-C_6H_4-]_n$	225*
Poly(ethylene adipate)	C	$HO-CH_2CH_2-OH$ + $HOOC-(CH_2)_4-COOH$	$[-OCH_2CH_2-O-C(=O)-(CH_2)_4-C(=O)-]_n$	−50
Poly(ethylene terephthalate) Dacron, Mylar PET	C	$HO-CH_2CH_2-OH$ + $HOOC-C_6H_4-COOH$	$[-OCH_2CH_2-O-C(=O)-C_6H_4-C(=O)-]_n$	69
Poly(dimethyl siloxane) Silicone rubber	I	$(CH_3)_2SiCl_2$	$[-O-Si(CH_3)_2-]_n$	−123
Natural rubber	–	$CH_2=CH-C(CH_3)=CH_2$	$[-CH_2-C(H)=C(CH_3)-CH_2-]_n$	−72

*Denotes melting point rather than T_g.

EXPERIENCE THIS

You can use spaghetti as a model for polymer crystallinity. Line up strands of uncooked spaghetti side by side. The resulting structure resembles figure 11.2c, the highly crystalline polymer form, with all its chains lined up in one direction.

To reproduce the semicrystalline polymer structure (figure 11.2b), take a bundle of spaghetti and pour boiling water over the center, but not the ends, so the sample sticks together in the middle. After the middle hardens, dip the ends into boiling water until they soften. The resulting structure is slightly crystalline.

Finally, place a bundle of spaghetti into boiling water and cook it for several minutes. Drain the water away and toss the cooked spaghetti onto a plate. The resulting disordered arrangement represents an amorphous polymer, shown in figure 11.2a.

If you wish, after you are satisfied you understand these concepts, add some sauce and cheese to your spaghetti. Bon appetit!

Table 11.2 lists the U.S. production, in metric tons, of the most common commercial plastics. Notice that the top five polymers are all thermoplastics, made from inexpensive, petroleum-derived monomers by addition polymerization (see chapter 9). In sixth place is the most common thermosetting polymer, the phenolic resins. These resins are made from inexpensive monomers by a condensation reaction (see chapter 10).

Fibers, or textiles, are the second type of polymer use. **Fibers** are thermoplastics with individual linear polymer chains held together by hydrogen bonds, polymer crystallinity, or both. Table 11.3 shows the production data for some common fibers. Noncellulosic fibers (those not made from plant fiber, or cellulose) make up 87 percent of the textiles produced in the United States. Following are cellulose fibers (those made from cellulose, a natural plant polymer) at 9 percent and inorganic fibers at 4 percent. As recently as 1970, cellulose fibers were the most widely produced, but synthetic polymers surged ahead despite the fact that cotton and the other cellulose fibers are made from less expensive raw materials.

The third polymer use is represented by elastomers. Elastomers are also called rubbers, after the foremost member of the elastomer class. **Elastomers** begin as linear or branched thermoplastics, but become lightly crosslinked during vulcanization, increasing their strength and stretchability. Table 11.4 shows the production figures for the main elastomers. Synthetic elastomers, derived almost exclusively from petroleum, make up about 61 percent of the production, but natural rubber remains the most produced single elastomer. Thermoplastic elastomers, a new group rapidly growing in importance, are described later in this chapter.

TABLE 11.2 U.S. commercial plastic production in 1993.

POLYMER	CLASS (TP, TS)[t]	METRIC TONS × 10^3	COST PER POUND ($)
Poly(ethylene), low-density (LDPE)	TP	21,047	0.36–0.40
Poly(vinyl chloride)	TP	19,045	0.37–0.50
Poly(propylene)	TP	15,580	0.45–0.60
Poly(ethylene), high-density (HDPE)	TP	14,132	0.43–0.50
Poly(styrene)	TP	2,130*	0.48–0.52
Phenolic resins	TS	1,350*	–
Poly(amides)	TP	1,088	–
ABS Copolymer	TP	1,040*	–
Urea-formaldehyde resins	TS	620*	–
Poly(carbonates)	TP	600	–
Poly(acetals)	TP	380	–
Poly(esters)	TS	280	–
Epoxy resins	TS	200*	–
Melamine-formaldehyde resins	TS	100*	–
Acrylates and methacrylates	TP	50*	1.00

SOURCE: Data from *Chemical Week*, December 1993.
[t] TP = thermoplastic; TS = thermosetting
*Estimated

TABLE 11.3 U.S. fiber production data. The classes are cellulosic (C), noncellulosic (NC), and inorganic (I).

FIBER	CLASS	METRIC TONS × 10^3	COST PER POUND ($)
Poly(esters)	NC	8,621	0.73
Nylons	NC	3,765	1.03
Acrylics	NC	2,357	0.78
Cotton	C	1,270	0.30
Glass	I	635	–
Poly(olefins)	NC	454	–
Rayon, viscose	C	191	–
Cellulose acetate	C	91	–
Carbon	I	2	20–30
Aramid	NC	<0.5	8–40

SOURCE: Data from *Chemical & Engineering News*, January 22, 1992 and various other sources.

TABLE 11.4 Worldwide (1985) and U.S. (1992) production of selected elastomers.

ELASTOMER	WORLDWIDE (METRIC TONS × 10^3)	U.S.
Natural rubber	4,131	889
Styrene-butadiene	2,632	801
Poly(butadiene)	1,060	410
Carboxylated styrene-butadiene	933	501
Ethylene-propylene copolymer	494	199
Poly(chloroprene)	259	75
Nitrile rubber	216	108
All other synthetic elastomers	931	376
Thermoplastic elastomers	–	286
Total synthetic rubber	6,525	2,756
Total rubber	10,656	3,645

SOURCE: Data from *Chemical & Engineering News*, March 21, 1988, p. 26 and January 27, 1992.

PRACTICE PROBLEM 11.1

What are the five most produced plastics in the United States? What types of plastic—thermoplastic or thermosetting—are they?

ANSWER As table 11.2 indicates, the five most produced plastics in the United States are low-density poly(ethylene), poly(vinyl chloride), poly(propylene), high-density poly(ethylene), and poly(styrene). They are all thermoplastics.

PRACTICE PROBLEM 11.2

Look around you for examples of elastomers (rubbery materials). How many can you find?

ANSWER A few possibilities include tires, hard-rubber combs, shoes, balls, faucet gaskets, refrigerator seals, rubber bands, and rubber-coated wire (clothes hangers, electrical wiring).

Preparation: Addition and Condensation

In chapter 9, we subdivided polymers into addition and condensation classes according to the reactions that occur in their preparation. Table 11.1 shows that nearly all addition polymers have "backbones" or chains composed entirely of C atoms. Functional groups do occur in addition polymers, but they are always pendantlike attachments to the backbone chain. For example, poly(vinyl acetate) has a pendant ester unit.

Table 11.1 also shows examples of polymers by condensation, including poly(ethylene terephthalate) (commonly known as Mylar or Dacron) and poly(dimethyl siloxane), also called silicone rubber. Condensation polymers often have noncarbon atoms in their backbone chains, as table 11.1 shows. Their functional groups may either hang from the chain or be part of the chain itself. In condensation reactions, some of the monomer atoms split off to form water or another byproduct.

ADDITION POLYMERS

Poly(olefins) and Vinyl Polymers

Now that we have examined polymer classification, let's look more closely at some specific addition polymers. Poly(olefins) are prepared by polymerizing olefins (alkenes). Their monomers have at least one C=C double bond, and their names carry the *ene* ending. The raw materials ordinarily originate from petroleum; some are derived from coal or biomass (living matter). The major poly(olefins) are poly(ethylene), poly(propylene), poly(butadiene), poly(chloroprene), and poly(isoprene).

Vinyl polymers contain the vinyl group ($CH_2{=}CH{-}$) or some derivative of it. They include poly(vinyl chloride), poly(vinylidene chloride), poly(styrene), poly(tetrafluoroethylene), poly(vinyl acetate), and poly(vinyl alcohol).

Poly(ethylene) and Poly(propylene)

Poly(ethylene), the simplest addition polymer, has been commercially available since 1941 and is the most common synthetic polymer in quantity and use. Figure 11.3a shows the polymerization of ethylene ($CH_2{=}CH_2$). As table 11.2 shows, the two main kinds of polyethylene produced and marketed are low-density (LDPE) and high-density (HDPE). LDPE, prepared by addition polymerization using a radical (a highly reactive atom with an unpaired electron), has a density of 0.93 g/mL; it is branched and has only modest crystallinity. HDPE is prepared using a surface catalyst and has a density close to 1.0 and very little branching; it is a moderately crystalline polymer.

These structural differences give LDPE and HDPE different properties. Because LDPE is branched, the carbon chains do not pack as closely together (making LDPE less dense). LDPE is thus more flexible then HDPE; it is used as plastic wrap for foods, grocery bags, and trash bags. By contrast, HDPE's unbranched chains can be arranged more closely, giving HDPE a higher density, and its more crystalline structure also gives it rigidity. HDPE is used to make tougher, more rigid products such as milk and bleach bottles.

The total PE produced annually in the United States exceeds 35 million tons, almost double that of PVC and nearly half the production of *all* thermoplastics put together. The reason is simple; PE is made from the least expensive monomer—ethylene ($CH_2{=}CH_2$). Table 11.5 summarizes some uses of PE. Since HDPE and LDPE differ markedly in properties, they have widely different uses. Soft drink bottle caps are usually made from crosslinked LDPE.

Poly(propylene), or **PP,** is of more recent vintage than PE; it was first marketed commercially in 1957. PP, which replaces one

TABLE 11.5 Poly(ethylene) uses (as percentages of total use).

USE	HDPE	LDPE
Molded products	57%	15%
Film (plastic wrap, storm windows, sheeting, and so on)	8%	55%
Wire coatings	0%	4%
Pipes, conduits	7%	0%
Exported (to other countries)	11%	10%
Miscellaneous	17%	16%

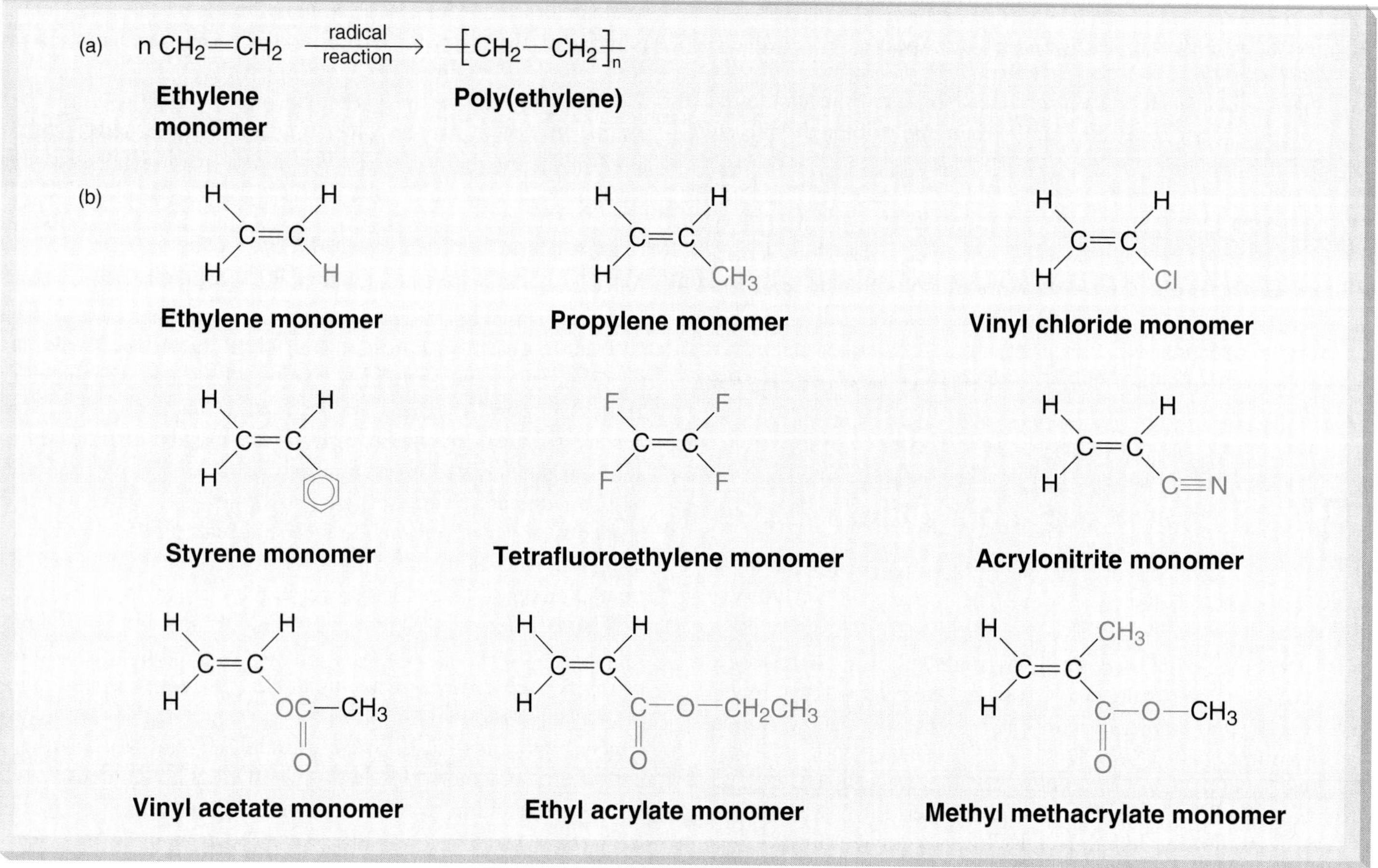

Figure 11.3 (a) The polymerization of ethylene. (b) The relationship between the poly(ethylene) monomer and other addition polymer monomers. All substitute other atoms or groups for the H atoms in ethylene.

of the PE's H atoms with a methyl (CH_3) unit (figure 11.3b), is the lightest major plastic (density = 0.88 g/mL). Table 11.6 summarizes polypropylene uses, including poly(olefin) fibers, storage battery cases, plastic film, automobile trim, and ropes. This polymer is prepared by addition polymerization.

Vinyl Polymers

Poly(vinyl chloride), or **PVC,** has been manufactured since 1927. Like PE and PP, PVC is an inexpensive polymer; it is made from monomeric vinyl chloride, a gas. Its Cl atom replaces one of the H atoms in poly(ethylene); thus, while the PE monomer is $CH_2{=}CH_2$, the PVC monomer is $CH_2{=}CH{-}Cl$. Although vinyl chloride monomer is a carcinogen (a cancer-causing chemical), the polymer contains no free monomers and is not carcinogenic.

One important use for PVC is in-home water piping. Most homes formerly had copper water pipes held together by solder, a lead alloy (although the more recent solders are lead-free). Hot water passing through the pipes extracts small amounts of the lead, an exceedingly poisonous material, (chapter 8), from these joints. PVC is "glued" together using a solvent that evaporates out of the system; it appears to be much safer. Another major PVC use is in garden hoses. PVC has a T_g of 80° C, which means it is a rigid plastic even in the hottest places on earth. Hoses are

TABLE 11.6 Poly(vinyl chloride), poly(propylene), and poly(styrene) uses (as percentages of total use).

Use	POLYMER		
	PVC	PP	PSt
Piping (water, electrical)	45	–	–
Fibers (clothing, carpets, rope, and so on)	–	35	–
Extruded (sheets)	25	–	35
Molded objects (gears, cases)	5	30	45
Expanded beads (cups, insulation)	–	–	15
Films (plastic wrap, coatings)	10	10	–
Exported and miscellaneous	15	25	5

made from a plasticized form of PVC (we will discuss plasticization later in this chapter); rainwear, handbags, and coated fabrics are also made from plasticized PVC. Some additional PVC uses are summarized in table 11.6.

Poly(styrene) or **PSt,** has been commercially available since 1937. It is a rigid plastic prepared by the free radical addition polymerization of a petrochemical—the liquid styrene monomer ($CH_2{=}CH{-}C_6H_5$). Table 11.6 features the main fabricated forms of PSt. Expandable PSt beads are compressed to form the foamed cups, egg cartons, hot drink cups, and packaging containers used in stores and fast food restaurants. Foamed PIP (plastic insulating panel) sheets, made from PSt, give outstanding insulation with very little weight. PSt is also used in housewares, furniture, electronics, construction, and toys.

Foamed PSt concerns environmentalists because it does not biodegrade, but foamed PSt is one of the best food packaging materials available and actually offers some advantages over the alternatives. For example, the alternatives, paper and cardboard, are less sanitary and also do not biodegrade in landfills; several landfill studies have shown that undegraded paper products may last more than twenty-five years. In addition, foamed PSt is recyclable, but it is difficult to recycle all waste paper or cardboard because some of the inks used on them are toxic and are difficult and expensive to remove. Plastics make up only 5 to 6 percent (weight or volume) of a typical landfill. Recycling plastics is an excellent practice that will grow in importance in the coming decades. (We'll look further at recycling later in this chapter.)

Poly(tetrafluoroethylene), or **PTFE,** is a polymer in which all hydrogen atoms in polyethylene, $(CH_2{=}CH_2)_n$, are replaced with fluorine atoms to create $(CF_2{=}CF_2)_n$. PTFE is widely used as a nonstick coating for cookware under the names of Teflon and PTFE. It is also used to produce electrical insulation for wires, cables, transformers, motors, generators, gaskets, and nonlubricated bearings. Many people use a white film of PTFE plastic to keep pipe threads from seizing and sticking. Drawer riders are often coated with PTFE to take advantage of its nonsticking, high-lubrication properties.

PTFE was first marketed in 1943 as Teflon, just five years after Roy Plunkett, a DuPont chemist, accidentally discovered it. Plunkett and his coworkers were studying various applications for fluorinated hydrocarbons (which DuPont named Freons). One day they found that a tetrafluoroethylene gas cylinder would not release gas, even though the net weight implied it was nearly full. After carefully checking for obstructions, they gingerly cut the cylinder in two behind concrete barriers. Plunkett then discovered the cylinder no longer contained gas—it was powder, later ascertained to be PTFE, which had polymerized from the tetrafluoroethylene gas. Further research revealed how to polymerize the monomer intentionally, and a new, useful plastic was born. Similar events happen often in science, but not all are utilized so quickly. Poly(styrene) was polymerized ninety-nine years before it became an industrial plastic.

Poly(vinyl acetate), or **PVAc,** has been manufactured since 1936, but not in as large a quantity as the top five poly(olefin) and vinyl polymers. Vinyl acetate monomer synthesis was shown in figure 9.16; most of the 670,000 tons of vinyl acetate produced annually is converted directly into the polymer (figure 11.4a). PVAc is used in latex (water-based) paints, in chewing gum, and as an intermediate for poly(vinyl alcohol) preparation (figure 11.4b). (Poly(vinyl alcohol), PVAl, cannot be made directly from a monomer—it must be made from PVAc.) PVAl is used as a thickening agent, a packaging film, an adhesive, and the glue on postage stamps. Many white glues utilized in homes and offices are mixtures of a PVAc emulsion polymer and PVAl.

Acrylates and Methacrylates

Along with poly(olefins) and vinyl polymers, acrylates and methacrylates are addition polymers (table 11.1). However, production of the acrylates and methacrylates is significantly lower

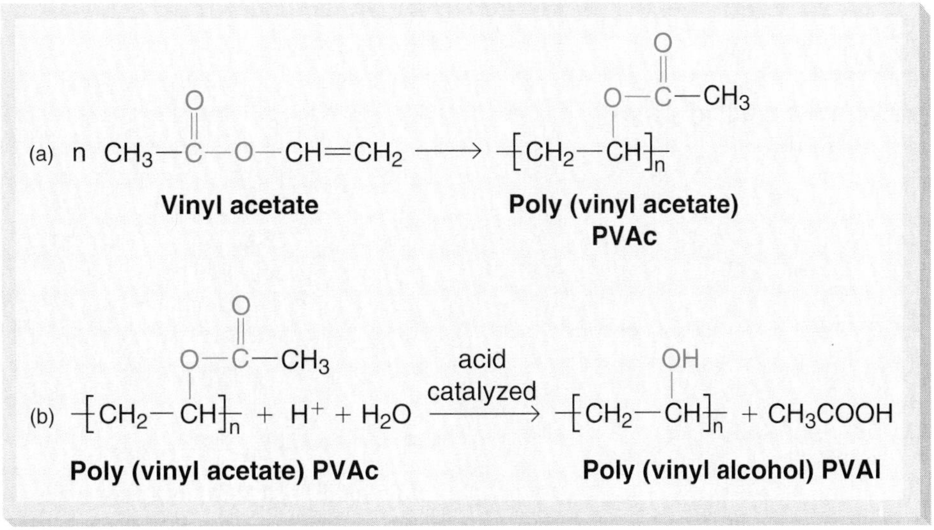

$$\text{(a)}\quad n\ CH_3{-}\overset{\overset{\displaystyle O}{\|}}{C}{-}O{-}CH{=}CH_2 \longrightarrow {-}\!\!\left[CH_2{-}\underset{}{\overset{\overset{\displaystyle O-\overset{\overset{\displaystyle O}{\|}}{C}-CH_3}{|}}{CH}}\right]_n$$

Vinyl acetate → **Poly (vinyl acetate) PVAc**

$$\text{(b)}\quad {-}\!\!\left[CH_2{-}\overset{\overset{\displaystyle O-\overset{\overset{\displaystyle O}{\|}}{C}-CH_3}{|}}{CH}\right]_n + H^+ + H_2O \xrightarrow{\text{acid catalyzed}} {-}\!\!\left[CH_2{-}\overset{\overset{\displaystyle OH}{|}}{CH}\right]_n + CH_3COOH$$

Poly (vinyl acetate) PVAc → **Poly (vinyl alcohol) PVAl**

Figure 11.4 **(a)** The polymerization of vinyl acetate. **(b)** Poly(vinyl acetate) serves as an intermediate for poly(vinyl alcohol) preparation.

than for olefin or vinyl polymers, mainly because of high monomer cost. **Poly(methyl methacrylate), PMMA,** became a commercial plastic in 1931, and the acrylates, such as poly(ethyl acrylate), or PEA, followed a few years later. The related poly(acrylonitrile), PAN, was introduced in 1950. PMMA has outstanding optical properties, rivaling those of optical glass (density = 2.5 g/mL), but it is much lighter (density about 1.1 g/mL) and has higher impact resistance. These optical attributes make PMMA useful in light-weight plastic spectacles and also in automobile taillights.

As figure 11.3b demonstrates, PEA, PMMA, and PAN are further variants of the simpler PE polymer. In the PEA monomer, an ethyl ester (ethyl = $—CH_2CH_3$, ester = —COO— plus other organic groups) replaces one of the ethylene H atoms. In the PMMA monomer, two H atoms are replaced—one by a methyl group (CH_3), one by a methyl ester ($—COO—CH_3$). In PAN's monomer, a nitrile (—C≡N) group takes the place of one H atom.

PMMA and acrylate copolymers are used in water-based paints, floor polishes, paper coatings, and skylights. They are also used as ion-exchange resins for water softening and other applications. Tough, rigid PMMA is used in aircraft windows and is the dominant polymer in dentures and hard contact lenses. The major polymer used in soft contact lenses is the closely related poly(2-hydroxyethyl methacrylate), PHEMA (figure 11.5).

Poly(acrylonitrile), or **PAN,** is used almost exclusively as a fiber; essentially all the acrylic fibers listed in table 11.3 are PAN or PAN copolymers. (A copolymer is a polymer made from two or more monomers.) PAN is made into carpets, sweaters, skirts, socks, slacks, and yarn.

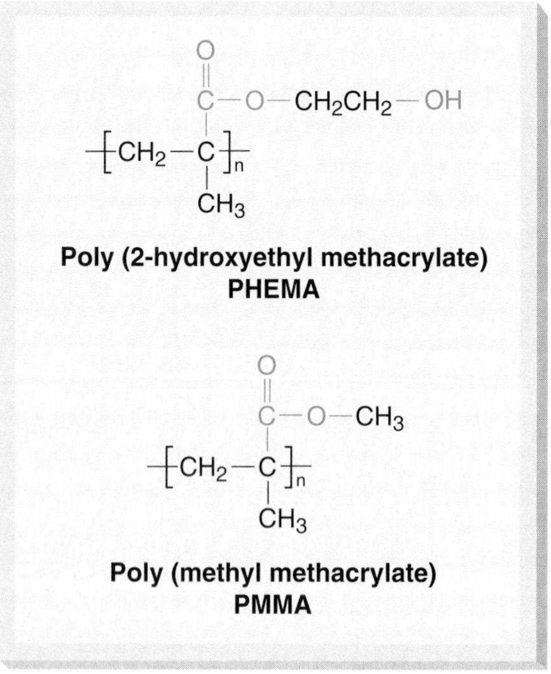

$$-\!\!\left[CH_2-\underset{CH_3}{\overset{C(=O)-O-CH_2CH_2-OH}{C}}\right]_n\!\!-$$

Poly (2-hydroxyethyl methacrylate)
PHEMA

$$-\!\!\left[CH_2-\underset{CH_3}{\overset{C(=O)-O-CH_3}{C}}\right]_n\!\!-$$

Poly (methyl methacrylate)
PMMA

Figure 11.5 Comparison of the repeat unit structures of PHEMA, used in soft contact lenses, and PMMA.

PRACTICE PROBLEM 11.3

Review tables 11.1, 11.5, and 11.6. Which addition polymers do you use regularly?

ANSWER Common objects made from poly(ethylene) include toys, cups, wire coating, milk bottles, and plastic wrap; from poly(propylene): fibers, storage battery cases, automobile trim, rope; poly(vinyl chloride): pipes, wire coatings, garden hoses, rainwear; poly(styrene): foam cups, egg cartons, furniture, electronics, toys; poly(tetrafluoroethylene): nonstick cookware, wire coatings, white film used to line pipe threads, gaskets; poly(vinyl acetate): latex (water-based) paints, chewing gum; acrylates: dentures, contact lenses, floor polishes, skylights; poly(acrylonitrile): carpets, sweaters, skirts, socks, slacks, yarn.

CONDENSATION POLYMERS

Condensation polymers account for 17.004 × 10^6 metric tons (17.89 percent) of synthetic plastics produced in the United States annually, compared to 75.835 × 10^6 metric tons (79.80 percent) addition polymers and 2.189 × 10^6 metric tons (2.31 percent) natural and inorganic polymers (excluding glass). The most common condensation polymers are the poly(esters), poly(amides), and poly(acetals), but the phenol-formaldehyde, urea-formaldehyde, melamine-formaldehyde resins and poly(carbonates) are also commercially important.

Poly(esters)

Counting both plastic and fiber applications, **poly(esters)** (chains of esters) comprise about 60 percent of the total condensation polymer production. Over 95 percent of this production is used in textiles, and the remainder goes into plastic film and bottles. The most important poly(ester) is poly(ethylene terephthalate), PET, whose synthesis is shown in figure 11.6. The main fiber trade names for PET are Dacron, Kodel, and Terylene, and Mylar is the most common trade name for the film or plastic version of this polymer. Mylar is used in the magnetic recording tapes for VCRs, audio players, and computers, in greenhouse constructions, and in home heat-sealers. PET is also molded into recyclable beverage bottles and is the main poly(ester) fiber in automobile tires.

Poly(amides)

Poly(amides) (chains containing amides) make up 26 percent of the total output of condensation polymers. About 95 percent of this output is used in textiles. The synthesis of nylon 66, one of the more common poly(amides), was shown in figure 10.24. The

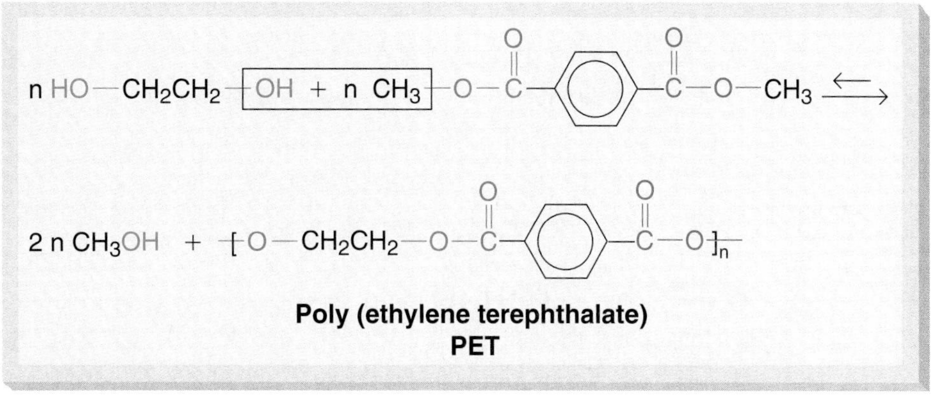

$$n\,HO-CH_2CH_2-OH + n\,CH_3-O-\overset{O}{\overset{\|}{C}}-C_6H_4-\overset{O}{\overset{\|}{C}}-O-CH_3 \rightleftarrows$$

$$2n\,CH_3OH + \left[O-CH_2CH_2-O-\overset{O}{\overset{\|}{C}}-C_6H_4-\overset{O}{\overset{\|}{C}}-O\right]_n$$

Poly (ethylene terephthalate)
PET

Figure 11.6 The synthesis of poly(ethylene terephthalate), or PET.

term **nylon** is now used generically to describe several distinctly different fibers; table 11.1 shows some of them.

Nylons are named by placing two digits after the word *nylon*. The first digit represents the number of C atoms in the diamine, and the second digit represents the number of C atoms in the organic diacid. Thus, nylon 66 is made from 1,6-diaminohexane and adipic acid (6 carbons), while nylon 610 is prepared from 1,6-diaminohexane and sebacic acid (10 carbons). Some nylon names contain only one number, such as nylon 6. The number still represents the number of C atoms, but the polymer is prepared by a ring-opening reaction using a single monomer (figure 11.7).

Aramids are poly(amides) with aromatic groups in their polymer backbone chain; trade names include Kevlar, and Nomex. The main uses for these polymers are to make bulletproof vests and other products with ultra-strong fibers.

Poly(carbonates)

Poly(carbonates) first entered the marketplace in 1957. Figure 11.8 shows the synthetic route for a typical example. These polymers have very high impact resistance, are transparent, and cost comparatively little because both monomers ($COCl_2$ and bisphenol A) are inexpensive. The major applications are in impact-resistant "glass" in windows and doors. Other uses include gears and bushings.

PRACTICE PROBLEM 11.4

a. What is the nylon name for a polymer made from an eight-carbon dicarboxylic acid and a six-carbon diamine?

b. What is the nylon name for a polymer made from a cyclic lactam (like ϵ-caprolactam in figure 11.7) containing eight C atoms and one N atom?

ANSWER *a.* Nylon 68, since the number of C atoms in the diamine becomes the first digit. *b.* Nylon 8; only the C atoms are represented in the nylon name.

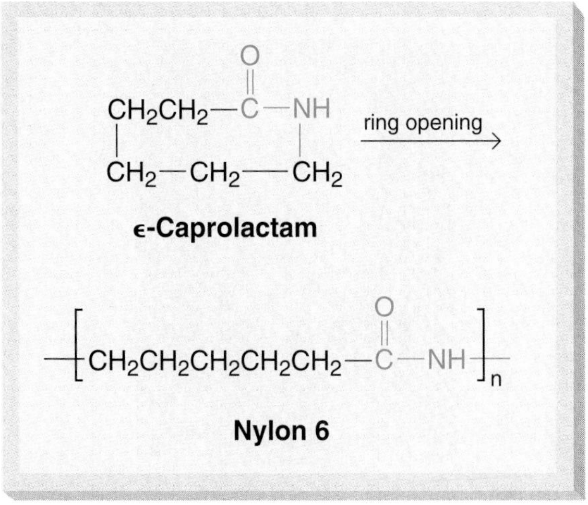

CH₂CH₂—C(=O)—NH, CH₂—CH₂—CH₂ (ring) —ring opening→

ϵ-Caprolactam

$$\left[CH_2CH_2CH_2CH_2CH_2-\overset{O}{\overset{\|}{C}}-NH\right]_n$$

Nylon 6

Figure 11.7 Nylon 6 is prepared by a ring-opening reaction using the monomer shown.

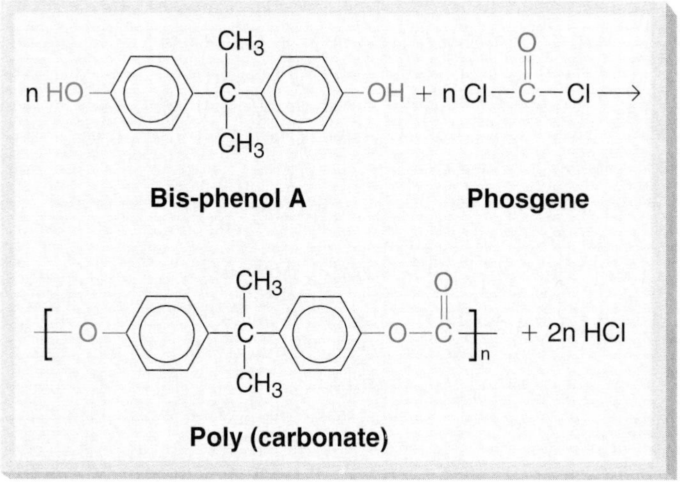

$$n\,HO-C_6H_4-C(CH_3)_2-C_6H_4-OH + n\,Cl-\overset{O}{\overset{\|}{C}}-Cl \longrightarrow$$

Bis-phenol A　　Phosgene

$$\left[O-C_6H_4-C(CH_3)_2-C_6H_4-O-\overset{O}{\overset{\|}{C}}\right]_n + 2n\,HCl$$

Poly (carbonate)

Figure 11.8 The synthesis of a typical polycarbonate.

$$n\ O{=}C{=}N{-}(CH_2)_6{-}N{=}C{=}O + n\ HO{-}(CH_2)_4{-}OH \longrightarrow$$

1,6-Hexane di-isocyanate **1,4-Butanediol**

$$\left[-\overset{H}{\underset{|}{N}}-\overset{O}{\overset{\|}{C}}-O-[CH_2]_4-O-\overset{O}{\overset{\|}{C}}-\overset{H}{\underset{|}{N}}-[CH_2]_6- \right]_n$$

Poly (urethane)

Figure 11.9 The synthesis of a typical poly(urethane).

Poly(urethanes)

Poly(urethanes), first introduced in 1943, are exceptionally strong elastomers with good abrasion and grease resistance. A typical synthesis is shown in figure 11.9. Although this reaction releases no small byproduct molecule, it is still classed as a condensation reaction since the di-isocyanate loses a small molecule during its synthesis. Table 11.7 summarizes some typical uses for poly(urethanes). The fiber form is known by the generic name spandex and the trade name Lycra.

One other class of polymer exists along with addition and condensation polymers—the semi-inorganic class (table 11.1). Semi-inorganic polymers are similar to other polymers, but they are not carbon-based. Box 11.1 discusses a commercially important member of this class: silicone rubber.

TABLE 11.7 Typical poly(urethane) uses.

FORM	USES
Fibers	swimsuits, stretch undergarments
Elastomers	small wheels, heel lifts
Foams	pillows, cushions, clothing insulation
Coatings	dance floors, synthetic varnish
Molded objects	bowling balls, artificial hearts

PRACTICE PROBLEM 11.5

Reexamine table 11.1. Which condensation polymers do you use regularly?

ANSWER Nylon: rugs, hosiery, clothing, fishing line, gears, toothbrushes; poly(carbonate): impact-resistant windows and doors, gears; poly(esters): fabrics, VCR tape and audio tapes, heat-sealing films, automobile tire cord, beverage bottles; poly(dimethyl siloxane), or silicon rubber: sealants, gaskets, implants, waxes, polishes, antifoaming additives, synthetic varnish.

Thermosetting Condensation Polymers

Thermosetting polymers comprise 14 percent of the annual plastic production. They are crosslinked, rigid, strong polymers, which allows them to perform many tasks at elevated temperatures which would cause thermoplastics to fail. The highly crosslinked rubber Ebonite, introduced in 1851 by Nelson Goodyear, was the first thermosetting plastic and is still used in bowling balls (see also chapter 9). Other examples of thermosetting condensation polymers include the phenolic, urea-formaldehyde, melamine-formaldehyde, and epoxy resins.

Phenolic Resins

The first totally synthetic commercial polymer, Bakelite, was introduced in 1907 by Dr. Leo Baekeland (Belgian, 1863–1944; figure 11.10) when he was in his early forties. The **phenolic resins** are made from phenol and formaldehyde (figure 11.11). Both chemicals have significant toxicity but can be used safely in an

Figure 11.10 Leo Baekeland made the first totally synthetic plastic in 1907. Earlier (1893), he developed the first photographic paper that was sensitive enough for printing by artificial light.

industrial setting. Note that the phenolic —OH group does not react during polymer formation. Since it is impossible to draw accurate structures for crosslinked polymers, the structure in figure 11.11 represents just a portion of the crosslinked network. The initial chemical reactions that lead to polymer formation can take place at the number 4 C atom and the equivalent numbers 2 and 5 of the aromatic ring. Eventually all of the C atoms in the ring become potential reaction sites, creating many possible structures for each polymer chain. Phenolics are marketed under several trade names, including Bakelite, Durite, and Hitanol.

Phenolic resins are normally prepared in three stages. The linear A-stage polymer is soluble in several solvents. The higher-molecular-weight B-stage resin is branched and is soluble in fewer solvents. The final C-stage is highly crosslinked and is insoluble in essentially every solvent. Phenolics are never completely colorless—they always have varying degrees of red, yellow, or brown color; this limits their uses to substances in which color is not important (table 11.8).

TABLE 11.8 Phenolic resin uses.

FORM	USES
Molded objects	electrical, radio, TV, and automotive parts, printed circuit boards, countertops
Adhesive	strong, water-resistant building trade and furniture glues
Wood modification	exterior plywood, chip or wafer board
Impregnated paper	dielectrics and transformers
Ion-exchange resins	water softening and purification systems
Other	NASA spacecraft nose-cones

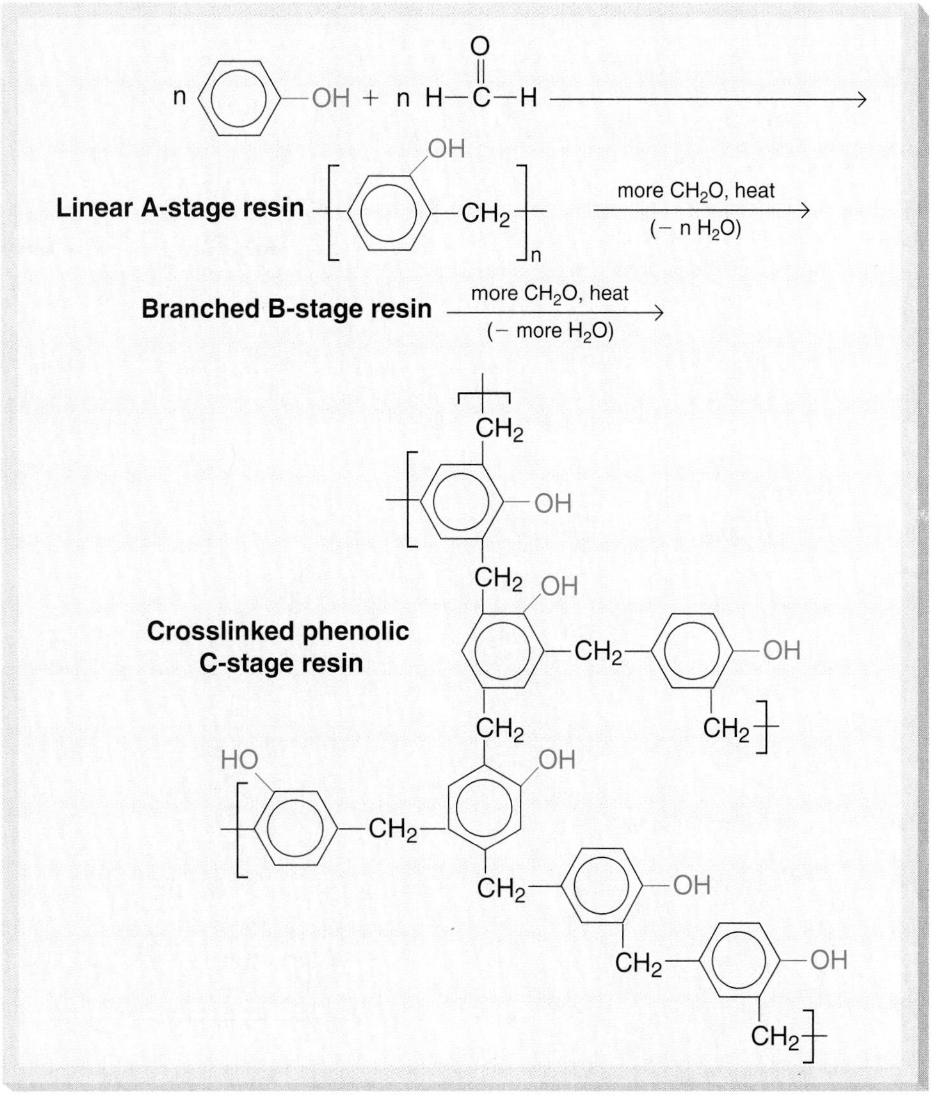

Figure 11.11 The preparation of a phenolic resin from phenol and formaldehyde.

BOX 11.1

SILICONES AND IMPLANTS

Poly(silicones), first introduced to the marketplace in 1943, are also called poly(siloxanes) or silicone rubber. Their synthesis is outlined in figure 1. Poly(silicones) come in a variety of forms, both liquid and solid, and serve a wide variety of uses—from protecting your car's finish to replacing damaged parts of the human body.

Poly(silicone) fluids are used as coolants because they are chemically inert and have stable physical properties. Low-molecular-weight liquid poly(silicones) are used in waxes and polishes, as release agents (chemicals that allow a substance to be removed from another surface or object), and as antifoaming additives for paper and textiles. You may have used a silicone-based wax on your car. Elastomeric (rubbery) poly(silicones) are used in gaskets, seals, and wire or cable insulation. Poly(silicone) resins are used as synthetic varnish, for encapsulating, and for impregnating other materials. The popular toy Silly Putty is an elastomeric poly(silicone).

Many poly(silicones) have proven useful in the field of medicine; highly purified silicone rubber has been used in surgical implants for many years. Figure 2 shows some of these uses, including (a) the hydrocephalus shunt. This simple piece of silicone rubber tubing drains excess fluid from the brain of a person with hydrocephalus (or of a postsurgical patient); such shunts save more than ten thousand lives annually in the United States. The Malecot catheter, is used to introduce medicines into the body or to remove samples of body fluids for chemical analysis.

Other poly(silicone) surgical implants replace severely injured or damaged body parts. Figure 2c, *left* shows an ear prosthesis that is used to replace an ear missing due to a birth defect, accident, or cancer. The chin implant, (2c, *top*) is used as a replacement for a portion of chin that was removed due to cancer. A tendon prosthesis (2c, *center*) replaces a damaged tendon and can restore normal motion in other parts of the body. Finger joint surgery using silicone implants (2c, *bottom*) restores hand mobility to more than 400,000 patients each year.

One silicone implant has achieved unenviable notoriety. More than two hundred thousand breast implants (figure 2d, *left*) are surgically implanted each year, although they have been the center of a major debate that began in 1992. A small number of doctors and researchers believe that breast implants pose a danger to the patient: they may rupture and leak silicone into the body, causing an adverse autoimmune response with arthritislike symptoms. Other medical experts believe these fears are groundless. Because it is very difficult to determine the presence or absence of an autoimmune response, a resolution to the debate may require more than a decade of additional studies. The 1994 studies using the largest numbers of women with silicone breast implants appear to show no negative autoimmune response—that is, the incidence of autoimmune responses is approximately the same whether the woman has a silicone polymer breast implant or not. An even larger study released in 1995 confirmed the absence of an increase in the autoimmune response. In August 1995, FDA director Dr. Kessler announced that "women with breast implants are not at a big risk" for autoimmune diseases.

It remains unclear why potential autoimmune problems appeared to be associated with breast implants. The testicular implant (figure 2d, *right*) uses an almost identical type and design of silicone polymer and causes no apparent immune problem. Similarly, the other medical uses of silicone polymers seem to elicit no autoimmune responses. In any event, a large settlement was reached in 1994 to compensate women with autoimmune problems that may have been induced by silicone breast implants.

Unfortunately, because of the litigation that produced this settlement, and possible future liability, Dow Corning, the dominant manufacturer, has stopped mak-

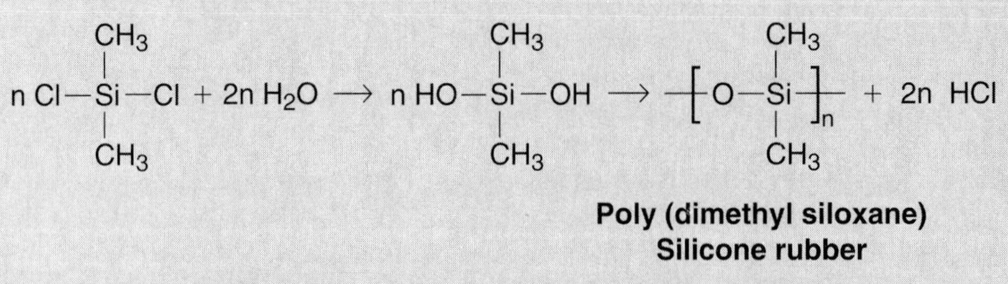

$$n\,Cl{-}\underset{CH_3}{\overset{CH_3}{\underset{|}{\overset{|}{Si}}}}{-}Cl + 2n\,H_2O \longrightarrow n\,HO{-}\underset{CH_3}{\overset{CH_3}{\underset{|}{\overset{|}{Si}}}}{-}OH \longrightarrow \left[O{-}\underset{CH_3}{\overset{CH_3}{\underset{|}{\overset{|}{Si}}}}\right]_n + 2n\,HCl$$

Poly (dimethyl siloxane)
Silicone rubber

Figure 1 The synthesis of a poly(silicone), or silicone rubber.

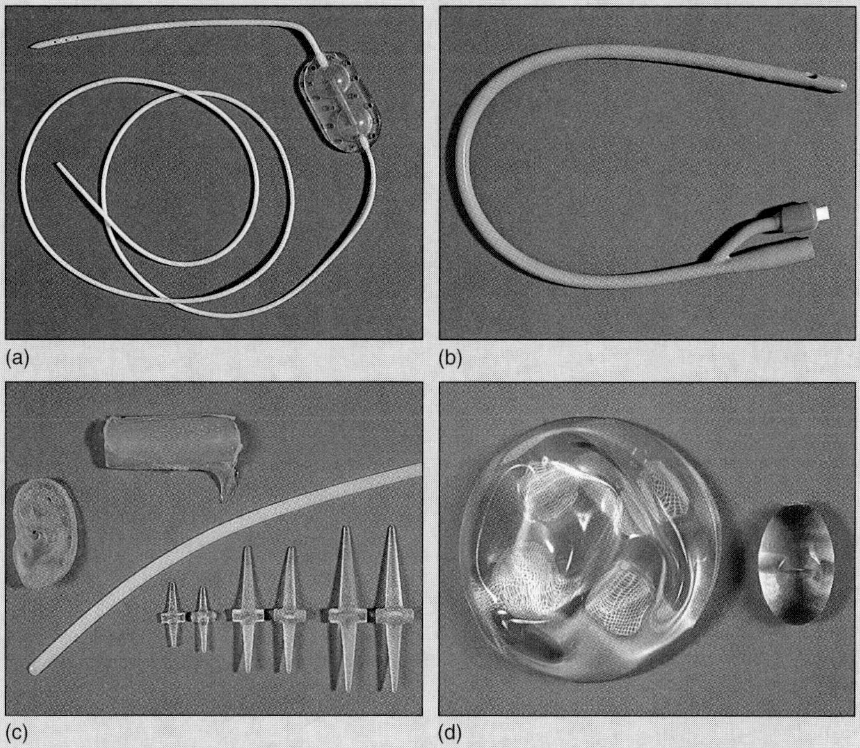

(a) (b)

(c) (d)

Figure 2 Examples of silicone implants. **(a)** Hydrocephalus shunt; **(b)** Malecot catheter; **(c)** ear prosthesis (*left*), chin implant (*top*), tendon prosthesis (*center*), finger joints of various sizes (*bottom*); **(d)** breast implant (*left*), testicular implant (*right*). All samples are from Dow Corning.

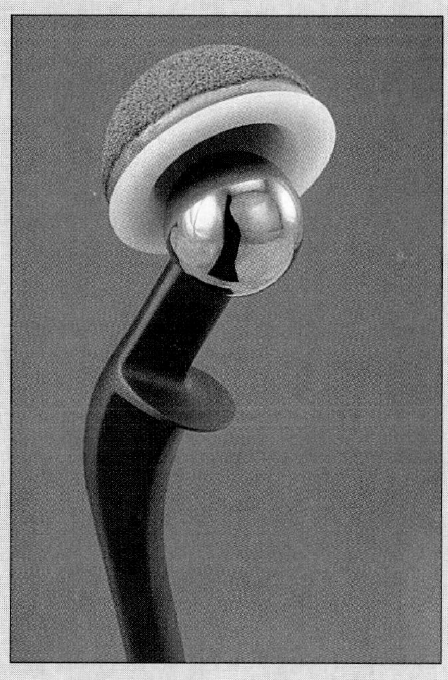

Figure 3 A typical hip prosthesis. The metal ball on the femur is inserted into a plastic socket made from high density poly-ethylene. Both the ball and the socket are cemented in place using poly(methyl methacrylate).

ing breast implants. This leaves the market to less experienced manufacturers, some based in foreign countries with less stringent quality control. Dow Corning continues to produce the other devices that improve the quality of life for millions.

New silicone rubber uses extend beyond mechanical tubing and structural prostheses. An artificial skin made from a silicone polymer and collagen (a natural polymer) has been developed by I. V. Yannas at M.I.T.; this is already proving a boon to burn patients and others with severe skin damage. Research suggests that similar materials may enhance nerve regeneration.

Other polymers find use as implants as well. A poly(ethylene) or poly(propylene) mesh is used to correct hernias. Hip and knee replacements contain several plastic materials. The usual hip prosthesis contains a metal ball attached to the femur (leg) and inserted into a high-density poly(ethylene) socket placed in the hip. Both the ball and the socket are cemented in place using poly(methyl methacrylate) (figure 3).

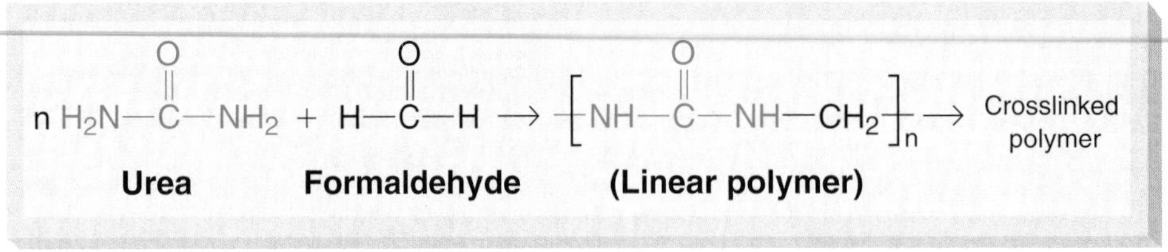

$$n\,H_2N-\overset{\overset{\large O}{\|}}{C}-NH_2 + H-\overset{\overset{\large O}{\|}}{C}-H \rightarrow \left[NH-\overset{\overset{\large O}{\|}}{C}-NH-CH_2\right]_n \rightarrow \text{Crosslinked polymer}$$

Urea **Formaldehyde** **(Linear polymer)**

Figure 11.12 Urea and formaldehyde react to produce a colorless, crosslinked polymer.

Urea-Formaldehyde

The reaction between formaldehyde and urea ultimately produces a colorless, crosslinked polymer (figure 11.12). The polymer solubility decreases rapidly with increased crosslinking. The urea-formaldehyde or UF resins were first introduced in 1929 and are marketed under several trade names, including Uformite and Beetle. UF resins are light colored and more water-sensitive than the phenolics. Most applications are similar to those for the phenolic resins; some other uses are summarized in table 11.9.

Urea-formaldehyde has come under attack because it sometimes releases formaldehyde vapors. Formaldehyde is toxic and can cause a wide range of allergic reactions in a very small number of people. When the UF polymerization is run properly, all the formaldehyde is consumed and none is left to cause such problems. However, there is an enormous difference between carrying out polymerization in a laboratory or factory, under carefully controlled conditions, and running a similar reaction at home, where control is extremely difficult if not impossible. This is unfortunate because the UF foams form an excellent thermal barrier. Polymers are among the best thermal barriers (table 11.10), but they cost more than mineral fibers or asbestos. (See box 6.2 for a discussion of asbestos and its problems.) UF foams are less expensive than foamed PSt (PIP boards) because they are synthesized from cheaper monomers; UF foams are also more flame resistant. Foamed poly(olefins) and poly(vinyl chloride), or PVC, are now available and promise quality insulation at lower cost.

Some of the materials in table 11.10 (for example, copper, graphite, and iron) are not insulators but are, instead, thermal conductors included in the table for comparison. Glass also conducts heat fairly well and is the reference point for table 11.10. Glass windows are often the major source of thermal energy loss in homes.

Melamine-Formaldehyde

The melamine formaldehyde (MF) polymers were first introduced in 1939. Their main trade name is Melmac, and they have an exceptionally light color. Although the MF resins are harder and more moisture resistant than the UF resins, the greater cost of the melamine monomer reduces their usage. The MF polymerization (figure 11.13) is similar to the polymerization of the phenolics.

TABLE 11.9 Urea-formaldehyde polymer uses.

FORM	USES
Molded objects	countertops, tabletops
Wood modification	interior plywood, particle board
Impregnated materials	crease-resistant fabrics
Foamed materials	home insulation

TABLE 11.10 Thermal barrier properties for some materials; higher values mean less conductivity and better insulation.

MATERIAL	BARRIER VALUE
UF foam	33
Polyurethane foam	33
Polystyrene foam	25
Polystyrene sheets	6
Poly(vinyl chloride) sheets	6
PVC + 35% plasticizer	6
Natural rubber	5
Mineral fibers	5
Asbestos	4
Nylon 66	4
Poly(tetrafluoroethylene)	4
Poly(ethylene), low-density	3
Poly(ethylene), high-density	2
Glass	1
Quartz	0.1
Iron	0.01
Graphite	0.007
Copper	0.0001

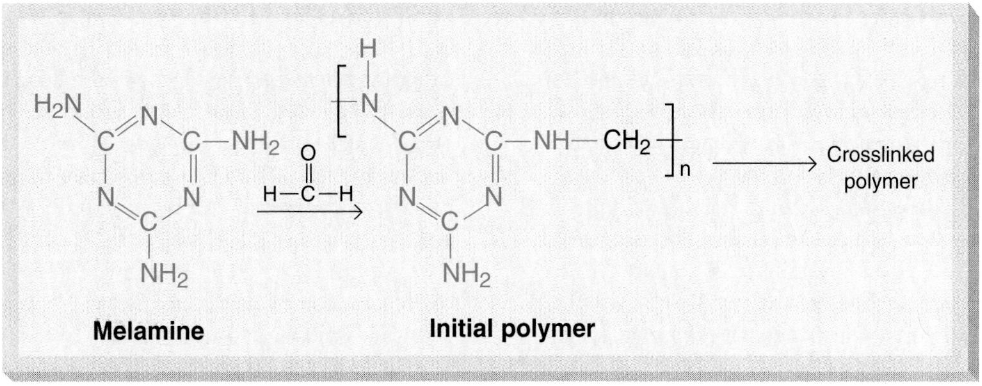

Figure 11.13 The polymerization of a melamine formaldehyde polymer.

Rapid crosslinking involves all three —NH_2 groups. The major use for MF polymers is in decorative plastic dinnerware.

Epoxy Resins

Epoxy resins were first placed on the market in 1947 and have become nearly indispensable in adhesives, laminating, and molding. Their chemistry is complicated because they come from a variety of starting monomers. The common epoxy cement you can purchase at any hardware store is usually a two-component mixture; one component contains a linear epoxy polymer and the other contains a poly(amine) that reacts with the epoxy groups to crosslink the system. The result is an extremely strong, excellent adhesive bonding agent. Figure 11.14 shows the reaction of epichlorohydrin with bis-phenol A to form a typical epoxy resin. The subsequent poly(amine) crosslinking reaction occurs with the free epoxide groups on the ends of the polymer chains.

COPOLYMERS—ADDING MORE VERSATILITY

Although a variety of polymers exist, their properties ranging from very flexible to very stiff, copolymers add even more versatility. Versatility is important for at least two reasons. First, certain

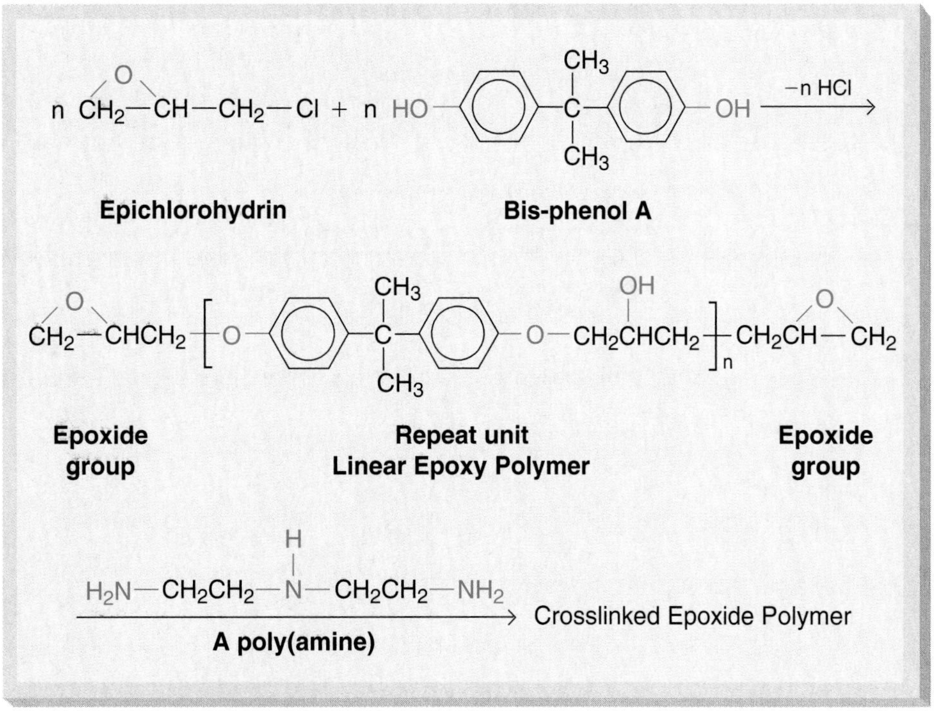

Figure 11.14 Epichlorohydrin and bis-phenol A react to form a typical linear epoxy polymer. Free epoxide groups on the ends of the polymer chain then can react with poly(amines) to form a crosslinked epoxide polymer.

applications cannot be met by any common plastic. Second, even when a plastic is available to meet a need, existing polymers do not always provide the needed properties at an affordable price. Mixing two polymers together seldom achieves the desired mix of properties because most polymer blends are incompatible and separate into layers. Copolymers provide an answer.

Part of figure 11.15 shows two separate polymer chains with the repeat units denoted as H or T, respectively. This type of polymer is a **homopolymer** because it contains only a single repeating monomer. If homopolymers are mixed together, the chains remain individual collections of either polymer H or polymer T. Let's consider two polymers—poly(vinyl chloride) (PVC) and poly(vinylidene chloride) ($PVCl_2$). Unmodified PVC has a T_g of +80° C and is a rigid plastic completely unsuitable for plastic food wrapping film. It also permits oxygen to penetrate and would not protect enclosed foods. However, PVC is inexpensive and has excellent tensile strength (it resists breaking when stretched)—important characteristics in a simple household plastic film. On the other hand, $PVCl_2$ has a low T_g of −17° C and is very resistant to oxygen penetration, making it potentially flexible and protective. Unfortunately, $PVCl_2$ is also a crystalline polymer, which makes it rigid at room temperature and useless as plastic film. Thus, each of these homopolymers is unsuitable for plastic wrapping film even though the desired cost, strength, and oxygen-penetration resistance exist as properties in one polymer or the other. What happens if we try to combine the two? A blend of the two individual polymers is still a rigid plastic with no useful oxygen-barrier properties because the monomer units remain in separate polymer chains and have no effect on each other. But what if we can find a way to mix the two monomers in a single chain—to create a **copolymer?** Chemists make copolymers to satisfy specific industrial or consumer needs. In copolymers, the monomer units intermix and modify each other's properties.

Chemists set to work on the plastic wrapping film problem and soon found that a copolymer formed from the PVC and $PVCl_2$ monomers has substantially reduced crystallinity. Eventually, they achieved the right mix; they discovered that a noncrystalline copolymer containing 20 to 30 percent vinyl chloride and 70 to 80 percent vinylidene chloride has a T_g below room temperature and is a flexible film with good oxygen resistance. Better yet, it is inexpensive! This copolymer is now marketed as Saran. Pliovic, a similar copolymer with a higher vinyl chloride level, is used for

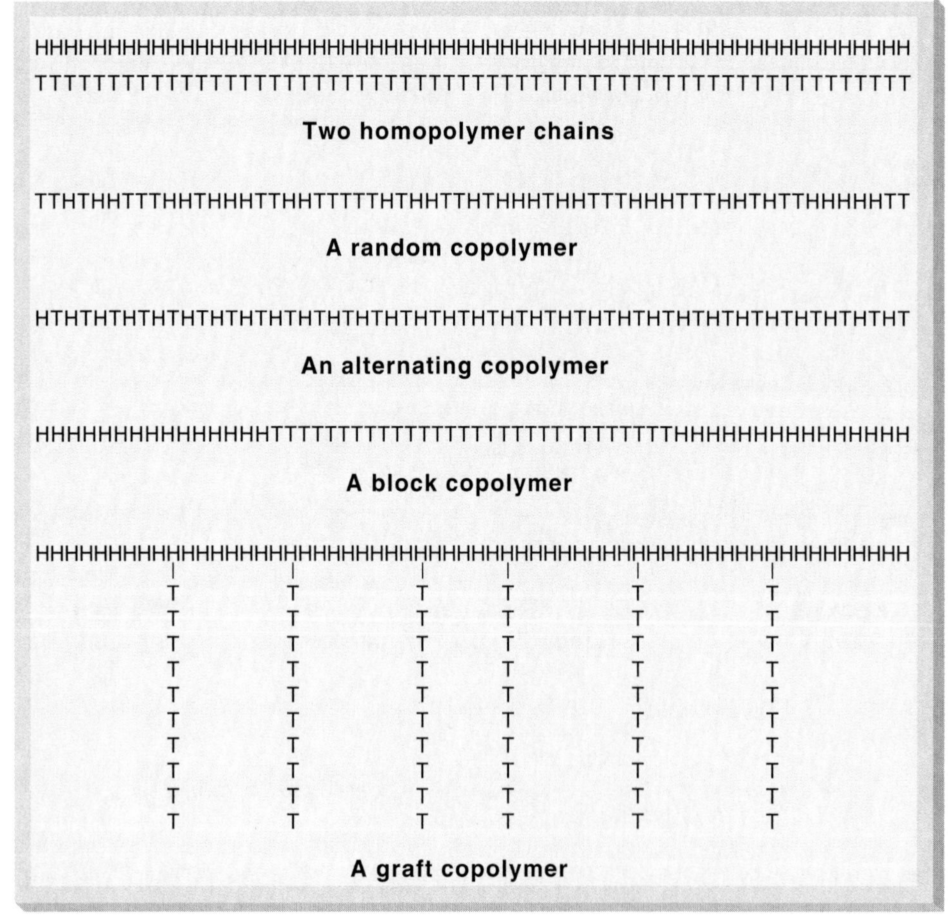

Figure 11.15 Different types of polymer and copolymer structures that could be formed from the monomers H and T.

coatings such as those on coated fabrics. Both copolymers were introduced in the 1930s.

In most copolymers, the monomers are arranged along the polymer chain to form a **random copolymer** (figure 11.15). For example, ethyl acrylate (EA) and methyl methacrylate (MMA) arrange themselves randomly in a copolymer. You could simulate this randomness by assigning each monomer one side of a coin and then flipping the coin, so that the "head" of the coin represents EA and the "tail" MMA. Figure 11.15 shows what this random copolymer might look like after 60 such coin tosses (creating a chain of sixty monomers) using the symbols H and T. You could repeat this procedure for the vinyl chloride and vinylidene chloride monomer pair. Most copolymers have a random arrangement.

Some monomer combinations do not create a random copolymer. Instead, they alternate regularly along the chain, as depicted in figure 11.15; such chains are called **alternating copolymers.** A pair of monomers may either combine in a random or alternating pattern, but not both.

Two other kinds of copolymers are shown in figure 11.15, **block copolymers** and **graft copolymers.** Chemists use special techniques to "force" the monomers into each of these unique arrangements. A block copolymer has long blocks, or sequences, of each monomer. A graft copolymer has a chain of one monomer type with shorter chains of the other monomer grafted onto it, much as the branch of one plant is grafted onto the rootstock of another.

Many useful copolymers exist. Copolymers of methyl methacrylate (MMA) and ethyl acrylate (EA) are used in water-based paints. Poly(methyl methacrylate) (PMMA) has a T_g of +105° C and is a stiff plastic, while poly(ethyl acrylate) (PEA) has a T_g of −24° C and is a tacky, flexible polymer. Neither homopolymer could function as a paint base because one is too hard and the other is too tacky, but copolymers of the two with T_g values below 15° C are suitable for paints. Other MMA-EA copolymers are used in floor polishes and paper coatings.

The Pluronic® surfactants (which we will discuss further in chapter 13) were the earliest commercially used block copolymers; they are synthesized from blocks of poly(ethylene oxide) and poly(propylene oxide) (figure 11.16). As the figure shows, the total number of repeat units in a Pluronic ABA block polymer is $x + y + z = n$. The ABA denotes the arrangement of monomers in each block. The spandex poly(urethane) fibers, made from poly(ether) or poly(ester) and di-isocyanate blocks (figure 11.9), are examples of an AB block copolymer (figure 11.16). Thermoplastic elastomer block copolymers of styrene and butadiene have been marketed since 1965, under the trade names Solprene (an AB block) and Kraton (an ABA block).

The ABS systems are the most common graft copolymers, with an annual production of over 907,000 metric tons. They are made from acrylonitrile, butadiene, and styrene; the name ABS is derived from the first initial of each of these monomers. ABS copolymers are used in automotive parts, telephones, luggage, refrigerator door liners, building panels, radiator grilles, business

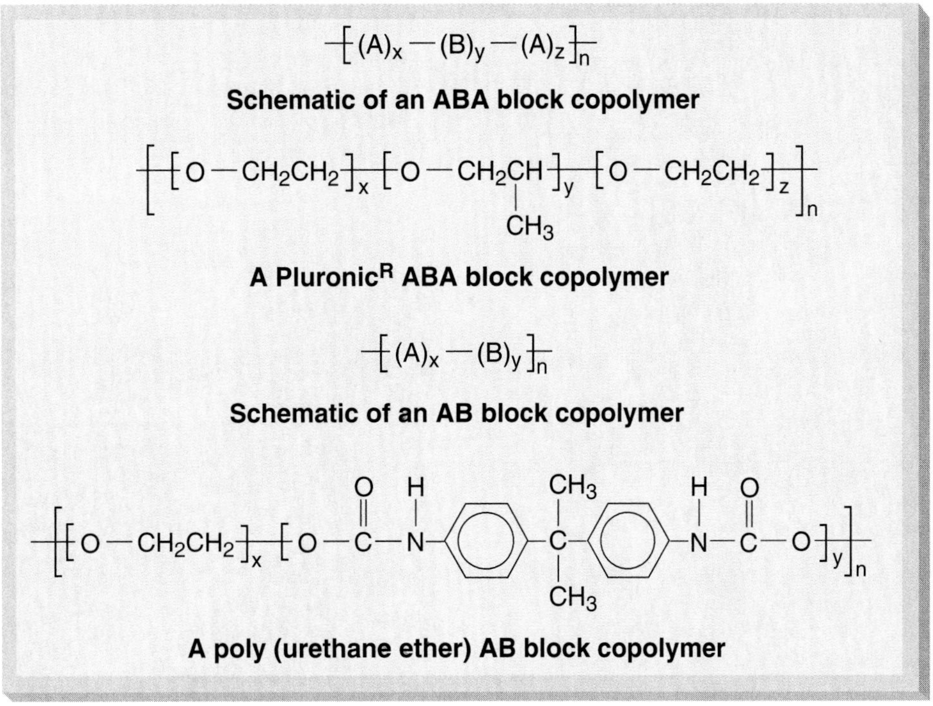

$$\left[(A)_x - (B)_y - (A)_z \right]_n$$

Schematic of an ABA block copolymer

$$\left[\left[O - CH_2CH_2 \right]_x \left[O - CH_2\underset{\displaystyle CH_3}{\underset{|}{C}}H \right]_y \left[O - CH_2CH_2 \right]_z \right]_n$$

A Pluronic[R] ABA block copolymer

$$\left[(A)_x - (B)_y \right]_n$$

Schematic of an AB block copolymer

$$\left[\left[O - CH_2CH_2 \right]_x \left[O - \overset{O}{\overset{\|}{C}} - \overset{H}{\overset{|}{N}} - C_6H_4 - \underset{CH_3}{\overset{CH_3}{C}} - C_6H_4 - \overset{H}{\overset{|}{N}} - \overset{O}{\overset{\|}{C}} - O \right]_y \right]_n$$

A poly (urethane ether) AB block copolymer

Figure 11.16 Schematic drawings and examples of block copolymers. The Pluronic® example is used as a surfactant in dishwasher detergents and the poly(urethane ether) is used in spandex garments. One version of the latter is used to make artificial hearts.

machines, and shoe soles. Numerous other graft copolymers are formed by attaching synthetic polymer chains onto natural polymers such as starch, cellulose, or proteins.

ELASTOMERIC POLYMERS

Rubber was the first widely used elastomer, but its availability was affected by wars and economic uncertainties. (The history and development of the rubber industry was discussed briefly in chapter 9.) Finding a suitable rubber substitute was not easy. The efforts to do so followed three main pathways: (1) duplicating natural rubber in the laboratory; (2) developing other elastomeric polymers; and (3) matching rubber's properties with elastomeric copolymers.

Poly(chloroprene)

Poly(chloroprene), the first commercially successful synthetic elastomer, was developed in 1931 at DuPont and received the trade name Neoprene. Wallace Hume Carothers (figure 11.17) invented poly(chloroprene). This elastomer is made by polymerizing monomeric chloroprene (IUPAC name = 2-chloro-1,3-butadiene) using a free radical catalyst (figure 11.18). The resulting polymer has many physical properties similar to those of natural rubber (table 11.11). In fact, poly(chloroprene) actually has some properties superior to rubber's; it has much better oil, grease, and solvent resistance, which accounts for its widespread use in gaskets, seals, oil hoses, and grease-resistant shoe heels and soles. Currently, poly(chloroprene) ranks fifth in synthetic rubber production. The Cl atom attached to the carbon chain confers some flame resistance, but the elastomer will burn. Poly(chloroprene) was the first successful attempt to duplicate natural rubber.

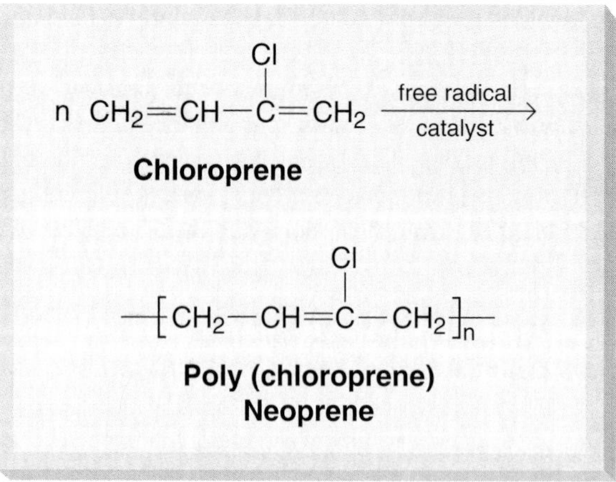

Figure 11.18 The preparation of poly(chloroprene), a synthetic rubber.

Figure 11.17 Wallace Hume Carothers (American, 1896–1937), demonstrating a piece of his new synthetic rubber at the December 1931, American Chemical Society meeting. His pioneering research led to many industrial polymers, including Neoprene, nylon, and Dacron.

Synthetic Poly(isoprene)

While Carothers tried to synthesize a rubberlike polymer in DuPont's lab, other researchers explored along similar lines. Although many early attempts were made to polymerize butadiene or isoprene monomers, these attempts were abject failures and led to useless materials. The key to both reactions turned out to be a special group of catalysts developed by Karl Ziegler and Giulio Natta (figure 11.19) in the early 1950s. (These two scientists subsequently shared the 1963 Nobel Prize for their research.)

(a)

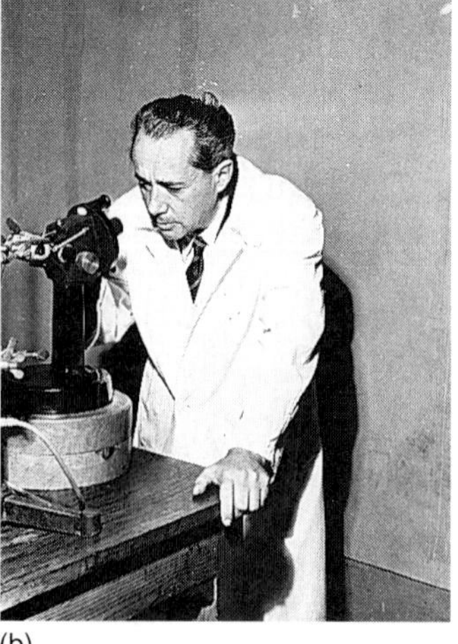
(b)

Figure 11.19 (a) Karl Ziegler (German) and (b) Giulio Natta (Italian) won the 1963 Nobel Prize in chemistry for their process used to make a synthetic version of natural rubber.

TABLE 11.11 The physical properties of poly(isoprene) and related elastomers.

POLYMER NAME	POLYMER STRUCTURE	T_g(°C)	T_m(°C)	DENSITY g/mL
Poly(butadiene) (free radical)	$-[CH_2-CH=CH-CH_2]_n-$ (irregular)	−63	none	0.99
Poly(chloroprene) (Neoprene)	$-[CH_2-CH=C(Cl)-CH_2]_n-$	−50	none	1.23
cis-poly(isoprene) (Natural rubber)	$-[CH_2-C(H)=C(CH_3)-CH_2]_n-$ (cis)	−72	+27	0.92
trans-poly(isoprene) (Gutta percha)	$-[CH_2-C(H)=C(CH_3)-CH_2]_n-$ (trans)	−53	+60	0.92
Isobutylene-isoprene copolymer (Butyl rubber)	$-[CH_2-C(CH_3)_2]_{0.97}-[CH_2-CH=C(CH_3)-CH_2]_{0.03}-$	−70	none	0.92

These catalysts were also used to prepare the previously unavailable high-density poly(ethylene) and high-molecular-weight poly(propylene).

EXPERIENCE THIS

Although both stretch, rubber has some unusual properties as compared to a metal spring. For example, place a rubber band near (but not between) your lips and stretch it. Immediately touch the rubber band to your lips and note its temperature. Next, let the rubber band relax back to its original size. Hold it to your lips again and notice the temperature change. Repeat the procedure with a spring—you won't find a temperature change.

For a more complicated experiment, find a small spring and a heavy rubber band. Cut the rubber band and tie a weight to one end. Attach an identical weight to the spring. Now fasten both the spring and the rubber band to a horizontal bar high enough to let the weights hang freely. Heat the spring and the rubber band with a hair dryer or suspend them over a hotplate. What happens? The spring stretches when heated, but the rubber band contracts.

This unusual behavior is virtually unique to rubber. It arises because rubber can partially crystallize when stretched or cooled but loses its crystalline character when heated. Metals do not behave this way.

Other Elastomers

Along with attempting to duplicate rubber in the lab, chemists worked on developing other elastomeric polymers. Poly(butadiene) has been synthesized since the 1920s, but the material prepared via free radical polymerization was a poor elastomer. With the advent of the Ziegler-Natta catalyst system, a satisfactory elastomeric form of cis-poly(butadiene) was developed in 1959. (See table 11.11 for the properties of the poly(butadiene) polymer.)

Poly(sulfide) elastomers, the first synthetic elastomers, were introduced in 1929 under the trade name Thiokol. These elastomers were developed by J. C. Patrick in a garage-scale operation that rapidly grew into a major corporation named Thiokol, now Morton-Thiokol. The polymers, prepared by reacting sodium polysulfide with dihaloalkanes, are called poly(alkylene polysulfide) (figure 11.20). Thiokol rubbers have outstanding oil and

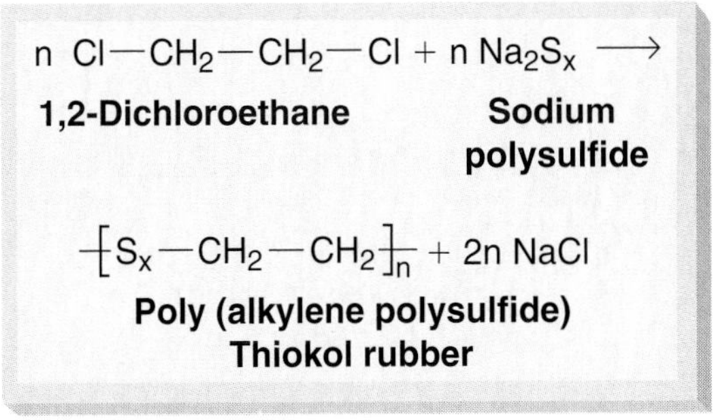

$$n\ Cl-CH_2-CH_2-Cl + n\ Na_2S_x \longrightarrow$$

1,2-Dichloroethane **Sodium polysulfide**

$$-[S_x-CH_2-CH_2]_n- + 2n\ NaCl$$

Poly (alkylene polysulfide)
Thiokol rubber

Figure 11.20 The synthesis of Thiokol rubber—a poly(sulfide) elastomer.

solvent resistance plus good weather resistance, but they are not as strong as most other synthetic elastomers and usually have an unpleasant sulfur smell. They are widely used in gaskets, gasoline hoses, and sealants. Recently, these polymers have also been used as a binder for solid rocket propellants, such as those used in space shuttles, because they burn well.

Most elastomers require vulcanization (see chapter 9) to strengthen them. Since sulfur vulcanization does not always work, metal oxide vulcanization is used for many elastomers. The individual elastomer molecules become connected or crosslinked by C—C bonds when vulcanized by a metal oxide or peroxide (figure 11.21). Crosslinked chains are much stronger than linear or branched chains.

Elastomeric Copolymers

Another path chemists took in developing a rubberlike compound, along with developing a synthetic rubber and synthesizing other elastomers, was to develop elastomeric copolymers. Prior to the development of the Ziegler-Natta catalyst system, several diene copolymers with a great practical value were prepared (figure 11.22). The first diene copolymer elastomers were the butadiene-styrene (Buna-S) and the butadiene-acrylonitrile (Buna-N) systems, marketed in 1937. The commercial impetus for these elastomers came during World War II when natural rubber supplies to the United States were cut off. Both elastomers contain 70 to 75 percent butadiene, although other combinations can be synthesized for specific uses.

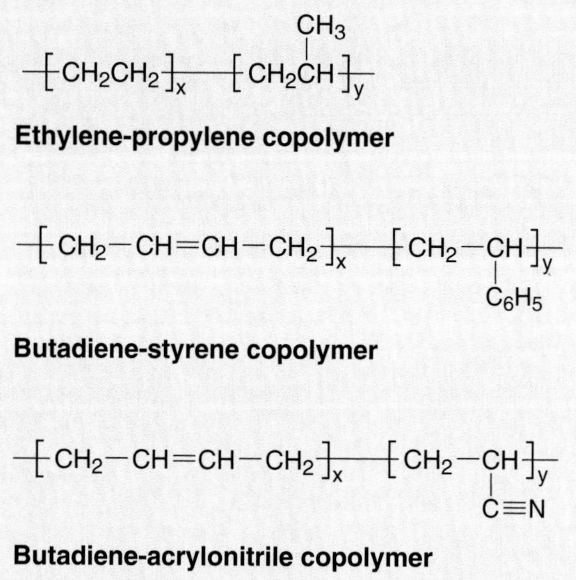

$$-\!\!\left[CH_2CH_2\right]_x\!\!-\!\!\left[CH_2CH(CH_3)\right]_y\!\!-$$

Ethylene-propylene copolymer

$$-\!\!\left[CH_2-CH=CH-CH_2\right]_x\!\!-\!\!\left[CH_2-CH(C_6H_5)\right]_y\!\!-$$

Butadiene-styrene copolymer

$$-\!\!\left[CH_2-CH=CH-CH_2\right]_x\!\!-\!\!\left[CH_2-CH(C\equiv N)\right]_y\!\!-$$

Butadiene-acrylonitrile copolymer

Figure 11.22 The repeat unit structures of some elastomeric copolymers. The subscripts *x* and *y* are the fraction of each monomer in the copolymer.

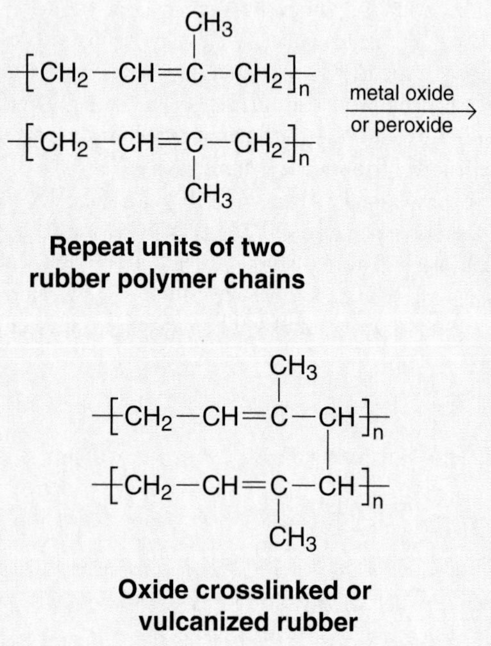

$$-\!\!\left[CH_2-CH=C(CH_3)-CH_2\right]_n\!\!- \quad -\!\!\left[CH_2-CH=C(CH_3)-CH_2\right]_n\!\!- \xrightarrow[\text{or peroxide}]{\text{metal oxide}}$$

Repeat units of two rubber polymer chains

$$-\!\!\left[CH_2-CH=C(CH_3)-CH\right]_n\!\!- \quad | \quad -\!\!\left[CH_2-CH=C(CH_3)-CH\right]_n\!\!-$$

Oxide crosslinked or vulcanized rubber

Figure 11.21 An example of metal oxide vulcanization.

The butadiene-styrene elastomer (B-St) is slightly inferior to natural rubber, but it is still the most widely produced synthetic elastomer because it is so inexpensive. Its main use is in automobile tires, but interestingly, B-St is not used in truck tires because the heat builds up too much and B-St's resilience is too low. Other uses for B-St elastomers include belting, molded goods, flooring, shoe soles, electrical insulation, and hoses.

The butadiene-acrylonitrile elastomer (B-AN) has much greater oil and grease resistance than natural rubber. This elastomer is used for gaskets, seals, fuel tank liners, shoe soles, O-rings (in the space shuttles), adhesives, and as a rocket fuel binder.

Elastomeric ethylene-propylene copolymers (EP) were introduced in 1960. Each homopolymer is crystalline, but the copolymer is amorphous. EP elastomers have outstanding oxidative resistance and are used in tires.

CHANGING THE PROPERTIES OF PLASTICS

In addition to synthesizing copolymers, a variety of ways exist to modify the properties of polymers to make them more useful or less expensive. We will only touch lightly on this complex area in which chemists routinely make use of additives such as plasticizers, pigments, and stabilizers.

Plasticizers

A **plasticizer** makes a polymer more flexible by lowering its T_g. Unplasticized PVC is a stiff plastic with a T_g of + 80° C, but when a plasticizer such as dioctyl phthalate or tricresyl phosphate (figure 11.23) is added to PVC, the polymer becomes soft and flexible. Most plasticizers are liquids with high boiling points and good solubility in the polymer. Water even acts as a plasticizer for many naturally occurring polymers in the human body; if it didn't, our soft body tissues would be relatively rigid and would break when bent. Fats and oils can also plasticize body tissues.

Plasticizers sometimes provide other benefits to a polymer—tricresyl phosphate, for example, confers some flame resistance. Plasticizers can also extend the usable temperature range for the inexpensive PVC used in garden hoses. A word of warning—the T_g of the blend is reduced, but not eliminated; thus, a plasticized garden hose left out in the cold winter weather becomes stiff and cracks when handled. Polychlorinated biphenyls (PCBs) were once used as plasticizers, but this practice ceased when chemists found PCBs were highly toxic.

Pigments and Fillers

Inexpensive **pigments** are added to plastics to impart color and reduce cost. Although essentially any inorganic salt can serve as a pigment, the less soluble salts withstand water extraction and retain color longer. **Fillers** are added to make the composition less expensive. Since they must be cheap, typical fillers include wood flour, wood fibers, carbon black, glass, sand, and chalk. Fillers sometimes provide added benefits to the polymer. For example, carbon black increases the strength of tires and decreases UV light degradation. The impact strength of nylon is increased by adding tiny glass spheres.

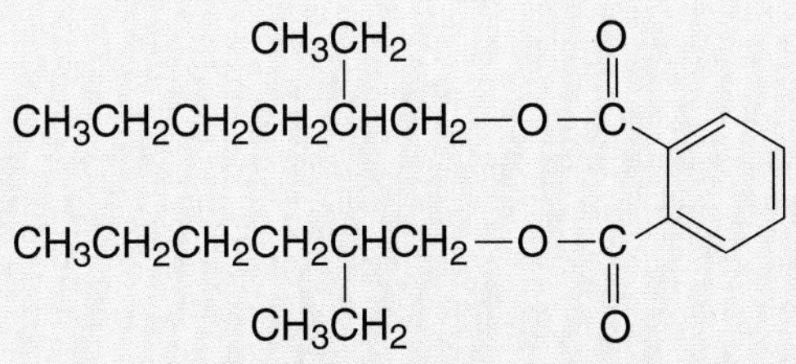

Dioctyl phthalate or Di-2-Ethylhexyl phthalate

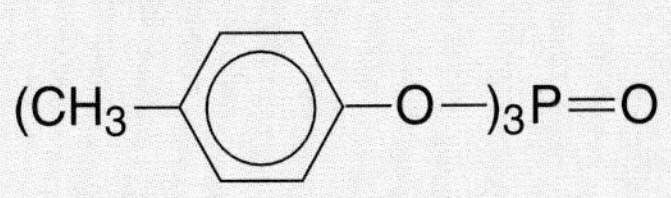

Tricresyl phosphate

Figure 11.23 Some common plasticizers.

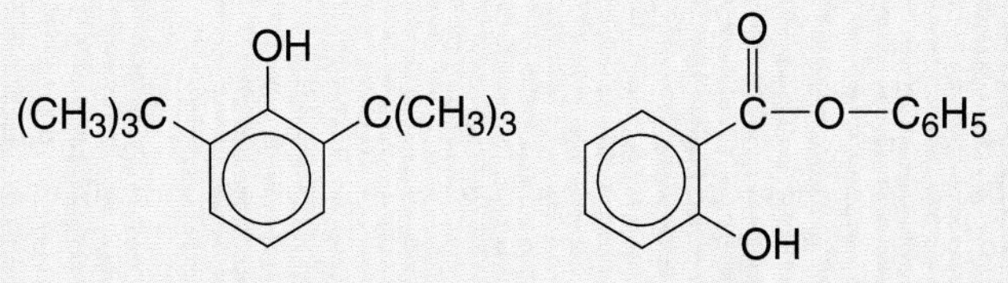

(a) **1,5-Di-*tert*-butyl phenol** (b) **Phenyl salicylate**

Figure 11.24 Some stabilizers used for plastics.

Stabilizers

Stabilizers include antioxidants, heat stabilizers, ultraviolet stabilizers, and flame retardants added to improve a specific polymer property. **Antioxidants** prevent oxidation. Poly(propylene) fails in long-term outdoor use because the H atoms on the tertiary carbons (the C atoms attached to three other C atoms; see the center CH in the PP structure in table 11.1) readily cleave from the chain, leading to decomposition. The antioxidant 1,5-di-*tert*-butylphenol (figure 11.24a) prevents this degradation.

Poly(vinyl chloride) or PVC, decomposes slightly when heated, losing HCl molecules. Chemists prevent this thermal degradation by adding lead, cadmium, magnesium, or barium salts. Alkyl tin compounds were once used, but this was discontinued because of their high toxicity. Although lead and cadmium salt toxicity is lower than that of alkyl tin compounds, Pb and Cd salts are not used in plumbing, since they might leach out and contaminate the household water supply with these heavy metals.

Many polymers, including the top five in table 11.2, degrade under the influence of ultraviolet light. The effect of UV on a plastic is somewhat similar to its effect on human skin (tanning), and similar chemicals are used for UV protection in suntan lotions and polymers. The main **UV stabilizers** for plastics are salicylate esters, such as phenyl salicylate (figure 11.24b).

Most polymers burn when placed in a flame and are potentially dangerous when used as building materials or furniture. It is hardly surprising that polymers burn, because they are carbon compounds. Flame retardants contain a halogen, phosphorus, or antimony atom. Examples include tricresyl phosphate, antimony oxide, and antimony chloride (figure 11.25).

Figure 11.25 Some additives used to impart flame resistance to plastics.

PAINTS AND COATINGS

In 1992, paint sales amounted to nearly 13 billion dollars, with more than 1.1×10^9 gallons sold in the United States (about 4.75 gallons per person). Paint falls into several classes—architectural coatings (55 percent), product coatings (29 percent), and special-purpose coatings (16 percent). The uses for each class appear in table 11.12.

Paints are dispersions of (1) the *binder,* which is a drying oil or polymer, (2) a solvent or thinner, and (3) pigments. The *vehicle* is the combination of (1) and (2). Common solvents are organic liquids, water, or a combination of both. The pigments are either *prime,* which color the paint, or *inert,* which make the paint last longer. Prime pigments include white titanium dioxide (TiO_2), brown or red iron oxides (Fe_2O_3 or Fe_3O_4), and green or blue phthalocyanines. Inert pigments include clay, talc, calcium carbonate ($CaCO_3$), and magnesium silicate. They are usually white.

There are two major kinds of paint—oil-based and latex. **Oil-based paints,** the older type, are a pigment suspension in a drying oil, such as linseed or tung oil. This drying oil polymerizes to produce an insoluble, crosslinked film that binds the pigments in place. Modern oil-based paints often contain a synthetic polymer replacing all or part of the drying oil.

Latex paints, also called water-based or emulsion paints, are an *aqueous polymer dispersion* with dispersed pigments. Latex paints have captured more than 50 percent of the market because they have less odor than oil-based paints, dry faster, and clean up easily with water. A typical formulation of a latex paint is shown in table 11.13. Acrylate, methacrylate, and vinyl acetate copolymers are the most common polymer components. Inexpensive styrene-butadiene copolymers are also used, but they discolor after long sunlight exposure. The better paints, latex or oil-based, usually contain appreciable amounts of titanium dioxide (TiO_2) because it produces a better shade of white and also allows better color definition with other pigments.

TABLE 11.12 Paint usage classification. The numbers in parentheses are the percentages of each category in 1993.

CLASSIFICATION	EXAMPLES
Architectural coatings (55%)	interior and exterior paints, undercoats, primers, sealers, stains
Product coatings (29%)	wood or metal furniture and fixtures, automobiles, machinery and equipment, factory-finished wood, coated sheets, strips and coils, appliances, coatings on film, paper, and foil; toys, sporting goods
Special purpose coatings (16%)	automotive and machinery refinishing, high-performance and bridge maintenance, traffic paint, roof coatings, aerosols, swimming pool coatings, arts and crafts

TABLE 11.13 A typical latex paint formulation.

COMPONENT	AMOUNT
polymer	10–20%
pigments	35–45%
coloring pigments	<5%
water and solvents	50–65%

Shellac is a natural polymer secreted on trees in India by the insect *Laccifer lacca.* The exudate is collected, washed, purified by melting and filtering, and then dissolved in ethanol. Shellac is used as a sealer under varnish or as a finish coating on wood, but the alcohol makes the solution highly flammable. A polymer-based synthetic shellac also exists.

A varnish is a natural or synthetic polymer dissolved in a drying oil-solvent blend; on drying, it forms a tough, crosslinked film. Varnishes are used on wood floors, doors, and furniture. Poly(urethanes) are used as a synthetic varnish with great toughness and abrasion resistance. An enamel is a varnish with added pigments (normally fewer than in oil-based paints). In an enamel, the polymers crosslink during drying. Enamels are used mainly to paint metal surfaces. A lacquer is a polymer solution with a small amount of added pigment. Lacquers contain no drying oils and do not crosslink, and they are thus less durable than enamels or varnishes. New environmental laws will drastically reduce the amount of organic solvent permitted in varnishes, enamels, and lacquers in the late 1990s.

EXPERIENCE THIS

Visit a hardware store, a paint store, or even your own home, and locate paints and coatings. Examine labels and try to classify each example under the headings in this section.

You should be able to find latex and oil-based paints in any paint store. Shellac and varnishes are also common, but you might need to visit a hobby section to locate lacquers. These compounds are commonly used to paint toy models.

Other Coatings

Floor polishes consist of waxes, such as carnauba (ester), beeswax (ester), and paraffins (alkanes) or low-molecular-weight PE (polyethylene), along with special acrylate, methacrylate, or styrene copolymers. Most writing paper or paperboard is coated with a small amount of a pigment plus a polymeric binder to improve the surface quality and permit printing or writing. (Ink spreads out on uncoated paper—try writing in ink on a paper towel!) Paper and paperboard use a tiny percentage of polymer, but with around 64 million metric tons of paper produced annually, this market is substantial. Ink itself is a pigment dispersion with a polymer binder.

EXPERIENCE THIS

Plastics are often blamed for occupying too much space in landfills, but this need not be the case if recycling is used. You will need two special materials for this experiment: (1) a bottle of acetone-type nail polish remover, and (2) at least two gallons (or eight liters) of polystyrene "worms" ("peanuts") or the brittle, white, foam plastic used in packaging. (If you cannot find the worms, you could use foamed PSt drinking cups, but break them into small pieces.) You will also need a large, wide-mouthed jar (about a gallon in volume); this can be either glass or recyclable plastic but must be see-through.

Place about 200 mL acetone in the gallon jar, and slowly add the PSt worms as you swirl the acetone. What happens? Continue adding the PSt worms until the container is filled or you run out. Keeping in mind that you started with about eight liters (two gallons) of worms, what volume of acetone plus worms would you expect at the end of the addition step? What is the actual volume? You can recover the plastic by repeatedly washing it with water and then allowing it to air dry. How does the final volume compare with the volume of worms at the start?

It is possible to recycle PSt by treating it with acetone, washing it, and allowing it to dry. When recycling PSt, the acetone is recovered and reused. Box 11.2 discusses polymer recycling in more detail.

BOX 11.2

POLYMER RECYCLING AND DISPOSAL

If you wander through a grocery store, you will see thousands of plastic bottles and jars filled with all kinds of household items including foods, beverages, and dishwashing detergent (figure 1). Visit the meat counter and you will see hundreds of pounds of meat wrapped in plastic. You can buy boxes of plastic bags or plastic film for home use, and as you check out, you can have your groceries packed into plastic bags. What happens to all this plastic? How much is there? Where does it go?

Most of the plastics used for packaging come from the "Big Five" listed in table 11.2 plus the polyester PET, which is commonly used in soft drink bottles. The production of these top five plastics totaled 71.9 million metric tons in 1993. Even though not all applications of these plastics end up in the grocery store, all of the uses eventually end. This means we must get rid of 1.43×10^{11} pounds of plastic each year (621.7 pounds per person)! Although many people express fears that scrap plastic will soon fill up our landfills, the waste plastic in a typical landfill is only about 6 percent by weight (figure 2); paper waste, in comparison, makes up 37 percent by weight.

When you examine landfill contents on a volume basis, however, the percentage of plastics is higher. There are several reasons for this. First, plastics have lower densities than paper, glass, or metals and can occupy a larger volume, pound for pound, unless they are compressed. Secondly, many disposed plastics are empty bottles and similar containers. A typical 2 liter soft drink bottle only contains 51 g PET, and only 51 mL of actual *plastic* in

Figure 1 Plastics in the grocery store.

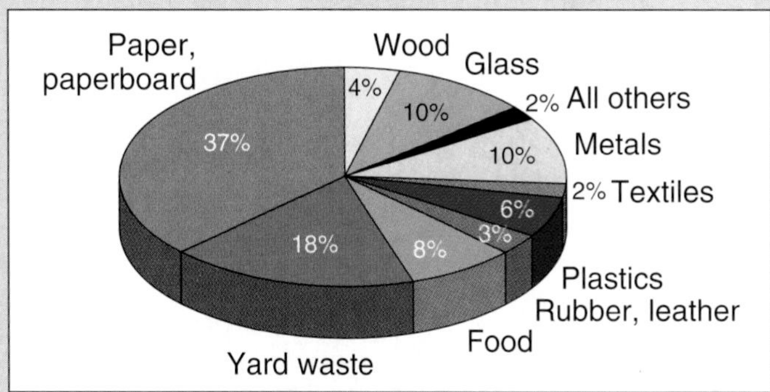

Figure 2 The composition of a typical landfill.

(Continued)

BOX 11.2—Continued

volume; however, the volume of the uncrushed bottle is 2,000 mL! Compressing the bottle reduces the volume—but recycling eliminates the plastic from the landfill entirely.

Fortunately, the "Big Five" thermoplastics of table 11.2 and the PET poly(ester), which makes up approximately 84 percent of the plastics entering landfills, are all recyclable. To take advantage of this, we must learn to think of discarded plastics as a valuable resource rather than trash. Thousands of communities have initiated recycling programs to alleviate the huge amount of trash dumped into landfills, and many such programs include plastics. (Figure 3 shows some of the plastics being recycled.) However, much more remains to be done—only about 2.5 percent of all plastics are now recycled. One factor that has hampered recycling is that the cost of the recycled plastic is greater than that of "virgin" plastic made from petrochemicals.

Incineration is another way to dispose of plastics such as PE, PP, and PSt that are mainly hydrocarbons and differ from fuel or heating oil only in molecular weight. These plastics could serve as raw materials for a substitute fuel and would release about the same amount of energy per kilogram as petroleum. Poly(esters) also burn, although their energy output is lower. Incinerating PVC requires special incinerators because it releases HCl (hydrochloric acid).

Which is a better solution—recycling or incineration? We should probably view incineration as a last resort for plastics since recycling them seems preferable in terms of conservation. Incineration may be a reasonable disposal method for plastics that are difficult to recycle, such as thermosetting plastics and vulcanized rubber, and might provide energy instead of wasted landfill space. But even some of these plastics are recyclable as fillers in other plastics. Actually, incineration is a better alternative than landfilling for paper and yard wastes.

A third alternative for plastics is to heat the polymers and degrade them back to the monomer, or to a lower-molecular-weight fuellike material, in a procedure called pyrolysis. In this manner, "virgin" plastics can be made from the monomers. At the present time, pyrolysis costs too much to be competitive, but that can change as improvements are made. Possibly, you may someday drive a car that runs on fuel obtained from scrapped plastics rather than from an oil field.

How to identify plastic:

Look for these triangle symbols on most container bottoms.

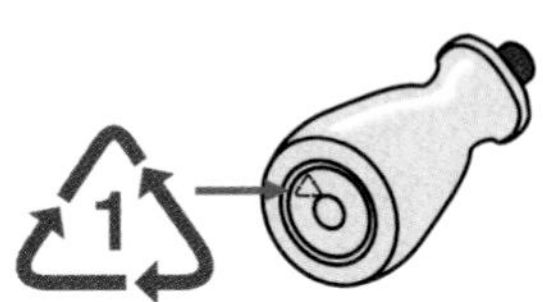

PET Poly(ethylene terephthalate)
Crystal clear. Look for the central dot left by the injection molding machine.

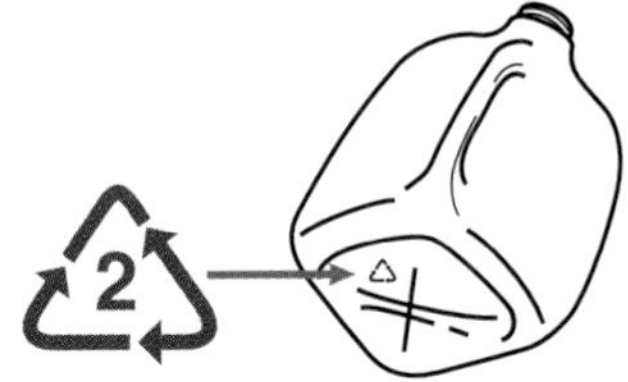

HDPE High-density poly(ethylene)
Cloudy white or opaque. Used extensively in milk jugs and soap containers.

PVC Poly(vinyl chloride)
Clear or slight blue tint. Look for the "ears" on the bottom seam. The seam will become cloudy when the bottle is squeezed.

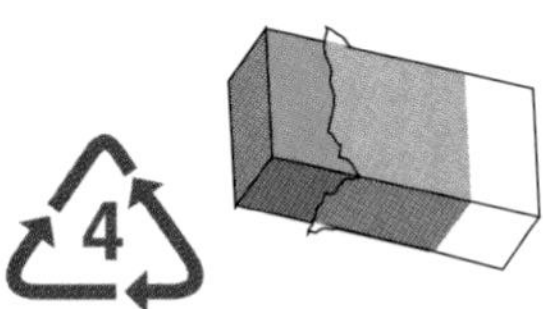

LDPE Low-density poly(ethylene) Used mostly in packaging wrap.

PP Poly(propylene)
Bottles have a seamed bottom. Most yogurt containers use this material.

PS Poly(styrene)
Eating utensils, plates, and clear glasses. Items tend to be brittle except when foamed, as in styrofoam.

Other
Mixed materials or materials that do not fit any of the previous six categories. Difficult to recycle.

Figure 3 Identification code for plastic recycling types.

SUMMARY

Polymers are large molecules made up of repeated units of a smaller molecule called a monomer. Synthetic polymers—generally known as plastics—have revolutionized twentieth-century life. With a variety of properties, including varying T_g, or glass transition temperatures, plastics are extremely versatile substances.

Polymers are classified in several ways—according to thermal characteristics, structure, use, or method of preparation. In terms of thermal behavior, synthetic polymers may be either thermoplastic—softening every time they are heated—or thermosetting—hardening when heated and never softening once they are formed. Most industrial polymers are thermoplastics.

Polymers are also divided according to structure—they can be either linear, branched, or crosslinked. Linear and branched polymers are usually thermoplastics and are generally soluble in solvents. They flow when pressure or heat is applied. Crosslinked polymers (usually thermosetting) are insoluble and do not flow under pressure or heat.

Yet another way to classify polymers is according to their use. The most common division is to label them as either plastics, fibers, or elastomers. Plastics are molded thermoplastic or thermosetting polymers. Most are amorphous (unstructured), but some are slightly or highly crystallized (structured). Fibers are thermoplastics with individual linear polymer chains held together by hydrogen bonds, crystallinity, or both. And elastomers, also called rubbers, begin as thermoplastics, but become slightly crosslinked, increasing their strength and elasticity, through vulcanization.

A final way to group polymers is according to their preparation. Addition polymers form as smaller molecules add to the growing polymer chain. Condensation polymers form as smaller molecules react, combining into a polymer chain and producing water or another compound as a byproduct.

Addition polymers themselves may be further subdivided into groups such as the poly(olefins), and vinyl polymers. Poly(olefins) are synthesized from olefins (alkenes). They include poly(ethylene) (PE) and poly(propylene) (PP). Vinyl polymers contain the vinyl group and include poly(vinyl chloride), or PVC, poly(styrene) (PSt), poly(tetrafluoroethylene) (PTFE), and poly(vinyl acetate) (PVAc). Methacrylates and acrylates are also addition polymers. They include poly(methyl methacrylate), or PMMA; poly(ethyl acrylate), or PEA; and poly(acrylonitrile) or PAN. These polymers lend themselves to all kinds of uses; PE is used for molded products and plastic films, PP in textiles and storage battery cases, and PVC in water piping, rainwear, and coated fabrics. PSt is used to make molded objects, extruded plastic items, plus styrofoam containers and insulation sheets, and PTFE serves as a nonstick cookware coating. PVAc is used in latex paints, chewing gum, and as an intermediate in poly(vinyl alcohol) preparation. Finally, PMMA is used in lightweight plastic eyeglass lenses and PAN in synthetic fibers.

Condensation polymers include poly(esters), poly(amides), and poly(acetals), along with the poly(carbonates), phenolic resins, and urea-formaldehyde and melamine-formaldehyde polymers. Poly(esters) are used mainly in the textile industry, in soft drink bottles, and plastic films; poly(amides), or nylons, have a wide range of uses including textiles and molded objects. Poly(carbonates) are used in impact-resistant window "glass." Phenolic resins are strong, insoluble polymers used in everything from electronics to furniture glue. Urea-formaldehyde (UF) polymers are used in countertops and home insulation. Melamine-formaldehyde (MF) polymers are hard and moisture-resistant; they are used to manufacture plastic dinnerware.

Copolymers—polymers that include two or more separate monomers arranged in a single chain—add even more versatility to plastics. They allow chemists to combine desirable characteristics from two distinct polymers in a single substance. Copolymers may be random, alternating, block, or graft copolymers in structure.

Chemists have developed at least five types of elastomeric polymers to imitate the behavior of natural rubber. Poly(butadiene), developed in the early 1920s, was of poor quality. In 1929, Patrick developed Thiokol, a poly(alkylene polysulfide). In 1931, less than two years later, Carothers invented poly(chloroprene), now marketed as Neoprene. About 1937, butadiene copolymers with either styrene or acrylonitrile were introduced as elastomers. The structure and properties of natural rubber, cis-poly(isoprene), were duplicated during the mid-1950s using a special catalyst Ziegler and Natta developed. Ethylene-propylene elastomeric copolymers entered the marketplace in the 1960s.

Even with all these versatile compounds, chemists continued to refine plastics. By adding plasticizers to lower T_g and increase flexibility, pigments to impart color, and stabilizers to prevent breakdown, they improved on plastic properties.

Finally, paints and coatings contain polymers, along with solvents, and pigments. Paints can be oil-based or latex (water-based). Other coatings include shellac, varnish, enamel, and lacquer. It truly seems that a synthetic polymer exists to fill almost any need.

KEY TERMS

alternating copolymer 289
amorphous 273
antioxidants 293
block copolymers 289
branched 273
copolymer 288
crosslinked 273
elastomer [rubber] 276
fibers [textiles] 276
fillers 293
glass transition temperature (T_g) 272
graft copolymer 289
homopolymer 288
latex paints 294
linear 273
nylon 281
oil-based paints 294
phenolic resins 282
pigments 293
plasticizers 293
plastics 273
poly(acrylonitrile) (PAN) 280
poly(amides) (nylon) 280
poly(esters) (Dacron, Mylar) 280
poly(ethylene) (PE) 277
poly(methyl methacrylate) (PMMA) 280
poly(propylene) (PP) 277
poly(silicones) (silicone rubber) 284
poly(styrene) (PSt) 279
poly(tetrafluoroethylene) (PTFE) 279
poly (vinyl acetate) (PVAc) 279
poly(vinyl chloride) (PVC) 278
random copolymer 289
stabilizers 293
thermoplastic 272
thermosetting 272
ultraviolet stabilizers 293

READINGS OF INTEREST

General

Alper, Joseph, and Nelson, Gordon L. 1989. *Polymeric materials: Chemistry for the future*. Washington, D.C.: American Chemical Society, (110 pp.) A short book, with many color photographs, that contains a good chapter on recycling. Easy reading for the non-chemist.

Seymour, R. E., and Carraher, C. E. Jr. 1989. *Giant molecules*. New York: John Wiley (314 pp.). This book starts with a brief survey of organic chemistry and then examines polymers, including the natural plant and animal polymers. Numerous applications are covered as well.

Gait, A. J., and Hancock, E. G. 1970. *Plastics and synthetic rubbers*. Oxford: Pergamon Press (302 pp.). A fairly easy-to-read book that discusses the manufacture and uses of plastics and elastomers. Most of the statistical data is for the United Kingdom, but you can still use it to trace the growth of the polymer industry.

Couzens, E. G., and Yarsley, V. E. 1968. *Plastics in the modern world*. Middlesex, England: Penguin Books (386 pp.). A very easy-to-read, inexpensive book that focuses almost entirely on the history, manufacture, and applications of plastics. (Chemistry is discussed in only about 20 pages.)

Specific Plastics or Applications

Mullin, Rick; Cottrill, Ken; Kemezis, Paul; and Howard, John. "Paints and coatings." *Chemical Week*, October 12, 1994, pp. 35–36. A business-oriented set of articles on paints and coatings. One discusses the automakers' market and another explores house paints.

Reisch, Marc S. "Paints and coatings." *Chemical & Engineering News*, October 3, 1994, pp. 44–46. New environmental regulations have spurred the development of new types of paints and coatings that contain smaller amounts of organic solvents. Sales continue to increase from the 1992 figure of $13 billion.

Coeyman, Marjorie; Young, Ian; Wood, Andrew; and Kemezis, Paul. "Plastics." *Chemical Week*, June 1, 1994, pp. 25–36. A series of business-oriented articles about plastics markets in the United States and Europe.

Reisch, Marc S. "Plastics in building and construction." *Chemical & Engineering News*, May 30, 1994, pp. 20–43. As building codes change, plastics have become common in such varied uses as plumbing, window frames, siding, and insulation. The current market is about $5.6 billion.

Kauffman, George B. "Rayon: The first semi-synthetic fiber product." *J. Chem. Ed.* vol. 70, no. 11 (November 1993), pp. 887–93. Includes the early history of this fabric, originally billed as artificial silk, and a list of rayon products.

Fariss, Robert H. "Fifty years of safer windshields." *Chemtech*, September 1993, pp. 38–43. Automobiles routinely use safety glass laminated with plasticized poly(vinyl butyral). This simple application, which has saved countless thousands of lives, has also proved useful in bullet-resistant glass, aircraft cockpits, and intruder-resistant windows.

Seymour, Raymond B., and Kauffman, George B. "The ubiquity and longevity of fibers." *J. Chem. Ed.* vol. 70, no. 6 (June 1993), pp. 449–450. A brief history of fibers and how they are made.

Stapleton, Constance. "The latex scare." Woman's Day, February 2, 1993, pp. 74–75. Latex polymers are used to make more than 40,000 items, including rubber gloves, balloons, stretch fabrics, condoms, bathing caps, rubber footwear, and water-based paints or glues. Some people are allergic to these items due to chemicals in the latex.

Seymour, Raymond B., and Kauffman, George B. "Elastomers. III: Thermoplastic elastomers." *J. Chem. Ed.* vol. 69, no. 12 (December 1992), pp. 967–970. This article tells what thermoplastic elastomers are and how they work. It also gives statistical data on the amounts of these materials manufactured in the United States, Japan, and Europe in 1966 and 1991, plus a projection to year 1996. (If you think chemistry is unimportant in the "real" world, look at these numbers and try to imagine the size of a pile of these elastomers.)

Seymour, Raymond B., and Kauffman, George B. "Polyurethanes: A class of modern versatile materials." *J. Chem. Ed.* vol. 69, no. 11 (November 1992), pp. 909–910. This paper covers some of the historical background on these useful polymers and examines foamed plastics in more detail than we covered in this text.

Reisch, Marc S. "Thermoplastic elastomers bring new vigor to rubber industry." *Chemical & Engineering News,* May 4, 1992, pp. 29–42. Learn how these materials are helping the rubber industry bounce back from hard times.

Morris, Gregory D. L. "Flame retardants." *Chemical Week,* April 15, 1992, pp. 26–28. A short report on a hot topic.

Seymour, Raymond B., and Kauffman, George B. "The rise and fall of celluloid." *J. Chem. Ed.* vol 69, no. 4 (April 1992), pp. 311–314. Celluloid was one of the first plastics, but you hardly see it any more. Why? This excellent article, complete with photographs of the celluloid inventor (John Wesley Hyatt), details the history and value of an old type of plastic.

Mullin, Rich; Roberts, Michael; Kirschner, Elisabeth; Kiesche, Elizabeth; Morris, Gregory; and Kemezis, Paul. "Adhesives and sealants—Innovating under pressure." *Chemical Week,* March 11, 1992, pp. 26–38. Up-to-date, business-oriented account on an important type of polymer not discussed much in this chapter. Lots of useful statistical data on who, what, when, and where for this industry.

Blohm, Margaret L.; Van Dort, Paul C.; and Pickett, James E. "A processible conduction polymer." *Chemtech,* February 1992, pp. 105–107. Polymers are normally insulators. This paper will introduce you to some that can conduct electricity.

Morris. Gregory D. L. "Fibers—First strand in recovery?" *Chemical Week,* January 22, 1992, pp. 20–22. Up-to-date report with good business data.

Konrad, Walecia. "The textile industry is looking threadbare." *Business Week,* September 16, 1991, pp. 114–118. Business-oriented article on current problems in the textile industry.

King, Roy H., and Reuben, Bryan G. "Vinyl chloride in Asia." *J. Chem. Ed.* vol 68, no. 6 (June 1991), pp. 480–482. This article discusses the radical polymerization reaction and shows a typical plant diagram.

Vogel, Shawna. "Smart kind." *Discovery,* April 1990, p. 26. Delves into the question of whether a humanlike skin can be designed that would permit a robot to feel objects.

Turner, G.P.A. 1988. *Introduction to paint chemistry and principles of paint technology.* 3d ed, New York: Chapman & Hall (252 pp.). If paint or painting is your interest, this is the book for you. It includes both the chemical reactions and more general information on paint and coatings.

Silicones and Breast Implants

"Breast implant study: No autoimmune disease link." *Chemical Week,* June 8, 1994, p. 64. A University of Michigan study of 377 women with scleroderma, an autoimmune disease, showed no link between the disease and silicone breast implants.

Reisch, Marc. "Breast implant agreement: $4 billion pledged to settle lawsuits." *Chemical & Engineering News,* February 21, 1994, pp. 4–5. The actual amount of money set aside to settle breast implant claims was $4.75 billion. But this does not appear to be the end of the problem; new suits and claims were still being filed near the end of 1995.

Finch, Steven. "Beyond breast implants." *Health,* July/August 1992, pp. 74–75. Very short, but has a good table showing various other medical applications for silicones.

Ember, Lois. "Dow Corning to quit making breast implants." *Chemical & Engineering News,* March 23, 1992, p. 6. Potential litigation is more than Dow Corning wants to handle.

Findlay, Steven. "New limits, more questions." *U.S. News & World Report,* March 2, 1992, p. 43. What should women who already have implants do? This article included some toll-free help phone numbers.

Ember, Lois. "Breast implants: Silicone effects in body to be probed." *Chemical & Engineering News,* March 2, 1992, pp. 4–5. Probably one of the more factual articles on this topic.

Byrne, John A. "Here's what to do next, Dow Corning." *Business Week,* February 24, 1992, p. 33. Advice for the corporate giant from the business community.

Fields, Suzanne. "Drop phony moralizing about breasts." *Insight,* February 17, 1992, pp. 18–19. This article asks why the moralizing takes place.

Ember, Lois. "Silicone breast implants: New Dow Corning chief to tackle crisis." *Chemical & Engineering News,* February 17, 1992, pp. 4–5. This article claims that "science supports the safety and effectiveness of breast implants."

Smart, Tim. "The man who sounded the silicone alarm—in 1976." *Business Week,* January 27, 1992, p. 34. Something was known about potential silicone problems many years ago.

Davis, Lisa. "Breast implants: Two million guinea pigs?" *In Health,* September/October 1991, pp. 30–34. Discusses what kinds of problems the implants might cause.

Gebelein, Charles G. 1982. "Prosthetic and biomedical devices." In *Kirk-Othmer encyclopedia of chemical technology,* vol. 19, pp. 275–313. New York: John Wiley. Although this encyclopedia is aimed at scientists, much of this article is readable to a nonscientist. Includes information on which plastics are used in implants and other medical devices. Also cites data on how many devices are used annually and provides some cost figures.

Rochow, Eugene G. 1987. *Silicon and silicones*. New York: Springer-Verlag (180 pp.). Good account of silicone polymer history and chemistry by a man who worked in this field for his entire career. Fairly easy reading for those who want to know more about these important polymers.

Recycling Plastics

Rotman, David, and Chynoweth, Emma. "Plastics recycling: Back to fuels and feedstocks." *Chemical Week*, March 2, 1994, pp. 20–22. If polymers are recycled back to fuels or to the original monomers, they have even more uses than recycled plastics. For one thing, any polymer made from these recovered chemicals would be "virgin" plastic. Perhaps that's one reason this procedure has become very popular in Europe.

Ward, Mike, and Kirschner, Elisabeth. "U.S. and Europe regulate plastics recycling." *Chemical Week*, March 2, 1994, p. 23. Regulation is done through taxes, mandates, and rules stipulating the recycled plastic content in a container.

Lipkin, R. "Waste plastic yields high-quality fuel oil." *Science News*, August 28, 1993, p. 134. At the moment, the cost of this process is too high to be practical, but further research may reduce costs. If so, you might drive your car, cook your food, or heat your home with fuel derived from scrap plastics.

Wood, Andrew, and Kiesche, Elizabeth S. "Plastics: Conventional recycling moves downstream." *Chemical Week*, August 25/September 1, 1993, pp. 32–34. Virgin plastic materials are usually less expensive than recycled plastics, and this has forced many recyclers out of business. Reconverting the plastics to monomers that can be repolymerized to virgin plastics seems to be a way around this dilemma.

Coeyman, Marjorie. "Novon serves up a new course of biodegradable polymers." *Chemical Week*, August 18, 1993, p. 12. These are plastics derived from starch (a natural polymer) that biodegrade fairly well. However, the price ranges from $2.00 to $2.40 per pound—higher than many synthetic plastics.

Thorsheim, Helen R., and Armstrong, David J. "Recycled plastics for food packaging." *Chemtech*, August 1993, pp. 55–58. Although no rules are yet in place, the FDA does plan to issue regulations concerning the use of recycled plastics for food packaging.

Young, John E. 1991. *Discarding the throwaway society. Worldwatch Institute Paper 101*. Washington, D.C.: Worldwatch Institute (45 pp.). How can we change our throwaway society? This pamphlet makes several suggestions. Not all are practical, but the booklet is well worth the short time it takes to read.

Wood, Andrew; Chynoweth, Emma; Rotman, David; and Kiesche, Elizabeth. "Plastics makers recycle for new growth." *Chemical Week*, December 18/25, 1991, pp. 28–43. Very good article with data for several different years showing changes in plastic making and recycling and comparing the United States with Europe and Japan. Which of these three regions do you think leads in recycling?

Babinchak, Steve. "Recycling must work." *Chemical & Engineering News, Chemtech*, December 1991, pp. 728–730. Describes plastics recycling process.

Stone, Robert F.; Sagar, Ambuj D.; and Ashford, Nicholas A. "Recycling the plastic package." *Technology Review*, July 1992, pp. 49–56. Good article on how plastic packaging is recycled and what the ultimate savings could be. Easy reading and possibly the best article listed in this area.

Reisch, Marc. "Dow sets big plastics recycling program." *Chemical & Engineering News*, June 22, 1992, p. 16. Not everyone is happy about this program—find out why.

Reisch, Marc. "Dow offers recycled resins for packaging." *Chemical & Engineering News*, June 1, 1992, p. 14. Curbside pickups are not much use unless the plastics are actually reused.

Strauss, Stephen. "The haze around environmental audits." *Technology Review*, April 1992, pp. 19–20. Which are better for our environment, paper cups or foam cups? What do we need to know to answer this question?

Kaminsky, Walter, and Rössler, Harald. "Olefins from wastes." *Chemtech*, February 1992, pp. 108–113. Olefins are alkenes, one of the major raw materials for making polymers. Aside from petrochemicals, where can we get olefins? How about waste polymers and rubber? This article discusses yet another form of recycling that needs to grow.

Grassy, John. "Bottle bills—Headed for a collision at curbside?" *Garbage*, January/February 1992, pp. 45–47. What effect might a tax on plastic bottles have on recycling?

PROBLEMS AND QUESTIONS

1. What were the main contributions of each of the following people to polymer chemistry? (You may need to consult chapters 9 and 10.)

a. Leo Baekeland
b. Wallace Hume Carothers
c. Benjamin Franklin Goodrich
d. Charles Goodyear
e. Nelson Goodyear
f. Giulio Natta
g. J. C. Patrick
h. Roy Plunkett
i. Hermann Staudinger
j. Karl Ziegler

2. Define each of the following terms.
   a. Amorphous
   b. Block copolymer
   c. Branched polymer
   d. Crosslinked polymer
   e. Elastomer
   f. Fiber
   g. Graft copolymer
   h. Linear polymer
   i. Plastic
   j. Plasticizer
   k. Stabilizer
   l. Thermoplastic
   m. Thermosetting

3. What are the most common uses for each of the following types of polymers? (You may need to consult chapters 9 and 10 for some of these materials.)
   a. Melamine formaldehyde (MF) polymers
   b. Nylon polymers
   c. Phenolic resins
   d. Poly(carbonate)
   e. Poly(ethylene)
   f. Poly(ethylene terephthalate)
   g. Poly(propylene)
   h. Poly(silicones)
   i. Poly(styrene)
   j. Poly(tetrafluoroethylene)
   k. Poly(vinyl acetate)
   l. Poly(vinyl chloride)
   m. Urea formaldehyde (UF) polymers

4. What are the five most common plastics?

5. What are the five most common fibers?

6. What are the five most common elastomers?

7. Determine the polymer class for each of the following repeat units (for example, polyester, acrylic, polyolefin, nylon, polyether, and so forth).

a. $\left[O-CH_2CH_2 \right]_n$

b. $\left[CH(C(=O)-OCH_3)-CH_2 \right]_n$

c. $\left[NH-(CH_2)_4-NH-C(=O)-(CH_2)_7-C(=O) \right]_n$

d. $\left[O-(CH_2)_8-O-C(=O)-(CH_2)_6-C(=O) \right]_n$

8. Draw repeat units for each of the following polymers.
   a. Poly(ethylene)
   b. Poly(styrene)
   c. Poly(vinyl chloride)
   d. Poly(propylene)

9. What is the difference between a plastic and an elastomer?

10. What are the two major types of reactions used to prepare polymers? How do they differ?

11. How do linear, branched, and crosslinked polymers differ?

12. Give some examples of common crosslinked polymers.

13. What are some examples of common linear polymers?

14. What are some uses for poly(ethylene)?

15. What are some uses for poly(vinyl chloride)?

16. What polymer is used as the adhesive on postage stamps?

17. What polymer is used in the common white glue found in most homes and businesses?

18. Give examples of common uses for each of the following polymers.
   a. Phenolics
   b. Poly(esters)
   c. Poly(amides)
   d. Epoxy resins
   e. Poly(carbonates)

19. What is poly(chloroprene), and how does it relate to natural rubber? Who invented this compound?

20. Who developed Thiokol elastomers?

21. What is a thermoplastic elastomer?

22. What do plasticizers do? Give some examples of plasticizers.

23. What are pigments? Give examples of prime and inert pigments.

24. What is the difference between a pigment and a filler?

25. What kinds of stabilizers are used in polymers and why? Give an example of each type of stabilizer.

26. What are the main ingredients in water- and oil-based paints?

27. What are the main ingredients in a varnish?

28. What are the main ingredients in an enamel?

29. Which plastics can be recycled most easily?

30. Which plastics can be incinerated most easily?

CRITICAL THINKING PROBLEMS

1. Read several articles related to silicone breast implants and make your own assessment of this problem. Four good articles are (1) Lisa Davis "Breast Implants: Two Million Guinea Pigs?" *In Health*, September/October 1991, pp. 30–34; (2) Lois Ember, "Breast Implants: Silicone Effects in Body to be Probed," *Chemical & Engineering News*, March 2, 1992, pp. 4–5; (3) Steven Finch, "Beyond Breast Implants," *Health*, July/August 1992, pp. 74–75; and (4) Suzanne Fields, "Drop Phony Moralizing about Breasts," *Insight*, February 17, 1992, pp. 18–19. In your analysis consider the following: How much responsibility should a manufacturer have for their products? Is this responsibility greater for implants than for paints or plastic dishes? Should individuals who had breast implants assume part of the responsibility? Finally, would this problem have been treated differently if it centered on a masculine prosthesis—say a silicone testicle?

2. Review the prices of the fibers in table 11.3 and then examine your clothing labels. Estimate how much of each kind of fiber you generally use. Why are the more expensive fibers so common?

3. There is little doubt that the U.S. textile industry has fallen on hard times. Read these articles on textiles: (1) Walecia Konrad, "The Textile Industry Is Looking Threadbare," *Business Week*, September 16, 1991, pp. 114–118; and (2) Gregory D. L. Morris, "Fibers—First Strand in Recovery?" *Chemical Week*, January 22, 1992, pp. 20–23. How can the textile industry be revitalized? Do you play any part in the problem or solution? (Consider the source of your clothing and the fibers used to make them before you answer.)

4. Review the PVC uses and try to find further examples of uses on the store shelves. Which applications probably utilize plasticized PVC?

HINT The T_g of PVC is 80° C. Plasticized PVC has a T_g value below room temperature, making it flexible.

5. Examine the structures in table 11.1 and try to estimate whether the acrylonitrile monomer is in the price range of styrene/vinyl chloride (about $0.50/pound) or the acrylates/methacrylates (about $1.00/pound). What is the basis for your guess?

6. Foam cups are an object of environmental scorn, but is this deserved criticism? Make your own assessment of this question after reading the article by Stephen Strauss in *Technology Review*, April 1992, pp. 19–20, entitled "The Haze Around Environmental Audits." Are paper cups really better than foam?

7. Most people agree that recycling plastics is a good idea, yet less than 5 percent of all plastics are recycled. Why do you think this is so, and what can be done to improve this rate? To aid you in your thinking, read the following background articles, (1) Steve Babinchak, "Recycling Must Work," *Chemtech*, December 1991, pp. 728–730; (2) Robert F. Stone, Ambuj D. Sagar, and Nicholas A. Ashford, "Recycling the Plastic Package," *Technology Review*, July 1992, pp. 49–56; and (3) Marc Reisch, "Dow Offers Recycled Resins for Packaging," *Chemical & Engineering News*, June 1, 1992, p. 14.

8. Thousands of tires are discarded in landfills each year. How would you feel if a tire incinerator was built in your community to provide low-cost electricity? Other than landfills, what other options exist?

9. What are the arguments for and against plastic incineration? What would be your reaction if someone proposed building an incinerator for medical plastics, such as syringes or tubing, in your community? Would this be a safer disposal method than using a landfill?

10. Besides phenol itself, other members of the phenol class react with formaldehyde, but these polymers have never become very popular. Why do you think this is so?

11. The use of plastics in the building trade, with a current market of $5.6 billion, is increasing. Read the following article: Marc S. Reisch,"Plastics in Building and Construction," *Chemical & Engineering News*, May 30, 1994, pp. 20–43. What are the advantages to using plastics in home, factory, and office construction?

12. There are several ways you can easily portray copolymers. For example, you could use two different colors of gumdrops, say black and orange, to represent the two monomers. Obtain a bag of gumdrops and make models of (a) random, (b) alternating, (c) block, and (d) graft copolymers. Can you suggest any other simple models?

CHAPTER 12

CHEMISTRY AND OUR WATER SUPPLY

OUTLINE

KEY IDEAS

1. How Pollution Poses Problems for Our Water Supply
2. The Different Potable Water Sources
3. Ground Water and the Types of Activities That Pollute It
4. The Composition of Landfills and How People Can Change It
5. How Drinking Water is Purified
6. Why Chlorination, or an Equivalent Process, is Essential for Health
7. What Happens to Waste Water
8. The Nature of Hard Water and How It Is Softened
9. Six Methods That Might Potentially Harvest Fresh Water from Seawater

"Water, water every where, nor any drop to drink."
SAMUEL COLERIDGE, *The Ancient Mariner*

Sapo Falls, Canaima National Park, Venezuela.

Water is a truly amazing compound. It is the only common compound that exists in all three states—solid, liquid and gas—at ordinary temperatures. Possibly the most unusual property of water is that the density of the solid form, ice, is lower than that of liquid water. (The solids for virtually all other chemicals have higher densities.) Water possesses unusually high boiling and freezing points and a large temperature range for its liquid state, abnormal features due primarily to hydrogen bonding (chapter 6). Water is also an excellent solvent that dissolves many different kinds of substances (see chapter 8), including pollutants. In this chapter, we will learn about our water supply, possible threats from pollution, and new sources of fresh water.

WATER—THE LIQUID OF LIFE

Water truly is "the liquid of life." If water behaved like other materials, ponds would freeze from the bottom up and all aquatic life would perish. Water is the major component in our body (approximately 70 percent), and it is difficult to imagine life existing without water. Even under the most severe desert conditions, the plants and animals contain water.

The water on our planet recycles from oceans, rivers, streams, lakes, and ponds into the clouds, and then returns back as rain, snow, or hail in the **water-vapor cycle.** Figure 12.1 diagrams the main features of this cycle. Take special note of the ground water since this is usually the most important source of drinking water. The water-vapor cycle has been known for many centuries; one of the older references appears in the book of Job in the Old Testament, conservatively dated between 2000 and 1000 B.C.: "He draws up the drops of water which distill as rain to the streams; the clouds pour down their moisture and abundant showers fall on mankind." This passage demonstrates that even ancient peoples understood the water-vapor cycle.

Although water is abundant on earth, only 0.7 percent is fresh water suitable for human consumption (figure 12.2). However, this is still more than 4×10^8 gallons for every person on our planet!

POLLUTION AND OUR WATER SUPPLY

Although enough fresh water exists to supply every person on earth, there are several problems involved in utilizing this fresh water supply. First, the supply is not always in the "right" places. A huge metropolitan area never has sufficient water on location, and thus will need to pipe water in from miles away. Southern California actually gets much of its water from the Colorado River, which touches only the southeast border of the state and runs through Colorado, Utah, Arizona, and finally into Mexico. Whose property is the water? If California does not have access to the river, four million people in San Diego, Los Angeles, and Long Beach will go thirsty. Yet if California tapped all the water of the Colorado River, the other three states would suffer, leaving 8.4 million people high and dry. The tension over this problem

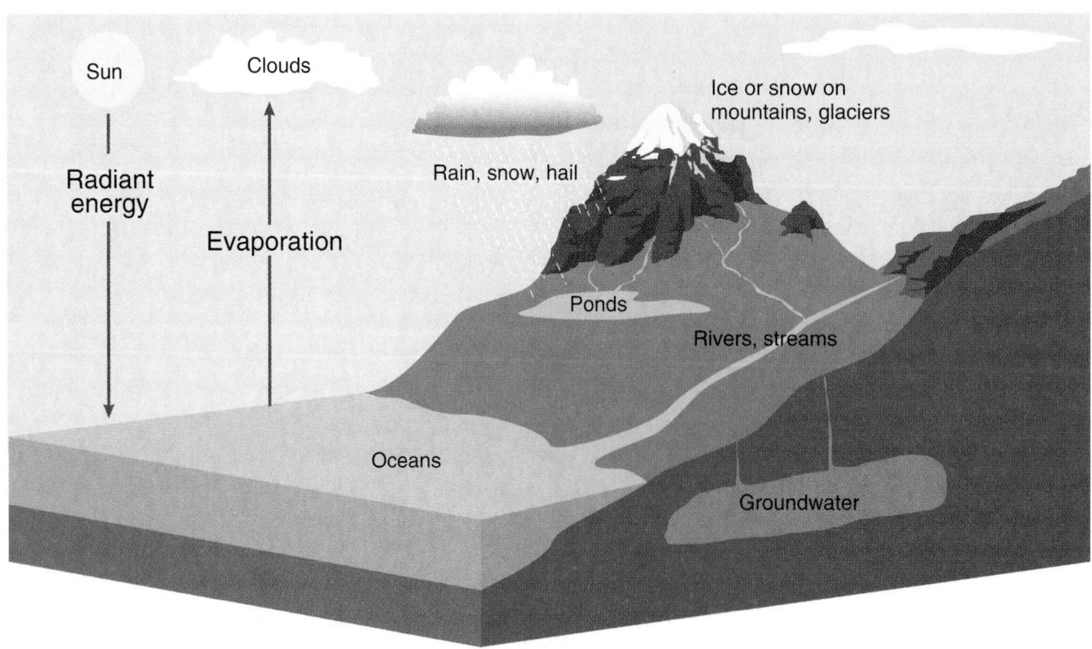

Figure 12.1 The water-vapor cycle.

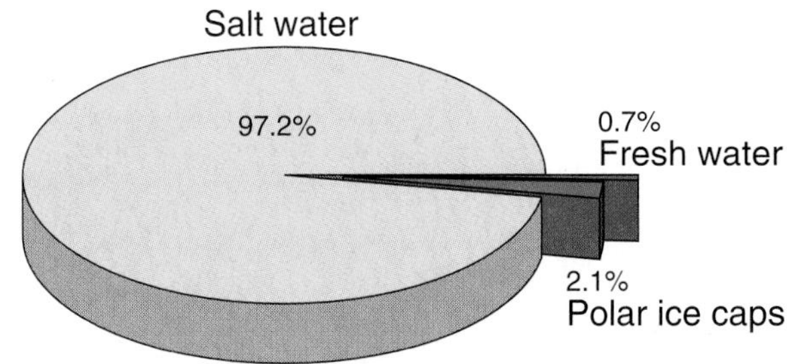

Figure 12.2 Earth's water supplies.

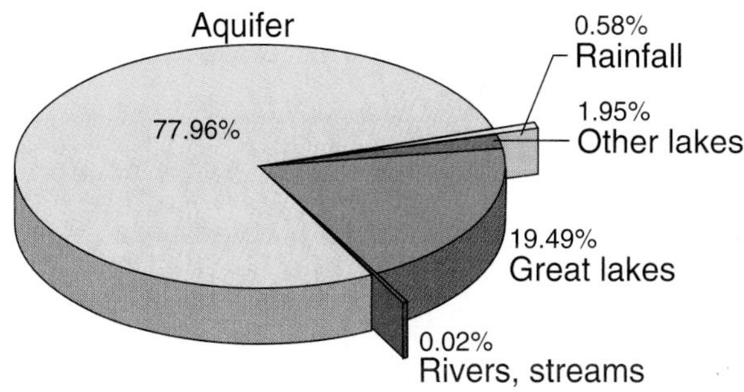

Figure 12.3 The estimated fresh water sources in the U.S.A.

has increased since 1990 when Arizona started withdrawing more water and California lost 25 percent of its supply. More recently, the federal government mandated that California desist from withdrawing more water from the Colorado River.

The second problem is that much of the fresh water is not suitable for human use. For example, as the mighty Mississippi River flows southward from Minnesota to the Gulf of Mexico, the river and its tributaries pass by hundreds of communities, most of which discharge waste material into the water. Consequently, each downstream town that draws water from the Mississippi River must purify its drinking water to remove the waste. This is never easy, always expensive, and sometimes not even possible with present treatment systems. At times, an industrial plant spill may be so bad that it forces downstream communities to shut down their water purification plants for several days, compelling them to obtain their water from other sources or to boil the water in each home.

Still other pollution problems exist. The Florida Everglades are polluted with mercury from both natural and agricultural sources. This poisons fish and makes them dangerous for human consumption. Similarly, the tap water in at least 20 (out of the 60 tested) major U.S. cities contains traces of cancer-causing chemicals that are impossible to remove by simple treatment procedures.

The third problem is that even simple water purification costs a lot of money. The days of crystal-clear streams or sparkling pure lakes are long past. Today, governments spend megabucks to obtain good quality water.

Groundwater

About half the people in the United States get their tap water from groundwater sources. People have used **groundwater** from wells and springs throughout recorded history. Today, about 75 percent of the U.S. cities supplement their water supplies with groundwater, and 95 percent of rural homes rely on this source.

Groundwater lies a few feet to several thousand feet under the Earth's surface. It is usually contained within pores between rock and soil particles and is part of the freshwater circulatory system (figure 12.1). Most groundwater is contained within zones called **aquifers** that lie within permeable rocks and soils; this is the most usable groundwater supply. Aquifers are replenished periodically by rainfall (or snow) and by seepage from surface water (streams, rivers, and lakes). An **aquitard** contains water below or within rocks or soils of low permeability, such as crystalline rocks, clays, or shales, where it is less accessible.

Figure 12.3 shows the estimated proportions of different sources of fresh water in the United States. The aquifer is clearly the largest source, but its level is reduced because people pump out water. When more water is removed than replenished, the water level drops and several problems occur, including a reduction in the available supply, land settling, and **saltwater intrusion,** which is a problem in both the interior and continental coastline areas. Figure 12.4 shows the groundwater supply status for the United States; levels have decreased significantly in many regions.

Sinkholes form when land settles into the partially depleted aquifers. In some parts of the San Joaquin Valley of California, the ground level has dropped as much as 29 feet because extensive amounts of groundwater were used for agriculture. Central Florida is also famous for its sinkholes. Saltwater intrusion also occurs under these conditions and is accentuated in times of drought. The Florida rainfall was unusually low in early 1989, and water usage restrictions were imposed to avert saltwater intrusion. Although the rainfall in subsequent years seemed normal, water restrictions continue in Florida as a prudent course of action.

Groundwater always contains dissolved salts leached from the rocks and soils. Water containing these salts is called **hard water** when 150 ppm or more of Ca^{2+} and Mg^{2+} ions are present (Ba^{2+} and Fe^{2+} ions are sometimes included in the hard water totals). Hard water does not normally pose a health problem. Good drinking water contains less than 500 ppm total dissolved salts and a pH of 6 to 9. Suspended matter, including microorganisms, can produce undesirable turbidity or haziness in water. The **turbidity** is eliminated when particles are removed by filtration. Some other natural constituents found in groundwater can produce a distinct taste, color, or odor. For example, iron compounds impart a yellowish color, and hydrogen sulfide gas (H_2S) imparts a foul smell akin to rotten eggs.

In addition to these natural pollutants, groundwater contamination occurs from (1) salt used to remove snow or ice from roads, (2) oil field brines, (3) acid drainage from mines, (4) injection wells, (5) agricultural runoff, (6) underground storage tank leaks,

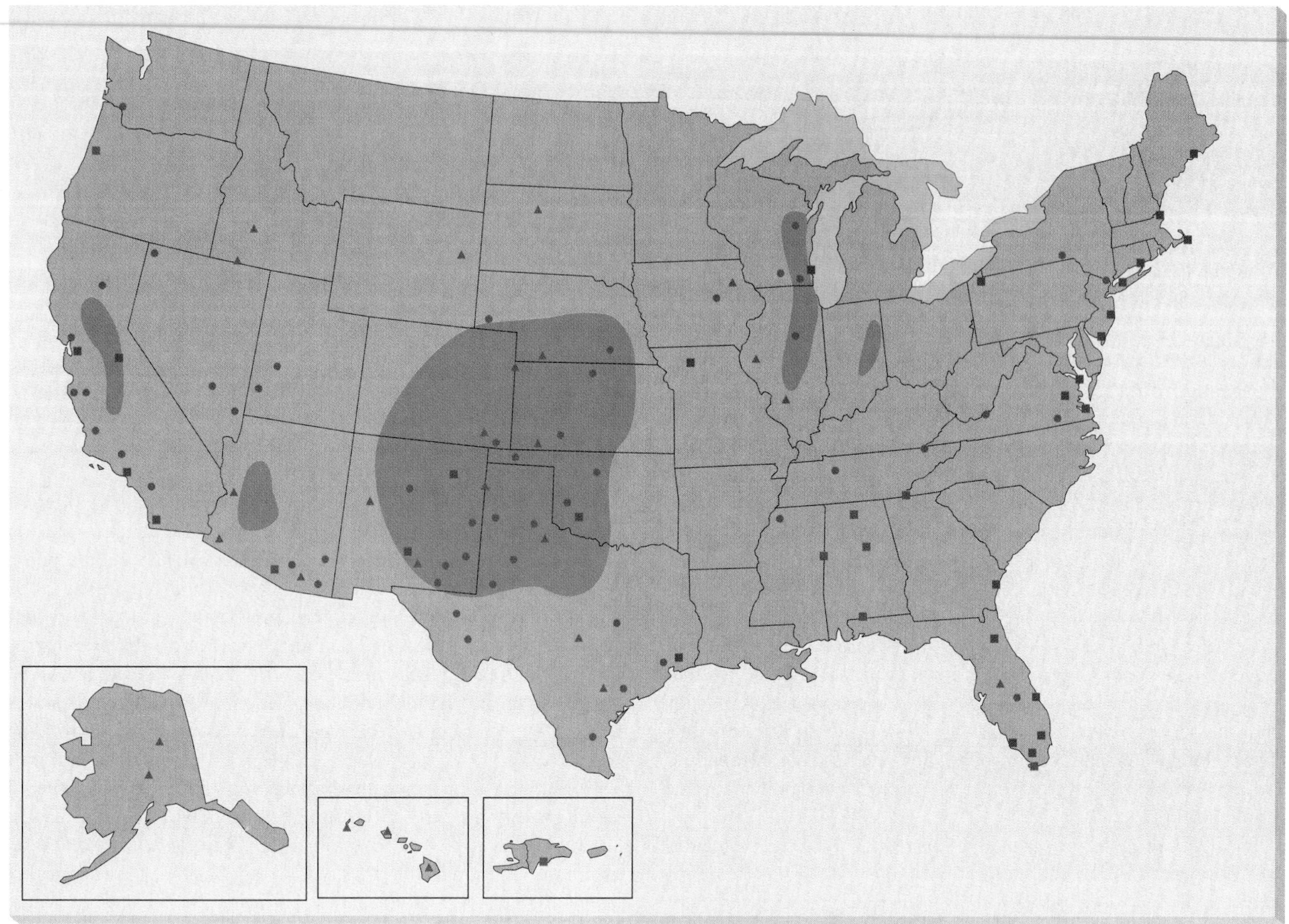

Area problem

- Area in which significant groundwater overdraft is occuring
- Area with less significant problem or no problem

Specific Problem (as identified by U.S. Federal and State/Regional study teams)

- ● Declining groundwater levels
- ▲ Diminished springflow and streamflow
- ■ Formation of fissures and subsidence
- ■ Saltwater intrusion into freshwater aquifers

Figure 12.4 The status of the groundwater supplies in various sections of the United States. In many areas, such as the Ogallala Aquifer (Nebraska, Kansas, Oklahoma and Texas), the removal rate is 40 times greater than the replenishing rate.

(7) landfills, and (8) industrial waste. The extent varies from region to region. We will discuss how each of these pollutants is removed during the water treatment process later in this chapter.

Salting the Roads

Ten million tons of salt (mostly NaCl), sometimes mixed with sand, are spread over northern U.S. roads during a typical winter to remove ice and snow and improve driving conditions. Salt is also used to clear sidewalks, driveways, and steps, but $CaCl_2$ is preferred for these areas since a smaller amount is needed. The use of these salts is considered an indisputable benefit since safe roads and walkways are important. But several byproducts of salt usage are not benefits, including groundwater pollution, loss of soil fertility, and automobile corrosion. Less polluting alternatives, such as shoveling, using ice scrapers, or using a snowblower are not as easy as using salt. Shoveling snow can even endanger the lives of some people.

PRACTICE PROBLEM 12.1

Using the data from figure 12.3 and the fact that the total volume of fresh water in the United States is about 5.76×10^{16} gallons, calculate how many gallons each source contains.

ANSWER To determine the proportions in gallons, you must multiply the percentage (as a decimal) of each source by the total volume. For the aquifer (groundwater), this is $0.7796 \times 5.76 \times 10^{16} = 4.49 \times 10^{16}$. The other results are: Great Lakes, 1.12×10^{16}; all other lakes and ponds, 1.12×10^{15}; rainfall, 3.34×10^{14}; and all the streams and rivers, 1.15×10^{13}. (If you need to, review the section on using exponentials for very large numbers in chapter 3.)

PRACTICE PROBLEM 12.2

What is the per capita amount of salt used for snow and ice removal in the United States? Assume our population is 230 million.

ANSWER Since ten million tons (20,000,000,000 pounds) of salt are used each year, a total of 2.0×10^{10} pounds are used for 2.3×10^{8} people. Dividing the first number by the second yields about 87 pounds/person.

Oil Field Brines

Brine (salt water) occurs naturally, under great pressure, in many sand or shale deposits of oil or gas; when the fuel is tapped, this brine is released. Oil or gas well drilling produces more than 4.6×10^{11} gallons of brine annually. Disposal of brine into the ground, streams, or ponds is now banned, but some leakage occurs during drilling. The only way to avoid this hazard is to leave the fuel in the ground, but then we cannot obtain necessary fuel supplies. Desalinating brine is another possibility, but it is an expensive process, and salt is more readily available from less costly sources.

Agricultural Chemicals

Farmers regularly apply fertilizers, soil conditioners, herbicides, and pesticides to increase yields and to deter crop destruction by pests. Inevitably, some of these chemicals find their way into the water supply. Some herbicides and pesticides are toxic. For example, organic mercury compounds are used to coat seeds to protect them against fungus, rot, and disease. However, mercury is a heavy metal that causes vision loss, muscle weakening, numbness, paralysis, and even coma or death in humans if levels are sufficiently high. (Heavy metals have a valence of two or greater and usually have an atomic weight above 100; the exact value varies with the metal.) Mercury has been detected in the Florida Everglades, coming mainly from agricultural (mostly sugarcane) runoff, but part also comes from natural sources, such as the mercury naturally present in the soil. The problem of mercury poisoning is not new; stores removed millions of cans of tuna from their shelves in the late 1970s because they contained 1 ppm mercury. Mercury entered the tuna's food chain and threatened to enter ours as well.

Inorganic fertilizers contain the nitrates and phosphates growing plants need. Water solutions of these fertilizers can drain into ponds and serve as food for algae. When algae grows rapidly, it leads to **eutrophication** (or algae bloom), a situation in which algae overrun the pond, covering it with a green, slimy blanket that inhibits the exchange of oxygen and produces a bad odor (figure 12.5). Eutrophication is actually the over-enrichment of water with nutrients—a result of biodegradation gone out of control. There are two main biodegradation pathways—aerobic and anaerobic. Oxygen is directly involved in the chemical reactions of the preferred aerobic pathway, but is excluded from the anaerobic pathway. Most aquatic decomposition follows an aerobic route, but the anaerobic pathway occurs if the dissolved O_2 supply is depleted. Eutrophication occurs when the lower layers of algae do not get enough sunlight and die. These dead algae decay and consume the dissolved O_2 in the water by an aerobic pathway, reducing the amount available for fish and other aquatic life and starting up an anaerobic pathway that produces foul-smelling byproducts. Eventually the pond dies and becomes a rotting basin of smelly garbage. Certain household detergents and municipal wastewater that contain phosphates can produce similar results, but phosphate additives are banned in many states. Ponds can gradually restore themselves after eutrophication; the time required is reduced by aeration (blowing air into the pond).

Figure 12.5 A lake that has undergone eutrophication. Note the heavy covering of algae and water plants.

Industrial Wastes

Industrial wastes present a more challenging problem because (1) many waste materials are not biodegradable and (2) the variety is enormous. Federal laws regulate the construction of hazardous waste landfills used mainly for industrially derived materials. Assessing the hazardous nature of industrial waste is a difficult exercise in risk appraisal, and decisions are sometimes based on incomplete data. Box 12.1 demonstrates these difficulties by examining the controversy over a family of chlorinated hydrocarbons known as dioxins.

All waste is chemical by nature, but not all pollution comes directly from chemical manufacturing. For example, a printing plant disposes of enormous amounts of solvents and water contaminated with heavy metal ions (from inks) each year. However, printing plants are not the only source of heavy metal contamination. On a typical college campus, the on-site print shop normally contributes 20 to 50 times as much pollution as the entire chemistry department. Art departments often take second place. Chemists are aware of the potential dangers of pollution, but printers and artists often are not. (However, artists and printers have become increasingly aware of these problems, and many now know how to safely dispose of or recycle their materials.) Chemists are developing alternative inks, often based on soybean oil, that are less toxic.

Landfills

Landfills have made the national news repeatedly over the past twenty years. Chances are that you have landfill problems in your community; most areas do. More than 2.6×10^{11} pounds (1.2×10^{11} kg) of solid wastes from household, industrial, and municipal sources are disposed of in landfills annually. In a simple landfill, trash is covered with dirt. When these landfills are not lined, as was typical in the past, soluble materials inevitably leach out as rainwater percolates through the soil around the waste. Plastic liners are now required in hazardous waste landfills, but these sometimes leak due to improper construction, insufficient thickness, or chemical attack. Figure 12.6 shows a diagram of a modern, lined landfill that does not leach undesired waste into the groundwater.

Figure 2 in box 11.2 showed the contents of a typical landfill; paper is the main culprit, with yard wastes (grass, leaves, and so on) in second place. Note, however, that many communities no longer allow residents to put grass and yard wastes into landfills. Plastics are in sixth place on the list of landfill wastes, but they get a lot of flack because most plastics are not biodegradable. Recent landfill studies revealed a startling picture. When twenty-five-year-old debris was unearthed, researchers found completely readable newspapers. They even found undecomposed food!

Landfill wastes are classified into three types: (1) solid wastes, (2) sewage sludge, and (3) industrial waste. Any of these types of wastes could lead to groundwater pollution in a poorly constructed landfill. Solid waste is mainly paper (see box 11.5, figure 2), which is presumed to be **biodegradable**—to break down naturally. However, recycling paper circumvents many tons of waste matter. Other solid wastes include recyclable plastics, metals, and glass. As previously noted, it takes twenty times more energy to produce aluminum from the ore than to reclaim it from recycled cans; the same is true for other metals, glass, and plastics. Thus, recycling saves both energy and valuable materials.

Recycling is important from at least one other standpoint—landfill space is limited. There are over 13,000 landfills in the United States, but many are already filled to capacity and others expect to close their doors in the next few years. Figure 12.7 shows the expected landfill availability for the next decade; the

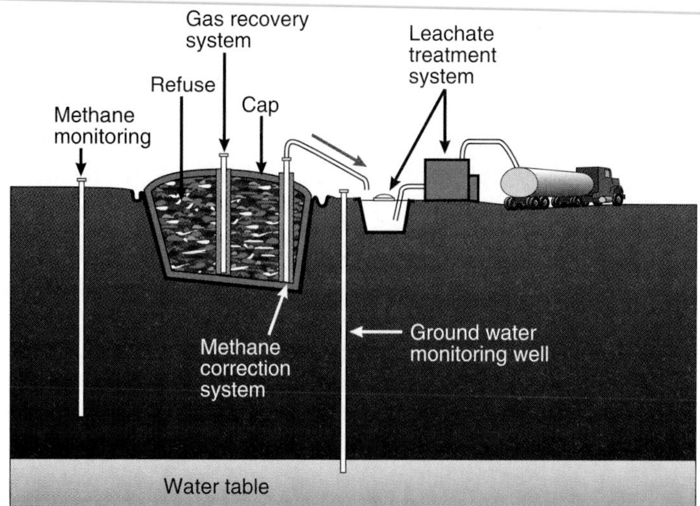

Figure 12.6 Diagram of a modern landfill. Unlike the older types of landfills, this landfill is sealed when filled.
SOURCE: Data from National Solid Wastes Management Association.

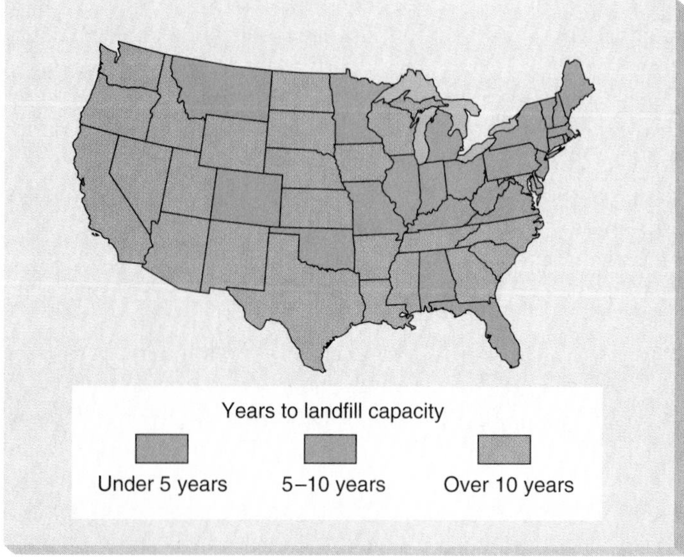

Figure 12.7 The expected availability of landfill space in the United States in the next decade.
SOURCE: Data from National Solid Wastes Management Association.

outlook is bleak, especially in the eastern United States. Shipping trash to other states is not usually viewed fondly by the recipients.

Very little decomposition occurs in a landfill because the debris is isolated from oxygen and microorganisms. If this trash (especially the paper, lawn, and food waste) were actually buried in your backyard, it would biodegrade in a few years. Microorganisms and oxygen are available in backyard dirt, but not in landfills; landfills contain almost no microorganisms, and air seldom penetrates into their deeper layers. The net result is that the presumably biodegradable materials (paper, food, and grass) become isolated from two necessary ingredients—microorganisms and oxygen—and remain undegraded. Landfills may become "happy hunting grounds" for future archaeologists, but they are certainly not good solid disposal sites today.

Sewage **sludge,** which comes from municipal sewage treatment plants, is composed of solid waste from kitchens, laundry water, human feces, soil erosion, vegetation debris, and other sources. Most of this sludge could be utilized as a fertilizer, alleviating inorganic fertilizer pollution problems and partially reducing the landfill crisis. Initially sludge has a bad odor, but this fades after a short time. Sludge is often priced higher than the corresponding inorganic fertilizers, making its use on gardens, lawns, and farms more expensive. This may change as sewage treatment plant managers develop better marketing skills, but some caution is necessary since sewage sludge can contain appreciable levels (0.01–0.1 percent) of heavy metals, such as Zn, Cu, Pb, and Cd. Some bags of sludge even carry warning labels: "For use on flower crops only."

Alternatives to Landfills

There are several alternatives to landfill use, including incinerators and recycling. Neither option will completely remove the need for future landfills, however, since certain materials will not burn and others are not readily recyclable. Moreover, toxic wastes will still require disposal sites, and landfills often seem the best option.

Incineration, or burning, produces very little air pollution if properly done. Modular incinerators can provide heat to generate electricity while producing only low levels of air pollution (figure 12.8a). Many of us think of incinerators as smoky, smelly trash burners, but the modern incinerator is different; figure 12.8b shows a photograph of a modular unit in operation. Notice that this unit produces no heavy, black cloud of smelly smoke exiting the chimneys. If one of these incinerators were a few blocks from your home, you would not notice any difference in air quality. However, trash trucks must come and go, so some prudence must be exercised in selecting a site.

There are many good economic and environmental reasons to recycle. If all paper were recycled, millions of trees could be saved annually. (Note, however, that many thousands of trees are planted just to become paper stock. In a sense, this is biomass-recycling.) Recycling metals like tin and aluminum has proven benefits, as discussed in earlier chapters, and salvaging plastics can save millions of barrels of crude oil each year. Glass is recyclable, and the recycled product is cheaper than making fresh glass from sand.

EXPERIENCE THIS

Place a screen or cloth over a container of tap water and allow the water to evaporate to dryness. If possible, repeat the process with ocean or river water, but filter each before setting it out to evaporate. After the water has evaporated, examine the container for residue. If you also used river or ocean water, did they have more or less residue than tap water? Why?

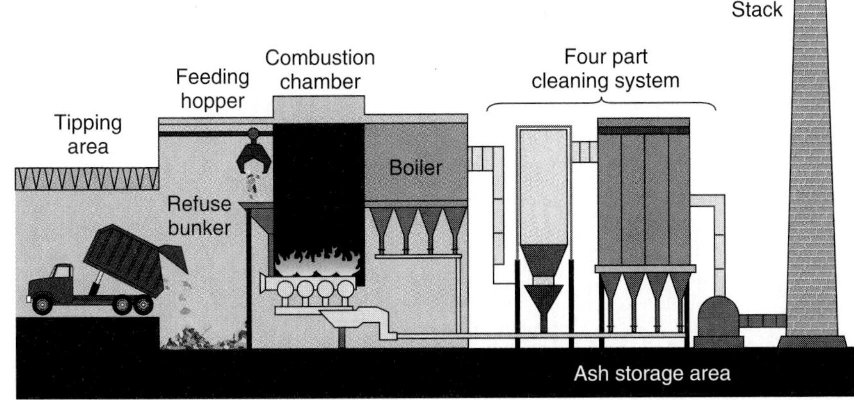

(a)

(b)

Figure 12.8 **(a)** Diagram of a modern incinerator showing the cleaning systems that prevent massive amounts of air pollution. **(b)** Unlike the old-fashioned versions, modern incinerators are a clean method of solid waste disposal, and can also give us a bonus when the energy is harvested from the incineration.

SOURCE: **(a)** is from William P. Cunningham and Barbara Woodworth Saigo, *Environmental Science*, 2d ed.

BOX 12.1

DIOXINS—JUST HOW TOXIC ARE THEY?

Nobody questions that dioxins are toxic; but a highly charged debate centers on exactly how dioxins affect human health and the environment. Dioxins are a family of at least 75 related chlorinated hydrocarbons. A typical example, abbreviated as TCDD, is shown in figure 1. Dioxins are produced by nature in forest fires, volcanic eruptions, and compost piles. They are also trace byproducts in some manufacturing operations and in medical, municipal, and hazardous waste incinerators. At the present time, researchers have pinpointed the sources for only about half of the dioxins in the environment.

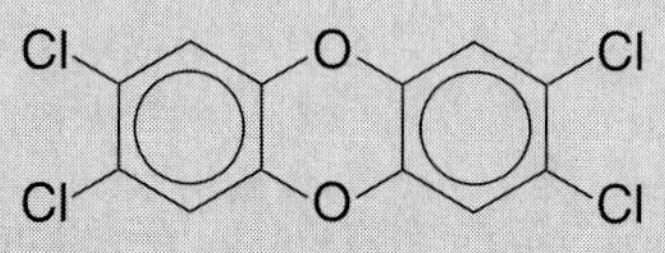

Figure 1 The structure of TCDD, a typical dioxin. The full chemical name is 2,3,7,8-tetrachloro-dibenzo-p-dioxin.

Since their detection in the late 1950s, dioxins have acquired the reputation of being among the most toxic synthetic chemicals known, with an LD_{50} value of 0.03 mg/kg. (Recall from chapter 10 that the lower the LD_{50} value, the greater the toxicity.) Table 1 compares this value to several other chemicals. As the table indicates, the dioxin TCDD is decidedly more toxic than many other herbicidal, medicinal, or common chemicals.

Dioxins began to make headlines in the 1980s, particularly as concerns mounted about the Agent Orange herbicide used as a defoliant in Vietnam during the Vietnam war (figure 2). Agent Orange consisted of two herbicides called 2,4-D and 2,4,5-T; it also contained dioxins as an impurity. Vietnam veterans alleged that these chemicals caused birth defects in their children and numerous health problems in the veterans themselves. Careful laboratory studies revealed that neither 2,4-D nor 2,4,5-T caused these birth defects when pure; scientists attributed the defects to the dioxins present in the Agent Orange. In 1985, the EPA banned 2,4,5-T, which sometimes contained dioxins.

Several studies, before and since, have examined the effects of dioxins on human health with varying conclusions. In 1976, some people were exposed to large doses of dioxins when a chemical plant exploded in Seveso, Italy. Scientists are keeping careful track of these people to determine what effects this large-scale exposure to dioxin may have; so far, the

TABLE 1 A comparison of the toxicities of defoliant chemicals used in Vietnam with some insecticides (*), some medicinal chemicals (**), and some other common chemicals (***). The data are expressed as LD_{50} values; higher LD_{50} values indicate less toxicity.

COMMON NAME	CHEMICAL NAME	ORAL LD_{50} (mg/kg)
Picloram	4-amino-3,5,6-trichloropicolinic acid	8,200
Aspirin**	acetylsalicylic acid	1,100†
Cacodylic acid	dimethylarsenic acid	830
2,4-D	(2,4-dichlorophenoxy)acetic acid	375
Cyclophosphamide**	an anticancer drug	350†
2,4,5-T	(2,4,5-tricholorphenoxy)acetic acid	300
Lidocaine**	a dental anesthetic	292†
Nicotine***	found in tobacco	230†
Caffeine***	stimulant found in coffee	127†
DDT*	1,1,1-trichloro-2,2-bis-(p-chloro-phenyl)ethane	87
Diazinon*	0,0-diethyl-0-(2-isopropyl-6-methyl-4-pyrimidinyl) phosphorothioate	66
Dioxin	2,3,7,8-tetrachlorodibenzo-p-dioxin (TCDD)	0.03

†These LD_{50} values were determined using mice; the rest were determined on rats.

(a)

(b)

Figure 2 More than 4.6 million acres were defoliated by chemical agents during the Vietnam war. **(a)** An aerial view of unsprayed forest near Saigon; **(b)** a forest area five years after being sprayed. (Both photographs were taken in 1970.)

only adverse reactions have been chloracne—a reversible skin-blistering condition—and some minor changes in enzyme structures. A 1987 study by the Centers for Disease Control (CDC) tested TCDD levels in Vietnam veterans exposed to Agent Orange, but the study reported no significant increases in TCDD blood levels in these men. An Air Force study did find higher levels in the 1,200 men who handled and sprayed Agent Orange, but uncovered no consistent evidence of harm. But author Fred A. Wilcox presented a decidedly different picture in his 1983 book entitled *Waiting for an Army to Die: The Tragedy of Agent Orange* (New York: Vintage Books). Wilcox reported that several foreign and American studies showed that dioxins such as TCDD were dangerous (although some of the studies appeared incomplete and used only a few animals or people to obtain data).

In May 1990, the Bush administration agreed to compensate veterans exposed to Agent Orange. Many believe that this decision, which will cost about $23 million per year, was primarily politically motivated since supporting scientific data seem questionable. However, there is little doubt that some Vietnam veterans suffered serious medical problems. Most of the redress efforts were directed toward Agent Orange, but the possibility remains that the disastrous effects were due to a combination of Agent Orange and other herbicides or to repeated exposure over a prolonged time.

Although a political resolution was reached, the scientific research continued. During the 1980s, the EPA lacked definitive human data on dioxin toxicity; to be safe, they adopted the worst case scenario. Their model assumed that even a single molecule of dioxin could be enough to produce birth defects or cancer in humans. On the basis of this assumption, 2,242 residents of Times Beach, Missouri were evacuated in 1983 because an oil spray applied to the streets for dust control contained dioxins. Dr. Vernon N. Houk of the CDC now states, "Given what we now know about this chemical's toxicity, it looks as though the evacuation was unnecessary. . . . It was based on the best scientific information we had at the time. . . . It turns out that we were in error."

By 1990, a conference on dioxin toxicity concluded that, rather than a single molecule being deadly, a specific level of dioxin was required to accumulate before any adverse effects could occur. (This is typical of chemical toxicity.) In early 1991, a study from the National Institute for Occupational Safety and Health (NIOSH) appeared in the January issue of the *New England Journal of Medicine*. This study tracked the health of 5,172 male maintenance and production workers at twelve U.S. plants that produced chemicals containing dioxin. The study showed that the overall death rate for these workers was about the same as for the general population, but those who were exposed to *500 times* more dioxin than the general public had a cancer rate almost 50 percent higher. (Some scientists question whether the cancer rate increases were due solely to dioxin exposure or whether they might also be due to other factors, such as exposure to other chemicals or cigarette smoking.)

The results of the NIOSH study and the earlier conference persuaded the EPA to restudy the dioxin situation. In September 1994, they issued a 2,000-page report that claimed that trace amounts of dioxins *might* cause cancer, disrupt regulatory hormones, cause reproductive and immune system disorders, and produce abnormal fetal development. Still, the EPA report stressed that the actual risks remain uncertain. All their conclusions are based on animal studies, and it is well known that different animal species show disparate levels of dioxin sensitivity and health effects from exposure. For example, less than 0.001 g TCDD will kill a guinea pig, but 300 times more TCDD per unit body weight is needed to kill a dog, and 5,000 times more is needed to kill a

(Continued)

BOX 12.1—Continued

hamster. As yet, low to moderate levels of TCDD do not appear to harm humans. Only the NIOSH study reports any definitive adverse effects, and then only at 500 times the normal exposure levels.

In October 1994, just a month after the EPA report was released, a report on the health risks of dioxins was released by the French Academy of Applied Sciences (CADAS in Paris). Their report stated that, "Contrary to popular opinion, there is no evidence to suggest that dioxins and their related compounds constitute a major risk to public health." Why is this conclusion so different from the EPA's? Pierre Fillet, the director of CADAS, says that the EPA report stressed the *potential* toxicological effects of dioxins—in speculative language that could inflame public opinion—while the CADAS report stressed the *known* toxicological effects and practical ways to minimize them. The CADAS report notes that "no fatal case of poisoning by these produces has ever been reported," and that "the only clearly established effect on human health is chloracne . . . [which] is not life threatening." Furthermore, CADAS notes that "exposure to dioxins is in decline." They state that dioxin emission levels can be decreased by using more sophisticated incineration methods for medical and municipal wastes, which are the sources of 95 percent of known dioxin emissions.

The final chapter in the dioxins story remains to be written. Clear evidence that dioxins are generally dangerous to people is still lacking. And yet, environmentalists and others continue to seek a ban on any activities or compounds that may release dioxins; as a result, any restrictions are likely to be based on factors other than purely scientific data.

But lack of scientific evidence of toxicity does not necessarily guarantee dioxins are safe. They are probably toxic to humans to some extent and, since they have no practical uses, dioxins should be eliminated whenever possible. The dire claims that dioxins are the most toxic or most cancer-causing chemicals known certainly appear exaggerated. But until we clearly understand the effects of dioxins on human health, it may be prudent to continue to seek ways to decrease our exposure.

EXPERIENCE THIS

Tally up the trash and garbage you generate during the next week. (To be fair, include any trash or garbage generated on your behalf by others. For example, if you eat out, count any table scraps or disposable containers in your week's total.) When the week ends, estimate either the volume or the weight of the trash you generated in just seven days. Multiply by 52 to determine your annual output; then multiply by 230,000,000 to get an estimate of the annual national trash tally. Your estimate will likely be too low, since you probably did not include your fair share of industrial or municipal waste.

Healthy Water

Fully oxygenated water contains 8 mg oxygen/L, but to be healthy, water need only contain 5 to 7 mg O_2/L. The amount of oxygen either living creatures or decaying organic matter consume is called the **biochemical oxygen demand,** or **BOD.** The BOD, expressed as mg O_2/L, gives us an idea of the amount of healthy water needed to offset the effects of oxygen-consuming contamination.

Untreated residential sewage has a BOD of 150 to 300 mg O_2/L; just one liter of sewage with a 300 mg O_2/L BOD consumes all the oxygen available in 50 liters of healthy water with around 6 mg O_2/L. Even worse, milk processing canning wastes have a BOD of about 6,000 mg O_2/L; one liter of this waste consumes 1,000 liters of water with 6 mg O_2/L. The wastewater from wood pulping, a major industry, has a 15,000 mg O_2/L BOD, consuming 2.5 times as much oxygen as the waste from the milk processing plant. BOD values climb even higher during eutrophication as decaying algae consume oxygen.

Consider a square pond 1.609 km on each side and 15.24 m deep (a mile square and 50 feet deep). This pond contains 3.95×10^{10} liters of water and 3.16×10^{11} mg O_2, if fully oxygenated at 8 mg O_2/L. This pond could therefore tolerate about 21 million liters of waste with a BOD of 15,000 before all the O_2 is consumed. This may sound like a lot of waste, but the pulping waste from the paper stock used in a large Sunday newspaper in a medium-sized community could soon wipe out the pond. Aerating the wastewater reduces the BOD, but this added expense could bankrupt a small manufacturer.

BOX 12.2

THE ATTACK OF THE NIMBYS

Landfills are becoming increasingly unpopular as people recognize the potential dangers, although some of the "dangers" have been exaggerated. Recycling can alleviate part, but not all, of this problem. Where can we locate these undesirable landfills?

Many Americans would answer, "Not in my backyard!" Nobody wants a landfill nearby. As a result, NIMBYs (an acronym taken from this response) march, picket, and do everything else they can to prevent landfilling in their areas. Unfortunately, everyone, including the NIMBY, generates trash, and this trash must end up somewhere. Figure 1 shows the world's largest landfill, located in the Netherlands. It takes in one million tons of assorted trash each year. Gazing at this picture makes it easy to sympathize with the NIMBYs, or even to become one.

NIMBYs do not usually want trash incinerators near them either. But NIMBYs often will not listen to the arguments or even consider the options available. The technology exists to construct safe incinerators and safe landfills. No matter how much we recycle, such facilities will remain essential in our society. Anyone can become a NIMBY by refusing to study the problems and learn about new technologies that may help solve them.

Figure 1 The world's largest compost pile, at the waste treatment facility in Wijister, the Netherlands. About 1,000,000 tons of trash are delivered here annually and about 100,000 tons of compost are sold for fertilizing farms and gardens, after removal of glass and metal.

Reducing trash production is another option that would help alleviate the crisis. This option is sightly less convenient for the individual—it might require, for example, using washable rather than disposable dishes. Nonetheless, even as individuals alter their behavior, the industrial toxic waste problem remains and makes many more landfills a necessity.

The political NIMTOO, which means "Not In My Term Of Office," is closely related to the NIMBY. In combination, the two frequently stifle all attempts to find solutions to the real problems. Safe landfills in close proximity to where the trash originates are necessary. Modern incinerators can not only help alleviate the trash problem, they can also provide much-needed energy. The decisions people must make on these topics will never please everyone; the key is to make the choices based on sound science.

PRACTICE PROBLEM 12.3

Great news! You are now the proud owner of a paper pulping plant, and that 3.95×10^{10} liter pond is now your property. Your pulp output is used exclusively for making disposable diapers, and you have cornered the market. This means that you will sell 510,000,000 kg wood pulp per year. Each kg wood pulp requires 20 L water to produce, and you run wastewater with a BOD of 15,000 mg O_2/L into your pond each day. How many days can you operate before your effluent water consumes all the O_2 in your pond? (Assume your plant operates 365 days a year.)

ANSWER Let's do this step by step. You produce a total of 5.1×10^8 kg wood pulp per year, using 20 liters of water per kg. This means you would produce 1.02×10^{10} liters of wastewater per year (the product of these two values). Each liter will consume 15,000 mg oxygen. That means your total annual oxygen consumption is the product of these two numbers, or $(1.02 \times 10^{10}) \times (1.5 \times 10^4) = 1.53 \times 10^{14}$ mg O_2/year. Divide this figure by 365 to determine the amount of oxygen consumed in one day: $(1.53 \times 10^{14}) \div (3.65 \times 10^2) = 4.19 \times 10^{11}$ mg O_2/day. Since your pond contains only 3.16×10^{11} mg oxygen, you cannot even make it through the first day of operation.

Are there other options? Consider this—the high BOD comes from waste material in the water. If you remove this waste, the BOD will drop to a more manageable value. You could set up a small water treatment plant beside your factory to clean up the water. You may need filters and chemicals to remove the waste, and some aeration may still prove necessary. All this will cost a bundle of money, but our society deems clean water desirable, and you won't be able to run your plant without some type of waste removal. How much you spend and how efficiently you clean up the water will depend on many complex social, economic, and scientific factors.

This is an example of a social as well as scientific problem: these two factors are often closely intertwined. The scientific method may help in the decision, but ultimately the matter depends on social as well as scientific considerations.

WATER TREATMENT PROCESSES

Water treatment is subdivided into several categories, including (1) municipal drinking water supplies, (2) softened water, (3) recovered wastewater, and (4) desalinated seawater. Each of these categories is examined in the following sections.

Municipal Drinking Water

Potable water (good-quality water) is essential to all societies. We pay attention to water quality in order to safeguard the public health. People still drink contaminated water infested with disease-causing microorganisms in many parts of the world. Figure 12.9 shows the devastating effects this water has on child mortality. Part *a* shows that countries with little access to clean water have high infant mortality rates. Part *b*, a plot of the data shown in *a*, makes the correlation even more clear. In Third World countries, educated leaders understand the link between polluted water and infant or adult deaths, but their citizens are often less aware of the problem. Almost the entire water supply of the Third World is of poor quality. Illness is common in these countries because drinking water is contaminated with sewage and loaded with virulent microorganisms.

In the more developed nations, microorganisms in water are destroyed by **chlorination** in municipal water plants. Chlorine kills microorganisms (germs) through an oxidation-reduction reaction:

$$\underset{\text{Oxidizing agent}}{Cl_2} + \underset{\text{Reducing agent}}{\text{Germs}} \longrightarrow \underset{\text{Reduced}}{2\,Cl^-} + \underset{\text{Oxidized}}{\text{Dead Germs}}$$

Other oxidizing agents, such as ozone, are used in many European communities in procedures such as **ozonolysis;** but chlorine is used more often.

Water-borne disease epidemics, such as typhoid fever and cholera, have caused widespread death throughout human history. Public health officials first associated epidemics with drinking water in the mid-1800s; they knew they needed to prevent

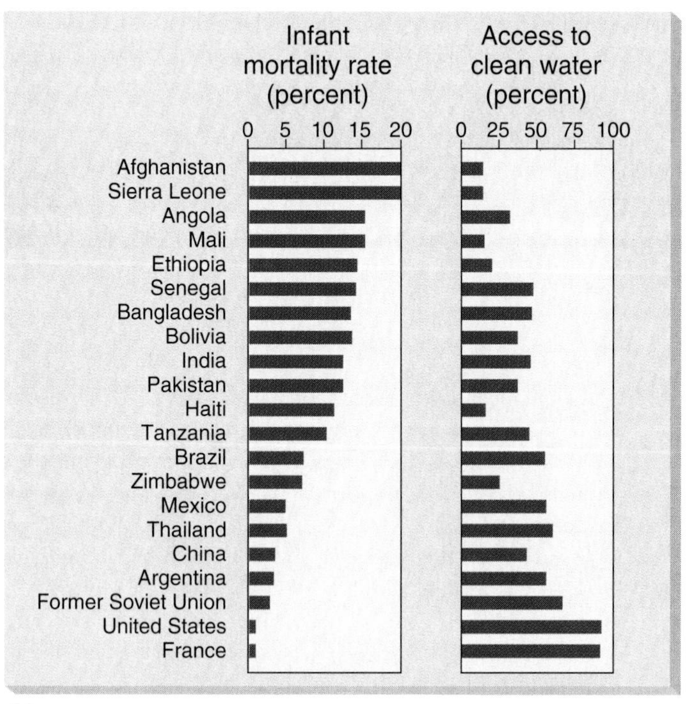

(a)

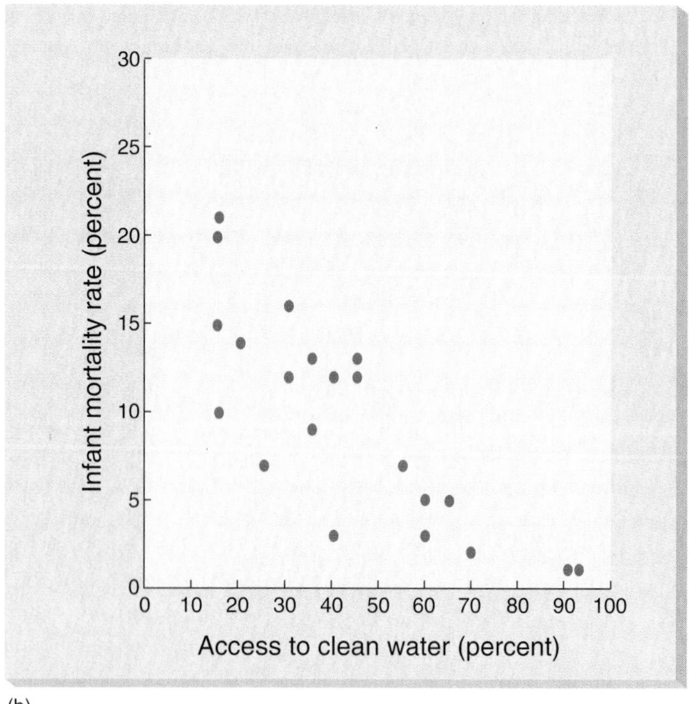

(b)

Figure 12.9 **(a)** The relationship between clean drinking water and infant mortality (1982). **(b)** A plot of the data of **(a)** showing that the infant mortality rate correlates with clean water.

the spread of diseases through water, but they did not know why the disease spread this way. Emergency water chlorination to control epidemics began as early as 1850, even though the bacterial causes were unknown. When water-borne bacteria were discovered in the 1870s, public health officials realized they were the cause of many diseases.

The first regular public water chlorination began in England in 1904 and at the Union Stockyards in Chicago in 1908. Jersey City holds the distinction as the first U.S. city to chlorinate water on a regular basis; the procedure spread rapidly between 1910 and 1915. The effectiveness of chlorination is evident from the marked decrease in typhoid deaths after chlorination was widely introduced in the years following 1910 (figure 12.10).

Water treatment was not the only practice that eliminated the scourge of typhoid. This disease spread primarily through contact with either infected water or milk; pasteurization eliminated the problem in milk. Typhoid was eliminated as a major health problem in many areas of the world years before scientists discovered a drug (chloromycetin or chloramphenicol) to treat it in the 1950s. It remains a major problem in Third World countries, and outbreaks even occur sometimes in the United States when people drink water from supposedly "clean" streams or ponds. Box 12.3 details a startling recent attack on chlorination.

Once the scourge of Europe, cholera was the most feared epidemic in the nineteenth century. Health officials believed cholera was under complete control as the twentieth century dawned. However, more than 391,000 cholera cases were reported in Central and South America in 1991, with about 4,000 deaths. Chlorination also controls this disease, but many Third World countries do not have chlorinated water.

Along with treating water to control or eliminate disease, municipal water is also treated to remove turbidity, color, taste, and some salts (for example, phosphates). Additional chemical treatments are now necessary because heavy metals are contaminating the groundwater and streams. Although turbidity may be removed by filtration, settling, or alum treatment, the latter is the most effective because small particles can pass through the filters. In alum treatment, alum (aluminum sulfate) is mixed with lime (calcium hydroxide) to form a floc (short for *flocculate*, which

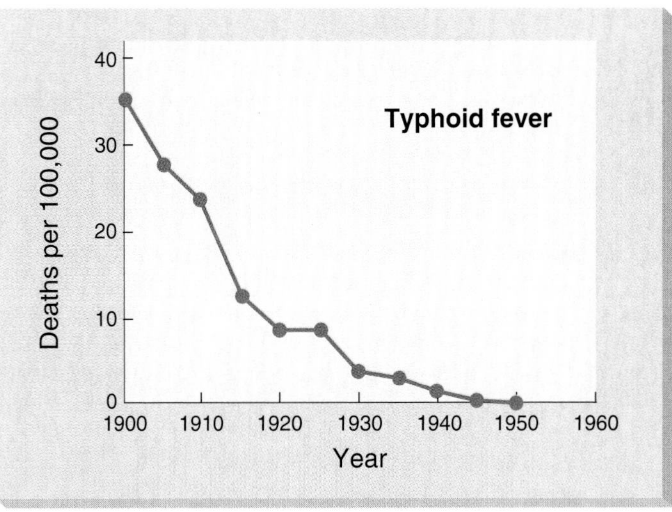

Figure 12.10 Typhoid fever death rates in the United States since 1900 (expressed as deaths per 100,000 people).

means to gather together in clumps) that traps the suspended matter that causes turbidity:

$$\underset{\text{Alum}}{Al_2(SO_4)_3} + \underset{\text{Lime}}{3\,Ca(OH)_2} \rightarrow \underset{\text{Aluminum hydroxide floc}}{2\,Al(OH)_3} + CaSO_4$$

Filtration then removes the floc and the suspended matter.

Taste, odor, and color are removed from water using activated carbon filters, chlorine dioxide, or potassium permanganate (a powerful oxidizing agent):

$$\underset{\text{Oxidizing agent}}{KMnO_4} + 2\,H_2O + \underset{\text{Reducing agent}}{\text{ODOR AGENT}} \rightarrow$$

$$\underset{\text{Brown solid}}{MnO_2} + 4\,KOH + \underset{\text{Oxidized material}}{\text{NON–ODOR AGENT}}$$

Most communities use more than one agent to improve drinking water quality; activated carbon is the most common. On the rare occasions when a community does not use such treatment, most people detect a bad taste or odor in the water.

Figure 12.11 diagrams a typical water treatment plant. In such a plant, most treatment methods are combined into a single

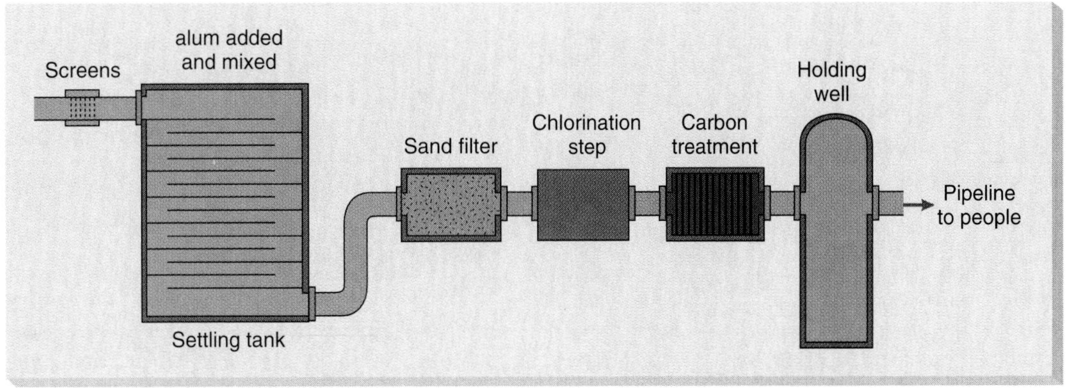

Figure 12.11 Diagram for a typical municipal drinking water treatment plant.

BOX 12.3

THE CHLORINE CHALLENGE

An astonishing movement began in the late 1980s and accelerated in the 1990s—the combined attempt by several environmental groups to ban all use of chlorine and chlorinated organic compounds from use in the United States and, eventually, the entire world. This was the first time in the history of chemistry, or the environmental movement, that all the compounds of a single element in the periodic table were simultaneously attacked.

Those who support a chlorine ban claim that all chlorinated compounds are potentially harmful to people. They point to such toxic compounds as dioxin (box 12.1) and DDT (box 1.2), both of which are already banned in the United States and most of the world. Joe Thornton of Greenpeace says, "There are no uses of chlorine that we regard as safe...no further organochlorine pollution should be permitted. This means phasing out the substance that is their root—chlorine, since whenever chlorine is used, organochlorines result." Greenpeace proposes a ten- to twenty-year transitional phaseout of all chlorine compounds. Gordon Durnil (the U.S. cochair of the International Joint Commission on the Great Lakes) states, "We decided you can't distinguish among different compounds of chlorine as to which is harmful and which is not. We decided we needed to look at chlorine as a class and decided because of the effects of dioxin, that use of chlorine as feedstock should be sunset." The American Public Health Association chimes in: "The only feasible and prudent approach to eliminating the release and discharge of chlorinated organic chemicals and consequent exposure is to avoid the use of chlorine and its compounds in manufacturing processes." Mary S. Wolff of the Mt. Sinai School of Medicine adds, "I don't think any scientist would be against a ban on chlorinated organics." These proclamations have had impact. In February 1994, President Clinton's Clean Water Act was released. It states that "the administration will develop a national strategy for substituting, reducing, or prohibiting the use of chlorine and chlorinated compounds."

Contrary to what the advocates claim, most scientists do not agree with this blanket assessment of chlorine risk. H. Leon Bradlow, an endocrinologist at Cornell University's Strang Cancer Prevention Center, calls a chlorine ban "an extremist position, and like most extremist positions, ridiculous. I think methylene chloride is a dandy solvent and would hate not to be able to use it." Philippe Shubik (a cancer researcher and toxicologist at Oxford University) flatly asserts, "Any scientifically based toxicologist finds that kind of general approach abhorrent."

The chemical industry is even more emphatic concerning a potential ban on chlorine usage. Joseph Walker, Chlorine Institute spokesperson, says, "You wouldn't have modern society as we have it." According to Brad Lienhart, managing director of the Chlorine Chemistry Council, "This is the most significant threat to chemistry that has ever been posed." Lienhart further notes that "a very high percentage" of chlorine compounds "are neither toxic, persistent, nor bioaccumulative." On the use of chlorine, W. Joseph Stearns, director of chlorine issues for Dow Chemical Company, says, "It is the single most important ingredient in modern [industrial] chemistry." And John Sesody (vice president and general manager of Elf Atochem, North America) adds, "It is such a valuable and useful molecule because it does so many things and is involved in so many end products." Most of the nation's governors, mayors, cities, counties, and state legislatures, who would be responsible for carrying out a chlorine ban, also oppose any blanket ban on the use of chlorine and chlorine products. Aside from causing the loss of jobs in their communities, it would prevent the chlorination of municipal drinking water supplies and put the public health at risk.

Global chlorine production is about 38 million tons per year, with about 12 million tons produced in the United States. More than 1.3 million jobs in the United States depend directly or indirectly on the use of chlorine. Figure 1 shows where this huge amount of chlorine goes in the United States; the attack on chlorine takes aim at all of these areas (Figure 2). Environmentalists claim that water purification produces trace quantities of compounds that can sometimes cause cancer. The use of chlorine in pulp and paper manufacturing does tend to produce small

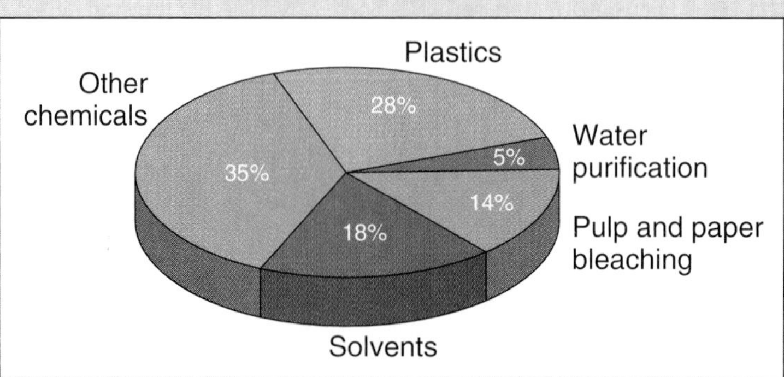

Figure 1 The percentages for the various industrial uses of chlorine.

Figure 2 Some "fruit" from the chlorine "tree." Note that this shows only a few of the more than fifteen thousand uses of chlorine compounds.

(Continued)

BOX 12.3—Continued

amounts of dioxins, and PVC plastic is made from vinyl chloride monomer, which can cause cancer (though the polymer does not cause cancer). Similarly, chlorinated solvents have been implicated as potential cancer-causing agents. The class called "other chemicals" in figure 1 includes various chlorine-containing herbicides and insecticides. Although some of these chemicals may indeed cause cancer, the concentrations in our environment are extremely low and the danger is considered insignificant by most cancer researchers. The level of dioxins paper manufacturing releases is also very low, and this problem is becoming less acute as the chlorine bleaching process is replaced by a process using chlorine dioxide, which does not produce dioxins.

Chlorine is used directly or indirectly in more than 15,000 products, with annual U.S. sales greater than $71 billion. Some of these uses are illustrated in figure 2. Chlorine and its derivatives are used in almost all modern industries. The chlorine "tree" shows that virtually every aspect of our modern lifestyle has come directly or indirectly from the use of this important chemical.

Chlorinated pharmaceutical chemicals are excluded from any ban, at least at the present time. However, all pharmaceuticals require the use of solvents, many chlorinated, for their manufacture. Melvin J. Visser, divisional vice president of Upjohn, notes that 31 pounds of source materials, including chlorinated solvents, are needed to produce one pound of a pharmaceutical agent.

Ban advocates claim that readily available substitute chemicals or processes can replace chlorine, but this is seldom true. For example, they claim that ozone can replace chlorine for drinking water disinfection. However, ozone tends to dissipate more rapidly and requires repeated treatment to maintain safe levels of bacteria and other infectious agents. (In other words, ozone does not guard against recontamination.) Ozonolysis is also more expensive than chlorination; switching to this process would cost at least $6 billion per year. In addition, ozone produces its own collection of potentially harmful products, including bromates, some of which are potential carcinogens.

In 1994, the Chemical Manufacturer's Association and the American Chemical Society asked CanTox, a Canadian consulting firm, to prepare an evaluation of chlorinated organic compounds. Their report evaluated the compounds in eight product and process categories: (1) chlorine, (2) drinking water, (3) incineration, (4) pesticides, (5) polychlorinated biphenyls (PCBs), (6) pulp and paper, (7) PVC and its monomer, and (8) solvents. In general, the report notes that far more benefits accrue from chlorinated organic compounds than problems. It stresses that the newer chlorine dioxide process will dramatically reduce the dioxins produced by the pulp and paper industry. It also notes that controlled high-temperature incineration in state-of-the-art facilities produces no significant quantities of dioxins.

Based on all the evidence, and on apparent benefits and risks, is it feasible or wise to ban all chlorine and chlorinated products? Keep in mind that there are more than 2,000 different *natural* organochlorine compounds in our environment. Some of these natural chemicals may prove harmful to people, while others will probably prove harmless. In any event, it is impossible to ban these natural organochlorine compounds.

How successful the attack on chlorine will be remains to be seen. These questions might well be best resolved using the scientific method; however, the game is not presently taking place on this playing field. The ultimate decisions may well come from the political arena. We can only hope decision makers will receive sufficient scientific input to make sound judgments.

multistep operation. An initial screening removes debris from the incoming water; then the alum treatment removes most of the remaining particulate matter at a sand filter. Chlorination occurs after these steps to reduce the amount of chlorine required; particulate matter generally reacts with Cl_2. After chlorination, the water is treated with activated carbon and other taste or odor removers and is finally stored in holding wells until needed. Figure 12.12 shows a photograph of an advanced potable water treatment plant.

The cost of municipal water varies greatly and depends on several factors, including (1) the availability of water, (2) the quality of the incoming water, (3) the age of the plant (plants cost less to build years ago), and (4) the level of purification (modern regulations stipulate higher purity). Actual costs in the United States are as low as $1.41 per 1,000 gallons and as high as $2.80 per 1,000 gallons. The average is about $2.18 per 1,000 gallons of water.

These costs may represent the expense of building new plants, installing new water lines in the community, or upgrading existing plants or lines. Most older water lines use leaded joints; because of the risks of lead contamination (chapter 8), many older lines are being replaced. Replacing old water lines, which usually requires digging under paved streets, is expensive, and communities find it challenging to obtain the money.

Figure 12.12 The new ultramodern Alan R. Thomas Water Purification Plant in Edgewater, Florida—an advanced water treatment plant that began operating in early 1993.

PRACTICE PROBLEM 12.4

Most communities use a sliding scale to price potable water. For example, the first 2,000 gallons might cost \$4.00 per 1,000 gallons and subsequent amounts only \$2.50 per 1,000 gallons.

a. Using these cost figures, what is your water bill if you use 8,000 gal/month?

b. On this basis, what is your average cost per 1,000 gallons?

ANSWER *a*. The cost is as follows:

2,000 gal × \$4.00/1,000 gal = \$8.00
6,000 gal × \$2.50/1,000 gal = \$15.00
\$8.00 + 15.00 = \$23.00 per month

b. Your average cost is \$23.00 ÷ 8,000 gallons = \$2.88/1,000 gallons. Do you think water rates should decrease if people use more water? What effect does this have on water conservation?

Wastewater Treatment

Most communities have wastewater treatment facilities to clean up their sewage rather than putting it directly into the streams, ponds, and groundwater. Several kinds of waste are removed in primary, secondary, and advanced treatments or stages. The **primary water treatment** is simply a screen that removes solid materials and settling tanks to remove the smaller particulate matter; still, this stage reduces the wastewater BOD from about 200 to about 140 mg/L. This primary treatment (figure 12.13a) removes about 50 to 75 percent of the particulates. Chlorination is sometimes run after the filtering and settling steps to eliminate most of the bacteria, but when a secondary stage follows immediately, the chlorination is deferred to a later stage.

Secondary water treatment is shown diagrammatically in figure 12.13b; the main treatment is a biological process using an activated sludge that consists of microorganisms and inorganic materials from previously processed wastewater. In this stage, 85 to 95 percent of the dissolved organic material and 85 to 95 percent of the remaining suspended matter is removed. After the secondary stage is complete, the BOD has decreased to an average value of 30 mg/L (15 percent of the initial level). The wastewater-activated sludge mixture is aerated and then sent to a settling tank where the solid waste collects. Part of this solid waste is used as a fertilizer, while the rest is recycled with more wastewater in the secondary treatment plant. Sludge is not widely used as a fertilizer for two main reasons: many people worry about its safety, and it costs more than inorganic fertilizers with the same nutrient levels. Sewage sludge disposal remains a problem.

Advanced water treatments are essentially chemical and are usually fairly expensive. They include carbon black (charcoal) treatment to reduce odors and improve taste, and alum—$Al_2(SO_4)_3$—treatment to remove phosphates via $AlPO_4$ precipitation. Phosphates are food for algae; their removal is important to avoid eutrophication. Advanced treatments remove 98 percent of the waste ingredients and cause the BOD to drop to an average value of 2 to 5 mg/L. Figure 12.14 shows modern primary and secondary wastewater treatment plant operations.

The removal of all waste and soluble impurities from sewage is virtually impossible, and the cost increases dramatically with each additional treatment stage. Very little treated wastewater is used directly for drinking since it is not sufficiently purified. After it passes through the soil into the aquifer, however, and after a potable water purification process, the treated wastewater is safe for drinking. Treated wastewater can be used for washing cars or irrigating crops; these activities do not require potable water, though the water must be free from sewage and bacteria. The cost for wastewater treatment approaches that for potable water purification. Cleaning up sewage is important, but expensive. If the decrease in disease is considered, wastewater treatment is a worthwhile public health step. If you are unconvinced of this, reexamine figures 12.9 and 12.10.

Industrial wastewater is more complex than household sewage because it contains numerous chemicals, many of which are not biodegradable and some of which are extremely toxic. The EPA requires the removal of most of these materials from the industry's effluents. The treatments are typically specific for each chemical and are beyond the scope of this book.

Box 12.4 focuses on yet another stage of water purification. In many homes, people make use of home water softening or filtering devices to improve their tap water.

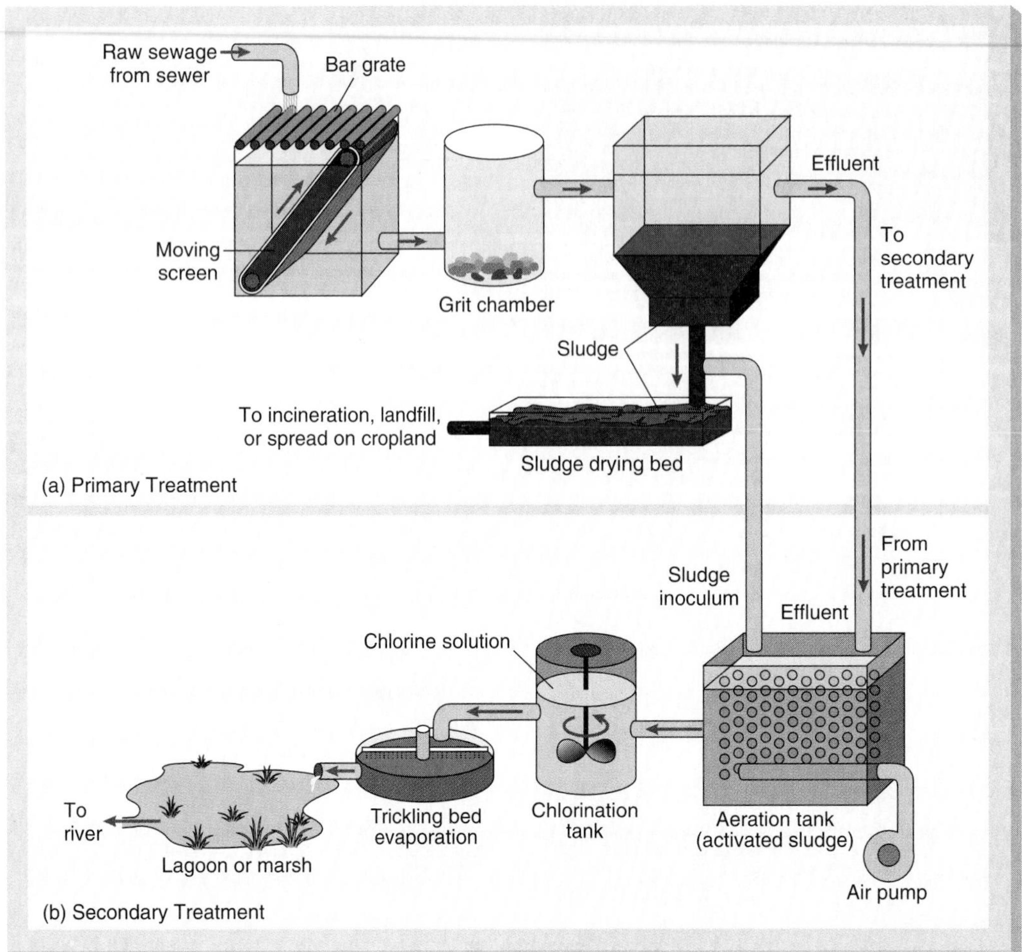

Figure 12.13 Diagrams of **(a)** a primary wastewater treatment facility that only removes the larger solids and particulate matter, and **(b)** a secondary wastewater treatment facility that removes most organic compounds and bacteria via chemical (chlorination) and biological (activated sludge) means.

(a)

(b)

Figure 12.14 **(a)** Primary and **(b)** secondary wastewater treatment plants. The primary treatment involves only physical settling of waste particles, while the secondary treatment utilizes chemical and biological processes.

FRESH WATER FROM THE SEA

Figure 12.2 showed that seawater makes up 97.2 percent of the total water on our planet. Seawater contains 35 g/L of dissolved NaCl and other salts that must be removed to make this water drinkable or potable. The process of removing these salts is called **desalination.** Not all regions in the United States would gain a direct benefit from desalination, but the coastal states will ultimately need to utilize this potentially important resource. Closely related to seawater desalination is the removal of salts from brackish (salty) water, which has 0.1 to 0.5 percent total salt content. The desalination of brackish water, sometimes called desalting, is possible virtually everywhere.

Desalination is not a totally new idea, although interest has grown greatly recently. The Egyptian, Hebrew, Persian, and Greek cultures explored desalination, usually by condensing the vapors from evaporated salt water. Greek sailors obtained fresh water from evaporated seawater as early as the fourth century B.C. In the 1700s, United States and British naval vessels were equipped with simple distillation devices for seawater desalination, and by 1900, land-based distillers were in use in many arid parts of the world that bordered on the sea.

Desalination is expensive because water requires 540 cal/g to evaporate or 80 cal/g to freeze, and energy costs money! At least six potential desalination methods exist: (1) distillation (including solar distillation), (2) reverse osmosis, (3) electrodialysis, (4) freezing, (5) ion exchange, and (6) hydrate formation. Table 12.1 compares the cost to prepare 1,000 gallons of potable water by each of these six methods against ordinary municipality-treated water. The distillation or reverse osmosis purification methods cost about the same, but they run roughly five times the cost of ordinary treated water; they are the only methods currently in common usage.

TABLE 12.1 Estimated cost for desalination compared with the cost of ordinary city-treated water (shown in the first two entries on the table).

METHOD	$/1,000 GALLONS
Older city water plants	0.50–1.50
Newer city water systems	2.00–2.50
Desalination of brackish water*	1.50–4.00
Desalination of seawater*	5.00–10.00
Solar distillation of seawater**	2.50–6.00
Electrodialysis	15.00–40.00
Freezing	30.00–50.00
Ion exchange	40.00–80.00
Hydrate formation	80.00–100.00

*Using reverse osmosis or distillation
**Distillation heated by solar energy

Solar distillation is estimated to cost half as much as ordinary distillation because the energy source (sunlight) is free. The remaining methods are considerably more expensive than either distillation or reverse osmosis. Most cost estimates are based on small-scale systems and could decrease markedly when applied to a large metropolitan water system. The cost for brackish water reclamation is less than for seawater because the salt content is much lower.

The cost of conventional water treatment will increase as water pollution continues; so will the demand for more water. As of 1991, more than one thousand desalination plants were operating in the United States, mostly to purify brackish water. Currently, more than one hundred desalting units operate in Florida alone, mostly using the reverse osmosis technology. The total capacity of these plants is more than 100 million gallons/day (mgd).

The majority of desalination plants are located in Third World locations, notably in the Middle East, where fresh water supplies are extremely scarce. The worldwide capacity exceeds 3.5 billion gallons/day. Many of these plants are small and experimental. Their costs are controlled partially by economic factors and should drop as more water is desalinated.

PRACTICE PROBLEM 12.5

How much water does the typical person in your community use each month? You may be able to obtain this information from your water department; if they do not have the average value available, ask them for the total volume used each month and divide by the number of people in your community. (You should be able to obtain this information from city hall or the local library.)

ANSWER Nationwide, per capita consumption ranges between 4,400 and 8,800 gal/month. How does your community compare?

Distillation

The **distillation** method is basically the same as that described in chapter 2; water is boiled and the condensate is collected. Water has a large heat of vaporization (540 cal/g); this means a lot of heat is needed to boil water to vapor. Seawater is usually heated by burning a fossil fuel, adding to air pollution, but nuclear or solar heating is presently possible. However, building a desalination plant beside a plant that generates "waste" heat (for example, an incinerator or a steel mill) can provide "gift" heat for distillation. A steel mill uses about 26,400 gallons of cooling water for each ton of steel produced. If the extra heat is transferred from

BOX 12.4

HOME WATER-TREATMENT DEVICES

Water softeners are one of the more common water-treatment devices installed in homes. Hard water contains mainly Ca^{++} and Mg^{++} ions. These ions cause no apparent health effects, but they do create problems in washing with ordinary soap. Soap is a mixture of Na^{+} salts of fatty acids with 14 to 20 C atoms in their carbon chains. When soap encounters Ca^{++} or Mg^{++} ions, it forms a water-insoluble product that shows up as a scum on the washed items. **Water softening** through ion exchange combats this problem; synthetic detergents are another tool used to overcome hard water problems. (The detergent approach will be discussed in chapter 13.)

Ion exchange attacks the hard water problem by removing the Ca^{++} or Mg^{++} cations and replacing them with Na^{+} cations. The ion-exchange process is run in homes in special tanks or cylinders that contain small beads of an **ion exchange polymer.** These tanks are connected to the incoming potable water line (figure 1). The ion-exchange polymers are small, crosslinked spheres, usually a styrene-divinylbenzene copolymer.

Water softening is normally limited to cation exchange. Although anion exchange is possible, it is cations that cause hard water problems. The Ca^{++} or Mg^{++} ions in the water flow past the beads and are swapped for Na^{+} ions, which do not precipitate soap. Eventually, all the Na^{+} ions are removed from the beads, and they must be replaced by backwashing the beads with a concentrated $Na^{+}Cl^{-}$ solution. The washing process removes the Ca^{++} or Mg^{++} ions from the beads and replaces them with Na^{+} ions. The excess salt washing solution is then discarded. The following equation summarizes the process:

$$\text{Polymer-}(Na^{+})_2 + Ca^{2+} \longrightarrow \text{Polymer-}Ca^{2+} + 2\ Na^{+}$$

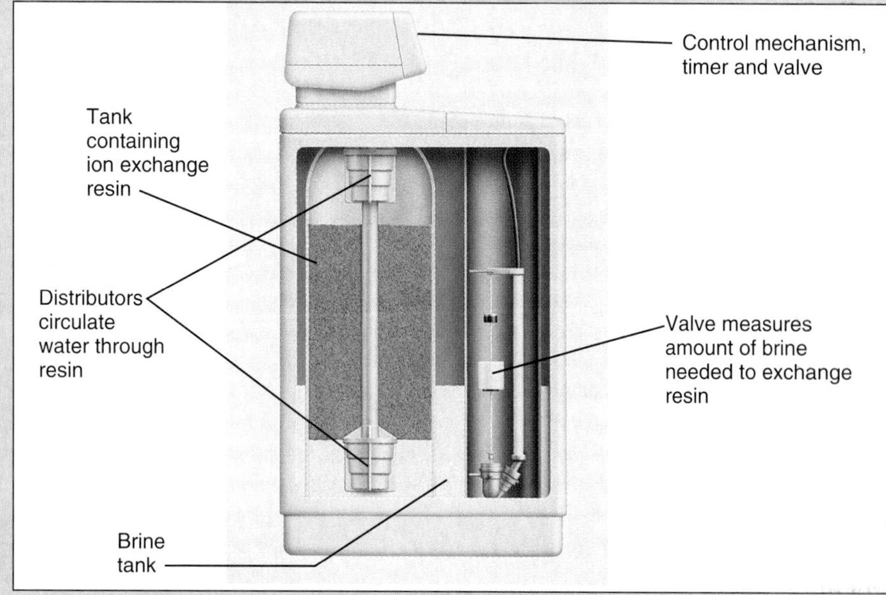

Figure 1 A typical home water softening unit.

Cation-exchanged water has a higher concentration of Na^{+} ions than ordinary drinking water. Anyone who must maintain a low sodium intake (for example, a person with a heart problem) should not drink softened water. Besides, only the water used for washing clothes or bathing needs to be softened in the first place. Ion-exchange treatment is also expensive and unnecessary for watering lawns or washing cars; ideally, part of the incoming water is diverted past the ion-exchange unit, avoiding the difficulties excessive Na^{+} ions can cause. Another problem with softened water is that it contains less total ions and can dissolve metal ions, such as lead, more readily. (See box 8.2 for information on the problems caused by lead ingestion.)

Besides water softening, several other options exist for home drinking water treatment. The next most common device is a carbon filter, normally attached to a faucet, that removes chlorine, radon, many organic compounds (including volatile organic compounds, or VOCs), most halogenated methanes, and some pesticides. The procedure is based on the adsorption of impurities onto the activated carbon in the filter, a physical process. Some carbon filters also contain a small quantity of adsorbed silver compounds to kill some bacteria via oxidation. This is a chemical process.

Distillation devices are relatively uncommon in homes (figure 2), but they are effective in removing lead, other metals, and some organic compounds. The distillation process also kills bacteria during heating. Its chief deficiency is that distillation seldom removes volatile organic compounds (VOCs) or totally halogenated methanes (THMs). It also increases metal pipe corrosion.

Another method used in homes is **reverse osmosis** purification (figure 3), which is described in detail later in this chapter.

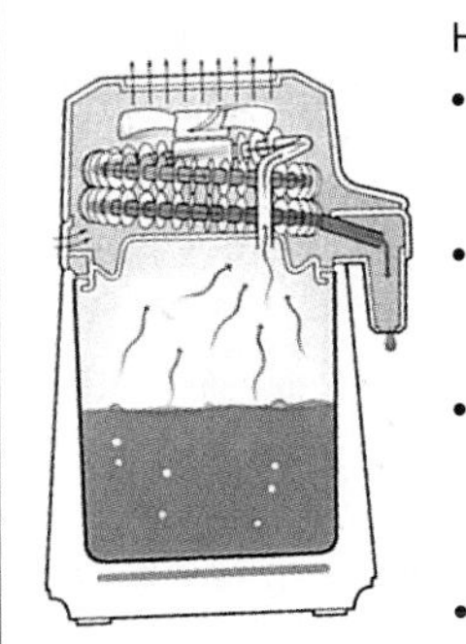

Figure 2 A home distillation unit.

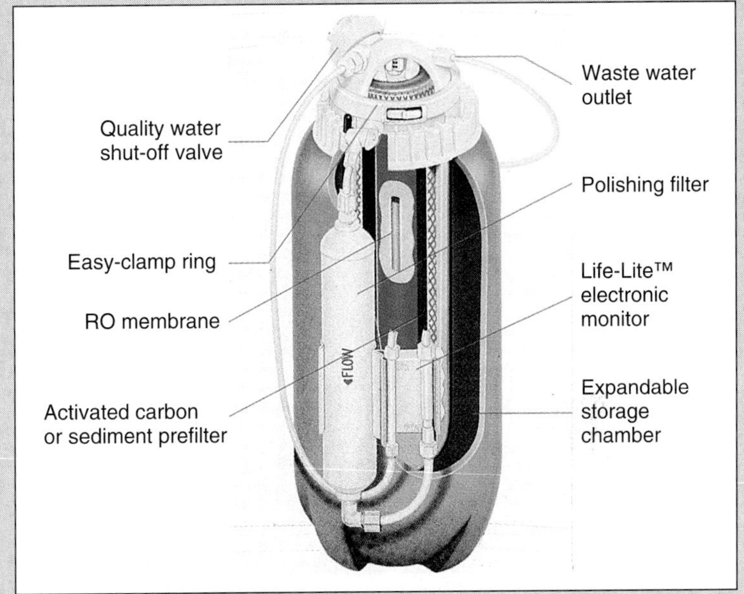

Figure 3 A home reverse osmosis (RO) unit.

RO devices remove lead, nitrates, radon, some THMs, most VOCs, and many pesticide residues. But these units are wasteful in that 75 to 90 percent of the water is lost during the process.

Home water treatment devices vary greatly in price. In general, carbon filters are the least expensive of the four systems, including water softeners.

the cooling water to incoming seawater, the amount of new heat required for distillation is greatly reduced. The incoming water can also be preheated by the heat released from condensed steam vapors, reducing energy costs by recycling part of the energy. Over 60 percent of the world's desalted water is prepared via distillation.

Solar Distillation

Solar distillation uses free energy from sunlight to distill water. The water used in current experimental units costs nearly $20.00 per 1,000 gallons, but this figure should decrease with large-scale production. A solar distillation system is basically a mini-greenhouse that harvests seawater instead of plants (figure 12.15). By trapping heat inside the system, a solar distiller raises the internal

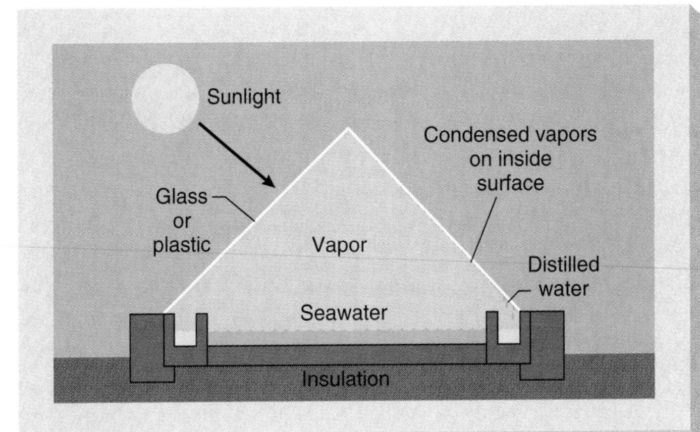

Figure 12.15 A solar seawater distillation unit.

temperature and evaporates the water more rapidly; the salts remain behind. The water vapor is cooled on transparent glass or plastic walls, often chilled with seawater. It condenses and runs down the surface to collection troughs.

Solar desalination systems occupy a large area (as much as 2,500 cubic meters) and only work effectively in regions with much interrupted sunshine. Solar distillation systems would work well in California and Florida, states blessed with long coastlines and an abundance of sunshine. To a lesser degree, solar distillation is useful in other states because the greenhouse effect will still work under partially sunny conditions, and a solar distillation process can desalt brackish water as well as seawater. The corrosive nature of salt water limits the usable materials; metals have a short lifetime, but glass and plastics last a long time in a solar distiller. Plastics also provide insulation. Small versions of solar stills are used in many Third World countries to provide potable water for individual homes. Similar small units are also used in the United States.

PRACTICE PROBLEM 12.6

How much energy is needed, per gram of water, to distill seawater via solar energy? How does this compare to distilling through conventional heating?

ANSWER Just when you thought this text would not pose trick questions, we spring this one on you. Boiling is boiling is boiling. Distillation raises the water temperature to the boiling point and converts liquid water into steam, a process that requires 540 cal/g. The source of the energy does not matter; the same amount of heat is required from sunshine as from a coal-fired burner. Only the cost of the heating is different.

Reverse Osmosis

Osmosis is a process in which water flows through a **semipermeable membrane** (figure 12.16). This membrane is usually a sheet or tube with a huge number of tiny channels or pores in it. The pores are so small that only certain molecules can fit through them; the membrane is thus selective by molecular size. When pure water is on both sides of a semipermeable membrane, as in figure 12.16a, the height of water is the same in both tubes or columns. This is an example of physical equilibrium—water flows equally fast in both directions through the membrane. When one chamber contains pure water and the other contains a salt solution, as in figure 12.16b, the salts are unable to flow through the membrane. Therefore, the water flows from the solution of lower salt concentration (pure water) into the solution with a higher concentration (brine). In effect, the water dilutes the more concentrated solution to try to reach equilibrium. The resulting height difference between the two tubes measures the osmotic

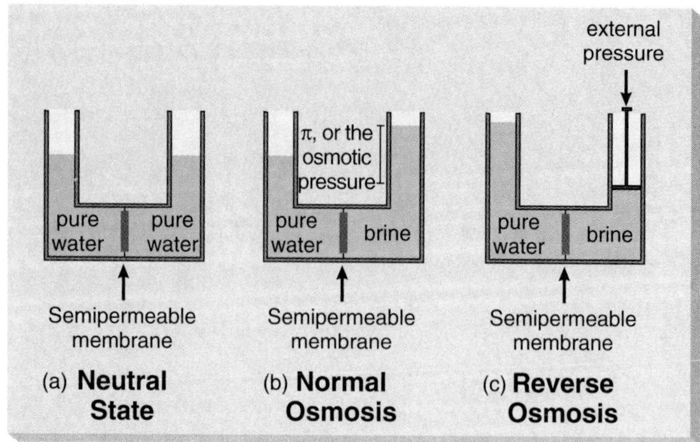

Figure 12.16 Osmosis and reverse osmosis.

pressure, which is represented by the symbol π (pi). The force opposing the osmotic pressure is the weight of the water rising in the tube. At equilibrium, the downward force from the weight exactly matches the upward force of the osmotic pressure, and no further visible change occurs.

You experience the effects of osmotic pressure whenever an insect bites you. The insect injects salts and formic acid into your body tissues, increasing the total salt concentration, changing the pH, and resulting in a stinging response. (Insect bite allergies are normally responses to proteins or other chemicals transmitted in insect bites.) Fluid flows through the body's semipermeable membranes to reduce the higher salt concentration around the location of the bite, resulting in a raised lump. Eventually the diluted salt solution flows away and the lump disappears.

PRACTICE PROBLEM 12.7

Aside from insect bites, what other examples of osmosis can you think of?

ANSWER A few other common osmosis examples include (1) the control of water flowing in and out of cells, (2) seeds rupturing during germination, (3) water flowing up a tree trunk (which is also partially controlled by capillary pressure), (4) controlled release of drugs, (5) wilting of plants in salt water, (6) polymer molecular weight determination, and (7) water purification.

Reverse osmosis (figure 12.16c) uses pressure to force water to flow backward through the semipermeable membrane (that is, from the more concentrated chamber into the less concentrated one). The downward force of the piston drives the water out from the salt solution, through the membrane, and into the pure water side. The salts remain behind because they can not penetrate the membrane. Membrane rupture, because of the high pressures employed, is the greatest problem with reverse osmosis. Membranes are usually made from cellulose acetate reinforced with a mesh made from a strong polymer such as poly(propylene).

This promising desalination method currently costs about \$5.00 to 10.00 per 1,000 gallons of seawater, less for brackish water. With brackish water, reverse osmosis recovers about 50 to 80 percent of the source water; with seawater, the recovery drops to 20 to 50 percent. Bottles of reverse-osmosis-purified water are available at many food stores, and home reverse osmosis units are available at lower cost than the better home distillation systems. These home units are mainly used for water purification rather than desalination. Home and industrial reverse osmosis systems use hollow fiber membranes that permit more rapid purification and reduce the cost. Hollow fiber membranes are also used in dialysis (artificial kidneys).

The city of Santa Barbara, California, built a seawater reverse osmosis desalination unit with a capacity of 6.6 mgd (millions of gallons per day) in 1992. The world's largest reverse osmosis plant, with a capacity of 78 mgd of fresh water from brackish water, operates in Yuma, Arizona. Many other cities will likely follow suit in the future.

Freezing

The ice from frozen seawater contains a lower salt concentration than the remaining liquid. If this ice is separated by filtration, remelted, and refrozen, the purity increases with each step. Eventually, good-quality water results, but it requires 80 cal/g to freeze water in each step. In addition, salt water is very corrosive and damages the purification equipment. In spite of these limitations, a few plants have been constructed that can produce more than 250,000 gallons of potable water daily by freezing and filtering seawater. Freezing is potentially less expensive than distillation because only 15 percent as much energy is required. Melted icebergs might supply some fresh water; in this case, nature does some of the purification, but the potential water supply is erratic.

PRACTICE PROBLEM 12.8

The energy required to obtain fresh water from seawater by freezing is only 15 percent of the energy needed in distillation. Why is this so? Where does the "extra" energy go in distillation?

ANSWER It takes 540 cal/g to convert water to steam, but only 80 cal/g to convert water to ice. Dividing 80 by 540 and multiplying by 100 gives you 15 percent. The "extra" energy in distillation causes the water vapor molecules to move into the gas phase. Both types of phase changes—freezing and boiling—require energy, but distillation requires more energy than freezing.

Electrodialysis

In **electrodialysis,** salt water is placed in a three-chambered electrolysis cell divided by two special semipermeable membranes (figure 12.17a). Each membrane allows either cations or anions to pass through but excludes the other. (These membranes are different from those used in reverse osmosis, which do not permit the passage of salts.) Initially, all three chambers contain Na^+ and Cl^- ions. When an electric current is applied, the Na^+ cations pass through the **cation-permeable membrane** and the Cl^- anions pass through the **anion-permeable membrane** because of the attraction toward the opposite charge. This leaves pure water in the center compartment (figure 12.17b).

Electrodialysis is expensive because the membranes are costly and the process requires a lot of electricity. It is used more often to treat brackish water than seawater; it recovers 80 to 90 percent of the source water.

Ion Exchange

Ion-exchange desalination is similar to water softening, except the polymers replace the seawater cations with H^+ ions and the seawater anions with OH^- ions. Two separate ion-exchange resins are necessary (figure 12.18). Although the H^+ and OH^- ions combine to make water, the amount produced represents only a small percentage of the total water purified. Ion-exchange desalination is very expensive because ion-exchange resins are costly and they require frequent reactivation. This method is used on small, private islands.

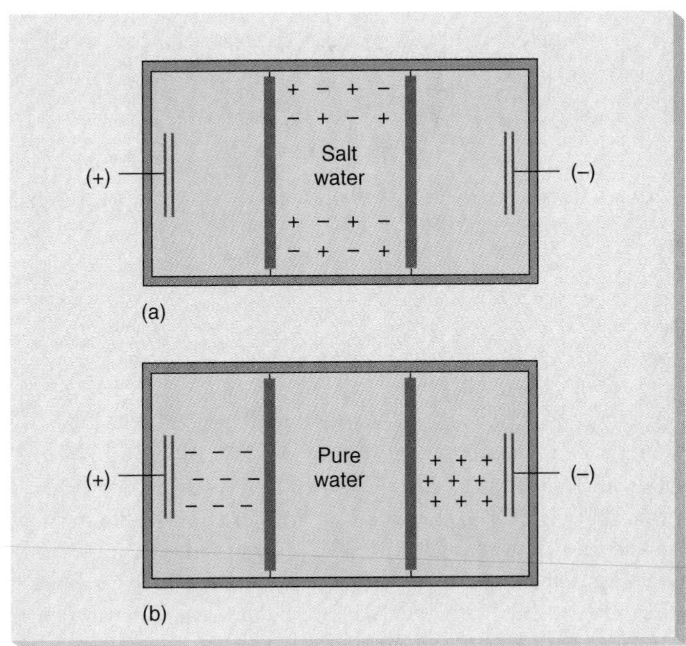

Figure 12.17 The electrodialysis of seawater. **(a)** Before the electrical current is applied. **(b)** After the current is applied. The cations and anions migrate through the membranes and out of the center section, leaving pure water.

$$\text{Polymer—SO}_3\text{H} + \text{Na}^+ \longrightarrow \text{Polymer—SO}_3\text{Na} + \text{H}^+$$

Cation exchange resin

$$\text{Polymer—}\overset{\text{R}}{\underset{\text{R}}{\text{N}^+}}\ \text{OH}^- + \text{Cl}^- \longrightarrow \text{Polymer—}\overset{\text{R}}{\underset{\text{R}}{\text{N}^+}}\ \text{Cl}^- + \text{OH}^-$$

Anion exchange resin

Net forward reaction: $H^+ + OH^- \longrightarrow H{-}OH$

Figure 12.18 Desalinization using ion-exchange resins.

Hydrate Formation

Certain hydrocarbons, such as propane, form a solid hydrate with water molecules when they are solidified in the presence of water. Salt ions are excluded from this solid because they are too large. Hydrates form when water molecules become attached to or entrapped within another substance. Electrostatic or hydrogen bonds generally form between water and the other material (although not with propane). After the hydrate is filtered out, the propane is evaporated, leaving pure water.

Even though the propane is recycled to reduce costs, the **hydrate formation** method of desalination is fairly expensive and not widely used. Propane is also highly flammable, posing the danger of exploding in the desalination plant. It seems unlikely that hydrate formation will ever be the method of choice.

Conclusions on Converting Seawater

Ultimately, many communities will need to convert seawater into fresh water to supply the people's needs. Future generations will probably look back with amazement at our time and wonder why governments even considered restricting population growth in states such as Florida when the solution to continuing water shortages was right on the coastlines. The sooner seawater is utilized on a large scale, the sooner many water supply problems will be solved. Meanwhile, we need to conserve water by reducing unnecessary uses such as lawn watering and car washing, and also by using household water more cautiously.

Many of the largest cities in the United States are within three hundred miles of the coastline. Most of these cities now pipe in fresh water, often from hundreds of miles away. Why not obtain water from the ocean? The barrier is the usual one—money. Solar seawater stills or reverse osmosis plants cost money, and tax revenues are difficult to raise. Current small-scale operations show that harvesting seawater is possible with present technology. Most municipal water systems are underwritten by taxes and bond issues; since this added cost is not as visible to the taxpayer, seawater desalination continues to lag in favor compared to municipal water treatment.

SUMMARY

Water is a unique substance essential to all forms of life. This precious natural resource recycles from the oceans, rivers, streams, lakes, and ponds into the clouds, and then returns as rain, snow, or hail in the water-vapor cycle. Although enough fresh water exists to supply every person on earth, several problems make it difficult to utilize: the water may not be located close to where it is needed; it may not be suitable for human use; and lastly, making it pure enough for human use may be an expensive and problematic undertaking.

About half the people in the United States get their drinking water from groundwater sources. Most ground water is contained in aquifers or less accessible aquitards. Rain and snowfall replenishes the aquifer. Ground water may be polluted by either natural or manmade pollutants. Natural contaminants include salt water, dissolved salts, and minerals. Human-produced pollutants include salt used to remove ice from roads and sidewalks, oil field brines, agricultural runoff, industrial wastes, and landfill materials. Salt and brine leave salts in the water, making it unfit for human consumption; agricultural chemicals can cause eutrophication, a situation in which algae overrun a pond or lake, depleting the oxygen supply in the water and producing foul odors. Industrial wastes—some of them toxic—accumulate in the water supply, and the contents of landfills can leach into the ground water. Most large companies have water treatment systems to reduce industrial waste products, and modern landfills have liners to prevent leakage.

Healthy water contains about 5 to 7 mg of oxygen per liter of water. Decaying organic matter and living creatures consume this oxygen, leaving a polluted body of water depleted; the measure of their consumption is called the biochemical oxygen demand, or BOD. Various wastes may have BODs of anywhere from 150 to 15,000 mg O_2/L, or even higher, and reducing their BODs can be an expensive and complex process. Several water treatment processes exist to help clean contaminants from the water.

Water may be treated through municipal treatment facilities, water softeners, wastewater recovery, or desalination. Municipal treatment plants often use chlorine to kill potentially disease-causing microorganisms. Primary wastewater treatment involves physically screening out solid materials and allowing them to settle in settling tanks; secondary water treatment uses activated sludge to further remove dissolved organic material and suspended matter; and advanced water treatment uses chemicals such as carbon black, alum, and oxidizing agents to eliminate still more impurities. Chlorination is normally performed at the end of the secondary water treatment stage. Home water softeners rely on ion exchange to replace Ca^{++} and Mg^{++} ions with Na^{+} ions. Other home water purifiers include distillation devices, reverse osmosis systems, and carbon filters.

The oceans are an abundant source of water, but seawater must be desalinated to become potable. Some desalination methods include distillation (including solar distillation), reverse osmosis, electrodialysis, freezing, ion exchange, and hydrate formation. Distillation involves heating liquid water until it vaporizes, then recollecting the condensate; solar distillation reduces the cost by using sunlight to heat the water. Reverse osmosis forces salt water through a semipermeable membrane; the salt cannot pass through the membrane, but the water can. Frozen seawater contains a lower salt concentration than the remaining liquid. If seawater is frozen, filtered, and refrozen, the purity increases with each step. In electrodialysis, an electric current attracts Na^{+} and Cl^{-} ions to opposite sides of a three-chambered tank; purified water remains in the center chamber. Ion-exchange desalination is similar to water softening, except the seawater cations are replaced with H^{+} and the seawater anions with OH^{-} ions. Finally, hydrate formation causes a solid hydrate to form, which can then be filtered out to remove impurities. At present, small-scale operations are using some of these methods to desalinate seawater, but because of the expense, desalination continues to lag in favor of traditional municipal water treatment.

KEY TERMS

advanced water treatment 319
anion-permeable membrane 325
aquifer 305
aquitard 305
biodegradable 308
biochemical oxygen demand (BOD) 312
cation-permeable membrane 325
chlorination 314
desalination 321
distillation 321
electrodialysis 325
eutrophication 307
groundwater 305
hard water 305
hydrate formation 326
incineration 309
ion-exchange polymers (resins) 322
osmosis 324
ozonolysis 314
potable water 314
primary water treatment 319
reverse osmosis 324
saltwater intrusion 305
secondary water treatment 319
semipermeable membrane 324
sludge 309
solar distillation 323
turbidity 305
water softening 322
water-vapor cycle 304

READINGS OF INTEREST

Water Availability

Koeppel, Gerard. "A struggle for water." *Invention & Technology*, Winter 1994, pp. 19–30. This article outlines New York City's struggle to establish its public water system. Even though the system was authorized in 1774, it did not provide a water supply until about seventy years later.

Hanson, Gayle. "Can a river run through the open market?" *Insight*, March 21, 1994, pp. 18–20. There are places in the United States where water is scarce, Some suggest it might be more effective to allow entrepreneurs to sell water to these regions.

Morrison, Micah. "Water war swamps Everglades." *Insight*, January 17, 1994, pp. 12–16. Balancing ecology and economic growth is not always easy.

Postel, Sandra. "The politics of water." *Worldwatch*, July/August 1993, pp. 10–18. Will water rights supplant oil and gold as a precious enough resource to fight a war for?

Sisson, Robert F. "Frost, nature's icing." *National Geographic* vol. 149, no. 3, March 1976, pp. 398–405. No chemistry, just some delightful photos of ice crystals and frost.

Water Pollution

"The lichens did it." *Discover,* April 1992, p. 17. Damage to Trajan's Column in Rome appears to be caused by lichens rather than acid rain, as was previously believed.

McDowell, Jeanne. "A race to rescue the salmon." *Time,* March 2, 1992, pp. 59–60. Chinooks and red-sockeyes are favorites of fisherman and restaurant-goers alike. These species cannot really swim away from pollution.

Allen, Herbert E.; Henderson, Mary Ann; and Haas, Charles N. "What's in that bottle of water?" *Chemtech,* December 1991, pp. 738–742. Many people use bottled water in the hope of improving the quality of their drinking water supply. But just how good is the bottled water?

Geiser, Ken. "The greening of industry: Making the transition to a sustainable economy." *Technology Review,* August/September 1991, pp. 65–72. Controlling potentially toxic wastes requires more than just passing a law against their release. Ultimately it must involve changing how our industries operate. Some changes can actually save money!

Marx, Wesley. "Can strip mining clean up its act?" *Reader's Digest,* March 1987, pp. 121–125. Not so far. Many operations are "hit and run."

Drinking Water Purification

Peaff, George. "Water treatment companies embrace expanding international markets." *Chemical & Engineering News,* November 14, 1994, pp. 15–23. Includes a listing of some of the major suppliers in this field.

Coeyman, Marjorie; Roberts, Michael; and Cottrill, Ken. "Water treatment." *Chemical Week,* May 11, 1994, pp. 35–50. A trio of articles about treating water and the nature of the chemicals used. One article considers new legislation in Europe.

Richardson, Susan D. "Scoping the chemicals in your drinking water." *Today's Chemist At Work,* March 1994, pp. 29–32. This article compares water treatment using chlorine, chlorine dioxide, and ozone.

Foran, Jeffery A., and Adler, Robert W. "Cleaner water, but not clean enough." *Issues In Science & Technology,* Winter 1993–4, pp. 33–39. This paper stresses the need for making the water supply even cleaner. One suggested method is to reduce industry's use of toxic chemicals; and another is to reclaim (or recycle) other chemicals. Very little consideration is given to the cost of these ideas, although some examples are cited.

World Resources Institute staff. Water. In *The 1992 information please environmental almanac,* pp. 85–106. Boston: Houghton Mifflin, 1992. A fairly comprehensive article about the sources and purification of water. Includes much useful data.

Kourik, Robert. "Even during droughts, cisterns deliver rain water." *Garbage,* July/August 1992, pp. 43–47. Rain water is potable almost everywhere, and much of the Third World obtains their potable water this way.

Kiesche, Elizabeth S.; Mullin, Rick; Begley, Ronald; and Roberts, Michael. "Water treatment chemicals and services." *Chemical Week,* May 13, 1992, pp. 32–47. Everything you might want to know about a $2.1 billion U.S. industry. Includes many topics not covered in this chapter.

Ember, Lois R. "Clean water act is sailing a choppy course to renewal." *Chemical & Engineering News,* February 17, 1992, pp. 18–24. Good review of earlier legislation and a look at where the future may be heading.

Carpenter, Betsy; Hedges, Stephen; Crabb, Charlene; Reilly, Mark; and Bounds, Mary C. "Is your water safe?" *U.S. News & World Report,* July 29, 1991, pp. 48–55. Certainly the water supply is safer than in the mid 19th century, but is it safe enough for the 21st century? This article considers some of the less obvious sources of pollution.

Church, Vernon M. "Purifying water with sunshine." *Popular Science,* August 1991, pp. 26–27. Reports on a study done at Solar Energy Research Institute, in Golden, Colorado, on detoxifying polluted groundwater.

Hall, Ross Hume. 1990. *Health and the global environment.* Cambridge, Mass.: Blackwell. Hall, a biochemist, argues that improving our environment makes sense from a health perspective. For some, this human-centered approach is more persuasive than one focused on owls or trees. (A short review of this book appears in *Chemical & Engineering News,* December 16, 1991, p. 25.)

Tunley, Roul. "Is your water safe?" *Reader's Digest,* November 1986, pp. 113–118. If you think your drinking water is unsafe, you may want to try some home remedies.

Cholera

Glass, R. I.; Libel, M.; and Branding-Bennett, A. D. "Epidemic cholera in the Americas." *Science,* vol. 256 (June 12, 1992), pp. 1524–1525. Not all cholera epidemics are ancient history; there were over 391,000 cases in the Americas in 1991 (though only 25 were in the United States).

Carey, Benedict. "The old German's double death wish." *Health,* May/June 1992, p. 16. The results of an argument about cholera that took place in 1892.

Diamond, Jared. "The return of cholera." *Discover,* February 1992, pp. 60–66. Just when you thought is was safe to drink the water, cholera returns. This is mainly a problem in Third World countries, but even in developed nations, such diseases may sometimes get out of control.

Wastewater Purification

Jewell, William J. "Resource-recovery wastewater treatment." *American Scientist,* vol. 82, July/August 1994, pp. 366–375. The best wastewater treatment procedures can exhaust a community's financial base and require isolated facilities. This article discusses some newer technologies that use living plants for wastewater treatment. Such facilities can also function as recreational areas.

Mullin, Rick, and Roberts, Michael. "Water treatment." *Chemical Week,* May 12, 1993, pp. 37–47. These three articles deal mostly with wastewater treatment. One compares the effectiveness of ozone and chlorine treatments (chlorine wins).

Winkler, Michael. "Sewage sludge treatments." *Chemistry & Industry,* April 5, 1993, pp. 237–240. Where sludge comes from, and where it goes.

Mackerron, Conrad; Rotman, David; Morris, Gregory D. L.; and Wilkerson, Sophie L. "Water treatment." *Chemical Week,* May 17, 1989, pp. 32–48. This business-oriented article, in considering water treatment, provides much useful data. Did you know, for example, that heavy industry uses more than twice as many water treatment chemicals as municipal water treatment plants do? Add in light industries, commercial, and institutional users, and the ratio surpasses 4:1.

Seawater

Some additional information on desalination can be obtained by contacting the National Water Supply Improvement Association (NWSIA), P.O. Box 102. St. Leonard, MD 20685.

Nuytten, Phil. "Money from the sea." *National Geographic* vol. 183, no. 1 (January 1993), pp. 109–117. Aside from providing a huge amount of mineral wealth and all the potential drinking water humans could ever need, the sea was actually a source of "money" for North American Indians for more than twenty-five hundred years.

Holloway, Margaret. "Abyssal proposal." *Scientific American,* February 1992, pp. 30–36. Why not dump sewage into the ocean? Surely it is big enough to accommodate all human-made garbage without a problem, right? Wrong! Read how life on the ocean floor is affected by pollution.

Kunzig, Robert. "Where the water goes." *Discover,* August 1991, pp. 26–27. How can scientists determine exactly where effluents end up in the ocean? What happens when they arrive? This article tells the story.

Trash and Landfills

Poore, Patricia. "Is garbage an environmental problem?" *Garbage,* November/December 1993, pp. 40–45. Just what role does garbage play in our legitimate concerns about the environment?

White, Peter T. "The fascinating world of trash." *National Geographic* vol. 163, no. 4 (April 1983), pp. 424–457. You might be surprised to find out what ends up in the trash—for example, gold and silver ingots culled from trash in Palo Alto, California. Thousands make their livings by trash-picking while millions continue to produce more fodder.

World Resources Institute staff. "Waste." *The 1992 information please environmental almanac.* Boston: Houghton Mifflin, 1992, pp. 107–128. Where waste comes from and where it goes. Good survey with many tables and figures.

Rathje, William, and Murphy, Cullen. 1992. *Rubbish! The archaeology of garbage.* New York: Harper/Collins. An entire book about garbage—and a good one, since it gives firsthand reports about myths and errors in our understanding.

Granfield, Mary. "Home sweet toxic home." *Money,* June 1992, pp. 124–139. Suppose that a dump (or should we say a landfill) was built near your new and expensive home, and that it started leaking toxic materials into the local groundwater. What would you do?

Rathje, William L. "From Tucson to Armenia: The first principle of waste." *Garbage,* March/April 1992, pp. 22–23. Includes a color photograph of an uneaten, undegraded T-bone steak excavated from the landfill where it lay buried since 1973.

Dunkel, Tom. "The big fill: N.Y. piles it on." *Insight,* December 23, 1991, pp. 14–16, 36–39. With over 8 million people, New York City produces a lot of trash. What do they do with it now, and what will they do in the future?

Rathje, William L. "Once and future landfills." *National Geographic* vol. 179, no. 5 (May 1991), pp. 116–134. Tells where trash now goes and how long it stays intact; shows undegraded newspapers buried more than twelve years, and explodes the myth that paper cups are better than poly(styrene) foam because they degrade.

Grossman, Dan, and Shulman, Seth. "Down in the dumps." *Discover,* April 1990, pp. 36–41. Some good photographs of intact trash salvaged from landfills after more than twenty years burial.

Superfund and Hazardous Waste

Kirschner, Elisabeth. "Love Canal settlement." *Chemical & Engineering News,* June 27, 1994, pp. 4–5. Between 1942 and 1953, chemical wastes were dumped into Love Canal. A black, oily fluid began to ooze from it in 1977, leading to an evacuation of the residents in five hundred houses during 1978. A $98 million legal settlement was reached in 1994, but the story has still not ended.

Hanson, David. "Industrial releases of toxics continue to decline." *Chemical & Engineering News,* April 25, 1994, p. 8. Toxic releases were down 6.6 percent from 1991 but still stood at 3.18 billion pounds, according to EPA figures.

Bartsch, Charles, and Munson, Richard. "Restoring contaminated industrial sites." *Issues in Science & Technology,* Spring 1994, pp. 74–78. The General Accounting Office estimates that 130,000 to 425,000 sites may need cleanup, at an estimated cost of 650 billion dollars.

Regan, Mary Beth; Weber, Joseph; Roush, Chris; and Kelly, Kevin. "Toxic turnabout." *Business Week,* April 25, 1994, pp. 30–31. Could there finally be light at the end of the tunnel for superfund?

Conner, Jesse R. "Stabilizing hazardous waste." *Chemtech,* December 1993, pp. 35–44. Some technical information on how to prevent hazardous waste from becoming a more serious threat.

Kirschner, Elisabeth; Pospisil, Ray; Begley, Ronald; Kemezis, Paul; and Chynoweth, Emma. "Hazardous waste." *Chemical Week,* August 18, 1993, pp. 23–32. A quintet of articles on a timely topic, plus a short piece on the Superfund, from a business aspect. Hazardous waste handling is emerging as a major new business. This set of articles contains a lot of information on incineration.

Chynoweth, Emma, and Anyadike, Nnamdi. "Europe agrees on hazardous waste burner rules, no recycling accord." *Chemical Week*, July 28, 1993, p. 11. Incineration is possible in Europe, if it is done properly.

DiChristina, Mariette. "Cleaning up deadly chemical weapons." *Popular Science*, July 1993, pp. 70–71, 88. Both the United States and the former Soviet Union have huge stockpiles of toxic chemical weapons—a total of more than 65,000 metric tons. Incineration is about the only method that can be used to destroy them.

Kirschner, Elisabeth, and Begley, Ronald. "EPA policy puts incinerators in limbo, tightens furnace requirements." *Chemical Week*, June 2, 1993, p. 9. Obtaining a permit for an incinerator can take more than a year; in the meantime, hazardous wastes continue to accumulate.

Hanson, Gayle. "Superfund supermass: Loaded for bear, EPA hits worms instead." *Insight*, May 3, 1993, pp. 7–9, 34–36. An interesting story about what happened at a superfund site near Burlington, Vermont.

Hanson, David. "Hazardous waste incineration presents legal, technical challenges." *Chemical & Engineering News*, March 29, 1993, pp. 7–14. Roadblocks come from many directions, including residents and activists. The article includes a good diagram of a modern incinerator.

Stover, Dawn. "Toxic avenger." *Popular Science*, July 1992, pp. 70–74, 93. Some microorganisms actually thrive on toxic chemicals. The process that makes use of them is called bioremediation.

Hong, Peter, and Galen, Michele. "The toxic mess called Superfund." *Business Week*, May 11, 1992, pp. 32–34. When both environmentalists and industry say the Superfund plan is a disaster, you know something is amiss. Only 6.75 percent of the 1,245 Superfund sites have been cleaned up—and at an average cost of over $132 million.

Noble, Charles. "Keeping OSHA's feet to the fire." *Technology Review*, February/March 1992, pp. 43–51. Safety standards can take many years to develop, as workers continue to be exposed to toxic chemicals. Between one and three thousand new chemical substances, some toxic, appear in the workplace every year, most are not regulated by OSHA.

Recycling

Scarlett, Lynn. "Recycling rubbish." *Reason*, May 1994, pp. 30–34. This article dwells mainly on the recycling program in Germany, perhaps the most extensive in the world. Unfortunately, much of the material collected is not, in fact, recycled. Over 100,000 tons of scrap plastics are stored in warehouses.

Kleiner, Art, and Dutton, Janis. "Time to dump plastics recycling?" *Garbage*, Spring 1994, pp. 44–51. Most plastics are not actually reused, even when they are collected for recycling. Less than 5 percent of the total sales of most plastics come from recycled products; only about 24 percent of the PET used comes from recycled PET.

Rockwell, Llewellyn H., Jr. "Recycling wastes time and money." *Insight*, April 18, 1994, p. 30. The title describes the viewpoint of this article.

Breen, Bill. "Is recycling succeeding?" *Garbage*, June/July 1993, pp. 36–43. Although the public demands more recycling and collections have increased in recent years, the markets for recycled materials have decreased. (Some materials, such as recycled aluminum, do have a strong market.)

Brady, Catherine. "Plastics recycling moves ahead." *Chemical Week*, March 29, 1989, pp. 20–21. This brief article lists some new companies, often spawned by industry giants, that are active in plastics recycling.

Goldbaum, Ellen. "A new wave of plastics recycling." *Chemical Week*, May 10, 1989, pp. 9, 13. A useful article that gives data on current plastic recycling efforts. Curbside recycling of plastics is useless if the plastics end up stockpiled somewhere. Unless some companies actually acquire and use these materials, the entire process is certain to abort.

The Paper Industry

Coy, Peter, and Sandler, Neal. "First Gutenberg, now Benny Landa?" *Business Week*, July 11, 1994, pp. 143–144. Describes a new method of color printing based on laser technology.

Begley, Ronald. "Chlorine takes a hit in EPA pulp and paper rule." *Chemical Week*, November 10, 1993, p. 9. The EPA wants the use of chlorine drastically reduced.

Regan, Mary Beth. "How much green in 'green' paper?" *Business Week*, November 1, 1993, pp. 60–61. The cost of recycled paper is still high.

Thayer, Ann M. "Paper chemicals." *Chemical & Engineering News*, November 1, 1993, pp. 28–41. Paper making is a multibillion dollar industry that uses million of dollars worth of chemicals, and one of the major chemicals is chlorine. Some changes are pending in this industry.

Ayres, Ed. "Making paper without trees." *Worldwatch*, September/October 1993, pp. 5–8. There are several other sources for paper besides wood pulp. They include straw, bagasse, bamboo, hemp, and kenaf.

Seidman, Ethan; Breen, Bill; and Packert, Bob. "An inside look at paper recycling." *Garbage*, September/October 1993, pp. 30–37. Most communities recycle paper, but where is the recycled product used? There is some demand, even though the cost is still much greater than for virgin paper.

Borchardt, John K. "Paper de-inking technology." *Chemistry & Industry*, April 19, 1993, pp. 273–276. The ink must be removed before recycled paper can be useful.

Chlorine Usage

Graff, Gordon. "The chlorine controversy." *Technology Review*, January 1995, pp. 54–60. A general ban on chlorine is considered unlikely.

Lucas, Allison, Begley, Ronald, and Roberts, Michael. "Health studies raise more questions in chlorine dispute." *Chemical Week*, December 21/28, 1994, pp. 26–27. Not all studies indicate the same magnitude of chlorine problem.

Hileman, Bette; Long, Janice R.; and Kirschner, Elisabeth. "Chlorine industry running flat out despite persistent health fears." *Chemical & Engineering News*, November 21, 1994, pp. 12–26. A full ban on chlorine in the United States or Canada appears unlikely. Nonetheless, recent studies show that *some* chemicals containing chlorine may cause health and environmental problems.

Amato, Ivan. "The crusade to ban chlorine." *Garbage*, Summer 1994, pp. 30–39. A chlorine ban would have devastating effects on industrial processes and products and could lead to a rise in the incidence of water-borne diseases.

Raloff, J. A. "The role of chlorine—And its future." *Science News*, January 22, 1994, p. 59. The U.S. chlorine industry does a $70 billion annual business. Chlorine is used to disinfect more than 98 percent of our drinking water systems, and it is also used to produce more than half of all commercially produced chemicals.

Cottrill, Ken; Roberts, Michael; Kirschner, Elisabeth; and Hunter, David. "Attacks on chlorine gather force." *Chemical Week*, November 3, 1993, pp. 28–31. A trio of short articles. The first tells how industry is mobilizing for the battle. The second details how PVC is the next target of the European stage of this battle, and the final article is a summary of a debate between the Chlorine Chemistry Council and Greenpeace.

Amato, Ivan. "The crusade against chlorine." *Science*, July 9, 1993, pp. 152–54. How chemistry, industry, and scientists are responding to the unprecedented attempt to ban the entire use of an element.

Dioxins

Roberts, Michael. "French Academy rebuffs EPA dioxin report." *Chemical Week*, October 5, 1994, p. 18. The French Academy claims that dioxin hazards are insignificant for humans.

Lucas, Allison. "EPA report heightens concerns about dioxin health risks." *Chemical Week*, September 21, 1994, p. 10. The EPA maintains that dioxins may be exceedingly toxic to humans.

Henderson, Rick. "Blinded by pseudoscience." *Reason*, June 1992, pp. 6–8. Not all environmental activities are well grounded in science. This article discusses some EPA foibles and fables.

Schmidt, Karen F. "Puzzling over a poison." *U.S. News & World Report*, April 6, 1992, pp. 60–61. Problems with dioxin seem to come and go. They appear again in this article, which concentrates on studies other than the Times Beach fiasco. Just what dioxin really does, however, remains at least a partial mystery.

Schmidt, Karen F. "Dioxin's other face." *Science News* vol. 141 (January 11, 1992), pp. 24–27. How safe or how dangerous is dioxin, considering what we now know about the Times Beach evacuation panic?

Felten, Eric. "The Times Beach fiasco." *Insight*, August 12, 1991, pp. 12–19. Times Beach, Missouri, was evacuated in 1983 on the supposition that it was contaminated with dioxin. It now appears this action was unwarranted.

Wilcox, Fred A. 1983. *Waiting for an army to die: The tragedy of Agent Orange*. New York: Vintage Books. An interesting account of the health problems Vietnam veterans experienced due to herbicides used for deforestation in Vietnam.

Problems and Questions

1. What are some of the unique properties or characteristics of water?

2. What is the water-vapor cycle? Why is it important?

3. What percentage of the earth's water supply is fresh water?

4. What is groundwater? Where is it found?

5. How much groundwater exists in the United States? What percentage of the population relies on it for their main source of potable water?

6. What is the difference between an aquifer and an aquitard?

7. List the main U.S. sources of fresh water in descending order of quantity.

8. What substances can cause discoloration in groundwater?

9. What substances might produce foul odors in groundwater or in other fresh water supplies?

10. How do oil fields affect groundwater?

11. What effects can agriculture have on groundwater supplies?

12. Define biodegradability.

13. List at least five items that biodegrade and five others that do not.

14. What is the maximum amount of oxygen present in water at ordinary temperatures? What level of oxygen is normally present in healthy ponds and streams?

15. What is eutrophication? What causes it?

16. How is biochemical oxygen demand (BOD) determined?

17. What effect does BOD have on water supplies?

18. How many liters of a 30,000 mg/L BOD wastewater could be added to a square pond that measures 300 m on each side and is 30 m deep? (Assume the water starts out fully oxygenated.)

19. How does agricultural runoff affect ponds?

20. How does using salt on roads or sidewalks add to groundwater contamination?

21. Why is calcium chloride more effective than sodium chloride in removing ice and snow?

22. Outline the basic features of a municipal drinking water treatment plant, and list the purification steps, both physical and chemical. Are there any steps that are both chemical *and* physical in nature?

23. What concentrations of salts are commonly found in ordinary drinking water? What is its pH?

24. What is turbidity? How is it removed from water?

25. Write an equation that shows how chlorination kills microorganisms. Show what is oxidized and what is reduced.

26. What chemicals are used to remove odor and color from drinking water?

27. Write an equation showing how odor-causing substances are removed from our drinking water. Label which chemicals are oxidized and reduced.

28. What types of devices enhance water quality in the home? How is chemistry involved in each device?

29. What is the difference between hard and soft water?

30. How can hard water be treated or softened?

31. What is ion exchange? What are the two major kinds of ion-exchange membranes?

32. What is involved in primary wastewater treatment? Are these steps chemical, physical, or both?

33. What is involved in secondary wastewater treatment? Are these steps chemical, physical, or both?

34. What differences exist between secondary and advanced wastewater treatments?

35. List the various processes by which fresh water can be obtained from seawater. Which process do you believe shows the greatest potential? Why?

36. Describe how the distillation process for harvesting fresh water from seawater works.

37. How does solar distillation differ from ordinary distillation?

38. What is osmosis? List at least five different examples of osmosis in action.

39. What is reverse osmosis, and how is it used?

40. What are hollow fiber membranes? Cite two uses for these membranes.

41. What are hydrates? How might they be used in water purification?

42. Ion-exchange resins can be used in both water softening and desalination. How do these two applications differ?

CRITICAL THINKING PROBLEMS

1. Water is essential for life. Can too much water prove harmful to plant life? Why?

2. Why does a person dehydrate more rapidly by drinking seawater than by not drinking any water at all?

3. Some possibly toxic chemicals were discovered to be leaking from drums buried many years ago at a chemical plant landfill. Would a survey of the current health problems in nearby communities reveal anything about potential problems caused by these chemicals? How should the survey questions be phrased to avoid bias? What are the limitations of such a study?

4. Read the following two articles: Conrad B. Mackerron, "EPA Under Attack on Groundwater Program," *Chemical Week*, May 10, 1989, p. 8; and Lois R. Ember, "Clean Water Act is Sailing a Choppy Course to Renewal," *Chemical & Engineering News*, February 17, 1992, pp. 18–24. Then read some recent news items on this same topic. Has the situation changed significantly over the past several years?

5. The Superfund was established to rid our country of toxic waste sites, but unfortunately, it has not succeeded. Read Peter Hong and Michele Galen, "The Toxic Mess Called Superfund," *Business Week*, May 11, 1992, pp. 32–34. Assume you have the decision-making power to correct this mishandled mess. What would you plan to do?

6. Reread box 12.1 plus the following three articles: Karen F. Schmidt, "Dioxin's Other Face," *Science News*, January 11, 1992, pp. 24–27; Karen F. Schmidt, "Puzzling Over a Poison," *U.S. News & World Report*, April 6, 1992, pp. 60–61; and Rick Henderson, "Blinded by Pseudoscience," *Reason*, June 1992, pp. 6–8. If possible, locate and read some articles about the use of Agent Orange during the Vietnam war. Make your own assessment on the dioxin problem. Were the actions at Times Beach, Missouri, based on factual information or misinformation? If the latter, how might such actions be avoided in the future? Suppose something similar happened near your hometown. What would you do about it?

7. Reread box 12.3 and at least three of the following articles: Gordon Graff, "The Chlorine Controversy," *Technology Review*, January 1995, pp. 54–60; Allison Lucas, Ronald Begley, and Michael Roberts, "Health Studies Raise More Questions in Chlorine Dispute," *Chemical Week*, December 21/28, 1994, pp. 26–27; Bette Hileman, Janice R. Long, and Elisabeth Kirschner, "Chlorine Industry Running Flat Out Despite Persistent Health Fears," *Chemical & Engineering News*, November 21, 1994, pp. 12–26; Ivan Amato, "The Crusade to Ban

Chlorine," *Garbage*, Summer 1994, pp. 30–39; and J. A. Raloff, "The Role of Chlorine—And Its Future," *Science News*, January 22, 1984, p. 59. Then answer the following questions:

a. What are the benefits and drawbacks society derives from chlorine?

b. Are substitute materials really available to replace most chlorine-containing compounds?

c. Is a total ban on chlorine a logical and reasonable course of action?

d. What effect, if any, would a ban on chlorine have on public health?

8. Solar desalination was discussed in this chapter. Read the article by Vernon M. Church in *Popular Science*, August 1991, pp. 26–27, that tells about "Purifying Water With Sunshine." Is this approach feasible for most areas of the United States?

9. Read the following three articles about cholera: Jared Diamond, "The Return of Cholera," *Discover*, February 1992, pp. 60–66; Benedict Carey, "The Old German's Double Death Wish," *Health*, May/June 1992, p. 16; and R. I. Glass, M. Libel, and A. D. Branding-Bennett, "Epidemic Cholera in the Americas," *Science*, June 12, 1992, pp. 1524–1525. Is it possible for a cholera epidemic to occur in the United States? What changes in our lifestyle might lead to one? What would be the consequences of such an epidemic? What would you do if it did happen?

10. How would you handle the trash problems in a large city such as New York? For background, read the article by Tom Dunkel in *Insight*, December 23, 1991, pp. 14–16, 36–39, called "The Big Fill: N.Y. Piles It On."

11. As noted in this chapter, more than 2.6×10^{11} pounds (1.2×10^{11} kg) of solid wastes from household, industrial, and municipal sources end up in U.S. landfills every year. What is the volume of this waste, assuming a density of 1.0 g/mL? What might be the total volume of trash after twenty-five years? How do you view the trash and landfill problem? Is it as drastic as some environmental groups claim, or is there more than enough potential landfill space available, as other groups suggest? (Remember that all projections are based on landfill areas now available. More sites could become available in the future.)

12. Landfills are likely to be with us for many years to come. What really happens in a landfill? Read these two articles by William L. Rathje, an expert on landfills: "Once and Future Landfills," *National Geographic*, vol. 179, no. 5, May 1991, pp. 116–134; *and* "From Tucson to Armenia: The First Principle of Waste," *Garbage*, March/April 1992, pp. 22–23. If you have time, also read the book by Rathje and Cullen Murphy called *Rubbish! The Archaeology of Garbage*, New York: Harper/Collins, 1992. Why do substances fail to biodegrade in a landfill? Can new techniques change this?

13. Recycling has been held up as a benefit to society with few or no risks or disadvantages, but not everyone agrees with this assessment. Read at least two of the following articles: Lynn Scarlett, "Recycling Rubbish," *Reason*, May 1994, pp. 30–34; Art Kleiner and Janis Dutton, "Time to Dump Plastics Recycling?" *Garbage*, Spring 1994, pp. 44–51; Llewellyn H. Rockwell, Jr., "Recycling Wastes Time and Money," *Insight*, April 18, 1994, p. 30; and Bill Breen, "Is Recycling Succeeding?" *Garbage*, June/July 1993, pp. 36–43. Then answer the following questions.

a. Are recycling programs working as well as claimed, or are they uneconomical drains on a community's resources?

b. Are all recycled materials reused, or are many stored aboveground, instead of in a landfill?

c. What recyclable materials actually find a market in our society?

d. Where do recycled plastics go?

e. How can recycling be made more economically feasible?

f. What alternatives exist, and are these more economical?

g. Will these alternatives cause massive landfill problems in the future?

14. Both chlorination and ozonolysis have their supporters as the method of choice in municipal water purification. Both oxidizing agents are also toxic. Ground level O_3 is a common air pollutant that causes severe eye, skin, and respiratory problems; Cl_2 does the same and produces trace amounts of cancer-causing agents in the water. Ozonolysis is a newer method that many believe is less toxic (though many studies concentrate on water purification and do not consider the potential damage of ground-level ozone; see chapter 18). Chlorination has a 90-year history as a water treatment and its side effects seem less severe than some claim. Weigh the potential use of each chemical and form at least a tentative opinion on which one appears safer for use in water purification.

15. Campaigns are sometimes directed at ending water chlorination for a variety of reasons, including cost (though costs are low), toxicity (which is high), and the fact that chlorination sometimes creates other hazardous chemicals which may cause cancer in laboratory animals (though at concentrations far greater than occur in the water). Suppose a campaign succeeded and chlorination was banned. Would the incidence of typhoid fever be likely to rise? Low levels of some cancer-causing chemicals do appear in the potable water supplies for many cities. How would you balance the risk of getting cancer against the potential of a typhoid fever epidemic?

HINT You would be comparing an almost certain increase in the death rate from typhoid against the *possibility* of an increase in cancer.)

16. How pure should wastewater be, in your opinion? Should it be potable, or is less purification adequate? Remember that *you* must pay the cost of the treatments in your water bills. Give the reasons for your answer.

HINT Some factors to consider include: (1) the potential of spreading disease, (2) the cost, (3) the fact that your wastewater eventually becomes someone's potable water source, and (4) the nondrinking uses for reclaimed water.)

17. Some people oppose water softening because it removes calcium and magnesium ions the body needs. What is your opinion on this? Why?

CHAPTER 13

SOAPS, DETERGENTS, AND CLEANING

OUTLINE

KEY IDEAS

1. The Structures that Make Molecules Behave as Surfactants
2. The Four Categories of Surfactant Molecules
3. The Chemical Nature of Anionic Surfactants
4. The Chemical Nature of Nonionic Surfactants
5. The Chemical Nature of Cationic Surfactants
6. How Surfactants Affect Detergency
7. The Main Constituents of a Laundry Detergent
8. What Phosphates Do in Laundry Detergents
9. What Can Replace Phosphates in Detergents
10. The Difference Between Standard and Compact Laundry Detergents
11. The Main Ingredients in Shampoos and Hair Rinses
12. The Main Ingredients in Toothpaste

"Cleanliness is, indeed, next to godliness."
JOHN WESLEY (1703–1791)

Interference patterns on soap bubbles.

Children may find it hard to keep clean, but most adults realize that cleanliness is a benefit and that soaps and detergents make cleanliness possible. In the old days, people made their own soaps and cleansers from fats and lye (NaOH). Today, a quick stroll through a supermarket reveals a staggering choice of cleansers. The shelves are stocked with solid bars, liquids, powders, and pellets; some blends produce lots of suds and some almost none; some are labeled environmentally friendly, and others carry poison symbols. Soap and detergent formulations exist to clean almost anything, from our bodies to our homes, cars, and factories. The total sales of soaps and detergents reached $14.9 billion in 1993 and remained at the same level in 1994. This constitutes about 14 percent of all supermarket sales in the United States.

SURFACTANTS

Soaps and synthetic detergents, the critical components of cleansers, are called **surface active agents** or **surfactants** because they reduce the surface tension of water and allow water to displace dirt or grease (see chapter 2). Surfactant production in the United States in 1993 totaled about 3.1×10^9 kg (6.9 billion pounds, or 3.14×10^6 tons).

Surfactants consist of molecules with both a polar and a nonpolar end. When these portions are balanced properly, the surfactant has good surface active properties. The polar portion is **hydrophilic** (water-loving) and the nonpolar, hydrocarbon end is **hydrophobic** (water-hating or fearing). The polar end dissolves in water, the nonpolar end in oily substances such as fats and lipids.

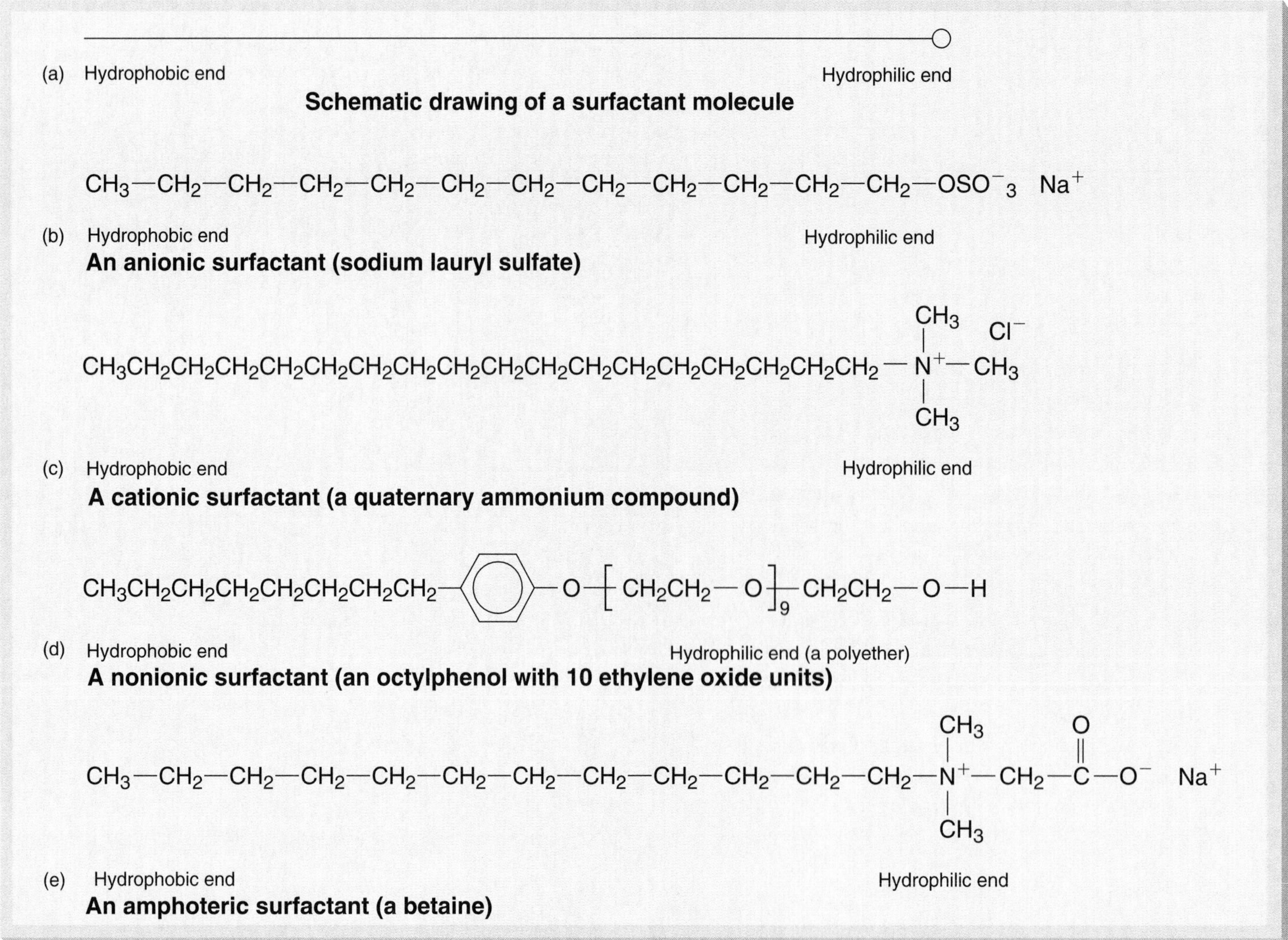

Figure 13.1 Schematic diagram of a surfactant and an example of each class. Notice that the charge attached to the hydrophobic portion in (a) and (b) determines whether the surfactant is anionic or cationic.

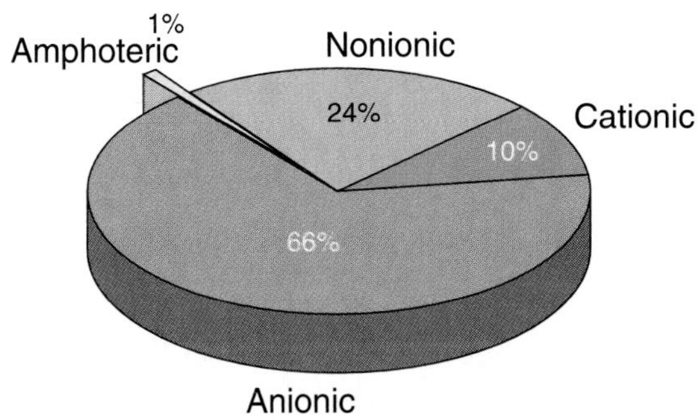

Figure 13.2 Distribution of surfactants among the four classes.

Chemists class surfactants as anionic, cationic, nonionic, or amphoteric, depending on the charge of the nonpolar part of the molecule (figure 13.1). The terms *anionic* and *cationic* describe a hydrophobic end with either a negative or positive charge, respectively. Most surfactants are **anionics** (figure 13.2) since they have a negative charge on the hydrocarbon portion of the molecule. In sodium lauryl sulfate, for example, a 12-carbon atom chain is the hydrophobic portion; this chain is attached to an anionic sulfate group (figure 13.1b). The Na^+ cation and the sulfate group form the hydrophilic portion. By contrast, when the hydrophobic portion is positively charged, the surfactant is **cationic** (figure 13.1c). **Nonionic** surfactants, as in figure 13.1d, have no charge; **amphoteric** surfactants contain both a positive and a negative charge on the hydrophobic part of the molecule (figure 13.1e).

Table 13.1 shows the structures and the 1992 production data for the main types of surfactants. About 1,200 different chemical compounds, made by at least 170 companies, are currently used as surfactants. Many of these 1,200 materials are represented by the classes listed in table 13.1. Most differ only in the nature of the hydrocarbon group they include, and many are mixtures of several hydrocarbon groups. Part of the reason anionics make up such a large share of the market is cost; anionics average about $0.30/pound, while nonionics run about $0.50/pound. Cationics cost about $1.00/pound, and amphoterics sell for about $1.07/pound.

Soap—The Old Standard

Soap, a carboxylic acid salt, remains one of the most commonly used surfactants. Figure 13.3 illustrates the preparation of soap from tristearin, an ester of glycerol containing three stearic acid units. Typical **fatty acids,** of which stearic acid is an example, are organic acids with between 12 and 22 carbon atoms. When they are treated with a strong base, such as NaOH, they form glycerol and sodium salts of the fatty acids (soap). The soap-forming reaction is called **saponification.** Glycerol, the reaction byproduct, is useful for a variety of purposes; therefore, very little glycerin is left in the soap. Commercially prepared soap is almost always made from a variable mixture of fatty acids.

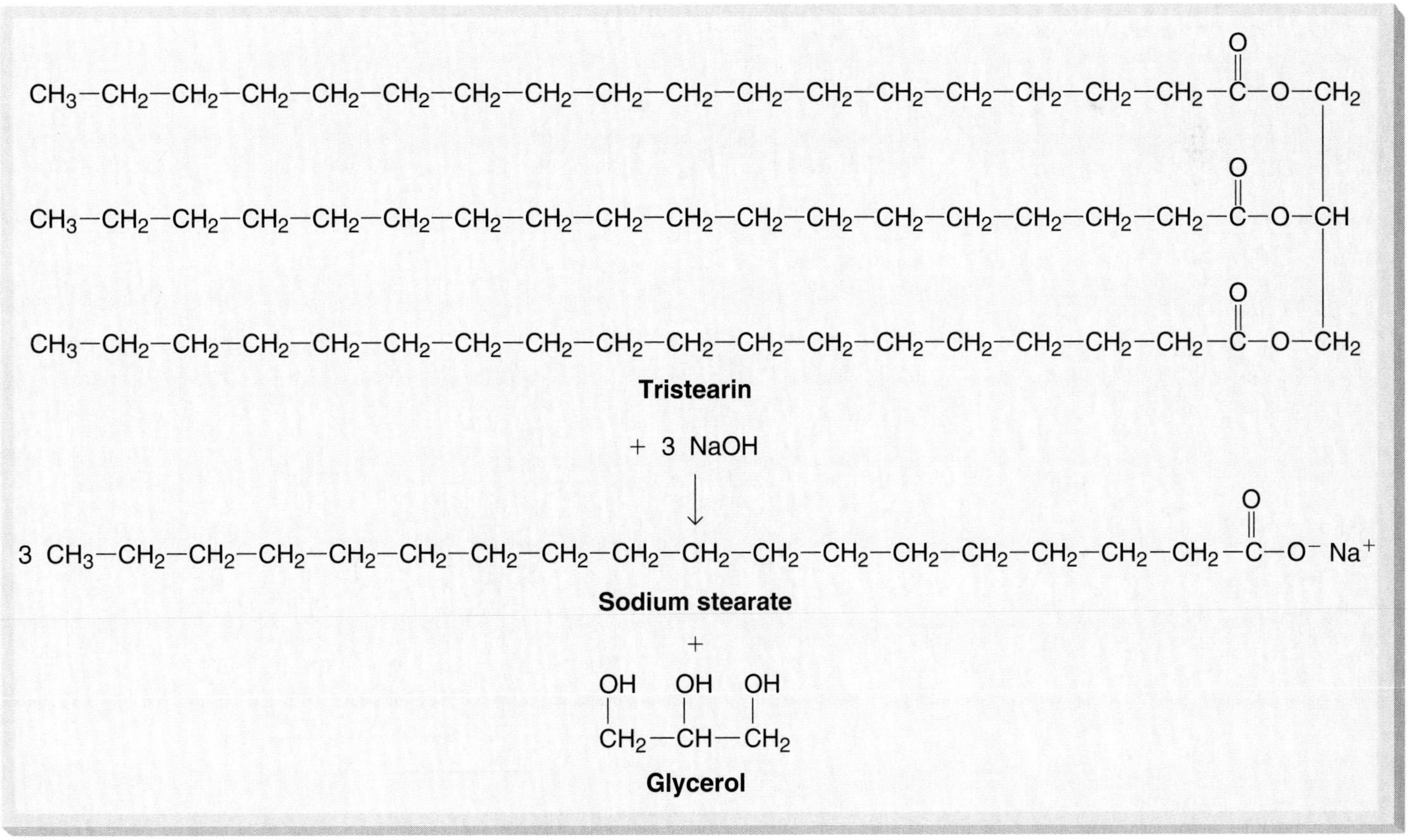

$$\begin{array}{l} CH_3-CH_2-CH_2-CH_2-CH_2-CH_2-CH_2-CH_2-CH_2-CH_2-CH_2-CH_2-CH_2-CH_2-CH_2-CH_2-CH_2-\overset{O}{\overset{\|}{C}}-O-CH_2 \\ CH_3-CH_2-CH_2-CH_2-CH_2-CH_2-CH_2-CH_2-CH_2-CH_2-CH_2-CH_2-CH_2-CH_2-CH_2-CH_2-CH_2-\overset{O}{\overset{\|}{C}}-O-CH \\ CH_3-CH_2-CH_2-CH_2-CH_2-CH_2-CH_2-CH_2-CH_2-CH_2-CH_2-CH_2-CH_2-CH_2-CH_2-CH_2-CH_2-\overset{O}{\overset{\|}{C}}-O-CH_2 \end{array}$$

Tristearin

$$+\ 3\ NaOH$$

$$\downarrow$$

$$3\ CH_3-CH_2-CH_2-CH_2-CH_2-CH_2-CH_2-CH_2-CH_2-CH_2-CH_2-CH_2-CH_2-CH_2-CH_2-CH_2-CH_2-\overset{O}{\overset{\|}{C}}-O^-\ Na^+$$

Sodium stearate

$$+$$

$$\begin{array}{ccccc} OH & & OH & & OH \\ | & & | & & | \\ CH_2 & - & CH & - & CH_2 \end{array}$$

Glycerol

Figure 13.3 The preparation of soap from tristearin.

TABLE 13.1 Surfactant structures and 1992 production data.

NAMES	STRUCTURES		AMOUNT PRODUCED ($\times 10^9$ lbs.)
ANIONICS			**4.52**
Carboxylic acid salts	$R{-}C({=}O){-}O^-Na^+$	$R = C_{11-17}$	0.90
Alkylbenzene sulfonates	$R{-}C_6H_4{-}SO_3^-Na^+$	$R = C_{10-13}$	2.67
Alkane sulfonates	$R{-}CH(CH_3){-}SO_3^-Na^+$	$R = C_{10-17}$	—[1]
α-Olefin sulfonates	$CH_3{-}(CH_2)_x{-}CH{=}CH{-}(CH_2)_y{-}SO_3^-Na^+$	$x + y = 9–15$	—[1]
Fatty alcohol sulfates	$R{-}CH_2{-}O{-}SO_3^-Na^+$	$R = C_{11-17}$	0.95
Fatty oil sulfonates	variable structures, but like FAS		—[2]
Oxo-alcohol ether sulfates	$R{-}CH_2{-}O{-}(CH_2{-}CH_2{-}O)_n{-}SO_3^-Na^+$	$R = C_{12-14}$; $n = 1–4$	—[2]
NONIONICS			**1.68**
Alkylphenol ethoxylates	$R{-}C_6H_4{-}O{-}(CH_2{-}CH_2{-}O)_n{-}H$	$R = C_{8-12}$; $n = 5–15$	1.22
Fatty or oxo-alcohol ethoxylates	$R{-}CH_2{-}O{-}(CH_2{-}CH_2{-}O)_n{-}H$	$R = C_{6-18}$; $n = 3–15$	—[3]
Ethylene oxide: propylene oxide copolymers	$H(OCH_2CH_2)_x{-}(OCH_2CH(CH_3))_y{-}(OCH_2CH_2)_z{-}OH$	$x = 2–60$; $y = 15–80$	0.49
Fatty alcohol polyglycol ethers	$RO{-}(CH_2{-}CH_2{-}O)_x{-}(CH_2{-}CH(CH_3){-}CH_2{-}O)_y{-}H$	$R = C_{8-18}$; $x = 3–6$; $y = 3–6$	—[4]
CATIONICS			**0.66**
Quaternary ammonion salts	$R{-}N^+(CH_3)_2{-}R'$	$R, R' = C_{16-18}$	0.33
Amine oxides	$R{-}N^+(OH)(CH_3){-}CH_3$	$R = C_{12-18}$	0.17
AMPHOTERICS			**0.05**
Alkyl betaines	$R{-}N^+(CH_3)_2{-}CH_2{-}C({=}O){-}O^-Na^+$	$R = C_{12-18}$	

Notes:
[1]Included in alkylbenzene sulfonate total
[2]Included in fatty alcohol sulfate total
[3]Included with alkylphenol derivatives
[4]Included with the ethylene:propylene oxides
SOURCES: Data from *Chemical & Engineering News*, January 25, 1988 and January 24, 1994; and *Chemical Week*, January 26, 1994.

When soap is used in hard water, it forms insoluble calcium, magnesium, or barium salts that interfere with the soap's cleaning action. This problem can be alleviated by water softening (chapter 12), but this is an expensive solution. Along with failing to clean in hard water, soap will *not* function in acidic solutions because it forms insoluble fatty acids that are not surfactants. Nor does soap work well in seawater because the high level of Na^+ ions in the water reduces its solubility.

EXPERIENCE THIS

You can demonstrate surface tension several ways. Place a loop of string on the surface of water in a dish; then touch a bar of soap to the water in the center of this loop. The loop stretches outward toward the sides of the dish because the surface tension was reduced inside the loop, and the higher outside tension pulls the string toward the edges. Now place a needle or metal paper clip on the surface of the water. It floats because the surface tension holds it up. When this surface tension is reduced with the addition of a little soap or detergent, the metal object promptly sinks. For one final demonstration, make a flat paper boat, as in the following diagram, with an open slot at the back end. Place a tiny chip of soap at the front end of this slot. The boat should be propelled across the water. Afterwards, try placing a small amount of liquid detergent in the slot. Does this work as well as the soap?

Detergents—Chemistry Creates New Surfactants

The problems with soap led chemists to develop new molecules patterned after soap's basic structure (figure 13.4a shows the chemical formula for soap). The first synthetic surfactants were developed commercially in the 1930s as fatty alcohol *sulfates* (figure 13.4b). Their synthesis is expressed as follows:

$$R{-}CH_2{-}OH + H_2SO_4 \rightarrow R{-}CH_2{-}O{-}SO_3{-}H + HOH$$

$$R{-}CH_2{-}O{-}SO_3{-}H + NaOH \rightarrow R{-}CH_2{-}O{-}SO_3^- Na^+$$

The alkylbenzene *sulfonates* (figure 13.4c and d), which followed a few years later, are now the most widely produced synthetic surfactants because the starting alkylbenzene materials are less expensive than fatty alcohols. Their synthesis is outlined in the following equation:

$$R{-}C_6H_4{-}H + H_2SO_4 + SO_3 \rightarrow R{-}C_6H_4{-}SO_3{-}H$$

$$R{-}C_6H_4{-}SO_3{-}H + NaOH \rightarrow R{-}C_6H_4{-}SO_3^- Na^+$$

Both of these surfactants—the fatty alcohol sulfates and the alkylbenzene sulfonates—are soluble in hard water and can work under acidic conditions or in seawater. The inexpensive alkylbenzene sulfonates, which are anionics, rapidly became the main surfactant used in powdered laundry detergents. Box 13.1 explains how these detergents work.

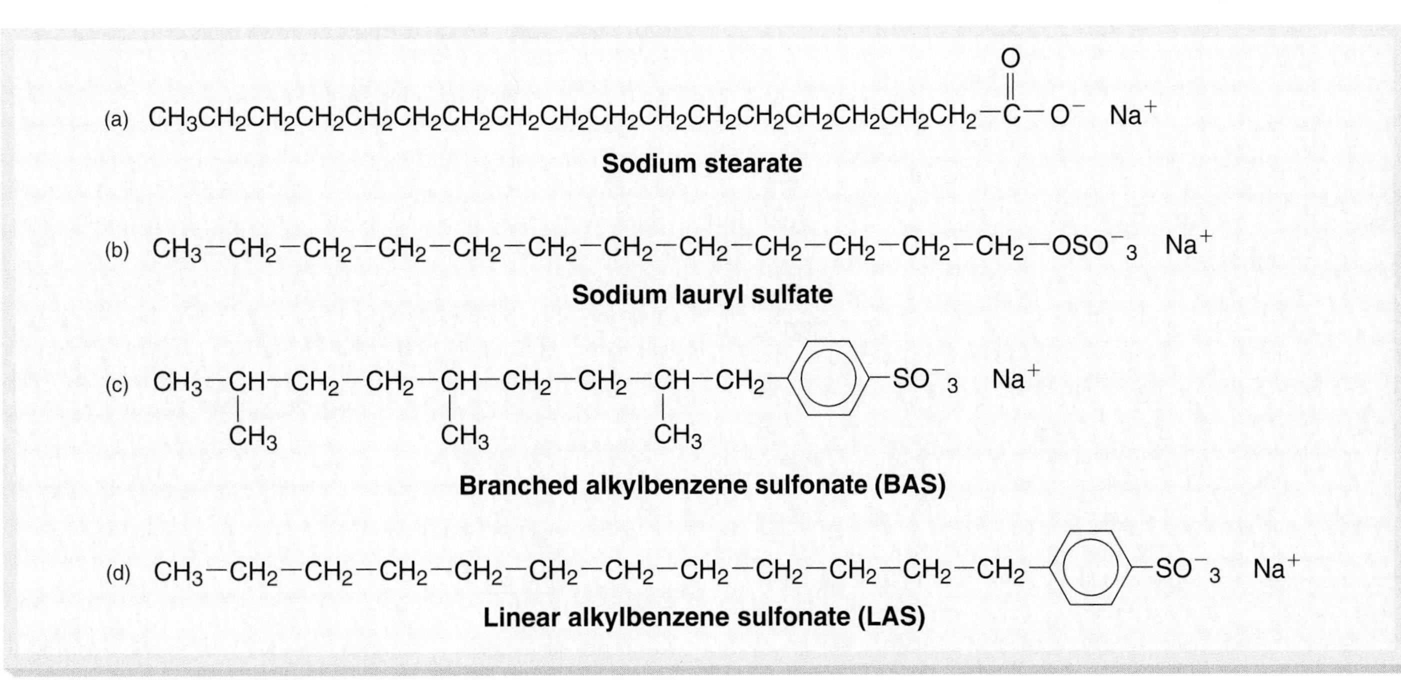

Figure 13.4 Structures of some anionic surfactants.

EXPERIENCE THIS

Soap has some useful properties that are unrelated to cleaning. For example, soap is a good lubricant; you can run a bar of soap on the edges of a drawer to make it slide more easily. In the same way, soap helps a hammered nail penetrate wood and assists screws rotating into nuts. If you make a paste from soap and water, you can use it to decorate windows or mirrors and easily wash it off, and soapy water is used to detect gas bubbles escaping through a leak. If you are a camper who cooks over an open fire, try rubbing the bottom of your pots with a bar of soap. This will keep soot from building up.

PRACTICE PROBLEM 13.1

Table 13.1 lists some surfactants called sulfonates and other called sulfates; the formation reactions appear in the section on detergents. What is the difference between a sulfate and a sulfonate?

ANSWER First, sulfates contain one more O atom than the sulfonates. In addition, the sulfate group is attached to the organic fragment via this O atom, whereas the sulfonate is attached by the S atom. This means the sulfates are *esters* of sulfuric acid; the sulfonates are not esters and are analogous to the carboxylic acids.

PRACTICE PROBLEM 13.2

Why are the newer surfactants more expensive than soap?

ANSWER Notice from figure 13.3 that soap formation (saponification) consists of a single, easy-to-run reaction on a natural fatty acid ester that is easy to obtain. Ordinary people with no knowledge of chemistry made soap in their homes not many years ago using fireplace ashes to provide the lye (NaOH) and common animals fats as the fatty acid esters. The blend was commonly called Grandma's lye soap. The reactions involved in the preparation of sulfates and sulfonates use more expensive chemicals and require two different steps. Multistep reactions are normally more difficult and more expensive to run. Grandma could not have made these surfactants, unless she happened to be a chemist.

PRACTICE PROBLEM 13.3

Examine table 13.1 carefully. Which part of each molecule would stick out of the water surface in a solution? Which part would be on the outside portion of a micelle, and which would penetrate an oil droplet or film?

ANSWER The hydrophobic hydrocarbon portion of each molecule will protrude from the water surface and will also be inside the micelle. The hydrophobic hydrocarbon tails penetrate the oil because the materials are similar chemically. The hydrophilic part will stick down into the water and be on the outside of a micelle.

EXPERIENCE THIS

Determine for yourself why synthetic detergents are popular. You will need the following materials: (1) an ordinary soap bar, (2) liquid dishwashing detergent, (3) shampoo, (4) calcium chloride ($CaCl_2$, which is used for melting ice and snow), (5) ordinary table salt (NaCl), (6) margarine, (7) some dark-colored dirt, (8) vinegar and (9) twelve 2×2-inch pieces of identical white cloth. Blend the margarine and dark dirt together to make a uniform paste. Apply some of this oily dirt to each of the twelve pieces of cloth, rubbing it in. Then prepare the following solutions: (A) 300 mL distilled water, (B) 300 mL tap water + 6 tablespoons $CaCl_2$, (C) 300 mL tap water + 6 tablespoons NaCl, (D) 200 mL tap water + 100 mL vinegar. Divide each solution into three equal parts and label the containers A-1, A-2, A-3 for the distilled water, B-1, B-2, B-3 for the $CaCl_2$ solution, and so on. Add small amounts of soap shavings to containers A-1, B-1, C-1, and D-1. Add about the same volume of dishwashing detergent to containers A-2, B-2, C-2, and D-2. Add an equal volume of shampoo to containers A-3, B-3, C-3, and D-3. Mix all the solutions thoroughly but try not to generate any foam. Note any precipitate or cloudiness that appears in these containers. Now wash a cloth square in each solution, using the same amount of time and effort cleaning each. (The easiest way to do this is to shake each container gently from side to side without generating foam.) Which detergents were effective in each solution? How did the soap work? Which worked better, dishwashing detergent or shampoo?

The Nonionic Solution

Nonionic surfactants offer an alternate method for cleaning in hard water; since they have no ions, nonionics are not affected by hard water ions (Ca^{++}, Mg^{++}, or Ba^{++}), by acidity, or by rela-

BOX 13.1

HOW DETERGENTS WORK

A surfactant molecule has a split personality—its hydrophilic portion loves water, while its hydrophobic part hates water. Because of this, the hydrophobic end tries to escape the water by sticking out of the solution, and the hydrophilic head remains in solution. This causes an accumulation of hydrophobic hydrocarbons on the surface, which reduces the surface tension.

Once the surface of the solution is saturated (filled) with surfactant molecules, the excess surfactant molecules collect to form spheres or droplets called **micelles** (figure 1). In these micelles, the surfactants arrange themselves so that their water-hating hydrocarbon tails converge in the center of the sphere and their water-loving hydrophilic heads radiate to the outside, like the seeds in a dandelion puff. The hydrocarbon tails then associate only with other hydrocarbon tails, while the hydrophilic heads (for example, the sodium carboxylate group in soap) surround and protect them from the water molecules. *Micelle* comes from a Latin word meaning crumb or grain. The tiny micelles are only about 0.00001 mm (10^{-8} meters), but they are the key to detergency and the main reservoir of surfactant in water solutions.

Recall that the polar, hydrophilic end of a surfactant molecule dissolves in water, while the nonpolar, hydrophobic end dissolves in oily substances. If some oily dirt is present in the solution, the surfactant molecules stick their nonpolar, hydrophobic tails into the oil droplets, leaving their polar, hydrophilic heads on the outside in the water phase. Everyone knows oil and water do not mix and tend to separate into layers. However, surfactant molecules surround or coat each oil droplet with a layer of polar heads, anchored in place by the nonpolar tails embedded in the oil droplet. These polar heads carry a negative charge (from the $—COO^-$ or $—SO_3^-$ groups on the surfactant molecules) that causes the oil droplets to repel each other. Rather than coalescing to form a layer on the surface, the oil now forms separate emulsion droplets. In other words, the oil becomes emulsified.

When an object soiled with oily or greasy dirt is washed, the surfactant pries off the oil in a process called **wetting.** In detergency, the surfactant wets the dirty materials and then emulsifies any oil or grease present. This emulsified oil is rinsed away in the wash water.

Not all dirt is oily. Dirt includes water-soluble or insoluble salts and other insoluble matter. Soluble salts (obviously) dissolve and are thus easily removed. Insoluble, non-oily matter is dislodged from a surface by the wetting action of the surfactant. Surfactant molecules then coat this dirt as they do oil droplets and suspend the insoluble dirt in the water. The suspended dirt is also rinsed away with the wash water.

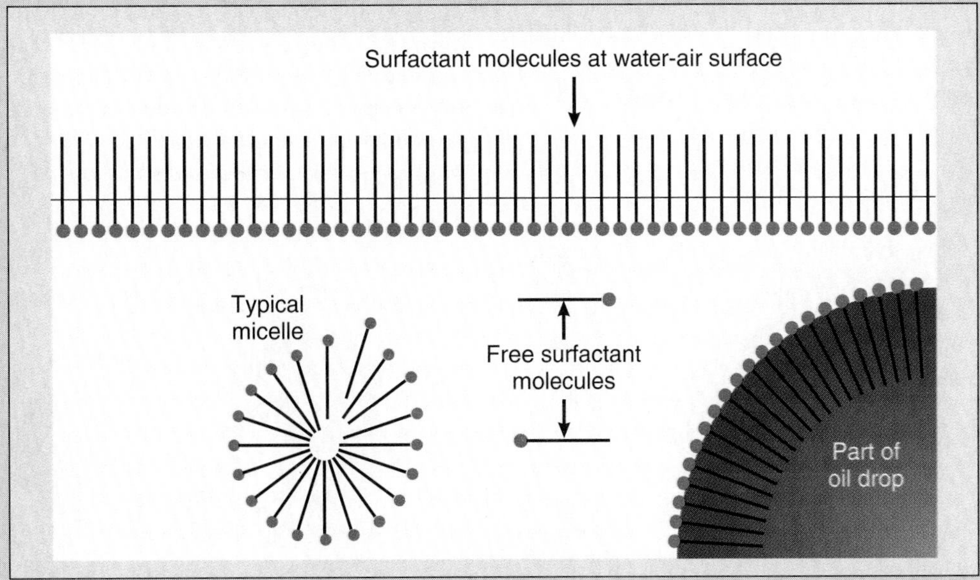

Figure 1 The detergency process, showing free surfactant molecules, surfactant molecules lined up at the air-water interface, a micelle, and surfactant adsorbed on the surface of an oil drop. In this figure, the —— portion is hydrophobic while the ● is the hydrophilic portion of the surfactant molecule.

tively high salt levels. Nonionics include a polyether chain derived from ethylene oxide:

$$C_8H_{17}—C_6H_4—OH + x\ CH_2—CH_2 \text{ (epoxide, O bridging) } \rightarrow C_8H_{17}—C_6H_4O\left[CH_2—CH_2—O\right]_x H$$

The polyether chain acts as the hydrophilic portion of the molecule, counterbalancing the **lipophilic** (fat- or oil-loving) alkylphenol portion. Water molecules are attracted to the ether linkages and draw the surfactant into the solution.

Nonionics do have their drawbacks. They cost 50¢/pound, about twice as much as some anionics. In addition, most nonionics are not biodegradable unless they are made from the more expensive linear alkylphenol molecules or fatty alcohols. However, nonionics dominate in liquid laundry detergents because most are liquid and formulate more readily with the other ingredients in liquid detergents than the solid anionic surfactants.

Cationics Tame Static

Although used less often than the anionics or nonionics, cationics have their own niche among surfactants. They dominate in antistatic applications such as hair conditioners and fabric softeners. Hair and fabrics usually possess a negative charge after washing, due largely to adsorbed anionics (anionics clinging to the hair or fabric surface). The positive charge of the quaternary ammonium ion in a cationic adsorbs strongly on negatively charged fabrics and hair and neutralizes this residual charge. The molecules line up with the positive charge toward the hair or fabric, leaving the hydrocarbon tails sticking outward, giving an antistatic feel.

Some cationics possess *germicidal* properties and are used as disinfectants, usually in hospitals or other health care settings. This cationic property is especially important in institutional cleaning since it aids in preventing the spread of disease, especially *Staphylococcus* bacteria.

Amphoterics

Amphoteric surfactants are mainly used in cosmetics and hair shampoos. Many people tolerate their higher prices more readily in such applications; they may be more willing to spend extra money for a better shampoo or cosmetic than for a laundry or dishwashing detergent. Amphoteric surfactants are usually more gentle on the skin and eyes than anionic surfactants. Several baby shampoos contain amphoteric surfactants for this reason.

Where Are These Surfactants Used?

Figure 13.5 depicts the main uses for the 7.84 billion pounds of anionic, nonionic, and cationic surfactants produced and used in a single year in North America. Household usage is the greatest portion, with nearly half the total. As figure 13.6 shows, about 74 percent of this household usage is for laundry detergents, with dishwashing accounting for another 22 percent; cleaning hard surfaces, such as cars, floors, and counters, consumed only 4 percent of the total. Industrial usage comes second to household usage (figure 13.5). Although industrial uses include some cleaning activities, most of the surfactant is consumed in industrial preparations—for example, in emulsifying or suspending a reactant or product. Institutional applications come third in the list of surfactant uses. These applications include cleaning and sanitizing floors, walls, dishes, and laundry. Oil field uses (for example, emulsifying petroleum inside the wells so that it can be pumped out more easily) make up about 4 percent of total surfactant usage. Personal care applications account for another 4 percent; hand or bath soap bars, shampoos, hair rinses, toothpaste, and some cosmetics fall into this category.

Table 13.2 shows that the less expensive anionics are the main type of surfactants used in household, industrial, institutional, and personal care areas. Nonionic surfactants, the second most used, are relatively widely used in the industrial and institutional cleaning areas. Cationic surfactants, a distant third, are used mostly in the industrial sector, and amphoterics are mainly used in shampoos, a personal care item. Box 13.2 discusses the environmental problems some earlier types of detergents caused and how chemists tackled those problems.

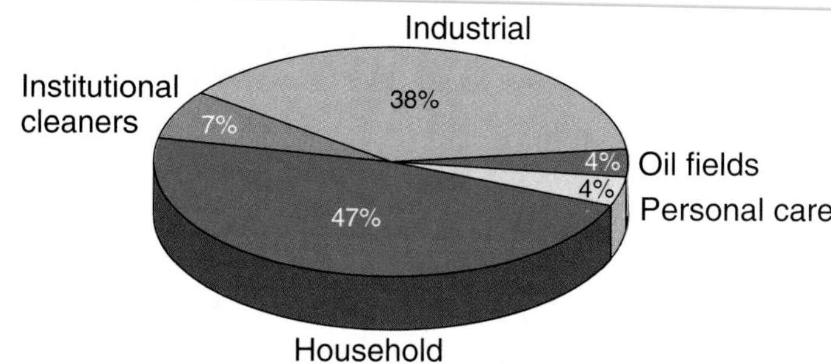

Figure 13.5 How 7.84 billion pounds of surfactants were used in a single year (1990) in North America.
SOURCE: Data from *Chemical & Engineering News*, January 25, 1993.

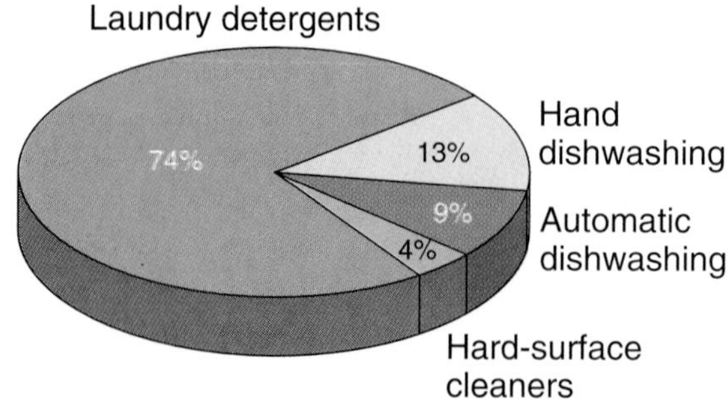

Figure 13.6 How 5.4 billion pounds of surfactant were consumed in households in North America during 1994.
SOURCE: Data from *Chemical Week*, January 25, 1995.

TABLE 13.2 Summary of surfactant uses in 1990. (The total use of amphoterics was only 30 million pounds, all in the personal care field.)

APPLICATION	AMOUNT USED (in millions of pounds and as a percentage)			
	Anionics	Nonionics	Cationics	Total
Industrial	1,075 (43)	875 (35)	550 (22)	2,500
Household	1,275 (67)	475 (25)	150 (8)	1,900
Institutional	192 (48)	168 (42)	40 (10)	400
Personal care	160 (64)	60 (24)	30 (12)	250
Grand totals	2,702 (54)	1,578 (31)	770 (15)	5,050

SOURCE: Data from *Chemical & Engineering News*, January 20, 1992.

DETERGENCY AND DETERGENTS

Detergency is the power or process of cleaning, and the chemicals used in cleaning are called by the class name **detergents.** This designation is sometimes applied, erroneously, to only the surfactant. The detergency process almost always involves surfactants blended into other detergents of various kinds. This section examines the detergency process and some specific detergent formulations.

HEAVY-DUTY LAUNDRY DETERGENTS

The U.S. heavy-duty laundry detergent market has undergone a marked change since the 1990 introduction of compact powdered detergents and 1992 debut of concentrated liquids. By the end of 1991, the U.S. market shares were fairly evenly divided between standard powders, standard liquids, and compact powdered laundry detergents. These ratios were expected to shift more toward the compacts and liquid concentrates, and away from the standard powders and liquids, by the year 2000 (figure 13.7).

Liquid laundry detergents have made little impact in Europe, where compacts currently hold a greater share of the market than in the United States (figure 13.8). However, the standard powder is still the most popular. In Japan, the situation is different. Compacts already comprised 80 percent of the Japanese powdered detergent market in 1990; by the end of 1992, compacts had virtually captured the entire Japanese market.

The driving force behind the rapid acceptance of compacts is the fact that they are phosphate-free and, presumably, better for the environment. Moreover, compacts are more concentrated and contain essentially no inert fillers; thus, they require less packaging, another environmental plus. Smaller packages also reduce shipping costs.

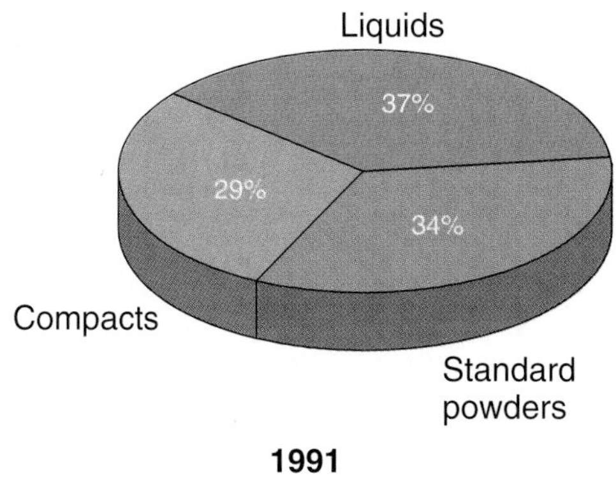

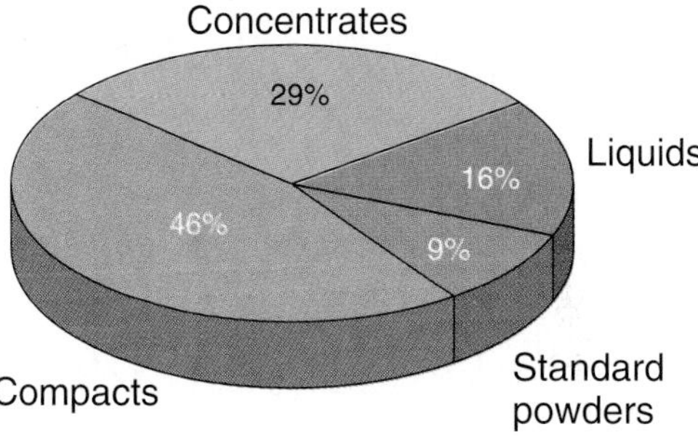

Figure 13.7 The United States market shares for the various types of detergents in 1991 and the projected market shares for 2000.
SOURCE: Data from *Chemical & Engineering News*, January 25, 1993.

Table 13.3 shows the composition of both the standard powdered laundry detergents and the newer compact powders. As you can see, each type contains many ingredients along with the surfactant. The two columns in the table are subdivided into groups according to use. In a sense, this subdivision also corresponds to the amount of chemicals used in each grouping.

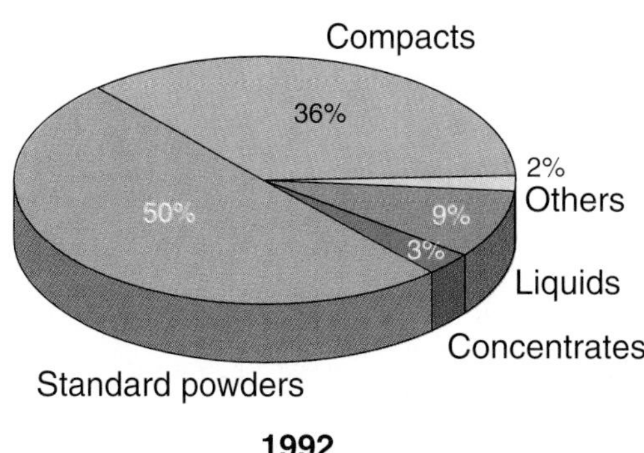

Figure 13.8 The European market shares for the various types of detergents in 1992.
SOURCE: Data from *Chemical Week*, January 26, 1994.

TABLE 13.3 Comparison of standard and compact heavy-duty laundry detergent.

INGREDIENT	TYPES OF CHEMICALS USED	PERCENT USED IN STANDARD	PERCENT USED IN COMPACT
Surface active agent	Anionic/nonionic	10–20	10–20
Builders	Sodium tripolyphosphate	20–25	0–5
	Zeolites	0–10	20–35
Alkali	Sodium carbonate	5–15	5–20
Corrosion inhibitor	Sodium silicates	2–6	2–6
Fillers	Sodium sulfate, etc.	0–20	—
Cobuilders	Poly(acrylic acids)	3–5	3–8
Bleach	Sodium perborate	0–25	0–20
Bleach activator	SNOBS, TAED*	0–3	0–8
Enzymes	Proteases, amylases, etc.	0.5–2	0.5–2
Fabric softeners	Quaternary ammonium salts	0.5–2	0.5–2
Optical brighteners	Stilbene derivatives	0.5–1	0.5–1
Antiredeposition	Na carboxymethylcellulose	0–1	0–1
Stabilizers	Various compounds	0.2–1	0.2–1
Antifoam agents	Silicones	0.1–4	0.1–2
Fragrances	Many	0.5–1	0.5–1
Dyes (for coloring)	Many	trace	trace
Water	Water	balance	trace
Bulk density		500–650 g/L	600–900 g/L

*SNOBS = sodium nonanoyloxybenzene sulfonate; TAED = tetraacetylethylenediamine
SOURCE: Data from *Chemical Week*, January 29, 1992, p. 40.

Surfactants

The first group is surfactant, which may make up as much as 20 percent of the total powder; mixtures of surfactants are now used in nearly all formulations. Originally, the alkylbenzene sulfonates used in laundry detergents were branched-chain molecules, but these surfactants caused environmental problems. Thus, for over thirty years, alkylbenzene sulfonates used in laundry detergents have always been the linear-chain, LAS type (see box 13.2). LAS surfactants are anionic; others are nonionic.

Inorganic Compounds

In the second group, which can make up as much as 50 percent of the powder, are inorganic compounds, including builders, alkalis, **anticorrosion agents,** and fillers. The alkalis make the detergent solution basic because detergents function better above pH 9. Anticorrosion chemicals help protect the metal in the washing machines from excessive corrosion or from dissolving in the detergent solution; silicates, which are also highly alkaline, are usually used for this purpose. Silicates also help improve the detergency; in other words, they act as builders. Sodium sulfate serves no purpose other than to fill the box; it is therefore called a filler. Fillers are seldom used in the compact powders.

Phosphate Builders

Builders, usually **phosphates,** enhance the efficiency of the detergency process. The major phosphate used in powdered detergents is sodium tripolyphosphate, or STPP, which has the chemical formula $Na_5P_3O_{10}$ and is an inexpensive white powder with low toxicity. Polyphosphates perform several tasks, including: (1) softening hard water (by binding Ca^{2+}, Ba^{2+}, or Mg^{2+} salts, i.e., sequestration); (2) making the solution alkaline; and (3) aiding detergency by suspending the removed dirt. When phosphates are not used in detergents, more surfactant is required, increasing the cost.

However, as noted in earlier chapters, phosphates can create environmental difficulties by serving as food for algae and microorganisms. There is no doubt that phosphates contribute to eutrophication, but whether detergent phosphates actually cause this problem is the subject of debate. About 30 percent of the phosphorus content of wastewater comes from detergents; the bulk comes from human waste, food waste, and fertilizers. By the end of 1992, fifteen states had banned phosphates and seven more had restricted their use, although no adequate STPP replacements exist. Phosphates are also banned in Japan, Korea, Switzerland, and Norway, and restricted in France, Germany, Italy, and several other European countries.

Whether these bans will persist remains to be seen. Sweden, for example, is reexamining its phosphate ban. The Swedes note that their wastewater treatment plants can remove 95 percent of phosphates, which would eliminate any biocontamination problem. They further note that the current substitutes (mostly zeolites) are not removed in wastewater treatment, and their environmental impact is unknown. In addition, zeolites are insoluble and increase sludge by 10 to 15 percent. Another substitute material, the poly(acrylic acids) may have even worse impact. Scientists do not yet know what these acids will do to the environment, and they are seldom biodegradable.

Polyphosphates work better as a builder than any other known substance. Sodium citrate is used in liquid laundry detergents because it is more soluble than the polyphosphates, but it is not as effective. Citrates are also more expensive than polyphosphates. Only about 36,500 tons of citrates were used in laundry detergents in 1992, compared to about 265 million tons of STPP, and citrates are essentially restricted to liquid formulations. More than a billion pounds (500,000 tons) of sodium carbonate (Na_2CO_3) are used in laundry detergent formulations, but sodium carbonate is not as good a builder as STPP; it mainly works to adjust pH.

The phosphate ban applies only to laundry detergents. Phosphates are still permitted in powdered automatic dishwasher detergents, since no adequate substitute currently exists. The insoluble zeolites cause excessive water spotting, and citrates do not clean effectively.

PRACTICE PROBLEM 13.4

What function does a builder perform in a laundry detergent? What chemicals serve as builders?

ANSWER A builder enhances, or builds, detergent activity by binding calcium or magnesium ions, increasing pH (a minor effect), and preventing redeposition of removed dirt. The chemicals used as builders are phosphates, especially STPP, but also include the zeolites, silicates, and Na_2CO_3 (sodium carbonate).

Zeolites

The two types of chemicals currently used most often to replace STPP are the zeolites and poly(carboxylates). The latter are water-soluble polymers, or copolymers, of acrylic or methacrylic acids and are more expensive than either STPP or zeolites.

Zeolites are insoluble sodium aluminosilicates with a formula of $2Na_2O \cdot 2Al_2O_3 \cdot 4SiO_2 \cdot 9H_2O$. They are ion exchangers that can soften water by removing Ca^{2+} ions, but they cannot remove Mg^{2+} ions. This limits their effectiveness to hard water that contains calcium but not magnesium ions. Zeolites do not enhance any other aspect of detergency, and they are more expensive than phosphates. They do not suspend removed dirt.

While phosphate sales have dropped greatly, zeolite sales have soared. During the mid-1980s, annual U.S. zeolite sales amounted to about 116 million pounds; this figure rose to approximately 814 million pounds by the end of 1992. Although several countries have switched from phosphates to zeolites, consumer response in the United States is less enthusiastic. When offered the option, consumers choose the phosphate-containing brands because they clean more effectively. But zeolites have gained ground in the compact detergent market since most of these formulations are based on zeolites, not STPP. Zeolites will likely dominate the compact detergents because they are more effective at absorbing the normally nonionic surfactants. Nonionics are playing an increasing role in laundry detergents.

We still need a cheap, effective chemical compound to replace the polyphosphates in laundry detergents. The phosphate ban—a legislative "solution" to the problem"—has not proven totally satisfactory. Little evidence shows that the change away from phosphates improves the quality of lakes and streams. Without phosphates, people could end up using even more surfactant, which costs more and feeds microorganisms and algae as well as phosphates do.

PRACTICE PROBLEM 13.5

What are the advantages and disadvantages of using zeolites in laundry detergent?

ANSWER Probably the chief advantage of zeolites is that they cannot serve as "food" for algae and will not lead to eutrophication. Other advantages include zeolite's ability to bind Ca^{2+} ions and to absorb (soak up) nonionic surfactants. The disadvantages of zeolites include higher costs, increased wastewater sludge, and less effective action as a builder.

BOX 13.2

FOAM-CAUSED ENVIRONMENTAL PROBLEMS

The original alkylbenzene sulfonates used in laundry and other detergents were called **branched-chain alkylbenzene sulfonates (BAS).** Though initially popular, these BAS did not biodegrade readily, while both soap and the fatty alcohol sulfates did. Eventually, this created a major environmental problem. Foam from the BAS surfactants began to accumulate in remote streams, waterfalls, and fountains as the branched alkylbenzene sulfonates entered into the waste flow patterns. The foam produced in wastewater treatment plants was even more problematic. Soap was always present in these bodies of water, but soap was biodegraded by natural microorganisms in the water. These microorganisms could not consume the BAS materials.

The BAS problem was caused by the specific structure of this surfactant. Microorganisms normally "eat" the chemicals in their environment in two-carbon-atom bites, starting from the exposed alkyl group. The BAS surfactants contained a branched side chain that did not permit microorganisms to take a two-carbon "bite;" instead, they would have to consume a three-carbon mouthful, which was not possible (figure 1). The undegraded BAS began to build up in the environment.

When BAS pollution became apparent, chemists began to seek ways to make new types of surfactants. They redesigned the entire process for making alkylbenzene sulfonates so that the final molecule was a ***linear* alkylbenzene sulfonate,** or **LAS,** instead of the branched BAS version (figure 2). This simple change made a great difference because the LAS molecules were biodegradable, and streams and waterfalls were no longer marred by heaps of billowy white foam.

Unfortunately, most environmental progress comes at a cost. The LAS process is more expensive than the BAS. Our environment is safer, but our detergents cost more. In this case, such a trade-off was necessary. Over the past thirty or more years, the biodegradable LAS surfactants, though more expensive, have been much better for the environment.

Figure 1 Branched alkylbenzene sulfonates (BAS) led to foaming problems in the environment.

Figure 2 Chemists developed linear alkylbenzene sulfonates (LAS) to solve the environmental problem.

Bleaches, Softeners, and Enzymes

The third grouping in table 13.3 includes bleaches, softeners, and enzymes. These substances are only used in small quantities in laundry detergents. Bleaches remove stains by an oxidation-reduction reaction, removing the color from the material that is causing the stain. Sodium perborate ($NaBO_3$) is the most common bleaching chemical added to powdered detergents. It works by converting some water molecules into hydrogen peroxide (H_2O_2), an oxidizing agent. **Perborates** release H_2O_2 only in very warm to hot water; this poses a problem in the U.S. market since we use lower washing temperatures than Europeans do. Activators are added to make perborates function more effectively at lower temperatures. In Europe, about 80 percent of laundry detergents contain perborate bleaches, but fewer do in the United States. The U.S. market is about 100 million pounds per year, compared with 1.1 billion in Europe.

While slightly less effective than chlorine bleaches, perborates are less harmful to fabrics than their chlorine-containing counterparts. The old chlorine standbys, sodium or calcium hypochlorite—$NaOCl$ or $Ca(OCl)_2$—can destroy fabrics on direct contact. When it is used, $NaOCl$ is added separately to the wash as a 5.25 percent aqueous solution, since the solid compound is unstable.

Solid chlorine bleaches, such as hydantoins or cyanurates (figure 13.9), are only used in powdered detergents, but they, too, are less popular than perborates because they harm fabrics. They also impart a hard-to-eliminate odor to the fabrics.

Enzymes, which we will discuss in more detail in Chapter 14, are polymeric protein catalysts whose names always end with the suffix *ase*. They were first introduced in detergents in the 1950s. Enzymes contribute to detergency by breaking down stain-causing chemicals and dirt that adhere strongly to fabrics. Their use was discontinued for many years because they were expensive, they were less effective than desired, and concerns arose that they might produce adverse skin reactions. Recently, chemists developed new enzymes and better formulation methods, and enzymes are reappearing in U.S. household laundry detergents. Almost all detergents in Japan contain enzymes. The most common laundry detergent enzymes are the proteases that hydrolyze proteins, but others include amylases (which hydrolyze starch), cellulases (which hydrolyze cellulose), and lipases (which hydrolyze fats or oils). One problem in using enzymes is that peroxides (such as the H_2O_2 perborates release) normally destroy them.

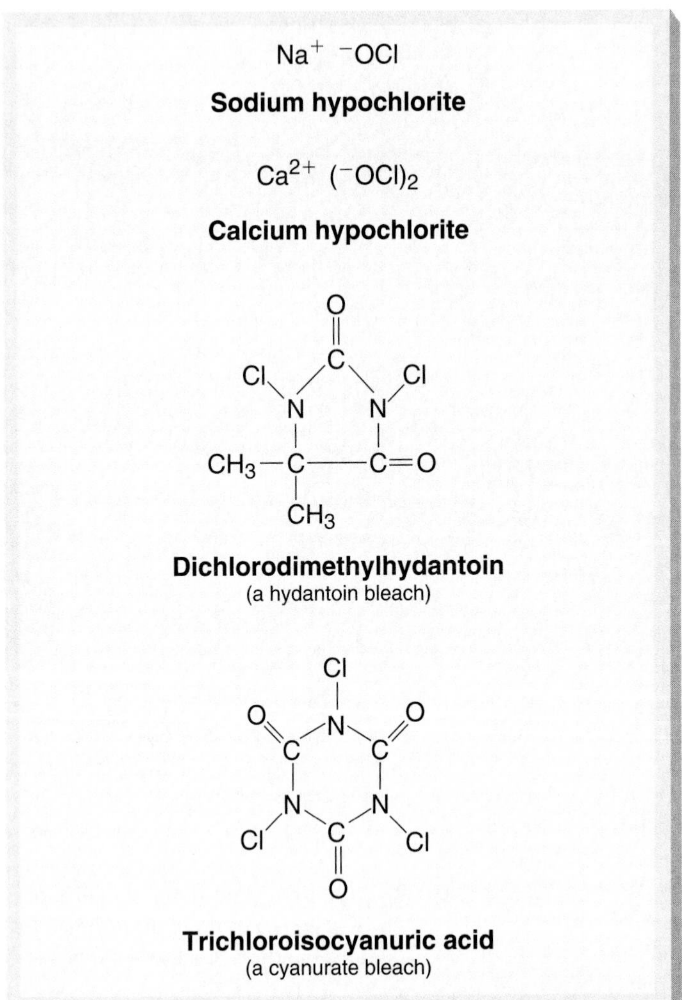

Figure 13.9 The structures of chlorine-containing chemicals that are added to detergents as bleaches.

Softeners make fabrics feel softer; cationics are used for softening purposes. Quaternary ammonium salts, common fabric softeners, are not used in formulations that contain anionic surfactants, since the oppositely charged ions combine and form a useless precipitate. The ammonium salts do work in formulations containing nonionics. Most people in the United States use a separate softener they add to the rinse cycle or during drying. The opposite is true in Europe—most detergents contain built-in softeners.

PRACTICE PROBLEM 13.6

What functions do bleaches, softeners, and enzymes serve in laundry detergents? What chemicals are used for these purposes?

ANSWER Bleaches, chemicals that act as oxidizing agents, oxidize dirt and stains to a colorless form. The main bleaching chemical used in laundry detergent formulations is sodium perborate. Softeners are cationic surfactants that make a fabric feel soft; quaternary ammonium salts serve this purpose. Finally, enzymes are polymeric catalysts that remove stains by hydrolysis reaction. Proteases are the most commonly used laundry detergent enzymes, followed by amylases, cellulases, and lipases.

Other Components

The last group of chemicals in table 13.3 includes several miscellaneous materials added in very small quantities to laundry detergents. **Optical brighteners** are chemicals that absorb UV light and emit blue light in its place, making the fabric appear a brighter white. Most optical brighteners are complex stilbene derivatives (figure 13.10). The double bonds of stilbene derivatives are conjugated or alternating, which means the electrons can circulate throughout the entire molecule. This allows the molecule to absorb UV energy and emit it in the blue region of the visible spectrum. Studies show that people perceive this bluish cast as whiter and cleaner; a yellowish color seems dirtier and less desirable.

Antiredeposition agents help prevent dirt from settling back on fabrics. The most common material used is sodium carboxymethyl-cellulose, or CMC. When antiredeposition agents are absent, fabrics do not remain as clean because some of the dirt redeposits. These expensive agents assist phosphates and surfactants in preventing redeposition.

Sometimes antifoaming agents are added to reduce the amount of foam the surfactant produces in automatic clothes washers. Fragrances are added to laundry detergents for esthetic reasons, and some detergents have added dyes to impart colors for the same purpose.

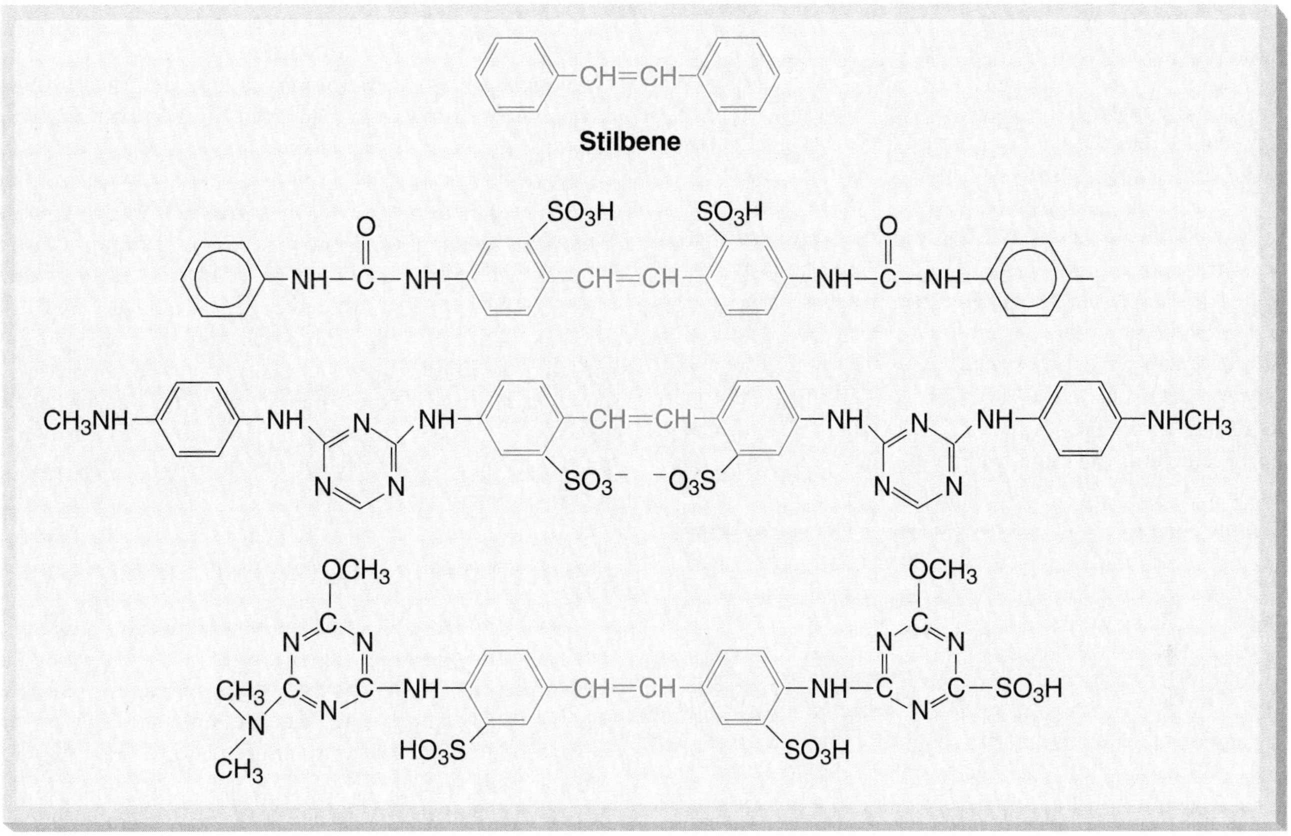

Figure 13.10 The chemical structure of stilbene and three optical brighteners that contain the stilbene unit as the central portion of the molecule.

PRACTICE PROBLEM 13.7

What functions do optical brighteners, antiredeposition agents, fragrances, and dyes serve in laundry detergents? What chemicals are used to achieve these functions?

ANSWER Optical brighteners impart an illusion of whiteness to fabrics by absorbing UV light and emitting blue light in its place. Optical brighteners are mainly stilbene derivatives. Antiredeposition agents, such as CMC, prevent removed dirt from resettling back onto a surface. Many kinds of chemicals are used as fragrances and dyes to impart pleasant odor and color to detergents.

COMPACT LAUNDRY DETERGENTS

Compact laundry detergents, sometimes called superconcentrates, were introduced in 1990 and comprised 29 percent of the U.S. market by the end of 1991. This growth occurred at the expense of liquid formulations and was spurred by the bans on phosphates. The main ingredient removed from compact powders is water, which occurs in hydrate form in the standard formulations.

Table 13.3 compares the typical compact laundry detergents against the standard powders. One conspicuous absence in compacts is the filler, which makes up as much as 20 percent of some standard powders. Compacts also replace most of the phosphates with zeolites and increase the carbonate levels. Zeolites absorb the liquid nonionics, keeping the powders free-flowing. Notice that more zeolite is used in compacts than STPP in the standard formulations; the surfactant percentages are *not* higher in compacts than in the standard powders, as might be expected from the term *superconcentrated*. Perborate bleaches are commonly used in both standards and compacts, but the levels of activator, which make the bleaches work at lower temperatures, are generally higher in compacts.

Even if phosphate bans are lifted, by no means a certain proposition, the compacts are expected to survive and prosper (figure 13.6). The idea of using less laundry detergent has special consumer appeal. Since compacts are designed around zeolites, that industry should also show growth, as should production of laundry enzymes. Less obvious pluses for compacts include (1) less packaging, producing less trash to clutter landfills, (2) reduced shipping costs, since more packages fit in a truck, and (3) the availability of refillable cartons. The compact liquids (concentrates), and even standard liquids, are now usually packaged in bottles made with 25 to 50 percent recycled plastics, as environmental concerns exert an increasing influence on detergent production and use (box 13.3).

PRACTICE PROBLEM 13.8

What are the major differences between compact and standard laundry detergents? Which type do you believe is best for the environment? Is this a simple comparison?

ANSWER Compact detergents contain less water than the standard powders and replace the phosphate builders with zeolites. If phosphate removal were the only important environmental factor, compact laundry detergents would therefore be "greener." If other factors are considered, such as sludge production and the effects of zeolites on the environment, the choice is less certain. Obviously, determining which detergents are more friendly to the environment is a complex task.

WASHING DISHES AND HANDS

Hand dishwashing detergents are usually 10 to 25 percent aqueous solutions of LAS, FAS, and nonionic surfactants with only small amounts of the inorganic salts typically found in laundry detergents. Unlike laundry detergents, these materials make direct contact with our skin and must have a lower pH to avoid irritation. The powdered formulations used in automatic dishwashers are mostly polyphosphates and silicates with only a small amount of surfactants (usually nonionic); they produce a very high pH. Zeolite-based cleansers do not work well in automatic dishwashers.

Most people wash themselves using toilet soap bars made from sodium salts of fatty acids. People have used these chemicals since ancient times. Bath soap bars have nearly neutral pH values; very high or low pH could harm skin. In addition to actual soap bars, some skin cleansing bars are made from synthetic surfactants. The earliest synthetic versions did not work as well as soap, but the newer bars are effective cleaners, especially in hard water.

Soap bars often contain added chemicals. Often a fragrance is added that rapidly dissipates from the skin. Deodorant soap bars contain a deodorizing chemical that is usually a mild antibacterial agent since body odor comes partially from bacterial sources. Triclocarban, also called triclosan (figure 13.11), is the most common deodorant chemical. Hexachlorophene, a phenol derivative, was once widely used, but its safety and toxicity were questioned and it was removed from the market.

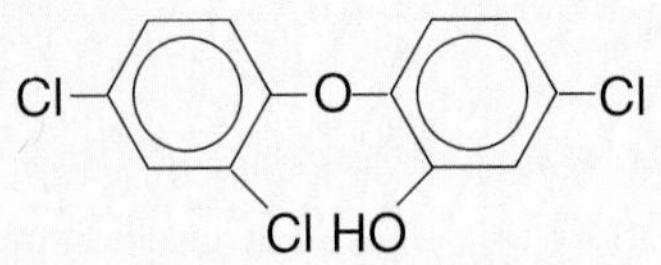

Cl–C6H3(Cl)–O–C6H3(OH)–Cl

2,4,4′-Trichloro-2′-hydroxydiphenyl ether
Triclosan

Figure 13.11 The chemical structure of triclosan.

Some soap bars contain a gritty material, such as pumice, to help remove heavy or greasy dirt from the skin. Others go the opposite direction and add emollients (skin softeners) such as glycerine, lanolin, or fat. Virtually all bath soap bars contain small amounts of perfumes, dyes, and glycerin. Some contain aloe, other herbal materials, or vitamins to further improve skin.

Liquid hand soaps are aqueous solutions of fatty acid salts, but more soluble K^+ or triethanolamine cations are used instead of the less soluble Na^+ salts. These salts are used in shaving creams, although synthetics now dominate the field. Synthetic surfactants, such as a triethanolamine lauryl sulfate, are also used in liquid hand cleansers. Transparent soap bars contain high alcohol levels (about 30 percent) that partially dissolves the soap, making it transparent.

SHAMPOOING HAIR AND BRUSHING TEETH

A casual stroll through the shampoo section of a supermarket or drug store reveals a staggering melange of brands, claims, and special ingredients. Some shampoos contain aloe or practically any herb you might desire. They also contain nearly any fragrance you might enjoy, from flowers to fruits or nuts. Some brands even include food items in their formulations, such as milk, sugar, honey, oats, or bran. Whether these "extras" confer any benefits is debatable, but they apparently appeal to consumers and give advertising departments something to do.

Shampoos clean hair; the word *shampoo* comes from the Hindu word *champo* meaning to massage or knead. Most shampoos were soap solutions before World War II, but then the picture changed drastically. Soap is an adequate shampoo in soft water, but in hard water it forms an undesirable, insoluble Ca^{2+} salt crud. Now the shampoo market is dominated by synthetics with sodium lauryl sulfate and related salts leading the pack.

The first shampoo was marketed in Germany during the 1890s. In the United States, John Breck introduced his first normal hair shampoo in 1930, and versions for oily and dry hair three years later. Breck actually developed the shampoo in the process of seeking a cure or preventative for baldness. In this goal, he was unsuccessful.

Shampoos

Shampoos have a neutral pH; hair is chemically similar to skin, and neither can tolerate high or low pH values. Most shampoos contain, along with surfactants, thickening agents that make the solutions more viscous, giving the impression that they are more concentrated. These thickening agents are natural alginates (from seaweed) or synthetic water-soluble polymers; they tend to discourage the use of excessive amounts of shampoo. Some shampoos also contain animal proteins that glue together "split ends" and coat the hair shafts, making the hair feel smoother and thicker. Finally, they may contain a clarifier, such as an alcohol, to make the solution appear clear.

BOX 13.3

HOW GREEN IS MY SURFACTANT?

In the past few decades, the concept of using environmentally safe materials has grown markedly popular in the consumer and industrial markets. So-called "green" products or processes supposedly keep our environment safe, healthy, and green. The green phenomenon has struck the surfactant and detergent market in several aspects, causing the development of linear alkylbenzene sulfonates (LAS) and the decreasing use of phosphates in laundry detergents.

One new development on the detergent front is the use of fatty alcohol sulfates in place of LAS. The **fatty alcohol sulfates (FAS)** are obtained from natural plant oil sources rather than petroleum, as LAS surfactants are. Some people believe the "natural" FAS surfactants are more environmentally friendly, but this is not necessarily true. Table 1 compares the laboratory biodegradability of LAS and FAS surfactants. Although these laboratory tests slightly favored the FAS surfactant, over thirty years of actual field experience clearly shows that LAS materials do biodegrade under actual use. Since the FAS also biodegrade, neither surfactant has a particular advantage on this basis—they are equally environmentally friendly. Remember, the value of experimental data depends on careful planning. In this case, the laboratory experiment evidently did not imitate the real-world environment closely enough.

However, we must consider other factors besides biodegradability. For example, consider raw material costs, atmospheric or water pollution, and the energy consumed during production or waste disposal. Table 1 shows this information for a fatty ester sulfonate (MES) derived from natural palm oil and for LAS derived from petroleum. Since the MES is chemically similar to the FAS surfactants, it is no surprise it fully biodegrades. However, the LAS surfactants are definitely more energy efficient. Chemists also consider which surfactant causes greater air or water pollution; again, the naturally-derived product loses the contest, producing more atmospheric and waterborne wastes. The palm-oil-derived MES surfactant also produces more solid waste, both in weight and volume. By most standards, the natural product comes in second best; hardly a strong selling point for protecting the environment.

What can we conclude? The lesson is that substances derived from natural or renewable sources are not always more environmentally friendly than petrochemically derived materials. We must consider a complex set of factors and risks. By the way, these studies were not merely industrial hype; they were conducted in cooperation with the Environmental Protection Agency.

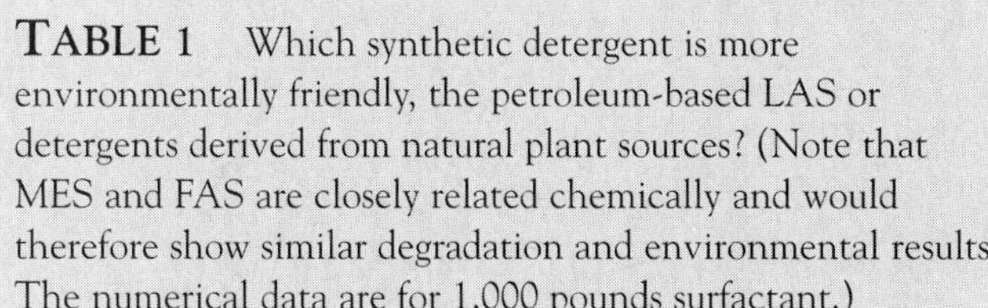

TABLE 1 Which synthetic detergent is more environmentally friendly, the petroleum-based LAS or detergents derived from natural plant sources? (Note that MES and FAS are closely related chemically and would therefore show similar degradation and environmental results. The numerical data are for 1,000 pounds surfactant.)

	SURFACTANT	
	LAS[1]	MES[2]
Aerobic degradation	95%	99^{+}%[3]
Anaerobic degradation	—[4]	—[5]
Energy for processing and transportation (10^6 btu)	13.2	17.2
Atmospheric wastes (lbs.)	41	91
Waterborne wastes (lbs.)	19	24
Industrial solid waste (lbs.)	124	191
(cubic feet)	2.5	3.8

[1]LAS is a linear alkylbenzene sulfonate, a petroleum derivative.
[2]MES is a methyl ester sulfonate of palm oil, a "natural" derivative.
[3]This data is for a fatty alcohol sulfate, such as lauryl sulfate, also a natural derivative.
[4]Does not degrade under laboratory conditions.
[5]Degrades under laboratory conditions.
SOURCE: Data from *Chemical Week*, January 29, 1992.

Figure 1 How good are "green" detergents?

TABLE 13.4 A typical shampoo formulation.

COMPONENT	EXAMPLE	PERCENT OF SHAMPOO
Detergent	ethanolamine lauryl sulfate	10–20
Thickener	sodium alginate	2–5
Clarifying solvent	ethanol	5–15
Protein	any soluble animal protein	0–3
Water		60–80
Perfume		trace
Colorant		trace

Table 13.4 shows a typical shampoo formulation. Actual formulations often contain two or more surfactants, and the exact nature of the surfactant used in a shampoo varies with the oiliness of the scalp. Generally, shampoos are marketed for normal, dry, or oily hair. These shampoos vary in the relative ratios of sodium lauryl sulfate (a surfactant) and a cocamide (figure 13.12). Children's shampoos usually contain amphoteric surfactants because they irritate the eyes less; amphoterics also appear in some adult shampoos.

Conditioners

Conditioners are solutions used on hair after shampooing to make hair softer and more manageable. Conditioners are cationic surfactant solutions, often quaternary ammonium salts or amine oxides. Cationics react with the anionic surfactant residues left on the hair after shampooing and form a precipitate that imparts a soft feel and reduces static. The effect is very similar to clothes treated with a fabric softener.

PRACTICE PROBLEM 13.9

In what ways does a shampoo differ from a laundry detergent?

ANSWER Shampoos are normally liquids that contain less surfactant than laundry detergents and almost no salts. The pH of a shampoo is neutral, about 7.0, whereas a laundry detergent solution will be over 9. And while laundry detergents contain a variety of builders, bleaches, and other chemicals to promote cleaning, shampoos contain thickeners, clarifiers, fragrances, and proteins to condition as well as clean the hair.

Toothpaste

Toothpaste—the oral equivalent of a shampoo—also has a wide range of ingredients and formulations. However, toothpaste contains large amounts of insoluble salts that the hair cannot tolerate. The Egyptians developed the first toothpaste about four thousand years ago from powdered pumice and vinegar. A lot has changed since then.

Although the basic toothpaste recipe (table 13.5) is simple, different commercial brands add everything from spices to mints

TABLE 13.5 A typical toothpaste formulation.

INGREDIENT	EXAMPLES	PERCENT OF TOOTHPASTE
Abrasive	calcium carbonate ($CaCO_3$)	40–50
	titanium dioxide (TiO_2)	
	alumina (Al_2O_3)	
	silica (SiO_2)	
Sweetener	glycerin	15–25
	sorbitol	
	saccharin	
Surfactant	sodium lauryl sulfate	3–5
	soap	
Fluoride	stannous fluoride (SnF_2)	1
	sodium fluoride (NaF)	
Thickener	sodium alginate	1
	cellulose derivatives	
Flavoring	peppermint oil	1
	wintergreen	
	strawberry	
	lime	
Water		20–40

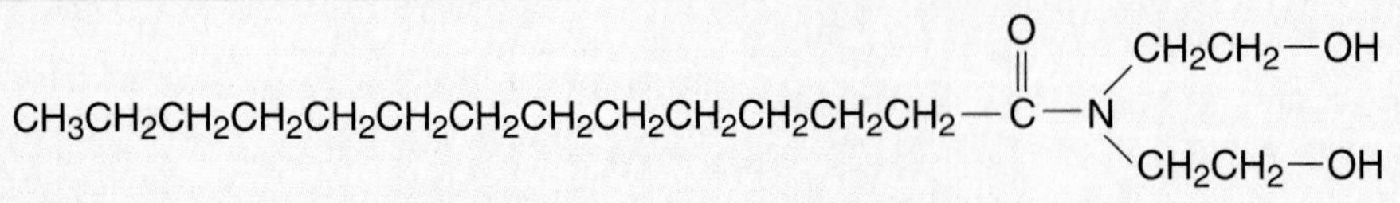

$$CH_3CH_2CH_2CH_2CH_2CH_2CH_2CH_2CH_2CH_2CH_2CH_2CH_2{-}\overset{O}{\overset{\|}{C}}{-}N(CH_2CH_2{-}OH)_2$$

Figure 13.12 The structure of a typical cocamide.

BOX 13.4

TOOTH WHITENERS

Research performed during the 1980s led to dozens of tooth bleaches and whiteners appearing on the home market in the 1990s. Prior to this time, only a dentist could whiten one's teeth, using solutions of hydrogen peroxide and protecting the patient's gums and other mouth tissues with rubber dams (shields). Similar treatments are now available for home use, but without the rubber dams. What are these home tooth whiteners, and are they as safe as the dentist's treatment?

Do-it-yourself whiteners contain 1 to 3 percent hydrogen peroxide, a much more dilute solution than the 35 percent solution a dentist typically uses. This means that any whitening action will be slower than in the dental chair. While the lower H_2O_2 concentration will reduce the chance of gum damage, the possibility still exists. The American Dental Association staunchly opposes home tooth whiteners, claiming that gum and nerve damage is too likely to result. However, the whitener manufacturers claim that the ADA is merely trying to protect a very lucrative business for its members. Dental office treatments can run between \$400 and \$2,000; home treatments run between \$3 and \$8 per tube.

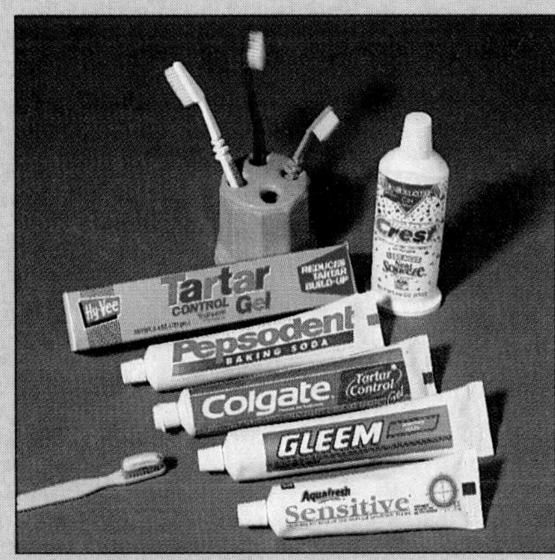

Figure 1 Tooth cleaners come in all sizes and types.

What is the FDA's position on this matter? According to Bradford Williams, an FDA enforcement officer, the FDA has not made tooth whiteners a priority. Williams notes, "If there were bodies in the street from this stuff, we wouldn't be sitting by."

Meanwhile, sales of home tooth whiteners had risen to \$150 million in 1994, and they are still rising. The safety questions remain unresolved.

to fruity flavors. Basically, toothpaste is a mixture of an abrasive with a surfactant, plus a few other minor ingredients. The most common abrasive is $CaCO_3$ (chalk), although titanium dioxide, alumina, and silica are also used. The surfactant is usually sodium lauryl sulfate (the main surfactant in shampoo), but sodium stearate is still used in a few brands; Japanese toothpaste normally contains soap. Many brands of toothpaste also add fluoride to reduce tooth decay. (See chapter 8 for more information on fluoridation). Flavoring agents and colorants run the gamut, presumably reflecting varied consumer tastes, or perhaps the ingenuity of the marketing and advertising departments. Toothpastes are usually sweetened, but not with typical sugars because these promote tooth decay. Instead, the sweeteners are glycerine, sorbitol (a natural sugarlike molecule), or an artificial sweetener. Tooth whiteners, as box 13.4 discusses, are a more recent entry in the market for tooth care products.

SUMMARY

Soaps and synthetic detergents, the essential ingredients of cleansers, are surface active agents, or surfactants. They reduce the surface tension of water and lift and surround dirt and grease molecules. Surfactants consist of molecules with both a polar and a nonpolar portion. The polar part is hydrophilic (attracted to water) and the nonpolar, hydrocarbon end is hydrophobic (repelled by water).

Chemists class surfactants as anionic, cationic, nonionic, or amphoteric. Anionics have negative charges on the hydrophobic portion of the molecule, while cationics have a positive charge. Nonionics have no charge, and amphoteric surfactants have both a positive and negative charge on the hydrophobic portion of the molecule.

Soap is prepared in a saponification reaction between fatty acids and a strong base. The products of the reaction are glycerol and sodium salts of the fatty acids (soap). Soap, the earliest organic surfactant, is not an effective cleanser in hard water, acidic solutions, and seawater. Newer surfactants, the fatty alcohol sulfates and alkylbenzene sulfonates, are soluble in hard water and can function in acidic solutions or seawater. These compounds are anionics. The alkylbenzene sulfonates are commonly used in powdered laundry detergents.

Nonionic surfactants also clean in hard water, salt water, and acidic solutions. They dominate in liquid laundry detergents, but are more expensive and less readily biodegradable than the anionics.

Cationics have amazing antistatic properties as well as some germicidal properties. They are used in shampoos and fabric softeners as well as disinfectant solutions. Amphoterics are mainly used in cosmetics and hair shampoos; they are gentle to the hair and skin and are often used in baby shampoos.

Almost 8 billion pounds of surfactants are produced in a single year in North America. Most are used in households, then in industry, institutions, oil fields, and in personal care. Anionics are the most widely used, followed by nonionics and cationics.

Detergency is the power or process of cleaning. Detergents contain surfactants blended with other compounds and are used to clean laundry, dishes, skin, hair, and teeth. Often, the surfactant molecules in a detergent form micelles; their hydrophobic tails converge in the center of a dirt or grease droplet, and their hydrophilic heads face to the outside, toward the water. This emulsifies the greasy dirt, or separates it into tiny drops that can then be washed away.

Laundry detergents are about 20 percent surfactants; the linear alkylbenzene sulfonates (LAS) are the most common surfactant in standard powders. Standard powders also contain phosphate builders, anticorrosion agents, bleaches, softeners, enzymes, optical brighteners, and fillers. Newer, compact powders contain the same types of ingredients, but they replace the phosphates, which have been banned in many states, with zeolites, and they contain very few fillers.

Hand dishwashing detergents are usually 10 to 25 percent aqueous solutions of various anionic, including LAS, and nonionic surfactants. These materials, which contact skin directly, must have a lower pH to prevent skin irritation. Bath soap bars are made from sodium salts of fatty acids (soaps) or synthetic surfactants; they also have a lower pH.

Shampoos also clean hair with fatty alcohol sulfates, such as sodium lauryl sulfate, combined with thickeners, fragrances, proteins, and other extras. Conditioners are cationic surfactants, solutions that impart softness and reduce static. Finally, toothpaste also contains sodium lauryl sulfate, as well as an abrasive, flavors, and in some cases, fluoride.

KEY TERMS

amphoteric surfactant 337
anionic surfactant 337
anticorrosion agent 344
antiredeposition agents 348
branched alkylbenzene sulfonate (BAS) 346
cationic surfactant 337
detergency 341
detergent 339
fatty acid 337
fatty alcohol sulfates (FAS) 351
hydrophilic 336
hydrophobic 336
linear alkylbenzene sulfonate (LAS) 346
lipophilic 342
micelle 341
nonionic surfactant 337
optical brighteners 348
perborate 347
phosphates 344
saponification 337
surface active agent 336
surfactant 336
wetting 341
zeolite 345

READINGS OF INTEREST

General

Ainsworth, Susan J. "Soaps and detergents." *Chemical & Engineering News*, January 23, 1995, pp. 30–53. This annual series is published every January with new and different information in each article.

Breskin, Ira; Westervelt, Robert; Kemezis, Paul; and Chynoweth, Emma. "Soaps and detergents." *Chemical Week*, January 25, 1995, pp. 38–52. An annual report on soaps and detergents that is more business oriented than the articles in *Chemical & Engineering News*.

Ainsworth, Susan J. "Soaps and detergents." *Chemical & Engineering News*, January 24, 1994, pp. 34–59. The 1994 edition of this annual series tells how industry now relies on suppliers to keep up with reformulation demands.

Coeymen, Marjorie; Mullin, Rick; Morris, Gregory D. L.; Kemezis, Paul; and Chynoweth, Emma. "Soaps and Detergents." *Chemical Week*, January 26, 1994, pp. 35–48. This annual series relies on reporters located at different locations, including overseas, for its data.

Thayer, Ann M. "Soaps and detergents." *Chemical & Engineering News*, January 25, 1993, pp. 26–27. This report from an annual series deals with compact powdered detergents and the new liquid compacts.

Mullin, Rick; Kagan, Andrew; Kemezis, Paul; Roberts, Michael; and Morris, Gregory D. L. "Soaps and detergents: New generation of compacts." *Chemical Week*, January 27, 1993, pp. 28–49. A business-oriented annual report. This article dwells on how compact laundry detergents have changed the field and discusses the prospect that some phosphate bans may end.

Ainsworth, Susan J. "Soaps and detergents." *Chemical & Engineering News*, January 20, 1992, pp. 27–63. Part of an annual series, published every January, this article contains quite a bit about actual surfactants.

Mullin, Rick; Wood, Andrew; Morris, Gregory D. L.; Chynoweth, Emma; and Roberts, Michael. "Soaps and detergents." *Chemical Week*, January 29, 1992, pp. 24–44. This business-oriented annual report introduces the compact laundry detergents and their impact on the detergent field.

Greek, Bruce F. "Detergent components become increasingly diverse, complex." *Chemical & Engineering News*, January 25, 1988. pp. 21–53. This particular report is a good basic article on detergent components and presents the essentials of detergent formulation.

"Green" Surfactants

Kirschner, Elisabeth, and Grund, Howard. "Green chemicals." *Chemical Week*, October 27, 1993, pp. 36–40. A trio of short articles on "green" chemicals that can be used in detergents.

Ishigami, Yutaka. "Biosurfactants face increasing interest." *Inform*, vol. 4, no. 10, October 1993, pp. 1156–1165. This chapter only touched on possible "natural" surfactants. This article describes a variety of other kinds of biosurfactants.

Kemezis, Paul, and Ward, Mike. "Specialty surfactants." *Chemical Week*, September 29, 1993, pp. 25–32. Several new surfactants have emerged as part of the drive toward more natural surfactants. At present, these materials have a limited market and a higher price than traditional surfactants. In this article, some questions are raised as to whether the added cost is worthwhile.

Wang, Penelope. "Going for the green." *Money*, September 1991, pp. 98–102. Interesting article about which detergents are "greener" and how you can become "green" yourself (without transforming into a frog).

Soap Bars

Reese, K. M. "The true story of the purity of Ivory Soap." *Chemical & Engineering News*, February 28, 1994, p. 60; "Ivory Soap saga hard to pin down conclusively," March 28, 1994, p. 68; "Definitive source claimed for Ivory's famous purity," April 18, 1994, p. 64. A trio of short articles on how the famous 99.44 percent pure slogan for Ivory soap originated.

Power, Christopher. "Everyone is bellying up to this bar." *Business Week*, January 27, 1992, p. 84. A new soap bar, of course. Article discusses and compares a new "soap star" with the old faithfuls.

Hager, Bruce. "Colgate: Oh what a difference a year can make." *Business Week*, March 23, 1992, pp. 90–91. What one of the soap giants is up to lately.

Toothpaste

Kluger, Jeffery. "Oh, rubbish." *Discover*, August 1994, pp. 30–35. "Trash archaeologists" are looking for information on the dental hygiene practices in the United States in recent years.

Reese, Ken. "From twig to tube: The search for a brighter smile." *Today's Chemist at Work*, February 1994, pp. 71–72. An easy-to-read account of the development of today's toothpaste formulas.

Gaffar, Abdul; Afflitto, John; and Nabi, Nuran. "Toothbrush chemistry." *Chemtech*, January 1993, pp. 38–42. Describes the chemistry behind the development of Colgate Total, a germicide-containing toothpaste claimed to be effective against gum disease, cavity formation, and plaque.

Problems and Questions

1. What makes a molecule behave as a surface active agent, or surfactant?

2. What is the name and structure of the oldest known organic surfactant? How is this chemical made?

3. Define hydrophilic and hydrophobic. How do these qualities relate to one another?

4. Define lipophilic. How does this quality relate to hydrophilic and hydrophobic?

5. What are the four categories of surfactants?

6. Which category of surfactant has the largest annual production?

7. What is the difference between an anionic and a cationic surfactant?

8. What are anionic surfactants? Give two examples.

9. What are the main uses for anionic surfactants?

10. What are cationic surfactants? Give two examples.

11. What are the main uses for cationic surfactants?

12. What are nonionic surfactants? Give two examples.

13. What are hydrophilic and hydrophobic portions in a typical nonionic surfactant?

14. What are the main uses for nonionic surfactants?

15. What is an amphoteric surfactant, and what is its main use?

16. What is the difference between a branched alkylbenzene sulfonate (BAS) and a linear alkylbenzene sulfonate (LAS)? Which compound is used in detergents today?

17. What is surface tension? How does it relate to surfactants?

18. What is detergency? How does the detergency process work?

19. What is a micelle? What is its function in the detergency process?

20. What are the differences between a surfactant and a detergent?

21. What are the main components in a standard powdered laundry detergent? What is the purpose of each component?

22. What are the differences between standard and compact laundry detergents?

23. What are the functions of a builder in a laundry detergent?

24. What is the main concern about using phosphates in laundry detergents?

25. Can phosphates still be used for any detergency applications in areas where they are banned? If so, what are these applications?

26. What chemicals have been considered as replacements for phosphates?

27. What are zeolites, and how are they used in laundry detergents? What are their advantages and disadvantages?

28. What do silicates do in laundry detergents?

29. Why do laundry detergents contain sodium carbonate?

30. Why are enzymes added to laundry detergents?

31. What kinds of bleaches are used in laundry detergents? How do they accomplish their function?

32. What kind of chemical has been used most effectively as a builder? Give a specific example of this type of chemical.

33. What is redeposition? What chemicals are used in laundry detergents to prevent this problem?

34. What do optical brighteners do? Give an example of a molecule that has this property.

35. What is the main type of surfactant used in bath soap bars?

36. Besides the surfactant, what other kinds of chemicals are added to bar soap?

37. What are the main ingredients in a typical shampoo?

38. What is the most commonly used surfactant in shampoos?

39. What is the major component in a hair conditioner? How does it work?

40. What are the main ingredients in a typical toothpaste?

41. What is the most common surfactant used in toothpaste in the United States? In Japan?

42. What types of sweeteners are added to toothpaste?

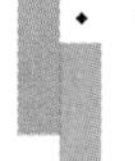

CRITICAL THINKING PROBLEMS

1. Read box 13.3 and the following articles: (1) Elisabeth Kirschner and Howard Grund, "Green Chemicals," *Chemical Week*, October 27, 1993, pp. 36–37; and (2) Penelope Wang, "Going for the Green," *Money*, September 1991, pp. 98–102. Then consider the following five surfactants: (a) sodium stearate (ordinary soap); (b) sodium alkylbenzene sulfonate, an anionic made from petrochemicals; (c) sodium lauryl sulfate, an anionic made from the lauryl alcohol derived from coconuts; (d) a typical quaternary surfactant; and (e) a nonionic alkylphenol ethoxylate made from petrochemicals. The structures and synthesis of these five materials are shown in table 13.1. Now try to list these five surfactants in order of decreasing friendliness to the environment. Which of these materials is the most environmentally friendly? Which is the least? (*Caution:* This is a tough question. You must consider many factors to assess environmental impact, and it is sometimes difficult to gather all the necessary information. Remember that you can form an opinion on available information and change it if future data warrant it. On this basis, make your tentative listing for these five surfactants.)

2. We are currently bombarded with surveys. Unfortunately, survey questions can be designed to induce almost any desired answer. Try asking at least 25 people the following three survey questions:

a. Would you like to use a laundry detergent that makes your clothes cleaner?

b. Would you like to have a cleaner environment?

c. Which would you prefer—(1) a cleaner environment regardless of the cost, or (2) a laundry detergent that makes your clothes cleaner and does essentially no harm to the environment?

Note that questions *a* and *b* are phrased to produce "yes" answers, and *c* is weighted toward option 2. Can you think of ways to ask the same questions that might change the results?

3. Using the articles in the Readings of Interest section, trace how the detergent market has changed during the past few years. If the projections for the future come true, how will they affect your lifestyle and the environment?

4. A better environment is a worthy goal. How can you contribute? Read the article by Penelope Wang, entitled "Going for the Green," in *Money*, September 1991, pp. 98–102. This article contains several suggestions, but are all of them scientifically sound? For example, are natural ingredients automatically less toxic than synthetic? Are the more expensive "green" products invariably better for the environment?

5. Review the information presented on the use of phosphates, zeolites, and poly(acrylic acids) as builders and the potential effects each has on the environment. Rank these materials in order of biodegradability and in order of effectiveness as a builder. Are these rankings the same? Which of these three chemicals are natural and which are synthetic? Taking as much information as possible into account, which chemical seems to be the safest detergent builder? (Consider the fact that more surfactant is needed with some builders than with others. Even if the surfactant is biodegradable, it might cause some environmental damage by reducing the BOD in the wastewater stream.)

CHAPTER 14

THE CHEMISTRY OF LIFE

OUTLINE

Classes of Biochemicals–The Stuff of Life
Polypeptides: Amino Acids and Proteins
Saccharides and Sugars—Not Just for Dessert
Nucleic Acids
Lipids
Hormones
Vitamins
Minerals
Birth Control
The Chemistry of Our Senses

BOX 14.1 Chiral Objects
BOX 14.2 Sickle Cell Disease
BOX 14.3 Antibodies, Antigens, and Allergies
BOX 14.4 Cystic Fibrosis
BOX 14.5 Cholesterol Testing
BOX 14.6 Biochemical Energy Production

KEY IDEAS

1. The Seven Major Groups of Biochemicals
2. The Relationship Between Amino Acids and Proteins
3. How Optical Activity Relates to Chemical Structure
4. The Four Levels of Protein Structure
5. The Nature and Function of Enzymes
6. The Chemical Nature of Saccharides
7. The Names of the Main Mono-, Di- and Polysaccharides
8. What Antibodies and Antigens Do
9. The Nature and Function of Nucleic Acids
10. The Classifications and Functions of Lipids
11. The Nature and Functions of Hormones
12. The Role Vitamins Play in Basic Biochemistry
13. How Minerals are Involved in Biochemistry
14. The Main Features of Energy Production in Living Systems
15. The Chemistry of Contraception
16. The Chemistry of our Senses

"Life is not a problem to be solved but a reality to be experienced."
SÖREN KIERKEGAARD (1813–1855)

Beta carotene seen through transmitted light microscope. (4×).

Biochemistry, the study of living systems, involves many complex biochemicals. Biochemistry can be complicated, but in this chapter, we will focus on an area most people can relate to—how biochemistry interacts with our lives. Table 14.1 divides the main biochemicals into seven groups. This chapter will examine each of these groups, then turn to the topics of biological energy production, the chemistry of our senses, and the chemistry of birth control. Since all life on this planet is chemical in nature, the study of biochemistry truly is the study of life itself.

CLASSES OF BIOCHEMICALS—THE STUFF OF LIFE

Biochemicals may be classified in several ways; we have divided them into seven main groups. The first group of biochemicals are the **polypeptides,** which include the proteins and enzymes. The name *protein* derives from the Greek word *proteios* (primary, or of first importance); this is appropriate, since these compounds are essential to life. Proteins are amino acid polymers that resemble the polyamides (nylons). In fact, Carothers developed nylon as a synthetic replacement for silk—a polypeptide (chapter 10).

PRACTICE PROBLEM 14.1

List the seven groups cited in table 14.1. What chemical structural feature or function do they have in common?

ANSWER The groups are polypeptides, saccharides, nucleic acids, lipids, hormones, vitamins, and minerals. These chemicals have no common structural feature. Their molecules are totally different from each other. They do, however, have a common function—at least in a broad sense, they are all involved in some biological processes.

POLYPEPTIDES: AMINO ACIDS AND PROTEINS

Amino acids are the building blocks of proteins and enzymes. They combine two organic functional groups in their molecule—a basic amine group (NH_2) and a carboxylic acid ($—COOH$). The normal amino acids have an α-carbon (or second carbon) attached to both the $—NH_2$ and $—COOH$ group. There are two ways to write their structure—the usual form, with neutral groups, and the zwitterion (charged) form, in which a proton transfers from the $—COOH$ to the $—NH_2$ to yield $—COO^-$ and $—NH_3^+$ groups; the zwitterion form is the usual state for amino acids. Table 14.2 shows both forms for some typical amino acids.

Twenty different amino acids are needed to form the proteins in our bodies and in our food (figure 14.1). Approximately ten of them are called essential amino acids because our bodies cannot synthesize them; an asterisk marks the essential amino acids in table 14.2 and figure 14.1. Each molecule in figure 14 has two sections; the colorless section contains the amino ($—NH_2$) and carboxylic acid ($—COOH$) groups plus the α-carbon, while the light purple section contains the R group that makes the amino acids different. Proteins, a rather monotonous breed, only form from a small, select group of amino acids. Even with this limitation, more protein molecules are possible than could be constructed from all the atoms in the entire universe.

TABLE 14.1 One method of grouping biochemicals.

GROUP	CHEMICAL NATURE	EXAMPLES OF USES AND FUNCTIONS
Polypeptides	polymers of amino acids	proteins and enzymes; found in muscles, nerves, other body structures
Saccharides	varies from monomers to polymers; $C_x(H_2O)_y$ formula	sugars; structural material in plants and insects
Nucleic acids	DNA, RNA polymers	controllers of heredity; protein synthesis
Lipids	fats, oils; nonpolymeric	cell membranes
Hormones	polypeptides, three other nonpolymeric classes	mediate body chemistry; control sexual processes
Vitamins	variable; nonpolymeric	essential to life; chemistry of vision; coenzymes
Minerals	inorganic, nonpolymeric	aid enzymes; found in bones, teeth

PRACTICE PROBLEM 14.2

Each amino acid in figure 14.1 has a —COOH group, an —NH_2 group, an H atom, and an α-carbon atom, but the groups that make up the rest of each molecule differ.

a. Which amino acid has two identical groups on the α-carbon atom?

b. Which amino acid contains two —NH_2 groups?

c. Which amino acids contain two —COOH groups?

d. Which amino acids contain an aromatic unit?

e. Which amino acid has the amine group contained in a ring?

f. Which amino acids contain an —OH group (other than in the —COOH unit)?

g. Which amino acid contains the —SH (thiol) group?

ANSWER *a.* Glycine. *b.* Lysine, arginine, glutamine, and asparagine (the last two actually contain amide, rather than amine groups). *c.* Glutamic acid, aspartic acid, and also glutamine and asparagine (as amides). *d.* Phenylalanine, tyrosine, and tryptophan. *e.* Proline. (Possibly you added histidine or tryptophan since these do contain amino groups in a ring, but the α-amino group (attached to the α-carbon) is not ring-bonded.) *f.* You can find —OH groups in serine, threonine, and tyrosine. *g.* Cysteine.

TABLE 14.2 Some typical amino acids, and the general formula for an amino acid, where R = an H or an organic group. Note that the α-carbon is attached to four different groups, except when R = H.

GENERAL FORMULA

$$\mathrm{R{-}C(NH_2)(H){-}COOH} \qquad \mathrm{R{-}C(\overset{+}{N}H_3)(H){-}COO^-}$$

Neutral formula — Zwitterion (charged) formula

NAME	ABBREVIATION	STRUCTURE OF R
Glycine	Gly	H—
Alanine	Ala	CH_3—
Phenylalanine*	Phe	$C_6H_5CH_2$—
Valine*	Val	$(CH_3)_2CH$—
Leucine*	Leu	$(CH_3)_2CHCH_2$—
Serine	Ser	HO—CH_2—
Glutamic acid	Glu	HOOC—CH_2CH_2—
Lysine*	Lys	H_2N—$CH_2CH_2CH_2CH_2$—
Cysteine	Cys	HS—CH_2—

*Denotes an essential amino acid

Optical Isomerization

Amino acids introduce a new kind of isomerization called **optical isomerization.** This occurs when a C atom has four different groups attached to it. Most amino acids contain four groups—H, COOH, NH_2, and R—attached to a C atom—and they usually occur in a tetrahedral shape. These amino acids may have isomers with the same molecular formula and tetrahedral structure, but with their functional groups in different places on that structure. Amino acids with such isomers are called **asymmetrical** or **chiral.** Box 14. 1 explains chiral molecules in more detail. A pair of chiral molecules is called an **enantiomer** pair.

EXPERIENCE THIS

Examine the following common objects and determine which are chiral; *(a)* a table knife, *(b)* a pencil, *(c)* a pair of gloves, *(d)* a shirt that buttons, and *(e)* a TV set.

ANSWER A table knife *(a)* is symmetrical (nonchiral) when it is viewed along its cutting edge. Likewise, a pencil *(b)* is symmetrical when viewed along its length or from either end. Gloves *(c)* are not symmetrical; a pair is chiral. A buttoned shirt *(d)* is also not symmetrical and this is chiral. A TV set *(e)* may seem symmetrical, but it usually has something different on one side, perhaps a speaker or a different knob.

Optical Activity

The two enantiomers of a chiral molecule have many identical physical and chemical properties, but they differ in at least one respect—the way they respond to plane polarized light.

Ordinary light waves move in all directions, but plane polarized light waves move only in one plane (figure 14.2). We encounter polarized light whenever light reflects from certain surfaces; thus, sunlight reflected off snow or sand produces polarized light. Polarized sunglasses reduce the glare from sand or snow because they block the polarized light reflected from these surfaces. When a beam of light passes through a polarizing lens, it emerges in one plane and is thus called **plane polarized light.** If plane polarized light is then passed through a solution containing one enantiomer of an asymmetric or chiral molecule, the enantiomer is thus **optically active.**

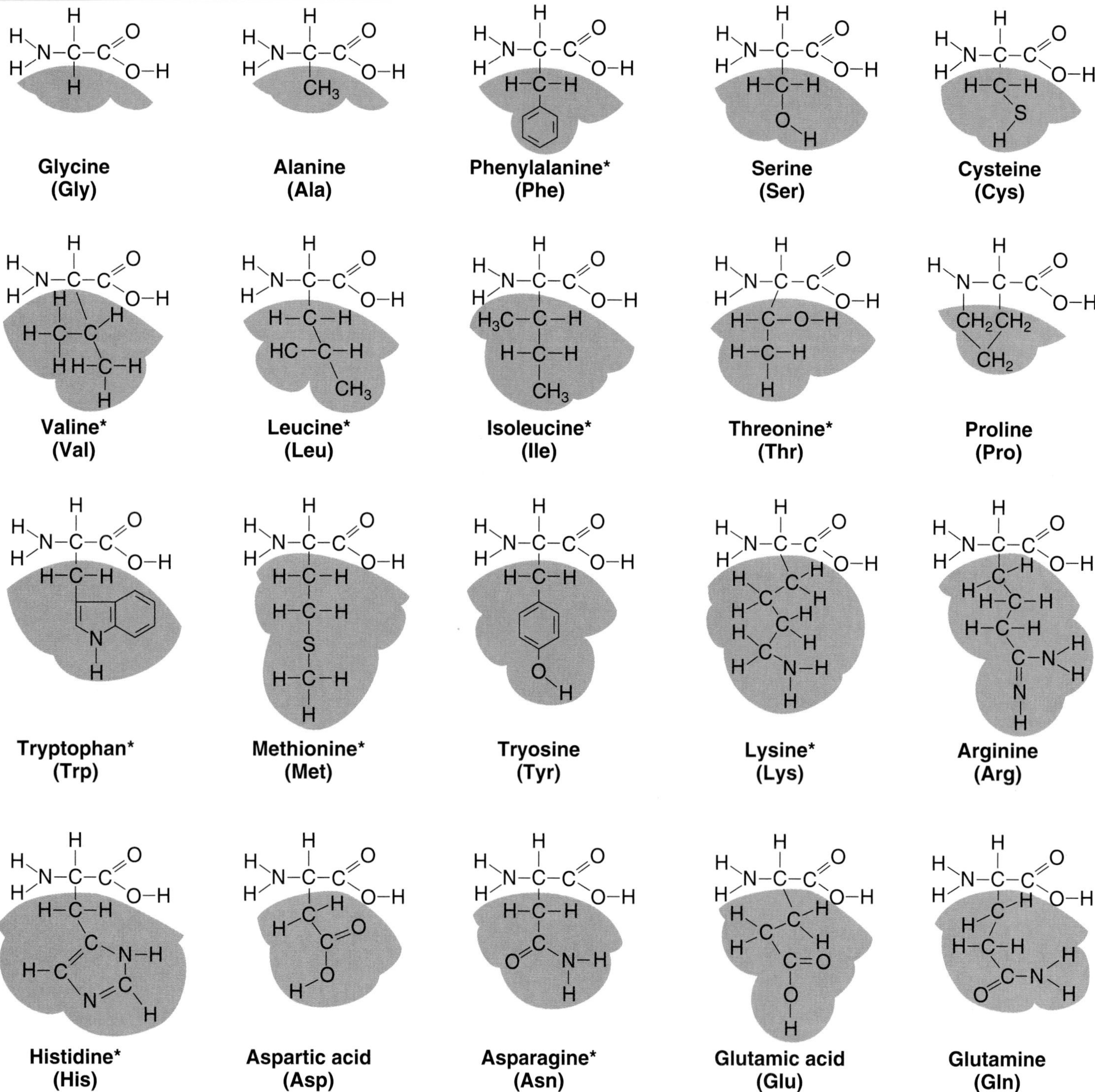

Figure 14.1 The structures of the twenty common amino acids and their three letter abbreviations. (*Essential amino acids.)

The two enantiomers of a chiral molecule rotate plane polarized light in opposite directions. If one rotates the plane 15 degrees clockwise, its mirror image would rotate the plane 15 degrees counterclockwise.

Chirality, optical isomerization, and optical activity are not restricted to amino acids or their polymers. Figure 14.3 shows typical chiral and nonchiral molecules. Notice that the three chiral molecules each have one C atom that bears four different groups.

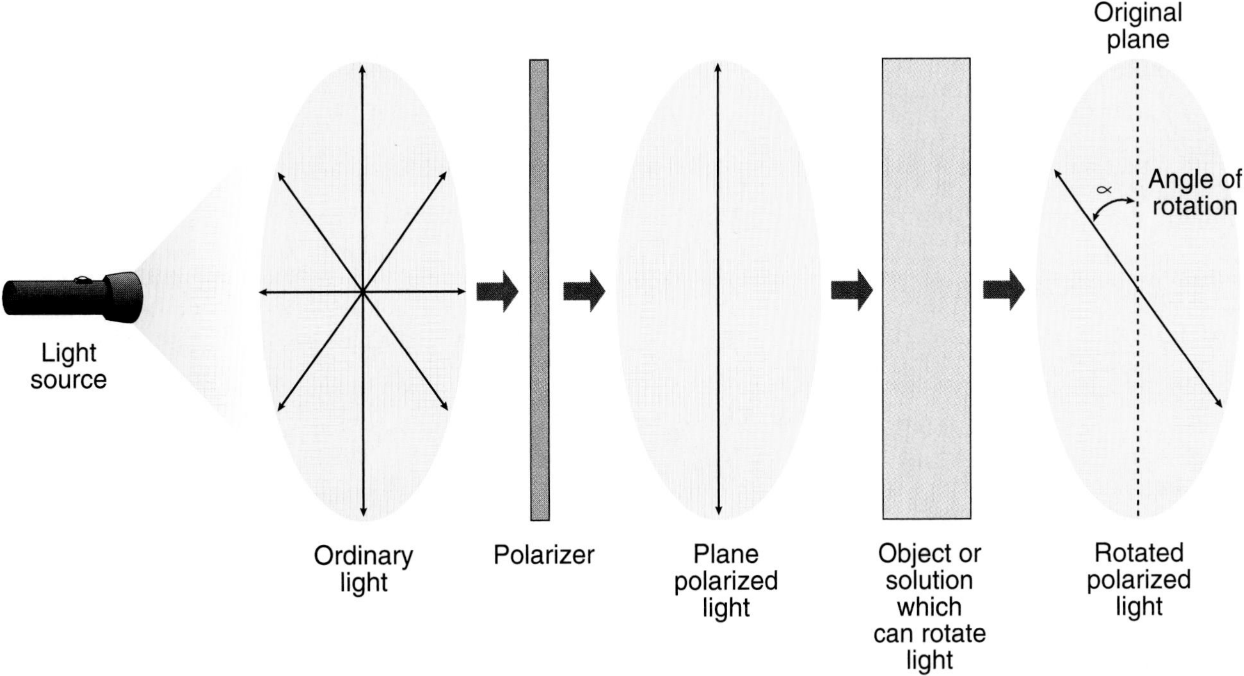

Figure 14.2 When plane polarized light passes through an object or solution that can cause rotation, the light motion shifts from the vertical to a horizontal direction.

Potentially, all these chiral molecules could rotate polarized light. On the other hand, the nonchiral (or symmetric) molecules do not have four different groups attached to their C atoms. The mirror images of each of these tetrahedrons can superimpose on one another. In a chiral molecule, the mirror images do not superimpose.

Chirality and optical activity are vitally important because all amino acids (except glycine) and many saccharide (carbohydrate) molecules contain at least one chiral carbon and are optically active. These molecules form the basis of life. The two isomers or enantiomers that exist in chiral molecules are called righthanded, or D-, and lefthanded, or L-isomers. In nature,

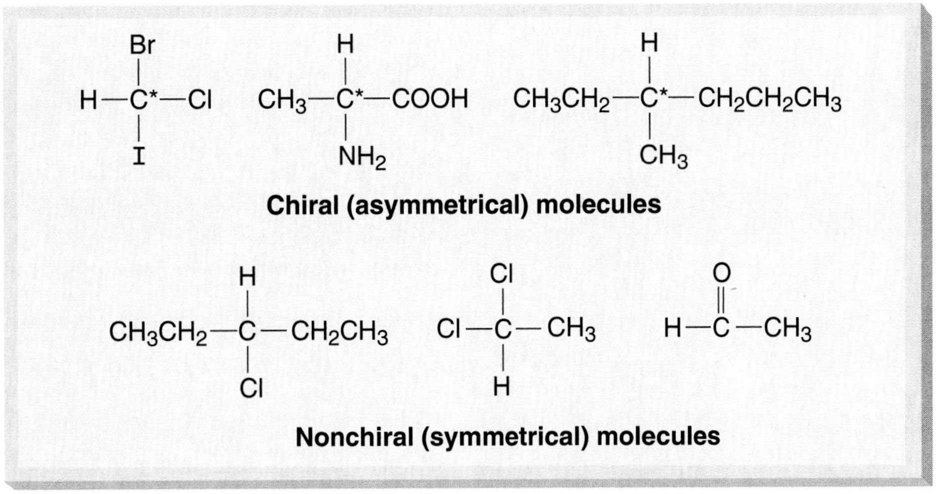

$H-C^*(Br)(I)-Cl$ $\quad$ $CH_3-C^*(H)(NH_2)-COOH$ $\quad$ $CH_3CH_2-C^*(H)(CH_3)-CH_2CH_2CH_3$

Chiral (asymmetrical) molecules

$CH_3CH_2-C(H)(Cl)-CH_2CH_3$ $\quad$ $Cl-C(Cl)(H)-CH_3$ $\quad$ $H-C(=O)-CH_3$

Nonchiral (symmetrical) molecules

Figure 14.3 Some typical chiral (asymmetric) and nonchiral (symmetrical) molecules. The * denotes a chiral carbon atom.

BOX 14.1

CHIRAL OBJECTS

Although you can arrange many molecules so that both sides look alike, the righthand side of an asymmetrical or chiral molecule looks different from the left. A **chiral** molecule comes in two versions, or **enantiomers;** each enantiomer is the *mirror image* of the other (figure 1).

You encounter many chiral objects daily—your hands are a good example. (In fact, the word *chiral* comes from the Greek word meaning "hand.") When you place your hands with both palms facing you, you cannot superimpose one hand on the other and match the equivalent fingers (that is, each thumb lines up with the little finger of the other hand). The mirror images of chiral objects do not superimpose.

Of course, not everything is chiral. An ordinary drinking glass is symmetrical; you could split it in two at any point and superimpose one piece over the other. Hence, it is not chiral. Some objects seem chiral until you examine them carefully. For example, a banana looks asymmetric if laid on its side, but it is fairly symmetrical if you held it with both ends pointing toward you. Likewise, a cup with a handle jutting out the side appears chiral (asymmetric), but if it is viewed with the handle turned toward you, its symmetrical nature becomes apparent.

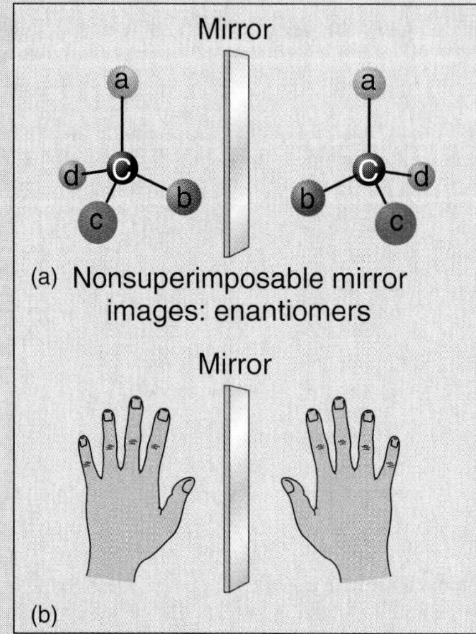

Figure 1 How chirality operates.

carbohydrates are usually D-isomers and amino acids are usually L-isomers. Though scientists can synthesize D-isomer amino acids and L-isomer sugars, these forms would not work in our metabolism. In fact, such "wrong-handed" molecules are frequently poisonous.

EXPERIENCE THIS

Collect some gumdrops and toothpicks and prepare three pairs of single C atom models. Make the C atom black in all three model pairs and arrange the four toothpicks in a tetrahedron. Use figure 1a in Box 14.1 to prepare a mirror-image pair for each of the three types of molecules.

In model pair 1, attach yellow, red, and green gumdrops to three toothpicks, then add another yellow gumdrop to the fourth toothpick. Model 1 corresponds to molecules with two identical and two different groups attached to the carbon atom. Model pair 2 uses three yellow gumdrops plus a red one on the toothpicks; this corresponds to a carbon atom attached to three identical groups. Finally, attach yellow, red, green, and white gumdrops to the toothpicks on each molecule in model pair 3; this pair will represent a molecule with four different groups attached to the carbon atom. Compare each model in the set with its mate; try to superimpose one over the other, matching the gumdrop colors. Which pair(s) are chiral and which are not?

ANSWER You can superimpose each colored gumdrop on top of the same color on the other model in pairs 1 and 2. Therefore, neither set is chiral. You cannot match up the four colored gumdrops in the two models in pair 3; this corresponds to a chiral molecule, with four different-colored gumdrops (different functional groups) attached to the carbon atom.

Amino Acid Polymers—Polypeptides and Proteins

Polypeptides and proteins are polymers of amino acids, but they are more complex than the polymers we encountered in chapters, 9, 10, and 11. A schematic drawing of a five-unit polypeptide chain is shown in figure 14.4; the R, R_a, R_b, and so on represent the various side chain groups present on the amino acids (the R

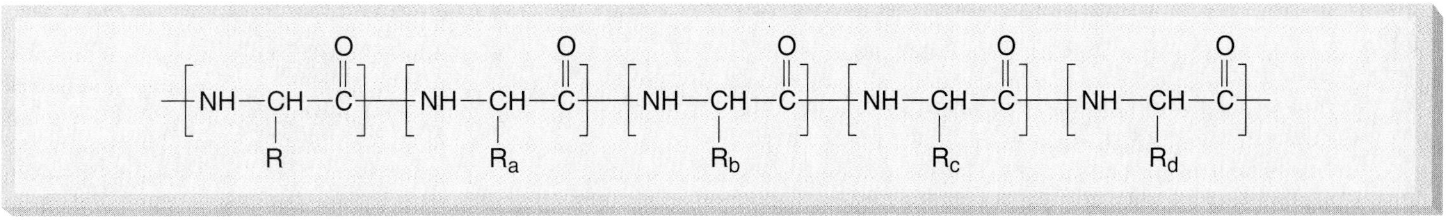

$$\left[-NH-\underset{R}{CH}-\overset{O}{\overset{\|}{C}}-\right]\left[-NH-\underset{R_a}{CH}-\overset{O}{\overset{\|}{C}}-\right]\left[-NH-\underset{R_b}{CH}-\overset{O}{\overset{\|}{C}}-\right]\left[-NH-\underset{R_c}{CH}-\overset{O}{\overset{\|}{C}}-\right]\left[-NH-\underset{R_d}{CH}-\overset{O}{\overset{\|}{C}}-\right]$$

Figure 14.4 A five-repeat-unit portion of a polypeptide, using general formulas for the amino acids. Each peptide unit is shown in a bracket.

groups in table 14.2). Notice that the nature of the side group changes with each repeat unit. This variability makes proteins much more complex than simple plastics and rubber. The primary repeat unit, called a **peptide** unit, contains an amide group attached to the carbon that bore it in the original amino acid. The peptide units are enclosed in brackets in figure 14.4. Polypeptides with molecular weights exceeding 10,000 are normally called proteins. An actual protein molecule is hundreds of times longer than the short sequence in the figure. Although poly(amino acids) can form by the following condensation reaction, the actual procedure is far more complex:

$$n\,H_2N-\overset{R}{\overset{|}{CH}}-\overset{O}{\overset{\|}{C}}-OH \rightarrow \left[-HN-\overset{R}{\overset{|}{CH}}-\overset{O}{\overset{\|}{C}}-\right]_n + n\,H_2O$$

When the amino acids glycine and alanine are combined, as in figure 14.5, two different dipeptides can form, depending on which amino acid loses the —OH group. Examine these structures carefully and you will see that you cannot match them up atom-by-atom; they cannot be superimposed for the simple reason that they are different compounds. If a specific dipeptide is desired, glycylalanine for example, it is necessary to apply special chemical techniques to avoid making both products at the same time.

The fact that two distinct dipeptides form from the same pair of amino acids means these dipeptides must have different names. Most of the time, chemists use the abbreviations shown in figure 14.1 or table 14.2 as a shorthand notation to depict polypeptides. Using Gly and Ala for glycine and alanine, respectively, the two formulas in figure 14.5 become Gly-Ala and Ala-Gly. It is not allowable to reverse these shorthand names; the free amine (NH_2) end must always come first. Violation of this firm rule could cause confusion between the shorthand formulas for two different compounds.

The **primary structure** of a polypeptide is the sequence of amino acids in its chain. The potential number of polypeptides becomes greater as more amino acids are reacted. For example, reacting glycine (Gly), alanine (Ala), and phenylalanine (Phe), yields six possible tripeptides (figure 14.6). Four amino acids provide 24 possible primary structures. Ten amino acids result in more than 3.6×10^6 possible structures, and 100 amino acids (five each of the twenty most common) produce more than 9.3×10^{157} combinations. If a protein has a molecular weight of a million, it contains more than 8,500 amino acids in its primary structure. There is not enough matter in the entire universe to make a single molecule of every possible arrangement of these 8,500 amino acids.

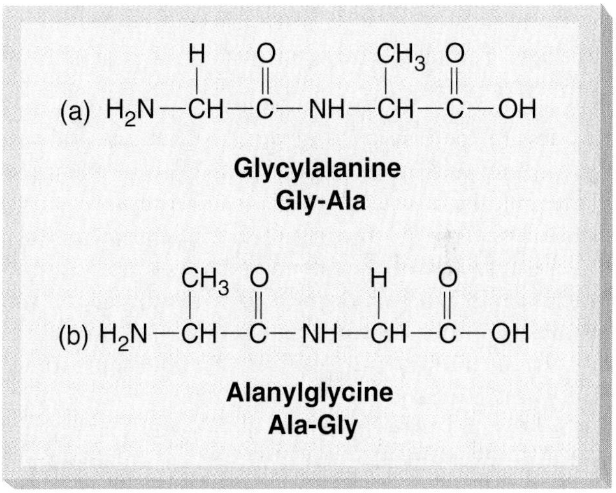

(a) $$H_2N-\overset{H}{\overset{|}{CH}}-\overset{O}{\overset{\|}{C}}-NH-\overset{CH_3}{\overset{|}{CH}}-\overset{O}{\overset{\|}{C}}-OH$$

Glycylalanine
Gly-Ala

(b) $$H_2N-\overset{CH_3}{\overset{|}{CH}}-\overset{O}{\overset{\|}{C}}-NH-\overset{H}{\overset{|}{CH}}-\overset{O}{\overset{\|}{C}}-OH$$

Alanylglycine
Ala-Gly

Figure 14.5 Two dipeptides prepared from the same pair of amino acids—glycine and alanine. The name and shorthand notation appears for each structure using the abbreviations from table 14.2.

Gly-Ala-Phe	Gly-Phe-Ala
Ala-Gly-Phe	Ala-Phe-Gly
Phe-Gly-Ala	Phe-Ala-Gly

Figure 14.6 The six possible tripeptides that can be prepared using one of each of the three amino acids glycine, alanine, and phenylalanine.

PRACTICE PROBLEM 14.3

Write a set of six tripeptides, as in figure 14.6, using the amino acids valine (Val), leucine (Leu), and serine (Ser). Are there any other possible combinations these amino acids can form?

ANSWER The six possible tripeptides are: Val-Leu-Ser, Val-Ser-Leu, Leu-Val-Ser, Leu-Ser-Val, Ser-Val-Leu and Ser-Leu-Val. There are no other possible combinations for these three amino acids.

Secondary, Tertiary, and Quaternary Protein Structures

In addition to the primary structure, proteins also have a secondary, tertiary, and quaternary structure. The **secondary structure** determines the protein's shape; it is a repetitive folded pattern arising from the interactions of the units in the polypeptide chain, mostly from hydrogen bonding. In some instances, covalent bonds aid in maintaining the shape. Insulin (figure 14.7) has an unambiguous primary structure that results in hydrogen bonding. It also contains a pair of —S—S— crosslinks that help maintain its specific shape.

Figure 14.8 illustrates the details of an **α-helix,** a major type of secondary protein structure, with (a) a schematic drawing, (b) a molecular model, and (c) a top view. A spiral staircase is actually a helix; perhaps the classic movie *The Spiral Staircase* should have been named *The Helical Staircase*. Another kind of secondary structure, the **β-pleated sheet,** appears in figure 14.9. Although the α-helix is more common, both of these structures occur often in proteins. Most muscle and nerve proteins are α-helices, but silk and some other fibers have the β-pleated sheet structure.

Collagen, which comprises almost half the total weight of protein in the human body, consists of a trio of α-helices intertwined to form a **tertiary structure.** This type of structure combines more than a single protein molecule, but these molecules are not chemically linked. Figure 14.10 shows the structure of a collagen-containing hair strand. The portion called the protofibril is a tertiary structure containing several α-helices. Aside from its occurrence in hair, collagen forms many of the structural features in the body, providing a scaffold for bone development and forming

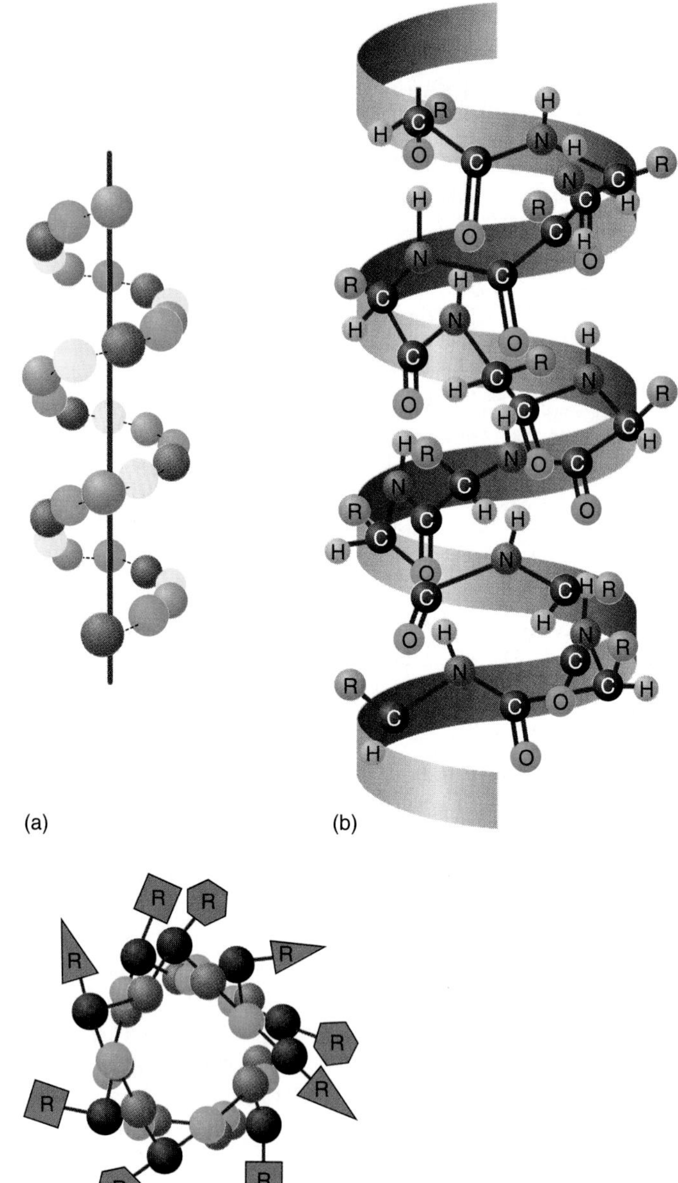

Figure 14.8 The α-helix, a protein secondary structure. **(a)** a schematic model, **(b)** a molecular model, and **(c)** a top view of the helix. In **(a)** the balls represent repeat units.

SOURCE: From Robert L. Caret, et al., *Foundations of Inorganic, Organic, & Biological Chemistry*.

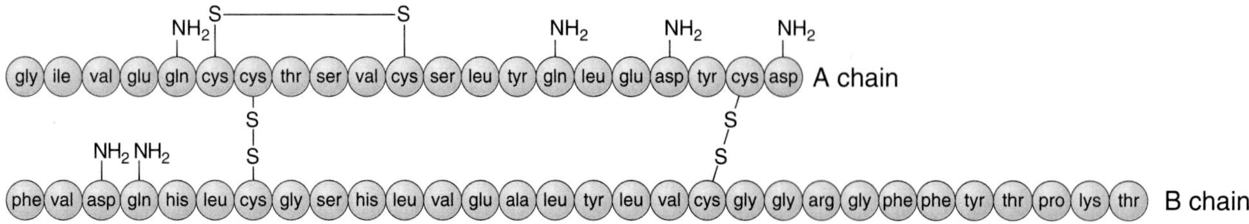

Figure 14.7 The chemical structure of insulin.

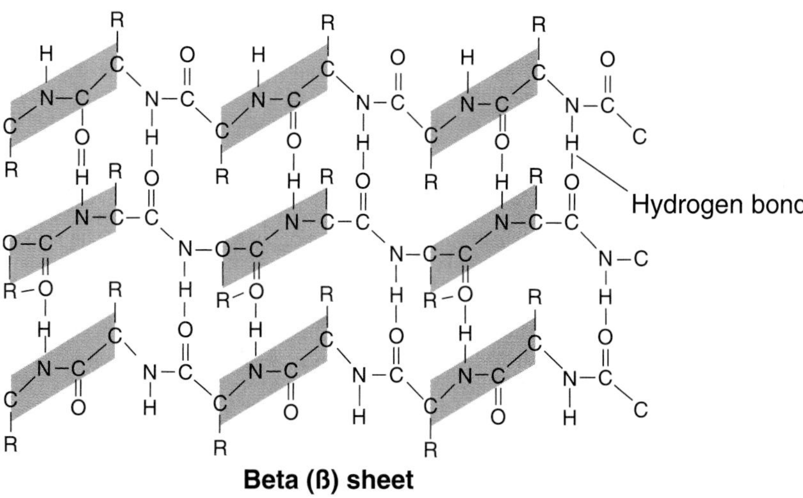

Figure 14.9 The β-pleated sheet, another protein secondary structure.

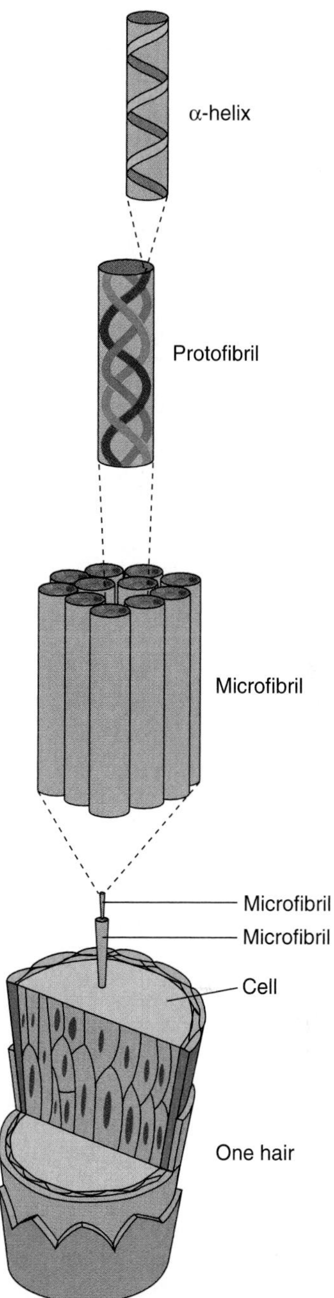

Figure 14.10 Hair, a collagen related structure formed from α-keratins.

SOURCE: From Robert L. Caret, et al., *Foundations of Inorganic, Organic, & Biological Chemistry*. Copyright © 1995 Times Mirror Higher Education Group, Inc., Dubuque, Iowa. All Rights Reserved. Reprinted by permission.

the major constituent in the skin and blood vessels. In skin, collagen occurs in a loosely woven arrangement that can expand in all directions; this allows the skin to stretch in all directions. Hair cannot do this. Blood vessels also have collagen arranged into loose helical tertiary structures to achieve high elastic strength.

Hemoglobin consists of two different proteins and four heme units responsible for conveying oxygen throughout the body. Each heme unit is a heterocyclic molecule (that is, a ring-shaped molecule that contains other atoms besides carbon) with an iron atom in the center (figure 14.11). In hemoglobin, the protein chains and the heme units fold together to form a **quaternary structure,** or a **multisubunit protein** that contains several proteins folded together with other units. Figure 14.12 shows the relationship between these four types of protein structures for the hemoglobin molecule. As box 14.2 reveals, a tiny change in the primary structure can have dire results.

PRACTICE PROBLEM 14.4

What is the main structural difference between the α-helix and the β-pleated sheets?

ANSWER The α-helix is a coil, like a spring, while the β-pleated sheet is flat.

PRACTICE PROBLEM 14.5

How do the four levels of protein structure differ from each other?

ANSWER The primary structure is the sequence of amino acids in the polymer chain, while the secondary structure defines the shape this molecule assumes as the primary units bond and the structure folds. The tertiary structure is the final form of a protein, consisting of more than one protein molecule that are folded together but not chemically linked. Finally, a quaternary structure has several different proteins and other subunits that fold together to form a multisubunit protein.

```
        CH3  CH=CH2
         |    |
         C====C
         |    |
    HC—C     C=CH
       ||  \\ /  |
CH3—C—C    N    C=C—CH3
    |    \ |2+ /
    |     N--Fe—N
    ||  //  |  \
CH2—C—C     N   C=C—CH=CH2
|      |  /  \   |
CH2   HC=C    C=CH
|         |    |
COOH      C====C
          |    |
         CH2  CH3
          |
         CH2
          |
         COOH
```

Heme group

Figure 14.11 The structure of the heme prosthetic group.

Enzymes—Natural Catalysts

Enzymes are polypeptides that act as catalysts, speeding reactions and allowing them to occur at lower temperatures. Most enzymes require at least one other material, such as a vitamin or a mineral, to achieve this effect.

Most enzymes operate by a "lock-and-key" mechanism (figure 14.13). In this mechanism, the reactant molecule (called the substrate) binds with the enzyme to form a complex; the complex site on the enzyme has a shape that relates to the reactant as a lock relates to the key. This specific shape requirement makes enzymes highly specific; enzymes usually react with only certain specific chemicals—those that fit into the enzyme's cavity. In figure 14.13, only the hexagonal end of the substrate fits into the enzyme; the pentagonal end does not fit. If both ends were the wrong shape, the molecule would not fit into the cavity of the enzyme and would not react any better than an incorrect key would open a lock. After the reactant fits into the enzyme, it reacts rapidly. Without an enzyme, the reaction might take hours, or might not occur at all. After the reaction is complete, the reactant exits and the enzyme is restored for further use.

Enzymes are important for cleaving proteins back to amino acids, cleaving saccharides to monosaccharides, and synthesizing body tissues from these smaller parts. Enzyme names normally end is *ase,* which makes them easy to identify. The first part of the name often comes from the substrate; thus, *maltase* hydrolyzes maltose and *lactase* hydrolyzes lactose. You encountered the term *amylase* (hydrolyzes starch), *cellulase* (hydrolyzes cellulose), and *lipase* (hydrolyzes fats or oils) in chapter 13.

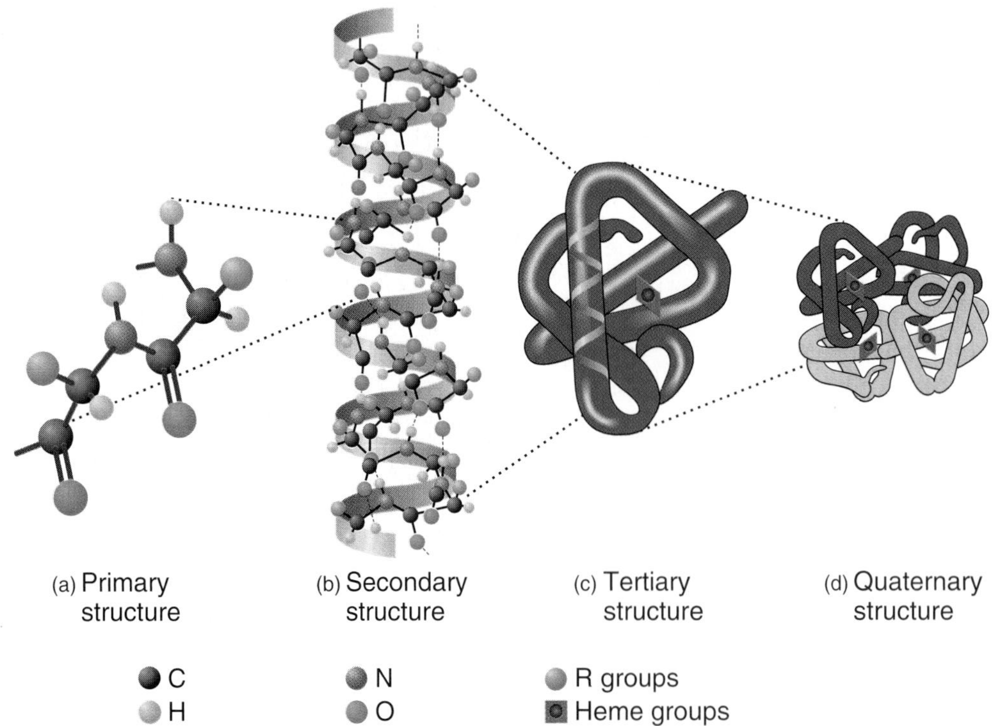

Figure 14.12 A summary of the four levels of protein structure, using hemoglobin as the example.

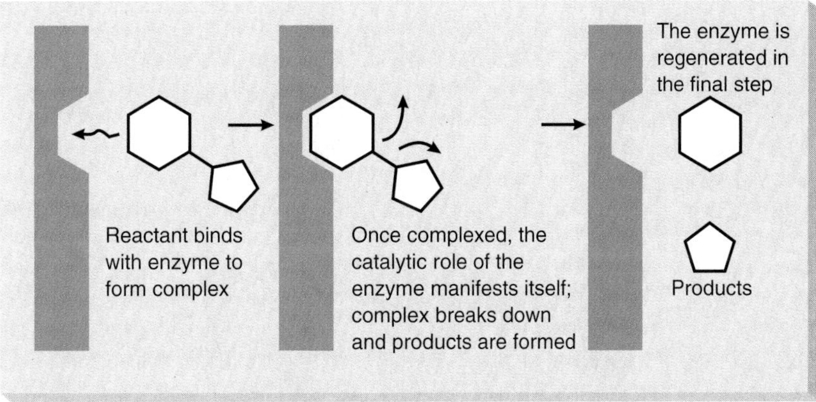

Figure 14.13 The "lock and key" analogy for enzyme catalysis.

PRACTICE PROBLEM 14.6

What other analogies, besides the lock-and-key mechanism, could be used to explain enzyme reactivity and specificity?

ANSWER Any combination of objects that fit together snugly could illustrate enzyme specificity; for example, a hand and glove, a foot and shoe, or even a nut and bolt. The main requirement is that there must be only a single way the pieces fit together. Sort of a jigsaw puzzle—there's another one!

SACCHARIDES AND SUGARS—NOT JUST FOR DESSERT

The second group of biochemicals, after the polypeptides, is the saccharides. **Saccharides,** also called carbohydrates, are very familiar to us. Cellulose (plant fiber), a polymer, is the most common saccharide on earth. Many other saccharides, including starch, are also polymeric.

The term *saccharide* comes from the Latin word *saccharon*, meaning sugar. The alternate name *carbohydrate* arose because many saccharides appear to be hydrates of carbon. Glucose ($C_6H_{12}O_6$), can be written as $C_6(H_2O)_6$, suggesting a hydrate of the six carbons. Similarly, sucrose ($C_{12}H_{22}O_{11}$) could be depicted as $C_{12}(H_2O)_{11}$, a hydrate of twelve carbons. No water molecules exist in either sugar, but the name has persisted. Table 14.3 lists some common saccharides; the names end in the suffix *ose*.

Saccharides are subdivided into classes according to the number of units in the molecule; **monosaccharides** have a single unit, **disaccharides** have two, and **polysaccharides** have many saccharide units. Monosaccharides are subdivided further according to the number of carbon atoms in the molecule. Thus, a three-C-atom monosaccharide is a triose, while a six-C-atom monosaccharide is a hexose. Monosaccharides are called simple sugars because they cannot hydrolyze to a smaller saccharide unit. A disaccharide, such as sucrose, hydrolyzes into two simple sugars—in the case of sucrose, into glucose and fructose. Cellulose or starch, polysaccharides, hydrolyze into many units of glucose.

TABLE 14.3 Some saccharide molecules.

NAME	CLASS	FORMULA
Ribose	mono-	$C_5H_{10}O_5$
Deoxyribose	mono-	$C_5H_9O_4$
Glucose	mono-	$C_6H_{12}O_6$
Fructose	mono-	$C_6H_{12}O_6$
Galactose	mono-	$C_6H_{12}O_6$
Sucrose	di-	$C_{12}H_{22}O_{11}$
Lactose	di-	$C_{12}H_{22}O_{11}$
Maltose	di-	$C_{12}H_{22}O_{11}$
Starch	poly-	$(C_6H_{10}O_5)_n$
Cellulose	poly-	$(C_6H_{10}O_5)_n$

PRACTICE PROBLEM 14.7

Examine table 14.3. Could chemists write the formula of either starch or cellulose as a hydrate of carbon? What would this formula look like?

ANSWER Since both starch and cellulose have the same general formula, $[C_6H_{10}O_5]_n$, the hydrate formula is the same: $[C_6(H_2O)_5]_n$, implying five water molecules attached to six carbons.

BOX 14.2

SICKLE CELL DISEASE

The shape of a protein determines its biological structure; thus, the primary structures are extremely specific. A change in the placement of a single amino acid can change the nature of a protein. For example, when a valine unit replaces a glutamic acid unit in the hemoglobin molecule, which contains 146 amino acid units, the red blood cells change from their normal rounded shape into sickle-shaped cells (figure 1). This form of hemoglobin is called hemoglobin S; the normal version is called hemoglobin A. People with hemoglobin S have a disease called sickle cell anemia.

Sickle cell anemia is a recessive genetic disease—that is, for someone to acquire the disease, he or she must inherit one sickling cell gene from each parent. A person with only one of the pair of genes has what is called the sickle cell trait, but will not suffer from any of the symptoms of the disease. Sickle cell anemia is found in all parts of the world where malaria is prevalent; it is therefore common in Africa and among people of African descent. Approximately 1 out of every 400 people of African ancestry in the United States has this disease, which is also moderately common among Hispanics (1 in 1,500 have the disease). Some scientists believe that hemoglobin S confers resistance to malaria, but that is of little benefit to people in the United States.

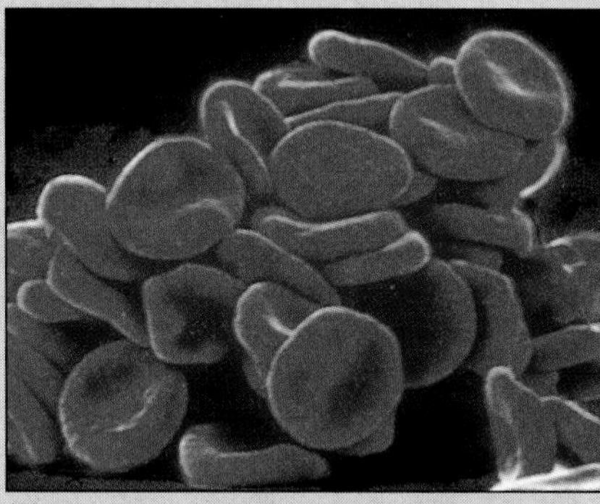
(a)

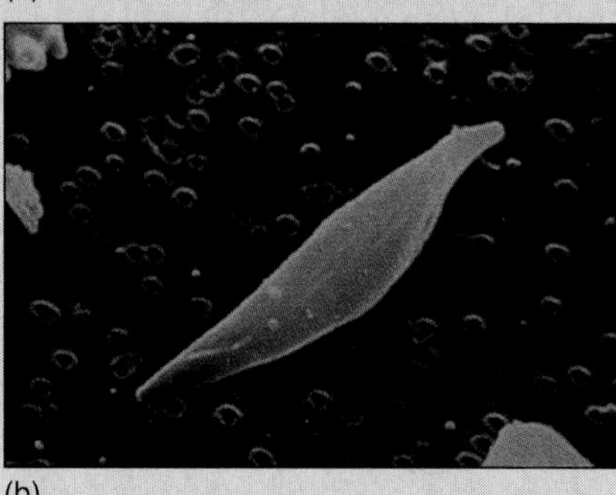
(b)

Figure 1 Normal **(a)** and sickled **(b)** red blood cells.

Sickled red blood cells do not transport oxygen as well as normal red blood cells. They also tend to block the smaller blood vessels (capillaries) because they do not flow through these small vessels as smoothly as the disk-shaped hemoglobin A cells. Accordingly, people with sickle cell disease experience physical weakness, brain damage, kidney damage, pain and stiffness in joints, rheumatism, and even death. There is no cure for sickle cell anemia. Treatment consists mainly of giving the patient extra fluids and additional oxygen when necessary.

In early 1995, physicians made the announcement that hydroxyurea, [$H_2N—CO—NH—OH$], a medicine used to treat leukemia, is also effective in treating sickle cell anemia. Hydroxyurea only treats the symptoms of sickle cell and is not a cure for the disease. Sickle cell anemia is preventable with genetic counseling since it is possible to test for the presence of the defective gene in the prospective parents. Presently, when a couple both carry the sickle cell trait, they are advised against having children since each of their offspring has a 1 in 4 chance of inheriting the disease. In the meantime, biotechnology researchers seek a cure for those who have sickle cell disease.

Monosaccharides

Some of the most important monosaccharides are glucose, fructose, ribose, and deoxyribose. Most monosaccharides exist in cyclic and open-chain forms that can convert from one to the other (figure 14.14). Notice that there are two cyclic and only one open-chain form of glucose. The difference between α- and β-glucose occurs at the carbon on the farthest right in each structure; in the α-form, the —OH group points downward, while in the β-form, it points upward. The higher saccharides consist of cyclic units rather than open chains. There are several chiral C atoms in each saccharide unit, and the cyclic form adds another chiral atom to the total. In glucose, for example, there are four chiral C atoms in the open-chain form, and the cyclic form has five.

Glucose and fructose, with six C atoms, are called hexoses; ribose and deoxyribose, with 5, are called pentoses. Monosaccharides have an aldehyde or a ketone unit in the open-chain form, and are therefore also called aldoses and ketoses, respectively. The C atoms at the end of the chain have a C=O bond and cannot be chiral. Thus, *aldohexoses* have six C atoms (including the terminal C=O) and four chiral atoms. *Ketohexoses* also have six C atoms (including the C=O) and normally have three chiral α-Glucose ⇌ Glucose (open-chain form) ⇌ β-Glucose

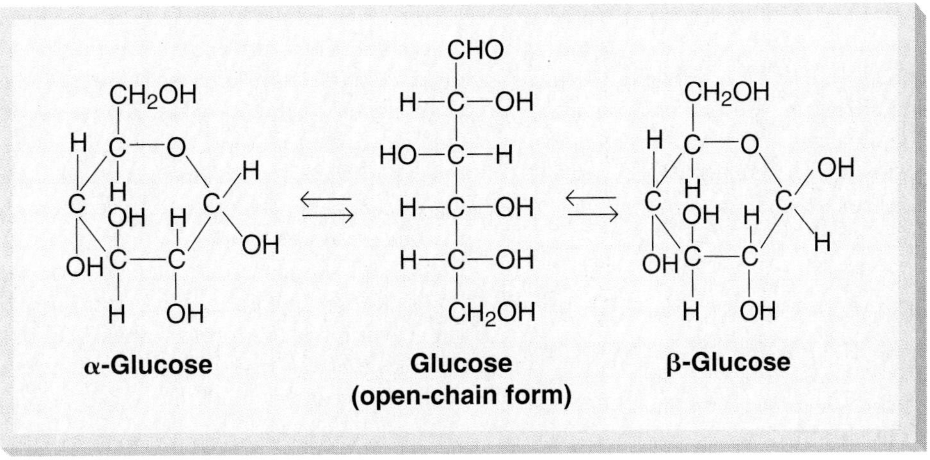

Figure 14.14 The equilibrium between the cyclic and open-chain forms of the glucose molecule.

atoms. The four central C atoms in the open-chain form of glucose (figure 14.14) are all chiral, and *each* of these four C atoms has an isomer with the —OH on the other side, giving a total of sixteen isomeric $C_6H_{12}O_6$ aldohexoses.

As you can see from figure 14.14, the —OH group on C atom number 5 in the open form of glucose is on the righthand side; monosaccharides with analogous placements of this —OH group are called D-sugars, from *dextro,* meaning right. As we previously mentioned, essentially all natural monosaccharides exist in the D-form, and all the important natural amino acids exist in the L-form. If the —OH on the fifth C atom lay on the left side, the sugars would be in the L-form, from *levo,* meaning left. Of the sixteen possible glucose aldohexose isomers, eight are D-isomers and eight are L-isomers.

PRACTICE PROBLEM 14.8

Using figure 14.14 as a guide, draw the open-chain structure for L-glucose.

ANSWER The only difference between L-glucose and D-glucose is that the —OH groups on each C atom are shifted to the opposite side of the molecule.

Glucose

Glucose is the most important monosaccharide. It is the building block (or repeat unit) for both starch and cellulose, and for several other di- or higher saccharides. Glucose is also called dextrose because it rotates polarized light to the right (clockwise). It occurs in the blood and our bodies use it for energy production. Glucose is only 74 percent as sweet as sucrose (ordinary table sugar). Natural sources include honey (which also contains fructose), but the major source is the hydrolysis of cellulose or starch in the following reaction:

$$[C_6H_{10}O_5]_n + n\,H_2O \rightarrow n\,C_6H_{12}O_6$$

Starch is readily hydrolyzed in our bodies, but cellulose is not because the human body lacks the proper enzyme. In other words, we can digest starch but not cellulose.

Fructose

Fructose, found in honey and fruit juices, is the most important ketohexose. It is also called levulose, because it rotates polarized light to the left (counterclockwise). Fructose is 1.73 times sweeter than sucrose and 2.3 times sweeter than glucose; this is why honey is so sweet.

Ribose and Deoxyribose

For a long time, the aldopentoses **ribose** and **deoxyribose** were mere laboratory curiosities even though they occurred in many types of plant, animal, and human tissues. Chemists ultimately learned that these two sugars are extremely important; ribose forms the basis for RNA, and deoxyribose the basis for DNA, the essential molecules of heredity.

Disaccharides

Disaccharides consist of two monosaccharide units. The most important disaccharide is **sucrose,** which forms when glucose and fructose combine. Sucrose is obtained from sugar cane or sugar beets. The average annual consumption of sucrose in the United States is about 65 pounds per person, including that present in prepared foods (*1993 Information Please Almanac*). (The use of other sugars increases the total sugar level to about 125 pounds per person.) Lactose, another disaccharide, is formed from glucose and galactose and occurs in milk; that is why it is called "milk sugar." Maltose, formed from two glucose units, is found in sprouting grains.

Disaccharides are enzymatically cleaved in our bodies to yield two monosaccharides. Strong acids can also produce this reaction, and sucrose is frequently converted into an invert sugar mixture of glucose and fructose this way. Invert sugar is 23 percent sweeter than sucrose—about as sweet as honey.

PRACTICE PROBLEM 14.9

Would you expect an invert sugar made from maltose to be sweeter than sucrose? Why or why not?

ANSWER No, for the simple reason that maltose is formed from two glucose units. The invert sugar is merely glucose, which is only about 74 percent as sweet as sucrose.

Polysaccharides

Polysaccharides consist of many saccharide units. Cellulose, the most common saccharide in the world, forms the structural components of trees and plants. Starch is the next most common polysaccharide; it is used as an energy reserve for plants. Both cellulose and starch are glucose polymers. In insects and shellfish, another polysaccharide called chitin forms the structural parts of the exoskeleton. Chitin is an aminosaccharide polymer—that is, it contains the $—NH_2$ unit in addition to —OH groups.

Starch

Starch (a mainstay of our food supply found in grains, potatoes, vegetables, and fruits) exists in two forms that differ in water solubility and structure: **amylose** is water soluble, and **amylopectin** is insoluble in water. Most plant starches contain about 20 percent amylose and 80 percent amylopectin. Both are polymers of glucose, but amylose is a **linear** polymer while amylopectin is a *branched* polymer (see figure 14.15, where the letter G represents the glucose repeat unit). **Glycogen** is a third starchlike polymer (figure 14.15). Made in the liver, it serves as the storage supply for small glucose surpluses. Glycogen is more highly branched than amylopectin and has shorter branches. The human body contains enzymes that can hydrolyze the α-linkages in starch or glycogen back to glucose, but our bodies cannot hydrolyze the β-linkage in cellulose (figure 14.16).

Chemists have long used starch in industrial and consumer applications. For example, starch is used as a binder to hold white pigments onto paper. This improves the paper's color and makes it more suitable for writing or printing. Starch is also used to reduce wrinkles in clothing by adding stiffness to fabrics.

PRACTICE PROBLEM 14.10

Which form of starch—amylose or amylopectin—might you use to make a wallpaper paste?

ANSWER Since this paste must be at least partially water-soluble to work as a glue, you must use amylose. However, you could probably tolerate a fair amount of amylopectin in the blend.

```
-G-G-G-G-G-G-G-G-G-G-G-G-G-G-G-G-G-G-G-G-G-G-G-G-G-G-G-G-G-G-G-G-G-G-G-G-
```

Amylose

```
G-G-G-G-G-G-G-G-G-G-G-G-G-G-G-G-G-G-G-G-G-G-G-G-G-G
    |                   |
    G-G-G-G-G-G         G-G-G-G-G-G-G-G-G-G-G-G-G
              |
      G-G-G-G-G-G-G-G-G-G-G-G-G-G-G
```

Amylopectin

```
          G-G-G-G-G-G-G-G-G
              |           |
        G-G-G-G     G-G-G-G-G-G-G-G       G-G-G-G
                    |                     |
G-G-G-G-G-G-G-G-G-G-G-G-G-G-G-G-G-G-G-G-G-G-G-G-G-G-G-G
      |                                 |
      G-G-G-G-G-G-G-G-G-G-G-G     G-G-G-G-G-G
            |       |                 |
            G-G-G   G-G-G-G     G-G-G-G
```

Glycogen

Figure 14.15 Amylose, amylopectin, and glycogen structures. The letter G denotes the glucose repeat unit.

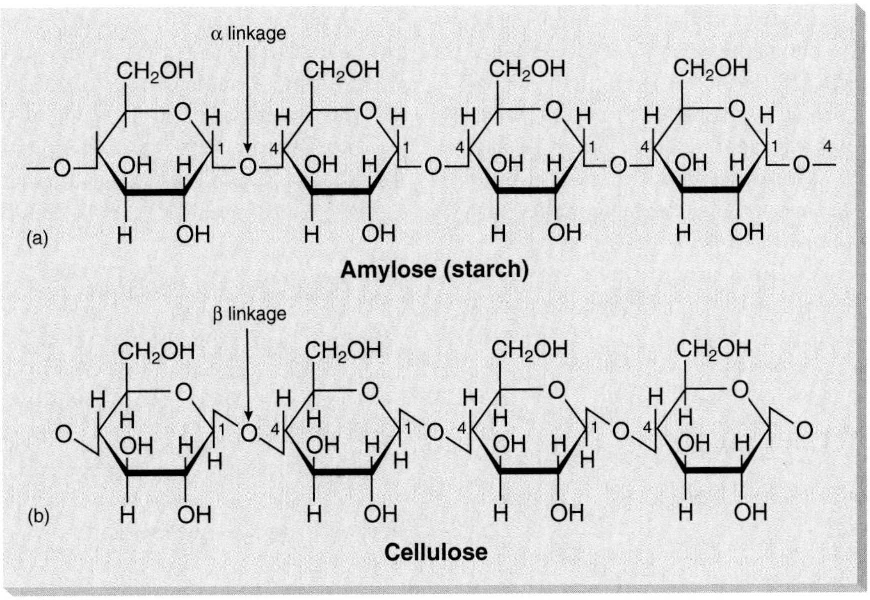

Figure 14.16 A comparison of the α-linkage in starch and the β-linkage in cellulose.

Cellulose

Cellulose is a linear polymer of glucose with the units jointed by β-linkages. Although humans cannot digest cellulose, some animals (cows for example) and insects (notably termites) do so quite well because bacteria in their digestive systems contain the proper enzyme. Many studies suggest that soluble fiber reduces the incidence of colon cancer; cellulose is insoluble, but it is a helpful dietary fiber.

People have used cellulose for centuries, with cotton and wood the most commonly used varieties. Cellulose is not moldable and is virtually useless as a plastic unless modified chemically. Inexpensive modified cellulosics were among the earliest commercial plastics and are still used in tonnage quantities today. The two most common examples are the cellulose nitrates and cellulose acetates (figure 14.17). The figure shows the triester derivatives of these two compounds, but mono- or di- derivatives are also made.

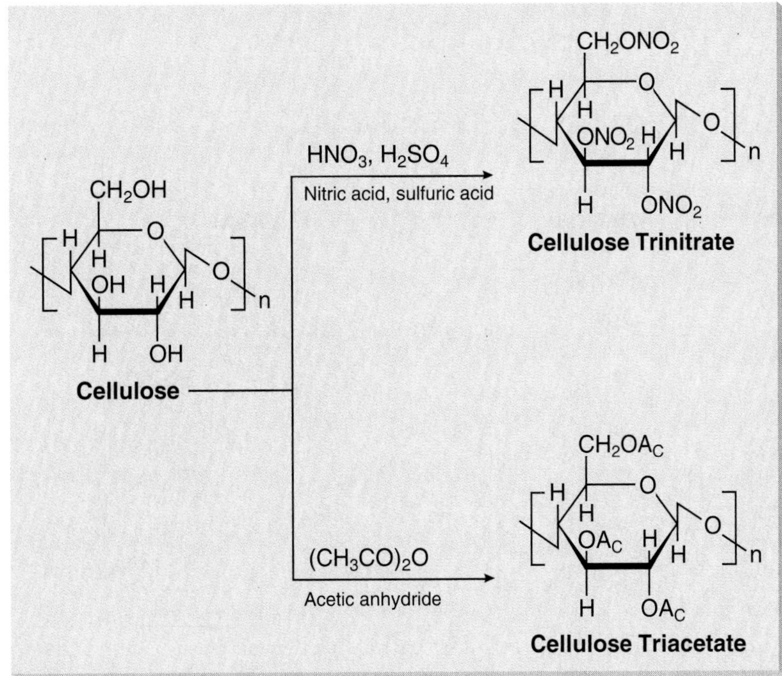

Figure 14.17 Two common examples of modified cellulosics: cellulose trinitrate and cellulose triacetate.

Cellulose trinitrate (also called nitrocellulose) is moldable or extrudable, but it is extremely flammable and it is outlawed for use in plastic toys or fabrics in the United States. The main use for cellulose nitrate is in explosives—guncotton, for example. Pyroxylin, a partially nitrated cellulose encountered as an ether-alcohol solution called collodion, is sometimes used to cover cuts or scrapes. Cellulose acetate is not extremely flammable and is used in textiles, films, lacquers, and photographic or magnetic tape.

Proteins and saccharides are essential to many biological functions. Box 14.3 outlines one—the way the human immune system functions.

PRACTICE PROBLEM 14.11

Look around your home and workplace—what common objects contain cellulose?

ANSWER Certainly you are likely to find cotton, wood, and paper in almost any setting. Many laundry detergents also contain a small amount of a cellulose derivative. Finally, do not overlook the contents of your kitchen cabinets and refrigerator. (You are eating plenty of fiber, aren't you?)

NUCLEIC ACIDS

The third main group of biochemicals, after the polypeptides and saccharides, is the nucleic acid polymers. **Nucleic acids,** though more complex than polysaccharides or polypeptides, are composed of three distinct units—a phosphate, a monosaccharide, and a nucleic base. The main examples are DNA and RNA. Figure 14.18 shows the two monosaccharides and five nucleic bases that occur in the nucleic acid polymers. The monosaccharide is

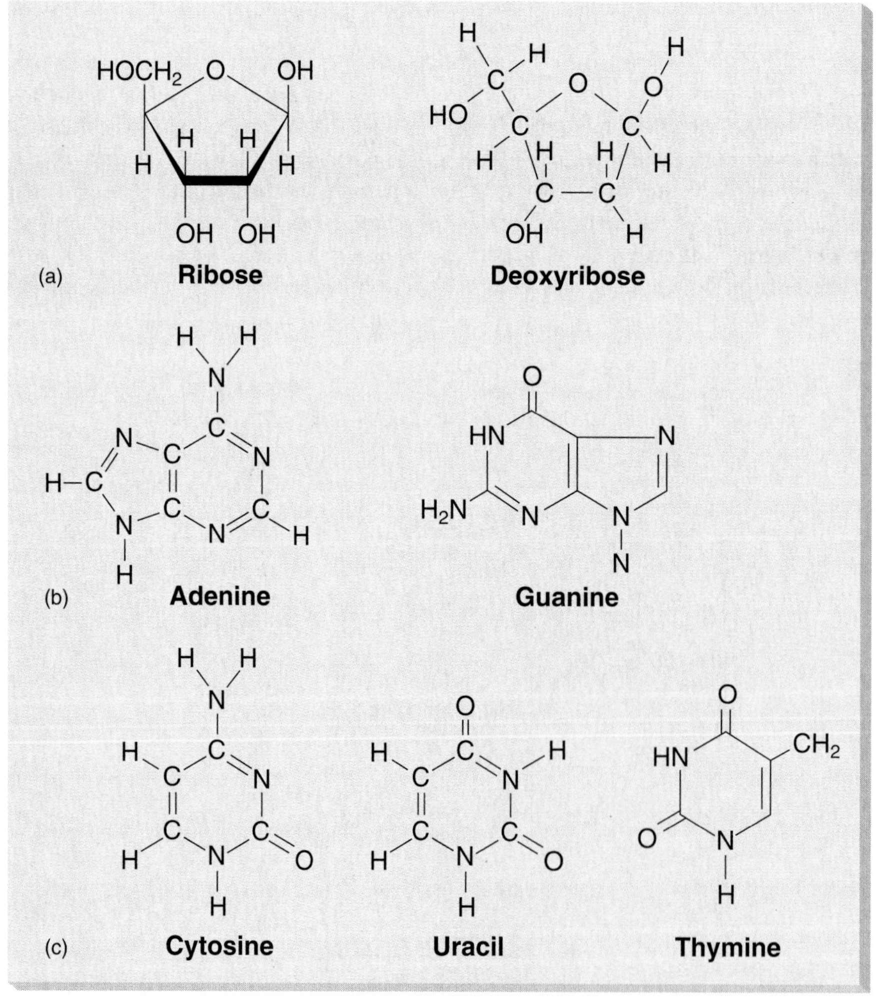

Figure 14.18 The organic building blocks of nucleic acids. **(a)** The two monosaccharides; **(b)** the two purines; **(c)** the three pyrimidines. The other building block, the phosphate group, is not shown.

ANTIBODIES, ANTIGENS, AND ALLERGIES

Antibodies, also called immunoglobins, are special proteins produced by the immune system in response to antigens. An **antigen,** the substance that triggers the body to produce antibodies, could be almost any compound, but proteins and saccharides are common examples. Each of our immune systems contains about 10^9 cells that have three major goals: (1) creating an explicit antibody for a specific antigen, (2) recalling antibodies previously synthesized for an antigen, and (3) distinguishing human biochemicals from others so our cells do not "eat" each other. When the immune system cannot accomplish this last objective, an often-fatal autoimmune reaction develops. Certain infections, including AIDS, induce an autoimmune response.

Figure 1 shows the structure of a typical antibody. It consists of four protein chains bound together with S—S linkages. The part of the antibody at the Y-shaped end is variable (V), while most of the protein structure is constant (C). The antigen fits into the Y-shaped cavity of the antibody.

Most antibodies are synthesized in the B-lymphocytes, specialized white blood cells, to combat the infections or allergic reactions antigens provoke. The major antibody is immunoglobin G (IgG), which combats microorganisms, but there are four other types: IgA, IgD, IgE, and IgM. IgA occurs in saliva and mucus; it also combats microorganisms. IgD is believed to moderate antibody synthesis, while IgM forms the vanguard of the attack against antigens. IgE is the antibody responsible for allergic reactions to dust, pollen, and similar allergens (figure 2).

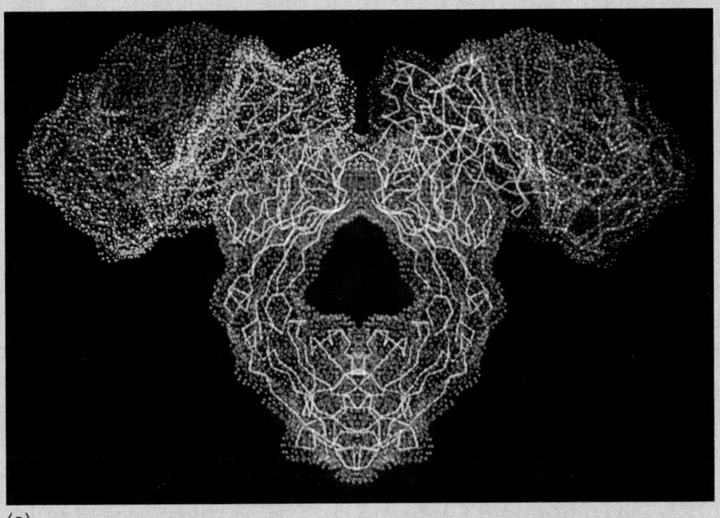
(a)

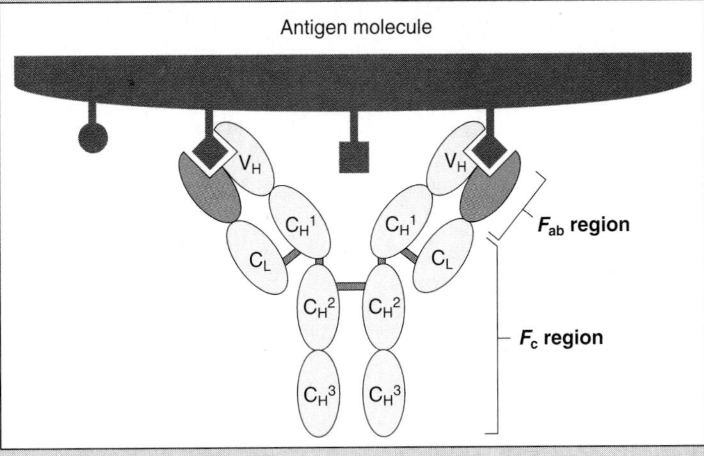

(b)

Figure 1 **(a)** A computer generated image of an antibody. **(b)** A schematic diagram of an antibody that shows where the antigen becomes attached at the end of the molecule.

SOURCE: From Kent M. Van De Graaff and Stuart Ira Fox, *Concepts of Human Anatomy and Physiology*, 4th ed. Copyright © 1995 Times Mirror Higher Education Group, Inc., Dubuque, Iowa. All Rights Reserved. Reprinted by permission.

Figure 2 The allergic response.

A special set of saccharide-based antigens in the blood bind to the surfaces of the red blood cells and give rise to the blood groups. The blood also contains antibodies in the plasma or serum. Unless these antigens and antibodies are suitably matched, they undergo antibody-antigen reactions, leading to massive blood coagulation. This is why people are blood-typed before donating or receiving blood; errors are almost always fatal.

ribose in RNA and deoxyribose in DNA. Both DNA and RNA contain the nucleic bases adenine (A), guanine (G), and cytosine (C); DNA also contains thymine (T), and RNA also contains uracil (U).

The nucleic acids play major roles in heredity, the regulation of protein synthesis, and overall body chemistry. Figure 14.19 shows a portion of a pair of DNA strands. The polymer "backbone" is made from phosphate and monosaccharide (deoxyribose) units, and the nucleic base is attached as a pendant group to the deoxyribose. The individual repeat units are called **nucleotides,** the polymers polynucleotides. The sugar and phosphate units are always the same, but the identity of the nucleic acid base changes along the chain. DNA exists as a pair of complementary chains held together by hydrogen bonds between the nucleic bases. The nucleic bases in each DNA strand are arranged so that an adenine (A) on one strand is always opposite a thymine (T) on the other, and a cytosine (C) is always opposite a guanine (G); this is called **base pairing** (figure 14.19). The two chains maintain a constant distance, permitting the formation of a double helix. Figure 14.20 outlines the relationship between DNA and **chromosomes.** A portion of a DNA chain comprises a gene, and a gene is part of a chromosome.

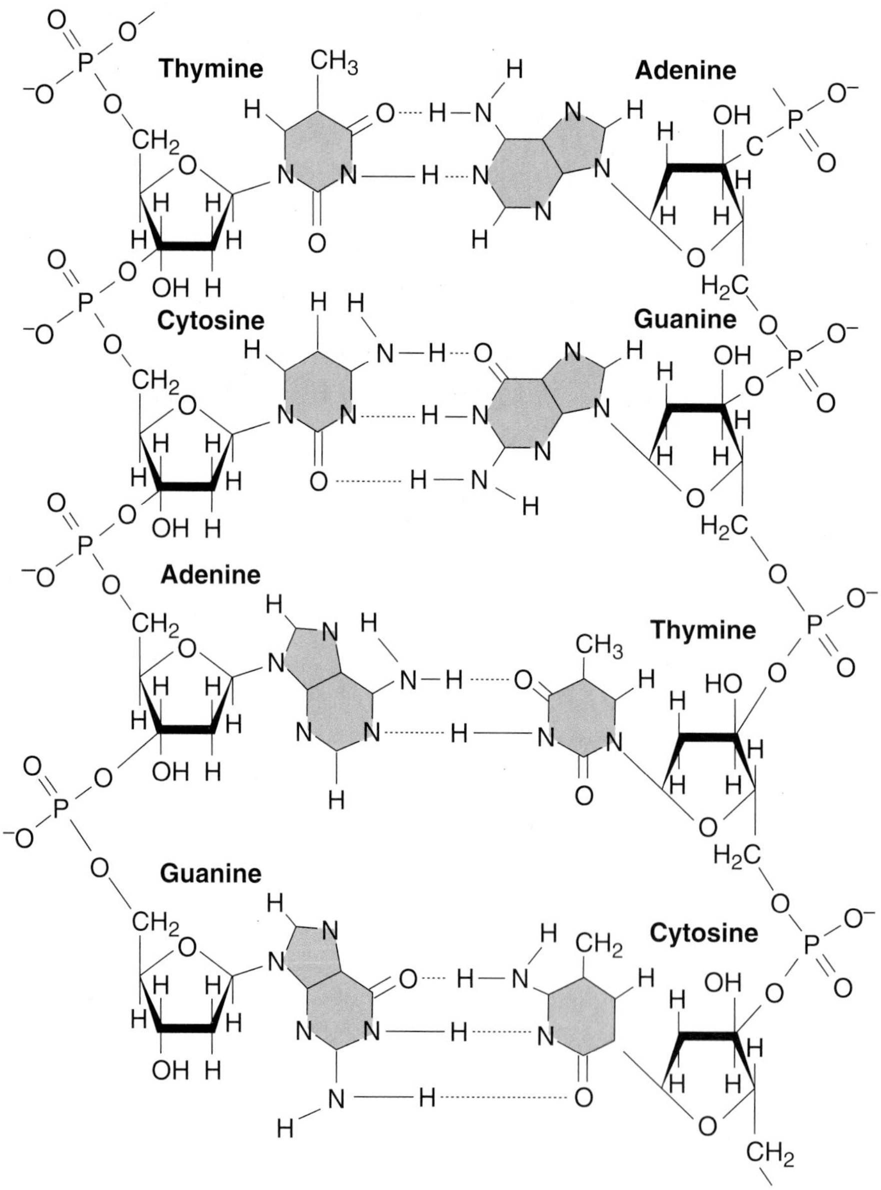

Figure 14.19 Two DNA strands. Note the chemical structure of the polymer and the hydrogen bonding between the two chains. The polymer backbone consists of phosphate and deoxyribose units with the nucleic base attached to the latter as a pendant group.

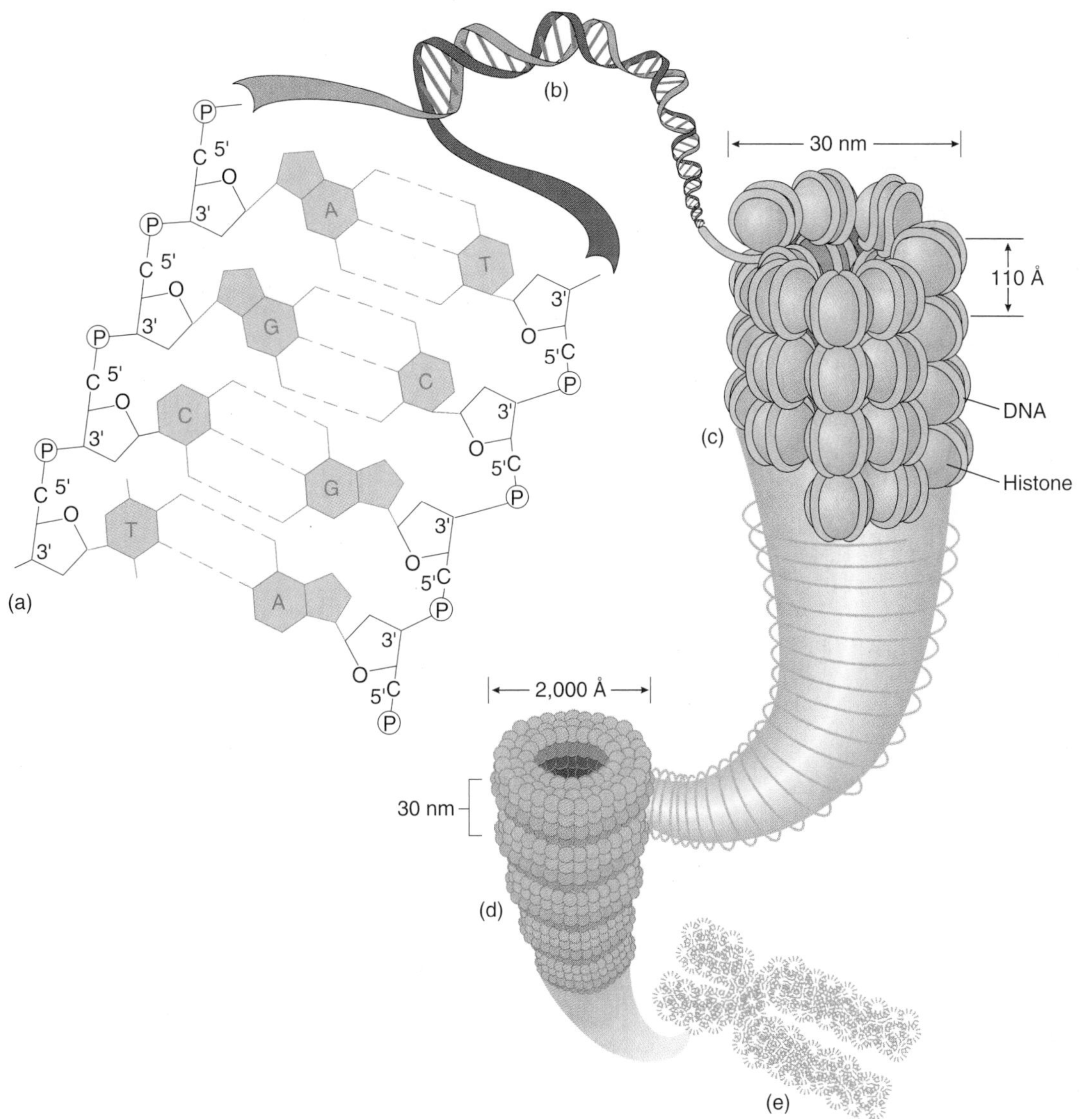

Figure 14.20 How DNA forms chromosomes. The two DNA strands form a duplex chain **(a)** with hydrogen bonds between the nucleic bases; this duplex chain forms the double helix **(b)**, which wraps around a small protein to form a hollow core structure **(c)**. This structure then coils to form a fiber **(d)** that makes up the chromosome **(e)**. Each chromosome contains many genes, which are segments of the DNA chain.

During sexual reproduction, sperm and egg each contribute one set of chromosomes or DNA chains, which then combine to form the pairs of chromosomes in an individual. The key step in this process is DNA **replication,** which ensures the faithful transmission of the genetic information in the chromosomes. During replication, the double helix unwinds and the bases in each separated chain attract nucleotides containing the complementary base units (figure 14.21). A new pair of DNA chains forms, each containing—if all goes well—the exact same bases in the exact same order as the original chain. Thus, the genetic material is faithfully replicated.

People are often tempted to think of DNA as a single chemical entity, like a salt or a simple organic molecule. This is not the case. The individualized nature of DNA makes each DNA molecule unique for every individual (except identical twins). This uniqueness has led to a technique called **DNA fingerprinting** that can determine whether two DNA samples came from the same person. This technique uses chromatographic methods, similar to those discussed in chapter 3, to examine the DNA. The resulting data is now used in criminal courts.

DNA has other related functions as well. Reproduction is an occasional operation, but protein molecules are constantly

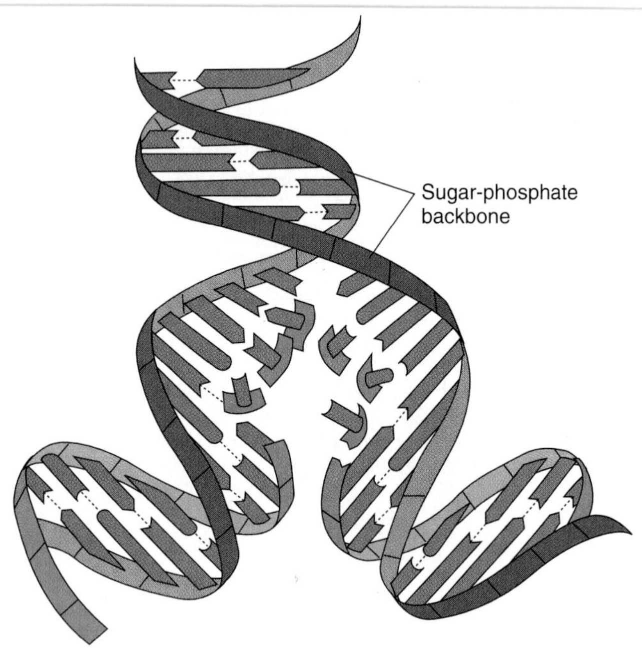

Figure 14.21 The replication of DNA. After the original chains unwind, a new pair of chains is formed from monomeric nucleotides.

synthesized on demand; DNA mediates this synthesis by partially unpairing to form short sequences of complementary messenger RNA (figure 14.22). Triplet nucleotide sequences, called **codons,** on the messenger RNA encode information for specific amino acids that line up along the RNA and then polymerize into a specific polypeptide. Each codon commands the incorporation of a specific amino acid into the growing polypeptide. AUG commands methionine, GGC commands glycine, UCC commands serine, and so on, until the required polypeptide is synthesized.

The simplest function of DNA or RNA occurs in the virus, which is merely a nucleic acid strand covered with a protein coat. Scientists have debated whether viruses are living organisms since the discovery of the tobacco mosaic virus in 1892. Heinz Fraenkel-Conrat (German 1910–) first elucidated the virus chemical structure in 1955, and the first complete structure of a virus was determined in 1978. There is no debate that viruses are infective agents that cause many diseases. AIDS, for example, is evidently caused by the HIV virus, and some other viruses can cause cancer.

Although DNA and RNA normally ensure the transmission of the genetic code through self-replication, **genetic engineering** allows geneticists to manipulate the nucleotide sequences in the nucleic acids. Scientists hope to cure genetic diseases, such as sickle cell anemia (box 14.2) or cystic fibrosis (box 14.4), in this manner. Meanwhile, chemists have already used genetic engineering to improve human life. They have cloned the gene responsible for making human insulin, a polypeptide hormone that regulates the blood glucose level, into *E. coli* bacteria, from

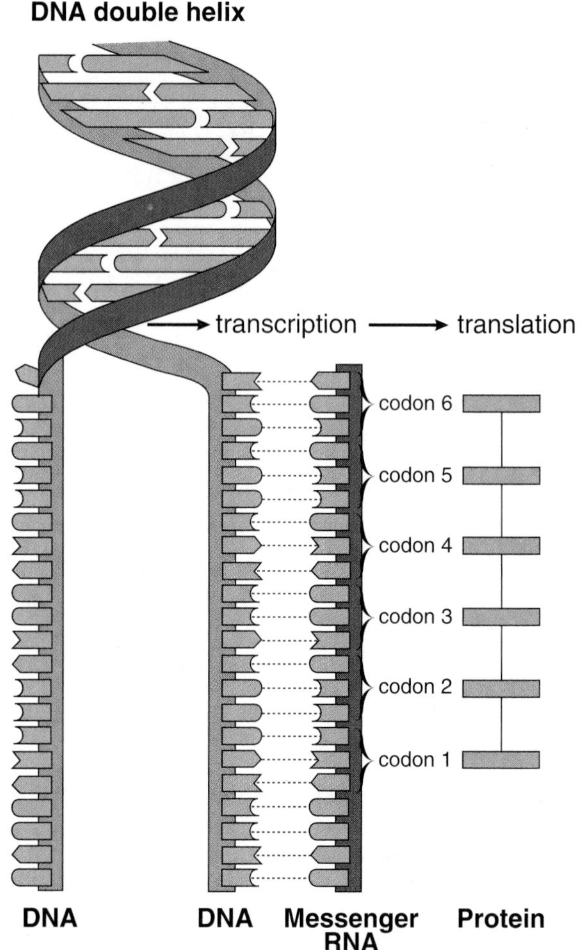

Figure 14.22 Messenger RNA forms from a partially unwound DNA strand (transcription). The codons on the RNA then lead to the formation of a specific polypeptide (translation).

which the insulin may then be isolated. This provides human insulin for diabetics to inject, since they lack this specific hormone. Genetic engineering has also led to the manufacture of several other hormonal polypeptides; it may well generate a revolution in the medical field.

PRACTICE PROBLEM 14.12

Two types of nucleic base molecules appear in figure 14.18. Those with two rings are purines, while those with only one ring are pyrimidines. Which nucleic bases fit into each class?

ANSWER Adenine and guanine are purines; cytosine, uracil, and thymine are pyrimidines.

BOX 14.4

CYSTIC FIBROSIS

Cystic fibrosis (CF) is a recessive genetic disease, meaning that the abnormal genes that cause it must be inherited from each parent. This disease affects about 1 in 2,000 people and is most common among Caucasians. It is uncommon among people of African descent and apparently nonexistent among Asians. Only 25 percent of the children of two CF gene carriers, who are unaffected by the disease, are born totally free of this gene. Another 25 percent have both sets of genes, and thus have cystic fibrosis, and the last 50 percent with one normal gene and one CF gene, become carriers.

Cystic fibrosis normally becomes symptomatic shortly after birth. Because of the abnormal genes, a child with CF constantly produces an extremely thick mucus in place of the normally thin and slippery substance. Normal mucus bathes and protects all internal body surfaces and helps transport the chemicals the body produces from one organ to another. Normal mucus also carries bacteria, dirt, and waste materials away from all parts of the body and permits these unwanted materials to be excreted.

The thickened CF mucus will not perform any of these actions. CF mucus obstructs the tiny passages in the lungs called bronchioles; this results in impaired breathing and increased susceptibility to respiratory infections such as pneumonia. Repeated infections often result in damaged lung tissue. The CF mucus also prevents the flow of the pancreatic enzymes to the duodenum (the part of the small intestine connecting to the stomach). This hinders fat digestion and absorption. Even though CF children may eat ravenously, they seldom gain weight or grow normally, and they often show signs of malnutrition.

The sweat exuded from CF patients is three to five times greater in salt (NaCl) content than normal sweat, and this feature has formed the basis of a CF screening test since 1954. More precise tests are necessary to confirm that the disease is cystic fibrosis. Children with CF experience discomfort and undergo frequent hospitalizations. They usually die before they reach their twenties.

There is no cure for CF at this time. The infections the disease causes are treated with antibiotics (see chapter 17), and the digestive problems are treated with pancreatic extracts. Scientists learned how cystic fibrosis produces its special combination of symptoms in the early 1980s and discovered the gene that causes this disease in 1989. Some have successfully prevented CF by removing the affected gene at an extremely early stage in embryotic development. Unfortunately, such high-tech treatments take years to put into widespread use. In the meanwhile, people with cystic fibrosis have many resources to help them stay as healthy as possible (figure 1).

Figure 1 A young cystic fibrosis patient on an outing.

PRACTICE PROBLEM 14.13

In what ways do the helices in proteins and nucleic acids differ? How are they similar?

ANSWER The main difference is in the nature of the repeating units that make up the chains. Proteins have amino acids, whereas nucleic acids form their chains from phosphate-sugar units with a pendant nucleic base. Both proteins and nucleic acids form a similar springlike coil or helix.

LIPIDS

Lipids are the fourth major group of biochemicals. A **lipid** is a biological molecule, containing carbon, oxygen, and hydrogen, that is soluble in organic solvents. Lipids include triglycerides—the fats and oils you encounter every day—plus several other kinds of chemicals such as steroids, prostaglandins, terpenes, waxes, and fatty acids. Many lipids are esters of fatty acids (that is, carboxylic acids with between 10 and 22 C atoms) with glycerin. The body uses them in cell membranes and in the synthesis of other important biochemicals, such as the prostaglandins. Certain waxes, such as lanolin, are esters of fatty acids and fatty alcohols and are used in cosmetics (chapter 16). An ester found in beeswax, used in polishes and cosmetics, has the formula $C_{35}H_{71}COO—C_{36}H_{73}$.

Fatty Acids

Fatty acids are subdivided into fats and saturated or unsaturated oils that differ mainly in the nature of the fatty acid. Saturated fatty acids contain only single C—C bonds; unsaturated fatty acids contain one or more double C=C bonds. Some typical fatty acids are listed in table 14.4. Notice that the saturated fatty acids are solids with melting points greater than 32°C, while the unsaturated fatty acids are liquids with melting points of 16°C or less. Fats contain high proportions of solid saturated fatty acids, such as stearic acid (18 C atoms) or palmitic acid (16 C atoms). The oils have high amounts of liquid unsaturated fatty acids, such as oleic, linoleic, and linolenic acids (18 C atoms in each case, but with one, two, or three C=C double bonds respectively). Fats are found principally in animals, while oils come mainly from plant sources, but exceptions exist; for example, palm and coconut oils are low in unsaturated fatty acids and high in saturated fatty acids, while fish oils are moderately high in unsaturated fatty acids. The double bond in the unsaturated fatty acids causes the molecules to fold back on itself, dramatically reducing its ability to pack tightly.

Fatty acids, especially unsaturated fatty acids, are important in body chemistry since they are involved in cell structure and in forming other lipids with important functions. Some unsaturated fatty acids (for example, linoleic and linolenic acids) are termed "essential" because the human body cannot synthesize them; they are the basic units for the synthesis of arachidonic acid, the key chemical in the synthesis of prostaglandins, leukotrienes, and thromboxanes, which have hormonelike activity. The prostaglandins are usually abbreviated as PGE_1, PGE_2, and PGF_1, and so on; the PG stands for prostaglandin, and the balance denotes the specific chemical compound. Some prostaglandin structures are shown in figure 14.23.

Prostaglandins

Prostaglandins are involved in an incredible array of biological activities that chemists break down into four categories. **Prostaglandins** (1) stimulate smooth muscle action and platelet aggregation; (2) regulate hormone synthesis and nerve transmission; (3) inhibit platelet aggregation, gastric secretion, and the

TABLE 14.4 The structures, sources, and uses of some typical fatty acids.

COMMON NAME	IUPAC NAME	M.P. (°C)	STRUCTURE	MAIN SOURCE	SOME USES
SATURATED					
Capric	n-decanoic	32	$CH_3(CH_2)_8COOH$	coconut oil, goats	fruit flavors, perfumes, surfactants
Lauric	n-dodecanoic	44	$CH_3(CH_2)_{10}COOH$	coconut oil	surfactants, cosmetics, insecticides
Myristic	n-tetradecanoic	54	$CH_3(CH_2)_{12}COOH$	coconut oil	soaps, cosmetics, flavors, perfumes
Palmitic	n-hexadecanoic	63	$CH_3(CH_2)_{14}COOH$	natural fats	soap, lube oil, waterproofing
Stearic	n-octadecanoic	70	$CH_3(CH_2)_{16}COOH$	animal fats	soap, polishes, lubricants
Arachidic	n-eicosanoic	77	$CH_3(CH_2)_{18}COOH$	peanuts, plants	lubricating greases, waxes
UNSATURATED					
Palmitoleic	cis-9-hexadecenoic	0	$CH_3(CH_2)_5CH{=}CH(CH_2)_7COOH$	marine oils	organic synthesis
Oleic	cis-9-octadecenoic	16	$CH_3(CH_2)_7CH{=}CH(CH_2)_7COOH$	olive oil	soaps, cosmetics, ointments, lubricants
Linoleic	cis, cis-9,12-octadecadienoic	5	$CH_3(CH_2)_4CH{=}CHCH_2CH{=}CH(CH_2)_7COOH$	linseed oil, safflower oil	soaps, coatings, margarine, medicine
Linolenic	cis,cis,cis-9,12,15-octadecatrienoic	−11	$CH_3CH_2(CH{=}CHCH_2)_2CH{=}CH(CH_2)_7COOH$	linseed oil, seed fats	drying oil (paints), medicine
Arachidonic	all cis-5,8,11,14-eicosatetraenoic	−50	$CH_3(CH_2)_4(CH{=}CHCH_2)_4(CH_2)_2COOH$	lecithin, liver	plastics, lubricants

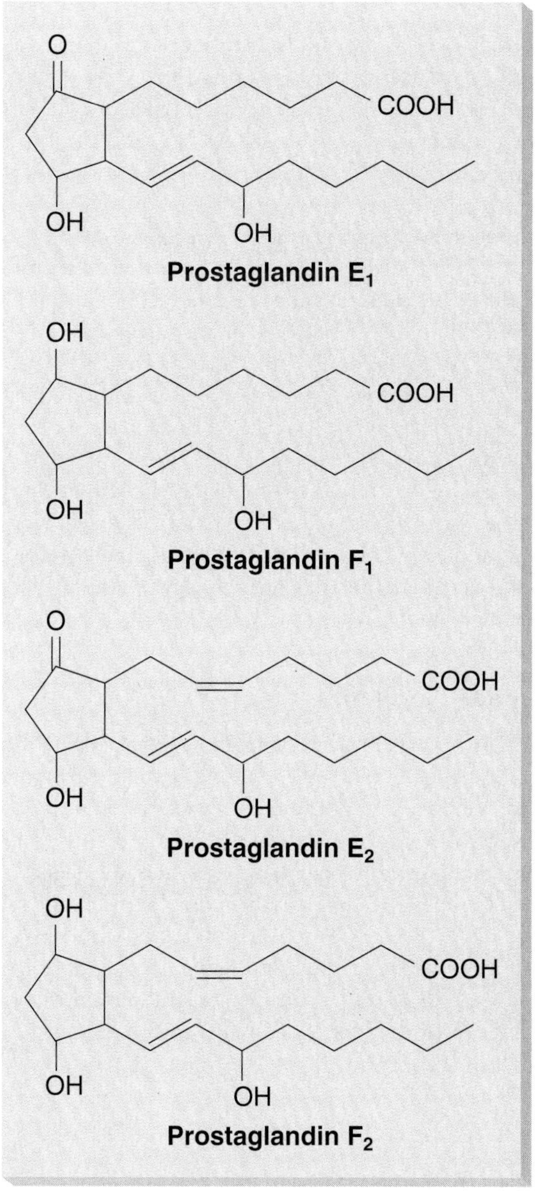

O COOH OH OH
Prostaglandin E_1

OH COOH OH OH
Prostaglandin F_1

O COOH OH OH
Prostaglandin E_2

OH COOH OH OH
Prostaglandin F_2

Figure 14.23 The structures for some prostaglandins.

activity of some lipases; and (4) mediate or sensitize the inflammatory response and pain. Certain prostaglandins oppose the actions of others. For example, different prostaglandins stimulate and inhibit platelet aggregation. Blood clotting is caused by platelet aggregation, which is promoted by thromboxane A_2, but prostaglandin I_2 (also called prostacyclin or PGI_2) inhibits clot formation. Likewise, PGI_2 causes blood vessel dilation, whereas thromboxane A_2 constricts blood vessels. PGE_2 stimulates smooth muscle contraction in the uterus and is used medically to induce labor; it is also used to induce abortions during the final trimester. Several PGs seem to promote the inflammatory response in the body with its associated pain; the analgesic aspirin is believed to block PG synthesis and alleviate pain. Blocking PG synthesis is not always desirable; PGs inhibit excess secretion of stomach acid and promote secretion of the mucous protective liner. If these PGs are blocked, stomach ulcers can develop. (In fact, this is one of aspirin's few negative side effects.) Leukotrienes are produced by some of the white blood cells and promote the bronchi constriction that characterizes asthma. Other PGs promote dilation of these bronchi.

Glycerides

The glycerides are subdivided into the neutral glycerides, which are esters of fatty acids, and the phosphoglycerides, which contain a phosphate group. The neutral glycerides have no charge, but the phosphoglycerides are anions. The neutral glycerides are used for energy storage, usually as a triglyceride; they do not form part of the cell membrane structure. Phosphoglycerides form part of the cell membrane and also serve as natural emulsifying agents. Lecithin, chemically known as phosphatidylcholine, is the best known phosphoglyceride and is found in egg yolks, soybeans, and other sources. Food processing facilities add lecithin as an emulsifying agent in ice cream, mayonnaise, and salad dressings. Any food product using egg yolks or soybeans contains lecithin that probably acts as an emulsifying agent.

Nonglyceride Lipid Subclasses

There are three nonglyceride lipid subclasses: sphingolipids, steroids, and waxes. Sphingolipids are phospholipids formed from sphingosine, an aminoalcohol. Sphingomyelin combines sphingosine, a phosphate group, choline, and a fatty acid to form a group of chemicals found in the brain and nerves. Approximately 25 percent of the human myelin sheath is composed of sphingolipids.

Complex Lipids: Lipoproteins and Glycolipids

The complex lipids—lipids combined with other substances—are subdivided into lipoproteins and glycolipids. Lipoproteins consist of a hydrophobic lipid core surrounded by an outer layer of proteins and hydrophilic lipids, such as phospholipids. In many ways, lipoproteins resemble the micelles we studied in chapter 13. Their function is also similar—to transport lipids throughout the body. Lipoproteins fall into four classes: chylomicrons, very low-density lipoproteins (VLDL), **low-density lipoproteins (LDL)** and **high-density lipoproteins (HDL).** Chylomicrons transport triglycerides from the intestines to other body tissues. The VLDL carry triglycerides synthesized in the liver to storage sites in the adipose (fatty) tissue.

LDL and HDL play a role in cardiovascular health since they are involved in cholesterol transport. The relationship between cholesterol and heart disease is complex and involves several factors. One factor seems to be the nature of the lipoproteins used to transport the cholesterol. LDL provides the main transport system for cholesterol, which the cells of the body take up via LDL receptors on the cells. (A receptor is a site designed to accept a particular chemical. It usually has a specific shape that limits chemicals that can enter the site.) When cholesterol is present inside a cell, it inhibits its own biosynthesis by limiting the number of LDL receptors synthesized. Some people have mutations in the gene that codes for LDL receptors. They accumulate cholesterol in the blood plasma, and the excess is deposited on the blood vessel walls. HDL transports cholesterol back to the liver and away from the cells or blood; it reduces the plasma concentration. In the liver, some cholesterol is converted to bile, which aids in fat digestion, and the balance is excreted.

Diets high in saturated fats tend to result in high levels of LDL and blood cholesterol; diets high in unsaturated fats (or oils) result in decreased blood cholesterol levels (box 14.5). This appears to reduce the incidence of atherosclerosis, or "hardening of the arteries."

PRACTICE PROBLEM 14.14

Summarize the relationship between dietary fats and cholesterol.

ANSWER Unsaturated fatty acids are found in edible oils derived from plant sources, whereas saturated fatty acids come mainly from animal products. The unsaturated lipids form HDL that transports excess cholesterol out of the bloodstream; the LDL from saturated fatty acids deposits the excess in the blood vessels. The accumulation of cholesterol could eventually cause cardiovascular disease.

PRACTICE PROBLEM 14.15

Is complete exclusion of dietary fats good for your health? Why or why not?

ANSWER Complete exclusion of fats from the diet is a recipe for disaster. As we have discussed, many essential biochemicals are derived from fatty acids. Limiting fat intake to a reasonable level makes sense, but excluding fats completely could prove disastrous to one's health.

HORMONES

Hormones are the fifth main group of biochemicals. These substances, which trigger important changes in body chemistry, are secreted by glands within the body when their unique activity is needed; they may not be obtained in the diet. There are four main classes of hormones—steroidal, polypeptide, thyroid, and catecholamine.

Steroids

The **steroids,** which are lipids that act as hormones, are cyclic compounds containing four fused rings; figure 14.24 shows some important examples. Cholesterol, the precursor of the steroidal hormones, occurs in most body tissues (10 percent of it occurs in the brain); the body will synthesize cholesterol if its supply drops too low.

Since the major sex hormones are steroidal, it is not surprising that most birth control chemicals mimic their structure (figure 14.24). Other steroidal hormones include cortisone, aldosterone, and digitoxin; table 14.5 outlines their sources and functions.

Anabolic steroids build up biostructures, while catabolic steroids break them down. Athletes take these natural hormones, or their synthetic analogs, to increase body mass, especially muscle structure. Although they achieve these results, there is a negative side. Anabolic steroid use increases the risk of cardiovascular disease, strokes, liver and kidney damage, impotence, infertility, and even cancer. Before using anabolic steroids, a person should consider the trade-offs. Is developing a "good body" worth dying young?

Polypeptide Hormones

Like proteins, polypeptide hormones have a specific sequence of amino acids in their polymer chain, a primary structure, and a specific shape. Like any hormone, they exert potent physiological effects; several sex-related hormones fall into this group. Table 14.6 lists examples of polypeptide hormones. These polypeptides are often of modest size and variable complexity; thus, vasopressin and oxytocin have only 9 repeat units; glucagon has 29 repeat units; HGH contains 191 amino acid units; MSH is a pair of polypeptides with 13 and 22 amino acids; and insulin contains 51 amino acids in two polypeptide chains linked together.

TABLE 14.5 The sources and functions of some steroidal hormones.

NAME	SOURCE	FUNCTION
Estradiol	ovaries	regulates menstrual cycle; secondary sexual characteristics
Progesterone	ovaries	promotes ovulation; maintains pregnancy
Testosterone	testicles	controls male secondary sexual characteristics
Cortisone	adrenal gland	has many effects on protein synthesis; inflammation; water balance
Aldosterone	adrenal gland	maintains body's salt balance
Digitoxin	foxglove plant	treats some forms of heart disease

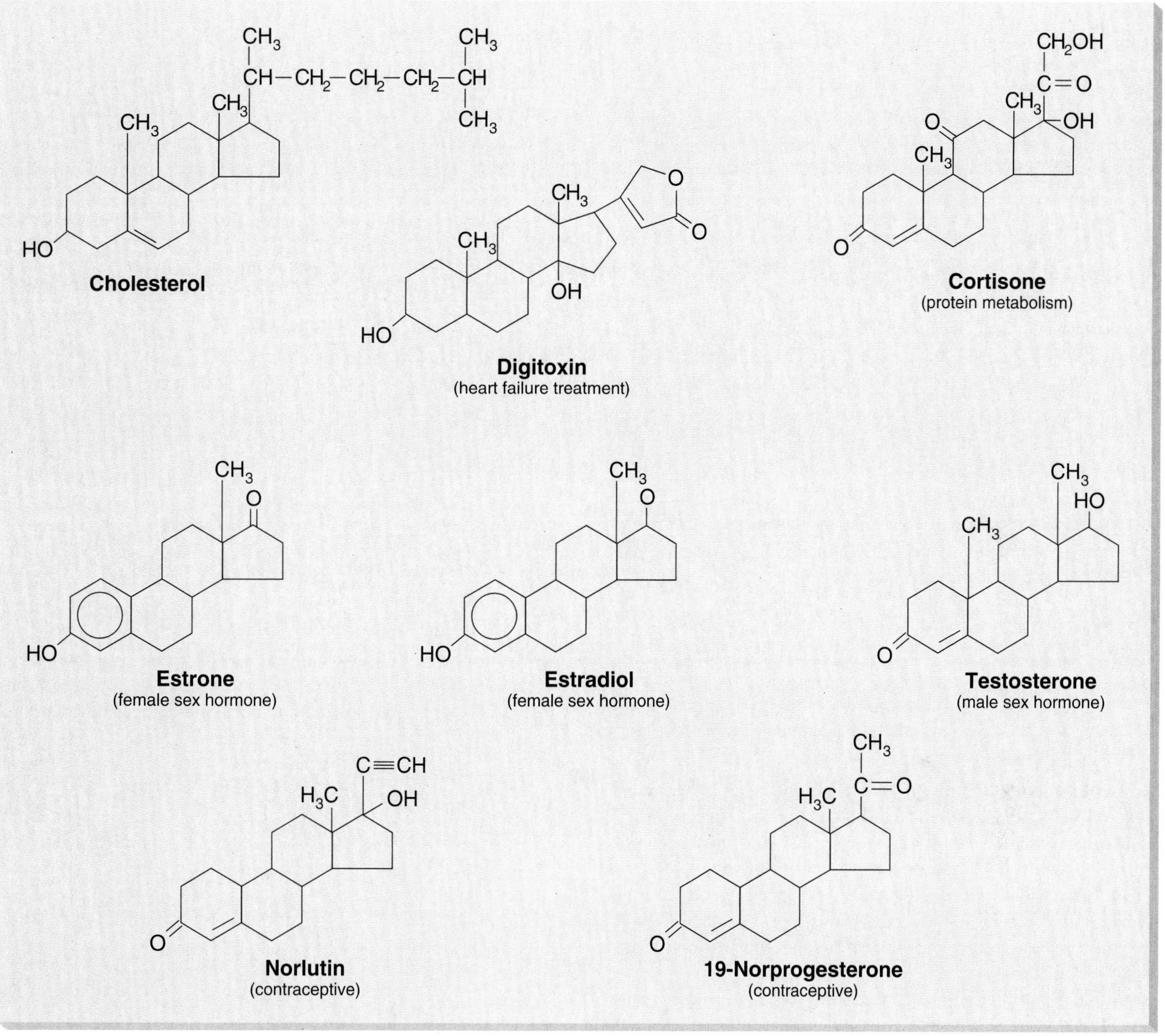

Cholesterol

Digitoxin
(heart failure treatment)

Cortisone
(protein metabolism)

Estrone
(female sex hormone)

Estradiol
(female sex hormone)

Testosterone
(male sex hormone)

Norlutin
(contraceptive)

19-Norprogesterone
(contraceptive)

Figure 14.24 Some steroidal hormones.

Insulin, figure 14.7, one of the most important polypeptide hormones, is secreted by the β-cells of the pancreas; it regulates blood sugar levels and glycogen storage. It also produces the mucopolysaccharides that make the mucous fluids slippery.

Human growth hormone (HGH) controls muscular and skeletal growth. HGH deficiency causes dwarfism, while HGH excess leads to various forms of giantism. HGH deficiency is correctable through hormone therapy—that is, by administering HGH.

Three hormones control the pigmentation level in the skin and hair: MSH, MIH, and MRH. The MIH and MRH regulate the amount of MSH, which actually controls the amount of melanin pigment. Many systems consist of three or more interacting hormones that regulate one another or that control another chemical (for example, Ca^{++}).

Some polypeptide hormones are essential to life, while others are not; however, all hormones are necessary for normal growth and development, and some are essential for reproduction. A large number of polypeptide hormones are involved in the reproductive cycle. Synthetic analogs have been considered as possible contraceptives, but polypeptides hydrolyze rapidly in the body and seldom show prolonged activity.

TABLE 14.6 The sources and functions of some polypeptide hormones.

POLYPEPTIDE HORMONE	ABBR.	SOURCE	FUNCTION
Insulin*		pancreas	controls blood glucose levels; deficiency leads to diabetes
Human growth hormone	HGH	pituitary	controls skeletal growth, visceral growth
Adrenocorticotrophic hormone*	ACTH	pituitary	mobilizes, oxidizes fatty acids
Corticotrophin-releasing hormone	CRH	pituitary	controls ACTH release
Melanocyte-stimulating hormone	MSH	pituitary	controls skin pigmentation; absence leads to albinism
MSH inhibiting hormone	MIH	pituitary	inhibits MSH release
MSH releasing hormone	MRH	pituitary	causes release of MSH
Thyroid-stimulating hormone*	TSH	pituitary	regulates body temperature
Thyroxine*		thyroid	regulates growth, metabolism
Parathyroid hormone*		parathyroid	controls Ca^{++} levels; affects bone strength
Thyrocalcitonin*		thyroid	opposes effect of parathyroid
Glucagon		pancreas	controls glucose metabolism
Oxytocin		pituitary	controls uterine contraction
Vasopressin		pituitary	controls blood pressure, smooth muscles
Leutinizing hormone	LH	pituitary	controls ovulation; steroid levels
Follicle-stimulating hormone	FSH	pituitary	triggers sperm, ova (egg) production
Prolactin		pituitary	stimulates milk production.

*Essential to life

Other Hormones

Two important hormones are derived from the phenolic amino acid tyrosine—epinephrine and norepinephrine (figure 14.25). Epinephrine, also called adrenaline, is secreted by the adrenal cortex. It controls blood circulation (and therefore, blood pressure), helps dilate the blood vessels, and increases heart output and rate. It affects most body organs and prepares them to respond to emergencies. Adrenaline also stimulates the production and release of ACTH.

Norepinephrine is chemically similar to epinephrine. The prefix *nor* denotes the absence of a methyl group on the amino group in norepinephrine. This hormone is secreted by the adrenal cortex and has functions similar to epinephrine's. Neither hormone is essential for life, but a deficiency shortens the life span.

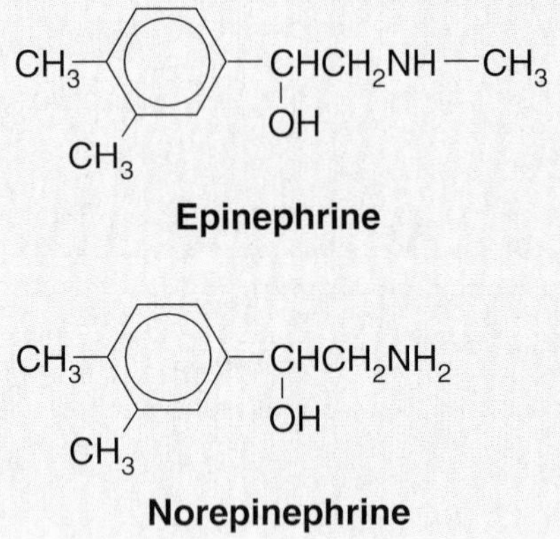

Figure 14.25 The structures of epinephrine (adrenaline) and norepinephrine.

PRACTICE PROBLEM 14.16

Which of the polypeptide hormones discussed in this section are regarded as essential to life?

ANSWER ACTH, TSH, thyroxine, parathyroid, insulin, and possibly thyrocalcitonin.

VITAMINS

The sixth group of biochemicals is vitamins. In the United States, nearly everyone knows vitamins are important for good health; a thriving vitamin supplement industry reminds us of it in the unlikely event that we forget. The human body cannot synthesize its own supply of vitamins. In fact, by definition, a **vitamin** is a trace organic substance obtained in animal or human diets. A

BOX 14.5

CHOLESTEROL TESTING

Cholesterol testing is becoming increasingly common as our society worries more about heart disease. Blood tests are often available at supermarkets, health fairs, and workplaces. The patient is expected to fast (not eat) for twelve hours prior to blood withdrawal; the blood is removed from a vein (figure 1). Then the blood is taken to a clinical or hospital laboratory for actual testing.

In the lab, the cells in the blood sample are removed, leaving blood serum. The serum is tested for its cholesterol content. Lab technicians use several methods for the actual determination of cholesterol levels, but most tests rely on the amount of color produced by treating the serum with sulfuric acid and some other chemicals. In the most common test, a solution of ferric chloride ($FeCl_3$) is added to the acidified blood serum. The resulting color is then compared with colors produced by known cholesterol concentrations and expressed as mg cholesterol per 100 mL serum. Another test involves extracting the cholesterol from the serum with various solvents and then conducting the color tests. These extraction procedures are more expensive, but they also provide more accurate measurements of the cholesterol levels since many other chemicals in the blood serum can also produce colors.

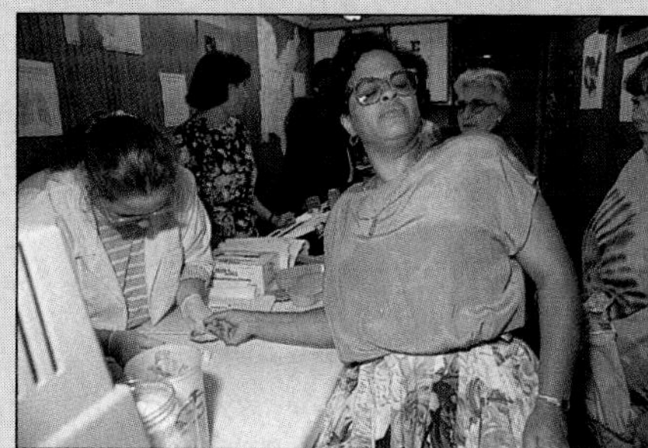

Figure 1 Blood being removed for cholesterol testing.

The designation of a "normal" cholesterol level has varied greatly over the years. A normal value falls somewhere between 150 and 280 mg/mL; many physicians encourage patients to aim for a level below 200 mg/mL. The average person consumes about 600 mg cholesterol daily. The main sources are foods of animal origin; foods from plant sources are usually cholesterol free. Many people now attempt to restrict their cholesterol intake. However, this substance is indispensable for the proper development of the brain and nervous system and for the production of the sex hormones. It is important not to reduce cholesterol consumption to an extreme degree.

The relationship between blood serum cholesterol levels and heart or cardiovascular disease remains uncertain. At this writing, the latest research suggests that high blood levels of both cholesterol and low-density lipoproteins (LDL) increase the risk of disease. Maintaining reasonable cholesterol levels and high levels of high-density lipoproteins (HDL) seems to be the healthiest combination.

good, balanced diet can supply daily vitamin needs, but multiple vitamin supplements, though unnecessary, are generally harmless. Vitamins are a remarkably mixed group of structures; chemicals were placed into this classification on the basis of function, not structure. We will discuss vitamins in more detail in chapter 15.

Vitamins affect the body's biochemistry in several ways. Some can function as a coenzyme (enabling an enzyme to function) or hormone (vitamin D). Others prevent a broad spectrum of diseases or afflictions, including poor night vision (A), beriberi (a muscular paralysis—B_1), several forms of dermatitis (B_2, C, niacin), scurvy (a skin disorder—C), rickets (a bone disorder—D), sterility (E), muscular dystrophy (E), and pellagra (niacin).

MINERALS

Minerals are the seventh and final group of biochemicals in our survey. **Minerals** are inorganic chemicals, usually ions, that are essential to health. For example, calcium and phosphorus for the teeth and bones, and iron is essential in the hemoglobin of the blood and transports the oxygen throughout the body. Table 14.7 lists some essential minerals and their important role in our physiology. When a mineral activates an enzyme to perform its function, the mineral is called a coenzyme or cofactor. Cofactors are often needed in trace amounts.

Sodium, potassium, and chloride ions deserve special emphasis because they regulate the electrolyte equilibrium in the blood and body tissues and help maintain proper fluid balance inside and outside the cells. This balance is controlled primarily by the Na^+/K^+ ion ratio, which ideally is 0.6. Potassium is the major cation within the cells; the 1.9 to 5.6 g/day requirement must be replenished since the ion is excreted in the urine. A K^+ deficiency is fairly common in people using a diuretic to control body fluids. The Na^+ (1-2 g/day) and Cl^- (1.7–5.1 g/day) requirements are usually easy to meet since most foods contain NaCl; in fact, the average Na^+ intake in the United States is about 10 g/day, between five and ten times the amount needed. Excess sodium intake may contribute to hypertension, or high blood pressure.

Phosphorus is another very important mineral. One of its functions involves energy production; box 14.6 explains this in more detail.

BOX 14.6

BIOCHEMICAL ENERGY PRODUCTION

With the knowledge you have already acquired about chemistry, you might guess that the energy produced in living systems comes from an oxidation-reduction reaction. This is fairly close to the truth, but there are some problems to overcome with oxidation reactions. For example, rapidly oxidizing glucose completely into water and carbon dioxide, produces a total of 686 Kcal/mole:

$$C_6H_{12}O_6 + 6\,O_2 \rightarrow 6\,H_2O + 6\,CO_2 + 686\text{ Kcal/mole}$$

If this much energy were released in a short time interval, the heat could destroy the surrounding cells. Something more

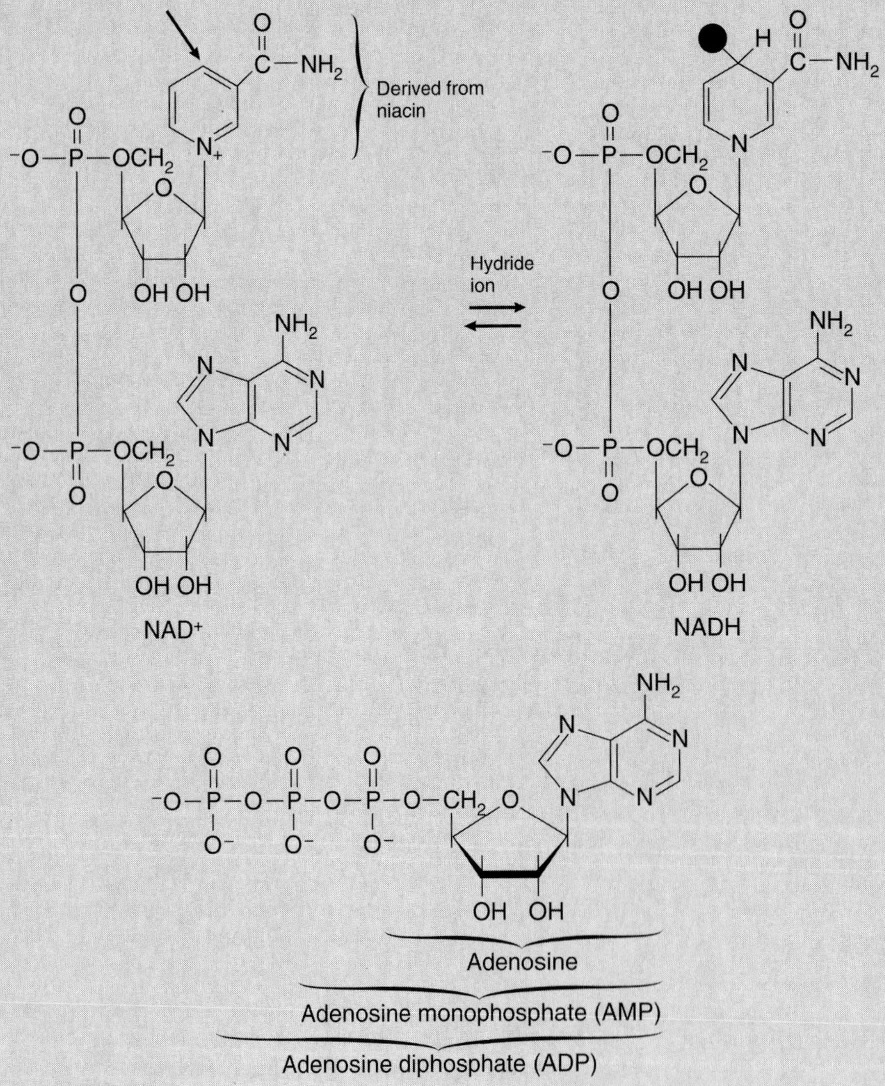

Derived from niacin

Hydride ion

NAD+

NADH

Adenosine

Adenosine monophosphate (AMP)

Adenosine diphosphate (ADP)

Figure 1 The structures of the adenosine phosphates and the nicotinamide adenine derivatives NAD^+ and NADH.

supple, more gentle, is needed to maintain life while providing energy. This is where the molecules AMP, ADP, ATP, NAD^+, and NADH fit into the picture. Figure 1 shows the structures of these adenosine phosphates and derivatives, which free the energy from glucose and other chemicals without destroying life.

Adenosine derivatives cause oxidation to proceed step by step in a series of simple reactions. One such series, glycolysis, is briefly outlined in figure 2. As the figure indicates, the reaction sequence releases two additional ATP and NADH molecules for every glucose molecule consumed. Various enzymes and coenzymes appear at essentially every reaction step. The glucose is converted to a series of six-C-atom open chains, then split into a pair of three-C-atom chains, then degraded further to pyruvic acid to be used in other reactions by living systems. **ATP (adenosine triphosphate)** is a high-energy phosphate and is the key energy molecule in the body; other energy-producing cycles that obtain energy from proteins, lipids, and the other saccharides also involve ATP.

Many energy production cycles exist in living systems. All are extremely efficient and most are intertwined with each other. In other words, one cycle cannot function adequately without the others. Life is hardly a series of random reactions.

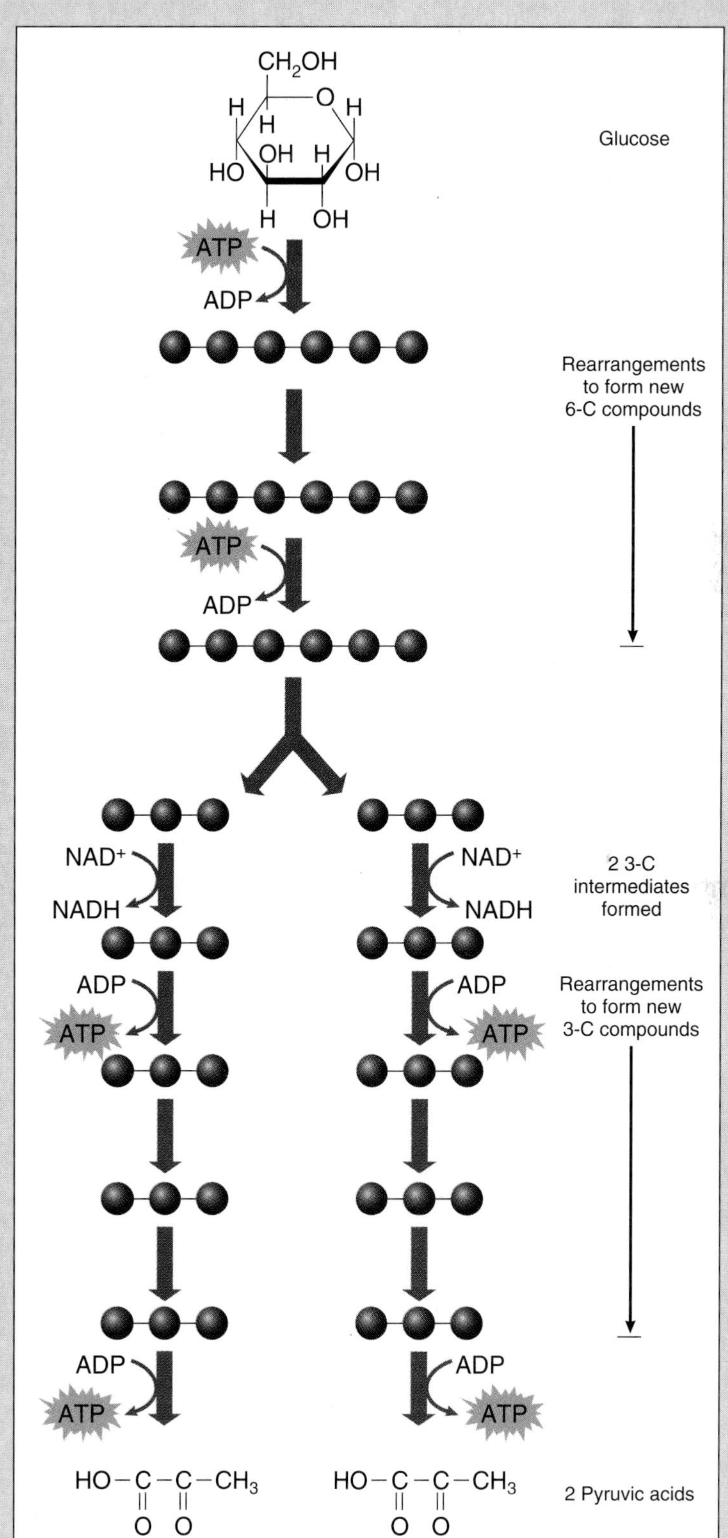

Figure 2 Glycolysis, one way energy is produced in the body. Glucose is rearranged to a 6-C atom chain and split into a pair of 3-C-atom chains that eventually yield pyruvic acid. In this process, 4 ATPs and 2 NADHs are produced.

TABLE 14.7 The function of minerals in the body.

MINERAL	SYMBOL	FUNCTION
METALS		
Sodium	Na^+	regulates fluid balance
Potassium	K^+	regulates fluid balance
Magnesium	Mg^{++}	contributes to cellular metabolism; nerve conductance; muscle function; bone development; required for enzyme activity
Calcium	Ca^{++}	forms bone, teeth
Iron	Fe^{++}	facilitates O_2 transport in hemoglobin
Copper	Cu^{++}	contributes to hemoglobin production; enzyme activation; cofactor for connective tissue protein synthesis
Chromium	Cr^{+++}	activates certain enzymes
Manganese	Mn^{++}	activates certain enzymes
Molybdenum	Mo	acts in enzyme synthesis
Zinc	Zn	acts in enzyme synthesis
Vanadium	V	regulates blood circulation
NONMETALS		
Chloride	Cl^-	regulates fluid balance
Iodine	I	contributes to thyroid hormone
Sulfur	S	is present in the amino acid cysteine found in insulin, hair, and skin
Phosphorus	P	is present in bones, DNA, and many other biochemicals; aids in energy production and conversion

PRACTICE PROBLEM 14.17

Many elements are considered essential minerals in trace quantities, including such diverse examples as selenium, silicon, tin, and nickel. Based on the information given in this section, make an educated guess concerning what functions these minor minerals might perform.

ANSWER Most likely these trace elements serve as cofactors that enable an enzyme to function. The concentration of enzymes in the body is never very high, so a cofactor is only needed in trace amounts.

EXPERIENCE THIS

Obtain two small, meat-free chicken bones. (A wishbone broken into two pieces is ideal for this experiment.) Place one bone in a sealed jar of water and place the other bone in a sealed jar of vinegar. After several days, remove the bones and rinse them with water. Does one bend more easily than the other? Why do you think this happens?

HINT Bones contain a lot of calcium. What effects do acids have on calcium salts, as we learned in chapter 8?

BIRTH CONTROL

As the number of people on our planet increases, there is a distinct possibility that the quality of life may decrease for most. Even now, many people have inadequate food, clothing, or shelter. Part of this comes from poor distribution of available supplies, either inadvertent or deliberate, but part seems to stem from overpopulation. One way to remedy this problem is through birth control, which can be a biochemical activity.

In the section on hormones, we noted that many hormones are steroids and that many birth control chemicals mimic these molecules. Several other birth control or **contraceptive** (anticonception) devices are also available. Attempts at birth control are not new. Ancient Egyptian records (dated between 2000 and 1950

TABLE 14.8 A summary of birth control methods, in order of increasing effectiveness.

METHOD	RANGE OF PREGNANCIES PER 100 FERTILE WOMEN/YEAR	AVERAGE NUMBER OF PREGNANCIES PER 100 FERTILE WOMEN/YEAR
None	40–80	60
Douche	34–61	48
Rhythm method	14–58	36
Spermicides	9–40	25
Diaphragm and jelly	11–28	20
Condoms	8–17	13
IUD (intrauterine device)	3–8	6
Pill (steroid)	0.03–2	1
Sterilization	0–0.003	0

SOURCE: From E. Peter Volpe, *Biology and Human Concerns*, 3rd ed. Copyright © 1983 Times Mirror Higher Education Group, Inc. Dubuque, Iowa. All rights reserved. Reprinted by permission.

B.C.) describe a tamponlike device, loaded with whatever chemicals were present in some powdered scrubs, that the woman inserted to prevent conception. Similar devices exist today, but the chemicals are purer and more controlled.

There are at least five requirements for creating a satisfactory contraceptive device. It must be (1) effective, (2) safe, (3) simple to use, (4) reversible (that is, after use of the device ceases, pregnancy must be possible), and (5) able to protect the user from disease. Table 14.8 lists some birth control methods arranged according to effectiveness. Many of these methods are not very effective. In fact, some methods are not much better than no birth control at all. To evaluate effectiveness, you must compare the results with the contraceptive against no contraception at one extreme and sterilization at the other. An ideal contraceptive would be as effective as sterilization, but potentially reversible, and as safe as the use of no contraceptive (ignoring AIDS and other venereal diseases). Not all contraceptive methods involve chemistry directly; those that do include douches, spermicides, diaphragms and jellies, condoms, IUDs, and the pill. Only the "natural" method of no contraception (the rhythm method) and sterilization do not involve chemistry directly. Figure 14.26 shows several modern contraceptive devices.

Douching offers little advantage over no contraception (48 percent pregnancy rate versus 60 percent). The rhythm, spermicide, and diaphragm-plus-jelly methods are about equally effective (or ineffective, depending on your point of view); on average, around 27 percent of the females using these methods become pregnant in any given year. It does not require great mathematical acumen to realize that populations will grow rapidly if 25 to 60 percent of all sexually active females of childbearing age have children annually. Even condoms, which many regard as almost infallible, have a 13 percent pregnancy rate. Only at the bottom of table 14.8 do the devices show real potency; IUDs and the pill have pregnancy rates of 6 percent and 1 percent, respectively. These numbers partially explain why these are such popular methods. Both are effective and reversible.

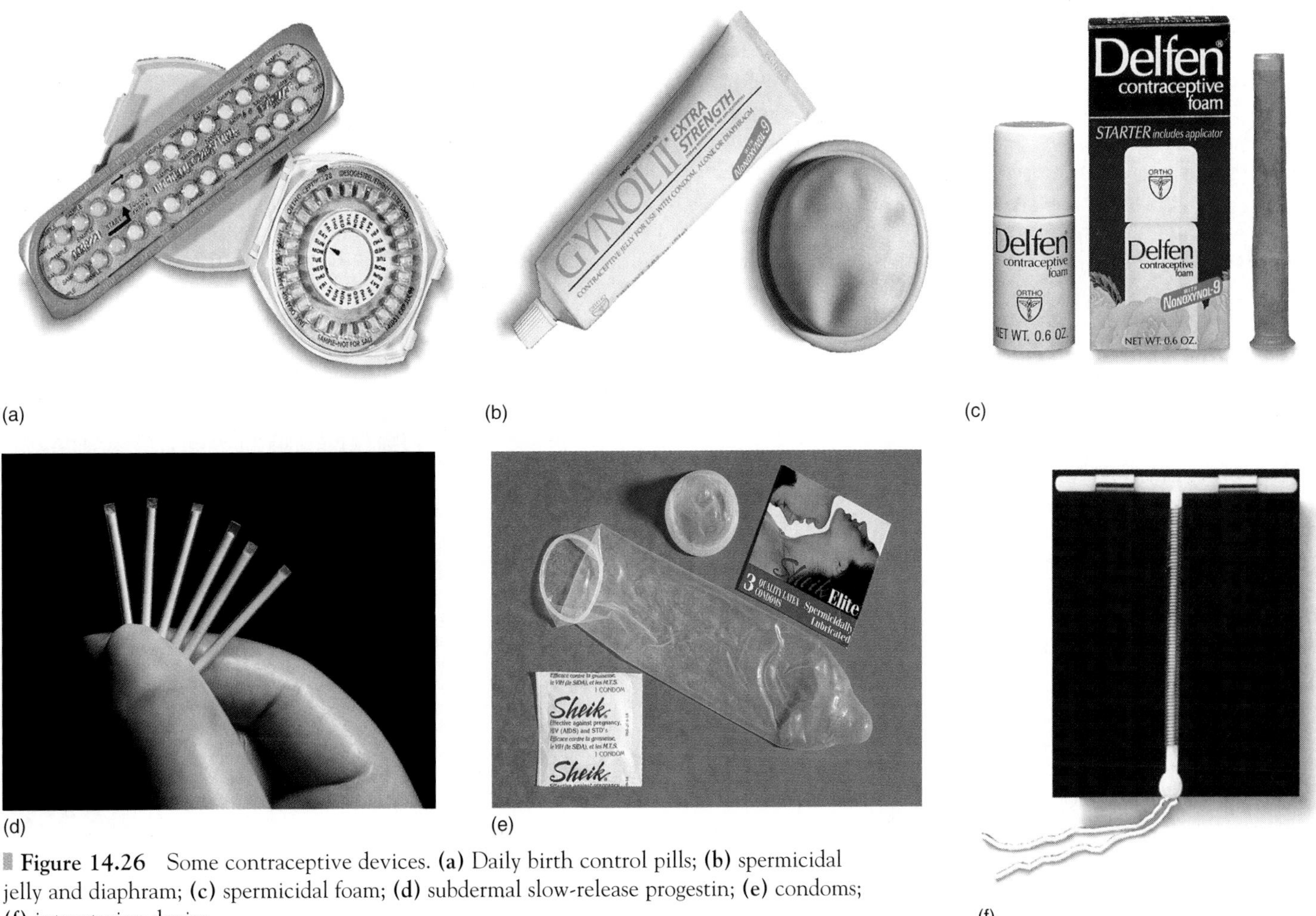

Figure 14.26 Some contraceptive devices. **(a)** Daily birth control pills; **(b)** spermicidal jelly and diaphram; **(c)** spermicidal foam; **(d)** subdermal slow-release progestin; **(e)** condoms; **(f)** intrauterine device.

Why do these methods work so well? To answer this question, we must understand what controls conception in the female. Of course, sperm and egg unite, but this process also depends on the actions of several steroid and polypeptide hormones. Progesterone and estradiol (figure 14.23) were singled out early in birth control research since these hormones help regulate the female menstrual cycle. Early researchers found that injected progesterone was an effective contraceptive agent. The more advantageous oral administration was not possible since progesterone is not sufficiently soluble.

After Adolf Butenandt (German, 1903–) determined the structure of progesterone in 1934, chemists knew what structure to mimic. (Butendandt received the 1939 Nobel Price in chemistry for his research on the sex hormones.) Hans Inhoffen synthesized ethisterone, the first oral contraceptive, in 1938, but this agent was effective only in fairly large doses and was never widely accepted. Carl Djerassi synthesized 19-norprogesterone in 1951; this molecule lacked one of the methyl groups in progesterone and was about six times more effective as a contraceptive, but it, too, required injection. Djerassi later developed an analog that could be taken orally. This analog was marketed as Norlutin and known chemically as 17-α-ethynl-19-norestosterone. Shortly after this, Frank Colton developed a similar compound that was marketed as Enovid. Other agents soon followed, until the birth control pill arrived on the scene. The pill induces a pseudopregnancy in the female, preventing the release of an egg and thereby averting a real pregnancy. (Pregnant women do not ovulate). These chemicals required extensive testing before they were released for general use. These studies were conducted in Puerto Rico by Gregory Pincus (American, 1903–67) and John Rock.

The original pill contained high levels (about 150 μg) of progestin (a generic name for synthetic steroids related to progesterone). It had undesirable side effects that actually led to some deaths, mostly from heart attacks or strokes. The pill, like pregnancy, increased blood clotting and caused an increase in cardiovascular problems. Even at that, the death rates from the pill were only 3 per 100,000 women, compared against 30 per 100,000 women in pregnancy. Today, the progestin level in birth control pills is about 3 μg and these problems are no longer serious. The early pill had a few other side effects, such as acne, hypertension, abnormal bleeding, weight gain, and upset stomach, but these were also reduced at the lower progestin levels. This means the pill is reasonably safe—one of the criteria for a satisfactory contraceptive. Since it is ingested orally, the pill is also easy to use, unless the user is very forgetful.

Norplant, an implantable device that contains a progestin, was introduced in 1990. This controlled-release device is implanted under the skin in the woman's upper arm. It gradually releases the progestin over a period of several years. The failure rate (that is, conception) is quoted as only 0.2 percent, rivaling the pill for effectiveness. Norplant is not the first controlled-release contraceptive device; Progestert, developed in the 1970s, is still used in Third World countries. One major obstacle to the use of the pill is the general lack of education in the less developed countries, making it difficult to convince users to take the pill faithfully. Implantable devices alleviate this problem. However, the Norplant device—as well as the pill, for that matter—provides absolutely no protection against sexually transmitted diseases, including AIDS.

The contraceptive pills and implantable devices mimic the female hormone system and are used by the woman. Why not a male pill? One oft-suggested reason is that men did the research; though this is possible, male contraceptives are also the topic of much research. There are two major obstacles to an effective male contraceptive, one statistical and the other biomedical. Conception only requires the union of one egg with one sperm cell. A female normally releases a single egg during the monthly cycle, but the male ejects millions of sperm cells during each act of intercourse. You need not be a statistician to realize that controlling the release of a single egg is much easier than controlling millions of sperm cells. If a male contraceptive reduced the number of viable sperm to just a few hundred, enough sperm could still remain to induce pregnancy. A sperm count this low will drastically reduce the probability of conception; it would be almost, but not quite, equivalent to sterilization. Still, the sheer numbers make an effective male contraceptive pill virtually unachievable. Another problem exists as well; several chemicals reduce sperm counts to very low values, even zero at times, but the effects are ordinarily irreversible. Such a contraceptive fails to meet the requirement of reversibility. However, research continues on a male chemical contraceptive pill.

Diethylstilbestrol (DES), which is similar to the steroid molecules, was once used as a potential birth control agent, but it also induces high levels of vaginal cancer in both the user and her female offspring. DES was also used as a growth promoter for cattle and poultry, but the FDA banned this use.

More recently, a new drug dubbed RU-486 has entered the picture and incited a great amount of controversy. RU-486 is Mifepristone, a progesterone derivative that terminates, or aborts, a pregnancy. Moral issues aside, RU-486 may be safer than a surgical abortion, but abortions carry greater risks than contraceptives. In addition, RU-486 requires that the doctor see the patient for several days and is hardly a quick, simple process. In August 1995, the *New England Journal of Medicine* reported on a study that showed that a combination of two readily available drugs, the anticancer drug methotrexate and the ulcer drug misoprostol, is about as effective as RU-486 for inducing abortion. Both drugs have FDA approval for their other uses.

Spermicides are chemicals that kill sperm cells. Although aimed at the male gamete, spermicides are used by the woman. Most spermicides will kill many sperm cells, but the odds against complete success are staggering; the failure rate averages 25 percent. The combination of a diaphragm and spermicide is more effective than the spermicide alone, but the failure rate still averages 20 percent.

Intrauterine devices (IUDs) are made of plastic or metal; plastic is now the favored material. Jack Lippes introduced the first IUD in 1961. The device was made of polyethylene and was dubbed the Lippes' Loop. Most subsequent IUDs used an inert plastic, but some used copper metal that aided contraception by releasing presumably harmless copper ions; most metallic IUDs are no longer in use. IUDs are fairly effective (6 percent pregnancy rate), but they do present several problems, including (1) increased and irregular bleeding, (2) increased and irregular spotting, and (3) an increase in vaginal and cervical infections. The main advantage is simplicity; once inserted, an IUD prevents conception for many years with little additional attention.

This leaves condoms to consider. Condoms are thin sheaths made of an elastomer and are subject to failure in several ways. The main cause of failure is leakage through pinhole-sized holes or through rips and tears that occur during use. Both problems contribute to the condom's 13 percent pregnancy rate. However, condoms are the *only* contraceptive device that guards against disease transmission. All other methods listed in table 14.8 (plus Norplant) provide absolutely no protection against sexually transmitted diseases, including AIDS.

Even condoms do not provide absolute protection against diseases; the rupture failure rates are estimated at about 10 to 20 percent. (The pregnancy rate is slightly lower because not all episodes of intercourse result in pregnancy.) From a practical standpoint, relying on a condom during sex with an infected person still leaves a 1 in 10 to 1 in 5 chance of contracting the disease. Very few people regard such odds as encouraging, especially when AIDS is the potential infection. Abstinence may seem old-fashioned, but it is the one sure protection from sexually transmitted diseases.

PRACTICE PROBLEM 14.18

Teratogenic substances cause birth defects. Alcoholic beverages, coffee, soft drinks, and cigarettes contain known potential teratogens such as ethanol, caffeine, and the polynuclear hydrocarbons. Should society restrict the use of such agents? Should restrictions be placed on females of childbearing age? Caffeine is claimed to affect sperm cells adversely, possibly leading to birth defects. Should males of childbearing age be restricted from drinking beverages that contain caffeine?

ANSWER Your answer will be your own opinion. Can you support it with some scientific evidence?

THE CHEMISTRY OF OUR SENSES

Although it may not be obvious, most of our senses involve chemistry at one point or another. For example, the sense of touch requires that nerve impulses travel to the brain for interpretation. Nerve impulses rely on chemicals to shift the signal from one neuron to another. The first neurotransmitter identified was the relatively simple chemical **acetylcholine,** $CH_3COO—CH_2CH_2N^+(CH_3)_3$, which controls the nerve signals with the help of acetylcholinesterase, an enzyme. Acetylcholine is stored in all nerve endings in membrane-bound bags called synaptic vesicles. When a nerve impulse arrives, an influx of Ca^{++} ions causes the synaptic vesicles to migrate to the nerve membranes adjacent to a muscle cell. Acetylcholine is released, and it diffuses across the space between the nerve and muscle cells and binds to a receptor site on the muscle cell. This opens pores through which Na^+ and K^+ ions flow, generating the nerve impulse and causing the muscle to contract. The process is stopped by acetylcholinesterase, which hydrolyzes acetylcholine.

Taste and smell are almost entirely chemical. In both senses, the nature of the stimulus is determined via direct chemical interactions. The quality of taste is confirmed using taste buds, sensory units in the tongue and mouth. Taste buds respond to four different kinds of taste: sweet, sour, salty, and bitter. The perception of flavor consists of varying combinations of these four kinds of taste. Most ionic compounds, such as NaCl, taste salty and stimulate the salt-specific taste buds. The sour taste usually arises from acidic substances. Sweet or bitter tastes are not as restricted; the taste buds perceive many chemicals as sweet, including sugar, artificial sweeteners, glycerine, and lead salts. These substances have no similarity in structure, shape, or chemical nature. (Some people suspect that the sweet taste of lead compounds causes children to eat lead-containing paint chips.) The bitter taste also arises from many different compounds. Temperature, texture, and odor blend into what people call taste.

How the sense of smell operates is still a mystery. Scientists know that odors impinge on cells that make up the olfactory epithelium in the nasal cavity and that molecular shape seems to play a major role in odor. Some evidence suggests that odors are detected via infrared, which might explain why this sense can detect quantities as small as one or two molecules.

Vision is, perhaps, the most intriguing sense of all. This sense depends on two types of cells found in the retina of the eye, the rods and cones. Rods are responsible for dim light vision, while the cones operate in bright light and also perceive color; this is why color is harder to discern at night. Vitamin A plays an important role in vision. It is converted to cis-11-retinal, an unsaturated aldehyde found in the retina, where it combines with a protein called opsin to form rhodopsin. Cis-11-retinal is also obtained from β-carotene, an orange pigment found in carrots and other yellow foods, plus tomatoes and spinach. When light strikes the rods, the cis-11-retinal is converted to the trans-isomer, changing the rhodopsin into metarhodopsin. This change in shape produces a nerve impulse that travels to the brain and is interpreted as an image. Metarhodopsin is converted back into rhodopsin for further use by a series of chemical reactions.

This series of reactions is powered by light and communicated to the brain via chemically modulated nerve impulses. Molecular shape is as important here as in the enzyme lock-and-key mechanism. Retinal contains five double bonds (four in a linear chain and the other in a ring), but only the double bond at one position changes back and forth from cis- to trans-; all the other double bonds remain trans- throughout the photochemical reactions. Cis-11-retinal fits snugly into the rhodopsin cavity and forms a chemical link to this protein through a pendant amino group (figure 14.27). When cis-11-retinal converts into trans-11-retinal, much of the molecule no longer fits snugly into the protein's cavity. The amine:aldehyde linkage is severed, and the trans-11-retinal is set free. Trans-11-retinal is then changed back to the cis-form, and the process repeats as often as necessary. The entire process requires less than a second; not all rods or cones are stimulated at the same time under normal conditions, so the sense of sight is not lost, even briefly.

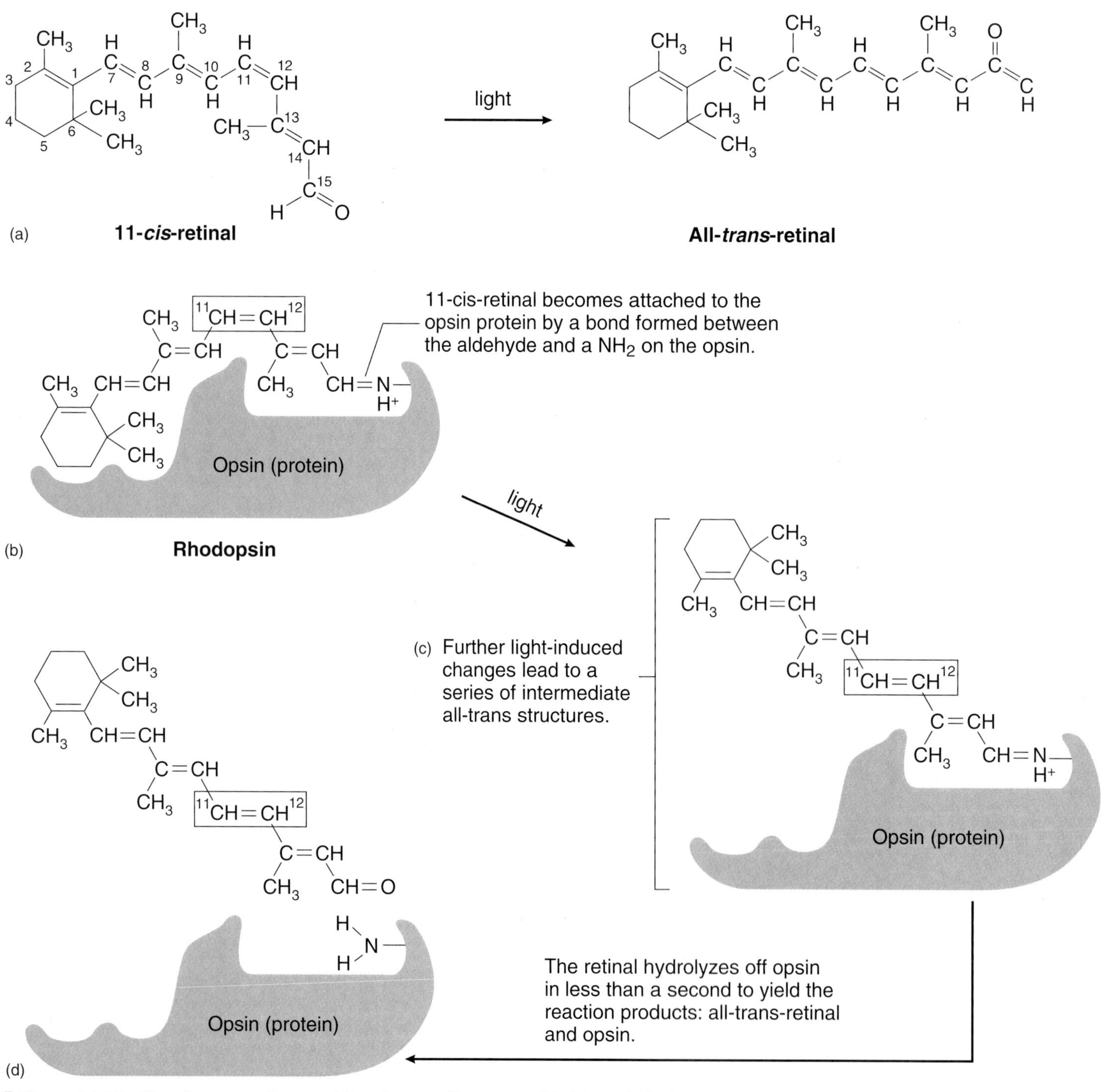

Figure 14.27 The chemistry of vision. **(a)** is the overall reaction; **(b)**, **(c)**, and **(d)** show details for the steps involved.

Chemistry is involved in hearing in much the same way as it is in touch. The motion of the ear drum (the tympanic membrane) is passed to the nerves in the back of the ear by a trio of tiny bones, where the nerve impulse is stimulated. This impulse is transmitted to the brain through the acetylcholine-acetylcholinesterase combination.

PRACTICE PROBLEM 14.19

Certain chemicals are acetylcholine inhibitors. How might such a chemical block the action of acetylcholine?

ANSWER Acetylcholine binds to a receptor site. If some other chemical binds there first, acetylcholine has no place to bind, and nerve transmission is blocked.

SUMMARY

Biochemistry is the study of life and living systems. Biochemicals, including polypeptides, saccharides, nucleic acids, lipids, hormones, vitamins, and minerals, form the chemical foundation for life.

Polypeptides include the proteins and enzymes. Twenty different amino acids form all the proteins in our bodies and in our food. Amino acids introduce a new kind of isomerization called optical isomerization. The isomers of an asymmetric or chiral molecule display this type of isomerization—they are mirror images of one another.

Proteins and polypeptides are polymers of amino acids. These polymers have primary, secondary, tertiary, and quaternary structures. The primary structure of a polypeptide is the sequence of amino acids in its chain; the secondary structure is a repeated folding pattern that arises from the interactions between the units in the polypeptide chain. The tertiary structure combines more than a single protein molecule, and the quaternary structure contains several proteins folded together with other units. Even a tiny change in the primary structure can have devastating effects on human health.

Enzymes are polypeptides that act as natural catalysts. They operate by a lock and key mechanism—one enzyme fitting one specific protein—to cleave proteins back to amino acids.

Saccharides, also called carbohydrates, are starches and sugars. They may be classified as monosaccharides (single unit in the molecule), disaccharides (two units), or polysaccharides (many units). Important monosaccharides include glucose, fructose, ribose, and deoxyribose. Disaccharides include sucrose (table sugar), lactose, and maltose, and polysaccharides include cellulose and starch.

Nucleic acids consist of three distinct units—a phosphate, a monosaccharide, and a nucleic base. The two most important nucleic acids—RNA and DNA—encode and transmit genetic information. Genetic engineering allows scientists to manipulate the sequence in a DNA chain. By doing so, they may be able to cure or prevent genetic diseases.

A lipid is a biological molecule that is soluble in organic solvents. Lipids include the triglycerides—fats and oils—plus steroids, prostaglandins, waxes, and fatty acids. Fatty acids may be saturated or unsaturated. Prostaglandins perform a variety of functions, and glycerides store energy and act as emulsifying agents. Nonglycerine lipids include sphingolipids, steroids, and waxes.

Complex lipids are either lipoproteins or glycolipids. Lipoproteins transport lipids throughout the body; two types, low- and high-density lipoproteins (LDL and HDL, respectively) appear to play a role in heart disease.

Hormones trigger important changes in body chemistry. They may be steroidal, polypeptide, thyroid, or catecholamine. Most sex hormones are steroidal, while polypeptide hormones control a variety of processes, including the manufacture of insulin, human growth and development, uterine contraction, skin and hair pigmentation, and others.

Vitamins are organic substances essential to good health and obtainable only through the diet or dietary supplements. Vitamins may act as coenzymes, enabling enzymes to function, or hormones. They prevent disease and keep eyes, bones, muscles, and skin healthy.

Minerals are inorganic substances also essential to health. Many are metals that help ensure proper fluid and blood circulation, develop strong bones and teeth, or activate enzymes. Phosphorus is essential to energy production.

Many birth control methods make use of biochemistry to prevent conception. A good contraceptive should be effective, safe, simple to use, reversible, and able to protect the user from disease. Present methods vary widely in these qualities.

Finally, our senses depend on biochemicals to function. Nerve impulses rely on acetylcholine, vision on vitamin A, cis-11-retinal, and trans-11-retinal. Taste and smell are almost entirely chemical in nature. We would not be able to sense the world around us—or even to sustain life—without a complex set of biochemicals that function in a delicate balance.

KEY TERMS

acetylcholine 389
adenosine triphosphate (ATP) 385
alpha (α) helix 364
amino acids 358
amylopectin 370
amylose 370
antibodies 373
antigens 373
asymmetric molecule 359
base pairing 374
beta (β) pleated sheet 364
chiral molecule 359
chromosome 374
codons 376
contraceptive 386
deoxyribose 369
disaccharides 367
enantiomer 359
enzyme 366
fructose 369
genetic engineering 376
glucose 369
glycogen 370
high-density lipoproteins (HDL) 379
hormones 380
lipids 377
low-density lipoproteins (LDL) 379
minerals 383
monosaccharides 367
multisubunit protein 365
nucleic acids 372
optical activity 359
optical isomerism 359
peptide unit 363
plane polarized light 359
polypeptides 358
polysaccharides 367
primary structure 363
prostaglandin 378
quaternary structure 365
replication 375
ribose 369
saccharides 367
secondary structure 364
steroid 380
sucrose 369
tertiary structure 364
vitamins 382

READINGS OF INTEREST

General

Ezzell, Carol. "Sticky situations: Picking apart the molecules that glue cells together." *Science News*, June 13, 1992, pp. 391–395. Ever wonder what holds the cells of your body together? This article explains.

Kohlwein, Sepp D. "Biological membranes: Function and assembly." *J. Chem. Ed.*, vol. 69, no. 1 (January 1992), pp. 3–9. This readable article describes what biological membranes are, how they are made, and what they do. The role of lipids is featured prominently.

Gore, Rick. "The awesome worlds within a cell." *National Geographic*, vol. 150, no. 3 (September 1976), pp. 355–395. From the workings of DNA to genetic engineering, this article transports you on an exciting journey. Learn more about the immune system, Tay-Sachs disease, photosynthesis, clones, carrots, and tomatoes.

Ponte, Lowell. "Killer diseases from the dawn of time." *Reader's Digest*, February 1985, pp. 128–132. Disease-causing agents even smaller than the virus may be the source of some unusual and incurable diseases. Apparently small polypeptide-like organisms, these infectious agents are immune to heat, radiation, and most drugs.

Amino Acids and Proteins

Copeland, Robert A. "Enzymes, the catalysts of life." *Today's Chemist at Work*, March 1994, pp. 51–53. This article describes the lock and key mechanism of enzyme activity along with several enzyme applications, such as tenderizing meat, dissolving blood clots, and preserving milk.

Linder, Maurine E., and Gilman, Alfred G. "G proteins." *Scientific American*, July 1992, pp. 56–65. These proteins on cell surfaces seem to make cell-to-cell communication possible. If so, G proteins enable life, awareness, and thinking.

Marx, Jean. "A new link in the brain's defenses." *Science*, vol. 256 (May 29, 1992), pp. 1278–1280. The link appears to be protease, like the blood-clotting agent thrombin, which is an amino acid polymer. This research may lead to an understanding of Alzheimer's disease.

Heiser, Terry L. "Enzyme activity: The ping-pong ball torture analogy." *J. Chem. Ed.*, vol. 69, no. 2 (February 1992), p. 137. Another simple analogy of how enzymes do their stuff.

Abel, Kenton B., and Halenz, Donald R. "Enzyme activity: A simple analogy." *J. Chem. Ed.*, vol. 69, no. 1 (January 1992), p. 9. Explains some enzyme activity features using children popping balloons with pins.

Optical Activity and Chirality

Jackson, W. Gregory. "Symmetry in automobile tires and the left-right problem." *J. Chem. Ed.*, vol. 69, no. 8 (August 1992), pp. 624–626. Are tires chiral or achiral? If you do not get bogged down in the "point group and characteristic elements" discussion, you ought to figure out the answer from this article.

Tavernier, Dirk. "The square knot and the granny knot." *J. Chem. Ed.*, vol. 69, no. 8 (August 1992), pp. 627–628. More information about everyday chiral objects, including some knotty examples.

Borman, Stu. "Mirror-image structures: Enzymes made using all D-amino acids." *Chemical & Engineering News*, June 8, 1992, pp. 4–5. Natural proteins and enzymes contain only L-amino acids. Here are the synthetic looking-glass versions.

Petsko, Gregory A. "On the other hand . . . " *Science*, vol. 256 (June 5, 1992), pp. 1403–1404. What if chiral polymers were made from the other kinds of molecules? That is, what if proteins were made from D-amino acids and polysaccharides from L-units? What shapes might they take?

Amoto, Ivan. "Looking glass chemistry." *Science,* vol. 256 (May 15, 1992), pp. 964–966. Chiral molecules do not react the same way in chemistry and biochemistry. This difference is exploited with chiral drugs.

Saccharides

Borman, Stu. "Centennial of first determination of glucose configuration honored." *Chemical & Engineering News,* June 8, 1992, pp. 25–26. This brief article shows several photographs of Emil Fisher, the pioneer in this field, and his laboratory. Discusses why we call the common molecule D-glucose rather than the opposite.

Lienhard, Gustav E.; Slot, Jan W.; James, David E.; and Mueckler, Mike M. "How cells absorb glucose." *Scientific American,* January 1992, pp. 86–91. Glucose is a major energy source, but how does it enter cells?

Antibodies, Antigens, and Allergies

Radetsky, Peter. "The end of allergies." *Longevity,* April 1994, pp. 36–37, 72–74. It is amazing how many things people are allergic to—this popularized article lists many of them. It also discusses some over-the-counter allergy treatments.

Brownlee, Shannon. "The cellular battlefield." *U.S. News & World Report,* March 28, 1994, pp. 66–68. This battlefield is where the war against allergies takes place, and this short article tells how some scientists are trying to win the battle.

Radetsky, Peter. "Of parasites and pollens." *Discover,* September 1993, p. 54–62. This article contains many good pictures of these allergy-causing entities.

Jaroff, Leon. "Allergies, nothing to sneeze at." *Time,* June 22, 1992, pp. 54–62. Good popular-style article with interesting diagrams of antigen-antibody interactions. Covers asthma as well.

Shine, Jerry. "Understanding the immune system." *Arthritis Today,* May-June 1995, p. 8. New research suggests a particular protein called "ku" may be the key to understanding the immune system.

Johnson, Howard M.; Russell, Jeffry K.; and Pontzer, Carol H. "Superantigens in human disease." *Scientific American,* April 1992, pp. 92–101. A superantibody might be nice, but superantigens spell real trouble! These proteins cause food poisoning, toxic shock, and possibly arthritis and some AIDS-related problems. Every piece of bad news is countered by good; awareness of a problem is the first step toward solving it.

Goodenough, Ursula W. "Deception by pathogens." *American Scientist,* vol. 79 (July/August 1991), pp. 344–355. Viruses and bacteria can be sneaky and attack by deceptive routes. This article has some excellent diagrams of antigen-antibody interactions plus outstanding photographs of the rhinovirus (which causes the common cold) and the influenza virus.

Jaret, Peter. "Our immune system, the wars within." *National Geographic,* vol. 169 no. 6 (June 1986), pp. 702–734. While AIDS has focused the public's attention onto the immune system, it is hardly unique in impairing the immune system, which battles all kinds of infectious microorganisms, cancer, allergens, and much more. This easy-to-read article contains a wealth of details on how this system operates.

Newman, Cathy. "Pollen, breath of life and sneezes." *National Geographic,* vol. 166, no. 4 (October 1984). pp. 490–520. Allergies are nothing to sneeze at, but could pollen have some useful purposes? This article contains excellent pollen photographs plus a splendid diagram on sneezing.

Nucleic Acids and Genes

Erickson, Deborah. "Hacking the genome." *Scientific American,* April 1992, pp. 128–137. A good article about mapping detailed descriptions of the nucleotides in human genes.

Carey, John; Hamilton, Joan; and McWilliams, Gary. "The gene doctors roll up their sleeves." *Business Week,* March 30, 1992, pp. 78–82. A business-oriented article about using gene therapy to cure diseases such as cystic fibrosis, cancer, AIDS, and many others.

Jaroff, Leon. "Making the best of a bad gene." *Time,* February 10, 1992, pp. 78–79. This gene is the cause of Huntington's disease. Here is what one person did after she realized she had this sequence of DNA.

Brownlee, Shannon. "Courtroom genetics." *U.S. News & World Report,* January 27, 1992, pp. 60–61. What DNA evidence can be used in court cases?

Montgomery, Geoffrey. "The ultimate medicine." *Discover,* March 1990, pp. 60–66. Recombinant DNA technology is being used to seek cures for genetic diseases. In many instances, this approach uses modified viruses as "good news" carriers to enable organs to function better, or even to perform functions they never did before.

McAuliffe, Karen. "Genes that predict disease." *Reader's Digest,* September 1987, pp. 17–24. The presence of certain genes predicts the probability that an individual may acquire common afflictions such as cancer or heart disease.

Cystic Fibrosis

Palca, Joe. "The promise of a cure." *Discover,* June 1994, pp. 76–86. Can a genetically engineered virus be used to combat cystic fibrosis? Some early results suggest this may be possible.

Findeis, Mark A. "Genes to the rescue." *Technology Review,* April 1994, pp. 47–53. Scientists are now trying to cure cystic fibrosis and other genetic diseases by transferring copies of normal genes into the patients.

Pennisi, E. "Gene, biochemical fixes sought for CF." *Science News,* October 23, 1993, p. 260. Cystic fibrosis has been linked to a protein defect. This article tells of efforts to correct this disease by administering protein to the patients.

Purvis, Andrew. "Laying siege to a deadly gene." *Time,* February 24, 1992, pp. 60–61. Breakthroughs in treating cystic fibrosis using DNA. In the early 1980s, scientists learned why cystic fibrosis caused its symptoms. Then, in 1989, scientists discovered the gene that causes this disease. Now scientists are on the verge of finding DNA treatments to cure cystic fibrosis.

Lipids

Hudnall, Marsha. "Clearing up misconceptions about cholesterol." *Arthritis Today*, May-June 1995, pp. 59–61. The latest words on cholesterol levels.

Lawn, Richard M. "Lipoprotein (a) in heart disease." *Scientific American*, June 1992, pp. 54–60. More information, vital to your health, about the cholesterol-lipid connection.

Quigley, Michael N. "The chemistry of olive oil." *J. Chem. Ed.*, vol. 69, no. 4 (April 1992), pp. 332–335. Olive oil has been valued since antiquity. Learn about its chemistry in soap, health, and elsewhere.

Lasic, Danilo. "Liposomes." *American Scientist*, vol. 80 (January/February 1992), pp. 20–21. These biological entities are sometimes called "little bags of fat," but that is not an entirely correct characterization. Liposomes are growing in importance in nonbiochemical areas and are now used in controlled release, cosmetics, and other applications.

Hormones

Murray, Mary. "Your sex hormones at 20, 30, 40." *Glamour*, October 1993, pp. 206–9, pp. 280–282. This woman-oriented article tells how the functions of these hormones change with age.

Ainscough, Eric W.; Brodie, Andrew M.; and Wallace, Anna L. "Ethylene—An unusual plant hormone." *J. Chem. Ed.*, vol. 69, no. 4 (April 1992), pp. 315–318. After seeing what sort of molecules usually are hormones, this may come as a shock: ethylene is a plant hormone (in a sense). Find out why.

Schrof, Joannie M. "Pumped up." *U.S. News & World Report*, June 1, 1992, pp. 55–63. The sorry story of anabolic steroid usage.

Rowan, Carl, and Mazie, David. "The mounting menace of steroids." *Reader's Digest*, February 1988. pp. 133–137. By most estimates, over a million Americans use anabolic steroids to make themselves bigger and stronger. They also "gain" the side effects of sterility, impotence, cancer and heart disease, plus possible early death.

Minerals

Ochiai, El-Ichiro. "Biomineralization." *J. Chem. Ed.*, vol. 68, no. 8 (August 1991), pp. 627–630. The manner in which minerals are involved in human biochemistry and nutrition.

Pennisi, Elizabeth. "Natureworks: Making minerals the biological way." *Science News*, May 16, 1992, pp. 328–331. How minerals are incorporated into living organisms, such as oysters.

Csintalan, B. P., and Senozan, N. M. "Copper precipitation in the human body: Wilson's disease." *J. Chem. Ed.*, vol. 68. no. 5 (May 1991), pp. 365–367. Wilson's disease, which involves improper utilization of copper in the human body, is relatively uncommon. This article gives information on the disease and describes how it is treated chemically.

Birth and Population Control

Ingrassia, Michelle; Springer, Karen; and Rosenberg, Debra. "Still fumbling in the dark." *Newsweek*, March 13, 1995, pp. 60–62. Why do so many women turn to sterilization when so many contraceptives are available?

Rubin, Rita. "Stopping this birth control can hurt." *U.S. News & World Report*, July 25, 1994, pp. 59–60. The method discussed is Norplant, which is not easy to remove once implanted.

Hanson, Gayle M. B. "'Morning after' pill has political side effects." *Insight*, July 4, 1994, pp. 15–17. An article about RU-486 and the ongoing controversy that surrounds it. Information is presented from both sides of the debate.

Criner, Lawrence. "'Safer sex' ads downplay risks." *Insight*, May 9, 1994, pp. 22–24. Also Funderburk, Patricia. "None, not safer, is the real answer." *Insight*, May 9, 1994, pp. 25–27. A pair of articles that claim that abstinence is the only method that can absolutely prevent unwanted pregnancy and avoid the dangers of sexually transmitted diseases, including AIDS.

Richman, Sheldon. "Forget the myth of overpopulation." *Insight*, December 20, 1993, pp. 20–22. The title tells what this author has in mind, but you will need to read the article to find out why.

Hardin, Garrett. "Limits to growth are nature's own." *Insight*, December 20, 1993, pp. 23–25. The growth this article focuses on is the expansion of the human population.

Franklin, Deborah. "The birth control bind." *Health*, July/August 1992, pp. 42–52. How effective are modern birth control methods? This article, largely anecdotal, claims they often fail.

Weber, Joseph, and Cuneo, Alice. "The war of the pill." *Business Week*, June 29, 1992, pp. 71, 74. Who makes birth control pills? This is a billion-dollar-per-year business!

Riddle, John M. and Estes, J. Worth. "Oral contraceptives in ancient and medieval times." *American Scientist*, vol. 80 (May/June 1992), pp. 226–233. More information on old-time contraception.

Berreby, David. "The numbers game." *Discover*, April 1990, pp. 42–49. This author asks whether population increases will harm or help humanity—by producing more geniuses and problem solvers.

Ehrlich, Paul R., and Ehrlich, Ann H. "Population, plenty, and poverty." *National Geographic*, vol. 174, no. 6 (December 1988), pp. 915–945. These authors discuss the potential problems of overpopulation. What role should technology play in resolving this dilemma?

The Senses and Neurotransmission

Feldman, Paul L; Griffith, Owen M.; and Stuehr, Dennis J. "The surprising life of nitric oxide." *Chemical & Engineering News*, December 20, 1993, pp. 26–38. This noncarbon compound is involved in such varied biochemical processes as transmission of nerve signals, blood pressure control, and the operation of the immune system.

Long, Michael E. "The sense of sight." *National Geographic*, vol. 182, no. 5 (November 1992), pp. 3–41. Sight is arguably our most important sense, but how does it work, and what can be done when something threatens it? Many diseases, including some that are contagious, can destroy eyesight.

Borman, Stu. "New light shed on mechanism of human color vision." *Chemical & Engineering News,* April 6, 1992, pp. 27–29. Why do we see in color? The explanation is still being researched.

Snyder, Solomon H., and Bredt, David S. "Biological roles of nitric oxide." *Scientific American,* May 1992, pp. 68–77. Nitric oxide plays a variety of roles in the body; some may be helpful, but others are dangerous. It can even act as a neurotransmitter.

Lancaster, Jack R., Jr.; "Nitric oxide in cells." *American Scientist,* vol. 80 (May/June 1992), pp. 248–259. As both messenger and destroyer, nitric oxide plays the role of a double agent in biochemistry.

Biotechnology

Kennedy, Dan. "Udder angst." *Garbage,* Summer 1994, pp. 40–45. How scientific is the fight against bovine growth hormone? For that matter, how scientific is the fight for it?

Tucker, William. "As health reform looms, the biotech industry cringes." *Insight,* July 18, 1994. pp. 6–9. Biotechnology companies fear that potential price controls on new drugs may not allow them to recoup development costs. This fear has already resulted in cutbacks on research into cancer, AIDS, and rare childhood diseases.

Elmer-Dewitt, Philip. "Fried gene tomatoes." *Time,* May 30, 1994, pp. 54–55. This short article explores new varieties of tomatoes, potatoes, rice, and other food crops emerging from biotechnology.

Thayer, Ann. "FDA gives go-ahead to bioengineered tomato." *Chemical & Engineering News,* May 23, 1994, pp. 7–8. At last, a tomato that can sit for weeks without rotting! "Flavr Savr," a genetically engineered food, has received FDA approval, but some activist groups plan to intensify their fight against it, threatening a national "tomato war."

Carey, John, and Smith, Geoffrey. "The next wonder drug may not be a drug." *Business Week,* May 9, 1994, pp. 84–86. The next wonder drug may be a gene that can treat a disease, such as cancer, cystic fibrosis, or AIDS. Much research is still required, but this approach looks very promising.

Borman, Stu. "Bioengineering makes plants disease resistant." *Chemical & Engineering News,* December 6, 1993, p. 9. Here is one way to reduce the need for pesticides.

Impoco, Jim. "Green genes." *U.S. News & World Report,* October 18, 1992, pp. 58–60. Biotechnology is a growing area of science that may become involved in many different markets, including medicines, diagnostics, and agriculture.

Campbell, Todd. "Nature's building blocks." *Popular Science,* October 1993, pp. 74–77. Scientists are studying ways to get new and valuable materials from such sources as beetles, spiders, and rat teeth.

Armstrong, Larry. "Supertomatoes—The old-fashioned way." *Business Week,* October 4, 1993, pp. 116, 121. Not everyone is trying to improve tomatoes via biotechnology.

Glanz, James. "Herman: The pharmaceutical industry's next star?" *R&D Magazine,* June 1992, pp. 36–42. Genetically engineered animals can produce medically useful proteins—a new benefit from chemistry via biotechnology.

Weaver, Robert F. "Beyond supermouse: Changing life's genetic blueprint." *National Geographic,* vol. 166, no. 6 (December 1984), pp. 818–847. The new world of biotechnology seems bizarre and frightening to many, but it also promises us many exciting benefits, such as improved medicine, healthier plants, and just maybe, the eradication of cancer. An easy-to-read article describes all this and more.

PROBLEMS AND QUESTIONS

1. What are the seven main classes of biochemicals?
2. Which biochemical classes are natural polymers?
3. What are amino acids?
4. How many amino acids commonly exist in proteins? Which are "essential," and why are they called essential?
5. What does an optically active molecule do?
6. What structural feature can cause optical isomerization in a chemical compound?
7. What is the meaning of the term chiral? What other term describes the same property?
8. Give examples of some biochemical compounds that exhibit optical isomerization.
9. What are the building blocks for proteins?
10. What is the special name for the repeat unit in a protein?
11. Which synthetic polymer class does the repeat unit in proteins resemble?
12. In what important manner does the repeating sequence in proteins differ from that found in typical synthetic polymers?
13. Describe briefly the four levels of structural organization in proteins. How do they differ from each other? Give an example for each of these four levels of protein structure.
14. Where are proteins found in our bodies?
15. What is the most common protein?
16. Using abbreviated names, draw all the possible primary sequences for a polypeptide containing one unit of each of the following amino acids: glycine (Gly), proline (Pro), tryptophan (Trp), and alanine (Ala).

17. What is the difference between an α-helix and a β-sheet structure?

18. What kinds of protein structures are in human hair?

19. To what class of biochemicals do the enzymes belong?

20. In what ways do enzymes affect the chemistry of living systems?

21. Briefly describe the lock and key mechanism for enzyme activity. What role does molecular shape play in this mechanism?

22. What part of the IUPAC name identifies an enzyme?

23. What function do catalysts perform?

24. What kinds of chemical compounds are antibodies? What is another name for antibodies?

25. What role do antibodies play in biochemistry?

26. What are antigens? How do they differ from antibodies?

27. What is another name for the saccharides?

28. What is the characteristic suffix used to denote a saccharide in the IUPAC system?

29. List several examples of monosaccharides.

30. List some examples of disaccharides.

31. List at least three examples of polysaccharides.

32. How do ribose and deoxyribose differ? In what class of polymers do these molecules occur?

33. What is the difference between amylose and amylopectin? How do these chemicals relate to glycogen in structure?

34. What is the most common natural polymer in the world?

35. What is the nature of the repeat unit in starches and cellulose?

36. What are the similarities and differences between cellulose and chitin?

37. List the following saccharides in order of increasing sweetness: fructose, glucose, invert sugar, and sucrose.

38. How is invert sugar prepared?

39. What are the names of the five nucleic bases?

40. What is the structure of the nucleic acids (DNA or RNA)? Show this answer in schematic form, indicating the structure of the repeat unit for either DNA or RNA.

41. What is another name for the repeat unit in a nucleic acid?

42. What are the main biochemical functions of the nucleic acids?

43. How does RNA differ from DNA?

44. What is a virus, from a chemical standpoint?

45. Lipids are made from what types of chemicals?

46. What are saturated and unsaturated fatty acids?

47. What are prostaglandins, and what are their functions?

48. What are phosphoglycerides? Name the most common example.

49. What are sphingolipids, and what important role do they play in the chemistry of the human body?

50. How do low-density lipoproteins (LDL) and high-density lipoproteins (HDL) differ in function?

51. What roles do LDL and HDL play in cholesterol metabolism?

52. What are the two major classes of hormones? Give the names of some important hormones from each class.

53. Cite some examples of how hormones function in living systems.

54. What structural feature do all steroids have in common?

55. What are anabolic steroids? How do they differ from other molecules in the steroid class?

56. What chemical functions do minerals perform in living systems?

57. What specific functions do sodium and potassium ions perform in human physiology?

58. What are the major molecules involved in energy production and transmission in living systems?

59. What essential mineral is present in AMP, ADP, and ATP?

60. What are the five main requirements for a satisfactory contraceptive?

61. Which contraceptives are the most effective in preventing pregnancy?

62. To what biochemical class do the chemicals used in birth control pills belong?

63. What is the chemical nature of RU-486, and what does it do?

64. What is a spermicide?

65. How does the sense of taste involve chemistry?

66. What are the key chemical compounds used in vision? How does this process operate at the chemical level?

CRITICAL THINKING PROBLEMS

1. Why is it essential to maintain a minimum level of water in the body?

2. There was an episode in the original *Star Trek* TV series in which a silicon-based life form called the Horta existed. Based on the periodic table and information in chapter 9, could living systems develop based on silicon instead of carbon? What limitations exist? Could such a life form exist here on earth?

3. Almost everyone is allergic to something. Read at least two of the following articles and then make two lists: one that shows how to avoid potential allergies, and a second that shows allergy treatment. How safe are the new treatments the articles discuss? Peter Radetsky, "The End of Allergies," *Longevity*, April 1994, pp. 36–37, 72–74; Shannon Brownlee, "The Cellular Battlefield," *U.S. News & World Report*, March 28, 1994, pp. 66–68; Peter Radetsky, "Of Parasites and Pollens," *Discover*, September 1993, pp. 54–62; Leon Jaroff, "Allergies, Nothing to Sneeze At," *Time*, June 22, 1992, pp. 54–62.

4. Read Robert F. Weaver, "Beyond Supermouse: Changing Life's Genetic Blueprint," *National Geographic*, vol. 166, no. 6 (December 1984), pp. 818–847, and John Carey, Joan Hamilton, and Gary McWilliams, "The Gene Doctors Roll Up Their Sleeves," *Business Week*, March 30, 1992, pp. 78–82. What are some potential uses for recombinant DNA? Should research in this field be restricted?

5. The Human Genome Project is trying to map the details of the human DNA structure. How does this structure relate to an individual? What individual differences exist that the Project is not likely to determine? How might the Project benefit humanity, and are there potential problems with gaining a detailed knowledge of the human genome? To read more about this project, see Deborah Erickson, "Hacking the Genome," *Scientific American*, April 1992, pp. 128–137.

6. Read the article by Shannon Brownlee entitled "Courtroom Genetics," in *U.S. News & World Report*, January 27, 1992, pp. 60–61. What limitations, if any, do you think should be placed on the use of DNA evidence in court cases? How might "DNA fingerprinting" aid in proving a criminal suspect's innocence or guilt?

7. What chemical factors are important in aging? Is a person's chronological age always the same as their biological age?

8. Why do you think contraceptive research and development has centered mainly on devices and drugs used by females?

9. Read Joseph Weber and Alice Cuneo, "The War of the Pill," *Business Week*, June 29, 1992, pp. 71, 74. Should restrictions or regulations be placed on the industries that manufacture contraceptives?

10. Some suggest that abstinence is the *only* true preventative for unwanted pregnancy and the threat of sexually transmitted diseases. Is this really a valid option for today's world? Read the following pair of articles and form your own opinion: Lawrence Criner, "Safer Sex" Ads Downplay Risks, *Insight*, May 9, 1994, pp. 22–24; and Patricia Funderburk, "None, Not Safer, Is the Real Answer," *Insight*, May 9, 1994, pp. 25–27.

11. Many believe that providing better education, new economic outlets for women, and readily available contraceptives may be the solution to the population problem. Read the following five articles on these ideas and form a tentative conclusion: (1) Eugene Linden, "More Power to Women, Fewer Mouths to Feed," *Time*, September 26, 1994, pp. 64–65; (2) Emily MacFarquhar, Stephen Budiansky, Betsy Carpenter, and Traci Watson, "Population Wars," *U.S. News & World Report*, September 12, 1994, pp. 54–65; (3) Emily T. Smith, Margot Cohen, and Elisabeth Malkin, "Too Many People?" *Business Week*, August 29, 1994, pp. 64–66; (4) Bryant Robey, Shea O. Rutstein, and Leo Morris, "The Fertility Decline in Developing Nations," *Scientific American*, vol. 269, no. 6 (December 1993), pp. 60–67; and (5) Gayle Hanson, "Norplant Joins War on Teen Pregnancy," *Insight*, March 8, 1993, pp. 6–11, 34–35.

12. Is our world on the verge of drastic overpopulation? Not everyone views this issue from the same vantage point. Read the first two articles and at least one of the others in the following list, then make your own decision on this question: Sheldon Richman, "Forget the Myth of Overpopulation," *Insight*, December 20, 1993, pp. 20–22; Garrett Hardin, "Limits to Growth Are Nature's Own," *Insight*, December 20, 1993, pp. 23–25; David Berreby, "The Numbers Game," *Discover*, April 1990, pp. 42–49; Paul R. Erlich and Ann H. Erhlich, "Population, Plenty, and Poverty," *National Geographic*, vol. 174, no. 6 (December 1988), pp. 915–45.

13. Just how important is population control for our future? There are many sides to this question and less agreement than you might expect. For example, some believe that population numbers are exaggerated and overpopulation is not as severe as others claim. (How would a poor country without many resources determine its population?) Some even go so far as to claim that overpopulation may be a benefit. Read several of the following articles on this topic and summarize the history of and present thinking on population control: (1) Barbara Ehrenreich, "The Bright Side of Overpopulation," *Time*, September 26, 1994, p. 86; (2) John Bongaarts, "Population Policy Options in the Developing World," *Science*, February 11, 1994, pp. 771–76; (3) Tim Stafford, "Are People the Problem?" *Christianity Today*, October 3, 1994, pp. 45–60; (4) Nicholas Eberstadt, "Population Policy: Ideology as Science," *First Things*, January 1994, pp. 30–38; (5) Gregory Benford, "The Designer Plague," *Reason*, January 1994, pp. 37–41; and (6) Midge Decter, "The Nine Lives of Population Control," *First Things*, December 1993, pp. 17–23.

14. The average life span has increased significantly in the past several centuries. Do you think there may be some ultimate limit on the maximum human life span? What problems exist in studying longevity? What problems might an increased life span cause for humanity? How could birth control fit into this picture? Read the article "Population, Plenty, and Poverty" by Paul R. Ehrlich and Ann H. Ehrlich, *National Geographic*, vol. 174, no. 6 (December 1988), pp. 915–945, for important background information.

15. Imagine that you are a UN population expert and have received the assignment to develop birth control options for a small, democratic, largely protestant, agricultural country. What would you recommend?

16. You are still a UN population expert, but now your assignment is to recommend birth control procedures for a small, largely Catholic country that is rapidly developing its industrial base. What are your recommendations?

17. The legacy of Thomas Malthus has endured into our generation. How valid are these ideas in today's world? Read some background articles, pro and con, and come to a tentative conclusion. Possibly the best summary on this topic is the book by Lester R. Brown and Hal Kane entitled *Full House: Reassessing the Earth's Population Carrying Capacity* (New York: W. W. Norton, 1994, 261 pp.) though Brown and Kane heavily favor the pro side. Some shorter articles include: (1) Samuel C. Florman, "Overpopulation Alarm," *Technology Review*, October 1994, p. 65; (2) Emily MacFarquhar, Stephen Budiansky, Betsy Carpenter, and Traci Watson, "Population Wars," *U.S. News & World Report*, September 12, 1994, pp. 54–65; (3) Emily T. Smith, Margot Cohen, and Elisabeth Malkin, "Too Many People?" *Business Week*, August 29, 1994, pp. 64–66; (4) John Bongaarts, "Can the Growing Human Population Feed Itself?" *Scientific American*, vol. 270, no. 3 (March 1994), pp. 36–42; and (5) Ronald Bailey, "Malthus for the Modern Man," *Reason*, April 1994, pp. 68–71. For more information, you might also consult the book *The Population Bomb* by Paul R. Ehrlich, a modern-day Malthus advocate.

18. Read the article entitled "Udder Angst," by Dan Kennedy, in *Garbage*, Summer 1994, pp. 40–45. Make your own appraisal of the merits and demerits of the bovine growth hormone.

19. Some biotech companies are backing off from developing new medical treatments because potential price controls may not allow them to recoup their development costs. How real is this concern? Read the following article and form your own opinion: William Tucker, "As Health Reform Looms, the Biotech Industry Cringes," *Insight*, July 18, 1994, pp. 6–9.

20. Why do many scientists consider the virus a living organism?

HINT Consider what characteristics are used to define life. One is the ability to reproduce. Do viruses reproduce? Can they survive without some other kinds of organisms? Can we?

21. Science can now do things that could change the basic nature of humankind, but are all these potential changes positive? What should be done to reduce or eliminate possible dangers while still allowing the development of useful and valuable technologies? Read some of the following articles and form a tentative opinion on this very difficult problem: (1) Shannon Brownlee, Garth G. Cook, and Viva Hardigg, "Tinkering with Destiny," *U.S. News & World Report*, August 22, 1994, pp. 59–67; (2) David R. Carlin, "For Luddite Humanism," *First Things*, April 1994, pp. 9–10; (3) John Horgan, "Eugenics Revisited," *Scientific American*, June 1993, pp. 122–133; (4) W. Warren Wager, "Tomorrow and Tomorrow and Tomorrow," *Technology Review*, April 1993, pp. 51–59; (5) Robert C. Post, "The Frailties and Beauties of Technological Creativity," *Invention & Technology*, Spring 1993, pp. 16–14; and (6) Gerald F. Joyce, "Directed Molecular Evolution," *Scientific American*, December 1992, pp. 90–97.

CHAPTER 15

FOOD AND AGRICULTURAL CHEMISTRY

OUTLINE

KEY IDEAS

1. The Basic Chemistry of Food Digestion
2. Food Classes and Their Recommended Daily Consumption
3. The Vitamins and How They Contribute to Health
4. The Categories of Food Additives
5. The Essence of Natural and Artificial Sweeteners
6. The Chemicals Used as Flavoring Agents
7. The Chemical Nature of Colorants
8. Chemical Food Preservation Methods
9. The Chemistry of Baking
10. The Nature and Use of Fertilizers
11. The Types of Chemicals Used as Pesticides or Herbicides
12. The Promise of Biotechnology

"An empty stomach is not a good political advisor."
ALBERT EINSTEIN (1879–1955)

Fruits and vegetables.

Chapter 14 explored some of the chemicals involved in life; many of these enter living systems as food or drink, or from the air. These chemicals then become part of one or more life cycles. In food chemistry, the overall cycles are digestion and respiration. These two cycles are interrelated.

Food may be found in the wild or deliberately grown. The most common food sources for humans are deliberately prepared plants or animals; the cultivation of such food sources is agriculture. This chapter examines some features of agricultural chemistry because it plays such a large role in food chemistry. Since food often contains small quantities of chemical additives, we will examine them, too.

THE CHEMISTRY OF FOOD DIGESTION

The saying "you are what you eat" is unequivocally true at the chemical level. Figure 15.1 summarizes the overall food digestion process. This process involves one basic chemical reaction—**hydrolysis,** in which the compounds are broken down into smaller units in the presence of water. Hydrolysis occurs with polypeptides (proteins), polysaccharides (carbohydrates), and lipids, the three main food chemical groups. Hydrolysis is the opposite of the condensation reactions that form proteins and polysaccharides. Vitamins and minerals aid in these reactions as cofactors or coenzymes.

The small hydrolysis products (amino acids, simple sugars, fatty acids, and glycerin) are either converted directly into energy, reassembled into new body tissues, or stored for future use as glycogen (in small amounts in the liver and muscle) or fatty tissue deposits (adipose tissue). Fat is an excellent storage medium with an energy content of 9 Kcal/g, which is more than either carbohydrates or proteins (both 4 Kcal/g). These values represent the amount of energy released by lipids, proteins, and carbohydrates during combustion. The **dietitian's calorie** is actually a kilocalorie from the chemist's standpoint. In this book, these two measures are distinguished by using Kcal for the chemical unit and Dcal for the dietetic unit.

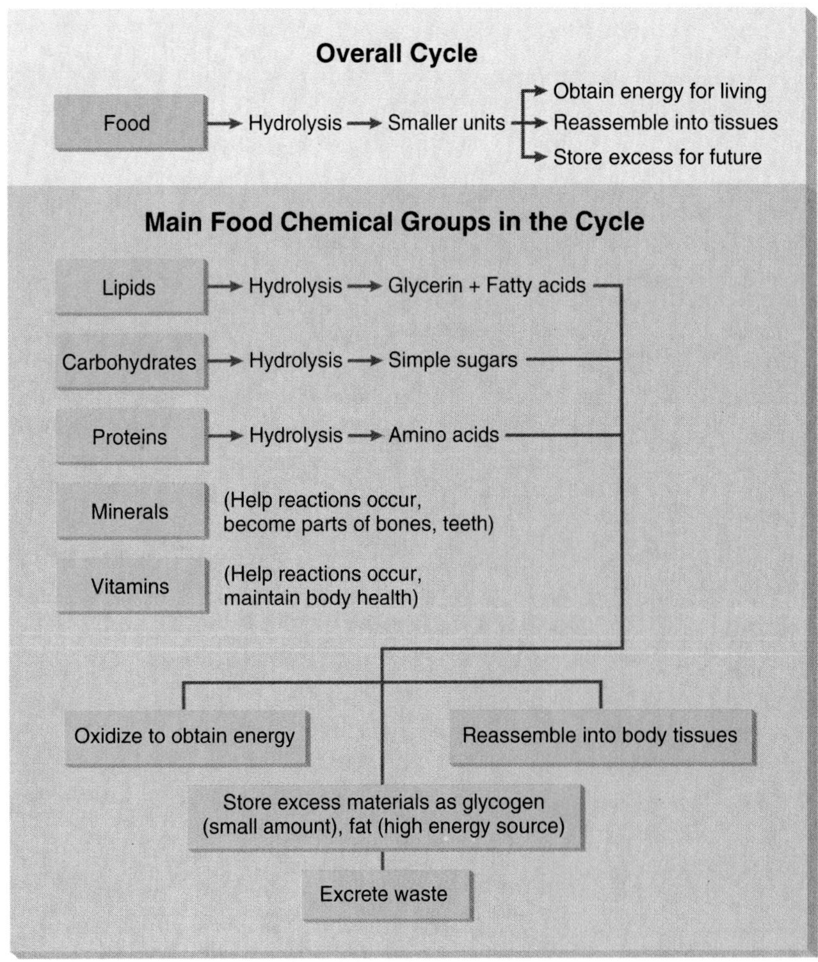

Figure 15.1 The food digestion cycle.

Protein Digestion

Although the basic reaction is the same for proteins, polysaccharides, and lipids, different enzymes are required for the hydrolysis of each type of food. These enzymes are called **hydrolases;** protein-hydrolyzing enzymes are called **proteases.** Some proteases can only cleave a specific amino acid, but others can cleave several; the digestion of a single protein could require dozens of enzymes to cleave the different amino acids in the polypeptide chain. After the proteins are cleaved back to amino acids, other enzymes are needed to reassemble them into body tissues. Meats, beans, nuts, and eggs provide protein, along with dairy products, such as milk, yogurt, and cheese.

Saccharide Digestion

Polysaccharides are usually made from a single saccharide unit and only require a single enzyme for hydrolysis. The specific enzyme required depends upon both the geometry and structure of the polymer. Thus, different enzymes hydrolyze the glucose α-linkages in amylose and β-linkages in cellulose. These saccharide-cleaving enzymes are named after the saccharide, with the suffix *-ose* replaced by an *ase*. (For example, lactase acts on lactose and maltase on maltose.)

Some humans possess lactase, the enzyme that splits lactose (milk sugar) into galactose and glucose. However, many people lack this enzyme and are unable to digest milk sugar and products that contain it, including cheese and ice cream. This leads to a condition called lactose intolerance that causes severe stomach cramps, bloating, and diarrhea. Lactose intolerance is fairly common among Asians, people of African descent (70 percent in the United States), and people with Jewish heritage; it usually affects women more than men, and is more common in adults than children. Moderately concentrated forms of lactase, available in pill form or added to milk products, enable an enzyme-deficient person to digest milk products. Whether lactose intolerance arises from genetic or environmental factors is not known.

In galactosemia, another genetic disease, the enzyme that converts galactose (a hydrolysis product of lactose) into glucose is missing. In an individual with galactosemia, a toxic, reduced form of galactose called dulcitol forms:

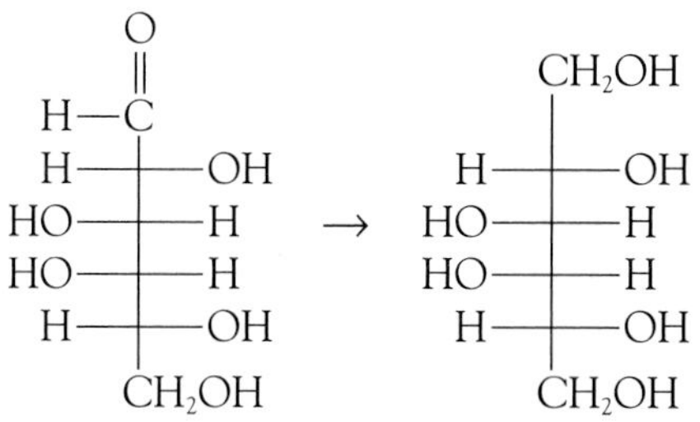

$$
\begin{array}{c}
\mathrm{H{-}C{=}O} \\
\mathrm{H{-}C{-}OH} \\
\mathrm{HO{-}C{-}H} \\
\mathrm{HO{-}C{-}H} \\
\mathrm{H{-}C{-}OH} \\
\mathrm{CH_2OH} \\
\text{Galactose}
\end{array}
\quad \longrightarrow \quad
\begin{array}{c}
\mathrm{CH_2OH} \\
\mathrm{H{-}C{-}OH} \\
\mathrm{HO{-}C{-}H} \\
\mathrm{HO{-}C{-}H} \\
\mathrm{H{-}C{-}OH} \\
\mathrm{CH_2OH} \\
\text{Dulcitol}
\end{array}
$$

Although not normally fatal, dulcitol does cause severe stomach upset. Grains, bread, pasta, rice, and fruits and vegetables provide polysaccharides.

Lipid Digestion

Lipids hydrolyze to fatty acids and glycerol (glycerine) using enzymes termed **lipases.** Individual fatty acids are classed as saturated, unsaturated, or polyunsaturated depending on the number of double bonds they contain. Table 15.1 shows the iodine number for some fatty acids, fats, and oils. The **iodine number,** expressed as grams iodine per 100 grams substance, is the amount of iodine that could add to the double bonds; the higher the iodine number, the more unsaturated the material. **Saturated fatty acids** contain no double bonds and have an iodine number of zero (0.0). **Unsaturated fatty acids,** with one C=C, have iodine numbers between 90 and 100. **Polyunsaturated fatty acids** (linoleic and linolenic) have two or three double bonds, with corresponding iodine numbers of 181 and 273. Solid fats have a high unsaturated fatty acid content, while most oils contain low amounts of saturated fat. There are exceptions; coconut and palm oils, for example, are much higher in saturated fats than other oils are.

Unsaturated fatty acids are considered better for human health; this means we are well advised to consume mostly oils with unsaturated or polyunsaturated fatty acids and iodine numbers ranging from 80 to about 200. Olive oil has been used for many thousands of years; peanut oil has about the same level of unsaturation and is much less expensive. Both olive and peanut oil have iodine numbers of about 80 to 100, but the currently preferred oils are safflower, canola (from rapeseed), sunflower, and soybean, with iodine numbers between 120 and 140. Fish oils have made a strong impact in recent years since scientists discovered that oils from cold-water fish seem to benefit the cardiovascular system. Highly unsaturated oils, with high iodine numbers, form the favored lipid-cholesterol combination, while solids such as butter and margarine can contribute to high levels of the undesirable type of cholesterol.

Where Does Cholesterol Fit?

Chapter 14 explained why diets high in polyunsaturated lipids result in the more healthy HDL cholesterol, whereas diets high in saturated fats yield the poorer LDL cholesterol. Neither diet totally prevents cholesterol formation since the body synthesizes this steroid when necessary, but diets moderately low in cholesterol are preferable. Prudent behavior would also include maintaining a diet high in the polyunsaturated lipids. Such a diet would include fish as the main protein choice and emphasize vegetables considerably more than the typical American diet does. While this may not guarantee low cholesterol and ward off all possible heart attacks, it makes a lot more sense from a chemical perspective than any other alternative.

PRACTICE PROBLEM 15.1

Coconut oil is most likely a liquid because the fatty acids it contains have low melting points. Why is palm oil, which is composed mainly of palmitic and lauric acids, also a liquid?

TABLE 15.1 The structures and iodine numbers for selected fatty acids, fats, and oils.

COMMON NAME	COMPOSITION OR STRUCTURE	M.P. (°C)	IODINE NUMBER
SATURATED FATTY ACIDS			
Capric	$CH_3(CH_2)_8COOH$	32	0
Lauric	$CH_3(CH_2)_{10}COOH$	44	0
Stearic	$CH_3(CH_2)_{16}COOH$	70	0
UNSATURATED FATTY ACIDS			
Oleic	$CH_3(CH_2)_7CH{=}CH(CH_2)_7COOH$	16	90
Linoleic	$CH_3(CH_2)_4CH{=}CHCH_2CH{=}CH(CH_2)_7COOH$	5	181
Linolenic	$CH_3CH_2(CH{=}CHCH_2)_2CH{=}CH(CH_2)_7COOH$	−11	273
FATS (SOLIDS)			
Butter	mostly butyric and caproic		17–35
Beef fat	mostly stearic, other saturated acids		35–42
Human	assorted		57–73
Margarine	partially hydrogenated unsaturated acids		70–90
OILS (LIQUIDS)			
Coconut	mostly caprylic and capric		6–10
Palm	mostly palmitic, some lauric		49–59
Olive	mostly oleic		79–88
Peanut	nearly 50 percent oleic		88–98
Corn	high in linoleic		111–128
Fish	high in linolenic		120–180
Soybean	high in linoleic		122–134
Safflower	high in linoleic		122–141
Canola	80 percent unsaturated acids (rapeseed oil)		125–135
Sunflower	high in linoleic		129–136

ANSWER There are two major reasons. First, mixtures have lower melting points than pure compounds; food oils are mixtures. Second, the melting points in table 15.1 are for the fatty acids, not their esters; esters melt at lower temperatures than the acids.

EXPERIENCE THIS

Collect as many cooking oils as you can find and place a small amount (about 15 mL, or a half ounce) of each in separate oven-safe dishes or aluminum pans. Carefully try to set fire to each. Do they burn? (They should!) Do those that burn give off much heat? Try to find some raw peanuts, or at least some that are not dry roasted, and see if you can set them on fire. Then try burning some animal fat and even a few small samples of meat and vegetables. You should find that most of these foods items will burn. What does this tell you about the conversion of foods into energy in the body? If the digestion process were not controlled by enzymes, so that it can operate at a lower temperature, what do you think would happen to your body?

THE CHEMISTRY OF NUTRITION

What is a chemically balanced, nutritious, diet? (Note that the term *diet* means a regulated program of eating and drinking and has little to do with weight reduction.) From a chemical perspective, a healthy diet must provide amino acids, fatty acids, and simple saccharides, plus vitamins, minerals, and fiber. Such a diet thus contains proteins, saccharides (preferably as polysaccharides), and lipids. The U.S. Department of Agriculture has created a food pyramid diagram that shows the ideal plan for healthy eating (figure 15.2). A person who bases his or her diet on the recommended number of food portions from each segment of the pyramid does not need to keep track of the details to maintain a healthy diet.

How much of each food group is needed? The total Kcal required daily for a 150-pound (68 kg) person, assuming moderate physical activity, is about 3,300. This person only requires 30 g (about an ounce) of protein daily, equivalent to about 105–120 Kcal or Dcal. To ingest this much protein, one needs less than two ounces, or about 57 grams, of meat each day. Although this amounts to about 4 percent protein in the diet, most Americans eat far more protein than their bodies require.

Nutritionists recommend that a diet also contain 65 to 80 percent polysaccharides, which means that 2,150 to 2,650 Kcal of a 3,300-Kcal diet should come from this source. The main polysaccharides are the starches amylose and amylopectin, plus cellulose. Along with the lignins, these also provide most of the 25 to 30 g of dietary fiber recommended daily. The daily polysaccharide intake should lie between 538 and 663 g, or about 19 to 23 ounces (1.2 to 1.5 pounds), including fiber.

The current lipid content in the American diet is nearly 40 percent of our caloric intake, and sometimes higher; the recommended levels are around 15 to 25 percent. This means that only 495 to 660 Kcal of a 3,300-Kcal diet should come from lipids, an amount equal to 55 to 73 g/day.

TABLE 15.2 Minimum recommended daily intake of proteins, polysaccharides, and lipids, in Kcal and grams for a 150-pound (68 Kg) person.

FOOD CLASS	Kcal	GRAMS
Proteins	105–120	30–57
Polysaccharides	2,150–2,650	538–663
Lipids	495–660	55–73
Total	2,750–3,430	623–793 22–28 oz 1.35–1.75 lbs.

These minimum requirements are summarized in table 15.2. You may be surprised at how low these values are, since most of us consume much more food than we truly need each day. Perhaps that is why 40 to 50 million Americans are on a weight-loss diet at any given time. Notice also how little protein we need, compared with polysaccharides or lipids. This is what constitutes a healthy, chemically balanced diet that should contain everything the body needs: amino acids, fatty acids, simple saccharides, fiber, minerals, and vitamins.

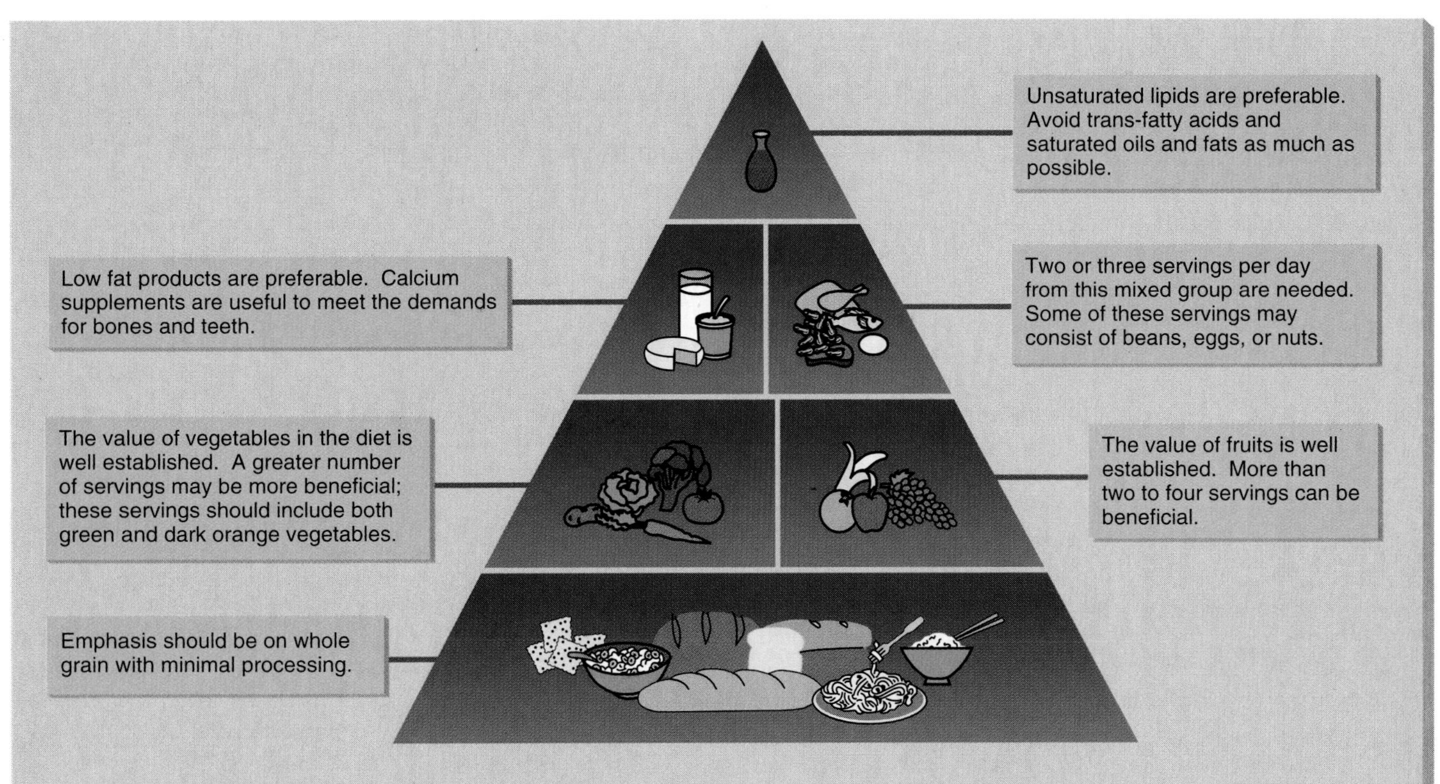

Figure 15.2 The food pyramid.

PRACTICE PROBLEM 15.2

Using table 15.2 as a guide, determine the minimum amounts of proteins, polysaccharides, and lipids that a person weighing 91 Kg (200 lbs.) must consume to maintain good health.

ANSWER The data in table 15.2 is based on a 68 Kg person; 91 Kg is 1.33 times 68 Kg, so each amount on the table must be multiplied by 1.33. Accordingly, a person weighing 91 Kg needs 40 to 76 g proteins, 717 to 884 g polysaccharides, and 73 to 97 g lipids each day, for a total of 830 to 1057 g (slightly more than two pounds of food).

The Role of Vitamins

Nearly everyone in the United States knows vitamins are important for good health, and a thriving multimillion dollar vitamin supplement industry is prepared to remind us if we forget. Vitamins are trace organic substances required in the diet; many are precursors of coenyzmes. A good, balanced diet will probably supply daily vitamin needs, but vitamin supplements are available to ensure one obtains the required amounts. From a pragmatic standpoint, it is healthier to take extra vitamins than to suffer a deficiency.

Vitamins are subdivided into water-soluble and lipid-soluble classes (figure 15.3). The **water-soluble** group includes the B-family (a mixed group that includes niacin and pyridoxal), biotin, ascorbic acid (or vitamin C), folic acid, and pantothenic acid. The **lipid-soluble** group includes vitamins A, D, E, and K. Some vitamin structures, such as niacin or ascorbic acid, are simple compounds, while others, such as vitamin B-12, are extremely complex.

Unlike the water-soluble vitamins that people excrete easily, lipid-soluble vitamins can deposit and accumulate in the fatty tissues of the body; lipids are present in most cells. High levels of lipid-soluble vitamins might exceed toxic limits and cause harm. This does not necessarily mean that water-soluble vitamins are always harmless; they just are not as readily stored in the body. Toxic vitamin levels are high, and a problem seldom occurs unless a person consumes massive quantities (box 15.1).

Table 15.3 lists the functions and sources of vitamins; it is divided into lipid- and water-soluble groups. A quick glance down the list reveals that vitamins function just about everywhere in the body. A good rule of thumb is that any water-soluble vitamin (except vitamin C) is probably a coenzyme—that is, without it some enzyme could not perform its function. Protein, lipid, nucleic acid, and saccharide synthesis and metabolism depend on vitamins. Obviously, the human body needs all these vitamins for optimal health, and as the table indicates, they come from a variety of sources. A balanced diet is essential in order to obtain an adequate supply of each vitamin.

Government scientists have run numerous studies on the nutritional status of people living in the United States, and the surveys consistently reveal vitamin intakes well below the RDAs. According to the 1985 Nationwide Food Consumption Survey, this deficiency is greater among the poor; the percentages of people taking in less than 70 percent of the RDA for a given vitamin include 26 to 51 percent for vitamins A, C, and B_6, and 12 to 17 percent for vitamins B_1, B_2, and B_{12}. Good dietary practices help ensure that an individual meets all vitamin and mineral requirements. If this is not possible, vitamin supplements are an inexpensive solution.

In December 1994, the World Bank reported that more than one billion people in the less developed countries have serious vitamin and nutrient deficiencies that could be remedied for a cost of one dollar per person. They reported that over 13 million people were blind due to a vitamin A deficiency; 6 of every 10 school children with this deficiency die. Over a billion people have an iodine deficiency; this deficiency kills 5 to 10 babies out of every 1,000 births and leaves many others mentally impaired, deaf, or mute. These problems can easily be corrected at a very modest cost, through vitamin supplements.

Many vitamins were discovered after scientists associated diseases with the lack of specific foods. For example, when sailors were out at sea for long periods, they developed scurvy, a skin disease. Aside from severe famine, scurvy is probably the most calamitous nutritional deficiency disease ever to strike humanity. Eventually, scientists discovered a lack of vitamin C in the diet caused scurvy (box 15.2).

Studies on pellagra (a skin disorder that leads to nerve disorders) led to the discovery of niacin, while investigations of beriberi (a muscular paralysis) led to the discovery of vitamin B_1 (thiamine). Rickets (a bone disorder) showed the need for vitamin D, and various vision problems led to the discovery of vitamins A and B_2. Lack of B_2 can also cause severe dermatitis. Vitamin E deficiency causes sterility and muscular dystrophy, while vitamin K deficiency produces improper blood clotting. Obviously, vitamins prevent some rather nasty problems.

PRACTICE PROBLEM 15.3

Using figure 15.3, answer the following questions.

a. Which vitamins contain a free —COOH group?

b. Which vitamins contain a benzene ring?

c. Which vitamins do *not* have an —OH group (COOH does not count as an —OH group.)

ANSWER *a*. Nicotinic acid, pantothenic acid, biotin, and folic acid; notice these are all water-soluble vitamins, due to the presence of this highly polar group. *b*. Vitamin K, vitamin B_2, folic acid; vitamin K is lipid-soluble, the other two are water-soluble. *c*. Vitamin K, nicotinic acid, nicotinamide, biotin; all except vitamin K are water-soluble.

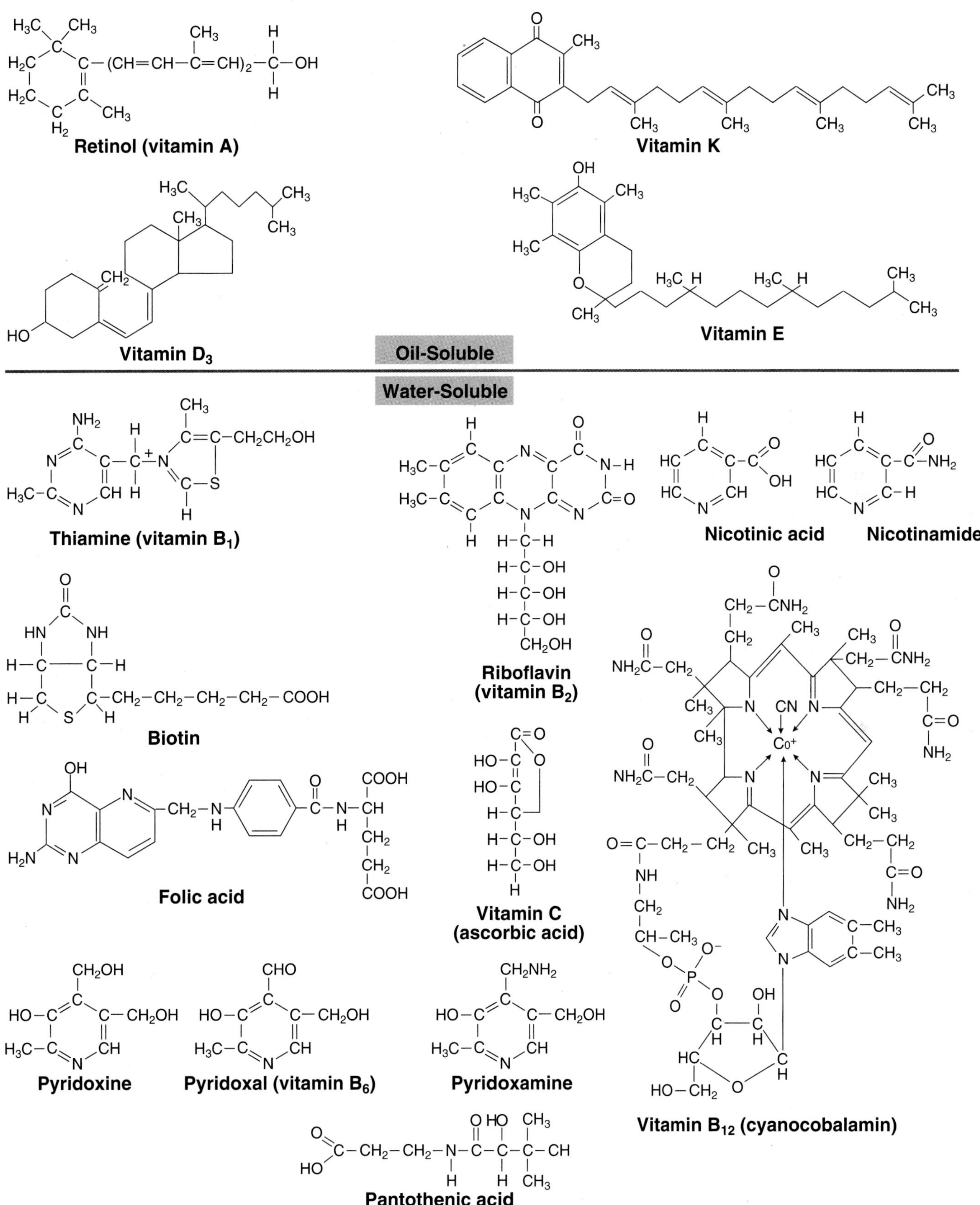

Figure 15.3 The structures of the vitamins divided into oil (lipid)-soluble and water-soluble groupings.

TABLE 15.3 The functions and sources of vitamins.

VITAMIN	FUNCTIONS	SOURCES
LIPID-SOLUBLE GROUP		
A	Promotes growth. Prevents night blindness. Aids normal immune system operation. Essential for vision. Promotes the health of the skin and hair. Has some anticancer activity.	Milk, eggs, liver, green and leafy vegetables, carrots, cheese, fish
D	Prevents rickets. Helps the body utilize calcium and phosphorus; hence, essential for healthy bones and teeth.	Milk, tuna, salmon, cod liver oil, egg yolks, halibut
E	Essential for reproduction. Maintains growth, acts as an antioxidant, absorbs unsaturated fatty acids, aids membrane structure, detoxifies. Vital for red blood cell function.	Beans, apples, olives, many vegetables, eggs, grains, nuts, unsaturated vegetable oils
K	Essential for normal blood clotting. Aids electron transport mechanisms, body growth.	Green and leafy vegetables, pork, beef, fish, nuts, eggs, grains
WATER-SOLUBLE GROUP		
B_1 (Thiamine)	Prevents beriberi. Essential for normal heart and nerve functions. Aids in energy production.	Fruit, wheat germ, rice bran, soybeans, beef, chicken, pork, nuts, fish, eggs, milk
B_2 (Riboflavin)	Develops and maintains body tissues, including skin, brain, blood. Aids fetal development. Contributes to redox reactions and degrades drugs and other chemicals.	Beef, chicken, pork, yeast, beans, nuts, grains, eggs, milk
B_6 (Pyridoxine)	Enables immune system operation; deficiency leads to decreased antibody levels. Essential for teeth, gums, red cells, nervous system; anticancer effects.	Liver, fish, nuts, grains, vegetables, bananas, grapes, beef, pork, eggs
B_{12}	Prevents anemia; promotes proper nerve function; promotes growth in children. Coenzyme for nucleic acid, protein and lipid synthesis.	Liver, egg yolk, clams, salmon, crabs, sardines, oysters, herring
Niacin	Prevents pellagra. Aids synthesis of NAD and NADP, hydrogen and electron transfer. Cofactor. Vital for nerve function and energy generation.	Peanuts, rice brain, liver, turkey, chicken, tuna, halibut, swordfish, yeast, beans, figs, dates, grains
C	Prevents scurvy. Antioxidant. Essential for teeth, gums, bones, blood vessels. Some anticancer activity. Aids iron absorption, wound healing. Contributes to polysaccharide and collagen synthesis.	Citrus fruits, broccoli, kale, cabbage, collards, cauliflower, spinach, strawberries, guavas
Biotin	Prevents anemia. Cofactor for carboxylation. Aids fatty acid synthesis. Maintains skin, hair, nerves.	Yeast, liver, wheat, rice, oats, barley, eggs, some fish, chicken, nuts
Folic acid	Prevents certain anemias. Aids in nucleic acid synthesis. Cofactor. Aids in choline synthesis, amino acid synthesis, and metabolism.	Liver, asparagus, spinach, wheat, bran, beans, yeast, mushrooms
Pantothenic acid	Contributes to fat, protein, and saccharide metabolism. Aids in synthesis of steroids and acetylcholine.	Beef, pork, chicken, lamb, eggs, herring, wheat germ, bran, peanuts, broccoli, soybeans, carrots, peas

PRACTICE PROBLEM 15.4

Most people are fairly familiar with vitamin C. Can you name some good sources for vitamin C?

ANSWER Citrus fruits (specifically lemons, limes, oranges), green vegetables and, to a lesser extent, other fresh vegetables.

PRACTICE PROBLEM 15.5

Assume a person took 30 g vitamin C, the toxic level, and that the $Vt_{1/2}$ (half-life) is one day. How long would it take for that 30 g to decrease to below 4 g? How long before the vitamin C level drops to the 60 mg RDA value? Would the person become sick?

BOX 15.1

RECOMMENDED DAILY ALLOWANCES

Deciding how much of a food or food component is optimal for health is sometimes difficult. Often, not enough data are available to determine exact amounts, but this is not the case for vitamins (figure 1). The minimum quantities needed to avoid gross vitamin deficiency are set as Recommended Daily Allowances (RDA). RDAs are expressed in milligrams, activity units, or in another suitable measure for the "typical" adult. The RDAs for children are smaller, the RDAs for larger adults greater. Since the RDAs offer minimal levels of needed vitamins, it is possible that certain people may require higher levels of a specific vitamin. Scientists have known from the early days of vitamin research that vitamins are toxic if used excessively, and they have set upper limits for most vitamins based on the appearance of toxic side effects. These upper limits are occasionally revised.

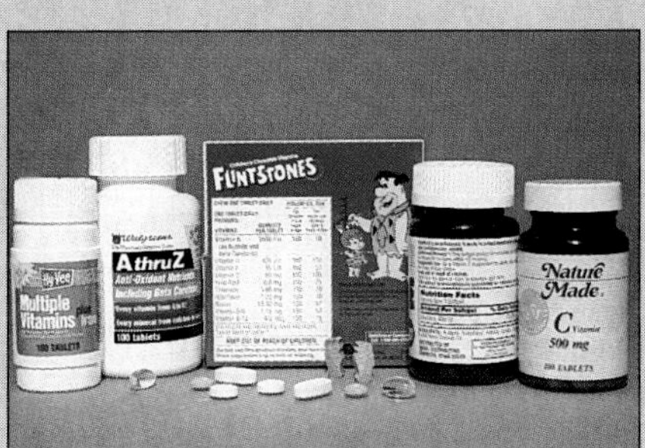

Figure 1 Vitamin pills.

Table 1 shows the current RDAs and upper limits for selected vitamins, along with the toxic side effects that occur when limits are exceeded. Since the RDAs establish lower limits, these two figures define a range to stay in for health and safety. You risk a deficiency if you fall below the RDA; if you surpass the upper limits, you risk toxic symptoms. As the table indicates, the toxic levels are at least ten times higher than the RDAs.

TABLE 1 The Recommended Daily Allowances (RDA), the upper (toxic) limits, and the toxic side effects of vitamins.

VITAMIN	RDA*	UPPER LIMIT	TOXIC EFFECTS
LIPID-SOLUBLE GROUP			
A	1,000 μg	20,000 μg	Irritability, nerve lesions, fatigue, insomnia, pain in bones and joints, loss of hair, jaundice, decreased blood clotting time
D	10 μg	100 μg	Nausea, thirst, diarrhea, joint pain, muscular weakness, calcification of soft tissues (blood vessels, muscle), resorption of bone
E	10 mg	high?	Possible increase in blood pressure
K	80 μg	unknown	Possible thrombosis, vomiting
WATER-SOLUBLE GROUP			
B_1	1.5 mg	15,000 mg	Possible edema, nervousness, sweating
B_2	1.8 mg	high?	Essentially nontoxic in humans
B_6	2.0 mg	200 g ?	Limited toxicity in humans, convulsions above about 300 g
B_{12}	2.0 μg	unknown	General lack of toxicity in humans
C	60 mg	10–30+ g	Normally none other than mild diarrhea, possible kidney stones with gout
Niacin	18 mg	1,500 mg	Burning, itchy skin, decreased serum cholesterol, increased pulse rate, and peripheral vasodilation
Biotin	300 μg	70 g	Essentially nontoxic in humans
Folic acid	200 μg	40,000+ mg	No reported toxicity in humans; renal damage and convulsions in mice
Pantothenic acid	15 μg	700+ g	Respiratory failure in mice; essentially nontoxic in humans

*For adult males weighing 160 pounds. Female and children RDAs are lower; pregnant or lactating female RDAs are higher. Note that not all units are the same.

(Continued)

BOX 15.1—Continued

Although it would take large amounts of most vitamins to produce toxic effects, lipid-soluble vitamins can accumulate in the body and could exceed toxic limits if exorbitant quantities are consumed regularly. Vitamins have an effective half-life in the body, similar to the radioactive half-lives of elements, so their concentrations continually decline. Vitamin half-lives ($Vt_{1/2}$) are determined directly through urinalysis, or by restricting intake and looking for signs of vitamin deficiency; the latter is less reliable since the symptoms require a long time to develop. Most ($Vt_{1/2}$) are not known precisely, but table 2 lists some estimates. Lipid-soluble vitamins (with the exception of vitamin E) often have high $Vt_{1/2}$ values, while water-soluble vitamins usually dissipate more rapidly. (The exceptions are vitamins B_{12} and biotin.)

TABLE 2 The estimated half-lives ($Vt_{1/2}$) of selected vitamins.

VITAMIN	ESTIMATED $Vt_{1/2}$
LIPID-SOLUBLE GROUP	
A	weeks to months
D	days to weeks
E	<1 day
K	<10 days
WATER-SOLUBLE GROUP	
Biotin	3–4 weeks
Niacin	a few days
B_1	a few days
B_2	a few days
B_6	a few days
B_{12}	a year
C	<1 day

ANSWER A $Vt_{1/2}$ works just like the half-life for nuclear decay. The 30 g level drops to 15 g after one day, 7.5 after 2 days, and 3.75 g (3,750 mg) after three days. It requires nine days to drop to the 60 mg RDA level. A person who ingested 30 g of vitamin C would probably have severe diarrhea for the first three days.

Reassembly of the Basic Units

Once the food chemicals are broken down into smaller units, the individual units either enter an energy-releasing cycle (as shown in figure 2 of box 14.6) or are used to rebuild body tissues. In some cases, enzymes must be available to convert one type of raw material into another, although humans cannot synthesize certain chemicals—for example, the essential amino acids (box 15.3). A balanced diet provided a balanced supply of the amino acids, simple saccharides, and fatty acids necessary for forming new or replacement tissues in the body. Naturally, food should also provide the necessary vitamins and minerals, or they must come from a supplement.

Classes of Foods

The best protein sources are meat, including fish and poultry, dairy products, and bean or peas; lesser sources include grains, nuts, and a few vegetables (table 15.4). In principle, people can obtain all the protein they need from grains and bean or nuts, but not all sources provide an equivalent amount of energy. The reason many foods in table 15.4 are low in usable energy is that they contain more fiber than protein; energy is consumed in digesting fiber, and part of this energy comes from the proteins. Rice, for example, has much fiber and starch. You would need to eat a lot of rice to take in 3,300 Kcal-day. (About 13.4 cups to be exact, if it is cooked white rice. Though 4.5 cups would provide enough raw protein, not all of it would be available as energy.) Proteins are not the only source of energy; most of our energy comes from the saccharides, with a significant amount from lipids as well.

The main polysaccharides in foods are the starches amylose and amylopectin, plus cellulose. **Carbohydrate** is a more common term for polysaccharide, but **polysaccharide** emphasizes the chemical aspects of food more accurately and emphasizes the polymers rather than the simple sugars. Sugar is sweet . . . but it is not an essential foodstuff. Fruit contains some simple sugars, notably glucose and fructose, and milk contains lactose, but our main concern is the polysaccharides.

TABLE 15.4 Percentages of protein and usable energy available from various food sources.

FOOD SOURCE	PERCENT PROTEIN	PERCENT USABLE ENERGY*
Beans (lima)	8	26
Beef	29	38
Eggs	13	50
Fish (tuna)	28	62
Milk (cow, 3.5 percent fat content)	4	22
Milk (cow, skimmed)	4	40
Peanuts	22	19
Peas	6	26
Pork	17	38
Poultry (chicken)	24	40
Rice	3	8
Soybeans	42	45
Sweet potatoes	2	4
Wheat flour	12	13

*Usable energy percentages from Charles B. Simone, *Cancer and Nutrition*. Garden City Park, N.Y.: Avery Publishing, 1992.

The main starch sources are grains, potatoes, beans, nuts, peas, most vegetables, and some fruits. Since these foodstuffs should comprise the central portion of our diet, most of our energy should come from them. One advantage polysaccharides have over simple sugars is that they release their energy over long time periods; sugars virtually dump their energy almost instantly upon ingestion. A candy bar may stifle hunger briefly, but a slice of bread quells hunger for a longer time.

Dietary fiber consists of cellulose, lignin, pectin, and various gums, all of which are nondigestible. The major cellulose sources are green vegetables, especially leafy ones, plus certain grains. Typical fiber sources include bran, carrots, oranges, apples, Brussels sprouts, spinach, wheat, rice, corn, beans, and cabbage, plus some leafy green vegetables and grains that provide cellulose.

Lipids have such a high caloric content (9 Kcal/g) that we need only 55 to 73 g daily. You easily obtain this much in one large hamburger, especially if it has added cheese. Keeping the lipid level below the recommended 25 percent upper limit is difficult if you eat much meat or cheese. The percentage of calories derived from lipids is usually under 15 percent for vegetables, normally unsaturated or polyunsaturated (table 15.5).

As the table shows, a small portion of beef or pork supplies more saturated fat than you need in one day. Poultry has less fat than beef or pork, but is not fat free. About 50 to 75 percent of the caloric content in a typical cut of beef is from fat; even lean cuts are 35 to 50 percent fat calories. Marbleized beef is higher

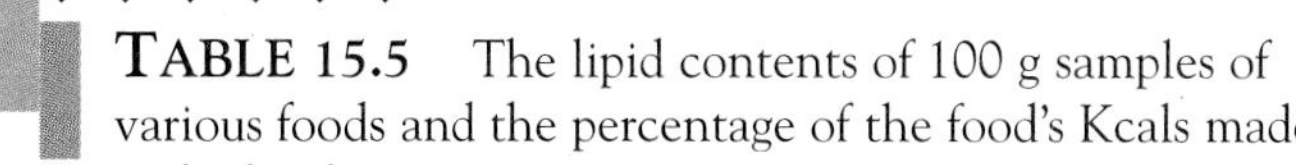

TABLE 15.5 The lipid contents of 100 g samples of various foods and the percentage of the food's Kcals made up by lipids.

FOOD ITEM	TOTAL Kcal	GRAMS LIPID	PERCENT LIPID (Kcal)	LIPID TYPE*
Bacon, regular, cured	678	68	90	S
Bacon, fat-free	125	4	26	S
Cheese, American	393	32	74	S
Cheese, American, light	250	14	51	S
Cheese, Swiss	393	29	66	S
Pork, center loin, roast	319	24	69	S
Beef, T-bone steak	324	25	68	S
Beef, T bone/lean	214	10	44	S
Beef, chuck roast	354	27	67	S
Ham, cured, roast	228	13	60	S
Turkey, dark/skin	197	10	47	S
Turkey, dark, no skin	185	7	34	S
Turkey, light, no skin	154	3	19	S
Chicken, light, skin	206	10	43	S
Chicken, light, no skin	170	4	22	S
Salmon, raw	142	6	37	U
Halibut, raw	109	2	19	U
Ocean perch, raw	94	1	11	U
Tuna, yellowfin, raw	106	1	10	U
Banana	91	0.4	4	U
Pinto beans, raw	288	0.9	3	U
Potato, baked, no butter	111	0	0	—
Peas, green, fresh	59	0	0	—
Apples, raw	28	0	0	—

*S = saturated; U = unsaturated or polyunsaturated

since the "marbled" lines are fat. Some cuts of pork are as high or higher than beef in proportion of fat. Between 75 and 90 percent of bacon's calories come from fat. (This is true of the special, higher-priced, lean bacons, as well as cheaper cuts; two major brands have 75 and 80 percent of their calories invested in fat.) Chicken gets 35 to 50 percent of its calories from fat, but peeling off the skin and cutting attached fat drops the value to 25 percent. Turkey white meats runs about 20 to 40 percent fat calorie content (without skin), while dark meat is in the 40 to 50 percent range.

Fish contains the more desirable polyunsaturated lipids and contains lower overall percentages of lipid-based calories. How you cook the fish may change these lipid levels, and their type, significantly. Fried fish has more fat, usually saturated; broiled fish is close to the raw levels.

BOX 15.2

THE VITAMIN C SAGA

In 1498, the famous Portuguese explorer Vasco da Gama wrote one of the earliest observations on scurvy after his fleet of ships rounded the Cape of Good Hope and docked for a month at the mouth of a river in south Africa (figure 1). The ship's log records that "many of our men fell ill here, their feet and hands swelling, and their gums growing over their teeth so that they could not eat." The most striking scurvy symptom is the appearance of purple spots scattered all over the person's body. These symptoms were apparently relieved after da Gama's crew obtained some oranges from a Moorish vessel. They put out to sea once more, and twelve weeks later the sailors redeveloped the same problems. This time about 30 of the 140 crew members died, and only 7 or 8 were healthy enough to navigate each ship. Fortunately, favorable winds blew the vessels toward Mitindy, where they docked and sent ashore for oranges; the sick crewmen believed their earlier cure was from the fruit's magical properties. The fleet eventually returned safely home, but with about half the original crew.

Figure 1 Sailing vessel of the type used by Vasco da Gama in 1496.

Did da Gama's crew really recognize that oranges (a vitamin C source) prevented scurvy? This point is debatable, since most ships after da Gama did not carry oranges or other citrus fruits, and passengers and crew often fell victim to scurvy. Scorbutic symptoms (that is, the symptoms of scurvy) were not restricted to seafaring vessels and sailors; land-based scurvy was quite common, although it was many years before physicians realized the two diseases were identical. In 1734, John Bachstrom noted that a scurvy epidemic that occurred during a siege of Thorn (Poland) in 1703 was not cured until the townspeople resumed eating fresh vegetables and greens. A similar outbreak occurred in 1719–20 during a military campaign against the Turks in Hungary. The soldiers did not recover from scurvy until they obtained fresh vegetables.

Though a dietary connection was suspected years before scurvy's true cause was diagnosed, it was 1747 before any experimental research linked scurvy with certain food deficiencies. James Lind (British, 1716–94, figure 2) isolated a dozen scorbutic sailors and gave pairs of them six different treatments, including (1) cider, (2) elixir vitriol (dilute sulfuric acid), (3) vinegar, (4) seawater, (5) oranges and lemons, and (6) a paste of garlic, mustard seed, balsam, radish root, and gum myrrh. Lind concluded "that oranges and lemons were the most effectual remedies for this distemper at sea," and that cider was second best.

Figure 2 James Lind, the British doctor who first determined that certain fruit juices could prevent scurvy.

However, Lind's most definitive conclusion was that sulfuric acid was ineffective for treating scurvy. Sulfuric acid was commonly used in the early days of medicine. Sailors probably tried it as a replacement for oranges or lemons because both fruits decomposed during storage, while sulfuric acid was stable. After Lind left the navy, he published a 400-page book on scurvy in 1753. His earliest account of a large scurvy outbreak was when the French fought the Saracens in Egypt during the winter of 1249–50. Nevertheless, it was many years before an antiscurvy food became a firm requirement on British ships.

The first antiscorbutic fruit the navy used was lemons, but they were soon replaced by limes, apparently on the basis of cost and availability. (Lemons have nearly double the amount of vitamin C.) Soon, the British sailors were called "limeys" because they regularly drank lime juice, which prevented them from becoming a "scurvy lot." They tried many other materials as well, but eventually returned to the use of lemons. Much later, in 1959, McCord estimated that two million sailors died from scurvy in the interval between 1500 and 1800.

Meanwhile, scurvy epidemics continued on land. An estimated ten thousand people died from scurvy during the 1849 California Gold Rush, far more than died from cholera. Amazingly, only one notation on scurvy appeared in a medical journal—a letter from a Thomas Logan, who noted that there was no difference between land-based scurvy and the sea version, and that the disease was not caused by climate. The letter suggested fresh meat and vegetables as cures. Logan did not mention citrus fruits.

A strange episode in the history of scurvy occurred during the early 1900s, shortly after the introduction of milk sterilization by boiling. This treatment not only killed bacteria, a positive result, it also destroyed vitamin C (which was still unknown). As a result, 15,000 babies died from scurvy each year in New York City alone.

Shortly thereafter, Axel Holst (Norwegian, 1860–1931) fed guinea pigs restricted diets that lacked vitamin C. The animals developed a scurvylike disease. Aided by Theodor Frölich, a pediatrician with an interest in infantile scurvy, Holst produced an experimental scurvy in 1907.

Subsequent studies used guinea pigs to determine which foods were antiscorbutic. In 1918, S. S. Zilva (German) began systematic studies; in 1928, he reported that lemon juice appeared to have a high antiscorbutic activity. Later in 1928, Albert Szent-Györgyi (Hungarian, 1893–1986, figure 3) isolated hexuronic acid (vitamin C) from lemon juice; in 1932, Waugh and Kink determined that this was the specific antiscorbutic agent in lemons. Norman Haworth (British, 1883–1950) elucidated hexuronic acid's structure in 1933, and Tadeus Reichstein (Polish, 1897–) synthesized the molecule later that year. Finally, in 1933, Haworth and Szent-Györgyi changed the name of the compound to ascorbic acid, meaning an antiscorbutic acid. Szent-Györgyi received the 1937 Nobel Prize for his discovery of vitamin C and for studies on respiration.

The research on vitamin C continues today. Recent arguments concern whether vitamin C can prevent or cure cancer or the common cold. The leading advocate of this position was Linus Pauling, (American, 1901–1994, figure 4). Pauling, a double Nobel Prize winner in chemistry and peace, recommended massive doses of 10 g/day. Although Pauling was considered one of the leading scientists of the late twentieth century, not all scientists agreed with his assessment of vitamin C; some of them note that Pauling's claims have not yet been rigorously proven. Some supporting data do exist for using vitamin C as an anticancer agent, since ascorbic acid prevents the formulation of nitrosamines, which are powerful carcinogens. Several studies show that vitamin C protects against human bladder cancer, possibly through nitrosamine prevention. Vitamin C also appears to inhibit the action of hyaluronidase, an enzyme that enables cancer cells to metastasize and spread the cancer throughout the body. Finally, vitamin C seems to protect against the effects of carcinogenic hydrocarbons, possibly because it is a good antioxidant and can prevent free radical reactions.

Many studies show that vitamin C has some antiviral activity and that it stimulates the immune system; either of these actions *might* prevent a cold. One reasonable concern in this matter is the safety of taking such massive doses over prolonged periods; thus, the vitamin C saga continues.

Figure 4 Linus Pauling (American, 1901–1994), a double Nobel Prize winner in chemistry and peace, did much research on vitamin C.

Figure 3 Albert Szent-Györgyi, who isolated vitamin C from lemon juice.

BOX 15.3

OBTAINING ESSENTIAL AMINO ACIDS

In principle, eating a balanced diet is easy in the United States; but obtaining adequate dietary protein is often difficult in Third World countries. Meat, the best source of protein, is too scarce, too expensive, or banned for religious reasons in many Third World nations; accordingly, it is often consumed once a week or less. How can people in the Third World obtain the nutrients they need? Is it possible to do so on a vegetarian diet? The answer is yes.

The human body must take in the essential amino acids that it cannot synthesize. If they are not obtained from meats, the amino acids must come from vegetable sources. Individual vegetables seldom provide a balanced or complete selection of amino acids, but an assortment can provide every essential amino acid from vegetarian sources alone. This normally requires a combination of beans, peanuts, or peas (protein sources), with rice, wheat, or corn (grains that also contain proteins). If dairy products (milk, cheese, and eggs) are added, it is much easier to obtain all the essential amino acids.

Many ethnic foods have traditionally utilized these types of combinations to achieve a good balance of amino acids. A mainstay of the Japanese diet is rice, and tofu, made from soybeans, complements the rice with the essential amino acids it lacks. Mexicans combine corn-based tortillas with beans to achieve the same result. Rice and peanuts are eaten in many parts of Africa, and Native Americans have long eaten succotash, a mixture of corn kernels and beans. Cajuns are known for their spicy bean and rice dishes, and Italians are famous for their pasta dishes, which combine wheat with cheese and small amounts of meat. Even an old-time favorite in the United States—the peanut-butter-and-jelly sandwich—combines peanuts and wheat.

Some ethnic dishes are healthy even though they seem unreasonable; the Eskimo diet is almost exclusively raw fish and meat, with an excessively high fat content and essentially no fiber. Nutritionists conjecture that the Eskimos obtain their vitamins from the meat, eaten fresh and uncooked.

American, cheddar, Swiss, Muenster, and many other cheeses are 65 to 85 percent fat. Low-fat versions still contain 50 to 60 percent fat. Cottage cheese, the weight-watcher's favorite, is lower at 10 to 40 percent. Cream cheese goes as high as 90 percent. Several brands of fat-free cheese are on the market; they contain, as their name implies, 0 percent fat.

PRACTICE PROBLEM 15.6

How much fish must you consume to obtain the same amount of usable energy as you would obtain in 100 g rice?

ANSWER Fish provides 62 percent usable calories, compared to 8 percent for rice. You would need only 8/62nds as much tuna fish as rice, or about 13 g.

EXPERIENCE THIS

Mix inexpensive clay-dough by adding 1-1/2 cups flour to 1/2 cup salt, then slowly adding 1/2 cup vegetable oil and 1/4 cup water. Knead this dough until it is uniform in feel. If it seems too sticky, add a little bit more flour. If it seems too stiff, add more vegetable oil. You can mold whatever you desire out of this dough and it will harden if you let it stand in the air for a few days. You can also heat it in the oven to speed the hardening process (try two hours at 250°F). After the dough dries, you can paint it or shellac it.

Flour consists of natural proteins and polysaccharide polymers (chapter 14). The polymer responsible for the hardening action in the dough is a protein called gluten. It is found in most kinds of cereal grains.

THE CHEMISTRY OF FOOD ADDITIVES

Some additional chemicals, such as pesticide residues or manufacturing process debris, are inadvertently present in foods; other chemicals are added deliberately. These deliberately added additives fall into several distinct categories: sweeteners, flavoring agents, colorants, preservatives, antimicrobial agents, and antioxidants.

Sweeteners, Natural and Artificial

Humans have used sweeteners for centuries; the use of honey, for example, goes back thousands of years. Today, the main sweetener in our foods is sucrose, commonly called table sugar (figure 15.4). This disaccharide, harvested from sugar cane or sugar beets, had a combined harvest of 58.3 billion tons of plants in the United States alone in 1991. The annual per capita consumption of refined sugar during the same year was 64.9 pounds (1993 *Information Please Almanac*); despite its huge harvest, the United States had to import over $713 million dollars of sugar to meet our "needs." The American sweet tooth is hardly satisfied with this amount, however. In 1991, a typical American also consumed 73.9 pounds of corn sweeteners and 16.4 pounds of ice cream. Not counting the ice cream, each of us, on the average, ate 0.38 pounds of sugar or corn sweetener *each day* in 1992; much of this sugar came from processed foods. (Just look at a box of breakfast cereal if you doubt this.) It is estimated that 15 to 25 percent of our calorie intake comes from sugar—calories dietitians refer to as "empty" because they contain no nutritive value.

Along with sweetening, sucrose also functions as a preservative in many foods, such as fruits, jellies, and jams. Sucrose draws water out from microorganisms via osmosis in what is called a hypertonic response. The dehydrated microorganisms die. (Salt acts in the same way to preserve meats and other foods.)

Artificial sweeteners, chemicals synthesized to use as sweeteners, help diminish the amount of sucrose in our foods. There are three reasons cited for using artificial sweeteners: (1) diabetics cannot tolerate sugar, (2) noncaloric sweeteners facilitate weight loss, and (3) artificial sweeteners do not promote tooth decay. Diabetics are unable to produce the insulin that controls blood sugar levels and must limit or totally eliminate sugar intake, so this is a legitimate reason to use artificial sweeteners. The second reason has a more hollow ring—Americans continue to be overweight even with the prevalence of artificial sweeteners. The third reason for using artificial sweeteners, preventing tooth decay, is also somewhat suspect since decay is greatly reduced simply by brushing one's teeth after eating. Toothpaste (chapter 13) normally contains no refined sugar; the most common sweetener in toothpaste is glycerine. Glycerine is sometimes added to foods as a moisturizing agent, or humectant, because it draws moisture out from the surrounding air. Glycerine is not used as an artificial sweetener in foods because it is only 60 percent as sweet as sucrose and is more expensive.

Table 15.6 compares the sweetness of several natural and artificial sweeteners, using sucrose as a reference point; most artificial sweeteners are much sweeter. Interestingly enough, **saccharin,** the oldest artificial sweetener, is still the sweetest. Natural sugars such as glucose, lactose, and maltose are less sweet than sucrose; fructose is substantially sweeter, and the sugar alcohol xylitol has about the same sweetness. Figure 15.4 shows the chemical structures for several of these sweeteners.

Artificial sweeteners include not only saccharin, but calcium cyclamate, aspartame, and Acesulfame K. Saccharin (figure 15.5) was first synthesized in 1879 after Constanin Fahlberg discovered a sweet taste on his hands after working with coal tar chemicals.

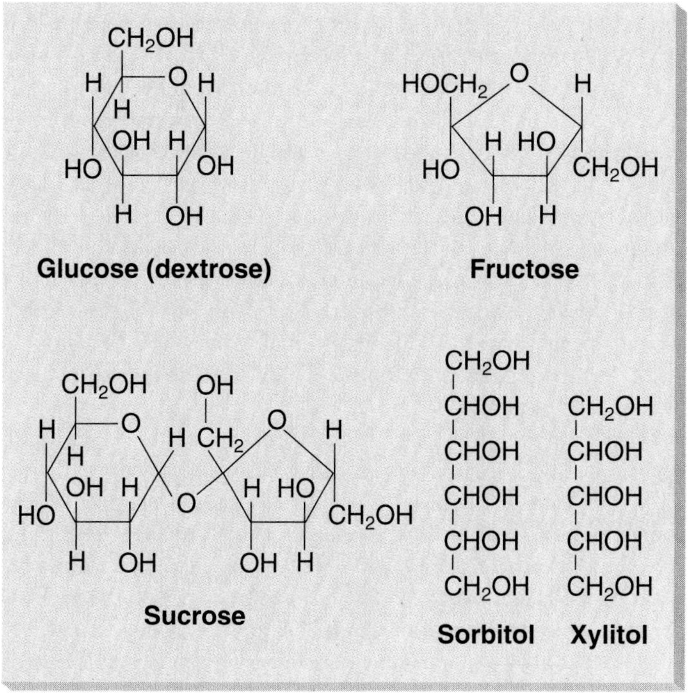

Glucose (dextrose)

Fructose

Sucrose

Sorbitol

Xylitol

Figure 15.4 The structures of some natural sweeteners.

TABLE 15.6 The relative sweetness for various chemical compounds (sucrose = 1).

COMPOUND	SWEETNESS
Lactose	0.16
Maltose	0.33
Glycerol (glycerine)	0.6
Sorbitol	0.6
Glucose	0.74
Xylitol	1.0
Sucrose	1.0
Fructose	1.73
Calcium cyclamate	30
Aspartame	160
Acesulfame K	200
Saccharin	500
P-4000	4,000

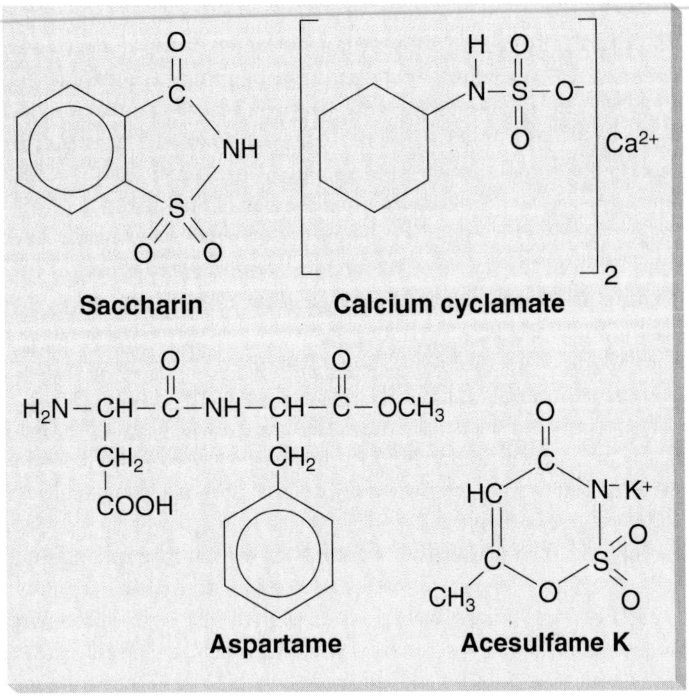

Saccharin

Calcium cyclamate

Aspartame

Acesulfame K

Figure 15.5 The structures of some artificial sweeteners.

Although saccharin does not resemble sugar chemically, it does produce a strong sweet taste, along with a fairly unpleasant, bitter aftertaste. Were it not for the fact that its sweetness is 500 times greater than sucrose, it is unlikely that saccharin would have become popular because of this aftertaste. However, because it is so sweet, very small amounts of saccharin may be used in beverages or foods, mitigating the bitterness.

Chemists continued researching new artificial sweeteners and found that several chemicals that did not have common structures nevertheless shared the property of sweetness (figures 15.4 and 15.5). Sorbitol and xylitol are reduced forms of monosaccharides that do not promote tooth decay; they are used in sugarless chewing gums. Sorbitol occurs naturally in many berries and in cherries, plums, pears, apples, and seaweed; it is made commercially by the reduction of glucose:

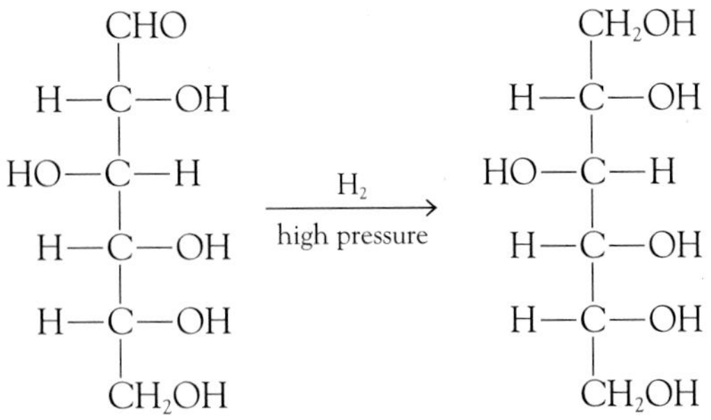

$$\begin{array}{c} CHO \\ | \\ H-C-OH \\ | \\ HO-C-H \\ | \\ H-C-OH \\ | \\ H-C-OH \\ | \\ CH_2OH \\ \text{Glucose} \end{array} \xrightarrow[\text{high pressure}]{H_2} \begin{array}{c} CH_2OH \\ | \\ H-C-OH \\ | \\ HO-C-H \\ | \\ H-C-OH \\ | \\ H-C-OH \\ | \\ CH_2OH \\ \text{Sorbitol} \end{array}$$

Although neither sorbitol nor xylitol causes tooth decay, both are metabolized by the body, releasing approximately 4 Kcal/g, just like glucose; both sugar alcohols also tend to cause diarrhea when used extensively, so although they are used in small amounts as humectants, they are not generally used as sugar replacements.

Calcium **cyclamate** has thirty times the sweetness of sucrose and no bitter aftertaste. It was widely used as an artificial sweetener in the 1960s. In 1969, animal studies indicated cyclamates could cause bladder cancer in mice, and the FDA (Food and Drug Administration) banned them from human consumption shortly afterwards. Cyclamates are not the only artificial sweeteners that may cause cancer, and researchers continue to test sweeteners as well as other food additives. Saccharin also induced bladder cancer in laboratory mice in 1977, but the FDA did not ban its use because at that time it was the only noncaloric artificial sweetener on the market, and diabetics needed it. Products containing saccharin now carry a warning label pointing out that the chemical causes cancer in laboratory animals. Interestingly, Canada has banned saccharin but permits cyclamates.

Cancer tests are run using massive amounts of the chemical on laboratory animals, and the validity of this procedure has been questioned. Do these massive doses over short periods correspond to consuming relatively small amounts of a chemical for many years? Often, the research dose equates to a person consuming 100 pounds per day of a specific food for over twenty years. In addition, do animal tests correspond to what happens with humans? Mice or rats are used because they are inexpensive, small, and short-lived. A typical test is run using thirty or more animals, with an equal number of control animals that do not receive the chemical. The virtue of using a short-lived animal is that the effect on life quickly becomes apparent. Massive doses are used because lower levels would cause the cancer (or other effects) in relatively few animals, requiring researchers to test thousands of animals to assay a chemical. But because of these limitations, the results are at best screening tests to show which chemicals *might* cause problems in humans.

In principle, test results should be reproducible, lending credence to the data. The tests leading to the cyclamate ban have been repeated several times since 1969, and the retesting did not show that cyclamates caused cancer in mice; on the other hand, saccharin has tested positive for inducing cancer in mice several times since 1977. Still, the ban on cyclamates remained in place, much to the dismay of the manufacturer, while saccharin remains on the market, partly because it has no history of inducing cancer in humans after over a century of actual usage since its discovery in 1879. Keeping saccharin on the market required a special decision by Congress.

There are very few other tests scientists can run to assay the potential cancer-inducing properties **(carcinogenicity)** of chemicals. The **Ames test,** developed by Bruce Ames, is a simple, quick, inexpensive screening test that measures mutations in the genetic makeup of the bacterium *Salmonella typhimurium*, responsible for a major variety of food poisoning. The Ames test therefore tests for mutagenicity instead of carcinogenicity. Although these are two distinct phenomena, many known carcinogens give a positive Ames test, so this procedure can work as a preliminary screening test. One issue the Ames test does not address is that cancer is sometimes caused by a chemical's metabolism products rather than the compound itself. We will further examine carcinogens in chapter 17.

A congressional action called the **Delaney Amendment** (or the Delaney Clause), signed into law in 1960, prohibits placing any natural or synthetic chemical into food if laboratory testing shows that the chemical is cancer-causing. In other words, any positive carcinogenicity test could eliminate a food additive from use even when there is no evidence that the agent causes cancer in human. In fact, even if medical studies strongly suggest that a chemical does *not* cause cancer in humans, the basis for removal of an additive from the market is not long-term medical studies on human beings, but laboratory tests that may use incredibly high levels of a chemical on rats or mice. New additives must pass this test as well.

The Delaney Amendment also states that the allowable amount of a laboratory-tested carcinogen in food is zero. This sounds like an excellent safeguard, but chemical analysis has improved significantly since 1960, and trace quantities completely unobservable back in the sixties are now easy to spot. Does this mean that additives considered safe in 1960 must now be banned? In general, the answer is yes, unless either the FDA or Congress takes special action. Such actions are rare.

The newer sweeteners **aspartame** and **Acesulfame K** have appeared in the past twenty years (figure 15.5). Both meet the requirements of the Delaney Amendment. The FDA approved aspartame, the methyl ester of a dipeptide formed from phenylalanine and aspartic acid, in 1981; in a 1974 attempt, toxicity questions caused its removal from the market. Aspartame is about 160 times sweeter than sucrose and has no bitter aftertaste; consumer acceptance was phenomenal, although some safety questions persist. Aspartame is an ester and can hydrolyze to give aspartic acid, phenylalanine, and methanol. Phenylalanine is a life-threatening chemical to people with a condition called phenylketonuria; the product labels warn of this danger.

Methanol, one of the other hydrolysis products of aspartame, is not classed as a carcinogen and does not violate the Delaney Amendment, but it does form the highly toxic formaldehyde when metabolized. Since very little aspartame (less than 1 g per serving) is needed for sweetening, the total amount of methanol produced is small in any food formulation. The chance of death from methanol produced is remote, since the usual lethal dose is 100 to 200 mL. You would need to ingest over 920 g of aspartame rapidly and have it all hydrolyze immediately, in order to reach the toxic level. It is unfortunate that aspartame was not created as an ethyl ester; the hydrolysis product would be the less toxic ethanol.

With aspartame, sold as Nutrasweet, Equal, and other tradenames, now found in so many different food items, questions concerning safety are legitimate. Along with the phenylalanine warning, the sweetener packages also list warnings not to use aspartame in cooking, due to hydrolysis. Aspartame is now used in some brands of baby food, and babies' chemical defense systems are less well developed than those of adults. Trace quantities of methanol might impair a baby for life.

The problem with aspartame is that the product has swept the market with extraordinary rapidity. If it is discovered later that this sweetener causes some health problems, a considerable portion of our population might be affected. And despite the widespread use of aspartame and other artificial sweeteners, the average weight of Americans has increased.

Acesulfame K is not structurally related to any other sweetener in the market. Unlike aspartame, Acesulfame K is usable in cooking since it is stable to hydrolysis. The FDA has approved Acesulfame K as an artificial sweetener.

The most desirable artificial sweetener is one that is safe and much sweeter than sucrose without an unpleasant aftertaste. Many health food stores promote fructose as a sucrose replacement, but fructose has the same number of calories as sucrose or glucose. People tend to consume more artificially sweetened food because they believe they are consuming less calories; this mistake would prove fattening with fructose. Aspartame also imparts 4 Kcal/g, as sucrose does, but only small amounts are used in sweetening. Saccharin, Acesulfame K, and the cyclamates are nonnutritive (noncaloric) sweeteners.

The reason certain chemicals impart a sweet taste remains unknown, although some researchers have pursued this question. Figure 15.6 shows a trio of structurally related compounds known as P-4000, Compound I, and Compound II that thousands of people taste-tested. Almost everyone perceived P-4000 as extremely sweet; its name comes from the observation that it is 4,000 times sweeter than sucrose. Compound I is generally tasteless, while Compound II is perceived by most people as decidedly bitter. None of these materials, including P-4000, is a candidate for a new artificial sweetener; they are too toxic.

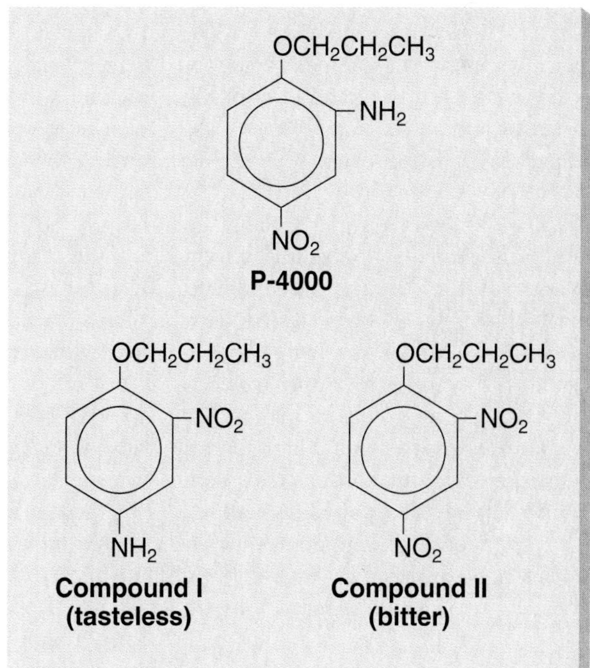

Figure 15.6 A trio of experimental compounds used to test people's responses to sweetness and bitterness.

In summary, several fairly safe artificial sweeteners now exist that diabetics find useful, but their value in controlling obesity is much more questionable. Artificial sweeteners seem safe if they are ingested at low levels, but their use in baby food seems indefensible.

Flavoring Agents: Adding Better Taste

Taste ranks high on the list of reasons why people eat certain foods. Many people regard certain foods as bland and add seasonings to enhance taste; this seems especially true of vegetables. Humans have sprinkled salt, the most widely used flavoring agent, on their food for many thousands of years, but too much salt upsets the Na^+/K^+ balance and promotes high blood pressure. Processed foods already contain plenty of salt; the typical American averages 10 g salt per day, which is 500 percent of the recommended 2 g/day level. Along with hypertension, health problems caused by excess salt intake include kidney disease, congestive heart disease, and toxemia during pregnancy. Foods such as pickles, dressings, canned vegetables, sauerkraut, and cured ham are high in salt content due to preparation methods. The salt shaker is one of the most dangerous antihealth weapons available.

Sugar is the next most commonly used flavor enhancer, but after sugar comes **monosodium glutamate (MSG),** the sodium salt of the amino acid glutamic acid. This additive is contained in most processed foods and is quite common in Asian foods. Some claim that MSG causes severe health problems, including brain lesions. These fears first cropped up in 1968 when a doctor experienced temporary numbness after eating Chinese food and early studies seemed to implicate MSG. However, studies that began in 1980 suggested that this conclusion was in error. In 1994, the largest study yet conducted (reported in the British journal *Food and Chemical Toxicology*) showed that MSG did not make people sick or cause any of the claimed symptoms. In effect, this result vindicates nearly two thousand years of MSG usage.

The FDA announced that MSG was safe for most people in September 1995. Before then, the FDA permitted MSG usage based on what is commonly called the "Grandfather Clause" to the Delaney Amendment. This clause grants a special **GRAS (Generally Regarded As Safe)** listing to materials used for many years prior to the Delaney Amendment. Most GRAS-listed chemicals have never been tested for long-term effects on humans; this is unlikely to change with the enormous expense associated with such testing. Who would pay to test red or black pepper? Both are mixtures of many chemicals with unknown effects, but they have been used for centuries without any apparent deleterious side effects other than "burned" tongues. Are they safe to use? Perhaps, but nobody is certain.

Other common flavor enhancers include vinegar and various spices, but none are used in large quantities and they probably pose no long-term health threats. Vinegar is merely a 5 percent aqueous solution of acetic acid plus a few dozen minor constituents. Scientists are only now discovering the total chemical contents of many spices.

Almost all ice creams contain added flavors. Some flavors are natural organic aldehydes, esters, and alcohols, and some are synthetic versions of the same chemicals. Aldehydes produce the primary taste in "natural" almond, cinnamon, and vanilla flavors (figure 10.11). These aldehydes are now manufactured synthetically from petroleum. While small amounts of these aldehydes appear to be safe, natural flavorings contain many additional chemicals for which little, if any, toxicological data exists. They are on the GRAS list because they have long existed in natural flavoring agents. Some health-food advocates claim that the natural flavoring blends are safer, but the opposite is more likely to be true. The synthetic flavoring versions contain a single chemical whose identity and effect are known, at least to some extent. But natural or synthetic, flavoring toxicity is seldom seen because such small quantities are used.

Many esters, an important flavoring group, possess fruit flavors (table 10.3). Amyl or isoamyl acetate is widely used as a synthetic banana flavor; some inexpensive banana ice cream brands never see real bananas unless they pass the fresh fruit department in the supermarket. Methyl salicylate is the main ingredient in wintergreen flavor; it is called "oil of wintergreen" for that reason.

Colorants: Pleasing to the Eye

Food has an esthetic appeal in addition to taste, and food colorings can enhance this image. Some food colors are natural, such as β-carotene from carrots or various dyes from cabbage, but many are synthetic compounds. The natural materials are as safe as the foods they come from; β-carotene is the precursor for vitamin A as well as a coloring agent. Not all synthetic dyes added to foods are as safe as this compound.

Why add colors to food? First, some natural food colors are not stable over long-term storage, especially in the presence of sunlight, so they fade into a bland-looking, strawlike shade. Would you eat strawberry preserves with a lifeless straw color? Possibly not, so many manufacturers combat this "defect" by adding a small amount of red dye that retains color indefinitely. Taste and appearance are not chemically related; the bland strawberries taste just like the red ones, as long as you do not see what you are eating.

The FDA has controlled food coloring agents since the Food and Drug Act was enacted in 1906. Most artificial food coloring compounds are derived from coal-tar chemicals (obtained from coal by pyrolysis), and they are generally azo-derivatives of naphthalene. (The azo group is —N=N—.) Figure 15.7 shows the structures of four such food coloring agents, along with β-carotene, a natural dye. The yellow dyes FD&C Yellow no. 3 and FD&C Yellow no. 4 are structurally similar to β-naphthylamine, a known carcinogen. If the bond to the —N=N— group breaks, the molecule could release free β-naphthylamine; traces of β-naphtylamine were observed in food colorings in the early fifties and both dyes were banned for food use. Laboratory studies eventually showed that either dye could cause bladder cancer in mice, apparently by hydrolysis in the stomach to release

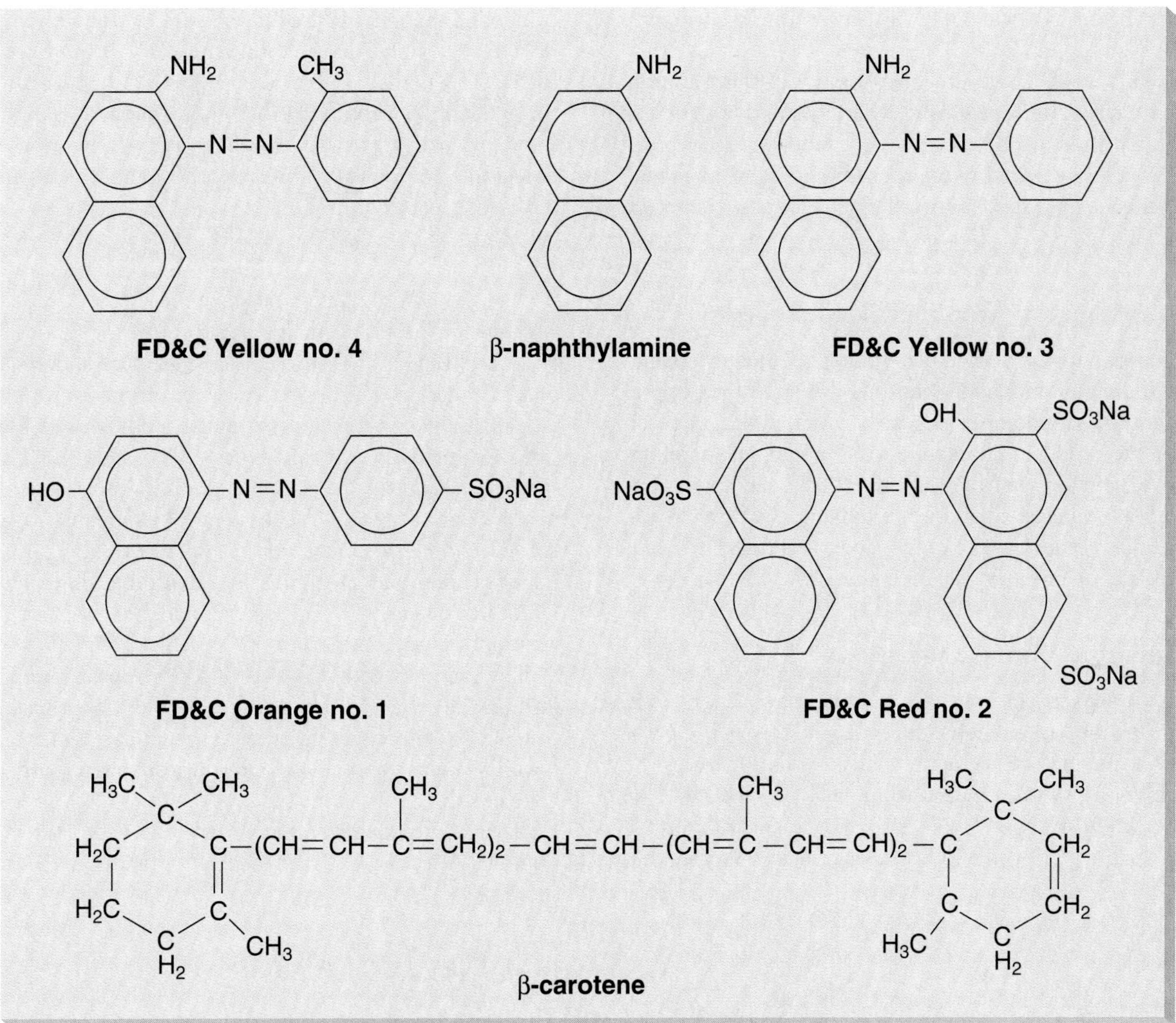

Figure 15.7 The structures of β-naphthylamine and four related FD&C food coloring compounds. β-carotene is added for comparison.

β-naphthylamine. The FD&C Orange no. 1 was banned in 1950, after causing severe gastrointestinal problems in children. FD&C Red no. 2 was banned in 1976, after tests showed it caused cancer in laboratory animals. These last two dyes do not contain the β-naphthylamine group, so the cancer must arise by another route.

A few artificial FD&C food colorants remain on the market; they fall under suspicion because their structures are similar to those of other dyes known to be carcinogenic. Ascribing guilt by association is normally undesirable, but it might be reasonable here. Unless a synthetic material has been tested and proved safe for human consumption, we are better off using only natural food colors.

Preservatives: Making Food Last

Everyone knows that food spoils. Several types of chemicals are added to delay food spoilage; people have used sugar and salt for centuries to preserve food. What other compounds do we use, and how safe are they?

Sodium and potassium nitrites ($NaNO_2$ and KNO_2) have long been used to retard spoilage in processed meats such as hot dogs, luncheon meats, and cured ham. Although nitrites are effective preservatives that even guard against botulism, they may react in the human body with secondary amines to form carcinogens called nitrosamines. No evidence exists that meats preserved with nitrites actually cause cancer, but the stomach acids can convert nitrites into nitrous acid, making the following reaction possible:

$$KNO_2 + HCl \longrightarrow \underset{\text{Nitrous acid}}{HNO_2} \xrightarrow{R_2N\text{–}H} \underset{\text{A nitrosamine (carcinogenic)}}{R_2N\text{–}N{=}O}$$

Some have attempted to relate this reaction to the incidence of cancer in countries where the inhabitants eat greater amounts of meat (such as the United States), but a relationship, if one exists, is tenuous at best.

Sulfur dioxide, used as a disinfectant on dried fruits, appears harmless. (SO_2 causes lung irritation when inhaled, but the gas vanishes shortly after disinfecting the fruit and poses no problem.) Sulfites were once used to maintain freshness at salad bars, but because they sometimes caused allergic reactions, they were banned in 1986.

Antimicrobial Agents: Warding Off Micropests

Antimicrobial agents—for example, sodium benzoate, sodium sorbate, and sodium propionate—retard spoilage by killing or preventing the growth of microbes such as bacteria, molds, or yeast. Sodium benzoate is used in nonalcoholic beverages, juices, jellies, jams, preserves, pickles, and salad dressings at a maximum concentration of 0.1 percent. Sodium propionate is used in breads, cheeses, chocolate products, pie crusts, and pie fillings; its maximum allowable concentration is 0.3 percent.

Antioxidants: Preventing Spoilage

Oxidation by atmospheric oxygen is a common cause of food spoilage. Storing food in an inert atmosphere alleviates the problem, but this is impractical for many foods. Oxidation poses the greatest problem with fats, which undergo a chain reaction. The chain begins when a fat molecule (R—H) reacts with an O_2 molecule to form a pair of free radicals, high-energy species that react rapidly:

$$\text{R—H} + O_2 \longrightarrow \underbrace{\text{R}\cdot + \text{HOO}\cdot}_{\text{Free radicals}}$$

The resulting alkyl radical (R•) then reacts with a second O_2 molecule to form a peroxyradical:

$$\text{R}\cdot + O_2 \longrightarrow \underbrace{\text{R—O—O}\cdot}_{\text{Peroxyradical}}$$

Next, the peroxyradical (ROO•) reacts with another fat molecule to form a hydroperoxide (ROOH) and the R•, which repeats the cycle:

$$\text{ROO}\cdot + \text{R—H} \longrightarrow \underbrace{\text{R—OOH}}_{\text{Hydroperoxide}} + \text{R}\cdot$$

Similarly, R—OOH splits apart to form the same radicals the first reaction formed—R• and HOO•:

$$\text{R—OOH} \longrightarrow \text{R}\cdot + \text{HOO}\cdot$$

These free radicals also continue the reaction chain.

In any free radical chain process, many molecules of a chemical react from each initial reaction. One reaction between a R—H molecule and an O_2 molecule triggers the reaction of thousands more R—H molecules; the fat goes rancid. Fats can seldom be isolated from O_2, so to preserve them it is necessary to stop the chain reaction. Molecules called radical traps do so. Three such compounds commonly used in food are **butylated hydroxyanisole (BHA), butylated hydroxytoluene (BHT),** and **propyl gallate** (figure 15.8); these substances are called **antioxidants** because they sidetrack the oxidation chain process. An antioxidant (Ar—H) reacts with R• to reform R—H, creating a relatively unreactive Ar• radical. This stops the chain process and prevents further fat oxidation:

$$\text{Ar—H} + \text{R}\cdot \longrightarrow \text{R—H} + \underbrace{\text{Ar}\cdot}_{\text{Stable radical}}$$

Research literature makes some surprising claims concerning BHA and BHT. While these molecules were undergoing tests for potential carcinogenic activity (which turned out negative), scientists discovered that rats fed massive amounts of BHT actually lived longer than controls fed no BHT. Does BHT extend the lifespan? Could this be a chemical fountain of youth? Many popular articles have voiced such claims, but chemists remain skeptical. The same chemical, BHT, also causes allergic reactions in many people, and pregnant mice fed large doses had offspring with birth defects. On the positive side, BHT seems to have some antitumor properties.

The amounts of BHT used in foods appear harmless. Rancid fats, on the other hand, are definitely bad for our health. Though the public sometimes bristles at chemical additives, the long chemical names you see on food labels are often there to safeguard your health.

Several vitamins also have antioxidant properties, including vitamins A, E, and C. Traces of these vitamins in foods also retard spoilage.

Miscellaneous Food Additives

The list of food additives seems never-ending; here are a few more short entries. Citric acid is added to food, usually as the Na^+ salt, to serve as a sequestrant, a molecule that binds metal atoms (chapter 13). Copper, iron, or nickel atoms or ions catalyze fat oxidation and interfere with antioxidants. The sequestrant removes these metals or ions from the game and lets the antioxidants do their job. Other sequestrants include ethylenediaminetetraacetic acid, or EDTA.

Buffers help maintain the pH in a specific range. One buffer frequently added to food is potassium hydrogen tartrate, $KHC_4H_4O_6$. This compound is also in baking powder, where it reacts with sodium bicarbonate to release CO_2 during baking, making the dough rise. When the pH of a food is too high, citric or lactic acids (called acidulants) help make the necessary adjustment.

Small amounts of magnesium silicate are added to table salt to reduce caking. The silicate binds moisture as water of hydration and the salt flows freely.

Stabilizers or thickeners are added to improve smoothness or increase the viscosity of foods. Carrageenin, a natural polymer isolated from seaweed, is one of the more common thickening

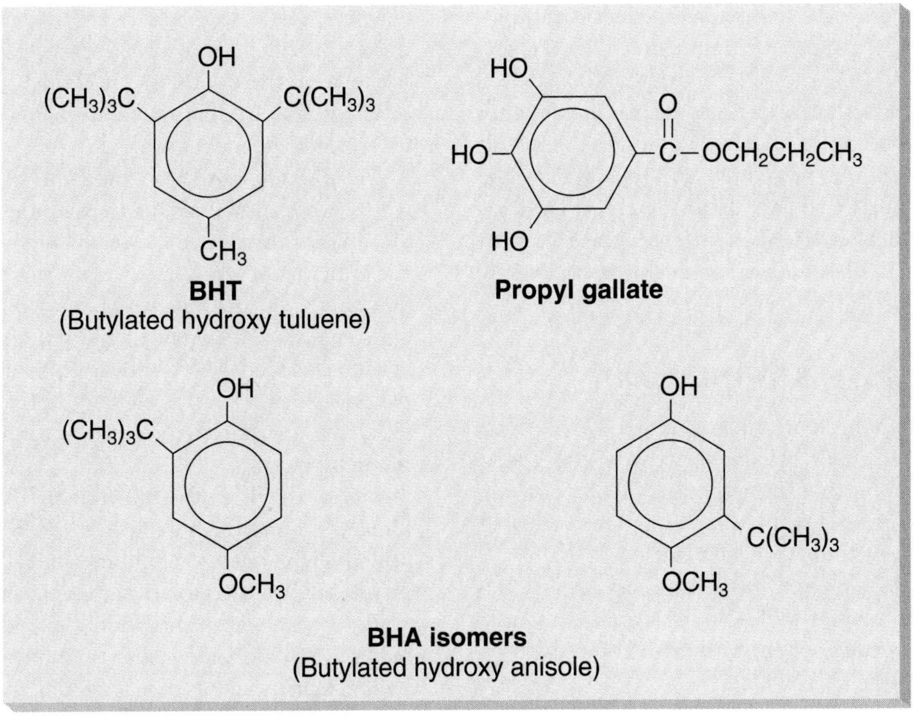

Figure 15.8 The structures of the antioxidants used in foods.

agents. Stabilizers or thickeners are often used in ice cream, salad dressings, cheese, whipped cream, and icings.

Emulsifiers are added to liquid mixtures to prevent phase separation. Lecithin is an emulsifier, as are some synthetic diglycerides of fatty acids and some sorbitol esters. The latter stabilize water in oil emulsions, an important activity in foods.

PRACTICE PROBLEM 15.7

Obtain some processed foods and examine the labels, singling out the chemicals present in minor amounts. Determine the function of each chemical.

ANSWER Answers will vary for this problem because different foods contain different additives, but you should expect to find combinations of colorants, antioxidants, antimicrobial agents, sweeteners, preservatives, and flavoring agents in most processed foods.

NOW YOU'RE COOKING

In an earlier chapter, we asked what happens when food (specifically, an egg) is cooked. You can now answer that question. In protein-based foods, the peptides lose their secondary or tertiary structures during cooking and become disorganized. Cooking also breaks down the proteins or polysaccharides into smaller chained polymers through hydrolysis and heat-induced cleavage. Other chemical reactions also occur with the functional groups located along the polymer chains, converting them into new molecules. The reactions that occur in cooking are one-way; that is, they do not reverse. In other words, food cooks when heated but does not "uncook" when cooled.

Baking bread provides a good example of kitchen chemistry. During baking, the protein gluten forms a crosslinked matrix that traps pockets of gaseous CO_2. This CO_2 may form in any of several reactions. The first and oldest is the yeast reaction, which releases CO_2 that causes the bread to rise. Without yeast, the bread would be unleavened, like a cracker:

$$C_6H_{12}O_6 \xrightarrow[\text{yeast}]{\text{zymase from}} 2CO_2 + 2C_2H_5\text{—}OH$$

Baking soda (sodium bicarbonate, $NaHCO_3$) was discovered later, but also works well as a leavening agent.

A second reaction that produces CO_2 comes from vinegar and sour milk. Do you know any cake recipes that call for sour milk? Chances are it forms the CO_2 during baking:

$$NaHCO_3 + H^+ \longrightarrow Na^+ + H_2O + CO_2$$

Finally, a third CO_2-producing reaction uses baking powder. Baking powder contains both potassium hydrogen tartrate and sodium bicarbonate, which react to form CO_2:

$$KHC_4H_4O_6 + NaHCO_3 \longrightarrow KNaC_4H_4O_6 + H_2O + CO_2$$

These chemical reactions have one common effect—they produce the CO_2 that becomes trapped within the bread or cake, giving it a desirable airiness or "lightness." A cake falls flat if the CO_2 escapes during baking. Modern cake or bread recipes are designed to prevent this mishap, but in the old days slamming a door while the bread or cake was in the oven was often all it took to flatten it.

Long before we can mix bread and place it in the oven, however, a variety of chemical reactions must take place to grow usable food products. Let's now turn to the chemistry of agriculture—raising food crops.

AGRICHEMISTRY—GROWING FOOD

Plants play the fundamental role in the food chain. They obtain their energy from the sun, their carbon supply from the atmosphere (as CO_2), and their minerals from the soil. Plants then produce saccharides, which provide both a structural framework (cellulose) and energy storage (starch) for the plants.

Saccharide formation is powered by the sun's energy and catalyzed by chlorophyll, a complex molecule structurally similar to hemoglobin, in a process called **photosynthesis.** The usual reaction product of photosynthesis is glucose.

$$6\ CO_2 + 6\ H_2O \xrightarrow{\text{catalyst}} \underset{\text{Glucose}}{C_6H_{12}O_6} + 6\ O_2$$

Other processes convert glucose to larger saccharide molecules; virtually each step in this sequence requires enzymes and minerals. If the minerals are missing from the soil, plant growth is stunted and the crop yield is low; in past times, agriculture was primitive and low crop yields were the norm. Then came fertilizers and a new day for agrichemistry.

Fertilizers

The earliest fertilizers were natural organic materials such as manure and other waste materials. Most likely some observant early farmer noticed that crops grew larger near sites where waste was dumped. This clever agriculturalist then intentionally planted food crops at these sites and was rewarded with a bountiful harvest. Early farmers began to collect waste materials, such as feces or food scraps, to make their crops grow better. This was the birth of organic fertilizers.

As communities and populations grew, it likely became more difficult to obtain enough manure for the crops. The harvest began to diminish. Eventually, another observant individual noticed that the harvest improved if certain crops were planted in the same field in alternate years. Crop rotation (including leaving a field fallow, or unplanted, for a year), combined with organic fertilizers, gradually improved overall harvest, providing enough food to feed the community.

Gradually, a science of farming developed. As chemistry developed into a science, chemists eventually realized that plants required certain minerals to grow well. (Box 15.4 focuses on one of the most famous agricultural chemists—George Washington Carver.) These scientists knew that manure and other organic wastes provided some of these necessary minerals. Could they improve on these substances? Perhaps, but first scientists needed to know exactly what nutrients plants required. That was a task for the analytical chemist.

Table 15.7 shows the outcome of this research—in other words, which minerals, or nutrients plants need. The nutrients are divided into three classes. Plants require large quantities of the primary nutrients, lesser quantities of the secondary nutrients, and only trace amounts of the micronutrients. Primary nutrients are somewhat analogous to the proteins and saccharides people consume as food. Plants, however, take these nutrients in as simple molecules or even as elements. Some secondary nutrients and all of the micronutrients serve as catalysts or cofactors to make the plant grow.

Notice that the nutrients in table 15.7 are different from those humans require. Plants have essentially no need for sodium and only a slight need for chlorine. In fact, NaCl is toxic to most plants; salt was used to destroy enemy farmlands in ancient times. Plants demand more phosphorus and potassium than humans, but require less calcium because they have no bones. On the other hand, plants need lots of magnesium, the key metal in the chlorophyll molecule.

The three main elements plants need, carbon, hydrogen, and oxygen, come from the atmosphere or from water. Nitrogen, phosphorus, and potassium come almost exclusively from the soil. Although the atmosphere contains abundant amounts of nitrogen, most plants cannot use this source directly. Leguminous plants (such as peas, beans, alfalfa, timothy, and clover) are able to "fix" elemental nitrogen from the atmosphere and deposit it into the soil in the form of a nitrate (via oxidation) or ammonia (via reduction). Legumes accomplish this through colonies of bacteria that grow around their roots. Nitrogen-fixing plants are an important factor in crop rotation; when a farm has animals, the rotating crop is often alfalfa, timothy, or clover, which not only replenish the nitrogen in the soil but provide feed for livestock.

TABLE 15.7 Nutrients needed for plant growth.

PRIMARY NUTRIENTS (large quantities)	SECONDARY NUTRIENTS (small quantities)	MICRONUTRIENTS (trace quantities)
carbon	calcium	boron
hydrogen	magnesium	copper
oxygen	sulfur	iron
nitrogen		manganese
phosphorus		zinc
potassium		molybdenum
		chlorine

BOX 15.4

GEORGE WASHINGTON CARVER

George Washington Carver (figure 1), one of the foremost agricultural chemists of all time, was born in 1860 to slaves on a plantation near Diamond Grove, Missouri. Carver's early eduction was hit-or-miss, but he did learn to read as a child. He later attended Simpson College in Iowa for two years and then entered Iowa Agricultural College in Ames, Iowa (now Iowa State University). Carver worked as an assistant botany instructor and was the director of the college greenhouse. He graduated with a Master's Degree in 1896.

In 1896, Booker T. Washington, founder and head of Tuskegee Institute in Alabama, invited Carver to become an instructor at the institute. Carver remained at Tuskegee the rest of his life. One of his first major achievements was to refine the concept of crop rotation. Noting that the harvest from the Alabama cotton fields was steadily dwindling, Carver concluded that the constant planting of this single crop for nearly two centuries had depleted the soil of essential minerals. He recommended alternating cotton with peanuts, clover, and peas to replenish the soil with both minerals and nitrogen.

Carver is perhaps most famous for developing more than 300 products from the peanut, including peanut butter, face cream, shampoo, and soap (figure 2). He found similar success with the sweet potato, pecans, and clay. Carver's research caused whole new industries to spring up all over the South. In 1916, the agriculturalist achieved special fame when he was elected a fellow of the Royal Society of Arts, Manufacturers, and Commerce of Great Britain. Carver received numerous other honors and awards for his pioneering research.

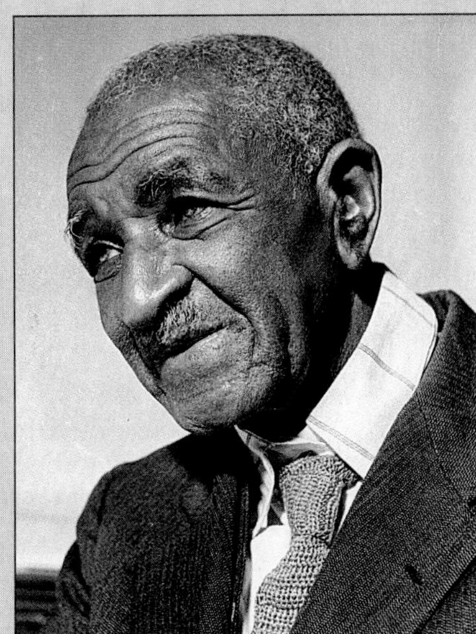

Figure 1 George Washington Carver (1860–1943)

Carver died of heart disease on January 5, 1943, at Tuskegee. A joint resolution of Congress later proclaimed January 5, 1946 George Washington Carver Day. Tuskegee Institute established a museum in Carver's honor, and the U.S. Government designated the farm where Carver was born a national landmark. From his early life as a slave, Carver rose to unparalleled achievement in agricultural research.

Figure 2 Some products made from peanuts.

The main chemical alternate for manure arose in 1908 when Fritz Haber (Polish-German, 1868–1934) discovered a direct way to fix atmospheric nitrogen. Haber received the 1918 Nobel Prize in chemistry for this discovery. The **Haber process** converts N_2 and H_2 directly into NH_3 through a combination of high temperature, high pressure, and a catalyst. Part of the ammonia is then oxidized to nitric acid, which combines with more ammonia to form ammonium nitrate, an excellent nitrogen fertilizer. The following equations summarize this process:

$$3\,H_2 + N_2 \xrightarrow[\text{high pressure catalyst}]{\text{high temperature}} \underset{\text{Ammonia}}{2\,NH_3}$$

$$2\,NH_3 + 4\,O_2 \longrightarrow \underset{\text{Nitric acid}}{2\,HNO_3} + 2\,H_2O$$

$$HNO_3 + NH_3 \longrightarrow \underset{\text{Ammonium nitrate}}{NH_4NO_3}$$

History sadly records that most of the early supplies of ammonium nitrate were used as explosives during World War I rather than as fertilizer to help feed the hungry. But the Haber process did form the basis of today's inorganic fertilizer industry.

Phosphorus is found as mineral deposits in various kinds of rocks. Calcium phosphate, one of the main mineral forms, is water-insoluble. Primitive farmers inadvertently added phosphates to the soil whenever they added fish or animal bones to their scrap piles.

Widespread use of phosphorus as a fertilizer required a specific chemical discovery. In 1842, John Bennett Lawes (English, 1814–1900) developed superphosphate fertilizer after treating phosphate rocks with sulfuric acid. (H. Köhler had converted bone phosphate into a soluble form with a similar sulfuric acid treatment in 1831—over a decade earlier—but his discovery went largely unnoticed.) The phosphorus comes from mined rocks, so it is a nonrenewable resource; the soluble phosphates migrate through the soil to rivers, and eventually end up in oceans. U.S. phosphate rock deposits are expected to be depleted before the turn of the century. When we reach this limit, we must either find another source of phosphorus or face severe crop reduction and potential famine.

Potassium abounds in nature, and its salts are mined worldwide. Since they are water-soluble, these salts require no special chemical conversion to prepare them for fertilizer use. The salts are usually mixed with others to obtain an optimal nitrogen, potassium, and phosphorus combination. Even though potassium resources seem adequate, the salts used in agriculture eventually end up in the ocean and become effectively unavailable. Perhaps the next revolution in the fertilizer industry will involve harvesting minerals from the sea. This may prove difficult, since NaCl, the major salt in seawater, is toxic to most plants.

Inorganic fertilizers are sold as combinations of ingredients with the percentages of N, P, and K listed in order. Thus, a 16-6-10 fertilizer contains 16 percent nitrogen, 6 percent phosphorus, and 10 percent potassium. Nitrogen, often the most abundant component in inorganic fertilizers, is necessary for the plant's leaf and stem development. Higher nitrogen levels promote green color in leaves and lawns. Phosphorus affects root, flower seed, and fruit development. For flowering plants or fruit trees, the ideal fertilizer has a high percentage of P and a lower percentage of N. This change in concentration prods the plants to emphasize flower or fruit development rather than leaf and stem growth. Potassium helps plants resist disease and probably plays a role in polysaccharide and polypeptide synthesis. Different plants need distinct **N-P-K blends,** which is why garden and farm supply stores carry such diverse mixtures.

The better fertilizers also contain small amounts of the secondary nutrients and micronutrients. The secondary nutrient Ca neutralizes acidic soils and plays a role in plant growth; Mg is an essential part of the chlorophyll molecule, and S forms portions of several amino acids and plays a role in protein synthesis.

Among micronutrients, iron, as the Fe^{++} ion, is essential because it catalyzes chlorophyll formation and functions as a cofactor. Lack of iron makes leaves yellow; iron must be applied separate from the phosphorus, or in the form of a chelate, since it forms an unusable complex with phosphates. Iron forms the insoluble hydroxide in basic soils and becomes unavailable; plant nutrients must be water-soluble. Copper, as Cu^{++}, is also an essential cofactor for chlorophyll synthesis.

The micronutrient boron is necessary for protein synthesis and seed production. Zinc and manganese aid plant growth, and Mn^{++} is a key ion in redox reactions. Legumes use molybdenum for nitrogen fixation and the reduction of nitrates. Chlorine plays a role similar to its role in humans—in water content control—but plants need much less Cl^- than we do.

PRACTICE PROBLEM 15.8

Visit the fertilizer section of a garden or farm supply store. What elements are present in some of the inorganic and organic fertilizers?

ANSWER Virtually all contain N-P-K, listed in that order, and some will contain a few other elements as well. The percentages will be higher in the inorganic fertilizers than in natural versions.

EXPERIENCE THIS

Obtain three small plants of about the same size in pots sufficiently large to allow them to grow for several weeks. Label the pots 1 through 3. Next, obtain three 1-gallon bottles of tap water and label these bottles as 1, 2, and 3. (Empty, clean, plastic milk bottles work well.) Add nothing to bottle 1. To bottle 2 add 1 teaspoon of each of the following: baking soda, Epsom salts, and household ammonia. To bottle 3 add 1 teaspoon baking soda, 1 tsp. Epsom salts, 1 tablespoon (tbs.) saltpeter (potassium nitrate) and 1 tbs. household ammonia. Place these plants in sunlight and water each on a regular basis, using the water from the bottle numbered the same as each plant. Do the observed differences in plant growth correspond to the additives in the water? Which solution(s) contain the chemicals used in fertilizers?

Pesticides and Herbicides

The annual crop loss to pests, weeds, and diseases costs more than $15 billion in the United States alone. To combat these problems, farmers use pesticides and herbicides. A **pesticide** is simply a chemical that kills pests, such as insects or vermin; a **herbicide** kills weeds and other unwanted plants. (Specific herbicides may be called fungicides or slimicides, but we will stick with the more generic term.) If farmers do not use these agricultural chemicals, it is virtually impossible to stay in business in today's farm market.

Pesticides

Crop damage is not the only problem insects and other pests cause. Pests carry many diseases, including the infamous bubonic plague of the 1300s. This plague, which fleas and rats spread, wiped out almost one fourth of the population of Europe. Malaria, spread by mosquitoes, has long been a scourge for humans, killing millions. The human loss from pests is impossible to calculate. Pesticides and related chemicals have saved countless lives; DDT alone, according to World Health Organization estimates, has saved at least 25 million human lives. However, DDT has also presented some problems because it is a **broad-spectrum** insecticide that affects many plant and animal species. (See box 1.2, "Risking the Cure.")

Researchers have developed several more specific, or **narrow-spectrum,** insecticides in recent decades. Many of the early examples were organophosphorus compounds, organic molecules that contain phosphorus (figure 15.9). Derived from chemicals originally developed as nerve gases, these insecticides are more toxic to humans than DDT and also accumulate in fatty tissues. However, they biodegrade relatively rapidly compared to DDT. The organophosphorus insecticides include malathion, parathion, diazinon, and chlorpyrifos. Malathion, which is less toxic than the others, was used in California to combat an infestation of white medflies that threatened to obliterate the grape crops. It was also used to combat a 1993 mosquito-transmitted horse virus epidemic in Florida.

Table 15.8 shows the oral or dermal (skin) toxicity of several insecticides and herbicides. As you examine this data, remember that chemicals with higher LD_{50} values are less toxic (an inverse relationship). Ethanol and aspirin are included in table 15.8 for comparison; both chemicals are considered safe for human consumption. Malathion is less toxic than DDT, while Parathion is far more toxic, with a LD_{50} of only 3 mg/kg.

Another group of narrow-spectrum insecticides are based on the organic carbamate group, which is midway between an ester and an amide. Carbaryl (Sevin), aldicarb (Temik) and carbofuran (Furadan) are examples of this group (figure 15.10). The carbamate group breaks down rapidly in the environment, preventing the chemical from accumulating. Probably the worst side effect of carbamates is that one of them, carbaryl, kills honeybees, a desirable insect. Carbaryl is more toxic than Malathion, but Aldicarb is extremely toxic with an LD_{50} of merely 1 mg/kg.

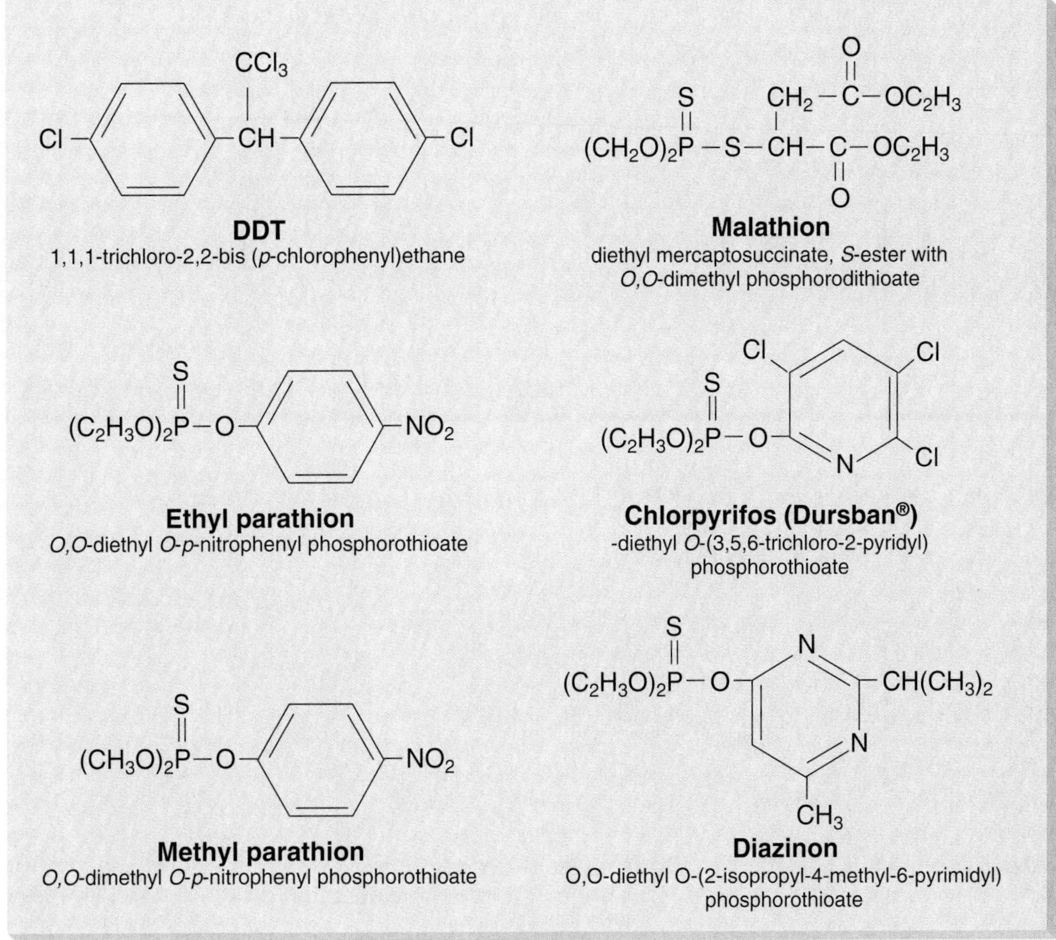

Figure 15.9 The structures of DDT and some organophosphorus insecticides.

TABLE 15.8 The toxicities of some insecticides and herbicides. Higher values are less toxic.

COMMON NAME	CHEMICAL NAME	ORAL LD_{50} (mg/kg)	DERMAL LD_{50} (mg/kg)
Alachlor**	2-chloro-2′,6′-diethyl-N-(methoxymethyl)acetanilide	1,200	—
Aldicarb*	2-methyl-2-(methylthio)propionaldehyde-O-(methylcarbamoyl)oxime	1	5
Allethrin*	cis,trans-(±)-2,2-dimethyl-3-(2-methyl-propenyl)cyclopropane-carboxylic acid ester with (±)-2-allyl-4-hydroxy-3-methyl-2-cyclopenten-1-one	680	11,200
Amitrole**	3-amino-s-triazole	1,100	—
Atrazine**	2-chloro-4-(ethylamino)-6-(isopropylamino)-s-triazine	3,080	—
Benefin**	N-butyl-N-ethyl-α,α,α-trifluoro-2,6-dinitro-p-toluidine	10,000	—
Butacide*	α-[2-(2-butoxy)ethoxy]-4,5-(methylene-dioxy)-2-propyltoluene	7,500	7,500
Carbaryl*	1-naphthyl methylcarbamate	307	2,000
Chlorpyrifos*	O-O-diethyl-O-(3,5,6-trichloro-2-pyridyl) phosphorothioate	97	2,000
DDT*	1,1,1-trichloro-2,2-bis-(p-chlorophenyl)ethane	87	1,931
Diazinon*	O,O-diethyl-O-(2-isopropyl-4-methyl-6-pyrimidyl) phosphorothioate	66	379
Dioxin	2,3,7,8-tetrachlorodibenzo-p-dioxin (TCDD)	0.03	—
Diuron**	3-(3,4-dichlorophenyl)-1,1-dimethylurea	3,400	—
Endrin*	1,2,3,4,10,10-hexachloro-6,7-epoxy-1,4,4a,5,6,7,8,8a-octahydro-1,4,-*endo*, *endo*-5,8-dimethanonaphthalene	3	12
Glyphosate**	N-(phosphonomethyl)glycine	4,320	
Lindane*	1,2,3,4,5,6-hexachlorocyclohexane, gamma isomer of not less than 99% purity	76	500
Malathion*	Diethyl mercaptosuccinate S-ester with O-O-dimethyl phosphorodithioate	885	4,000
Monuron**	3-(p-chlorophenyl)-1,1-dimethylurea	3,600	—
Paraquat**	1,1′-dimethyl-4,4′-bipyridinium ion	150	—
Parathion*	O,O-diethyl-O-(p-nitrophenyl) phosphorothioate	3	4
Propanil**	3′,4′-dichloropropionalide	1,384	—
Pyrethrins*	mixture of pyrethrins & cinerins	200	1,800
2,4-D**	(2,4-dichlorophenoxy)acetic acid	375	—
2,4,5-T**	(2,4,5-trichlorophenoxy)acetic acid	300	—
For comparison			
	Aspirin	1,500	—
	Ethanol	10,600	—

*Insecticide
**Pesticide
SOURCE: From *Pesticides Theory and Application* by Ware. Copyright © 1978 by W.H. Freeman and Company. Used with permission.

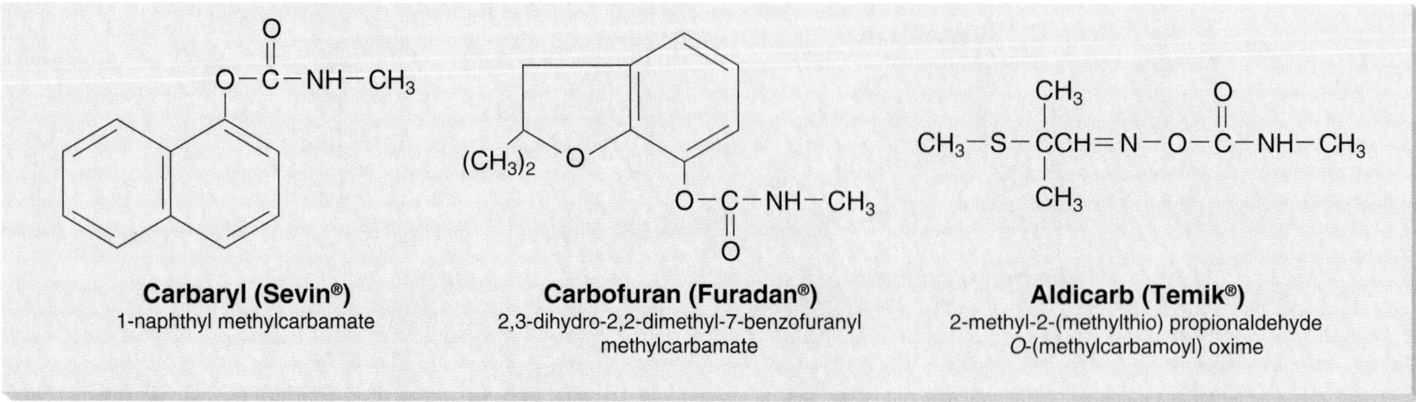

Figure 15.10 The structures of some carbamate insecticides.

Some other insecticides include butacide, endrin, and lindane, plus various pyrethrin derivatives (natural insecticides found in certain plants). Their toxicities are listed in table 15.8. In addition to these chemicals, as box 15.5 details, scientists have been partially successful in developing alternatives to insecticides.

PRACTICE PROBLEM 15.9

How do a broad-spectrum and a narrow-spectrum insecticide differ? Is there any difference in the biodegradability of the insecticide? Is there any difference in toxicity to humans?

ANSWER Broad-spectrum insecticides kill many different types of insects, while narrow-spectrum versions are usually fatal to only a few species. Biodegradability and toxicity vary; both broad- and narrow-spectrum insecticides may be harmless or dangerous, biodegradable or long-lasting, because these phenomena are independent of the insecticide's specificity.

Herbicides

Herbicides are chemicals that kill unwanted plants, often defined as weeds. These chemicals may be either broad- or narrow-spectrum. Glyphosphate (figure 15.11), the active component in herbicides known as Kleenup and Roundup, kills virtually every plant it contacts.

Although many chemicals were tried and discarded as herbicides over the centuries, the first effective compound was 2,4-D (an abbreviation of the chemical name 2,4-dichlorophenoxyacetic acid), which works against newly emerging broad-leafed plants. The chemical, 2,4,5-T is a relative of 2,4-D with one more chlorine atom in the molecule; it is effective against woody plants and causes defoliation. However, 2,4,5-T was often contaminated with dioxins (see box 12.1). In 1985, the U.S. Environmental Protection Agency (EPA) banned all used of 2,4,5-T and any other agent that might produce dioxins. The original 2,4-D is still used.

Figure 15.12 shows herbicide structures. Typical examples include atrazine, a triazine molecule effective for controlling weeds in corn fields; monuron, a urea derivative used as a preemergent herbicide; allidochlor (or CDAA), a chlorinated amide selective for grasses; benefin, a nitroaniline derivative that is a preemergent herbicide; and terbacil, a uracil derivative that controls grass and broad-leaf plants. Table 15.8 lists some of these herbicides and their toxicity levels. Notice that many herbicides are less toxic than most insecticides—some are even less toxic than aspirin.

$$HO-\overset{\overset{O}{\|}}{C}-CH_2-NH_2-CH_2-\underset{\underset{OH}{|}}{\overset{\overset{O}{\|}}{P}}-OH$$

Figure 15.11 Glyphosate, a broad-spectrum herbicide.

PRACTICE PROBLEM 15.10

Among the more common insecticides and herbicides available for home use are benefin, butacide, carbaryl, chlorpyrifos, diazinon, diuron, glyphosate, lindane, malathion, and pyrethrins. Rank these chemicals in order of increasing toxicity.

ANSWER (Lowest to highest): benefin, butacide, glyphosate, diuron, malathion, carbaryl, pyrethrins, chlorpyrifos, lindane, diazinon.

BOX 15.5

ALTERNATIVES TO INSECTICIDES

For years, chemists have researched for alternative methods of insect control. Their efforts have centered on limiting the insects' ability to reproduce, using natural chemicals called pheromones in traps, preventing insects from reaching sexual maturity, and controlling populations through specific predators and diseases.

Because the basic problem in insect control is too many insects (figure 1), one approach is to keep insects from reproducing. Scientists developed two methods based on this approach. In the simpler method, huge numbers of radiation-sterilized male insects were released into the environment to mate with the females. Many insects mate only once in a lifetime, so if a female mates with a sterilized male, the female produces no offspring. The total number of insects decreased dramatically when researchers tried this method with the bothersome screwworm fly. However, this approach does not work well with most other insects since females prefer to mate with unsterilized males.

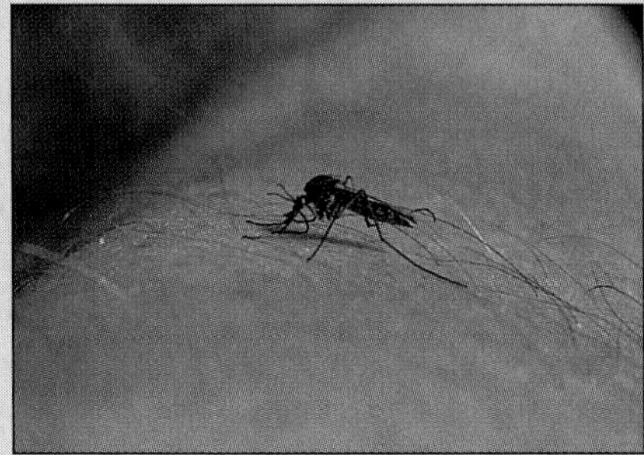

Figure 1 Too many insects.

The second approach used insect pheromones, chemicals that insects excrete in minute amounts to send special messages (such as alarms) or to attract a mate. Chemists wanted to synthesize the sexual pheromones and place them in traps, so that the captured insect, normally a male, would be unable to mate. Although it was difficult to determine the sexual attractant from among the thousands of chemicals an insect carries, chemists were nevertheless able to isolate and synthesize the pheromones. Figure 2 shows the structures of insect pheromones. Traps containing such chemicals are now available at most garden and farm centers. Unfortunately, scientists have still not analyzed the pheromones for most insects.

Like humans, insects have a vast array of hormones that control the processes in their bodies. The juvenile growth hormone keeps insects in the youthful pupae stage of development. Eventually other hormones take over and the insect matures into the bothersome adult form. Chemists correctly reasoned that if the juvenile growth hormone were continually applied, the insects would not mature. Determining which chemicals caused this effect is difficult, but chemists have succeeded in several cases. The structure of a mosquito juvenile growth hormone is shown in figure 3. These highly selective chemicals do not harm other species of insects or animals.

Other approaches to insect control include using predators and using diseases that only affect specific insects. Some consider the latter method potentially dangerous since viruses and bacteria can mutate. Nobody knows whether these mutations would be safe, or whether they would plague other insects, possibly even humans. Predator insects are somewhat more benign; most people like ladybugs and the famous praying mantis. Undoubtedly, humans will continue to search for ways to control insect populations.

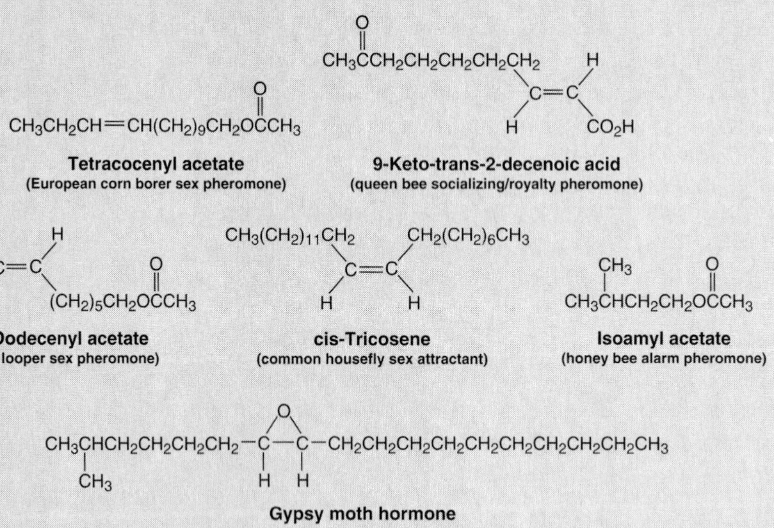

$CH_3CH_2CH{=}CH(CH_2)_9CH_2OCCH_3$ (C=O on the ester carbon)

Tetracocenyl acetate
(European corn borer sex pheromone)

$CH_3C(=O)CH_2CH_2CH_2CH_2CH_2$, H / C=C / H, CO_2H

9-Keto-trans-2-decenoic acid
(queen bee socializing/royalty pheromone)

H, H / C=C / $CH_3(CH_2)_3$, $(CH_2)_5CH_2OCCH_3$ (C=O)

cis-7-Dodecenyl acetate
(cabbage looper sex pheromone)

$CH_3(CH_2)_{11}CH_2$, $CH_2(CH_2)_6CH_3$ / C=C / H, H

cis-Tricosene
(common housefly sex attractant)

$CH_3CH(CH_3)CH_2CH_2OCCH_3$ (C=O)

Isoamyl acetate
(honey bee alarm pheromone)

$CH_3CH(CH_3)CH_2CH_2CH_2CH_2{-}CH(O)CH{-}CH_2CH_2CH_2CH_2CH_2CH_2CH_2CH_2CH_2CH_3$ (epoxide; H, H on ring carbons)

Gypsy moth hormone

Figure 2 The structures of some insect pheromones.

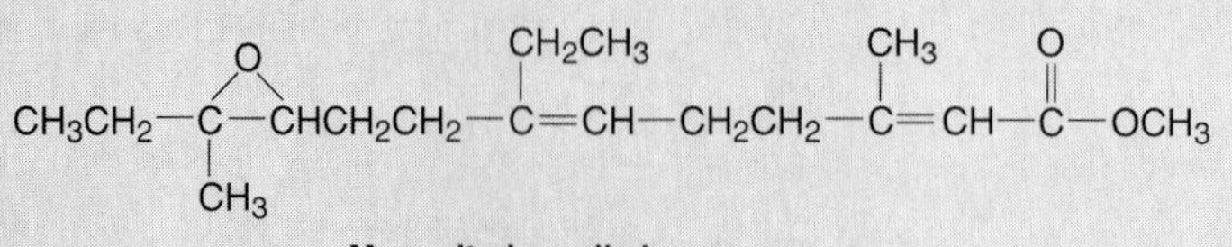

$CH_3CH_2{-}C(CH_3)(O)CH{-}CH_2CH_2{-}C(CH_2CH_3){=}CH{-}CH_2CH_2{-}C(CH_3){=}CH{-}C(=O){-}OCH_3$ (epoxide)

Mosquito juvenile hormone

Figure 3 The mosquito juvenile hormone.

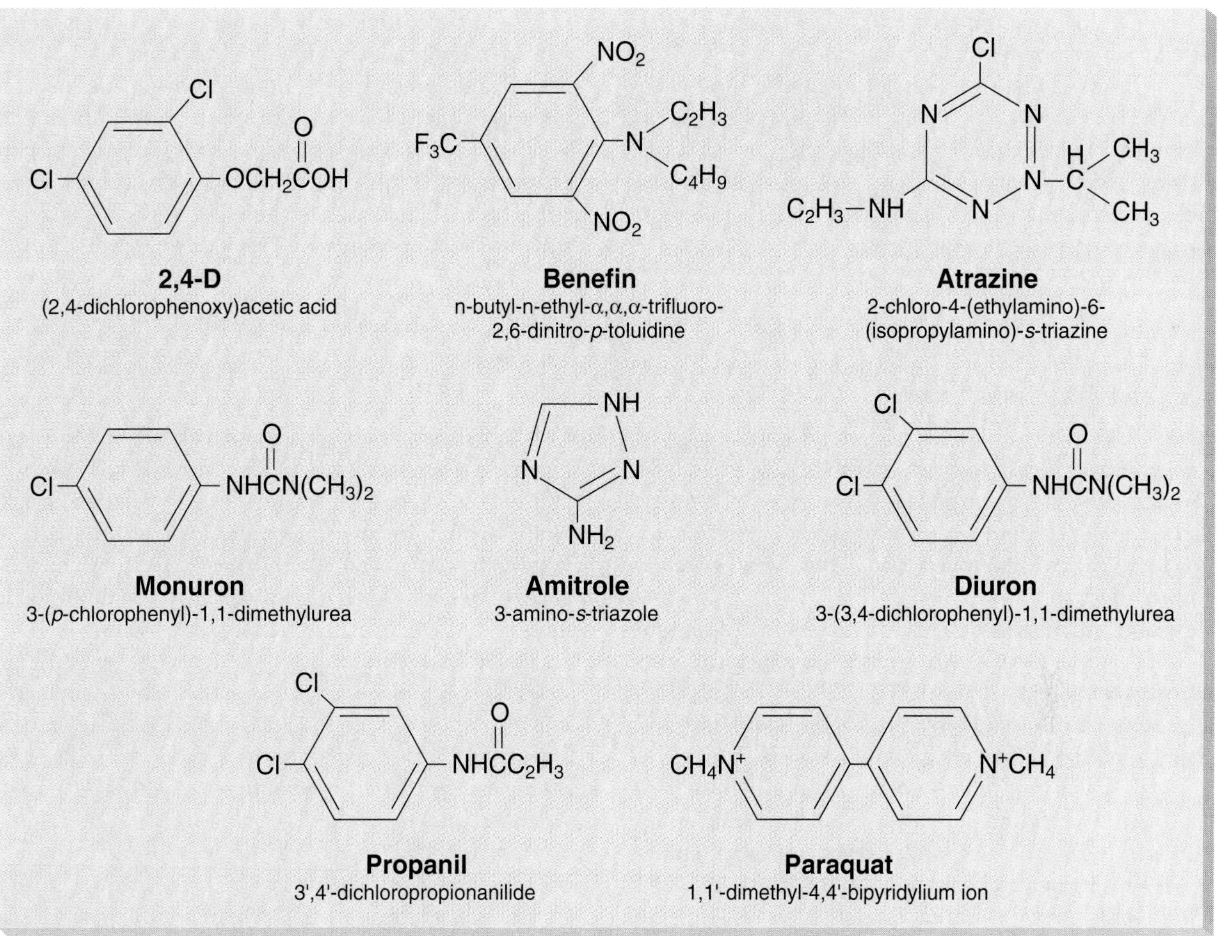

Figure 15.12 The structures of some herbicides.

BIOTECHNOLOGY—CHEMISTRY IN DISGUISE

The successful Green Revolution of the 1950s and 1960s, which stressed using fertilizers, petrochemical-based herbicides, and large amounts of energy to promote the growth of certain grains to alleviate human hunger, was almost entirely dependent on inorganic fertilizers, pesticides, herbicides, and petroleum. This technology doubled the world's wheat yield, but not without drawbacks. The crops are designed for a single harvest, using efficient machinery, but when entire fields are planted with a single crop, they become more susceptible to damage caused by pests, disease, and adverse weather. In addition, the high cost of this intensive form of farming has driven many smaller farms out of existence in both the United States and the Third World.

Many scientists believe the answer to feeding humanity and improving our overall lifestyle may come from **biotechnology,** a fusion of chemistry and biology. Biotechnology utilizes living systems for human benefit. This sometimes means finding improved ways to grow crops, developing new or modified crops, or using known materials in new ways. Several new and fully biodegradable polymers from renewable resources should appear on the market by the end of the century.

Scientists can now modify the genetic makeup of plants and animals. One agricultural innovation involves a genetic change in the bacteria *Pseudomonas syringue*, which lives in most plants and serves as a nucleus for ice formation. Ice nucleation arises from a protein, but the genetically modified bacteria cannot synthesize this protein and cannot serve as ice nucleation sites. The overall result is that the plants treated with genetically altered bacteria are more frost-resistant, permitting longer growing seasons for crops such as tomatoes, corn, strawberries, and potatoes. This also increases the amount of food produced by one field. During 1994, Flavr Savr, a genetically modified tomato that does not soften as quickly, entered the marketplace. Other biotech developments were to follow rapidly.

Another potential biotechnological innovation may be the development of food crops, such as corn or wheat, that fix their own nitrogen supply, as the legumes do. This would drastically reduce (or eliminate) the need for inorganic fertilizers. Biotechnological improvements in the amino acid composition of some corn and rice crops seem to provide a built-in herbicide resistance.

Genetic engineering may improve milk yields by 30 percent through the addition of bovine (cow) growth hormone (BGH) to the cow's diet. Ultimately, scientists might manipulate the cow's genes so that the cow produces more of this natural hormone on its own. (Not everyone is enthused about these changes, however, and many have protested emphatically.) Other possible genetic changes include increasing the proportion of lean meat in animals—making beef (or pork) less fatty.

Biotechnology is hardly limited to food or plastic production. Already scientists have modified bacteria so they can produce insulin and human growth hormones. Diabetics benefit from this human insulin; the older forms were derived from cows or pigs. Similarly, doctors now administer HGH to patients who lack adequate output of this hormone and are very small in stature. In either case, biotechnology provides a necessary protein whose structure and function were previously determined by chemists.

SUMMARY

Many chemicals essential to life enter living systems as food or drink to take part in the digestive cycle. Digestion, nutrition, food additives, and cooking are important aspects of food chemistry. Agrichemistry and biotechnology—which apply chemical and genetic knowledge to grow improved foods—are also key factors in the chemistry of foods.

Food digestion involves one essential chemical reaction: hydrolysis, which breaks foods down into smaller chemical units in the presence of water. Hydrolysis breaks down polypeptides (proteins), polysaccharides (carbohydrates), and lipids (fats and oils), the three main food chemical groups, into smaller units.

Proteins break down into amino acids with the help of proteases, or protein-hydrolyzing enzymes. Similarly, saccharide-cleaving enzymes break polysaccharides down into saccharides (simple sugars), and lipases break lipids down into fatty acids and glycerol. Fatty acids may be either saturated or unsaturated; polyunsaturated fats are more healthful in the diet than saturated fats. In fact, high consumption of saturated fats is alleged to be one of the culprits in cholesterol-related heart disease.

What is a chemically balanced, nutritious diet? Humans need a diet that is 65 to 80 percent polysaccharides, 5 to 10 percent polypeptides, and 15 to 25 percent lipids, mostly unsaturated lipids. A healthy diet must also include a variety of essential vitamins and minerals.

Vitamins may be either water- or lipid-soluble. The water-soluble vitamins include B vitamins, niacin, vitamin C, biotin, folic acid, and pantothenic acid. The lipid-soluble vitamins include A, D, E, and K. All of these vitamins are essential to good health, but in extremely high doses, they can be toxic—especially the lipid-soluble vitamins, which can accumulate in fatty tissues.

Vitamins aid in a variety of body functions, including growth and development and the functioning of the immune system and the nervous system—virtually every organ in the body depends on vitamins. Hydrolyzed foods either provide energy, help rebuild body tissues, or are stored for future use. Meats, beans, nuts, and dairy products provide proteins; breads, grains, pastas, vegetables, and some fruits supply carbohydrates; and meats, fish, oils, and dairy products provide lipids. All of these foods provide different vitamins.

Food additives are chemicals we deliberately add to foods during processing. They include sweeteners, flavoring and coloring agents, preservatives, antimicrobial agents, and antioxidants. Sweeteners may be natural or artificial (laboratory-synthesized). Natural sweeteners include glucose, fructose, lactose, maltose, and sucrose, among others. Artificial sweeteners, which are generally sweeter in small amounts, include saccharin, the cyclamates, aspartame, and Acesulfame K. Neither natural nor artificial sweeteners add significant nutritive value to foods—for the most part, they just make the food taste better.

Flavoring agents are also designed to enhance food taste. These agents include salt, sugar, monosodium glutamate, and various aldehydes and esters. Food colorings make foods look more appetizing. Some are natural food products, such as β-carotene; others are synthetic. Preservatives may also be natural (such as sugar and salt) or synthetic (such as nitrites and sulfur dioxide).

Antimicrobial agents kill or prevent the growth of microorganisms, and antioxidants slow or block the oxidation reactions that can cause foods to spoil. Other additives act as sequestrants, buffers, thickeners, or emulsifiers. Some of these additives—both natural and artificial—appear to be potentially cancer-causing; research continues to determine which are safe for human consumption, and in what amounts.

Cooking and baking help break down the structures of polypeptides and polysaccharides before digestion begins. Baking causes the formation of "pockets" of carbon dioxide, making baked goods light and airy.

Before we can cook or bake food, we must apply agrichemistry to produce the food from plants or animals. Plants produce saccharides, such as glucose, through photosynthesis. Agrichemical techniques such as fertilizing and rotating crops can increase the efficiency of this process by providing plants with needed minerals and other nutrients, including nitrogen, phosphorus, and potassium, through the soil. Pesticides and herbicides further aid plant growth by eliminating insects that can harm plants and weeds that compete for precious nutrients. Scientists continue to search for insecticide and herbicide formulations that can enhance plant harvests without endangering human health.

Biotechnology combines chemistry and biology to utilize living systems for human benefit. Scientists can now genetically engineer plants and animals to improve food yields and reduce the incidence of disease, as well as to produce biochemical products that treat human illnesses and conditions. In the future, we will undoubtedly see more and more biotechnological breakthroughs applied to food and agricultural chemistry.

KEY TERMS

Acesulfame K 415
Ames test 414
antimicrobial agent 418
antioxidant 418
artificial sweetener 413
aspartame 415
biotechnology 427
broad-spectrum 423
butylated hydroxyanisole (BHA) 418
butylated hydroxytoluene (BHT) 418
carcinogenicity 414
cyclamate 414
Delaney Amendment 415
GRAS list (Generally Regarded As Safe) 416
Haber process 421
herbicide 422
hydrolase 401
hydrolysis 400
iodine number 401
lipase 401
lipid-soluble vitamins 404
monosodium glutamate (MSG) 416
narrow-spectrum 423
N-P-K fertilizer blends 422
pesticide 422
photosynthesis 420
polyunsaturated fatty acid 401
propyl gallate 418
protease 401
saccharin 413
saturated fatty acid 401
unsaturated fatty acid 401
water-soluble vitamins 404

READINGS OF INTEREST

General

Brownlee, Shannon, and Barnett, Robert. "A loaf of bread, a glass of wine." *U.S. News & World Report*, July 4, 1994, pp. 62–63. There is more than one food pyramid; here is the Mediterranean version.

Volti, Rudi. "How we got frozen food." *Invention & Technology*, Spring 1994, pp. 47–56. The earliest attempt to develop frozen food led to the death of Sir Francis Bacon in 1626. How did this industry rise?

Troxell, Terry. "Government, chemistry, and food." *Chemtech*, April 1994, pp. 54–59. A survey of the interactions between these three entities.

Whelan, Elizabeth. "Proposed food safety laws are starved for scientific merit." *Insight*, November 1, 1993, pp. 30–31. This article's author is the president of the American Council on Science and Health.

Marshall, Douglas L., and Wiese Lehigh, Patti L. "Nobody's nose knows." *Chemtech*, October 1993, pp. 38–42. How do you tell if fish is still good to eat? Some tests are being developed to determine how.

Speights, Robert M.; Perna, Peter J.; and Downing, Steven L. "Do those calories really count?" *Chemtech*, July 1993, pp. 54–59. Of course they do, but there is more to dieting than counting calories.

Raloff, Janet. "Paring protein." *Science News*, November 21, 1992, pp. 346–47. It appears that low-protein diets may actually slow aging.

Moser, Penny Ward. "There's always room for . . ." *Health*, May/June 1992, pp. 20–25. If you watch TV commercials, you already know what the next word is. But do you know what gelatin really is?

Holmes, Hannah. "Eating low on the food chain." *Garbage*, January/February 1992, pp. 32–37. This article advocates eating a diet containing many more vegetables than the average American eats. It is similar in content to the original USDA food-groups pyramid.

Hapgood, Fred. "The prodigious soybean." *National Geographic*, vol. 172, no. 1 (July 1987), pp. 66–96. Once called the "yellow jewel" or the "great treasure" by the Chinese, the humble soybean now forms the food base for billions of people around the world.

Bode, Richard. "This little bean is big business." *Reader's Digest*, March, 1985, pp. 127–130. A summary of the $12+ billion coffee bean market.

Lipids and Cholesterol

Lemonick, Michael D. "Are we ready for fat-free fat?" *Time*, January 8, 1996, pp. 52–61. Some up-to-date information about Olestra, a non-digestable fat-like material. Also has some information about previous "fake foods," including cream-free cream and coffee substitutes.

Raloff, J. "Margarine is anything but a marginal fat." *Science News*, May 21, 1994, p. 325. In fact, margarine ranks higher than most foods, including butter, in fat content.

Adler, Tina. "Designer fats." *Science News*, May 7, 1994, pp. 296–97. Food makers are designing new fat substitutes.

Davies, H. Maelor, and Flider, Frank J. "Designer oils." *Chemtech*, April 1994, pp. 33–37. How genetic engineering is improving plants and producing more useful oils.

Rotman, David. "National Starch's fake fats." *Chemical Week*, January 15, 1992, p. 15. Several fat substitutes are now either hitting the market or are poised to enter. Some are made from starch.

Teagle, Philip L. 1991. *Understanding your cholesterol*, New York: Academic Press. If you really want an easy-to-read and accurate report on the cholesterol debate, this may be the book for you. It is written by a medical school professor using language no more difficult than the writing in this text. This is probably your best single, simple source for information on this "hot" topic.

Hamilton, Joan. "Cholesterol management for all seasons." *Business Week*, December 23, 1991, pp. 96–97. Not much glitz, just some straightforward recommendations on controlling cholesterol through a healthy diet.

Jandacek, R. J. "The development of Olestra, a noncaloric substitute for dietary fat." *J. Chem. Ed.*, vol. 68, no. 6 (June 1991), pp. 476–479. Olestra, a sucrose octaester than can be used in cooking and as a fat substitute, is noncaloric because the body does not digest it. This article gives historical background on this compound.

Ross, Irwin. "A new drug that fights cholesterol." *Reader's Digest*, December 1987, pp. 91–94. How the drug lovastatin was developed.

Vitamins

Richardson, Sarah. "The war on radicals." *Discover*, July 1994, pp. 27–28. Chemical free radicals are usually harmful inside our bodies, but vitamins can counteract them. But are all vitamins equally advantageous?

Nowak, Rachel. "Beta-carotene: Helpful or harmful?" *Science*, vol. 264 (April 22, 1994), pp. 500–501. The answer depends on whether you smoke. If you do, you are out of luck.

Fackelmann, K. A. "Vitamin A-like drug may ward off cancers." *Science News*, May 30, 1992, p. 358. If vitamins work, why not synthesize drugs patterned after them?

Silberner, Joanne. "Health Guide: Building the best defense." *U.S. News & World Report*, May 4, 1992, pp. 62–70. This "best defense" is vitamins, in multivitamin tablets if necessary, and occasional aspirin tablets. At last count, 35 percent of all Americans take vitamin and mineral supplements, and 20 percent take an aspirin daily.

Fackelmann, K. A. "Vitamin D: Too much of a useful thing." *Science News*, May 2, 1992, p. 295. A woeful tale about milk with too much vitamin D added; instead of the 400 international units the RDA calls for, this batch had 232,565.

Toufexis, Anastasia. "The new scoop on vitamins." *Time*, April 6, 1992, pp. 54–59. This article strongly advocates adding more vitamins to our diets, with multivitamin tablets the logical source. It also contains an interesting table showing some possible benefits various vitamins may provide in combating cancer, heart disease, and even pregnancy problems.

Raloff, Janet. "Garden-variety tonic for stress." *Science News*, February 8, 1992, pp. 94–95. This article claims that vitamins from fresh vegetables may help fight the effects of stress. Fiber may help also.

Raloff, J. "Vitamin C protects blood from radicals." *Science News*, August 26, 1989, p. 133. Vitamin C may help protect our bodies from dangerous radicals—chemical radicals, not political.

Findlay, Steven. "Squaring off over vitamins." *U.S. News & World Report*, April 10, 1989. pp. 62–64. Claims and counterclaims abound in the continuing controversy over the quantity of each vitamin we need.

Carpenter, Kenneth J. 1986. *The history of scurvy and vitamin C*. New York: Cambridge University Press. This is the complete story of scurvy and vitamin C. You may also want to read the 1981 *Pellagra* by the same author.

Stroetzel, Donald S. "Do you need extra vitamins?" *Reader's Digest*, April 1986, pp. 77–80. This articles says "no"—or perhaps a qualified "yes."

Kutsky, Roman J. 1973 *Handbook of vitamins and hormones*. New York: Van Nostrand Reinhold. This handbook contains most of the hard data on vitamins and hormones. Some numbers have changed since 1973 (RDAs for example), but most of these data are still up to date.

Cancer Prevention and Health Aspects of Food

Willett, Walter C. "Diet and health: What should we eat?" *Science*, vol. 264 (April 22, 1994), pp. 532–537. An easy-to-read article that discusses most foods and their relationship to many diseases.

Smith, Trevor. "Eating your way to health." *Today's Chemist at Work*, March 1994, pp. 54–57. This article provides a low-fat meal plan that can help you lose weight while maintaining good health.

Schmidl, Mary K. "Food products for medical purposes." *Trends in Food Science & Technology*, June 1993, pp. 163–168. This article deals with the use of food products in medicine rather than with nutrition itself.

Fackelmann, K. A. "Cancer protection from fruits and veggies." *Science News*, June 5, 1993, p. 357. When Mom and Dad told you to eat your veggies, they had the right idea.

Williams, Christine M. "Diet and cancer prevention." *Chemistry & Industry*, April 19, 1993, pp. 280–283. We are learning more about the food-cancer prevention connection every year. This article contains a lot of information on vitamins.

Simone, Charles B. 1992. *Cancer and nutrition*, Garden City, N.Y.: Avery Publishing. Written by a medical doctor, this highly readable book covers most aspects of the cancer-nutrition connection and gives solid evidence for each position it takes.

Pennisi, Elizabeth. "Hairy harvest." *Science News*, May 30, 1992, pp. 366–367. This short paper, subtitled "Bacteria turn roots into chemical factories," tells how some potential antileukemia drugs are harvested from the common periwinkle plant.

Raloff, J. "Pectin helps fight cancer's spread." *Science News*, March 21, 1992, p. 180. Pectin is a polysaccharide used in home canning.

Jaret, Peter. "Bet on broccoli." *In Health*, September/October 1991, pp. 58–62. Interestingly enough, this veggie contains several natural chemicals that seem to prevent cancer. All this plus good fiber.

Jacobson, Michael F.; Lefferts, Lisa Y.; and Garland, Anne Witte. 1991. *Safe food: Eating wisely in a risky world*. Los Angeles: Living Planet Press. As the title implies, this book discusses many potential problems with the food supply, including additives, bugs, drugs, pesticides, and "organic" food suppliers. Not a completely balanced account, but still worth reading.

McCourt, Richard. "Some like it hot." *Discover*, August 1991, pp. 48–52. Capsaicin, isolated from hot red peppers, has desirable pain-relieving properties. But keep it out of your eyes.

Csintalan, R. P., and Senozan, N. M. "Copper precipitation in the human body: Wilson's disease." *J. Chem. Ed.*, vol. 68, no. 5 (May 1991), pp. 365–367. Wilson's disease involves the body's improper utilization of copper. This article gives some information on the disease itself and describes how it is treated chemically.

Quillin, Patrick. 1989. *Healing nutrients*. New York: Vintage Press. A well-documented and readable book on how nutrition relates to almost every aspect of health.

McConnell, Carol and Malcolm. "Ancient secrets of modern nutrition." *Reader's Digest*, September 1986, pp. 37–44. Not everything we hear today about nutrition and health is new information. This article tells about combating cancer and cardiovascular disease with careful food selection. While the old-timers may not have known why some foods worked, they were often correct in their choices.

Hager, Tom. "Take fish to heart." *Reader's Digest*, August, 1985, pp. 127–130. In case you haven't heard, fish contain desirable polyunsaturated lipids and plenty of vitamins.

Burgess, Anthony. "The glory of garlic." *Reader's Digest*, 1984, pp. 185–190. Used for centuries to season food, garlic also has some valuable medical properties. Just remember your mouth rinse.

Pesticides and Fertilizers

Kirschner, Elisabeth M. "Agricultural chemical producers rebound from floods of 1993." *Chemical & Engineering News*, October 3, 1994, pp. 13–16. The 1993 market was $6.8 billion. About 68 percent was spent on herbicides and 21 percent on insecticides.

Roberts, Michael, and Begley, Ron. "Pesticide producers innovate under recessionary pressure." *Chemical Week*, September 14, 1994, pp. 27–28. A business-oriented article that contains an excellent graphic showing the numbers of pesticides and various types introduced between 1983 and 1992. This number has dwindled over time because testing is so expensive.

Cottrill, Ken. "Recovery in fertilizers exceeds expectations." *Chemical Week*, September 14, 1994, p. 29. U.S. fertilizer production capacity exceeded 40 billion tons in 1994.

Schmidkofer, Regina M. "The growing specialty fertilizer market." *Chemtech*, March 1994, pp. 54–57. Fertilizer markets are as varied as the clientele of farms, landscaping, nurseries, and homeowners can make them.

Gianessi, Leonard. "The quixotic quest for chemical-free farming." *Chemtech*, December 1993, pp. 48–53. *Also* Gianessi, Leonard. "The quixotic quest for chemical-free farming." *Issues in Science and Technology*, Fall 1993, pp. 29–36. A pair of articles with the same title by the same author. The thrust of both is that pesticide bans are rare because viable alternatives seldom exist. Those that do are often as hazardous as the pesticides they attempt to replace.

Begley, Ronald; Plishner, Emily S.; Roberts, Michael; and Young, Ian. "Farm chemicals." *Chemical Week*, September 8, 1993, pp. 21–25. Five short, business-oriented articles about pesticides and fertilizers. One is entitled "Is the Organic Grass Greener?"

Shen, Samuel K., and Dowd, Patrick F. "Detoxifying enzymes and insect symbionts." *J. Chem. Ed.*, vol. 69, no. 10 (October 1992), pp. 792–797. Although some claim that synthetic chemicals cause all toxicity problems, the truth is that the natural environment is extremely toxic. How does nature handle this problem?

Weber, Peter. "A place for pesticides?" *Worldwatch*, May/June 1992, pp. 18–25. A brief overview of pesticides and some of the emerging problems with resistant bugs.

Kourik, Robert. "Flower power." *Garbage*, May/June 1992, pp. 26–31. A primer on natural means to control insects by using flowers and predator bugs rather than synthetic pesticides.

Hanson, David J. "Dupont, EPA investigate possible role of fungicide in crop deaths." *Chemical & Engineering News*, April 6, 1992, pp. 15–17. Two sides of the story on Benlate, a fungicide.

Stone, Richard. "A biopesticide tree begins to blossom." *Science*, vol. 255 (February 28, 1992), pp. 1070–1071. The neem tree contains several chemicals that fight off insects. Other of the chemicals may even improve human health.

Mackerron, Conrad B. "Alar panel subject of inquiry." *Chemical Week*, June 7, 1989, pp. 16–18. Was there a conflict of interest for some members of this panel?

Department of Governmental Relations and Science Policy, American Chemical Society. 1987. *Pesticides*. Washington, D.C.: American Chemical Society. You can obtain a free copy of this very informative paper by writing or calling the A.C.S. at 1155 Sixteenth Street N.W., Washington, D.C. 20036; (202) 872-8725.

Higdon, Hal. "New tricks outwit our insect enemies." *National Geographic*, vol. 142, no. 3 (September 1972), pp. 380–399. This article discusses such "tricks" as sterilization, infecting insects with specific diseases, chemical lures, and growth-retarding hormones.

Sweeteners

Pennisi, Elisabeth. "Valentine bind." *Science News*, February 15, 1992, pp. 110–111. What makes something taste sweet? This article presents some of the latest findings on the nature of sweetness in all kinds of molecules.

Benson, Susan. "Has sugar taken too many lumps?" *In Health*, September/October 1991, pp. 20–21. Sugar has been used as a sweetener for centuries, but most other sweeteners are fairly new. Can we use sugar safely?

Ross, Walter S. "Artificial sweeteners: Are they safe?" *Reader's Digest*, December 1985, pp. 140–144. How safe are artificial sweeteners? A part of the story on testing these now common chemicals.

Other Food Additives

Pospisil, Ray. "Low fat driving market for ingredients." *Chemical Week*, June 22, 1994, pp. 27–28. Lowering the fat content in foods makes a host of additives necessary to maintain taste.

Raloff, Janet. "Not-so-hot hot dogs." *Science News*, April 23, 1994, pp. 264–265, 269. What kinds of risks do you encounter because of the way meats are processed?

Coeyman, Marjorie. "Food additives: Catering to the health conscious." *Chemical Week*, June 23, 1993, pp. 21–23. The market for additives was estimated at more than 3.47 billion dollars for 1996. A good, quick survey of what kinds of additives are used.

Anderson, Denise A. "A matter of taste." *Chemtech*, August 1993, pp. 45–49. How can you make food taste better and yet have a longer shelf life? The answer is additives, many of which come from natural sources.

Bird E. W., and Sturtevant, Floyd. "Extraction of FD&C dyes from common food sources." *J. Chem. Ed.*, vol. 69, no. 12 (December 1992), pp. 966–968. This article explains how dyes are extracted from foods.

Iyengar Radha, and McEvily, Arthur J. "Antibrowning agents: Alternatives to the use of sulfites in foods." *Trends in Food Science & Technology*, vol. 3 (March 1992), pp. 60–63. Sulfites have long been used to keep foods from turning brown. While effective, they sometimes produce adverse health reactions. This paper discusses the chemical nature of browning and suggests several alternatives to sulfites.

Raloff, Janet. "Cancer-fighting food additives." *Science News*, February 15, 1992, pp. 104–106. These additives are both natural and synthetic. They include BHT and BHA, conjugated linoleic acid (CLA, which is similar to normal linoleic acid), tannic acid, myristicin (from parsley oil), quercetin (from onions and garlic), ellagic acid (from many fruits), green tea, and soybeans.

Dawes, Lisa. "There's wood in your ice cream." *Hippocrates*, July/August 1989, pp. 18–20. Cellulose finds many food applications, not only as fiber, but in modified forms as thickeners and to improve shelf life.

Biotechnology

Wrubel, Roger. "The promise and problems of herbicide-resistant crops." *Technology Review*, May/June 1994, pp. 56–61. Herbicides do not affect these bioengineered crops. Thus, these plants could allow better weed control and enhanced yields.

Adler, T. "Tomato biotechnology heads for the market." *Science News*, May 28, 1994, p. 342. A new FDA-approved tomato that softens slowly.

Beck, Charles I., and Ulrich, Thomas. "Biotechnology in the food industry." *Bio/Technology*, vol 11 (August 1993), pp. 895–902. This article is mostly a table of foodstuffs that biotechnology is improving, including apples, corn, and potatoes. Easy and valuable reading.

Fox, Michael W. 1992. *Superpigs and wondercorn*. New York: Lyons & Burford. This readable book presents a fairly negative view of biotechnology. While acknowledging that biotechnology may benefit the medical field and a few other areas, the author questions the way this science operates.

Good, Mary L. 1988. *Biotechnology and materials science*. Washington, D.C.: American Chemical Society. A good biotechnology survey; well illustrated and moderately easy reading. Presents a positive view of biotechnology.

PROBLEMS AND QUESTIONS

1. Outline the basic steps in metabolism (digestion). Write simplified chemical equations that describe the breakdown of the major kinds of food materials.

2. What is the main chemical reaction in the digestive process?

3. What type of enzyme catalyzes the hydrolysis of proteins?

4. What are the minimum daily requirements for proteins and lipids?

5. What foods are the best protein sources?

6. How much protein would a person weighing 82 kg (180 lbs.) need per day?

7. What are the relative energy contents of proteins, polysaccharides, and lipids?

8. What is the minimum recommended amount of dietary fiber content, in grams, and what chemical groups supply this fiber?

9. How many grams of polysaccharides should a 100-kg (220-lb) person eat each day?

10. Why can't humans digest cellulose?

11. What effect does a lactase deficiency have on the human body?

12. What is the recommended upper limit for total lipids in the diet of a 68-kg (150-lb.) person?

13. What types of enzymes promote the hydrolysis of lipids?

14. What are unsaturated lipids, and how does saturation relate to the iodine number?

15. Which are more beneficial in the human diet—fats or oils?

16. What is iodine number? What is its significance in nutrition?

17. What general differences exist between the iodine numbers of fats and those of most oils?

18. What would the iodine number be for an equal-weight mixture of stearic, oleic, and linoleic acids?

19. What would the iodine number be for oil formed from equal weights of oleic and linoleic acid?

20. Are human fats composed mainly of unsaturated or saturated lipids?

21. What are vitamins?

22. List the lipid- and water-soluble vitamins.

23. Do water- or lipid-soluble vitamins persist longer in the body?

24. Which vitamins contain a free carboxylic acid group?

25. Which vitamins contain a hydroyxl (or phenol) group?

26. What disorder led to the discovery of vitamin A?

27. What disease led to the discovery of vitamin B_1?

28. What disease led to the discovery of vitamin C?

29. Who determined the structure of vitamin C?

30. What disease led to the discovery of vitamin D?

31. Which vitamin deficiency can lead to sterility?

32. Which vitamin deficiency can result in poor blood clotting?

33. What disease led to the discovery of niacin?

34. What is an RDA?

35. If four times the RDA for vitamin E were consumed, how long would it require before the body contained only the RDA level?

36. Which effects are usually more severe—those caused by an excess or a deficiency of a vitamin?

37. According to the 1985 Nationwide Food Consumption Survey, do the poorer members of our society have vitamin deficiencies or an excess supply?

38. What is the ratio between the upper limit and RDA values for each of the following vitamins: A, B_1, C, D, niacin, biotin, and folic acid?

39. What are some natural sweeteners?

40. How do we determine how safe food additives are?

41. Which artificial sweeteners are currently approved for use in the United States?

42. Aside from being used as a sweetener, what other uses does sucrose have?

43. List three reasons why people want to use artificial sweeteners.

44. What are the advantages and disadvantages of aspartame, saccharin, and Acesulfame K?

45. Why were cyclamates removed from the U.S. market? Is their use permitted in any other country?

46. What kinds of chemicals are added to foods as flavoring agents?

47. What is the GRAS listing?

48. Why are additional colorants sometimes added to foods?

49. What antimicrobial agents are commonly added to foods?

50. What kinds of compounds are added to foods as antioxidants?

51. Why does the density of bread decrease during baking?

52. What are the primary nutrients plants need? From what sources do plants obtain these nutrients?

53. What is the Haber process? What is it used for?

54. Who developed the first superphosphate fertilizer?

55. What do the numbers 20-5-10 represent on a fertilizer package?

56. What are the symbols for the secondary plant nutrients?

57. Why do plants require magnesium?

58. In the United States, what is the annual value of the crops lost to pest, weeds, and diseases?

59. What is the difference between a broad-spectrum and a narrow-spectrum insecticide?

60. What was the first major, broad-spectrum insecticide? Who discovered it?

61. What are some common organophosphorus insecticides?

62. Are the carbamate insecticides more or less persistent in the environment?

63. What alternative methods of insect control exist?

64. What are pheromones?

65. What does a juvenile growth hormone do?

66. List some examples of the classes of compounds used as herbicides.

67. In general, which are more toxic, insecticides or herbicides?

68. What is the most common broad-spectrum herbicide?

69. Why was DDT banned, and under what conditions can it still be used?

70. Which was the first major broad-spectrum herbicide? How is it related to dioxin?

71. What were the main goals and methods of the Green Revolution?

72. List some recent advances in biotechnology.

CRITICAL THINKING PROBLEMS

1. Can our food possibly supply all the vitamins we need in today's stress-filled world? In other words, are vitamin supplements useful? Read at least three of the following articles and make your own assessment: Steven Findlay, "Squaring Off Over Vitamins," *U.S. News & World Report*, April 10, 1989, pp. 62–64; Donald S. Stroetzel, "Do You Need Extra Vitamins?" *Reader's Digest*, April 1986, pp. 77–80; Anastasia Toufexis, "The New Scoop on Vitamins," *Time*, April 6, 1992, pp. 54–59; Joanne Silberner, "Health Guide: Building the Best Defense," *U.S. News & World Report*, May 4, 1992, pp. 62–70; Trevor Smith, "Vitamins: Worthy Food Supplements or Gigantic Rip-Offs?" *Today's Chemist at Work*, May 1994, pp. 43–44, 62.

2. No vitamin gets more press and is ascribed more miracles than vitamin C, but are these claims valid? Make your own decision after reading the following three articles: P. Long, "The Power of Vitamin C," *Health*, October 1992, pp. 67–72; J. Raloff, "Vitamin C Protects Blood From Radicals," *Science News*, August 26, 1989, p. 133; S. Findlay, "Squaring Off Over Vitamins," *U.S. News & World Report*, April 10, 1989, pp. 62–64.

3. Why is vitamin A necessary for good eyesight. Could a deficiency lead to blindness?

HINT Review the section on vision in chapter 14. Are there any similarities between the chemicals of vision and vitamin A?

4. Popular magazines often report studies claiming marvelous, almost miraculous, health effects from certain foods and food supplements. Do the studies seem scientifically valid? Read at least three of the following articles to frame your answers: Peter Jaret, "Bet on Broccoli," *In Health*, September/October 1991, pp. 58–62; Carol and Malcolm McConnell, "Ancient Secrets of Modern Nutrition." *Reader's Digest*, September 1986, pp. 37–44; Tom Hager, "Take Fish to Heart," *Reader's Digest*, August 1985, pp. 127–130; Anthony Burgess, "The Glory of Garlic," *Reader's Digest*, 1984, pp. 185–190; Richard McCourt, "Some Like It Hot," *Discover*, August 1991, pp. 48–52; Walter C. Willett, "Diet and Health: What Should We Eat?" *Science*, vol. 264 (April 22, 1994), pp. 532–537; Trevor Smith, "Eating Your Way to Health," *Today's Chemist at Work*, March 1994, pp. 54–57; Mary K. Schmidl, "Food Products for Medical Purposes," *Trends in Food Science & Technology*, June 1993, pp. 163–168.

5. Some animals eat meat, some eat plants, and others consume both. What would happen if all animals (including humans) were to convert to completely plant diets?

6. Read at least two of the following articles on fake fats: R. J. Jandacek, "The Development of Olestra, a Noncaloric Substitute for Dietary Fat," *J. Chem. Ed.*, vol. 68., no. 6 (June 1991), pp. 476-479; David Rotman, "National Starch's Fake Fats," *Chemical Week*, January 15, 1992, p. 15; Tina Adler, "Designer Fats," *Science News*, May 7, 1994, pp. 296–297; H. Maelor Davies, and Frank. J. Flider, "Designer Oils," *Chemtech*, April 1994, pp. 33–37. Fake fats are gradually catching on in the marketplace because they remove the calorie punch of fats. Do you think such materials are safe? Would you use them?

7. Vegetarianism has never really caught on in the United States. What are some of the virtues and problems in not eating meat? The article by Hannah Holmes entitled "Eating Low on the Food Chain," *Garbage*, January/February 1992, pp. 32–37, will supplement the information this chapter contains.

8. This chapter discussed several reasons for using artificial sweeteners. How many of these reasons are really valid when the overall health and diet of the American public is considered?

9. This chapter mentioned the GRAS list. What problems do you think might crop up with such a listing of chemicals? How might the older chemicals on this list be tested, if it were deemed necessary? What would make the testing process extremely expensive?

10. The queen honeybee secretes a chemical that prevents ovary development in other female bees, making the queen the only bee able to produce offspring. What potential problems exist with such an action? Is this chemical more likely to belong to the hormone, pheromone, or prostaglandin class?

11. Biotechnology has developed new types of plants such as frost-resistant tomatoes and fruits that do not quickly rot. Read T. Adler, "Tomato Biotechnology Heads for Market," *Science News*, May 28, 1994, p. 342. How would you evaluate the potential of plant biotechnology to produce better food sources for humans? Could this new branch of science help alleviate famine and food shortages? Could it improve nutrition?

12. Does the Haber process use a renewable source?

HINT Where do the nitrogen and hydrogen come from?

13. Biotechnology has developed some vegetables that are resistant to naturally occurring insect pests. Do you think this is a good idea? What problems might this present?

14. Biotechnology has developed herbicide-resistant plants. For more details, read Roger Wrubel, "The Promise and Problems of Herbicide-Resistant Crops," *Technology Review*, May/June 1994, pp. 56–61. Why is this innovation expected to *increase* the use of herbicides?

15. The broad-spectrum herbicide Agent Orange has been accused of causing birth defects in the children of Vietnam War veterans. In what ways would the chemicals have to affect the human body to cause such defects? How might such claims be proved or disproved?

HINT Review box 12.1 and the section on birth control in chapter 14.

16. You have just acquired the advertising account for a food company marketing a "super" mayonnaise made from oils with high iodine numbers. Devise an advertising campaign that will explain this in such a way that consumers will understand and want to buy this new wonder product.

17. As the newly appointed Special Science Attache to the Secretary of Agriculture, your first assignment is to perform a risk/benefit analysis of a pesticide. (Pick any one you wish.) You can make up any cost figures you desire (within reason), but you must assess the costs of using the pesticide against the crop loss if the chemical is not used. You must also weigh the potential pollution hazards and human health risks against potential food shortages if the pesticide is not used.

18. You are now a Special Science Attache for the Director of the Food and Drug Administration (FDA). Your assignment is to perform a risk/benefit analysis for a new red food coloring. Assume the cost is \$25 per pound, but assume the amount used in any food is 0.001 percent, based on the total weight of the food. What kinds of information will you need to make your decision, and how long will it take to obtain this information? Project the risks and benefits assuming three different levels of toxicity for the new red dye. What effect might it have on your decision if the red dye causes cancer in some laboratory mice when fed to them at high amounts per body weight? Is there any legal basis for making any of these decisions, or must they be made without this kind of guideline?

19. *Chemophobia* is the fear of chemistry or anything chemical. As this book has shown, chemistry is intimately involved in all parts of life, including in everything we use. Many times we may hear that a product is "completely natural" and "contains no chemicals." How can we evaluate these claims from a scientific perspective? What part might chemophobia play in the prevalence of such claims?

CHAPTER 16

COSMETIC CHEMISTRY—THE SCIENCE OF BEAUTY

OUTLINE

KEY IDEAS

1. The History of the Cosmetics Industry
2. The Types of Chemicals Used in Cosmetics
3. Which Chemicals Are Used in Different Kinds of Skin Care Products, Including Sunscreen Formulations
4. The Nature of Retin A and its Role in Cosmetics
5. The Chemicals Used in Decorative Cosmetics, Such as Lipstick, Nail Polish, and Eye Makeup
6. How Chemicals Control Body Odor
7. Which Chemicals Are Used on Hair
8. The Chemistry of the Three Classes of Hair Dyes
9. The Chemistry of the Permanent Wave
10. A Chemical Method to Treat Male Pattern Baldness

"Beauty is eternity gazing at itself in a mirror."
KAHLIL GIBRAN (1883–1931)

False color scanning electron micrograph of human hair "split ends."

Humans have been preoccupied with beauty for countless centuries. Ancient literature (for example, records left by the Egyptians) is full of assorted cosmetic recipes. However, only in the last few centuries have chemists achieved success in developing safe, effective, cosmetics for widespread use (box 16.1).

COSMETICS—LOOKING GOOD, SMELLING GOOD

What, exactly, are cosmetics? The United States Food, Drug, and Cosmetic Act of 1938 defines **cosmetics** as "articles intended to be rubbed, poured, sprinkled, or sprayed on, introduced into, or otherwise applied to the human body or any part thereof, for cleansing, beautifying, promoting attractiveness, or altering the appearance." Cosmetics then, are chemical formulations used for beautifying the external human body. Some common cosmetics (such as antiperspirants) are also classed as drugs because they effect changes in body functions. In fact, for this reason, sunscreening formulations are now being reevaluated to determine whether they should also be classed as drugs.

Cosmetic applications may be divided into the groups summarized in table 16.1. We will discuss each of these groups in the remainder of this chapter.

TABLE 16.1 The main types of cosmetics and some examples of each.

TYPE	EXAMPLES
Skin care	emollients, creams, suntan lotion
Decoration	lipstick, face powder, eye makeup, nail polish
Odor control	deodorants, perfumes, aftershave lotions, colognes
Hair applications	permanents, colorants, bleaches, sprays, depilatories

SKIN CARE—FOR THAT SOFTER, SMOOTHER FEELING

The skin (figure 16.1) is the largest organ of the body. It serves many important functions, including temperature regulation, fluid maintenance, and disease control. Cosmetics are used on the

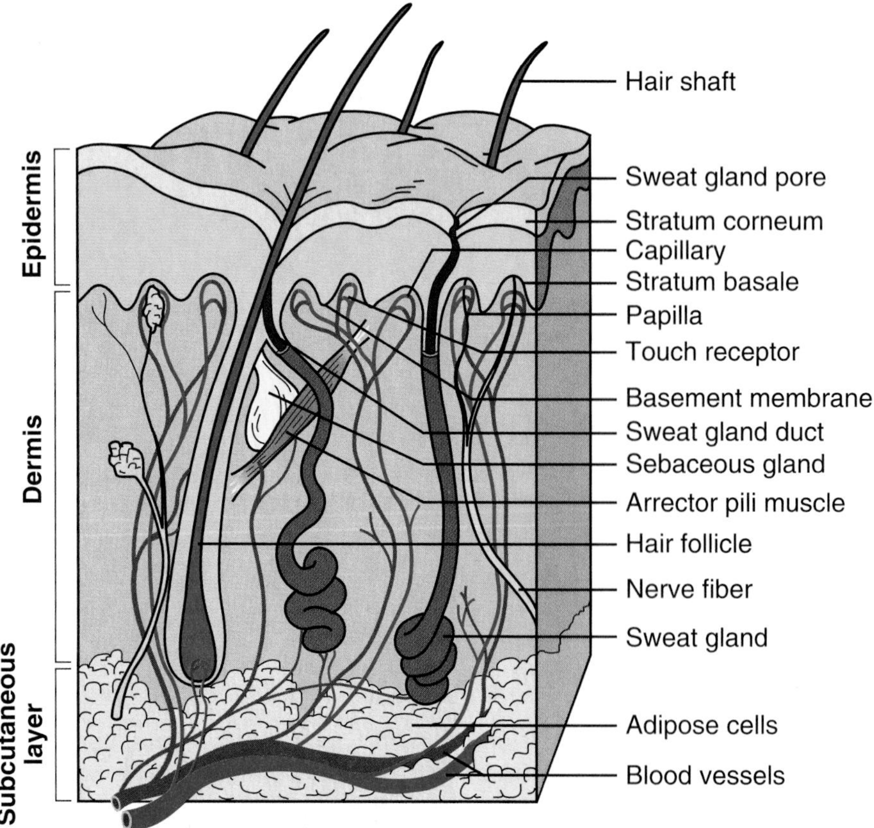

Figure 16.1 The structure of the skin.

BOX 16.1

THE RISE OF THE COSMETICS INDUSTRY

The birth of cosmetics dates back to the dawn of civilization. Archaeologists estimate that cosmetics existed as long ago as 6,000 B.C.; beauty shops and perfumeries existed in ancient Egypt by about 4,000 B.C. Eye makeup was used not only in Egypt, it was also popular among the ancient Hebrews and Phoenicians (figure 1). Rouge was in common use among the Greeks by 4,000 B.C.—in fact, soldiers going forth into battle wore it regularly.

These ancient cosmetics were seldom made from safe materials. Many were derived from arsenic salts or poisonous plants. No one knows how many users of these substances died from poisoning.

As the centuries passed, not all societies embraced the use of cosmetics. At times, rejection of cosmetics was based on moral or religious grounds, but the threat of death by poisoning may well have swayed some from using these products. Not too long ago, cosmetics were considered respectable only for those in the theater. That notion has changed considerably, propelled in part by the rise of the modern cosmetic industry and safe cosmetics.

Much of this growth occurred in France and spread to other countries.

Figure 1 Eye makeup in ancient times.

Industry leaders included names still easily spotted by strolling the aisles of a cosmetics department: Coty, Rubenstein, Arden, and Revlon. Naturally, these companies employed chemists to develop safer materials with qualities customers sought.

The cosmetic industry is considered relatively recession-proof. In poor economic times, people may not purchase cars, but they still buy and use cosmetics. From 1986 to 1993, very little overall change occurred in the total U.S. cosmetics market, which varied between 18 and 24 billion dollars annually. U.S. sales of cosmetics and toiletries in 1987 were $24 billion, and additional sales in Western Europe ($19 billion), Japan ($14 billion), and the rest of the world ($12 to $15 billion) brought the grand total to $69 to 72 billion. The chemicals used to manufacture cosmetics cost an estimated $7 billion, or roughly 10 percent of total sales. (Advertisement and marketing costs consume the lion's share of this fortune.)

outer skin layer, or epidermis, which consists of the corneal layer (mostly dry, dead skin cells) and the living cells directly beneath. The skin consists of proteins and other natural polymers, with about 10 percent moisture. Skin treatments try to either soften the corneal layer or add more moisture. Nearly 20 billion dollars are spent annually on skin care.

The most common skin softening treatment merely coats the skin with mineral oil or **petroleum jelly** (a semisolid alkane with 20 or more carbons). The March 1992 issue of the *Journal of the American Academy of Dermatology* reports that petroleum-jelly-based cosmetics are effective in assisting the repair of damaged skin. Vaseline is the best known brand name, but generic petroleum jellies are just as effective. This effectiveness is endorsed to the tune of $3.04 billion in annual sales.

The first skin softeners and moisturizers were developed in the dry, arid region of the Near East around 3,000 B.C. Presumably, their main goal was to prevent the skin from becoming too dry. Most of these concoctions were based on plant oils and even contained some perfume. Galen, a second-century physician, used similar mixtures on his patients. Cosmetic skin care formulations often contain colorants or perfumes for esthetic reasons, but the main ingredients are shown in figure 16.2

Emollients are oil- or fat-based compounds that also soften skin, but not all work well. The most effective preparations contain lanolin, cetyl alcohol, or cocoa butter. Lanolin is an ester found in sheep's wool. It is derived from cholesterol (an alcohol with the formula $C_{27}H_{45}OH$) and fatty acids such as stearic acid ($C_{17}H_{35}COOH$). Impure grades of cetyl alcohol ($C_{16}H_{33}OH$) are useful in emollients because they melt near body temperatures (37°C). Cetyl alcohol was originally obtained from sperm whales, but it is now made by the reduction of palmitic acid, a plant source. Cocoa butter, another emollient, is a mixture of esters obtained from cacao beans. It also melts near body temperature. Other materials included in emollients are beeswax and various

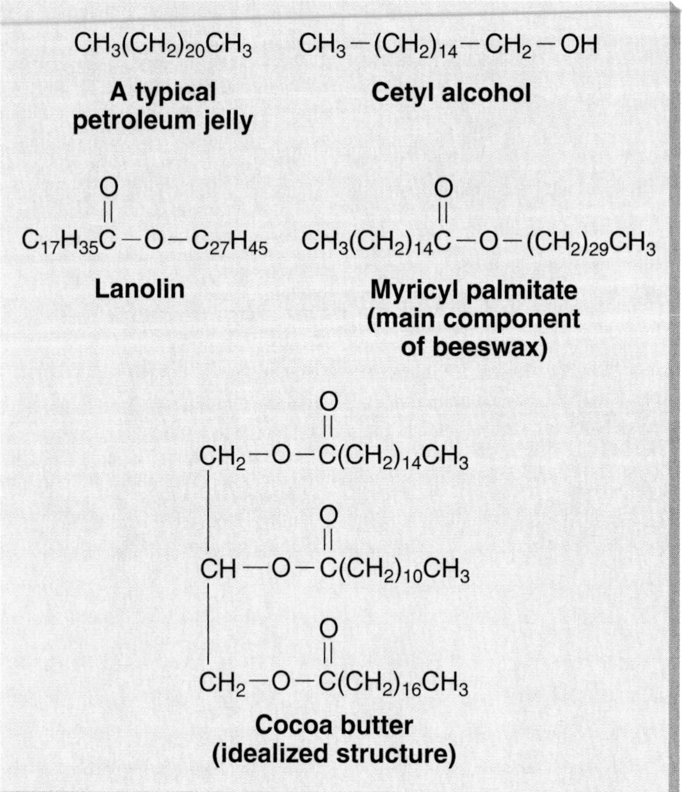

Figure 16.2 Some chemicals used in skin lotions and ointments.

plant waxes, which are used primarily to adjust the viscosity or stiffness of the formulation.

Skin creams are oil-in-water emulsions. Their main purposes are to soften and moisten the skin at the same time. Creams always contain soap or another surfactant to stabilize the emulsion. Other ingredients found in creams include vitamins (such as vitamin E) and herbs or plant extracts (such as aloe vera), but the effectiveness of such additives is debatable. Nevertheless, the sales for the vitamin or herbal creams increased greatly during the 1990s due to the consumer's desire for "natural" products.

Cold creams are used mainly to clean the skin and remove cosmetics that are not water-soluble. Table 16.2 shows a typical cold cream formulation. **Vanishing creams** are suspensions of stearic acid in water that are stabilized by a soap, such as potassium stearate. They seem to vanish when rubbed on the skin, but they actually spread out in a thin layer over the surface. Vanishing creams have some emollient properties.

TABLE 16.2 A typical cold cream formulation.

COMPONENT	PERCENT
Mineral or vegetable oil	35–60
Beeswax (or other wax)	10–15
Lanolin	5–15
Water (perfumed)	30–50

PRACTICE PROBLEM 16.1

Could petroleum jelly be added into an emollient formulation?

ANSWER Yes. Emollients are oil or fat-based formulations, and petroleum jelly readily blends into such a mixture. This could reduce the cost since petroleum jelly is less expensive than many ingredients in emollients.

PRACTICE PROBLEM 16.2

Could petroleum jelly be added into a cold cream or other cream formulations?

ANSWER Yes; since these creams are oil-in-water emulsions, petroleum jelly could be incorporated. In fact, table 16.2 shows a formulation that contains 35 to 60 percent mineral oil. However, the higher viscosity of petroleum jelly limits the amount that could be blended into a cream without making the mixture too stiff to spread on the skin.

SUNTAN LOTIONS—WARRIORS IN THE BATTLE AGAINST SKIN CANCER

Suntan lotions come in dozens of forms, including creams, oils, and lotions. Traditionally, people used these formulations to promote tanning rather than **prevent** it. Contrary to popular belief, however, suntan lotions do not promote tanning. A few suntan preparations, such as Coppertone QT, contain dihydroxyacetone (DHA), a colorless chemical that reacts with the skin to form a yellow-brown color that can last for a week or so. However, DHA provides no protection against sunburn. Suntan lotions have always contained various oils or lanolin to soften the skin both during and after sun exposure in an emollientlike action. Most suntan lotions now contain chemical **sunscreens** to block out ultraviolet (UV) light and help prevent skin cancer (box 16.2).

The most common sunscreen chemicals are **para-aminobenzoic acid (PABA),** PABA esters, benzophenones, cinnamates, salicylates, and anthranilates (figure 16.3). Chemists have used PABA and its esters as sunscreens for many years because they

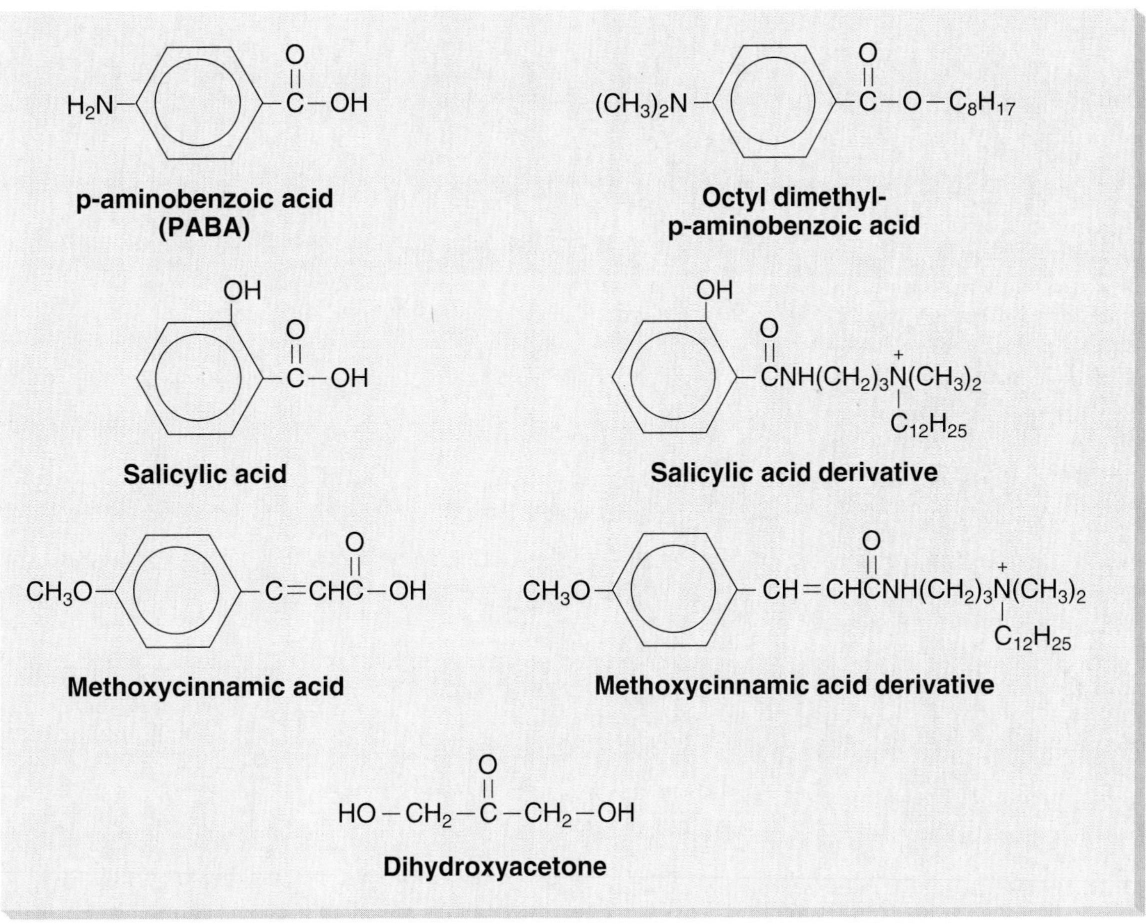

Figure 16.3 Some chemicals used as sunscreens.

effectively screen out ultraviolet (UV) rays. However, the current trend is away from PABA esters due to potential toxicity and allergic reactions in some people. This toxicity seems surprising since PABA is used naturally in the human body for folic acid synthesis. The favored chemical in sunscreens now seems to be a methoxycinnamate derivative, but cosmetic companies are also examining benzophenones and other cinnamates.

New sunscreens should appear on the market before the year 2000, possibly including one containing **melanin,** the natural polymer that imparts color to skin and hair. This seems a logical choice since the skin's melanin level increases after exposure to the sun's rays to prevent further absorption of UV light. Natural melanin is a dark brown-black material, but sunscreens use a light tan version that functions just like unmodified melanin.

Most topical sunscreens absorb the medium UV light with wavelengths between 290 and 320 nm (nm = nanometer = 10^{-9} m), which is called the UV-B region and is the major cause of sunburn. However, most photosensitive reactions occur in the 320 to 400 nm range, which is known as the UV-A region; only the benzophenones and anthranilates are effective in blocking rays in this range. Penetration below the surface of the skin is greater for UV-A than for UV-B. There is some uncertainty about which range actually causes **melanoma,** or skin cancer, so the prudent choice is to protect in both ranges. (Melanin absorbs UV in both ranges.) The ideal sunscreen may contain more than one active agent. For example, some products may contain both UV-A and UV-B blockers plus additional skin care compounds (see box 16.3 for a discussion of one called Retin A.)

UV protection is a separate concern from the **Sun Protection Factor (SPF)** given on most products. The SPF is the ratio of time required to tan or burn with the sunscreen compared to the time required without it. An SPF of 4 therefore means you must be exposed 4 times as long to tan or burn as much as you would without the sunscreen. An SPF of 2 is considered minimal and a value of 15 or above is considered good protection. Some brands have SPF values as high as 50, but many scientists do not believe ratings beyond 15 are realistic. (In other words, you get nothing extra for your money by buying a 50 SPF sunscreen as compared to a 15 SPF sunscreen.) In early 1993, the FDA actually advised against these higher SPF formulations, on the basis that the higher values triggered more allergic reactions.

Another thing to consider in a sunscreen is **substantivity,** or how long the sunscreen lasts under the stress of exercise, sweating, and swimming. PABA esters are generally the most substantive,

BOX 16.2

SUNTANNING AND MELANOMA

Most modern suntan lotions and creams contain a screening agent, a chemical that absorbs the sun's ultraviolet rays. Such suntan products, called sunscreens, are designed to reduce the risk of melanoma. **Melanoma** (figure 1) is a virulent form of skin cancer that appears to be linked to severe sunburn, especially for people with light complexion, blue eyes, naturally blond or red hair, and freckles.

The occurrence of melanoma, as figure 2 indicates, is increasing in terms of actual total deaths (upper line) as well as deaths per 100,000 people (lower line). (Use caution in interpreting figure 2; although these lines appear linear, they actually aren't. If they were, no melanoma cases would have occurred prior to 1870, but the history of this cancer goes back to prehistoric times.

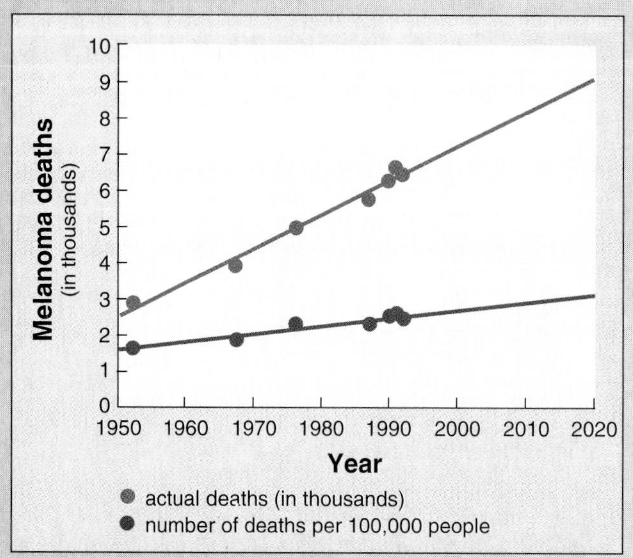

Figure 2 Deaths due to melanoma. The upper line shows the actual number of deaths in thousands of people and the bottom line shows the number of deaths per 100,000 people.

Melanoma is 100 percent curable if diagnosed early, but it can spread to other organs if neglected. When it spreads, the cure rate drops to only 15 to 20 percent. Prevention is the best way to deal with melanoma; one means of prevention is to use sunscreens, but another is to avoid overexposure to the sun's rays. Approximately 600,000 new cases of nonmelanoma skin cancer are reported each year. In 1992, 32,000 new cases of melanoma were reported, with 6,400 deaths resulting from skin cancer. (By comparison, lung cancer, closely linked with smoking, resulted in 146,000 of the 520,000 total cancer deaths in 1992.) Melanoma is dangerous, but it accounted for less than 1.2 percent of all cancer deaths in 1992. Some believe the thinning of the ozone layer may lead to even more melanoma deaths. Though most forms of skin cancer are not fatal, this deadly cancer is one to avoid—by avoiding sun exposure and by using sunscreens when you must be out in the sun.

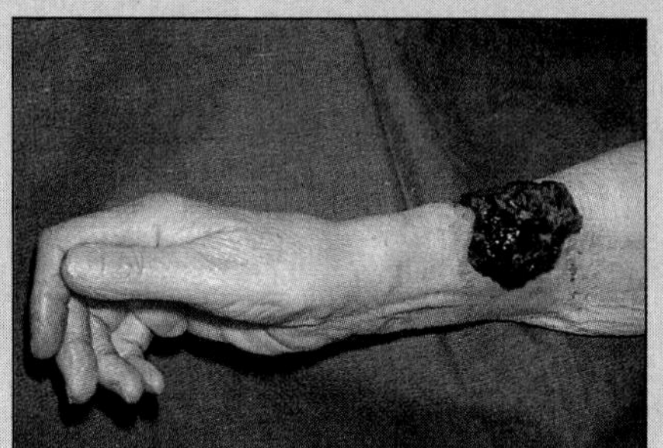

Figure 1 A section of skin showing a malignant melanoma.

and octyl-dimethyl-PABA is the most commonly used ester. Formulation into an oily lotion or cream increases the suncreen's staying power.

Sunscreens are not without problems. As we previously mentioned, PABA causes allergic reactions in some people; it can also stain clothing. Benzophenones and cinnamates occasionally cause a skin rash, or cause the rash on exposure to sunlight. However, these reactions are fairly rare and are far less dangerous than overexposure to UV rays, which could lead to fatal skin cancer.

PRACTICE PROBLEM 16.3

Suppose a person found that a sunscreen allowed them to remain in the sun for eight hours with the same degree of sunburn they acquired in only one hour of exposure without it. What is the SPF value for this location?

ANSWER The sunscreen allowed 8 times as much exposure for the same degree of burning; hence, the SPF is 8.

BOX 16.3

RETIN A—A NEW WRINKLE IN WRINKLE REMOVAL?

Since humans first caught sight of their images reflected in pools of water, they seem to have been concerned with finding a "fountain of youth." Surely there must be some way to wash away the effects of aging. In 1513, Ponce de Leon explored the wilds of Florida, searching in vain for the legendary fountain. Of course, he failed. Today the multibillion dollar cosmetics industry has taken up the crusade. **Retin A** is making a splash in today's cosmetic market as a promising modern-day equivalent of this fabled fountain.

Retin A, or retinoic acid (figure 1), is a vitamin A derivative that belongs to a class of chemicals called tretinoins. It was first marketed in 1972 as an acne treatment. At one time, Retin A was available only by prescription, but you can now purchase it over-the-counter. Retin A accelerates the shedding of dead skin cells, reduces pore blockage, and decreases the oily buildup that usually causes acne. It also dilates blood vessels beneath the skin and creates new collagen, the fibrous protein that gives skin its smooth texture and elasticity. Aging and prolonged sun exposure reduce this elasticity, causing wrinkles.

A scientific study showed that when thirty Caucasian patients thirty-five to seventy years old applied Retin A to their forearms for sixteen weeks, wrinkles and roughness disappeared (*Journal of the American Medical Association*, January 22, 1988). The doctors called the results "dramatic." A four-year University of Arizona study involving 280 patients suggested that some tretinoins can treat cancers—for example, cervical cancer. Retin A's potential as a skin cancer treatment was a major news item in early 1993. Because of its versatility, this cosmetic now treads a fine line between being a beauty care product and a medical treatment.

In the meantime, Japanese cosmetic giant Shiseido has introduced a skin treatment that contains the sodium salt of **hyaluronic acid** (figure 1), a natural polysaccharide obtained from rooster combs. Shiseido claims that treating one's skin with this product for two minutes a day will give "radiant, healthy, younger-looking skin"—at a cost of fifty dollars an ounce. Liposomes, which resemble natural cells, are another potential cosmetic chemists are exploring. They claim that liposomes can blend with natural cell components, restoring fluidity and vitality to skin cells.

We probably will never find a true "fountain of youth" in rooster combs or vitamin A-derived chemicals, but chemists may be on the verge of a new era in cosmetic skin treatment. Is Retin A really an anti-aging compound? Will it prove to have unexpected side effects? Chemists, doctors, and consumers do not yet know.

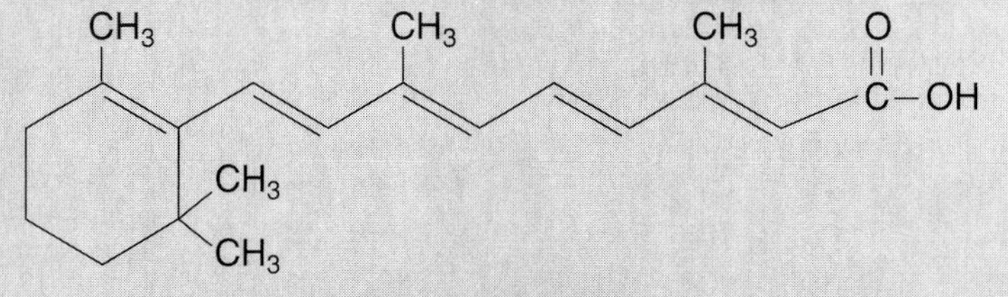

Retinoic acid (Retin A)

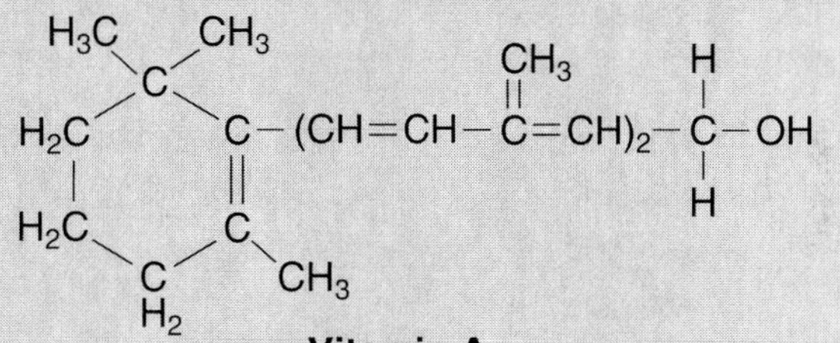

Vitamin A

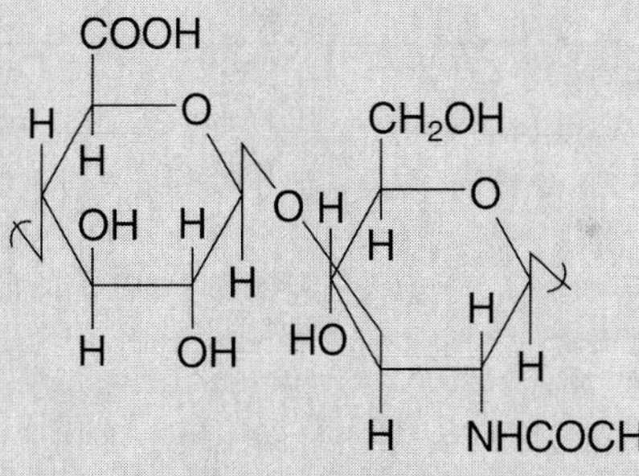

$$H_2C-C(CH_3)_2-C-(CH{=}CH-C(CH_3){=}CH)_2-CH_2-OH$$

Hyaluronic acid (a segment of a polymeric compound)

Figure 1 Some chemicals used in wrinkle-removing formulations; vitamin A is included for comparison.

PRACTICE PROBLEM 16.4

Would the addition of petroleum jelly to a sunscreen of SPF 15 improve its action?

ANSWER There would be no improvement in the SPF value, but the sunscreen might remain on the skin longer, or acquire greater substantivity, since petroleum jelly could keep the sunscreen from washing off. Hydrocarbon oils or jellies are added to some sunscreens for this purpose. Other agents used in sunscreens include vegetable oils, animal fats (including lanolin), and silicon oils. An increase in substantivity makes a sunscreen more effective and also acts as an emollient.

Figure 16.4 Applying lipstick for beauty.

DECORATIVE COSMETICS

Cosmetics are used to promote beauty, not health. Let's look into the staggering array of lipsticks, polishes, powders, eye cosmetics, and creams that people hope will improve their looks without causing harm.

Lipstick and Lip Balm

The lips are the most sensitive and exposed area of the skin; they are especially susceptible to dryness and its effects. Cosmetic companies have therefore developed **lipsticks** and **lip balms** to protect the lips from the effects of moisture loss. Table 16.3 shows a typical formulation. The ingredients are similar to those used in emollients, though lipsticks have significantly higher levels of dyes than skin creams; lip balms, which make up a much smaller market, are generally uncolored. Several other colorants are red dyes that are often similar in structure to the FD&C dyes discussed in chapter 15 (see figure 15.7).

A major problem in designing a suitable lipstick is balancing stiffness against ease of application (figure 16.4). Oils tend to increase fluidity, while waxes tend to make the lipstick stiffer. If the blend of oils and fats is too stiff, the lipstick may break during application. If the blend is too fluid, the lipstick will run. The formula must also balance ease of removal and substantivity; a lipstick should come off easily when washed, but not as easily when kissing or eating.

TABLE 16.3 A typical lipstick or lip balm recipe.

COMPONENT	PERCENT OF TOTAL COMPOSITION
Castor oil, sesame oil, fats, or mineral oils	45–55
Lanolin or cetyl alcohol	12–25
Beeswax or carnauba wax	18–33
Dyes	4–8
Perfume	trace

PRACTICE PROBLEM 16.5

Why do you think a lipstick or lip balm protects the skin?

ANSWER These cosmetic formulations coat the skin with a thin layer of an oily substance that retards evaporation of moisture from the skin. It is moisture evaporation that causes the skin to dry or chap.

Nail Polish

Although nails are designed to protect the fingers and toes (figure 16.5), the cosmetic industry views the nails as perfectly designed for aesthetic applications. The Egyptians were the first to paint nails, they were using red stain from henna by 3,000 B.C. By 600 B.C., the Chinese were using egg whites, gum arabic, and beeswax to formulate nail lacquers. Their favorite colors were red and black.

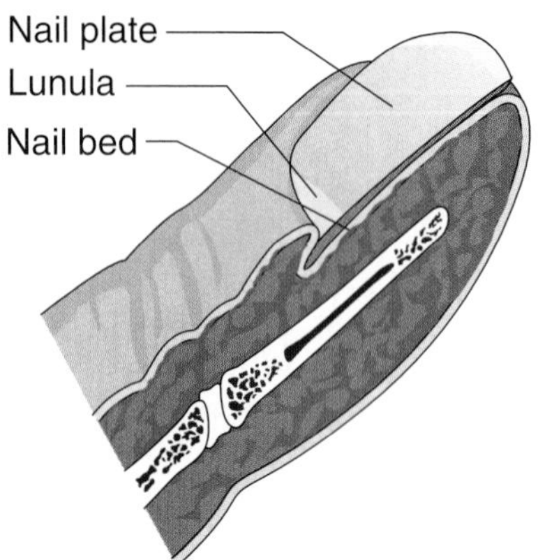

Figure 16.5 The structure of the nail.

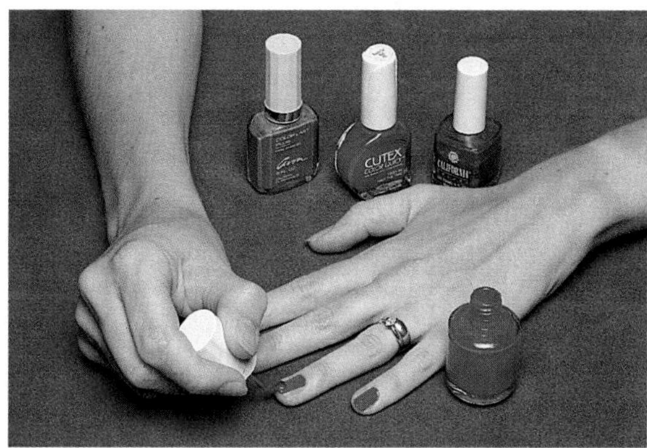

Figure 16.6 Applying nail polish to beautify the nails.

Modern **nail polish** is still a lacquer used to paint the nails, but this lacquer is made from a polymer, a plasticizer, a solvent, and a dye (figure 16.6). Table 16.4 shows a typical nail polish formula. Nitrocellulose is the most common polymer, but nylon and acrylates are also used. The plasticizer makes the final film more flexible, to prevent chipping. Mixed solvents are often used because they dissolve the polymer more effectively.

Nail polish removers are solvents that remove the lacquer film; the most common solvents are acetone and ethyl acetate, which is similar to the amyl acetate listed in table 16.4. A trace of perfume is added to mask the solvent odor, and oils are added to make the polish remover feel smooth. Ethyl acetate, though more expensive than acetone, has increased in popularity because people find its odor less objectionable.

PRACTICE PROBLEM 16.6

Does nail polish protect the fingernails? Why or why not?

ANSWER Most studies suggest the opposite. Nail polish keeps air from penetrating the nails and reaching the tissues beneath. This may reduce the health of these tissues and possibly harm the nails themselves. Nail polish is designed for decoration, not protection.

Face Powder

Face powders are used to cover blemishes or hide oily spots on the face. The main ingredients in most powders are inexpensive, inert organic compounds, such as chalk ($CaCO_3$) or talc ($3MgO{\cdot}4SiO_2{\cdot}H_2O$) —table 16.5. Some face powders contain an astringent, such as zinc oxide (ZnO), that reacts with oil secretions to make them more easily absorbed. Soap binder helps hold the powder onto the face. The amount of dye or pigment varies with the specific type of face powder. The pigment level is increased when face powders are used for blushers or rouge.

TABLE 16.4 A typical nail polish formulation.

COMPONENT	TYPICAL EXAMPLE	PERCENT OF TOTAL COMPOSITION
Polymer	nitrocellulose	12–20
Solvents	acetone	40–50
	amyl acetate	25–35
Plasticizer	butyl stearate	4–6
	rosin*	4–6
Colorants		trace
Perfumes		trace

*A fatty ester mixture

EXPERIENCE THIS

1. Visit a cosmetics counter and write down the ingredients for several creams, emollients, and lotions. Are there any similarities between these classes of cosmetics?
2. Determine which chemicals are used as sunscreening agents in several lotions and creams. How many different chemicals did you find? The location of a chemical in the listing should reflect the amount in the formulation—the first ingredient usually has the highest percentage. Do sunscreen ingredients seem to vary with the SPF?
3. Make a list of the ingredients in several different brands of lipstick. Do the brands differ much?
4. Examine at least ten different brands of nail polish and make note of what polymers and solvents they contain. Do nail polishes vary much between brands?
5. Make a list of the main ingredients in at least eight brands of face powder. Do you see more similarities or differences in these eight?

TABLE 16.5 A typical face powder formulation.

COMPONENT	EXAMPLE	PERCENT OF TOTAL COMPOSITION
Absorbent	talc	60–70
	powdered chalk	10–15
Astringent	zinc oxide	15–20
Binder	zinc stearate or other soap	5–8
Perfumes and dyes		trace

Eye Makeup

Eye makeup is produced by the cosmetic industry for purely aesthetic purposes. This class of cosmetics includes several different products, including eyebrow pencils, mascara, and eyeshade, or eye shadow. Eyebrow pencils are similar in composition to lipstick, but they contain different colorants (table 16.6). Many contain the same chemicals used by the ancient Egyptians to enhance beauty. Are all these chemicals safe? Probably not. Most of these colorants are known poisons when ingested, and most eye makeups warn the user against getting the material into the eye.

Mascara (table 16.7) is used to darken and "lengthen" eyelashes. It contains dark coloring agents suspended in a mixture of soap, fats, oils, and waxes. Mascara usually contains high levels of soap, which causes the mascara to run. The remainder is a waxy material containing large amounts of the types of pigments shown in table 16.6. Mascara is usually designed for easy soap-and-water removal, since you cannot scrub your eyelashes the same way you scrub your face or hands.

Eye shadow, also a suspension of colorants in an oil-wax-fat base, is featured in table 16.8. Note that there is little or no soap in eye shadow and that the wax mixture is softer (mostly petroleum jelly) than in mascara. The colorants are similar to those used in mascara, but the pigment level in eye shadow is significantly higher than in mascara.

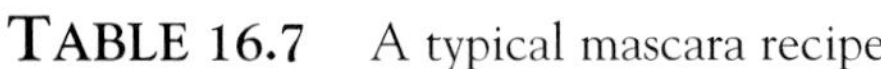

TABLE 16.7 A typical mascara recipe.

COMPONENT	PERCENT OF TOTAL COMPOSITION
Soap	50
Wax and/or paraffin	40
Lanolin or cetyl alcohol	5
Colorant	5

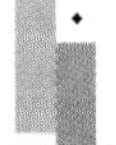

TABLE 16.8 A typical eye shadow formulation.

COMPONENT	PERCENT OF TOTAL COMPOSITION
Petroleum jelly	60
Lanolin	6
Fats, waxes (coca butter, beeswax)	10
Zinc oxide and colorants	24

PRACTICE PROBLEM 16.7

Eye shadow rarely contains soap, whereas mascara usually does. Why?

ANSWER Eye shadow is used on the skin surfaces around the eyes, while mascara is applied to the eyelashes. It is fairly easy to wash skin with soap and water or use a cold cream, but this is more difficult with eyelashes. Soap makes the mascara come off more easily, without scrubbing.

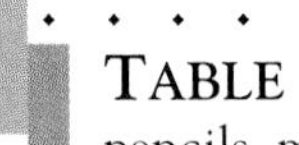

TABLE 16.6 Some of the colorants used in eyebrow pencils, mascara, and eye shadows.

COMPOUND	FORMULA	COLOR
Titanium dioxide	TiO_2	white
Bismuth oxychloride	$BiOCl$	white
Iron oxides	FeO_x	reds and yellows
Bronze powder	an alloy	reddish-yellow
Chromic oxide	Cr_2O_3	dark green
Ultramarine	organic dye	blue
Manganese salts	varied	violet
Aluminum powder	Al metal	silver

ODOR CONTROL—SMELLING THE ROSES

Perspiring is a normal body function that helps regulate body temperature and remove waste products. The more than two million sweat glands in the skin (see figure 16.1) secrete amines, low-molecular-weight fatty acids (for example, caproic acid), and sulfur compounds with foul smells. (Sweat also contains salts, such as NaCl, but these have almost no odor.) We alleviate body odor by bathing with soap or by using additional agents to mask, change, or prevent odor. Three categories of odor-controlling agents exist: (1) deodorants, (2) antiperspirants, and (3) perfumes, colognes, and aftershave lotions.

Deodorants are chemicals that attempt to mask or absorb body odor. They come in the form of powders, sticks, liquids, and creams. The absorbent powders are mainly composed of talc or calcium carbonate. Sometimes they also contain zinc peroxide to oxidize fatty acids and other odor-causing compounds into compounds with less odor. Stick deodorants are similar in formulation to lipsticks, but they contain less colorant. Deodorant creams are like cold cream in texture, but while some liquid and cream formulations also exist, powders and sticks dominate the field.

Antiperspirants are similar to deodorants, but they also add a chemical that prevents the formation of sweat. The most commonly used preventive chemical is **aluminum chlorohydrate,** $Al_2(OH)_5Cl \cdot 2H_2O$, an astringent that constricts the sweat gland pores, thereby restricting perspiration. If one were to use aluminum chlorhydrate on large portions of the body over long periods of time, this chemical could be harmful. However, it is fairly safe when used only under the arms, though a few people develop a rash from antiperspirant use.

The common ground for **perfumes, colognes,** and **aftershave lotions** is the presence of a fragrance (table 16.9). Perfumes contain the most fragrance and also contain **fixatives,** substances that prevent rapid evaporation of the aroma agent. Cosmetic fragrances are usually pleasant smelling esters, alcohols, aldehydes, or ketones (figure 16.7). The fixatives usually have unpleasant odors; civetone and muscone are cyclic ketones from the civet cat and the musk deer, respectively, and indole is obtained from feces. Ambergris (basically sperm whale vomit) and castor, obtained from certain species of beavers, are also used as fixatives. Perhaps someone once poured a perfume over one of these foul-smelling agents to reduce its odor and discovered that the fixative actually enhanced and prolonged the good odor. In any event, fixatives,

TABLE 16.9 The overall composition of perfumes, colognes, and aftershave lotions.

	PERCENTAGES WHEN THE FORMULATION IS:		
CHEMICAL	PERFUME	COLOGNE	AFTERSHAVE LOTION
Fragrance	10–25	1–3	<1
Fixative	2–5	0	0
Ethanol	70–88	77–84	59
Water	nil	15–20	40

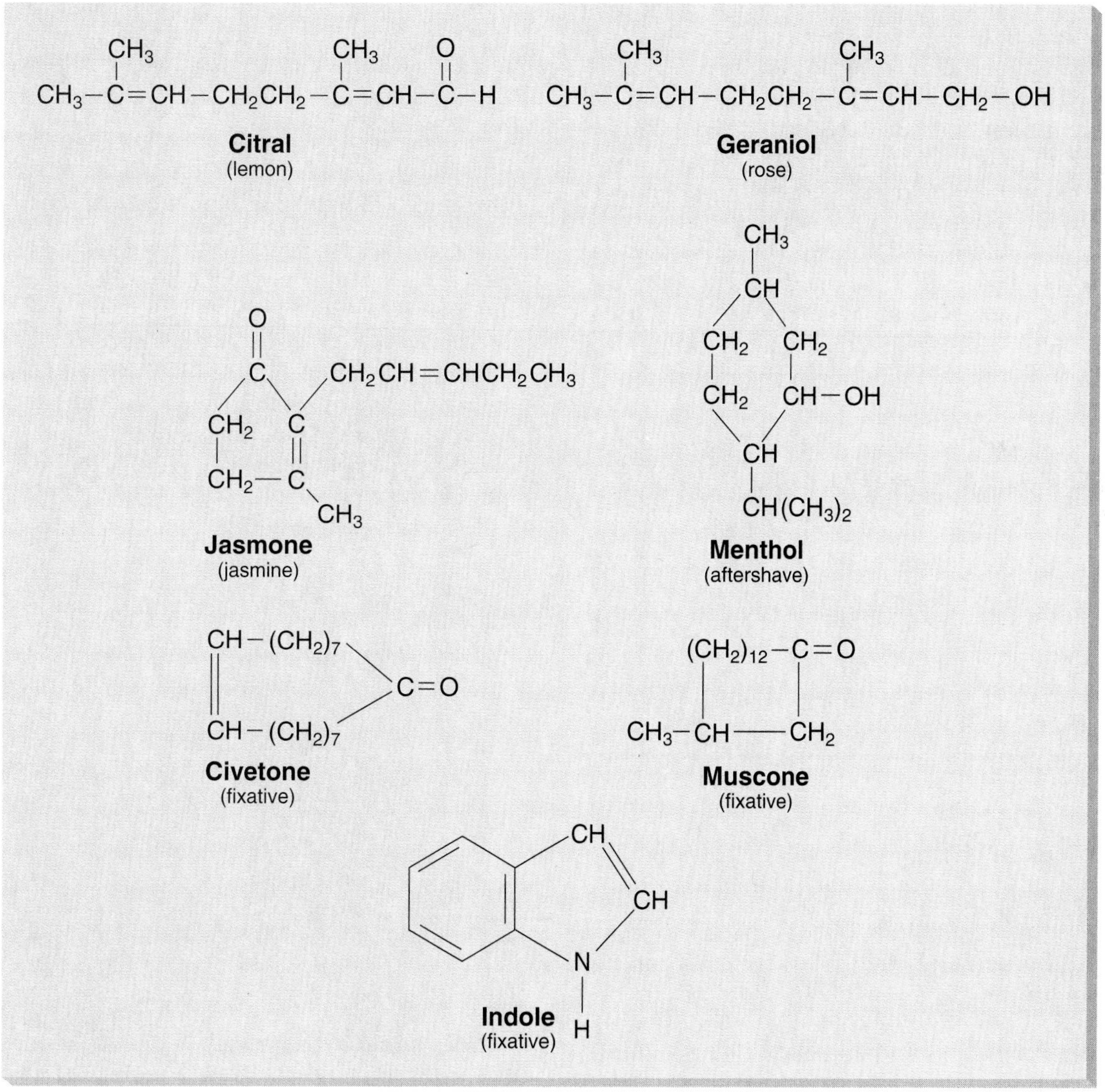

Citral
(lemon)

Geraniol
(rose)

Jasmone
(jasmine)

Menthol
(aftershave)

Civetone
(fixative)

Muscone
(fixative)

Indole
(fixative)

Figure 16.7 Some chemicals used in perfumes and deodorants. The odor, type of compound, or specific use appears beneath the chemical structure.

now synthesized in the laboratory, remain an essential part of the perfume industry.

Generally, perfumes contain several chemicals with similar odors, but each has a different volatility so that the basic odor continues long after the initial odor-causing agent evaporates. These discrete ingredients, called "notes," are rated as "top notes" (most volatile), "middle notes," and "final notes" (least volatile). Matching these odors is difficult. Some specialized instrument analysis techniques such as infrared spectroscopy or gas chromatography can help, but the final "analysis" usually takes place when an expert smells the perfume. (These individuals, called "noses," are generally well paid and well treated. If a nose gets sick and his or her sense of smell is impaired, an entire cosmetics plant could end up on hold.)

Colognes are diluted perfumes, which makes them less expensive. Aftershave lotions contain even less perfume; they are basically solvent mixtures of water and ethanol, which makes them cost even less than colognes. Menthol (a solid alcohol) is often added to aftershave to give it a cooling effect.

One major problem looming for companies that produce nail polishes and other cosmetics such as perfumes, deodorants, and antiperspirants is the potential impact of new regulations in California that limit the VOC (volatile organic compounds) to under 80 percent in 1993, under 55 percent by 1995, and 0 percent by 2000. Critics claim VOCs contribute to smog. Several other states are poised to enact similar rules, and since many cosmetic formulations cannot meet these stipulations, cosmetic manufacturers must redesign their formulations. Whether these products will remain effective after redesign is uncertain.

PRACTICE PROBLEM 16.8

What is the major difference between a deodorant and an antiperspirant? Could these functions be combined in one cosmetic?

ANSWER Antiperspirants prevent sweat formation, while deodorants mask sweat odor. Many cosmetics on the market combine these two functions.

HAIR CHEMISTRY

Like skin, hair is composed mostly of protein molecules, which are α-amino acid polymers (chapter 14). The major hair protein is keratin, which is composed of 16 to 18 percent cystine, an α-amino acid with an —SH group. Skin keratin contains only 2.3 to 3.8 percent cystine. Cystine content is important because cystine forms the **disulfide (S—S) crosslinks** between different protein chains in the hair, giving it durability and shape. Skin is more flexible or elastic than hair because it has fewer of these crosslinks.

We use many different kinds of chemicals on hair, including hair sprays, bleaches, hair removers, hair dyes, permanents, and hair straighteners.

Hair Sprays

Hair sprays are simple solutions of about 4 percent polymer (resin) in a volatile solvent. The ingredients in hair sprays include the polymer, a plasticizer, and silicone oils dissolved in the solvent. The polymer helps hold the hair in place; the polymer most commonly used for this purpose is poly(vinyl pyrrolidone) or its copolymers (figure 16.8). Plasticizers make the plastic softer and more pliable, and silicone oils impart sheen.

Hair sprays are often sold in aerosol cans. Years ago, the propellant in these aerosols was a chlorofluorocarbon (CFC), but the United States banned CFCs in 1978 after chemists discovered these compounds may contribute to the destruction of the ozone layer (see chapter 19). Manufacturers came up with two solutions: in some cases, they changed to a pump-style bottle, and in others they continued to use aerosols but with non-CFC formulations that do not harm the ozone layer.

Bleaches—What Your Hairdresser Doesn't Know

Hair contains two major and two minor pigments embedded within the protein chains of the hair fibers. The major pigments are **melanin,** a brown-black polymeric pigment also responsible for skin coloration, and **phaeomelanin,** an iron-containing red pigment derived from melanin. The two minor pigments are the yellow pigment **oxymelanin** and the brown pigment **brown melanin.** All of these natural pigments are polymeric, but their exact structures are unknown. The same pigments also occur in the skin and eyes.

The relative proportions of these pigments determine an individual's hair color. For example, high melanin content gives an individual black or dark brown hair, while low melanin and high phaeomelanin content results in red or blond hair color, depending on the proportions. Oxymelanin is also present in blond hair;

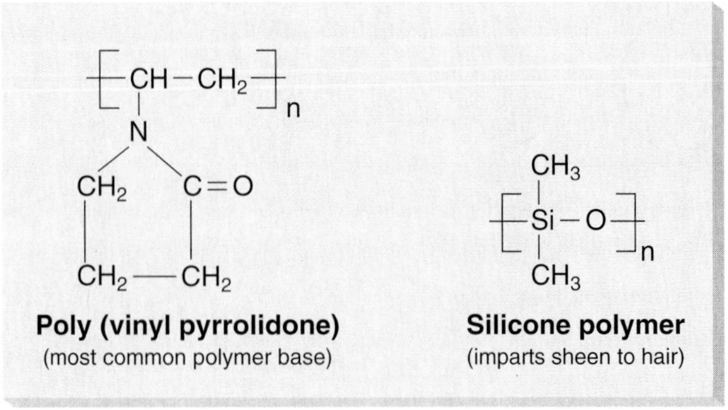

Figure 16.8 Some polymers used in hair sprays.

this pigment is formed, to some extent, by the oxidation of melanin during bleaching. Brown hair color comes from various pigment combinations but is mainly attributable to brown melanin. White or gray hair contains very few or no pigments. Albinos also lack pigment in the hair, skin, and eyes.

Bleaching is an oxidation-reduction reaction that uses hydrogen peroxide (H_2O_2) to convert the hair pigments to colorless forms. These forms are lower-molecular-weight melanin degradation products. Hair bleaches usually contain ammonia to adjust the pH above 9, which makes the oxidation more efficient.

Along with removing color, bleaching weakens the hair and makes it more brittle because the peroxide also reacts with the hair proteins and reduces their molecular weight. Other bleaching side reactions crosslink the smaller protein fragments, making them more brittle and weaker. Animal protein conditioners are added to soften hair after bleaching.

The following equations summarize the two main reactions in the bleaching process—degrading melanin and reducing the protein molecular weight. In the second equation, subscripts m + p = n.

$$\text{Melanin} + H_2O_2\ (\text{pH} > 9) \longrightarrow \text{Degraded Melanin}$$

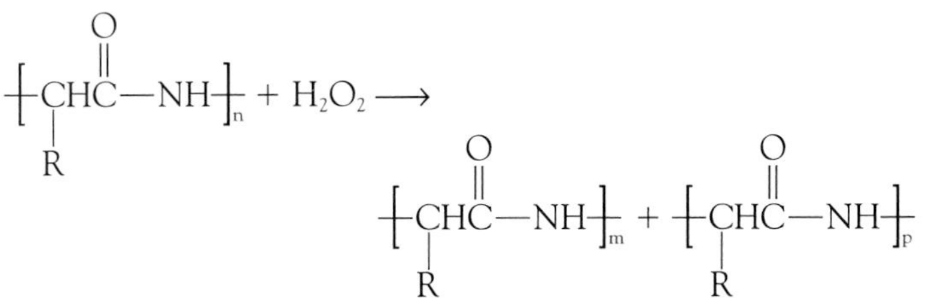

$$\left[\begin{array}{c}\text{O}\\ \|\\ \text{CHC—NH}\\ |\\ \text{R}\end{array}\right]_n + H_2O_2 \longrightarrow \left[\begin{array}{c}\text{O}\\ \|\\ \text{CHC—NH}\\ |\\ \text{R}\end{array}\right]_m + \left[\begin{array}{c}\text{O}\\ \|\\ \text{CHC—NH}\\ |\\ \text{R}\end{array}\right]_p$$

Hairdressers can use other bleaching chemicals in place of H_2O_2. Alternatives include the more expensive perborates (for example, $NaBO_3 \cdot 4H_2O$), which release H_2O_2 by reacting with water, and chlorine-based bleaches, which are less commonly used because they may discolor the hair and leave it with a greenish tinge.

Hair Dyes—Why Be Satisfied with What Mother Nature Provided?

Hair coloring is not a new cosmetic process; it occurred as far back as the first century A.D. Interestingly, in the early days, many people chose colors not seen in nature. For example, the Saxons dyed their hair and beards blue, green, or orange. In the 1790s, blue was a favorite hair color in the French courts, but pink, violet, yellow, and white were also popular. Most of the dyes came from various plants or inorganic salt powders; many were fairly toxic. French chemist Eugène Schueller first developed modern hair coloring in 1909; similar dyes are still in use today.

Millions of people recolor their hair again and again. Some do this to restore color lost with aging, while others simply want a different hue. The cosmetic industry provides an amazing assortment of hair dyes and colorants. They may be classified in three distinct categories: (1) metallic dyes, (2) temporary colorants, and (3) semipermanent dyes.

Metallic Dyes

Metallic dyes are the simplest types of hair dyes. Usually these dyes are a dispersion of sulfur in an aqueous lead acetate solution (figure 16.9). The dyeing process involves the reaction of lead ions with both the elemental sulfur from the dye and sulfur atoms on the hair protein chains. The resulting compound, black PbS, adheres to the hair surfaces and penetrates the hair follicles, darkening the hair. Small amounts of PbS wash off the hair strands during shampooing, so this method is not permanent. However, PbS is very water-insoluble and this dyeing procedure seems longer-lasting than other methods. Hair dyed with a lead acetate-sulfur mixture *darkens* in sunlight or in a chlorinated pool; the other classes of hair dye lose color under these conditions. This happens because either strong light (in a photochemical oxidation reaction) or a chemical oxidizing agent, such as chlorine, accelerates PbS formation.

Temporary Colorants

Temporary colorants are wash-in or shampoo-in dyes that coat the surface of the hair to change its color. These colorants are easily removed by repeated shampooing, or even by rinsing in ordinary water or pool water. Many of these dyes are coal-tar chemicals present in their oxidized colored forms (see next section).

Semipermanent Dyes

Semipermanent dyes last longer than the temporary versions because they partially penetrate into the hair fibers. The dye is applied in a reduced form that oxidizes to the colored form after penetrating the hair fibers, usually by alkaline H_2O_2. The reduced form easily penetrates the hair, but the oxidized form can't, and it becomes entrapped within the hair fibers. This makes the color more durable.

The most commonly used semipermanent dyes are derivatives of p-phenylenediamine, which is black in the oxidized form; a yellow dye is made from p-aminodiphenylamine sulfonic acid. Chemists can create a full palette of colors from combinations of these two dyes and a few related ones. (Recall that the enormous

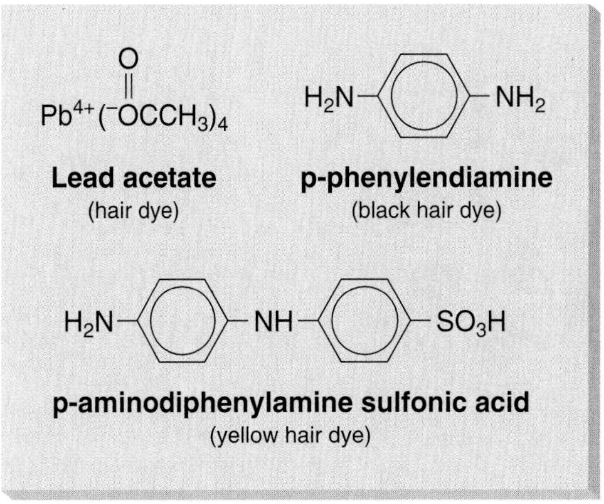

$Pb^{4+}(^{-}OCCH_3)_4$ (with C=O)

Lead acetate
(hair dye)

$H_2N-C_6H_4-NH_2$

p-phenylendiamine
(black hair dye)

$H_2N-C_6H_4-NH-C_6H_4-SO_3H$

p-aminodiphenylamine sulfonic acid
(yellow hair dye)

Figure 16.9 The structures of some chemicals used in hair dyes.

variety of natural hair colors come from only four pigments.) The dye is supplied in a reduced form (that is, the amino compound), usually with sodium sulfite (Na_2SO_3), a reducing agent, added to maintain this reduced state. The dye is later oxidized to nitro groups:

$$R{-}NH_2 + 3\,H_2O_2 \longrightarrow R{-}NO_2 + 4\,H_2O$$

Ammonia is added to increase the pH above 9, which makes oxidation occur more readily. As in bleaching, H_2O_2 reacts with the hair proteins and makes them brittle and weak. Animal proteins are regularly included in dyeing formulations to counteract this.

Some people raise safety questions concerning these "coal-tar" derived dyes. The term *coal-tar dye* comes from the original source of these chemicals—coal tar, the material formed when coal is pyrolyzed to coke. Coal-tar contains many important organic molecules. Although these compounds are now obtained from petroleum, the name persists. There is little doubt that most of these chemicals are toxic when ingested; however, hair dyeing is strictly an external matter. The real question is whether these chemicals may cause cancer in either the cosmetologist or his or her clients. Although there were several media scares on this subject in 1993, the American Cancer Society released the results of a study in the February 1994 issue of the *Journal of the National Cancer Institute* that showed no credible link between hair dye use and cancer.

How Hair Dyes Operate Chemically

Each hair dye class operates in a different chemical manner; this is illustrated in figure 16.10. Naturally colored hair fibers contain "pockets" filled with the melanin compounds. When hair turns gray or white, the melanin is no longer present in these pockets, so they become empty voids. Temporary colorants merely sit on the surface of the hair fiber and coat the outside, while semipermanent dyes penetrate into the void spaces of the hair fibers in a manner similar to melanin. Metallic dyes do both; some of the lead compound reacts after entering the void spaces, while other molecules react on the surface of the hair. If the PbS color (dark brown or black) is the desired shade, this method probably works the best. It also causes less damage to the hair since these chemicals do not oxidize the hair proteins.

The longer-lasting dyes are those that enter the void spaces in the hair. When pigmented hair is being recolored, rather than restored from gray or white to the person's natural shade, there are very few void spaces the dye can penetrate into. If the new color is a lighter shade, a bleaching process forms some voids as melanin is oxidized. The nonmetallic dyes (that is, the "coal-tar" dyes) undergo further reactions with any oxidizing or reducing agent in or on the hair. This causes color changes, frequently to an orange shade, that limits how long the desired color will last. PbS is less prone to these reactions and can endure longer. Still, since new hair growth is the person's natural color, redyeing is necessary.

PRACTICE PROBLEM 16.9

What similarities and differences exist between the chemicals used in the three main types of hair dyes?

ANSWER The chemicals have several similarities: (1) all impart color; (2) most are water-insoluble; (3) all adsorb onto the hair; and (4) two types penetrate into the hair. The differences include: (1) metallic dyes are inorganic salts, while the other types are organic coal-tar derivatives; (2) permanent dyes start out as reduced forms that are oxidized; and (3) temporary dyes are always in the oxidized form.

Depilatories

A **depilatory** is a cosmetic used to remove hair by chemical action. The most common chemical depilatories are Na_2S, CaS, SrS, and calcium thioglycolate (figure 16.11). These chemicals remove hair by disrupting the disulfide crosslinks and partially dissolving the hair. They usually attack the skin at the same time, because the chemical compositions of skin and hair are similar. Emollients are added to alleviate the skin damage problem; how-

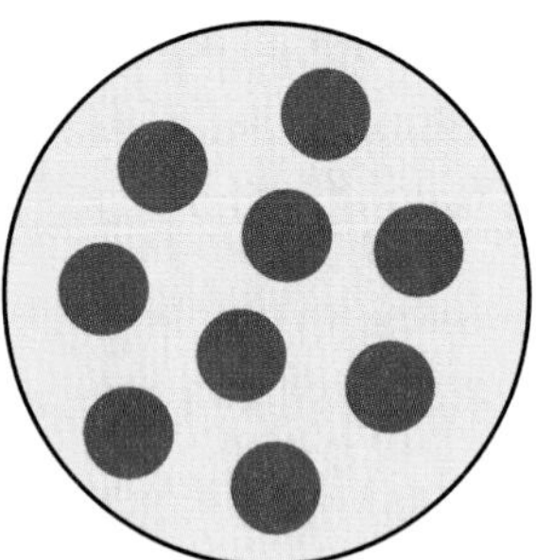

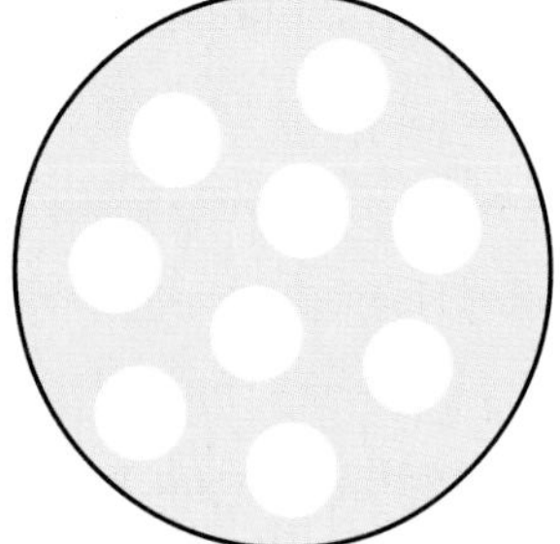

Figure 16.10 The chemistry of dye interaction with hair. Hair strands are shown in cross-section.

$$\left(H{-}S{-}CH_2{-}\overset{\displaystyle O}{\overset{\|}{C}}{-}O^-\right)_2Ca^{2+}$$

Calcium thioglycolate
(depilatories)

$$H{-}S{-}CH_2{-}\overset{\displaystyle O}{\overset{\|}{C}}{-}OH$$

Thioglycolic acid
(permanent wave solutions)

Figure 16.11 Some chemicals used in depilatories and permanent wave solutions.

ever, frequent depilatory use can still damage the skin. Most depilatories are alkaline (with a pH above 9) because their chemical activity works best under alkaline conditions. Table 16.10 shows a typical formulation.

Alternative hair removal methods include shaving or plucking the hair and electrolysis. In electrolysis, a permanent hair removal process, a tiny electric current "zaps" individual hair follicles to stop hair production.

Permanents and Straighteners

Permanent waves make straight hair curly, while hair straighteners do just the opposite. In fact, both systems operate the same way chemically, and both are examples of oxidation-reduction reactions. In both waving and straightening, the disulfide linkages (RS—SR) between hair fibers are broken, the hair is reshaped as desired, and the crosslinks are then re-formed. The most common agent used to reduce the disulfide linkages to thiol (R—SH) groups is thioglycolic acid (figure 16.9). This is the waving lotion. The usual oxidizing agent (or oxidizing solution) is hydrogen peroxide (H_2O_2), but sodium perborate ($NaBO_2{\cdot}H_2O_2$) or potassium bromate ($KBrO_3$) are sometimes used. When mixed with water, the perborates and bromates react to generate H_2O_2.

Table 16.11 shows a typical waving and straightener lotion and oxidizing solution. Many products contain added animal proteins (from animal bones) to coat or condition the hair after the waving or straightening process because the hydrogen peroxide (H_2O_2) also reacts with the hair, partially damaging it and making it brittle. Usually both the waving lotion and the oxidizing formulations contain a water-soluble polymer as a thickener for ease of application. These are often natural polymers, such as alginates or carrageenin, obtained from seaweed. In the absence of this thickener, the mixture might run down the user's face, neck, and

TABLE 16.10 A typical depilatory formulation.

COMPONENT	EXAMPLE	PERCENT OF TOTAL COMPOSITION
Depilatory agent	calcium thioglycolate	7–8
Filler	calcium carbonate	20–22
pH adjuster	calcium hydroxide	1–1.5
Skin conditioner	cetyl alcohol or mineral oil	6–8
Detergent	sodium lauryl sulfate	0.5–1.0
Solvent	water	60–66

TABLE 16.11 A typical permanent wave (or straightener) formulation.

COMPONENT	EXAMPLE	PERCENT OF TOTAL COMPOSITION
WAVING LOTION		
Reducing agent	thioglycolic acid	5–7
pH adjuster	ammonia	2–3
Solvent	water	90–93
OXIDIZING SOLUTION		
Oxidizing agent	hydrogen peroxide	3–5
pH adjuster	ammonia	2–3
Solvent	water	92–95

SOURCE: By permission of Johnny Hart and Creators Syndicate, Inc.

body. The thickened solution also gives the illusion of higher concentration. Ammonia is added to adjust the pH.

The actual chemistry for a permanent wave is shown in figure 16.12, which depicts the process from two viewpoints. The view on the left is a schematic diagram showing how pairs of protein chains are tied together into a particular shape by S—S crosslinks. The view on the right shows the actual chemical reactions for a pair of S—S crosslinks on a protein chain.

The process begins when alkaline **thioglycolic acid** is applied to the hair. As the acid severs S—S linkages (step 1), many individual protein chains become disconnected from each other. While these bonds are still severed, the hair is wrapped around a roller to impart a new shape to the strands. Next, the rolled hair is treated with H_2O_2 (at a high pH) to form new S—S linkages (step 2). This realignment of protein molecules forces the hair into a new spatial arrangement. The hair is now wavy.

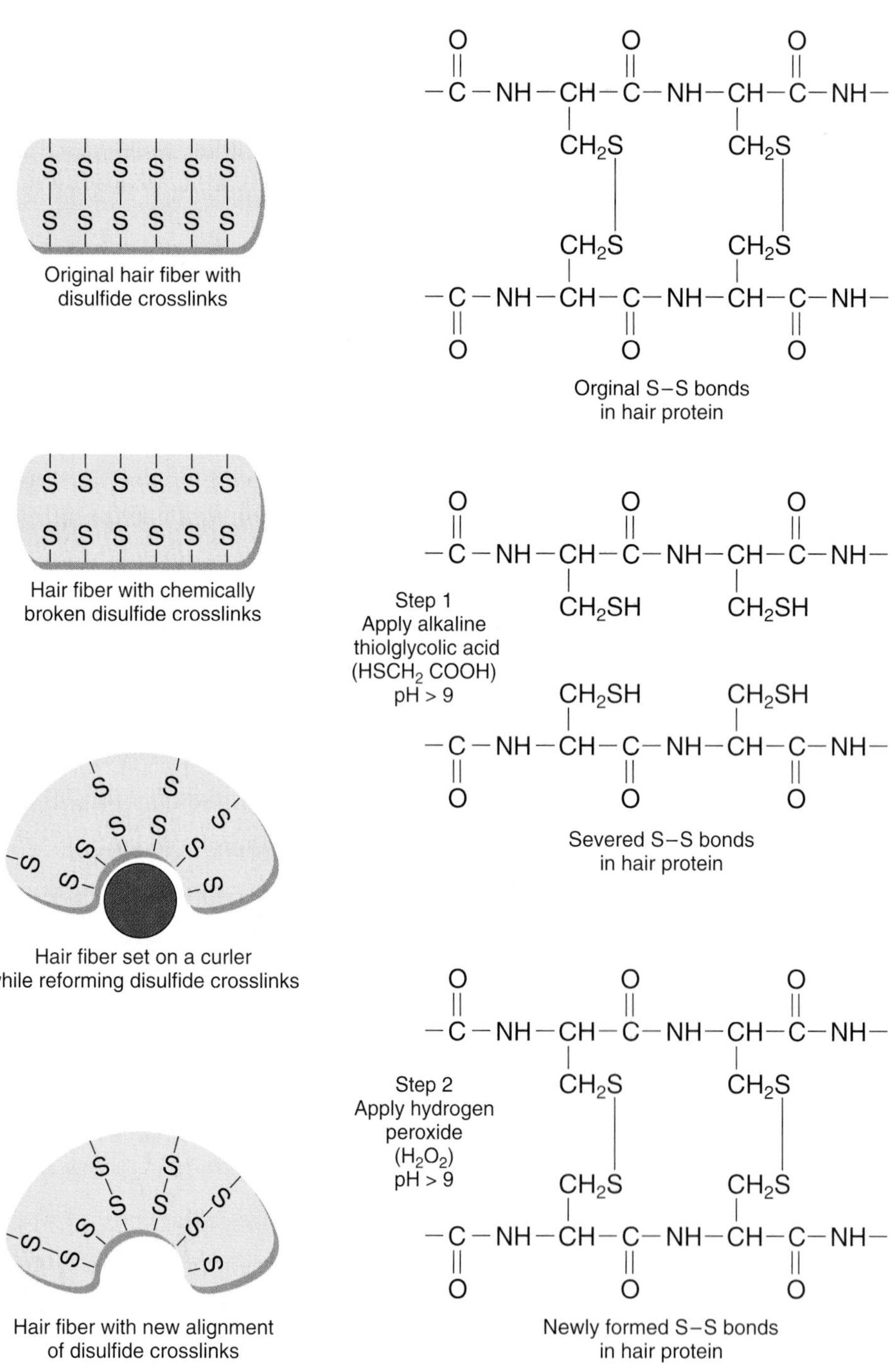

Figure 16.12 The chemistry of the permanent wave.

Although this process is called a "permanent" wave, the fundamental nature of the hair remains unchanged. New hair will grow in straight, and the process will have to be repeated endlessly to maintain the wave. The term *permanent* arose because the curls do last much longer than curls set by merely wetting the hair and twisting it on rollers. A hair straightener works in a similar way, except the hair is stretched after the reduction step to make it straight.

Other hair treatments do not color, wave, or straighten hair; they restore it. Box 16.4 discusses minoxidil, a drug used to treat high blood pressure that has an unusual side effect—the stimulation of new hair growth.

PRACTICE PROBLEM 16.10

Why don't cosmetic chemists simply mix the ingredients in table 16.11 together to wave or straighten hair in a single step?

ANSWER Mixing would cause the thioglycolic acid to react with the H_2O_2 instead of cleaving the S—S linkages. In addition, the H_2O_2 would not be available to oxidize the —SH units and form the new S—S patterns; to put it plainly, the mixture would no longer function as a waving solution.

BOX 16.4

GROWING NEW HAIR—THE MINOXIDIL STORY

Male pattern baldness (figure 1), the typical receding male hairline, can hardly be classed as a disease. Yet the most promising treatment for this cosmetic problem—a drug called minoxidil—is produced by a pharmaceutical company. Approximately 35 million men in the United States have male pattern baldness. Interestingly, about 20 million women have essentially the same problem. About 50 percent of these women and 40 percent of the men are seeking treatment for this condition.

Minoxidil (figure 2) is a drug developed for treating high blood pressure, but scientists discovered it had a curious side effect. Since drug side effects are quite common and hair loss is one common side effect, few would be surprised if minoxidil caused hair loss. Instead, scientists found minoxidil caused new hair growth. The first report that linked minoxidil with hair growth was mentioned in 1980 in a brief letter to the editor of the prestigious *New England Journal of Medicine*.

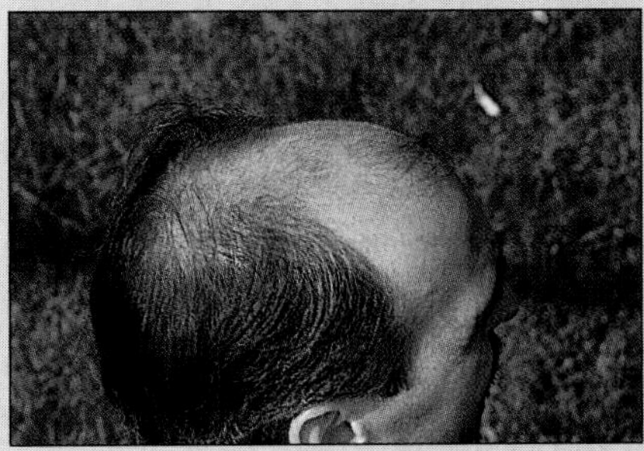

Figure 1 Male pattern baldness.

Unfortunately, hair growth is not minoxidil's only side effect. Some patients experience an increased heartbeat, dizziness, or fainting after oral or topical administration of the drug. Nonetheless, the Upjohn Company is marketing this drug in a 2 percent solution as a hair restorative under the trade name Rogaine. This is the first hair restorative to gain FDA approval; doctors promptly wrote 300,000 prescriptions for it in 1990. Over 2 million people have started using the topical minoxidil solution. The potential market is large and lucrative. The current price for a year's treatment runs about \$800 to \$1,000. Naturally, drug companies are looking for other drugs that also can grow hair.

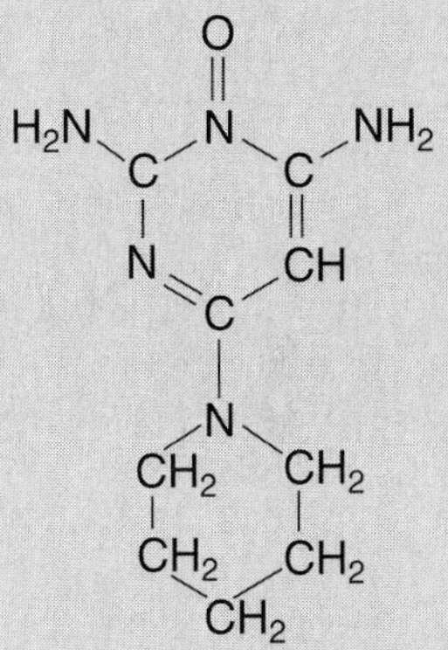

Figure 2 Minoxidil, the drug sold under the brand name Rogaine, restores hair.

In a Canadian clinical study, about 75 percent of the men using topical minoxidil grew new hair, albeit a thin fuzz, on their scalps. Around 8 percent grew enough hair to cover bald spots. Is this the long-hoped-for panacea for baldness? Only time will tell.

Meanwhile, Rogaine received FDA approval for over-the-counter sales in February, 1996.

EXPERIENCE THIS

1. Visit the cosmetics section in a store and list some of the ingredients found in the different types of eye makeup. Make separate columns for eye pencils, mascaras, and eyeshadows. What ingredients are common to all three types of eye makeup? Which are found only in a single type?
2. Examine at least seven deodorants and an equal number of antiperspirants. What chemicals are used in both? What chemicals are unique to one or the other?
3. What chemicals are found in the hair sprays at your local cosmetic counter? Are polymers present in every hair spray?
4. Examine some wash-in (temporary) hair dyes as well as some permanent dyes. How do these formulations differ? Are similar dyes used in both types?
5. Examine several different brands of permanent wave solutions. Are there more similarities or differences between brands?

SUMMARY

Humans have been preoccupied with beauty for countless centuries, but only in the last few hundred years have chemists been able to develop safe, effective cosmetics. The main types of cosmetics are used for skin care, decorative purposes, odor control, and hair.

The skin, the largest organ of the body, helps regulate body temperature, maintain proper fluid balance, and protect the body from disease. Skin care products act on the corneal layer of the epidermis to soften and add moisture. Common skin care treatments include petroleum jelly, emollients, and creams. All contain lipids, mostly oils. Sunscreens (most containing PABA) protect the skin from the harmful effects of the sun's ultraviolet rays and help prevent skin cancer.

Decorative cosmetics include lipsticks and lip balms, nail polishes, powders, eye cosmetics, and creams. All are designed to enhance beauty rather than health. Lipsticks and lip balms, which contain blends of oils, waxes, alcohol, dyes, and perfumes, do help protect the lips by slowing moisture evaporation. Nail polish contains a polymer, a plasticizer, a solvent, and a dye. Although advertisers may claim that nail polishes protect the fingernails, they may actually prevent needed air from penetrating the nails and reaching the tissues underneath. Face powders contain inert ionic compounds such as chalk or talc that adhere to blemishes and oily spots. Powdered cosmetics such as rouges and eye shadows also contain pigments. Eye makeups serve no protective purpose and are used strictly for aesthetic reasons.

Odor control cosmetics include deodorants, antiperspirants, and fragrances such as perfumes, colognes, and aftershave lotions. Deodorants contain chemicals such as calcium carbonate and zinc peroxide to mask or adsorb odors, and antiperspirants add astringent chemicals such as aluminum chlorohydrate to constrict sweat glands and reduce perspiration. Perfumes, colognes, and aftershaves contain fragrances such as esters, aldehydes, and ketones to impart a pleasant odor.

The chemicals we use on hair include sprays, bleaches, hair removers, dyes, and permanents and straighteners. Hair sprays contain a polymer to hold hair in place, a plasticizer to keep it soft and pliable, and silicone oils to impart sheen. Bleaches act on hair pigments melanin, phaeomelanin, oxymelanin, and brown melanin. The bleaching process involves using hydrogen peroxide in an oxidation-reduction reaction to break down these pigments and leave the hair colorless. Hair coloring masks or replaces the hair's color through metallic dyes, temporary coloring agents, or semipermanent dyes. Temporary colorants coat the surface of the hair fiber with color; semipermanent dyes penetrate void spaces in the hair fiber that lack pigment; and metallic dyes do both.

Depilatories remove hair by chemical action. The most common chemicals in depilatories are Na_2S, CaS, SrS, and calcium thioglycolate. These chemicals work by breaking the disulfide crosslinks that give hair its shape and partially dissolving the hair.

Permanents make straight hair wavy, while straighteners do the opposite. Both operate the same way chemically. A waving lotion severs the disulfide crosslinks in the hair, leaving it "shapeless." Then the hair is rolled and an oxidizing solution is applied to form new disulfide (S—S) linkages. These linkages will reflect the rolled or curled position of the hair, making the hair wavy. In either case—whether the individual wants to straighten or curl his or her hair—chemistry enables the desired result in a safe, effective way.

KEY TERMS

aftershave lotion 445
aluminum chlorohydrate ($Al_2(OH)_5Cl{\cdot}2H_2O$) 444
antiperspirant 444
brown melanin 446
cold cream 438
cologne 445
cosmetics 436
disulfide (S—S) crosslinks 446
deodorant 444
depilatory 448
emollient 437
eye shadow (eyeshade) 444
face powder 443
fixatives 445
hair spray 446
lip balm 442
lipstick 442
melanin 446
melanoma 439
mascara 444
metallic hair dye 447
minoxidil 451
nail polish 443
nail polish remover 443
oxymelanin 446
para-aminobenzoic acid (PABA) 438
perfume 445
permanent wave 449
petroleum jelly 437
phaeomelanin 446
Retin A 439
semipermanent hair dye 447
skin cream 438
substantivity 439
sun protection factor (SPF) 439
sunscreen 438
suntan lotions 438
thioglycolic acid 450
vanishing cream 438

READINGS OF INTEREST

General

Begoun, Paula. 1991. *Don't go to the cosmetics counter without me.* Seattle:Beginning Press. *Also* 1988. *Blue eyeshadow should still be illegal.* Seattle:Beginning Press. Two good, easy-to-read discussions of cosmetics. The 1991 book includes Begoun's evaluations of the cosmetics lines of most major manufacturers. The 1988 book contains a lot of information on cosmetics in general, but very little chemistry (and no cosmetic or chemical formulas). Both books are available as paperbacks.

Ainsworth, Susan J. "Personal care products." *Chemical & Engineering News*, April 4, 1994, pp. 38–49. This version of an annual series dwells mainly on the trend toward "natural" cosmetics. The controversy over animal testing is also discussed.

Ainsworth, Susan J. "Cosmetics." *Chemical & Engineering News*, April 26, 1993, pp. 36–48. Part of an annual series on this topic. Among the subjects this article discusses are animal testing, new products, the trend toward natural ingredients, and the latest statistics on cosmetics markets.

Coeyman, Marjorie. "Cosmetics: Working harder to stay in the game." *Chemical Week*, February 3, 1993, pp. 36–40. Business-oriented article on cosmetics.

Weber, Joseph. "A big company that works." *Business Week*, May 4, 1992, pp. 124–132. A case study on how Johnson & Johnson manages its profitable cosmetics and medicinal business.

Ainsworth, Susan J. "Cosmetic suppliers rally to meet consumer concerns." *Chemical & Engineering News*, April 20, 1992, pp. 31–42. Many basic facts about the cosmetics industry, including the size of various markets. Part of an annual series on this topic.

Fentem, Julia, and Balls, Michael. "In vitro alternatives to toxicity testing in animals." *Chemistry & Industry* (London), March 17, 1992, pp. 207–211. Product safety is impossible without testing; what options exist other than testing animals?

"Skin deep." *Discover*, August 1991, p. 16. This short article describes the use of Testskin, a semisynthetic skin, for cosmetics testing.

Layman, Patricia L. "Cosmetics." *Chemical & Engineering News*, April 4, 1988, pp. 21–41. This summary deals mainly with sunscreens, Retin A, and minoxidil. Part of an annual series on this topic.

Skin Care and Retin A

Jaroff, Leon. "New age therapy." *Time*, January 23, 1995, p. 52. Is it possible that a hormone called DHEA can retard the aging process?

Wills, Christopher. "The skin we're in." *Discover*, November 1994, pp. 74–81. Melanin and skin color.

Byrd, Veronica, and Zellner, Wendy. "The Avon lady of the Amazon." *Business Week*, October 24, 1994, pp. 93–96. Cosmetics are spreading around the globe; now Avon is calling in the Amazon.

Repinski, Karyn. "Safe face?" *Longevity*, September 1994, pp. 36–40. This author questions the safety of the glycolic acid peel and Retin A.

Brink, Susan. "The dewrinkling of America." *U.S. News & World Report*, June 6, 1994, p. 79. A short article about alpha-hydroxy acids, one of the latest fads in wrinkle prevention.

Sawahat, Lesa. "Not just another pretty facial." *Longevity*, April 1994, pp. 28–32, 70. This article contains some information on alpha-hydroxy acids, but mostly it discusses nonchemical treatments.

Marshall, Eliot. "A new wrinkle in the Retin A dispute." *Science*, vol. 256, May 1, 1992, pp. 607–608. Who should profit most from this potentially billion-dollar cosmetic/drug?

Van Pelt, Dina. "Drug's uses more than skin-deep." *Insight*, January 23, 1989, p. 56. Includes some before and after photographs of Retin A treatments.

Weiss, Rick. "Wrestling with wrinkles." *Science News*, September 24, 1988, pp. 200–202. Good article on how skin ages and what Retin A does to slow the process.

"Beauty bonus: Choosing the right foundation." *Ladies Home Journal*, August 1988, p. 28. Compares various skin treatments from a woman's perspective.

Henderson, Nancy. "Saving your skin." *Changing Times*, July 1988, pp. 59–63. Keeping that youthful look with Retin A, chemical peels, and moisturizers.

"Lipsticks." *Consumer Reports*, February 1988, pp. 75–80. Information on the chemicals in lipsticks and comparative costs and ratings of various brands.

Moser, Penny Ward. "Search for the wrinkle-free face." *Reader's Digest*, February 1988, pp. 111–114. Nontechnical summary of some dermatology research on Retin A.

Ostro, Marc J. "Liposomes." *Scientific American*, January 1987, pp. 102–111. Liposomes are gaining attention as a cosmetics additive. This article tells you what they are and how they are used in medicine.

"Skin care: What you need to know." *Reader's Digest*, May 1986, pp. 143–148. This brief article highlights skin care items ranging from acne treatments and Accutane to moisturizers and sunscreens.

Sunscreens

Pennisi, E. "Visible UV-A light tied to skin cancer." *Science News*, July 24, 1993, p. 53. Sunscreens attempt to provide protection from UV light. This article implies that visible light may also induce skin cancer.

Roach Mary. "Sun struck." *Health*, May/June 1992, pp. 41–50. A good article about the relationship between sunburn and skin cancer. It includes an evaluation of the potential increases in skin cancer if the ozone layer decreases significantly.

Anselmi, Cecilia. "Staying on the surface." *Chemtech*, February 1992, pp. 99–104. Sunscreens work on the surface of the skin; this article tells us which compounds work best and how scientists test them.

Ezzell, C. "Waving a red flag against melanoma." *Science News*, December 14, 1991, p. 388. Melanoma kills one-fifth of its victims within five years. Is it possible to develop a vaccine against it?

Groves, David. Skin cancer. "The sun's revenge." *McCalls*, August 1988, pp. 74–76. Describes skin cancer in some detail.

Perfumes and Deodorants

Buchholz, Fredric L. "Keeping dry with superabsorbent polymers." *Chemtech*, September 1994, pp. 38–43. The same polymers used in stay-dry diapers can also be used in cosmetics to maintain underarm dryness.

Layman, Patricia L. "Flavors and fragrances business marked by growth, mergers." *Chemical & Engineering News*, May 30, 1994, pp. 10–11. Cosmetics is the second largest market for fragrances—household products rank number one.

Ramsey, Joy. "Pampering your senses with aromatherapy." *Body, Mind, & Spirit*, March/April 1994, pp. 18–23. Aromatherapy, with its roots in ancient Egypt, regained popularity in the early 1990s.

Dunkel, Tom. "It's time to wake up and smell the rosemary." *Insight*, June 15, 1992, pp. 14–16, 36–38. Can aromas relieve stress? Aromatherapy advocates say they can.

Black, Pamela J. "No one's sniffing at aroma research now." *Business Week*, December 22, 1991, pp. 82–83. This article considers the possible relationship between odors and illnesses. It also discusses, in a nontechnical manner, recent discoveries on how the sense of smell operates.

Wilkinson, Sophie L. "Only a rose smells as sweet as a rose." *Chemical Week*, August 24, 1988. pp. 64–66B. Good article on the kinds of scents isolated from roses. The amount of scent differs greatly between fresh and picked flowers.

Hirshlag, Jennifer. "Making scents for the brain." *Chemical Week*, August 24, 1988. pp. 64B–64C. Aromatherapy is not a new stress-relief technique, but it grew in popularity during the 1990s.

Gibbons, Boyd. "The intimate sense of smell." *National Geographic*, vol. 170, no. 3 (September 1986), pp. 324–360. Perhaps you have wondered how you can tell one odor from another? Or how you remember the smell of something you sniffed many years ago? In fact, how do we detect odors in the first place?

Gilbert, Avery N., and Wysocki, Charles J. "Smell survey results." *National Geographic*, vol. 172, no. 4 (October 1987), pp. 514–525. Follow-up on a 1986 article on smell.

Hair Chemistry

Frackelmann, K. A. "Mixed news on hair dyes and cancer risk." *Science News*, February 5, 1994, p. 86. American Cancer Society study collected information on 573,369 women and found no significant link between cancer and hair dyes. A few women who use hair dyes, however, may be at risk.

Rheim, Rex. "The bald truth about minoxidil." *Business Week*, November 28, 1988, p. 186. What minoxidil does, and what it costs to restore hair this way.

Fox, Marisa. "Sun-safe summer hair." *Health*, June 1988, pp. 49–51, 74. Hair care in the summer time.

Wellborn, Stanley N. "Searching for their roots." *U.S. News & World Report*, October 27, 1986, p. 66. An early account of the development of Rogaine (a minoxidil solution). Includes some before and after photographs.

Powitt, A. H. 1985. *Hair structure and chemistry simplified*. Bronx, N.Y.: Milady Publishing. This brief, relatively nontechnical book contains just about anything you might want to know about hair chemistry.

PROBLEMS AND QUESTIONS

1. When did humans start using cosmetics? When did the modern industry begin?

2. What parts of the skin do cosmetic formulations treat?

3. What is the main type of chemical used for cosmetic skin treatment?

4. What is an emollient?

5. What are the major components of an emollient?

6. What are cosmetic creams made of?

7. What is a cold cream? What does it do?

8. What property gives vanishing cream its name?

9. What is the main purpose for using a suntan lotion? How does this differ from its purpose in the past?

10. What are the major components of a suntan lotion?

11. What are sunscreens?

12. What is the most common chemical sunscreen in use?

13. What is melanoma? Is it dangerous? Is the incidence of melanoma increasing or decreasing? How can it be prevented?

14. What is the sun protection factor (SPF)? How is it determined?

15. What is the SPF for a tanning lotion that allows the same amount of burning in 36 hours of exposure that would normally occur in 3 hours exposure?

16. What is substantivity? How does it affect sunscreen formulations?

17. What is Retin A and what does it do (or what is it claimed to do)?

18. What other kinds of chemicals have been developed in an attempt to impart the same benefits as Retin A?

19. What are the major components of a lipstick?

20. How do lipsticks and lip balms differ?

21. What are the major components of nail polish?

22. What are the major components of nail polish remover?

23. Do nail polishes protect the nails? Why or why not?

24. What are the major components of face powder?

25. How do mascara and eye shadow differ?

26. What are the major components in mascara?

27. What are the major components in eye shadow?

28. Are the dyes used in eye shadows and mascara dangerous or completely harmless?

29. What is an antiperspirant?

30. What is the main chemical used in antiperspirants? How does it work?

31. What does a deodorant do?

32. What kinds of chemicals are used in deodorants?

33. How do deodorants and antiperspirants differ?

34. What are the major components in perfumes?

35. How do perfumes, colognes, and aftershave lotions differ?

36. What kinds of chemicals are used in a cologne.

37. What do the terms *top note*, *middle note*, and *final note* mean in perfume design?

38. How are perfume odors matched?

39. What is the basic composition of the hair? How does it differ from that of the skin?

40. What are the major components in hair sprays?

41. How much harm do hair sprays do to the ozone layer?

42. What does a depilatory do?

43. What are the major components of a depilatory?

44. What are the names of some natural hair coloring materials?

45. What are the three main classes of hair dyes?

46. What is hair bleach, from a chemical standpoint?

47. How do each of the three types of hair dyes work?

48. What are the major components of metallic hair dyes?

49. What types of chemicals are used in temporary hair colorants?

50. What types of chemicals are used in semi-permanent hair dyes?

51. What is the chemical difference between the substances used in temporary colorants and in semipermanent hair dyes?

52. What is the main type of chemical reaction that takes place when a hair dye is used?

53. What is a coal-tar derivative?

54. What kind of chemical bonds control the hair's shape?

55. What are the two main components of a permanent wave mix?

56. How do permanent waves work?

57. What unusual side effect does minoxidil have?

58. What percentage of the adult population (both male and female) have male pattern baldness?

59. What was the original use for the drug minoxidil?

CRITICAL THINKING PROBLEMS

1. There is a current trend toward using "natural" ingredients in cosmetics. For background, read the following two articles by Susan J. Ainsworth, "Personal Care Products," *Chemical & Engineering News*, April 4, 1994, pp. 38–49; and "Cosmetics," *Chemical & Engineering News*, April 26, 1993, pp. 36–48. In what ways are so-called natural cosmetics superior or inferior to the older types? Many of these "natural" cosmetics consist mainly of synthetic chemicals. Why do you think this is true? Is it wise to assume that natural cosmetics are safer than the versions that rely completely on synthetic chemicals?

2. Every spring break, thousands of college students travel south for some fun in the sun. What are the chances that some of these students will eventually contract skin cancer as an undesired bonus from repeatedly making the trip? For background, read at least two of the following articles on skin cancer: E. Pennisi, "Visible, UV-A Light Tied to Skin Cancer," *Science News*, July 24, 1993, p. 53; Mary Roach, "Sun Struck," *Health*, May/June 1992, pp. 41–50; David Groves, "Skin Cancer: The Sun's Revenge," *McCalls*, August 1988, pp. 74–76; and C. Ezzell, "Waving a Red Flag Against Melanoma," *Science News*, December 14, 1991, p. 388. How could a potential skin cancer disaster be prevented?

3. Modern suntan lotions normally contain sunscreens. How do these agents protect a person from the harmful portion of the sun's rays. How are they tested? Could any material that absorbs ultraviolet light be used as a sunscreen? For more background, read the following two articles: Cecilia Anselmi, "Staying on the Surface," *Chemtech*, February 1992, pp. 99–104. Patricia L. Layman, "Cosmetics." *Chemical & Engineering News*, April 4, 1988, pp. 21–41.

4. How safe is Retin A? Would you consider using this compound on your skin to remove wrinkles? For more background, read at least three of the following six articles: Patricia Layman, "Cosmetics," *Chemical & Engineering News*, April 4, 1988, pp. 21–41; Nancy Henderson, "Saving Your Skin," *Changing Times*, July 1988, pp. 59–63; Penny Ward Moser, "Search for the Wrinkle-Free Face," *Reader's Digest*, February 1988, pp. 111–114; Eliot Marshall, "A New Wrinkle in the Retin A Dispute," *Science*, vol. 256, May 1, 1992, pp. 607–608; Rick Weiss, "Wrestling with Wrinkles," *Science News*, September 24, 1988, pp. 200–202; Dina Van Pelt, "Drug's Uses More than Skin-Deep," *Insight*, January 23, 1989, p. 56.

5. The sense of smell, briefly discussed in chapter 14, is important in several areas of cosmetics, including odor prevention and perfumes. Read the following two articles, which deal with how this sense operates: Boyd Gibbons, "The Intimate Sense of Smell," *National Geographic*, vol. 170, no. 3 (September 1986), pp. 324–360; Avery N. Gilbert and Charles J. Wysocki, "Smell Survey Results," *National Geographic*, vol. 172, no. 4 (October 1987), pp. 514–525. (This survey was a followup to the 1986 article.) Recently one trend has been to use special aromas to treat disease or enhance the work place. For background on this, read Tom Dunkel, "It's Time to Wake Up and Smell the Rosemary." *Insight*, June 15, 1992, pp. 14–16, 36–38; and Pamela J. Black, "No One's Sniffing At Aroma Research Now." *Business Week*, December 22, 1991, pp. 82–83. Does it seem reasonable to you that aromas could aid in treating disease or creating a better place to work? What is the basis of your opinion?

6. Male pattern baldness affects about 35 million males and 20 million females in the United States. (Perhaps the problem should not have a gender-specific name. Can you think of another?) The drug minoxidil is now available to treat this problem, but is this a safe therapy? For additional background, read the following trio of articles: Patricia L. Layman, "Cosmetics," *Chemical & Engineering News*, April 4, 1988, pp. 21–41; Stanley N. Wellborn, "Searching for Their Roots," *U.S. News & World Report*, October 27, 1986, p. 66; and Rex Rheim, "The Bald Truth about Minoxidil," *Business Week*, November 28, 1988, p. 186. Since any drug presents potential hazards and side effects, is using this drug a better solution for baldness than letting nature take its course, or wearing a hair piece?

7. Using animals to test cosmetics is an important part of the industry. What are the differences between using animals to test cosmetics safety and using animals to test drug safety? What problems would be raised by eliminating animal testing of cosmetics? For background, read this trio of articles: Julia Fentem and Michael Balls, "In Vitro Alternatives to Toxicity Testing in Animals," *Chemistry & Industry* (London), March 17, 1992, pp. 207–211; "Skin Deep" (no author credited), *Discover*, August 1991, p. 16; and Susan J. Ainsworth, "Personal Care Products," *Chemical & Engineering News*, April 4, 1994, pp. 38–49. Consider the fact that essentially the same chemicals have been used in cosmetics for many years, while drugs are often totally new formulations. Will the nonanimal tests discussed in these articles provide definitive answers on drug safety as well as the safety of cosmetics?

8. A crop of newer face peels (which are brushed over the surface of the face and then peeled away when dry) use the more gentle alpha-hydroxy acids. Are these really safe? And do they improve the quality of the skin? Four articles can provide a starting point for your evaluation: Karyn Repinski, "Safe Face," *Longevity*, September 1994, pp. 36–40; Susan Brink, "The Dewrinkling of America," *U.S. News & World Report*, June 6, 1994, p. 79; Lesa Sawahat, "Not Just Another Pretty Facial," *Longevity*, April 1994, pp. 28–32, 70; Nancy Henderson, "Saving Your Skin," *Changing Times*, July 1988, pp. 59–63.

9. How safe are hair dyes? Read K. A. Frackelmann, "Mixed News on Hair Dyes and Cancer Risks," *Science News*, February 5, 1994, p. 86, and make your own decision based on the evidence.

10. Aromatherapy advocates claim aromatherapy can treat diseases and other problems through the use of various aromas. But is this treatment method effective? Is it safe? Read the following three articles for additional background: Joy Ramsey, "Pampering Your Senses with Aromatherapy," *Body, Mind, & Spirit*, March/April 1994, pp. 18–23; Tom Dunkel, "It's Time to Wake Up and Smell the Rosemary," *Insight*, June 15, 1992, pp. 14–16, 36–38; and Jennifer Hirshlag, "Making Scents for the Brain," *Chemical Week*, August 24, 1988; pp. 64B–64C.

CHAPTER 17

CHEMISTRY IN HEALTH AND MEDICINE

OUTLINE

KEY IDEAS

1. The Positive Effects Chemistry has on Public Health
2. The Nature and Examples of Analgesics and Antipyretics
3. The Meaning of the Term LD_{50}
4. The Nature of Chemotherapy
5. How Bacteriostatic and Bactericidal Agents Differ
6. The Names of Some Important Antibacterial Groups
7. How do the Terms Carcinogen and Carcinogenic Differ
8. The Main Classes of Anticancer Drugs and Some Examples
9. The Nature and Treatment of AIDS
10. Alzheimer's Disease and its Treatment
11. The FDA Recommendations for the Best Cold Medicines
12. The Differences Between Herbal (Folk) Medicine and Quackery
13. How Polymers are Used in Medicine
14. The Best Way to Dispose of Medical Waste

"Confront disease at its onset."
PERSIUS (34–62 A.D.)

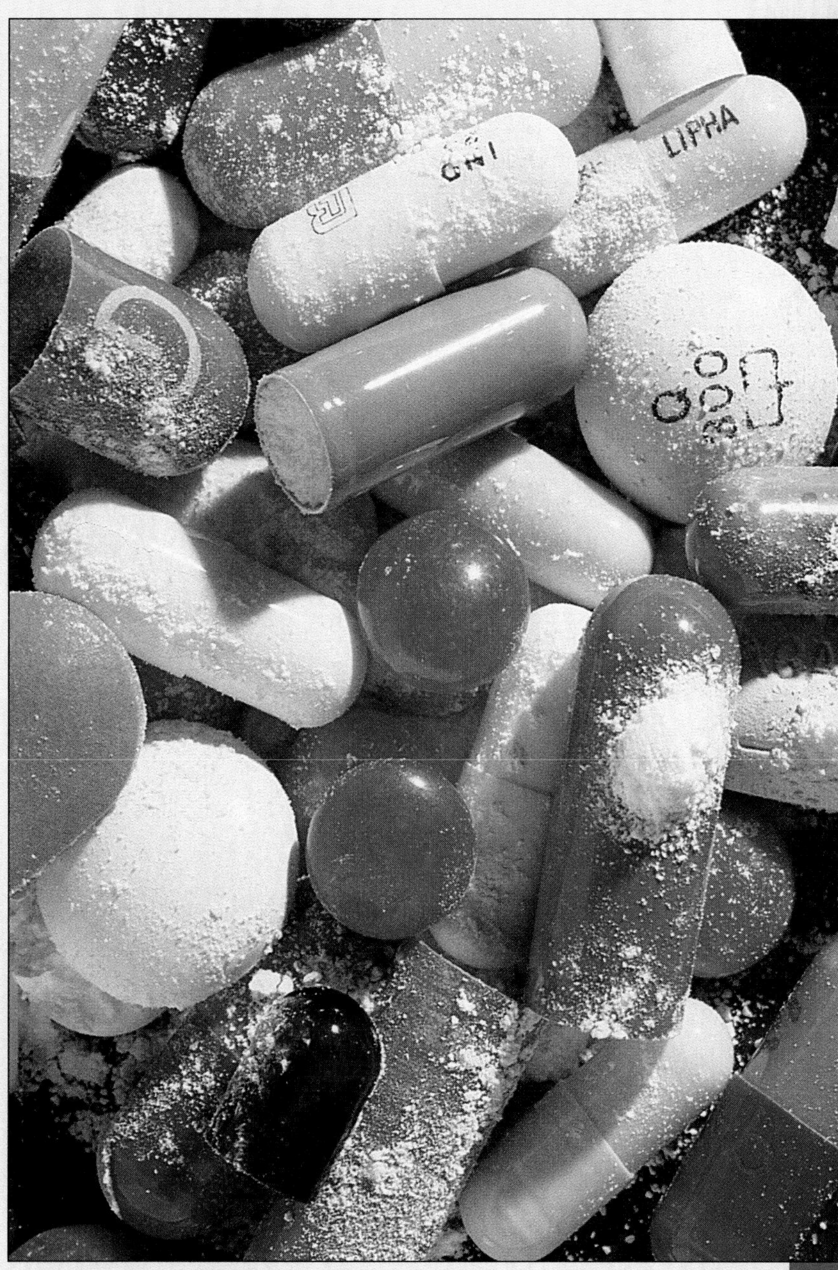

Prescription drugs: pills, tablets, and capsules.

Iatrochemistry (chapter 3) unsuccessfully tried to link chemistry and medicine. Although these early chemists worked in vain, their twentieth century counterparts have seen astonishing results as they apply chemistry to medicine and public health, wiping out one disease after another. Figure 17.1 shows the leading disease-related causes of death for the years 1900 and 1993. In 1900, 61.5 percent of all disease-caused deaths were due to infectious disease; this figure declined to a mere 7.1 percent by 1993. Chapter 12 discussed the dramatic decrease of typhoid fever deaths after water chlorination became commonplace. Chapter 13 explained how effective surfactants gave us the means to create a clean personal environment. This chapter will examine more of the many contributions chemistry has made to medicine and health.

PRACTICE PROBLEM 17.1

Examine figure 17.1 and answer the following questions.

a. What were the six leading causes of death from disease in 1993?

b. Which disease shows the greatest percentage increase between 1900 and 1993?

c. What diseases that killed in 1900 were either wiped out or treatable by 1993? Why?

ANSWER *a.* The leading causes of death in 1993 were, in order, heart disease, cancer, cardiovascular disease, degenerative lung disease, pneumonia and influenza, and diabetes. AIDS is in seventh place. *b.* To determine percentage increases, divide the 1993 percentage by the 1900 percentage and multiply the result by 100. According to this formula, AIDS actually increased the most; since it didn't exist in 1900, its increase is infinite. Aside from AIDS, cancer shows the greatest increase at 417 percent ($100 \times 31.3/7.5$); heart disease increased 258 percent and diabetes 246 percent over this period. Pneumonia decreased to 24 percent of the 1900 values. *c.* Infectious diseases, except pneumonia and AIDS, were wiped out or no longer killers by 1993. Drugs have eliminated most infectious diseases in the United States.

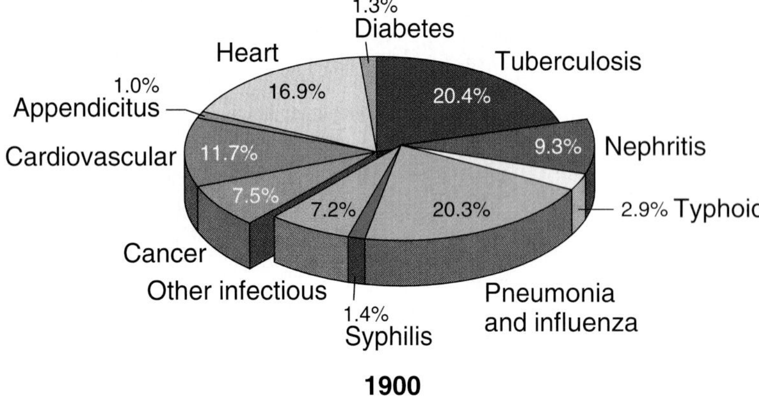

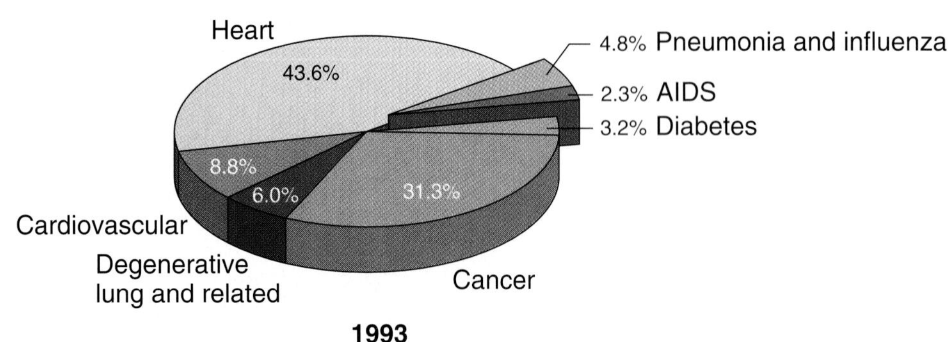

Figure 17.1 The percentages of U.S. deaths from disease in 1900 and in 1993; the offsets show infectious disease deaths.
SOURCE: Data from Department of Health & Human Services, National Center for Health Statistics.

ANALGESICS—WHEN YOU DON'T HAVE TIME FOR THE PAIN

Throughout most of history, pain has been a constant of human life. Pain has one survival advantage—it warns that something is wrong. Nonetheless, humans have long sought ways to reduce or eliminate pain. Chemists still have not attained the pinnacle of achievement in this quest.

This odyssey began thousands of years ago. From antiquity, people knew that alcohol and opium could reduce pain. In the fifth century B.C., Hippocrates recommended willow bark for pain relief, and centuries later—about 1763—English clergyman Edward Stone observed that a willow bark extract could reduce fever. But it wasn't until almost a hundred years later that the true beginnings of pain **chemotherapy** (the chemical treatment of disease) came with the 1860 isolation of salicylic acid from willow bark (figure 17.2). As Hippocrates and Stone had claimed, salicylic acid not only reduced fever, making it an **antipyretic,** it also relieved pain and thus was an **analgesic.** Unfortunately, salicylic acid has a very sour taste and causes stomach irritation.

In 1875, chemists found that the sodium salt of salicylic acid (figure 17.2) tasted less bitter but still caused stomach problems. They continued to experiment with salicylic acid derivatives, and in 1886, they introduced another chemical, salol, a phenyl ester (figure 17.2). Salol was somewhat of an improvement because it did not hydrolyze into salicylic acid and phenol until it reached the intestines, which meant it bypassed the stomach irritation problem. But by about 1900, doctors no longer recommended salol as a general analgesic because an overdose could produce phenol poisoning. Today, salol still finds limited use.

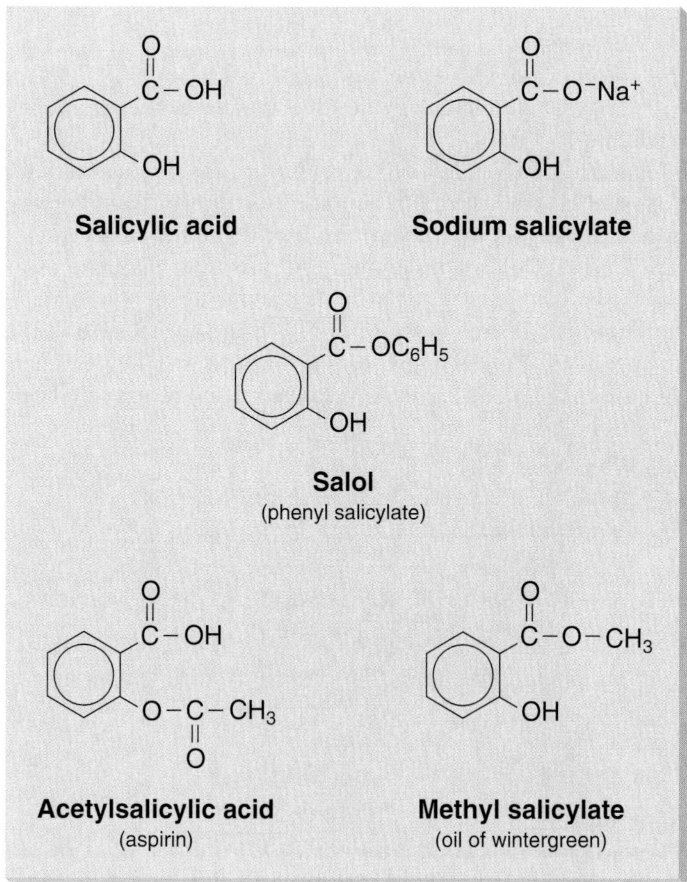

Figure 17.2 Some analgesics derived from salicylic acid.

At the same time sodium salicylate and salol became available as analgesics, Charles-Frederick Gerhardt (1816–1856) was experimenting with what was to become the most important analgesic—**acetylsalicylic acid.** This compound, first synthesized in 1853, was marketed in 1899 as aspirin. Aspirin is now the most commonly used analgesic with an annual U.S. usage exceeding 43,000,000,000 tablets (179 per person). Acetylsalicylic acid is both antipyretic and analgesic. It provides these benefits without drowsiness, euphoria, or other side effects common to some analgesics.

This does not mean aspirin is the perfect analgesic. It still causes stomach irritation in some users and rashes or other allergy symptoms in others. Aspirin is also dangerous for children suffering from flu or chicken pox because it is associated with the occurrence of a potentially fatal disease called Reye's syndrome. Furthermore, it must be stored out of the reach of children who might consume an overdose. Aspirin is the number one cause of poisoning in children.

Nevertheless, the utility of aspirin grows with each passing year. Doctors now know that aspirin reduces the risk of a first heart attack by about 47 percent; it also seems effective in reducing the incidence of certain types of stroke. It is definitely useful for relieving the joint pain of arthritis and rheumatism. Aspirin cannot cure a cold, but it does relieve the fever and pain caused by a cold. Large doses of aspirin even seem to slow eye cataract formation for up to ten years. Aspirin can prevent toxemia in pregnancy and avert vision loss in some people with diabetes. What was once an outstanding pain killer now seems to be an all-purpose medicine.

Can another chemical challenge such a wonder drug? The answer is yes. Aspirin works by reducing the body's prostaglandin levels (chapter 14). but prostaglandins play an important role in healing and body maintenance. In fact, prostaglandins cause the pain associated with fever, swelling, and menstruation. Since aspirin reduces prostaglandin levels, it reduces pain. However, too much pain relief is not a good idea; if prostaglandin levels are too low, we can develop an ulcer or suffer from excessive bleeding. Other drugs can offer pain relief without these dangers, though they have their own side effects. Let's look at several.

Analgesic Alternatives

Alternatives to aspirin include acetaminophen, ibuprofen, and naproxen. Among these, the most popular is **acetaminophen** (figure 17.3), a derivative of acetanilide marketed under the trade name Tylenol. Gerhardt discovered acetaminophen in 1852, one year before he discovered aspirin. Acetaminophen reduces pain and fever about as well as aspirin, but it does not reduce the

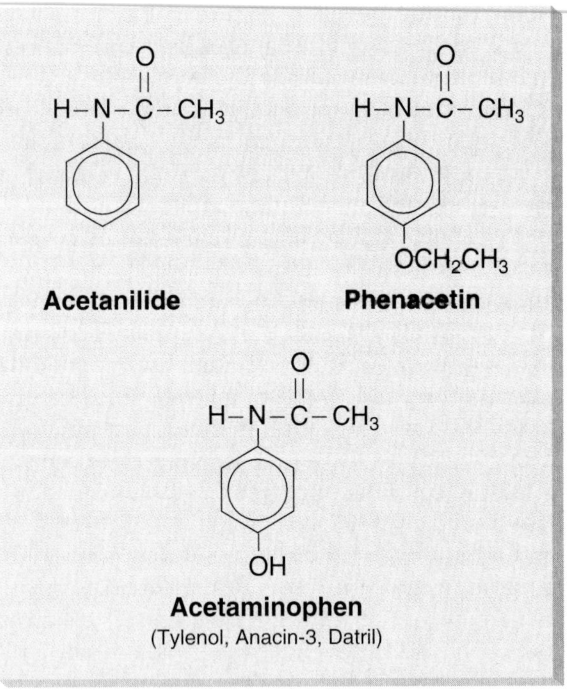

Acetanilide

Phenacetin

Acetaminophen
(Tylenol, Anacin-3, Datril)

Figure 17.3 Some derivatives of acetanilide used as analgesics. Only acetaminophen is still in use as an analgesic.

TABLE 17.1 The toxicities for some analgesics. Ethanol, caffeine, and sodium cyanide are included for comparison

COMPOUND	LD_{50} (mg/kg) (tested orally in mice except as noted)
Ethanol	10,600
Phenacetin	1,650[1]
Ibuprofen (Advil, Nuprin, Motrin)	1,255
Naproxen (Aleve)	1,234
Acetylsalicylic acid (aspirin)	1,100
Methyl Salicylate	887[1]
Acetanilide	800[1]
Sodium salicylate	780[1] (i.p.)[2]
Phenol	530[1]
Salicylic acid	500 (i.v.)[2]
Acetaminophen (Tylenol, Anacin-3, Datril)	338
Caffeine	127
Sodium cyanide	15[1]

SOURCE: Data from *Merck Index of Chemicals and Biologicals*, 11th ed.
[1] Tested in rats
[2] i.v. = intravenous; i.p. = intraperitoneal

inflammation caused by arthritis and rheumatism. Acetaminophen is more expensive than aspirin and is considerably more toxic (table 17.1).

Recall that **LD_{50}** values, the measure of toxicity, represent the amount of a chemical that causes deaths of 50 percent of the laboratory animals in a toxicity test. This amount is usually expressed as mg of the substance per kg of body weight (mg/kg). The test animals are normally mice or rats because they are small and have a fairly short life span; often toxic effects do not appear until a drug is used for most of an animal's life span. The *higher* the LD_{50} value, the safer the chemical.

Relating LD_{50} values to human poisoning is not easy, but table 17.2 gives a general indication of what to expect in a healthy person. As the two tables indicate, it would take many aspirin or acetaminophen tablets to kill you, but some damage often occurs at lower doses. For a 60-kg (132-pound) person, the potential lethal dosages are 75 g ibuprofen, 74 g naproxen, 66 g aspirin, 47 g methyl salicylate, and 20 g acetaminophen. Typical adult tablets contain only 0.25 to 0.5 g of analgesic, so you would need to ingest at least 132 aspirin or 40 acetaminophen tablets (0.5 g, or 500 mg each) to reach toxic levels. Acetaminophen is thus less safe than aspirin, ibuprofen, or naproxen—but all four are safer than caffeine, a naturally occurring stimulant in coffee and tea that is now added to many soft drinks.

Excessive use of acetaminophen has been linked to kidney and liver damage. As few as 50 extra-strength tablets (taken in one dose) can cause death from liver failure. Acetaminophen is sold under such names as Tylenol, Anacin-3, and Datril and is also available in generic versions. It is considered safe for relieving pain and fever in children because it does not cause Reye's syndrome.

Another acetanilide derivative, phenacetin, was once used for the relief of premenstrual pain; this was discontinued because phenacetin causes liver damage and is believed to be a carcinogen. Acetanilide's use as an analgesic also ceased because of its slightly burning taste, water solubility, and toxicity.

Ibuprofen, a third pain reliever, became available over the counter in 1984. After many years as a prescription drug, it is now sold under the brand name Advil, Nuprin, and Motrin, and as a

TABLE 17.2 Estimated relationship between LD_{50} and lethal human dose.

LD_{50} (mg/kg)	PROBABLE LETHAL HUMAN DOSE*
<5	<0.3 g
5–50	0.3–3 g
50–500	3–30 g
500–5,000	30–300 g
5,000–15,000	300–900 g

*Based on an oral dose at 60 kg (132 pounds) body weight

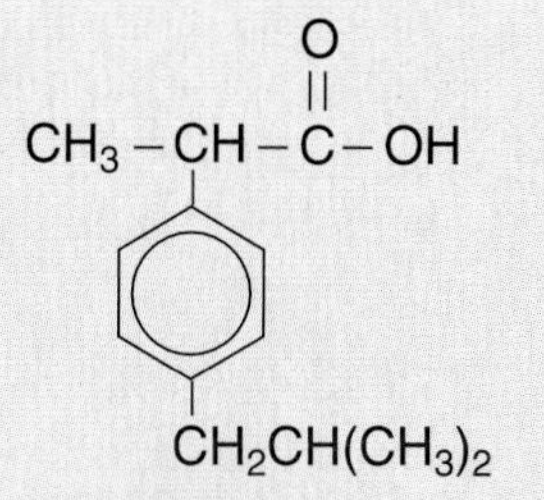

Ibuprofen
(Advil, Nuprin, Motrin)
(1984)

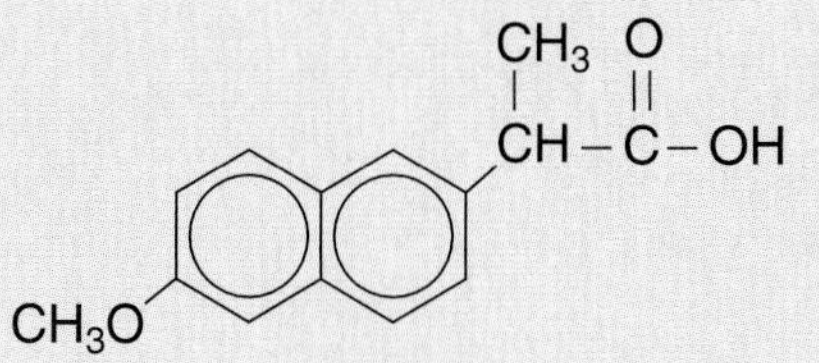

Naproxen
(Aleve)
(1994)

Figure 17.4 Two chemicals that became over-the-counter analgesics in the years listed.

generic drug. Ibuprofen (figure 17.4) is not a salicylic acid derivative, but it behaves similarly and causes comparable allergic reactions in people sensitive to aspirin. Also like aspirin, ibuprofen relieves inflammation due to arthritis and rheumatism; it even relieves inflammation from gout (for which aspirin is ineffective) and is also effective in providing menstrual pain relief. Ibuprofen seems to have more side effects than aspirin or acetaminophen. However, its higher LD_{50} value means it is slightly less toxic than aspirin and much less toxic than acetaminophen.

A fourth over-the-counter analgesic, **naproxen** (figure 17.4), was approved by the FDA in early 1994. Naproxen is antiinflammatory as well as analgesic and antipyretic. Sold under the trade name Aleve, it also has an LD_{50} value greater than that of aspirin.

Methyl salicylate (oil of wintergreen) is the most common topical analgesic (figure 17.2). It is used in rubbing creams and balms but is not ingested. This compound has a slightly different structure than aspirin, but it has similar analgesic properties. A related chemical, the triethanolamine ester of salicylic acid, does not have the strong odor of the methyl ester and is a popular topical analgesic. One recently developed topical analgesic comes from a hot pepper extract called capsaicin (figure 17.5). This product is marketed as Satogesic, Axsain, and other brand names. Capsaicin is claimed to alleviate arthritis pain and the severe pain of shingles.

More potent analgesics are available by prescription, but most are addictive. Their structures appear in figure 17.6. Morphine, a very effective (and addictive) pain killer, is used to relieve severe pain, as a patient with terminal cancer might experience. Codeine is only 10 percent as effective as morphine, but it can be taken orally; morphine requires injection. Meperidine or Demerol is another powerful, addictive analgesic. Methadone is an addictive synthetic narcotic used to treat heroin addiction, but heroin is not used medically to manage pain. In general, these addictive analgesics are more toxic than the over-the-counter drugs (table 17.3).

TABLE 17.3 The toxicities of some addictive chemicals; most are narcotics and are classed as controlled substances. Ethanol, caffeine, and sodium cyanide are included for comparison.

COMPOUND	LD_{50} (mg/kg) (tested orally in mice except as noted)
Ethanol	10,600
Codeine hydrochloride	300
Nicotine	230
Morphine hydrochloride	226 (i.v.)[2]
Meperidine (Demerol)	170[1]
Caffeine	127
Cocaine	18[1] (i.v.)[2]
Sodium cyanide	15[1]
Heroin	1.4 (i.v.)[2]

SOURCE: Data from *Merck Index of Chemicals and Biologicals*, 11th ed.
[1] Tested in rats
[2] i.v. = intravenous; i.p. = intraperitoneal

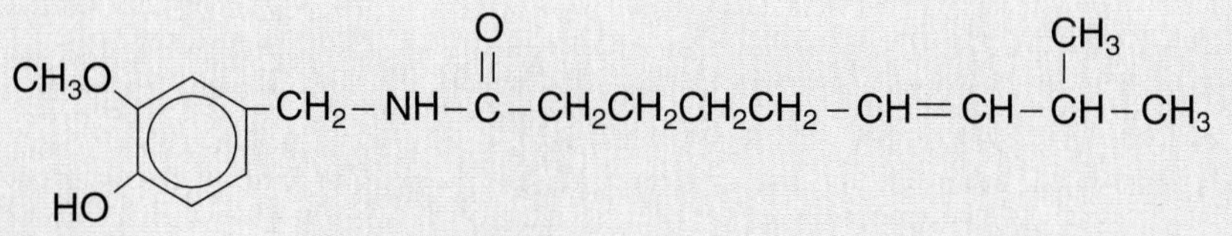

Figure 17.5 The chemical structure of capsaicin, an analgesic isolated from cayenne peppers.

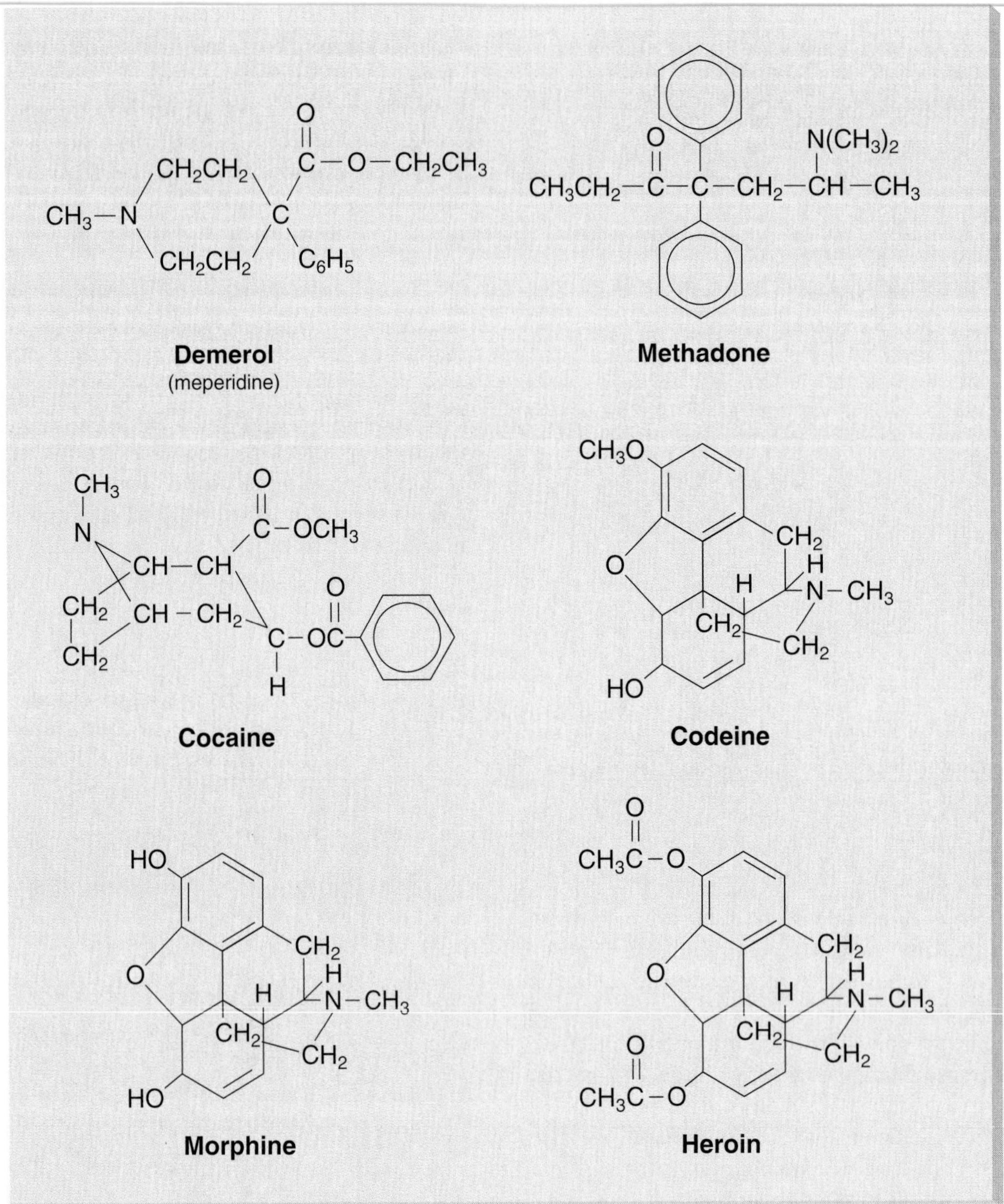

Figure 17.6 Addictive drugs sometimes used as analgesics.

The human body produces its own pain relievers as well. These natural pain relievers, known as **enkephalins,** are small polypeptides made in the brain (figure 17.7). Some enkephalins are a thousand times as effective at relieving pain as morphine. Chemists have synthesized versions with over a *million* times the analgesic activity of morphine, and with less addictive effect. Whether these synthetic enkephalins will ever become prescription drugs remains to be seen. Because the body quickly hydrolyzes them, the duration of the analgesic effect they provide is short.

LD_{50} values are known for all drugs and for many other chemicals. Examples of LD_{50} values for several drugs appear in table 17.4. Other drugs known as anesthetics also relieve pain; box 17.1 discusses them in more detail.

EXPERIENCE THIS

Visit a large analgesic section at a local drug store or supermarket. Make a list of the names of the analgesics available and write down the active ingredients each one contains. Be certain to include generic or store brands in this tally. Also write down the cost and unit size for each brand examined. Later, answer the following questions.

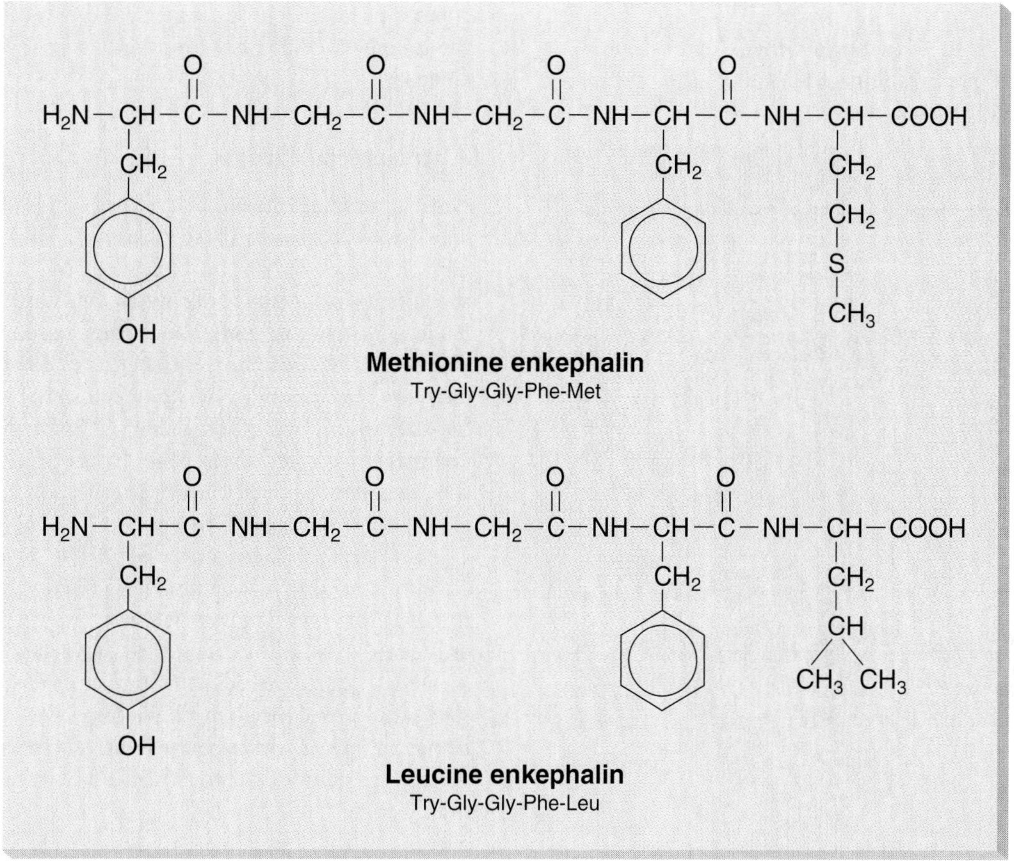

Methionine enkephalin
Try-Gly-Gly-Phe-Met

Leucine enkephalin
Try-Gly-Gly-Phe-Leu

Figure 17.7 The structures for two enkephalins, the addictive analgesics produced in the brain.

a. How many different chemicals are used in these analgesics?

b. Which chemicals are actually analgesics (as opposed to other ingredients)?

c. If possible make an educated guess at the purpose for adding other chemicals (caffeine).

d. How much do the costs of different brands of the same analgesic vary?

e. Aside from cost, what other factors might you consider in selecting a particular analgesic?

ANSWER *a.* Tetracycline is the least toxic and cisplatin is the most toxic. *b.* The LD_{50} value for aspirin is 1,100. Only those drugs with higher LD_{50} values are less toxic; these include tetracycline and sulfanilamide. (Penicillin is close in toxicity to aspirin.) *c.* Cyclophosphamide, an anticancer drug, has about the same LD_{50} value as acetaminophen, but cisplatin, another anticancer drug, is even more toxic; *d.* Ibuprofen and naproxen are similar in toxicity to aspirin. They are less toxic than three of these drugs, more toxic than tetracycline, and sulfanilamide.

PRACTICE PROBLEM 17.2

Consider the LD_{50} data for the following five drugs: tetracycline, sulfanilamide, penicillin V, cyclophosphamide, and cisplatin. Referring to tables 17.1 and 17.4, answer the following questions:

a. Which of these five drugs is least toxic? Which is most toxic?

b. Which of these drugs are less toxic than aspirin?

c. Which drugs are more toxic than acetaminophen?

d. How do ibuprofen and naproxen compare with these five drugs in toxicity?

PRACTICE PROBLEM 17.3

Refer to tables 17.3 and 17.4 to determine *(a)* which of the drugs in Practice Problem 17.2 are more toxic than nicotine and *(b)* which are more toxic than cocaine.

ANSWER Any drug with an LD_{50} lower than the 230 value for nicotine is more toxic; thus, only cisplatin is more toxic than nicotine—something to remember if you smoke. Cisplatin is also the only drug of these five that is more toxic than cocaine.

TABLE 17.4 The toxicities of other chemicals used in medicine. Ethanol, caffeine, and sodium cyanide are included for comparison.

COMPOUND	LD_{50} (mg/kg) (tested orally in mice except as noted)
Ethanol	10,600
Tetracycline	6,443[1]
Sulfanilamide	2,000[2]
Penicillin V	>1,040[1]
Diphenhydramine	500[1]
Cyclophosphamide	350
Lidocaine	292
Phenobarbital	162[1]
6-mercaptopurine	157
Caffeine	127
Formaldehyde (37% in water)	80[1]
Sodium pentothal	78 (i.v.)[4]
Procaine	45 (i.v.)[4]
Sodium cyanide	15[1]
Methotrexate	14 (i.v.)[1,4]
Cisplatin	9.7 (i.v.)[3,4]

SOURCE: Data from *Merck Index of Chemicals and Biologicals*, 11th ed.
[1]Tested in rats
[2]Tested in dogs
[3]Tested in guinea pigs
[4]i.v. = intravenous; i.p. = intraperitoneal

PRACTICE PROBLEM 17.4

Refer to tables 17.2 and 17.4 to determine the probable lethal human dose for each of the five drugs listed in Practice Problem 17.2.

ANSWER Tetracycline 300–900 g; sulfanilamide 30–300 g; penicillin V 30–300 g; cyclophosphamide 3–30 g; cisplatin 0.3–3 g.

ANTIBIOTICS—CHEMICALS THAT FIGHT DISEASE

Today we take access to needed medications for granted, but the development of effective chemical agents for treating infectious diseases is a comparatively recent event. We now have drugs that combat bacteria, viruses, cancer, AIDS, Alzheimer's disease, and the common cold. New drugs and new techniques continue to emerge.

Antibacterial Drugs

Antibacterial agents either destroy bacteria **(bactericidal)** or prevent bacterial growth **(bacteriostatic)**. They work by eliminating enough bacteria that the body's defense network can eradicate the infectious agents. Folk medicine, dating back to antiquity, made use of herbs, garlic, and other natural substances to treat diseases. Paracelsus (sixteenth century) recommended using mercury compounds and other toxic substances to treat disease, but his suggestions were not widely adopted, probably because the chemicals were too dangerous. In the mid-1800s, people knew that elemental iodine killed bacteria; however, too much iodine is harmful to most body tissues.

Paul Ehrlich (German, 1854–1915) developed the first true antibiotic agent in 1907; it was the red dye trypan, and he used it to treat African "sleeping sickness." In 1909, Ehrlich developed an arsenic-containing drug called Salvarsan 606 (figure 17.8) that seemed to cure the sexually transmitted disease syphilis. This story was popularized in the movie *The Magic Bullet*, a term Ehrlich coined to describe the relatively specific action of his drug on the spirochete bacteria that cause syphilis. Salvarsan 606 is highly toxic (LD_{100} = 140 mg/kg i.v. in rats, meaning that 100 percent of the animals die at this dose), but it was the drug of choice to treat syphilis for many years because there was nothing better. Ehrlich won the 1908 Nobel Prize in medicine and is considered the founder of chemotherapy, the use of chemicals to treat disease.

The next major attack in the battle against bacteria did not occur until the 1930s. This is surprising, considering Ehrlich's pioneering research and the fact that the penicillin family was discovered in 1929. **Sulfanilamide** (figure 17.9), the prototype **sulfa drug,** was first synthesized in 1908, but its drug potential was not discovered for many years. In 1932, Gerhard Domagk (Polish, 1895–1964) showed that a related compound, a red dye called

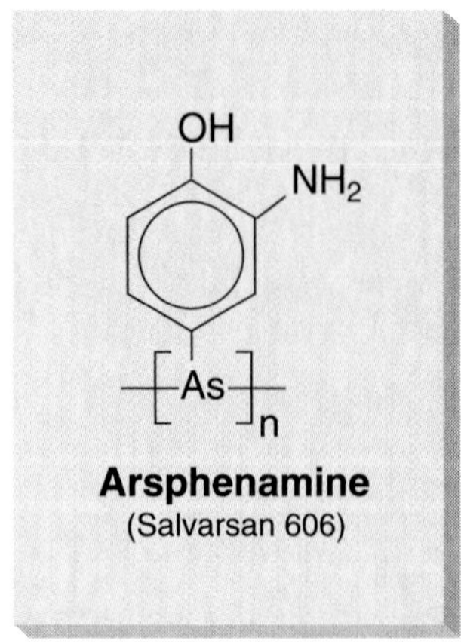

Figure 17.8 Salvarsan 606.

BOX 17.1

ANESTHETIC AGENTS

Like analgesics, **anesthetics** relieve pain, but unlike analgesics, they produce a complete or partial lack of feeling. Modern surgery is almost always performed with the patient under an anesthetic. Anesthetics may be local or general; local anesthetics operate on localized nerve centers, and the person remains conscious; while general anesthetics act on the brain centers and render the person unconscious. The first example of a chemical anesthetic appears to be Henry Hickman's 1824 use of carbon dioxide as a general anesthetic for an animal.

Nitrous oxide (figure 1) was discovered in 1800 by Humphrey Davy (English, 1778–1829). Davy suggested its use as a general anesthetic. In 1844, Horace Wells (American, 1815–1848) became the first person to use nitrous oxide as an anesthetic, in dentistry. Subsequently, nitrous oxide proved to be fairly dangerous and to cause brain damage.

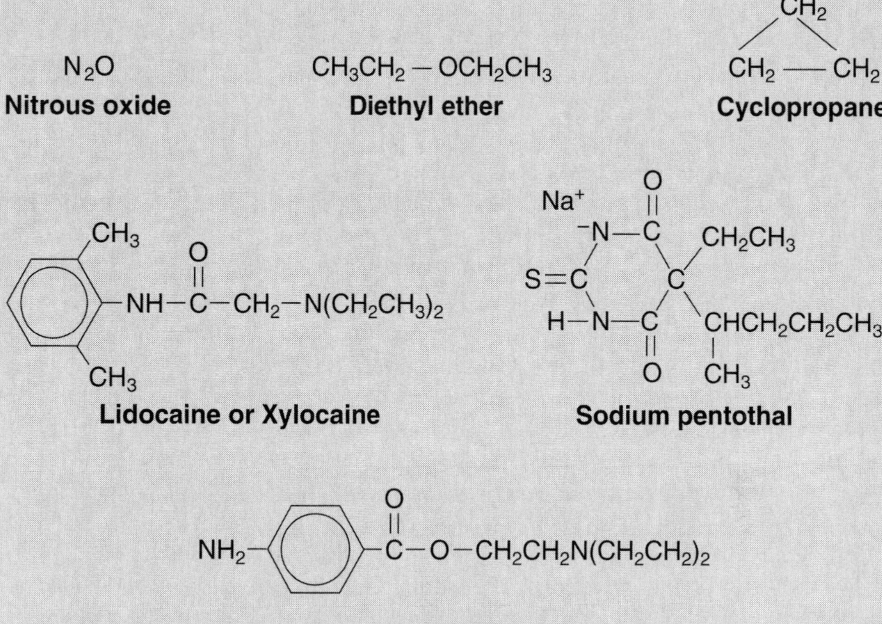

Figure 1 Some chemicals used as anesthetics.

Charles Thomas Jackson (American, 1805–1880) first discovered diethyl ether's anesthetic properties in 1841. By 1842, Crawford Long (American, 1815–1878) used it to perform surgery; however, Long did not publish his results until 1849. Meanwhile, in 1844, Jackson suggested the use of ether to deaden pain to American dentist William Morton (1819–1868). Morton actually used ether as an anesthetic in 1846. The use of ether soon became widespread (figure 2). However, ether is extremely flammable—it forms explosive mixtures with air—and it also frequently produces nausea. Because of these hazards, diethyl ether is seldom used in the United States, though it is still widely used in China.

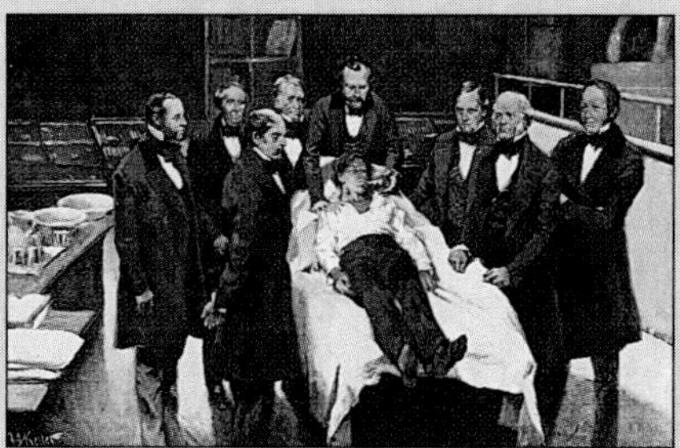

Figure 2 The early uses of ether as an anesthetic.

Chloroform ($CHCl_3$) was discovered in 1831 by Samuel Guthrie (American, 1782–1848), but no one used it as an anesthetic until 1846. At that time, James Simpson (Scottish, 1811–1870) concluded that chloroform was a better anesthetic than either nitrous oxide or ether and began to use it to relieve the pain of childbirth for his patients. Though chloroform was used as an anesthetic for many years because it is less flammable than ether, doctors now know that $CHCl_3$ causes liver damage and that the effective anesthetic dose is rather close to the lethal dose.

Cyclopropane (C_3H_6) was used as an anesthetic for a short time, but it forms an even more potent explosive mixture with air than ether does, so its use has been discontinued. The highly flammable gas acetylene ($CH{\equiv}CH$) was also used briefly as an anesthetic in 1924. Some fluorinated compounds, such as halothane ($C_2HBrClF_3$) or halopropane ($C_3H_3BrF_4$), were tested as anesthetics but appeared to cause miscarriages and other side effects. Modern surgery relies on combinations of chemicals such as sodium pentothal, an intravenous anesthetic.

Your dentist might use a local anesthetic, such as procaine, (Novocain) or lidocaine (Xylocaine), when working on a sensitive spot in your mouth. Novocain was first introduced in 1905 by Albert Einhorn. In the old days, before these pain-relieving chemicals were available, patients literally had to "bite the bullet," use alcohol, or take an addictive drug to undergo surgery or dental work. Modern anesthetics, used judiciously, are certainly preferable.

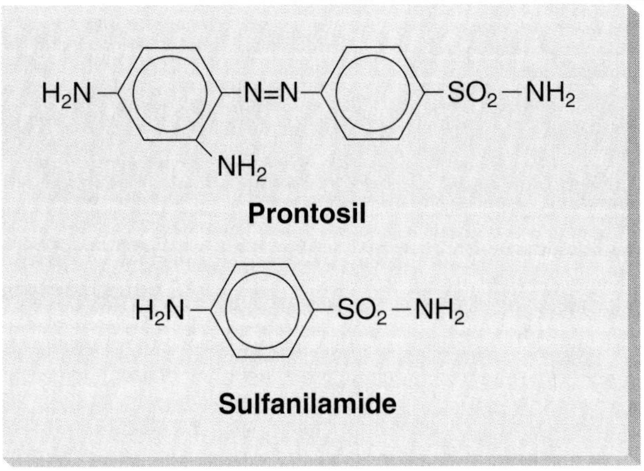

Prontosil

Sulfanilamide

Figure 17.9 The sulfa drugs.

Prontosil (figure 17.9), was effective in treating streptococcal infections in animals. Soon afterwards, his daughter contracted a dangerous streptococcal infection. In desperation, Domagk gave her Prontosil and effected a cure; he won the Nobel Prize in medicine in 1939.

By the end of World War II, several other sulfa drugs had been developed. Sulfa drugs are bacteriostatic and structurally similar to p-aminobenzoic acid (PABA, the key ingredient in many sunscreens). Sulfa drugs insert themselves into the bacteria's folic acid production pathway in the place of PABA. Folic acid is essential for bacterial growth, but the pseudopathway with a sulfa drug in it does not work. Thus, the bacteria can no longer grow and reproduce, and the body's natural defense systems can then control the more limited number of bacteria, winning the battle against the disease.

The story of **penicillin** is interesting. In 1928, Alexander Fleming (Scottish, 1881–1955) discovered penicillin when a germ culture accidentally became moldy. Before he tossed away the errant experiment, Fleming noticed that there was a circular area with no bacterial growth around each mold spore. Fleming pursued his chance discovery and published a report on the bactericidal activity of the penicillin molds. He noted that the penicillin killed some bacteria and not others, and that it did not harm white blood cells. It is interesting to note that John Tyndall (Irish, 1820–1893) had also seen the same lack of bacterial growth around mold spores in the mid-1860s, but Tyndall did not pursue this observation.

Fleming never isolated the active agent in penicillin; Howard Florey (Australian, 1898–1968) and Ernst Chain (German, 1906–1979) picked up Fleming's unfinished work in 1939 and completed that task. Both men were seeking new antibacterial agents to use on wounds inflicted during World War II, and they settled on penicillin in 1940. Florey and Chain found they could isolate and manufacture the active mold ingredient by growing the mold in large tanks. The Fleming, Florey, and Chain trio won the 1945 Nobel Prize for this work. Penicillin has saved thousands of lives and is still one of the most widely used antibiotics.

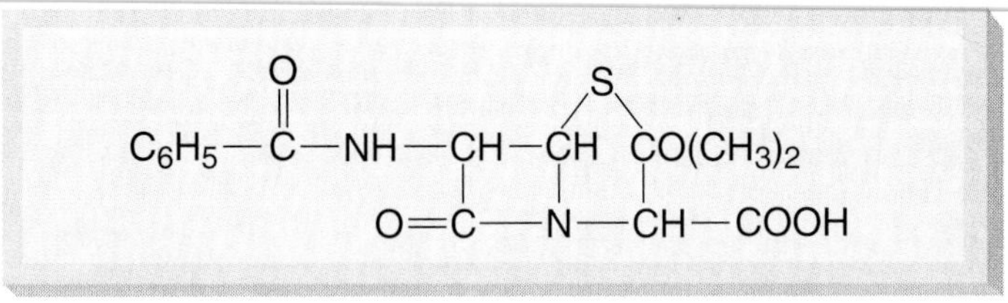

Figure 17.10 The chemical structure of penicillin G.

Penicillin's structure (figure 17.10) was determined in 1949 by Dorothy Hodgkin (English 1910–). Hodgkin later won the 1964 Nobel Prize for her 1956 delineation of the structure of Vitamin B_{12}. The penicillin molecule contained structural features chemists could not synthesize at that time, but new synthetic techniques were eventually developed, largely by John Sheehan (American, 1916–1992). As a result, many different penicillin versions are now made synthetically. The penicillins are bactericidal; they work by preventing cell wall formation in bacteria, causing the insides to spill out. A fair number of people are allergic to penicillin and must use other drugs.

Chemists have developed numerous other antibiotic agents, including streptomycin, erythromycin, and the tetracycline family. Drugs with names ending in *mycin* were first derived from various microorganisms in the soil. The first successful researcher in this area was René Jules Dubos (French, 1901–1982). In 1939, Dubos was deliberately seeking antibacterial agents in soil samples and found gramicidin and tyrocidin. In 1943, Selman Waksman (Russian-American, 1888–1973) discovered streptomycin in a similar manner. Unlike the previous agents, which were only effective against gram-positive bacteria, streptomycin worked against gram-negative bacteria. Waksman (1941) coined the term **antibiotic** for substances that kill bacteria without injuring other forms of life. In 1944, Benjamin Duggar (American, 1872–1956) discovered aureomycin, the first of the tetracycline family, in soil samples. The tetracycline family (figure 17.11) are broad-spectrum antibiotics—that is, they kill a broad range of bacteria. Numerous other antibiotic agents have been discovered in plants, soils, and animals; the search continues for newer and better drugs.

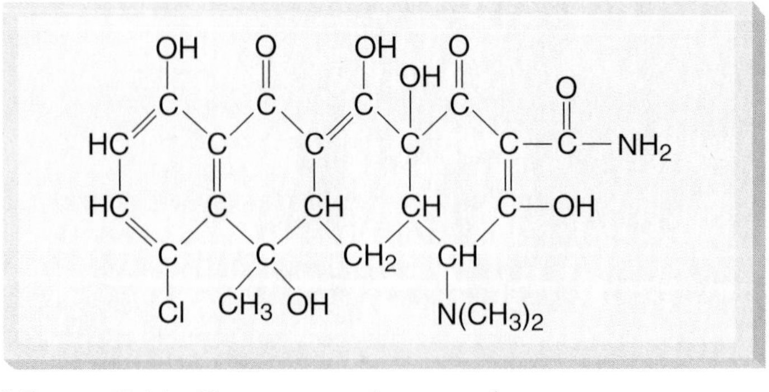

Figure 17.11 The structure of a tetracycline.

THE BATTLE AGAINST CANCER

Today, in industrialized countries, the major afflictions are not infectious diseases but heart and cardiovascular disease or cancer. The very word *cancer* strikes fear in many; at any given time, this dreaded disease affects about half the families in the United States, and the incidence of cancer is showing a steady increase (figure 17.12). (Use caution in interpreting the data in figure 17.12. The increase appears linear over time, but that cannot be true since the straight line implies there were no cancer cases prior to 1870, and cancer has existed for thousands of years.) A major problem in the war against cancer is that we are fighting several different enemies. Although cancer always involves the abnormally rapid division and replication of body cells, it is actually many loosely related diseases with many causes. There is no all-purpose cure for cancer, but several forms are increasingly curable. (See also the discussion on melanoma in box 16.2.)

Cancer prevention is now considered a priority, and this means avoiding certain activities or materials. Tobacco is clearly linked to cancer and probably causes more cancers than any other single substance; it is estimated that as many as 85 percent of lung cancer cases are related to tobacco use. Studies also appear to link certain cancers to lifestyle and diet. Some experts estimate that 35 percent of all cancers may be diet related, and that maintaining a healthy diet may aid in resisting cancer; studies show that adding fiber and reducing saturated fat seems to combat gastrointestinal cancer, for example. Broccoli, cauliflower, cabbage, and other foods contain a natural cancer-averting chemical called sulforaphane. On the other hand, naturally occurring substances such as the radioactive gas radon, asbestos, coal dust, and lint have been implicated in causing cancer. "Natural" is not always "better."

800
700
600
500
400
300
200
100
0
Cancer deaths
1960
1970
1980
1990
2000
2010
2020
Year
= number deaths in thousands
= death rate per 100,000 people

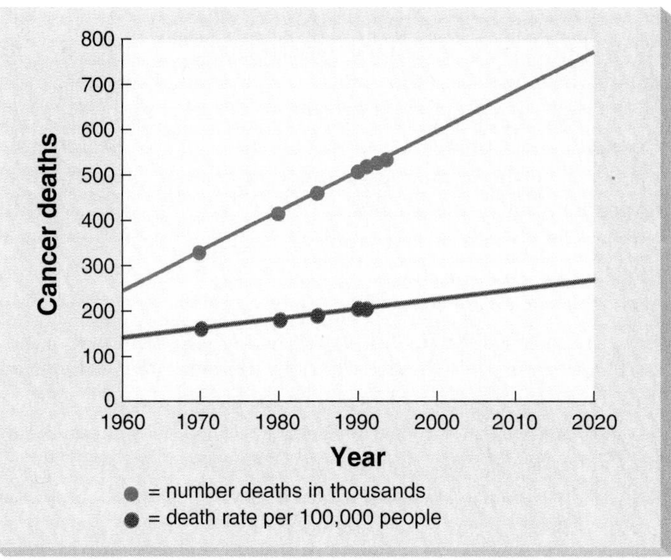

Figure 17.12 The increase in U.S. cancer deaths from 1970 to 1993. The upper line shows the total number of deaths in thousands; the lower line shows the death rate per 100,000 people. SOURCE: *The American Almanac*, 1993–1994.

Carcinogens are chemicals that cause cancer. Figure 17.13 shows a few examples of carcinogens; as the figure demonstrates, they have an incredible assortment of structures. The oldest

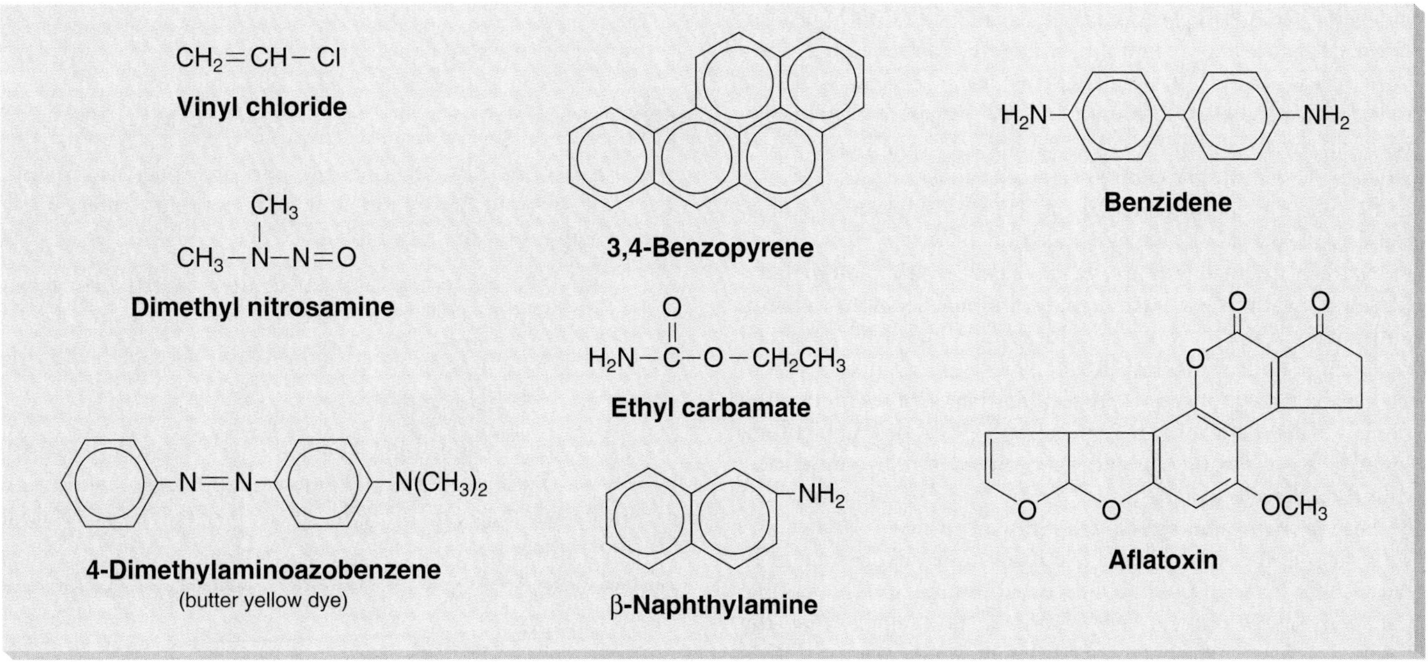

CH2=CH–Cl
Vinyl chloride
CH3
CH3–N–N=O
Dimethyl nitrosamine
N=N
N(CH3)2
4-Dimethylaminoazobenzene
(butter yellow dye)
3,4-Benzopyrene
O
H2N–C–O–CH2CH3
Ethyl carbamate
NH2
β-Naphthylamine
H2N
NH2
Benzidene
O
O
O
O
O
OCH3
Aflatoxin

Figure 17.13 The structures of several carcinogens (cancer-causing chemicals).

known carcinogen is 3,4-benzopyrene, which was first isolated from chimney soot. Chimney sweeps, usually small boys, frequently developed scrotum cancer—a cancer that was rare in the rest of the population. Scientists asked, what are chimney sweeps exposed to that the rest of society avoids? Careful research eventually linked the scrotum cancer with soot; then laboratory tests on mice revealed that specific compounds in the soot caused the cancer. This was the first study that linked specific chemicals directly to cancer. Later studies revealed that 3,4-benzopyrene also occurs in cigarette smoke.

The carcinogens benzidine and β-naphthylamine (figure 17.13), once used in the dye industry, were credited with causing an elevated number of bladder cancer cases among dye factory workers compared to the general population. These chemicals are no longer used in dye manufacture. Another dye called "butter yellow" (4-dimethylaminoazobenzene) was used to color margarine before chemists discovered it was a carcinogen; it is no longer used in foods.

Of the nearly two hundred nitroso-compounds tested, only a few were *not* carcinogenic. Unfortunately, nitroso-compounds can form when amines react with potassium nitrite (KNO_2), a chemical used to prevent meat spoilage. The use of KNO_2 has been restricted in the United States for this reason.

The most potent carcinogens yet discovered are the aflatoxins, excreted by molds growing on grains, sweet potatoes, and peanuts. The FDA restricts the aflatoxin level to less than 20 ppb. Food and candy companies must painstakingly test for aflatoxins before using peanuts in their products. All things considered, it is nearly impossible to avoid encountering some carcinogens in our daily lives. Fortunately, most of these encounters do not lead to cancer. Wariness can keep you reasonably safe.

Chemists remain uncertain about what makes a chemical carcinogenic. Some believe the chemical interacts with DNA, but chemists are only now beginning to understand such interactions. Other causes for cancer, besides chemical carcinogenesis, include radiation and viruses. It is unlikely that all cases of chemically induced cancer arise in the same manner and equally unlikely we will develop a universal cure; research continues on finding causes and treatments for specific cancers. (Box 17.2 describes animal testing to learn about human disease.)

Anticancer chemicals may be classified as either **antimetabolites** or **alkylating agents** (though a few agents do not fit well into either class). Since cancer is a disease of abnormally rapid cell replication, both antimetabolites and alkylating agents attempt to slow or halt the replication process. Antimetabolites imitate the normal components of DNA and interfere with cell metabolism, usually by preventing the cell from dividing and forming new cells. Cancer cells require more RNA and DNA than normal cells because of their rapid reproduction rate. This means that cancer cells take up the available nucleic bases more rapidly than ordinary cells. If an antimetabolite is substituted for the normal nucleic bases, the cancer cells take it up more rapidly than the healthy cells; since antimetabolites do not produce "good" DNA, the cancer cells cannot reproduce.

At present, it is very difficult, if not impossible, to create drugs that are specific for cancerous cells rather than healthy cells. Many times a number of healthy cells are damaged in chemotherapy treatments. Still, this is preferable to leaving the cancer unchecked. The main antimetabolites used in chemotherapy are 5-fluorouracil, 6-mercaptopurine, and methotrexate (figure 17.14).

Alkylating agents appear to react with DNA and change its basic structure by adding an alkyl group (figure 17.15). Alkylating agents were discovered when doctors noticed that the deadly mustard gas used in World War I sometimes inhibited or reduced cancer growth. Patients exposed to these poisonous gases often showed a reduction in lung tumor size. Was there a relationship? If so, the compounds in mustard gas might prove useful in cancer treatment. Researchers developed milder derivatives containing nitrogen instead of sulfur. Cyclophosphamide is now the preferred alkylating agent because it affects healthy tissues less than other agents do. Most alkylating agents can also cause cancer, probably because they react with healthy DNA and convert it into an abnormal form.

Cis-dichlorodiamino platinum, called cisplatin or cis-DDP, also interacts with DNA (figure 17.16). However, instead of adding an alkyl group, cisplatin produces crosslinks between DNA strands and thereby prevents cancer cell replication.

New anticancer agents appear annually. Among the newer waves are some biotechnology products, including the polypeptides interleukin-2 (IL-2), colony-stimulating factor (CSF), and erythropoietin (EPO). Interferon (another polypeptide) also shows promise. All of these polypeptides are extremely complicated in structure and often highly specific. Some are also moderately toxic.

PRACTICE PROBLEM 17.5

Review the LD_{50} data for the anticancer agents cisplatin, cyclophosphamide, and 6-mercaptopurine (table 17.4). Then answer the following questions:

a. Which of these anticancer drugs is the least toxic? Which is the most toxic?

b. Are any less toxic than nicotine (see table 17.3) or caffeine?

ANSWER Cyclophosphamide is the least toxic, cisplatin the most. Both cyclophosphamide and 6-mercaptopurine are less toxic than caffeine. Only cyclophosphamide is less toxic than nicotine.

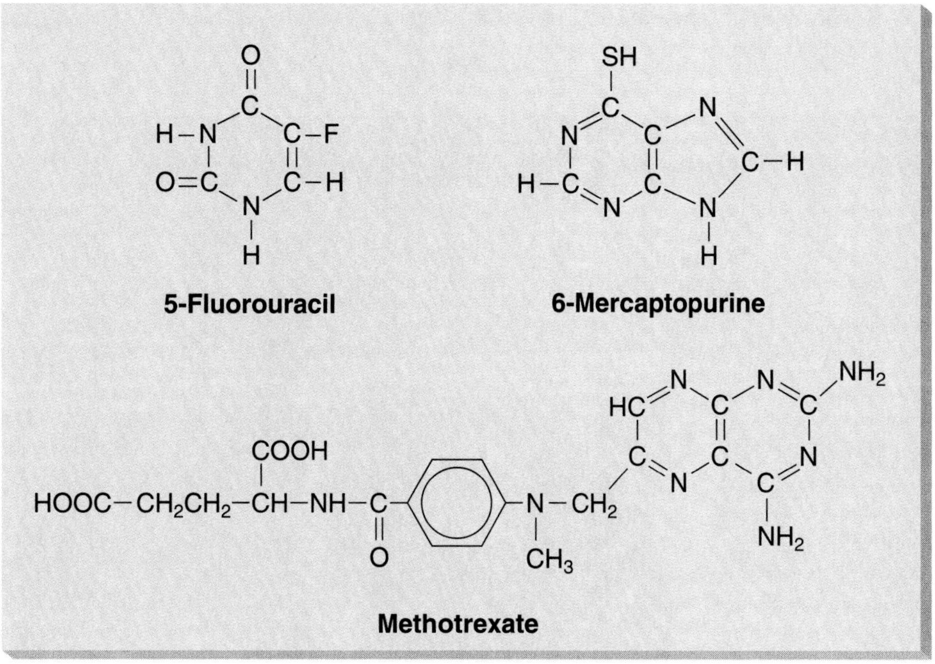

5-Fluorouracil

6-Mercaptopurine

$HOOC-CH_2CH_2-CH(COOH)-NH-C(=O)-C_6H_4-N(CH_3)-CH_2-$(pteridine ring, NH_2, NH_2)

Methotrexate

Figure 17.14 Three antimetabolites used to treat cancer.

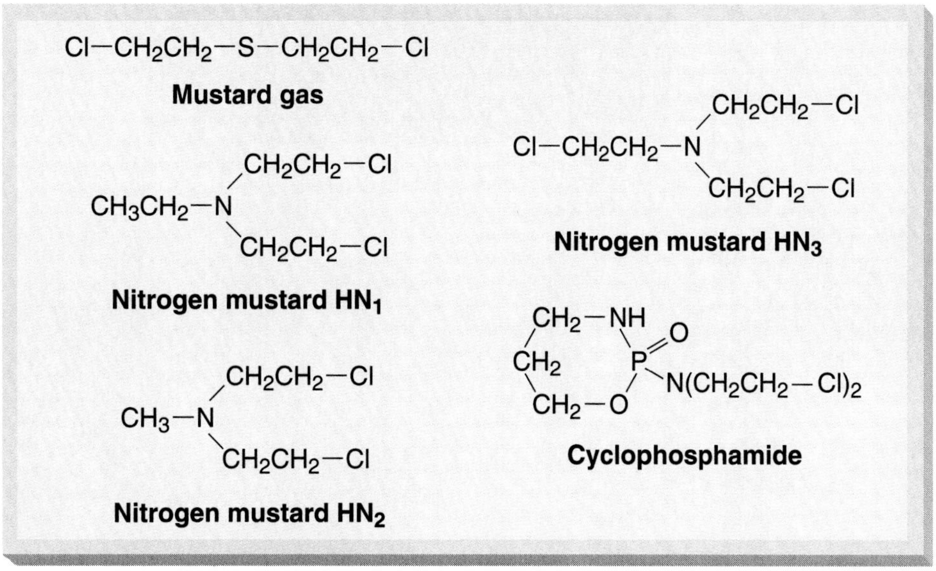

$Cl-CH_2CH_2-S-CH_2CH_2-Cl$

Mustard gas

$CH_3CH_2-N(CH_2CH_2-Cl)_2$

Nitrogen mustard HN_1

$CH_3-N(CH_2CH_2-Cl)_2$

Nitrogen mustard HN_2

$Cl-CH_2CH_2-N(CH_2CH_2-Cl)_2$

Nitrogen mustard HN_3

CH_2-NH, CH_2, CH_2-O, $P(=O)$, $N(CH_2CH_2-Cl)_2$

Cyclophosphamide

Figure 17.15 Some alkylating agents used to treat cancer.

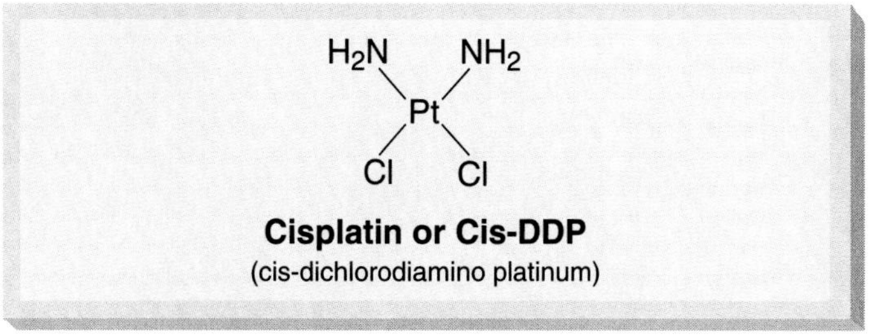

H_2N NH_2 Pt Cl Cl

Cisplatin or Cis-DDP

(cis-dichlorodiamino platinum)

Figure 17.16 Cisplatin, an anticancer agent that appears to work by interacting directly with DNA, but is not an alkylating agent.

BOX 17.2

ANIMAL TESTING AND HUMAN HEALTH

Chemists have screened over 22,000 of more than 50,000 common chemicals as potential carcinogens. Screening tests are very difficult and costly procedures—some studies of a single compound may cost millions of dollars and require several years of animal tests. The standard testing procedure is to feed large amounts of the chemical to laboratory animals—usually mice or rats—on a regular basis and examine them for tumors after several months or years (figure 1). The animals must be fed and housed in clean facilities and carefully monitored during the test period, or the data are meaningless.

Serious scientists are very careful with laboratory animals because they recognize their crucial contribution to our knowledge of the disease process. However, some animal rights activists, believing laboratory animals are mistreated, break into laboratories, destroy equipment and research records, and set infected animals loose—even those infected with incurable diseases. These activists have increased the costs of performing medical research by causing researchers to divert research funds to security operations; this slows progress toward curing diseases and ultimately increases medical care costs and prolongs human suffering (and animal suffering, since many medical discoveries are used to treat animals as well as humans). Some groups want to ban animal testing altogether. If there were an effective alternative, scientists might agree, but this is often not the case.

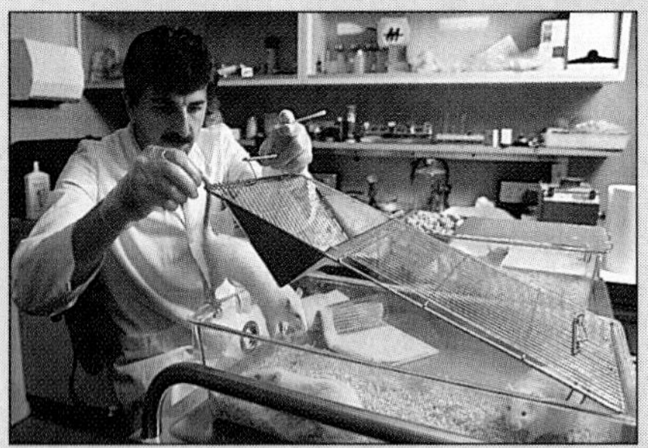

Figure 1 A laboratory animal colony.

Figure 2 Perhaps scientists will someday rely on computers to simulate animal research.

Two suggested alternative techniques include running computer simulations and testing tissues or cultures rather than live animals. Unfortunately, these techniques are not always effective. Although most scientists would undoubtedly prefer running computer simulations (figure 2), considering the funds and time needed to obtain reliable animal data, it is virtually impossible to create the software to run such simulations without first gaining data from animal testing. (In fact, the few existing programs were written based on data from animal testing.) Perhaps someday a good computer simulation will exist, but this is not a viable option at present.

A few screening tests do exist that use tissues or cultures instead of animals—for example, the HeLa cell research as shown in figure 3. But most scientists are reluctant to conclude that a compound is or is not carcinogenic in humans based on such tests; it they are wrong, the cost may be measured in human lives. Even "helpful" chemicals—drugs to treat cancer or AIDS, for example—may have deadly side effects that do not show up in a laboratory culture. At least for now, these tests require animals.

Figure 3 A scientist doing *HeLa* cell research. *HeLa* cells were the first human cell line used for cancer research (1952) and were obtained from the cervix of Henrietta Lacks. *HeLa* cells, which were named after her, thrive in the laboratory and are used in viral, genetic, and anticancer research. In this example, a medium containing a poliovirus vaccine will be used to kill the *HeLa* cells.

THE FIGHT AGAINST A MODERN-DAY PLAGUE—AIDS

Just a few decades ago, nobody had ever heard of **AIDS,** an acronym for **Acquired Immune Deficiency Syndrome.** AIDS, which destroys the body's disease defense mechanisms, is a viral disease caused by **HIV (Human Immunodeficiency Virus).** Unlike other viral diseases (such as influenza, which is spread through an airborne virus), AIDS is spread in a limited number of ways, most of which involve the exchange of body fluids. AIDS transmission vehicles may be summarized into four groups: (1) engaging in sexual activity with an HIV-infected person, (2) receiving a blood transfusion containing HIV, (3) sharing hypodermic needles (drug use) with an HIV-infected person, and (4) transfer of body fluids from a person infected with HIV. People who engage in such behaviors are at high risk for contracting AIDS.

Rumors persist that AIDS also spreads in other ways, such as through sneezing or mosquito bites, but no compelling evidence supports these alternate routes. A person not in one of the four high-risk groups has a risk of less than 1 in a million of getting AIDS. The chances of being killed by a terrorist while traveling abroad is about 1 in 700,000; the risk of dying in an automobile accident is about 1 in 5,300. Still AIDS is a serious health problem throughout the world.

The question of how AIDS originated has never been totally settled, but the most probable original source for the HIV virus appears to be the simian T-lymphotropic virus III (STLV-III) that affects green monkeys in Africa. The crossover of a virus between species is uncommon, but not unheard of. One factor that may have facilitated this crossover is the practice of some Africans to eat green monkeys. (Box 17.3 discusses viral transmission and treatment in more detail.)

When did this crossover occur, and how rapid was the spread of the human form of the infection? Stored African blood samples, tested in the 1980s, revealed that HIV already existed in Central and Eastern Africa during the late 1950s. The earliest serological (blood) proof for HIV infection was found in the area surrounding Lake Victoria, bordered by Kenya, Tanzania, and Uganda. By the mid-1970s, AIDS was an epidemic in Africa. AIDS decimated entire villages in Uganda, leaving virtually the entire adult population dead. The HIV infection spread rapidly as the society in this region changed from a rural tribal system to a more modern society, leading to changes in the traditional values of monogamy. In recent years, the AIDS problem has worsened in Africa, with 60 percent of all worldwide AIDS cases now recorded in a continent that contains only 12 percent of the world's population. Pessimistic estimates claim that 33 percent of South Africa could be infected by 2010, with 6 million deaths.

AIDS next spread to the United States, the Caribbean, Europe, and Asia. By the late 1970s, it had become a global epidemic—a pandemic—in just one third of a single human generation. The HIV virus probably arrived in the United States in the late 1960s but remained undetected until the early 1980s, since its early spread in the United States was primarily in homosexual communities. In the meantime, AIDS remained a heterosexual problem in most of the world, though it went largely undetected as a specific disease. The next major group to contract AIDS was intravenous drug users sharing HIV-contaminated needles. Infected blood supplies also spread the disease (box 17.4).

By 1987, 10,000 people were infected worldwide, and the World Health Organization predicted this number would increase to 100,000,000 by 1991. However, these estimates were not borne out; the number of HIV-infected people totaled 11 to 14 million in 1992 (figure 17.17). In 1992, at the time the number of AIDS cases was estimated at 11.8 million worldwide, health organizations projected a rise to 17.5 million by 1995. The lowest expected percentage increase in new HIV infections was in North America (28 percent), while the greatest increases were expected in Asia (81 to 95 percent). These reductions in the more advanced countries presumably arise from changes in behavior prodded by AIDS education.

Currently, the highest percentage increases in AIDS transmission in the United States are among heterosexuals; earlier increases occurred almost exclusively among homosexuals and intravenous drug users. (Heterosexual contact was always the main transmission mode in Africa.) In 1993, only 9,279 cases out of 103,500 AIDS cases were heterosexually transmitted (9 percent of AIDS cases and 0.0036 percent of the U.S. population). However, by the year 2000, heterosexual transmission may become the predominant mode (figure 17.18). Still, AIDS is not running rampant among the heterosexual population. Most people can avoid infection by restricting themselves to a monogamous relationship or reduce the risk by using condoms.

Researchers are working on two main approaches to dealing with AIDS: (1) developing an AIDS vaccine and (2) treating the disease with chemical agents. As box 17.3 discusses, vaccines are typically made from modified bacteria or viruses rendered nonvirulent. Whether an HIV vaccine is a viable goal remains to be seen, but while such a vaccine would reduce the spread of AIDS, it would not cure an already infected patient. In addition, difficult ethical decisions crop up in testing a potential HIV vaccine since AIDS is an incurable fatal disease and administering the virus might either vaccinate or infect. In 1993, researchers successfully tested an AIDS vaccine in monkeys, but at the Ninth International Conference on AIDS in Berlin (June 1993), scientists stated that an effective human AIDS vaccine was unlikely in the near future. This opinion was reaffirmed at the 1994 conference in Japan. In fact, researchers at these conferences do not believe any cure for AIDS is on the immediate horizon.

Chemical agents able to combat AIDS are a rare breed; only a few drugs are currently FDA-approved for AIDS treatment. These drugs include AZT, DDC, and DDI (see figure 17.19 for their full names and structures). Several experimental drugs are undergoing development, but this normally requires many years and extensive animal testing. However, in December 1995 the FDA approved saquinavir (trade name Invirase) in only 97 days. Unlike other AIDS drugs that block DNA formation, the protease inhibitor

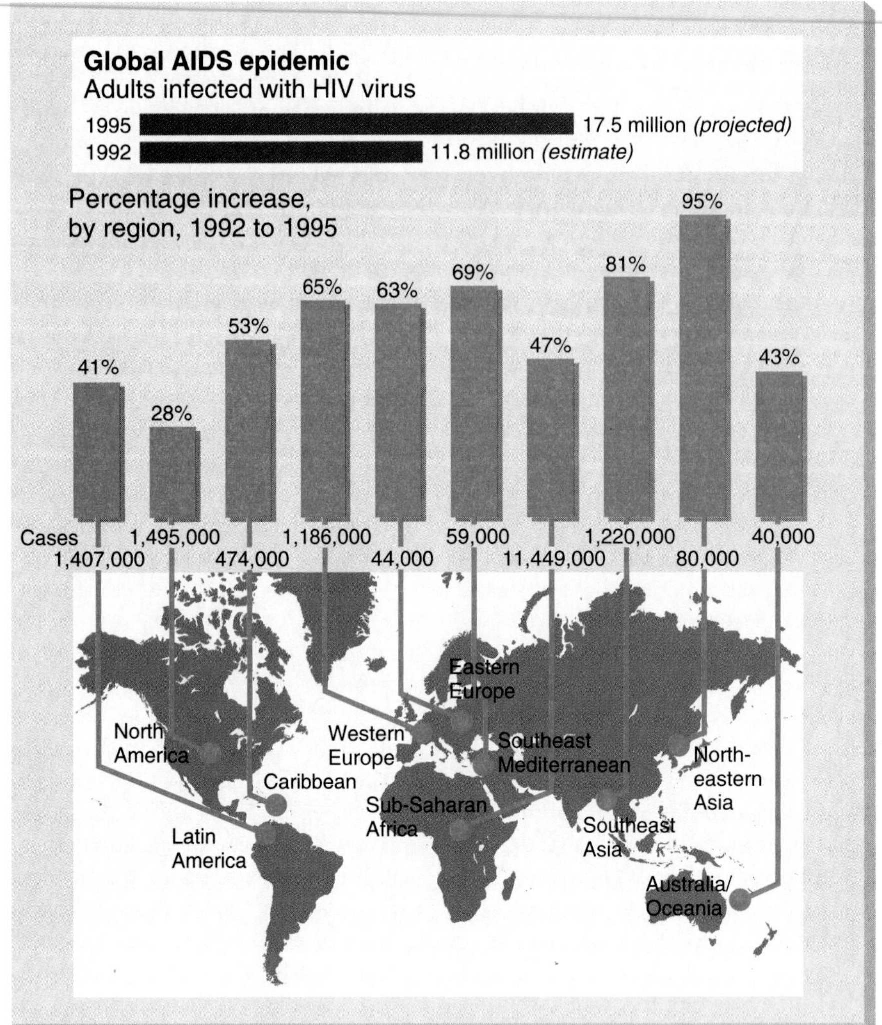

Figure 17.17 Projected AIDS and HIV infection by region.
SOURCE: Data from Harvard University.

saquinavir inhibits an enzyme that is essential in the final stages of HIV replication.

AZT, DDC, and DDI contain a nucleic base (chapter 14) or analog. Such an agent disrupts the virus molecule by causing it to take up the wrong kind of nucleic base. AZT prolongs the lives of AIDS victims, especially when treatment begins in the early stages of the disease, but it is very toxic and causes severe anemia, necessitating frequent blood transfusions. Research reports at the ninth AIDS Conference were less than enthusiastic regarding AZT effectiveness; some claimed no enhanced survival benefits. DDC appears less toxic than AZT, but the available data are limited. The more recently developed DDI seems even more gentle, but nobody yet knows if it is effective. In any case, though these drugs may prolong the patient's life, the patient remains uncured and infectious.

Interleukin-2 (IL-2), a T-cell growth-inducing polypetide, tested unsuccessfully against AIDS, but it *may* enhance AZT effectiveness when the two are used in combination. α-Interferon, another polypeptide, also shows some promise in combination with AZT. French researchers report that cyclosporin, a highly toxic drug normally used to prevent organ transplant rejection, controls AIDS in its early stages; and GLQ223, a Chinese drug extracted from a plant in the cucumber family, is another new anti-AIDS agent. Some synthetic nucleic base polymers show promise in treating AIDS, though not in curing the disease.

Scientists do not always know exactly why these drugs work. They believe that AZT, DDC, and DDI interfere with virus reproduction (although this was questioned at the ninth AIDS Conference). GLQ223 seems to work differently; it cuts apart the ribosomes—large particles containing RNA—in cells infected

15.0
10.0
5.0
0
Percentage
1980
1985
1990
1995
2000
Year

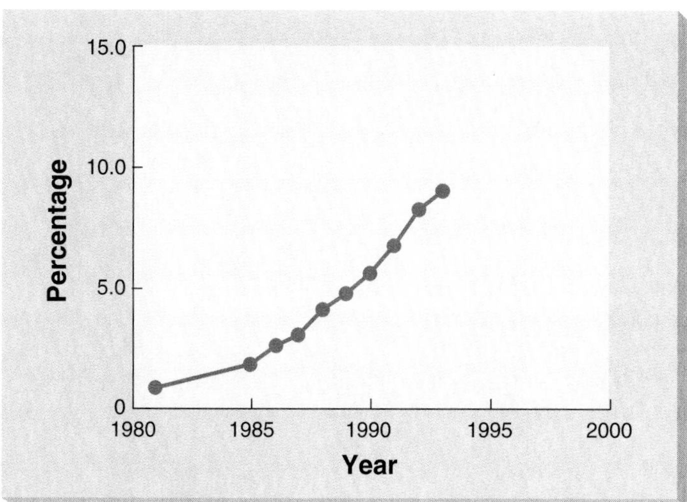

Figure 17.18 The percentage of AIDS cases attributed to heterosexual transmission in the United States between 1981 and 1993.

SOURCE: Data from The Centers for Disease Control and Prevention.

with the AIDS virus and kills them. GLQ223 does not appear to have this effect in normal cells, although no one understands why.

The economic costs of AIDS have steadily risen over the years. The 1991 medical care costs for AIDS patients approximated \$32,000 per patient in the United States, \$22,000 in Western Europe, \$2,000 in Latin America, and \$393 in sub-Sahara Africa. Also, by 1991, the annual amount of money spent on AIDS research surpassed 1.6 billion dollars, eclipsing the 1.5 billion dollars spent on cancer and far surpassing the 1.0 billion dollars spent on heart disease, stroke, and cardiovascular diseases combined. Annual deaths due to cancer and heart disease are forty-five times greater than deaths from AIDS; in 1992 there were 521,000 deaths from cancer (figure 17.12)—15.5 times the 33,590 AIDS deaths. Some have questioned the wisdom of investing so much funding into AIDS research at the expense of research on other afflictions. This seems an especially valid question in light of the fact that AIDS is largely a preventable disease.

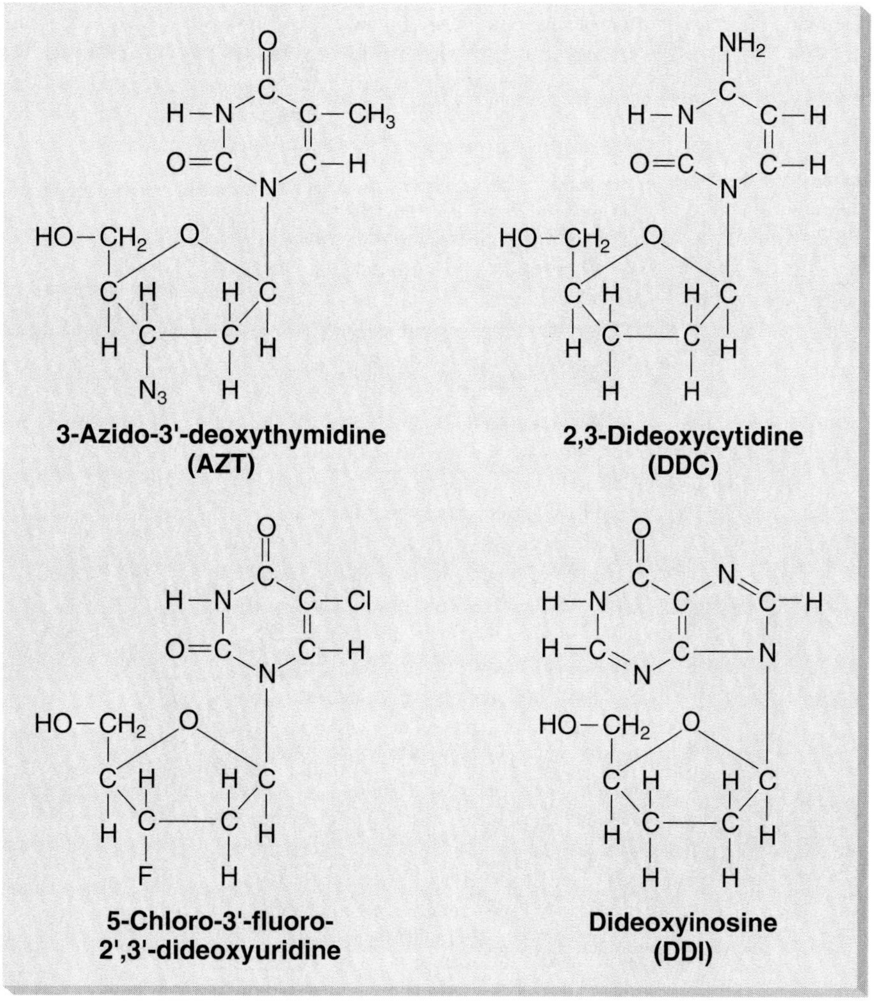

3-Azido-3'-deoxythymidine
(AZT)
2,3-Dideoxycytidine
(DDC)
5-Chloro-3'-fluoro-
2',3'-dideoxyuridine
Dideoxyinosine
(DDI)

Figure 17.19 Some AIDS drugs.

BOX 17.3

VIRUSES AND VACCINES

Viruses and vaccines are not new concepts to most of us. But what are they, and how did scientists first learn about them?

A **virus** is a nucleic acid molecule (DNA or RNA) with a protein coating. Almost from the beginning, scientists regarded viruses as falling at the edge of living matter. Unlike bacteria, viruses are unable to reproduce without entering another living cell. Since they invade body cells, it is difficult to attack and cure them.

In 1892, Dmitri Ivanovsky (Russian, 1864–1920) showed for the first time that viruses existed. Later that year, tobacco mosaic disease was attributed to a virus too small to see (figure 1). Subsequently, the tobacco mosaic virus became the first example of a virus to be isolated as a pure crystalline solid when Wendell Stanley (American, 1904–1971) did so in 1935. A year later, Stanley isolated the viral nucleic acids that caused the disease. He shared the 1946 Nobel Prize in chemistry for this research.

In the meantime, other scientists identified several viruses and viral diseases. In 1898, foot and mouth disease became the first identified example of an animal viral disease. Many additional viruses and related diseases (such as polio, measles, small pox, and influenza) were uncovered later. In 1955, Heinz Fraenkel-Conrat (German, 1910–) showed that viruses consist of the noninfective protein coat and an infective nucleic acid core. Francis Peyton Rous (American, 1879–1970) discovered in 1911 that some animal cancers are caused by a virus. (Rous shared the 1966 Nobel Prize in medicine for this research.) In 1966, Daniel Gajdusek (American, 1923–) succeeded in transferring kuru, a viral disease of the central nervous system, from humans to chimpanzees in the first interspecies exchange of a virus. (HIV, the virus that causes AIDS, is an extremely virulent virus that appears to have crossed over species lines.) Gajdusek shared the 1976 Nobel Prize in medicine for his research.

Figure 1 The tobacco mosaic virus.

As scientists discovered viruses, they also learned how they function. Once a virus invades a cell, its genetic material (DNA or RNA) commandeers some of the cell's functions and causes the cell to produce new chains of viral nucleic acid. Each of these nucleic acids then acquires a protein coat and exits the cell as a virus. The new virus enters other cells and the viral production process continues.

Although both RNA and DNA are natural polymers with specific nucleic bases attached as side groups, DNA and RNA viruses operate differently. **DNA viruses** *do not* insert into the host cell's genes, although they reproduce using the cellular machinery. When a DNA viral attack ends (or is cured), the virus is no longer active in the patient. A typical DNA example is a polio virus. **RNA viruses,** on the other hand, take control of the cell's chemistry by actually entering the host's genes. Such viruses are called **retroviruses** because they retrofit the cell's DNA to a new use; HIV is a retrovirus that causes *permanent* genetic mutation in the cells. RNA viruses thus persist in patients for the remainder of their lives.

Although doctors are unable to cure viral diseases, they can treat the symptoms and they can prevent an individual from contracting certain viral diseases through vaccines. A vaccine is a preventative medicine derived from a modified bacteria or from a virus rendered nonvirulent. Vaccines develop immunity to a specific disease.

Edward Jenner (English 1749–1823) developed the first vaccine in 1796 (figure 2). Jenner observed that milkmaids who had had cowpox seemed to have developed some immunity against smallpox, an especially virulent disease that killed

Figure 2 Edward Jenner (1749–1823), the Englishman who developed the first vaccine.

many thousands each year. Jenner isolated fluid from cowpox pustules (blisters) and inoculated some people who were at great risk for smallpox with the fluid. (Vaccinia was the technical name for cowpox, hence the term *vaccine* for this fluid.) The cowpox disease was similar to smallpox but not as deadly. Nonetheless, the cowpox vaccine was effective in preventing smallpox infection; we now know that this is because it produces antibodies similar to those produced by smallpox.

Although the Royal Society rejected Jenner's new vaccine in 1797, this was only a temporary setback. The smallpox vaccine was first used in the United States in 1800. The vaccine eventually led to the total eradication of this viral disease; the last recorded case of smallpox occurred in Somalia in 1979. The only remaining specimen of the smallpox virus in the entire world is a sample kept in cold storage at the Centers for Disease Control and Prevention in Atlanta.

There are three ways to create a vaccine. The first way is to find a similar infectious disease that is not virulent and develop the vaccine from this source—as Jenner did using cowpox in place of smallpox. Unfortunately, the causative agents for most diseases are not amenable to this approach.

The second method, now the preferred approach, is to develop a weakened, nonvirulent form of the bacterium or virus that causes the disease. This procedure requires scientists to grow many generations of the bacterium or virus in order to engineer a safe, noninfectious version—called an attenuated bacterium or virus—that can no longer cause serious illness. The final modified, living bacterium or virus is then grown and used as the vaccine. Examples of successful vaccines of this type include those for polio, German measles, and yellow fever.

Not all bacteria or viruses can be easily grown in the lab. For these infectious agents, the third approach is most suitable—using extracts from dead bacteria. Examples of vaccines made by using this approach are those that prevent whooping cough and cholera.

Research scientists are currently exploring one more method to develop a vaccine. This possible fourth method uses synthetic polymers as the vaccine base. We don't yet know just how successful this approach will be. Vaccines have revolutionized preventative medicine (figure 3). The last reported case of smallpox in the United States occurred in 1949. Polio disappeared even more swiftly: in 1955, the year the polio vaccine was introduced, there were 30,000 reported cases in the United States; a decade later there were less than 80 cases, and the last reported case was in 1983. The World Health Organization (WHO) expects polio to join smallpox as an eradicated disease by 2005. Similarly, cholera, diphtheria, and tetanus are extremely rare. Whooping cough, measles, German measles, and mumps are now under control in the United States, although most of them killed or crippled many children as recently as forty years ago.

In March 1995, a new vaccine was introduced to prevent chicken pox. Most people consider chicken pox a nuisance disease, but it infected 134,722 Americans in 1993, hospitalized 9,000, and killed 90. Japanese researchers had developed a vaccine in 1974, but American researchers were unable to duplicate their success. The chicken pox virus was extremely difficult to grow under laboratory conditions, making a vaccine almost impossible to develop. However the Japanese shared their vaccine with researchers at Merck & Company in 1981. Testing and large-scale development eventually produced an effective vaccine. With this new vaccine, the incidence of chicken pox is expected to decrease dramatically. As an added benefit, scientists believe this vaccine may also prevent shingles, a painful disease of the nervous system that the chicken pox virus produces in adults.

Figure 3 Vaccination in action.

At this time, no one knows how to create a vaccine for a retrovirus such as HIV. AIDS remains a fatal viral disease with no cure or preventative on the horizon. Only a handful of chemical agents work against any virus; doctors ordinarily treat the symptoms and depend on the body's immune system to overcome the virus. With AIDS, this approach is doomed to failure because HIV destroys the T-cells in the human immune system, rendering the system inoperative. The battle against cancer (sometimes virus induced) has not yet been won after spending billions of dollars and decades of research. At present, it appears that the battle against AIDS will also be long and hard-fought.

BOX 17.4

HOW SAFE IS THE BLOOD SUPPLY?

Health experts have known for some time that the HIV virus may be transmitted through contaminated blood. Accordingly, hospitals and blood banks routinely screen blood for the viral antibodies (figure 1). Is our blood supply safe now that these organizations test the blood?

The answer may not be as positive as we would wish. When Romania emerged from behind the Iron Curtain, the world discovered thousands of children with AIDS via blood transfusions given to infants to enhance their health. The blood supply was infected with AIDS, and the transfusions, instead of improving health, destroyed it. More recently, hundreds in France were infected with HIV-tainted blood because the people responsible for testing did not check the blood carefully. More frightening was the fact that the French government tried to conceal this information and let the disease spread unchecked. The U.S. Red Cross recalled blood plasma from ten states in 1992 because donors were unscreened for possible HIV infection, and similar events were reported in Canada in 1993 and in Germany even more recently. During the 1980s, before blood was tested, many people, mostly hemophiliacs, were infected with AIDS via blood transfusions. Today, infection through a transfusion is much more unlikely—but not impossible.

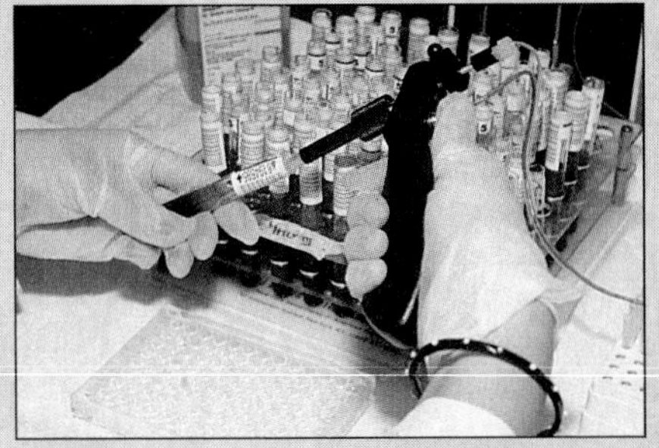

Figure 1 Blood banks now routinely test for AIDS antibodies.

HIV-infected people still donate blood. Medical personnel test for the presence of AIDS antibodies in the blood; whenever AIDS antibodies are discovered, the blood is destroyed. Current tests do not determine the presence of the actual HIV virus—just the antibodies responding to the virus. These antibodies do not develop until between six weeks to a year after HIV infection; theoretically, blood donated prior to antibody formation could still contain the HIV virus and escape detection. Researchers are trying to develop a direct test for the virus; until this test is perfected, no one can be absolutely certain the blood supply is clean. The prudent course of action is to donate your own blood in case of a future surgery or emergency procedure, or to choose donors you are certain are not HIV-infected.

Numerous articles in popular science magazines and the newspapers have reported on the development of artificial blood. A replacement blood that is not obtained from humans or animals should be free from any infectious agents. It could be produced in a nearly limitless supply and stored indefinitely. The synthetic blood substitutes chemists have experimented with thus far carry only oxygen and carbon dioxide. All the other functions of blood are excluded from these synthetic materials. Is it likely that a complete artificial blood will be developed in the near future? Why do many doctors feel an artificial blood would be valuable? Obviously, one reason is to establish a truly safe blood supply.

CHEMISTRY AND ALZHEIMER'S DISEASE

Alzheimer's disease is a progressive, irreversible, brain disorder that affects more than 20 million people worldwide. First described by Alois Alzheimer in 1906, this disease affects about 4 million Americans, causing over 100,000 deaths annually; it is the fourth leading cause of death after heart disease, cancer, and stroke. Most victims are over sixty-five years of age, but Alzheimer's does strike people in their forties or fifties and cases have been reported in people as young as 32. The main symptoms include gradual memory loss, especially short-term memory, inability to perform routine tasks, such as dressing or eating; judgment impairment; learning difficulties; disorientation; and personality changes. There is no known cure and the victim eventually becomes totally dependent on others. Alzheimer's is a devastating disease because the patient can no longer function normally and no longer remembers friends and family.

The cause of Alzheimer's disease is uncertain. However, recent research suggests that a specific protein called β-amyloid is the culprit. This sticky protein seems to form masses called plaques on the brain's nerve endings, and these plaques interfere with normal brain functioning. The disordered plaques occur principally in the brain centers that control memory, emotions, and cognition. Some suggest that β-amyloid forms channels to allow excess Ca^{++}—which is toxic—to enter the nerve cells. Others suggest that β-amyloid short-circuits the nerve impulses. Scientists do know that a shortage of acetylcholine, one of the chemicals that transmits nerve signals (chapter 14), leads to an increase in the amount of β-amyloid present, at least in laboratory-grown brain cells.

Some scientists suspect that a genetic cause underlies Alzheimer's; in fact, they have observed similarities between Alzheimer's and Down syndrome patients who live past age fifty that seem to support this idea. In 1993, researchers discovered additional genetic evidence for a version of Alzheimer's disease commonly afflicting patients older than sixty-five. It is still uncertain, however, whether these discoveries will lead to any cure in the foreseeable future.

At one time, some scientists believed aluminum, mostly in the ionic form, was the cause of Alzheimer's, and some "health store" literature still makes this claim, but the scientific community has discredited this view. Aluminum does accumulate in plaques of many Alzheimer's patients, but the plaque formation seems to occur first. The current question many researchers are focusing on is whether β-amyloid causes plaque formation or whether it, like aluminum, later deposits in the plaques.

No effective drug treatments exist for Alzheimer's. **Tacrine (THA)** (figure 17.20), the main drug used for treating Alzheimer's, was first synthesized in 1931 for other medical uses. Tacrine first received FDA approval in mid-1994, even though clinical trials on Alzheimer's patients dated from before 1986. Tacrine, which affects the neural system, restores mental alertness to some patients but does not reverse the effects of Alzheimer's disease. Because nerve impulse transmission becomes dysfunctional in Alzheimer's patients, it makes sense to try chemicals that affect the neural system. Other scientists are seeking drugs to prevent β-amyloid formation.

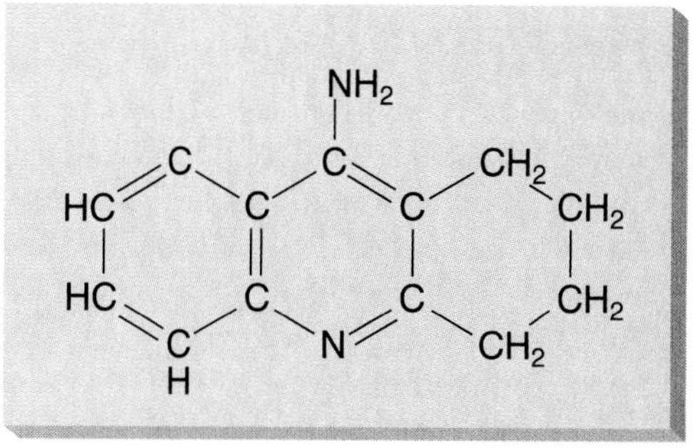

Figure 17.20 The structure of Tacrine (THA), the drug used to treat Alzheimer's disease.

PRACTICE PROBLEM 17.6

Which afflicts more people, AIDS or Alzheimer's?

ANSWER Recent data shows over 20 million people in the world have Alzheimer's disease, compared to 11 to 14 million people with HIV.

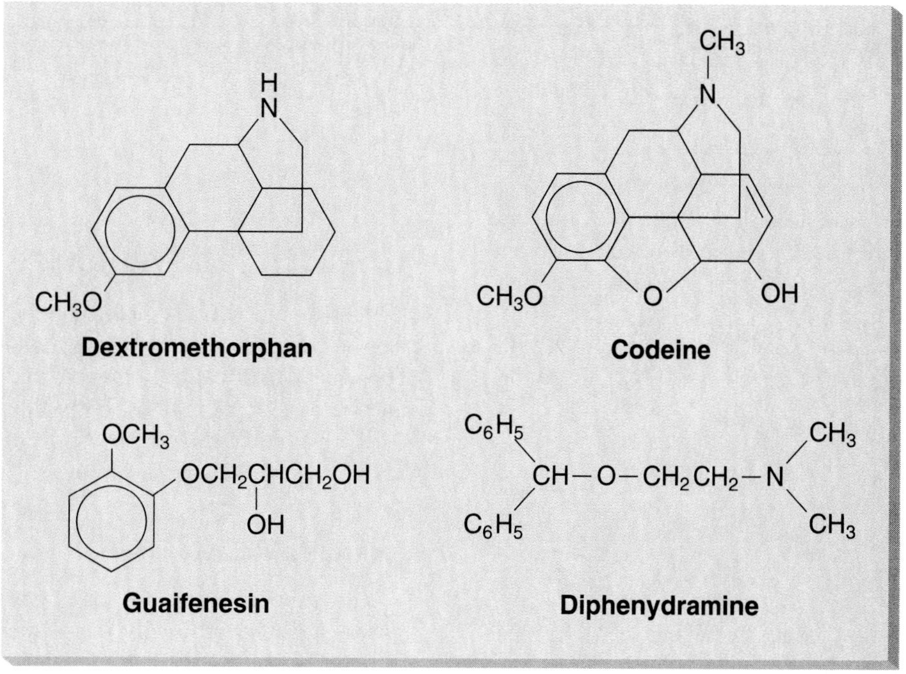

Figure 17.21 The chemical structures of several drugs used to treat the common cold.

"CURING" THE COMMON COLD

Not all diseases are as severe as cancer, AIDS, and Alzheimer's. Doctors estimate that about 7.7 million people will get colds this winter and about 2.7 million more will have one in the spring. There is no cure for the common cold; all you can do is treat the symptoms and wait until your body fights off this annoyance. While you wait, you usually want something to make you feel better; a quick trip to a drug store reveals a staggering assortment of alleged remedies. Which one is best? Is there any real difference between all these potions and elixirs?

The FDA put together a report on exactly this question in February 1989. This report notes that two medicines *seem* more effective in fighting the symptoms of the common cold than any others: dextromethorphan and guaifenesin (figure 17.21). **Dextromethorphan** is the most effective **cough suppressant** and **guaifenesin** is the *only* drug the FDA found effective as an **expectorant**—an agent that helps bring up mucus by decreasing its viscosity or thickness. Dextromethorphan is a less addictive chemical modification of codeine (figure 17.21), which also works as a cough suppressant; both dextromethorphan and codeine are classed as narcotics.

People have tried many chemicals over the long history of combating the common cold. Expectorant "losers" include chloroform, turpentine, iodine, creosote, camphor, and pine tar extracts. Since the 1989 FDA edict, only guaifenesin is approved to be labeled as an expectorant. In August 1987, the FDA listed five approved cough suppressants: camphor, menthol, chlophedianol, dextromethorphan, and codeine; the most effective were the last two. However, codeine may make you drowsy, nauseous, or constipated, and it is highly addictive. This leaves dextromethorphan as the suppressant drug of choice.

Some other agents are also useful in relieving cold symptoms; an antipyretic reduces fever and analgesics relieve the headaches and body aches that frequently accompany colds. An antihistamine such as diphenhydramine (figure 17.21) can help relieve stuffiness, although it may make you sleepy. But the best drugs available when you have a cold, according to the FDA, are dextromethorphan and guaifenesin. (Of course, you would use only one or the other since one suppresses coughs and the other helps you cough and bring up mucus.)

Some have maintained, both in centuries past and today, that herbal medicines are agents for fighting disease. Box 17.5 explores this claim.

PRACTICE PROBLEM 17.7

Let's make some toxicity comparisons between chemicals used as insecticides, herbicides, and medicine. Which are generally most toxic? Are any medicines more toxic than any insecticides?

INSECTICIDES	LD_{50} (mg/kg)
Aldicarb	1
DDT	87
Chlorpyrifos	97
Endin	3
Diazinon	66
HERBICIDES	**LD_{50} (mg/kg)**
2,4-D	375
Glyphosate	4320
Amitrole	1100
Diuron	3400
Dioxin	0.03
MEDICINES	**LD_{50} (mg/kg)**
Ibuprofen	1255
Acetaminophen	338
Lidocaine	292
Phenobarbitol	162
Cocaine	18

ANSWER Insecticides, with the lowest LD_{50} values, are generally most toxic. However, cocaine is more toxic than any of the insecticides except aldicarb and endin. It is also more toxic than any of the herbicides other than dioxin; in fact, most of the medicinal drugs on the list are more toxic than many of the herbicides.

PLASTICS IN MEDICINE

Polymer chemistry has some applications in medicine that are not often recognized. Dentists, ophthalmologists, orthopedic surgeons, and other medical professionals routinely use polymers in timed release drugs, dialysis, surgical procedures, artificial hearts, blood vessel replacements, eye applications, and dental uses. Let's look at each of these applications.

Controlled Release Drugs

Why are timed release drugs sometimes more effective in treating a disease? Almost all involve a polymer as one of their key components. The polymer regulates the flow of the drug, maintaining the optimal level to treat the disease and yet avoid undesirable side effects.

BOX 17.5

HERBAL MEDICINE

Herbal medicine poses an interesting set of chemical questions: Exactly what is an "herbal medicine" and how do we know whether it is safe?

In the broadest sense, herbal medicines are simply derived from plants. Historically, the earliest medicinal agents were derived from herbs; the salicylates and various antibacterial agents were herbal extracts. In fact, thousands of medicinal agents have been discovered in plant or animal sources, and pharmaceutical companies continue to search for such substances today. One of the arguments for preserving the Amazon Rain Forest is that it might be a source of new and valuable medicines (figure 1).

Although herbal medicine has a time-honored history, this does not mean all herbal agents are safe. *All* drug agents are toxic when the dose is too high, and many "natural" compounds are among the most poisonous known to humankind. (Who has not heard of poison hemlock?) Even some herbs used as home remedies for centuries might be characterized as potentially dangerous substances.

Figure 1 The Amazon Rain Forest, a potential source of valuable medicines.

The real problem in herbal medicine today is the way "health food" advocates make blithe, casual endorsements of unknown and untested substances. Suppose you visited a health food store and salespeople recommended Herb Q (a fictitious concoction) for some serious disease. They tell you that Herb Q has been used by the South American native tribes for hundreds of years to treat many ailments. There are several questions you should ask before jumping on the Herb Q bandwagon. First, does Herb Q actually have any medicinal value, especially for your condition? Some herbs are harmless even if ineffective, but some are dangerous. Could Herb Q be toxic or have bad side effects?

A second question to ask: what kind of health do the natives have who used Herb Q for decades or longer? Imagine that the natives have an average life span of thirty-eight years. Would Herb Q affect us any differently since our life span is much longer and we could take Herb Q much longer? Recall that afflictions such as heart disease, cancer, and Alzheimer's occur mainly in the older population. Could the natives avoid these ailments not because they use Herb Q but because of their shorter life span? And even if the natives are truly healthier, did Herb Q confer this good health? Heart disease and other ailments have increased in Third World countries after they adopted the American habit of eating fatty foods. How does this affect your evaluation of Herb Q's benefits?

Finally, remember that chemical agents have varied effects on human biochemistry. Before rushing out to try the latest health food fad, read some good books or articles on the topic. The better ones warn of potential side effects and are less likely to endorse untried herbs. Even if Herb Q only gives the South Americans more pep, you might ask why this happens. Matè tea, a South American herbal product, peps up most people for the simple reason that it contains a lot of caffeine. So does ordinary tea, which is usually less expensive (figure 2).

Figure 2 A variety of popular teas including traditional "black tea" with caffeine and some herbal teas without caffeine.

Figure 17.22 shows how a controlled release drug operates. All drugs have an optimum concentration range for effective activity; below this range the drug is ineffective, and above it the drug may have toxic side effects. In the ordinary administration of medication, the initial dose raises the body's drug level toward the upper portion of the effective range; this gradually declines and falls below the effective limit. In **controlled release,** as figure 17.22 shows, the same drug is released continuously, keeping the dosage level within the effective region for longer time periods. Controlled release substantially reduces the side effects that can arise when the drug's level exceeds the upper limit.

At least six different controlled release techniques exist. **Erodable devices** enclose the drug inside a substance, usually a polymer, that erodes or wears away due to chemical activity in the

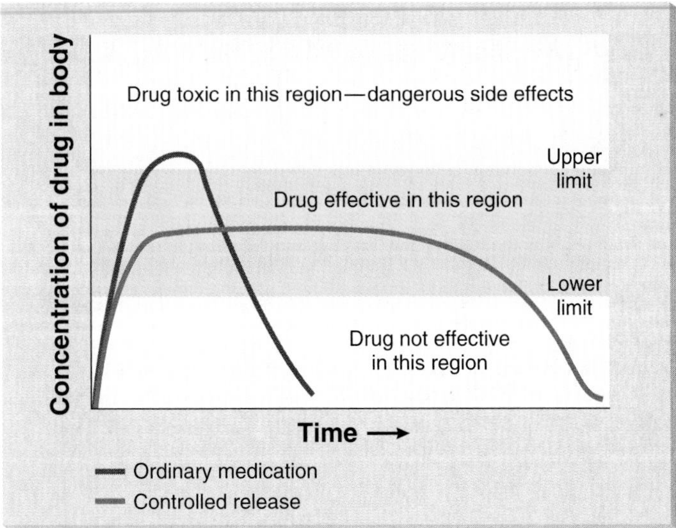

Figure 17.22 Controlled release systems operate by maintaining the drug concentration in the effective range. This is accomplished by releasing the drug continuously for long time periods.

body and releases the drug for a few hours to several days. Many over-the-counter vitamins and cold, headache, and allergy medications utilize this simple method. In **reservoir** and **monolithic devices,** the drug is enclosed within a nonerodable polymer and slowly diffuses out into the body fluids; these devices can operate effectively for periods as long as two years. Norplant, a contraceptive device, falls into this category. **Transdermal devices** administer the drug through the skin using a polymeric membrane to control the delivery. Nitroglycerin patches are placed on the chest to treat angina pectoris (heart pain), patches that contain drugs to alleviate motion sickness are placed behind the ear near the semicircular canals, and nicotine patches are placed on the arm to break the smoking habit. Another controlled release device, **microcapsules,** are tiny capsules molded or compressed into a caplet (a cross between a capsule and a tablet). And finally, **polymeric drugs** contain an active drug unit as part of the polymer.

PRACTICE PROBLEM 17.8

Visit your local store or supermarket. How many controlled or timed release formulations are available?

ANSWER You will find many. Common over-the-counter items include cold remedies, vitamins, allergy medications, and analgesics.

Dialysis

More than one hundred thousand people undergo dialysis every year because their kidneys are defective. The **dialysis** unit, sometimes called an artificial kidney, consists of thousands of hollow polymer fibers in a plastic case. As the patient's blood flows through these hollow fibers, the body's waste products wash away into an aqueous solution called dialyzing fluid. The polymers in the dialysis fibers are usually cellulose derivatives (a natural polymer) or poly(acrylonitrile).

Surgical Procedures

Surgeons have used silicone polymers in implants for many years (see box 11.1). Examples of these implants include breast replacements, a testicular prosthesis, finger joints, an ear prosthesis, a chin implant, a tendon prosthesis, and various forms of tubing, including the hydrocephalus shunt.

The importance of these polymer products is evident from the sheer number of people they have helped. Finger joint surgery restores hand mobility to more than four hundred thousand patients each year. The hydrocephalus shunt drains excess body fluids away from the brain; these shunts save over ten thousand lives annually. An artificial skin made of silicone polymer and collagen is used to treat some of the more than one hundred thousand people who suffer severe skin damage each year from fires and accidents. Research suggests that similar materials may facilitate nerve regeneration, which could restore the use of impaired arms or legs.

Hip and knee replacements make use of several plastic materials. The usual hip prosthesis consists of a metal ball attached to the femur (leg bone) and inserted into a high-density poly(ethylene) socket placed in the hip. Both ball and socket are cemented in place using poly(methyl methacrylate).

Artificial Hearts and Cardiovascular Devices

Artificial hearts, studied for over forty years, leaped into the public spotlight with the 1982 permanent implantation into Barney Clark. An early artificial heart was built from glass by Charles Lindbergh, better known for his solo airplane flight across the Atlantic Ocean in 1927. Artificial hearts are now primarily constructed using segmented poly(ester urethane urea)s or PEUUs, an exceptionally strong elastomer. Most plastics are incompatible with blood, but the PEUUs used in artificial heart devices, such as the Jarvik heart, have moderate blood compatibility and are able to withstand the rigorous pumping action needed. Figure 17.23 shows the basic PEUU structure, a Jarvik heart, some replacement blood vessels made from Dacron and Teflon, and a balloon heart-assist device made from PEUU. Well over one hundred thousand coronary bypass procedures each year replace blood vessels with artificial blood vessels.

Eye Applications

We can trace the use of plastics in the eye back to World War II, when it was observed that fighter pilots had small particles of poly(methyl methacrylate) (pMMA) embedded in their eyes but

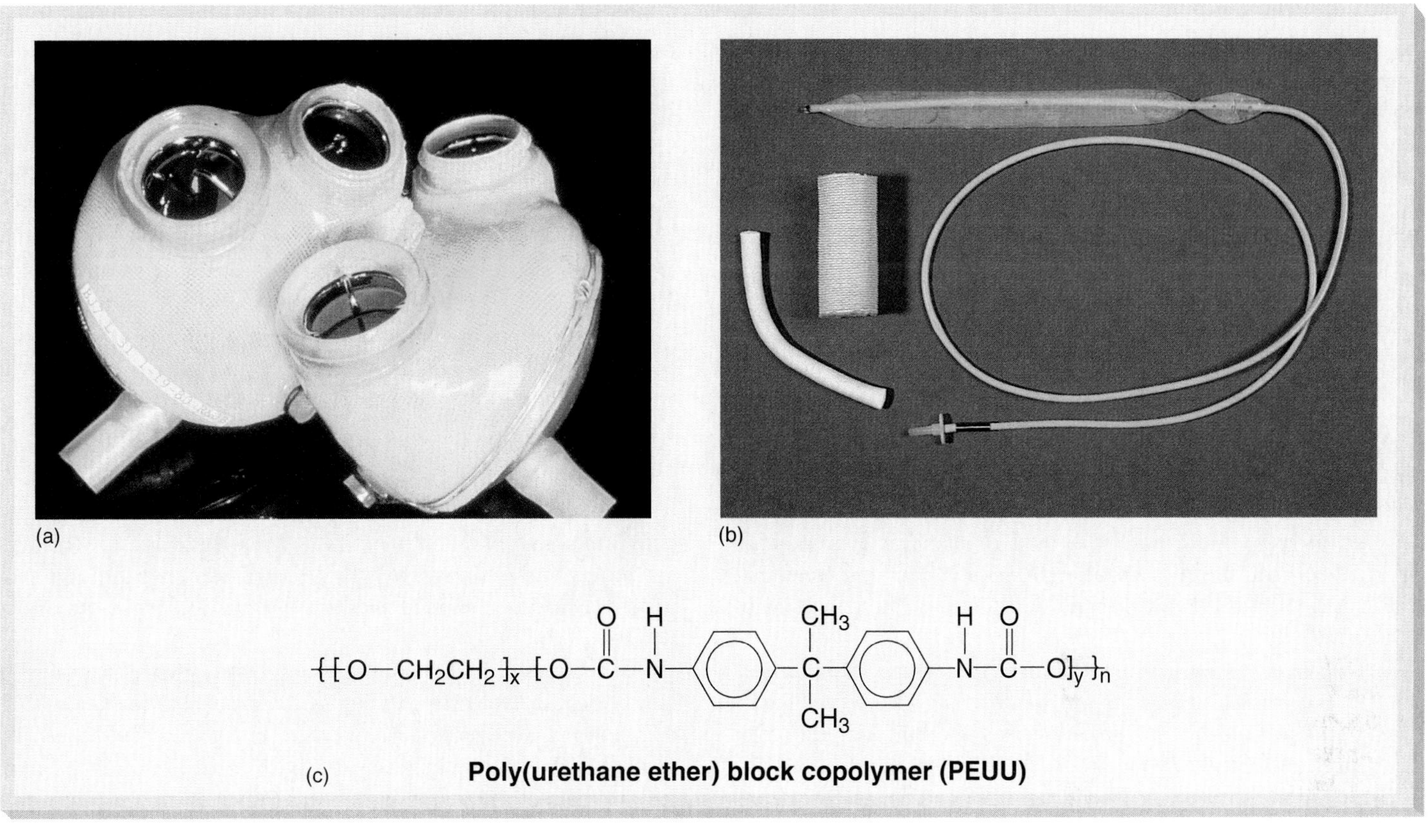

(a)

(b)

$$\text{-[-[O-CH_2CH_2]_x-[O-C(=O)-N(H)-C_6H_4-C(CH_3)_2-C_6H_4-N(H)-C(=O)-O]_y-]_n}$$

(c) **Poly(urethane ether) block copolymer (PEUU)**

Figure 17.23 The structure of the PEUU polymers and some cardiovascular devices made from plastics. **(a)** The Jarvik total artificial heart. **(b)** From left: Dacron aorta replacement; intraaortic balloon heart-assist device made from poly(urethane); replacement blood vessels made from poly(tetrafluoroethylene). **(c)** PEUU structure.

showed no signs of irritation. This observation led optical companies to use pMMA in contact lenses and intraocular lens replacements. Subsequent research led to soft lenses in 1963 and extended-wear lenses in the 1980s; these lenses are made of copolymers of acrylates or methacrylates.

Dental Uses

Several decades ago, front tooth dental fillings were made from silicates that only survived a few months before falling out. Acrylate fillings arrived on the scene about 1949 and replaced these unsatisfactory silicates within a few years. Modern dental fillings, made from composites of poly(methyl methacrylate)—pMMA—with special inorganic fillers, can last many years. Dentures are also made from pMMA. Other polymers are used to make mouth molds, root canals, and jaw bone replacements.

MEDICAL WASTE

Medical waste presents a special, almost unique, challenge in hazardous waste disposal. Medical waste includes used syringes, needles, bandages, cotton swabs, dressings, test specimens, miscellaneous chemicals, and various forms of biological tissue. Medical waste is potentially infected with disease-causing organisms that create the possibility of spreading disease.

Normal hazardous waste can also cause health problems—that's why it's considered hazardous. Unlike some medical waste, however, ordinary hazardous waste is not self-reproducing. Even if this waste leaks from a landfill, the original quantity does not increase. By contrast, many types of medical waste multiply; infectious agents such as bacteria, viruses, fungi, or parasites are capable of reproducing themselves under suitable conditions. These potentially dangerous infectious agents could produce epidemics and endanger human lives. Many ordinary landfills leak (chapter 12). Even without leaks, vermin plague these sites. Most vermin will not bother ordinary hazardous waste since it is toxic to them, but some do get into medical waste. Biological specimens can serve as food for vermin, infecting them and spreading the disease throughout a community. Placing medical waste into landfills is unwise.

Putting these materials in an open environment, such as a beach or an ocean, is equally problematic. Nor is recycling a reasonable option; who would use recycled bandages or swabs, even if they were assured all infectious agents were removed? Still, since medical wastes are inevitable, how can we dispose of them safely? Incineration is practically the only sure method, since infectious agents will not survive fire. Medical waste incineration poses fewer problems than hazardous waste incineration because the latter can contain heavy metals, while most medical waste is organic, with essentially no heavy metal content. Yet great opposition usually arises whenever medical waste incineration is proposed. Part of this resistance is spurred by fear, but part is due to resistance to any form of waste disposal under the claim of protecting the environment. Our environment is not protected if medical wastes accumulate. Only incineration can prevent that.

PRACTICE PROBLEM 17.9

What would you do if someone planned to build a medical waste incinerator in your neighborhood?

ANSWER Answers will vary. The important point is to act intelligently rather than emotionally. If you decide you would oppose the incinerator, what alternative waste disposal method would you propose?

SUMMARY

Chemistry is an integral part of modern medicine. Natural and synthetic chemicals relieve pain, fight disease, help us understand the causes of illness, and provide devices that improve human health and life.

From ancient times, human beings have sought pain relief. One of the earliest analgesics was willow bark; chemists discovered that the salicylic acid in the bark had both analgesic and antipyretic properties. Eventually, Gerhardt synthesized acetylsalicylic acid, or aspirin. This "wonder drug" proved useful in treating pain, fever, inflammation, and a number of other disease-related symptoms. Alternatives such as acetaminophen, ibuprofen, and naproxen have joined aspirin as modern-day analgesics.

Antibacterial agents are chemicals that fight the bacteria that can cause infectious diseases. These chemicals may be either bactericidal or bacteriostatic. Penicillin, sulfanilamide, and other antibiotics kill a broad range of bacteria. They have drastically reduced the number of deaths attributed to infectious disease.

Cancer is a disease in which body cells divide and replicate abnormally rapidly. Cancer prevention studies link cancers to diet and lifestyle. Cancer is treated with antimetabolites and alkylating agents. Both interfere with cell replication to prevent the rapid proliferation of cancer cells.

AIDS has been called a modern-day plague. This disease, linked to the Human Immunodeficiency Virus (HIV), has infected over 10 million people worldwide. AIDS is an RNA virus, or retrovirus, that causes permanent genetic mutations in body cells, as opposed to a DNA virus that uses the cellular machinery without changing the genes. Researchers are trying to treat AIDS with chemical agents as well as to develop an AIDS vaccine.

Alzheimer's disease is a progressive brain disorder that affects mostly people over fifty years of age. A protein called β-amyloid, combined with a genetic predisposition, may be the underlying cause. Scientists are experimenting with tacrine and other chemicals that affect the neural system in an effort to treat and prevent Alzheimer's.

The common cold is a viral illness that affects nearly everyone now and then. Though people have tried dozens of chemicals to relieve cold symptoms, only a few have proven effective in laboratory tests. Since no one yet knows how to cure a virus, we have not conquered the common cold.

Plastics play a number of roles in modern medicine. Polymers enable timed release drugs to operate at optimal levels over longer periods of time; they are also essential in artificial kidneys, which are filled with hollow polymer fibers, as well as in prostheses and in the artificial heart. Many of us wear plastic contact lenses or depend on pMMA fillings to ensure dental health.

On the other side of the coin, medical waste presents a challenge in hazardous waste disposal. Incineration is the safest disposal method. But many people resist incineration, mistakenly believing other disposal methods are safer. Few would argue, however, that the benefits chemistry has offered modern medicine make the waste disposal challenge a risk worth taking.

KEY TERMS

acetaminophen (Tylenol) 459
acetylsalicylic acid (aspirin) 459
AIDS (Acquired Immune Deficiency Syndrome) 471
alkylating agents 468
Alzheimer's disease 476
analgesic 459
anesthetic 465
antibacterial agent 464
antibiotic 466
antimetabolites 468
antipyretic 459
bactericidal 464
bacteriostatic 464
carcinogen 467
chemotherapy 459
controlled release 478
cough suppressant 478
dextromethorphan 478
dialysis 480
enkephalins 462
erodable device 479
expectorant 478

READINGS OF INTEREST

Pharmaceutical Agents

Pharmaceutical Manufacturers Association. *The role of prescription medicine in healthcare.* Washington, D.C.: Pharmaceutical Manufacturers. Good brochure about drugs. You can request a free copy by writing to the association at 1100 Fifteenth Street N.W., Washington, D.C. 20005.

Podolsky, Doug, and Loeb, Penny. "Dangerous drugs." *U.S. News & World Report,* January 9, 1995, pp. 48–54. Drug side effects are deadly to some patients.

Bugg, Charles E; Carson, William H.; and Montgomery, John A. "Drugs by design." *Scientific American,* vol. 269, no. 6 (December 1993), pp. 92–98. Computers may aid in developing new drugs by relating structure and function.

Wigzell, Hans. "The immune system as a therapeutic agent." *Scientific American,* vol. 269, no. 3 (September 1993), pp. 127–135. It is possible to use the immune system therapeutically; vaccines are one example of this technique.

Beardsley, Tim. "Blood money." *Scientific American,* vol. 269, no. 2 (August 1993), pp. 115–117. Do pharmaceutical firms make too much profit?

Alper, Joseph. "Ulcers as an infectious disease." *Science,* April 9, 1993, pp. 159–160. This was initially a surprising concept, but good evidence supports the idea that some ulcers are caused by bacterial infection and are treatable with antibiotic drugs.

Hirschhorn, Norbert, and Greenough, William B. III. "Progress in oral rehydration therapy." *Scientific American,* vol. 4, no. 1, special issue (1993), pp. 126–132. Each year, this therapy—which consists of feeding a simple electrolyte solution to children dehydrated from diarrhea—saves the lives of a million children. New compounds are being developed to improve that number.

Oliwenstein, Lori, "New victories in an old war." *Discover,* June 1992, pp. 16–17. The malaria war rages on, with more fatalities in one year than AIDS has caused in fourteen.

Gorman, Christine. "Can drug firms be trusted?" *Time,* February 10, 1992, pp. 42–46. The article is mainly concerned with drug testing data.

Cox, Francis E. G. "Malaria control: Challenges for chemotherapy and vaccination." *Chemistry & Industry,* August 5, 1991, pp. 533–536. More than a million people die from malaria each year and the older drugs are no longer effective. This article tells about some new approaches.

Analgesics

Rubin, Rita. "One more pill for pain." *U.S. News & World Report,* August 1, 1994, pp. 62–64. This short article compares the four over-the-counter analgesics now available.

Leutwyler, Kristen. "Something to chew on." *Scientific American,* vol. 270, no. 1 (January 1994), p. 24. Salicylic acid, the analgesic that comes from willow tree bark, is the subject of this article.

Brink, Susan. "Everybody's miracle drug." *U.S. News & World Report,* September 13, 1993, p. 77. The "miracle" drug is aspirin.

Melzack, Ronald. "The tragedy of needless pain." *Scientific American,* vol. 4, no. 1 special issue (1993), pp. 45–51. Contrary to popular belief, says the author, morphine taken solely to control pain is not addictive. Yet patients worldwide continue to suffer unnecessary agony.

Musto, David F. "Opium, cocaine, and marijuana in American history." *Scientific American,* vol. 4, no. 1, special issue (1993), pp. 30–37. Over the past two hundred years, Americans have twice accepted and then vehemently rejected drugs. Understanding these dramatic historical swings gives perspective to our current reaction to drug use.

Matthews, Robert. "A low-fat theory of anesthesia." *Science,* January 10, 1992, pp. 156–157. One theory about how anesthetics render people unconscious.

Stone, Judith. "The novocaine mutiny." *Discovery,* March 1990, p. 36. A mini-article on painless dentistry.

Whote, Peter T. "Coca: An ancient Indian herb turns deadly." *National Geographic,* vol. 175, no. 1 (January 1989), pp. 2–47. The story of cocaine, an import from South America, in today's world. This article answers the questions of whether Coca-Cola contained cocaine before 1903.

Animal Testing

Breen, Bill. "Why we need animal testing." *Garbage,* April/May, 1993, pp. 38–45. This magazine concentrates on environmental issues. The author asserts that animal testing is necessary to study the impact of contaminants on the environment and human health.

Ritvo, Harriet. "Toward a more peaceable kingdom." *Technology Review,* February/March, 1992, pp. 55–61. A good article that describes the animal rights movement from a historical perspective.

"Sigma Xi statement on the use of animals in research." *American Scientist,* January/February 1992, pp. 73–76. Find out what scientists think about animal research in this carefully reasoned article written by representatives of a leading honorary research society.

Singer, Peter. 1975. *Animal liberation*, New York: Avon. The other side of the story about animal testing, including a section on becoming a vegetarian.

Cancer

Cowley, Geoffrey. "Surviving against all odds." *Newsweek*, March 13, 1995, p. 63. Some cancer patients seem to survive much longer than expected or even to overcome their cancers. Why?

Rubin, Rita. "The new promise of cancer vaccines." *U.S. News & World Report*, March 6, 1995, pp. 87–88. There is a possibility researchers may be able to develop a vaccine that could prevent some cancers.

Gross, Neil, et al. "Quiet strides in the war on cancer." *Business Week*, February 6, 1995, pp. 150–152. Part of the battle is shifting to a study of the immune system in order to develop a vaccine.

Gorman, Christine. "How to starve a tumor." *Time*, January 9, 1995, p. 60. Cutting off the blood supply may be one way to kill a tumor.

Brink, Susan. "A different kind of cancer risk." *U.S. News & World Report*, January 9, 1995, pp. 58–61. Some research suggests that certain drugs will cause cancer to grow even though they do not cause the cancer itself.

Gorman, Christine. "Do abortions raise the risk of breast cancer?" *Time*, November 7, 1994, p. 61. A seven-year study suggests that the answer is yes, but more research is needed to affirm or refute this claim.

Garnick, Marc B. "The dilemmas of prostate cancer." *Scientific American*, vol. 270, no. 4 (April 1994), pp. 72–81. Do the risks of aggressively treating early prostate cancer outweigh the benefits? This question is one of several unresolved issues concerning prostate cancer.

Raloff, Janet. "Manhood's cancer." *Science News*, February 26, 1994, pp. 138–140. Testicular cancer does not restrict itself to older men. Its incidence is growing nationwide.

Beardsley, Tim. "A war not won." *Scientific American*, vol 270, no. 1 (January 1994), pp. 130–138. This ongoing war is against cancer. An interesting table on page 134 shows that most kinds of cancer were less frequently seen in 1990 than in 1973.

Beardsley. Tim. "A Joycean mutation." *Scientific American*, vol. 269, no. 6 (December 1993), p. 47. A possible new mechanism for cancer.

Mann, Charles C. "The prostate cancer dilemma." *Atlantic Monthly*, November 1993, pp. 102–118. A huge screening program is underway to test for prostate cancer. Surgery is not the only treatment option.

Beardsley, Tim. "Sharks do get cancer." *Scientific American*, vol. 269, no. 4 (October 1993), pp. 24, 27. Cartilage therapy will not work if it relies on the misinformation that sharks do not get cancer; they do, even in their cartilage.

Passeri, Daniel R., and Spiegel, Jack. "Killing cancer without surgery." *Chemtech*, July 1993, pp. 24–28. An antibody can guide anticancer drugs to tumor cells without harming other cells.

Raloff, Janet. "EcoCancers." *Science News*, July 3, 1993, pp. 10–13. This article asks whether environmental factors are causing a breast cancer epidemic.

Marshall, Eliot; Roberts, Leslie; Marx, Jean; Henderson, Maureen; and Lippman, Marc E. "The politics of breast cancer." *Science*, vol. 259, January 29, 1993, pp. 616–632. Six articles on an important topic that spans both science and politics.

Rosenberg, Steven A. "Adoptive immunotherapy for cancer." *Scientific American*, vol. 4, no. 1 special issue (1993), pp. 94–101. Also called cell-transfer therapy, this technique is one of a new class of approaches designed to strengthen the innate ability of the immune system to fight cancer.

Kartner, Norbert, and Ling, Victor. "Multidrug resistance in cancer." *Scientific American*, vol. 4, no. 1 special issue (1993), pp. 110–117. An ancient pump protein that flushes toxins out of cells may be to blame when cancer chemotherapy fails. The identification of this protein offers hope that multidrug-resistant cancers may be made vulnerable again.

Liotta, Lance A. "Cancer cell invasion and metastasis." *Scientific American*, vol. 4, no. 1 special issue (1993), pp. 142–150. The most life-threatening aspect of cancer is the undetected spread of tumor cells throughout the body. Improved understanding of how these cells invade tissues is leading to new treatments.

Dunn, Don. "Prostate cancer: How to thwart a killer." *Business Week*, March 16, 1992, pp. 132–133. Prostate cancer is to men as breast cancer is to women—a deadly killer.

Radetsky, Peter. "Stopping the crab." *Discover*, March 1993, pp. 32–33. Brief history of cancer and a quick overview of the disease.

Neuman, Elena. "Cancer: The issue feminists forgot." *Insight*, February 24, 1992, pp. 6–11, 32–34. Breast cancer is a political issue that promises to grow in importance during the remainder of the 1990s.

Rosenberg, Steven A., and Barry, John M. 1992. *The transformed cell*. New York: Putnam & Sons. A readable, up-to-date account of the nature of cancer and its treatments.

Bloch, Annette, and Bloch, Richard. 1985. *Fighting cancer*. Kansas City, Mo.: R. A. Bloch Cancer Foundation. A very readable book with lots of information on cancer chemotherapy written by a cancer survivor.

Weaver, Robert F. "The cancer puzzle." *National Geographic*, vol. 150, no. 3 (September 1976), pp. 396–399. What causes cancer?

AIDS

Frackelmann, Kathleen. "Staying alive: Scientists study people who outwit the AIDS virus." *Science News*, March 18, 1995, pp. 172–174. Some AIDS patients remain alive years longer than expected. Why?

Solomon, Goody L. "Women with AIDS, fastest rising category, are also most invisible." *Insight*, March 13, 1995, p. 33. Women are increasingly joining the ranks of those with AIDS—yet no one seems to notice.

Gorman, Christine. "Salk vaccine for AIDS." *Time*, February 6, 1995, p. 53. Dr. Jonas Salk (who invented the polio vaccine) tried to develop an AIDS vaccine, but most experts did not expect him to achieve success.

Cowley, Geoffrey. "HIV's raw aggression." *Newsweek,* January 23, 1995, p. 58. HIV slowly overwhelms the immune system's T-cells.

Watson, Traci. "New understanding of the AIDS virus." *U.S. News & World Report,* January 23, 1995, p. 61. Some new insights on how the AIDS virus attacks the immune system.

Brownlee, Shannon. "Can a hormone fight AIDS?" *U.S. News & World Report,* November 21, 1994, pp. 88–89. Only time will tell, but hormones can do many other things.

Carey, John. "AIDS: Maybe there isn't a magic bullet." *Business Week,* October 24, 1994, pp. 108–109. As early hopes of a quick cure fade, the emphasis is shifting toward a methodical approach.

Essex, Max. "Confronting the AIDS vaccine challenge." *Technology Review,* October 1994, pp. 22–29. Some expect to have an AIDS vaccine on the market within five years.

Des Jarlais, Don C., and Friedman, Samuel R. "AIDS and the use of injected drugs." *Scientific American,* vol. 270, no. 2 (February 1994), pp. 82–88. This article focuses on regions where drug injection has led to the prevalence of AIDS. The authors advocate changes in governmental drug policies.

Piel, Gerald. "AIDS and population control." *Scientific American,* vol. 270, no. 2 (February 1994), p. 124. The author discusses a controversial claim—that AIDS arrived in time to slow down global population increases.

Steinman, Lawrence. "Autoimmune systems." *Scientific American,* vol. 269, no. 3 (September 1993), pp. 107–114. Many autoimmune diseases, including multiple sclerosis, Crohn's disease, myasthenia gravis, psoriasis, rheumatoid arthritis, and lupus.

Greene, Warner C. "AIDS and the immune system." *Scientific American,* vol. 269, no. 3 (September 1993), pp. 99–105. How HIV uses the immune system to self-replicate, thereby wreaking havoc on the system.

Baum, Rudy M. "AIDS: Scientific progress but no cure in sight." *Chemical & Engineering News,* July 5, 1993, pp. 20–27. A summary report on the ninth AIDS conference.

Science, May 28, 1993, pp. 1253–1293. This issue contains several articles on AIDS, including some on current research, vaccines, therapies, and social issues.

Tarantola, Daniel, and Mann, Jonathan. "Coming to terms with the AIDS pandemic." *Issues in Science & Technology,* Spring 1993, pp. 41–48.

Volker, Eugene J. "An attack on the AIDS virus: Inhibition of the HIV-1 protease." *J. Chem. Ed,* vol. 70, no. 1 (January 1993), pp. 3–9. This article tells about the virus and how to destroy it.

Anderson, Roy M., and May, Robert M. "Understanding the AIDS pandemic." *Scientific American;* vol. 4, no. 1, special issue (1993). pp. 86–92. Mathematical models help reveal how the AIDS virus infects individuals and communities. They sometimes produce results that upset simple intuition.

Aral, Sevgi O., and Holmes, King K. "Sexually transmitted diseases in the AIDS era." *Scientific American,* vol. 4, no. 1, special issue (1993), pp. 118–125. Gonorrhea, syphilis, and other infections still exact a terrible toll. Social conditions help fuel the new epidemics, and only a combination of social and health programs can defeat them.

Root-Bernstein, Robert. 1993. *Rethinking AIDS: The tragic cost of premature consensus.* New York: Free Press. Not everyone agrees with the general consensus that HIV causes AIDS. This book presents another viewpoint and evidence to support it.

Scientific American, special issue, October 1988. A single-topic issue devoted to AIDS. (This was also released in book form.) Individual articles discuss the origin and spread of the disease, the nature of the disease from both the clinical and cellular standpoints, therapies, vaccines, and the social aspects of AIDS. Because of the rapid advances in AIDS research, some of the information is out of date. However, the historical sections, the sections on epidemiology, and the sections on the sociological aspects of AIDS are outstanding.

Chemical & Engineering News, November 23, 1987. This special issue of *Chemical & Engineering News* contains several articles on AIDS topics, including some on the spread of the disease, the nature of the virus, the search for a vaccine, HIV detection tests, AIDS therapy, and social issues.

McNamee, Lawrence J., and McNamee, Brian F. 1988. *AIDS, the nation's first politically protected disease.* La Habra, Calif.: National Medical Legal Publishing House. A pair of doctors write pointedly on the social and political nature of AIDS and its treatments. Definitely controversial!

Alzheimer's Disease

Brownlee, Shannon. "Hopeful hunt for an Alzheimer's cure." *U.S. News & World Report,* November 21, 1994, p. 89. With a new test and a new theory about its cause, an Alzheimer's cure may come soon.

Seligmann, Jean, and Springen, Karen. "Progress on Alzheimer's." *Newsweek,* November 21, 1994, p. 80. An eyedrop test could confirm the diagnosis of Alzheimer's even though a cure is not yet available.

Dorfman, Andrea. "An eye on Alzheimer's." *Time,* November 21, 1994, p. 89. A new eyedrop test may be able to show the presence of Alzheimer's in living patients.

Pennisi, Elizabeth. "A molecular whodunit." *Science News,* January 1, 1994, pp. 8–11. A summary of recent information on Alzheimer's disease. This is probably the best up-to-date summary of the chemistry of Alzheimer's.

Beardsley, Tim. "Unravelling Alzheimer's." *Scientific American,* vol. 269, no. 5 (November 1993), pp. 28–31. Researchers have found a gene that appears to cause the release of the blood protein apolipoprotein E; some believe this protein causes 80 percent of Alzheimer's cases.

Freundlich, Naomi. "Quietly closing in on Alzheimer's." *Business Week,* May 3, 1993, pp. 112–113. Easy-to-read account of recent findings on Alzheimer's.

Selkoe, Dennis J. "Amyloid protein and Alzheimer's disease." *Scientific American,* vol. 4, no. 1 special issue (1993), pp. 54–61. When this protein accumulates excessively in the brain, Alzheimer's disease may result. Understanding how the amyloid fragment forms could be the key to an effective treatment.

Fackelmann, Kathy A. "Anatomy of Alzheimer's." *Science News*, December 5, 1992, pp. 394–396. Includes a brief report on Tacrine.

Ezzell, Carol. "Alzheimer's alchemy." *Science News*, March 7, 1992, pp. 152–153. How a brain protein is converted into β-amyloid.

Lushin, Guy. 1990. *The Living death: Alzheimer's in America*. Potomac Publishing Company. This 110-page booklet was prepared for the National Foundation for Medical Research. It will tell you everything you might want to know about the disease itself, but contains very little on the chemistry of Alzheimer's.

Other Diseases

Cooke, Robert. "A plague on all our houses." *Popular Science*, January 1996, pp 50–56. Author warns that we should prepare for outbreaks of some new and several old diseases. Often adequate medical countermeasures do not exist for these new threats, and the treatments for the older diseases are not always manufactured at this time.

Watson, Traci. "Hope for fighting stroke." *U.S. News & World Report*, February 20, 1995, p. 67. Some new drugs are on the market to combat stroke.

Kaberia, Kirimi. "New malaria vaccine shows some promise." *Insight*, December 12, 1994, p. 32. Many strains of malaria are resistant to medications. This vaccine may provide a way out of this dilemma.

Price, Joyce. "Drug-resistant TB continues to plague cities." *Insight*, November 28, 1994, pp. 32–33. TB was nearly eliminated in the United States, but new strains do not always respond to medical treatment.

Adler, Tina. "Desert Storm's medical quandary." *Science News*, June 18, 1994, pp. 394–395. Thousands of Desert Storm veterans seem to have contracted a mysterious ailment during the conflict.

Nowak, Rachel. "Flesh-eating bacteria: Not new but still worrisome." *Science*, June 17, 1994, p. 1665. No doubt you have heard about this strain of bacteria. Here is more information on it.

Beardsley, Tim. "Molecular mischief." *Scientific American*, vol. 270, no. 3 (March 1994), pp. 20–21. A possible cause for schizophrenia.

Revkin, Andrews. "Hunting down Huntington's." *Discover*, December 1993, pp. 101–108. A look at people who have this genetic disease.

Gutin, Jo Ann C. "The infection unto death." *Discover*, November 1993, pp. 115–120. Sepsis, a poisoning of the body by bacterial products is on the rise in today's hospitals.

Paul, William E. "Infectious diseases and the immune system." *Scientific American*, vol. 269, no 3 (September 1993), pp. 91–97. This article discusses various viral diseases.

Aldhous, Peter. "Malaria: Focus on mosquito genes." *Science*, July 30, 1993, pp. 546–548. Could malaria be controlled by manipulating mosquito genes?

Lawn, Richard M. "Lipoprotein(a) in heart disease." *Scientific American*, vol. 4, no. 1, special issue (1993), pp. 12–18. A remarkable protein that transports cholesterol and binds with blood clots can raise the risk of heart attack. Comparisons between it and other blood proteins may explain why.

Houston, Charles S. "Mountain sickness." *Scientific American*, vol. 4, no. 1, special issue (1993). pp. 150–155. The varieties and subtle symptoms of this potentially lethal disorder humble many who scale the summits. But the problem is often preventable.

Rosenthal, Elisabeth. "Return of consumption." *Discovery*, June 1990, pp. 80–83. Once nearly eliminated, tuberculosis has made a terrifying comeback. Worse yet, many of the medicines formerly used to treat TB are no longer effective.

Disease Carriers and Transmission

Brownlee, Shannon. "The disease fighters." *U.S. News & World Report*, March 27, 1995, pp. 48–58. Any time a new disease threatens to develop into an epidemic, the Center for Disease Control is on the scene to check everything out.

Levins, Richard, et al. "The emergence of new diseases." *American Scientist*, January/February 1994, pp. 52–60. As humans have become more mobile, diseases once restricted to narrow regions of the world travel, too. How can we combat this problem?

Hughes, James M., et al. "Hantavirus pulmonary syndrome: An emerging infectious disease." *Science*, November 5, 1993, pp. 850–851. A bit of history on how this disease spread.

Dunkel, Tom. "U.S. unready for viral invasion." *Insight*, October 11, 1993, pp. 16–19. If you believe we are ready, you had better read this article. Killer pandemics are not a thing of the past; AIDS proved that.

Gibbons, Ann. "Where are new diseases born?" *Science*, August 6, 1993, pp. 680–681. A good question. Some exit the forest when the forest is cleared away.

Jaret, Peter. "The disease detectives." *National Geographic*, vol. 179. no. 1 (January 1991), pp. 114–140. This article starts with cholera and moves on to other diseases, such as Lyme disease, measles, AIDS, and possible harm from electromagnetic fields, showing how researchers learn about them. It even discusses the Asian tiger mosquito that can transmit viruses to humans.

Duplaix, Nicole. "Fleas: The lethal leapers." *National Geographic*, vol. 173, no. 5 (May 1988). pp. 672–694. This article covers the three major pandemics spread by these insects, each of which killed millions of people.

Gerster, Georg. "Tsetse-fly of the deadly sleep." *National Geographic*, vol. 170. no. 6 (December 1986), pp. 814–833. This is still a deadly blight for 50 million people in Africa; 20,000 fall victim to sleeping sickness each year.

Roach, Mary. "Secrets of the shamans." *Discover*, November 1993, pp. 58–65. Tribal medicine men have always used herbal medicines, though not always with success.

Wagner, Betsy. "Nature's tropical medicine chest." *U.S. News & World Report*, November 1, 1993, p. 77. This article discusses several chemicals, including some from an orange African frog.

Sotheeswaren, S. "Herbal medicine: The scientific evidence." *J. Chem. Ed*, vol. 69, no. 6 (June 1992), pp. 444–446. Some people swear by herbal medicines, other swear at them. Who is correct? It turns out there is plenty of evidence to support using some herbal medicines, but not all.

Mowrey, Daniel B. 1986. *The scientific validation of herbal medicine*. New Canaan Conn.: Keats Publishing. Possibly the best single source of information on this topic.

Steiner, Richard P., ed. 1986. *Folk medicine—The art and the science*. Washington, D.C.: American Chemical Society. Everything you might ask about folk medicines from many parts of the world, including a 5,000-year study on the medical benefits of garlic.

Aikman, Lonnell. "Nature's gifts to medicine." *National Geographic*, vol. 146, no. 3 (September 1974), pp. 420–440. All sorts of helpful potions and deadly poisons exist in nature. This is a brief account about how we discovered some of the good and bad.

New Medical Techniques

Leutwyler, Kristin. "Optical tomography." *Scientific American*, vol. 270, no. 1 (January 1994), pp. 147–149. How ordinary light can be a noninvasive imaging tool to diagnose disease.

Fitzgerald, Karen. "Magnetic apprehensions." *Scientific American*, vol. 269, no. 4 (October 1993), pp. 106–107. There are now new high-speed MRI machines, but are they safe for patients?

Piel, Jonathan. "Medicine." *Scientific American*, vol. 4, no. 1, special issue (1993), pp. 4–10. A powerful healing technology seeks institutions and regulations that can deliver it to those who need it.

Montgomery, Geoffrey. "The ultimate medicine." *Discover*, March 1990, pp. 60–68. Curing diseases via genetic manipulation.

Sochurek, Howard. "Medicine's new vision." *National Geographic*, vol. 171, no. 1 (January 1987), pp. 2–41. CAT (Computed Axial Tomography), MRI (Magnetic Resonance Imaging), sonography, PET (Positron Emission Tomography), SPECT (Single Proton Emission Computed Tomography), radioisotope imaging, and many other modern medical techniques cut their eyeteeth in chemistry and physics laboratories.

Plastics and Medicine

Edelson, Edward. "Robo surgeons." *Popular Science*, April 1995, pp. 62–65, 90. Plastics have long been used in surgery. Now robots are performing some of the operations that use them.

Fitzgibbons, Stella Jones. "Making artificial organs work." *Technology Review*, August/September 1994, pp. 32–40. Good review of the artificial kidney, lung, liver, and heart.

Gibbs, W. Wayt. "Deliverance." *Scientific American*, vol. 269, no. 6 (December 1993), pp. 30–33. A report on advances in developing an artificial liver featuring hollow fibers.

Gibbs W. Wayt. "More fun than a root canal." *Scientific American*, vol. 269, no. 5 (November 1993), p. 106. An alternative dentistry technique that removes only the infected portion of a tooth and covers the rest with a composite that allows more dentin to regenerate and cover the nerves.

Gunther, Judith Anne. "Treating disease without drugs." *Popular Science*, May 1993, pp. 78–60, 104–116. Implants of natural cells encased within a plastic membrane are rapidly becoming a major method of treating some diseases, such as diabetes and Parkinson's.

Gebelein, Charles G. 1982. "Prosthetic and biomedical devices." In *Kirk-Othmer encyclopedia of chemical technology*, vol. 19, pp. 275–313. New York: Wiley. Although this encyclopedia was designed for scientists, a nonscientist could understand much of this article. It includes data on how many devices are used annually and some cost figures.

Gebelein, Charles G. 1985. "Medical applications of polymers." In *Applied polymer science*, R. W. Tess and G. W. Poehlein, eds., pp. 535–556. Washington, D.C.: American Chemical Society. A short review of polymers used in the medical field.

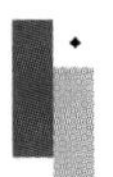

PROBLEMS AND QUESTIONS

1. What contributions did each of the following people make to the chemistry of medicine?

- **a.** Chain
- **b.** Domagk
- **c.** Dubos
- **d.** Duggar
- **e.** Ehrlich
- **f.** Fleming
- **g.** Florey
- **h.** Hodgkin
- **i.** Sheehan
- **j.** Waksman

2. How have the causes of death changed between 1900 and 1992? How did chemistry help bring about these changes?

3. What is an analgesic?

4. What source provided the first analgesics?

5. List the major analgesics used today. If you can, show their chemical formulas.

6. What are the potential side effects of analgesics?

7. What does an antipyretic do?

8. What kind of chemical is most commonly used as a topical analgesic (for application on the skin)?

9. What is the source, chemical nature, and activity of the enkephalins?

10. How does anesthetic activity differ from analgesic activity?

11. What chemicals have been used as anesthetics? Which are still used?

12. What is the LD_{50} value, and how is it determined?

13. What do LD_{50} numbers mean in relation to human toxicity?

14. Which is more toxic, a compound with an LD_{50} of 100 mg/kg or one with an LD_{50} of 1,000 mg/kg?

15. Which would be more toxic—3 g caffeine or 3 g ibuprofen?

16. Assuming one cup of regular coffee contains 100 mg caffeine, how many cups of coffee must you drink to imbibe a dose equivalent to caffeine's LD_{50}? (*Hint:* Review tables 17.1 and 17.2)

17. Which are usually more toxic, medicinal drugs or insecticides?

18. What is the difference between bacteriostatic and bactericidal agents?

19. What are the main chemical features (or formulas) for the following medicinal drug agents: (a) sulfa drugs, (b) penicillin, and (c) tetracycline?

20. Why is Paul Ehrlich considered the founder of chemotherapy?

21. What did Ehrlich mean by the term *magic bullet?*

22. What structural detail makes sulfa drugs work as antibacterial agents?

23. How does penicillin kill bacteria?

24. What does the suffix *mycin* tell us about the source of a drug?

25. What are some of the causes of cancer? Which are chemical?

26. What does the term *carcinogen* mean? Give some examples of carcinogens?

27. What was the first identified example of chemical carcinogenesis?

28. What are aflatoxins? What affliction can they cause?

29. What are the major classes of anticancer agents?

30. Which class of anticancer agents does cyclophosphamide belong to?

31. Which class of anticancer agents does 5-fluorouracil belong to?

32. What is the nature of the agent which causes AIDS? How does this differ from other infectious agents?

33. What are the main chemicals used to treat AIDS? Into what class of compounds do these anti-AIDS drug fit?

34. What is a vaccine?

35. How do retroviruses differ from regular viruses?

36. How long must an AIDS patient receive drug therapy before he or she is cured?

37. What is currently believed to be the cause of Alzheimer's disease?

38. Is aluminum still believed to cause Alzheimer's disease?

39. What is the main chemical used to treat Alzheimer's disease?

40. What does a cough suppressant do?

41. What chemical does the FDA consider the best cough suppressant?

42. What does an expectorant do?

43. What chemical does the FDA consider the only effective expectorant?

44. What is herbal (folk) medicine? Has it ever been reliable in treating diseases?

45. Give examples of some of the ways plastics are used in medicine.

46. How does a controlled release drug system operate? How does this differ from ordinary medication?

47. List the six methods used in the controlled release of drugs.

48. What is dialysis? How does it operate from a chemical standpoint?

49. What plastics are used in artificial hip replacements?

50. What are some medical uses of poly(urethanes) or PEUUs?

51. What polymers are commonly used in replacement of blood vessels?

52. What are the major polymers used in eye care? How are they used?

53. What polymers are commonly used in dentistry?

54. What is the most effective method for the disposal of medical waste?

55. Why shouldn't medical waste be placed into a landfill?

CRITICAL THINKING PROBLEMS

1. Can the biochemical agents in plants be used to treat human disease? What are some advantages and disadvantages of this approach? Two good background articles you might want to read are Lonnelle Aikman, "Nature's Gifts to Medicine," *National Geographic*, vol. 146, no. 3 (September 1974), pp. 420–440; and Mary Roach, "Secrets of the Shamans," *Discover*, November 1993, pp. 58–65.

2. Suppose a chemical agent was developed that could extend the life span to an average of 200 years, but about 1 in 1,000 people died from the treatment. Would you take this drug? How would you evaluate the relative risks and benefits?

3. Drug research requires testing to determine the safety and effectiveness of various drugs. Most of this research uses insects, mice, or rats. The campaigns against animal research tend to concentrate on

dogs, cats, and rabbits. Why do you think this is so? For more information, two useful background articles: Bill Breen, "Why We Need Animal Testing." *Garbage*, April/May, 1993, pp. 38–45; and Harriet Ritvo, "Toward a More Peaceable Kingdom," *Technology Review*, February/March, 1992, pp. 55–61.

4. Occasionally, reports in the popular press describe the results of medical studies performed on a limited number of people. Sometimes these studies are restricted to narrow age groups or to one gender, race, or ethnic group. Many medical studies use data amassed solely from examinations of athletes. What problems arise in extending the conclusions of such studies to the general population? How do such studies stack up in relation to scientific method? Are there limitations on the validity of the conclusions in these studies? The article called "The Disease Detectives," by Peter Jaret in *National Geographic*, vol. 179, no. 1 (January 1991), pp. 114–140, can provide some useful insights on these questions.

5. Everyone knows drugs have side effects. However, some drugs seem unusually dangerous for some people. Read Doug Podolsky and Penny Loeb, "Dangerous Drugs," *U.S. News & World Report*, January 9, 1995, pp. 48–54; and Susan Brink, "A Different Kind of Cancer Risk," *U.S. News & World Report*, January 9, 1995, pp. 56–61. Should such dangerous drugs be allowed on the market? (Remember that they effectively treat some diseases.) What kind of warning label should each drug carry? Would you take one of these drugs if the alternative was possible death?

6. Claims are made almost daily that herbs can heal various diseases. Some herbal remedies have proven to be effective medical treatments. How can you be sure that the latest herbal fad is safe and effective? What tests would you recommend to determine the safety of an herb, and how would you determine whether an herb will work in treating a particular affliction? Two good, short articles on herbal medicine are S. Sotheeswaren, "Herbal Medicine: The Scientific Evidence," *J. Chem. Ed*, vol. 69, no. 6 (June 1992), pp. 444–446; and Lonnelle Aikman, "Nature's Gifts to Medicine," *National Geographic*, vol. 146, no. 3 (September 1974), pp. 420–440.

7. A placebo is an inert substance given to patients that provides no benefits or harm. Its effects are usually measured against a drug researchers are testing. In some instances, the patient reports a positive effect from the placebo; sometimes the patient even goes into remission from a disease. How could you determine whether this spontaneous remission is a placebo effect or whether it is unrelated to the placebo? How can spontaneous remission be explained?

8. Which over-the-counter analgesic do you like best? Review the information in this chapter and read "One More Pill for Pain" by Rita Rubin in *U.S. News & World Report*, August 1, 1994, pp. 62–64.

9. Some cancer cells resist chemotherapy and radiation treatments because they are hidden deep within the mass of a tumor. These cancer cells can cause a new attack of cancerous growth at a later date. What steps can doctors take to avert this medical disaster? Read Robert F. Weaver, "The Cancer Puzzle," *National Geographic*, vol. 150, no. 3 (September 1976), pp. 396–399 for more information.

10. Some tumor cells can be grown in culture. How might this type of cancer cell help us research anticancer drugs? For more background on cancer, read Peter Radetsky, "Stopping the Crab," *Discover*, March 1992, pp. 32–33.

11. Information on breast cancer frequently appears in the newspapers or on TV, but you don't often hear about prostate cancer. Why? Some helpful articles on the subject include Elena Neuman, "Cancer: The Issue Feminists Forgot," *Insight*, February 24, 1992, pp. 6–11, 32–34; Don Dunn, "Prostate Cancer: How to Thwart a Killer," *Business Week*, March 16, 1992, pp. 132–133; and the six articles in *Science*, vol. 259 (January 29, 1993), pp. 616–632, collectively called "The Politics of Breast Cancer."

12. A study released in late 1994 suggested a link between breast cancer and abortions. For background, read Christine Gorman, "Do Abortions Raise the Risk of Breast Cancer?" *Time*, November 7, 1994, p. 61. Do you think this is a reasonable question? If additional studies provided the same information, what solutions or prevention methods might you propose?

13. Cancer vaccines are a fairly new concept. Read the following articles and form your own opinion on the viability of these programs: Rita Rubin, "The New Promise of Cancer Vaccines," *U.S. News & World Report*, March 6, 1995, pp. 87–88; and Neil Gross, et al., "Quiet Strides in the War on Cancer." *Business Week*, February 6, 1995, pp. 150–152. Are scientists more likely to develop a cancer vaccine or an AIDS vaccine? Why?

14. The battle against cancer has been waged for a long time. Are we winning? Has the battlefield changed (that is, are some forms of cancer more curable than others)? Do there seem to be new causes for cancer that did not exist years ago? For background, read Tim Beardsley, "A War Not Won," *Scientific American*, vol. 270, no. 1 (January 1994), pp. 130–138; Janet Raloff, "EcoCancers," *Science News*, July 3, 1993, pp. 10–13; and Peter Radetsky, "Stopping the Crab," *Discover*, March 1992, pp. 32–33.

15. The presence of HIV can be detected in a person's blood long before AIDS symptoms develop. (a) Should the person and their family be informed of the results of the test? (b) Should the person's doctor, dentist, and coworkers be informed? (c) Should any restrictions be placed on the person to restrain the spread of the HIV to others? (d) Should AIDS patients be quarantined? (e) Should scientists develop a required HIV virus detection for all blood supplies? You can find some useful background in the articles on AIDS in *Scientific American*, October 1988, *Chemical & Engineering News*, November 23, 1987, and *Science*, May 28, 1993, pp. 1253–1293.

16. Why do many scientists consider it unlikely that an AIDS vaccine will be developed in the near future? In what ways would an AIDS vaccine be of limited usefulness, at least for a while? You can find some valuable information on this topic in the collection of articles in *Science*, May 28, 1993, pp. 1253–1293; plus Rudy M. Baum, "AIDS: Scientific Progress, But No Cure in Sight," *Chemical & Engineering News*, July 5, 1993, pp. 20–27; Christine Gorman, Salk Vaccine for AIDS," *Time*, February 6, 1995, p. 53; and Max Essex, "Confronting the AIDS Vaccine Challenge," *Technology Review*, October 1994, pp. 22–29.

17. Imagine you have been appointed an advisor to the FDA. Your first assignment is to make a study of the risks and benefits of an experimental AIDS treatment to be given to a twenty-year-old woman who contracted AIDS from a blood transfusion. What kind of information will you need, and how will you get it? What other factors would you use in making your decision? Would your decision change if the patient was fifty years old? What if the patient had contracted the disease from intravenous drug usage?

18. In your opinion, should money be spent on AIDS research at the expense of research on other diseases?

19. Alzheimer's disease normally afflicts a person over age fifty. Some recent studies seem to relate this disease to genetic defects that could be identified in younger people. What are the advantages and disadvantages of gaining such information? You can find more background on Alzheimer's in Naomi Freundlich, "Quietly Closing in on Alzheimer's," *Business Week*, May 3, 1993, pp. 112–113; Kathy A. Fackelmann, "Anatomy of Alzheimer's," *Science News*, December 5, 1992, pp. 394–96; and Elizabeth Pennisi, "A Molecular Whodunit," *Science News*, January 1, 1994, pp. 8–11.

20. Some patients who later develop Alzheimer's disease (or Parkinson's disease) lose a part of their sense of smell before the onset of the disease. Could this observation be used to diagnose these diseases, or is more information necessary? An article that might provide some useful information is Carol Ezzell, "Alzheimer's Alchemy," *Science News*, March 7, 1992, pp. 152–153.

21. Malaria is still a major disease that causes more than one million deaths annually—more than died from AIDS in fourteen years. Many of the older drugs are no longer effective against malaria. For background, read the following four articles: Francis E. G. Cox, "Malaria Control: Challenges for Chemotherapy and Vaccination," *Chemistry & Industry*, August 5, 1991, pp. 533–536; Lori Oliwenstein, "New Victories in an Old War," *Discover*, June 1992, pp. 16–17; Kirimi Kaberia, "New Malaria Vaccine Shows Some Promise," *Insight*, December 12, 1994, p. 32; and Peter Aldhous, "Malaria: Focus on Mosquito Genes," *Science*, July 30, 1993, pp. 546–548. Are any of these problems likely to eliminate malaria? Should malaria research receive more funding? Should DDT be used more extensively to eradicate the mosquitoes that carry malaria?

22. Why has tuberculosis once again become a dangerous disease in the United States? What can we do to prevent this type of problem from happening again? For background, read Joyce Price, "Drug-Resistant TB Continues to Plague Cities," *Insight*, November 28, 1994, pp. 32–33; Elisabeth Rosenthal, "Return of Consumption," *Discover*, June 1990, pp. 80–83; and Richard Levins, et al., "The Emergence of New Diseases," *American Scientist*, January/February 1994, pp. 52–60.

23. How likely is it that the United States will have major epidemics of new diseases in the near future? What can we do to lessen this possibility? For background, read the Levins, et al. article mentioned in question 22; Shannon Brownlee, "The Disease Fighters," *U.S. News & World Report*, March 27, 1995, pp. 48–58; Tom Dunkel, "U.S. Unready for Viral Invasion." *Insight*, October 11, 1993, pp. 16–19; and Ann Gibbons, "Where Are New Diseases Born?" *Science*, August 6, 1993, pp. 680–681.

24. What are the problems in developing artificial organs? What role does chemistry play in this process? For background, read Stella Jones Fitzgibbons, "Making Artificial Organs Work," *Technology Review*, August/September 1994, pp. 32–40; W. Wayt Gibbs, "Deliverance," *Scientific American*, vol. 269, no. 6 (December 1993), pp. 30–33; Judith Anne Gunther, "Treating Disease Without Drugs," *Popular Science*, May 1993, pp. 78–60, 104–106; and Charles G. Gebelein, "Prosthetic and Biomedical Devices," in *Kirk-Othmer Encyclopedia of Chemical Technology*, vol. 19, pp. 275–313 (New York: Wiley, 1982).

25. Tacrine helps certain Alzheimer's patients but not others. In your opinion, should any drug be allowed on the market if it is not effective for *all* patients with a particular disease? Why or why not? (This argument kept Tacrine unavailable for many years.)

26. Now that you have studied both contraception (chapter 14) and AIDS (this chapter), what is your opinion of the use of Norplant to prevent teen pregnancy? A good starting reference is: G. Hanson, "Norplant Joins War on Teen Pregnancy," *Insight*, March 8, 1993, pp. 6–11, 34–35.

27. Fetal tissue research is very controversial. Are the objections valid, or do they unfairly restrict scientific research? Read at least two of the following trio of articles and make your own assessment of this matter. (1) Cheryl Wetzstein, "Research on Embryos Fuels Controversy," *Insight*, January 2, 1995, p. 33; (2) Christine Gorman, "Brave New Embryos," *Time*, August 29, 1994, pp. 60–61; and (3) Jon Cohen, "New Fight Over Fetal Tissue Grafts," *Science*, February 4, 1994, pp. 600–601.

28. Scientists have studied the human brain for many decades, but recent data have shed new light on the power of our brains. Read several of the following articles and summarize this information. (The first article provides the most information.) (1) Joannie M. Schrof, "Brain Power," *U.S. News & World Report*, November 28, 1994, pp. 88–97; (2) Bennett Daviss, "Brain Powered," *Discover*, May 1994, pp. 58–65; (3) B. Bower, "Scientists Peer into the Mind's Psi," *Science News*, January 25, 1994, p. 68; and (4) "What Is Consciousness?" (no author listed), *Discover*, November, 1992, pp. 95–106.

29. Weight gain is common after people stop smoking. Why do you think this happens? What can be done to prevent it?

HINT Research the nature of nicotine.

30. Some people believe that the human mind can control pain through the release of enkephalins in the brain. Research enkephalins. What is your opinion of this idea. How could you test whether enkephalins have this effect?

31. Since many antibiotics are derived from molds, would it be a good idea to rub an infected sore with a mold found growing around the home? Why or why not?

HINT Try to find out what other substances molds produce besides antibiotics.

32. Problems always crop up in finding people to test a medical treatment. Ehrlich encountered such difficulties when he tested the early chemotherapeutic agents on syphilis victims. Testing vaccines poses even more complications; Salk and Sabin encountered many obstacles in testing their polio vaccines. (a) Why is testing a vaccine more problematic than testing a drug? (b) What special problems exist in testing an AIDS vaccine?

33. Could medical problems arise if a heavy smoker used a cough suppressant for smoker's hack?

HINT Coughing is important to remove anything lodged in the lungs, including particles from the smoke.

C H A P T E R 18

AIR POLLUTION, ENERGY, AND FUELS

OUTLINE

KEY IDEAS

1. The Nature and Composition of the Atmosphere
2. The Nature of the Two Main Types of Pollution
3. The Chemicals Involved in Industrial Pollution
4. Several Ways to Reduce Industrial Air Pollution
5. Particulate Matter Sources and Removal
6. The Chemicals Involved in Photochemical Air Pollution
7. Several Ways to Reduce Photochemical Air Pollution
8. The Relationship Between Chemistry, Fuels, Energy, and Pollution
9. The Promise and Hazards of Nuclear Energy
10. Alcohols and Related Compounds as Fuels
11. The Feasibility of Hydrogen as a Fuel
12. Some Nonchemical Energy Sources
13. How to Harness Solar Energy
14. Storing and Transporting Energy
15. Whether Any Form of Energy is Absolutely Safe

"Pollution is nothing but the resources we are not harvesting."
R. BUCKMINSTER FULLER (1895–1983)

Photovoltaic panel.

Most people realize that air is essential to life, and that good air quality is necessary for a good quality of life. Reports on air pollution appear nearly daily in the news. In response, you may either become excessively alarmed or dismiss the whole problem out of desperation or boredom. Neither approach is useful. Instead, we need to understand the types and causes of air pollution. The relationship between pollution and fuels is critical to any rational consideration of pollution problems.

AIR—A GASEOUS OCEAN FOR LIFE

The "ocean" of air we live in is a homogeneous gas mixture (a solution) that obeys the gas laws chapter 2 described (for example, Boyle's and Charles' Laws). Figure 18.1 shows the typical composition of dry air; air's water content can vary so widely that it is omitted. This combination of nitrogen, oxygen, argon, carbon dioxide, and other gases makes up the troposphere, the layer of the atmosphere that extends to about 10 km above the surface of the planet (figure 18.2). This is the air we breathe.

Not all air quality problems are attributable to human activities. A single volcanic eruption spews many more pollutants into the air than the worst smokestacks bellowing out industrial pollution. The 1980 eruption of Mount St. Helens in the state of Washington belched out more than 3 km^3 of dust and ash that settled across most of North America (figure 18.3). This output pales by comparison with the 1815 eruption of Tambora (in Indonesia), which expelled 175 km^3 debris and caused temperatures to drop worldwide so that 1815 became the year with no summer. Volcanoes emit SO_2, HCl, and CO_2 during eruptions and even during the dormant periods between. Likewise, lightning discharges generate many tons of nitrogen oxide compounds (NO_x) annually. In this chapter, our main concern will be human-caused pollutants, which humans can control; lightning and volcanoes are beyond our influence.

Table 18.1 summarizes the major air pollutants and their sources. Some materials are classed as **primary pollutants,** while others (for example, ozone) are **secondary pollutants** derived from a chemical reaction involving a primary pollutant.

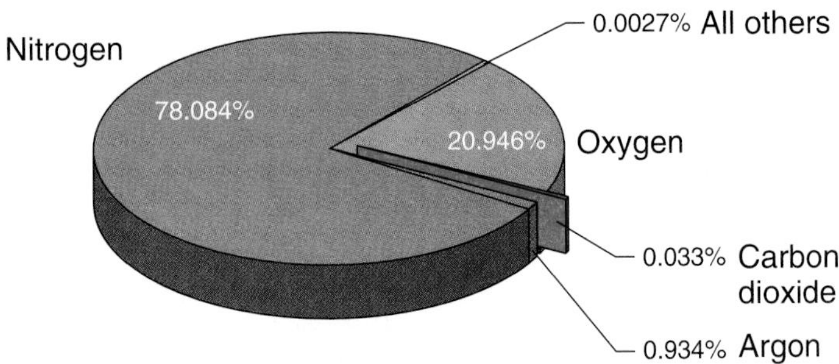

Figure 18.1 The chemical composition of dry air.

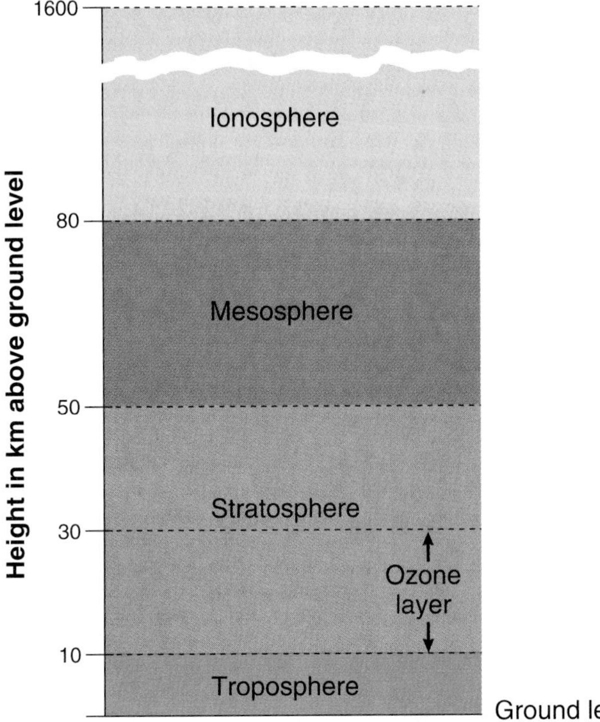

Figure 18.2 The physical structure of the atmosphere.

PRACTICE PROBLEM 18.1

List the materials cited as examples of natural air pollution. Is it reasonable to say something causes air pollution if it is completely natural?

ANSWER The items mentioned are dust and ash (particulate matter) and the gases SO_2, CO_2, HCl, and NO_x. The fact that a compound is produced in nature does not mean it is not a pollutant. In fact, humanmade pollutants may normally be present in low concentrations (below 100 ppm), while natural pollutants may on occasion be released in huge (and dangerous) quantities.

PRACTICE PROBLEM 18.2

What is the main difference between primary and secondary pollutants?

Figure 18.3 The 1980 eruption of Mount St. Helens poured out over 3 km^3 of dust and ash into the atmosphere.

ANSWER The difference is the source. Primary pollutants are released directly into the atmosphere, whereas secondary pollutants are formed by a chemical reaction, generally with a primary pollutant.

TYPES OF AIR POLLUTION

Humanmade air pollution is divided into two basic types: the **industrial (London) type,** and the **photochemical (Los Angeles) type.** Even though there are similarities between them, these two types of air pollution are sufficiently different to require separate study.

Industrial (London) Type Pollution

Industrial air pollution, due primarily to industrial emissions, was first described in detail in London, England; consequently, it is also called "London Type" air pollution. This type of pollution follows industrialization wherever it develops, including the United States (figure 18.4).

The two major pollutants in industrial type pollution are gaseous **sulfur dioxide (SO_2),** and **particulate matter,** including dust, soot, and fly ash. As we discussed in chapter 8, SO_2 is the major cause of acid rain; soot and ash makes neighborhoods dirty. Both are byproducts of the Industrial Revolution and remain the hallmarks of industrial type pollution. Many years ago, in some cities, it was necessary to turn on the streetlights on a winter day in order to see. In those days, factory smokestacks belched a continuous stream of soot and fly ash into the air, along with invisible polluting gases. As Charles Dickens described in his fictional

TABLE 18.1 The main air pollutants and their sources.

POLLUTANT	FORMULA	SOURCES
Carbon monoxide	CO	Incomplete human-generated or natural combustion of carbon-containing fuels.
Hydrocarbons	C_nH_{2n+2}	Residue from petroleum combustion; cattle also release methane into the atmosphere.
Particulate matter	—	Incomplete combustion of fuels; farming, construction, industry, demolition, dust and sand storms; volcanic eruptions.
Sulfur dioxide	SO_2	Combustion of soft coal or other sulfur-containing fuels; smelting of sulfur-containing metal ores. Most anthropogenic SO_2 comes from power plants or industrial operations.
Nitrogen oxides	NO_x	Burning of any fuel containing nitrogen, or fuel combustion at high temperatures, such as in an internal combustion engine. Lightning and sparking motors also produce some NO_x, but most anthropogenic NO_x comes from transportation sources.
Ozone	O_3	Reactions between NO_x, sunlight, and O_2 (secondary pollutant).
Aldehydes	R—CHO	Reactions between O_3 and hydrocarbons (secondary pollutant).
Peroxynitrates	RCO_3NO_2	Reactions between aldehydes and NO_x (secondary pollutant). Also called PANs, for peroxyacylnitrates.

Figure 18.4 A factory scene in Donora, Pennsylvania—the location of a major industrial air pollution crisis in October 1948.

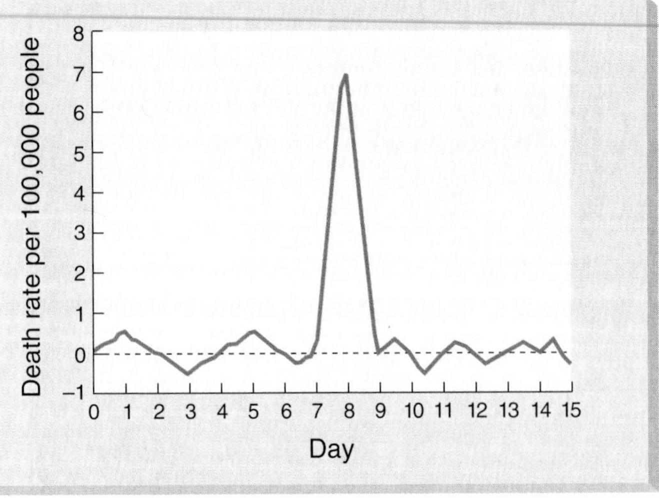

Figure 18.5 A plot of the death rates against time for a hypothetical community. Note that this death rate increased markedly at Day 8; the increase represents "excess deaths" due to some unusual circumstance.

book *Hard Times*: "Coketown lay shrouded in a haze of its own, which appeared impervious to the sun's rays." Although many western nations have attacked the problem, particulate matter continues to be a major concern in Eastern Europe, the former U.S.S.R., China, and southern Asia.

Major industrial air pollution problems are less prevalent today in modernized countries, such as the United States, because strict air pollution control laws were enacted years ago. An impelling force behind these laws was the observation that air pollution crises cause human deaths. Table 18.2 summarizes the history of this problem from 1930 to the mid-1960s, when the United States enacted its first air pollution legislation. The term *excess deaths*, as used in the table, means deaths over and above the number normally expected (figure 18.5).

In the cases cited in table 18.2, public health officials could find no evidence of obvious causes of death. Ultimately the death rate increases were associated with abnormally high air pollution levels, usually aligned with a **thermal inversion** weather pattern that trapped the pollutants in the community (figure 18.6). Thermal inversion normally requires a valley or nearby mountain range. As a cold air layer moves over the valley, it behaves like a stopper on a bottle, holding the warmer air and its pollutants down in the valley. The pollutant levels increase beyond the normal range and people with respiratory or heart problems are affected. Some at-risk people die prematurely, giving the sort of sudden surge appearing on Day 8 in figure 18.5. Most of the communities in table 18.2 are in geographical regions suitable for thermal inversions.

TABLE 18.2 Some major air pollution crises. SO_2 and particulate matter come mainly from industrial pollution; NO_x comes mainly from photochemical pollution.

LOCATION	DATE	POLLUTANTS	EXCESS DEATHS
Meuse Valley, Belgium	December 1930	SO_2, NO_x	60
Donora, PA	October 1948	SO_2, NO_x, P*	20
London, England	November/December 1948	SO_2, P	700–800
London, England	December 1952	SO_2, P	4,000
London, England	January 1956	SO_2, P	1,000
London, England	December 1962	SO_2	750
Osaka, Japan	December 1962	SO_2, NO_x	60
New York, NY	January/February 1963	SO_2, NO_x, P	200–400
New York, NY	February/March 1964	SO_2, NO_x, P	168
New York, NY	Winter, 1966	SO_2, NO_x	80

*P = Particulate matter

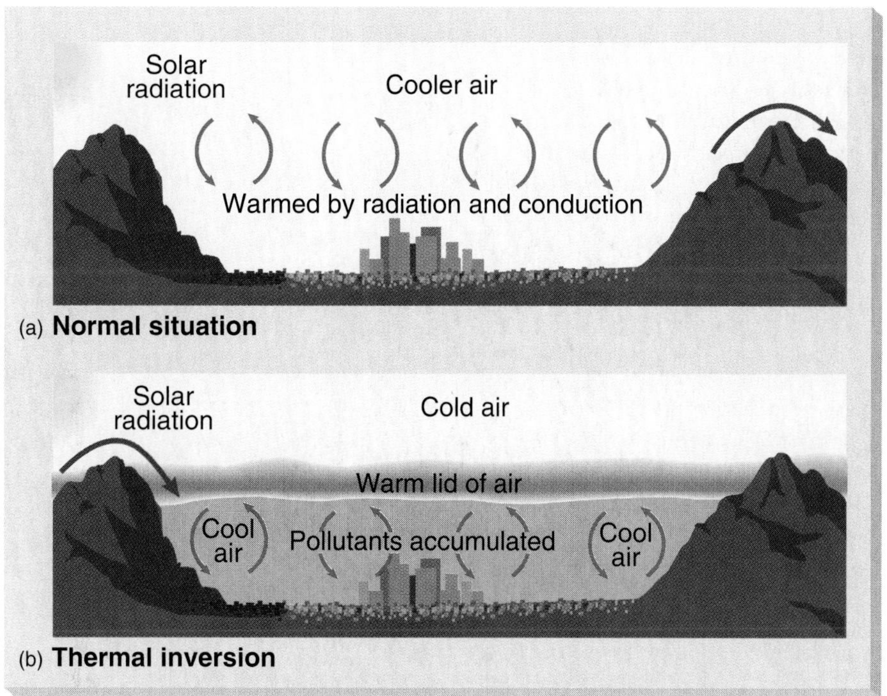

Figure 18.6 The thermal inversion pattern. **(a)** In the normal pattern, the warmer air rises upward in the atmosphere, carrying its pollutants away from ground level. **(b)** In a thermal inversion, a layer of cooler air acts as a blanket or trap that holds the warmer air, and its pollutants, at ground level. This weather pattern generally only develops in a valley or similar terrain.

SOURCE: From Eldon D. Enger and Bradley F. Smith, *Environmental Science: A Study of Interrelationships*, 5th ed. Copyright © 1995 Times Mirror Higher Education Group, Inc., Dubuque, Iowa. All Rights Reserved. Reprinted by permission.

Because air pollution crises can be so deadly, people have long tried to prevent them. The first "clean air act" may have been when King Edward I of England made his 1306 proclamation banning the burning of "sea coles," a form of coal found along seashores. However, until air pollution was related directly to human health, very few actual laws were enacted to reduce emissions. Following the infamous 1952 "Black Fog" disaster in London, which killed four thousand, the British Parliament enacted a series of stricter emission regulations. Perhaps these regulations helped reduce the excess deaths in later crises, as table 18.2 shows. Similar laws were enacted elsewhere, including the United States in 1970.

We now rarely have air pollution crises that cause large numbers of excess deaths. Still, clean air laws did not solve the problem completely; table 18.3 lists some cities where less lethal air pollution crises occur fairly regularly. Figure 18.7 emphasizes the global aspect of air pollution; winds eventually carry pollution from one part of the world to another. Rainfall removes some pollutants, reducing their spread, but massive transfers of pollutants are well documented, especially for particulate matter.

Sulfur Dioxide

As we previously mentioned, the main pollutants in industrial type pollution are SO_2 and particulates. Sulfur dioxide arises mainly from two industrial processes: burning a sulfur-containing fuel, which is almost always soft or **bituminous coal,** or smelting a metal sulfide ore to form the metal. Bituminous (soft) coal is the most common coal resource in the United States and is used in power plants to heat boilers to generate electricity and in manufacturing plants for heat and machinery operation. Smelting plants manufacture useful metals from sulfide ores. The following equations summarize the way these two processes—burning coal and smelting metal sulfide ore—release SO_2 as a byproduct:

$$S \text{ (in coal)} + O_2 \rightarrow SO_2$$

$$CuS + O_2 \rightarrow Cu + SO_2$$

The 1990 Clean Air Act mandates the reduction of SO_2 emissions through three methods: (1) using sulfur-free coal, such as **anthracite;** (2) removing sulfur from coal before burning; and (3) removing SO_2 from combustion products prior to release into the atmosphere. The first method is problematic. In many regions, the local coal is bituminous (soft); importing another type is expensive. Moreover, fewer lodes of anthracite coal remain in the United States, which also makes this form of coal more expensive, and local miners fear they would lose their livelihood. The 1990 Act includes a \$250 billion soft-coal worker retraining fund; but few major industries offer other types of work in Appalachia.

TABLE 18.3 Some representative cities (listed alphabetically) with frequent air crises. The main pollutants are SO_2, NO_x, and particulates. "Usual" implies that the problem occurs most of the time, while "often" implies the problem is a regular occurrence but not present most of the time.

LOCATION	FREQUENCY	POLLUTANTS
Athens, Greece	Usual	NO_x
Berlin, Germany	Often	SO_2, NO_x, P*
Beijing, China	Usual	SO_2, P
Budapest, Hungary	Usual	NO_x
Denver, CO	Often	SO_2, NO_x
Hamburg, Germany	Often	SO_2, NO_x, P
Hamilton, Ontario	Often	NO_x
Houston, TX	Often	NO_x
Los Angeles, CA	Usual	NO_x
Mexico City, Mexico	Usual	SO_2, P
Milan, Italy	Often	SO_2, NO_x, P
New Delhi, India	Often	P
New York, NY	Often	NO_x
Paris, France	Often	NO_x
Rome, Italy	Usual	NO_x
Santiago, Chile	Often	SO_2, NO_x, P
Sáo Paulo, Brazil	Usual	SO_2, NO_x, P
Seoul, Korea	Usual	SO_2, P
Shenyang, China	Usual	SO_2, P
Sydney, Australia	Often	NO_x
Teheran, Iran	Usual	SO_2
Tel Aviv, Israel	Often	NO_x
Tokyo, Japan	Often	SO_2, NO_x, P

*P = particulate matter

SOURCE: Jon Bowermaster, *A Citizen's Guide to Environmental Action*. 1990: Alfred Knopf, Inc., New York.

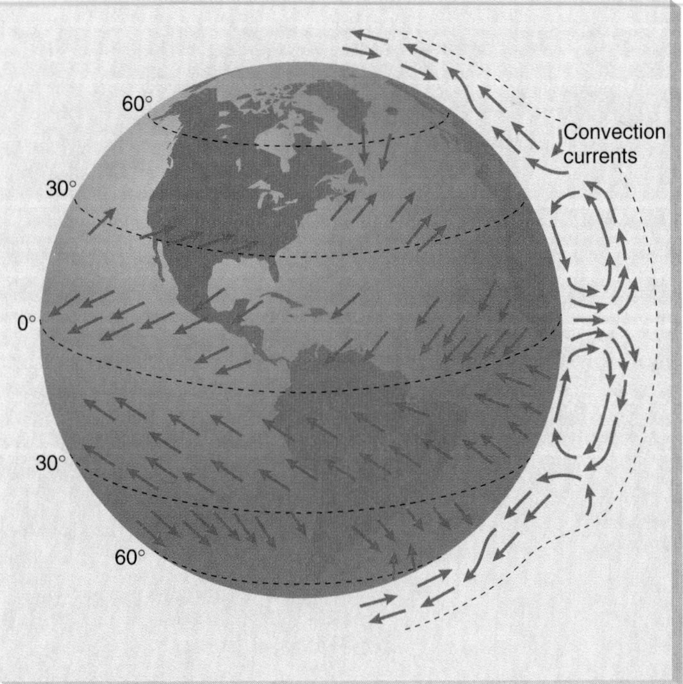

Figure 18.7 Global wind patterns eventually transfer pollutants from one region to another. This two-dimensional figure only shows part of the world, but the wind patterns are similar elsewhere, and air pollution is a global problem.

SOURCE: From Eldon D. Enger and Bradley F. Smith, *Environmental Science: A Study of Interrelationships*, 5th ed. Copyright © 1995 Times Mirror Higher Education Group, Inc., Dubuque, Iowa. All Rights Reserved. Reprinted by permission.

Alternate fuels are sometimes considered to replace coal. Both natural gas and petroleum, though more expensive, would reduce SO_2 emissions because they are normally sulfur-free. Ethanol, methanol, wood, or biomass (plant matter) could provide SO_2-free heat, but these substances are also either more expensive than coal or are less effective fuels on the basis of energy output per pound; they are already partially oxidized (see chapter 7). Nuclear power is relatively cheap and sulfur free and would reduce SO_2 emissions.

The second SO_2 reduction method, removing sulfur from coal before burning, also presents some difficulties. The sulfur in soft coal is present in covalently bonded compounds, making its removal moderately difficult. Although some removal methods are under development, this added step will undoubtedly increase the cost of the treated coal, and this cost will be passed on to the customer.

The third method for reducing SO_2 emissions is in use, at least to some degree. Industrial or power plants can trap the generated SO_2 in **scrubbers** that either dissolve the SO_2 in water or react it with another chemical. If it reacts with another substance, the SO_2 is converted into a potentially useful product; perhaps sulfurous acid, sodium sulfite, or calcium sulfite:

$$SO_2 + H_2O \rightarrow \underset{\text{Sulfurous acid}}{H_2SO_3}$$

$$SO_2 + 2NaHCO_3 \rightarrow \underset{\text{Sodium sulfite}}{Na_2SO_3} + CO_2 + H_2O$$

$$SO_2 + CaCO_3 \rightarrow \underset{\text{Calcium sulfite}}{CaSO_3} + CO_2$$

Other scrubbing reactions are also possible. But while these chemicals have valuable uses (for example, Na_2SO_3 is used in photography and bleaching), power plants are not chemical distributors and disposal is sometimes a problem.

The amount of sulfur released into the world's atmosphere annually, from all sources, is about 182 million metric tons (figure 18.8). Among these SO_2 sources, only coal burning and ore smelting are anthropogenic, or humanmade. Though these two activities account for only 25 percent of the worldwide total, this proportion rises to nearly 90 percent in urban areas. The United States (21.1 million metric tons) and China (12.9 million metric tons) release the largest amounts; together, they emit nearly 75 percent of the 45.5 million metric tons of anthropogenic sulfur. (Box 18.1 explores the energy pattern differences between nations.) Ultimately, about 75 percent of all the sulfur in the atmosphere comes from natural sources.

EXPERIENCE THIS

Are there power plants in your community that might emit SO_2? Observe these companies several times at different times of day. Are there noticeable differences in the emissions at different times?

Particulate Matter

Particulates are released when coal burns unless they are trapped or removed from the effluent smokestack gases. Particulate matter consists of two main substances: (1) soot formed from incomplete combustion, and (2) small amounts of inorganic minerals, present in most coals, which exit as a fine powdery ash called fly ash. Particulate matter can contaminate an entire community; because it is visible, it is a more obvious symptom of industrial pollution than invisible SO_2 gas.

While coal is a major source of particulate pollution, other fuels can produce particulates as the result of incomplete combustion. Particulate matter also normally includes dust from construction, demolition, and farming; in fact, dust is one of the causes of the dire pollution problems in Eastern Europe, the former U.S.S.R., China, and much of South Asia. Natives of these regions frequently wear face masks to reduce their intake of particulates. The U.S. Armed Forces seldom allow a serviceperson's family to move to an American base in these regions if the family includes members who are asthmatic.

Figure 18.9 shows the level of particulates per square mile for several countries. The United States and Western Europe may have lower levels than you expected; their industries have cleaned up much of the dirty, sooty pollution from the early days of the Industrial Revolution. This is not true in Eastern Europe, as Poland illustrates. China is also high in particulates; this is likely to increase as industrialization continues.

The best method for removing particulates from effluent smokestack gases is to use an **electrostatic precipitator** (figure 18.10a). In this device, a positive electrical charge is added to the smokestack chimney walls, or to electrodes placed in the gas stream, and a negative charge is imparted to the particles. As effluents rise up the chimney, the negatively charged particles are attracted to the walls or electrodes, and fly ash settles into a chamber at the chimney bottom. The recovered fly ash is used for valuable purposes, such as a tire filler, rather than polluting the community.

Alternate methods of fly ash removal include bag filtration (figure 18.10b) and the cyclone collector (figure 18.10c). In **bag filtration,** the effluents pass through a porous filter, usually shaped as a bag, which entraps the particles. Filtration slows down the overall gas emission release rate; this is undesirable since it increases costs. The **cyclone collector** accomplishes the same end result more rapidly by forcing the particulates out of the gas stream by centrifugal action.

Finally, particulates can contribute to excess deaths for several reasons. First, particulate matter can irritate and inflict damage on the lungs or other tissues. In addition, particulates have large

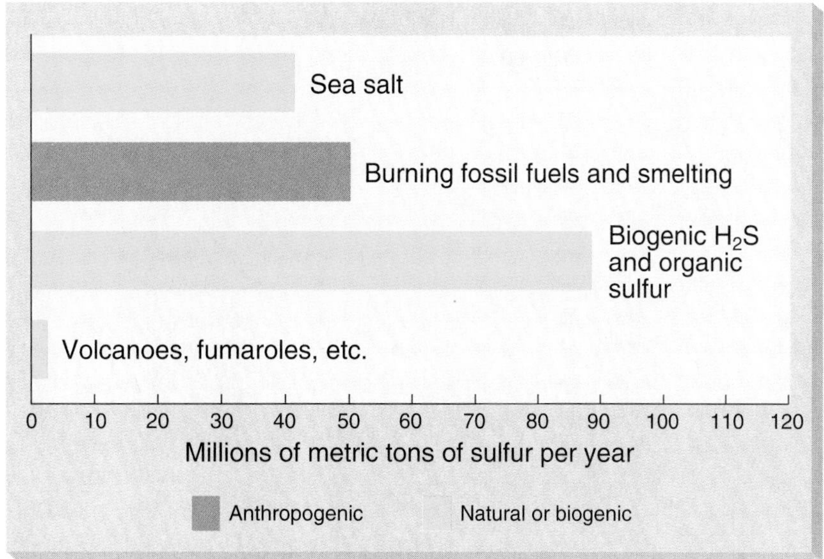

Figure 18.8 Annual amounts of sulfur released into the atmosphere worldwide.
SOURCE: From William P. Cunningham and Barbara Woodworth Saigo, *Environmental Science*, 2d ed.

BOX 18.1

WORLDWIDE ENERGY PRODUCTION AND CONSUMPTION

Air pollution is directly linked to energy production and consumption. Many humanmade SO_2 emissions come from electricity-producing power plants, while NO_x emissions arise from automobile use. Even if all the industrialized nations reduce these emissions below 1940 levels, the emerging nations will more than make up for these reductions. Although it is possible that these emerging nations will utilize modern technology to avoid the emission overloads the industrial nations experienced, the fact remains that these underdeveloped nations comprise over 60 percent of the world's population. This almost guarantees increased air pollution as these nations become able to produce more energy.

Unfortunately, we have few realistic options since virtually all sources of energy are ultimately chemical in nature (chapter 5). As figure 1 illustrates, petroleum, coal, natural gas, and biomass—all chemical energy sources—provide 88 percent of available energy worldwide. Hydroelectric energy generation, limited by geography, supplies only about 6 percent of the energy consumed, and nuclear sources account for the remaining 6 percent. More unusual energy sources, such as wind, solar, or geothermal sources, provide less than 0.01 percent of all the energy the world consumes.

Nuclear power generation actually provides proportionately more energy in North America (7.5 percent) than worldwide (6 percent, including North America), but most Third World countries have yet to tap into this resource. Some European countries rely more on nuclear power than the United States does, but the total energy demands are greater in the United States. France generates about 73 percent of its electricity via nuclear reactors, while the United States generates about 22 percent. However, the United States generates more gigawatts of electricity via nuclear reactors (99.8 in 1991) than any other country; France is second with 56.9 gigawatts. (A gigawatt is a trillion watts, or 10^9 watts. You purchase electricity in kilowatts, and a gigawatt is a billion kilowatts.) Belgium, Sweden, Hungary, and the Republic of Korea generate about 50 percent of their electricity through nuclear reactors, while Canada and the former U.S.S.R. generate 17 percent and 13 percent, respectively.

Some forms of chemical energy generation yield less SO_2 or NO_x emissions than others, but additional costs and lower fuel efficiency may make these forms impractical at present. Interestingly, overall energy utilization patterns exist in almost a direct relationship with the Gross National Product (GNP) (figure 2), but per capita energy consumption is not always directly related to the standard of living. Qatar and the United Arab Emirates have high GNPs and high levels of per capita energy consumption, yet many of their people are poor. Similarly, Switzerland consumes less energy than the United States but has a higher standard of living. However, for most countries the standard of living generally improves as per capita energy consumption rises.

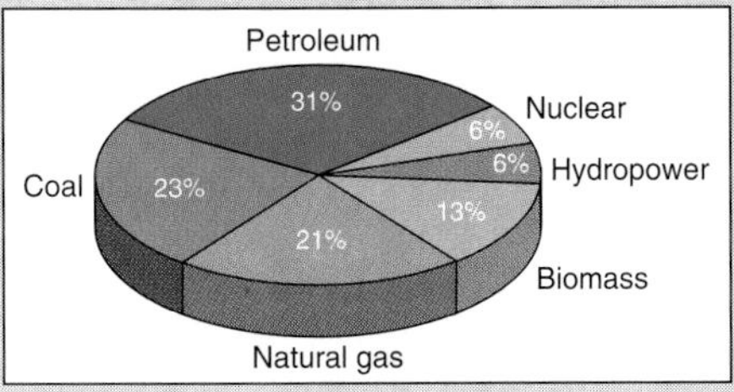

Figure 1 Worldwide consumption of energy in 1992.

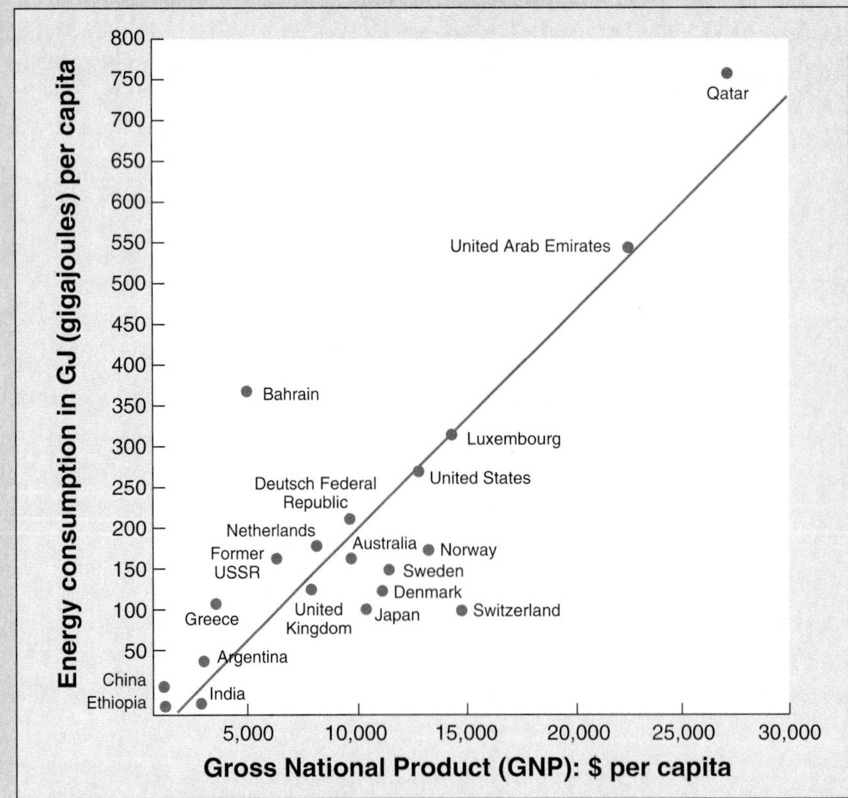

Figure 2 The relationship between Gross National Product (GNP) and the per capita consumption of energy.

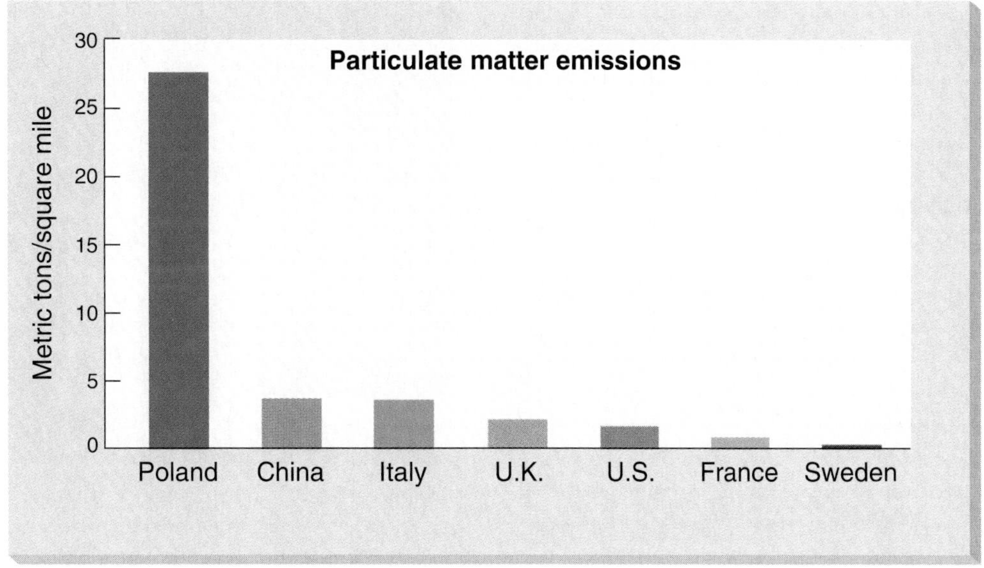

Figure 18.9 The particulate matter emissions in metric tons per square mile for seven countries.
SOURCE: Data from *Global Environment Monitoring Systems*, 1984 and *The World Almanac*, 1993.

surface areas with many crevices and can adsorb pollutants on their surfaces or absorb gaseous pollutants into these crevices. When these particles enter the lungs, the pollutants are released at a higher concentration than present in the atmosphere. This could exacerbate a health problem in the sick individual and could even be fatal.

PRACTICE PROBLEM 18.3

List the ways particulates are removed from smokestack emissions. Why must this be a separate step from the removal of SO_2?

ANSWER Electrostatic precipitators, cyclone collectors, and bag filters are the methods used for particulate removal. None of these devices removes SO_2 or other gaseous pollutants, so the gases must be removed separately.

Photochemical (Los Angeles) Type Pollution

Unlike industrial (London Type) pollution, photochemical (Los Angeles) type pollution contains essentially no SO_2. Instead, the culprits are **nitrogen oxides** (for example, N_2O, NO_2, and N_2O_4, all abbreviated as **NO_x**), ozone, hydrocarbons, aldehydes, and peroxynitrates. The main source for these pollutants is automobile

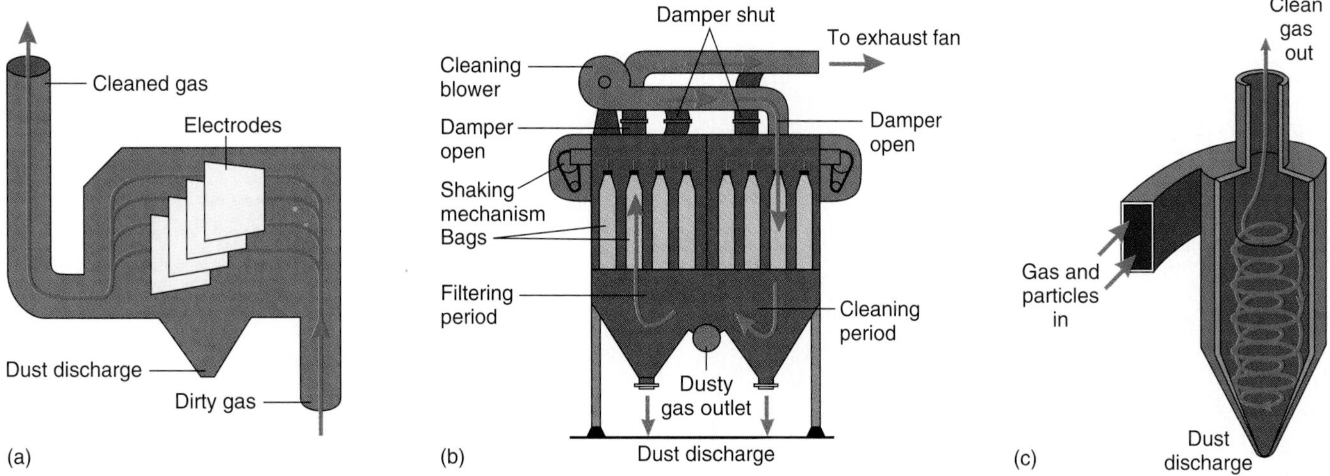

Figure 18.10 Devices used to remove particulate matter from effluent gas streams. **(a)** Electrostatic precipitator. **(b)** Typical bag filter. **(c)** Basic cyclone collector.
SOURCE: From William P. Cunningham and Barbara Woodworth Saigo, *Environmental Science*, 2d ed.

emissions; very little photochemical pollution comes from industries or power plants. Because a haze or fog often combines with these pollutants, they form what is called photochemical smog (figure 18.11).

Photochemical pollutants may arise from natural sources; for example, NO_x forms from electrical discharges, including those in thunderstorms. Power plants also emit some NO_x in addition to SO_2, though the latter is their main problem. But the major NO_x source is the automobile (figure 18.12). This was first detected in California because it has the highest ratio of cars to people. Since cars are omnipresent over most of the globe, this newer pollution form has spread worldwide, but it is predictably more prevalent in affluent societies. A quick review of table 18.3 reveals that most cities that have frequent problems with NO_x are in the United States, Europe, Japan, or another developed nation. It is an eerie feeling to watch this haze from inside a landing airplane when you realize you will soon walk through this brownish ground-level cloud.

The chemical processes that create photochemical pollution are sun-powered—hence the prefix *photo* (light). Nitrogen dioxide, or NO_2, is a reddish-brown gas that is toxic when concentrated; the maximum safety limit for NO_2 is 5 ppm, but even at this level NO_2 causes eye and lung irritation. NO_2 causes the distinctive brownish haze typical of photochemical smog. It forms when nitrogen and oxygen combine at high temperatures:

Figure 18.12 The automobile, the main cause of photochemical smog, produces carbon monoxide, nitrogen oxides (NO_x) and hydrocarbons as primary pollutants. These substances then react under the influence of sunlight to generate the secondary pollutants: ozone, aldehydes, and peroxynitrates.

(a)

(b)

Figure 18.11 The effects of photochemical smog. **(a)** Los Angeles skyline on a clear day and **(b)** under a blanket of photochemical smog.

$$2\ N_2 + O_2 + \text{high temperature} \longrightarrow 2\ NO_2$$

Once NO_2 is present in the air, it reacts with atmospheric water to form nitric and nitrous acids, both of which contribute to acid rain:

$$2\ NO_2 + H_2O \longrightarrow \underset{\text{Nitric acid}}{HNO_3} + \underset{\text{Nitrous acid}}{HNO_2}$$

Nitrogen dioxide has many uses, including the production of HNO_3 and rocket fuel.

While irritating, NO_x are not the main problems in photochemical type pollution. NO_2 interacts with sunlight to form NO and an O atom, which in turn reacts with an O_2 molecule to form ozone, O_3:

$$NO_2 + \text{sunlight} \longrightarrow NO + O$$

$$O + O_2 \longrightarrow O_3$$

Ozone is extremely toxic; its safety limit is only 0.08 ppm. An oxidizing agent, O_3 is sometimes used (instead of Cl_2) to purify water and is frequently employed in industrial syntheses. In the ozone layer of the stratosphere (see chapter 19), O_3 protects us against excessive ultraviolet radiation. Loose in the atmosphere at ground level, however, this chemical is an unequivocal menace.

Ozone is a bluish-colored gas with a sharp, pungent odor, in contrast to the colorless and odorless O_2 allotrope. Most people can detect ozone in concentrations as low as 0.01 ppm; it is the odor we usually associate with electrical sparking and the aftermath of a lightning storm. Ozone generators were once commonly used in doctor's offices and hospitals because O_3 oxidized bacteria and disinfected the area, but this usage was discontinued because of the potential health hazards.

There is little doubt that ground-level ozone concentrations have increased although we have less data from previous years than desired. Ozone measurements made at the Montsouris Observatory near Paris, France from 1876 to 1910 were between 5 and 15 ppb, while recent measurements at Arkona, East Germany showed levels of 15 to 25 ppb from 1956 to 1984 (figure 18.13). These data suggest there is nearly twice as much ozone now as at the end of the last century. (Of course, the data were collected at different locations; though combining such data presents problems, sometimes scientists must use data that are less than ideal.) Since O_3 is not a normal industrial pollution product, this increase is most likely due to photochemical pollution. The observed levels of about 20 ppb are below the safety limit of 0.08 ppm (80 ppb), but photochemical smog levels periodically rise above 20 ppb.

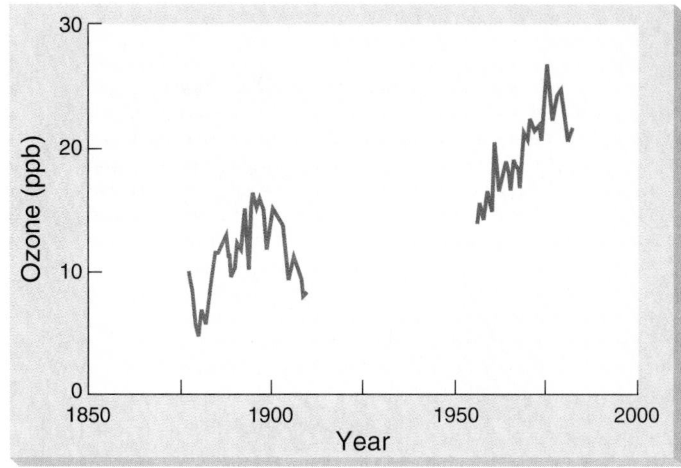

Figure 18.13 The annual mean ozone concentrations at Montsouris Observatory near Paris (1876–1910) and Arkona, East Germany (1956–84).

SOURCE: From A. Volz and D. Kley, "Ozone Measurements in the 19th Century: An Evaluation of the Montsouris Series," in *Nature*, 1988 332:240–242. Copyright © 1988 Macmillan Magazines Limited. Reprinted by permission.

Both NO_x and O_3 are less annoying than the peroxynitrates, often called **PANs** (for **peroxyacylnitrates**). The first step in the formation of these obnoxious compounds is the reaction of O_3 with a hydrocarbon (most likely from unburnt gasoline) to form an aldehyde, which then reacts with O_2 and NO_2 to form the peroxynitrates:

$$\underset{\text{Hydrocarbons}}{O_3 + R{-}H} \rightarrow \underset{\text{Aldehydes}}{R{-}\overset{\overset{\displaystyle O}{\|}}{C}{-}H} + O_2$$

$$R{-}\overset{\overset{\displaystyle O}{\|}}{C}{-}H + O_2 + NO_2 \rightarrow \underset{\text{Peroxynitrates (PAN)}}{R{-}\overset{\overset{\displaystyle O}{\|}}{C}{-}O{-}O{-}NO_2}$$

One example of a peroxynitrate, peroxybenzoyl-nitrate (C_6H_5—CO—OO—NO_2), irritates the eyes 100 times worse than formaldehyde. Concentrations as low as 0.02 ppm (20 ppb) cause moderate to severe eye irritation (the tolerance limit for formaldehyde is 2 ppm). The lower molecular weight peroxyacetyl-nitrate (CH_3—CO—OO—NO_2) is even more volatile and irritating.

Aldehydes are also highly irritating and sometimes toxic. Aqueous formaldehyde solutions, called formalin, were once used in wick-deodorizer devices to reduce airborne odors in homes, but chemists learned that formaldehyde worked by reducing the sensitivity of the olfactory nerves that enable us to detect odors. Subsequently, formaldehyde was removed from the wick devices. Chemists may want to make their homes smell nice, but not by damaging their sense of smell!

How can we avert photochemical type pollution? If you examine the equations in this section, you will see that the foundational step is the formation of NO_2 (or NO_x in general). If NO_2 is eliminated, the formation of O_3, aldehydes, and peroxynitrates will also be eradicated. The formation of NO_x from N_2 and O_2 requires high temperatures, such as occur in either lightning discharges or the internal combustion engine. If we can reverse this reaction, we can eliminate the photochemical pollution problem at its source. A catalyst enables the reaction:

$$2\,NO_2 \xrightarrow[\text{Catalyst}]{V_2O_5} N_2 + O_2$$

The catalyst in the equation is vanadium oxide, an oxide of a semiprecious metal. V_2O_5 is a red-to-yellow powder that serves as a catalyst for a variety of chemical reactions. In this case, V_2O_5 is used to coat pellets that are then encased in a metallic cylinder. This cylinder, called a **catalytic converter,** is then placed in an automotive exhaust system (figure 18.14). Once the NO_2 decomposes to the elements, N_2 and O_2, they cannot recombine because the high temperatures required for the reaction are unavailable in the exhaust.

The reaction that forms NO_2 from N_2 and O_2 is an endothermic reaction (requires heat), while the reaction that decomposes NO_2 is an exothermic reaction (emits heat). Like any catalyst, V_2O_5 makes the reaction run faster without being consumed during the reaction. Catalysts operate by providing a reaction surface or by direct involvement in the actual chemical reaction pathway. Even when providing a surface, the catalyst may be directly involved in the chemical reaction, but this reaction is reversed and the original catalyst remains when the reaction is completed. A single catalyst molecule might interact thousands of times during a chemical reaction, but it reverts to its original form at the end of the reaction.

Catalysts often react with more than one kind of chemical agent. Sometimes these other reactions are not easily reversible and they "poison" the catalyst by changing it into a compound that lacks catalytic activity. Coating a catalyst surface with another agent also prevents the catalytic activity. Tetraethyl-lead (used in leaded gasoline) reacts with the V_2O_5 catalyst, coating it and rendering it inactive. This is one reason modern automobiles

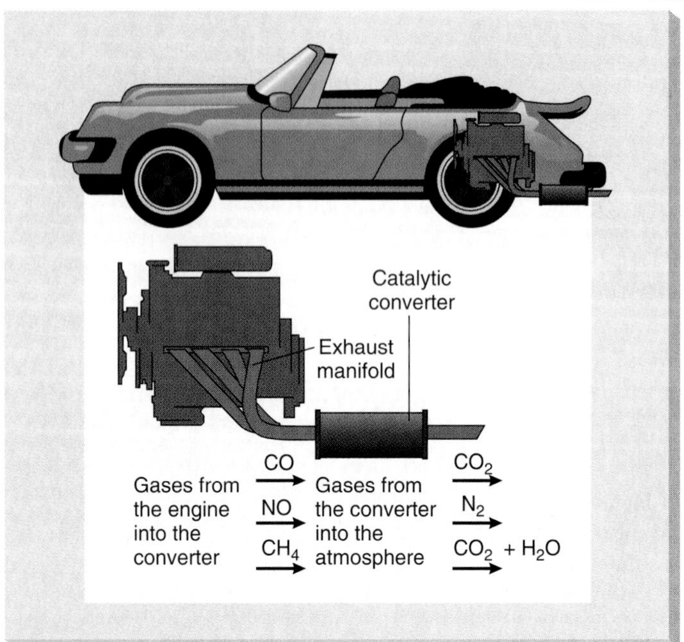

Figure 18.14 A catalytic converter in a Porsche.
SOURCE: From *Chemistry: The Central Science*, 5/E by Brown/LeMay/Bursten, © 1991. Reprinted by permission of Prentice-Hall, Inc. Upper Saddle River, NJ.

require unleaded gasoline; another is that the use of unleaded gasoline reduces the amount of poisonous lead in the atmosphere (see box 8.2). By inventing the catalytic converter and removing lead from gasoline, chemists helped solve two pollution problems.

When catalytic converters first came on the market, motorists sometimes ignored the instructions and used the cheaper leaded gasoline. Others assumed they could get better gas mileage if they bypassed the converter entirely, but this is not true. The first group blocked their catalytic converters with lead from the gasoline; this reduced their gas mileage and added to photochemical pollution. Fortunately, leaded gasoline is now difficult to find, so the first problem is unusual; unfortunately, the second still occurs, increasing NO_2 pollution.

Catalytic converters are not 100 percent perfect; some NO_2 is not reconverted to N_2 and O_2. This is especially true after a catalytic converter has been in use for several years. This means that people should change their catalytic converters periodically, but most do not. Eventually, periodic converter replacement may be mandated by law because photochemical pollution cannot be controlled without effective catalytic converters. In the meantime, more research is needed to develop better devices. Box 18.2 discusses the impetus to develop a "green" automobile.

Alternate fuels, such as ethanol or methanol, have lower combustion temperatures and produce less NO_x, but the lower combustion temperatures also reduce engine efficiency. Moreover, incomplete combustion of either alcohol leads directly to undesirable aldehydes. The photochemical pollution problem is unlikely to vanish in the near future because the automobile has increasingly displaced mass transportation (figure 18.15). The convenience of the automobile caused this change; a reversal of the trend seems improbable, even if desirable from an environmental point of view.

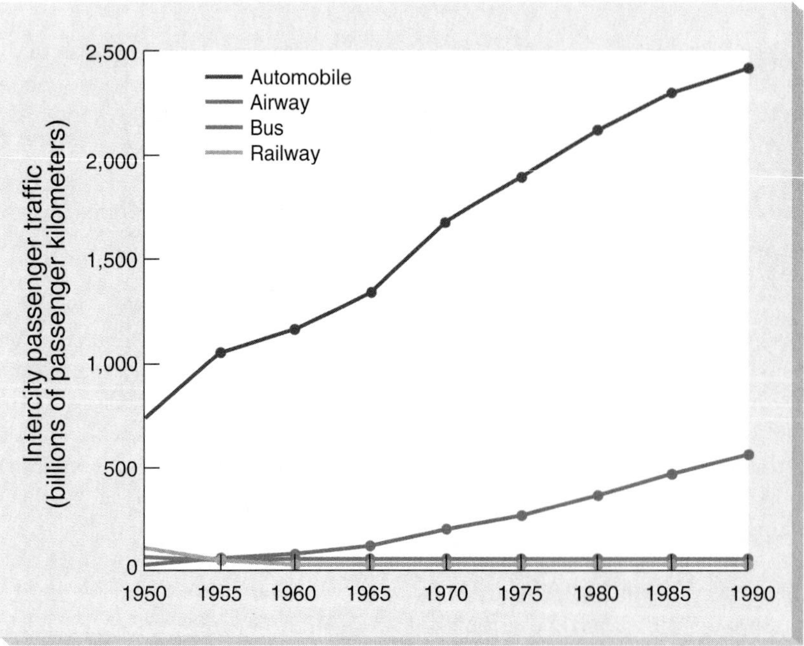

Figure 18.15 The decline of mass transportation in favor of the automobile from 1950 to 1990.
SOURCE: From Eldon D. Enger and Bradley F. Smith, *Environmental Science: A Study of Interrelationships*, 5th ed. Copyright © 1995 Times Mirror Higher Education Group, Inc., Dubuque, Iowa. All Rights Reserved. Reprinted by permission.

PRACTICE PROBLEM 18.4

Since the 1876–1910 era predated widespread car usage, what was the probable cause of the high O_3 concentrations reported at the Montsouris Observatory near Paris, France?

ANSWER No one knows. Low levels of O_3 probably came from natural sources such as lightning, while excess amounts most likely came from unidentified human sources.

PRACTICE PROBLEM 18.5

When was the last time you had your catalytic converter checked? Is is possible that you are contributing to photochemical smog even if you try to conscientiously prevent air pollution?

ANSWER If you are like most of us, and you do not live in a state that requires regular emission tests, your catalytic converter probably no longer works as efficiently as it originally did. You may well be contributing to photochemical air pollution.

OVERCOMING AIR POLLUTION

Hardly anyone seriously questions the need to reduce pollution, but air pollution reduction is expensive and requires compliance from essentially all the population. Since air pollution is worldwide in scope, compliance must eventually become worldwide in extent. The United States has made much progress in air pollution reduction; figure 18.16 shows the atmospheric levels of six pollutants from 1940 to 1984 as a percentage of the 1940 levels. Lead (Pb) and suspended particulate matter (SPM) have fallen to 25 percent of the 1940 values, and even greater changes are likely as a result of the 1990 Clean Air Act. The long-term levels of SO_2, CO, and VOC (volatile organic compounds) initially rose, peaked around the year 1970, and then dropped back to around the 1940 levels. However, the NO_x levels continued to rise during this period, although they seemed to level off after 1970 to about 300 percent of the 1940 levels. Since NO_x is the source of photochemical pollution, our cities will not become smog-free until this pollutant is reduced.

Some change should occur because of the 1990 Clean Air Act, which mandates a 60 percent NO_x reduction in all new cars by 1996. Still, this expensive change will leave NO_x levels at about 120 percent of the 1940 levels. Is the cost worthwhile? This is difficult to answer; it is often tough to analyze benefits and risks in a truly scientific manner. Sadly, a large portion of the actual antipollution expenditures are now drained in unproductive legal battles while pollution continues unabated.

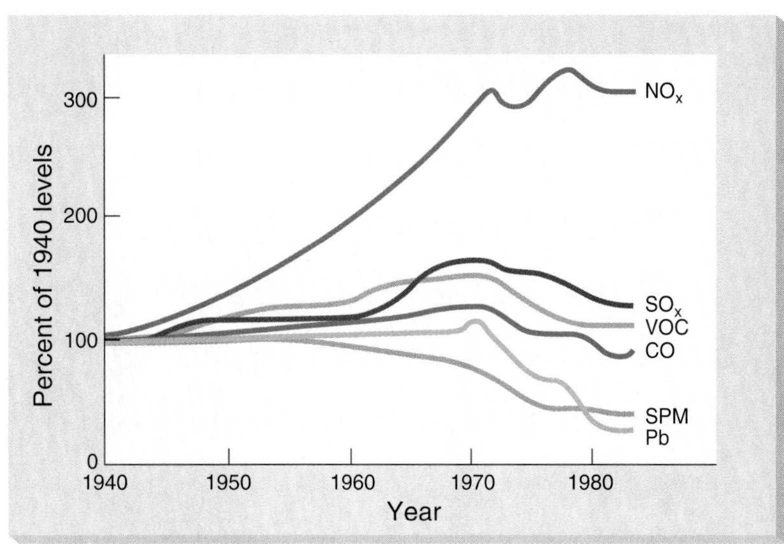

NO_x = nitrogen oxides (1940 level: 6.8 x 10^6 metric tons/yr)
SO_x = sulfur oxides (1940 level: 18.0 x 10^6 metric tons/yr)
VOC = volatile organic compounds (1940 level: 18.5 x 10^6 metric tons/yr)
SPM = suspended particulate matter (1940 level: 22.8 x 10^6 metric tons/yr)
Pb = lead (1940 level: c.230 x 10^3 metric tons/yr)
CO = carbon monoxide (1940 level: 81.6 x 10^6 metric tons/yr)

Figure 18.16 Air pollution trends in the United States from 1940 to 1984. The data are expressed as a percentage of the 1940 emission levels.
SOURCE: Data from Environmental Protection Agency.

BOX 18.2

"GREEN" AUTOMOBILES

The development of pollution-free automobiles would dramatically reduce air pollution, since the automobile is a major culprit in poisoning the atmosphere with pollutants. Automobile manufacturers have tried several approaches to reduce the levels of NO_x our vehicles emit, and as a result, the "green" cars of the future will likely weigh less and get better gas mileage—perhaps up to 80 mpg. They may rely on combinations of electric drive motors, fuel cells (see also box 18.4), and special flywheels. When will these redesigned cars hit the auto showrooms, and how well will they perform? As one writer advises, "Don't believe a thing you read until you've taken a test ride in Minneapolis in February." (J. Baldwin, *Garbage*, June/July 1993, pp. 24–29.)

Electric automobiles are, at present, the best bet for a low-polluting car. These vehicles are not an entirely new idea. Battery-driven cars were used in the early days of the horseless carriage but soon lost ground to their gasoline-powered cousins; their main drawbacks were slow speeds and the limited distance they could travel between battery charges. Today, the few modern electric automobiles in use have top speeds over 60 mph and can travel up to 165 miles between charges. Newer types of batteries have extended the travel range. These cars are sleek looking, which may be essential for a car that costs much more than a comparable gasoline-powered car (figure 1). However, governmental air quality regulations are pushing electric cars to the forefront. For example, the California Air Resources Board has mandated that 2 percent of all cars sold in that state in 1998 must have zero emissions. The percentage is to rise to 10 percent by 2003.

The electric car models are not available in the open market at the time of this writing and are not expected to arrive here until the late 1990s. Batteries and

Figure 1 An electric van.

recharging are the limiting factors. Two types of batteries offer distinct advantages and drawbacks: the lead-acid battery and the sodium-sulfur battery. The lead-acid battery is more powerful (300 watts/kg versus the sodium-sulfur's 175 W/kg) and is the most reliable at this time. It also costs less ($200 versus the sodium-sulfur's $1,200). But a bevy of high-powered, lead-acid storage batteries would weigh a whopping 1,100 pounds (compared to 780 pounds for a similar collection of sodium-sulfur batteries) and yield only 50 watt-hours/kg (Wh/kg) energy (compared to the sodium-sulfur's 140 Wh/kg). Sodium-sulfur batteries also allow more mileage between charges—about 100 to 200 miles against the lead-acid battery's 70 to 90.

Neither battery competes well with a gasoline-powered car, which can travel 400 or more miles between refills. Driving a steady 55 mph, you can complete a 400-mile trip in 7 hours and 16 minutes in a gasoline-powered car. With a sodium-sulfur battery that allows 200 miles between recharges, you need an additional 5 to 7 hours waiting time while the battery is made ready for the rest of the trip. A lead-acid battery, at 90 miles/recharge, would need to be serviced five times during the 400-mile trip. This would add an additional 15 hours minimum, parlaying your trip to over 22 hours—more than three times as long as it would take with a gasoline engine.

Some versions of the "green" car combine the electric engine with a hydrogen-fueled **fuel cell.** This permits the driver to recharge the battery as he or she drives and greatly extends the car's range. These vehicles are not quite zero emission, but their main effluent is water vapor. The limiting factor for these cars is the high cost of the fuel cells.

Solar cars offer another low-emissions option, but their size, to date, is incredibly small (figure 2). Annual solar car races have shown that these vehicles have a long range (3,200 miles or more) and can travel at high speeds (as fast as 90 mph). Obviously they will not run at night, or even when the sun disappears behind the clouds. Solar cars have been "available" for many years, but mostly as a do-it-yourself car.

Research is in progress on a car that produces solar-generated hydrogen (from water), which is then used in a fuel cell. Though the work is progressing, it is very unlikely that such a vehicle will arrive in automobile showrooms in the near future. Nonetheless, these new vehicles do promise a lower-pollution future. The question is, how long can we wait?

Figure 2 A solar powered car.

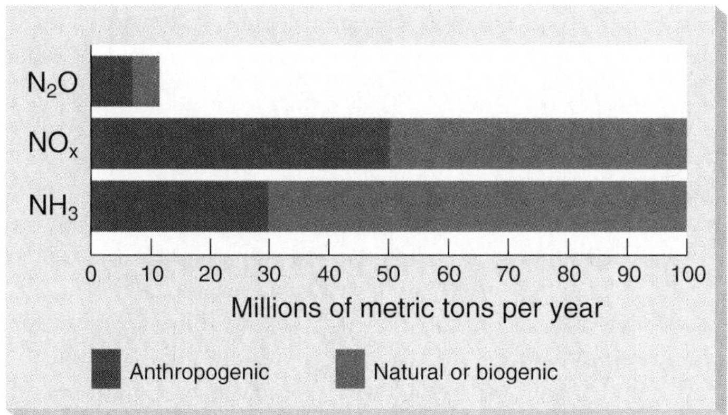

Figure 18.17 The annual amounts of nitrogen compounds released into the atmosphere.

SOURCE: From William P. Cunningham and Barbara Woodworth Saigo, *Environmental Science*, 2d ed. Copyright © 1992 Times Mirror Higher Education Group, Inc., Dubuque, Iowa. All Rights Reserved. Reprinted by permission.

Moreover, while legislation and research may help reduce human-generated emissions, no Clean Air Act will clean up natural pollutants. Figure 18.17 compares the annual release of anthropogenic and natural nitrogen compounds into the atmosphere. The NO_x comes in approximately equal amounts from human and natural sources, but about two-thirds of NH_3 comes from natural sources. Remember too, that pollutants move throughout the world (figure 18.7). It is unrealistic to expect the elimination of all humanmade NO_x if only a few nations attempt to reduce their output.

Achieving zero pollution is undoubtedly a worthy goal, but it is not realistic because of natural pollutants and because each incremental decrease in pollutant levels requires enormously larger sums of money. Fortunately, many chemicals that are dangerous at high concentrations are innocuous at low levels; natural levels of NO_x, for example, are part of our natural environment and even serve as a plant fertilizer. Perhaps a more realistic goal is to determine safe levels and work to contain pollutants within those levels.

NUCLEAR ENERGY—A POSSIBLE ANSWER?

Nuclear energy is a repugnant concept to many people; no other energy source in human history has had as many detractors. Part, but not all, of this bad reputation is deserved. Let's briefly examine the nuclear energy record in the United States.

Nuclear energy was developed in the 1950s as a safe, clean alternative to fossil fuel electricity generation. Despite its initial promise, no nuclear reactors have been ordered since 1978. Many that were in the construction process were abandoned at costs of as high as $25 billion.

One major nuclear reactor accident has occurred in the United States—at Three Mile Island (TMI), near Harrisburg, Pennsylvania on March 28, 1979. This event must not be confused with the accident at Chernobyl, which occurred in the former U.S.S.R. during 1986, because the reactor design and the results of the accidents were totally different. The Three Mile Island design included a shell made from several feet of reinforced concrete and a steel casing, both of which were missing in the Chernobyl reactor. As figure 18.18 shows, the TMI reactor used water as both a coolant and a moderator in what is known as a pressurized water nuclear reactor (PWR). In such a reactor, pressurized water is heated by the nuclear reaction; the heated water then transfers this heat to convert more water in a boiler into steam, and the

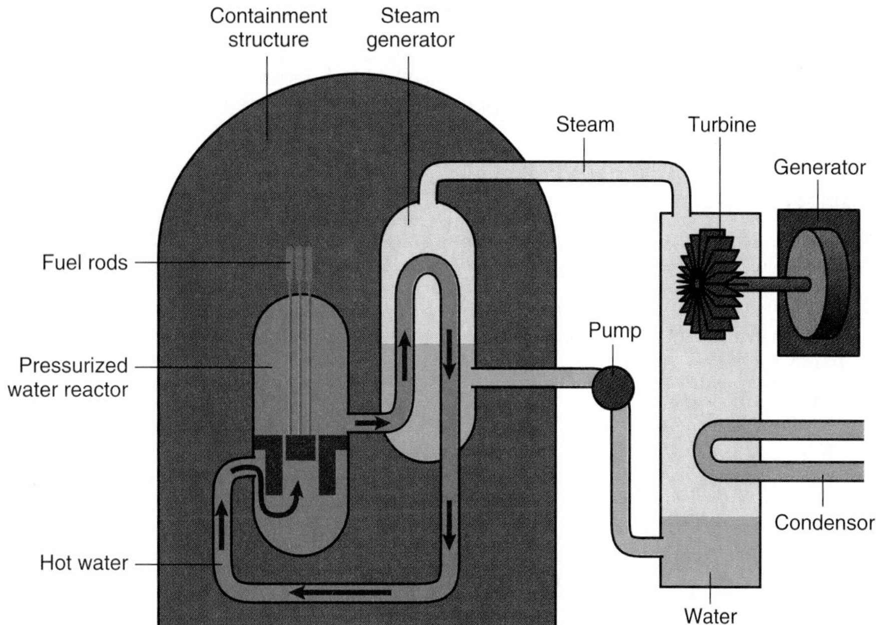

Figure 18.18 Diagram of the pressurized water nuclear reactor (PWR) used to generate electricity in 70 percent of the nuclear power plants in the United States.

SOURCE: Reprinted by permission of Northern States Power Company, Minneapolis, MN.

steam drives a turbine to generate electricity. The design is almost identical to a coal-fired generator, except the fuel is nuclear.

A nuclear reactor operates by splitting an atom, usually U-235, producing several neutrons that split more atoms enabling a chain reaction to occur (chapter 5). These reactions may yield different products, as the following equations show:

$$^{235}_{92}U + ^{1}_{0}n \longrightarrow ^{103}_{42}Mo + ^{131}_{50}Sn + 2^{1}_{0}n$$

$$^{235}_{92}U + ^{1}_{0}n \longrightarrow ^{90}_{38}Sr + ^{143}_{54}Xe + 3^{1}_{0}n$$

$$^{235}_{92}U + ^{1}_{0}n \longrightarrow ^{92}_{36}Kr + ^{141}_{56}Ba + 3^{1}_{0}n$$

To cleave the U-235 atom, the additional neutrons must move at slow speeds, within what is called the thermal range, or they will bounce off the U-235 with no **nuclear fission** (splitting) occurring. Several materials, including water, beryllium, and graphite, can slow the neutrons down to the correct range; these materials are called **moderators.** A similar process that completely removes neutrons from the reactor uses a **blocker;** typical blockers include boron steel, cadmium, and graphite rods.

Fission releases an enormous amount of heat, so it is essential to control the heat formation rate. Moderators assist by controlling how many thermal neutrons are available to react at any given time. Blockers also control the heat by removing excess neutrons from the reactor and therefore breaking the reaction chain. In the TMI incident, a water pump failed and the coolant-moderator water ceased to flow through the reactor. Although the production of thermal neutrons also terminated, the reaction continued to run for a short time using the neutrons already present. Since there was no coolant, the heat built up, causing a partial meltdown in the reactor core. To make matters worse, a human operator mistakenly prevented the use of readily available emergency cooling.

A core meltdown is bad news, but a nuclear explosion is not possible in a reactor. The amount of fissionable U-235 is too low (about 3 percent) to form the critical mass needed for a nuclear bomb (see chapter 5). The meltdown does ruin the reactor and could allow radioactive gases or water to leak into the surrounding environs. Almost no radioactive gases leaked from TMI, and nearly a million gallons of low-level radioactive water was restrained in the containment shell. No one was killed or injured during the TMI accident. The total amount of α-radiation downwind from TMI was determined to be 0.004 percent of the annual dosage from completely natural sources. The early reports of radioactive I-125 leakage were proven false, although antinuclear groups still cite these reports and claim huge numbers of cancer cases arose from the accident. Follow-up studies made six years later by the state of Pennsylvania showed no detectable increase in cancer rates within a twenty-mile radius of TMI.

TMI was the most serious nuclear accident in the United States. With no loss of life or injuries, or even serious radioactive leakage, it seems puzzling that most Americans view TMI as a major catastrophe. To the contrary, TMI shows how well American reactors are constructed—and additional safeguards were installed after TMI to provide added protection. The biggest problem connected with TMI appears to be the release of accurate follow-up information. Perhaps officials were worried about causing a panic, but the overall picture did not emerge until long after the TMI event, and part of the story may still be untold.

Chernobyl was different. This accident resulted in 31 deaths, more than 200 serious injuries, and the evacuation of over 135,000 people from their homes. Huge amounts of radioactive material were released into the atmosphere and drifted around the globe. Chernobyl was moderated only by graphite (instead of water, as at TMI), and when problems developed the nuclear reaction continued unabated until the reactor blew apart because the water coolant created too much pressure for the reactor to withstand. Since the reactor had no containment shell, the resultant meltdown spread radioactivity over a large land mass. Human error also added to the problems of Chernobyl. The lesson is not that nuclear power itself is too dangerous, but that poor reactor designs and inadequate construction can lead to dangerous situations. Studies released in late 1994 indicate that the results of the Chernobyl mishap were even worse now than they appeared in 1986.

Since TMI, the United States has improved the monitoring of nuclear power plants. Most units have computerized backup systems and better trained human operators. In 1979, the Nuclear Regulatory Commission authorized a study to determine the safety of and potential hazards in nuclear power plants. The results compiled in the Rasmussen report, concluded that if the United States had 200 nuclear reactors in operation, the chance of a serious radioactive leak would be less than one time in 10,000 years. Not all scientists agree with the conclusions of the Rasmussen report, and dissenters believe the number of potential radioactive leaks are at a much higher level. The Rasmussen report was authorized by a group whose jobs depended on a favorable outlook from the study. While this is not unusual, it does suggest some caution is necessary in evaluating the group's conclusions.

Still, by any measure, many established energy-related industries can be just as dangerous or much more so. At least 90,000 coal mining deaths have been reported this century. About 260 people still die each year in coal mines, and thousands more die from black lung disease. These numbers dwarf those attributed to the Chernobyl disaster. Herbert Inhaber (*Science*, February 23, 1979, p. 718) estimated the number of person-days lost per megawatt year at 3,000 for coal-fired reactors, 2,000 for oil-fired reactors, and 10 for nuclear reactors. If these estimates are even remotely accurate, nuclear reactors are several hundred times safer than their fossil fuel counterparts.

Fusion—Hot and Cold

Along with nuclear fission, nuclear fusion is another possible way to obtain energy from the atom (see chapter 5). **Nuclear fusion,** the fusion or joining of atoms, is the process used in the hydrogen bomb and the basis of energy generation on the sun. The goal of fusion is to fuse hydrogen atoms together to form helium and thereby release an enormous amount of energy. One possible reaction sequence appears in the following equations:

$$^{2}_{1}H + ^{3}_{1}H \rightarrow ^{4}_{2}He + ^{1}_{0}n$$

$$^{6}_{3}Li + ^{1}_{0}n \rightarrow ^{4}_{2}He + ^{3}_{1}H$$

The major difficulty in nuclear fusion is confining the reaction within a device that can then transfer the energy into electricity generation; most materials vaporize at temperatures well below those necessary for nuclear fusion (most research has been done at 45,000,000°C). Strong magnetic fields have shown some success as containment "vessels," and a fusion reaction self-sustained for almost a second in 1992. This may sound trivial, but it is well worth working on; the potential for fusion is greater than for fission since the raw materials and waste products are less harmful. Our oceans contain over 10^{13} tons of deuterium and could potentially supply all human energy needs for a billion years if fusion power could be harnessed.

As box 1.4 noted, some researchers claimed successful room-temperature ("cold") fusion in March 1989. After the initial reports a flurry of cold fusion research took place, but enthusiasm waned when many scientists were unable to reproduce the original results. The cold fusion question remains unsolved; some vigorous detractors claim outright fraud, but some countries are avidly pursuing the research. Japan, for example, has created several major research laboratories to seek ways to make this nonreproducible process work. The Japanese firmly believe they will develop a successful cold fusion before 2025. At this time, neither hot nor cold fusion actually works well enough to generate electricity. Many projections put such applications at least thirty years in the future.

Limits to Nuclear Power

Three major problems confront the nuclear power industry: (1) public perception of the safety of the industry, (2) disposal of nuclear waste products, and (3) fears of nuclear proliferation. The first two problems impose limitations on the growth of this industry in the United States, but they are solvable. Many of the public perceptions of nuclear industry dangers have been proven false; a concerted campaign might well convince the public of the industry's safety. (For an example of public fears concerning radioactivity exposure, see box 18.3 on radon.)

The nuclear waste disposal problem is also solvable—in fact, much of this solution has existed for several decades. A process called glassification (vitrification) incorporates radioactive waste into the chemical structure of a glass that is then made into beads. Once the beads are encased within polyethylene and then buried inside a labeled lead container, the chances are extremely slim that radioactive materials will leach out. Where to bury these containers presents a problem because very few people want nuclear waste in their own backyard (the NIMBY syndrome). Safe nuclear waste transportation carriers also already exist—in fact, these carriers are safer than those used for standard chemicals or food. Most of these marches, rallies, and campaigns against nuclear waste disposal do not differ substantially from those against nuclear power generation. Is it possible that the net result of these activities has been the denial of the safest form of power generation to the American public?

The U.S. government has been less helpful in this matter than you might expect. The cleanup record at government nuclear sites has been dismal at best. In fact, most government-controlled waste disposal sites, such as on military bases, are among the worst examples of pollution in the entire country. It is one thing to advocate a clean environment and quite another to take the necessary measures to ensure quality and safety.

Another source of concern in nuclear waste disposal is safety over long periods of time. Many companies buried toxic waste materials using the best technology available in the 1950s and 1960s. We now realize that this technology was inadequate. Is it possible that the same could happen with disposed nuclear waste? After all, thirty years is only a small fraction of the half life for many of these nuclear materials. Will they still be safely buried one hundred years from now? Unfortunately, there may be no way to answer this question without waiting and observing for many more years. Besides, who will monitor these disposal sites? Both government and industry have built up poor records in this regard.

"Hey, Myerson...How goes the cold fusion experiment?"

SOURCE: © 1993; Reprinted courtesy of Bunny Hoest and *Parade Magazine*.

OTHER ENERGY SOURCES

Other possible energy sources include non-fossil fuels, wind or hydroelectric power, and solar energy. Some are limited to specific geographical regions, while others have other types of limitations.

Alcohols as Fuels

Only two alcohols have any practical potential as fuels—methanol and ethanol, both of which were discussed in chapters 9 and 10. All the others are too expensive. The main objection to the use of either alcohol is that these chemicals are already partially oxidized and thus provide less energy per mass than coal or fuel oil. Both are used in automobiles to a limited extent; in fact, as we have mentioned before, Brazil has used alcohol extensively. Both alcohols are too costly for large-scale fuel usage unless

BOX 18.3

HOW RISKY IS RADON?

Radon, a colorless, odorless, radioactive gas, continues to make headlines even though this problem is far less serious than media releases suggest (see chapter 5). People have been exposed to natural sources of radiation, such as radon, throughout human existence; artificial sources entered the picture during the past one hundred fifty years. Figure 1, which shows the sources of radiation, reveals that natural sources comprise 82 percent of our annual exposure; radon is the largest natural source.

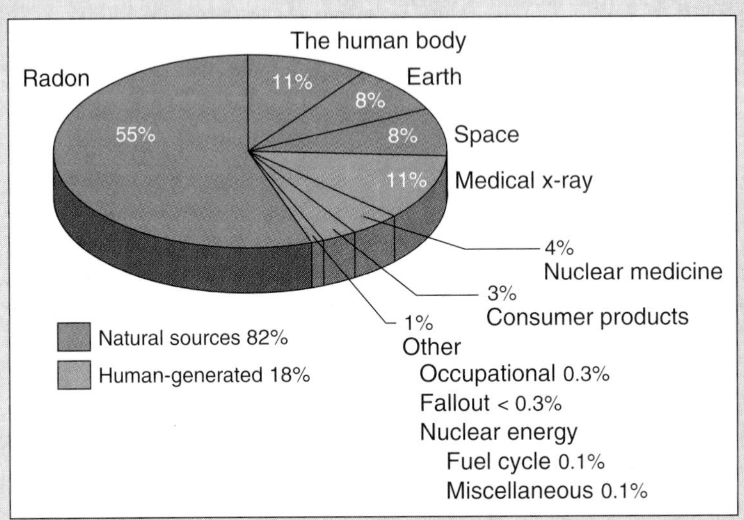

Figure 1 Sources of natural (82 percent) and artificial (18 percent) radiation. Radon is a natural radiation source.
SOURCE: Data from National Council on Radiation Protection and Measurements.

The potential dangers of radon are best considered by analyzing the results of a massive study that the High Background Radiation Research Group did in China in 1980. They compared the results of high and low background radiation exposure using 140,000 people, approximately equally divided in two otherwise identical samples. Although the higher-exposure group was exposed to more than double the amount of radiation as the lower-exposure group, no adverse health effects were observed. In addition, many of the Chinese families had lived in the high- or low-exposure regions for as long as sixteen generations without any signs of genetic problems (*Science*, August 22, 1980, pp. 877–880).

Most radiation experts believe radon poses no serious threats to the American public and that the scare is strictly a media event. Regardless, the EPA states it is unsafe to dwell in a house that contains 4 picocuries, or pCi, of radiation per liter of air in the basement. (The high-radiation group in China was exposed to 1700 to 2000 pCi.) The half-life of radon is four days or less, depending on the isotope, which means the material reaches the 20-half-life level in less than 80 days. While there is no doubt that radon can harm human health, the real question is whether the levels encountered in most places in the United States are likely to do so. Uranium miners have been harmed by radon at radiation levels above 12,000 pCi, but these results were confounded by the fact that many of these miners also smoked. Below 12,000 pCi no harm seemed to accrue to the miners.

Claims that 1 to 5 percent of those living in houses with 4 pCi radon radiation per liter air will develop cancer from this source seem unreasonable in light of the available data. The highest background radiation levels in the United States occur in North Dakota, where people also have the highest average longevity in the country. If, as claimed, it costs $500 to $5,000 per house to protect against radon, the total cost would run about $100 billion, with cleanup costs in schools estimated at $10,000 each. This seems a huge price to pay for a dubious benefit.

government subsidizes their use, but that merely spreads the cost over the larger population base.

One factor too often neglected in discussions on using methanol and ethanol as fuels is that incomplete oxidation of alcohols produces aldehydes (chapter 10). Aldehydes are one of the main pollutants in photochemical smog. No doubt engineers can design engines that generate fewer aldehydes, but complete prevention appears unlikely in the near future. The use of these oxygenated fuels in smaller motors, such as lawn mowers or chain saws, has been very unsatisfactory. If pollution abatement is the goal in switching to alcohol fuels, the trade could end up as a mere exchange of waste products. Moreover, oxygenated fuels do not reduce the amount of carbon dioxide released into the atmosphere.

Methyl t-butyl ether (MTBE) is an alcohol derivative added to gasoline to promote smoother burning. The toxicity of MTBE is under question. When it was put into widespread use during late 1994 and early 1995, a large number of complaints arose regarding its effects on public health. Many people complained about burning eyes and severe headaches. This health question should have been resolved before MTBE was put into large-scale use.

Finally, the total necessary fuel volume is beyond the production capacity for these alcohols. There is just not enough biomass available to convert to alcohol fuels, even if the food supply and forests were drastically reduced. True, Brazil uses alcohol fuels, but the number of cars and the total energy demands are far lower there than in the United States. In any event, these alcohol fuels are strictly for vehicles; much more fuel is consumed in electricity generation or manufacturing.

Hydrogen—An Almost Limitless Resource

Hydrogen, frequently singled out as a potential fuel, has many valuable characteristics. First, an almost limitless supply of hydrogen is potentially available in the waters of the world. Second, the combustion product is water, ordinarily not considered a pollutant.

There are downsides to using hydrogen gas as a fuel. First, it requires a lot of energy to electrolyze water and generate hydrogen. Second, H_2 gas requires heavy, pressurized tanks for safe handling. (The use of metal hydrides could alleviate part of the weight problem, but these systems generally use platinum, a very expensive metal.) Third, and probably most important, hydrogen and oxygen mixtures are explosive. Still, H_2 could become an important fuel in the future. The problems in handling and avoiding explosions should not be much more difficult to solve than similar problems in using propane or natural gas; plus, hydrogen is less polluting. Obtaining hydrogen at a low cost is probably the major obstacle for widespread fuel utilization.

If solar energy could be used to electrolyze water, the raw material cost for hydrogen would decrease markedly. This procedure would also free solar power from its dependency on direct sunlight by transferring the solar energy into a readily transportable form of chemical energy. At this time, solar-derived heat energy is stored briefly in rocks or other thermal "wells," and solar-derived electricity recharges batteries. Solar hydrogen production would provide an even more accessible form of stored solar energy. Several research groups are actively pursuing this goal and such systems may appear on the market during the 1990s.

Box 18.4 discusses another device that may become important in energy production and storage in the near future: the fuel cell.

Nonchemical Energy Sources

Hydroelectric plants are common in the United States and other parts of the world, but this power source is limited to locations with rapidly flowing water. In the same manner, **geothermal power** generation can provide many megawatts of energy from the heat of the earth, but works effectively in only a few places. Although the wind blows almost everywhere, wind power generation requires a steady stream. However, modern wind generators are far more efficient than the older windmills and can produce many megawatts of energy when set in the proper locations (figure 18.19). The bottom line on nonchemical energy sources is this: Even if we used all these devices in the United States, we would still need fossil fuel or nuclear power plants to supply our energy demands.

SOLAR ENERGY—SOMEWHERE THE SUN IS SHINING

"Somewhere the sun is shining, somewhere the skies are bright..." states a well-known line from the epic sports poem "Casey at the Bat." The sun does indeed shine somewhere at all times and could thus potentially be harnessed for power. The total amount of solar energy reaching our planet is a staggering 340 watts/square meter at the outer limits of the atmosphere. This amount is reduced by about 50 percent as it travels through the atmosphere, but that still leaves nearly 170 W/m^2 over the entire surface of the planet. That is approximately eight thousand times more energy than *all* the commercial energy used by *all* people on Earth! If we harness this vast supply, we could meet nearly all of our energy demands. There are several ways to harness energy from the sun, including (1) solar heating (or cooling), (2) photovoltaic conversion (solar-generated electricity), and (3) solar-generated hydrogen.

Solar energy is often called "free energy," a characterization that is partially true; you do not pay for sunshine. However, if you want to collect solar energy, you must invest in the equipment needed for the particular kind of solar energy you plan to use. Solar energy is not entirely free.

Solar energy use has limitations. You can only harvest solar energy during the daytime and must somehow store this energy for use at night, when the need is often greater. Second, some regions of the world have a frequent cloud covering that renders solar energy collection ineffective, or at least significantly reduces the uptake. Third, solar energy is inadequate as a heat source in the colder climates; backup systems are necessary.

A fourth limitation is that solar energy striking any surface is diffuse as well as variable in intensity. Although 170 watts/m^2 may

Figure 18.19 Modern wind-powered generators can produce many megawatts of electricity in suitable locations.

BOX 18.4

FUEL CELLS

A **fuel cell** is a device that converts the energy of an oxidation-reduction reaction directly into electricity. The first fuel cell was devised in 1839 by Sir William Grove (born in Swansea, Wales in 1811 and died in London in 1896). The concept lay dormant until the 1960s and 1970s, when the space program put it to work. The main reason fuel cells were ignored for so long was that the devices, in the forms known before the mid-twentieth century, were far too expensive for routine use: about \$100,000 per kilowatt (kW). However, this high cost made little difference to the Gemini, Apollo, and Space Shuttle programs, and they further developed and refined the fuel cell to meet their needs.

Fuel cells resemble batteries in that both work through oxidation-reduction reactions that occur at their electrodes. The distinction between fuel cells and batteries is that the fuel cell does not have much storage space for fuels, so they must be stored outside, while the fuel for a battery is contained within the unit. The oldest type of usable fuel cell, and the one used in the space program, is an alkaline cell that directly converts the energy of the reaction between hydrogen and oxygen gases into electricity. The cell also harvests drinking water for the astronauts. A typical alkaline fuel cell is diagrammed in figure 1. In this fuel cell, electrons are released at the anode and taken up at the cathode. Both electrodes are permeable platinum tubes that allow the chemical reactions to occur at their surfaces. The anode and cathode reactions are shown in the following equations:

Anode reaction:

$$2H_2 + 4\,OH^- \rightarrow 4\,H_2O + 4\,e^-$$

Cathode reaction:

$$O_2 + 2\,H_2O + 4\,e^- \rightarrow 4\,OH^-$$

Net reaction:

$$2\,H_2 + O_2 \rightarrow 4\,H_2O$$

One of the main advantages of the fuel cell is its efficiency. The conversion of energy from the H_2-O_2 reaction into electricity is usually about 60 percent efficient, and some fuel cells operate at greater than 80 percent efficiency. The alternate procedure, in which the two gases are burned and the resulting heat is converted into steam that drives a turbine, is less than 30 percent efficient. Another powerful advantage of the fuel cells is that they do not produce any nitrogen oxide emissions. This could reduce NO_x pollution by 50 to 90 percent if electricity production was handled entirely by fuel cells. Some fuel cells use fuels other than hydrogen, including hydrocarbons such as gasoline. Although this type of fuel cell will produce carbon dioxide, the total emissions would be only about 40 to 60 percent of the present levels emitted from coal-fired power plants.

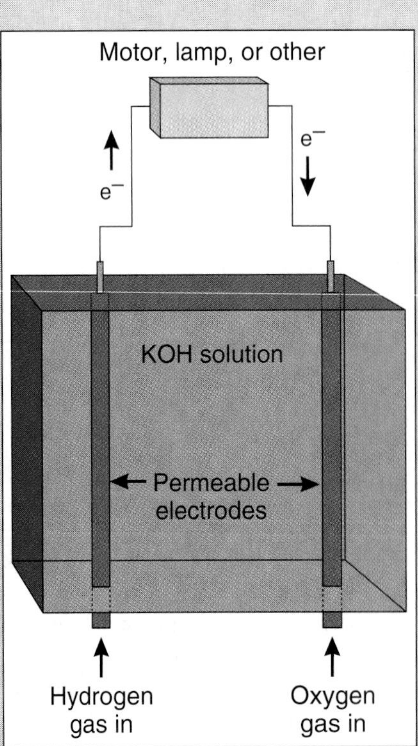

Figure 1 Diagram of a typical alkaline fuel cell.

Table 1 lists the five main types of fuel cells and their advantages and disadvantages. The most widely used type is based on phosphoric acid. These cells have been used to make large electricity-generating plants that are so quiet and pollution-free that they can be placed right in a shopping center and go unnoticed. The Japanese used an 11-megawatt phosphoric acid fuel cell to supply all the electricity for four thousand households in Tokyo.

Many believe that the polymeric version, which actually uses polymer membranes, will ultimately be the most useful type of fuel cell. Joseph Maceda (vice-president of H Power) notes that his company is working on a 200-watt unit the size of an adult's fist; he calls it a "Power Brick." Maceda foresees a kilowatt unit, no larger than a microwave, that could supply all the electricity, heat, and hot water for a single family home.

At present, the building costs for a large fuel cell run about \$3,500/kW compared against \$1,000 to 2,000/kW for a steam turbine powered by fossil fuels. On the other hand, fuel cell power stations can be placed wherever needed. This eliminates high-tension transmission lines that cost around \$10 million per mile and distribution lines that run about \$50,000 per mile. With this in mind, the actual cost is much closer to that of a traditional generator. Convenience of location and reduced pollution will probably make fuel cells a major item in the economy of the future.

TABLE 1 Comparison of five types of fuel cells.

FUEL CELL TYPE	ADVANTAGES	DISADVANTAGES	TYPICAL USES
Alkaline	70% efficiency; nonpolluting fuel	Most expensive system	Shuttle
Phosphoric acid	Most developed technology; easy to upscale	Only 40% efficiency	Anything; batteries, motors
Molten carbonates	50 to 60% efficiency; uses natural gas; best for cogeneration	Operating temperature is 650°C	Cars, buses, industrial
Solid oxides	Uses natural gas; reformer unnecessary	Operating temperature is 1,000°C; hard to make; requires two days to warm up	Industrial; high-power systems
Solid polymers	Low operating temperature; handles changing power demands readily; quick starting	Expensive system; needs much more testing	Cars, buses, large batteries

sound like a lot of energy when you consider the number of square meters in the Earth's atmosphere, the amount in one square meter would not light three 60-watt lightbulbs. If your home requires an average of 11,000 watts each day, you would need to collect 65 meters square, assuming you could convert 100 percent of the solar energy into electricity. Most photovoltaic cells are closer to 15 percent efficient, so the size would increase to 431 m^2, or 21 meters on each side. (Box 18.5 discusses photovoltaic electricity generation.) An enormous-sized collector would be necessary for a single home. A factory would require hundreds of times more energy.

Still, solar energy is the best unharnessed resource available. Why use wind power when you can harness the source of the wind? We can certainly hope to learn new ways to tap into this natural resource.

Solar Heating

The most direct way to capture solar energy is to heat something. Solar heating is subdivided into passive and active solar systems. A passive system collects solar energy without using other energy sources or energy-powered devices; an active system relies on pumps and collectors.

A typical **passive solar heating system,** as figure 18.20 shows, can be built into any style home, but it is essential that the main sunlight-collection windows face the sun (south and/or west). This limits the use of passive solar heating in a previously constructed home, but is a factor to consider in a new home; a passive system could save hundreds of dollars during a cold winter, even with cloud cover part of the time. Good insulation (both roof and walls) is absolutely essential to such a system. The only power needed in a passive system is a small fan to move the warm air through the storage area.

An **active solar heating system** requires pumps and collectors mounted on the roof or near the building (figure 18.21). The sun's energy heats water in the collectors, and the heated water is circulated throughout the building. Similar systems generate hot water for washing and cooking, but it is not possible to use the

"It's called 'fire'...it'll replace inefficient solar heating."

SOURCE: © 1991; Reprinted courtesy of Bunny Hoest and *Parade Magazine*.

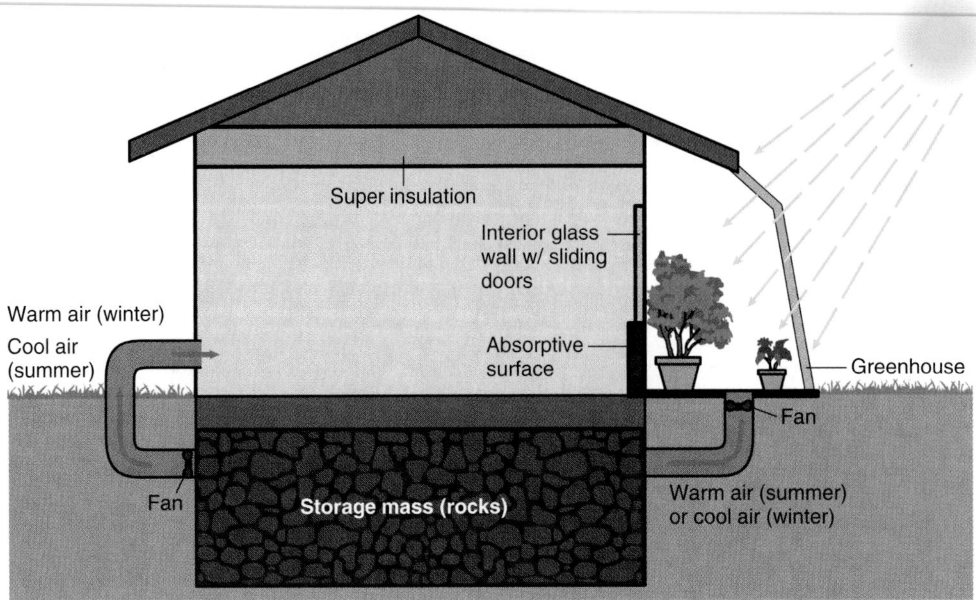

Figure 18.20 A home with a passive solar heating system. The sunlight enters a greenhouse or similar room and, ideally, moves through a storage area (rocks in this example) before heating the house. The overhanging roof prevents excessive sunlight from entering the building during the summer. In principle, this system is reversible for partial summer time cooling.
SOURCE: From William P. Cunningham and Barbara Woodworth Saigo, *Environmental Science*, 2d ed.

same system for both hot water and heating since the removal of the wash water renders the system ineffective for heating the building. Solar systems are also used to heat swimming pools.

Solar heating involves chemistry in several ways. First, all insulation materials are chemical in nature. Second, active systems use collector panels that typically consist of copper tubes within an insulated box with a reflective surface (often aluminum film) to amplify the heating effect. Swimming pool solar systems generally use a series of black plastic tubes (usually polypropylene) mounted nearby (on the roof for an indoor pool). The black

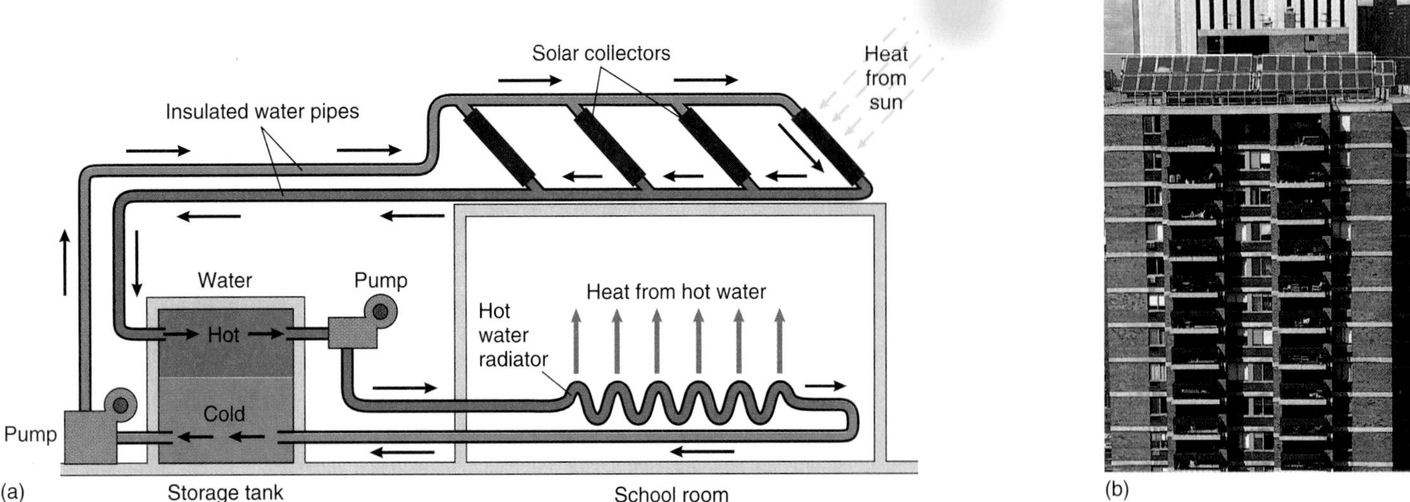

Figure 18.21 An active solar heating system. **(a)** Diagram showing how such a system operates. **(b)** A solar water heating system on the roof of an apartment complex in New York City.
SOURCE: **(a)** is from *Solar Energy: A Biased Guide International Library of Ecology Series*.

BOX 18.5

PHOTOVOLTAIC ELECTRICITY GENERATION

Photovoltaic cells convert solar radiation directly into electricity. A. E. Becquerel (better known for his research in radioactivity, chapter 5) observed the photovoltaic effect in 1839, but commercial development did not begin until the mid-1970s. At that time, the Bell Laboratories discovered that highly purified silicon wafers effectively changed sunlight into electricity. They developed photovoltaic cells (*photo* means light, *voltaic* electricity) containing ultrapure silicon wafers.

The ultrapure silicon was made using a technique called zone refining. In zone refining, a heated coil is passed along a column of material to heat only a small zone. The impurities are more soluble in this small molten portion of the material, so they move along this liquid portion, leaving a very pure material behind as a solid. Silicon must be of high purity to work in a photovoltaic cell. The development of zone refining meant that silicon, the second most abundant element on Earth, could be used to produce electricity from sunlight.

The early silicon photovoltaic cells were too expensive for widespread usage. The radio in the 1958 space vehicle Vanguard I was powered by six small silicon cells with a peak output cost of $2,000/watt. As research continued, the cost dropped to about $100/watt in 1970; it was expected to drop below $1/watt in the mid-1990s, which could make solar electricity competitive with other sources. At present, solar electricity costs about $0.25 to 0.50/kilowatt hour (kWh). Most photovoltaic cells are approximately 15 percent efficient and will last thirty years or longer. At a unit cost of $1/watt, the electricity from these cells would cost about $0.08/kilowatt-hour. Coal-fired electricity generation runs about $0.05/kWh, and some projections expect fossil-fueled electric power to increase to as much as $0.30/kWh; with nuclear-generated electricity as much as three times higher. At some point in time, solar-generated electricity will become competitive with these traditional methods. Photovoltaic cells are already in use for electricity generation in isolated places where other sources are unavailable (figure 1).

Continuing photovoltaic cell research is aimed at improving efficiency. The original cells were less than 1 percent efficient, which means that 99 percent of the incident sunlight was *not* converted into electricity. With the present rate at around 15 percent, a goal of 30 percent efficiency seems reasonable within the next few years. Even at the low efficiencies now available, photovoltaic devices are on the market for home and commercial uses such as outdoor lighting, and photovoltaic-powered automobiles compete in cross-country races every year. This method of electricity generation is safe, nonpolluting, and waste-free.

Figure 1 Photovoltaic cells mounted on a roof in Osaka, Japan provide an economical source of electricity.

color increases the heat absorption and protects the plastic against degradation from constant exposure to the sun. Finally, solar systems rely on chemical antifreeze compounds added to the water when a system is built in a cold-weather region; this prevents the pipes from rupturing at freezing temperatures.

A potential problem that may prevent the widespread use of solar energy use is the availability of copper and other necessary materials. Copper is used in numerous applications, including in electrical wiring, and the available resources are limited. Some estimates suggest that the world's copper supply is insufficient to permit every home in the United States to convert to solar energy for either heating or hot water. However, there is no reason to assume that a copper substitute cannot be developed; solar pool heaters use plastics. Some plastics may not work well in hot water solar systems because the water temperatures are too high, but thermally stable plastics now exist.

EXPERIENCE THIS

You can construct a simple solar heater by lining a salad bowel with aluminum foil. Keep the foil as wrinkle-free as possible. Line this solar reflector up to reflect the sunlight, and it will concentrate the sun's energy into a small area. Place your hand above the reflector—but be ready to remove your hand quickly! Using this basic technique, you could design a solar cooker to heat water or cook hot dogs. The curve of the foil is crucial to success; with a little trial and error, you can arrange the foil correctly.

PRACTICE PROBLEM 18.6

How can you use solar energy in your home?

ANSWER Assuming enough sunshine is available, most homes could use solar hot water systems. In fact, San Diego requires a solar hot water system be installed in new homes. You could also use passive solar heating if you have windows that face the sun for much of the day. You will need good insulation, and a backup heater is absolutely mandatory.

ENERGY STORAGE, TRANSPORT, AND CONSERVATION

Energy storage, transport, and conservation are as important as developing new energy sources. Chemical energy is the only energy form that is readily transportable and storable with no loss of total energy. This form of energy is contained in fuels such as petroleum, coal, and natural gas, and in the foods we consume. Batteries are chemical devices that release electricity on demand. In principle, microwave energy is readily transmitted, but the energy that forms this radiation is derived from chemical or nuclear sources. In short, chemical energy is also the only form that can be stored for long time periods without loss.

With the exception of light (electromagnetic radiation), no energy form is more mobile than electricity. Excepting light, electricity is also the cleanest form of energy, unless the pollution from the electricity-generating source is considered. Since most electricity is now generated using fossil fuels, the indirect pollution from electricity is high. One way to avoid this is to use generation methods that do not rely on fossil fuels; this probably means more extensive use of nuclear power. Supplementing the present generation plants with solar units will help, but total energy demands will quickly exceed the capacity a solar system can handle. Fossil fuel or nuclear electric power-generating plants are likely to remain the chief sources of electricity for many years into the future.

Energy Conservation

Conservation, as opposed to discovering a new fuel or an alternate energy source, can reduce the amount of energy needed for any given operation. For example, an automobile that gets 26 miles per gallon uses half as much fuel as one that chugs along at 13 mpg. And a car getting 52 mpg uses only half as much fuel as that one at 26 mpg.

Similarly, better insulation reduces the amount of energy needed to heat or air condition a home or building. Fiberglass, blown paper fibers, and foamed polystyrene boards (commonly called PIP for polystyrene insulation panel) have been used to insulate for many years. Asbestos, once used as a nonflammable insulation, was later deemed unsafe. Newer insulation materials include closed-cell polyurethane that is blown in as a foam using either CO_2 or HFC134a as the blowing agent, and recycled poly(ethylene terephthalate) (PET) fibers. Other foamed plastics,

Figure 18.22 A geodesic domed house in Florida.

such as foamed polyethylene or foamed poly(vinyl chloride) will undoubtedly become insulation materials in the future.

Geodesic domed houses, originally conceived by R. Buckminster Fuller, are another energy-saving device. The house in figure 18.22 has concrete walls covering foot-thick blocks of foamed polystyrene. The windows are also insulated. Annual heating and cooling costs are very low; the owners claim under $100 annually. The house is also nearly soundproof.

Conservation not only saves energy, it also saves a bundle of money. Recycling is one form of conservation that saves both energy and money; recycling aluminum from cans eliminates the need to mine it from ores. It is unfortunate that some antipollution or conservation measures, in contrast, require greater amounts of energy, defeating the purpose of the measures.

ENERGY AND POLLUTION

Energy and pollution are inextricably bound together. We cannot use energy without producing some pollution. Electricity, the cleanest energy form available on a large scale, has hidden pollution costs at the generating sites. Furthermore, electrical transmission requires wiring and polymeric insulation, each of which cause some pollution during their production. Wires require refining a metal ore, often through a dirty process such as copper smelting; polymers come mainly from petroleum, which also contributes to pollution.

Solar energy has a well deserved reputation as a clean energy form, but it, too, causes indirect pollution. Solar heating requires good insulation, and insulation manufacturing produces pollution. Active solar heating requires pumps and lots of copper tubing; copper is refined at smelters that emit copious quantities of SO_2. Production of the ultrapure silicon used in photovoltaic cells also generates pollutants.

Where does this leave those of us who truly desire a clean and safe environment? Like most areas of life, energy use requires us to compromise between how much pollution we can tolerate and how much energy we consume. Energy-caused pollution increases in a direct relationship with consumption, but switching to a less polluting energy source could reduce pollution. Although nuclear power may pollute less, it still presents a long-term quandary: how to deal with nuclear waste disposal. Energy is never free in the real world, and part of its costs is paid in air pollution.

IS ANY FORM OF ENERGY COMPLETELY SAFE?

In most cases, the answer is no, there is no completely safe type of energy. However, certain forms of energy are safer than others. Solar energy is probably the safest form, although pollution is associated with manufacturing the required equipment. Electricity is moderately safe, but electricity generation is hazardous. Thousands have died mining fossil fuels, and in the early 1990s, a war was fought over petroleum (among other issues); fossil fuels also pollute. Nuclear power involves certain risks as well, although nuclear power plant explosions are unlikely in the United States.

Some yearn for the "good old days" prior to the Industrial Revolution when the Earth was a pristine planet filled with crystal clear streams and blanketed with majestic forests. However, those clear streams were filled with typhoid and there was never enough food in those lush forests. The average life span was just under forty years and people labored twelve to sixteen hours a day to meet subsistence needs. That is why we purify water and cultivate land once covered with tall trees, and why we used modern technology to improve life. We were never promised a risk-free world. Humans have brains to think and reason with. Part of the reasoning process involves evaluating how much risk is acceptable in order to live a healthier, longer, more comfortable life.

SUMMARY

Good air quality is essential to a good quality of life. Dry air consists of nitrogen, oxygen, carbon dioxide, and tiny amounts of other gases. Both natural events and human activities add other chemicals to that mixture; some are primary pollutants themselves, and others react with other substances to form secondary pollutants.

Human-caused air pollution may be either industrial (London) type or photochemical (Los Angeles) type pollution. The London type is primarily caused by industrial emissions; the main pollutants of this type are sulfur dioxide (SO_2) and particulate matter. At times, these pollutants are trapped over a region by a thermal inversion weather pattern. The resulting air quality crisis may cause excess deaths over and above the expected number of deaths for that region.

Sulfur dioxide arises from the burning of bitumious (soft) coal and from smelting metal sulfide ores. Some industries use scrubbers to remove the SO_2 from effluent gases before they escape into the atmosphere. Still, 75 percent of the sulfur in the atmosphere comes from natural sources. Particulates such as soot, fly ash, and dust are another London type pollutant. Electrostatic precipitators, bag filters, and cyclone collectors help remove particulate matter from gaseous emissions.

Photochemical (Los Angeles) type pollution, or smog, contains little SO_2, but high levels of nitrogen oxides (NO_x). Automobile emissions are a major source of NO_x. Other photochemical pollutants include ozone (O_3) and peroxyacylnitrates (PANs). Catalytic converters in cars help reduce NO_x emissions through a catalyst, V_2O_5, that breaks NO_x down into N_2 and O_2.

Reducing and controlling air pollution is a challenging task. The United States has done a fairly good job of containing SO_2, CO, and VOC levels and has greatly reduced lead and suspended particulate matter levels. NO_x levels, however, have continued to climb. Scientists are working on developing "green" automobiles and alternative fuels to help alleviate this problem.

Other energy or fuel sources include nuclear and solar energy as well as alcohols, hydrogen, wind power, and geothermal power. Nuclear energy is clean and relatively safe. Its main drawback is the production of nuclear waste that must be safely disposed of. Presently, nuclear energy is produced through nuclear fission (splitting). Some scientists hope to develop a fusion (joining) process as well.

The only alcohols with potential as fuels are methanol and ethanol. Both are already partially oxidized (making them somewhat inefficient), and both produce aldehydes that can pollute the air. Moreover, not enough biomass exists to convert to alcohol fuels to meet human energy requirements.

Hydrogen is an almost limitless resource if we can "harvest" it from water. Presently, it takes so much energy to electrolyze water that the process is too expensive, but researchers are working on solar electrolysis to reduce costs.

Geothermal and wind power are viable energy sources only in certain locations. Because of this limitation, they are inadequate to supply all our energy needs.

Solar energy is another nearly limitless energy source. It is relatively nonpolluting as well (manufacturing the equipment creates some pollution). Passive and active solar heating systems are already in use. Some solar automobiles have also been invented, though they are not yet commercially available. Photovoltaic cells convert sunlight directly into electricity.

Energy must not only be harnessed to do work, it must be stored, transported, and conserved. Chemical energy is the only form that can be stored and transported without energy loss, and electricity is the cleanest, most mobile type. Conservation reduces the amount of energy needed to do a particular task; increasing a car's gas mileage and installing building insulation help conserve energy.

Energy and pollution are inextricably connected—we cannot use energy without producing some pollution. The key to making wise energy-usage decisions is to balance the risks against the benefits of a healthier, more comfortable life.

KEY TERMS

active solar heating system 511
anthracite (hard) coal 495
bag filtration 497
bituminous (soft) coal 495
blocker 506
catalytic converter 501
cyclone collector 497
electrostatic precipitator 497
fuel cell 510
geothermal power 509
Industrial (London) type pollution 493
moderator 506
nitrogen oxides (NO_x) 499
nuclear fission 506
nuclear fusion 506
particulate matter 493
passive solar heating system 511
peroxyacylnitrates (PANs) 501
photochemical (Los Angeles) type pollution 499
photovoltaic cells 513
primary pollutants 492
scrubbers 496
secondary pollutants 492
sulfur dioxide (SO_2) 493
thermal inversion 494

READINGS OF INTEREST

Petroleum, Gas, Coal, and Electric Power

Engardio, Pete, and Einhorn, Bruce. "Beijing's brownout." *Business Week*, August 29, 1994, pp. 43–44. China desperately needs more electricity-generating plants, but foreign investors do not like government-imposed profit limits.

Woodruff, David, and Engardio, Pete. "Plugging into the power surge abroad." *Business Week*, August 15, 1994, pp. 100–102. The demand for electricity is greatest in the emerging nations. However, profits are not attractive in all these countries.

Cullen, Robert. "The true cost of coal." *Atlantic Monthly*, December 1993, pp. 38–52. There are hidden costs in mining coal, including the environmental damage the mines cause.

Flavin, Christopher, and Kane, Hal. "Coal use levels off." *World Watch*, November/December 1993, p. 40. On the worldwide scene, the use of coal is decreasing, but it is still the main fuel in some nations.

Roodman, David Malin. "Power brokers: Managing demand for electricity." *World Watch*, November/December 1993, pp. 22–29. Demand-side management programs attempt to balance power demand, costs, and environmental concerns.

Lenssen, Nicholas. "A new path for the Third World." *Technology Review*, October 1993, pp. 43–51. In Third World countries, the per capita energy consumption is less than 10 percent of that in developed nations. The fuels needed for the future are biomass (including wood), oil, and coal.

Ruth, Lawrence A. "There is coal in our future." *Chemtech*, June 1993, pp. 33–39. New coal technologies offer higher thermal efficiency and reduced emissions.

Impoco, Jim. "Fueling the future." *U.S. News & World Report*, May 17, 1993, pp. 54–56. This article suggests that natural gas will become more important in the mid-1990s.

Britton, Peter. "Cleaning up coal: The 21st century imperative." *Popular Science*, April 1993, pp. 78–83. Emerging technologies could make coal into a cleaner fuel.

Lenssen, Nicholas. "All the coal in China." *World Watch*, March/April 1993, pp. 22–29. China currently uses a greater percentage of coal than any other nation, and this will likely increase with more industrialization.

Woodbury, Richard. "The great energy bust." *Time*, March 16, 1992, pp. 50–51. This article suggests that the U.S. oil and gas industry may be disappearing in favor of imports.

Martin, Amy. "A petrochemical primer." *Garbage*, September/October 1991, pp. 36–41. A good, easy-to-read article about the petrochemical industry.

Other Fuels

Tenebaum, David. "Tapping the fire down below." *Technology Review*, January 1995, pp. 38–47. Geothermal energy remains a major potential source of energy, but has yet to be widely utilized.

Peaff, George. "Methanol's transformation to commodity status stretches supply." *Chemical & Engineering News*, October 24, 1994, pp. 13–15. The increased demand for methanol was projected to use up 94 to 95 percent of the supply by 1996. Prices normally increase with increased demand.

Peaff, George. "Court ruling spurs continued debate over gasoline oxygenates." *Chemical & Engineering News*, September 26, 1994, pp.

8–13. A good review of the various oxygenated fuels with some comparisons against gasoline.

Morris, Gregory D. L. "Methanol moves into a new era." *Chemical Week*, August 31/September 7, 1994, pp. 36–37. As the price and demand for methanol increase, so do plans for building new methanol conversion plants.

Lipkin, Richard. "Firing up fuel cells." *Science News*, November 13, 1993, pp. 314–318. Fuel cells are beginning to come into use in buses and cars.

Raloff, J. "Digging up cleaner-burning cooking fuels." *Science News*, November 13, 1993, p. 309. Roots are potential sources for fuel in countries that use biomass for cooking and heating (over half the world).

Mayersohn, Norman S. "The outlook for hydrogen." *Popular Science*, October 1993, pp. 66–71, 111. This article describes the latest technologies for "mining" hydrogen from water; storage systems; and ways to use this fuel. Includes solar-driven hydrogen production.

Skerrett, P. J. "Fuel cell update." *Popular Science*, June 1993, pp. 88–91, 120–121. Fuel cells are able to convert fuels directly to electricity at efficiencies as high as 70 percent. The best fossil-fueled generators are only about 35 percent efficient.

Rotman, David. "Effects of oxygenated fuels are questioned at ACS meeting." *Chemical Week*, April 7, 1993, p. 9. Alcohol fuels seem to promote the formation of more aldehydes and peroxynitrates. Switching to these fuels may not be worthwhile.

McCosh, Dan, and Brown, Stuart F. "The alternate fuel follies." *Popular Science*, July 1992, pp. 54–59. An interesting comparison of the effects of various kinds of automobile fuel on the environment. Methanol or ethanol usually produce less hydrocarbon emissions than gasoline, but the CO_2 levels are about the same. In a properly tuned car, the hydrocarbon emissions are greatly reduced no matter what the fuel.

Starr, Chauncey, Searl, Milton F., and Alpert, Sy. "Energy sources: A realistic outlook." *Science*, vol. 256 (May 15, 1992), pp. 981–991. This article projects a 400 percent increase in global energy consumption by the mid-2000s due mainly to population increases and economic growth in Third World countries. Alternate sources of energy must be developed to meet this increased demand.

Pollution and Fuels

Hanson, David. "Energy department examines its own chemical waste problem." *Chemical & Engineering News*, January 30, 1995, p. 20. This is potentially good news for those who think the "pot has been calling the kettle black." The DOE record on waste is atrocious.

Carey, John. "We can fight smog without breaking the bank." *Business Week*, October 3, 1994, pp. 128–129. Six suggestions for low-cost smog reduction, including emphasizing NO_x rather than VOC reduction.

Kirschner, Elisabeth M. "Self-incrimination remains major problem with environmental audits." *Chemical & Engineering News*, August 22, 1994, p. 13–16. Audits expose the companies to potential prosecution using evidence they provide themselves.

Monastersky, R. "Fouling the air: Not just a modern problem." *Science News*, March 26, 1994, p. 198. Lead has contaminated the air for at least 2,600 years.

Budiansky, Stephen. "The doomsday myths." *U.S. News & World Report*, December 13, 1993, pp. 81–91. Many doomsday projections are overstated. Two myths are the number of species now extinct because of harvesting wood from the forests and the amount of tropical rainforest destroyed each year.

Adler, T. "Health effects of smog: Worse than thought." *Science News*, November 20, 1993, p. 326. Ground-level ozone appears more problematic than we once believed.

Grossman, Daniel, and Shulman, Seth. "A case of nerves." *Discover*, November 1993, pp. 67–75. How would you feel wandering through a chemical weapons plant?

Multiple authors. "Environment and the economy." *Science*, June 25, 1993, pp. 1883–1901. A series of thirteen short articles.

Kildow, Judith Tegger. "Keeping the oceans oil-free." *Technology Review*, April 1993, pp. 42–49. Can oil be transported over the oceans safely? Much doubt exists on this issue.

Bolch, Ben, and Lyons, Harold. 1993. *Apocalypse not*. Washington, D.C.: Cato Institute. Chapters 5 (A multibillion-dollar radon scare), 8 (Acid rain) and 9 (The safest fuel) relate to information in this chapter.

Ray, Dixy Lee. 1993. *Environmental overkill: What happened to common sense?* Washington, D.C.: Regnery Gateway. This book presents a counterview to environmentalism. Chapters 1 (The future according to Rio) and 5 (Urban air pollution) relate to this chapter.

Boly, William. "Smog city wants to make this perfectly clear." *Health*, April 1992, pp. 54–64. Los Angeles is among the smoggiest cities in the world, but this article tells about some rather drastic measures they are planning to alleviate the haze—measures such as restricting the use of lawnmowers, hair sprays, and charcoal fires.

Karplus, Walter J. 1992. *The heavens are falling: The scientific prediction of catastrophes in our time*. New York: Plenum Press. Chapters 7 (Nuclear radiation) and 8 (Air pollution: Acid rain) relate to information in this chapter.

Holloway, Marguerite. Soiled shores. *Scientific American*, October 1991, pp. 102–116. Describes the cleanup technologies used following the *Exxon Valdez* oil spill.

Canby, Thomas Y. "After the storm." *National Geographic*, vol. 180, no. 2 (August 1991), pp. 2–35. Desert storm, that is. The aftermath of one of the worst ecological disasters ever caused by humankind.

French, Hilary F. "Clearing the air: A global agenda." *Worldwatch Institute Paper 94*, January 1990. This 54-page pamphlet (including notes) probably tells all the bad news you could want to hear about dirty air.

Ray, Dixy Lee. 1990. *Trashing the planet*. New York: Harper Perennial. Chapters 5 (Acid Rain) and 8–11 (nuclear topics) relate to material in this chapter.

Lee, Douglas B. "Tragedy in Alaska waters." *National Geographic*, vol. 176, no. 2 (August 1989), pp. 260–63. A short, but not sweet, article on the Alaskan oil spill.

McLoughin, Merrill; Carpenter, Betsy; Cook, William J.; and Plattner, Andy. "Our dirty air." *U.S. News & World Report*, June 12, 1989, pp. 48–54. While this article seems almost an advertisement urging passage of the 1990 Clean Air Act, it does show how pollution is a global problem centered mainly in urban areas.

Grove, Noel. "Air: An atmosphere of uncertainty." *National Geographic*, vol. 171, no. 4 (April 1987), pp. 502–537. In the unlikely event that you doubt that air pollution is for real, this article will enlighten you. It covers smog, industrial air pollution, ozone (both upper- and lower-level problems), global warming, and even in-home air pollution.

Boyle, Robert H., and Boyle, R. Alexander. 1983. *Acid rain*. New York: Nick Lyons Books. Somewhat dated, but still an interesting account.

Nuclear Power

Jacobs, Michael. "Nuclear fusion project raises fiscal and military concerns." *Insight*, March 6, 1995, p. 32. Some believe that studying nuclear fusion is just another way to continue the nuclear arms race.

Hileman, Bette. "Stored plutonium found to pose major hazards." *Chemical & Engineering News*, December 12, 1994, pp. 8–9. This problem is occurring at Department of Energy sites.

Hileman, Bette. "Alternative disposal methods for excess U.S. weapons plutonium proposed." *Chemical & Engineering News*, December 5, 1994, pp. 21–22. One disposal method is to make plutonium into glass logs for storage. Another method is to use plutonium oxides with uranium oxides in a fuel mixture. Either way, the nuclear waste will eventually need disposal.

Illman, Deborah. "Expedited action recommended for spent nuclear fuel at Hanford." *Chemical & Engineering News*, November 28, 1994. pp. 29–30. More than 2,100 metric tons (nearly 80 percent of the DOE's total spent fuel supply) is stored in a pair of obsolete concrete water basins built in Washington state in 1951. These basins were meant to be temporary and were to be removed from service within twenty years. A change is long overdue.

Anderson, Christopher. "Fusion research at the crossroads." *Science*, April 29, 1994, pp. 648–651. Funding cutbacks over the past decade have reduced high-temperature fusion (Tokamaks) efforts considerably. Most Tokamaks now operating are in Europe or Japan.

Travis, John. "Inside look confirms more radiation." *Science*, February 11, 1994, p. 750. The results of the Chernobyl mishap look worse now than they did in 1986.

Taubes, Gary. "No easy way to shackle the nuclear demon." *Science*, February 4, 1994, pp. 629–631. This particular demon is plutonium disposal. The article has an interesting diagram for ultimate disposal, but locating a site is not simple.

Stone, Richard. "New radon study: No smoking gun." *Science*, vol. 263 (January 28, 1994), p. 465. Radon is deemed only a minor cause of lung cancer in the general population; its effects are swamped by the damage that comes from smoking.

Mann, Charles C. "Radiation: Balancing the record." *Science*, January 28, 1994, pp. 470–473. No doubt you have heard about experiments that deliberately exposed people to radiation. Here is more information on this touchy topic, presented in a nonsensational manner.

Dagani, Ron. "Latest cold fusion results fail to win over skeptics." *Chemical & Engineering News*, June 14, 1993, pp. 38–41. This paper briefly reviews the history of cold fusion studies, but it seems to dwell more on skeptics' comments than believers' claims. Intriguing reading includes provocative comments of a skeptic who claims the Japanese cold fusion workers are inadequately informed on this topic.

Ackland, Len. "Radiation risks revisited." *Technology Review*, February/March 1993, pp. 56–61. Everyone knows that large doses of radiation can cause cancer, but what about low dosage levels? This article reports on a study that indicates even low doses can cause an increase in the rate of cancer cell formation.

Ahearne, John F. "The future of nuclear power." *American Scientist*, vol. 81 (January/February 1993), pp. 24–35. This is possibly the best short article on the future of nuclear power. It suggests that America will only choose nuclear power if electricity demands increase and the cost of nuclear plants are kept within reasonable limits.

Kahn, Patricia, and Marshall, Eliot. "A grisly archive of key cancer data." *Science*, vol. 259 (January 22, 1993), pp. 448–451. This article discusses high-exposure data on East German uranium workers and also provides some much-needed insight into the long-term effects of low-level radiation.

Grossman, Dan, and Shulman, Seth. "Doing their low-level best." *Garbage*, December 1992/January 1993, pp. 32–37. Considers the obstacles inherent in solving any problem when everyone is trying to outshout everyone else.

Pasternak, Douglas, and Cary, Peter. "A $200 billion scandal." *U.S. News & World Report*, December 14, 1992, pp. 34–47. This article is about the cleanup programs at U.S. nuclear weapons plants. It is difficult to imagine how ineptly this program has been run and how much fraud is involved. This piece will not give you extra confidence in the ability of our legislators to deal with similar pollution problems.

Atwood, Charles H. "How much radon is too much?" *J. Chem. Ed*, vol. 69, no. 5 (May 1992), pp. 351–55. Good overview of the radon controversy.

White, David C.; Andrews, Clinton J.; and Stauffer, Nancy W. "The new team: Electricity sources without carbon dioxide." *Technology Review*, January 1992, pp. 42–50. This article stresses the need to find new power sources and quit arguing over how much global temperatures might be rising. Nuclear power is the main example the authors discuss.

Miller, Peter. "A comeback for electric power? Our electric future." *National Geographic*, vol. 180, no. 2 (August 1991), pp. 60–89. We definitely need more energy for tomorrow's world, but where will we get it? This article suggests nuclear power as the most likely source, but it also considers wind power, incinerating trash, and energy conservation.

Clark, Roy W. "What ever happened to cold fusion?" *J. Chem. Ed*, vol 68, no. 4 (April 1991), pp. 277–279. This article provides background on cold fusion, both before and after 1989. It accepts cold fusion as a verified phenomenon, although not in the sense Pons and Fleischmann had hoped. (This will hardly be the last word on this topic since some scientists are still intensively studying cold fusion.)

Wilson, Richard; Gale, Robert Peter; von Hippel, Frank; Lee, William S.; Winston, Donald C.; Lloyd, Marilyn; and Lovins, Amory B. "Forum Chernobyl." *Issues in Science & Technology*, Winter 1987, pp. 6–13. A series of letters on the use of nuclear energy and the Chernobyl disaster. These letters are worth reading just to see how different people view this problem. The general conclusion is that nuclear energy must be pursued because it presents less risk than the alternatives.

Carter, Luther J. "Nuclear imperatives and public trust: Dealing with radioactive waste." *Issues in Science & Technology*, Winter 1987, pp. 46–61. This article should have been subtitled "NIMBYs in Action." Learn how the solution to a problem has been hamstrung by legal maneuvering, outright hostility, and inept government actions. Unlike France, the United States has made nuclear energy a political issue.

Solar Energy

Linden, Eugene. "A sunny forecast." *Time*, November 7, 1995, pp. 66–67. Predicts more solar energy and less pollution.

Tyson, Peter. "Solar ovens heat up in the tropics." *Technology Review*, May/June 1994, pp. 16–17. Using a solar oven for cooking, instead of a smoky wood fire, reduces health problems.

Brown, Stuart F. "The eternal airplane." *Popular Science*, April 1994, pp. 70–75. A solar-powered airplane that can fly on and on.

Langreth, Robert. "Eco resort." *Popular Science*, March 1994, pp. 66–67, 87. Describes a luxury resort, made from trash, which uses solar and wind energy.

Borman, Stu. "Amorphous silicon film offers cheaper solar energy." *Chemical & Engineering News*, January 1994, pp. 7–8. This new technology is expected to reduce the cost of photovoltaic-generated electricity from \$0.25 to 0.50/kWh (kilowatt hour) to \$0.16/kWh. Eventually, the cost should drop to about \$0.12/kWh.

Lipkin, R. "Thin-film solar cells boost efficiency." *Science News*, December 4, 1993, p. 374. Reports 15.9 percent efficiency for a polycrystalline thin-film photovoltaic cell. The theoretical limit is about 23 percent.

Regan, Mary Beth. "The sun shines brighter on alternative energy." *Business Week*, November 8, 1993, pp. 94–95. A good comparison of the costs of electricity generation from alternative energy sources. Standard solar costs about twice as much as coal, and photovoltaic is about six times greater.

Zweibel, Ken. "Thin-film photovoltaic cells." *American Scientist*, vol. 51, (July/August 1993), pp. 362–369. A good review of photovoltaics, one of the main benefits derived from the space program.

El-Haggar, S. M., and Awn, A. A. "Optimum conditions for a solar still and its use for a greenhouse using the nutrient film technique." *Desalination* vol. 94 (1993), pp. 55–68. Describes some work done in Egypt that converts salty water into potable water.

Wilhelm, John L. "Solar energy, the ultimate powerhouse." *National Geographic*, vol. 149, no. 3 (March 1976), pp. 381–397. Covers the basic concepts behind solar energy uses, many of which are still valid today. Depicts a great vision yet to happen.

Miscellaneous Energy Sources

Turnbull, Andy. "Pond-size battery." *Popular Science*, April 1995, p. 34. Some Canadians are turning acid-waste ponds into gigantic batteries. The cost is lower than merely neutralizing the acid and you obtain energy as a bonus.

McGowan, Jon G. "Tilting toward windmills." *Technology Review*, July 1993, pp. 38–46. Good article on the latest windmill designs. Deflates several myths on wind power availability, cost, and reliability.

Thompson, Dick. "Breezing into the future." *Time*, January 13, 1992, pp. 48–49. Short article on the future prospects of wind power.

Hamilton, Roger. "Can we harness the wind?" *National Geographic*, vol. 148, no. 6 (December 1975), pp. 812–829. People have harvested the wind for centuries, but this article probes newer, more efficient ways to capture the wind's raging energy.

Energy Conservation

Wardell, Charles. "The revolution behind your walls." *Popular Science*, April 1995, p. 43. Short article on some new, very effective forms of insulation.

Cary, Peter. "The asbestos panic attack." *Business Week*, February 20, 1995, pp. 61–63. Billions of dollars have been spent on removing asbestos from schools even though it presented very little risk to students. This paper has an interesting table of comparative risks for different kinds of activities.

Langreth, Robert. "The \$30 million refrigerator." *Popular Science*, January 1994, pp. 65–67, 87. An improved refrigerator that uses better insulation and less energy.

Douglas, Carole. "Making houses out of trash." *World Watch*, December 1993, pp. 30–32. This "trash" is a virtual lumber made from scrap wood, paper, and cardboard.

Chaikin, Andrew. "The light stuff." *Popular Science*, February 1993, pp. 72–74, 100–101. Fascinating article about aerogels and SEAgels, the lightest forms of insulation yet available. They could revolutionize the insulation industry.

Harris, Tom. "The asbestos mess." *Garbage*, December 1992/January 1993, pp. 44–49. Recent information suggests that the alleged asbestos problem was a media-generated issue and that asbestos was seldom a real danger to people—unless it was removed from buildings.

Knauer, Gene. "The return of the geodesic dome." *The Futurist*, January/February 1992, pp. 29–32. These dome structures are potentially the most energy-efficient buildings in existence, but they are seldom built. Those that do exist tell an amazing story of energy conservation.

Buderi, Robert; Smith, Emily T.; Shao, Mario; Smith, Geoffrey; and Hong, Peter. "Conservation power." *Business Week*, September 16, 1991, pp. 86–92. While the government debates regulations, thousands of businesses are already putting energy conservation programs into operation.

Improved Automobiles

Mayersohn, Norman S. "The greening of the diesel." *Popular Science*, July 1994, pp. 46–51. Diesel cars have often caused more air pollution than gasoline counterparts. New technology may change that.

McCosh, Dan. "Emerging technologies for the supercar." *Popular Science*, June 1994, pp. 95–101. Cars of the future will be lighter weight and get better gas mileage—up to 80 mpg. They may use combinations of electric-drive motors, fuel cells, and special flywheels. How soon will they hit the auto showrooms? No one knows.

Williams, Robert H. "The clean machine." *Technology Review*, April 1994, pp. 20–30. Fuel cells could revolutionize automobiles and vastly reduce their pollution within the next twenty years.

Woodruff, David; Armstrong, Larry; and Carey, John. "Electric cars: Will they work? And who will buy them?" *Business Week*, May 30, 1994, pp. 104–114. Electric cars could do much to alleviate photochemical smog. Their main drawbacks have been cost and distance between charges. Newer battery designs and types extend the range, but prices remain high.

Silber, Kenneth. "Electric cars shift into second gear." *Insight*, May 9, 1994, pp. 18–21. Government regulations are pushing electric cars to the forefront.

Carey, John. "Magnetic field of dreams." *Business Week*, April 18, 1994, pp. 118–21. Giant magnetoresistance devices may help control automobile engines, brakes, and suspensions.

McCosh, Dan. "We drive the world's best electric car." *Popular Science*, January 1994, pp. 52–58. This article includes a table that compares several brands of electric cars. Their driving range is now up to 165 miles between charges, and their top speeds may exceed 60 mph. Most are not yet available on the open market, however.

Kelly, Kevin; Miller, Karen Lowrey; and Woodruff, David. "The rising rumble of American diesels." *Business Week*, September 6, 1993, pp. 84–86. These diesels have much-improved emission and noise controls. They are beginning to make an impact on the market, both domestic and foreign.

Flavin, Christopher. "Jump start the new automobile revolution." *World Watch*, July/August 1993, pp. 27–33. This piece considers both electric and solar cars.

Baldwin, J. "Green cars." *Garbage*, June/July 1993, pp. 24–29. An update on modifications in the works. This article advises not to "believe a thing you read until you've taken a test ride in Minneapolis in February."

Fischetti, Mark. "Here comes the electric car—It's sporty, aggressive, and clean." *Smithsonian*, April 1992, pp. 34–43. You could also add quiet to this list of virtues, except for a slight whine.

PROBLEMS AND QUESTIONS

1. What are the four main chemicals present in dry air?
2. What are the two major classes of air pollution?
3. What chemicals are found in industrial type air pollution?
4. How can sulfur dioxide be removed from gaseous industrial emissions?
5. What is a scrubber?
6. What are the sources of particulate matter?
7. Name three methods that remove particulates from gaseous emissions.
8. What chemicals are found in photochemical type air pollution?
9. What chemical is present in industrial type air pollution that seldom occurs in photochemical type pollution?
10. What is the difference between a primary and a secondary pollutant?
11. Which of the following chemicals are primary and which are secondary air pollutants: NO_x, O_3, CO, hydrocarbons, aldehydes, and peroxynitrates?
12. What chemical is common in photochemical air pollution but rare in industrial air pollution?
13. What percentage of the NO_x in the air comes from natural sources?
14. What are the sources of natural NO_x?
15. What percentage of the SO_2 in the air comes from natural sources?
16. What are sources of natural SO_2?
17. What device is used to reduce photochemical air pollution?
18. What does a catalytic converter do, and how does it work?
19. What types of energy generation are essentially nonchemical?
20. List several forms of chemically generated energy (for example, fuels).
21. How much does the percentage of electricity generated using nuclear power in the United States differ from that in France?
22. What are the potential dangers of nuclear power-generating plants?
23. How does the safety record of nuclear energy compare with the safety record of coal mining and electrical energy generation from coal?
24. What methods are used to dispose of nuclear waste?

25. How can we prevent leaching of chemicals from nuclear waste?

26. What is the difference between nuclear fission and fusion? Which is now used for electrical power generation?

27. What alcohols are most often used as fuels?

28. What are the limitations on the use of alcohol fuels?

29. What are the advantages and disadvantages of hydrogen as a fuel?

30. What are the advantages and disadvantages of solar energy?

31. What are the two methods of solar heating?

32. What is photovoltaic solar conversion, and how does it work?

33. Is it possible for wind power or geothermal energy sources to provide sufficient energy for all human needs?

34. What energy form is the easiest to transport?

35. What energy form is the easiest to store?

36. Is there a relationship between energy generation and pollution?

37. Which type of energy-generating procedure is the least polluting?

38. Which form of large-scale energy generation is the safest?

39. What types of energy generation can supply the huge amounts of energy that modern, industrialized societies consume?

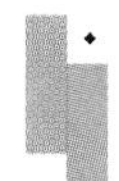

CRITICAL THINKING PROBLEMS

1. Are temperature and wind factors part of the quality of the environment? If so, can they be changed to improve conditions? Why or why not?

2. Massive efforts and enormous amounts of money were expended to clean up the effects of the *Exxon Valdez* oil spill. Based on the results, were these expenditures justified? Could some other approach have worked better? If so, what approach would you suggest? Three useful articles on this subject are: Douglas B. Lee, "Tragedy in Alaska Waters." *National Geographic*, vol. 176, no. 2 (August 1989), pp. 260–63; Amy Martin, "A Petrochemical Primer," *Garbage*, September/October 1991, pp. 36–41; and Judith Tegger Kildow, "Keeping the Oceans Oil-Free," *Technology Review*, April 1993, pp. 42–49.

3. Air pollution poses real problems for humanity. What are some things you can do about this problem? Some useful background can be found in the following two articles: William Boly, "Smog City Wants to Make This Perfectly Clear," *Health*, April 1992, pp. 54–64; and Noel Grove, "Air: An Atmosphere of Uncertainty." *National Geographic*, vol. 171, no. 4 (April 1987), pp. 502–537.

4. If a company has excellent emission controls, they can sell their "pollution rights" to another company, who can then emit more pollutants. What do you think of this idea, which finds its basis in the 1990 Clean Air Act?

5. Would there be more particulate matter per cubic foot in the air over (1) a city, (2) a farming community, or (3) the middle of the ocean? (These particles might include dust, soot, or salt.)

6. Asbestos, a fibrous silicate, is known to cause certain types of lung cancer. At one time the use of asbestos was mandated in public buildings and recommended for private homes because it is a superior, nonflammable insulation material. Asbestos only causes problems when the particles are airborne. Should the removal of asbestos from public buildings be required even if tests show no asbestos in the air? If so, who should pay the bill? Should asbestos manufacture be banned? You can find valuable background information in the following articles: Tom Harris, "The Asbestos Mess," *Garbage*, December/January 1993, pp. 44–49; and Peter Cary, "The Asbestos Panic Attack," *Business Week*, February 20, 1995, pp. 61–63.

7. Some believe nuclear power will someday be the main electricity-generation source in the United States. Read the following three articles: John F. Ahearne, "The Future of Nuclear Power," *American Scientist*, vol. 81 (January/February 1993), pp. 24–35; Peter Miller, "A Comeback for Nuclear Power? Our Electric Future," *National Geographic*, vol. 180, no. 2 (August 1991), pp. 60–89; and David C. White, Clinton J. Andrews, and Nancy W. Stauffer, "The New Team: Electricity Sources Without Carbon Dioxide," *Technology Review*, January 1992, pp. 42–50. What do you think about nuclear power, and what is the basis of your opinion?

8. Not all victims of atomic radiation exposure die immediately. Some live for many years and die from cancer or leukemia. How would you attempt to explore whether death from cancer is related to high radiation exposure? Two helpful articles are Patricia Kahn and Eliot Marshall, "A Grisly Archive of Key Cancer Data," *Science*, vol. 259 (January 22, 1993), pp. 448–451; and Richard Wilson, Robert Peter Gale, Frank von Hippel, William S. Lee, Donald C. Winston, Marilyn Lloyd, and Amory B. Lovins, "Forum: Chernobyl," *Issues in Science & Technology*, Winter 1987, pp. 6–13.

9. Fears about radon continue to surface. Are they justified? Form your own opinion by reading box 18.3 and at least one of the following articles: Richard Stone, "New Radon Study: No Smoking Gun," *Science*, vol. 263 (January 28, 1994), p. 465; Len Ackland, "Radiation Risks Revisited," *Technology Review*, February/March 1993, pp. 56–61; and Charles H. Atwood, "How Much Radon Is Too Much?" *J. Chem. Ed.*, vol. 69, no. 5 (May 1992), pp. 351–353.

10. Is fusion likely to be a major energy source in the near future? A good article for additional background is Christopher Anderson, "Fusion Research at the Crossroads," *Science*, April 29, 1994, pp. 648–651.

11. Cold fusion has a problematic history. Even scientists do not agree on whether this phenomenon is real. Read the following two articles on this topic and form a tentative opinion on cold fusion: Roy W. Clark, "What Ever Happened to Cold Fusion?" *J. Chem. Ed.*, vol. 68, no. 4 (April 1991), pp. 277–279; and Ron Dagani, "Latest Cold Fusion Results Fail to Win Over Skeptics," *Chemical &*

Engineering News, June 14, 1993, pp. 38–41. You may want to change your opinion as more information becomes available in the next ten years.

12. Many people believe that coal should provide a greater portion of our energy needs. Read some of the following articles and form your own opinion on this matter: Robert Cullen, "The True Cost of Coal," *Atlantic Monthly*, December 1993, pp. 38–52; Lawrence A. Ruth, "There is Coal in Our Future," *Chemtech*, June 1993, pp. 33–39; and Peter Britton, "Cleaning Up Coal: The 21st Century Imperative," *Popular Science*, April 1993, pp. 78–83.

13. The use of oxygenated alternative fuels has created a debate even hotter than the fuels themselves. Read the following two articles and form your own opinion on this topic: George Peaff, "Court Ruling Spurs Continued Debate Over Gasoline Oxygenates," *Chemical & Engineering News*, September 26, 1994, pp. 8–13; and Dan McCosh and Stuart F. Brown, "The Alternate Fuel Follies," *Popular Science*, July 1992, pp. 54–59.

14. Do you believe that solar energy will ever become an important energy source? Reread box 18.5 and at least two of the following articles: John L. Wilhelm, "Solar Energy, the Ultimate Powerhouse," *National Geographic*, vol. 149, no. 3 (March 1976), pp. 381–397; Eugene Linden, "A Sunny Forecast," *Time*, November 7, 1995, pp. 66–67; and Mary Beth Regan, "The Sun Shines Brighter on Alternative Energy," *Business Week*, November 8, 1993, pp. 94–95.

15. Many people feel fuel cells will solve most of our energy problems. Form your own opinion after reading box 18.4 and at least one of the following two articles: P. J. Skerrett, "Fuel Cell Update," *Popular Science*, June 1993, pp. 88–91, 120–121; and Richard Lipkin, "Firing Up Fuel Cells," *Science News*, November 13, 1993, pp. 314–318.

16. How practical is geothermal energy likely to be in the future? Read the following article to help form your opinion: David Tenebaum, "Tapping the Fire Down Below," *Technology Review*, January 1995, pp. 38–47.

17. How practical is wind power likely to be as an energy source in the United States? Three good background articles on this subject are: Jon G. McGowan, "Tilting Toward Windmills," *Technology Review*, July 1993, pp. 38–46; Dick Thompson, "Breezing Into the Future," *Time*, January 13, 1992, pp. 48–49; and Roger Hamilton, "Can We Harness the Wind?" *National Geographic*, vol. 148, no. 6 (December 1975), pp. 812–829.

18. Geodesic dome houses were mentioned only briefly in this chapter. Are they really a good energy conservation method? For further background, read Gene Knauer, "The Return of the Geodesic Dome," *The Futurist*, January/February 1992, pp. 29–32.

19. So-called "green" cars will probably be highly important in the future. What are their advantages and disadvantages? For background, read box 18.2 and at least one of the following articles: Dan McCosh, "Emerging Technologies for the Supercar," *Popular Science*, June 1994, pp. 95–101; Robert H. Williams, "The Clean Machine," *Technology Review*, April 1994, pp. 20–30; David Woodruff, Larry Armstrong, and John Carey, "Electric Cars: Will They Work? And Who Will Buy Them?" *Business Week*, May 30, 1994, pp. 104–114; Kenneth Silber, "Electric Cars Shift Into Second Gear." *Insight*, May 9, 1994; pp. 18–21; and J. Baldwin, "Green Cars," *Garbage*, June/July 1993, pp. 24–29.

20. Research a city's history of air pollution crises. How often are air pollution crisis alerts issued? How are the crises resolved? Is there any way to prevent them? What would you recommend to avert crises?

HINT Many large metropolitan areas are located in valleys subject to thermal inversion patterns. The only practical way to prevent or alleviate air pollution crises is probably to reduce emissions drastically when weather patterns threaten a thermal inversion. How could you restrict emissions?

21. Do you think a ban on sulfur-containing coal is reasonable? What would you suggest in its place?

22. Imagine you now have the responsibility and the power to control air pollution. Based on the data presented in this chapter, what changes would you put into effect?

23. You have now learned about nuclear electricity generation (and nuclear waste) and learned how it compares with fossil fuel-powered plants. Imagine you have the authority to decide which type of electricity-generating plants will be built in your community. What factors would you weigh? What might you decide?

24. How important is it to learn from past mistakes that caused environmental damage? Should there be special punitive penalties for such errors, or should the emphasis be on correcting past errors and preventing future problems of the same kind? What about situations where the damage was inadvertent or arose because laws changed after the damage was done? And what can we do about foreign countries who pollute the world's air excessively? Some articles that deal with these issues include: (1) Wade Roush, "Learning from Technological Disasters," *Technology Review*, August/September 1993; pp. 50–57; (2) Tom Waters, "Ecoglasnost," *Discovery*, April 1990, pp. 50–53; and (3) Rick Henderson, "Crimes Against Nature," *Reason*, December 1993, pp. 18–24.

25. Economic and regulatory changes have a profound effect on science and technology. Some believe that major changes in these factors have adversely affected technological development in the United States. Is this true? Has America shifted its policies to the detriment of future development? Read at least two of the following articles and come to your own conclusion: (1) Joan O'C. Hamilton, "Biotech," *Business Week*, September 26, 1994, pp. 84–92; (2) Neil Gross, John Carey, and Joseph Weber, "Who Says Science Has to Pay Off Fast?" *Business Week*, March 21, 1994, pp. 110–111; (3) John Carey, et al., "Could America Afford the Transistor Today?" *Business Week*, March 7, 1994, pp. 80–84; (4) Roger A. Sheldon, "Consider the Environmental Quotient," *Chemtech*, March 1994, pp. 38–47; and (5) "Forecast 94," *Chemical Week*, January 5/12, 1994, pp. 34–51.

CHAPTER 19

CHEMISTRY AND OUR ATMOSPHERE

OUTLINE

KEY IDEAS

1. Two Chemical Problems that May Affect Our Atmosphere
2. The Nature of the Greenhouse Effect
3. The Main Greenhouse Gas
4. Global Warming and the Chemicals Claimed to Cause It
5. Databases and the Long-Term Temperatures of the Earth
6. The Sources of Atmospheric Carbon Dioxide
7. How Scientists Estimate Past Atmospheric Levels of CO_2
8. Whether Global Warming Is a Real Phenomenon
9. The Nature and Variability of the Ozone Layer
10. What Chemicals (Besides CFCs) Cause Ozone Depletion
11. CFC Uses and Alternative Chemicals
12. The Nature of the Antarctic Ozone Hole
13. How the Concentration of the Ozone Layer Changes
14. Whether There Is a Connection Between the Ozone Layer, UV Radiation, and Skin Cancer
15. Whether the Ozone Layer is Really Being Depleted

"Facts do not cease to exist because they are ignored."
ALDOUS HUXLEY (1894–1963)

March marigolds along small stream in Whitefish Dunes State Park, Door County, Wisconsin.

In chapter 18, we discussed air pollution problems and solutions. Chemicals can affect the atmosphere and, perhaps, our quality of life in at least two other ways as well: they may cause global warming and deplete the ozone layer. Each of these problems could have a major impact on your lifestyle (figure 19.1).

Both global warming and ozone layer depletion are regular topics of discussion on television and in the popular press. Chemicals seem to play a role in each problem; unfortunately, the chemistry-related details seldom appear in the late news nor in the popular press. Both problems are deeply intertwined with politics and subject to the spread of misinformation.

You have learned enough about the scientific method and critical thinking to apply these principles to these political and environmental problems. Accordingly, in this chapter, we will seek to answer two specific questions: (1) is global warming occurring because of human activities, and (2) is the ozone layer being depleted because of human activities? We will explore a lot of data on each of these two problems. Study this information carefully and try to reach conclusions based on the facts.

QUESTION ONE: IS GLOBAL WARMING OCCURRING?

Not everyone agrees that global warming is actually occurring. To try to determine whether it is and how to deal with it, we must understand what global warming is and its relationship to the greenhouse effect; then we must study the relationship between temperature and the greenhouse gases, and, finally, assess the impact of global warming. Let's begin by discussing why global warming might occur.

Figure 19.1 Global warming and ozone layer depletion will affect your life at the beach and elsewhere.

Gaining a Background—Global Warming and the Greenhouse Effect

To discuss the question of global warming, it is first necessary to distinguish between global warming and the greenhouse effect. Global warming depends on the greenhouse effect, but the greenhouse effect has been operating throughout much of the history of our planet. Global warming, if it is happening, is apparently a recent phenomenon.

The Greenhouse Effect

Sometimes it seems that scarcely a day passes without someone claiming that an imminent, protracted **global warming** is underway because of certain gases in the atmosphere. These gases are claimed to cause a **greenhouse effect,** or a buildup of heat energy in the atmosphere, that could actuate catastrophic climatic changes around the world. The implicated gases are CO_2, the chlorofluorocarbons (CFCs), CH_4, and NO_x. Are these claims true, and if so, what can we do about it?

There is no doubt that the greenhouse effect is a real phenomenon, and that this phenomenon is unique to Earth. Earth is the only planet in the solar system with the right combination of air, water, and temperature to sustain life. The fact that we have an atmosphere means that part of the incoming solar radiation is held on the surface of the planet, rather than reflected back into space. Much of this "trapped" radiation is in the infrared (IR) region of the electromagnetic spectrum, which we perceive as

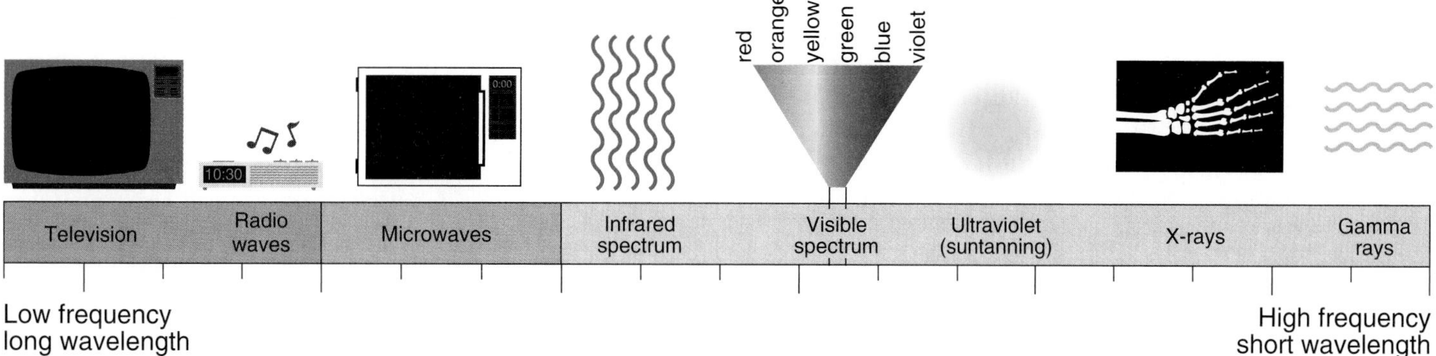

Figure 19.2 The electromagnetic spectrum.

SOURCE: From Ricki Lewis, *Life*.

heat (figure 19.2). The atmosphere acts somewhat like the glass in a greenhouse, allowing solar radiation to enter but trapping some of it as heat.

Figure 19.3 depicts the balance between the incoming and outgoing energy radiation for the Earth. The numbers in parentheses are the percentages of the total energy part of this cycle. The part of the cycle in which energy is converted into infrared and held within our atmosphere represents the greenhouse effect.

If the atmosphere of Mars were similar to Earth's, Mars would have an average temperature many degrees warmer than its −23°C (−10°F). Similarly, if Earth lacked an atmosphere, it would radiate away most of the incoming light and heat, and our

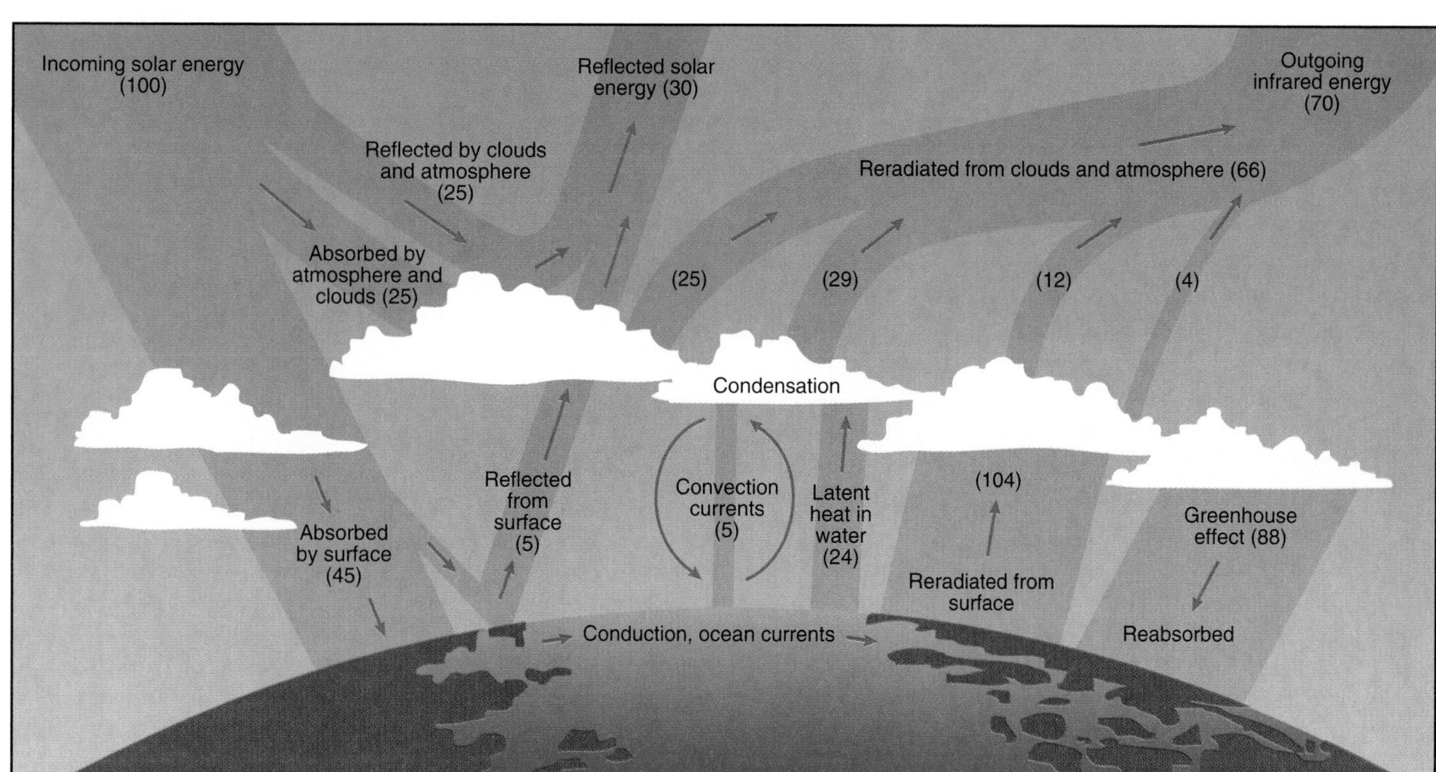

Figure 19.3 The estimated total incoming and outgoing solar energy balance for the Earth. The numbers denote the percentage of solar energy for each pathway. The greenhouse effect is illustrated on the right. More energy (104 percent) is radiated from the surface back into space than arrives; this is due to a net energy increase via the greenhouse effect, which causes 88 percent to be reabsorbed while the other 16 percent reflects back into space.

SOURCE: From William P. Cunningham and Barbara Woodworth Saigo, *Environmental Science*, 2d ed.

surface temperature would drop close to that of Mars. The dry atmosphere traps enough heat, via kinetic motion, to raise the average global temperature to around −15°C (5°F). However, the average temperature on Earth is much higher—about 15.5°C (60°F). This is because Earth has an atmosphere that contains up to 4 percent water vapor (40,000 ppm), and H_2O absorbs IR radiation extremely well. The absorbed energy gives our planet its livable temperature range. Water also creates a cloud layer that can retain heat. Without this cloud cover, most of our planet would resemble the polar regions, buried under dense layers of ice.

PRACTICE PROBLEM 19.1

What does the greenhouse effect do to the temperature of the Earth? What is the main gas that causes this effect?

ANSWER The greenhouse effect raises the temperature by nearly 30°C (from −15°C to 15.5°C). H_2O vapor is the main greenhouse gas.

Global Warming

While the greenhouse effect is a normal, natural phenomenon caused by the presence of water in the Earth's atmosphere, global warming is attributed to human activities. Other atmospheric gases, such as **carbon dioxide (CO_2), chlorofluorocarbons (CFCs), methane (CH_4), nitrogen oxides (NO_x),** and **ozone (O_3),** absorb IR radiation to a lesser extent than water (box 19.1). The global warming debate centers exclusively on these gases and totally ignores the effects of water. Of these gases, CO_2 is present to the greatest extent in our atmosphere (table 19.1); its concentration is about 330–350 ppm (compared to 40,000 ppm for water), and it is claimed to be responsible for at least 50 to 60 percent of the potential heat buildup in the global warming theory. The CO_2 in the atmosphere comes from both natural and human-controlled sources (for example, from both breathing and burning carbon-containing fuels).

PRACTICE PROBLEM 19.2

Why is H_2O the main greenhouse gas, and why is CO_2 far ahead of any other greenhouse gas?

ANSWER Water absorbs IR energy better than any other gas and is present in the atmosphere at a much greater concentration. While the amount of CO_2 in the air is low, the other greenhouse gases are at even lower concentrations. Even if a gas absorbs IR fairly efficiently, its total absorption will be low if its concentration is low.

TABLE 19.1 The composition of dry air showing the concentrations of the greenhouse gases (*). Water, the main greenhouse gas, can vary between nearly zero and 40,000 ppm.

GAS	PERCENTAGE (by volume)	PPM
Nitrogen	78.084	780,840
Oxygen	20.946	209,460
Argon	0.934	9,340
Carbon dioxide*	0.033	330
Neon	0.0018	18
Helium	0.0005	5
Methane*	0.0002	2
Krypton	0.0001	1
Xenon	0.0001	1
Hydrogen	0.00005	0.5
Nitrous oxide*	0.00005	0.5
CFCs, all types	—	<0.001
All others	0.0002	2

CFCs are believed to have a direct effect on the ozone layer. All CFCs come from human activities. Some environmentalists claim that the low atmospheric levels of CFCs cause 20 to 25 percent of global warming. However, that value is merely a guess; recent estimates place the potential warming from CFCs at much lower values. Realistically, we might expect CO_2 to contribute about ten thousand times as much to the greenhouse effect as CFCs, based on relative concentrations and IR absorption.

We discussed the nitrogen oxides (NO_x) in chapter 18 under the topic of air pollution. These compounds are claimed to contribute 5 to 10 percent of total global warming, with N_2O identified as the chief troublemaker. As we previously noted, NO_x compounds come roughly equally from natural and human-derived sources. The estimates for N_2O warming seem rather generous considering the low levels of this gas in the atmosphere (table 19.1). The atmospheric methane levels are four times as high as N_2O, yet methane is assumed to make only a slightly higher contribution (less than 14 percent) to overall global warming (and this also seems a very generous estimate for its warming potential). Methane comes from peat bogs, rice paddies, and cattle—all natural sources; there are no major human-derived methane emissions (although humans eat rice and cattle and thus indirectly encourage methane production).

In summary, some global warming theorists claim that certain gases comprising less than 0.034 percent of the dry atmosphere cause significant atmospheric warming. They also project that this will cause catastrophic warming within the next century

BOX 19.1

HOW GREENHOUSE GASES WORK

Not all molecules or chemical bonds are effective in absorbing infrared (IR) radiation, but those that do, including water vapor, warm the atmosphere as greenhouse gases. Exactly how does the greenhouse effect take place? Why do some compounds absorb IR more than others?

The ability to absorb IR depends on several factors, including (1) the relative polarity difference between the atoms in a bond (see chapter 6), (2) the presence of multiple bonds, (3) the molecular shape, and (4) the presence of unshared electrons (which has only a minor effect). The electronegativity (or polarity) differences for the greenhouse gas atoms are 1.4 for H_2O; 1.5 to 0.5 (depending on the specific bonds) for CFCs; 1.0 for CO_2; 0.5 for N_2O; and 0.4 for CH_4. In addition, CO_2 has multiple bonds and water has a nonlinear structure that accentuates its polarity. Thus, as we might expect, water absorbs IR effectively, as does CO_2. The more a molecule absorbs IR energy, the greater its molecular motion. This motion results in temperature increases.

The total greenhouse contribution of a particular type of molecule depends on: (1) how much IR energy a molecule absorbs and (2) the concentration of the molecule in the atmosphere. At 40,000 ppm, H_2O completely overshadows all other greenhouse gases. Even if CO_2 absorbed as much IR energy as H_2O, the warming effect of H_2O would be over a hundred times greater because its concentration is over a hundred times greater. Since water actually absorbs IR energy better than CO_2, the total warming from H_2O is even greater. By this same reasoning, if CO_2 and the CFCs absorb exactly the same amount of IR energy per molecule, the CO_2 would warm 330,000 times as much because the concentrations differ to that extent. CFCs do absorb more IR than CO_2, but not nearly enough to overshadow the concentration factor. Methane absorbs a smaller amount of IR than the other gases listed and its concentration in the air is small. The net effect of all these factors is that H_2O is the most powerful greenhouse gas, followed by CO_2, CFCs, methane, and N_2O.

unless we control emissions. To evaluate these claims, we will first determine the sources of the greenhouse gases, excluding water, and then assess data projecting long-term temperature trends for the Earth.

The Greenhouse Gases

The **greenhouse gases** claimed to cause global warming (CO_2, CH_4, N_2O, and CFCs) are minor components of the atmosphere; figure 19.4 shows the total amount, in millions of tons, of each of these chemicals in the Earth's atmosphere. Some of these figures are educated guesses at the actual amounts. The total amount, for example, of CFCs as listed in figure 19.4 was estimated as follows: In the mid-1970s (the peak years), 400×10^6 kg of CFCs were released into the atmosphere. Assuming the same amount was released during each of the past twenty years, and assuming that none degraded, there could be as many as 8.8 million tons of CFCs lodged in the atmosphere. These assumptions are actually on the high side for two reasons: (1) CFC production and emission decreased sharply after the 1970s, and (2) CFCs do biodegrade. For our purposes, however, it is better to err on the high side than to underestimate.

From examining the figure, it is immediately apparent that CO_2 dwarfs all other greenhouse gases—its total mass is 440 times greater than CH_4, the next highest. The other gases do not have

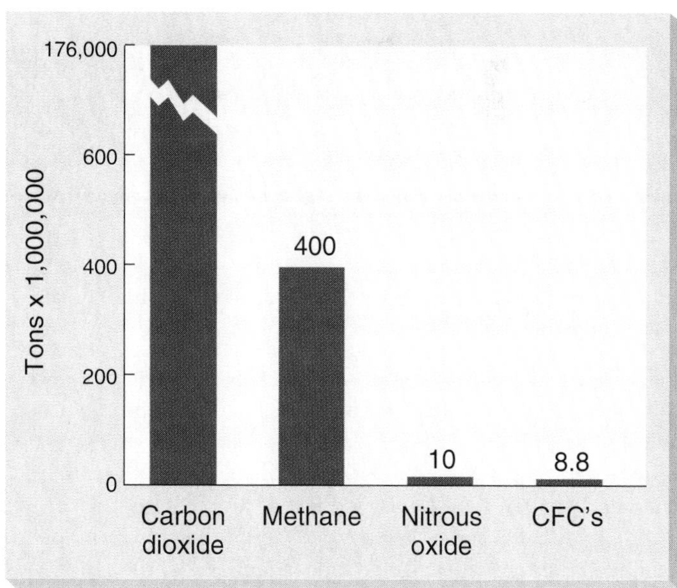

Figure 19.4 Estimated amounts of CO_2, CH_4, N_2O, and CFCs in the atmosphere in millions of tons.

SOURCE: Data from Renew America Project, *Chemtech*, August 1989, and Money, et al., in *Science*, 288:926, 1987.

extraordinary heat-absorbing properties, so they would not likely have the great warming effects some attribute to them. For this reason, we will limit the ensuing discussion to the greenhouse effects of CO_2. (This is not a completely new idea, incidentally. In 1896, Svante Arrhenius proposed that atmospheric levels of CO_2 may control temperatures and that the ice ages occurred because these levels had been reduced by some unknown process.)

A certain amount of caution is necessary in interpreting the claims regarding greenhouse gases. For example, according to an early 1995 report, termite flatulence accounts for as much as 20 percent of the world's methane. According to this study, termites produce about 176×10^9 pounds of methane every year. Since there are about 800×10^9 pounds total methane in the atmosphere (400 million tons $\times$ 2,000 = 800 billion pounds), termites alone could generate this amount in less than five years. However, if we examined the news report carefully, it would become evident that the researchers did not really know how much methane termites generate. Their estimate was aimed at acquiring support for their studies on termite populations. By tying their estimates to global warming, literally a "hot" research topic, they may have believed their program would receive more funding.

PRACTICE PROBLEM 19.3

Why were the CFCs dismissed from this discussion of global warming, even though some claim as much as 25 percent of the warming effect is due to CFCs?

ANSWER The total atmospheric levels and IR absorption tendencies of CFCs are so low that the alleged warming percentages seem unreasonable. Of course, if new data come to light, you can reevaluate CFCs accordingly.

Studying the Relationship between Temperature and the Greenhouse Effect

Before we go any further in our discussion, we need to try to ascertain whether global warming is a real phenomenon. The mere fact that certain scientists make claims or devise theories does not prove that these claims and theories are correct. Recall from chapter 3 that many scientists believed the phlogiston theory; it wasn't until Lavoisier gathered evidence for the oxidation theory that the older theory began to fade. Similarly, in the late 1960s and early 1970s, some scientists alleged a new form of water they called polywater or anomalous water, but careful studies showed this material did not exist; scientists continue to disagree on whether cold fusion is for real. Is global warming an actual occurrence or an invalid idea? Without studying temperature and greenhouse gas data, it would be impossible to tell.

Global Temperature Data

Climatology is a challenging science because the factors that influence weather patterns, especially from a long-range global view, are not fully understood. Everyone is familiar with the difficulties that face local weather forecasters. Assessing global warming trends expands these difficulties many steps beyond the present knowledge of weather and climate. Nonetheless, extrapolation beyond the known facts is a common practice in science. Scientists study whatever data exists and make reasonable predictions based on these data.

Long-term predictions of drastic climatic change go back two decades or more. In the April 28, 1975 issue of *Newsweek*, on page 64, an article states: "There are ominous signs that Earth's weather patterns have begun to change dramatically and that these changes portend a drastic decline in food production—with serious political implications for just about every nation on Earth." And later in the same article: "The evidence in support of these predictions has now begun to accumulate so massively that meteorologists are hard-pressed to keep up with it." The article then describes major changes in future weather patterns and expresses skepticism that government leaders will take the necessary precautions, such as stockpiling food, to avert disaster. The article concludes with the statement, "The longer the planners delay, the more difficult will they find it to cope with climatic change once the results become grim reality." Perhaps the most intriguing feature of this 1975 article is that the predicted regions of drought and flooding are almost exactly the same as those in today's projections. Would you be surprised to learn that this article was entitled "The Cooling World" and described a coming ice-age? What dramatic change has occurred in the past twenty or more years to change the prediction from a cooling to a warming world, with the same attendant problems? We can begin to try to answer these questions by exploring past data on temperature changes and trends.

EXPERIENCE THIS

In an Experience This in chapter 1, you began to record the daily high and low temperatures in your locality. Obtain a sheet of graph paper and label the horizontal side with the dates on which you collected the temperature readings, including any dates for which you have no data. Label the vertical scale from −25°F (bottom) to +105°F (top). (°C would be even better, if you want to convert all your °F temperatures.) Now place marks for the high (o) and low (x) temperatures at the correct locations on the vertical scale for each day's record. Connect the o's and x's to make lines. Save this graph to use in answering question 14 at the end of this chapter.

HINT You should expect two lines with many jagged ups and downs.

Past Temperature Data

There are at least six databases (sets of data) available for analysis of a trend toward or against global warming:

Data Set 1: Daily temperature readings. Everyone knows daily temperature values fluctuate. Your graph from the last Experience This exercise should show this fluctuation. However, short time-span data are normally considered useless for appraising long-term effects. In addition, most of the data are restricted to local areas and have limited application to the planet as a whole.

Data Set 2: In 1988, the Department of Agriculture released the average temperatures for the entire United States from 1885 to 1988. This database averaged the temperatures over the cities, rural areas, and forest environs and revealed an overall temperature increase of less than 0.1°C. These data were limited to the United States, but preliminary data from Europe also indicate an absence of major temperature increases. A large amount of data recently became available from the former U.S.S.R. These also show negligible temperature changes over similar time spans.

Data Set 3: Satellites have monitored the temperature of our planet for over a decade. These global measurements show less than a 0.1°C change; this change was a *decrease*. The problem with this database is that the time frame is too short, but measurements are continuing.

Data Set 4: Since 1856, all ocean-faring vessels have had to take global ocean surface temperature measurements. The oceans comprise about 70 percent of the Earth's surface, so this set (figure 19.5) may be the most reliable single database available. The measurements cover the entire planet and go back about 140 years.

Data Set 5: During 1993, additional long-term climatic data were deduced from two tree-ring studies. One study (*Science*, March 5, 1993) examined the annual temperatures in the Sierra Nevada Mountains over a 2,000-year period, while the other (*Science*, May 21, 1993) determined a 3,620-year temperature record through South America tree rings. Neither study reported evidence of significant temperature increases. Some scientists have raised questions about the reliability of this type of data. Larger tree rings could mean the temperature was higher that year, but they might also mean that some other factor, such as CO_2 level, promoted more rapid tree growth. (The tree-ring method is also used to determine CO_2 levels, but it is not obvious how to separate the two effects. If we use tree-ring size to estimate past CO_2 levels, we must assume temperatures and CO_2 levels are directly related, and this would be cyclic reasoning—depending on one unproven conclusion as the basis for another.

Data Set 6: Some scientists have studied the sediments on the ocean floor and estimated ancient temperatures from them. Potentially these records could go back more than 800,000 years; it would be difficult to find data covering a longer time frame.

Figure 19.5 is arranged so that the horizontal zero line is the overall average temperature (in degrees Celsius) during the entire time period. Values extend above and below this average; the temperature change varied only 0.7°C total. Figure 19.5 also shows that Earth's temperature seems to follow a recurring twenty-two-year cycle (presumably related to solar activity, which is known to follow an eleven-year cycle). In addition, the average temperature is farther below the zero line at about 1908 than at any other time. The highest temperatures seem to appear in the

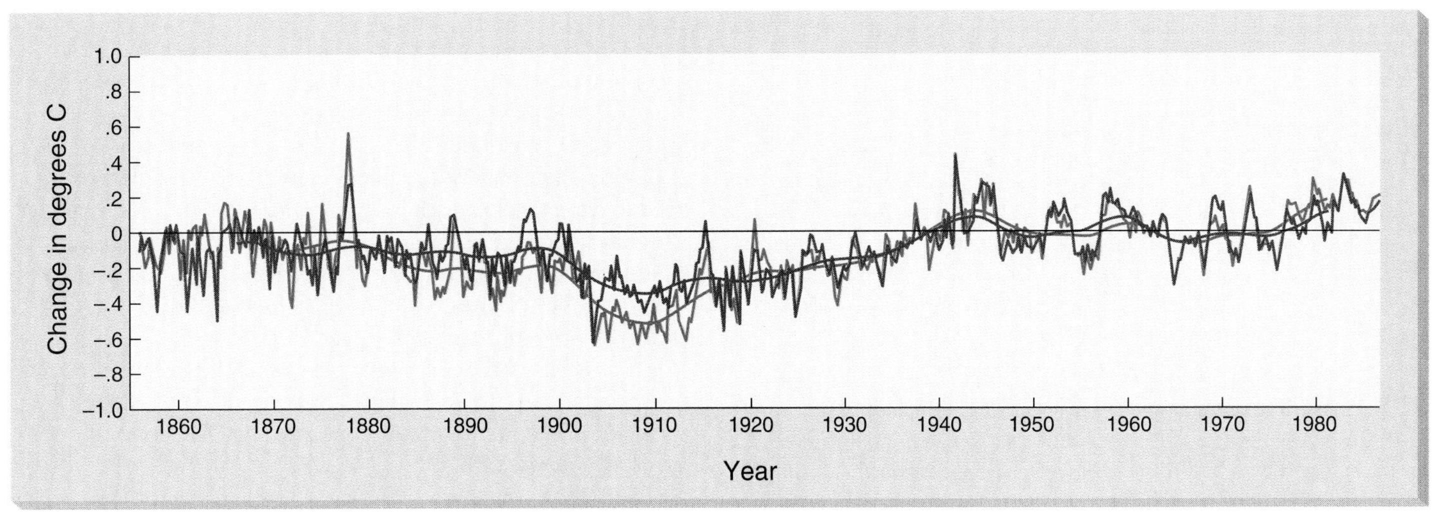

Figure 19.5 The average ocean surface temperatures from 1856 to 1988 (data set 4). The blue line is the sea surface temperature and the red line is the night air temperature.
SOURCE: *Technology Review*, November–December 1989, page 80.

1980s, although the late 1870s and the 1940s values are close. Do these data show definite evidence of global warming?

These data describe past temperatures; scientists make future predictions by extrapolation, usually with a computer, but not all their predictions are identical. Some computer simulations show no indication of global warming, while others predict a catastrophic increase. This happens because different computer programs utilize factors such as past temperatures, cloud effects, and ocean effects in completely different ways. Scientists are uncertain how to factor in the effects of clouds, but clouds definitely affect temperature. (Clouds maintain heat in an area at night, but cool that same area during the daytime.) Figure 19.6 shows how several computer projections might look, all starting from the data in figure 19.5. As you can see, the predictions vary widely.

Note also from figure 19.5 that there is a definite temperature dip around 1908. If this dip is ignored, the Earth's average temperature appears nearly constant. This suggests that perhaps the Earth was cooler than we would normally expect in the early part of this century. Why might this have happened?

In 1883, the volcano Krakatoa erupted catastrophically on an island in Indonesia. Two-thirds of the island was destroyed when Krakatoa erupted, and over 36,000 people died because of the resulting tidal waves that occurred as far away as Cape Horn. The eruption spread billions of tons of dust and ash around the globe, and they remained suspended in the atmosphere for several years. This increased concentration of particulate matter reflected much of the incoming solar radiation back out into space and cooled the planet. (Even in this post-Krakatoa period, the twenty-two year high-low temperature cycle is observable, but the pattern is suppressed.) After the dust settled, mainly thanks to rainfall, the planet's temperature slowly returned to its normal pattern. But it did not return to the average value until around 1940. The June 9, 1991, Mt. Pinatubo volcanic eruption also spewed ashes and dust around the globe, but because of Pinatubo's location and global wind patterns, the debris spread more rapidly into the northern latitudes and then settled out of the air more quickly as well.

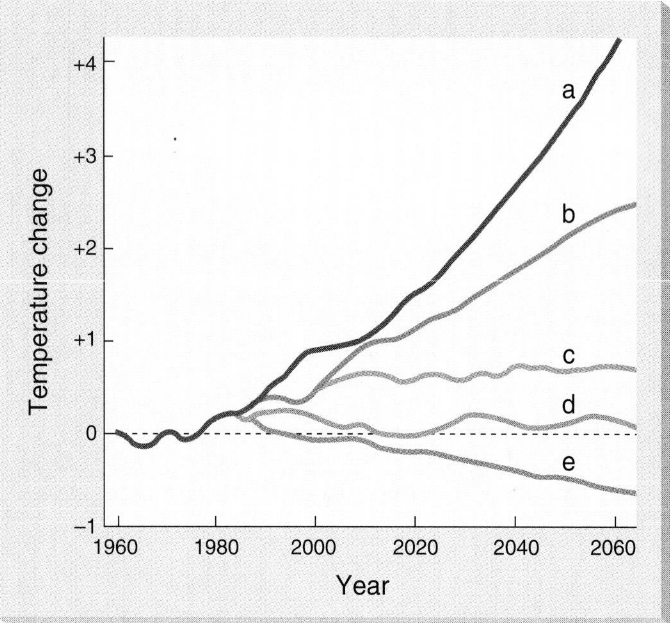

Figure 19.6 Computer projections of future temperatures on the Earth using the figure 19.5 temperatures as the starting point. Five scenarios are shown: the magnitude of each depends on the assumptions made. Scenarios **(a)**, **(b)**, and **(c)** show projected temperature increases or global warming. Scenario **(d)** shows very little change and seems in agreement with past trends. Scenario **(e)** predicts global cooling.

If we examined data from this century only, and if these data included the tail-end of the Krakatoa temperature dip, we would see a definite gradual increase of nearly 1°C in the Earth's temperature. Some computer simulations start their database at the turn of the century, including this data. But if we use the entire record shown in figure 19.5, essentially canceling out the dip, we would find only a slight temperature rise (possibly 0.2°C) over the past 140 years. Which view is correct?

These types of complications introduce an enormous amount of uncertainty into the question of global warming. Most scientists are reluctant to agree with James Hansen of NASA, who claimed he was "99 percent confident" of global warming when he addressed a congressional subcommittee on June 23, 1988—during an unusually hot summer. Stephen Schneider, another global warming theorist, stated (in *Discover,* October 1989, page 47) that ". . . scientists should consider stretching the truth to get some broad-based support, to capture the public's imagination. That, of course, entails getting loads of media coverage. So we have to offer up scary scenarios, make simplified dramatic statements, and make little mention of any doubts we might have . . . Each of us has to decide what the right balance is between being effective and being honest."

You might wonder why a credible scientist would make such a statement. The answer is simple: money. Grants are funded according to a combination of scientific merit and national need. Grants, scientific papers, and patents allow a scientist to advance in his or her career, especially a scientist who works in academia or in a research institution. Schneider was arguing that scientists need to depict global warming as a major national need that merits funding. The researchers working on termite flatulence were aiming for the same goal. This type of behavior is not peculiar to scientists; with a little effort, you can spot other examples in almost all areas of human endeavor.

As a result of these types of statements, many press releases treat global warming as a fact, even though not all scientists agree. A 1992 Gallup poll of the American Geophysical Union and the American Meteorological Society showed that 90 percent of their member scientists considered climate change research an "emerging science" and 70 percent considered the underlying research for global warming as "fair to poor." Only 41 percent believed there was "scientific evidence" for global warming, although 66 percent believed that some human-impelled climatic change (not necessarily global warming) may one day confront us. It seems significant that much of the media emphasis on global warming came during hot spells, as in the summer of 1988; we heard little about global warming during the winters of 1992–93 and 1993–94, when many areas of the United States recorded all-time record lows (and snows). These decreases were generally attributed to a "Pinatubo effect"—the idea that a layer of volcanic dust spread around the globe and reflected sunlight, back out to space. The global warming claims were "proven" mainly by temperatures from data set 1—daily temperature values, the least useful for making such long-term predictions.

This does not mean that global warming is impossible. Perhaps humans could cause global warming by flooding the atmosphere with greenhouse gases such as CO_2. Are we doing this? We will examine that possibility briefly in the next section.

The Global Concentration of CO_2

Our main considerations are restricted to CO_2 because global warming proponents claim that this greenhouse gas is responsible for at least 50 to 60 percent of the temperature increases related to global warming. (No one claims that water, the chief greenhouse gas, will cause catastrophic warming.) Scientists have measured atmospheric CO_2 levels since 1958 at Mauna Loa Observatory in Hawaii, on the rim of a volcano (figure 19.7). During a thirty-year period, the CO_2 levels rose from 315 ppm (1958) to 355 ppm (1988), a 12.7 percent increase. Is this change significant enough to produce global warming? Caution is essential in answering this question. First, because the measurements are from the rim of a volcano, they may not reflect global CO_2 increases; the volcano may emit more CO_2 than would be expected elsewhere. Second, assume the maximum 0.7°C temperature change shown in figure 19.5 was due solely to a 12.7 percent CO_2 increase. If this were the case, CO_2 would not have the dramatic effects some attribute to it. If the average temperature on Earth is about 15.5°C (or 288.5K), a 0.7° increase is only a 0.24 percent change ($0.7 \div 288.5 = 0.0024 \times 100 = 0.24$). This means only huge increases in CO_2 concentrations would effect significant warming.

Contrary to what the more flamboyant press releases indicate, there is no agreement on the global warming question. Until better computer models emerge, in which clouds, oceans, forests, and other factors are considered correctly, scientists cannot make relatively accurate predictions, much less make predictions with "99 percent confidence." More data are needed, probably from the data set 3 satellites, the most accurate global sources, before we can test models scientifically.

Still, an increase of nearly 13 percent in CO_2 concentrations seems too high to dismiss blithely. Is this a real increase, or an artifact of the measurement or of the specific time interval? Could this increase be cyclic? We need a longer time scale to be sure. Unfortunately, long-term data do not exist because the predecessors of meteorologists did not make these measurements. Only time will tell whether CO_2 is increasing in the Earth's atmosphere and affecting temperatures.

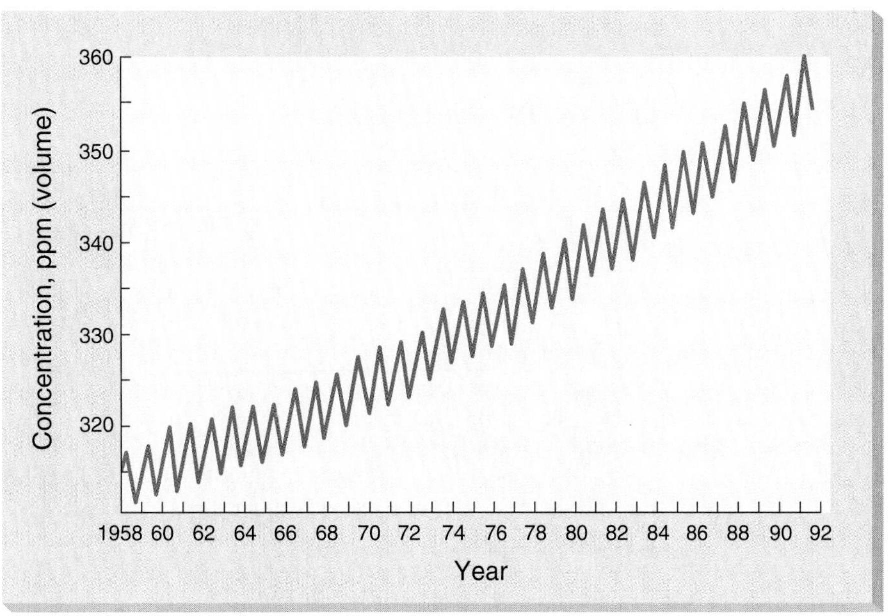

Figure 19.7 The atmospheric CO_2 levels measured between 1958 and 1992 at the Mauna Loa Observatory, located on the rim of a volcano in Hawaii.
SOURCE: C. D. Keeling et al., Scripps Institution of Oceanography, and National Oceanic & Atmosphere Administration.

Past CO_2 Concentrations

Considering the history of chemistry, it is not surprising that we have such a short record of atmospheric CO_2 measurements. Although the existence of CO_2 was known in the middle ages, no one understood its composition at that time. It was impossible to determine CO_2 concentration in the atmosphere until scientists developed accurate analytical procedures, beginning in the 1700s, and routine atmospheric analysis was not developed until well into the twentieth century. In short, chemists do not have much long-term data showing CO_2 levels during the course of human history, much less over the entire history of Earth. Unfortunately, this information is crucial if scientists are to determine whether a relationship exists between CO_2 and average global temperature.

Scientists are able to glean some information on CO_2 levels through indirect evidence from tree rings and ice-core air bubbles. The **ice-core analysis** is considered the more reliable (box 19.2).

The CO_2 levels obtained from ice cores (where air bubbles are trapped) are summarized in figure 19.8. From the interglacial period (about 130,000 years ago) until about 5000 B.C., the ice core CO_2 levels decreased from 300 ppm to around 230 ppm. Then, about 5,000 years ago, they rose to 280 ppm. The CO_2 level remained around 280 ppm from that time until the mid-1700s, then began to rise after the start of the Industrial Revolution, about 1770. Since then, the level shows a continuous rise to the current value of about 350 ppm at the Mauna Loa volcano. The shape of this curve, as figure 19.8 shows, suggests that the CO_2 level is continuing to rise. This raises two questions. First, how much of this CO_2 increase is due to the fact that current measurements are taken at a volcano, which is known to emit various gases, including CO_2? Perhaps the CO_2 levels should be measured in an area remote from volcanoes to get true values for atmospheric CO_2. Second, if the increase is real, could it eventually lead to global warming—even if global warming has not yet begun? If the answer is yes, then controlling CO_2 emissions becomes important.

Drawing Conclusions and Taking Action

Once we understand how greenhouse gases and the greenhouse effect may cause global warming, and we study past and present data on temperature and CO_2 trends, we come to perhaps the hardest task: deciding whether global warming is for real and how to respond if it is.

Should We Restrict Carbon Dioxide Emission?

Some people suggest that humans must immediately begin reducing CO_2 emissions or our planet will suffer dreadful consequences. Before we make any decision concerning CO_2, it is important to determine where CO_2 comes from. Table 19.2 gives a detailed summary of the sources of CO_2. Part of this information is also shown in figure 19.9.

BOX 19.2

ICE CORES AND CARBON DIOXIDE

When scientists analyze ice cores to determine ancient levels of CO_2, they assume that the deeper a sample lies inside a column of ice, the older it is. Ice always contains entrapped air, presumably entrapped when the ice froze. While this seems reasonable, several problems can distort the analysis. First, ice columns can become inverted because ice is more mobile than rock. This mandates caution in setting the dates for any given ice core. The best way to prevent errors is to use multiple core drillings to reduce or eliminate the possibility of basing dates on such inversions.

A second problem is that scientists must assume the amount of ice formation is moderately constant over a long time period in order to set the age of the sample. In the short-term, this assumption is invalid; winters vary in intensity in different regions and clearly vary from one year to another. The existence of past ice ages also argues against a long-term constancy of ice deposition. Nonetheless, ice-core analysis is used to estimate past CO_2 levels because it provides the best data available. After the core is removed and cut into sections and the ice is melted, the amount of CO_2 is determined at different depths in the ice core, which correspond to different time intervals.

Other factors pose additional uncertainties. Gases diffuse out of solid ice, and not all components of the entrapped air diffuse at the same rate. For example, CO_2 diffuses more rapidly through some plastics than either O_2 or N_2, a valuable property in the heart-lung machines that remove CO_2 from the blood. If diffusion occurs in ice cores, then the values researchers obtain will not necessarily represent the actual past concentrations. Ice-core data, however, remain the best available data for long-term CO_2 values. (See *American Scientist*, July/August 1990, pages 310–326 for a full discussion of ice-core data.)

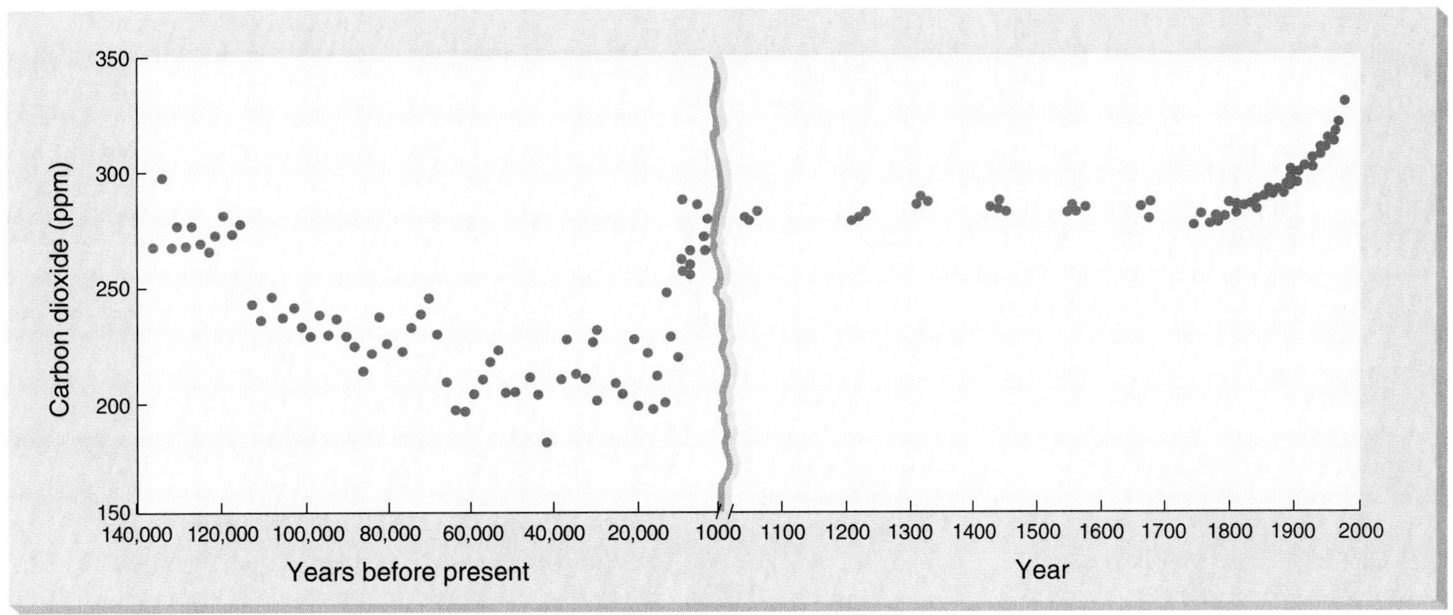

Figure 19.8 The atmospheric CO_2 levels over a 140,000-year time period. The data since 1956 are from the Mauna Loa Observatory (see figure 19.7); the older data are from air bubbles entrapped in ice columns. Note the change in the time scale at the year 1000; also note that the rise in CO_2 levels starting in the 1800s parallel the Industrial Revolution and the burning of fossil fuels.

SOURCE: From W. M. Port, et al., in *American Scientist*, July–August 1990, 78:310–326. Copyright © 1990 Sigma Xi-Scientific Research Society. Reprinted by permission.

TABLE 19.2 Sources of CO_2 in the atmosphere today.

SOURCE	TONS × 10^6	PERCENT OF TOTAL
NATURAL SOURCES		
Release from the oceans	105,000	59.66
Release from the land	64,000	36.36
Natural Sources Subtotal	169,000	96.02
HUMAN-DERIVED		
Electric power plants	1,670	0.95
Transportation	1,590	0.90
Industrial/commercial	1,375	0.78
Residential	365	0.21
Fossil Fuel Use Subtotal	5,000	2.84
Land conversion (deforestation, etc.)	2,000	1.14
Human-Derived Sources Subtotal	7,000	3.98
Grand Total	176,000	100.00

SOURCE: Data from Renew America Project and *Chemtech*, August 1989.

Figure 19.9 shows that 96.1 percent of all atmospheric CO_2 comes from natural sources and only 2.8 percent comes from fossil fuel use. The 1.1 percent from deforestation is not entirely human-derived; part of this comes from natural, lightning-induced forest fires. Table 19.2 gives more details concerning the specific sources of the 2.8 percent fossil fuel-generated CO_2. This CO_2 comes about evenly from electric power plants (0.95 percent), transportation (0.90 percent) and all other sources (0.99 percent), including industrial or commercial uses (0.78 percent) and residential uses (0.21 percent). We can do so little to reduce

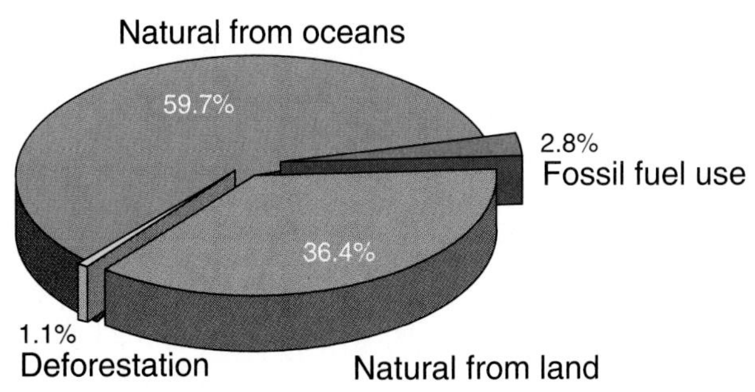

Figure 19.9 The sources of CO_2 in the atmosphere. (Data from table 19.2.)

CO_2 from natural sources; only the human-generated CO_2, about 7 billion tons per year, is reducible. Seven billion tons, or 14 trillion pounds, may sound like a lot of CO_2, but it is dwarfed by the amount that comes from natural sources.

Environmental activists advocate reducing human-generated CO_2 to avert global warming. Even assuming the entire 12.7 percent increase in the CO_2 level is from human-generated sources, which seems reasonable, this increase has not produced any appreciable global warming in the 220 years since the Industrial Revolution began (figure 19.5). Moreover, if the human-generated sources listed in table 19.2 are to be reduced, we need to know just where these reductions should occur. Should we cut back in heating homes? Curtail factory jobs? Cut back on electricity generation? Restrict transportation? While few people want to undergo the dramatic scenarios global warming proponents paint, even fewer favor major reductions in the activities that produce human-generated CO_2. Most might argue to pare back usage by conserving, but few want to return to the hardships of pre-Industrial Revolution life.

Still, we can reduce CO_2 production by taking simple steps to conserve energy and can even save money in the process. Mandated reductions of industrial CO_2 (less than 1 percent of the annual total output) are enormously expensive (estimated at \$450 billion/year). On the other hand, better, more efficient industrial processes that produce lower CO_2 emissions could ultimately save rather than cost money. If we spend the available money on expensive equipment aimed solely at CO_2 reduction, we may not have enough funds left to develop these more efficient processes. Continued research, then, may benefit society by improving efficiency, reducing CO_2 emissions, and saving money; this seems a wiser course than focusing strictly on reducing CO_2 output.

Another wise step might be to consider CO_2 from a global perspective. To control CO_2 emissions, it is necessary to control them worldwide. While developing nations might ignore or reject emission mandates, they might welcome economic aid that helps them install the latest, most efficient technology—reducing CO_2 emissions in the process. Both of these steps—developing new technologies and making them available worldwide—could also help reduce other, more troublesome greenhouse gases, such as N_2O. If this also averts global warming, we gain an added bonus.

The Impact of Global Warming

Not all meteorologists believe global warming is imminent. Some also question what effects global warming would have if it does occur. For example, some meteorologists believe global warming would not produce flooding along the coastal areas, as many global warming scenarios predict. Whether an increase in the average temperature on Earth would actually be catastrophic is debatable. See also Box 19.3 for natural controls over global warming.

Remember, too, that meteorology is a relatively young science, and long-term climatology is one of its younger offspring. Scientists are still trying to learn how to predict reliable short-term weather patterns; for long-term predictions, their accuracy is low, and computer simulations are only as accurate as the assumptions they are based on.

Furthermore, inferring future trends in the absence of hard data from the past is extremely difficult. Often, the data are unrevealing. For example, scientists have tried to deduce a long-term pattern for temperatures on Earth by analyzing ocean floor sediments. These data for an 85,000-year time period show enormous temperature fluctuations (figure 19.10). How can we apply this to future trends? The truth is, our best attempts to determine whether global warming is occurring and what its impact would be are woefully uncertain. The most prudent route may be to take reasonable precautions to control CO_2 emissions as we continue to gather data and apply the scientific method to test the global warming hypothesis.

The Scientific Method and Global Warming

While the scientific method works, it takes time. Global warming is one example of an issue in which it is important to apply the scientific method before forming faulty conclusions or taking rash actions. Some people attempt to make decisions according to public pressure rather than by a careful consideration of factual information. Even worse, the public is sometimes fed false, incomplete, or misleading information to promote the beliefs of certain groups. While this might be expected in a democratic society that protects free speech, it makes it difficult to make decisions based purely on scientific principles.

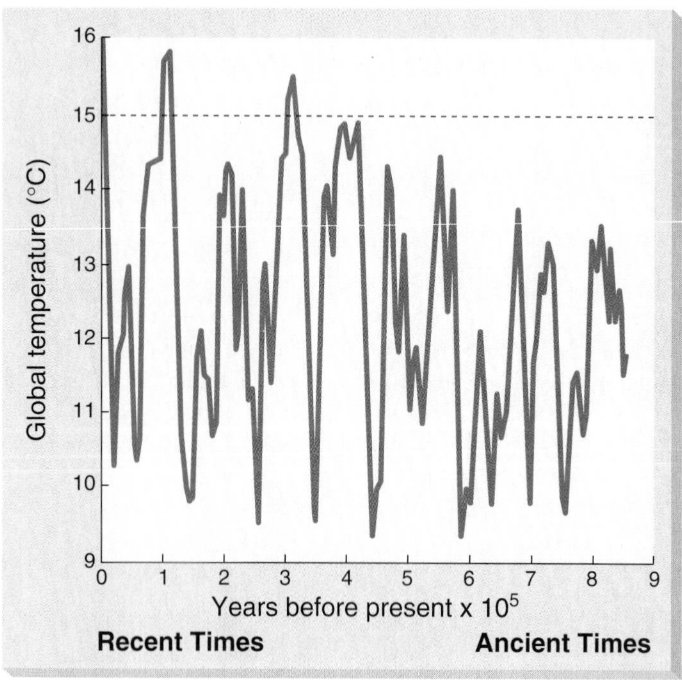

Figure 19.10 The average global temperature of the Earth for the past 850,000 years as reconstructed from ocean floor sediments. These fluctuations are called Milankovitch cycles.

SOURCE: Copyright 1992 in *The Heated Debate: Greenhouse Predictions Versus Climate Reality* by Robert C. Balling Jr. (page 6). Reprinted by permission of Pacific Research Institute for Public Policy, San Francisco, CA.

BOX 19.3

NATURAL CONTROLS OVER GLOBAL WARMING

Some meteorologists point out that many natural controls could prevent extensive global warming. Increased temperatures would increase the water evaporation rate and promote more cloud formation and rain. Thus, inland areas might receive more rain and some arid regions might become fertile, offsetting potential negative effects on food production. Increased cloud formation would also reduce the amount of sunlight that hits the surface of the Earth. Everyone is familiar with the cooling effects of a daytime cloud cover. At night, clouds work in the opposite way, containing heat against the land surfaces. In other words, clouds can either reduce or increase temperature. How clouds would affect global warming is one problem yet to be solved in the computer simulations.

Another major climate-controlling force is volcanic activity. A volcanic eruption spews enormous amounts of particulate matter into the atmosphere, and these particulates reflect incoming sunlight back out into space. The Mt. Pinatubo eruption caused many parts of the world to experience the coldest weather in recent history during the winters of 1992 through 1994. The Krakatoa eruption caused an even more drastic temperature lowering in the late 1800s and early 1900s. No doubt this volcano effect has occurred throughout our planet's history.

During the Industrial Revolution, massive amounts of human-made particulate matter were released into urban atmospheres. We don't know whether this was sufficient to lower temperatures, but this is certainly a possibility. If this particulate matter did produce a cooling effect, then our efforts at cleaning up the air may have caused temperatures to rise. This does not mean we should return to the days of smoky, sooty cities; it just shows how factors sometimes interrelate in unexpected ways.

Another factor that could affect global warming is the release of **sulfuric acid aerosols,** tiny droplets of dilute H_2SO_4 that forms when atmospheric SO_2 reacts with moisture. These aerosols form a haze that reflects part of the incoming solar radiation back out to space. Some scientists believe this factor could avert global warming. Since most atmospheric SO_2 comes from natural sources, this is ultimately another natural control Earth uses to regulate its temperature.

Finally, oceans and rocks are massive reservoirs for CO_2. Some scientists believe these reservoirs are nearly exceeding their limits, making them unable to store CO_2 as quickly as it is produced. If this is true, perhaps the present apparent CO_2 increase is simply a lag in storage time that started during the Industrial Revolution and will eventually correct itself as more rock forms. These and many other self-regulating systems in nature may prevent global warming from occurring.

The question of global warming is unlikely to be resolved quickly. *Science*, the official organ of the American Association for the Advancement of Science, ran a "news" report in its August 3, 1990 issue entitled "New Greenhouse Report Puts Down Dissenters." The article's subtitle was "An International Panel Assessing Greenhouse Warming Pointedly Denies the Validity of Objections Raised by a Prominent Minority." The following excerpts point out the controversy and confusion that surrounds global warming:

> . . . The authors of the report decided that one oft-cited piece of contrary evidence—the absence of a warming trend over the contiguous 48 states during nearly a century—was so trivial that they ignored it. "The U.S.A. is not an average place," notes Folland, "just as the U.K. isn't."

This comment blithely dismisses the fact that the United States constitutes about 10 percent of the total land area of the earth. These scientists decided to "ignore" U.S. data because it is not "average," whatever that may mean. More recent data from both Europe and the former U.S.S.R. disclose the same results for an even larger land mass. How would the authors of the greenhouse report view this data?

Consider also their view of temperature measurements gained via satellite:

> The authors did give space to a discussion of satellite temperature measurements that failed to show a warming trend. But they gave the observations a decidedly different twist. The authors . . . claim that the satellite results in fact buttress their conclusions that temperature measurements at the earth's surface over the past century reveal a warming . . . "I was surprised how closely the [satellite data and surface measurements] matched. It gave me more confidence . . ." stated Thomas Karl (National Climatic Data Center).

It is difficult to understand how this scientist gained confidence in the idea of global warming from "proof" gained from data that fails to show any warming trend.

Eventually, the scientific community will resolve the issue. They have resolved hundreds of previous controversies using the scientific method, but it takes time. Political and economic issues cloud the question more than a little; perhaps what we need most is a well-informed public that clamors for conclusions based purely on scientific evidence.

QUESTION TWO: IS THE OZONE LAYER DEPLETING?

Sydney Chapman (British, 1888–1970) first hypothesized the ozone layer in 1930. Figure 18.2 showed that the ozone layer is located about 10 to 30 km (6 to 18 miles) above the surface of the planet. It contains a modest and variable amount of ozone (O_3), an allotrope of oxygen. The ozone layer is not a discrete band; it is an ill-defined, variable region in the lower stratosphere.

The thickness of the ozone layer and its exact O_3 concentration vary greatly with location, season, and time. It is not unusual to see a 40 to 50 percent variation in ozone levels over a few days in the northern latitudes; the levels vary less in the equatorial regions. Any area of ozone reduction is usually called an ozone hole. The ozone concentrations are greater at the middle and upper latitudes than at the equator, due mainly to wind current circulation. Figure 19.11 shows the O_3 levels for three regions from 1989 to early 1993. (The United States is located within the 30°N to 50°N region, where O_3 levels are the highest.) Although seasonal changes do not occur at the same time in each region, decreases normally occur in the spring. In addition, the ozone levels are generally lower in the Southern Hemisphere.

The weather, the jet stream, and solar activity all affect ozone levels. Scientists have calculated that if all the ozone in this layer were compressed together at normal atmospheric pressure and placed on the surface of the Earth, it would form a layer only 3 mm (1/8 inch) thick. To summarize, then, the three main points to remember about the ozone layer are: (1) the thickness varies greatly from one season to another, and even daily; (2) it contains only a small concentration of ozone; and (3) the ozone concentration varies with latitude and is the lowest at the equator.

In chapter 18 you learned that ozone is a component in photochemical (Los Angeles) type air pollution and is dangerous at ground level. However, at 6 to 18 miles above the surface of the Earth, ozone plays an important and beneficial role. Ozone interacts with the ultraviolet (UV) portion of the incoming solar radiation (about 5 percent of the total radiation). It markedly decreases the UV level that reaches the surface of our planet. It is estimated that for every 1 percent decrease in the ozone (O_3) layer, the UV light reaching the surface increases by 2 percent.

The following equations describe the UV quenching process:

$$O_2 + \text{UV light} \rightarrow 2\ O\cdot$$

$$O_2 + O\cdot \rightarrow O_3$$

$$O_3 + \text{UV light} \rightarrow O_2 + O\cdot$$

The process starts in the first equation, when a pair of O atoms form by breaking the covalent bond in an oxygen (O_2) molecule. The energy necessary for this rupture comes from UV radiation, which has a higher energy content than visible light (see chapter 3). These unbonded O atoms are highly reactive, although at a low concentration. In the next equation, one of the unbonded oxygen atoms reacts with an O_2 molecule to form the ozone (O_3) molecule. Then, as the third equation shows, more UV radiation breaks the O_3 molecule apart, forming an O atom and an O_2 molecule. Notice that the first and second equations form O_3, while the third shows O_3 destruction; UV light is absorbed in each.

This combination of O_3-forming and destroying reactions can continue thousands of times in a chain reaction before two unbonded oxygen atoms happen to encounter each other and form an O_2 molecule. The net result is the production of several

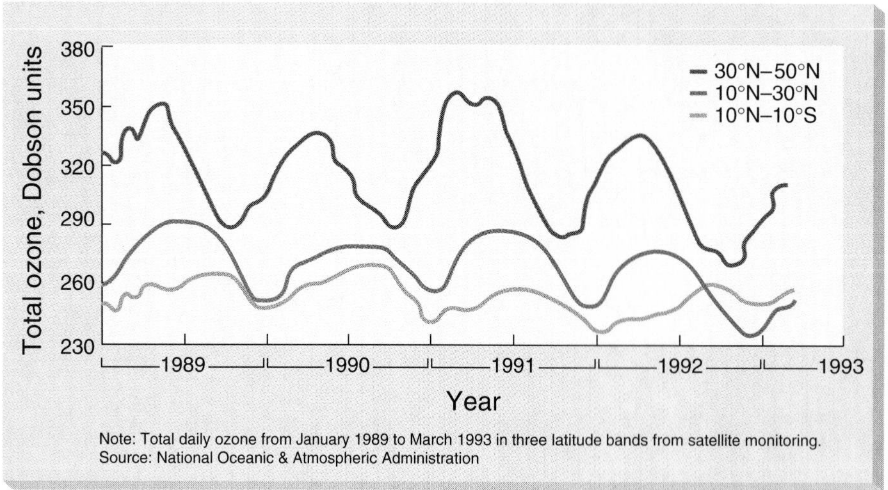

Figure 19.11 The stratospheric ozone concentrations for three different latitudes from 1989 to early 1993. The United States lies between 30°N and 50°N.
SOURCE: *Chemical & Engineering News*, May 24, 1993. Copyright 1993 American Chemical Society, Washington, DC. Reprinted by permission.

tons of ozone across the globe every second. This means that although some molecules of ozone decompose, other ozone molecules form each day.

This ozone-centered chain reaction removes thousands of UV photons from the incoming solar radiation; as a result, the solar radiation striking the Earth's surface contains fewer UV photons than the radiation entering the upper stratosphere. Some doctors estimate (but cannot prove or state with certainty) that a 15 percent increase in skin cancer deaths could occur if the ozone layer disappeared to any appreciable (usually unspecified) extent. Some evidence suggests that chlorofluorocarbons, or CFCs, destroy this ozone layer, allowing more UV radiation to reach the Earth.

PRACTICE PROBLEM 19.4

Before proceeding further, be certain you understand the nature of the ozone layer. What is it? How thick is it? How much ozone forms each day? Does the ozone layer ever vary in thickness or concentration?

ANSWER The ozone layer is an ill-defined, highly variable region of the stratosphere. It contains samll amounts of O_3, which varies from season to season by 40 to 50 percent in concentration. Several tons of O_3 are formed daily.

Ozone Layer Depletion

Many scientists now assume that chlorofluorocarbons (CFCs) have a deleterious effect on the ozone layer. CFCs were developed for several important uses, but they seem to interfere with the ozone layer via the following reactions:

$$CF_2Cl_2 + \text{UV light} \rightarrow \cdot CF_2Cl + \cdot Cl$$

$$\cdot Cl + O_3 \rightarrow \cdot ClO + O_2$$

$$\cdot ClO + O\cdot \rightarrow \cdot Cl + O_2$$

This reaction sequence was first reported in 1974 by F. Sherwood Rowland and Mario Molina of the University of California at Irvine. It is one of several processes that deplete ozone. In the first equation, a CFC (CF_2Cl_2 in this example) interacts with incoming solar UV radiation. This breaks a Cl—C covalent bond to produce a pair of free radicals, each of which has an unpaired electron and is extremely reactive. At first glance, this seems fine because it rids the atmosphere of more UV light.

However, complications arise because the newly formed •Cl atom (which has an unpaired electron and is a free radical) reacts directly with ozone to form an O_2 molecule and a second highly reactive free radical, •ClO. This •ClO radical then reacts directly with an O atom (also a free radical) to form an O_2 molecule and re-form the •Cl radical. The net result is the formation of two O_2 molecules—one from the O_3 (see the second equation), and the other from an oxygen atom (see the third equation)—and the restoration of the •Cl free radical. This free radical chain process can repeat itself many thousands of times, destroying many thousands of ozone molecules. The ultimate result is that more UV photons pass through the upper atmosphere to Earth.

CFCs do not readily biodegrade in the atmosphere and might linger for as long as fifty to a hundred years. Most CFCs do biodegrade in contact with soil microorganisms. Since CFCs are more dense than air and clearly settle toward the ground when poured from a container, you might expect that they would fall to settle into the soil and then biodegrade. However, gases mix with each other well, and most CFCs are gases at room temperature. This means that they can mix with other gases in the air and remain there. How many CFC molecules remain in the atmosphere? Trace amounts of CFCs were found in the ozone layer in the past few years, but rather significant quantities of CFCs were also detected in the soil, showing that not all of these materials go into the atmosphere, much less the stratosphere. All things considered, CFCs seem unlikely to build up in significant amounts in the stratosphere, but chain reactions only require small quantities.

Furthermore, laboratory experiments have shown the CFCs can interact with UV light to start the chain reaction shown earlier. Although the concentrations used in the lab were much higher than those found in the stratosphere, this seems to be impressive evidence supporting the role of CFCs in ozone depletion. (Molina and Rowland shared the 1995 Nobel Prize in chemistry, with Paul Crutzen of the Max Planck Institute, for their work. Crutzen studied the effects of NO_x on the ozone layer, which is discussed in the following paragraphs.)

Other chemical agents also participate in ozone-depleting reactions, including nitric oxide (NO). The following equations demonstrate these reactions:

$$NO + O_3 \rightarrow NO_2 + O_2$$

$$NO_2 + O\cdot \rightarrow NO + O_2$$

These reactions are especially problematic because NO arises naturally from lightning discharges. But in the same vein, we can reasonably assume that this reaction sequence has always taken place and that the long-term concentration of ozone in the upper atmosphere is the result of an equilibrium between O_3 and NO. Some NO is also formed by supersonic transport planes (SST) that fly in the upper atmosphere, and U.S. development of these planes was blocked in the 1960s on the basis that SSTs might harm the ozone layer. As history records, France went ahead and developed their version, the Concorde, and later studies revealed little, if any, ozone depletion from SST emissions.

Reactions also take place that stop ozone-depleting chains. For example, the •ClO can react with NO_2 (nitrogen dioxide) to form chlorine nitrate, which is relatively inert. When this happens, the chain reactions between •Cl and O_3 and between •ClO and O• stop, and no further ozone depletion occurs until new Cl• radicals form:

$$\cdot ClO + NO_2 \rightarrow ClONO_2$$

Hydrochloric acid, HCl, also can form the •Cl radical under certain conditions and start an ozone-depleting chain:

$$HCl + ClONO_2 \rightarrow Cl_2 + HNO_3$$

$$H_2O + ClONO_2 \rightarrow HOCl + HNO_3$$

$$Cl_2 + UV \rightarrow 2\ Cl\cdot$$

$$HOCl + UV \rightarrow HO\cdot + Cl\cdot$$

$$NaCl + H_2SO_4 \rightarrow Na_2SO_4 + HCl$$

It is possible for HCl to form indirectly by the reaction of naturally occurring chlorides (for example, NaCl) with the sulfuric acid emitted from volcanic or human activity (see the fifth equation). In addition, volcanoes directly emit large quantities of HCl during eruptions; shortly after the eruption of Mt. Pinatubo in June 1991, the ozone levels decreased sharply. (Oddly enough, when they did, some environmentalists redoubled their efforts to reduce CFC use, even though CFCs did not increase during that time frame.)

Some reports in recent popular literature claim that the largest source of chlorine in the atmosphere comes from NaCl evaporating out of the ocean. (It is never completely clear whether these reports actually mean chloride or chlorine, since the authors sometimes use these terms interchangeably.) This is impossible. NaCl is an ionic solid. When water evaporates, it leaves the salt behind, not the reverse. Salt does not evaporate from seawater. On the other hand, some dry NaCl, in the form of a fine powder, does enter the atmosphere by salt spray evaporation and could be carried by the wind for several miles. If some of this NaCl interacts with **sulfuric acid (H_2SO_4) aerosols,** HCl will form. Some scientists claim that almost all of this NaCl or HCl returns to the ocean or ground very quickly, by rainfall; however, there is no proof that rainfall results in a 100 percent return of the HCl. If this HCl moves into the ozone layer, it can serve as another source of chlorine free radicals (•Cl) and start ozone-depleting chains; these are *natural* sequences.

The current total atmospheric chlorine concentration is estimated to be about 3 ppb (not ppm), and is expected to increase to about 4.5 ppb by the year 2000. CFC concentrations are only 0.8 ppb (figure 19.12), but other halocarbons raise the total to about 1 ppb. The other sources of Cl are apparently natural, which means that CFCs contribute only about 27 percent of the total chlorine in the atmosphere. The other human-derived chlorine sources shown in figure 19.12 are under consideration for elimination, but they only bring the anthropogenic contribution to 33 percent of the total. Anthropogenic Cl-containing materials have been singled out because it is possible to reduce them; this is not true for either the natural Cl sources or the other ozone-depleting materials. If all nations cooperate by eliminating CFCs, or at least restricting their open-system usage, the 4.5 ppb level projected for the year 2000 is expected to drop below the present 3 ppb value by the year 2050. Even if this cooperative effort fails, the value will likely stabilize around 4.5 ppb.

The hydroxyl free radical (HO•) (produced in the fourth equation just shown) can also lead to ozone loss through the following reactions:

$$HO\cdot + O_3 \rightarrow HOO\cdot + O_2$$

$$HOO\cdot + O_3 \rightarrow HO\cdot + 2O_2$$

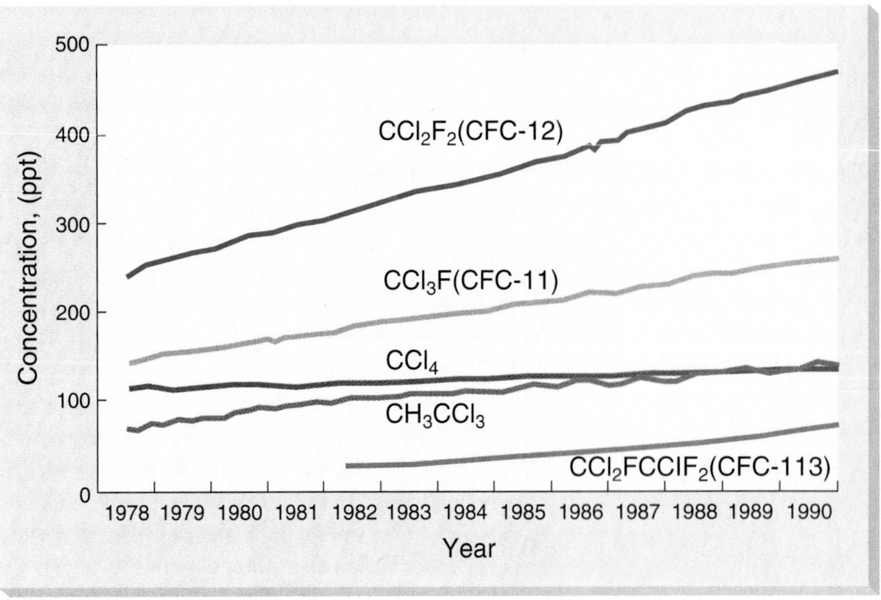

Figure 19.12 The atmospheric concentrations of CFCs and some other halocarbons. Note that the concentration is expressed in ppt (parts per trillion).

SOURCE: *Chemical & Engineering News*, April 17, 1992. Copyright 1992 American Chemical Society, Washington, DC. Reprinted by permission.

These two reactions also form a chain reaction, since HO• is regenerated in the second to supply the first. Until recently, scientists have considered the NO_x reactions, shown earlier in the chapter, as the major ozone-depletion pathways, but many now believe the HO• sequences may be more important.

In summary, several other materials besides CFCs are potential ozone-destroying agents, including HO• and NO_x. There are also other possible sources of Cl atoms that could form •Cl radicals. Thus, CFCs are not the only cause of ozone depletion. Many of the other causes are natural.

PRACTICE PROBLEM 19.5

Make a list of the major materials that can destroy ozone and keep it for future reference. Which substances are natural and which are human-made?

ANSWER The main culprits include the CFCs, the •Cl and HO• free radicals, NO_x, and substances such as HCl and NaCl that can form the •Cl atom. The only human-made chemicals are the halocarbons, including the CFCs.

CFC Usage

Figure 19.13 depicts chlorofluorocarbon (CFC) usage in different regions of the world, and box 19.4 outlines some of these uses. The main industrialized areas of the world—the United States, Canada, Western Europe, and Asia—use 85 percent of all CFCs. Third World countries, including Eastern Europe, use only 15 percent. Africa and Latin America lag far behind, with 4 percent total. This difference is possibly due to the lack of refrigeration and air conditioning in Third World countries. It is common to see food markets with meat spread out on open counters, occasionally using ice as a coolant.

Very few chemical compounds possess the right combination of physical and chemical properties, including nonflammability,

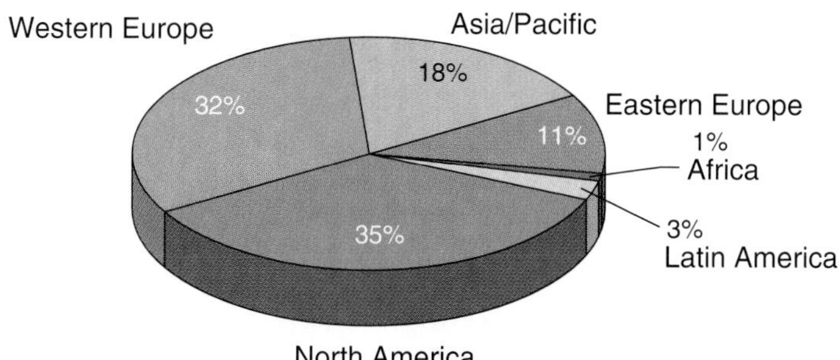

Figure 19.13 Regional CFC usage for 1986.
SOURCE: Data from R. Caplan, *Our Earth, Ourselves*. New York, Bantam Books, 1990.

to work effectively in cooling processes. The structures of the proposed CFC substitutes are shown in figure 1 of box 19.4. Note, however, that these structures are chemically similar to the CFCs; this means some are also likely to be banned. Moreover, these substitutes are less efficient coolants than CFCs, which means they will require more energy; most are also more costly, are flammable, and will not work properly unless refrigerators and air conditioning units are redesigned. If refrigeration became impossible, increased food spoilage could cause widespread illness, which some estimate could cause 20 to 40 million deaths per year. In spite of all this, U.S. production of CFCs was scheduled to cease by 1995, leaving no truly adequate replacement.

The prospect of increased refrigeration usage apparently accelerated the CFC ban. According to a 1989 statement by William Reilly, the former EPA chief, "The prospects of seeing countries moving forward with major development plans involving, as we heard in China, a proposal for 300 million new refrigerators possibly based on CFCs makes very clear that we must engage them in the process." The process he is talking about is banning CFCs.

Most CFC uses make important contributions to modern life. Air conditioning sounds like a luxury, but many elderly people require it for health and survival, especially in warm weather climates. (During one heat spell in July 1995, over 400 deaths were attributed to the heat in the city of Chicago alone.) The extensive industrialization in the southern United States was made possible through air conditioning, and as Third World countries become industrialized, they will want air conditioning and refrigeration. It is difficult for most Americans to realize that refrigerators, much less air conditioners, are rare commodities in the Third World.

Foamed plastics are widely used for insulation, providing significant reductions in energy and fuel consumption. Most foamed plastics can be made without using CFCs, however. Virtually all poly(styrene) foam is already made without CFCs.

The solvent uses of CFCs are difficult to replace because CFCs are excellent solvents that evaporate very cleanly. This is essential in the microchips and electronic components that enable computers and other devices to function. Without proper cleaning, these devices become unreliable.

Fire extinguishers require a nonflammable propellant, and Halons, CFC-like molecules, are used worldwide for this application. Halons are scheduled for phase-out by the year 2000 or earlier. None of the proposed replacements have the necessary nonflammable characteristics of the Halons. Unfortunately, we cannot ban fires by international accord.

The main source of CFC release is still from aerosols or solvent use (open systems), which make up 43 percent of the total global CFC use. This figure drops to only 15 percent in the United States. Most U.S. uses are contained in closed systems, (48 percent), and an additional 37 percent are in medical aerosols and fire extinguishers, which are closed much of the time. Yet oddly enough, most environmental activity against CFCs occurs in the United States, where the problem is the least worrisome. Possibly these groups are misinformed; some still avidly campaign against

BOX 19.4

USES FOR CFCS

Chlorofluorocarbons, or **CFCs,** are small organic chemical compounds that contain covalently bonded chlorine and fluorine atoms. Thomas Midgley, then at the Frigidaire division of General Motors, first developed these molecules in 1930 for use in refrigeration. Prior to this discovery, the chemicals used in refrigeration were ammonia, sulfur dioxide, and methyl chloride; these toxic chemicals caused many people to die during leakages. (In 1929, for example, over one hundred people died in a Cleveland, Ohio hospital from a refrigerant leak.) CFCs are nontoxic, nonflammable, and excellent refrigerants. Figure 1 shows the structures of the CFCs and some potential substitute materials.

CFCs have several major uses, some of which are open to the atmosphere and some of which are involved in a closed system:

1. *Refrigeration and air conditioning*. Both uses involve closed systems; CFCs normally do not enter the atmosphere except during rare equipment failures. In automobile air conditioners, vibrations result in more frequent failure rates.

2. *Aerosol propellants*. This CFC use was banned in 1978 in the United States by the EPA, although CFC-based aerosols are still used in other countries. The use of CFCs in medical devices, such as inhalants, was not banned. Recent agreements should restrict this aerosol usage entirely before the turn of the century. Aerosols are an open system, releasing CFCs into the atmosphere.

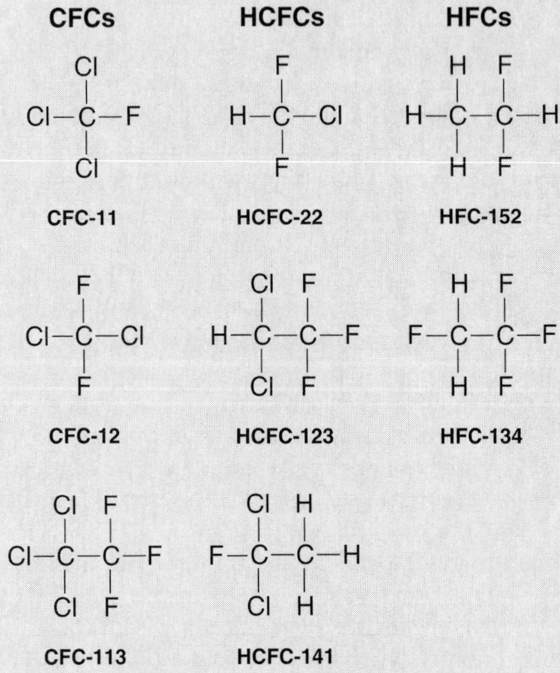

Figure 1 The structures of the CFCs and some potential substitutes.

3. *Solvents*. CFCs are used to clean microchips and electronic parts because they are excellent solvents that evaporate cleanly. They are also used in some dry cleaning operations. These are open systems that allow the CFCs to get into the atmosphere. It is often possible to redesign these processes and recycle the CFCs, making these solvents part of closed systems.

4. *Foamed plastics*. CFCs are used mostly in the preparation of some insulation plastics (but not foamed poly(styrene)). Plastics preparation is effectively a closed system that releases very few CFCs into the atmosphere.

5. *Fire extinguishers*. These are closed systems until they are used; they become open systems when used to fight fires. Fire extinguishers currently depend on the nonflammability of the **Halon molecules,** which are similar to the CFCs and are scheduled for banning.

6. *Medical uses*. Inhalants, medical sprays, and similar applications are mainly aerosol sprays that depend both upon the nontoxic nature of CFCs and their nonflammability. (A flammable inhalant could result in death.) CFCs are also used to clean surgical equipment. These medical applications are open systems during use, though they are generally closed systems before use.

Figure 2 shows the percentage of CFCs used in each application, both in the United States and throughout the world. The open systems are set apart as darker sectors. The total percentage of open systems is lower in the United States than in the world as a whole; the total percentage of open system uses in the United States is 15 percent, compared with 43 percent globally. (These numbers increase to 32 and 50 percent if fire extinguishers and medical uses are included.) As these figures indicate, the United States has led the world in CFC control. While the quickest way to reduce atmospheric CFC levels would be to change open systems into closed ones, this option unfortunately does not exist under the latest international agreements.

CFC refrigeration/air conditioning and foam uses are about the same proportions of the total in the United States and globally. Aerosols are used much more extensively in the rest of the world, however. The United States limits aerosol CFCs to medical applications. Finding a completely suitable replacement is not easy. Some proposed substitutes, such as the **fluorohydrocarbons (HFCs)** are flammable. The **fluorocarbons** are nonflammable, however, and may make good CFC replacements.

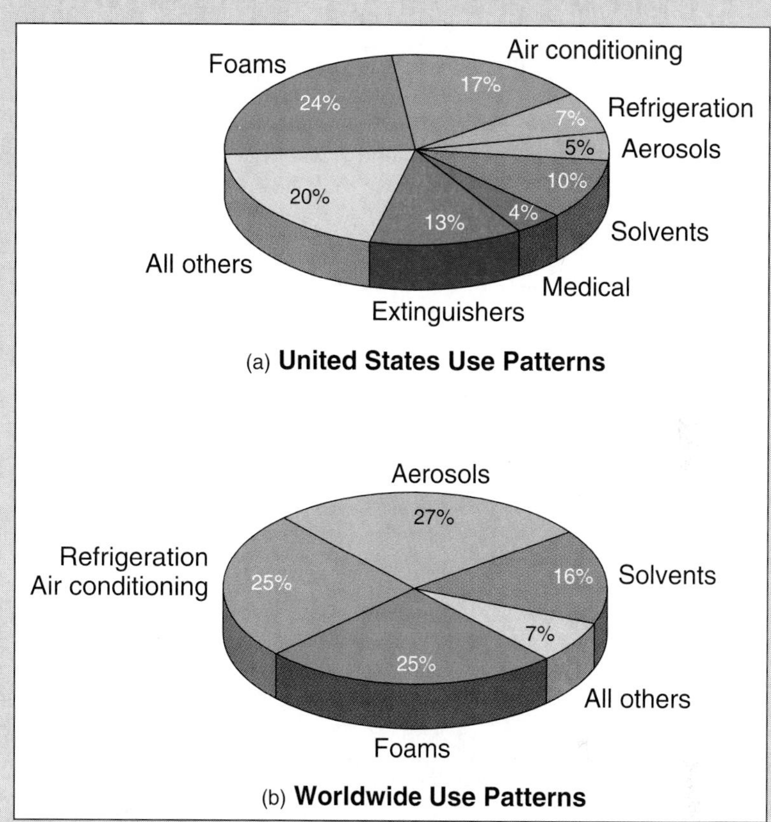

Figure 2 Percentage use patterns for CFCs **(a)** In the United States and **(b)** worldwide. Solvents and aerosols are open systems.
SOURCE: Data from Environmental Protection Agency.

deodorant aerosols, claiming that they destroy the ozone layer, in spite of the fact that the United States banned such systems in 1978.

There are several other types of propellants we can use in aerosols, including hydrocarbons, air, and nitrogen. Hydrocarbons are flammable, act as photochemical type pollutants, and often impart an unpleasant odor to the spray. Nitrogen aerosols are nonflammable and odorless; their use is expected to increase in the United States. Switching from CFCs to hydrocarbons or nitrogen is more difficult than you might expect because each specific gas requires a different type of can and nozzle. The new technology is now in use.

The Antarctic Ozone Hole

Antagonism against CFCs increased greatly after the discovery of the ozone hole in 1985. Analysis of the NASA Nimbus-7 weather satellite data revealed an extremely diminished amount of ozone (a "hole") in the atmosphere over the Antarctic. Subsequent reexamination of the Nimbus-7 records showed that this hole existed as early as 1979. (The Nimbus was launched in 1978, so the 1979 data are the oldest available.) As scientists monitored this Antarctic ozone hole, they observed that the size increased in the ensuing years, up to 1988, and then seemed to diminish. Sometimes this hole becomes smaller, sometimes larger, but the reasons are not clearly understood. Figure 19.14 shows how this ozone hole changed over a period of several years. Possibly the ozone hole is a natural phenomenon that follows a cyclic pattern, but more study is necessary before we can be sure.

The ozone hole forms during the Antarctic spring, after the sunless winter when the temperatures drop to −80°C. During the Antarctic winter, no new ozone is formed in the stratosphere, but any depletion mechanism that does not require sunlight continues to operate. For this reason, the ozone levels decrease and an ozone hole forms. Once the sunlight returns, the ozone levels return to normal ranges. At first it was surprising that this hole occurred only over the Antarctic, but later analysis showed that the air circulation patterns made this region more susceptible to hole formation than the Arctic. Some scientists predicted that a similar hole would develop over the Arctic, and false new reports claimed that an ozone hole was forming over the United States in 1992. Remember, an ozone hole is not a place with no ozone present—it is an area of reduced ozone concentration. For this reason, a person could proclaim a hole has formed in any region where the ozone levels have appreciably decreased, but such fluctuations are common (look back at figure 19.11). Studies of the ozone layer levels since 1957 indicate a decrease beginning about 1970, although this decrease is partially seasonal.

Scientists do not fully understand why there is an ozone hole in the Antarctic, but they have proposed several theories. CFC-induced ozone depletion was one of the earlier proposals, even though this route requires UV light to operate. Volcanic activity and sunspot activity were other theories. The earliest ozone hole observations showed high levels of NO_x on the surface of ice crystals present in the extremely cold atmosphere (−80°C); some researchers then proposed that the depletion was due to excess NO (nitric oxide). Since most scientists still believe that the vast majority of ozone depletion occurs through NO reactions, these original conclusions were not surprising. Later, others claimed the problem was due to CFCs, and the researchers looked for the presence of chlorine (not CFCs) in the ozone hole data. Once they found chlorine, the ozone hole was fully attributed to CFCs, the ozone scapegoat.

Volcanic activity increases the amounts of NO_x and HCl in the atmosphere, and there are many volcanoes in Antarctica, including Mt. Erebus. HCl could serve as a source of chlorine radicals, but this requires UV light; increased solar activity might produce higher levels of stratospheric NO_x, which might also come from volcanic activity and lightning. Such compounds, as we have seen, can destroy ozone. The NO_x are stable compounds that can migrate over the globe, propelled by air currents. The

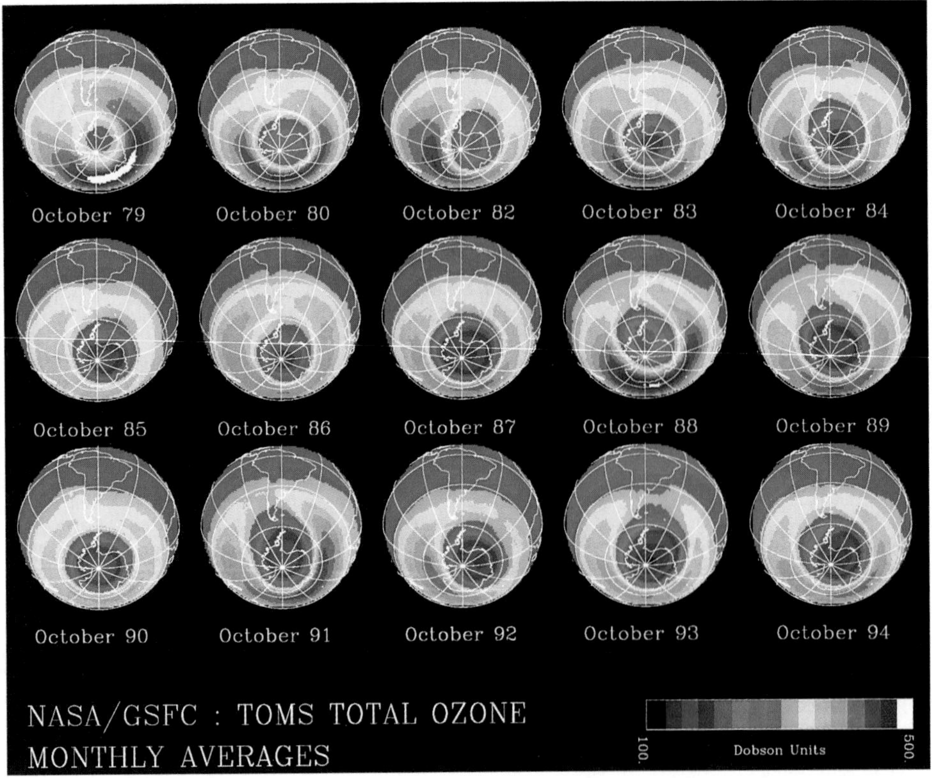

Figure 19.14 The Antarctic ozone hole as seen during the spring for each of the years shown. Notice that the size of this hole varies, decreasing in one year and increasing in later years. The size also varies with the season.

Cl• radical, on the other hand, is unstable and must be used wherever they are formed by UV-induced reactions; the radical is too unstable to move around the globe on air currents.

Scientists know far less about these problems than we need to know. Worse yet, prevailing attitudes make careful examination and assessment extremely difficult. One scientist bitterly described the situation as doing "science in a goldfish bowl." It is almost impossible to practice good science (or good politics) when every new discovery is used to proclaim disaster. Nevertheless, measurements suggest there is more chlorine, NO_x, and H_2SO_4 in the Antarctic atmosphere than in other parts of the stratosphere, and perhaps more than was previously present there. The sources of these compounds remain uncertain, but they could come from anywhere, traveling through natural air circulation patterns, and, perhaps be at least partially due to solar effects.

One significant fact usually omitted from ozone hole discussions is that this is not a new phenomenon. British scientist G. M. B. Dobson first observed the Antarctic hole in 1956–57, and P. Rigaud and B. Leroy observed it in 1958. These two French scientists reexamined their data in 1990, concluded it was correct, and then republished their results, showing that the ozone hole existed back in the 1950s. In other words, the Antarctic ozone hole existed before CFCs could have migrated into the stratosphere, and appears completely natural.

The Concentration of the Ozone Layer

Determining how serious the ozone layer problem actually is depends on obtaining stratospheric ozone concentration data. Ozone was discovered in 1840 by Christian Schönbein (a German chemist who invented guncotton in 1845). Semiregular ground level measurements of ozone were being made as early as 1876, but we do not have even this much data for stratospheric ozone. G. M. B. Dobson, the British physicist that the standard Dobson unit measurement was named after, made many observations on the ozone layer during the 1950s using balloons, but his data were erratic.

With the advent of the space age, scientists were able to study the upper atmosphere more completely using spectroscopic techniques and photography. Often the results were transmitted to Earth. Satellite readings began in 1979. At best, the combination of balloon and satellite measurements give us only about forty years of data, and this is not enough to answer questions regarding possible cyclic behavior patterns in the ozone layer.

Figure 19.15 shows the O_3 concentration changes for three locations from 1956 through the early 1990s. Great variations in the O_3 levels occurred in North America, Europe, and the Far East, but all three regions seem to show a steady total decrease of about 2.5 to 5 percent beginning about 1970. Some scientists believe these changes are due to natural variations in solar activity, but other scientists attribute the decreases mainly to human-produced CFCs. Until we gather long-term ozone data, it may not be possible to determine the cause(s) of this decrease.

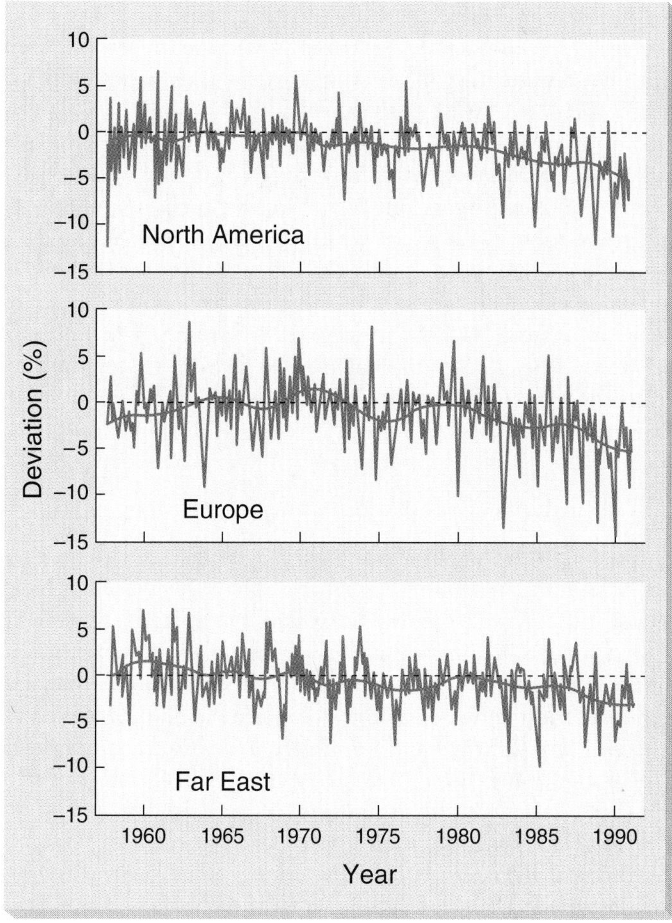

Figure 19.15 The changes in the stratospheric ozone levels from 1957 through the early 1990s over North America, Europe, and the Far East. The zero line represents the average during the entire period.

SOURCE: From Richard Stolarski, et al., "Measured Trends in Stratospheric Ozone" in *Science*, 256, 17 April 1992: 342–349. Copyright © 1992 American Association for the Advancement of Science. Reprinted by permission.

Past Ozone Concentrations

Scientists use ice-core data to estimate past CO_2 concentrations, but the stratosphere contains no ice columns. Is there any way to determine ozone layer concentrations for past eras? To date, there are no proposals for acquiring this information and it appears likely that scientists will never acquire it.

This means scientists are stuck with the undesirable and unwelcome dilemma of making decisions based on very limited data. Few scientists like this prospect, though politicians are used to it; in the absence of long-term data, they often decide with whatever information they have, and scientists may need to do the same on the ozone depletion question. Some data suggest that CFCs may cause ozone depletion, but the proportion of depletion attributable to CFCs must have been consistently reduced in recent years because CFC usage (especially open-system usage) has significantly declined. CFC depletion of ozone was estimated at 18 percent in 1980, but dropped to only 7 percent by 1984 and to merely 2 percent in the 1993 models. Current models predict a 5 percent decline in ozone over the next hundred years, or about a 0.05 percent decrease per year. This is rather surprising, since the data from figure 19.15 suggest a 5 percent decline since 1970.

We must remember that CFC-induced depletion of the ozone layer remains an assumption, even if some data make it appear to be a reasonable assumption. The political climate surrounding this topic is driving us to reduce atmospheric CFC levels as much and as quickly as possible. This is unfortunate, since CFCs are excellent chemicals for the uses box 19.4 discusses. The proposed substitutes (box 19.5) leave much to be desired. Yet, the rush to judgment seems unstoppable, and CFCs and their benefits to people lost.

UV, Skin Cancer, and the Ozone Layer

The main purpose in eliminating CFCs is to maintain a safe level of UV light impinging on our planet's surface. Many studies have shown that UV can contribute to skin cancer and a variety of other problems, including cataract formation and possibly lowering the ability of the immune system to fight disease. Crop damage and harm to the lower forms of marine life are also cited as potential problems, but some evidence points to the contrary. If none of these effects were possible, it seems unlikely many people would be alarmed that the ozone layer is diminishing.

Figure 19.15 suggests there has been a 5 percent decrease in the ozone layer since 1970. How does this correlate with the amount of UV light striking the surface? If the theories are correct, UV radiation should have increased by about 10 percent during this time. However, no study has reported a large global increase in UV light; in fact, almost all studies show a *decrease* in the UV radiation reaching the Earth's surface. In any event, the extra UV exposure caused by a 5 percent decrease in the ozone level is equivalent to moving fifty miles further south.

In all the brouhaha surrounding ozone depletion, the fact that UV radiation has some benefits is totally ignored. UV increases the formation of vitamin D in the human body and this, in turn, reduces the risk of osteoporosis and rickets. UV radiation is also an effective sterilizing agent for water or foods. However, it is also true that too much UV exposure can be harmful.

Estimates of the potential increase in skin cancer range all over the map, but the value of 15 percent is commonly quoted. The United Nations Environmental Program predicts a 26 percent global increase in nonmelanoma skin cancers if the ozone level drops 10 percent. The EPA predicts a 3 percent increase in nonmelanoma skin cancers for every 1 percent decrease in the ozone layer. With an ozone level decrease of 5 percent since 1970, there should be, by these estimates, a 15 percent increase in this form of skin cancer from this factor alone. Although the incidence of skin cancer did increase during this period, there was no increase in UV levels, the presumed causative agent.

Not all pronouncements on ozone depletion and skin cancer make sense. For example, William Reilly, the former EPA chief, declared on April 4, 1991 that with an ozone depletion of 5 percent and its resulting increases in UV radiation, "over the next 50 years about 12 million Americans will develop skin cancer and 200,000 of them will die." This works out to 240,000 skin cancer cases and 4,000 deaths per year; currently, 600,000 new cases of nonmelanoma skin cancer are reported annually, and 6,700 people died from all forms of skin cancer in 1992. Perhaps Reilly meant that these numbers reflected the expected increases over the present numbers; if so, he was predicting a 28 percent annual increase in skin cancer cases and a 33 percent increase in deaths from skin cancer. Most people would agree this is a frightening prospect. However, not all scientists agree with Reilly. For example, Dr. Frederick Urback of Temple University agrees that skin cancer rates are increasing, but adds, "the increases are due to people spending more time outside, not more UV." (Quoted in *Reason*, June 1992.)

There is no dispute that increased UV exposure will increase the chances of getting skin cancer. But even if UV levels increased, cancer is avoidable if one uses common sense and stays out of direct sunlight from 11:00 A.M. to 2:00 P.M., when the UV intensity is at its highest level. Sunscreens are useful as well. While it might be sensible, wearing long sleeves and slacks to the beach is a suggestion that most people, young or old, are unlikely to follow. But consider this: Even if the ozone layer *increased* by 10 percent and UV levels *dropped* by 20 percent, you could still get skin cancer from staying in the sun longer and soaking up more photons. CFCs have very little to do with our society's predilection for sun exposure.

Intriguingly, UV levels actually seem to be falling. John Delouisi (of the National Oceanic and Atmospheric Administration) reports an average surface reduction of 8 percent in the period 1973 to 1985 (*Reason*, June 1992). Shaw Li (also at NOAA) notes that the incidence of UV in the midlatitude Northern Hemisphere (including the United States) declined 5 to 18 percent during the twentieth century. Li believes this decrease is due to increased cloud cover and the haze formed from industrial pollutants. If this is so, then sulfuric acid aerosols might reduce the incoming amounts of UV light and behave like an artificial ozone layer in the lower atmosphere. Interestingly enough, some scientists believe these aerosols also prevent any signs of global warming due to increased CO_2.

One danger in associating skin cancer so closely with UV and ozone depletion is that other possible causative factors may be ignored. Box 16.2, figure 2 clearly shows that melanoma deaths are increasing even as UV levels decrease. Presumably, other factors besides UV are causing melanoma.

PRACTICE PROBLEM 19.6

Some claim there is no definite evidence that CFCs actually destroy the ozone layer. Review the information in this chapter and assess this statement. Is there any data, in this chapter or from other sources, that actually prove CFCs are instrumental in destroying ozone?

ALTERNATIVES TO CHLOROFLUOROCARBONS

When scientists first suspected CFCs might play a role in ozone depletion, a logical first step might have been to remove all open systems using CFCs (43 percent globally). Unfortunately, this was not done, and subsequent political actions rendered such steps useless. Now the focus is on developing CFC substitutes.

To date, all major CFC substitutes are variations on the CFC theme (see figure 1 in box 19.4). The few non-CFC-related compounds under development have not yet been studied extensively enough to ascertain whether they will or will not affect the ozone layer. Many of these CFC-related materials (for example, alkyl nitrates) may be poor choices in the first place because similar agents are fairly toxic.

The major CFC alternatives and the three most common CFCs are listed in table 1. This table also summarizes the estimated lifetimes for these compounds in the atmosphere, their ozone depleting potential, and their global warming potential. The lifetime estimates assume that CFCs do not biodegrade and that all degradation occurs from UV cleavage of CFCs. This is now known to be untrue; CFCs do biodegrade, at least in the soil.

The ozone depletion and global warming potentials are estimates made by people who believe these agents contribute to these presumed diasters. In each case, they make the estimates by assigning CFC-11 (CCl_3F) and CFC-12 (CCl_2F_2) the value of 1.0 and then deciding what relative effect the other chemicals should presumably have. At best, these values are guesses since nobody actually knows how long CFC-11 or CFC-12 will last in the atmosphere. The estimate of 108 years might be a good one for CFC-12, for example, but we have little, if any, data to support this number. Since the other lifetime guesses are based on CFC-12, they are equally uncertain.

Most of the CFC-like molecules in table 1 contain hydrogen atoms. These **HCFCs** (or **hydrochlorofluorocarbons**) have shorter presumed atmospheric survival times than the CFCs. Their potential impact on both ozone depletion and global warming is estimated to be significantly lower. Using HCFCs might help alleviate both environmental problems to a large degree while still permitting us to use most of the applications that rely on CFCs. However, many of these applications will require new and expensive modifications to use HCFCs.

HCFCs are more expensive than CFCs and require new manufacturing plants, but this may be a good, reasonable tradeoff for now. However, some environmental activist groups totally oppose this change and insist on a complete withdrawal of both CFCs and HCFCs, without proposing alternatives for essential uses, such as refrigeration, air conditioning, insulation, or fire fighting. In any event, HCFC use is unlikely to last for more than a few years; these compounds are not suitable for any applications near flames because most HCFCs are flammable, and in any event, they are scheduled for phase-out by the year 2000. This leaves the **HFCs (hydrofluorocarbons),** which are five to ten times more costly than CFCs and are flammable. Switching to HFCs would be expensive for consumers, although the manufacturer (DuPont) would make billions. Unless the present mandates are reversed, this is just about the only long-term option—just about, but not quite.

Hydrocarbons such as propane and butane were used as coolants in the past and could be once again. They would surely be less expensive then HCFCs or HFCs, but hydrocarbons are extremely flammable and form explosive mixtures with air. A leak in an automobile's air conditioning unit when such a blend is used might remove the car (and driver) from the road permanently. Is this preferable to using CFCs?

TABLE 1 Chlorofluorocarbons (CFCs) and their alternatives.

NAME	STRUCTURE	LIFETIME[1] (in years)	OZONE DEPLETION POTENTIAL[2]	GLOBAL WARMING POTENTIAL[3]	MAJOR USES
CFC-11	CCl_3F	64	1.0	0.32	Refrigeration, foamed plastics
CFC-12	CCl_2F_2	108	1.0	1.0	Refrigeration, aerosols
CFC-113	CCl_3CF_3	88	0.8	0.5	Solvent cleaning
HCFC-22	$HCClF_2$	22	0.05	0.07	Air conditioning
HCFC-123	$HCCl_2CF_3$	2	0.01	0.01	Foamed plastics
HCFC-141	$FCCl_2CH_3$	10	0.03	0.05	Foamed plastics
HCF-152	CH_3CHF_2	2	0	<0.1	Refrigeration
HFC-134	H_2CFCF_3	8	0	<0.1	Refrigeration

[1]Estimated atmospheric survival time in years; [2]Based on data using CFC-11 as the standard value of 1.0;
[3]Based on data using CFC-12 as the standard value of 1.0
SOURCE: Data from *Popular Science*, July 1990.

ANSWER Although most scientists will concede that CFCs *could* deplete ozone, the majority of atmospheric scientists believe NO_x is the primary culprit in ozone depletion. Some now think the •OH route may also play a large role. Only a relatively small percentage of atmospheric scientists express *certainty* that CFCs are the major ozone-depleting chemicals in our environment. As meteorologist Melvin Shapiro notes, "This is about money. If there were no dollars attached to this game, you'd see it played in a very different way. It would be played on intellect and integrity. When you say that the ozone threat is a scam, you're not only attacking people's integrity, you're going after their pocketbook as well. It's money, purely money." (Quoted in *Insight*, April 6, 1992.)

Shapiro's statement is not true of every scientist who believes the ozone layer is diminishing. While some do seek power, money, or both, some are honestly trying to work for the betterment of others. Still, as NASA scientist Linwood Callis points out, "CFCs come in a very poor last place as the cause for lower levels of global ozone." He calculates that fully "73 percent of the global [ozone] declines between 1979 and 1985 are due to natural effects related to solar variability." (*Reason*, June 1992.)

The Scientific Method and Ozone Layer Depletion

The question of ozone depletion has not always been investigated in a purely scientific manner and has certainly not been free from irrationality. Each study raised new questions and a considerable amount of dispute, often extremely bitter. There are no signs this will cease. Working in a "goldfish bowl" does not promote good science, which requires long period of time for introspection and weighing evidence. When society presses for immediate conclusions, and these conclusions must support a particular view in order to win grants or earn prestige, science suffers even more. Eventually, however, the scientific method will prevail.

We can hope the HCFCs and HFCs will give scientists a breathing spell so they can perform the real science needed to solve this problem. Perhaps the apparent decrease in the ozone layer is actually a cyclic variation in stratospheric ozone concentrations; perhaps it is related to solar activity. The only way to find out is to separate the science from the sensationalism.

EPILOGUE

Global warming and ozone depletion are not the first possibly spurious issues to surface in the past few decades. In 1972, a group called the Club of Rome published a book entitled *The Limits to Growth*. In this book, the Club warned that humanity could exhaust the oil supply in twenty years or less. This may indeed prove true someday; but we have rolled past the twenty-year mark in petroleum-powered vehicles. Hundreds of people warned that biotechnology would create new plagues and possibly destroy humankind. We are now beginning to harvest the benefits of new drugs, animals, and plants from biotechnology.

What might have happened in 1972 if a strong group of activists decided that all further use of petroleum should be banned? Gasoline is not the only important product we obtain from petroleum; this versatile resource also ends up in plastics, medicines, cosmetics, detergents, computers...and the list goes on. How would our world be different without petroleum products? This scenario points out how important it is to carefully apply the scientific method before making such decisions.

SUMMARY

Global warming and ozone depletion are environmental concerns that have become prominent in the media and in the public eye. Is either phenomenon actually occurring, and if so, what should we do about them?

Scientists who believe in global warming connect it to the greenhouse effect, a buildup of heat energy in the atmosphere due to the presence of greenhouse gases that absorb infrared (IR) radiation. There is no doubt the greenhouse effect really occurs and has occurred throughout history on Earth. Water vapor is the main greenhouse gas; the greenhouse effect raises the temperature on Earth enough to allow life to exist.

While the greenhouse effect is considered a natural part of the environment, global warming is thought to be a recent problem caused by human activity. Along with water, CO_2, CH_4, N_2O, and CFCs are the main greenhouse gases. A particular molecule's contribution to the greenhouse effect depends on the molecule's ability to absorb IR plus the molecule's total concentration in the pollution. After water, CO_2 is the main greenhouse gas. Those who believe in global warming think an increase in CO_2 concentration causes the retention of more infrared radiation (heat) in the atmosphere—and thus causes global warming.

To assess global warming, it is necessary to study temperature and CO_2 levels over long periods of time. Unfortunately, it is difficult to obtain much past data. The most reliable temperature data we have is relatively recent (within the last 150 years); these data indicate cyclic temperature variations within a relatively narrow range. Computer simulations of future temperature trends vary widely depending on assumptions about cloud, ocean, and other effects.

Recent CO_2 measurements taken from the rim of Mauna Loa volcano indicate a 12 to 13 percent increase in CO_2 concentrations. Scientists have yet to determine whether this increase is global, whether it stems from natural or human-produced sources, and whether it raises legitimate concerns about accompanying global warming. Ice-core analysis indicates fluctuations in past CO_2 concentrations; since the Industrial Revolution began in the late 1700s, the concentration has been steadily rising. Still, over 96 percent of atmospheric CO_2 comes from natural sources we cannot control. Whether we need (or even are able to) control the rise in CO_2 concentrations is a matter requiring further study. A prudent course might be to develop more efficient devices that emit less CO_2, as well as to share new low-CO_2 technologies with developing nations.

The bottom line on global warming: We need to continue to study global warming, gathering further data on temperature and CO_2 concentrations and assessing the potential effects of temperature increases. In the process, it is important to apply the scientific method in analyzing global warming data, claims, and predictions.

The same precautions apply to the question of ozone depletion. The Earth is surrounded by a layer of ozone that varies in thickness, or ozone concentration, depending on season, location, and time. An ozone hole, or area of low concentration, was discovered over Antarctica in 1956–57. Some environmentalists are concerned about this hole; they believe it warns that continued ozone depletion will allow too much ultraviolet (UV) radiation to reach the Earth's surface. Many believe chlorofluorocarbons (CFCs) are the primary cause of ozone depletion.

CFCs are used in refrigerators and air conditioners, as solvents, in fire extinguishers, and in medical and other applications. They react with UV light to produce highly reactive chlorine free radicals. These free radicals then react with ozone (O_3) to produce •ClO and O_2, starting a chain reaction that reduces the amount of ozone in the atmosphere. Other chemicals also participate in ozone-depleting reactions, including NO and HCl. In fact, many scientists now believe NO is the primary chemical causing ozone depletion, even though CFCs have been banned to alleviate the problem. Other scientists believe hydroxy free radicals (HO•) are the agents most responsible for ozone depletion. CFC replacements such as HCFCs (hydrochlorofluorocarbons) and HFCs (hydrofluorocarbons) lack nonflammability and other properties important in CFC applications.

One of the main concerns about ozone depletion is that excess UV exposure may cause skin cancer. The incidence of nonmelanoma skin cancer has increased in recent years, but, paradoxically, UV levels have decreased. Scientists studying the relationships between CFC use, ozone levels, and skin cancer need to continue to apply the scientific method and ignore the sensationalism that surrounds the issue. Only then can we learn the truth about ozone depletion and global warming.

Key Terms

Antarctic ozone hole 541
carbon dioxide (CO_2) 526
CFCs (chlorofluorocarbons) 526
climatology 528
fluorocarbons 541
fluorohydrocarbons 541
global warming 524
greenhouse effect 524
greenhouse gases 527
Halon molecules 540
HFCs (hydrofluorocarbons) 545
HCFCs (hydrochlorofluorocarbons) 545
ice-core analysis 532
methane (CH_4) 526
nitrogen oxides (NO_x) 526
ozone (O_3) 536
sulfuric acid (H_2SO_4) aerosols 538

Readings of Interest

General

Nordhaus, William D. "Expert opinion on climatic change." *American Scientist*, January/February 1994, pp. 45–51. The range of disagreement on this topic is staggering. Of course, these experts come from the physical, natural, and social sciences and disagreement often exists between these disparate fields.

Budiansky, Stephen. "The doomsday myths." *U.S. News & World Report*, December 13, 1993, pp. 81–91. Many times environmentalists' cases are overstated. The analyses of the ozone hole and global warming, among other "doomsday myths," are worth reading.

Monastersky, Richard. "The long view of weather." *Science News*, November 20, 1993, pp. 328–330. Efforts are now underway to predict the weather several seasons in advance.

Ray, Dixy Lee. 1993. *Environmental overkill: Whatever happened to common sense?* Washington, D.C.: Regnery Gateway. Definitely not an environmentalism text, but instead a reasoned attack on the movement. Chapters 2 (Global warming), 3 (Stratospheric ozone and the "Hole"), and 4 (The ozone and ultraviolet rays) relate to information in this chapter.

Bolch, Ben, and Lyons, Harold. 1993. *Apocalypse not*. Washington, D.C.: Cato Institute. Chapters 6 (Greenhouse effect: What is the real issue?) and 7 (The missing ozone controversy) relate to information covered in this chapter. These authors are decidedly skeptical about global warming and ozone depletion.

Balling, Robert C., Jr. 1992. *The heated debate: Greenhouse predictions versus climate reality*. San Francisco, Calif.: Pacific Research Institute. An entire book debunking global warming.

Karplus, Walter J. 1992. *The heavens are falling: The scientific prediction of catastrophes in our time*. New York: Plenum Press. Chapters 5 (Ozone layer depletion: The hole in the sky) and 6 (Climate modification: The greenhouse effect) relate to this chapter.

Matthews, Samuel W. "Is our world warming?" *National Geographic*, vol. 178, no. 4 (October 1990), pp. 66–69. A very good question, but this article not only considers potential global warming, and some of the possible consequences, it also discusses ozone depletion, another potential problem, and a possible solution—the use of solar energy.

Belton, Keith B. 1990. *Global climate change*. You can obtain a free copy of this short booklet by writing to the Department of Government Relations and Science Policy, American Chemical Society, 1155 Sixteenth Street, N.W., Washington, D.C., 20036. It provides a very good introduction to the terms, theories, and problems involved in climate change.

Fishman, Jack, and Kalish, Robert. 1990. *Global alert: The ozone pollution crisis*. New York: Plenum Press. In spite of its title, this book deals with ground-level ozone, the ozone layer, global warming, and air pollution, all of which the authors believe are major problems requiring immediate action.

Ray, Dixy Lee. 1990. *Trashing the planet*. New York: Harper Perennial. Chapters 1 (Who speaks for science?) and 4 (Greenhouse Earth) relate to material in this chapter.

Oppenheimer, Michael, and Boyle, Robert H. 1990. *Dead heat*. New York: Basic Books. This book definitely supports the idea that global warming is already in progress and something must be done immediately. It is easy reading and presents an intriguingly good idea in chapter 11, "The last war: Beating swords into solar cells."

Global Warming

Hileman, Bette. "Climate observations substantiate global warming models" *Chemical & Engineering News*, November 27, 1995, pp. 18–23. Some data seem to support global warming models.

Flavin, Christopher, and Tunali, Odil. "Getting warmer: Looking for a way out of the climate impasse." *World Watch*. March/April 1995, pp. 10–19. The authors lament that nothing positive has happened in the years since the signing of the Climate Change Treaty.

Monastersky, R. "Climate summit: Slippery slopes ahead." *Science News*, March 25, 1995, p. 183. The 1992 Climate Change Treaty requires only the industrialized nations to attempt to limit CO_2. These countries want this limitation extended to developing nations, but those nations are resisting. They say they need more, not less, industrialization to reduce poverty.

Taubes, Gary. "Is a warmer climate wilting the forests of the north?" *Science*, March 17, 1995, p. 1595. Warmer climates usually promote tree growth, but some recent research suggests that this may not always be true.

Balling, Robert C., Jr., and Cerveny, Randall S. "Influence of lunar phase on daily global temperature." *Science*, March 10, 1995, pp. 1481–1483. On the average, a full moon (compared to a new moon) increases the daily temperature by 0.02°C. This may not seem like much, but it could have some effect on our analysis of global warming.

Roemmich, Dean, and McGowan, John. "Climatic warming and the decline of zooplankton in the California current." *Science*, March 3, 1995, pp. 1324–1326. The biomass of this plankton has decreased 80 percent since 1951, presumably due to warmer surface waters.

Smith, Emily T. "Global warming: The debate heats up." *Business Week*, February 27, 1995, pp. 119–120. This article notes the lack of compelling scientific evidence to support global warming. Should CO_2 restrictions be put into place? Expect a business-oriented answer.

Graham, Nicholas E. "Simulation of recent global temperature trends." *Science*, February 3, 1995, pp. 666–671. The most recent portion of the global temperature record (1970–present) can be reproduced by models based on the observed ocean temperatures. The author thinks this may be due to an increase in atmospheric CO_2 levels.

Kerr, Richard A. "Darker clouds promise brighter future for climate models;" also Cess, R. D., et al., "Absorption of solar radiation by clouds: Observations versus models," and Ramanathan, V., et al., "Warm pool heat budget and shortwave cloud forcing: A missing physics?" *Science*, January 27, 1995, pp. 454, 496–499, 499–503. These studies indicate that clouds not only reflect 30 percent of UV radiation away from the Earth's surface, but that clouds also absorb 15 percent of this radiation.

Oerlemans, Johannes. "Quantifying global warming from the retreat of glaciers." *Science*, April 8, 1994, pp. 243–245. This article claims that the glacial retreat can be explained by a warming trend of 0.66°C per century.

Foukai, Peter. "Stellar luminosity variations and global warming." *Science*, April 8, 1994, pp. 238–239. This article dwells mainly on whether the sun is highly variable in its light output, which would have a major effect on climate. The author believes the variability is only slight.

Powell, Corey S. "Cold confusion." *Scientific American*, vol 270, no. 3 (March 1994), pp. 22–28. An assault on the link between CO_2 and climate.

Guilderson, Thomas P., et al. "Tropical temperature variations since 20,000 years ago: Modulating interhemispheric climate change." *Science*, February 4, 1994, pp. 663–665. Coral growth depends on ocean temperatures, and the Barbados corals indicate temperatures were about 5°C cooler 19,000 years ago than they are today.

Schneider, Stephen H. "Detecting climatic change signals: Are there any fingerprints?" *Science*, January 21, 1994, pp. 341–347. This author is a firm believer in the reality of global warming. He believes the trend now is different from the many global warming and cooling trends of the past several thousand years.

Holloway, Marguerite. "Core questions." *Scientific American*, vol. 269, no. 6 (December 1993), pp. 34–36. Cores drilled into the Greenland ice sheet indicate an abruptly changing climate over a 250,000-year period.

Monastersky, R. "Ancient ice reveals wild climate shifts." *Science News*, July 17, 1993, p. 36. Greenland ice-core records suggest great temperature fluctuations over the past 250,000 years.

Stauffer, Bernhard. "The Greenland ice core project." *Science*, June 18, 1993, pp. 1766–1767. How scientists are trying to learn about the gases and particles present in the atmosphere hundreds of years ago.

Pollack, Henry N., and Chapman, David S. "Underground records of changing climate." *Scientific American*, June 1993, pp. 44–50. This paper describes attempts to determine the temperatures of the distant past using boreholes in the continental rock. The data, while difficult to interpret, suggest cycles of warmer and colder temperatures in the past.

Lara, Antonio, and Villalba, Ricardo. "A 3,620-year temperature record from Fitzroya cupressoides tree rings in southern South America." *Science*, vol. 260 (May 21, 1993), pp. 1104–1106. This 3,620-year record does not reveal any evidence of significant changes in global temperatures.

Tomkin, Jocelyn. "Hot air." *Reason*, March 1993, pp. 52–55. Here is one paper that states that water is the main greenhouse gas. The author notes that the effect of CO_2 is so minor that if its concentration doubled, the change in infrared blocking (which would cause warming) would rise by only 3 percent. Strangely, this information never makes the newscasts.

Rubin, Charles T., and Landy, Marc K. "Global warming." *Garbage*, February/March 1993, pp. 24–29. While not siding for or against global warming, the authors of this paper are critical of the manner in which the prospects of global disaster are presented. They are particularly critical of those who claim a consensus when none exists. They wisely ask exactly what such a consensus would prove even if it did exist?

Michaels, Patrick. "Beware of misleading facts in reports of global warming." *Insight*, February 8, 1993, p. 20. The author asks why we only hear gloom-and-doom predictions when all the evidence seems to point in the opposite direction.

Easterbrook, Gregg. "A house of cards." *Newsweek*, June 1, 1992, pp. 24–33. Everyone knows what happens to a house of cards, but did you know that a major scientific society stated that global warming had already begun? Perhaps you missed the announcement—it occurred in the 1930s. The predictions changed to global cooling in the 1970s, and now have swung back again to global warming. Do CFCs cause global warming as some claim with certainty? The latest information suggests that the answer is no—they have almost no effect.

Kerr, Richard. "Greenhouse science survives critics." *Science*, vol. 256 (May 22, 1992), pp. 1138–1140. This is a summary report on the views of both dissenters and advocates of global warming. The author says greenhouse science "survives" although most of the views reported seem skeptical of global warming.

Hileman, Bette. "Web of interactions makes it difficult to untangle global warming data." *Chemical & Engineering News*, April 27, 1992, pp. 7–19. A good, balanced report on this tricky problem, with many intriguing figures. One estimates the average global temperatures over the past 10,000 years, showing a variation of less than 2°C. Another tries to relate global temperatures to the concentrations of CO_2 and CH_4 over a 160,000-year time frame. Current changes in CO_2 and CFC levels are illustrated and a variety of computer simulations are depicted as well.

Sanderson, George F. "Climate change." *The Futurist*, March/April 1992, pp. 34–38. This article attempts to assess the effects on human health a global rise in temperature might produce.

Stone, Peter H. "Forecast cloudy: The limits of global warming models." *Technology Review*, February/March 1992, pp. 32–40. This paper considers how clouds should be placed into the computer modeling simulations of climate changes. At present, the effects of clouds are unknown, but that could change in fifteen to twenty years.

Hileman, Bette. "Role of methane in global warming continues to perplex scientists." *Chemical & Engineering News*, February 10, 1992, pp. 26–28. A good paper for those who think there is a consensus on how CH_4 changes the climate. To sum up—scientists do not yet know whether or not it has much effect.

Matthews, Samuel W. "Is our world warming?" *National Geographic*, vol. 178, no. 4 (October 1990), pp. 66–99. This article not only considers potential global warming and some of its possible consequences, but it also considers ozone depletion and contains some discussion of solar energy utilization.

Gushee, David E. "Global climate change." *Chemtech*, August 1989, pp. 470–479. This author assumes global warming from human activity is already in progress, then measures this contention against tables of data showing that most of the claimed greenhouse gases come from natural sources.

Monastersky, Richard. "Cloudy concerns: Will clouds prevent or promote a drastic global warming?" *Science News*, August 1989, pp. 106–110. A good, easy-to-read discussion of this topic. The net result—who knows?

Monsastersky, Richard. "Global change: The scientific challenge." *Science News*, April 8, 1989, pp. 216–221; also "Looking for Mr. Greenhouse," *Science News*, April 15, 1989, pp. 233–235. A two-part series showing how models are used to predict climate changes and some comparisons with real data.

Carbon Dioxide

Zimmer, Carl. "Good news and bad news." *Discover*, May 1994, pp. 26–28. Greenhouses gas concentrations are not increasing as rapidly as in the past. However, the authors don't expect this to last.

Leutwyler, Kristin. "No global warming?" *Scientific American*, vol. 270, no. 2 (February 1994), p. 24. Mauna Lao CO_2 emissions have declined since 1990.

Normile, Dennis. "Science gets the CO_2 out." *Popular Science*, February 1994, pp. 65–70. This article discusses various ways to capture CO_2, instead of releasing it into the atmosphere. Some are quite expensive.

Zimmer, Carl. "The case of the missing carbon." *Discover*, December 1993, pp. 38–39. About half the CO_2 emitted into the atmosphere vanishes to an unknown place.

Sarmiento, Jorge L. "Ocean carbon cycle." *Chemical & Engineering News*, May 31, 1993, pp. 30–43. The oceans are the largest CO_2 reservoir; this paper describes how this works and discusses some of the implications for global temperatures.

Sundquist, Eric T. "The global carbon dioxide budget." *Science*, vol. 259 (February 12, 1993), pp. 934–941. This article suggests that the current CO_2 levels in the atmosphere are comparable to those in the most recent ice age. Some claim the ocean is a rapid CO_2 sink now, but was not during the recent ice age.

Kerr, Richard A. "Fugitive carbon dioxide: It's not hiding in the ocean." *Science*, vol. 256 (April 3, 1992), p. 35. Uncertainty exists on where human-generated CO_2 ends up; while this article says it's not in the ocean, the next article listed (from the same *Science* issue) concludes that the oceans are the main reservoir.

Quay, P. D.; Tilbrook, B.; and Wong, C. S. "Ocean uptake of fossil fuel CO_2: Carbon-13 evidence." *Science*, vol. 256 (April 3, 1992), pp. 74–79. This article presents evidence that the oceans are the main storage receptacle for human-generated carbon dioxide.

Kauppi, Pekka F.; Mielikäinen, Kari; and Kullervo. "Biomass and carbon budget of European forests, 1971 to 1990." *Science*, vol 256 (April 5, 1992), pp. 70–74. While some trees have died, this study reveals that the total biomass has increased during this time period. This has implications concerning the effects of pollution and the control of possible global warming. An increase in biomass also means some of the atmospheric CO_2 is ending up in trees and plants.

Raynaud, D.; Jouzel, J.; Barnola, J. M.; Chappellaz, J.; Delmas, R. J.; and Lorius, C. "The ice record of greenhouse gases." *Science*, vol. 259 (February 12, 1992), pp. 926–934. Describes studies to determine CO_2, CH_4, and N_2O levels from air bubbles located at various depths in ice cores. This study suggests that current atmospheric concentrations are higher than at any time in the past thousand years, or longer.

Bazzaz, Fakhri A., and Fajer, Eric D. "Plant life in a CO_2 rich world." *Scientific American*, January 1992, pp. 68–74. Higher levels of CO_2 are not always beneficial for plants.

Weiner, Jonathan. "Glacier bubbles are telling us what was in ice age air." *Smithsonian*, 1991, pp. 78–87. How scientists determine the CO_2 content of ancient air bubbles.

Soviero, Marcelle. "Exploring Earth's ancient climate." *Popular Science*, August 1991, pp. 70–73, 88. This short article discusses how scientists determine both climate and CO_2 from ice cores.

Ice Ages

Monastersky, Richard. "Staggering through the ice ages." *Science News*, July 30, 1994, pp. 74–76. It is clear that the Earth's climate has undergone severe fluctuations in the past. What is unclear is why these oscillations occurred.

Kerr, Richard A. "The whole world had a case of the ice age shivers." *Science*, vol. 262 (December 24, 1993), pp. 1972–73; also "How ice age climate got the shakes," *Science*, vol. 260 (May 14, 1993), pp. 890–892. It hardly seems worth studying about the Ice Age if we are heading to global warming. On the other hand, some information suggests another ice age is more likely than global warming.

Oliwenstein, Lori. "Cold comfort." *Discover*, August 1992, pp. 18–21. The basic thrust of this article is that global warming will lead to another ice age.

Ozone Layer Depletion

Hanson, D. R., and Lovejoy, E. R. "The reaction of $ClONO_2$ with submicrometer sulfuric acid aerosol." *Science*, March 3, 1995, pp. 1326–1328. These aerosols provide a ready site for the reactions of natural chemicals involved in ozone depletion. The article also notes that both $ClONO_2$ and HCl are relatively stable gaseous reservoirs of chlorine in the stratosphere.

Ainsworth, Susan. "Stratospheric ozone: Supersonic jets may pose lower risk." *Chemical & Engineering News*, October 24, 1994, pp. 6–7. This risk prompted the United States not to develop supersonic jets in the 1960s. It now appears the SST caused no real problems with the ozone layer.

Rensberger, Boyce. "A reader's guide to the ozone controversy." *Skeptical Inquirer*, vol. 18 (Fall 1994), pp. 488–497. Good, popular-style article on the ozone controversy, covering most aspects of the debate.

"Ozone depletion: 20 years after the alarm." *Chemical & Engineering News*, August 15, 1994, pp. 8–13. The first concerns about possible CFC depletion of the ozone layer arose in 1974. This is an overview of what has happened since.

Cicerone, Ralph J. "Fires, atmospheric chemistry, and the ozone layer." *Science*, March 4, 1994, pp. 1243–1244. Biomass burning can release fairly large amounts of methyl bromide that might cause ozone depletion.

Poore, Patricia, and O'Donnell, Bill. "Ozone." *Garbage*, September/October 1993, pp. 24–29. If you thought the only heated debates were over global warming, you need to read this article.

Zurer, Pamela S. "Researchers lack data on trends in UV radiation at Earth's surface." *Chemical & Engineering News*, July 26, 1993, pp. 35–37. Very few places, outside of the Antarctic, have made regular long-term observations of the amount of UV hitting the Earth's surface. Those that have report *decreases* in the amount of UV reaching the surface, contrary to the predictions based on ozone depletion.

Hogan, James P., plus response by Pohl, Frederik. "O-zone politics—They call this science?" *Omni*, June 1993, pp. 34–42, 91. A debate between a science writer (Hogan) and an outstanding science fiction writer (Pohl). This easy-to-read article reveals the political aspects of the ozone debate.

Zurer, Pamela S. "Ozone depletion's recurring surprises challenge atmospheric scientists." *Chemical & Engineering News*, May 24, 1993, pp. 8–18. Good article with several excellent figures. Read this one if you believe the ozone problem is fully understood.

Kerr, Richard A. "Ozone takes a nose dive after the eruption of Mt. Pinatubo." *Science*, vol. 260 (April 23, 1993), pp. 490–491. If you want proof that ozone is affected by natural phenomena, this article provides it. For more details see Gleason, et al., pp. 523–526, in the same issue.

Monsastersky, R. "Northern hemisphere ozone hits record low." *Science News*, March 20, 1993, pp. 180–181. Apparently seeing lower-than-average results two years in a row is proof, to some, of a long-term decrease in ozone levels—even if these changes are less severe than the usual seasonal variability. See also *Science News*, April 24, 1993, p. 260.

Ashley, Steven. "Ozone drone." *Popular Science*, July 1992, pp. 60–64. Scientists use an occupied power glider to gather ozone data.

Horgan, John. "Volcanic disruption." *Scientific American*, March 1992, pp. 28–29. Mt. Pinatubo's continuing legacy and its drastic effect on the ozone layer.

The Ozone Hole

Newman, Paul A. "Antarctic total ozone in 1958." *Science*, April 22, 1994, pp. 543–546. This paper states, "There is no credible evidence for an ozone hole in 1958." The author notes that a springtime dip in ozone levels over the Antarctic is a natural event; much of his argument seems to depend on what data he was willing to accept.

Vogel, Shawana. "Under the ozone hole." *Earth*, January 1993, pp. 30–35. This article deals with the potential harm increased UV light might have on living organisms. The main organisms considered are phytoplankton, tiny creatures that lie at the bottom of the food chain; some have predicted a 40 percent decline in phytoplankton due to increased UV exposure, but the actual probability seems closer to 2 to 4 percent.

Bailey, Ronald. "The hole story." *Reason*, June 1992, pp. 24–31. Possibly the best magazine article on this subject, this paper covers the entire ozone layer question, not just the ozone hole. It also discusses the CFC ban and potential replacements.

Lemonick, Michael D. "The ozone vanishes." *Time*, February 17, 1992, pp. 60–63. The true nature of an ozone hole is unclear, although the effects of presumed UV increases are detailed.

Monastersky, R. "Northern ozone hole deemed likely." *Science News*, February 8, 1992, p. 84. This panic-ridden prediction of an ozone hole over the northern hemisphere appeared almost everywhere and scared millions. Although this article perpetuates this story, it does at least point out that the ozone hole was a prediction, not an absolute certainty. It also contains some interesting information on ozone depletion.

CFCs and Substitutes

Roberts, Michael, and Sissel, Kara. "Dawn of the post-CFC era." *Chemical Week*, January 17, 1996, pp. 18–20. Article has a good timeline graphic showing the expected phase-out times for various CFC-like compounds. The market for the HCFCs seems to be flourishing.

Metzger, Alain. "CFCs depleting ozone or cash?" *Insight*, December 12, 1994, p. 32. More than three thousand scientists, including seventy-two who have won Nobel Prizes, have signed documents disputing the claim that CFCs deplete the ozone layer. Economists claim the CFC ban will cost Americans at least $200 billion.

Zurer, Pamela. "Europe just five weeks from phaseout of CFCs." *Chemical & Engineering News*, November 28, 1994, pp. 30–31. This article describes a rather unstable situation with some successes and many points of concern, especially in refrigeration.

Freemantle, Michael. "Pressure mounts as search for Halon replacements reaches critical phase." *Chemical & Engineering News*, September 29, 1994, pp. 29–32. Halons, used as propellants in fire extinguishers and other fire-fighting devices, are among the CFC-like chemicals being banned. No suitable, nonflammable replacements existed as the ban date grew nearer.

Kirschner, Elisabeth. "Producers of CFC alternatives gear up for 1996 phaseout." *Chemical & Engineering News*, July 4, 1994, pp. 12–13. Several companies have entered the CFC replacement market, so there should be some in supply by the time the ban hits. However, most alternatives do not work as CFCs do. Refrigerators, air conditioners, and other devices will require modifications.

Ravishankara, A. R., et al. "Do hydrofluorocarbons destroy stratospheric ozone?" *Science*, January 7, 1994, pp. 71–75. HFCs are to replace HCFCs after they are banned. The authors conclude that the ozone depletion potential from these compounds is negligibly small.

Zurer, Pamela S. "Looming ban on production of CFCs, Halons spurs switch to substitutes." *Chemical & Engineering News*, November 15, 1993, pp. 12–18. Some may not realize that the main CFC replacement chemicals, the HCFCs, are also scheduled for phase-out in several stages starting with aerosol usage in 1994. Total production phase-out begins in 2003.

Zurer, Pamela. "U.S. seeks exemptions from ban on CFCs." *Chemical & Engineering News*, October 11, 1993, p. 7. *Also*, "CFC users seek exemptions from ban on ozone-depleting substances." *Chemical & Engineering News*, August 1993, pp. 15–16. The exemption most likely to be granted is for inhalers such as those used by people with asthma. Automobile air conditioning units and other medical uses besides inhalers may also be exempted; no safe replacements are available.

Weber, Joseph. "Quick, save the ozone." *Business Week*, May 17, 1993, pp. 78–79. The search for CFC substitutes continues as the time draws near for their replacement. The HCFCs remain the main substitution candidates.

Glanz, James. "CFC replacement technologies: Help is on the way." *R&D Magazine*, December 1992, pp. 28–32. Details the almost mad scramble by a variety of industries to find a CFC replacement that will work in their specific application. This is not always easy since the HCFCs or HFCs are less efficient coolants and may also be flammable.

Martin, Amy. "Why we need air conditioning." *Garbage*, July/August 1992, pp. 22–28. The "green" advice is to turn off the air conditioner because it is unnecessary, wastes energy, and destroys ozone. This Houston-based writer disputes these contentions, telling of people becoming sick from spoiled food, suffering from continuing heat-induced rashes, and dying from fevers before air conditioning. She recommends that those in northern climates spend the winter in unheated houses to develop empathy for those who need air conditioning.

Zurer, Pamela S. "Industry, consumers prepare for compliance with pending CFC ban." *Chemical & Engineering News*, June 22, 1992, pp. 7–13. This is a fairly extensive discussion of the HCFCs and the likelihood that they, too, will be banned by the year 2000, leaving no safe material available for coolant use.

Stone, Richard. "Warm reception for substitute coolant." *Science*, April 3, 1992, p. 22. While many praise HFC-134a as an ozone safe material, others claim it will increase global warming.

Ellis, Jeffrey R. "Ozone depleting chemicals: Can the health-care industries survive without them?" *Medical Device & Diagnosis Industry*, March 1992, pp. 68–72. Health care systems use about 45 million pounds of CFC and related chemicals each year in a variety of therapeutic applications. The nonflammable nature of the CFCs made them particularly suitable for medical usage, but some replacements lack this property.

Cahan, Vicky. "Just when the ozone war looked winnable..." *Business Week*, June 12, 1989, p. 56. The EPA is also seeking to ban several other chemicals used in many industries.

Volcanoes and Other Natural Effects

Simkin, Tom. "Distant effects of volcanism—How big and how often? *Science*, May 13, 1994, pp. 913–914. Greenland ice-core data indicate that the effects of volcanic eruptions spread worldwide. Researchers are trying to determine how often this happens and the extent of the impact.

Jin, Fei-Fei, et al. "El Niño on the devil's staircase: Annual subharmonic steps to chaos." *Science*, April 1, 1994, pp. 70–72. El Niño, an irregularly occurring warming of the equatorial surface waters, has a profound effect on the weather. This paper attempts to explain its erratic occurrence and behavior.

Tziperman, Eli, et al. "El Niño chaos: Overlapping of resonance between the seasonal cycle and the Pacific Ocean-atmosphere oscillator." *Science*, April 1, 1994, pp. 72–74. Chaos theory can sometimes explain erratic events. These authors suggest that El Niño follows a low-order chaos pattern.

Kerr, Richard A. "Did Pinatubo send climate warming gases into a dither?" *Science*, March 18, 1994, p. 1562. It sure looks as if it did.

Zimmer, Carl. "Pinatubo lives." *Discover*, January 1994, pp. 65–66. How volcanic eruptions destroy ozone.

Kerr, Richard A. "Volcanoes may warm locally while cooling globally." *Science*, May 28, 1993, p. 1232. The cooling comes from particulates and from SO_2 aerosols.

Tabazadeh, A., and Turco, R. P. "Stratospheric chlorine injection by volcanic eruptions: HCl scavenging and implications for ozone." *Science*, vol. 260 (May 21, 1993), pp. 1082–1085. While noting that the amounts of chlorine emitted by a volcanic eruption as HCl vastly exceed the chlorine from all human sources, this article claims that nearly all of this HCl is removed by water in the atmosphere. It is unclear whether this removal mechanism applies at the − 80°C temperatures in the Antarctic, especially with the local volcanoes.

Minnis, P.; Harrison, E. F.; Stowe, L. L.; Gibson, G. G.; Denn, F. M.; Doeling, D. R.; and Smith, W. L., Jr. Radioactive climate forcing by the Mount Pinatubo eruption." *Science*, vol. 259 (March 3, 1993), pp. 1411–1415. The 1991 volcanic eruption had a definite cooling effect on the climate.

Monastersky, Richard. "Fire beneath the ice." *Science News*, February 13, 1993, pp. 104–107. There are several active volcanoes in Antarctica, including Mt. Erebus, but their full impact on global climate is obscure. They could have effects on global warming and the ozone layer.

Grove, Noel. "Volcanoes, crucibles of creation." *National Geographic*, vol. 182, no. 6 (December 1992), pp. 5–41. Volcanoes definitely have an effect on our climate and have caused devastation many times during human history. This profusely illustrated article also covers many other facets of volcanoes.

Monastersky, Richard. "A star in the greenhouse." *Science News*, October 24, 1992, pp. 282–285. The star is, of course, the sun. The question is how much effect does the sun have on Earth's climate, especially on temperature? Scientists are fairly certain that the brightness of the sun varies, directly affecting the amount of light that hits our planet, but they are unsure if this variation is enough to cause major climatic changes.

Carey, Steven; Sigurdsson, Haraldur; and Mandeville, Charles. "Fire and water at Krakatoa." *Earth*, March 1992, pp. 26–35. This eruption was one of the most severe in recent human history. The article shows pictures of this volcano today and discusses the 1883 eruption.

Monastersky, R. "Pinatubo's impact spreads around the globe." *Science News*, August 31, 1991, pp. 132–133. The eruption caused worldwide cooling.

Sulfur Dioxide Aerosols, the Temperature, and the Ozone Layer

Tolbert, Margaret A. "Sulfate aerosols and polar stratospheric cloud formation." *Science*, April 22, 1994, pp. 527–528. Chemical reactions on polar stratospheric cloud surfaces play an important role in Antarctic ozone hole formation. It is now believed these clouds form and grow on stratospheric sulfate aerosols, which are formed as a consequence of volcanic activity; this may explain why volcanic activity increases ozone depletion.

Charlson, Robert J., and Wigley, Tom M. L. "Sulfate aerosol and climatic change." *Scientific American*, vol. 270, no. 2 (February 1994), pp. 48–57. Sulfate aerosol droplets reflect solar radiation back into space and mask global warming tendencies. This article seems only to consider industrial sources and to ignore volcanic activity. They do consider plankton emissions.

Berreby, David. "The parasol effect." *Discover*, July 1993, pp. 44–50. Paper likens the SO_2-derived aerosols to a parasol in that both block out the sun's rays.

Kiehl, J. T., and Briegleb, B. P. "The relative roles of sulfate aerosols and greenhouse gases in climate forcing." *Science*, vol. 260 (April 16, 1993), pp. 311–314. These aerosols derived from SO_2 counteract the greenhouse effect and reduce the Earth's temperature. They accomplish this mainly by reflecting incoming solar radiation back out into space.

Monastersky, Richard. "Haze clouds the greenhouse." *Science News*, April 11, 1992, pp. 232–233. An easy-to-read account on how SO_2 aerosols may counteract global warming.

Kerr, Richard A. "Pollutant haze cools the greenhouse." *Science*, vol. 255 (February 7, 1992), pp. 682–683. Somehow this "review" concludes that only human-generated SO_2 interferes with the greenhouse effect, even though over 70 percent of SO_2 comes from natural sources.

Charlson, R. J.; Schwartz, S. E.; Hales, J. M.; Cess, R. D.; Coakley, J. A., Jr.; Hansen, J. E.; and Hofmann, D. J. "Climate forcing by anthropogenic aerosols." *Science*, vol. 255 (January 24, 1992), pp. 423–433. Human-generated SO_2 appears to counteract the effects of human-generated CO_2. However, most atmospheric SO_2 comes from natural sources; the authors ignore these sources.

PROBLEMS AND QUESTIONS

1. What is the greenhouse effect?

2. What is the major greenhouse gas in our atmosphere?

3. What does the greenhouse effect do to the temperatures of the Earth?

4. What is global warming, and how does it differ from the greenhouse effect?

5. Is global warming a problem related to the troposphere or the stratosphere?

6. Do all geologists and meteorologists agree with the theory of global warming?

7. What gases are considered the potential causes of global warming?

8. Which gases implicated in global warming occur naturally?

9. Which sources of CO_2 exist in the atmosphere?

10. Have the atmospheric concentrations of CO_2 changed over the past fifty years?

11. How can scientists deduce past levels of atmospheric CO_2?

12. What percentage of atmospheric CO_2 is human-generated?

13. What kinds of databases show Earth's temperatures in the past?

14. Do past temperatures show definite evidence of global warming?

15. Does SO_2 from industrial pollution increase global warming?

16. Did the Mt. Pinatubo eruption have any effect on the average temperature on Earth?

17. Have other volcanoes had the same effect as Mt. Pinatubo?

18. What effect, if any, might particulate matter have on the greenhouse effect?

19. What are the sources of sulfuric acid aerosols?

20. Do sulfuric acid aerosols have any effect on possible global warming?

21. On a long-term basis, perhaps over 800,000 years, has the temperature of the Earth remained fairly constant or fluctuated?

22. Where is the ozone layer located?

23. Does the ozone layer concentration ever vary?

24. Does the ozone layer concentration vary with latitude?

25. Does the thickness of the ozone layer vary with latitude?

26. Does the thickness of the ozone layer vary with time or season?

27. Where is the ozone layer thickest, and where is it thinnest?

28. What is the presumed problem with the ozone layer?

29. Which chemicals are claimed to deplete ozone?

30. Which ozone-depleting chemical do atmospheric scientists generally believe to the major cause of ozone loss?

31. What chemical species do atmospheric scientists consider to be the second most potent ozone-depleting agent?

32. Do any of the ozone-depleting chemicals enter the atmosphere from natural sources? If so, what percentage of each chemical is natural, and what percentage is human-generated?

33. What can we do to reduce or eliminate ozone depletion?

34. What are CFCs (chlorofluorocarbons) used for?

35. Are there any conditions under which CFCs biodegrade?

36. What makes CFCs difficult to replace?

37. What is the chemical nature of the current CFC replacements?

38. What are the assumed effects of CFC replacements on global warming?

39. What are the assumed effects of CFC replacements on ozone depletion?

40. Are HCFCs and HFCs more or less biodegradable than CFCs?

41. What properties do HCFCs have that might make them unsuitable replacements for CFCs?

42. What is an ozone hole?

43. When was the Antarctic ozone hole first observed?

44. What causes the ozone hole in the Antarctic?

45. Are ozone holes likely to form over the United States? Why or why not?

46. Is the thickness of the ozone layer greater over the United States or over the Antarctic?

47. Is it possible to determine ozone levels for time periods one hundred or more years ago?

48. Did the Mt. Pinatubo eruption have any effect on the ozone layer?

49. Why is it believed that a reduction in the ozone layer could be harmful to people?

50. What changes, if any, have been observed in the average O_3 levels in the ozone layer since 1970?

51. What is the percentage change in ground-level ultraviolet (UV) radiation since 1970?

52. Does increased UV exposure increase the possibility of getting skin cancer?

53. Has an increase in the incidence of skin cancer occurred in the past twenty years?

CRITICAL THINKING PROBLEMS

1. Read several of the following articles. Based on the data presented in the articles and in this chapter, is global warming a real problem? Patrick Michaels, "Beware of Misleading Facts in Reports of Global Warming," *Insight*, February 8, 1993, p. 20; Gregg Easterbrook, "A House of Cards," *Newsweek*, June 1, 1992, pp. 24–33; Richard A. Kerr. "Greenhouse Science Survives Critics," *Science*, vol. 256 (May 22, 1992), pp. 1138–1140; Bette Hileman, "Web of Interactions Make It Difficult to Untangle Global Warming Data," *Chemical & Engineering News*, April 27, 1992; pp. 7–19; Christopher Flavin and Odil Tunali, "Getting Warmer: Looking for a Way Out of the Climate Impasse," *World Watch*, March/April 1995, pp. 10–19; and Emily T. Smith, "Global Warming: The Debate Heats Up," *Business Week*, February 27, 1995, pp. 119–120.

2. What do long-term temperature records show regarding the stability of the temperature on Earth? Along with the information in this chapter, review some of the following articles: Marguerite Holloway, "Core Questions," *Scientific American*, vol. 269, no. 6 (December 1993), pp. 34–46; Henry N. Pollack and David S. Chapman, "Underground Records of Changing Climate," *Scientific American*, June 1993, pp. 44–50; Antonio Lara and Ricardo Villalba, "A 3,620-Year Temperature Record from Fitzroya cupressoides Tree Rings in Southern South America," *Science*, vol. 260 (May 21, 1993) pp. 1104–1106; and Bette Hileman, "Web of Interactions Makes it Difficult to Untangle Global Warming Data," *Chemical & Engineering News*, April 27, 1992, pp. 7–19.

3. If you lived in Siberia, would you be unhappy at the prospect of global warming, or would you view it as a positive? What other considerations besides temperature would figure into your assessment?

4. Assume that the amount of atmospheric CO_2 has increased 12 to 13 percent. Are there natural reservoirs where this material can go and thereby avoid any possibility of contributing to global warming? Some good background information can be found in the following articles: Jorge L. Sarmiento, "Ocean Carbon Cycle," *Chemical & Engineering News*, May 31, 1993, pp. 30–43; Eric T. Sundquist, "The Global Carbon Dioxide Budget," *Science*, vol. 259 (February 12, 1993), pp. 934–941; Richard A. Kerr, "Fugitive Carbon Dioxide: It's Not Hiding in the Oceans," *Science*, vol. 256 (April 3, 1992), p. 35; P. D. Quay, B. Tilbrook, and C. S. Wong, "Oceanic Uptake of Fossil Fuel CO_2: Carbon-13 Evidence," *Science*, vol. 256 (April 3, 1992), pp. 74–79; and P. E. Kauppi, et al., "Biomass and Carbon Budget of European Forests, 1971 to 1990," *Science*, vol. 256 (April 3, 1992), pp. 70–74. Try to read at least two of these articles. You will find that not all these authors agree; this makes the question even more intriguing.

5. Governmental studies released in 1994 suggest that massive efforts to alleviate acid ran were not only unsuccessful, but may have contributed to global warming. How do you perceive reports of this nature, and what do you think should be done to tackle these problems? Read several of the following background papers: Robert J. Charlson and Tom M. L. Wigley, "Sulfate Aerosol and Climatic Change," *Scientific American*, vol. 270, no. 2 (February 1994), pp. 48–57; J. T. Kiehl and B. P. Briegleb, "The Relative Roles of Sulfate Aerosols and Greenhouse Gases in Climate Forcing," *Science*, vol. 260 (April 16, 1993), pp. 311–314; Richard Monastersky, "Haze Clouds the Greenhouse," *Science News*, April 11, 1992, pp. 232–233; Richard A. Kerr, "Pollutant Haze Cools the Greenhouse," *Science*, vol. 255 (February 7, 1992), pp. 682–683; R. J. Charlson, S. E. Schwartz, J. M. Hales, R. D. Cess, J. A. Coakley, Jr., J. E. Hansen, and D. J. Hofmann, "Climate Forcing by Anthropogenic Aerosols," *Science*, vol. 255 (January 24, 1992), pp. 423–433; Margaret A. Tolbert, "Sulfate Aerosols and Polar Stratospheric Cloud Formation," *Science*, vol. 264 (April 22, 1994), pp. 527–528; and David Berreby, "The Parasol Effect," *Discover*, July 1993, pp. 44–50.

6. What effects do volcanoes have on global warming and ozone depletion? Read several of the following helpful articles: A. Tabazadeh and R. P. Turco, "Stratospheric Chlorine Injection by Volcanic Eruptions: HCl Scavenging and Implications for Ozone," *Science*, vol. 260 (May 21, 1993), pp. 1082–1085; P. Minnis, E. F. Harrison, L. L. Stowe, G. G. Gibson, F. M. Denn, D. R. Doeling, and W. L. Smith, Jr., "Radiative Climate Forcing by the Mount Pinatubo Eruption," *Science*, vol. 259 (March 3, 1993), pp. 1411–1415; R. Monastersky, "Pinatubo's Impact Spreads Around the Globe," *Science*

News, August 31, 1991; pp. 132–133; Richard A. Kerr, "Ozone Takes a Nose Dive After the Eruption of Mt. Pinatubo," *Science*, vol. 260 (April 23, 1993), pp. 490–491; Carl Zimmer, "Pinatubo Lives," *Discover*, January 1994, pp. 65–66; and Richard A. Kerr. "Did Pinatubo Send Climate Warming Gases Into a Dither?" *Science*, March 18, 1994, p. 1562.

7. Based on information from this chapter, estimate the percentage of ozone depletion attributable to natural sources.

8. Many areas of the ozone controversy were not covered in this chapter. Read several of the following articles and form your opinion on ozone depletion: Susan Ainsworth, "Stratospheric Ozone: Supersonic Jets May Pose Lower Risk," *Chemical & Engineering News*, October 24, 1994, pp. 6–7; Boyce Rensberger, "A Reader's Guide to the Ozone Controversy," *Skeptical Inquirer*, vol. 18 (Fall 1994), pp. 488–94; "Ozone Depletion: 20 Years After the Alarm," *Chemical & Engineering News*, August 15, 1994, pp. 8–13; Patricia Poore and Bill O'Donnell, "Ozone," *Garbage*, September/October 1993, pp. 24–29; Pamela S. Zurer, "Ozone Depletion's Recurring Surprises Challenge Atmospheric Scientists," *Chemical & Engineering News*, May 24, 1993, pp. 8–18; and Richard A. Kerr, "Ozone Takes a Nose Dive After the Eruption of Mt. Pinatubo," *Science*, vol. 260 (April 23, 1993), pp. 490–491.

9. Is there any proven relationship between the ozone layer and ground-level UV radiation? Read some of the following articles and form an opinion: Ronald Bailey, "The Hole Story," *Reason*, June 1992, pp. 24–31; Michael D. Lemonick, "The Ozone Vanishes," *Time*, February 17, 1992, pp. 60–63; R. Monastersky, "Northern Ozone Hole Deemed Likely," *Science News*, February 8, 1992, p. 84; and Pamela S. Zurer, "Researchers Lack Data on Trends In UV Radiation at Earth's Surface," *Chemical & Engineering News*, July 26, 1993, pp. 35–37.

10. Is reaching a consensus a worthwhile goal in making a scientific decision? Three useful articles that demonstrate this issue are: Charles T. Rubin and Marc K. Landy, "Global Warming," *Garbage*, February/March 1993, pp. 24–29; Boyce Rensberger, "A Reader's Guide to the Ozone Controversy," *Skeptical Inquirer*, vol. 18 (Fall 1994), pp. 488–497; and Patricia Poore and Bill O'Donnell, "Ozone," *Garbage*, September/October 1993, pp. 24–29.

11. If only 2 percent of all scientists adhere to some specific idea, should their ideas receive as much public exposure as the ideas of the other 98 percent of scientists? Can you think of any examples when this has occurred? What if the 2 percent turn out to be right? Some good insights on this are found in: Ronald Bailey, "The Hole Story," *Reason*, June 1992, pp. 24–31; Ben Bolch and Harold Lyons, *Apocalypse Not*, Washington, D.C.: Cato Institute, 1993; Robert C. Balling, Jr., *The Heated Debate: Greenhouse Predictions Versus Climate Reality*, San Francisco: Pacific Research Institute, 1992; Walter J. Karplus, *The Heavens Are Falling: The Scientific Prediction of Catastrophes in Our Time*, New York: Plenum Press, 1992; and Dixy Lee Ray, *Environmental Overkill: Whatever Happened to Common Sense?* Washington, D.C.: Regnery Gateway, 1993.

12. Should people in Africa be restricted from using CFCs since they have very few refrigerators or air conditioners on this continent? Should the development and use of these technologies be prevented in African nations? What if such restrictions might lead to an increase in deaths due to food poisoning?

13. CFCs are being replaced, but how good are the alternate materials? Read some of the following to find out: Joseph Weber, "Quick, Save the Ozone," *Business Week*, May 17, 1993, pp. 78–79; James Glanz, "CFC Replacement Technologies: Help Is on the Way," *R&D Magazine*, December 1992, pp. 28–32; Pamela S. Zurer, "Industry, Consumers Prepare for Compliance with Pending CFC Ban," *Chemical & Engineering News*, June 22, 1992, pp. 7–13; Richard Stone, "Warm Reception for Substitute Coolant," *Science*, April 3, 1992, p. 22; Jeffrey R. Ellis, "Ozone-Depleting Chemicals: Can the Health-Care Industries Survive without Them?" *Medical Device & Diagnosis Industry*, March 1992, pp. 68–72; Vicky Cahan, "Just When the Ozone War Looked Winnable..." *Business Week*, June 12, 1989, p. 56; and Amy Martin, "Why We Need Air Conditioning," *Garbage*, July/August 1992, pp. 22–28.

14. Look at the graph of the high and low daily temperatures you made in one of the chapter 19 Experience This exercises. This shows you the trends in local temperature for the time period studied. Could you use these data to decide whether the region's average temperature had increased by 1°F more than the past?

15. Assume you have the authority to put CO_2 emissions changes into effect. Analyze the information and decide what course you will pursue.

16. Make your own assessment of global warming. Reexamine the information presented in the chapter and review any other data you find. Be certain to examine all data as critically and as logically as you can.

17. Considering the uses of CFCs and the fact that many chemicals may produce ozone depletion, do you think a total CFC ban was a good idea? Try to make your own estimate of the risk-benefit ratio for this ban.

18. Review the information on the ozone hole and CFCs and decide whether you believe CFCs are causing an ozone-depletion problem.

19. Make your own assessment of whether ozone depletion is a real problem. In this evaluation, include answers to the following questions:

a. Is the level of the stratospheric ozone really decreasing?
b. Once CFCs are removed, will all ozone depletion cease?
c. What natural sources lead to ozone depletion?
d. What information suggests that the Antarctic ozone hole might be a natural cyclic phenomenon?
e. Has the scientific method been properly applied to this problem?
f. Is the problem mainly due to human-generated or natural chemicals?

20. The practice of lying or at least stretching the truth seems endemic in modern society. Is this practice good, bad, or neutral? How does it affect science and human activities in general? Read at least two of the following articles on this subject: (1) "Deceit for a Good Cause Is Deceit Nonetheless," *Garbage*, February/March 1993, p. 66; (2) P. S. Zurer, "Ethical Issues Underlying Responsible Conduct of Science Explored," *Chemical & Engineering News*, March 15,

1993, pp. 7–10; (3) Rushworth M. Kidder, "Ethics, a Matter of Survival," *The Futurist*, March/April 1992, pp. 10–12; and (4) Ronald Bailey, "Political Science," *Reason*, December 1993, pp. 61–63.

21. Many questions regarding ethics, poor quality research, and even fraud have arisen in recent years in regard to science. Who is at fault here—scientists, government, or society as a whole? Are there ways to eliminate or alleviate these difficulties? The following papers present some views on this issue. Read at least two: (1) Leslie Alan Horvitz, "Can Scientists Police Themselves?" *Insight*, May 2, 1994, pp. 6–11; (2) The Carnegie Commission on Science, Technology, and Government, "Political Influence vs. Good Science," *Chemtech*, May 1994, pp. 8–14; (3) Ronald Hoffman, "How Should Chemists Think?" *Scientific American*, February 1993, pp. 63–73; and (4) Rushworth M. Kidder, "Ethics, a Matter of Survival," *The Futurist*, March/April 1992, pp. 10–12.

22. If you ask several people about the Endangered Species Act, you are certain to find some who like the law and some who don't. What is your opinion on this topic, and is it based on the scientific method? A few references provide a good starting point on evaluating this issue: (1) T. Dunkel, "For One Law, It's a Jungle Out There," *Insight*, May 10, 1993, pp. 4–9, 34; (2) E. O. Wilson, "The Diversity of Life," *Discover*, September 1992, pp. 47–68; (3) E. O. Wilson, "Toward Renewed Reverence for Life," *Technology Review*, November/December 1992, pp. 72–73; (4) Betsy Carpenter, "What Price Dolphin?" *U.S. News & World Report*, June 13, 1994, pp. 71–74; (5) Jeff A. Taylor, "Species Argument," *Reason*, January 1994, pp. 52–53; and Betsy Carpenter, "The Best-Laid Plans," *U.S. News & World Report*, October 4, 1994, pp. 89–93.

23. What are some of the problems we encounter in maintaining forests, including rain forests? Summarize your ideas after reading several of the following articles plus any others you find on your own: (1) Thomas O'Neill, "New Sensors Eye the Rain Forest," *National Geographic*, September 1993, pp. 118–130; (2) Billy Goodman, "Drugs and People Threaten Diversity in Andean Forests," *Science*, July 16, 1993, p. 293; (3) William S. Ellis, "Brazil's Imperiled Rain Forest," *National Geographic*, vol. 174, no. 6 (December 1988), pp. 772–799; and (4) Rowe Findley, "Old-Growth Forests: Will We Save Our Own?" *National Geographic*, vol. 178, no. 3 (September 1990), pp. 106–136.

24. Whether you are for or against some of the extremes a few environmental groups advocate, it is useful to know about the origins of environmentalism. Read several of the following articles and then summarize this history: (1) Edward C. Krug, "Environmentalism," *Chronicles*, June 1993, pp. 44–46; (2) Thomas F. Homer-Dixon, Jeffrey H. Bourwell, and George W. Rathjens, "Environmental Change and Violent Conflict," *Scientific American*, February 1993, pp. 38–45; and (3) Richard H. Grove, "Origins of Western Environmentalism," *Scientific American*, July 1992, pp. 42–47.

25. Briefly consider the actions of environmental fringe groups. Are their ideas sound and are their actions warranted? Can a better environment be achieved through other means than those such groups advocate? Some good background materials to use in considering this question include: (1) Rick Henderson, "Crimes Against Nature," *Reason*, December 1993, pp. 18–24; (2) Robert Miniter, "Are the Greens Losing Their Grip?" *Insight*, December 27, 1993, pp. 6–12; (3) Edward C. Krug, "Environmentalism," *Chronicles*, June 1993, pp. 44–46; and (4) Gregory Benford, "The Designer Plague," *Reason*, January 1994, pp. 37–41.

26. Some claim that sensationalism tends to dominate news releases. Is this true? Read at least two of the following articles, and any others you can find, and come to your own tentative conclusion: (1) David Link, "Storyville: Turning Every Issue into a Drama is Warping Public Policy," *Reason*, January 1995, pp. 35–43; (2) Michael Fumento, "Shock Journalism," *Reason*, January 1995, pp. 23–29; and (3) Robert Root Bernstein, "Future Imperfect," *Discover*, November 1993, pp. 42–47.

27. Read the following short editorial: D. E. Koshland, Jr., "The Great Overcoat Scare," *Science*, vol. 259 (March 25, 1993), p. 1807. This is a brief parable on fake issues and their costs to us all. Analyze the article and decide how you will evaluate and respond to the media hype over the next fake issue that comes along.

28. Economics affect science and scientific study; no one can escape economic considerations in the real world. Some claim that foreign inroads into American markets have caused Americans to lose jobs. How valid is this claim? If valid, what can be done about it? Is the overall world view more or less important than one localized to a single nation? And does this apply differently to the scientific world than to other endeavors? Read several of the following five articles and form your own opinion: (1) Paul R. Krugman and Robert Z. Lawrence, "Trade, Jobs, and Wages," *Scientific American*, vol. 270, no. 4 (April 1994), pp. 44–49; (2) Jagdish Bhagwati, "The Case for Free Trade," *Scientific American*, vol. 269, no. 5 (November 1993), pp. 42–49; (3) Herman E., Daly, "The Perils of Free Trade," *Scientific American*, vol. 269, no. 5 (November 1993), pp. 50–57; (4) Vernon L. Smith and Arlington W. Williams, "Experimental Market Economics," *Scientific American*, December 1992, pp. 116–121; and (5) Amartya Sen, "The Economics of Life and Death," *Scientific American*, May 1993, pp. 40–47.

29. Read the following article: Loren McIntyre, "Last Days of Eden," *National Geographic*, vol. 174, no. 6 (December 1988), pp. 800–817. This article tells about the Rondônia's Urueu-Wau-Wau Indians, a primitive tribe who dwell deep in Brazil's rain forest. Should we leave them to their primitive lifestyle—knowing most of them will die young—or lead them into the twenty-first century to "improve" their lives? Would you choose the type of life the Urueu-Wau-Wau Indians live for yourself, your children, and your grandchildren? Why or why not?

THE METRIC SYSTEM

There are at least two main reasons why you need to know the basics of the metric system. Perhaps the most important, and certainly the most immediate, reason is that the metric system is used in this book. You will need to understand how this system operates in order to grasp most of the numerical information in this text. Almost everything that is done in science uses the metric system. Chemistry is a science; you will need to know enough about this rather simple measurement system in order to understand what is being said in this chemistry book.

The second reason to learn the metric system is that nearly all of the world has already adopted this measurement system in place of the English system, which is still used in the United States. Even England now uses the metric system. Although you may use English measurement every day in the United States, that cannot be the only factor to consider. If you lived or worked anywhere else in this world, you would almost certainly use the metric system for every measurement, and think nothing further about it. Our main problem in the U.S.A. is that the metric system is unfamiliar. Nevertheless, there are many advantages to the metric system which makes it far easier to use than the English system that's more familiar to us. Why bother with the metric system? Well, your future career may demand that you understand this system. Perhaps now is the best time to learn the basics.

The major measuring sub-systems are: length, area, volume, weight, and time. In the English system, we use the units of inch, foot, yard, and mile (plus several less common units such as rod and furlong) for measurements of length. There is no simple, constant numerical relationship between these length units. Thus, there are 12 inches to a foot, 3 feet to a yard and 5,280 feet to a mile. With each conversion from one length unit to another we must use a completely different factor. Sometimes this is hard to do in your head. Try determining how many miles in 316,382 feet! (The correct answer is 59.92 miles.) Most of us would not attempt to do this conversion while driving down a highway.

Area measurement is not any better: area units include square inches, square feet, square miles, and acres! It's not easy converting 684,395 square inches to square feet. (The answer is 4,752.74 because the conversion factor is 144.) Even though the term is familiar, most people have no idea how large an acre is.

Volume measurements are just as bad. Consider our mixed up collection of ounces, pints, quarts, and gallons. And if that's not enough, add barrels and gills to the mix. None of these have a common conversion factor: thus, there are 16 ounces to a pint, 2 pints to a quart, and 4 quarts to a gallon. If you want to befuddle your friends ask them how many gallons are in a barrel? The lack of knowledge in this area might seem surprising since crude oil is sold by the barrel. (The answer is 42.) If your great knowledge impresses them, spring this question on them: "How many gills are there in a quart?" The answer is 8, but most likely your friends (perhaps former friends by this time) will associate gills with fish rather than a liquid volume measurement.

Weight is also weighty. To be more accurate, however, we should say mass since that is the term commonly used in science, but the problem still remains massive. The English system has units of grains, ounces, pounds, and tons, (both long and short), plus drams, pennyweights, stones, and so forth. There are even two different weighing systems called the troy and the avoirdupois, which use different conversion factors. There are either 12 (troy) or 16 (avoirdupois) ounces in a pound, depending on which system is used. Lots of people ask the question, "which weighs more, a pound of feathers or a pound of gold?" The answer is feathers because a troy ounce (the unit in which gold is weighed) is the same size as an avoirdupois ounce (the unit used for feathers), but there are only 12 troy ounces to a pound. If you still have friends left, you might ask them "How many pounds are in a stone?" You'll probably not amaze them with the correct answer of 14.

With all these complex problems, you might wonder why *anyone* wouldn't prefer the simpler metric system. All the conversion factors are 10 or a multiple of ten (e.g., 100, 1,000, and so on). What could be easier to remember or use? Even the names of the sub-system units, table A1, are less complicated; there are only three different names (meter, liter, and gram).

Table A2 lists the more common prefixes that you will need: micro-, centi-, milli-, kilo-, mega-. Many of these prefixes are already familiar to you in your daily life. A cent is 0.01 dollar (or a centi-dollar). A mil (a common taxation base) is 0.001 or 1/1000th of a dollar. We sometimes call a million dollars ($1,000,000) a megabuck. Machinists have used the micron (μ) on their micrometers for many years, as one millionth of a meter. We buy our electricity in kilowatts (1,000 watt units).

Table A3 lists a few selected factors for converting metric

TABLE A1 Main metric system units.

SUB-SYSTEM	NAME OF UNIT	SYMBOL
Length	meter	m
	centimeter	cm
Area	square meter	m^2
	square centimeter	cm^2
Volume	cubic centimeter	cm^3
	liter	L
	milliliter	mL
Mass	gram	g
	kilogram	Kg

TABLE A2 The prefixes used in the metric system.

PREFIX	NUMERICAL VALUE
micro	0.000001
milli	0.001
centi	0.01
deci	0.1
—	1.0
deka	10
kilo	1,000
mega	1,000,000

TABLE A3 Common metric-English system conversion factors.

LENGTH		
1 inch (in)	=	2.54 centimeter (cm)
1 yard	=	0.914 meter (m)
39.37 inches (in)	=	1.0 meter (m)
1 mile (mi)	=	1.609 kilometer (km)
VOLUME		
1 ounce (oz)	=	29.57 milliliters (mL)
1 quart (qt)	=	0.946 liter (L)
1.06 quart (qt)	=	1.0 liter (L)
1 gallon	=	3.78 liters (L)
MASS (WEIGHT)		
1 ounce (oz)	=	28.35 grams (g)
1 pound (lb)	=	453.6 grams (g)
2.20 pounds (lb)	=	1.0 kilogram (kg)
1 ton (tn)	=	907.2 kilograms (kg)

values to English (or the reverse). We will use the metric system in most places in this book, but we will rarely convert from the metric system to the English units. The conversion factors shown in table A3 give a rough idea how metric units relate to some of our English units. Thus, a liter is slightly larger than a quart, and a kilometer is much shorter than a mile. A 100 mile trip would be 160.9 kilometers. A speed of 65 miles per hour would be 104.6 km/hour, which certainly sounds faster. (The time units are the same in both systems.) A meter is about 3.37 inches longer than a yard. Weight is a little more difficult to grasp because the gram is very small. A nickel weighs about 5 grams and a penny (pre-1975) weighs about 3 grams. This is why many mass values are given in kilograms. Thus, a 150 pound person weighs 68.04 kg. If that person is also 5 foot 10 inches tall, he or she is 177.8 cm or 1.778 meters tall.

We need to note one last thing regarding measurements. Whenever we make measurements we do not always get exactly the same value. For example, if you weighed each egg in two dozen eggs, in grams, you would probably not find any two eggs with exactly the same mass. Which value is correct? Obviously all of the numbers are right for the individual egg, but what does a "typical egg" weigh? We can find this out by weighing each egg, adding up the numbers, and dividing by the number of eggs. This gives us an average value. We could also get this average value by weighing all the eggs at the same time and then dividing by the number of eggs weighed. (This essentially adds up the masses of the invidivual eggs.)

We quickly realize that eggs, or just about anything else, are not exactly identical in mass, length, volume, and so forth. In science, this limitation is one form of experimental error. If we want to know how much an egg weighs, we must realize that all eggs are not alike and take an average value. Likewise, if we want to know how much material we would get in a chemical reaction, we must determine the average, and talk about that instead of an individual result. The same thing applies to any finding in science; one individual result might not really represent what will happen all the time.

This limitation is not always obvious or clear when medical, science, or chemistry results or conclusions are described by the news media. This is especially true in television or radio news where everything must be presented in sound-bites a few seconds long. You need to recognize this limitation and reserve judgment until more information becomes available.

Here are a few problems to help you better relate metric to English measurements. The answers are given following all of the questions.

PROBLEMS

1. (a) Convert 9,286 g to kg. (b) Now convert 2,432 ounces to pounds. Which was easier to do?

2. (a) Convert 86,400 meters to kilometers. (b) Next, convert 283,800 feet to miles. Which was easier to do?

3. Finally, convert the pounds in 1(a) to kg and the miles in 2(b) to km. Which is easier in each case?

4. (a) What would be the metric dimensions for a person with a chest-waist-hip size of 34-24-34 inches? (b) What would the measurements be if the person had measurements of 44-32-38?

5. Measure (or convert) your height, weight, and waist size to the metric system.

6. (a) How far, in km, is Los Angeles from New York City if the distance is 2,825 miles? (b) Likewise, how far, in km, is Chicago from Miami (1,188 miles)?

Answers

1. (a) 9.286 kg; (b) 152 pounds (avoirdupois ounces). You decide which was easier for yourself.

2. (a) 86.400 km; (b) 53.75 miles

3. 69.09 kg; 86.48 km

4. (a) 86-61-86, in cm; (b) 112-81-96, in cm

5. You have to do this one by yourself.

6. (a) 4,545 km; (b) 1,911 km

EXPONENTIALS AND SCIENTIFIC NOTATION

The concepts of exponentials and scientific notation were introduced in chapter 3. This appendix will provide some additional information on these two topics, both of which are important because scientists describe large and small numbers using exponentials and scientific notation. The main thing for you to realize is that an exponential is simply a number, even though it appears somewhat different than the usual numbers you have used all your life. Although a full discussion of this topic is beyond the scope of this book, the basic features presented here should be sufficient for your understanding.

It would be virtually impossible to express many of the exceptionally huge or exceedingly small numbers used in science without resorting to exponentials and scientific notation. For example, the mole, which tells how many molecules are in a quantity corresponding to the molecular weight of that compound, is a remarkably large 602,000,000,000,000,000,000,000. Writing out such a number takes up a lot of space, can lead to errors by omitting some zeros, and can be very confusing. Expressed as the scientific notation form, this number becomes 6.02×10^{23}, which is more compact; the number remains extremely huge but it is much easier to write. The trick is to understand what the terms mean.

Scientific notation consists of a simple decimal number, like you have used for many years, combined with an exponential, which is a number expressed in a different manner. An exponential number is a number that has been raised to a power, denoted by a superscript. For example, 2^2 is two raised to the second power, or squared. This means that the number two is multiplied by itself one time, or $2 \times 2 = 4$. The exponential number 2^5 means that five twos are to be multiplied, or $2 \times 2 \times 2 \times 2 \times 2 = 32$. In other words, $2^5 = 32$. An exponent is a shorthand notation for the process of multiplying the five twos.

However, it is more convenient and much simpler to use powers of ten for an exponential. Thus, one thousand is 1,000 or 10^3, which means we multiply three tens and get 1,000. Likewise, a million is 1,000,000 or 10^6, wherein six tens are multiplied. A billion is 10^9 or the result of multiplying nine tens to get 1,000,000,000. A trillion is 10^{12} or the product from multiplying ten twelve times to get 1,000,000,000,000. By now it should be evident that the exponential notation is more compact than writing out the full number. If you read the newspapers carefully, you will see that numbers like a billion or a trillion are seldom written out explicitly. The name is used as an abbreviation for these large numbers. Thus, you might see a congressional allocation of 35 billion dollars for some project, rather than \$35,000,000,000. This could just as readily be written as $\$35 \times 10^9$, or \$35 times a billion. This abbreviation is almost exactly what we use in scientific notation. The only difference is that scientific notation normally uses the number expressed as a decimal. In our example of 35 billion, this would become $3.5 \times 10^1 \times 10^9$. Exponents are always multiplied by adding them together, that is $1 + 9 = 10$. Thus, 35 billion becomes 3.5×10^{10}, or the number 3.5 multiplied by ten billion (10,000,000,000).

We often forget how big numbers like a million, billion, or trillion really are because we hear about them in news daily. For example, a single stealth bomber costs about 400 million dollars. The national debt is over three trillion dollars and has an interest of more than 139 billion dollars annually. Just how big is a billion? Consider this. If you were to spend exactly one dollar per minute, no more and no less, for 16 hours a day, allowing a third of the day to rest, eat, and so forth, you would spend \$350,400 dollars in a year. You would need 2,853.9 years to get rid of this billion dollars. If you had started spending that money in the year 1 A.D., there was no year zero, you would have the tidy sum of \$299,771,920 left at the end of 1997! Expressed in scientific notation that would amount to almost $\$3 \times 10^8$, still quite a tidy sum! You could continue spending at the same rate for another 856.9 years, assuming you didn't get any interest on the money. Imagine what might happen if you put the bulk of this money into a simple CD (certificate of deposit).

We can also express numbers less than one as an exponential by writing a negative exponent, which means we performed a repeated division. Thus, the quantity of one thousandth is 0.001, obtained by dividing ten into one three times. In other words, $1 \div 10 = 0.1$; $0.1 \div 10 = 0.01$; $0.01 \div 10 = 0.001$, and we can express this as $1 \div 10^3$ or 1×10^{-3}, meaning we divided 1 by 1,000. A penny is one-hundredth of a dollar or 10^{-2} dollars. A millionth of an inch is 0.000,001 or 10^{-6}. Parts per billion (ppb), a term often used by the Environmental Protection Agency, means you divide the number one by a billion in order to write it in the usual fashion. One part per billion is 0.000,000,001 or 10^{-9}. If you had three parts per billion, you would express this in scientific notation as 3×10^{-9}. You can see that this is easier to write than 0.000,000,003.

You can also see that the general rules for writing large or small numbers is the same in scientific notation. You write a combination of a decimal and an exponential number. Thus, 7.3×10^{-6} inches is 0.000,0073, or 7.3 millionths of an inch, in the same way, 9.3×10^{7} miles is 93 million miles, or 93,000,000 miles.

In the same manner, multiplication or division of exponentials is done by either adding the exponents, in multiplication, or by subtracting them, in division. If we multiply 10^9 by 10^6 we just add $9 + 6 = 15$ to get 10^{15}. If we were to divide 10^9 by 10^6, we would subtract the number being divided by the number dividing it; $9 - 6 = 3$, or 10^3. The same rules apply if we divide 10^6 by 10^9; $6 - 9 = -3$, or 10^{-3}.

When we are using scientific notation, the same rules apply for the exponential portion of the number; the decimal portion is multiplied or divided separately. If we multiply 3.2×10^5 by 6.4×10^7, we add the exponents to get 10^{12} and multiply 3.2 $\times$ 6.4 to get 20.48; this gives us 20.48×10^{12}. We're not quite finished here since our regular number can only have a single number to the left of the decimal point. Accordingly, we make this $20.48 \times 10^1 \times 10^{12}$ and then combine the exponents to get the final 2.048×10^{13}. If you are uncertain of this answer, write out the numbers, do the multiplication longhand and then convert to scientific notation. You will find the same product.

If we divide 3.2×10^5 by 6.4×10^7, we divide the regular numbers as usual; $3.2 \div 6.4 = 0.5$, or 5.0×10^{-1}. We next divide the exponents by subtracting the second from the first, $5 - 7 = -2$, or 10^{-2}. Now we combine the parts to get $5.0 \times 10^{-1} \times 10^{-2}$, or 5.0×10^{-3}, or 0.005. If you remain unsure of this answer or method, convert to ordinary numbers, divide longhand, and then convert to scientific notation. You will discover that you will get the same result.

Did you notice that in multiplying, the exponents are added together while the decimals are multiplied in the usual manner? In division, one exponent is subtracted from the other while the decimals are divided in the usual manner. At the end of this process, all numbers are converted into the correct, or proper, scientific notation.

We can illustrate this principle further by determining the mass of a single molecule of water. Water has a molecular mass of 18.02 g/mole. We know that a mole of water contains 602,000,000,000,000,000,000,000 water molecules/mole. As we do this operation, we are dividing grams per mole by molecules per mole; that will result in grams per molecule, which is what we are after.

If we divide mass of a mole of water by the number of molecules in a mole, the first step in our operation looks like this in scientific notation:

$$\frac{18.02 \text{ g/mole}}{6.023 \times 10^{23}} \text{ molecules/mole}$$

We now separate the parts to get:

$$18.02 \div 6.023 = 2.99$$

and:

$$1 \div 10^{23} = 10^0 \div 10^{23}$$

$$0 - 23 = -23, \text{ or } 10^{-23}$$

We now combine to get the answer 2.99×10^{-23}, or about 3×10^{-23}.

If you write this out correctly, you will see that a single molecule of water has a mass of 0.000,000,000,000,000,000,000,030 grams. This may seem to be a very small mass for a molecule of the most important compound on Earth, but that is the correct number. I believe you will agree that both these very large and very small numbers are difficult to write. (And certainly to write them correctly!) For this reason we normally use exponential numbers and scientific notation in chemistry.

Since they are actually numbers, we can add or subtract numbers in scientific notation. Whenever you do this, it is very important to make certain you line up the numbers in the same manner as you would do with "regular" numbers. For example, when you add 3,000 and 200 you must line up the numbers at the decimal point, or implied decimal point in this instance. If you don't do this you will get an incorrect answer. When they are lined up properly, the arrangement looks like this:

$$\begin{array}{r} 3{,}000 \\ +\ \ 200 \\ \hline 3{,}200 \end{array} \quad \text{or} \quad 3.2 \times 10^3$$

The arrangement never looks like this:

$$\begin{array}{l} 3{,}000 \\ +200 \\ \hline 5{,}000 \end{array}$$

That would be incorrect!

If we did this same addition in scientific notation, we would be adding 3×10^3 and 2×10^2. In order to do this correctly, the exponents must be the same during the addition (or subtraction) step. Usually it's easier to match the lower exponent by pulling out powers of ten into the decimal number of the larger one. Thus, 3×10^3 is expressed as 30×10^2 and then added to 2×10^2 to get 32×10^2. That answer is then converted to the correct scientific notation of 3.2×10^3.

If we subtracted 200 from 3,000 we have the same problem. The only correct arrangement is:

$$\begin{array}{r} 3{,}000 \\ -\ \ 200 \\ \hline 2{,}800 \end{array} \quad \text{or} \quad 2.8 \times 10^3$$

Again we convert the larger number to 30×10^2 and subtract 2×10^2 from it. The answer is 28×10^2, which we convert to 2.8×10^3.

Did you notice that the exponents remain the same during addition or subtraction? Only the decimal portion is involved in the operation, once the exponents are made the same. This is different from what you do in multiplication or division.

Here are a few numerical problems for you to practice operations using scientific notation.

PROBLEMS

1. Multiply 4.7×10^3 with 9.2×10^7
2. Divide 9.2×10^7 by 4.7×10^3
3. Multiply 5.5×10^{-3} with 7.2×10^4
4. Divide 5.5×10^{-3} by 7.2×10^4
5. Divide 7.2×10^4 by 5.5×10^{-3}
6. Add 4.8×10^6 to 3.7×10^5
7. Subtract 3.7×10^5 from 4.8×10^6

Answers

1. 4.32×10^{11}
2. 1.96×10^4
3. 3.96×10^2
4. 7.64×10^{-8}
5. 1.30×10^7
6. 5.17×10^6
7. 4.43×10^6

APPENDIX C

EQUATIONS AND THE MATHEMATICS OF CHEMISTRY

There are many things that a practicing chemist must be able to do that a non-scientist does not need to understand in detail. For that reason, these topics are not covered extensively in this text. One of these is the topic of stoichiometry, which is the measurement of the exact amounts of materials involved in a chemical reaction. In short, this is the mathematics of chemical equations.

It's not that this subject is too difficult for the non-scientist. Rather, it's that stoichiometry is just not that useful unless you are planning to determine the quantities of chemicals involved in a reaction. For example, we could ask how many grams of carbon dioxide gas are released when you react 100.00 g chalk with an excess of hydrochloric acid? For a non-scientist it's quite sufficient to know that chalk will react with hydrochloric acid to produce carbon dioxide; the other products, by the way, are calcium chloride and water.

Yet, we would be remiss not to at least show how this works, back here in an appendix so it won't get in the road of the main features of this book. This way, you can see what's involved, without getting very involved in the mathematics of chemical equations, and spend most of your time on the more down-to-earth, practical aspects of chemistry covered in this book. With that understanding, let's examine that question of the chalk and hydrochloric acid in more detail and see how many grams of CO_2 we can get from 100.00 g chalk, $CaCO_3$. The first step is to write an equation showing all the reactants and products in this reaction:

$$CaCO_3 + HCl \longrightarrow CaCl_2 + CO_2 + H_2O$$

The next step is to balance this equation:

$$CaCO_3 + 2\ HCl \longrightarrow CaCl_2 + CO_2 + H_2O$$

The third step is to calculate the molecular mass for each reactant and product; for convenience, these are placed underneath each chemical:

	$CaCO_3$	+ 2 HCl ⟶	$CaCl_2$	+ CO_2	+ H_2O
Grams/mole	100.09	36.46	110.98	44.01	18.02

Since certain chemicals require more than one mole in this reaction, we must multiply these molecular mass values by the number of moles used to get the grams of each reactant and product in a stoichiometric reaction, that is one in which all chemicals are present in exactly the amounts required by the balanced equation:

	$CaCO_3$	+ 2 HCl ⟶	$CaCl_2$	+ CO_2	+ H_2O
Grams in reaction	100.09	72.92	110.98	44.01	18.02

Notice that there are exactly 173.01 g total reactants and exactly 173.01 g total products. You will always have the same quantities of reactants and products in a stoichiometric reaction.

Now, if we run this reaction using exactly 100.09 g $CaCO_3$, chalk, and have at least 72.92 g HCl, we would get exactly 44.01 g CO_2 as a product. Of course 100.09 g is slightly more than 100.00 g, which was our original question so this cannot be the correct answer for a reaction using 100.00 g $CaCO_3$. To get that answer, we must determine how many moles of reactant we would use in 100.00 g $CaCO_3$, which will also tell us how many moles of CO_2 we will obtain.

Let's run our reaction this way, but still use the original amount of HCl; that poses no problem since our original question used excess HCl. We next calculate the number of moles of $CaCO_3$ we used by dividing 100.00 by 100.09; this gives us 0.999 and this is also the number of moles of CO_2 and H_2O we will get from this reaction:

	$CaCO_3$	+ 2 HCl ⟶	$CaCl_2$	+ CO_2	+ H_2O
Grams in reaction	100.09	72.92	?	?	?
Moles used or obtained	0.999	2.000	0.999	0.999	0.999

A simple calculation, multiplying the molecular masses of each product by 0.999, gives us the amount of grams that we will obtain for each product, starting with 100 g $CaCO_3$:

	$CaCO_3$	+ 2 HCl ⟶	$CaCl_2$	+ CO_2	+ H_2O
Grams in reaction	100.00	72.92	110.88	43.97	18.00
Moles used or obtained	0.999	2.000	0.999	0.999	0.999

Thus, the answer to our original question is 43.97 g CO_2.

Notice that the total mass is not the same for the reactants and products. These are 173.01 total grams reactants but only 172.85 g products. Why is this so? The answer is that we didn't include the excess HCl in our calculations. If we add this into the total "products" we get the same amounts for reactants and product, 173.01 g.

Let's do one more example. How many grams of silver chloride, AgCl, would we get by reacting 100.00 g silver nitrate, $AgNO_3$, with an excess of sodium chloride, NaCl? In this case, we don't care how much sodium nitrate is formed, nor is the exact amount of NaCl used of interest. The balanced equation is shown below with the molecular masses beneath.

	$AgNO_3$ +	NaCl ⟶	AgCl +	$NaNO_3$
Grams/mole	169.88	58.44	143.32	85.00
Grams used	100.00	excess	? g	—

Next, we divide 100.00 g $AgNO_3$ by 169.88 g/mole to calculate how many moles are present; it is 0.589, which will also be the number of moles of AgCl product.

	$AgNO_3$ +	NaCl ⟶	AgCl +	$NaNO_3$
Grams/mole	169.88	58.44	143.32	85.00
Grams used	100.00	excess	? g	—
Moles	0.589	excess	0.589	—

Now all we need to do is to multiply 143.32 g/mole, the molecular mass of AgCl, by 0.589 to get 84.42 g AgCl as the product. Since the amount of $NaNO_3$ is not important, we don't need to calculate it, but there would also be 0.589 moles of it produced. (That's 50.07 g if you're interested.)

As you can see, stoichiometric calculations are not really that hard. But a detailed study of such reactions is more appropriate, and essential, for the chemistry student. Non-science students generally prefer to see how chemistry enters into their life, and that is the main thrust of this text.

GLOSSARY

A

The numbers denote the page(s) where the term is discussed in more detail. The term is often in **bold** letters in text, is sometimes a section heading, or otherwise of sufficient importance to list here.

Acesulfame K (415) a nonnutritive, noncaloric, artificial sweetener
acetaminophen [Tylenol] (459) an analgesic derivative of acetanilide
acetic acid (260) a carboxylic acid that occurs in vinegar
acetone (255) a ketone; an important solvent
acetylcholine (389) a chemical compound that behaves as a messenger to transmit a message from one nerve cell to another, or from nerve cells to muscle cells
acetylsalicylic acid [aspirin] (459) a derivative of salicylic acid that is the most commonly used analgesic and antipyretic agent
achiral (359) nonchiral; having some symmetry
acid (195) a compound with a labile proton (H^+), that is, a proton that is readily able to separate from the rest of the molecule
acid rain (203) rainwater with a pH value between about 1.8 and 5.5
activation energy (183) an additional amount of energy that is required in order to make a reaction occur
active solar heating system (501) a solar heating system that uses pumps and other devices to move the heat
adenosine triphosphate [ATP] (385) the primary energy storage and transport molecule used by cells in metabolism; a nucleotide composed of the purine adenine, the sugar ribose, and a trio of phosphoryl groups
addition polymer (234) a polymer formed by an addition reaction
addition polymerization (234) a polymerization reaction in which the individual units add to each other without the loss of any atoms or groups
addition reaction (218) a chemical reaction in which other molecules add to a double or triple bond; similar to a combination reaction
additive (230) a chemical added to gasoline to increase the octane number; (292) a chemical added to a polymer to modify the properties
adhesive (279) a glue
ADP (385) adenosine diphosphate
advanced water treatment (319) an essentially chemical method of water treatment that can remove up to 98 percent of the waste ingredients and reduce the BOD average value to 2–5 mg/L
aerosol propellants (539) the chemicals used to propel materials out of an aerosol container
aflatoxin (467) one of the most potent carcinogens known, found on moldy peanut shells
aftershave lotion (445) a perfume-like lotion applied after shaving; composed mainly of ethanol and water and a very small amount of fragrant chemical agent
AIDS [Acquired Immune Deficiency Syndrome] (471) an affliction affecting the body's disease defense mechanisms; it is a viral disease caused by HIV (Human Immunodeficiency Virus)
alchemy age (48) a time period lasting roughly from 300 to 1600 A.D. during which alchemists worked on the transmutation of metals (e.g., lead into gold) and the search for the grand elixir or the philosopher's stone
alcohols (246) compounds that contain a hydroxyl group (OH) that is not attached to an aromatic ring
aldehydes (252) compounds that contain the $-\overset{\overset{\displaystyle O}{\|}}{C}-H$ group
aldosterone (380) a steroidal hormone secreted by the adrenal gland
alkali metals [IA] (87) elements in the first column of the periodic chart
alkaline (198) compounds with a pH above pH 7.0 are alkaline (or basic)
alkaline earths [IIA] (87) elements in the second column of the periodic chart
alkanes (211) a family of organic compounds containing only carbon and hydrogen atoms; also called paraffins and saturated hydrocarbons
alkenes (218) hydrocarbons that contain a carbon-carbon double bond (C=C); also called unsaturated hydrocarbons
alkyl groups (214) an alkane from which one H-atom is removed
alkylating agents (468) anticancer drugs that appear to react with the DNA and change its basic shape or structure by adding an alkyl group
alkynes (220) hydrocarbons that contain a carbon–carbon triple bond (C≡C); also called unsaturated hydrocarbons
allotrope (57) a different form of the same element
allotropy (57) the ability of an element to exist in two or more forms
alloys (21) solutions of a substance in a metal
alpha (α) helix (364) a right-handed, coiled secondary structure that normally maintains its shape by hydrogen bonds
alpha (α) particles (90, 119) a form of atomic radiation that consists of doubly-charged helium atoms
alternating copolymers (289) copolymers in which a pair of monomer units alternate along the backbone chain in a regular manner
alum treatment (315) a water treatment in which a floc formed from an aluminum sulfate:calcium hydroxide mixture traps the suspended matter causing turbidity which is then removed by filtration
aluminum chlorohydrate ($Al_2(OH)_5Cl \cdot 2H_2O$) (444) a chemical used as an antiperspirant
Alzheimer's disease (476) a progressive, irreversible, brain damage disorder that affects more than 20 million people worldwide
Ames test (414) a screening test that measures mutations in the genetic makeup of the bacterium *Salmonella typhimurium* and serves as a screening test for carcinogenicity
amides (258) compounds that contain the $-\overset{\overset{\displaystyle O}{\|}}{C}-N$ unit
amines (255) compounds that contain the $-NH_2$, $-NH$, or $-N$ group
amino acids (358) the subunits of proteins; molecules containing both the carboxylic acid and the amine functional groups
amorphous (273) a solid that lacks a crystalline structure arrangement of its molecules
amorphous solids (40) solids that lack a definite structure; solids that are not crystalline
AMP (385) adenosine monophosphate
amphetamine (256) a powerful and addictive central nervous system stimulant; known in "street drugs" as an "upper"
amphoteric surfactant (337) a surfactant that contains both a positive and a negative charge on the hydrophobic part of the molecule

amplitude (99) the height of a wave; the number of photons in a wave
amylopectin (370) a highly branched, and insoluble, polymer of glucose; a form of starch
amylose (370) a linear polymer of glucose; a form of starch
anabolic steroids (380) steroids that build up biostructures and are sometimes used by athletes to increase body mass; they have severe side effects
analgesic (459) a chemical agent that reduces pain
analysis (69) determining what components are present in a material; taking materials apart to determine their composition; the opposite of synthesis
anesthetic (250, 465) a chemical that produces a complete or partial lack of feeling
anion (146) a negative ion; an atom, or group of atoms, which has gained at least one electron
anion exchange resin (325) a crosslinked polymer that can exchange anions; in desalination, they exchange H^+ for Na^+
anion-permeable membrane (325) a semipermeable membrane that allows anions to pass through
anionic surfactant (337) a surfactant that contains a negative charge on the hydrophobic part of the molecule
anode (92) a positively-charged electrode or terminal
antacids (201) weak bases used to neutralize stomach acid
Antarctic ozone hole (541) a reduction in the concentration of the ozone over the Antarctic that varies from season to season and from year to year, but is most pronounced in the spring, following the sunless Antarctic winter season
anthracite (hard) coal (495) a sulfur-free coal that is more expensive than bituminous (soft) coal
antibacterial agents (464) drug agents that either destroy bacteria (bactericidal) or prevent bacteria growth (bacteriostatic)
antibiotics (466) substances that kill bacteria without injuring other forms of life
antibody (373) a specific molecule produced by the immune system, usually a glycoprotein, in response to the invasion of infectious agents (476) an agent developed by the body in response to infection
anticorrosion agents (344) chemicals that are added to detergent formulations to protect the metals in the washing machines from excessive corrosion or from dissolving in the detergent solution
antifreeze (248) a chemical, such as ethylene glycol or methanol, that is added to water to prevent freezing
antigen (373) any substance that can stimulate or activate the immune system, usually a foreign protein or a large polysaccharide
antimetabolite (468) an anticancer agent that imitates the normal components of DNA and interferes with cell metabolism, usually by preventing the cells from dividing and forming new cells
antimicrobial agents (418) chemical agents that retard spoilage by killing or preventing growth of bacteria, molds, or yeast
antioxidants (293) compounds that prevent or retard oxidation; (418) a chemical agent that prevents fats or oils becoming rancid by interrupting a free radical oxidation chain reaction
antiperspirant (444) a chemical agent that prevents or restricts perspiration
antipyretic (459) a drug agent that reduces fever
antiredeposition agents (348) chemicals that are added to detergent formulations to prevent removed dirt or soil from redepositing on the items washed; the most commonly used material is sodium carboxymethyl cellulose, CMC
aqueous solution (21) a solution of some substance in water
aquifer (305) natural water reservoirs, or retention volumes, that lie within permeable rocks and soils and are the most usable ground water supply; aquifers are recharged periodically by rainfall (or snow) and by seepage from surface water (streams, rivers, lakes)
aquitard (305) a natural ground water reservoir that has the water contained below or within rocks or soils of low permeability, such as crystalline rocks, clays, or shales, where it is less accessible
aromatic hydrocarbons (225) unsaturated cyclic hydrocarbons that do not undergo addition reactions; usually contain one or more benzene rings
artificial heart (289, 480) replacement heart made mainly from polymers
artificial sweetener (413) a nonnutritive or noncaloric sweetener
aspartame (415) an artificial sweetener that is the methyl ester of a dipeptide formed from phenylalanine and aspartic acid
asymmetric molecule (359) a chiral molecule; usually a carbon atom that has four different attached groups
atmospheric pressure (30) the pressure of air, which can be expressed in units of torr (or mmHg, the preferred unit in chemistry), inches mercury, or atmospheres
atoms (21, 56) the smallest subunits of matter; the basic building blocks of elements
atomic mass (weight) (67, 87) the mass (weight) in grams of a mole of atoms, or 6 x 10^{23} atoms of an element
atomic mass unit (amu) (64) the mass, in grams, of a mole of atoms
atomic number (97) the number of protons (or electrons) in an element; it is used to arrange the elements in the periodic table
ATP *see* adenosine triphosphate
autoimmune response (373) a reaction of the body's immune system against the body's own tissues
Avogadro's hypothesis (64) an assumption that equal volumes of gases contain equal numbers of atoms or molecules
Avogadro's number (64) the number of atoms or molecules in a mole, or approximately 6.0 x 10^{23} atoms or molecules
AZT, DCC, DDI [AIDS drugs] (471) drugs used to treat AIDS

B

bactericidal (464) drug agents that destroy (kill) bacteria
bacteriostatic (464) drug agents that prevent bacteria growth
bag filtration (497) a system used to remove particulate matter from exhaust gases via filtration
barometer (30) a device that measures air pressure
BAS *see* branched alkylbenzene sulfonate
base (195) a compound that will accept a proton
base pairing (374) base pairs that are held together with hydrogen bonds in DNA, such as adenine and thymine or cytosine and guanine
beeswax (437) wax from the honeycomb of bees; mainly myricyl palmitate, an ester of a fatty acid with a fatty alcohol
benefit (5) an advantage, good thing, or profit obtained from something
benefit-risk ratio (6) the ratio of the benefits from some item divided by the risk of using that item
beta (ß) particles (119) electrons; a form of radiation
beta (ß) rays (90) a form of atomic radiation consisting of electrons
beta (ß) pleated sheet (364) a protein secondary structure wherein the protein chains are arranged in the form of a sheet
binder (294) a drying oil or polymer in paints; a substance that binds or holds something together or onto something
binding energy (128) the energy that holds subatomic particles together in the nucleus
biodegradable (308) substances that degrade in the natural environment
biological oxygen demand (BOD) (312) the amount of oxygen consumed by decaying organic matter
biomass (496) material from plants
biotechnology (427) the utilization of living systems for the benefit of humans
bituminous (soft) coal (495) a soft coal containing sulfur; the most common coal found in the United States
bleaches (347) chemicals that remove colored stains by an oxidation-reduction reaction, changing the colored material into a colorless form; (446) an oxidation-reduction reaction using hydrogen peroxide (H_2O_2) that converts hair pigments to lower molecular weight, colorless, melanin degradation products
bleaching (446) the use of a bleach on the hair; *see also* bleaches
block copolymer (289) a copolymer that has long runs, or sequences, of each monomer
blocker (506) an agent that completely removes neutrons from a nuclear reactor
blood-alcohol level (247) the amount of ethanol in the blood
Bohr atom (102) an atomic model in which the protons and neutrons are localized in the nucleus, and the electrons are located in a series of specific orbits about this nucleus
Boyle's Law (32) an inverse relationship that shows that the gas volume decreases as the pressure increases
branched polymers (273) polymers that have branch chains coming off of the backbone chain in an irregular manner

branched alkylbenzene sulfonate [BAS] (346) a synthetic surfactant that did not biodegrade readily and caused some environmental problem due to its foam; a sulfonate surfactant that contains a branched hydrocarbon chain attached to its benzene ring
broad-spectrum (423) a nonspecific chemical agent that is effective against many kinds of insects, pests, or plants; the opposite to narrow-spectrum in its activity
brown melanin (446) a natural, brown-colored, polymeric pigment found in the hair and other parts of the body
butylated hydroxyanisole [BHA] (418) a chemical used as an antioxidant
butylated hydroxytoluene [BHT] (418) a chemical used as an antioxidant

C

calcium hydroxyapatite (196) the structural chemical compound that forms teeth and bones
calcium thioglycolate [$Ca(HSCH_2COO)_2$] (448) a chemical used in depilatories
canal ray (93) a beam of positively-charged particles that flows from the anode to the cathode in a cathode-ray tube; the mass of these particles varies with the gas present
carbon dioxide (526) a natural atmospheric gas claimed to be causing global warming
carbon filters (315, 322) water filters that contain activated carbon, which removes many unpleasant tastes or odors
carcinogen (467) any agent, especially chemical, that causes or induces cancer
carcinogenesis (467) the process of causing cancer
carbohydrate (246, 408) a common name for a saccharide; *see* saccharide
carbon cycle (129) the nuclear reaction sequence, found in the sun, that converts hydrogen atoms into helium atoms with the simultaneous emission of energy
carbonyl group (258) compounds that contain the C=O group
carboxyl group (258) the group containing the combination

$$-\overset{\displaystyle O}{\overset{\|}{C}}-O$$

carboxylic acid (258) a compound that contains the $-\overset{O}{\overset{\|}{C}}-OH$ group
carcinogenicity (414) the tendency, or ability, of any chemical or other kind of agent to induce cancer
cardiovascular devices (480) artificial devices, usually polymeric, that are used in the heart or blood vessels
carnauba wax (437) exudate from the leaves of a wax palm; the hardest and most expensive commercial wax
catalysis (184) the use of a catalyst in a chemical reaction
catalyst (184) an agent that makes a reaction run better, but still retains its own identity after the reaction is complete
catalytic converter (501) a device used in automotive exhaust systems to convert nitrogen oxides back to nitrogen and oxygen
catalytic cracking (232) a process used in the petroleum industry in which larger molecules are split into smaller ones
catalytic reforming (232) a process used in the petroleum industry in which smaller molecules are combined to form larger molecules or more useful isomers of molecules
catenation (210) the ability of an element to combine readily with many other atoms of itself
cation (146) a positive ion; an atom, or group of atoms, which has lost at least one electron
cation exchange resin (326) a crosslinked polymer that can exchange cations; in desalination, they exchange OH^- for Cl^-
cation-permeable membrane (325) a semipermeable membrane that allows the passage of cations through it
cationic surfactant (337) a surfactant that contains a positive charge on the hydrophobic part of the molecule
cathode (92) a negatively-charged electrode or terminal
cathode ray tube (92) an evacuated tube that permits the flow of a beam of electrons, called cathode rays, from a cathode to an anode
Celsius temperature scale (29) a temperature scale based on the values of 0°C and 100°C for the freezing and boiling points of water
cellulose (371) a polymer of glucose that contains a β-linkage between the repeat units; the most abundant polymer in the world
centi (14) one hundredth; 0.01 times; abbreviation is c
cetyl alcohol (437) a mixture of fatty alcohols with about 16 carbon atoms that melts at skin temperature; originally obtained from sperm whales, it is now made by the reduction of palmitic acid, a plant product
CFCs (chlorofluorocarbons) (526) halocarbons used in refrigeration, air conditioning, and other uses that are believed to cause a decrease in the concentration of ozone in the stratosphere
Charles' Law (34) a direct relationship that shows that the gas volume increases as the temperature increases
chemical bond (140) the attractive force holding atoms together as molecules in chemical compounds or molecular forms of elements
chemical change (76) alteration in the chemical nature of a material
chemical energy (115, 220) the potential (stored) energy in the chemical bonds of compounds that can be released during chemical reactions
chemical equations (66) a shorthand method for describing a chemical reaction in which formulas of the reactants are written on the left of a reaction arrow and formulas of the products are written on the right side
chemical formula (65) a summary of the composition of a compound (or element) in which subscripts on the symbols of the different atoms tell how many of each kind of atom is present
chemical reaction (66) a change in which the composition of one or more substances is changed into different compositions
chemiluminescence (116) the production of light during a chemical reaction with the production of little or no heat
chemistry (2) the science that deals with matter and the energy that causes changes in matter
chiral molecules (359) molecules that can exist as mirror images; an asymmetric molecule, usually a carbon atom with four different attached groups
chirality (360) possessing units that are chiral
chitin (370) a polysaccharide used in the structural exoskeletons of insects and shellfish
chlorination (314) a water purification method that uses elemental chlorine to kill microorganisms
chlorophyll (420) a complex plant pigment, structurally similar to hemoglobin, that catalyzes photosynthesis
chromatography (73) an analytical technique that purifies materials by flowing a solution (liquid or gas) over a stationary phase that is a solid or a liquid
chromosome (374) a threadlike structure in the cell's nucleus that carries the genetic information; a DNA molecule with an attached protein that is coiled into a short, compact unit
chronic endemic dental fluorosis (196) a condition in which teeth become mottled and more brittle
cis- (219) a term denoting that two similar units lie on the same side of a double bond
class formula (212) a formula that describes the chemical composition of all organic molecules in a homologous family, such as C_nH_{2n+2} for alkanes
climatology (528) a branch of meteorology that studies long-term weather phenomena
coal-tar (447) chemicals obtained from coal by pyrolysis (destructive distillation)
coatings (294) a covering layer; paint; enamel; lacquer
cocoa butter (437) a wax-like extract from the cacao bean that melts at 30–35°C, near skin temperature; it has a chocolate-like taste and consists mainly of fatty acid glycerides; also known as theobroma oil
codens (376) a group of nucleotides that specifies, and commands, the addition of a particular amino acid into a growing peptide chain
cold cream (438) a cosmetic that is used mainly to remove other cosmetics that are not water soluble and for cleaning the skin
collagen (441) the most common protein in the body; it contributes mechanical stability to the skin, bones, and tendons
colognes (445) diluted perfumes that contain 1–3 percent fragrance in a solvent that is mostly ethanol

colorants (416) food colorings, usually added for esthetic appeal
combination reaction (170) a reaction in which two or more chemical substances, either elements or compounds, combine to form a new compound
combustion (50) the process of burning; oxidation
compact or **condensed formulas** (163) chemical formulas in which many of the actual single bonds are not shown, although they are assumed to exist, however, the presence of any double or triple bonds is shown
compact laundry detergent (349) a more concentrated laundry detergent that contains essentially no inert fillers
complementary arrangements (374) the arrangement of pairs of DNA strands that are held together by hydrogen bonds; *see* also base pair
compound (21, 60) a pure substance that cannot be separated into other components by physical methods, and contains at least two different kinds of atoms, but can be separated into elements (or other compounds) by chemical methods
compressibility (26) a property describing what happens when a material is placed under pressure
concentration (72, 183) the amount of solute dissolved in the solvent; normally expressed as g/L or g/g solvent
condensation polymers (234) polymers formed by a condensation reaction, for example a poly(ester)
condensation polymerization (263) a polymerization reaction in which a small molecule is split out from the monomer(s) during polymer formation
condensation reaction (262) a chemical reaction in which a small molecule splits out during the course of the reaction
contraceptive (386) a device or chemical agent designed to prevent conception
controlled release (478) systems that regulate the flow of a drug and help maintain the correct level required to treat the disease, thereby reducing undesirable side effects
copolymer (288) a polymer formed from two or more monomer units; in other words, a polymer containing two or more different kinds of repeat units
copolymerization (288) the process used to make copolymers
cortisone (380) a steroid used to treat rheumatoid arthritis, asthma, and several other diseases; secreted by the adrenal gland
cosmetics (436) chemical formulations that are used for beautifying the external parts of the human body; defined by the United States Food, Drug, and Cosmetic Act of 1938 as "articles intended to be rubbed, poured, sprinkled or sprayed on, introduced into, or otherwise applied to the human body or any part thereof, for cleansing, beautifying, promoting attractiveness or altering the appearance."
cough suppressant (478) a drug agent that can suppress coughing
covalent bonds (140) chemical bonds formed by the sharing of pairs of electrons
covalent compounds (89) compounds whose chemical bonds are formed by sharing electrons
creams (438) oil in water emulsions, whose main purposes are to soften and moisten the skin
critical mass (129) the minimum amount of fissionable material necessary to produce an explosive chain reaction, as in an atomic bomb
crosslinked (235, 273) a polymer in which the backbone chains are chemically bonded together to form a larger, nonlinear structure
crosslinking (235) chemically bonding together the backbone chains of a polymer to form a larger, nonlinear structure, as in rubber vulcanization
crystalline solids (40) solids that have a precise structural arrangement for their atoms or molecules
cyclamate (414) a nonnutritive artificial sweetener now banned in the United States but not in Canada
cycloalkanes (216) alkanes that exist in the form of rings
cyclone collector (497) a system used to remove particulate matter from exhaust gases via centrifugation

D

Dacron (263) a poly(ester) fiber (textile) formed from ethylene glycol and terephthalic acid, or one of its esters; *see* also Mylar and polyester
Dalton's atomic theory (54) a theory consisting of a series of simple assumptions that explains both the law of the conservation of matter and the law of definite composition
data (8) an assumed or conceded fact from which further conclusions are reached; (singular is datum)
decanting (71) a method for separating a mixture of a liquid and a solid by pouring off the liquid and leaving the solid behind
decomposition reaction (170) a chemical reaction in which a chemical compound breaks up into two or more substances
Delaney Amendment (415) a law, enacted in 1960, that prohibits placing any chemical, natural or synthetic, into food if the chemical is shown to be cancer-causing in laboratory testing
Delrin (252) a poly(formaldehyde) thermoplastic
density (252) the ratio of the mass (m) to the volume (V)
dental (481) pertaining to the teeth
dental caries (196) cavities; holes formed in the teeth
depressant (247) that which retards any function
desalination (321) the removal of salt and other dissolved matter from ocean water or brines
detergency (38, 341) the process of cleaning, usually using a detergent containing a surface active agent, or surfactant
detergent (339) a chemical formulation used in cleaning; a word used as a synonym for surfactant
deodorant (444) chemicals that attempt to mask or adsorb body odor
deoxyribose (369) a five carbon monosaccharide; a component of DNA
depilatory (448) a cosmetic used to remove hair by chemical action
dextro (369) a term meaning right; the opposite of levo
dextrose (369) another name for glucose
dextromethorphan (478) a cough suppressant that is a less addictive chemical modification of codeine, which is also a cough suppressant
dialysis (480) a method used to separate substances in solution form that is used to treat people with defective kidneys
dielectric effect (157) another name for the dipole effect; the dielectric constant is a measurement of magnitude of the dielectric effect in a specific compound
dienes (236) hydrocarbons that contain two carbon–carbon double bonds (C=C)
diethylstilbestrol [DES] (388) a steroid-like molecule that was once used as a potential birth control agent; its use was discontinued because it also induced high levels of vaginal cancer
dietitian's calorie (400) equal to a kilocalorie
digestion (400) the process in which the compounds are broken down into smaller units by the basic chemical reaction of hydrolysis
digitoxin (380) a steroidal compound isolated from the foxglove plant that is used to treat some forms of heart disease
dipole (157) molecules that are simultaneously attracted toward both electrodes due to a polarized bond
dipole effect (157) the attraction between polar molecules
disaccharide (367) a saccharide composed of two monosaccharide units, joined via an oxygen bridge, such as sucrose
displacement reaction (171) a reaction in which one chemical agent displaces another one from a chemical compound
distillation (70) a purification technique that separates materials on the basis of boiling points; (321) a water purification method involving heating the water to the boiling point, cooling its vapors, and collecting the pure water
disulfide (S—S) linkages (446) linkages that form between cystine units in protein molecules, giving toughness and shape to the skin and hair
double displacement reaction (171) a chemical reaction in which two elements or groups trade places in a pair of chemical compounds
dyes (257, 348, 442, 447) a chemical additive that imparts color

E

efficiency (114) the ratio of the work accomplished to the amount of energy inputted
elastomers (236, 276) polymers that have extensibility and elastic recovery (i.e., rubber-like properties)
electrical energy (114) energy derived from the motion of electrons

electromagnetic fields [EMFs] (117) radiation ordinarily limited to the electromagnetic spectrum that is in the radio or microwave portions

electrons (92) negatively-charged subatomic particles with a mass of 0.00055 amu or 19.11×10^{-28} g (approximately 1/1833rd that of a hydrogen atom) and a charge of -1, or -1.60×10^{-19} coulombs

electron microscope (103) a microscope that utilizes the wave character of the electron and can "see" objects much smaller than can be detected by visible light

electrostatic precipitator (497) a system used to remove particulate matter from exhaust gases by electrostatic precipitation

electrodialysis (325) an electrochemical method of desalination that uses ion-selective semipermeable membranes

electronegativity [EN] (151) the attraction of the nucleus of an element for electrons

element (21, 56) a pure substance that consists of only a single kind of atom and cannot be separated into other components by physical or chemical methods

emission spectra (98) spectra produced by heating a material and causing it to release electromagnetic radiation

emollient (350, 437) chemical blends that are used to soften skin, such as glycerine, lanolin, or fat

emulsifier (419) food additives that prevent phase separation in liquid mixtures

emulsion (39, 341) a dispersion or suspension of one liquid in another

enamel (294) a varnish with added pigments in which the polymers crosslink during drying; used mainly to paint metal surfaces

enantiomer (359) stereoisomers that are mirror images of each other

endothermic reaction (76, 182) a chemical reaction that requires the addition of heat or energy in order to run

energy levels (100) the energy associated with the outermost electrons in an atom; the same number as the period in the periodic chart

enkephalins (462) natural, polypeptide analgesics that are made in the brain

Environmental Protection Agency [EPA] (310) a federal agency established to protect the environment

enthalpy [heat of reaction] (182) the amount of heat released or consumed by a reaction; also called the heat of reaction

entropy (182) a thermodynamic term used to describe the degree of randomness or disorder in a system; entropy generally increases with an increase in temperature

enzyme (347, 366) a protein that functions as a catalyst; a protein catalyst used in detergent blends to remove stains

epoxy resins (287) polymeric materials that contain a free epoxy group that can be cross-linked to form a rigid polymer; useful as adhesives

equilibrium (38, 261) a condition of equal balance between two opposing forces; a set of circumstances when things are going at the same rate, at the same time, in forward and reverse directions; a state in which no net change occurs with the passage of time

erodable device (479) a controlled release device that works by having part of the material eroded away by hydrolysis or other biochemical activity

essential amino acid (358) amino acids that cannot be synthesized within the human body

esters (260) compounds with the general formula $R{-}\overset{\overset{\large O}{\|}}{C}{-}OR'$

estradiol (388) a steroidal hormone, produced by the ovaries, that regulates the menstrual cycle and controls secondary sexual characteristics

ethanol (247) a two carbon alcohol; an important solvent and chemical intermediate; used in alcoholic beverages; ethyl alcohol; commonly known as grain alcohol

ethers (250) compounds with the general formula R—O—R, where R is an alkyl, alkenyl, or aromatic unit

ethyl alcohol (247) *see* ethanol

ethylene glycol (248) a glycol used as an antifreeze

eutrophication (307) the enrichment of water with nutrients from biodegradation, fertilizer runoff, or other sources, that leads to excessive algae bloom

evaporation (41) the process in which a liquid is converted to its gas state

excess deaths (494) deaths occurring during a period of abnormally high rates of death

exothermic reaction (76, 182) a chemical reaction that liberates heat or energy when it is run

expectorant (478) an agent that helps bring up mucus by decreasing its viscosity

experiment (10) a new set of events (including a new chemical reaction) or a different way of viewing old data that can test the predictions of a theory

exponential (77) a shorthand way of writing a number in which the number is expressed as another number raised to some power; usually involves the number ten raised to some power; e.g., $1{,}000{,}000 = 10^6$.

extensive property (26) a property that depends on the amount (extent) of material present

eye shadow [eyeshade] (444) a suspension of colorants in an oil-wax-fat base used as a cosmetic to shade around the eyes

F

face powder (443) cosmetics that are used to cover blemishes or hide oily spots on the face

fact (8) a statement or action that can be observed or shown to be true

Fahrenheit temperature scale (28) a temperature scale based on the values of 32°F and 212°F for the freezing and boiling points of water

families (87) the vertical columns in the periodic chart wherein the elements possess similar chemical properties

fatty acid (337) organic acids with 12–22 carbon atoms

fatty alcohol sulfate [FAS] (351) anionic surfactants obtained from natural plant oil sources

fertilizer (420) chemical mixtures that provide some of the minerals needed by plants

fibers (textiles) (276) polymers used to make fabrics and the like; linear thermoplastics that have the individual polymer chains held together by hydrogen bonds, polymer crystallinity, or both

fillers (293) compounds that are added to make a polymeric composition less expensive; (344) anything added to take up space or weight

filtering (71) a method for separating liquids from solids by passing the mixture through porous filter paper or cloth

final note (446) the least volatile fragrance in a perfume mixture

first law of thermodynamics (112) a law stating that energy cannot be created nor destroyed, but that it can be changed in form or type

fixatives (445) a substance that prevents the too rapid evaporation of the fragrance agent in perfumes

flame resistance (293) the ability to resist flames or fire

flame retardants (293) chemical agents that confer full or partial flame resistance on another substance

flavoring agent (416) chemicals added to impart a special taste to food

floc (315) short for flocculate; a precipitate that forms when aluminum sulfate solutions are mixed with calcium hydroxide solutions

fluoridation (196) the addition of fluoride ions to a drinking water supply to reduce caries formation

fluorocarbons (541) potential replacements for the chlorofluorocarbons that only contain fluorine and carbon

fluorohydrocarbons (541) potential replacements for the chlorofluorocarbons that contain only fluorine, hydrogen, and carbon

food additive (412) chemicals added to impart some desirable characteristic to foods

food irradiation (130) the process of radiating food with gamma (γ) radiation to destroy microbial life

fraction (70) a portion collected in a restricted temperature range during distillation

fragrance (348, 445) an aroma agent, an odor agent, such as a perfume

frequency (74, 99) the number of times a wave passes a specific point in a specific time interval

frequency of collision factor (183) a factor that relates to the number of collisions that occur between atoms or molecules in a given time interval

fructose (369) a hexaketose; a six carbon monosaccharide that contains a ketone unit

fuel cell (510) a device that converts the energy of an oxidation-reduction reaction directly into electricity

functional groups (244) atoms other than carbon or hydrogen, normally polar units, that allow organic molecules to have more reactions and applications
fungicides (257) chemicals that kill fungi

G

gamma (γ) radiation (119) a form of true radiation consisting of very short wavelength, or high frequency, electromagnetic radiation
gamma (γ) rays (90) true electromagnetic radiation that is released from some elements
gasohol (251) a gasoline-alcohol mixture
genetic engineering (376) manipulating the nucleotide sequence in DNA or RNA
geometrical isomers (219) isomers that arise because of the presence of a double bond or a cyclic unit in the molecule
geothermal energy (509) heat energy found in the ground
germicidal (250, 342) pertaining to killing germs
germicide (314) a chemical agent that destroys (kills) germs (bacteria)
glass transition temperature, T_g (272) the temperature at which a polymer becomes soft or flexible rather than hard or rigid
global warming (524) the theory that the average temperature of the Earth is increasing
global wind patterns (496) the pattern wind currents display over the entire planet
glucose (369) an aldohexose; a six carbon monosaccharide with an aldehyde functional unit; also called dextrose
glycerol (248) a three carbon chain with three hydroxyl groups, commonly used in cosmetics
glycogen (370) a highly branched polymer of glucose that is stored in the liver and muscles of animals
glycolipids (379) a complex lipid containing a saccharide group
graft copolymer (289) a copolymer that has a chain of one monomer type with shorter chains of the other monomer grafted onto it, much like a branch of one plant is grafted onto the rootstock of another
grain alcohol (247) *see* ethanol; ethyl alcohol
gram (14) a metric unit of mass; abbreviation is g
GRAS list [Generally Regarded As Safe] (416) a "grandfather clause" that grants a special listing to materials used for many years prior to the Delaney Amendment
greenhouse effect (524) that part of the solar energy cycle in which energy is converted into infrared radiation, held within our atmosphere, and warms the planet
greenhouse gases (527) gases that promote the greenhouse effect; the main such gas is water vapor with carbon dioxide being the second most important
ground state (102) the lowest energy form or state for an atom or molecule
groundwater (305) water that lies a few feet to several thousand feet under the earth's surface; it is the source of potable water for over 75 percent of the people in the United States
groups (87) a family; a column in the periodic chart wherein the elements have similar chemical properties
guaifenesin (478) a drug agent used as an expectorant

H

Haber process (421) a process used to manufacture ammonia directly from nitrogen and hydrogen
half-life (121) $t_{1/2}$, the length of time required for half of a radioactive sample to disintegrate naturally
halogenated methanes [THMs] (322) methane molecules that contain halogen atoms
halogens [VIIA] (87) elements in the seventh column of the periodic chart
Halon molecules (540) a halocarbon that is used in fire extinguishers
hair dyes (447) chemical formulations used to change the color of hair
hair permanent (449) chemical formulations used to make hair wavy
hair rinses (352) a hair conditioner, usually a cationic surfactant
hair spray (446) simple solutions of about 4 percent polymer (resin) in a volatile solvent that are sprayed onto the hair
hair straightener (449) chemical formulations used to straighten very curly hair
hard water (305) water that contains 150 ppm or more of Ca^{2+} and Mg^{2+} ions; Ba^{2+} and Fe^{2+} ions are sometimes included in the hard water totals
HCFCs [hydrochlorofluorocarbons] (545) replacements for chlorofluorocarbons that contain hydrogen atoms attached to the carbon atoms, in addition to the chlorine and fluorine atoms
HDPE (276) high density poly(ethylene)
heat (27) a measure of the total kinetic energy in a material
heating diagram (curve) (22) a diagram relating the temperature and the states of matter, showing the melting (freezing) and boiling (condensation) points
heat stabilizers (293) enable polymer compositions to resist heat degradation
heavy metals (197) metallic elements that have an atomic mass greater than about 200 amu and are usually poisonous to humans and animals
herbal medicine (479) the use of herbs as medical agents
herbicide (257, 422) chemicals that kill unwanted plants, often defined as weeds
heterogeneous matter (20) matter that does not appear to be the same throughout the sample
HFCs [hydrofluorocarbons] (545) halocarbons that contain hydrogen and fluorine; replacements for chlorofluorocarbons
high-density lipoproteins [HDL] (379) a lipoprotein, found in the blood, that transports cholesterol away from cells and back to the liver; this is considered the better form of lipoproteins
HIV [Human Immunodeficiency Virus] (471) the virus that causes AIDS
hollow fiber membranes (325) fibers used in dialysis and reverse osmosis
homogeneous matter (20) matter that appears to be the same throughout the sample
homologous series (212) a series of hydrocarbons differing from each other by a $—CH_2—$ group
homopolymer (288) a polymer that contains only one kind of repeat unit
hormone (380) a class of chemicals secreted by glands within the body, and not obtained from the diet, that cause important changes to occur in body chemistry; there are four main classes—steroidal, polypeptide, thyroid, and catecholamine
hybrid bond (222) bonds that involve combinations of s- and p-orbitals
hybridization (222) the process of forming hybrid bonds
hybrid orbital (222) an orbital formed by the combination of an s-orbital with one or more p-orbitals
hydrate formation (326) a desalination method that separates seawater from its salts by forming hydrates that are essentially free of salts
hydrocarbons (210) compounds composed solely of carbon and hydrogen
hydroelectric energy generation (509) electrical energy derived from flowing water, such as a dam or waterfall
hydrophile-lipophile balance (336) the balance in properties, in surfactants, between the attraction to water and lipids
hydrogen bond (158) relatively weak attractions in which an H-atom bonded to a N, O, or F atom is attracted to another N, O, or F atom in another molecule
hydrolase (401) an enzyme that catalyzes hydrolysis reactions
hydrolysis (400) the chemical reaction of a molecule with water in which the molecule becomes broken down into smaller segments
hydronium ion (199) a H_3O^+ ion
hydrophilic (336) the water-loving portion of a surfactant; they are the more polar portion of the surfactants and frequently are ionic
hydrophobic (336) the water-fearing, or hating, portion of a surfactant; this is the nonpolar portion of the molecule and is frequently the hydrocarbon portion of the surfactant
hydroxyl group (246) the covalently bonded —OH group
hypothesis (9) a set of assumptions conditionally accepted as a basis for reasoning or investigation
hyaluronic acid (441) a natural polysaccharide obtained from rooster combs; found naturally in the eyeball, synovial fluid, and other parts of the human body

I

iatrochemistry age (49) a time period, between approximately 1500 to 1700 A.D., when chemistry was placed in the service of medicine
ibuprofen [Advil, Nuprin, Motrin] (460) an over-the-counter analgesic and antiinflammatory agent
ice core analysis (532) a study of drilled ice cores to determine the possible carbon dioxide levels of the past
immiscible liquid (39) a liquid that doesn't dissolve in a second liquid
immunoglobin (373) special proteins produced by the immune system in response to antigens; an antibody
incineration (296, 309) a landfilling alternative that can be done safely with modern technology and is probably the safest way for the disposal of medical waste; a furnace or apparatus for reducing any substance to ashes; a method of waste disposal
independent chemistry age (49) a period starting about 1650 A.D. during which chemistry was freed from both alchemy and medicine and became an independent science
inert (284, 294) inactive; devoid of active chemical properties; in paint and coatings, pigments that make the paint last longer
indicator (198) a chemical substance that changes color in the presence of an acid or a base
Industrial Revolution (532) a period of time, centering about 1770, wherein industrialization growth increased extensively
industrial (London) type pollution (493) air pollution characterized by the presence of sulfur dioxide and particulate matter
inner transition elements (87) a collection of fourteen (14) vertical groups of elements that come out from the transition element series between elements 57 and 72. Their incoming electrons enter into f-orbitals
insulators (40) materials that do not conduct electricity or heat very well
intensive property (36) a property that depends only on the purity and chemical nature of the material, and is independent of the amount present
intensity (99) the relative strength or power of a light wave
invert sugar (370) the sugar produced by the hydrolysis of a disaccharide, usually sucrose
intrauterine device [IUD] (388) a device made of plastic or metal that is inserted into the uterus to prevent conception
ion-exchange polymers [resins] (322) crosslinked polymer beads that are used to remove ions (e.g., Ca^{2+}, Mg^{2+}) from hard water; they are also sometimes used to desalinate seawater
ionic bond (146) chemical bonds formed by the transfer of electrons
ionization potential (152) a measure of the energy required to remove an electron from an atom or ion
iodine number (401) a measure of unsaturation used for fats and oils; the amount of iodine that could add to the double bonds, expressed as grams iodine per 100 grams substance
ion (89) fragments of atoms, or compounds, that have either gained or lost electrons and acquired an electrical charge
ionic compound (89) a compound formed via the transfer of electrons from one atom to another
ionic solid (40) a solid consisting of charged particles called ions that are arranged into a crystalline lattice, and which conduct electricity in the molten state or when dissolved in water
isomers (213) compounds with the same molecular formula but having a different structure; several forms of isomerization exist; *see also* structural, geometrical, and optical isomers
isotopes (3, 98) an atom that has a slightly different atomic mass value than other atoms of the same element; an element with a slightly different atomic mass, but with the identical atomic number
IUPAC system (211) a standardized system of nomenclature; letters stand for International Union of Pure and Applied Chemistry

J

juvenile growth hormone (426) insect hormones that control the maturation to the adult form

K

Kelvin temperature scale (29) a temperature scale that runs from "absolute zero" (−273°C); water freezes at 273K and boils at 373K; sometimes called the absolute temperature scale
keratin (446) fibrous proteins that are present in the skin, hair (fur), and nails of people and most land animals
ketones (254) compounds with the general formula $R{-}\overset{\displaystyle O}{\overset{\|}{C}}{-}R'$, where R is an alkyl, alkenyl, or aromatic group
kilo (14) 1,000 times; abbreviation is k
kinetic energy (112) the energy of motion
kinetic molecular theory of gases (37) a theory that explains the behavior of gases based on assumptions about the motion of atoms or molecules

L

lacquer (294) a polymer solution with a small amount of added pigment; they contain no drying oils and do not crosslink
lactose (369) a disaccharide formed from glucose and galactose
LDPE (276) low density poly (ethylene)
landfill (308) a place used to bury waste materials
lanolin (437) an ester found in sheep's wool and made from cholesterol (an alcohol with the formula $C_{27}H_{45}OH$) and fatty acids such as stearic acid, $C_{17}H_{35}COOH$
LAS *see* linear alkylbenzene sulfonate
latex paints (294) paints made from an aqueous polymer dispersion with disperssed pigments; also called water-based or emulsion paints
law (8) an assembly of data or facts that always seems to be true
law of conservation of energy (112) a law stating that energy cannot be created nor destroyed, but that it can be changed in form or type
law of conversation of mass (52) a law stating that there is no loss in the total amount of matter during a chemical (or physical) reaction
law of conservation of matter and energy (116) a law that states that the total amount of mass and energy is constant, although either can be converted into the other
law of definite proportions (50) every compound has a specific chemical composition
law of multiple proportions (56) several distinct compounds can be made by different combinations of the same kinds of atoms
LD_{50} (249, 460) numbers that represent the amount of a chemical, usually expressed as mg/kg or g/kg body weight, that causes the deaths of 50 percent of the laboratory animals in a toxicity test
lead poisoning (197) lead (Pb) is a very poisonous chemical and can cause nerve damage and learning disabilities; the Public Health Service limits the permissible level in drinking water to only 15 ppb (0.0000015%);.
lecithin (379) also known as phosphatidylcholine; it is found in egg yolks, soybeans, and other sources and is used as an emulsifying agent in ice cream, mayonnaise, and salad dressings
Left-Handed LEGO (179) a mnemonic for oxidation; it stands for Lose Hydrogen, Lose Electrons, Gain Oxygen
legumes (420) plants, such as peas and beans, that can fix the elemental nitrogen from the atmosphere and place it into the soil in the form of a nitrate (via oxidation) or ammonia (via reduction)
leukotrienes (379) lipids with hormone-like properties; they are produced by some white blood cells and promote the bronchi constriction that is characteristic of asthma
levo (369) a term meaning left; the opposite of dextro
Lewis electron dot structure (144) structures that depict the number of electrons in the outermost shell of an atom; used to describe changes in this number after bond formation
light energy (115) electromagnetic radiation
lignin (409) one of the non-digestible components of dietary fiber, along with cellulose, pectin, and various gums
line structural formula (160) chemical structures wherein the bonds are represented by lines
linear polymers (235, 273) polymers in which the backbone chains do not have branching or crosslinking

linear alkylbenzene sulfonate [LAS] (346) alkylbenzene sulfonates that possess a linear hydrocarbon chain attached to the benzene ring and can biodegrade
lipase (401) an enzyme that hydrolyzes the bond between glycerol and the fatty acids in a fat or oil
lip balm (442) cosmetics, generally uncolored, that are applied to the lips to prevent the evaporation of moisture and the resultant dryness
lipids (377) simple lipids are the fats and oils
lipid soluble vitamins (404) vitamins that are soluble in fats or oils
lipophilic (342) the oil-loving portion of a surfactant, usually the hydrocarbon tail portion
lipoproteins (379) molecules consisting of a hydrophobic lipid core surrounded by an outer layer of proteins and hydrophilic lipids, such as phospholipids
lipstick (442) a cosmetic used to enhance the color of the lips and also to protect them against dryness
liter (14) a metric unit of volume; abbreviation is L
lock-and-key mechanism (366) a mechanism proposed for enzyme activity to explain enzyme specificity in which the reactant molecule (the substrate) binds with the enzyme to form a complex; the complex site on the enzyme has a shape that relates to the reactant like a lock relates to the key
low-density lipoproteins [LDL] (379) a lipoprotein, found in the blood, that transports cholesterol away from the liver and to the cells; this is considered to be a risky form of lipoprotein

M

macromolecule (234) a giant molecule; another term for a polymer
macromolecular solids (40) a lattice consisting of long chains or rings of atoms held together by chemical bonds; they usually do not conduct electricity
magic bullet (464) a term coined by Paul Ehrlich to describe the relatively specific action of his drug
main group elements (87) the eight families of elements (from IA through VIIIA); the incoming electrons in these elements enter into either s- or p-orbitals
malleable (40) a metal property that means they can be beaten or pressed into a definite shape
maltose (370) a disaccharide formed from two glucose units
matter (20) that which occupies space and has mass
mascara (444) a cosmetic that is used to darken eyelashes
mechanical energy (114) the energy produced by machinery, such as motors; a form of kinetic energy
mega (14) one million times, 1,000,000 times; abbreviation is M
melamine-formaldehyde polymers (286) crosslinked condensation polymers prepared from formaldehyde and melamine
melanin (446) a brown-black, polymeric pigment found in the skin, hair, and other parts of the body
melanoma (439) a virulent form of skin cancer that appears linked to severe sunburn, especially for people with light complexions, blue eyes, naturally blond or red hair, and freckles
Mendeleev inversions (97) places in the periodic table wherein the heavier element of a pair has a lower atomic number
mercury poisoning (197) mercury is a heavy metal that sometimes poisons the environment or people
metal (40, 89, 140) solids that consist of atoms arranged in a definite pattern called a crystal lattice; elements that are good conductors of electricity and heat, have luster and are malleable; metals form ionic compounds in which they exist as positively charged ions (cations); they are the largest group of elements in the periodic chart
metallic hair dye (447) a hair dye that contains lead acetate and a suspension of powdered sulfur
metal bonding (140) metals consist of spherical atoms arranged as compactly as possible in a crystalline solid; this explains why metals have high density. Scientists assume that the outermost electrons in each metal atom are shared equally by *all* the metal atoms, which is why they can conduct electricity.
metalloids (89) an element that lies near the borderline between metals and nonmetals in the periodic chart; they possess properties of both bordering classes
meteorology (534) the science of weather
meter (14) metric unit of length; abbreviation is m
methane (526) the gas CH_4, which is presumed to play a role in potential global warming
methanol (247) wood alcohol; methyl alcohol
methyl alcohol (247) *see* methanol
micelle (341) an arrangement of surfactant molecules in which the hydrocarbon tails are packed close together and the water-loving, hydrophilic parts are placed on the outside
micro (14) one millionth times, 0.000,001 times; abbreviation is μ
microcapsule (480) a small capsule used in controlled release systems
middle note (446) a perfume component with medium volatility
Milankovitch cycles (534) a theory that predicts cycles for the incoming solar radiation of 21,000, 41,000, and 100,000 years
milli (14) one thousandth times; 0.001; abbreviation is m
mineral (383) an inorganic chemical, usually an ion, which is essential to our health
mineral oil (437) a liquid alkane mixture
minoxidil (451) a drug, originally developed for treating high blood pressure, that can be used to grow hair; *see also* Rogaine
mirror image (362) the image that forms in a mirror; one chiral molecule is the mirror image of another similar chiral molecule
moderator (506) a material that controls the speed of the neutrons in a nuclear reactor and slows them down to the correct use range
molarity [moles/L] (194) the number of moles of a substance present in 1,000 mL total solution
mole (64) a collection or pile of atoms or molecules; more specifically, an Avogadro's number (6.02×10^{23}) of molecules or atoms; the quantity of a gas that occupies 22.4 liters at 0.0°C and 1.0 atmosphere pressure, the standard conditions for gas measurements
molecular mass [weight] (67) the mass (weight) in grams of an Avogadro's number of molecules; the mass in grams of a mole of molecules
molecular solids (40) solids that have molecules (or atoms) arranged in a crystalline lattice but do not conduct electricity
molecules (21, 60) the smallest subdivision units of compounds or some elements; collections of atoms that are held together by some force
moles/liter (194) *see* molarity
monolithic device (480) a type of controlled release system
monomer (234) a single unit that is used to form a polymer
monosaccharides (367) saccharides that have only a single unit
monosodium glutamate [MSG] (416) the sodium salt of the amino acid glutamic acid, which is used as a flavor enhancer
multiple bonds (150, 226) double or triple covalent bonds
multiple covalent bonds (150) covalent bonds involving more than a single pair of shared electrons
multisubunit protein (365) *see* quaternary structure
Mylar (263) a poly(ester) plastic prepared from terephthalic acid (or one of its esters) and ethylene glycol; *see also* Dacron and polyester

N

N-P-K fertilizer blends (422) fertilizers that list the nitrogen, phosphorus, and potassium content, in that order
nail polish (443) a cosmetic lacquer applied to the finger or toenails
nail polish remover (443) a solvent that is used to remove nail polish
naproxen [Aleve] (461) an over-the-counter analgesic agent
narrow-spectrum (423) a chemical agent that is fairly specific in its biological activity, as in a narrow-spectrum herbicide; the opposite of broad-spectrum regarding the range of activity
network polymer (235) a crosslinked polymer, as in vulcanized rubber
neutron (94) an uncharged (neutral) subatomic particle with a mass of 1.0087 (1.675×10^{-24} g), or about the same weight as a proton or a hydrogen atom
neurotransmitter (389) chemicals, such as acetylcholine, that control the transmission of nerve signals
neutralization (198) the reaction between an acid and a base to

produce a salt; water is also often produced during neutralization
NIMBY (313) an acronym for Not In My Back Yard
nitrogen oxides [NO_x] (499, 526) oxides of nitrogen that arise from both human and natural sources and are a form of air pollution; NO_x compounds can lead to the destruction of the ozone layer
noble (inert) gases [VIIIA] (87) elements in the last (eighth) column of the periodic chart
nomenclature (211) any system of naming compounds
nonaqueous solutions (21) solutions that do not contain water
nonbonding electron pairs (160) pairs of electron that are not involved in bond formation, though they do influence the shape of a molecule
nonionic surfactant (337) a surfactant that contains no charges on the hydrophobic or hydrophilic parts of the molecule
nonmetals (87) elements that are poor conductors of electricity or heat, often brittle, and form covalent compounds or ionic compounds in which they are negatively charged ions (anions); they are located in the upper right-hand side of the periodic chart
nonnutritive (415) has no nutritional value; noncaloric
nuclear chain blocker (506) an agent that completely removes neutrons from a nuclear reactor
nuclear chain moderator (506) an agent that controls the speed of neutrons and slows them down to the correct range
nuclear energy (116) the energy derived from reactions involving the nucleus of atoms
nuclear fission (125, 506) a nuclear reaction in which energy is released as an atomic nucleus is split into two other nuclei, plus several neutrons
nuclear fusion (129, 506) a nuclear reaction in which energy is released as two atomic nuclei are fused together to form one larger nucleus
nuclear medicine (129) the use of nuclear radiation to diagnose or treat disease
nuclear power generation (505) the generation of electricity, or other forms of power, using nuclear fission; potentially, nuclear fusion may also be used for this purpose
nuclear waste disposal (507) the process or methods for disposing of nuclear waste products in a safe manner
nucleic acids (372) very high molecular weight polymers that carry genetic information and regulate many aspects of body chemistry, such as protein synthesis; typical examples are deoxyribonucleic acid (DNA) and ribonucleic acid (RNA)
nucleus (96) the central part of an atom that contains all the protons and neutrons, in about one trillionth of the total atomic volume
nylon (281) a poly(amide)
Nylon 66 (264) a poly(amide) formed from adipic acid and hexamethylene diamine; it can be a plastic or a fiber
Nylon day (265) May 15, 1940, the day nylon stockings were introduced in the marketplace

O

octane number (230) numbers that are based on the combustion of poorly burning n-heptane (octane number of 0) and 2,2,4-trimethylpentane (octane number of 100), a good burning fuel that gives no knocking
octaves (84) a concept developed by John A. Newlands that claimed that certain properties recur with every seventh element
octet rule (141) a rule stating that the outermost shell of electrons, the valence electrons, will normally consist of eight electrons for every atom in a compound.
oil-based paints (294) paints that contains a drying oil, such as linseed or tung oil, which forms a crosslinked film during the drying stage; some modern oil-based paints contain synthetic polymers in place of all or part of the drying oil
oligomers (273) a polymer chain that only contains a few units
opsin (389) the protein portion of the visual pigment rhodopsin
optical activity (359) the ability to rotate plane polarized light
optical brighteners (348) chemicals that absorb UV light and emit blue light in its place, giving a more white appearance to a fabric; most optical brighteners are complex stilbene derivatives
optical isomerization (359) molecules that exist as enantiomers and have the ability to rotate plane polarized light
optical isomers (359) stereoisomers that exist as nonsuperimposable mirror images and can rotate plane polarized light
orbital (104) a volume of space within the atomic volume in which a specific electron is probably localized
orbit (100) the specific path an electron travels about the nucleus in the Bohr atom
organic mercury compounds (307) compounds that are often used as herbicides; they are fairly toxic
orientation (183) the alignment of molecules; chemical reactions do not occur unless the molecules are properly aligned; probability factor
osmosis (324) a process in which water flows through a semipermeable membrane, which includes synthetic polymers and cell membranes
outermost energy level (102, 141) the energy of the outermost electrons in an element; the same number as the period or the principal quantum number
outermost shell (141) valence shell of an element
oxidation (173) a chemical reaction that loses electrons, gains oxygen atoms, or loses hydrogen atoms
oxidation-reduction (redox) reaction (173) paired chemical reactions that always involve an oxidation reaction and a reduction reaction
oxidation theory (51) the modern theory that uses the element oxygen to explain combustion; it was developed by Lavoisier
oxymelanin (446) a yellow, polymeric pigment found in the skin, hair, and other parts of the body
ozone (536) an allotropic form of elemental oxygen that is present in the ozone layer, an ill-defined layer that is located about 10–30 km (6–18 miles) above the surface of our planet and contains a modest and variable amount of ozone (O_3)
ozone layer (536) a layer located about 10–30 km (6–18 miles) above the surface of our planet that contains a modest and variable amount of ozone (O_3), an allotrope of oxygen; it is not a discrete band but rather an ill-defined region in the lower stratosphere
ozone layer depletion (536) decreases in the stratospheric ozone concentration
ozonolysis (314) a water purification method that uses ozone to kill microorganisms

P

paint (294) a film, layer, or coat of a pigment applied to the surface of an object; usually contains a polymer or a drying oil
para-aminobenzoic acid [PABA] (438) a chemical compound, or an ester derivative, that is used in many sunscreening cosmetics because it effectively absorbs ultraviolet light
paraffins (211) another name for alkanes; saturated hydrocarbons
partial charge (153) fractions of a charge that appear to occur on atoms in a polarized covalent bond
particulate matter (493) small particles, such as from dust, soot, and fly ash, that are present in air pollution
parts per billion [ppb] (195) how many parts of a substance are present in a billion parts total mixture
parts per million [ppm] (194) how many parts of a substance are present in a million parts total mixture
passive solar heating system (511) solar heating systems in which the flow of heat occurs without the use of pumps and other large machinery
penicillin (466) a bactericidal agent that works by preventing cell wall formation in bacteria
peptide unit (363) the bond linking two amino acid molecules; it is an amide-like unit formed from the reaction of the carboxylic acid of one amino acid with the amine group on the other
perborate (347) an oxidizing agent that releases hydrogen peroxide when it comes into contact with water, and is added to detergent formulations to work as a bleach
perfume (445) a cosmetic that contains 10 to 25 percent fragrance and 2 to 5 percent fixative in ethanol solvent
periodic chart (table) (85) a listing of all the chemical elements by atomic number and chemical behavior
periods (87) the horizontal rows in the periodic chart; the same as the energy levels for the outermost electrons

permanent wave (449) chemical formulations used to make hair wavy
peroxyacylnitrates [PANs] (501) secondary pollutants that arise from a series of chemical reactions starting with the reaction of ozone with hydrocarbons
pesticide (422) any agent that can kill pests, such as insects, rodents, and the like
petroleum (228) an inflammable, oily, liquid mixture of numerous hydrocarbons found in many scattered subterranean deposits
petroleum jelly (437) a semisolid alkane with 20+ carbons that is used in many kinds of cosmetics
pH (202) a measure of the concentration of H^+ in a solution
phaeomelanin (446) an iron-containing, red pigment derived from melanin
phenol (250) an aromatic compound that contains the covalently bonded —OH group
phenolic resins (282) polymers formed from phenol and formaldehyde
pheromones (426) chemicals excreted in minute amounts by insects, and other animals, to define a trail, send a special message such as an alarm, or to attract a mate
phlogiston theory (50) an early theory attempting to explain combustion and some other chemical processes by the transfer of a substance called phlogiston
phosphates (344) a chemical compound, usually sodium tripolyphosphate (STPP), that is added to detergent formulations as a builder to enhance detergency performance
phosphoglycerides (379) fatty acid-glycerine esters that contain a phosphate group
photochemical (Los Angeles) type pollution (499) a form of air pollution characterized by the presence of some combination of nitrogen oxides, peroxyacylnitrates, ozone, and hydrocarbons
photoelectric effect (103) the emission of electrons caused when some minimum frequency of electromagnetic radiation strikes a metallic surface
photon (101) the subunit of light or electromagnetic radiation
photon's energy (103) the characteristic amount of energy associated with a photon (or quantum) of a specific frequency
photosynthesis (420) the formation of saccharides powered by the sun's energy and catalyzed by chlorophyll, a complex molecule structurally similar to hemoglobin
photovoltaic cells (513) devices that directly convert sunlight into electricity
primary pollutants (492) pollutants that are released directly
physical change (76) alternation in the physical state or form of a material without changing the chemical nature of the material
pi (π) bond (218) bonds that are formed from p-orbitals that are not used to form hybrid orbitals; π-bonds normally lie at right angles to the other bonds in the molecule
pigments (293) compounds added to a polymer to impart color
plane polarized light (359) light waves that move in only one plane
plasticized (279, 293) made to become plastic or pliable
plasticizers (293) compounds that are added to polymers to increase their flow or flexibility characteristics
plastics (273) polymers that are capable of being molded
polarized bonds (153) bonds that occur when there is a fairly large difference in electronegativity between the atoms in a covalent bond
polarized covalent bonds (153) bonds that occur when there is a fairly large difference in electronegativity between the atoms in a covalent bond
poly(acrylonitrile) [PAN] (280) a thermoplastic polymer made from acrylonitrile via addition polymerization
poly(amides) (280) condensation polymers that contain the amide group as part of the repeat unit; also called nylon
poly(carbonates) (281) thermoplastics made by a condensation reaction between bis–phenol A and phosgene
poly(carboxylates) (345) water soluble polymers, or copolymers, of acrylic or methacrylic acids that are added to detergent blends as a builder
poly(esters) (280) condensation polymers that contain the ester group as part of the repeat unit; *see also* Dacron and Mylar
poly(ester urethane urea) [PEUU] (480) a poly(urethane) used in artificial hearts and textiles
poly(ethylene) or PE (235, 277) an addition polymer produced from ethylene monomer; it is the most common synthetic polymer
poly(ethylene terephthalate) [PET] (263) a poly(ester) that is made from terephthalic acid (or one of its esters) and ethylene glycol; *see also* Dacron or Mylar
poly(hexamethylene adipamide) (264) a poly(amide) formed from adipic acid and hexamethylene diamine; *see also* Nylon 66
polymer (234) a chemical compound consisting of a large number of atoms arranged in the form of a chain; also called a macromolecule
polymeric drug (480) a drug agent in polymer form
poly(methyl methacrylate) [PMMA] (280) an addition polymer made from methyl methacrylate
poly(oxymethylene) (252) poly(formaldehyde); Delrin
polypeptides (358) polymers of amino acids
polypeptide hormone (380) a polypeptide that exhibits hormonal activity
poly(propylene) [PP] (277) an addition polymer made from propylene
polysaccharides (367, 408) polymers formed from a monosaccharide unit; examples include starch and cellulose
poly(silicones) [silicone rubber] (284) condensation polymers containing the —Si—O— linkage; the most common silicone polymer has a pair of methyl groups attached to the Si atom
poly(styrene) [PSt] (279) an addition polymer made from styrene monomer
poly(tetrafluoroethylene) [PTFE] (279) addition polymers that resemble poly(ethylene), except all the hydrogen atoms are replaced with fluorine atoms; commonly known as Teflon
polyunsaturated fatty acid (401) a fatty acid that contains more than one carbon–carbon double bond (C=C)
poly(vinyl acetate) [PVAc] (279) an addition polymer made from vinyl acetate monomer; contains a pendant ester unit
poly(vinyl chloride) [PVC] (278) an addition polymer made from vinyl chloride monomer; second most commonly manufactured polymer
poly(vinyl pyrrolidone) (446) a synthetic, water-soluble polymer that is used in hair sprays and other cosmetics
potable water (314) water of a sufficient quality to be drinkable
potential energy (112) the energy a material has because of its position or its internal chemical energy
preservative (417) a chemical that extends the lifetime of a food
primary structure (363) the linear sequence of amino acids in a polypeptide or protein chain
prime pigments (294) pigments that give color to paint
primary water treatment (319) a wastewater treatment method that uses screens and settling tanks to remove the smaller particulate matter; it removes about 50–75 percent of the particulates and reduces the BOD from about 200 to about 140 mg/L
probability factor (183) *see* orientation
product(s) (66) the compound(s) or element(s) formed by a chemical reaction
progesterone (380) a steroidal hormone secreted by the ovaries that promotes ovulation and maintains pregnancy; both this compound and its derivatives have been used as contraceptives
progestin (388) a generic term for synthetic steroids related to progesterone
propyl gallate (418) a chemical compound used as an antioxidant in foods
proof value (61) a measure of the amount of ethanol in an aqueous solution; its numerical value is twice the percent value
prostaglandin (378) a family of hormone-like substances derived from arachidonic acid, a 20 carbon fatty acid; they regulate many body functions and are prepared in many body tissues
protease (401) an enzyme that hydrolyzes the peptide bonds between amino acid units in a protein or polypeptide
proton (94) a subatomic particle with a mass of 1.0073 amu, 1.673×10^{-24} g, (or nearly the same mass as a hydrogen atom) and a charge of +1, or $+1.60 \times 10^{-19}$ coulombs
PTFE (279) poly(tetrafluoroethylene); Teflon

pure substances (21) homogeneous substances that consist of a single kind of material, such as compounds or elements
PVC (278) poly(vinyl chloride)
pyrolysis (59) destructive distillation; heating a substance, in the absence of air, and causing decomposition of that substance

Q

quantum [plural, quanta] (101) the subdivision units of energy
quantum number (101) an integer (whole number) used to describe the amount of energy associated with a subatomic particle; the principal quantum number is identical to the period number or the energy level for the outermost electrons in an element
quantum theory (101) a theory that claims that energy is not continuous, but exists as minuscule energy packets called quanta
quaternary structure (365) an aggregation of more than one folded peptide chain to form a functional protein

R

radioactivity (90, 119) the emission of particles of radiation from an element; the spontaneous decay of one atomic nucleus into another with the emission of some form of radiation
radioactive decay series (119) a nuclear serial sequence in which one element is transmuted into another, which, in turn, transmutes into yet another element
random copolymers (289) copolymers in which a pair of monomer units alternate along the backbone chain in an irregular, or random, manner
reactant(s) (66) the compound(s) or element(s) that undergo a chemical reaction
recommended daily allowances [RDA] (407) the recommended amounts of vitamin and minerals for daily consumption
recrystallization (71) a purification technique that separates materials on the basis of solubility differences at various temperatures
recycling (295, 308) the process of reusing a material or preparing that material for reuse
redox *see* oxidation-reduction
reduction (173) a chemical reaction that gains electrons, loses oxygen atoms, or gains hydrogen atoms
repeat unit (234) a collection of atoms regularly occurring in the polymer chain; a substance consisting of many units
replication (375) the process in which DNA unwinds and the bases on each chain attract monomeric nucleotides containing the complementary base units
reservoir device (480) a type of controlled release system
resonance (226) a concept in which electrons circulate about stationary atoms; this concept is used to explain the lack of certain kinds of isomers in aromatic compounds
Retin A (439) retinoic acid, which was first marketed in 1972 as an acne treatment, and is now also an over-the-counter cosmetic agent used to remove or reduce the amount of wrinkles in the skin
retinal (389) the aldehyde form of vitamin A that binds to opsin to form the visual pigment rhodopsin
retrovirus (474) a virus that retrofits a person's DNA to a new use; HIV is a retrovirus
reverse osmosis (324) a water purification method in which saline water, from seawater or a brackish source, is forced through a semipermeable membrane to effect removal of the dissolved salts
reversible reaction (261) a chemical reaction that can run in the opposite direction; in other words, the reactants and products can be reversed
rhodopsin (389) the visual pigment
ribose (369) a five carbon monosaccharide; a component of RNA
risk (5) an injury, hazard, or danger
Rogaine (451) trade name for minoxidil solutions used to treat male pattern baldness; *see also* minoxidil
rubber (236) a polymer obtained from the *Hevea brasiliensis* tree and numerous other plants; a general name for an elastomer
Rutherford's atomic model (96) an atomic model in which all the protons and neutrons, about 99.98% of the mass, are located in the nucleus and the electrons roam freely in the remainder of the atom's volume

S

saccharide (367) organic compounds composed chiefly of carbon, hydrogen, and oxygen that are also known as carbohydrates; usually the H:O ratio is 2:1; main examples are the sugars, starches, and cellulose
saccharin (413) a nonnutritive, noncaloric, artificial sweetener
salt (147, 199) the reaction product between an acid and a base; salts are always ionic compounds
saltwater intrusion (305) the flow of seawater into aquifers
saponification (337) the chemical reaction that forms soap
saturated fatty acids (378, 401) fatty acids that do not contain multiple carbon–carbon bonds
saturated hydrocarbon (211) an organic compound that contains only carbon and hydrogen and that has no multiple bonds present
science (2) a system of knowledge that develops, organizes, and confirms facts and theories by observation, testing, and experimentation
scientific method (10) a technique or way of viewing information, theories, hypotheses, and laws
scientific notation (78) a shorthand method for expressing large or small numbers in which a decimal number is multiplied by a power of ten, expressed as an exponential; e.g. $3.2 \times 10^3 = 3{,}200$.
scrubbers (496) devices designed to remove chemical contaminants, especially sulfur dioxide, from exhaust gas streams
second law of thermodynamics (112) a law that states heat spontaneously flows from a body at a higher temperature to one at a lower temperature
secondary pollutants (492) pollutants that are derived from other pollutants
secondary structure (364) the folding of the primary protein structure into an α-helix or a β-pleated sheet; such structures are maintained by hydrogen bonding
secondary water treatment (319) a biological process for treating wastewater that uses an activated sludge consisting of microorganisms and inorganic materials from previous wastewater; it removes 85–95 percent of the dissolved organic material and 85–95 percent of the remaining suspended matter and decreases the BOD to an average value of 30 mg/L
selective adsorption (73) preferential adsorption of a substance
semipermanent hair dye (447) hair dyes that last longer than the temporary versions because they partially penetrate into the hair fibers; they are applied as a reduced form that is later oxidized to the colored form
semipermeable membrane (324) a membrane that will not permit the passage of compounds exceeding a specific size but will allow free passage of smaller compounds; useful in water purification
separation (70) any procedure used to isolate one substance from another; this could include distillation, filtration, chromatography and so forth
sequestration (344, 418) binding metal ions with some substance
sewage treatment plants (309) wastewater treatment plants
shampoo (350) a surfactant blend used to wash the hair
shellac (294) a natural polymer secreted on trees in India by the insect *Laccifer lacca*
sigma (σ) bond (218) a bond that always forms on a straight line between the two atoms involved
silicates (344) chemicals that are added to detergent blends that reduce or retard the corrosion or dissolution of metals from washing machines; they are also highly alkaline and help adjust the pH; (156) chemical components of clays and ceramics
single displacement (171) chemical reaction in which an element replaces another element in a chemical compound to give a new compound and a new free element
skeletal fluorosis (196) a bone disorder that can lead to arthritis-like pain and immobility
skin cream (438) a cosmetic that is applied to soften or moisten the skin
sludge (309) solid waste that originally comes from kitchens, laundry water, human feces, soil erosion, vegetation debris, and other sources; this is not to be confused with an activated sludge that is used in secondary wastewater treatment
softener (342) a chemical that makes fabrics feel softer; usually a cationic surfactant

solar distillation (323) a desalination method that uses the energy from the sun for the purification of seawater via distillation
solar energy (509) energy from the sun
solar heating (511) heating buildings, and the like, directly using solar energy
solubility (71, 192) the measure of how much solute can dissolve in a specific amount of solvent; a description telling how much of one substance dissolves in another
solute (71, 193) the portion of a solution that is dissolved in the solvent
solvent (71, 193) the portion of a solution that dissolves the solute
solutions (21) homogeneous substances that contain more than one kind of material
sorbitol (414) a sugar alcohol sometimes used as a sweetener; although nonnutritive, it does have the same amount of calories per gram as glucose
sp-orbital (222) a hybrid orbital formed by combining an s-orbital with one p-orbital; its bond angle is 180°
sp^2-orbital (222) a hybrid orbital formed by combining an s-orbital with two p-orbitals; its bond angle is 120°
sp^3-orbital (222) a hybrid orbital formed by combining an s-orbital with three p-orbitals; its bond angle is 109.5°
spectroscopy (74) an analytical technique in which materials interact with light, or radiant energy, and the amount of absorbed or emitted energy is measured
spectrum [plural, spectra] (98) an image formed by radiant energy directed through an instrument (called a spectroscope) and observed on some kind of a recording or photographic device; they are arranged in the order of some characteristic, such as wavelength or frequency
spermicide (388) chemicals that kill sperm cells
sphingolipids (379) phospholipids containing the aminoalcohol sphingosine instead of glycerine
spreading (38) the movement of one liquid over another, as oil spreads over water
spontaneous reaction (182) a reaction that occurs without an outside influx of energy
stability (210) the ability of a molecule to resist destruction from outside forces, such as air or water
stabilizers (293, 418) compounds added to polymers to protect against various types of degradation, such as thermal or photochemical degradation; thickeners; agents added to improve smoothness or increase the viscosity of foods
standard laundry detergent (343) a "traditional" detergent blend that contains surfactants, builders, alkalis, anti-corrosion chemicals, and fillers, plus small amounts of other additives
steroid (380) a lipid derived from cholesterol that is composed of three six-membered rings and one five membered ring; they include many sex hormones and some anti-inflammatory compounds
stimulant (257) anything that increases activity
stoichiometry (53) the measurement of the exact amounts of materials involved in a chemical reaction
stratosphere (536) a region of the atmosphere located about 30 to 50 kilometers above ground level
strong acids (199) acids that dissociate nearly completely
strong bases (199) bases that dissociate nearly completely
structural isomer (213) molecules that have the same chemical formula but have a different structure
sublimation (24) the direct conversion of a solid into a gas
substantivity (439) a measure of how long a sunscreen lasts under the stress of exercise, sweating, and swimming
sucrose (369) a disaccharide formed from the monosaccharides glucose and fructose
sulfa drug (464) a bacteriostatic drug agent
sulfanilamide (464) an example of a sulfa drug
sulfuric acid aerosols (536) droplets of water containing sulfuric acid, from the reaction of atmospheric sulfur oxides with water
sulfur dioxide [SO_2] (493) an air pollutant present in the industrial (London) type air pollution; it is derived both from natural and human sources, the latter is derived mainly from smelters and the burning of sulfur-containing fuels, such as bituminous coal
sun protection factor [SPF] (439) the ratio of the time required to give the same amount of tanning or burning with and without the sunscreen
sunscreen (438) a chemical that blocks out ultraviolet (UV) light and helps prevent skin cancer
suntan lotion (438) a cosmetic lotion that aids in the suntanning process
superconcentrates (349) another name for compact laundry detergents
superimposition (362) placing the units of one object directly over the corresponding units of another object;
superphosphate (422) a fertilizer component obtained by treating phosphate rocks with sulfuric acid
supersonic transport planes (537) planes that fly in the upper atmosphere
superstructure (364) a structure that contains several intertwined molecules, such as a tertiary structure
surface active agent (38, 336) a chemical that collects at a surface and changes the nature of the surface; a chemical compound that reduces the surface tension of water and aids in the detergency process; also called surfactant
surface area (185) the total area available on which a surface reaction can occur
surface tension (38) a liquid state property which is a force that causes liquids to rise in a tube; surface active agents usually reduce the value of surface tension
surfactant (336) *see* surface active agent
symbol (87) a one or two letter representation of an element that is used in the periodic table and in chemical equations
synthetic dyes (257) chemicals that are artificial dyes
synthesis (69) the building up of compounds from atoms or other compounds; making new materials by combining or changing matter; the opposite of analysis

T

tacrine [THA] (477) a drug approved for the treatment of Alzheimer's disease
technology (3) the application of science for some practical use
temperature (27) a measure of the average kinetic energy in a sample
temporary hair colorant (447) a hair dye that is applied to the surface of the hair for a temporary coloring treatment
terephthalic acid (263):

HOOC—⟨benzene ring⟩—COOH

testosterone (380) steroidal hormone secreted in the testicles; it controls male secondary sexual characteristics
tertiary structure (364) the globular, three dimensional structure of a protein that results from folding the regions of secondary structure
tetrahedral (161) having the shape of a tetrahedron
tetrahedron (161) a three-dimensional geometrical figure that consists of four equilateral triangles
theory (9) similar to a hypothesis but covering a broader range of data; i.e., a broad set of assumptions conditionally accepted as a basis for reasoning or investigation
thermal energy (114) heat energy; energy due to the motion of atoms or molecules
thermal inversion (494) a weather pattern in which pollutants are trapped in a community that is usually in a valley
thermodynamics (114) the science that studies the movement of heat
thermoplastic (272) polymers that soften when heated and are able to be molded into new shapes
thermoplastic elastomers (289) block copolymers of styrene and butadiene, with the trade names of Solprene (an AB block) and Kraton (an ABA block)
thermosetting (272) polymers that crosslink when heated and are no longer capable of being molded into a new shape
thermosetting condensation polymers (282) polymers that crosslink on heating; such as the phenolics, urea-formaldehyde resins, and so forth
thickener (419) *see* stabilizer
thinner (294) a solvent used to reduce the viscosity of a paint
thioglycolic acid (450) a chemical that is used in hair permanents and depilatories
Thomson atomic model (94) a model of the atom in which the electrons and protons were fairly uniformly distributed in the atomic volume. originally the electrons were dispersed uniformly in a sea of positive electricity

three-dimensional formulas [shape formulas] (163) structures drawn to depict their three-dimensional shape
thromboxanes (379) lipids secreted by blood platelets that promote clotting and stimulate blood vessel dilation
toilet soap bar (350) a soap bar used mainly for cleaning the body
toothpaste (352) a blend of surfactant and abrasive inorganic chemicals that is used to clean teeth
top notes (446) the most volatile fragrance in a perfume mixture
toxicity (249) poisonous
trans- (219) a term denoting that two similar units lie on the opposite side of a double bond
transdermal devices (480) controlled release devices that allow a drug to transverse the skin
transition elements (87) a collection of ten (10) vertical groups of elements that fit into the periodic table between elements 20 and 31; the incoming electrons enter into d-orbitals
triads (84) a concept developed by Doebereiner in which sets of three elements always have the middle member with an atomic mass, melting point, and boiling point falling between the other two elements; triad elements also form similar chemical compounds
trigonal (162) a structure that has the shape of a triangle
turbidity (305) the haziness in water due to the presence of particulate matter
typhoid fever (314) an infectious fever caused by the typhoid bacillus; characterized by diarrhea and other symptoms

U

uncertainty principle (104) a concept developed by Heisenberg that says you cannot determine both the exact location and the exact speed (actually momentum) of any particle with certainty
unsaturated fatty acids (378, 401) fatty acids that contain one carbon-carbon double bond
ultraviolet stabilizers (293) compounds that protect plastics against degradation due to ultraviolet light
urea-formaldehyde polymers (286) crosslinked condensation polymers

V

vaccine (474) a substance used for inoculation that produces immunity and protects against a disease
vacuum distillation (41) a purification technique in which a liquid is distilled at a reduced pressure and boiling point
valence (145) the number of electrons transferred or shared by an atom in forming bonds
valence factor (145) normally the same value as the valence number of the other element in a chemical compound; the number of atoms required in the compound's formula
valence shell electron pair repulsion [VSEPR] (161) the basic idea in this approach is that the electrons in bonds arrange themselves so that they are as far apart as possible
vanadium oxide [V_2O_5] (502) a catalyst used in automotive catalytic converters
vanishing cream (438) suspensions of stearic acid in water that are stabilized by a soap; they appear to vanish when rubbed on the skin
vapor pressure (41) the equilibrium pressure created by the gas form of a liquid when it evaporates in a closed container
varnish (294) a natural or synthetic polymer dissolved in a drying oil-solvent blend that forms tough crosslinked films on drying
vehicle (294) in paints and coatings, a combination of a drying oil or polymer (the binder) and a solvent or thinner
virus (474) a nucleic acid molecule (DNA or RNA) with a protein coating
viscosity (231) the property of liquids that describes their ability to resist flow
vital force (210) a force derived from a living organism, which scientists once believed was necessary in order to form organic molecules
vitamins (382, 401) an organic compound that is required in the diet in small amounts; vitamins are involved in the synthesis of coenzymes, vision, calcium metabolism, and blood clotting
volatile organic compounds (VOCs) (322, 466) normally low molecular weight liquids used as solvents
vulcanization (236) a chemical process in which rubber molecules are converted from a linear arrangement into a crosslinked arrangement

W

wastewater treatment (319) the process of removing waste materials, such as sewage, from water
water ionization constant, k_w (200) the number that results when the moles of undissociated water is combined with the equilibrium constant for water's dissociation reaction; it has a value of 10^{-14}
water softening (322) the removal of Ca^{2+} and other hard water ions by ion exchange
water soluble vitamins (404) vitamins that are soluble in water
water treatment plant (314) a plant that produces potable (drinkable) water
water-vapor cycle (304) the natural process wherein the water on our planet recycles from the oceans, rivers, streams, lakes, and ponds into the clouds, and then returns back as rain, snow, or hail
wavelength (74, 99) the distance between equivalent points in a uniform wave
wave mechanics model of the atom (103) an atomic theory in which the electrons exist in the form of waves, called orbitals, within the entire atomic volume, while the protons and neutrons are localized in the nucleus
weak acids (199) acids that hardly dissociate at all
weak bases (199) bases that hardly dissociate at all
wettability (39) the ability of one material to wet another
wetting (341) the ability of one substance, usually a liquid or a solution, to make contact with (i.e., wet) a surface
wind power-generation (509) energy derived directly from the movement of the wind
wood alcohol (247) *see* methanol; methyl alcohol

X

xylitol (414) a sugar alcohol sometimes used as a sweetener; although nonnutritive, it does have the same amount of calories per gram as glucose

Y

yield (262) amount of product produced in a chemical reaction

Z

zeolites (345) insoluble sodium aluminosilicates with an empirical formula of $2Na_2O{\cdot}2Al_2O_3{\cdot}4SiO_2{\cdot}9H_2O$ that are used as replacements for phosphates
zone refining (513) a purification technique wherein a heated coil passes along a column of material heating only a small zone; the impurities are more soluble in the molten portion of the material and move along with this liquid portion leaving a very pure material behind

ANSWERS TO SELECTED PROBLEMS, QUESTIONS, AND CRITICAL THINKING PROBLEMS

ANSWERS TO ODD NUMBERED PROBLEMS AND QUESTIONS

Chapter 1

1. See glossary for definition; examples of events or items studied by chemists include the current questions of global warming and ozone layer depletion, developing new energy sources, and developing new materials.

3. Thousands of benefits that we receive from chemical technology exist and here are a dozen examples: better food quality and safety, better home insulation materials, better treatments for disease, computer chips, high quality drinking water, improved automobile fuels, latex paints, magnetic tape, new textiles, photography, plastics, and safer cosmetics. You can probably add another dozen of your own choice.

5. The steps are outlines in figure 1.3. First data is gathered and then this is organized. Next, a hypothesis is formulated and used to make predictions. Experiments are then run to test these predictions. If the test is positive, the hypothesis is tentatively accepted. In the event of negative experimental results, the hypothesis is rejected or revised. This overall process is continually repeated.

7. The main prefixes are: micro, milli, centi, kilo, and mega.

9. The average diameter is 23.56, which we round off to 23.6 mm. The range of sizes differs only by 0.8 mm, and none of the diameters are far from the average value. We conclude that the quarters are uniform.

Chapter 2

1. Milk, mayonnaise, air, dirt, soft drinks

3. See the glossary for the definitions. Some examples are: **(a)** dirt; **(b)** water; **(c)** soft drink; **(d)** table salt; **(e)** gold; **(f)** sugar; **(g)** hydrogen; **(h)** brass.

5. Extensive properties depend on how much material is present, while intensive properties are independent on the amount of material.

7. 2.4 g/mL

9. In a direct relationship, one property increases as a second property increases. In an inverse relationship, one property decreases as the other property increases.

11. **(a)** 116.6°F; **(b)** 98.6°F; **(c)** 149°F

13. **(a)** 134°C; **(b)** −230°C; **(c)** 1,227°C

15. Charles' law states that the volume of a gas increases directly with the temperature, expressed in degrees K. One form of the Charles' law equation is: $V = kT$

17. Inverse relationship; 22.3 L

19. 15.2 L

21. See the glossary for the definitions.

23. The equilibrium pressure created by the gas form of a liquid when it evaporates in a closed container.

25. See the glossary for the definitions.

Chapter 3

1. Early Greek, 450 B.C. to 65 B.C.; some Greeks talked about atoms but they were little understood. Alchemy age, 300 to 1200 A.D.; virtually no development of the atomic theory. Iatrochemistry age, started around 1200 A.D. and lasted until about 1700 A.D.; some rudimentary atomic theory development but no basic understanding of the difference between elements and compounds. Independent Chemistry Age, from about 1700 A.D. to the present; this is when atomic theory developed.

3. See the glossary for the definitions.

5. Copper, which does not occur free in nature, has been known since antiquity, so elements have been isolated from the ore since then.

7. **(a)** 3,200,000,000,000; **(b)** 0.0048; **(c)** 730,000;
(d) 0.0000000091; **(e)** 34,700;
(f) 602,000,000,000,000,000,000,000;
(g) 0.000000000000000000000303; **(h)** 0.0000511

9. To a large extent, chemistry is an applied science and these operations are instrumental in achieving the goals of preparing new materials or modifying existing ones.

11. Isolating vitamins from food substances.

13. A liquid, such as water, is heated to the boiling point and converted to the gaseous state, which is then condensed back to the liquid form. Any dissolved solids remain behind during distillation, resulting in a more pure liquid.

15. An impure solid is dissolved in as little hot liquid as possible. After this solution cools, some of the solid comes out of solution and is collected in a filtering step. This solid is more pure than the original solid. If necessary, the recrystallization is repeated until sufficient purity is achieved.

17. A sample is heated to glowing, or burned in a flame, and the emitted light is passed through a prism. Bands of light appear that are unique for the elements in that material, allowing these elements to be identified.

19. In any chemical reaction, the total mass of the reactants is exactly equal to the total mass of the products.

21. Every compound has a specific chemical composition.

23. **(a)** definite proportions; **(b)** definite proportions; **(c)** both laws; **(d)** both laws; **(e)** both laws; **(f)** definite proportions

25. The assumption that all atoms of a specific element are identical and have the same unique atomic mass was partially disproved by the existence of isotopes. Dalton himself disproved his assumption that the ratio of atoms in a molecule is constant and can be expressed as simple whole numbers by discovering the law of multiple proportions. Dalton also believed atoms were indivisible; the existence of subatomic particles required modification of this assumption.

27. **(a)** $2\ Ca + O_2 \rightarrow 2\ CaO$
(b) $S_8 + 8\ O_2 \rightarrow 8\ SO_2$
(c) $N_2 + 2\ O_2 \rightarrow 2\ NO_2$

29. Buckyballs are an allotropic form of the element carbon. Probably their most unusual feature is the fact that they exist as hollow cages of 32 or more carbon atoms. They also possess some useful and intriguing electronic properties that may enable them to function as insulators, conductors, semiconductors, and superconductors.

Chapter 4

1. See the glossary for the definitions.

3. Most of the elements in the periodic table are metals; the nonmetals lie in the upper right hand corner; the noble gases are shown as a separate column in figure 4.1.

5. Families are the vertical columns and periods are the horizontal rows.

7. Family IA, in the first column from the left.

9. Family VIIA, in the seventh column from the left.

11. Family IVA, in the fourth column from the left.

13. Family VIA, in the sixth column from the left.

15. Only sets **(d)**, **(f)**, and **(g)**.

17. **(a)** Arrhenius showed ions exist in solids, implying atoms lost something to form the ions; **(b)** Becquerel demonstrated that radioactivity came out from atoms, implying they might be divisible; **(c)** Chadwick discovered the neutron, a subatomic particle; **(d)** Curie demonstrated that radioactivity came out from atoms, implying they might be divisible; **(e)** Goldstein showed canal rays were positively-charged particles; **(f)** Millikan determined the charge on the atom; **(g)** Rutherford showed that most of the mass of the atom was located in a tiny portion of the total volume; **(h)** Thomson proposed an atomic theory that showed subatomic particles; **(i)** Wien showed that the mass of the canal rays varied with the mass of the gases in the tube.

All these scientists showed that the atom was actually divisible, in contradiction to Dalton's atomic theory.

19. Thomson deflected a beam of beta-rays toward a positive plate.

21. Thomson's first model had the electrons dispersed in a matrix of positive electrical charge. His later model had electron-proton pairs uniformly dispersed throughout the volume of the atom.

23. The nucleus of the gold atom was about $1/10^{12}$ the size of the atom.

25. In almost all cases the atomic mass increases as the atomic number increases. The few exceptions are called Mendeleev inversions.

27. Isotopes are atoms of the same element that differ slightly in mass due to differing numbers of neutrons. Hydrogen has three isotopes, as does carbon. Chlorine has two isotopes. Some elements, such as uranium, have many isotopes.

29. **(a)** Bohr developed the idea that electrons moved in orbits outside the nucleus; **(b)** de Broglie stated that electrons existed as both particles and waves; **(c)** Einstein, best known for the theory of relativity, also developed the photoelectric effect theory, which tells us that the energy of a photon varies with its frequency; **(d)** Heisenberg's uncertainty principle states that you cannot know the exact location and speed of a moving object or particle; **(e)** Schrödinger developed the theory of wave mechanics that is used in modern atomic structural theory.

31. **(a)** Amplitude is the height of a wave and corresponds to the number of photons in a wave; **(b)** Frequency is the number of times per second a complete wave moves past a fixed point; **(c)** Wavelength is the distance between equivalent points in a pair of waves.

33. Microwave radiation, while very useful for heating, has much lower energies than ultraviolet radiation.

35. Spectral lines are at definite frequencies and, therefore, have definite amounts of energy for each line. They relate to the Bohr atom in that they describe the allowable, or permitted, amounts of energy for the jumps of electrons from one orbit to another. Thus, they set the diameter of these orbits.

37. Their shapes are complicated and the details are not important in this text.

39. Main groups fill either s or p orbitals. Transition elements fill d orbitals, while inner transition elements fill f orbitals.

Chapter 5

1. See the glossary for the definitions of these terms. Some examples are: **(a)** waterfalls have potential energy that can be converted to electricity; **(b)** you use kinetic energy whenever you ride in a car; **(c)** motors use mechanical energy; **(d)** we heat our food and homes with thermal energy; **(e)** electricity has thousands of uses in our society; **(f)** we see using light energy; **(g)** nuclear energy is used to generate electricity; **(h)** fuels are forms of chemical energy.

3. Chemical energy is burned and converted into mechanical energy in a motor that drives a generator to make the electricity.

5. Take aluminum cans as an example. Energy is required to separate aluminum metal from the ore, but less energy is needed to convert already existing aluminum scrap into new and useful objects. Basically, this is a conservation of useful energy. The total mass is also conserved, but that's true even when the materials are not recycled.

7. You surely use thermal (heating or cooling), mechanical (automobile, bus, or train), chemical (fuels, food, batteries), electricity, light, and nuclear (electricity generation). Better insulation saves energy, as does recycling.

9. Cold light. Trinkets sold at fair, carnivals, and other places often contain chemicals that emit chemiluminescence, or cold light. Similar systems are used for emergency lighting. And, as box 5.2 depicts, some insects and other animals use chemiluminescence to attract mates.

11. A beta (β) particle is an electron. An alpha (α) particle is a helium nucleus bearing two positive charges.

13. Pa has 91 protons and 144 neutrons, U-235 has 92 protons and 143 neutrons, U-238 has 92 protons and 146 neutrons, Np has 93 protons and 146 neutrons, Pu-235 has 94 protons and 141 neutrons, Pu-239 has 94 protons and 145 neutrons; **(a)** The pairs of U and Pu atoms are isotopes; **(b)** The Pa and Np are not isotopes; **(c)** Pa-235, U-235, and Pu-235 have the same mass number–235; U-239 and Pu-239 also have identical mass numbers; **(d)** The pairs of U and Pu atoms have the same number of protons; **(e)** U-238 and Np have 146 neutrons; **(f)** The pairs of U and Pu atoms have the same number of electrons.

15. $^{222}_{86}Rn \rightarrow ^{4}_{2}He + ^{218}_{84}Po$

17. $^{223}_{87}Fr \rightarrow ^{0}_{-1}e + ^{223}_{88}Ra$

$^{223}_{88}Ra \rightarrow ^{4}_{2}He + ^{219}_{86}Rn$

19. Evaluating thyroid function, blood circulation studies, bone mineralization, detection of skin cancer, and many others.

Chapter 6

1. See the glossary for the definitions of these terms.

3. The metal atoms are arranged in a crystalline lattice and the outermost electrons roam freely through the metal.

5. **(a)** ·Ċ̣· **(b)** ·Ṇ̇: **(c)** Na· **(d)** :Ọ̇: **(e)** :Ḥ̈e: **(f)** :N̈ẹ: **(g)** :C̈ḷ: **(h)** Ca· **(i)** ·Ȧl·

7. **(a)** SBr_2 **(b)** NCl_3 **(c)** H_2Te **(d)** PCl_3

9. **(a)** :B̈ṛ:S̤̈:B̈ṛ: **(b)** :C̈ḷ:N̈:C̈ḷ: with :C̈ḷ: below N **(c)** H:T̈ẹ:H **(d)** :C̈ḷ:P̈:C̈ḷ: with :C̈ḷ: below P

11. **(a)** Na^+ :Ö:$^{2-}$ Na^+ **(b)** :B̈ṛ:$^-$ Ca^{2+} :B̈ṛ:$^-$ **(c)** K^+ :F̈:$^-$ **(d)** Ba^{2+} :Ö:$^-$

13. See the glossary for the definitions of these terms.

15. Electronegativity is the attraction of the nucleus of an element for electrons. Ionization potential measures the energy required to remove an electron from an atom or ion. Ionic bonds tend to form between elements showing greater differences in electronegativity. When the differences are small, the bonds tend to be covalent.

17. Properties such as boiling points showed abnormally high values in compounds such as water. This was explained by the formation of many hydrogen bonds that effectively linked a large number of water molecules together into a cluster.

19. **(a)** covalent **(b)** ionic **(c)** ionic **(d)** covalent **(e)** covalent **(f)**covalent

21. **(a)** metallic; **(b)** covalent; **(c)** ionic; **(d)** metallic for each metal in the alloy; **(e)** ionic; **(f)** covalent; **(g)** metallic for each metal in the alloy; **(h)** ionic; **(i)** covalent; **(j)** metallic for each metal in the alloy; **(k)** ionic; **(l)** covalent; **(m)** metallic.

Chapter 7

1. See the glossary for the definitions of these terms.

3. **(a)** double displacement; **(b)** double displacement; **(c)** combination; **(d)** decomposition; **(e)** combination; **(f)** single displacement

5. Use table 7.3 to determine whether or not a reaction is spontaneous.

(a) $Ca + ZnCl_2 \rightarrow CaCl_2 + Zn$ (spontaneous)
(b) $3\ Mg + Fe_2O_3 \rightarrow 2\ Fe + 3\ MgO$ (spontaneous)
(c) $4\ K + SnO_2 \rightarrow 2\ K_2O + Sn$ (spontaneous)
(d) $2\ Fe + 3\ HgS \rightarrow Fe_2S_3 + 3\ Hg$ (spontaneous)

7. The rate of the reaction doubles.

9. The rate of the reaction decreases eightfold; in other words, it runs only 1/8 as fast.

11. The sawdust burns 11.76 times faster. The sawdust has a surface area of 2.5 cm^2 for each of its 2,000 particles for a total surface area of 5,000 cm^2. The log only had a total surface area of 425 cm^2. Dividing 5,000 by 425 shows that the sawdust will burn 11.76 times faster.

Chapter 8

1. See the glossary for the definitions of these terms.

3. The percentages, ppm, and ppb for each gas are: O_2: 13.33%; 133,300; 133,000,000; N_2: 80%; 800,000; 800,000,000; CO_2: 5%; 50,000; 50,000,000; SO_2: 1.67%; 16,700; 16,700,000

5. **(a)** insoluble; **(b)** soluble; **(c)** soluble; **(d)** insoluble; **(e)** insoluble; **(f)** insoluble; **(g)** soluble; **(h)** insoluble; **(i)** soluble; **(j)** insoluble

7. Mercury is the main other heavy metal in water supplies.

9. This NaOH solution would be 0.30 M and would be basic since it supplies OH^- ions to the solution.

11. The solution is 0.10 M and is acidic since there are H^+ ions present from the H_2SO_4.

13. The water ionization constant is used in pH calculations.

15. Mostly sulfur dioxide, which comes from both natural and human sources. In the latter category, the main sources of SO_2 are smelters and power plants that burn bituminous (soft) coal that contains sulfur. Some of the acid rain is also caused by various nitrogen oxides.

17. The human-generated portion of the sulfur dioxide can be reduced by burning anthracite (hard) coal, instead of bituminous (soft) coal, and by adding scrubbers to gas emission pathways to remove the SO_2. There is almost nothing that we can do to reduce the natural SO_2 emissions.

Chapter 9

1. (a) Kekulé concluded that the benzene molecule consisted of alternate C═C and C—C bonds; **(b)** Wöhler synthesized the organic compound urea from inorganic ammonium cyanate, showing that a vital force was not required to make organic compounds; **(c)** Goodyear invented the vulcanization of rubber; **(d)** Goodrich produced rubber fire hoses using vulcanization to improve the elasticity.

3. Catenation and the stability of the resulting bonds.

5. (a) alkane; n-octane; **(b)** alkane; 2-methylhexane; **(c)** alkene; 1-pentene; **(d)** alkyne; 2-butyne; **(e)** alkene; cis-3-hexene; **(f)** alkene; 4-methyl-1-pentene; **(g)** alkane; 2-methyl-hexane; **(h)** aromatic; methylbenzene or toluene

7. (a) methyl; **(b)** ethyl; **(c)** propyl

9. (a) $CO_2 + H_2O$; **(b)** $CH_3CHBrCHBrCH_3$; **(c)** $CH_3CCl_2CHCl_2$; **(d)** $C_6H_5{-}Br$

11. Polymers, chemical intermediates, and fuels, especially automotive.

13. A fuel is a chemical in a reduced form. Good fuels are fully reduced whereas poor fuels already contain some oxygen atoms.

15. In order of increasing fuel value (from lowest to highest): wood, methanol, peat, bituminous coal, anthracite coal, gasoline.

17. It raised the effective octane number and improved combustion.

19. The oil contains a polymer that is coiled into a ball at low temperatures and has little effect on the oil's viscosity. As the temperature increases, the polymer opens up to form stretched-out chains which increase the oil's viscosity.

21. Poly(ethylene)

23. Goodyear

25. It becomes crosslinked, more elastic, and less tacky.

Chapter 10

1. See the glossary for these definitions.

3. Use table 10.1 to find these answers.

5. Alcohols, phenols, carboxylic acids, and amines

7. (a) alcohol; **(b)** alcohol; **(c)** aldehyde; **(d)** ketone; **(e)** amine (primary); **(f)** ketone; **(g)** amine (secondary); **(h)** aldehyde; **(i)** carboxylic acid; **(j)** ester; **(k)** amide; **(l)** carboxylic acid

9.

(a) $CH_3CH_2CH_2{-}\overset{\overset{\displaystyle O}{\|}}{C}{-}OH$

(b) $CH_3\overset{\overset{\displaystyle OH}{|}}{C}HCH_2CH_3$

(c) $CH_3CH_2CH_2{-}\overset{\overset{\displaystyle O}{\|}}{C}{-}OCH_2CH_3$

(d) $CH_3CH_2\overset{\overset{\displaystyle OH}{|}}{C}H{-}\overset{\overset{\displaystyle CH_3}{|}}{C}HCH_3$

(e) $H_2N{-}\overset{\overset{\displaystyle CH_3}{|}}{C}HCH_3$

(f) $CH_3{-}\overset{\overset{\displaystyle O}{\|}}{C}{-}CH_2CH_3$

(g) $CH_3CH_2CH_2CH_2CH_2\overset{\overset{\displaystyle O}{\|}}{C}H$

(h) $CH_3CH_2CH_2\overset{\overset{\displaystyle O}{\|}}{C}CH_2CH_3$

(i) $CH_3CH_2\overset{\overset{\displaystyle NH_2}{|}}{C}HCH_3$

(j) $CH_3{-}O{-}CH_2CH_3$

11. (a) methanol: a solvent for shellac, nitrocellulose, and ethyl cellulose, a temporary antifreeze, plasticizers, a fuel; **(b)** ethanol: alcoholic beverages such as beer, wine, and whiskey, an industrial and medicine solvent, a reagent for preparing other chemicals; **(c)** 2-propanol: rubbing alcohol, cleaning electronic components and tape heads; **(d)** ethylene glycol: permanent automobile antifreeze and summer coolant, brake fluids, cosmetics, printing inks, and lacquers, a solvent for polyesters and cellulose esters ; **(e)** glycerol: cosmetics, soaps, hydraulic fluids, printing inks, and a lubricant

13. A chemical that produces a complete or partial lack of feeling.

15. Diethyl ether

17. (a) a solvent for paint removers, nail polish, and nail polish remover; to clean and dry precision equipment parts; **(b)** cleaning fluid and as a solvent in paint removers, lacquers and coatings, and lube oil dewaxing; **(c)** a solvent, in medicine and organic synthesis.

19. Butanoic acid (butyric acid)

21. Formic acid

23. (a) the manufacture of dyes, insecticides, perfumes, medicines, and plastics; **(b)** found in vinegar, used in the manufacture of cellulose acetate, vinyl acetate, drugs, dyes, insecticides, photographic chemicals and food additives; **(c)** in perfumes and flavors and in the manufacture of emulsifying agents, pharmaceuticals and disinfectants; **(d)** soap making, lubricants, pharmaceuticals, cosmetics, rubber compounding, coatings, shoe polish, cosmetics, and toothpaste; **(e)** food additive, used in the manufacture of perfumes, flavors, medicines, plasticizers, dentifrices, and polymers.

25. Addition polymers do not split off a small unit, such as water, during formation and have the same empirical formula for the monomer as for the repeat unit. Neither of these features are found in condensation polymers.

27. In decomposing garbage and dead bodies.

29. **(a)** aniline: dyes, drugs, photography, herbicides, and fungicides; **(b)** amphetamine: central nervous system stimulant, in nasal sprays; **(c)** 1,6-diaminohexane: used to make nylon-66.

31. Esters are used in perfumes and polymer synthesis.

Chapter 11

1. **(a)** Baekeland invented Bakelite, the first totally synthetic polymer; **(b)** Carothers developed poly(isoprene), poly(esters), and poly(amides); **(c)** Goodrich developed rubber hoses using the vulcanization process; **(d)** Charles Goodyear invented vulcanization; **(e)** Nelson Goodyear developed hard rubber by using high amounts of sulfur during vulcanization **(f)** Natta codeveloped a special groups of catalysts that were used to make synthetic rubbers with structures almost identical to natural rubber; these catalyst systems were also used to prepare several other polymer systems; **(g)** Patrick developed the poly(sulfide) elastomers, the first synthetic elastomers; **(h)** Plunkett discovered poly(tetrafluoroethylene); **(i)** Staudinger developed the field of polymer chemistry and also discovered how to polymerize formaldehyde; **(j)** Ziegler codeveloped a special group of catalysts that were used to make synthetic rubbers with structures almost identical to natural rubber; these catalyst systems were also used to prepare several other polymer systems.

3. **(a)** Melamine-formaldehyde (Delrin): plastic dishware; **(b)** nylon: textiles, plastic gears, fishing line; **(c)** phenolic resins: molded objects, adhesives, exterior plywood, impregnated paper, ion-exchange resins, space craft nose-cones; **(d)** poly(carbonates): impact-resistant windows and doors, gears; **(e)** poly(ethylene): toys, cups, wire coating, milk bottles, plastic wrap; **(f)** poly(ethylene terephthalate): fabrics, VCR tape and audio tapes, heat-sealing films, automobile tire cord, beverage bottles; **(g)** poly(propylene): fibers, storage battery cases, automobile trim, rope; **(h)** poly(silicones): sealants, gaskets, implants, waxes, polishes, anti-foaming additives, and synthetic varnish; **(i)** poly(styrene): foamed cups, egg cartons, furniture, electronics, toys; **(j)** poly(tetrafluoroethylene): nonstick cookware, wire coatings, white film used to line pipe threads, gaskets; **(k)** poly(vinyl acetate): latex (water based) paints, chewing gum; acrylates: dentures, contact lenses, floor polishes, skylights; **(l)** poly(vinyl chloride): pipes, wire coatings, garden hoses, rainwear; **(m)** urea-formaldehyde polymers: molded objects, interior plywood, crease-resistant fabrics, foamed insulation

5. Poly(esters), nylons, acrylics, cotton, glass

7. **(a)** poly(ether); **(b)** methacrylate; **(c)** poly(amide); **(d)** poly(ester)

9. Plastics are moldable but they do not stretch well; elastomers have rubbery characteristics and stretch well.

11. The differences lie in their molecular structure: linear polymers have chains that are unbranched, branched polymers contain random branches off of the polymer chains, and crosslinked polymers have the chains linked together into a network.

13. Virtually any addition polymer can be made in the form of linear chains.

15. Poly(vinyl chloride) uses include: pipes, wire coatings, garden hoses, and rainwear.

17. Poly(vinyl acetate)

19. Chloroprene is the common name for 2-chlorobutadiene and is the monomer used to prepare poly(chloroprene), better known as Neoprene, which was invented by Carothers.

21. Thermoplastic elastomers are block copolymers of styrene and butadiene.

23. Pigments are solid chemicals added to plastics to impart color and to reduce cost. Prime pigments impart color to paints and include white titanium dioxide (TiO_2), brown or red iron oxides (Fe_2O_3 or Fe_3O_4), and green or blue phthalocyanines. Inert pigments include clay, talc, calcium carbonate ($CaCO_3$), and magnesium silicate.

25. Stabilizers include: antioxidants that prevent oxidation (example: 1,5-di-*tert*-butylphenol), heat stabilizers avert thermal degradation (examples: lead, cadmium, magnesium, or barium salts), ultraviolet stabilizers prevent the degradation of polymers under the influence of ultraviolet light (examples: salicylate esters), and flame retardants that reduce the flammability of polymers (examples: tricresyl phosphate, antimony oxide, or antimony chloride).

27. A varnish is a natural or synthetic polymer dissolved in a drying oil-solvent blend.

29. Almost all plastics are recyclable, but the usual ones recycled include poly(ethylene), poly(propylene), poly(ethylene terephthalate), and poly(vinyl chloride).

Chapter 12

1. Water has high boiling and freezing points, a density of the solid (ice) that is less than that of the liquid, and a polar structure that dissolves many inorganic compounds.

3. Only 0.7 percent of the Earth's total water supply is fresh water.

5. There are 4.49×10^{16} gallons of ground water in the United States and about half the people in the United States get their tap water from groundwater sources.

7. In descending order; groundwater, the Great Lakes, all other lakes and ponds, rainfall, and all the streams and rivers.

9. Hydrogen sulfide gas (H_2S) imparts a foul smell akin to rotten eggs.

11. Fertilizer, herbicide, and insecticide runoffs can enter the groundwater.

13. Under ordinary conditions the following products will biodegrade: food, dead trees, paper products, plant matter (biomass), dead animals. The following materials will not biodegrade under normal conditions: glass, most plastics, aluminum metal, industrial wastes, petroleum.

15. Eutrophication is the enrichment of water with nutrients from biodegradation, fertilizer runoff, or other sources, that leads to excessive algae bloom.

17. High BOD values that can cause fish kills unless the oxygen in the water is replaced.

19. The dissolved phosphates and nitrate support aquatic life and will deplete the oxygen supply and raise the effective BOD of the water.

21. The ice or snow removing chemicals operate by lowering the freezing (melting) point of the solid water. This depends on the total number of ions present. Calcium chloride produces three ions whereas sodium chloride only produces two ions and is less effective. Also, calcium chloride releases a lot of heat as it dissolves in the melting water and sodium chloride does not.

23. Drinking water contains less than 500 ppm total dissolved salts and a has a pH of 6–9.

25.

Cl_2	+ Germs	$\longrightarrow$	$2\ Cl^-$	+ Dead germs
Oxidizing agent	Reducing agent		Reduced	Oxidized

27.

$KMnO_4$ + $2\ H_2O$ +	Odor agent	$\longrightarrow$ MnO_2	+ 4 KOH +	Non-odor agent
Oxidizing agent	Reducing agent	Brown solid		Oxidized material

29. Hard water contains calcium and magnesium ions, which are replaced with sodium ions during water softening.

31. The two types of ion-exchange resins are anion-exchange and cation-exchange resins. Both are polymeric beads (spheres) that swap sodium cations for calcium or magnesium cations.

33. In secondary wastewater treatments, the water is treated with an activated sludge and then chlorinated. Both of these steps are chemical.

35. Seawater can be desalinated by the following treatments: distillation (including solar heated), reverse osmosis, electrodialysis, freezing, ion exchange, and hydrate formation. Reverse osmosis is generally considered the most promising method.

37. Solar distillation uses sunlight as the power source to heat the water.

39. Reverse osmosis uses pressure to force water to flow backwards through a semipermeable polymeric membrane (i.e., from the more concentrated chamber into the less concentrated one). The dissolved salts remain behind because they cannot penetrate the membrane, and pure water is collected.

41. Hydrate formation is a desalination method that separates seawater from its salts by forming hydrates that are essentially free of salts. Hydrates form when water molecules become attached to or entrapped within another substance. They generally involve electrostatic or hydrogen bond formation between water and the other material.

Chapter 13

1. One part of the molecule is attracted to water while the other part is repelled by the water.

3. Hydrophilic denotes the water-loving portion of a surfactant, which is the more polar portion of the surfactant and is frequently ionic. Hydrophobic denotes the water-fearing, or hating, portion of a surfactant; this is the nonpolar portion of the molecule and is frequently the hydrocarbon portion of the surfactant.

5. Anionic, cationic, nonionic, and amphoteric.

7. The name denotes which portion of the surfactant molecule is the hydrophobic part. Thus, the hydrophobic portion bears a negative charge in anionics and a positive charge in cationics.

9. Exclusive of soap, about 47 percent of the anionics are used in household applications, with another 40 percent used in various industrial applications. Only about 6 percent are used in personal care applications.

11. Cationic surfactants are used as hair conditioners and fabric softeners.

13. The hydrophobic portion in a nonionic surfactant varies a bit, but the usual ones are hydrocarbons. In nearly all cases, however, the hydrophilic portion is a polymeric chain formed from ethylene oxide units, or an ethoxylate.

15. Amphoteric surfactants have both a negative and a positive charge on the hydrophobic portion of the molecule. They are mainly used in shampoos, especially ones designed for children.

17. Surface tension is a liquid state property wherein a force causes liquids to rise in a tube against the pull of gravity; surface active agents usually reduce the value of surface tension.

19. Micelles are an arrangement of surfactant molecules in which the hydrophobic hydrocarbon tails are packed close together and the water-loving hydrophilic parts are placed on the outside in the water. Micelles serve as a source of surfactant in aqueous solutions of a surface active agent, and also serve as initial sites to hold oily dirt.

21. The answer is given as component (uses): surfactant (lowers surface tension and removes dirt); builder (enhances detergency and softens the water); alkali (adjusts pH and helps as a builder); bleach (oxidizes colored stains); optical brighteners (imparts a bluish appearance to washed fabrics); antiredeposition agents (prevents dirt from resettling back on fabrics or other surfaces); fillers (take up space).

23. They soften the water and also enhance the detergency action.

25. Yes, in automatic dishwashing detergents.

27. Zeolites are insoluble sodium aluminosilicates that have an empirical formula of $2Na_2O \cdot 2Al_2O_3 \cdot 4SiO_2 \cdot 9H_2O$; they are used as phosphate replacements. Other than being phosphate free and unable to promote eutrophication, their main advantage is that they remove calcium ions (but not magnesium ions). On the debit side, compared against the phosphates, zeolites do not aid in the detergency process, do not help suspend dirt, are more expensive, and increase wastewater sludge.

29. Sodium carbonate is added for pH adjustment.

31. The main bleaches in laundry detergents are perborates. They work by reacting with the water to form hydrogen peroxide.

33. Redeposition is dirt resettling back on a fabric or other surface after it was removed during the cleaning process. Sodium carboxymethyl-cellulose (CMC) is the main anti-redeposition agent.

35. Soap remains the main surfactant in bath soap bars, although various synthetic anionic surfactants are also used.

37. Surfactant, a thickener, ethanol, some protein in certain shampoos, water, and traces of perfumes and dyes.

39. These are mainly quaternary ammonium salts, or cationic surfactants, and they work by interacting with trace deposits of anionic surfactants on the hair to form a "softening" precipitate.

41. In the United States it's sodium lauryl sulfate, but the Japanese use ordinary soap.

Chapter 14

1. The seven groups are polypeptides, saccharides, nucleic acids, lipids, hormones, vitamins, and minerals.

3. These building blocks of polypeptides contain both an amino group and the carboxylic acid group.

5. Rotate plane polarized light

7. Chiral is derived from a Greek word meaning hand.

9. Amino acids

11. Poly(amides)

13. The primary structure is the sequence of amino acids in the protein; any protein is an example. The secondary structure is the folding of the primary protein structure into an α-helix or a β-pleated sheet; such structures are maintained by hydrogen bonding; silk is an example of a β-sheet. The tertiary structure is the three dimensional structure of a protein that results from folding the regions of secondary structure; collagen is one example. The quaternary structure is an aggregation of more than one folded peptide chain to form a functional protein. Hemoglobin is a good example.

15. Collagen

17. An α-helix is a right-handed, coiled secondary structure that normally maintains its shape by hydrogen bonds. A β-pleated sheet is a protein secondary structure wherein the protein chains are arranged in the form of a sheet.

19. Polypeptide

21. In the lock-and-key mechanism proposed for enzyme activity to explain enzyme specificity, the reactant molecule (the substrate) binds with the enzyme to form a complex that has a shape relating to the reactant like a lock relates to the key.

23. A catalyst is an agent that makes a reaction run better, but still retains its own identity after the reaction is complete.

25. They react with and deactivate antigens.

27. Carbohydrate

29. Fructose, glucose, galactose

31. Cellulose, amylopectin, amylose, glycogen

33. Amylose mainly consists of a linear chain whereas amylopectin is a highly branched chain. The repeat unit is glucose in each case.

35. Glucose

37. From less sweet to sweetest: glucose, sucrose, invert sugar, fructose.

39. Adenine, guanine, cytosine, uracil, and thymine.

41. Nucleotide

43. RNA contains ribose while DNA contains deoxyribose. Also RNA contains uracil instead of the thymine found in DNA.

45. Fatty acids

47. Prostaglandins are a family of hormone-like substances derived from arachidonic acid, a 20 carbon fatty acid; they regulate many body functions and are prepared in many body tissues.

49. Sphingolipids are phospholipids containing the aminoalcohol sphingosine instead of glycerine.

51. HDLs transport cholesterol away from cells and back to the liver; this is considered the better form of lipoprotein. LDLs transport cholesterol away from the liver and to the cells; this is considered to be a risky form of lipoprotein.

53. Hormones control most sexual and physiological functions in the body. See tables 14.5 and 14.6 for specific examples.

55. Anabolic steroids are synthetic steroids that build up biostructures and are sometimes used by athletes to increase body mass; they have severe side effects.

57. Sodium and potassium ions regulate fluid flow and balance.

59. Phosphorus

61. The steroidal pill, intrauterine device, and the condom.

63. RU-486 is mifepristone, a progesterone derivative, that terminates, or aborts, a pregnancy.

65. The nature of the sense of taste stimulus is determined via direct chemical interactions. Taste is confirmed using taste buds, sensory units in the tongue and mouth. The taste buds respond to four different kinds of taste: sweet, sour, salty, and bitter.

Chapter 15

1. The basic step in metabolism (digestion) is the hydrolysis of proteins, saccharides, and lipids, as outlined in figure 15.1. Schematic word equations for each of these reactions are:

Protein + $H_2O \longrightarrow$ Amino acids
Complex saccharides + $H_2O \longrightarrow$ Simple sugars
Lipids + $H_2O \longrightarrow$ Fatty acids + Glycerine

3. Proteases

5. Meats, beans, peas, eggs, nuts

7. Lipids contain about 9 Kcal/g while both proteins and polysaccharides are about 4 Kcal/g.

9. Between 789 and 972 grams

11. Lactase is the enzyme that splits lactose (milk sugar) into galactose and glucose. When this enzyme is not present its absence leads to a condition called lactose intolerance, with symptoms of severe stomach cramps, bloating, and diarrhea.

13. Lipases

15. Oils

17. Oils have much greater iodine numbers.

19. 135

21. Vitamins are trace organic substances that are required in the diet and many are precursors of coenzymes.

23. Oil soluble vitamins persist longer in the body.

25. Not counting carboxylic acids, all the water soluble vitamins except biotin contain a hydroxyl group.

27. Beriberi, a muscular paralysis

29. Albert Szent-Györgyi

31. Vitamin E deficiency leads to sterility.

33. Pellagra, a skin disorder that leads to nerve disorders

35. The $t_{1/2}$ for vitamin E is less than one day, so the time required to return from four times the RDA value to "normal" would be two days.

37. Definitely a deficiency

39. Sucrose, fructose, glucose

41. Saccharin, aspartame, and Acesulfame K

43. The three reasons cited for using artificial sweeteners are: (1) diabetics cannot tolerate sugar, (2) noncaloric sweeteners will permit weight loss, and (3) sugar promotes tooth decay.

45. Some animal tests suggested they might cause cancer. Canada still permits the use of cyclamates.

47. The GRAS list is a "Grandfather Clause," which grants a special GRAS (**G**enerally **R**egarded **A**s **S**afe) listing to materials used for many years prior to the Delaney Amendment.

49. Sodium benzoate, sodium sorbate, and sodium propionate

51. Carbon dioxide gas is released during baking and this forms tiny pockets of gas within the bread.

53. The Haber process prepares ammonia from atmospheric nitrogen and hydrogen. This is the starting point for inorganic fertilizers.

55. These numbers represent the nitrogen, phosphorus, and potassium content of a fertilizer.

57. Magnesium is the key metal in the chlorophyll molecule.

59. Broad-spectrum insecticides kill a wide variety of insects while narrow-spectrum are more selective.

61. Malathion, parathion, chlorpyrifos (Dursban), and diazinon.

63. Radiation sterilization of male insects; the use of insect pheromones; and the use of hormones such as the juvenile growth hormone.

65. Controls development from the larval to the pupal stage.

67. Insecticides are more toxic.

69. It entered the food chain and threatened several species of birds with extinction.

71. The goal of the Green Revolution was to increase crop yields. It was almost entirely dependent on inorganic fertilizers, pesticides, herbicides and petroleum.

Chapter 16

1. Apparently before as long ago as 6,000 B.C. The modern cosmetic industry dates from the mid-1800s.

3. Petroleum jelly, in skin softeners

5. Lanolin, cetyl alcohol, and/or cocoa butter

7. Cold cream is a cosmetic that is used mainly to remove other cosmetics that are not water soluble and for cleaning the skin.

9. The main purpose at the present is to avoid or reduce sunburn. In the past the main purpose was to enhance suntanning.

11. Sunscreens are chemicals that block out ultraviolet (UV) light and help prevent skin cancer.

13. Melanoma is a virulent form of skin cancer that appears linked to severe sunburn, especially for people with light complexions, blue eyes, naturally blond or red hair, and freckles.

15. The SPF is the ratio of 36 to 3, or 12.

17. Retin A, retinoic acid, was first marketed in 1972 as an acne treatment, and is now also an over-the-counter cosmetic agent used to remove or reduce the amount of wrinkles in the skin.

19. Oils, such as mineral, castor, sesame, or a fat; lanolin or cetyl alcohol; beeswax; traces of dyes and perfume.

21. A polymer (such as nitrocellulose), a solvent (such as acetone), a plasticizer, and traces of colorants and perfumes.

23. That's debatable.

25. Mascara usually contains a large amount of soap and can be removed by washing with water. Eye shadow is mainly petroleum jelly and waxes, with a fair amount of zinc oxide and colorants and is not as readily removed by washing.

27. Petroleum jelly, lanolin, fats and/or waxes, zinc oxide, and colorants.

29. An antiperspirant is a chemical agent that prevents or restricts perspiration.

31. Deodorants are chemicals that attempt to mask or adsorb body odor.

33. The intent is different. Deodorants try to adsorb or mask odors while antiperspirants attempt to reduce perspiration. These activities are often combined into a single product.

35. Mainly in the amounts of fragrance. Perfumes have 10 to 25 percent, colognes have 1 to 3 percent, and aftershaves are less than 1 percent.

37. These terms describe the evaporative nature of fragrance chemicals used in a perfume so that the odor lasts longer. Top notes are the most volatile fragrance in a perfume mixture; final notes are the least volatile.

39. Keratin, which is also the main component of the skin, but the extent of crosslinking is greater in the hair than in the skin.

41. None. Chlorofluorocarbons (CFCs) were removed from hair sprays many years ago in the United States.

43. Calcium thioglycolate is the main active ingredient in most depilatories, but sodium, calcium, or strontium sulfide are also used. Emollients are also added.

45. Metallic, temporary, and semipermanent.

47. In metallic hair dyes, lead acetate reacts with dispersed sulfur and the sulfur present in the hair proteins to form lead sulfide, a brown-black pigment, which both coats the hair fibers and enters within the empty pores. Temporary hair dyes are colorants derived from coal-tar that coat the outside of the hair fibers and are readily removable with shampooing. Permanent hair dyes also used coal-tar colorants but these are initially in a reduced state which allows them to enter into the void spaces in the hair where they are oxidized to the final color.

49. Many are oxidized coal-tar derivatives that already have their final color present.

51. When the dyeing chemical is a coal-tar derivative, it is in an oxidized form for the temporary hair dyes, but with the semipermanent hair dyes the initial form of the dye is a reduced state that is oxidized after application to the hair.

53. Coal-tar derivatives are chemicals obtained from coal by pyrolysis (destructive distillation).

55. Thioglycolic acid and hydrogen peroxide, but the two are kept in separate containers until they are used on the hair.

57. It grows hair on people, both male and female, that have male pattern baldness.

59. Minoxidil is a drug that was originally developed for treating high blood pressure and is now used to grow hair in certain people with male pattern baldness. This latter usage is now approved by the FDA.

Chapter 17

1. **(a)** Chain and Florey developed penicillin, based on Fleming's early work; **(b)** Domagk found that a red dye called Prontosil was effective in treating streptococcal infections in animals and shortly afterwards used this chemical to save his daughter's life; **(c)** Dubos looked for antibacterial agents in soil samples and found gramicidin and tyrocidin; **(d)** Duggar discovered aureomycin, the first of the tetracycline family, in soil samples; **(e)** Ehrlich developed the first true antibiotic agent in 1907—the red dye trypan used to treat African Sleeping Sickness and later developed an arsenic-containing drug called Salvarsan 606, which was a cure for the sexually transmitted disease syphilis; he is considered the "father of chemotherapy;" **(f)** Fleming discovered penicillin; **(g)** Florey and Chain developed penicillin, based on Fleming's early work; **(h)** Hodgkin determined the structure of penicillin and later determined the structure of Vitamin B_{12}; **(i)** Sheehan developed the new techniques needed to synthesize penicillin; **(j)** Waksman discovered streptomycin and coined the term antibiotic for substances that killed bacteria without injuring other forms of life.

3. An analgesic is a chemical agent that reduces pain.

5. Acetylsalicylic acid (figure 17.2), acetaminophen (figure 17.3), ibuprofen (figure 17.4), and naproxen (figure 17.4).

7. An antipyretic is a drug agent that reduces fever.

9. Enkephalins are natural, polypeptide analgesics that are made in the brain.

11. Diethyl ether, chloroform, nitrous oxide, cyclopropane, lidocaine, novocaine, and sodium pentothal. Only sodium pentothal and nitrous oxide are still used in the United States, but the others are still used in Third World countries.

13. The LD_{50} is an estimate of the toxicity of a chemical for humans.

15. Caffeine

17. Insecticides

19. See figures 17.9, 17.10, and 17.11 for the structures.

21. A magic bullet is a term that describes a drug with a high specificity of activity.

23. It disrupts the cell walls.

25. Cancer has many causes, including smoking, viruses, certain chemicals, radon, and ultraviolet light exposure.

27. Soot, causing scrotum cancer in chimney-sweeps.

29. Antimetabolites and alkylating agents.

31. Antimetabolite

33. The main AIDS treatment drugs are compounds that are mimics of nucleic bases, and include 3-azido-3′-deoxythymidine (AZT), dideoxyinosine (DDI), and 2,3-dideoxycytidine (DDC). Several new drugs were added to these in early 1996 and these are protease inhibitors, designed to block the protease enzymes that are vital in the final stages of HIV replication.

35. A retrovirus is a virus that retrofits a person's DNA to a new use; HIV is a retrovirus.

37. Recent research suggests that a specific protein called β-amyloid is the culprit.

39. Tacrine, or THA.

41. Dextromethorphan is the most effective cough suppressant.

43. Guaifenesin is the *only* drug agent the FDA claims is effective as an expectorant.

45. Controlled release systems for drugs, artificial organs, dialysis membranes, contact lenses, dental fillings and dentures.

47. The six different controlled-release techniques are: (1) Erodable devices, (2) reservoir devices, (3) monolithic devices, (4) transdermal devices, (5) microcapsules, and (6) polymeric drugs.

49. High density poly(ethylene) and poly(methyl methacrylate).

51. Poly(ethylene terephthalate) or Dacron, and poly(tetrafluoroethylene) or Teflon.

53. Poly(methyl methacrylate)

55. Medical waste is potentially infected with disease-causing organisms that create the possibility of spreading disease. In a landfill, this danger could spread to the neighboring communities, carried by vermin.

Chapter 18

1. Nitrogen, oxygen, argon, and carbon dioxide.

3. Mainly sulfur dioxide and particulate matter.

5. Scrubbers are devices designed to remove chemical contaminants, especially sulfur dioxide, from exhaust gas streams.

7. Electrostatic precipitors, bag filters, and cyclone collectors.

9. Sulfur dioxide

11. Using P to denote primary and S to denote secondary, the answers are: NO_2, P; O_3, S; CO, P; hydrocarbon, P; aldehydes, S; peroxynitrates, S.

13. About half of it.

15. Around 75 percent.

17. The catalytic converter.

19. Nuclear and solar, though both use materials made by chemical processes.

21. France generates about 73 percent of its electricity via nuclear reactors compared with only 22 percent for the United States.

23. Nuclear energy generation is safer that coal fired energy generation when everything is considered.

25. The glassified pellets can be sealed inside poly(ethylene) and then buried in lead containers.

27. Ethanol and methanol.

29. One advantage is that there is virtually a limitless supply available in the waters of the Earth. A second advantage is that the waste material after combustion is water. The disadvantages include: it takes much energy to electrolyze water to generate hydrogen; second, H_2 gas requires heavy, pressurized tanks for safe handling; third, and probably most important, hydrogen and oxygen mixtures are explosive.

31. Passive and active solar heating systems. The former only uses small fans to move the heat while the latter uses more machinery.

33. No, there are only a limited number of places where the wind blows enough or geothermal energy is large enough to utilize for power generation. Still, both methods can supply some of the power for humanity and need to be utilized.

35. Chemical

37. Solar

39. Unfortunately, this answer brings us back to those systems that can produce the massive amounts of energy demanded by our increasingly industrialized society. They are nuclear and coal-powered electrical energy generation. Until the other systems become more developed, humanity will need to rely on these standbys.

Chapter 19

1. The greenhouse effect is that part of the solar energy cycle in which visible light energy is converted into infrared radiation, held within our atmosphere, and warms the planet.

3. The greenhouse effect has raised the temperature of the Earth into a range that favors life.

5. Troposphere

7. Carbon dioxide, chlorofluorocarbons, methane, and nitrogen oxides.

9. Carbon dioxide is released naturally from life forms on land and in the oceans; approximately 96.02 percent of the atmospheric CO_2 comes from these sources. Definite human sources include: electric power plants (0.95 percent), transportation (0.90 percent), industrial/commercial (0.78 percent), and residential (0.21 percent), for a total of 2.84 percent. Land conversion (deforestation, etc.) is responsible for another 1.14 percent and, if added to the human sources, this gives a total of 3.98 percent of the atmospheric carbon dioxide as being due to human activities.

11. This is difficult, but two techniques have been tried. They are studies of tree ring sizes and analysis of the gaseous contents of ice cores. The latter values are considered the better estimates.

13. There are at least six data sets showing the past temperatures of the Earth and these are: (1) daily temperature readings; (2) the average temperatures for the *entire* United States from 1885 to 1988, released in 1988 by the Department of Agriculture; (3) satellite monitored temperatures of the planet; (4) global ocean temperature measurements have been made by all ocean-faring vessels since 1856; (5) tree ring studies; and (6) studies of the sediments on the ocean floor can yield estimated ancient temperatures.

15. Some now believe that sulfuric acid aerosols may act like a cloud and retain atmospheric heat at night. If this is true, then the answer is yes. On the other hand, sulfuric acid aerosols could reflect the incoming solar radiation and lower the temperatures. More study is necessary.

17. Yes

19. Any source of sulfur dioxide can ultimately cause sulfuric acid aerosols. These sources include volcanic eruptions, smelting, and the burning of bituminous (soft) coal that contains sulfur.

21. There have be huge fluctuations.

23. The ozone layer varies daily and is not constant everywhere over the globe. It is not unusual to see a 40 to 50 percent variation in ozone levels over a period of a few days in the northern latitudes; the variation is less in the equatorial regions.

25. The thickness of the ozone layer and the exact O_3 concentration varies greatly with location (latitude), season, and time.

27. The ozone concentrations are greater at the middle and upper latitudes than at the equator and the ozone levels are also lower in the Southern Hemisphere.

29. Chlorofluorocarbons (CFCs) are the most cited culprits, but other possible ozone-depleting chemicals include: the hydroxyl (HO•) radical and NO_x (nitrogen oxides) compounds. There are also other possible sources of Cl atoms besides the CFCs. Some of these causes are natural.

31. Currently, the number two depleting species is believed to be the nitrogen oxides (NO_x).

33. About the only thing that we can do is restrict chlorofluorocarbon (CFC) emissions into the atmosphere, since most other ozone-depleting chemicals come from natural sources that are beyond our control. Whether banning the CFCs is a better solution that requiring the uses to be in closed systems only is debatable but the international mandates have chosen the banning route.

35. Yes

37. Most replacements have added hydrogen atoms to the molecules to give increased biodegradability.

39. They are believed to be less.

41. HCFCs are flammable.

43. The Antarctic ozone hole was first observed by G. M. B. Dobson in 1956-57 and was also observed by P. Rigaud and B. Leroy in 1958.

45. Since an ozone hole is a region of diminished ozone concentration, which occurs almost on a daily basis, the formation of ozone holes anywhere, including over the United States, is to be expected; it is a natural phenomenon.

47. Unfortunately no.

49. The ozone layer reduces the amount of ultraviolet light that reaches the surface of the Earth. If the amount of this UV increases, it is expected that the incident of skin cancer will also increase.

51. Although there has been a 2-5 percent decrease in the concentration of ozone in the stratosphere, there have not been the expected corresponding increases in ultraviolet light at the surface of the Earth. In fact, the data suggests that the UV levels have decreased.

53. Yes, but most observers believe that this was due to an increased amount of exposure to the sun during sunbathing and the like, and not to an increase in the total amount of ultraviolet light reaching the surface of the Earth.

SUGGESTED ANSWERS TO ODD NUMBERED CRITICAL THINKING PROBLEMS

Many critical thinking problems are opinion questions and have no set answers. In some cases, additional hints are provided but such questions have no answer provided.

Chapter 1

1. Imagine yourself floating about freely, and everything else doing the same thing. Could you even pour out a drink of water? Would you be able to visit a friend?

3. The number of people killed every year, in the United States, by automobile accidents far exceeds the total number of murders.

5. With a benefit value of 80 percent and a risk value of 90 percent, the benefit-risk ratio is $80 \div 90 = 0.89$. That's the easy part of this question; the hard part is forming your opinion. The identity of this chemical is ethyl alcohol, which is found in all alcoholic beverages.

7. Sorry to say, most "wonder cures" leave much to be desired. Maybe we need to wonder more about the claims. What did you think of those listed in this article? Watch for new ones coming out every year.

9. At least you have 100 people in this trial, which is 99 more than on the TV show in question 8. Note that these results are negative as far as the curative affects of grapefruit are concerned, but it's still harmless so you can eat as much as you desire without fearing for your health.

11. Sometimes scientists have been incorrect in their initial conclusions and later research changes their opinions. On the other hand, some scientists persist in studying something long beyond the time others feel it's worthwhile. Call it stubborn if you wish, but scientists are people.

Chapter 2

1. It's easier to smell something at higher temperatures; see the hint for the explanation.

3. As the alcohol levels in the blood increase, the individual's control over any motor function, including driving, decreases—an inverse relationship.

5. Indeed it is! Tires heat up during use but the volume is restricted; therefore, the pressure increases and if the tire has a weak point, the gas will rupture the tire to relieve the increased pressure. This is Charles' law in action.

7. Air is affected by gravity and tends to concentrate (settle) slightly to lower levels. Since there is less air on a mountain top than at sea level, breathing to obtain the same amount of air is more difficult.

9. Both salt and sugar dissolve in water and solutions, by their very nature, are homogeneous. The fact that the original mixture was actually heterogeneous has no effect in this case. On the other hand, sand does not dissolve in water so when a sand-salt mixture is placed in water the result is a homogeneous solution of salt with the sand sinking to the bottom as a heterogeneous phase.

11. It's true all the time.

13. Unlikely. Life is assumed to have started in the oceans and filling them with ice would probably preclude the development of life. Even if life started on land, life needs liquid water to function.

15. The nanotech machines are usually made using chemicals.

Chapter 3

1. No doubt our world would be different, but it's impossible to give any accurate answer here. This is the stuff of alternate timeline histories in science fiction.

3. This is not truly a good fit with the scientific method, although there is much evidence that certain types of chemicals (e.g., foods, vitamins, and medicines) can help us live longer.

5. Assuming you followed the hints given, you already know how energy is used to prepare new chemicals. Likewise, chemistry is involved in the generation of electricity and light. Benefits include more reliable sources of electricity and better intrusion warning systems.

7. Density could settle the issue, and actually did do so at times. Gold has a much higher density than the less expensive metals.

9. Boyle surely must have wondered why gases decreased in volume when the pressure was increased. Squeezing corpuscles closer together could account for this result.

11. This must illustrate the law of multiple proportions.

13. Density is one way. Pentane and benzene have different density values and you could estimate the amounts of each in a mixture from measurements of the mixture's density. Spectroscopic methods would work also but you do not know enough about spectroscopy to explain the details of how this might be done.

15. Obviously the answer is yes, women have been hampered in pursuing scientific careers in the past. The extent of this problem is less in the United States today, but many other regions of the world still have the same problems as in the past.

Chapter 4

1. The energy of microwave radiation is much lower than that of gamma rays. Gamma rays can harm cells or genetic materials in many ways. If a living creature is placed within a microwave oven, they can be harmed by the heat, but will not otherwise suffer damage.

3. Liquids and gases do not have a crystalline lattice, which was essential for Rutherford's α-particle studies.

5. The theory was modified and then retested. The modified version worked.

7. Our theories would probably have stagnated.

9. Metalloids are used in transistors, diodes, and computer chips. All of these inventions have changed our lifestyles.

Chapter 5

1. Higher engine temperatures increase efficiency, which increases gas mileage.

3. This is an opinion question, but check the data before answering.

5. The average would be 18.8 deaths per 100,000 people. The risk per 100,000 for the cited diseases are: less than 10 per 100,000 for AIDS; 4 per 100,000 for skin cancer; and 360.3 for heart disease. Note, however, that this does not mean second-hand smoke is harmless. Remember that this is an unnecessary cause of death that has little to do with the person's medical background or unusual risks they may have taken.

7. The twenty half life (20 $t_{1/2}$) period would be 161.2 days, or about 5.3 months. You should answer the part about nuclear safety for yourself, but do note that anti-nuclear spokespersons still claim large numbers of cancer cases, or other forms of disease, in spite of medical reports to the contrary.

9. Although it may not seem that way, this is an opinion question. One thing is certain, however. Something better must surely be possible for nuclear waste storage than what's been done for plutonium.

11. The source of the radon will be natural emissions, probably from rocks beneath your house. If the amount worries you, then by all means have your house vented so that the gas is exhausted daily. This will, of course, add to your heating and cooling bills since venting does mean you now have new places for warm or cool air to exit.

13. For a brief period of time all anti-nuclear activists would be happy, but then the sun would go out and we would all perish.

15. This is an opinion question that is "loaded" with emotional issues. It's your opinion.

Chapter 6

1. It would probably caused problems for him, but Mendeleev might still have arrived at the same conclusion based on the majority of data available to him. Exceptions do sometimes occur and they are treated as such in creating theories. His critics, however, would probably have used this information to lambaste his idea, and Mendeleev also.

3. This is an opinion question.

5. This is an opinion question.

7. It's radioactive, even though it's nonreactive.

9. All these devices require very pure chemicals. Some will require the development of new chemical compounds.

Chapter 7

1. Wheat flour has a greater surface area, per gram, and will burn faster than wheat in the form of grains.

3. Ordinarily the iron would oxidize and the buried tank would quickly become useless. However, magnesium is more reactive than iron and, since they are wired together, the electrons are lost from the magnesium, which oxidizes instead of the iron.

5. You could design a chamber to hold your fish and wrap the outside with cloth soaked with seawater. As the seawater evaporates, the inside of your chamber should cool somewhat. Another thing to bear in mind is that cooked fish decompose slower than raw ones. Finally, if you can tolerate the salt, keep the fish moist with seawater. Salt can act as a preservative. Keep your water purification running, however, because salty fish will make you very thirsty.

7. Iron is more reactive than tin, but it is protected from the oxidizing environment by the less reactive layer of tin. Once this layer is fully reacted, or is ruptured, the iron reacts rapidly.

9. Space ventures are extremely expensive and most of this high cost is underwritten by our government. Whenever a company wants to have experiments run on a Space Shuttle, they must pay part of the cost. Most companies cannot afford this, but many such experiments have already been run. This will most likely increase in the future, especially if the cost per experiment decreases. After all, companies need to make a profit or they will eventually go out of business.

11. Hydrogen is extremely flammable and forms an explosive mixture with air. There are safer ways to chlorinate a swimming pool.

13. It's an oxidation reaction. You can reduce the amount by keeping the food in a refrigerator since reactions proceed slower at lower temperatures. Placing the food inside a container or wrapper to exclude oxygen will also reduce spoilage.

15. Refrigerators reduce food spoilage by providing a storage place with a lower temperature. Food spoilage is nothing more than a collection of chemical reactions and, like all chemical reactions, will run slower at cooler temperatures.

Chapter 8

1. Whether or not you would eat the fish is an opinion question. The best way to control such water pollution is to prevent the heavy metals from entering the water; millions of dollars are spent each year on devices that avert this form of contamination. As is always the case, this extra cost is passed on to the consumer. Cleaning up polluted waters is even more of a problem.

3. For one thing, removal of ionic fluorides would be extremely difficult and hopelessly expensive. Most anti-fluoridationists concentrate on keeping fluoride ions from being added to drinking water supplies because this is easier and less expensive than removal of fluoride ions. If the problem is real, keeping fluorides out of water supplies would seem a logical first step.

5. Chalk is a weak base and will react with an acid. Chalk will not react with another base. You need to use the dilute acid to remove the chalk.

7. Hard water deposits are calcium carbonate and the like. These are decomposed by acids but not by bases.

9. This is definitely an opinion question. Two things to consider follow. Would the removal of these smaller amounts have a significant impact on the environment? Was there damage due to something like acid rain centuries ago, perhaps called by another name?

11. This is definitely an opinion question, but one worth mulling over. Is there one correct, or best, way to define pollutant? Can a natural product be classed as a pollutant? (Be careful here—any answer you give will be considered wrong by someone else.)

13. This is definitely an opinion question with six parts.

Chapter 9

1. The amount of oil works out to be 2.76×10^9 quarts, or 6.90×10^8 gallons. By comparison, the 1988 oil spill from the *Exxon Valdez* at Prince William Sound was 11 million gallons and the 1991 Persian Gulf war's multiple spills totaled around 250 million gallons. The waste oil from cars is 690 million gallons.

3. There is little doubt that the two programs are contradictory. The matter of ethics is more debatable. If the tobacco subsidies were discontinued, many tobacco farmers might go out of business and become unemployed. There are other options to becoming unemployed, such as raising other crops or converting the farm land to housing or industrial projects. The (former) tobacco farmers could still have a productive life, perhaps more productive than before. Whether cigarette smoking, or even the sales of cigarettes, is justified should be your carefully weighed opinion, based on as much data as you can amass.

5. This is basically an opinion question with no set answer, but the articles cited, and parts of this chapter, provide much insight to the questions posed. It might be worthwhile to separate petroleum uses when making your judgments. Remember that petroleum is not only used as a fuel, it is also used in manufacturing plastics, dyes, medicines, and many other useful commodities.

7. The C—O single bond would be more flexible that the C═O double bond.

9. Two products might be expected since there are two atoms at the double bond that could acquire the —OH unit. They are:

(a) $CH_3-\underset{\displaystyle |}{\overset{\displaystyle OH}{CH}}-CH_3$ and

(b) $CH_3-CH_2-CH_2-OH$

The answer to (*b*) has already been given and chemists can, indeed, control which product is formed. There are special conditions used to make either of these two structural isomers.

11. Density measures the mass in a given volume. If the molecules are packed closer together, there would be more mass per unit volume. The irregular branching in low density poly(ethylene) prevents this close packing and results in a lower density.

Chapter 10

1. This is to be your opinion, use the facts wisely.

3. Examine the basic chemistry shown in the following equation:

$$CH_3-\overset{\displaystyle O}{\overset{\|}{C}}-OH + HO-CH_2CH_3 \rightleftarrows CH_3-\overset{\displaystyle O}{\overset{\|}{C}}-O-CH_2CH_3 + H_2O$$

If either the acetic acid or the ethanol concentrations were increased, the equilibrium would shift toward products; that's two ways. Now if we removed either the ethyl acetate or the water, the equilibrium would again shift to the right; that's two more ways.

5. This is a good question to probe and form your own opinion. But be careful what is considered evidence on either side of this debate. Some of this is debatable at best.

7. This is yet another opinion question. As you read the recommended article, and possibly research this topic, remember that some accounts on either side will probably be highly biased.

9. Make your list and draw your own conclusions. You may have a major problem finding substitutes for many of these household chemicals, at least substitutes that do a good job. Many of these chemicals have found their way into our homes because they do a good job.

11. Glycerine is used in many lotions; alcohols are used for rubbing and cleaning the body; acetone is used in nail polish and nail polish remover; sodium stearate is used as a toilet soap; esters are used in perfumes.

13. This is an opinion question. Be aware, however, that Brazil has curtailed much of their use of ethanol. The aldehyde problem is a major concern since this could lead to the formation of other, more irritating chemical agents. This actually occurs in the photochemical air pollution that is discussed in detail in chapter 18.

15. The best method is to distill off the ethyl acetate, which is the lowest boiling component of the equilibrium mixture.

17. Adipic acid does react with 1,6-diaminohexane to form a poly(salt). When this poly(salt) is recrystallized, you obtain an exact stoichiometric mixture of the adipic acid and the diamine, which leads to the formation of high molecular weight polymers.

Chapter 11

1. This is still an ongoing problem, and will likely remain as such for many years to come. The question is primarily an opinion one, however, and has no set answer. Weigh the evidence carefully before reaching your decision.

3. This is an opinion question.

5. Acrylonitrile monomer bears a strong resemblance to the acrylates so it would probably be in their price range.

7. This is an opinion question.

9. Some plastics could serve as a substitute fuel and would release about the same amount of energy per kilogram as petroleum. Incineration reduces the amount of plastics in landfills. We should view incineration as a last resort for plastics since it is more valuable to recycle them. However, incineration may be a reasonable disposal method for plastics that are difficult to recycle, such as thermosetting plastics and vulcanized rubber, and might provide energy instead of wasted landfill space.

11. Among the advantages are: better insulation characteristics, lower cost, impact resistant windows and doors, better plumbing, window frames, and siding materials.

Chapter 12

1. Certainly! Floods destroy much vegetation each year and watering plants too much can kill them by causing root rot.

3. This would be possible if enough time had elapsed to allow the effects of the toxic chemicals to become manifest in the population. Taking the survey does pose problems because many ways the questions could be phrased would suggest answers indicating the toxins caused these health problems even if this were not really the case. Among the limitations of such a study are the difficulty of separating the effect of the toxic chemicals from many other possible causes for the diseases. Sometimes simple solutions do not exist. Furthermore, even if the survey actually proved cause and effect, further work would be needed to correct the problems. Finding those responsible and getting them to rectify the situation is seldom easy and could plunge the community into a lengthy legal battle. On the other hand, finding money to clean up the problem can be even more difficult. Consider the results of the Superfund in this context.

5. This is an opinion question, and a very difficult one. The first thing you need to do, however, is to amass data on the problem. Only then can you make a reasonable decision. You might ask why the government can't seem to do this, considering all their resources. Could this be another case of pork-barreling?

7. **(a)** Look at the use data in box 12.3, figure 1, and at the varied fruit from the "chlorine tree" in figure 2 of this box. How many of these uses appear in your life each day? Most likely it's many dozens. **(b)** Alternate materials to replace chlorine-containing chemicals seldom exist. When they do, they are often either more expensive or less effective, or both. **(c)** Total bans on anything are seldom the wisest course of action. This is especially true when no alternative solutions exist, as is the case here. **(d)** Look up the data on typhoid fever, figure 12.10, and see how effective water chlorination was in controlling this disease. Any chlorine ban is almost certain to have an adverse effect on public health.

9. Whether cholera could make inroads in the United States is debatable, but if health habits change cholera could pose a major threat. One change that could propel cholera to the forefront might be the cessation of water chlorination. Cholera wiped out many thousands in major plagues years ago and still has that potential. Like other diseases, many forms of cholera cannot be treated with our current medical agents. Unless new ones are developed we could be partially defenseless. What you would do depends on your actions both now and later. That part of this question is opinion.

11. The volume would be 1.2×10^{11} mL. This would grow to 3×10^{12} in the next quarter century. The rest of this question is your opinion. Make it wisely, after all, it's your future too.

13. **(a)** Obviously not since less than five percent of those materials that are meant to be recycled are actually recycled. All these programs cost money and virtually none are self-sustained. **(b)** Aluminum and paper (to an extent) are re-used; many other materials are not. Stories about plastics and other recyclable materials being stored in above ground storage buildings or open areas have now been made public. **(c)** Aluminum and, to a degree, paper. **(d)** Unfortunately, plastics too often end up in above-ground storage. Sometimes the reuse of plastics becomes impaled on governmental regulations. **(e)** Possibly the only way recycling will become economical is for it to become profitable. Some private companies do make a profit from recycling, but too often this profit is subsidized by communities. Yet, many of the recyclable materials have a significant value and should lead to a profitable business, without governmental support. Perhaps we should be asking why this isn't happening more? **(f)** Landfilling does not seem to be an option but incineration does. **(g)** Not if the alternative approach is incineration.

15. Part of this is an opinion question, but if you look at figure 12.10 you almost have to conclude that eliminating chlorination will bring a disastrous return of typhoid fever. For the comparision of potential cancer against typhoid fever you need to consider several things. First, typhoid fever will kill people quickly and cancer takes many years to develop. Second, not every exposure to potential cancer-causing agents leads to actual cancers. Third, most estimates of the hazard of these trace amounts of potential cancer-causing agents are most likely grossly overestimated. Since these trace amounts have been around a fairly long time, we should assume that some of their effects already are present in the annual totals of new cancer cases and the deaths due to cancer. You could try to make your own guesstimate of the number of such cases for the most recent year you can locate data and compare that with your estimate of deaths that a resurgent typhoid fever might cause. Remember, express these estimates as the number of cases (or deaths) per 100,000 people.

17. Another opinion question. This hint might be useful: people obtain their calcium or magnesium, which are definitely essential for health, from either their foods or mineral supplements. They do not get it from water.

Chapter 13

1. The order of decreasing environmental friendliness is: **a, c, d, b** (assuming a linear alkyl group), **e.**

3. This is a do-it-yourself literature search and an opinion question.

5. This is a do-it-yourself ranking and an opinion question.

Chapter 14

1. The body operates using an aqueous-based chemistry system.

3. This is a literature search and opinion question.

5. Every individual human being, except identical twins, has a unique DNA structure. However, there are certain features we all share in common and the knowledge of some of these may help combat diseases that have a genetic component, such as cancer and Alzheimer's. Read more about this and form your own opinion.

7. Aging does occur at different rates for different people and some have a much younger appearing body than others of comparable age. All the factors involved in aging are not known but a good diet and exercise seem important in maintaining a "youthful" appearance. One major theory claims that aging is due, at least partially, to crosslinking of the proteins in the body, causing them to lose flexibility and operating efficiency. This theory suggests that antioxidants may help delay aging. This theory remains to be proved.

9. This is an opinion question.

11. This is an opinion question.

13. After your literature research, this is still an opinion question.

15. This is an opinion question and is a slightly different situation than the question in problem number 16. How important is birth control in an agricultural country where children may be needed to bring in the crops and tend the herds?

17. If you are not familiar with the legacy of Malthus, here it is in a nutshell. Thomas Malthus (English, 1766–1834) proclaimed that population increases were greater than the increases in providing food and clothing, and unless the birth rate was controlled, poverty and war would serve as a natural means of population constraint. He further argued that in the battle between sex and restraint, the moral argument seems to lose more often than better food production could possibly handle. Malthus' followers concluded that the population increased geometrically (i.e., 1, 2, 4, 8...) while food production increased arithmetically (i.e., 1, 2, 3, 4...). The longer the game goes on the worse is the result; you cannot possibly win in this race. This will at least give you a small amount of background for this question. This is still an opinion question.

19. This could become a serious problem if it happens. Read and form your own opinion on the relative merits for each side of the issue.

21. This is an opinion question, but it would be unreasonable to assume that all possible or potential changes would be good for humanity. Try to decide where this line should be drawn, and why.

Chapter 15

1. The answer to this question is debatable, but many claim that we do not get sufficient vitamins in the usual diet. Perhaps this is a weakness of the style of diet we use and, perhaps, this could be changed into one that does supply all our necessary vitamins. Read the articles and form your own opinion.

3. Vitamin A prevents night blindness and is essential for vision. The complete lack of this chemical could prevent the formation of the visual chemicals in the eye. That would result in blindness.

5. It is possible to survive on a vegetarian diet. One possible undesirable consequence of the total change mentioned here is that the "natural" controls over animal populations would disappear.

7. People can survive quite well on a vegetarian diet, though some care is necessary in order to acquire all the necessary nutrients. Many ethnic foods are primarily vegetarian and have worked well for centuries. If, however, you strongly like the taste of meat you will probably be dissatisfied with a vegetarian diet. On the other hand, if you do not want to see any animals killed for food, a vegetarian diet is almost mandatory.

9. The GRAS list (**G**enerally **R**egarded **A**s **S**afe) contains food items that have been used for centuries without any pronounced health problems, for the general population. Still, some chemicals on the GRAS list are probably harmful in high dosages and this is often unknown. The testing of such chemicals, like any other chemical, would require animal testing, which is very expensive. It's unlikely that anyone, except the government, would be willing to pay for such tests. At this time, even the government is not planning to test the materials on the GRAS list.

11. To an extent, the proof is in the tasting. If it does not taste good the "improved" plant products will not sell well. There is a legitimate concern for safety in any new food product and these biotech "wonders" are not immune to these considerations. Not to worry! Any new biotech product, food or otherwise, is greeted with hostility from some groups. Safety can be assayed by feeding the foods to animals that act like humans in their response to the particular foodstuff. (It wouldn't do to feed new meats to rabbits or carrots to tigers.) Assuming the foods met these strict requirements, they could be a blessing to humanity and could help alleviate famine and food shortages. Unless the foods were modified greatly, the nutritional value should be the same as for regular versions of the same food that had not deteriorated. Some biotech foods do not deteriorate as quickly as the older version. That alone could help avert famines and improve the food supply.

13. Once you realize that the annual crop loss in the United States is more than 15 billion dollars, the idea of growing insect-resistant crops looks ideal. Assuming there are no chemicals that appear in such crops that would be toxic or harmful to people, this is a good idea. Presumably any such insect-repelling chemicals would disappear soon after harvesting or cooking. If not, then humans would be consuming variable amounts of the insect-repelling agents. To a degree we already do this with insecticide residues remaining on the crops. Will this be any different? Nobody knows for certain except for those that have already been tested. As far as problems are concerned, some insects will likely develop resistance to whatever the plant produces to keep them away. Whether these are hardier insects remains to be seen, but most likely they will be very similar to those eating the crops today.

15. As box 12.1 showed, proving the connection between Agent Orange and birth defects has not been easy. Sometimes the changes take many years to manifest themselves in the veterans or their offspring. In most of these cases, cause and effect must be inferred rather than demonstrated directly and human experimentation is out of the question. Extensive studies need to be made of those exposed to Agent Orange and follow-ups on the veterans and their children is essential. Some believe that the government has been less than honest in their approach to this problem. The fact remains, however, that most veterans did not suffer severe damage after exposure to Agent Orange. Why some veterans have responded differently is unknown. Some have tried to link dioxin to this problem but many other studies have failed to show the severe genetic damage claimed by others. Almost forgotten in this problem is the fact that there were several chemicals used in defoliating the Vietnamese forests. Perhaps one of these, or some combination, is the real culprit. Perhaps total exposure, which is largely unknown, was the problem. This issue remains essentially unsolved; a similar problem seems to be surfacing involving Gulf War veterans.

17. This is a do-it-yourself research and opinion question. Have fun.

19. Basically, this is an opinion question.

Chapter 16

1. Many, but not all, natural ingredients used in cosmetics have the advantage of being biodegradable while some synthetic ingredients do not biodegrade. It is an odd fact that many so-called "natural" cosmetics contain many synthetic compounds because these ingredients work better than the alternative natural ingredients. It would be unreasonable to assume that all-natural cosmetics, whatever that may mean, are safer than those that do not make such a claim. Perhaps it is worth noting that most cosmetics contain substances, such as petroleum jelly, mineral oil, and lanolin, that can easily be classified as "natural."

3. Sunscreens are tested on animals before use on humans. Even before that, whether these chemicals can work as sunscreens can be estimated from their interaction with ultraviolet light, which can be measured with a spectroscope. Ultimately these chemicals are field tested on humans. Not all potential sunscreening chemicals would be safe for human use; some are too toxic or too irritating to the skin.

5. This is an opinion question based on knowledge you can obtain in the recommended readings. As a bonus, note that the effectiveness of aroma medical treatments is considered suspect. You will need to read the articles to find out why. However, aromas have been used for centuries to make a working or living place more pleasant.

7. Drug testing usually involves new chemicals that might become useful and important drugs. Cosmetic testing, on the other hand, normally involves chemicals that have been used for many years and have had many animal testing trials. Consider this as you derive your answer to this opinion question.

9. Condemning hair dyeing sometimes seems like an annual ritual but there seems to be little evidence that it is cancer causing or harmful, at least when done properly. There's more than that in the recommended article, so read and then make your opinion.

Chapter 17

1. Herbal medicines have a long history of use and many important drugs are obtained from plants. Several examples were given in this chapter and many others can be found in the two articles cited. Obviously the answer to the first question is yes, since that has already happened. It's your task to list the advantages and disadvantages of this approach.

3. Mice are now common pets; rats are also used as pets but have not really caught on that well. Insects are in most homes, but not usually as pets. Could this have any influence on whether people protest animal testing using insects?

5. This is a loaded opinion question. As a hint, let's rephrase this as "should a drug that is harmful to a few be disallowed for treating people whose only hope for a cure depends on this drug?" Almost all drugs already carry warning labels telling the potential user about the possible dangers and warning to seek medical advice before using if uncertain. This is still an opinion question. How do you answer that last question?

7. Spontaneous remission often cannot be explained, but the placebo effect is assumed due to the patient's belief that what was taken was an effective drug that would be of help. The difference between the placebo effect and spontaneous remission is subtle but real. The latter sometimes occurs in the absence of anything considered as medical treatment. Faith healing would fit into this category. The placebo effect requires some form of presumed medical treatment.

9. Cancer remains a difficult disease many years after the beginning of the so-called war on cancer. Perhaps the new tumor-specific drugs may help in this particular battleground. Some of these are an offshoot of the polymeric drugs presented in this chapter. Others involve liposomes and cloned antibodies, recent developments from biotechnology.

11. This is a good question whether you are male or female. Somehow the idea that one sex should receive preferential treatment for a generalized disease like cancer does not set well with most people. Read the articles for background on this touchy issue.

13. You should form your own opinion on the programs discussed in these articles. Here is a hint on the last question, however. Vaccines are used to prevent diseases (see box 17.3) and not to treat them. In addition, the cancer-causing viruses are normally of the DNA type, whereas HIV is a retrovirus. No vaccines have yet been developed against retroviruses but that is possible. If they are developed, they will help *prevent* the disease, not cure it.

15. These are opinion questions but some hints may be appropriate. **(a)** In many states it is illegal to inform others that a person has AIDS. **(b)** The doctor may already know, if the doctor recommended the testing. Otherwise none of the medical personnel will find out in states that forbid release of this information. **(c)** None are allowed and some laws forbid this. **(d)** This is not legal. **(e)** Absolutely! Sorry, but that was more than a hint. All we can now detect is the antibodies to HIV and a blood supply could have the virus and no antibodies. This has happened far too often and the solution is a matter of public safety. A possible blood test for the AIDS virus was announced in early 1996.

17. Most of this question is of the personal opinion type. All current AIDS treatments merely prolong life; none have been proven to cure the disease. Weigh your options carefully here. Is a twenty year old person more worthy of medical help than a fifty year old one? Is it truly important *how* the disease was contracted, as far as receiving treatment is concerned? Is there a limit to the total amount of money that can be used to treat one disease, probably at the expense of reducing expenditures on all other diseases?

19. Pre-knowledge can assist in planning, but how do you make plans for an incurable disease? Some people would prefer not to know that they have a potential tendency toward a deadly disease, others want to know so they can plan for their families. One thing that can be done is to support research in these other ailments, such as Alzheimer's. At the moment there is no cure but that can change. In addition, maybe scientists can learn how to prevent the development of Alzheimer's and similar diseases. That would be even better than a cure for most people.

21. It's easy to forget about malaria when you live in a country that does not have this disease as a major problem, but malaria is making a comeback in many Third World countries. As you read these articles, note that the new treatments are promising but not yet proven effective. It costs money to do the necessary research and that's not always easy to get. As far as DDT use is concerned, it is still permissible to use this insecticide to avert a plague.

23. Some firmly believe these disastrous events are almost certain in our future. Fortunately the United States has good medical research and disease prevention programs, but this may not guarantee success with new strains of old diseases and brand new diseases like Ebola. What do you think can be done to avert disasters?

25. This is an opinion question.

27. This is an opinion question.

29. This is an opinion question, but here is another hint: most people try to replace the smoking habit with food.

31. This is an opinion question, but most people would not rub mold on a cut. Remember, the molds produce other chemical agents besides antibiotics and these might not be beneficial.

33. This is an opinion question.

Chapter 18

1. Temperature and the wind are most assuredly part of the environment. Unfortunately, we cannot control either one, at least at this time, so we cannot use either to improve environmental conditions.

3. This is a literature research and opinion question. Smog City is, of course, Los Angeles. Their plans for improving their air quality could set a new standard for everywhere.

5. This is a bit tricky. It depends, in part, on when the measurements are being made. Farming communities could have a lot of dust during the harvest season if it's dry. Cities can have lots of particulate matter or very little depending on the recent weather conditions. The salt level in the air over the ocean is probably fairly constant for any given time of the year. So what is the answer? It depends on weather conditions. That may surprise you since cities are normally cited as the worst places for particulate matter, but that's not always true.

7. This is an opinion question but do your literature search before making your final decision. There was also additional information on this topic in chapter 5.

9. This is an opinion question.

11. Cold fusion is one of those enigmas in current science. It may turn out to be a real, and useful, phenomenon or it may be another scientific mistake. Some believe it was a hoax, but the scientists involved did not seem to gain anything from it. There have been what appear to be reliable studies on a cold fusion process, but that process does not show much promise for producing massive amounts of energy. This is one of those questions that will take many years to resolve; in the meanwhile, someone just might figure out how to make a useful cold fusion apparatus and change the entire picture of energy production. For now, however, this is an opinion question; don't be afraid to change your opinion if more data shows that you were wrong.

13. This is an ongoing debate and there will certainly be more recent data that can be cited here. This is also an opinion question, so look up what you can find before setting your decision in stone.

15. Fuel cells certainly sound exciting, as well as promising.

17. Large scale power generation from the wind is not likely to become a major energy source because there are not that many places where the wind blows reasonably constantly. Optimistic claims aside, wind power generation is unlikely to become a major factor in most parts of the world. For sites where wind power generation is feasible, this resource could be useful and should be augmented. Any additional source of energy, no matter how small, can help enhance the overall energy generation.

19. At the present time the electric car is priced out of the reach of the average consumer, even if they might like to purchase one. The same thing is true for other "green cars." This could change if the cars begin to be mass-produced, but that depends, in part, on consumer acceptance. Sounds like a Catch 22 situation, doesn't it?

21. This is an opinion question. While the most recent air quality legislation virtually mandates this action, such rules are always subject to change as new people, sometimes with vastly different ideas, assume office. The question is whether this is a good idea? Should alternatives, such as sulfur removal and improved scrubbers, be pursued instead?

23. Definitely a pair of opinion questions, but ones that require a lot of review and literature searching. Review the papers cited in the Readings of Interest section before making your decisions.

25. Many people believe the United States is fast losing its cutting edge in new technology developments, but is this true? One additional source of pertinent information is annual almanacs or yearbooks. Many of these list recent inventions and scientific discoveries. You can check the older issues and then compare the data to see whether there has been a change in the technological acumen for the United States. If you do find this to be true, what can be done about it? Or should anything be done on this matter? Isn't technological development a world concern?

Chapter 19

1. This is an opinion question.

3. Most Siberians would probably welcome global warming with open arms, and open fur-lined coats. One thing you should consider, if living in Siberia, is whether this change will help you raise your crops. Most likely it would.

5. Hopefully you greet such reports with concern. As far as what can be done to tackle these problems that is an opinion question for you to answer. Be certain to read some of the cited articles to get enough background to form a reasonable opinion.

7. This is more difficult than it appears. Since most ozone depletion is believed due to non-CFC use, it seems reasonable to conclude that most of this depletion is from natural causes. This is one of those things for which we really need more data.

9. The relationship between ultraviolet exposure and skin cancer appears well established, but this is not the case for any relationship between the ozone layer levels and the amount of ultraviolet radiation that reaches the Earth's surface. There has been a 5 percent decrease in the ozone layer since 1970 but the ultraviolet levels have not shown a corresponding increase. According to theory, the UV reaching the surface should have increased by about 10 percent during this time. Do read some of these articles and form your own opinion.

11. This depends on what you mean by public exposure. Scientists usually can get a hearing on their data or theories by their peers, but sometimes these theories are rejected. Even data is frequently ignored by other scientists for long time periods. The ultimate strength of the scientific method is that truth will prevail in the long run. It may take decades or longer, but the scientific method works. If public exposure means press releases or time on the newscasts, then that's another matter entirely. Many of the unusual viewpoints of a few scientists get far more uncritical news display than they deserve. It is, of course, difficult for news people to know what claims are correct and what are spurious, but surely some middle ground exists. Listen to the varied, and often contradictory, health reports on any local TV news show.

13. This is an opinion question.

15. This is an opinion question.

17. This is an opinion question.

19. This is an opinion question. All necessary information for answering **a** to **f** is provided in this chapter. Track it down and use it.

21. This is an opinion question, but note that money and reputation-making often play big roles in these problems.

23. This is a reading research and opinion question.

25. This is an opinion question. Ask yourself if harmful activities are a good means to effect change. After all, aren't similar things done in other realms of human activity?

27. This is a read and analyze opinion question. Have some fun doing it.

29. This is an opinion question. Be sure to tell why you formed this opinion, regardless of your choices.

CREDITS

CHAPTER 1

Opener: © David M. Dennis/Tom Stack & Associates; **p. 2 (left):** © Michael Newman/PhotoEdit; **p. 2 (right):** Courtesy Zenith Electronics; **p. 3 (top):** AP/Wide World Photos; **p. 3 (bottom):** © Gary Conner/PhotoEdit; **Box 1 fig.1:** Courtesy of Tupperware U.S., Inc.; **1.1:** © Tony Freeman/PhotoEdit; **p. 6:** AP/Wide World Photos; **1.2:** Bettmann Archive; **Box 1.5:** AP/Wide World Photos

CHAPTER 2

Opener: © Spencer Swanger/Tom Stack & Associates; **p. 28:** Bettmann Archive; **p. 32:** Chemical Heritage Foundation Pictorial Collection; **p. 34, Box 2.4 fig.1:** Bettmann Archive; **Box 2.4 fig.2:** North Wind Picture Archives; **2.11 a–c:** Corel Photo CD; **2.17:** © Yoav Levy/Phototake

CHAPTER 3

Opener: © Charles Walker Collection/Stock Montage, Inc.; **p. 48:** © E.R. Degginger; **p. 51:** Chemical Heritage Foundation Pictorial Collection; **3.1:** © Yoav Levy/Phototake; **Box 3 fig.1:** Bettmann Archive; **p. 54:** Chemical Heritage Foundation Pictorial Collection; **3.4:** Egyptian Museum, Cairo; **p. 58:** The Nobel Foundation, Courtesy of Chemical Heritage Foundation Pictorial Collection; **3.5a:** © Russ Lappa/Photo Researchers, Inc.; **3.5b:** © Len Lessin/Peter Arnold, Inc.; **3.5c:** © V. Fleming/SPL/Photo Researchers, Inc.; **3.5d:** © Yoav Levy/Phototake; **3.6:** UPI/Bettmann; **p. 63:** Chemical Heritage Foundation Pictorial Collection; **3.9 a–c, 3.10 a–c, 3.12, Box 3 fig.3:** © Times Mirror Higher Education Group, Inc./Bob Coyle, photographer; **3.13:** © Yoav Levy/Phototake; **3.14 a–d:** © Times Mirror Higher Education Group, Inc./Bob Coyle, photographer; **3.15:** © 1994 North Wind Pictures

CHAPTER 4

Opener: © Science Source/Photo Researchers, Inc.; **p. 84:** Bettmann Archive; **p. 85:** Courtesy of Edgar Fahs Smith Collection, University of Pennsylvania Libraries; **p. 86:** Bettmann Archive; **p. 92:** Chemical Heritage Foundation Pictorial Collection; **Box 4 fig.1; p. 95, p. 97, p. 100:** Bettmann Archive; **4.15:** Centers for Disease Control, Atlanta; **Box 4 fig.2:** Bettmann Archive

CHAPTER 5

Opener: © Wm L. Wantland/Tom Stack & Associates; **5.2:** © Jeff Greenberg/PhotoEdit; **Box 5 fig.2:** © E.R. Degginger; **Box 5 fig.3 a–f:** © Times Mirror Higher Education Group, Inc./Bob Coyle, photographer; **Box 5.4 fig.1:** © Bob Daemmrich/The Image Works; **Box 5.4 fig.2:** © Bruce & Kenneth Zuckerman/West Semitic Research/ PhotoEdit; **Box 5.4 fig.3:** NASA, **Box 5.5:** © Times Mirror Higher Education Group, Inc./
Bob Coyle, photographer; **5.9:** © Joe Sohm/The Image Works; **5.10 a–c:** © Southern Illinois University/Visuals Unlimited; **Box 5.6 fig.1:** Courtesy of FOOD TECHnology Service, Inc.; **Box 5.6 fig.2:** International Atomic Energy Agency, Public Information Products

CHAPTER 6

Opener: © J. Pickerell/The Image Works; **p. 142 (left):** Courtesy of ALCOA (Aluminum Company of America); **p. 142 (right):** Bettmann Archive; **p. 143:** Courtesy of Bancroft Library Archives, University of California, Berkeley; **Box 6.2:** from *Chemical Engineering News*, Courtesy of American Chemical Society; **Box 6.3 fig.2,3:** © Times Mirror Higher Education Group, Inc./Bob Coyle, photographer; **Box 6.3 fig.4:** © Doug Sherman/Geofile; **Box 6.4 fig.1:** © Gianni Tortoli/Photo Researchers, Inc.; **Box 6.4 fig.2:** © 1995 Cleveland Museum of Art, Purchase John L. Severance Fund, #57.61; **Box 6.4 fig.3a:** © Allen Green/Photo Researchers, Inc.; **Box 6.4 fig.3b:** NASA/SS/Photo Researchers, Inc.; **Box 6.5 fig.1:** © Times Mirror Higher Education Group, Inc./Bob Coyle, photographer; **Box 6.5 fig.2:** © Fujifotos/The Image Works

CHAPTER 7

Opener: © E.R. Degginger; **7.3 a,b; 7.4:** © Times Mirror Higher Education Group, Inc./Bob Coyle, photographer; **Box 7.1:** Monique Salaber/NASA/The Image Works; **Box 7.2 fig.1a:** © Jim Zerschling/Photo Researchers, Inc.; **Box 7.2 fig.1b:** © Times Mirror Higher Education Group, Inc./Louis Rosenstock, photographer; **Box 7.2 fig.2–4:** © Times Mirror Higher Education Group, Inc./Bob Coyle, photographer; **Box 7.3:** © Lee Snider/The Image Works; **Box 7.4 fig.4:** © Times Mirror Higher Education Group, Inc./Bob Coyle, photographer

CHAPTER 8

Opener: © E.R. Degginger; **Box 8.3, 8.2:** © Times Mirror Higher Education Group, Inc./Bob Coyle, photographer; **Box 8.4 fig.1:** © M.A. Cunningham; **Box 8.4 fig.2:** © John D. Cunningham/Visuals Unlimited

CHAPTER 9

Opener: © Joe Sohm/The Image Works; **9.1, 9.17:** Bettmann Archive; **9.25:** AP/Wide World Photos; **9.31:** © Paolo Koch/Photo Researchers, Inc.; **Box 9.4 fig.1:** North Wind Picture Archives

CHAPTER 10

Opener: © Bob Daemmrich/The Image Works; **Box 10.1:** © Tom McHugh/Photo Researchers, Inc.; **10.6:** © Times Mirror Higher Education Group, Inc./Bob Coyle, photographer; **Box 10.3 fig.1:** © Tom Pix/Peter Arnold, Inc.; **Box 10.4 fig.1:** Courtesy of the National Portrait Gallery of London; **Box 10.4 fig.2:** © Esbin-Anderson/The Image Works; **10.17:** © Andrew Syred/SPL/Photo Researchers, Inc.; **10.23:** © Times Mirror Higher Education Group, Inc./Bob Coyle, photographer

CHAPTER 11

Opener: © Adam Hart-David/SPL/Photo Researchers, Inc.; **11.1:** © Michael Greenlar/The Image Works; **p. 273:** Chemical Heritage Foundation Pictorial Collection; **Box 11.1 fig.2 a–d:** © Times Mirror Higher Education Group, Inc./Bob Coyle, photographer; **Box 11.1 fig.3:** © SIU/Photo Researchers, Inc.; **11.10:** Chemical Heritage Foundation Pictorial Collection; **11.17:** AP/Wide World Photos; **11.19a:** AP/Wide World Photos; **11.19b:** AP/Wide World Photos; **Box 11.2 fig.1:** © Times Mirror Higher Education Group, Inc./Bob Coyle, photographer

CHAPTER 12

Opener: © Rob Crandall/The Image Works; **12.5:** © W. Hill, Jr./The Image Works; **Box 12.1 fig.2 a,b:** AP/Wide World Photos; **Box 12.2:** © Louis Psihoyos/Contact Press Images, Inc.; **12.8b:** © Archives/The Image Works; **12.12:** Courtesy of Alan R. Thomas Water Purification Plant, Edgewater, FL; **12.14 a,b:** © Tom Pantages; **Box 12.4 fig.1–3:** Courtesy of EcoWater Systems, Inc.

CHAPTER 13

Opener: © Jeremy Burgess/SPL/Photo Researchers, Inc.; **Box 13.3:** © David Young-Wolff/PhotoEdit; **Box 13.4:** © Times Mirror Higher Education Group, Inc./Bob Coyle, photographer

CHAPTER 14

Opener: © Michael W. Davidson/Photo Researchers, Inc.; **Box 14.2 a,b:** © Jackie Lewin/SPL/Photo Researchers, Inc; **Box 14.3 fig.1a:** Courtesy Arthur J. Olson, Ph.D., The Scripps Research Institute, Molecular Graphics Laboratory; **Box 14.3 fig.2:** © Dennis MacDonald/PhotoEdit; **Box 14.4 fig.1:** Courtesy of the Cystic Fibrosis Foundation; **Box 14.5 fig.1:** © David Wells/The Image Works; **14.26 a–c:** © Times Mirror Higher Education Group, Inc./Bob Coyle, photographer; **14.26d:** © Hank Morgan/Photo Researchers, Inc.; **14.26e:** © Times Mirror Higher Education Group, Inc./Bob Coyle, photographer; **14.26f:** Courtesy of Gyno Pharma, Inc.

CHAPTER 15

Opener: © Matthew Klein/Photo Researchers, Inc.; **Box 15.1 fig.1:** © Times Mirror Higher Education Group, Inc./Bob Coyle, photographer; **Box 15.2 fig.1:** Historical Pictures/Stock Montage, Inc.; **Box 15.2 fig.2:** Bettmann Archive; **Box 15.2 fig.3:** AP/Wide World Photos; **Box 15.2 fig.4:** Chemical Heritage Foundation Pictorial Collection; **Box 15.4 fig.1:** AP/Wide World Photos; **Box 15.4 fig.2:** © Tom Pantages; **Box 15.5 fig.1:** © John Eastcott/YVA Momatiuk/The Image Works

CHAPTER 16

Opener: © CNRI/SPL/Photo Researchers, Inc.; **Box 16.1 fig.1:** Bettmann Archive; **Box 16.2 fig.1:** © BioPhoto Associates/Photo Researchers, Inc.; **16.4:** © Tony Arruza/The Image Works; **16.6, Box 16.4 fig.1:** © Times Mirror Higher Education Group, Inc./Bob Coyle, photographer

CHAPTER 17

Opener: © Garry Watson/SPL/Photo Researchers, Inc.; **Box 17.1 fig.2:** North Wind Picture Archives; **Box 17.2 fig.1:** © Larry Mangino/The Image Works; **Box 17.2 fig.2:** © Will & Deni McIntyre/Photo Researchers, Inc.; **Box 17.2 fig.3:** © James King-Holmes/SPL/Photo Researchers, Inc.; **Box 17.3 fig.1:** © Biology Media/Photo Researchers, Inc.; **Box 17.3 fig.2:** North Wind Picture Archives; **Box 17.3 fig.3:** © Michael Newman/PhotoEdit; **Box 17.4 fig.1:** © Leonard Lessin/Peter Arnold, Inc.; **Box 17.5 fig.1:** © Luiz C. Marigo/Peter Arnold, Inc.; **Box 17.5 fig.2:** © Times Mirror Higher Education Group, Inc./Bob Coyle, photographer; **17.23a:** NIH/Science Source/Photo Researchers, Inc.; **17.23b:** © Times Mirror Higher Education Group, Inc./Bob Coyle, photographer

CHAPTER 18

Opener: © Joe Sohm/Image Works; **18.3:** © David Weintraub/Photo Researchers, Inc.; **18.4:** © John D. Cunningham/Visuals Unlimited; **18.11 a,b:** © Peggy Yoram Kahana/Peter Arnold, Inc.; **18.12:** © Mark Antman/The Image Works; **Box 18.2 fig.1:** © Joe Sohm/The Image Works; **Box 18.2 fig.2:** © Charles Gebelein; **18.19:** © Inga Spence/Tom Stack & Associates; **18.21b:** © Mark Antman/Phototake, Inc.; **Box 18.5:** © Takeshi Takahara/Photo Researchers, Inc.; **18.22:** © Charles Gebelein

CHAPTER 19

Opener: © Doug Sherman/Geofile; **19.1:** © Charles Gatewood/The Image Works; **19.14:** NASA/Goddard Space Flight Center

INDEX

NOTE: Page numbers followed by lowercase f indicate illustrations; while page references followed by lowercase t indicate tables.

A

B

C

D

N

O

R

S

X

Z